Soft Computing

Soft Computing

Editors

Rui Araújo
António Pedro Aguiar
Nuno Lau
Rodrigo Ventura
João Fabro

MDPI

Basel • Beijing • Wuhan • Barcelona • Belgrade • Novi Sad • Cluj • Manchester

Editors

Rui Araújo
University of Coimbra
Coimbra
Portugal

António Pedro Aguiar
University of Porto
Porto
Portugal

Nuno Lau
University of Aveiro
Aveiro
Portugal

Rodrigo Ventura
Instituto Superior Técnico
Universidade de Lisboa
Lisbon
Portugal

João Fabro
Federal University of
Technology-Paraná (UTFPR)
Curitiba, Paraná
Brazil

Editorial Office
MDPI AG
Grosspeteranlage 5
4052 Basel, Switzerland

This is a reprint of articles from the Topic published online in the open access journals *Applied Sciences* (ISSN 2076-3417), *Algorithms* (ISSN 1999-4893), *Information* (ISSN 2078-2489), and *Mathematics* (ISSN 2227-7390) (available at: https://www.mdpi.com/topics/Soft_Computing).

For citation purposes, cite each article independently as indicated on the article page online and as indicated below:

Lastname, A.A.; Lastname, B.B. Article Title. *Journal Name* **Year**, *Volume Number*, Page Range.

ISBN 978-3-7258-1891-4 (Hbk)
ISBN 978-3-7258-1892-1 (PDF)
doi.org/10.3390/books978-3-7258-1892-1

Contents

Preface

The aim of this reprint is to provide a collection of relevant recent research in the field of soft computing, involving works on different methodologies, techniques and algorithms in soft computing, machine learning, computational intelligence, and optimization. This reprint is addressed to scientists and engineers, offering open access to a variety of innovative ideas and developments, which may further motivate new research endeavors and developments in related areas. We would like to thank the authors of all the included works and the Assistant Editors and topic specialists from MDPI who were involved.

Rui Araújo, António Pedro Aguiar, Nuno Lau, Rodrigo Ventura, and João Fabro
Editors

information

MDPI

Article

Regularized Mislevy-Wu Model for Handling Nonignorable Missing Item Responses

Alexander Robitzsch [1,2]

1 Department of Educational Measurement and Data Science, IPN—Leibniz Institute for Science and Mathematics Education, Olshausenstraße 62, 24118 Kiel, Germany; robitzsch@leibniz-ipn.de
2 Centre for International Student Assessment (ZIB), Olshausenstraße 62, 24118 Kiel, Germany

Abstract: Missing item responses are frequently found in educational large-scale assessment studies. In this article, the Mislevy-Wu item response model is applied for handling nonignorable missing item responses. This model allows that the missingness of an item depends on the item itself and a further latent variable. However, with low to moderate amounts of missing item responses, model parameters for the missingness mechanism are difficult to estimate. Hence, regularized estimation using a fused ridge penalty is applied to the Mislevy-Wu model to stabilize estimation. The fused ridge penalty function is separately defined for multiple-choice and constructed response items because previous research indicated that the missingness mechanisms strongly differed for the two item types. In a simulation study, it turned out that regularized estimation improves the stability of item parameter estimation. The method is also illustrated using international data from the progress in international reading literacy study (PIRLS) 2011 data.

Keywords: Mislevy-Wu model; missing data; nonignorable missingness; missing not at random; item response model; regularized estimation

Citation: Robitzsch, A. Regularized Mislevy-Wu Model for Handling Nonignorable Missing Item Responses. *Information* **2023**, *14*, 368. https://doi.org/10.3390/info14070368

Academic Editor: Heming Jia

Received: 17 May 2023
Revised: 21 June 2023
Accepted: 26 June 2023
Published: 28 June 2023

1. Introduction

In educational large-scale assessment (LSA) studies [1,2], such as the progress in international reading literacy study (PIRLS; [3]), the trends in international mathematics and science study (TIMSS; [4]), or the programme for international student assessment (PISA; [5]), students' abilities are assessed using cognitive test items. Often, however, students do not respond to specific items leading to missing item responses [6]. It is not obvious how item nonresponse [7] should be treated in the computation of values of abilities (i.e., values of the latent trait) in item response theory (IRT) models [8–10] that are used as scaling models.

Researchers frequently argue for applying complex IRT models that model missing item responses in order to avoid biased item parameters [6,11]. If students omit items, the most obvious option would be treating them as either wrong or missing, which effectively means removing them from the estimation. In the latter case, missing item responses are simply ignored. Slightly more complex treatments assume that missing item responses can be ignored when conditioning on further latent variables (i.e., latent ignorability; [12]). However, it has been shown that these kinds of models do not adequately fit typical LSA datasets [13]. Recently, the Mislevy-Wu (MW) model received some attention [7,13–16] that relaxes the strict assumption that missingness on item responses should either be treated as wrong or latent ignorable. However, the MW model tends to produce unstable parameter estimates if the missingness parameters are estimated item-specific. To circumvent this issue, this paper proposes a regularized estimation approach to the MW model to stabilize parameter estimation.

The decision of how to score missing item responses in LSA studies is a delicate one. On the one hand, students might omit item responses because of a lack of motivation.

On the other hand, students could simply not know the correct answer and therefore do not deliver an item response. Even if students have only low motivation to respond to an item in LSA studies, it could be generally questioned whether item omissions should not be scored as wrong. Treating item omission as wrong might induce a strategy for students to only respond to those items that they do know with sufficient confidence. Introducing different item selection (or solution) strategies would undoubtedly impact the interpretation of results in an LSA study. Hence, one can conclude that the decision on how to score item responses is not (mainly) a statistical one [17]. Nevertheless, we use the regularized MW model in this paper to explore potential causes of missing item responses and the consequences of different missing data treatments.

The rest of the article is structured as follows. The regularized MW model is introduced in Section 2. Section 3 presents results from a simulation study that investigates the performance of regularized estimation of the MW model. In Section 4, an empirical example involving PIRLS 2011 data is provided. Finally, the article closes with a discussion in Section 5.

2. Mislevy-Wu Model

In this section, we review missing data terminology and introduce the regularized MW model for handling nonignorable missing item responses.

A vector of item responses for person p is denoted by $\boldsymbol{X}_p$. In the presence of missing values, we decompose $\boldsymbol{X}_p$ into $\boldsymbol{X}_p = (\boldsymbol{X}_{\text{obs},p}, \boldsymbol{X}_{\text{mis},p})$, where $\boldsymbol{X}_{\text{obs},p}$ denotes the observed and $\boldsymbol{X}_{\text{mis},p}$ the missing item responses. Let $\boldsymbol{R}_p$ denote the vector of response indicators whose values are 1 if an item is observed and 0 if it is missing. We can factorize the joint distribution of $\boldsymbol{X}_p$ and $\boldsymbol{R}_p$ as

$$\mathrm{P}(\boldsymbol{R}_p, \boldsymbol{X}_{\text{obs},p}, \boldsymbol{X}_{\text{mis},p}) = \mathrm{P}(\boldsymbol{R}_p | \boldsymbol{X}_{\text{obs},p}, \boldsymbol{X}_{\text{mis},p}) \mathrm{P}(\boldsymbol{X}_{\text{obs},p}, \boldsymbol{X}_{\text{mis},p}) \quad (1)$$

Missing data literature distinguishes different missingness mechanisms regarding the assumptions of the conditional distribution $\mathrm{P}(\boldsymbol{R}_p | \boldsymbol{X}_{\text{obs},p}, \boldsymbol{X}_{\text{mis},p})$ (see [18,19]). The most important distinction is between missing at random (MAR; [20]) and missing not at random (MNAR). MAR holds if

$$\mathrm{P}(\boldsymbol{R}_p | \boldsymbol{X}_{\text{obs},p}, \boldsymbol{X}_{\text{mis},p}) = \mathrm{P}(\boldsymbol{R}_p | \boldsymbol{X}_{\text{obs},p}). \quad (2)$$

If (2) is violated, the missing data are MNAR. Based on the MAR assumption in (2), we can integrate out the missing data $\boldsymbol{X}_{\text{mis},p}$ and obtain

$$\int \mathrm{P}(\boldsymbol{R}_p | \boldsymbol{X}_{\text{obs},p}, \boldsymbol{X}_{\text{mis},p}) \mathrm{P}(\boldsymbol{X}_{\text{obs},p}, \boldsymbol{X}_{\text{mis},p}) \, \mathrm{d}\boldsymbol{X}_{\text{mis},p} = \mathrm{P}(\boldsymbol{R}_p | \boldsymbol{X}_{\text{obs},p}) \int \mathrm{P}(\boldsymbol{X}_{\text{obs},p}, \boldsymbol{X}_{\text{mis},p}) \, \mathrm{d}\boldsymbol{X}_{\text{mis},p} \quad (3)$$

The crucial point is that the factor $\mathrm{P}(\boldsymbol{R}_p | \boldsymbol{X}_{\text{obs},p})$ does not depend on missing data $\boldsymbol{X}_{\text{mis},p}$, which is the reason why likelihood-based inference can rely on the observed data by parametrizing the distribution $\int \mathrm{P}(\boldsymbol{X}_{\text{obs},p}, \boldsymbol{X}_{\text{mis},p}) \, \mathrm{d}\boldsymbol{X}_{\text{mis},p}$. If the model parameters of the two factors are distinct [18], missing data are denoted as ignorable. Hence, we also label the MAR assumption (2) as manifest ignorability (MI).

Latent ignorability (LI; [21–27]) is one of the weakest nonignorable missingness mechanisms. LI weakens the assumption of ignorability for MAR data. In this case, the existence of a latent variable $\boldsymbol{\eta}_p$ is assumed. The dimension of $\boldsymbol{\eta}_p$ is typically much lower than the dimension of $\boldsymbol{X}_p$. LI is formally defined as (see [27])

$$\mathrm{P}(\boldsymbol{R}_p | \boldsymbol{X}_{\text{obs},p}, \boldsymbol{X}_{\text{mis},p}, \boldsymbol{\eta}_p) = \mathrm{P}(\boldsymbol{R}_p | \boldsymbol{X}_{\text{obs},p}, \boldsymbol{\eta}_p). \quad (4)$$

That is, the probability of missing item responses depends on observed item responses and the latent variable $\boldsymbol{\eta}_p$, but not the unknown missing item responses $\boldsymbol{X}_{\text{mis},p}$ itself. By integrating out $\boldsymbol{X}_{\text{mis},p}$, we obtain

$$\int P(\boldsymbol{R}_p|\boldsymbol{X}_{\text{obs},p}, \boldsymbol{X}_{\text{mis},p}, \boldsymbol{\eta}_p)P(\boldsymbol{X}_{\text{obs},p}, \boldsymbol{X}_{\text{mis},p}|\boldsymbol{\eta}_p)\, d\boldsymbol{X}_{\text{mis},p}\, d\boldsymbol{\eta}_p = \int P(\boldsymbol{R}_p|\boldsymbol{X}_{\text{obs},p}, \boldsymbol{\eta}_p)P(\boldsymbol{X}_{\text{obs},p}, \boldsymbol{X}_{\text{mis},p}|\boldsymbol{\eta}_p)\, d\boldsymbol{X}_{\text{mis},p}\, d\boldsymbol{\eta}_p \quad (5)$$

Specification (4) is also known as a shared-parameter model [28,29]. In most applications, conditional independence of item responses X_{pi} and response indicators R_{pi} conditional on $\boldsymbol{\eta}_p$ is assumed [27]. In this case, Equation (5) simplifies to

$$\int P(\boldsymbol{R}_p|\boldsymbol{X}_{\text{obs},p}, \boldsymbol{X}_{\text{mis},p}, \boldsymbol{\eta}_p)P(\boldsymbol{X}_{\text{obs},p}, \boldsymbol{X}_{\text{mis},p}|\boldsymbol{\eta}_p)\, d\boldsymbol{X}_{\text{mis},p}\, d\boldsymbol{\eta}_p = \int \prod_{i=1}^{I} [P(R_{pi} = r_{pi}|\boldsymbol{\eta}_p)P(X_{pi} = x_{pi}|\boldsymbol{\eta}_p)^{r_{pi}}]\, d\boldsymbol{\eta}_p\,. \quad (6)$$

In the rest of this paper, it is assumed that the latent variable $\boldsymbol{\eta}_p$ consists of a latent ability θ_p and a latent response propensity ξ_p. The latent response propensity ξ_p is a unidimensional latent variable that represents the dimensional structure of the response indicators $\boldsymbol{R}_p$.

The IRT model of interest follows a two-parameter logistic (2PL) model [30]:

$$P(X_{pi} = 1|\theta_p) = \Psi(a_i(\theta_p - b_i))\,, \quad (7)$$

where Ψ is the logistic link function, a_i represents item discriminations, and b_i represents item difficulties.

Regularized Mislevy-Wu Model

For allowing nonignorable missing item responses, the conditional distribution $P(R_{pi} = 1|X_{pi} = x, \theta_p, \xi_p)$ must be specified. The conditional probability of a missing item response in the MW model [13,14,16,31–33] is defined as

$$P(R_{pi} = 1, X_{pi} = x|\theta_p, \xi_p) = \Psi(\xi_p - \beta_i - \rho_i x) \text{ for } x = 0, 1\,. \quad (8)$$

The total probability of a missing item response is given by

$$P(R_{pi} = 0|\theta_p, \xi_p) = P(R_{pi} = 1, X_{pi} = \text{NA}|\theta_p, \xi_p) = \sum_{x=0}^{1} P(R_{pi} = 1, X_{pi} = x|\theta_p, \xi_p)\,. \quad (9)$$

By combining (8) and (9), we get

$$P(X_{pi} = x, R_{pi} = r|\theta_p, \xi_p) = \begin{cases} [1 - \Psi(a_i(\theta_p - b_i))]\Psi(\xi_p - \beta_i) & \text{if } x = 0 \text{ and } r = 1, \\ \Psi(a_i(\theta_p - b_i))\Psi(\xi_p - \beta_i - \rho_i) & \text{if } x = 1 \text{ and } r = 1, \\ \Psi(a_i(\theta_p - b_i)\Psi(\xi_p - \beta_i - \rho_i) + [1 - \Psi(a_i(\theta_p - b_i))]\Psi(\xi_p - \beta_i) & \text{if } x = \text{NA and } r = 0. \end{cases} \quad (10)$$

Note that the model defined in Equation (10) can be interpreted as an IRT model for a variable U_{pi} that has three categories: Category 0 (observed incorrect): $X_{pi} = 0$, $R_{pi} = 1$, Category 1 (observed correct): $X_{pi} = 1$, $R_{pi} = 1$, and Category 2 (missing item response): $X_{pi} = \text{NA}$, $R_{pi} = 0$ (see [34,35]). The marginal distribution $P(X_{pi} = x|\theta_p, \xi_p)$ in (10) follows the 2PL model (7). The conditional probabilities for response indicators R_{pi} are modeled with parameters β_i and δ_i. The parameter β_i parametrizes the item-specific proportion of missing item responses, while the parameter δ_i quantifies the dependence of the responding to item i conditional on the true but possibly unobserved item response X_{pi}. It has been pointed out in [13,33] that the MW model (10) contains the special cases of treating missing item responses as latent ignorable and as wrong as two extreme cases. Moreover, the simulation studies in [13,14] demonstrated that a common δ parameter that is constant across items could be consistently estimated.

Figure 1 graphically displays the MW model. Note the dependency of response indicators R_i from items X_i.

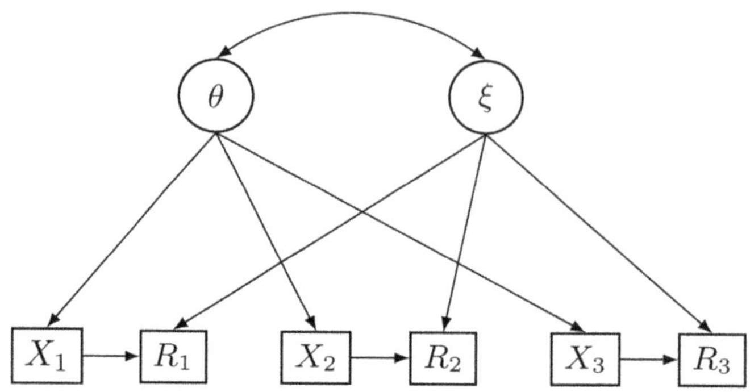

Figure 1. Graphical representation of the Mislevy-Wu model with three items X_1, X_2, and X_3, and their corresponding response indicators R_1, R_2, and R_3, and latent ability θ and latent response propensity ξ.

In this article, a bivariate normal distribution for (θ_p, ξ_p) is assumed, where $\mathrm{SD}(\theta_p)$ is fixed to one, and $\mathrm{SD}(\xi_p)$, as well as $\mathrm{Cor}(\theta_p, \xi_p)$, are estimated (see [36,37] for more complex distributions). LI in general and the model of Holman and Glas [12] in particular is obtained in the MW model by fixing all δ_i parameters equal to zero. If one fixes all δ_i parameters in the MW model at a sufficiently small (negative) value, such as -9.99, students with missing item responses are scored as incorrect. Moreover, if one fixes all δ_i parameters to zero and sets the correlation of θ and ξ to zero, MI (i.e., MAR) is obtained. Hence, LI can be tested against MI. Moreover, the MW model is more general than the LI model because the latter model only models the dependence of missingness on item i from ξ, but not the item itself.

The MW model can be estimated with maximum likelihood (ML). By denoting all item parameters by $\gamma = (\gamma_1, \ldots, \gamma_I)$ and distribution parameters by $\boldsymbol{\alpha}$, the log-likelihood function is given by

$$l(\gamma, \boldsymbol{\alpha}; \boldsymbol{X}, \boldsymbol{R}) = \sum_{p=1}^{N} \log \int_{-\infty}^{\infty} \prod_{i=1}^{I} [\mathrm{P}(X_{pi} = x_{pi}, R_{pi} = r_{pi} | \theta, \xi; \gamma_i)] f(\theta, \xi; \boldsymbol{\alpha}) \, \mathrm{d}\theta \, \mathrm{d}\xi \,, \tag{11}$$

where $\boldsymbol{X} = (x_{pi})_{pi}$ and $\boldsymbol{R} = (r_{pi})_{pi}$ denote the datasets of item responses and response indicators. The item-specific parameters are given as $\gamma_i = (a_i, b_i, \beta_i, \delta_i)$. The log-likelihood function can be numerically maximized to obtain item parameter estimates $\hat{\gamma}$ and distribution parameters $\hat{\boldsymbol{\alpha}}$. In IRT software, the expectation-maximization algorithm is frequently utilized [38,39].

In our experience, estimating item-specific δ_i parameters in the MW model can become quite unstable. Moreover, it has been shown that average δ_i parameters typically strongly differ between constructed response (CR) and multiple-choice (MC) items, because the omission of CR items is more associated with the true but not fully observed item response, while omissions of MC items are only weakly associated with true item responses [13]. For stabilizing the estimation of δ_i parameters in ML, we propose to employ regularized ML estimation with fused ridge-type penalty functions [40].

Let $\mathcal{I}_{\mathrm{MC}} \subset \mathcal{I}$ and $\mathcal{I}_{\mathrm{CR}} \subset \mathcal{I}$ be distinct integer sets of multiple-choice and constructed response items, respectively, where $\mathcal{I} = \{1, \ldots, I\}$. The fused ridge penalty function $\mathcal{P}$ for the MW model is defined by

$$\mathcal{P}(\gamma; \lambda) = \lambda \left[\sum_{i,j \in \mathcal{I}_{\mathrm{CR}}} (\delta_i - \delta_j)^2 + \sum_{i,j \in \mathcal{I}_{\mathrm{MC}}} (\delta_i - \delta_j)^2 \right] , \tag{12}$$

where λ is a fixed regularization parameter. In regularized ML estimation, one maximizes the penalized log-likelihood function l_{pen} defined by

$$l_{pen}(\gamma, \boldsymbol{\alpha}; \lambda, \boldsymbol{X}, \boldsymbol{R}) = l(\gamma, \boldsymbol{\alpha}; \boldsymbol{X}, \boldsymbol{R}) - \mathcal{P}(\gamma; \lambda) \ . \quad (13)$$

Using the penalty function in (12) implies normal priors for δ_i with means for CR and MC items, respectively, and a common variance [40]. Importantly, by only considering differences in pairs of item parameters δ_i, the item-type specific means of δ_i are not explicitly estimated. The MW model (10) applied with regularized ML estimation using the fitting function (13) is also called the regularized MW model.

The maximization of l_{pen} involves the unknown regularization parameter λ. The k-fold cross-validation approach is used for obtaining the optimal regularization parameter λ_{opt}. The dataset is divided into k groups, and the parameters of the regularized MW model are estimated on $k-1$ folds leaving one fold out to evaluate the cross-validation error. This is performed by leaving one fold out in turn and for each value of the regularization parameter λ. In this article, the error was evaluated using the negative log-likelihood function value [40]. The cross-validation error is calculated as $\sum_{h=1}^{k} l(\hat{\gamma}_{(-h)}, \hat{\boldsymbol{\alpha}}_{(-h)}; \boldsymbol{X}_h, \boldsymbol{R}_h)$, where $\hat{\gamma}_{(-u)}$ and $\hat{\boldsymbol{\alpha}}_{(-h)}$ are the vector of item parameter and distribution estimates obtained by excluding the hth group of data. Moreover, $\boldsymbol{X}_h$ and $\boldsymbol{R}_h$ denote the datasets of item responses and response indicators in the hth part of the data, respectively. In cross-validation, the log-likelihood function is predicted on the part of the data that has not been used for parameter estimation [41]. The smallest cross-validated log-likelihood value determines the optimal regularization parameter λ_{opt}. In practice, $k = 5$ or $k = 10$ is frequently chosen.

3. Simulation Study

3.1. Method

In this simulation study that studies the performance of the regularized MW model, we fixed the number of items to $I = 20$ and fixed item parameters a_i, b_i and δ_i throughout all replications. To mimic real-data situations, we assumed that the first ten items C01, ..., C10 were CR, while the last ten items M11, ..., M20 were MC. On average, the missing proportion of CR items was larger than for MC items. The δ_i parameters were varied according to two data-generating models DGM1 and DGM2. In DGM1, the missing proportions were 0.112 for MC items and 0.153 for CR items. In DGM2, a higher missing proportion of 0.341 for CR items was assumed while retaining the missing proportion for MC items at 0.112. The item parameters used in the simulation study can be found in Table A1 in Appendix A (see also the directory "*Simulation Study*" https://osf.io/5pd28 (accessed on 21 June 2023)).

We chose sample sizes $N = 1000$ and $N = 2500$. We did not opt for smaller sample sizes because we think that estimating an IRT model for response indicators requires sufficiently large sample sizes. Hence, the MW model is more suitable for LSA studies than for small-scale studies.

A bivariate normal distribution was simulated for the ability variable θ and the response propensity ξ. The standard deviation of θ was set to 1, while the standard deviation of ξ was fixed at 2. Moreover, the correlation of θ and ξ was fixed at 0.5 when simulating the data.

The regularized MW model was estimated for a fixed sequence of values for the regularization parameter λ. A grid of 21 regularization parameters was chosen: 0.000010, 0.000018, 0.000034, 0.000062, 0.000113, 0.000207, 0.000379, 0.000695, 0.001274, 0.002336, 0.004281, 0.007848, 0.014384, 0.026367, 0.048329, 0.088587, 0.162378, 0.297635, 0.545560, 1.0, and 10,000. Values between 0.000010 and 1.0 were equidistantly chosen on a logarithmic scale. In k-fold cross-validation, $k = 5$ folds were used. In the MW model, the estimated distribution parameters $\boldsymbol{\alpha}$ consisted of the variance of ξ and the covariance of θ and ξ.

In total, 2500 replications were conducted in each simulation condition. We assessed the performance of parameter estimates by bias and root mean square error (RMSE).

To provide simple summary statistics, we averaged absolute biases and RMSE values across items for the same item parameter groups (i.e., the a, b, β, and δ parameters).

The statistical software R [42] was employed for all parts of the simulation. The estimation of the regularized MW model was carried out using the `sirt::xxirt()` function in the sirt package [43]. Replication material can be found in the directory *"Simulation Study"* at https://osf.io/5pd28 (accessed on 21 June 2023).

3.2. Results

In Figure 2, the average absolute bias (red dashed line) and the average RMSE (solid black line) for item parameter groups δ, β, a, and b are displayed for the DGM2. It can be seen that for $N = 1000$, the minimum average RMSE for δ parameters is obtained for a λ value that is substantially larger than the optimal regularization parameter λ_{opt} obtained with k-fold cross-validated log-likelihood estimation. Interestingly, biases in all parameters became relevant for sufficiently large regularization parameters. Hence, the search for an optimal λ parameter regarding RMSE reflects a bias-variance tradeoff. However, it should be emphasized that for a broad range of sufficiently small λ values, the bias and RMSE for item discriminations a_i, and item difficulties b_i were almost unaffected by the choice of λ.

In Table 1, average absolute bias and average RMSE are displayed for the optimal regularization parameter λ_{opt}, and fixed regularization parameters 10^{-5}, 0.0263665, and 10^5. It can be seen that $\lambda = 0.0263665$ strongly outperformed the other λ choices in terms of RMSE for the δ_i parameters. However, a nonnegligible bias in δ_i and β_i parameters was introduced by using this regularization parameter. Nevertheless, inducing too much regularization could stabilize estimated item parameters for the response indicators, while the target parameters a_i and b_i were almost unaffected by the choice of λ. Hence, one could generally conclude that the MW model should be utilized to estimate the missing response mechanism flexibly. The regularization technique is only applied for stabilizing parameter estimates without introducing relevant bias in target item parameter estimates.

Table 1. Simulation Study: Average absolute bias (Bias) and average absolute RMSE of estimated item parameters of the regularized Mislevy-Wu model as a function of sample size N and different choices of the regularization parameter λ for two data-generating models (DGM) DGM1 and DGM2.

DGM	Par	N	Bias for $\lambda =$				RMSE for $\lambda =$			
			10^{-5}	λ_{opt}	0.0263665	10^5	10^{-5}	λ_{opt}	0.0263665	10^5
DGM1	δ_i	1000	0.123	0.076	0.304	0.754	0.915	0.845	0.638	0.875
		2500	0.067	0.045	0.190	0.757	0.563	0.540	0.451	0.812
	β_i	1000	0.063	0.068	0.092	0.221	0.313	0.309	0.285	0.329
		2500	0.028	0.031	0.052	0.221	0.189	0.189	0.186	0.277
	a_i	1000	0.008	0.008	0.011	0.028	0.154	0.154	0.154	0.156
		2500	0.002	0.002	0.005	0.028	0.096	0.096	0.096	0.102
	b_i	1000	0.016	0.018	0.028	0.064	0.146	0.146	0.144	0.154
		2500	0.008	0.010	0.017	0.062	0.089	0.090	0.090	0.109
DGM2	δ_i	1000	0.063	0.063	0.184	0.750	0.687	0.631	0.518	0.896
		2500	0.018	0.028	0.090	0.750	0.382	0.374	0.349	0.822
	β_i	1000	0.038	0.052	0.080	0.266	0.311	0.299	0.280	0.376
		2500	0.017	0.023	0.039	0.267	0.188	0.187	0.183	0.323
	a_i	1000	0.012	0.012	0.015	0.043	0.172	0.172	0.171	0.178
		2500	0.004	0.004	0.006	0.041	0.105	0.105	0.105	0.118
	b_i	1000	0.013	0.020	0.032	0.108	0.165	0.164	0.161	0.201
		2500	0.005	0.008	0.015	0.105	0.100	0.100	0.099	0.155

Note. Par = item parameter group; λ_{opt} = optimal regularization parameter selected with cross-validated log-likelihood.

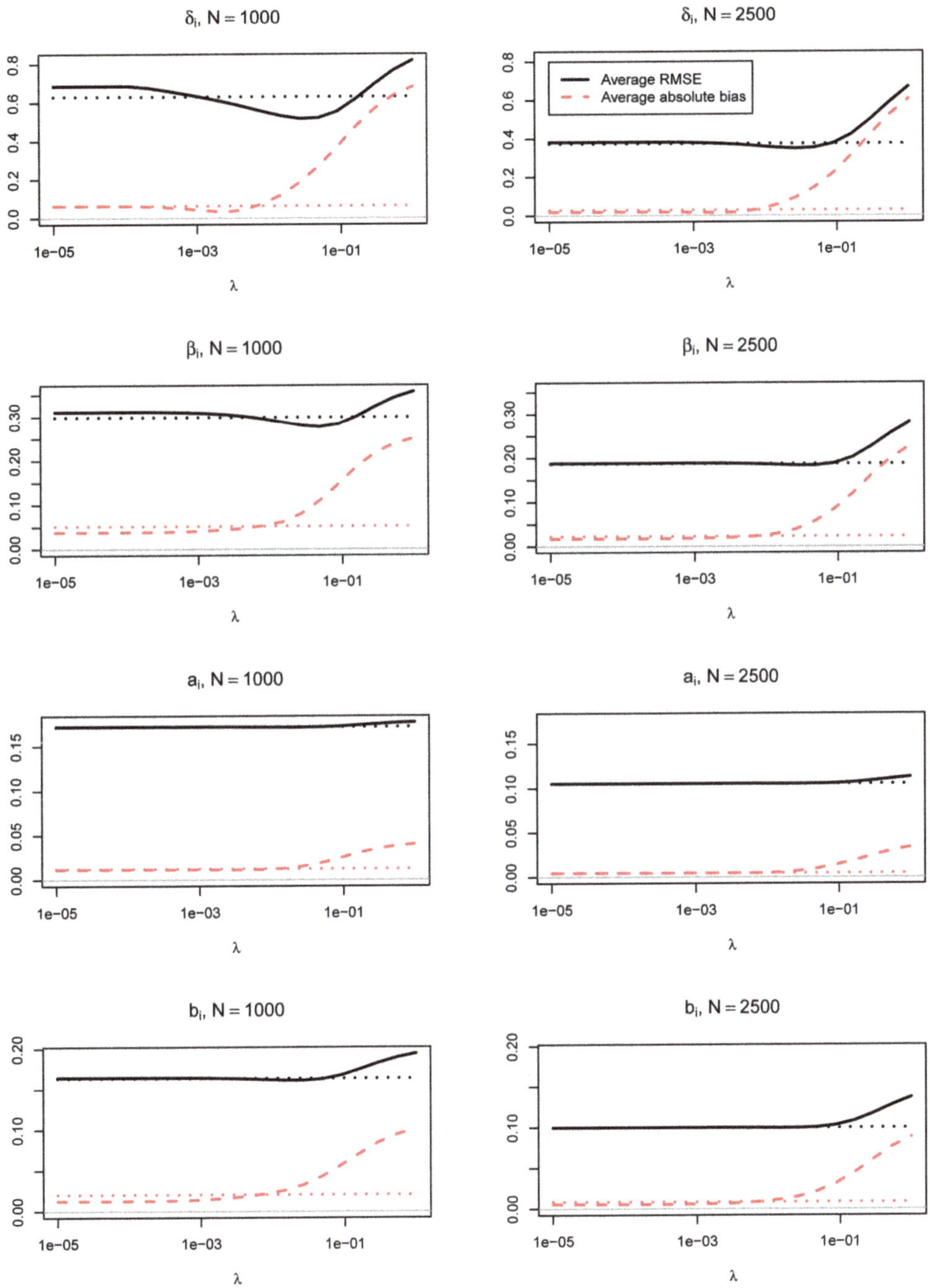

Figure 2. Average absolute bias and average root mean square error (RMSE) for item parameter groups of the regularized Mislevy-Wu model for data-generating model DGM2 as a function of sample size N and the regularization parameter λ. RMSE and bias values for the optimal regularization parameter λ_{opt} selected with cross-validated log-likelihood are displayed with dotted lines.

4. Empirical Example

4.1. Method

In the following analysis, item responses of booklet 13 in PIRLS 2011 (i.e., the "PIRLS Reader") consisting of 35 items (20 CR items and 15 MC items with four response alternatives) were used. For this booklet, item responses of 968 Austrian (AUT), 809 German (DEU), 901 French (FRA), and 802 Dutch (NLD) students were available. The resulting dataset is used for illustrative purposes in this section. For ease of presentation, all polytomous items were dichotomized, where only the highest scores were recoded as correct. The dataset has been made available as `data.pirlsmissing` in the R [42] package sirt [43].

Descriptive analyses showed that the average proportion of missing item responses varied considerably between items and countries (AUT: 0.112, DEU: 0.079, FRA: 0.136, NLD: 0.027). For MC items, the average rate of missing item responses was 0.023 ($SD = 0.016$). For CR items, the average rate of missing item responses was substantially larger ($M = 0.141$, $SD = 0.070$).

We estimated the nonregularized MW model with freely estimated δ_i parameters and compared this model to constrained alternatives. In the LI model [12], all δ_i parameters were fixed to zero. In the WR model, all missing item responses are treated as incorrect, which was implemented by fixing all δ_i parameters to -9.99 setting the response probabilities effectively to zero for students who do not know the item. Finally, in the MI model, we fixed all δ_i parameters to zero and fixed the correlation of θ and ξ to zero. Model comparisons were conducted based on the Akaike information criterion (AIC) and the Bayesian information criterion (BIC).

The MW model was also estimated using regularized estimation. The optimal regularization parameter λ_{opt} was selected by minimizing the negative cross-validated log-likelihood value. The sequence of the regularization parameter λ was selected between 10^{-8} and 10,000, equidistantly spaced on a logarithmic scale.

The R [42] package sirt [43] using the `sirt::xxirt()` was employed for fitting the IRT models. Replication material can be found in the directory "*Empirical Example*" at https://osf.io/5pd28 (accessed on 21 June 2023).

4.2. Results

In Table 2, model comparisons of the four nonregularized models are displayed. The most general MW model turned out to be the best-fitting model in terms of AIC and BIC. In line with [13], the WR model (i.e., treating missing item responses as incorrect) outperformed the LI model (i.e., treating missing item responses as missing). The standard deviation of ξ slightly varies across models, being smallest when treating missing item responses as wrong in model MW. Also note that the correlation of θ and ξ was practically identical for the models LI, WR, and MW.

Table 2. PIRLS Reader 2011: Model comparisons.

Model	#npars	AIC	BIC	SD(ξ)	Cor(θ,ξ)	δ_i
MI	106	162,192	162,844	2.41	0 ‡	0 ‡
LI	107	161,796	162,454	2.38	0.41	0 ‡
WR	107	161,414	162,073	2.29	0.41	−9.99 ‡
MW	142	**161,086**	**161,960**	2.34	0.41	est

Note. #npars = number of estimated model parameters; MI = manifest ignorabiity; LI = latent ignorability; WR = treating missing item responses as wrong (i.e., 0); MW = Mislevy-Wu model; ‡ = fixed model parameter; est = estimated model parameters; Entries with the least AIC or BIC are printed in bold font, respectively.

In Table 3, estimated item parameters of the regularized Mislevy-Wu model are displayed. Notably, the average δ parameters for CR items ($M = -2.01, Med = -1.65$, $SD = 1.44$) were lower than MC items ($M = 0.60, Med = 0.03, SD = 2.06$). Moreover, the average β_i parameter was larger for CR items ($M = -2.70, Med = -2.89, SD = 0.85$)

than for MC items ($M = -6.55, Med = -5.75, SD = 1.84$), reflecting that the missing proportions for CR items were larger than for MC items.

Table 3. PIRLS Reader 2011: Estimated item parameters of the regularized Mislevy-Wu model.

					a_i				b_i				β_i	δ_i
Item	**Type**	**Freq0**	**Freq1**	**FreqNA**	**MI**	**LI**	**WR**	**MW**	**MI**	**LI**	**WR**	**MW**	**MW**	**MW**
R31G02C	CR	0.28	0.64	0.08	0.86	0.85	0.89	0.91	−1.04	−1.04	−0.81	−0.80	−3.03	−4.45
R31G04C	CR	0.58	0.21	0.20	1.05	1.04	1.06	1.05	1.23	1.25	1.43	1.36	−2.34	−1.40
R31G08CZ	CR	0.41	0.40	0.19	1.27	1.26	1.34	1.23	0.18	0.20	0.48	0.30	−2.01	−0.90
R31G08CA	CR	0.40	0.37	0.23	1.76	1.74	1.82	1.74	1.32	1.35	1.44	1.39	−1.99	−0.76
R31G08CB	CR	0.61	0.14	0.25	1.45	1.44	1.51	1.41	0.09	0.11	0.31	0.17	−2.41	−0.79
R31G10C	CR	0.48	0.40	0.13	1.13	1.13	1.20	1.17	0.28	0.29	0.40	0.35	−3.08	−1.42
R31G12C	CR	0.51	0.29	0.20	0.49	0.48	0.63	0.49	1.24	1.26	1.47	1.25	−2.53	−0.04
R31G13CZ	CR	0.17	0.66	0.17	2.43	2.46	3.31	3.18	−0.55	−0.53	−0.27	−0.33	−1.55	−2.81
R31G13CA	CR	0.23	0.58	0.20	2.11	2.12	2.85	2.72	−0.36	−0.34	−0.11	−0.14	−1.35	−3.40
R31G13CB	CR	0.26	0.52	0.22	2.19	2.19	2.75	2.68	−0.12	−0.10	0.07	0.05	−1.54	−3.21
R31G13CC	CR	0.32	0.46	0.22	3.58	3.67	4.88	5.07	−0.76	−0.74	−0.51	−0.57	−1.69	−2.69
R31P02C	CR	0.23	0.73	0.04	0.81	0.81	0.79	0.81	−1.61	−1.62	−1.52	−1.48	−3.77	−3.88
R31P03C	CR	0.16	0.79	0.06	1.26	1.25	1.24	1.29	−1.57	−1.57	−1.42	−1.38	−2.88	−4.36
R31P05C	CR	0.45	0.48	0.08	1.08	1.09	1.07	1.02	−0.05	−0.05	0.04	−0.11	−4.51	0.59
R31P06C	CR	0.19	0.76	0.04	1.40	1.39	1.34	1.37	−1.28	−1.28	−1.23	−1.24	−3.85	−2.43
R31P07C	CR	0.19	0.74	0.07	1.74	1.74	1.67	1.74	−1.08	−1.08	−0.98	−0.98	−2.90	−3.12
R31P09C	CR	0.14	0.80	0.06	1.25	1.26	1.37	1.33	−1.71	−1.68	−1.40	−1.55	−3.40	−1.83
R31P14C	CR	0.33	0.54	0.13	1.06	1.06	1.15	1.07	−0.48	−0.47	−0.24	−0.39	−2.92	−1.22
R31P15C	CR	0.51	0.36	0.13	0.52	0.53	0.63	0.53	0.74	0.75	0.92	0.81	−3.19	−0.55
R31P16C	CR	0.49	0.38	0.13	0.76	0.75	0.86	0.80	0.48	0.49	0.62	0.57	−3.03	−1.48
R31G01M	MC	0.18	0.81	0.01	1.11	1.15	1.13	1.15	−1.66	−1.63	−1.66	−1.68	−10.51	4.20
R31G03M	MC	0.26	0.73	0.01	1.19	1.18	1.04	1.10	−1.11	−1.11	−1.24	−1.23	−7.52	1.24
R31G05M	MC	0.42	0.56	0.02	0.84	0.85	0.81	0.80	−0.41	−0.40	−0.42	−0.50	−7.56	2.18
R31G06M	MC	0.27	0.72	0.01	0.98	0.97	0.84	0.90	−1.20	−1.22	−1.38	−1.30	−5.72	−2.94
R31G07M	MC	0.40	0.58	0.02	1.04	1.04	0.95	0.99	−0.41	−0.41	−0.45	−0.45	−5.64	−0.82
R31G09M	MC	0.38	0.60	0.02	0.69	0.69	0.65	0.65	−0.71	−0.71	−0.74	−0.78	−6.06	0.08
R31G11M	MC	0.36	0.62	0.02	1.25	1.26	1.23	1.24	−0.54	−0.54	−0.56	−0.58	−5.81	0.03
R31G14M	MC	0.33	0.60	0.07	1.16	1.17	1.07	1.09	−0.65	−0.64	−0.53	−0.79	−5.41	1.68
R31P01M	MC	0.25	0.74	0.01	1.06	1.06	0.97	0.99	−1.24	−1.24	−1.33	−1.37	−9.84	3.88
R31P04M	MC	0.53	0.46	0.01	0.76	0.77	0.75	0.74	0.19	0.19	0.17	0.12	−8.56	3.12
R31P08M	MC	0.19	0.79	0.02	1.19	1.21	1.13	1.16	−1.52	−1.51	−1.53	−1.58	−5.75	−0.01
R31P10M	MC	0.11	0.87	0.03	1.88	1.89	1.74	1.81	−1.66	−1.65	−1.65	−1.69	−4.80	−1.29
R31P11M	MC	0.27	0.70	0.03	1.07	1.07	1.04	1.04	−1.05	−1.05	−1.05	−1.09	−5.18	−0.78
R31P12M	MC	0.33	0.64	0.03	1.02	1.03	1.02	1.00	−0.77	−0.76	−0.75	−0.80	−5.33	−0.32
R31P13M	MC	0.09	0.88	0.03	1.59	1.60	1.49	1.54	−1.97	−1.96	−1.90	−1.99	−4.53	−1.28

Note. Type = item format; CR = constructed response item; MC = multiple-choice item; MI = manifest ignorabiity; LI = latent ignorability; WR = treating missing item responses as wrong (i.e., 0); MW = Mislevy-Wu model.

It is also noteworthy that item discriminations a_i hardly varied between the MI and LI model (model MI for CR items: $M = 1.41, Med = 1.25, SD = 0.74$; model LI for CR items: $M = 1.41, Med = 1.25, SD = 0.75$). However, item discriminations a_i were larger for models WR ($M = 1.62, Med = 1.29, SD = 1.06$) and MW ($M = 1.58, Med = 1.26, SD = 1.09$). Similarly, item difficulties b_i did not show practical differences between MI and LI models (model MI for CR items: $M = -0.25, Med = -0.24, SD = 0.96$; model LI for CR items: $M = -0.24, Med = -0.22, SD = 0.97$). In line with expectations, the MW model (CR items: $M = -0.14, Med = -0.12, SD = 0.93$) and the WR model resulted in larger item difficulties (CR items: $M = -0.07, Med = -0.03, SD = 0.97$). The pattern was similar for MC items but less pronounced because the missing proportion rates were smaller for MC items compared to CR items.

In Figure 3, the negative cross-validated log-likelihood value is displayed as a function of the regularization parameter λ. For sufficiently small λ values, there is almost no difference in cross-validated log-likelihood values. The optimal regularization parameter was estimated as $\lambda_{\text{opt}} = 0.0004342$.

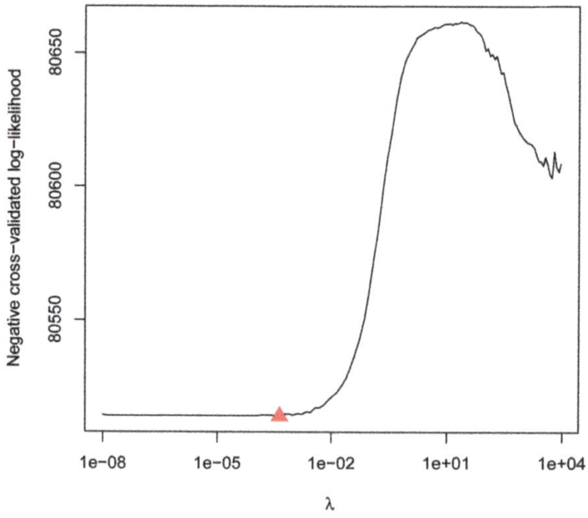

Figure 3. Negative cross-validated log-likelihood value as a function of the regularization parameter λ. The optimal regularization parameter λ_{opt} selected with cross-validated log-likelihood is displayed with a red triangle.

In Figure 4, estimated δ_i item parameters are displayed as a function of the regularization parameter λ for CR and MC items, respectively. With increasing λ parameters, item parameters are fused to item-format-specific parameters. The fused values were $\delta_i = -3.18$ for CR items and $\delta_i = -0.79$ for MC items. This result indicated that missing item responses for CR items are more likely associated with a wrong item response than for MC items. Notably, the fused δ_i parameters were both negative.

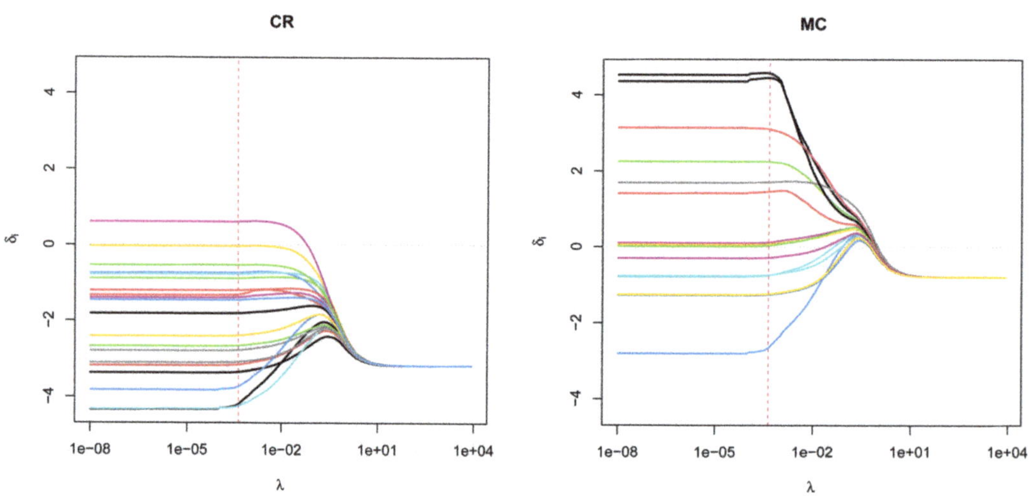

Figure 4. Curves of item parameter estimates δ_i are shown as a function of the regularization parameter λ for constructed response (CR) items (left panel) and multiple-choice (MC) items (right panel). The optimal regularization parameter λ_{opt} selected with cross-validated log-likelihood is displayed with a red dashed line.

In Figure 5, estimated β_i item parameters are displayed as a function of the regularization parameter λ for CR and MC items, respectively. The regularization of the δ_i also affected the estimated β_i parameters, particularly for MC items.

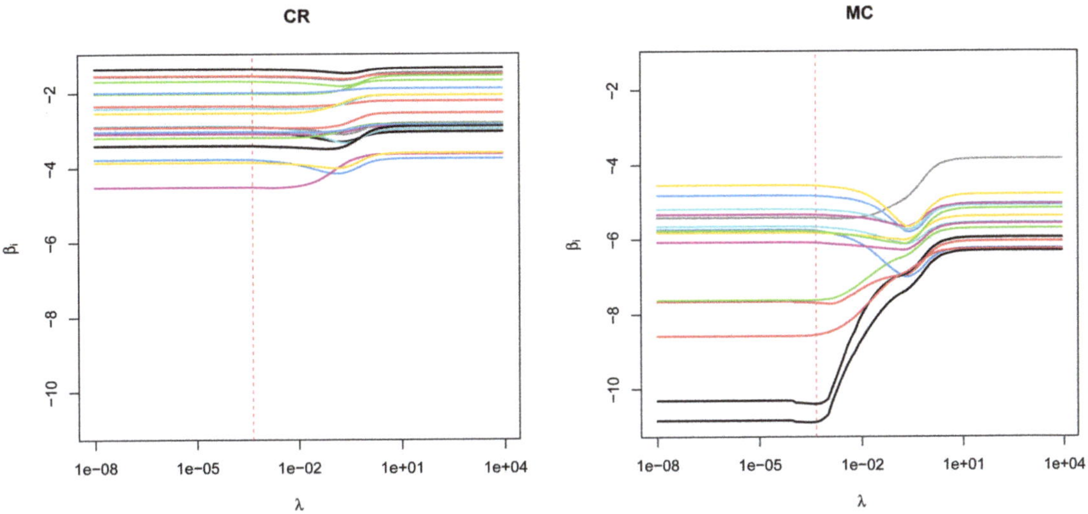

Figure 5. Curves of item parameter estimates β_i as a function of the regularization parameter λ for constructed response (CR) items (left panel) and multiple-choice (MC) items (right panel). The optimal regularization parameter λ_{opt} selected with cross-validated log-likelihood is displayed with a red dashed line.

Finally, Figures 6 and 7 display the a_i and b_i parameters as a function of the regularization parameter λ. The target item parameters are hardly affected for small values of the regularization parameter λ, but show some variation for λ parameters larger than 10^{-2}.

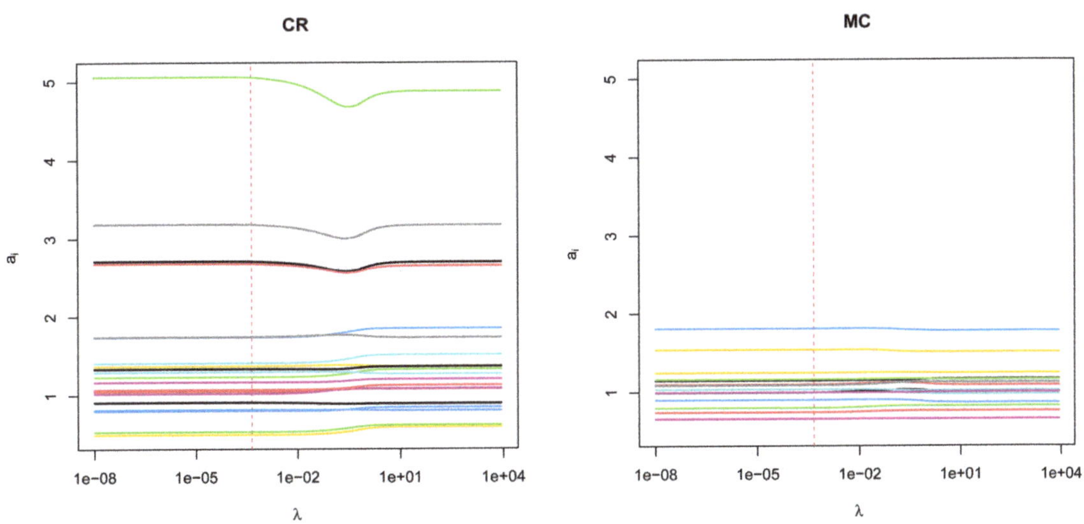

Figure 6. Curves of item parameter estimates a_i as a function of the regularization parameter λ for constructed response (CR) items (left panel) and multiple-choice (MC) items (right panel). The optimal regularization parameter λ_{opt} selected with cross-validated log-likelihood is displayed with a red dashed line.

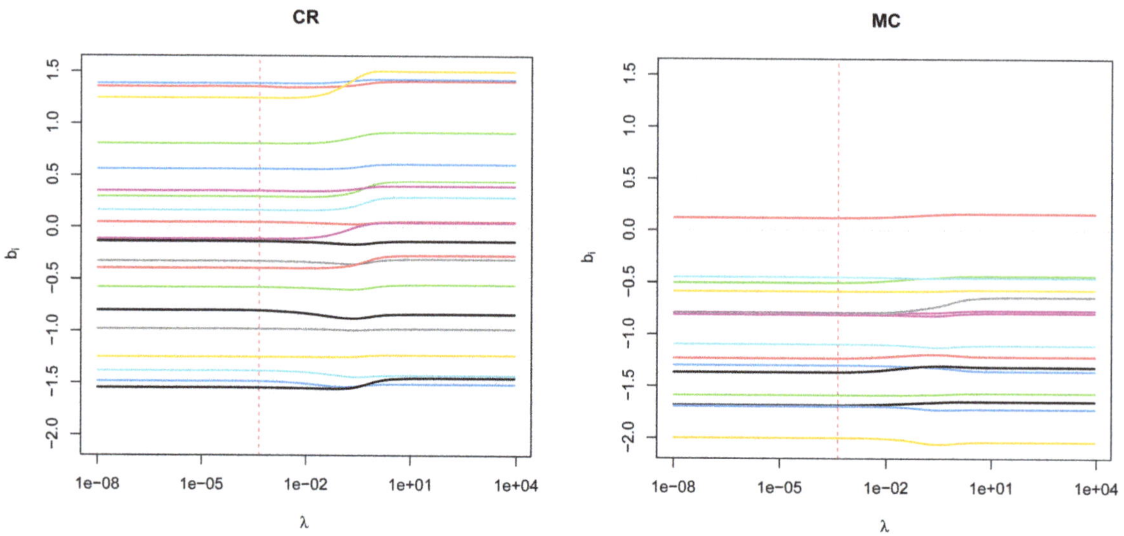

Figure 7. Curves of item parameter estimates b_i as a function of the regularization parameter λ for constructed response (CR) items (left panel) and multiple-choice (MC) items (right panel). The optimal regularization parameter λ_{opt} selected with cross-validated log-likelihood is displayed with a red dashed line.

5. Discussion

In this article, we proposed a regularization estimation approach to the Mislevy-Wu model. This approach allows sufficiently complex missingness mechanisms as well as estimation in moderate sample sizes such as $N = 1000$. Interestingly, the most stable item parameter estimates in terms of RMSE were obtained for values of the regularization parameters that were larger than the one obtained by k-fold cross-validation based on the log-likelihood function value.

To further stabilize estimation, the fused ridge penalty function could also involve the β_i parameters because they are also difficult to estimate for items with low missing proportion rates or in moderate sample sizes.

It has been shown in the PIRLS 2011 application that the Mislevy-Wu model outperformed all other estimation approaches. Omissions on constructed response items were strongly associated with true item responses. This implies that students who do not know an item likely do not respond to it [33]. In contrast, multiple-choice items were only weakly associated with true but non-fully observed item responses. Given these findings, it seems plausible in large-scale assessment studies to score omitted constructed response items as wrong while treating multiple-choice as fractionally correct in a pseudo-likelihood estimation approach [44]. In the latter case, a multiple-choice item with K_i answer alternative is scored with $1/K_i$.

It could be generally argued that constructed response items are omitted more to a lack of knowledge than multiple-choice items. In this sense, as argued by an anonymous reviewer, omissions on constructed response items are likely missing not at random data. In contrast, multiple-choice items could be regarded as missing at random data. In practice, the tendency to omit items can be associated with person traits [45].

The Mislevy-Wu model can be easily extended to item response models for polytomous items. For dichotomous items, the dependence of response indicators R_i from true item responses X_i is modeled by the item parameter δ_i. For polytomous items scored between 0 and K_i, K_i parameters $\delta_{i,k}$ ($k = 1, \ldots, K_i$) that differentially weigh the impact of item category k on the response indicator can be identified from the data.

The Mislevy-Wu model can be extended to include covariates for predicting the latent ability θ_p and the latent response propensity ξ_p [46]. In a latent regression model [47,48], the estimated item parameters could be fixed, and IRT packages such as TAM [49] could be utilized for estimation. Such an approach could also be applied in providing plausible values [50] in LSA studies as realizations of the latent ability θ_p that can be used for secondary analysis. In this sense, the Mislevy-Wu model can be implemented in operational practice when scaling item responses in LSA studies such as PISA, PIRLS, or TIMSS [51].

Missing item responses are typically classified into omitted and not-reached item responses [52]. In this article, we only investigated omitted item responses within a test. For speeded tests, it might be preferable not to score not-reached item responses as wrong. However, large-scale assessment studies like PIRLS are not strongly speeded such that there is only a low prevalence of not-reached items.

The Mislevy-Wu model follows a model-based strategy in which the missingness mechanism for the response indicators is simultaneously modeled with the item response model (e.g., 2PL model) of interest. It might be beneficial to weaken the assumption of a unidimensional ability variable θ and unidimensional response propensity variable ξ and to estimate multidimensional variables with an exploratory loading structure [35] in an imputation model. In this case, the imputation model is more complex than the intentionally misspecified analysis model [17,53]. Certainly, such an estimation approach would need even larger sample sizes, and regularized estimation could also be applied to the exploratory loading structure.

Although modeling missingness mechanisms in educational studies now receive wide attention, only in rare cases, the dependence of item omissions from the item itself is considered a viable alternative (e.g., see [6]). This is unfortunate because we empirically demonstrated that there are several studies in which treating constructed response items as wrong [13] instead of latent ignorable (i.e., as missing; [6]) resulted in superior model fit. The Mislevy-Wu model contains these two extreme scoring treatments as particular constrained models and also parameterizes processes that are a mixture of both. Hence, if missing item responses should be modeled in large-scale assessment studies, there is no excuse for neglecting the Mislevy-Wu model from the preferred psychometrician's toolkit.

Funding: This research received no external funding.

Data Availability Statement: The PIRLS 2011 dataset is available at https://timssandpirls.bc.edu/pirls2011/international-database.html (accessed on 21 June 2023). The part of the dataset used in this article can be accessed as the R object `data.pirlsmissing` in the R package sirt [43].

Conflicts of Interest: The authors declare no conflict of interest.

Abbreviations

The following abbreviations are used in this manuscript:

2PL	two-parameter logistic
AIC	Akaike information criterion
BIC	Bayesian information criterion
CR	constructed response
DGM	data-generating model
IRT	item response theory
LSA	large-scale assessment
LI	latent ignorability
MAR	missing at random
MC	multiple-choice
MI	manifest ignorability
ML	maximum likelihood
MNAR	missing not at random
MW	Mislevy-Wu

PIRLS progress in international reading literacy study
RMSE root mean square error

Appendix A. Item Parameters Used in the Simulation Study

In Table A1, data-generating item parameters for the simulation study are displayed.

Table A1. Simulation Study: Data-generating item parameters of the Mislevy-Wu model.

Item	Type	a_i	b_i	β_i		δ_i
				DGM1	DGM2	
C01	CR	1.7	1.4	−1.7	0.3	−2.0
C02	CR	1.2	0.4	−2.7	−0.7	−1.7
C03	CR	0.5	1.3	−2.2	−0.2	−3.6
C04	CR	2.2	−0.6	−1.4	0.6	−2.9
C05	CR	2.7	−0.3	−1.3	0.7	−3.1
C06	CR	2.8	−0.1	−1.2	0.8	−3.8
C07	CR	1.3	−1.4	−2.5	−0.5	−4.8
C08	CR	1.3	−1.5	−1.8	0.2	−2.0
C09	CR	1.1	−0.4	−2.5	−0.5	−1.3
C10	CR	0.5	0.8	−2.4	−0.4	−0.6
M11	MC	0.9	−1.3	−3.2	−3.2	0.5
M12	MC	1.0	−0.4	−3.4	−3.4	−0.8
M13	MC	0.7	−0.8	−3.6	−3.6	−0.3
M14	MC	1.2	−0.6	−3.7	−3.7	−0.2
M15	MC	1.1	−0.8	−2.8	−2.8	0.4
M16	MC	1.2	−1.6	−2.8	−2.8	−0.3
M17	MC	1.8	−1.7	−2.8	−2.8	−1.1
M18	MC	1.0	−1.1	−3.4	−3.4	−1.0
M19	MC	1.0	−0.8	−2.8	−2.8	−0.5
M20	MC	1.5	−2.0	−1.8	−1.8	−0.9

Note. DGM = data-generating model; Type = item format; CR = constructed response item; MC = multiple-choice item.

References

1. Lietz, P.; Cresswell, J.C.; Rust, K.F.; Adams, R.J., Eds. *Implementation of Large-Scale Education Assessments*; Wiley: New York, NY, USA, 2017. [CrossRef]
2. Rutkowski, L.; von Davier, M.; Rutkowski, D. (Eds.) *A Handbook of International Large-Scale Assessment: Background, Technical Issues, and Methods of Data Analysis*; Chapman Hall/CRC Press: London, UK, 2013. [CrossRef]
3. Foy, P.; Yin, L. Scaling the PIRLS 2016 achievement data. In *Methods and Procedures in PIRLS 2016*; Martin, M.O., Mullis, I.V., Hooper, M., Eds.; IEA: Chestnut Hill, MA, USA 2017.
4. Foy, P.; Yin, L. Scaling the TIMSS 2015 achievement data. In *Methods and Procedures in TIMSS 2015*; Martin, M.O., Mullis, I.V., Hooper, M., Eds.; IEA: Chestnut Hill, MA, USA 2016.
5. OECD. *PISA 2018. Technical Report*; OECD: Paris, France, 2020. Available online: https://bit.ly/3zWbidA (accessed on 21 June 2023).
6. Pohl, S.; Ulitzsch, E.; von Davier, M. Reframing rankings in educational assessments. *Science* **2021**, *372*, 338–340. [CrossRef] [PubMed]
7. Mislevy, R.J. Missing responses in item response modeling. In *Handbook of Item Response Theory, Vol. 2: Statistical Tools*; van der Linden, W.J., Ed.; CRC Press: Boca Raton, FL, USA, 2016; pp. 171–194. [CrossRef]
8. Bock, R.D.; Moustaki, I. Item response theory in a general framework. In *Handbook of Statistics, Vol. 26: Psychometrics*; Rao, C.R.; Sinharay, S., Eds.; Elsevier: Amsterdam, The Netherlands, 2007; pp. 469–513. [CrossRef]
9. van der Linden, W.J.; Hambleton, R.K. (Eds.). *Handbook of Modern Item Response Theory*; Springer: New York, NY, USA, 1997. [CrossRef]
10. van der Linden, W.J. Unidimensional logistic response models. In *Handbook of Item Response Theory, Volume 1: Models*; van der Linden, W.J., Ed.; CRC Press: Boca Raton, FL, USA, 2016; pp. 11–30. [CrossRef]
11. Rose, N.; von Davier, M.; Nagengast, B. Modeling omitted and not-reached items in IRT models. *Psychometrika* **2017**, *82*, 795–819. [CrossRef] [PubMed]
12. Holman, R.; Glas, C.A.W. Modelling non-ignorable missing-data mechanisms with item response theory models. *Brit. J. Math. Stat. Psychol.* **2005**, *58*, 1–17. [CrossRef]

13. Robitzsch, A. On the treatment of missing item responses in educational large-scale assessment data: An illustrative simulation study and a case study using PISA 2018 mathematics data. *Eur. J. Investig. Health Psychol. Educ.* **2021**, *11*, 1653–1687. [CrossRef]
14. Guo, J.; Xu, X. An IRT-based model for omitted and not-reached items. *arXiv* **2019**, arXiv:1904.03767.
15. Mislevy, R.J.; Wu, P.K. *Missing Responses and IRT Ability Estimation: Omits, Choice, Time Limits, and Adaptive Testing*; (Research Report No. RR-96-30); Educational Testing Service: Princeton, NJ, USA, 1996. [CrossRef]
16. Rosas, G.; Shomer, Y.; Haptonstahl, S.R. No news is news: Nonignorable nonresponse in roll-call data analysis. *Am. J. Political Sci.* **2015**, *59*, 511–528. [CrossRef]
17. Robitzsch, A.; Lüdtke, O. Some thoughts on analytical choices in the scaling model for test scores in international large-scale assessment studies. *Meas. Instrum. Soc. Sci.* **2022**, *4*, 9. [CrossRef]
18. Little, R.J.A.; Rubin, D.B. *Statistical Analysis with Missing Data*; Wiley: New York, NY, USA, 2002. [CrossRef]
19. Rubin, D.B. Inference and missing data. *Biometrika* **1976**, *63*, 581–592. [CrossRef]
20. Seaman, S.; Galati, J.; Jackson, D.; Carlin, J. What is meant by "missing at random"? *Stat. Sci.* **2013**, *28*, 257–268. [CrossRef]
21. Frangakis, C.E.; Rubin, D.B. Addressing complications of intention-to-treat analysis in the combined presence of all-or-none treatment-noncompliance and subsequent missing outcomes. *Biometrika* **1999**, *86*, 365–379. [CrossRef]
22. Harel, O.; Schafer, J.L. Partial and latent ignorability in missing-data problems. *Biometrika* **2009**, *96*, 37–50. [CrossRef]
23. Beesley, L.J.; Taylor, J.M.G.; Little, R.J.A. Sequential imputation for models with latent variables assuming latent ignorability. *Aust. N. Z. J. Stat.* **2019**, *61*, 213–233. [CrossRef]
24. Debeer, D.; Janssen, R.; De Boeck, P. Modeling skipped and not-reached items using IRTrees. *J. Educ. Meas.* **2017**, *54*, 333–363. [CrossRef]
25. Glas, C.A.W.; Pimentel, J.L.; Lamers, S.M.A. Nonignorable data in IRT models: Polytomous responses and response propensity models with covariates. *Psychol. Test Assess. Model.* **2015**, *57*, 523–541.
26. Bartolucci, F.; Montanari, G.E.; Pandolfi, S. Latent ignorability and item selection for nursing home case-mix evaluation. *J. Classif.* **2018**, *35*, 172–193. [CrossRef]
27. Kuha, J.; Katsikatsou, M.; Moustaki, I. Latent variable modelling with non-ignorable item nonresponse: Multigroup response propensity models for cross-national analysis. *J. R. Stat. Soc. Ser. A Stat. Soc.* **2018**, *181*, 1169–1192. [CrossRef]
28. Albert, P.S.; Follmann, D.A. Shared-parameter models. In *Longitudinal Data Analysis*; Fitzmaurice, G., Davidian, M., Verbeke, G., Molenberghs, G., Eds.; Chapman and Hall/CRC: Boca Raton, FL, USA, 2008; pp. 447–466. [CrossRef]
29. Little, R.J. Selection and pattern-mixture models. In *Longitudinal Data Analysis*; Fitzmaurice, G., Davidian, M., Verbeke, G., Molenberghs, G., Eds.; Chapman and Hall/CRC: Boca Raton, FL, USA, 2008; pp. 409–431. [CrossRef]
30. Birnbaum, A. Some latent trait models and their use in inferring an examinee's ability. In *Statistical Theories of Mental Test Scores*; Lord, F.M., Novick, M.R., Eds.; MIT Press: Reading, MA, USA, 1968; pp. 397–479.
31. Deribo, T.; Kroehne, U.; Goldhammer, F. Model-based treatment of rapid guessing. *J. Educ. Meas.* **2021**, *58*, 281–303. [CrossRef]
32. Robitzsch, A.; Lüdtke, O. An item response model for omitted responses in performance tests. Personal communication, 2017.
33. Robitzsch, A. Nonignorable consequences of (partially) ignoring missing item responses: Students omit (constructed response) items due to a lack of knowledge. *Knowledge* **2023**, *3*, 215–231. [CrossRef]
34. Kreitchmann, R.S.; Abad, F.J.; Ponsoda, V. A two-dimensional multiple-choice model accounting for omissions. *Front. Psychol.* **2018**, *9*, 2540. [CrossRef]
35. Rose, N.; von Davier, M.; Nagengast, B. Commonalities and differences in IRT-based methods for nonignorable item nonresponses. *Psych. Test Assess. Model.* **2015**, *57*, 472–498.
36. Köhler, C.; Pohl, S.; Carstensen, C.H. Taking the missing propensity into account when estimating competence scores: Evaluation of item response theory models for nonignorable omissions. *Educ. Psychol. Meas.* **2015**, *75*, 850–874. [CrossRef] [PubMed]
37. Xu, X.; von Davier, M. *Fitting the Structured General Diagnostic Model to NAEP Data*; (Research Report No. RR-08-28); Educational Testing Service: Princeton, NJ, USA, 2008. [CrossRef]
38. Aitkin, M. Expectation maximization algorithm and extensions. In *Handbook of Item Response Theory, Vol. 2: Statistical Tools*; van der Linden, W.J., Ed.; CRC Press: Boca Raton, FL, USA, 2016; pp. 217–236. [CrossRef]
39. Hanson, B. IRT Parameter Estimation Using the EM Algorithm. 2000. Technical Report. Available online: https://bit.ly/3i4pOdg (accessed on 21 June 2023).
40. Battauz, M. Regularized estimation of the four-parameter logistic model. *Psych* **2020**, *2*, 269–278. [CrossRef]
41. Bates, S.; Hastie, T.; Tibshirani, R. Cross-validation: What does it estimate and how well does it do it? *J. Am. Stat. Assoc.* **2023**. [CrossRef]
42. R Core Team. *R: A Language and Environment for Statistical Computing*; R Core Team: Vienna, Austria, 2023. Available online: https://www.R-project.org/ (accessed on 15 March 2023).
43. Robitzsch, A. sirt: Supplementary Item Response Theory Models. R Package Version 3.13-151. 2023. Available online: https://github.com/alexanderrobitzsch/sirt (accessed on 23 April 2023).
44. Lord, F.M. Estimation of latent ability and item parameters when there are omitted responses. *Psychometrika* **1974**, *39*, 247–264. [CrossRef]
45. Hitt, C.; Trivitt, J.; Cheng, A. When you say nothing at all: The predictive power of student effort on surveys. *Econ. Educ. Rev.* **2016**, *52*, 105–119. [CrossRef]

46. Köhler, C.; Pohl, S.; Carstensen, C.H. Investigating mechanisms for missing responses in competence tests. *Psych. Test Assess. Model.* **2015**, *57*, 499–522.
47. Mislevy, R.J. Randomization-based inference about latent variables from complex samples. *Psychometrika* **1991**, *56*, 177–196. [CrossRef]
48. von Davier, M.; Sinharay, S. Analytics in international large-scale assessments: Item response theory and population models. In *A handbook of International Large-Scale Assessment: Background, Technical Issues, and Methods of Data Analysis*; Rutkowski, L., von Davier, M., Rutkowski, D., Eds.; Chapman Hall/CRC Press: London, UK, 2013; pp. 155–174. [CrossRef]
49. Robitzsch, A.; Kiefer, T.; Wu, M. TAM: Test Analysis Modules. R Package Version 4.1-4. 2022. Available online: https://CRAN.R-project.org/package=TAM (accessed on 28 August 2022).
50. Wu, M. The role of plausible values in large-scale surveys. *Stud. Educ. Eval.* **2005**, *31*, 114–128. [CrossRef]
51. von Davier, M. Omitted response treatment using a modified Laplace smoothing for approximate Bayesian inference in item response theory. *PsyArXiv* **2023**. [CrossRef]
52. Gorgun, G.; Bulut, O. A polytomous scoring approach to handle not-reached items in low-stakes assessments. *Educ. Psychol. Meas.* **2021**, *81*, 847–871. [CrossRef]
53. Robitzsch, A. On the choice of the item response model for scaling PISA data: Model selection based on information criteria and quantifying model uncertainty. *Entropy* **2022**, *24*, 760. [CrossRef]

applied sciences

MDPI

Article

Load Forecasting Based on LVMD-DBFCM Load Curve Clustering and the CNN-IVIA-BLSTM Model

Linjing Hu [1], Jiachen Wang [1,*], Zhaoze Guo [2] and Tengda Zheng [1]

1 College of Electric Power, Inner Mongolia University of Technology, Hohhot 010051, China; hulinjing666@163.com (L.H.); www_15838898917@163.com (T.Z.)
2 The Sixth Research Institute of China Aerospace Science and Industry Corporation, Hohhot 010010, China; guozhaoze111@163.com
* Correspondence: wangjiachen103@163.com

Abstract: Power load forecasting plays an important role in power systems, and the accuracy of load forecasting is of vital importance to power system planning as well as economic efficiency. Power load data are nonsmooth, nonlinear time-series and "noisy" data. Traditional load forecasting has low accuracy and curves not fitting the load variation. It is not well predicted by a single forecasting model. In this paper, we propose a novel model based on the combination of data mining and deep learning to improve the prediction accuracy. First, data preprocessing is performed. Second, identification and correction of anomalous data, normalization of continuous sequences, and one-hot encoding of discrete sequences are performed. The load data are decomposed and denoised using the double decomposition modal (LVMD) strategy, the load curves are clustered using the double weighted fuzzy C-means (DBFCM) algorithm, and the typical curves obtained are used as load patterns. In addition, data feature analysis is performed. A convolutional neural network (CNN) is used to extract data features. A bidirectional long short-term memory (BLSTM) network is used for prediction, in which the number of hidden layer neurons, the number of training epochs, the learning rate, the regularization coefficient, and other relevant parameters in the BLSTM network are optimized using the influenza virus immunity optimization algorithm (IVIA). Finally, the historical data of City H from 1 January 2016 to 31 December 2018, are used for load forecasting. The experimental results show that the novel model based on LVMD-DBFCM load c1urve clustering combined with CNN-IVIA-BLSTM proposed in this paper has an error of only 2% for electric load forecasting.

Keywords: power load forecasting; LVMD; DBFCM; CNN; BLSTM; IVIA

Citation: Hu, L.; Wang, J.; Guo, Z.; Zheng, T. Load Forecasting Based on LVMD-DBFCM Load Curve Clustering and the CNN-IVIA-BLSTM Model. *Appl. Sci.* **2023**, *13*, 7332. https://doi.org/10.3390/app13127332

Academic Editor: Seung-Hoon Yoo

Received: 27 April 2023
Revised: 12 June 2023
Accepted: 15 June 2023
Published: 20 June 2023

1. Introduction

As quality of life has improved, the demand for power energy has also risen, and thus, the requirements for power generation and transmission and the use of electricity have increased [1]. Power load forecasting applies machine learning methods to mine the key factors affecting load from historical data, such as weather and time data, to build load forecasting models [2]. Load forecasting is the basis for ensuring the balance of power supply and demand. Accurate forecasting results can reduce the pressure on transmission and distribution links and facilitate the optimal scheduling of power transmission links. They can effectively reduce power generation costs and improve economic and social benefits [3,4].

Traditional forecasting methods use the overall historical load numbers as their basis [5–9]. The trend extrapolation predicts future loads based on the historical trends of the predictor variables. It predicts near future loads with high accuracy, but error gradually increases with time. The regression analysis method predicts future load changes by establishing a functional relationship between variables. It has high fitting ability but has a large error in predicting future changes. Time series extrapolation (the autoregressive moving average model) approximates a nonlinear relationship with a linear equation [10]. The grey forecasting method performs correlation analysis by identifying the degree of dissimilarity in development trends between system factors. Traditional forecasting models have a relatively simple structure and require highly accurate historical data, resulting in forecasting results with poor accuracy.

In recent decades, deep learning has made important breakthroughs in the field of artificial intelligence. In 1957, Frank Rosenblatt invented an artificial neural network to solve complex problems by means of a multilayer perceptron (MLP). Deep neural networks (DNNs) are currently a popular topic in load prediction research, and recurrent neural networks (RNNs) have advantages in memory parameter sharing and Turing completeness [11]. Long short-term memory (LSTM) networks overcome the problem that RNNs are prone to gradient disappearance/explosion, but their ability to correct errors is weak [12]. M. Schuster proposed a BLSTM network that can make full use of the hidden layer history state and has stronger robustness [13]. Yet, BLSTM networks have many parameters, and without proper optimization, the model may be overfitted or slow to train, resulting in inefficiency. The number of hidden layer neurons, the number of training epochs, the learning rate, and the regularization coefficient in the BLSTM network are optimized using the IVIA. It can improve the model's efficiency. The challenges of the machine learning algorithms to load forecasting for power systems, with complex, nonsmooth, nonlinear time-series and "noisy" data, will decrease the accuracy of the algorithm, and a single prediction model has difficulty meeting the accuracy requirements. Integrating a hybrid preprocessing method can significantly improve the prediction accuracy [14–17].

In summary, to improve the accuracy and meet the needs of practical problems, this paper proposes a new model based on data mining and deep learning. It considers not only historical load data but also weather information, date types, real-time electricity prices, etc. Historical data load, weather information, and real-time electricity prices are normalized, and the date and holiday information are expressed through One-Hot encoding. The DBFCM algorithm is applied to load curve clustering to overcome the problems of traditional C-means clustering and fuzzy C-means clustering in which data are clustered into small classes, inaccurate classification is performed, and the "uniform effect of the cluster size" degrades performance. First, the LVMD is used to decompose and denoise the input data to improve the continuity and stability of the data. Second, the DBFCM algorithm is used to perform load curve clustering. Third, feature fusion extraction is performed using CNN. In addition, the extracted feature vectors are used as the input of IVIA-BLSTM. The IVIA is proposed to optimize the parameters in the BLSTM network. Finally, this paper proposes a new load forecasting model based on the LVMD-DBFCM algorithm and CNN-IVIA-BLSTM.

The remainder of this article is organized as follows. Section 2 describes the IVIA algorithm, Section 3 presents the methodology used in this paper, Section 4 is a case study of this paper, and Section 5 concludes the paper.

2. Influenza Virus Immunity Optimization Algorithm

The IVIA is a new metaheuristic optimization algorithm that is inspired by the way influenza viruses spread in a population and the process of population immunity. The process of an individual being infected is divided into three states: uninfected, infected, and immune, and the population has herd immunity when the number of immune individuals in the population exceeds 80% of the total population. This section examines the IVIA with 23 sets of standard test functions and compares it with seven optimization algorithms proposed in recent years. They are turbulent flow of water-based optimization (TFWO) [18], golden eagle optimization (GEO) [19], the parasitism–predation algorithm (PPA) [20], the rat swarm optimizer (RSO) [21], gray wolf optimization (GWO) [22], particle swarm optimization (PSO) [23], and whale optimization algorithm (WOA) [24]. The results of the tested functions show that the IVIA converges faster with higher accuracy, optimizes the function better, and finds the optimal value of the function in fewer iterations. Then, the IVIA is applied to the neural network BLSTM training of the power load forecasting model to optimize the number of hidden layer neurons, the number of training epochs, the learning rate and the regularization coefficient, and other relevant parameters to improve the accuracy of the forecasting model.

2.1. Biological Characteristics

The influenza virus is very contagious, spreads quickly, and has many routes of transmission [25]. The spread of the virus in the population can be achieved through droplet transmission and physical contact, and the rate of spread is related to the social distance between people [26]. The vast majority of the population can recover on their own after infection, and those who recover produce the corresponding antibodies in their bodies and have immune functions. A small number of people with underlying diseases, that is, elderly people or those with weak resistance, are at risk. When the number of immune individuals in the population exceeds 80%, the population has herd immunity, thus preventing the next round of disease transmission. A schematic diagram of the transmission process of the influenza virus in a population is shown in Figure 1.

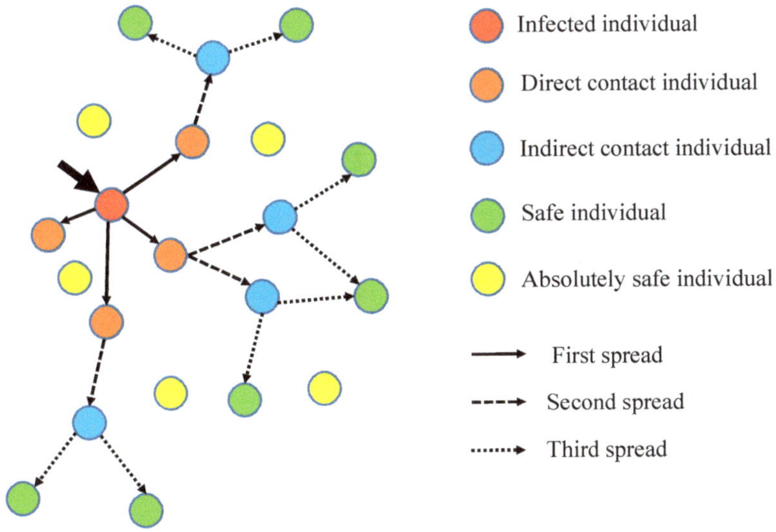

Figure 1. Diagram of the process of influenza virus population transmission.

The population is divided into infected individuals, highly susceptible individuals, susceptible individuals, safe individuals, and absolutely safe individuals.

Extremely susceptible individuals are direct contacts of infected individuals, who are socially close to infected individuals and are infected by getting influenza virus from infected individuals; susceptible individuals are indirect contacts of infected individuals who are socially distant from the infected individuals, may be infected through third parties as vectors, and have a higher risk of infection; safe individuals are not in contact with infected individuals or other individuals, and because the influenza virus is transmitted as a vector, through a change in social distance, safe individuals may become susceptible individuals or infected individuals; absolutely safe individuals have immune functions through vaccination or healing after infection and are not susceptible to infection, regardless of the contact distance of infected individuals. In this paper, the influenza virus immunization algorithm is proposed through the transmission of the influenza virus in a population and the immunization mechanism. To map the spread process of the influenza virus with the optimization algorithm, the following assumptions are made:

1. The initial number of infected individuals is small, and they are randomly distributed in the population. The category to which an individual currently belongs and the corresponding update formula are determined based on the distance between the individual and the infected individual;
2. The overall number of the population is kept constant, and absolutely safe individuals randomly move to different locations in the population. The number and location of other individuals change dynamically with the number of iterations;
3. Individuals in the population have three states: infected, uninfected, and immune. Infected individuals have immunity after infecting other individuals and will not be infected in the following process. A population is immune when 80% or more of the individuals in the population are immune;
4. Individuals are infected by contracting virus cells from an infected person, and the process of virus cell exchange is shown in Figure 2. The influenza virus cells and the diseased cells are taken as the smallest units within an individual to represent the dimension of the optimization problem.

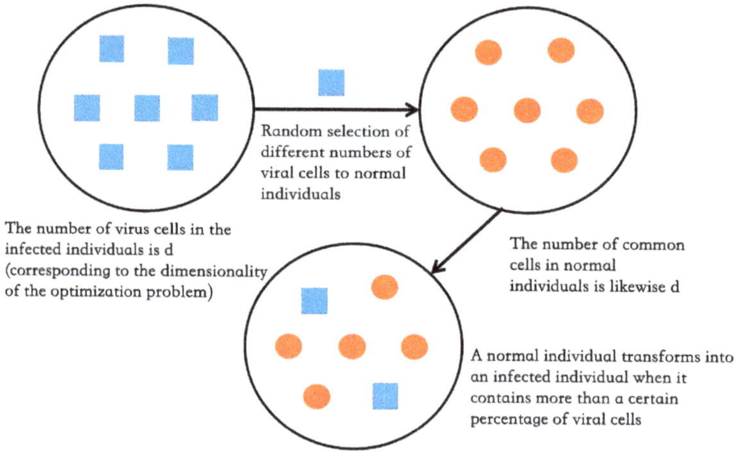

Figure 2. Diagram of the virus cell exchange process.

Immunized individuals: The individuals who are categorized as immunized are protected against the virus, and they are not affected by infected individuals.

Figure 3 shows the proportion of immune individuals to the entire population with the 23 sets of standard test functions run by IVIA. From the figure, except for functions F4, F7, F9, F16, F21, and F23, all the remaining 17 tested functions stop the iterative search process when the population immunity rate reaches 0.8. In this paper, the minimum population immunity rate for IVIA is set at 0.8, i.e., the search for optimization stops when the number of immunized individuals in the population exceeds 80% of the total. It ensures that the algorithm has high optimization results.

In this paper, we propose the IVIA based on the transmission and immunization process of the influenza virus. The influenza virus spreads rapidly in a population, and the growth curve of the cumulative number of infections is shown in Figure 4. Curve ① indicates that the number of infected individuals in the population increases rapidly within 2~3 weeks, and curves ② and ③ indicate that the number of infected individuals peaks at 7~8 weeks and then gradually starts to decline. Curve ④ indicates that the number of infected individuals gradually decreases and the number of immunized individuals increases near the end of this influenza outbreak. The IVIA has a good optimization capability at the beginning of the iteration. Figure 4 is a schematic diagram summarized according to the epidemic transmission rules; the website of the epidemic transmission rules is shown below.

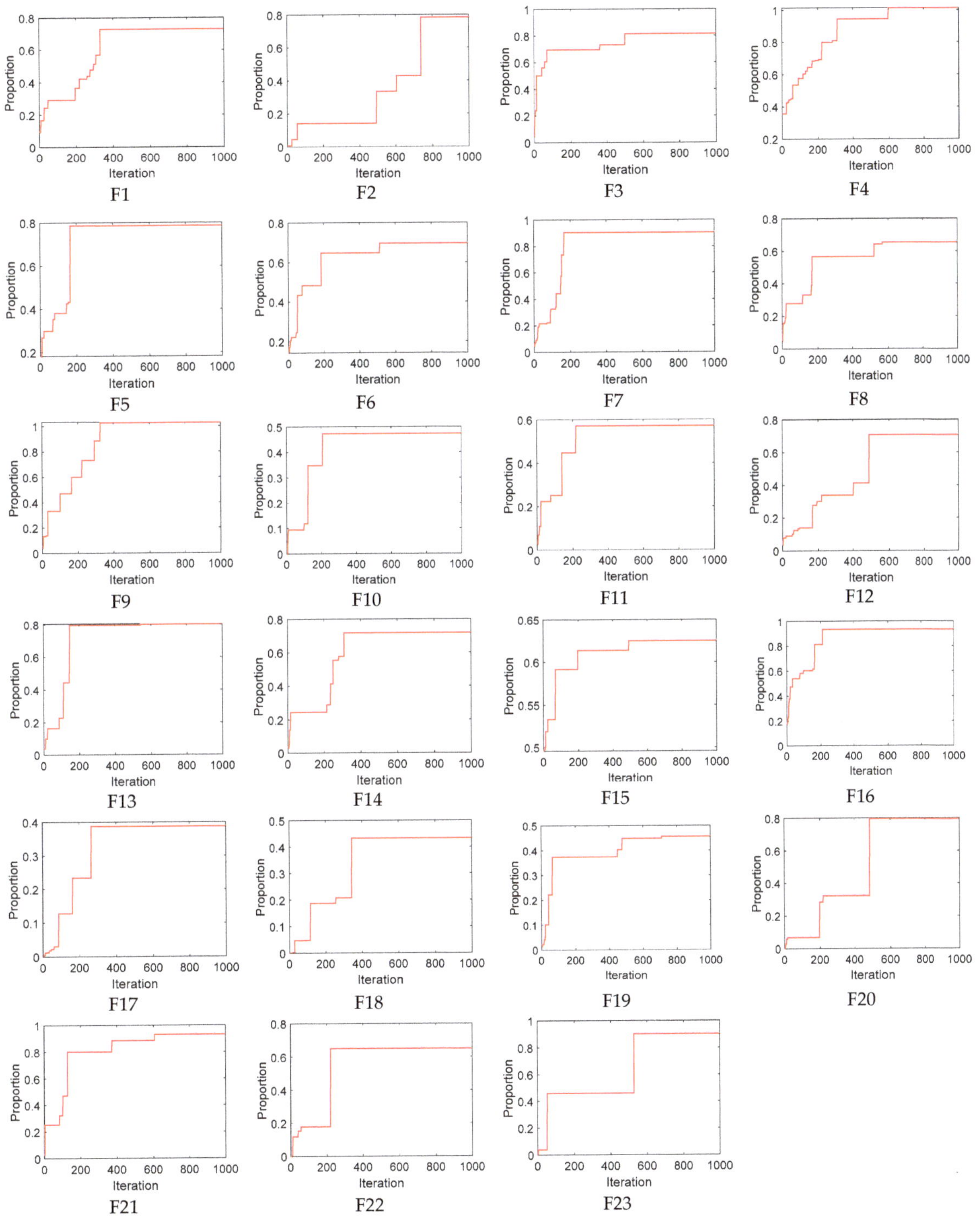

Figure 3. The proportion of immune individuals in the whole population.

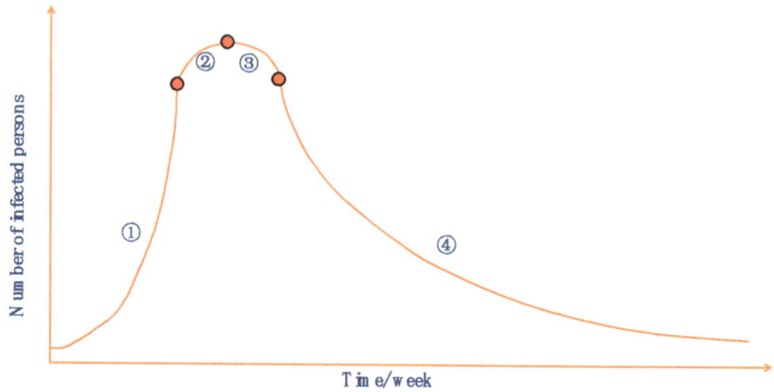

Figure 4. Growth curve of the number of infected people.

https://www.zhzyw.com/crb/gm/08521151042985F4176E7H4KEC0B47A.html (accessed on 12 June 2023).

2.2. Mathematical Model

The process of influenza virus transmission in a population is mapped with the influenza virus immunization algorithm. The population is represented by the matrix X, as shown in Equation (1), where n is the population size and the number of individuals, and d is the number of cells invaded by the influenza virus into the human body, corresponding to the dimensionality of the objective function. The initialization process of the algorithm randomly generates an $n \times d$ dimension matrix X, which sets the initial number of infected individuals to x.

$$X = \begin{bmatrix} x_{11} & x_{12} & \cdots & x_{1d} \\ x_{21} & x_{22} & \cdots & x_{2d} \\ \vdots & \vdots & \ddots & \vdots \\ x_{n1} & x_{n2} & \cdots & x_{nd} \end{bmatrix} \tag{1}$$

The fitness values of different individuals in the population are represented by Equation (2). Each row in the matrix is represented as the fitness value of the current individual. The individual corresponding to the optimal fitness value is the current optimal solution of the objective function.

$$F(x) = \begin{bmatrix} f(x_{11}, x_{12}, \cdots, x_{1d}) \\ f(x_{21}, x_{22}, \cdots, x_{2d}) \\ \vdots \\ f(x_{n1}, x_{n2}, \cdots, x_{nd}) \end{bmatrix} \tag{2}$$

The distance between individuals is represented by C, as shown in Equation (3). $x = \{x_1, x_2, \ldots, x_d\}$ is the position of individual x, $y = \{y_1, y_2, \ldots, y_d\}$ is the position of individual y, and d is the dimension of the optimization problem.

$$C = \sqrt{(x_1 - y_1)^2 + (x_2 - y_2)^2 + \ldots + (x_d - y_d)^2} \tag{3}$$

L is the maximum safety distance, as shown in Equation (4). $(L_{\min}, L_{\max})$ represents the range of values for individuals.

$$L = \frac{1}{2}\sqrt{L_{\min}^2 + L_{\max}^2} \tag{4}$$

The virus spreading process corresponds to the iterative optimization-seeking process in the algorithm. It is assumed that the safe distance between other individuals and the infected person when the influenza virus spreads in the population is L. The iterative formulas for individuals in different cases are shown in Equations (5) to (9).

- Infected individuals are updated by Equation (5):

$$x_{i,j}(t+1) = x_{i,j}(t)\left(\exp\left(\frac{N_{\max}-n}{N_{\max}}-1\right)\times\frac{1}{d-1}\right)^{q} \tag{5}$$

When the contact distance between individuals in the population and infected individuals is $C < 0.2L$, the individuals are moved into the infected state. The position of infected individuals is updated by Equation (5). $N_{\max}$ is the maximum number of iterations, n is the current number of iterations, d is the dimensionality of the optimization problem, and q is used to regulate the individual position update rate. The larger the value is of q, the faster the position update rate and the higher the accuracy of the optimization search.

- Extremely susceptible individuals are updated by Equation (6):

$$\begin{gathered} x_{i,j}(t+1) = x_{i,j}(t) + \alpha\left|x_{i,j}(t) - x_{i,q}(t)\right| \\ \alpha = \tfrac{1}{d}\sum_{p=1}^{d} rand(-1,1) \end{gathered} \tag{6}$$

When the contact distance between individuals in the population and infected individuals is $0.2L < C < 0.5L$, the position is updated by Equation (6). $x_{i,q}(t)$ is the random acquisition of viral cells by highly susceptible individuals; α represents update coefficient at the contact distance of $0.2L < C < 0.5L$. When the number of exchanged cells exceeds a quarter of the total number of cells, the highly susceptible individuals are transformed into infected individuals. Cell exchange occurs only once during each iteration, and the number of exchanges is random.

- Susceptible individuals are updated by Equation (7):

$$\begin{gathered} x_{i,j}(t+1) = x_{i,j}(t) + \beta\left|x_{i,j}(t) - x_{i,v}(t)\right| \\ \beta = \exp\left(\tfrac{C-L}{\varepsilon\times N_{\max}}\right) \\ \varepsilon = rand(0,1) \end{gathered} \tag{7}$$

When the contact distance between individuals in the population and infected individuals is $0.5L < C < 0.8L$, the position is updated by Equation (7). $x_{i,v}(t)$ is a random acquisition of viral cells by highly susceptible individuals, β represents update coefficient at the contact distance of $0.5L < C < 0.8L$, ε is a random number in the range 0~1. When the number of exchanged cells has exceeded one-third of the total number of cells, the highly susceptible individuals are transformed into infected individuals.

- Secure individuals updated by Equation (8):

$$\begin{gathered} x_{i,j}(t+1) = x_{i,j}(t) + \gamma\left|x_{i,j}(t) - x_{i,w}(t)\right| \\ f(x_{i,w}) = \operatorname{argmin} f(x_{i,j}) \\ \gamma = rand(-1,1) \end{gathered} \tag{8}$$

When the contact distance between individuals in the population and infected individuals is $0.8L < C < L$, the position is updated by Equation (8). γ is a random number in the range -1~1. There is no cellular exchange between safe individuals and infected individuals. However, as the number of infected individuals increases, the contact distance between safe individuals and infected individuals may decrease, and safe individuals may become susceptible.

- Absolutely safe individuals updated by Equation (9):

$$\begin{aligned} x_{i,j}(t+1) &= \lambda x_{i,j}(t) \\ \lambda &= \gamma \times \left(\frac{\sqrt{2}-1}{2}\right) L \end{aligned} \tag{9}$$

Absolutely safe individuals are immune, and the contact distance between individuals and infected individuals is $C > L$; λ represents the update coefficient at the contact distance of $C > L$. They randomly change their positions during the iterative process and can be used to escape the current search region when the algorithm falls into local optima.

Stop criterion IVIA: The algorithm terminates when the number of immune individuals in the population exceeds more than 80% of the population or when the maximum number of iterations is reached. Therefore, the individual with the highest population immunity corresponds to the optimal solution of the optimization problem, and the corresponding function value is the optimal value of the optimization problem. If the problem to be solved requires higher accuracy, the population immunity rate in the Algorithm 1 can be increased. The pseudo-code of IVIA is presented below.

Algorithm 1 IVIA pseudo-code

Input:
n: The number of people
$N_{\max}$: The maximum number of iterations
m: Initial number of infected individuals (usually set to 1)
L: Maximum safe contact distance
R: Herd immunity ratio
C: The distance between individuals
lb,*ub*: Search boundary
Population initialization
Setting parameters
while ($t < N$)
    Calculate fitness value and sort
    for $i = 1{:}n$
        if ($C < 0.2L$) then
            Using Equation (5) to update the location of the infected individuals
            Record the current status of the individuals
        else if ($0.2L < C < 0.5L$) then
            Using Equation (6) to update the location of highly susceptible individuals
            Record the current status of the individuals
        else if ($0.5L < C < 0.8L$) then
            Using Equation (7) to update the location of susceptible individuals
            Record the current status of the individuals
        else if ($0.8L < C < L$) then
            Using Formula (8) to update the position of a safe individuals
            Record the current status of the individual
        else
            Using Equation (9) to update the position of an absolutely safe individuals
        end if
        Recalculate fitness value
    end for
    Calculate immunity rate
    $t = t + 1$
end while

2.3. Algorithm Testing

To test the actual optimization effect of the proposed IVIA, it was evaluated in this paper on 23 sets of standard test functions. The optimization results were compared with those of turbulent flow of water-based optimization (TFWO), golden eagle optimization (GEO), the parasitism–predation algorithm (PPA), the rat swarm optimizer (RSO), gray wolf optimization (GWO), particle swarm optimization (PSO), and the whale optimization algorithm (WOA).

Table 1 shows the expression of the test function, the dimension of the optimization problem, and the overall search range, and the last column shows the minimum value of the test function expectation, which is the theoretical optimum. The IVIA population size n is 30, the maximum number of iterations $N_{\max}$ is 1000, the maximum safe contact distance L is $0.5ub$, and the population immunity rate R is 80% during the test period. The number of dimensions of the test function and the parameters, such as search intervals lb and ub, are set according to Table 1. The overall and iteration numbers of the other optimization functions are the same as those of the IVIA.

Table 1. Standard test function.

Standard	Dimension	Search Space	Minimum
$F_1(x) = \sum_{i=1}^{n} x_i^2$	30	[−100, 100]	0
$F_2(x) = \sum_{i=1}^{n} \lvert x_i \rvert + \prod_{i=1}^{n} \lvert x_i \rvert$	30	[−10, 10]	0
$F_3(x) = \sum_{i=1}^{n} \left(\sum_{j=1}^{n} x_j \right)^2$	30	[−100, 100]	0
$F_4(x) = \max_i\{\lvert x_i \rvert, 1 \le i \le n\}$	30	[−100, 100]	0
$F_5(x) = \sum_{i=1}^{n-1} \left[100(x_{i+1} - x_i^2) + (x_i - 1)^2 \right]$	30	[−30, 30]	0
$F_6(x) = \sum_{i=1}^{n} ([x_i + 0.5])^2$	30	[−100, 100]	0
$F_7(x) = \sum_{i=1}^{n} i x_i^4 + random[0, 1)$	30	[−128, 128]	0
$F_8(x) = \sum_{i=1}^{n} -x_i \sin\left(\sqrt{\lvert x_i \rvert} \right)$	30	[−500, 500]	−12,569.5
$F_9(x) = \sum_{i=1}^{n} \left[x_i^2 - 10\cos(2\pi x_i) + 10 \right]$	30	[−5.12, 5.12]	0
$F_{10}(x) = -20\exp\left(-0.2\sqrt{\frac{1}{n}\sum_{i=1}^{n} x_i^2} \right) - \exp\left(\frac{1}{n}\sum_{i=1}^{n} \cos(2\pi x_i) \right) + 20 + e$	30	[−32, 32]	0
$F_{11}(x) = \frac{1}{4000}\sum_{i=1}^{n} x_i^2 - \prod_{i=1}^{n} \cos\left(\frac{x_i}{\sqrt{i}} \right) + 1$	30	[−600, 600]	0
$F_{12}(x) = \frac{\pi}{n}\left\{ 10\sin(\pi y_1) + \sum_{i=1}^{n-1} (y_i - 1)^2 \left[1 + 10\sin^2(\pi y_{i+1}) \right] + (y_n - 1)^2 \right\} + \sum_{i=1}^{n} u(x_i, 10, 100, 4)$ $y_i = 1 + \frac{x_i + 1}{4} u(x_i, a, k, m) = \begin{cases} k(x_i - a)^m & x_i > a \\ 0 & -a < x_i < a \\ k(-x_i - a)^m & x_i < a \end{cases}$	30	[−50, 50]	0
$F_{13}(x) = 0.1\left\{ \sin^2(3\pi x_i) + \sum_{i=1}^{n} \begin{matrix} (x_i - 1)^2 \left[1 + \sin^2(3\pi x_i + 1) \right] + \\ (x_n - 1)^2 \left[1 + \sin^2(2\pi x_n) \right] \end{matrix} \right\} + \sum_{i=1}^{n} u(x_i, 5, 100, 4)$	30	[−50, 50]	0
$F_{14}(x) = \left[\frac{1}{500} + \sum_{j=1}^{25} \frac{1}{j + \sum_{i=1}^{2} (x_i - a_{ij})^6} \right]^{-1}$	2	[−65.56, 65.56]	1

Table 1. *Cont.*

Standard	Dimension	Search Space	Minimum
$F_{15}(x)=\sum\limits_{i=1}^{11}\left[a_i-\frac{x_1(b_i^2+b_i x_2)}{b_i^2+b_i x_3+x_4}\right]^2$	4	[−5, 5]	0.0003075
$F_{16}(x)=4x_1^2-2.1x_1^4+\frac{1}{3}x_1^6+x_1x_2-4x_2^2+4x_2^4$	2	[−5, 5]	−1.0316
$F_{17}(x)=\left(x_2-\frac{5.1}{4\pi^2}x_1^2+\frac{5}{\pi}x_1-6\right)^2+10\left(1-\frac{1}{8\pi}\right)\cos x_1+10$	2	[−5, 5]	0.398
$F_{18}(x)=\left[1+(x_1+x_2+1)^2(19-14x_1+3x_1^2-14x_2+6x_1x_2+3x_2^2)\right]\times\left[30+(2x_1-3x_2)^2(18-32x_1+12x_1^2+48x_2-36x_1x_2+27x_2^2)\right]$	2	[−2, 2]	3
$F_{19}(x)=-\sum\limits_{i=1}^{4}c_i\exp\left(-\sum\limits_{j=1}^{3}a_{ij}(x_j-p_{ij})^2\right)$	3	[1, 3]	−3.86
$F_{20}(x)=-\sum\limits_{i=1}^{4}c_i\exp\left(-\sum\limits_{j=1}^{6}a_{ij}(x_j-p_{ij})^2\right)$	6	[0, 1]	−3.32
$F_{21}(x)=-\sum\limits_{i=1}^{10}\left(\sum\limits_{j=1}^{5}\left(x_j-C_{ji}\right)^2+\beta_i\right)^{-1}$ $\beta=\frac{1}{10}(1,\ 2,\ 2,\ 4,\ 4,\ 6,\ 3,\ 7,\ 5,\ 5)$ $C=\begin{bmatrix}4.0 & 1.0 & 8.0 & 6.0 & 3.0 & 2.0 & 5.0 & 8.0 & 6.0 & 7.0\\ 4.0 & 1.0 & 8.0 & 6.0 & 7.0 & 9.0 & 3.0 & 1.0 & 2.0 & 3.6\\ 4.0 & 1.0 & 8.0 & 6.0 & 3.0 & 2.0 & 5.0 & 8.0 & 6.0 & 7.0\\ 4.0 & 1.0 & 8.0 & 6.0 & 7.0 & 9.0 & 3.0 & 1.0 & 2.0 & 3.6\end{bmatrix}$	4	[0, 10]	−10.1532
$F_{22}(x)=-\sum\limits_{i=1}^{7}\left(\sum\limits_{j=1}^{5}\left(x_j-C_{ji}\right)^2+\beta_i\right)^{-1}$	4	[0, 10]	−10.4028
$F_{23}(x)=-\sum\limits_{i=1}^{10}\left(\sum\limits_{j=1}^{4}\left(x_j-C_{ji}\right)^2+\beta_i\right)^{-1}$	4	[0, 10]	−10.5364

The viability of the proposed IVIA is tested using 23 well-known benchmark functions with different sizes and complexity. Figure 5 shows the three-dimensional graphs of the standard test functions and their corresponding test results. The convergence of the IVIA is very good for both single-peaked and multipeaked functions.

From the test results in Table 2, the accuracy of the IVIA for function finding is much higher than that of several other optimization algorithms, and the theoretical optimal value can be found for some functions. IVIA has fewer iterations, a shorter running time, and faster convergence compared to the other seven algorithms.

For comparative evaluation, the proposed IVIA is compared against seven well-established comparative methods using the same benchmark functions. IVIA optimizes best in F1, F2, F3, F4, F5, F6, F7, F8, converges fastest in F11, F12, F13, F14, F15, F17, and it has the same effect as other algorithms in F18, F19, F20, F21, F22, F23. The comparative results show the effectiveness of IVIA.

In summary, the IVIA is based on the characteristics of fast convergence, better robustness, and good optimization search. In this paper, the IVIA is selected to optimize the relevant parameters, such as the number of neurons in the hidden layer, the number of training times, the learning rate, and the regularization coefficients in the BLSTM network.

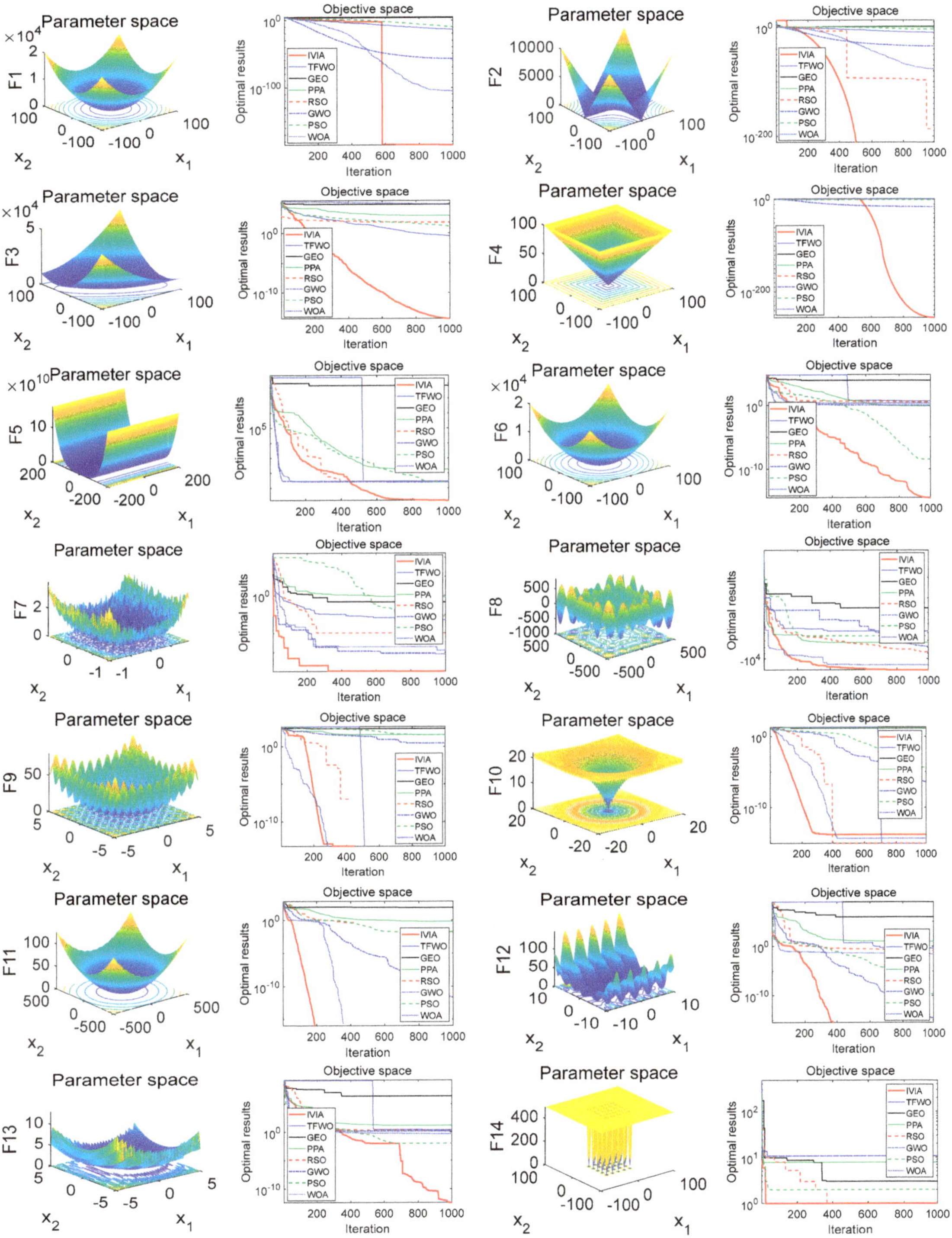

Figure 5. *Cont.*

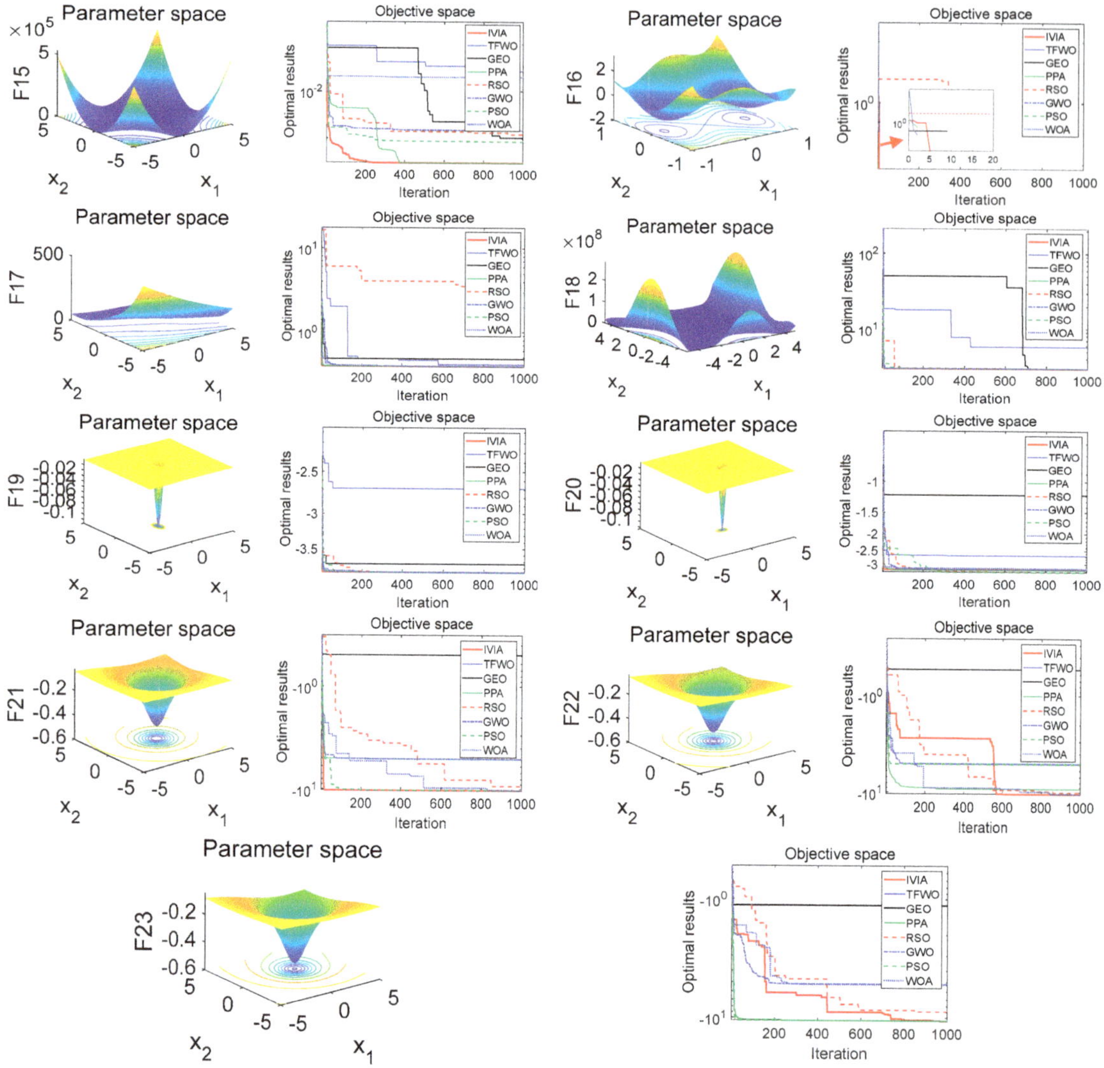

Figure 5. Standard test function optimization result.

Table 2. Results of the testing functions optimized by different algorithms.

F		IVIA	TFWO	GEO	PPA	RSO	GWO	PSO	WOA
	optimal	**0.000×10^{0}**	0.000×10^{0}	5.988×10^{-6}	0.000×10^{0}	0.000×10^{0}	0.000×10^{0}	8.428×10^{-5}	0.000×10^{0}
F1	average	**1.820×10^{-248}**	4.900×10^{-1}	2.136×10^{-1}	1.598×10^{-6}	9.458×10^{-102}	1.985×10^{-47}	6.789×10^{-12}	9.486×10^{-71}
	worst	**1.245×10^{-162}**	3.500×10^{-11}	5.121×10^{-1}	2.185×10^{-1}	1.169×10^{-13}	8.942×10^{-10}	8.628×10^{-18}	6.839×10^{-27}
	optimal	**0.000×10^{0}**	0.000×10^{0}	0.000×10^{0}	0.000×10^{0}	0.000×10^{0}	6.843×10^{-36}	6.880×10^{-10}	1.642×10^{-104}
F2	average	**9.840×10^{-53}**	1.249×10^{-8}	4.878×10^{-6}	9.785×10^{-7}	1.854×10^{-36}	6.842×10^{-13}	6.874×10^{-7}	4.859×10^{-47}
	worst	**2.350×10^{-30}**	1.895×10^{-6}	1.236×10^{-4}	8.545×10^{-5}	1.561×10^{-18}	6.842×10^{-6}	6.420×10^{-4}	6.465×10^{-28}
	optimal	**0.000×10^{0}**	2.855×10^{-1}	1.565×10^{1}	1.585×10^{-7}	1.265×10^{-6}	6.872×10^{-3}	4.265×10^{-2}	1.063×10^{3}
F3	average	**4.185×10^{-11}**	1.522×10^{0}	5.016×10^{2}	1.855×10^{-5}	1.295×10^{-6}	3.000×10^{1}	2.326×10^{1}	3.064×10^{3}
	worst	**1.855×10^{-8}**	5.489×10^{0}	1.000×10^{5}	1.855×10^{-3}	1.486×10^{-3}	1.014×10^{2}	6.019×10^{1}	1.036×10^{4}
	optimal	**9.459×10^{-77}**	1.125×10^{0}	2.153×10^{0}	2.126×10^{0}	3.165×10^{-3}	2.894×10^{-24}	8.486×10^{-2}	3.492×10^{-1}
F4	average	**5.895×10^{-34}**	2.459×10^{0}	4.153×10^{0}	3.124×10^{0}	2.989×10^{0}	1.987×10^{-14}	1.425×10^{0}	9.482×10^{-1}
	worst	**2.349×10^{-19}**	3.157×10^{0}	8.166×10^{0}	5.156×10^{0}	6.362×10^{0}	6.850×10^{-7}	6.895×10^{0}	3.843×10^{0}
	optimal	**9.646×10^{-10}**	2.360×10^{-1}	1.465×10^{1}	1.946×10^{0}	1.235×10^{0}	2.169×10^{0}	9.892×10^{-1}	1.979×10^{0}
F5	average	**5.419×10^{-8}**	6.523×10^{0}	5.988×10^{2}	1.249×10^{1}	1.565×10^{1}	1.249×10^{1}	1.468×10^{1}	9.162×10^{0}
	worst	**1.242×10^{-2}**	2.104×10^{1}	1.275×10^{4}	1.991×10^{1}	1.035×10^{2}	5.616×10^{1}	2.317×10^{1}	3.216×10^{1}
	optimal	**2.355×10^{-18}**	3.985×10^{-4}	8.852×10^{-2}	1.289×10^{-3}	1.965×10^{-6}	3.687×10^{-1}	7.985×10^{-10}	6.485×10^{-2}
F6	average	**9.475×10^{-10}**	9.125×10^{-1}	8.049×10^{-2}	1.597×10^{-1}	1.547×10^{-1}	9.006×10^{-1}	7.864×10^{-7}	9.717×10^{-1}
	worst	**9.855×10^{-5}**	1.836×10^{0}	3.598×10^{0}	1.099×10^{0}	6.218×10^{-1}	2.663×10^{0}	6.843×10^{-2}	2.097×10^{0}
	optimal	**1.968×10^{-7}**	0.000×10^{0}	3.942×10^{-2}	3.975×10^{-6}	4.885×10^{-3}	3.946×10^{-3}	9.847×10^{-5}	3.550×10^{-5}
F7	average	**2.495×10^{-3}**	1.698×10^{-1}	3.965×10^{-1}	3.735×10^{-1}	1.125×10^{-2}	8.972×10^{-1}	8.885×10^{-1}	4.165×10^{-1}
	worst	**6.515×10^{-1}**	1.317×10^{0}	1.569×10^{0}	1.032×10^{0}	3.339×10^{-1}	1.398×10^{0}	1.763×10^{0}	1.223×10^{0}
	optimal	**-1.256×10^{4}**	-1.024×10^{4}	-5.663×10^{3}	-1.124×10^{4}	-1.035×10^{4}	-9.176×10^{3}	-1.034×10^{4}	-1.224×10^{4}
F8	average	**-1.156×10^{4}**	-9.563×10^{3}	-4.983×10^{3}	-8.832×10^{3}	-8.865×10^{3}	-5.863×10^{3}	-6.856×10^{3}	-8.936×10^{3}
	worst	**-1.055×10^{4}**	-7.934×10^{3}	-3.527×10^{3}	-7.295×10^{3}	-5.235×10^{3}	-4.330×10^{3}	-5.845×10^{3}	-7.532×10^{3}
	optimal	**0.000×10^{0}**	4.946×10^{-6}	9.533×10^{-7}	0.000×10^{0}	1.968×10^{-22}	6.427×10^{-3}	6.848×10^{-8}	7.842×10^{-28}
F9	average	**5.649×10^{-5}**	1.568×10^{-3}	3.798×10^{-3}	1.765×10^{-3}	1.765×10^{-6}	3.499×10^{-5}	3.492×10^{-4}	9.787×10^{-9}
	worst	**3.476×10^{-2}**	1.065×10^{0}	9.833×10^{-1}	3.177×10^{-2}	1.685×10^{-1}	1.297×10^{0}	1.379×10^{-1}	3.790×10^{-3}
	optimal	**9.852×10^{-21}**	4.682×10^{-11}	1.649×10^{-4}	6.842×10^{-8}	9.845×10^{-6}	5.400×10^{-16}	6.463×10^{-5}	6.842×10^{-15}
F10	average	**1.855×10^{-8}**	4.190×10^{-3}	1.515×10^{-1}	6.850×10^{-3}	9.815×10^{-2}	8.943×10^{-3}	3.163×10^{-3}	8.463×10^{-8}
	worst	**1.642×10^{-5}**	1.684×10^{-1}	1.331×10^{0}	1.146×10^{0}	3.855×10^{-1}	6.315×10^{-1}	9.198×10^{-1}	6.318×10^{-3}
	optimal	**5.989×10^{-9}**	9.847×10^{-5}	5.646×10^{-3}	1.685×10^{-7}	9.946×10^{-5}	3.489×10^{-2}	3.965×10^{-4}	3.787×10^{-9}
F11	average	**3.500×10^{-8}**	3.490×10^{-1}	3.654×10^{1}	4.846×10^{-4}	1.654×10^{-3}	3.326×10^{-1}	1.038×10^{0}	9.779×10^{-5}
	worst	**1.985×10^{-4}**	2.165×10^{0}	1.068×10^{2}	3.168×10^{0}	5.486×10^{-1}	1.235×10^{0}	1.380×10^{0}	3.480×10^{-3}
	optimal	**5.648×10^{-12}**	2.165×10^{-6}	6.846×10^{1}	6.546×10^{-7}	6.468×10^{-7}	4.893×10^{-8}	4.896×10^{-13}	1.687×10^{-3}

Table 2. *Cont.*

F		IVIA	TFWO	GEO	PPA	RSO	GWO	PSO	WOA
F12	average	**6.942×10^{-8}**	1.656×10^{-3}	6.845×10^{2}	8.646×10^{-5}	6.579×10^{-4}	7.893×10^{8}	4.987×10^{-10}	6.447×10^{-1}
	worst	**1.685×10^{-4}**	9.924×10^{-2}	1.007×10^{4}	1.634×10^{-1}	6.846×10^{-2}	3.589×10^{-3}	4.983×10^{-7}	1.349×10^{0}
	optimal	**6.115×10^{-8}**	4.216×10^{-5}	2.647×10^{1}	5.242×10^{-2}	5.464×10^{-4}	4.685×10^{-4}	4.165×10^{-7}	4.198×10^{-8}
F13	average	**4.359×10^{-20}**	6.410×10^{-3}	1.464×10^{2}	5.825×10^{0}	1.351×10^{-2}	3.199×10^{-3}	4.147×10^{-4}	4.168×10^{-6}
	worst	**4.642×10^{-8}**	1.416×10^{0}	1.001×10^{4}	2.353×10^{1}	3.861×10^{0}	2.056×10^{0}	1.325×10^{0}	9.650×10^{-3}
	optimal	**1.000×10^{0}**	1.000×10^{0}	1.001×10^{0}	1.024×10^{0}	1.115×10^{0}	1.259×10^{0}	1.000×10^{0}	1.000×10^{0}
F14	average	**1.001×10^{0}**	1.674×10^{0}	1.009×10^{0}	2.196×10^{0}	1.351×10^{0}	1.950×10^{0}	1.350×10^{0}	1.268×10^{0}
	worst	**1.007×10^{0}**	2.419×10^{0}	2.486×10^{0}	1.117×10^{1}	2.652×10^{0}	1.286×10^{1}	2.149×10^{0}	1.635×10^{8}
	optimal	**3.000×10^{-4}**	5.419×10^{-4}	1.684×10^{-3}	6.554×10^{-4}	3.948×10^{-4}	9.157×10^{-4}	9.165×10^{-3}	3.917×10^{-4}
F15	average	**3.425×10^{-4}**	8.646×10^{-3}	1.682×10^{-3}	6.834×10^{-3}	6.422×10^{-2}	2.925×10^{-3}	6.117×10^{-3}	9.164×10^{-3}
	worst	**6.453×10^{-3}**	6.550×10^{-1}	1.954×10^{-1}	9.846×10^{-2}	9.648×10^{-1}	1.896×10^{-1}	1.430×10^{-2}	9.168×10^{-2}
	optimal	**-1.032×10^{0}**	-9.186×10^{-1}	-1.021×10^{0}	-1.001×10^{0}	-9.425×10^{-1}	-9.168×10^{-1}	-8.250×10^{-2}	-1.000×10^{0}
F16	average	**-1.000×10^{0}**	-8.634×10^{-1}	-9.648×10^{-1}	-9.545×10^{-1}	-8.217×10^{-1}	-2.198×10^{-1}	-1.625×10^{-8}	-5.491×10^{-1}
	worst	**4.618×10^{-1}**	-5.462×10^{-1}	-6.139×10^{-1}	-8.316×10^{-1}	-7.717×10^{-1}	1.198×10^{-1}	1.032×10^{0}	-1.493×10^{-1}
	optimal	**3.981×10^{-3}**	4.856×10^{-2}	4.404×10^{-1}	4.040×10^{-1}	3.990×10^{-2}	4.299×10^{-1}	4.399×10^{-1}	3.992×10^{-1}
F17	average	**4.040×10^{-2}**	4.000×10^{-2}	6.004×10^{-1}	6.040×10^{-1}	8.036×10^{-1}	6.498×10^{-1}	9.413×10^{-1}	1.065×10^{0}
	worst	**5.033×10^{-1}**	9.911×10^{-1}	1.032×10^{0}	2.030×10^{0}	1.603×10^{0}	2.064×10^{0}	2.319×10^{0}	1.916×10^{0}
	optimal	**3.000×10^{0}**	3.005×10^{0}	3.016×10^{0}	3.149×10^{0}	3.001×10^{0}	3.198×10^{0}	3.616×10^{0}	3.264×10^{0}
F18	average	**3.015×10^{0}**	3.147×10^{0}	3.656×10^{0}	3.648×10^{0}	3.004×10^{0}	3.616×10^{0}	4.086×10^{0}	3.949×10^{0}
	worst	**3.146×10^{0}**	3.142×10^{0}	4.411×10^{0}	4.245×10^{0}	3.133×10^{0}	4.691×10^{0}	4.620×10^{0}	5.017×10^{0}
	optimal	**-3.860×10^{0}**	-3.861×10^{0}	-3.195×10^{0}	-3.619×10^{0}	-3.812×10^{0}	-3.852×10^{0}	-3.801×10^{0}	-3.859×10^{0}
F19	average	**-3.677×10^{0}**	-3.852×10^{0}	-2.870×10^{0}	-3.595×10^{0}	-3.816×10^{0}	-3.742×10^{0}	-3.550×10^{0}	-3.726×10^{0}
	worst	**-3.165×10^{0}**	-3.832×10^{0}	-8.456×10^{-1}	-3.455×10^{0}	-3.795×10^{0}	-3.516×10^{0}	-2.949×10^{0}	-3.030×10^{0}
	optimal	**-3.320×10^{0}**	-3.146×10^{0}	-3.307×10^{0}	-3.032×10^{0}	-2.919×10^{0}	-1.982×10^{0}	-2.098×10^{0}	-3.156×10^{0}
F20	average	**-2.996×10^{0}**	-2.920×10^{0}	-3.019×10^{0}	-2.533×10^{0}	-1.398×10^{0}	-1.089×10^{0}	-9.497×10^{-1}	-1.979×10^{0}
	worst	**-2.649×10^{0}**	-2.520×10^{0}	-2.120×10^{0}	1.298×10^{0}	-5.945×10^{-1}	1.685×10^{-1}	-6.430×10^{-2}	-1.189×10^{0}
	optimal	**-1.015×10^{1}**	-1.007×10^{1}	-1.002×10^{1}	-1.015×10^{1}	-8.103×10^{0}	-1.002×10^{1}	-9.162×10^{0}	-1.006×10^{1}
F21	average	**-9.122×10^{0}**	-9.004×10^{0}	-9.654×10^{0}	-1.012×10^{1}	-4.137×10^{0}	-8.981×10^{0}	-7.198×10^{0}	-9.682×10^{0}
	worst	**-7.596×10^{0}**	-5.642×10^{0}	-8.514×10^{0}	-9.632×10^{0}	-6.517×10^{-1}	-1.192×10^{0}	-1.192×10^{0}	-8.336×10^{0}
	optimal	**-1.040×10^{1}**	-1.032×10^{1}	-1.035×10^{1}	-1.040×10^{1}	-1.039×10^{1}	-1.032×10^{1}	-9.198×10^{0}	-1.006×10^{1}
F22	average	**-1.015×10^{1}**	-8.250×10^{0}	-1.015×10^{1}	-5.365×10^{0}	-3.349×10^{0}	-9.195×10^{0}	-6.927×10^{0}	-8.198×10^{0}
	worst	**-1.002×10^{1}**	-4.820×10^{0}	-9.671×10^{0}	-2.389×10^{-1}	8.924×10^{-2}	-4.928×10^{0}	-1.162×10^{0}	-1.265×10^{-1}
	optimal	**-1.052×10^{1}**	-9.525×10^{0}	-1.051×10^{1}	-8.449×10^{0}	-1.046×10^{1}	-1.000×10^{1}	-9.198×10^{0}	-9.062×10^{0}
F23	average	**-1.025×10^{1}**	-6.489×10^{0}	-1.050×10^{1}	-4.168×10^{0}	-6.265×10^{0}	-5.984×10^{0}	-3.198×10^{-1}	-6.894×10^{0}
	worst	**-9.219×10^{0}**	-1.220×10^{0}	-9.647×10^{0}	-1.608×10^{-1}	-2.487×10^{-1}	-1.963×10^{-1}	1.625×10^{-2}	-1.820×10^{0}

3. Methodology

In this paper, we propose a method for power load forecasting based on LVMD-DBFCM load curve clustering, and the CNN-IVIA-BLSTM model is shown in Figure 6. First, data preprocessing is performed, and the influences of load are divided into continuous and discrete series. After normalizing the continuous sequences, they include load data, meteorological data (temperature, wind speed, and humidity), and economic factors. One-hot encoding is applied to discrete sequences, and the discrete sequences include the data types of workdays, weekends, and holidays. Load data decomposition and denoising with LVMD are performed on the preprocessed data. Then, the power load curve is clustered using the DBFCM algorithm. The processed data are extracted with a 1D CNN for feature extraction and finally, predicted with the IVIA-BLSTM model.

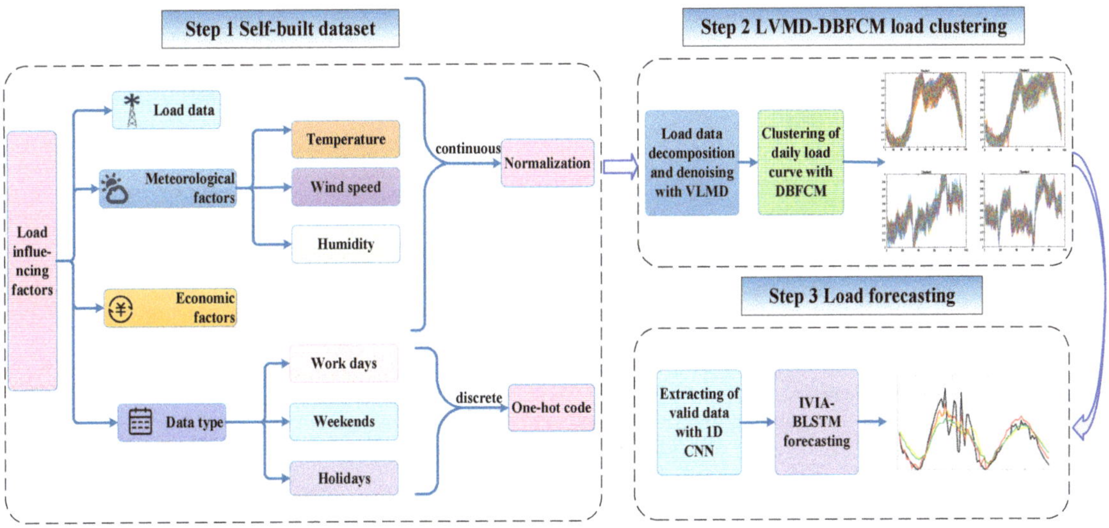

Figure 6. The overall structure design of this paper.

3.1. Self-Built Dataset

We expect to study the electrical load of a particular city or region where such datasets are rare. Data from the internet are collected and organized to produce an HS dataset. The following cities are replaced by H City because the electricity data relate to the economic performance of the country. The dataset includes 1096 days of electricity load data from 1 January 2016 to 31 December 2018, in H City. The dataset was divided; 80% is used as the training set and 20% is used as the test set.

The effects of the three factors of temperature, humidity, and wind speed on the power load are as follows. In summer, the temperature is high, and as air-conditioners usage increases, the power load increases. When the relative humidity is in the sensitive range 40~95%, the meteorological sensitive load varies significantly with the relative humidity. The load decreases with the increase of relative humidity. When the wind speed is in the sensitive range 2~6 m/s, the weather sensitive load changes significantly with the wind speed. The load increases with the increase of wind speed [27].

3.1.1. Dataset Visualization

The overall power load data from 2016–2018 are presented in Figure 7. From the box-whisker plot, the difficulty of prediction is reflected by the high number of data outliers in the morning from 6:00 to 8:00 and in the afternoon from 17:00 to 18:00 and 19:45 to 22:00. In summary, the importance of the data preprocessing and clustering and prediction models that follow in this paper is illustrated.

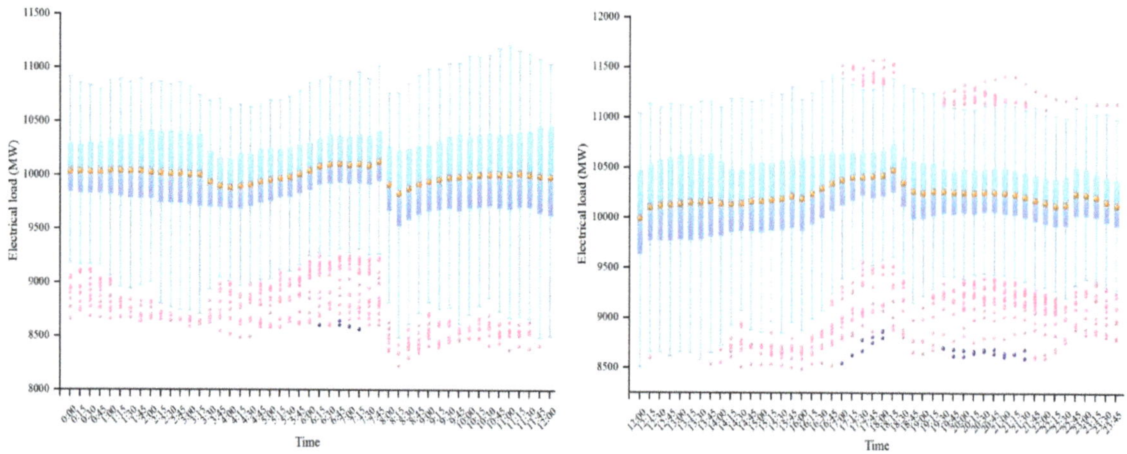

(a) (b)

Figure 7. Distribution of whole load data in the dataset for 15 min. (**a**) Data from 00:00 to 12:00; (**b**) Data from 12:00 to 23:45. Box-whisker plot, the dots replace the abnormal value, the middle line of the box is he median of the data, which represents the average level of the sample data.

3.1.2. Data Preprocessing

- Abnormal data recognition and correction

Data preprocessing is first performed to resolve data outliers before clustering and prediction is performed [28].

Missing data: Lagrange interpolation is used to handle missing values.

Data duplication: Remove duplicate datasets by similarity.

Data mutation: 3 criteria.

- Data normalization

Normalized values for the clustering process can reduce the adverse effect on the clustering effect in the data domain due to the weight of the distance occupied between different attributes of the initial values. Normalization can eliminate the influence of the size of the load data volume on the distance in the clustering analysis, and thus, the information on the load pattern is highlighted [29].

Data normalization: Uniform metrics on data. *Z* Score normalization is shown in Equation (10). After normalization, the load data are normalized to the interval [−1, 1].

$$Z_X = \frac{x_i - \overline{X}}{\sigma} \tag{10}$$

Data preprocessing: The equation for normalizing continuous data is given in Equation (11). Using One Hot encoding for discrete sequences, each state has unique register bits; only one bit is 1, and the rest are 0.

$$MinMax_x = \frac{x_i - x_{\min}}{x_{\max} - x_{\min}} \tag{11}$$

3.2. LVMD-DBFCM Imbalanced Data Clustering

The goal of load clustering is to mine the typical daily load model [30].

Step 1. De-noising. The load data are decomposed and denoised using the LVMD (in Section 3.2.1).

Step 2. Clustering. The load data are clustered using the DBFCM (in Section 3.2.2).

Step 3. Clustering evaluation. The load data clustering effect are analyzed using the SSE evaluation index (in Section 3.2.3).

3.2.1. LVMD Double Decomposition Modal Strategy

- VMD Variational Mode Decomposition

VMD [31] is an iterative process that is used to search for the optimal solution of a variational model and to determine the mode $u_k(t)$ and its corresponding central frequency w_k and bandwidth. VMD has good denoising capabilities. VMD denoising are as follows:

1. Initialize parameters $\hat{u}_k^1$, w_k^1, $\hat{\lambda}^1$, and let $n = 0$.
2. $n = n + 1$, update w_k and u_k.
3. $k = k + 1$, repeat the previous step until $k = K$.
4. Update λ, and the equation is shown in Equation (12).
5. Repeat steps 2–5 until the end, when the condition of Equation (13) is satisfied.
6. The component that contains the minimum time information after decomposition is judged and considered to be random noise that is eliminated. Then, the data are reconstructed by the remaining modal component pairs, and the reconstructed data are the load data that do not contain noise.

$$\hat{\lambda}^{n+1}(\omega) = \hat{\lambda}^n(\omega) + \tau\left[\hat{f}(\omega) - \sum_k \hat{u}_k^{n+1}(\omega)\right] \tag{12}$$

$$\sum_k \left\|\hat{u}_k^{n+1} - \hat{u}_k^n\right\|_2^2 / \|\hat{u}_k^n\|_2^2 < e \tag{13}$$

- LMD Local Mean Decomposition

LMD is a new time–frequency analysis method that adaptively decomposes complex signals into a finite sum of product function (PF) components. Each of these PF components is actually a single component of the AM-FM information. LMD has good feature decomposition abilities [32,33]. The LMD principle is as follows.

1. First, find all the extreme points contained in the data series $x(t)$, assuming that the distribution of extreme points is $\{n_1, n_2, n_3, \cdots\}$, and then calculate the mean m_i and envelope a_i of the adjacent extreme points according to Equations (14) and (15).

$$m_i = \frac{n_i + n_{i+1}}{2} \tag{14}$$

$$a_i = \frac{|n_i - n_{i+1}|}{2} \tag{15}$$

The local mean function curve $m_{11}(t)$ and the envelope function curve $a_{11}(t)$ are obtained by connecting m_i and a_i with line segments and smoothing, respectively, and the sliding average formula is given in Equation (16):

$$Y_s(i) = \frac{1}{2R+1}(Y(i+R) + Y(i+R-1) + \cdots + Y(i-R)) \tag{16}$$

where $Y(i)$ is the sequence to be smoothed, $2R + 1$ is the sliding span, and R is the distance from $Y(i)$ to the starting point of the sliding process.

2. $m_{11}(t)$ is removed from the original signal $x(t)$ to obtain $h_{11}(t)$, and $s_{11}(t)$ is obtained by demodulating $h_{11}(t)$ using Equations (17) and (18):

$$h_{11}(t) = x(t) - m_{11}(t) \tag{17}$$

$$s_{11}(t) = \frac{h_{11}(t)}{a_{11}(t)} \tag{18}$$

When $a_{12}(t) = 1$, $s_{11}(t)$ decomposes into a pure FM function; if $a_{12}(t) \neq 1$, repeat the above process until $a_{1n}(t) = 1$ and $s_{1(n-1)}(t)$ is the corresponding pure FM function; see Equation (19) for details:

$$h_{1n}(t) = s_{1(n-1)}(t) - m_{1n}(t) \tag{19}$$

3. The PF component of the envelope signal is the product of the envelope signal and the pure FM signal, and the above steps are repeated after stripping the PF component from the original signal to obtain the new signal until the residual component is a monotonic function. The final result of LMD is shown in Equation (20):

$$x(t) = \sum_{p=1}^{k} PF_p(t) \tag{20}$$

where the last component is a monotonic function.

LMD decomposes the original signal into multiple high-frequency and low-frequency components without distorting the original signal in the decomposition process. If the sliding step of the sliding average algorithm is not chosen properly in the first step, the decomposition result will be greatly affected. In this paper, the original sliding average process is replaced by the three-time Hermite interpolation method to improve the decomposition accuracy of the LMD algorithm while reducing the overshoot and undershoot of the envelope. That is, after obtaining the extreme value points, the Hermite interpolation method is used to form the upper and lower envelopes, and then the other steps continue to apply the LMD without changes.

3.2.2. DBFCM Double Weights Fuzzy C-Means Algorithm

- FCM Fuzzy C-means Algorithm

FCM is a soft clustering algorithm and is unlike traditional hard clustering (HCM) algorithms, which allows the same object to belong to the same cluster [34]. By optimizing the objective function to obtain the affiliation of each sample point to all class centers, the affiliation range is [0, 1] to determine the class of the sample points to achieve the purpose of automatic classification of sample data [35].

The objective function and constraints *s.t.* of the FCM algorithm are given in Equation (21):

$$\begin{gathered} \min_{u_{ij}, c_i} J(u_{ij}, c_i) = \sum_{i=1}^{K} \sum_{j=1}^{N} u_{ij}^{m} \left\| x_j - c_i \right\|^2 \\ s.t. \sum_{i=1}^{K} u_{ij} = 1, \qquad j = 1, 2, \ldots, N \end{gathered} \tag{21}$$

where u_{ij} denotes that the sample belongs to the cluster affiliation value, m represents the fuzziness, K is the number of clusters, c_i is the i-th clustering center, and $\left\| x_j - c_i \right\|^2$ is the 2 parity of the Euclidean distance from each data point to the cluster center.

The clustering center and affiliation update equations are shown in Equations (22) to (23):

$$u_{ij} = \frac{1}{\sum_{l=1}^{K} \left(\frac{\left\| x_j - c_i \right\|}{\left\| x_j - c_l \right\|} \right)^{\frac{2}{m-1}}} \tag{22}$$

$$c_i = \frac{\sum_{j=1}^{N} u_{ij}^{m} x_j}{\sum_{j=1}^{N} u_{ij}^{m}} \tag{23}$$

- DBFCM Double Weights Fuzzy C-means Algorithm

Traditional FCM clustering is inaccurate in classification for holidays in small categories. From the date type, the power load data are unbalanced data. FCM has a "uniform effect of cluster size" issue, which affects the recognition effect and causes the algorithm to not recognize the holiday load model implied in the historical dataset [36,37]. In this paper, the DBFCM algorithm is proposed to improve the clustering performance of imbalanced data. The objective function of DBFCM utilizes the clustering volume as the weight and the weighted Euclidean distance as the metric distance.

1. The affiliation matrix defines the class volume, which is then introduced as a constraint *s.t.* into the traditional FCM algorithm objective function, as shown in Equation (24):

$$J_{DBFCM} = \sum_{i=1}^{K}\sum_{j=1}^{N} \frac{u_{ij}^{m}\left\|x_j - c_i\right\|^2}{v_j} \tag{24}$$

where v_j is the volume of the j-th class; see Equation (25). Constraint *s.t.* is shown in Equation (26):

$$v_j = \frac{\sum_{i=1}^{N} u_{ij}}{N} \tag{25}$$

$$s.t.\sum_{i=1}^{K} u_{ij} = 1, \quad j = 1,\ 2,\ \ldots,\ N \tag{26}$$

The objective function of DBFCM uses the clustering volume as weights, which can balance the volume of each class in the clustering process. Thus, it can compensate for the unequal interactions between classes and improve the clustering performance of traditional algorithms for unbalanced data.

$$\frac{\partial J_{DBFCM}}{\partial c_i} = \frac{\partial \sum_{i=1}^{N} u_{ij}^{m}\left\|x_j - c_i\right\|^2 / v_j}{\partial c_{\mathrm{i}}} \tag{27}$$

The derivative of Equation (27) with respect to the affiliation degree is constantly positive. Therefore, the Lagrange multiplier method is used to solve the affiliation degree and clustering center by using a greedy strategy and introducing constraint variables. The Lagrangian equation is given in Equation (28):

$$L = \sum_{j=1}^{K}\sum_{i=1}^{N} u_{ij}^{m}\left\|x_j - c_i\right\|^2 / v_j - \sum_{i=1}^{N} \lambda_i\Big(\sum_{i=1}^{K} u_{ij} - 1\Big) \tag{28}$$

The partial derivative of Equation (28) is found by setting the partial derivative = 0; see Equation (29):

$$\frac{\partial L}{\partial u_{ij}} = m u_{ij}^{m-1}\left\|x_j - c_i\right\|^2 / v_j - \lambda_j = 0 \tag{29}$$

The updated equations for the affiliation u_{ij} as well as the clustering center c_i are given in Equations (30) to (31):

$$u_{ij} = \frac{\dfrac{v_j}{\left\|x_j - c_i\right\|^{\frac{2}{m-1}}}}{\sum_{q=1}^{K} \dfrac{v_q}{\left\|x_j - c_i\right\|^{\frac{2}{m-1}}}} \tag{30}$$

$$c_i = \frac{\sum_{j=1}^{N} u_{ij}^m x_j}{\sum_{j=1}^{N} u_{ij}^m} \tag{31}$$

The DBFCM algorithm is more biased towards small classes and avoids the problem of the FCM uniformity effect when dealing with the same sample Euclidean clustering.

2. The load curve is characteristically weighted by introducing a weighting factor w to the traditional Euclidean distance d_{ij}. The DBFCM is used to perform cluster analysis on the decomposed and reconstructed load curves of LVMD to improve the optimal number of clusters.

The degree of influence of each dimensional feature of the analyzed sample on the classification result is assumed to be w. The feature weight coefficients of each dimension $w = \{w_1, w_2, \cdots, w_n\}$ are introduced to the Euclidean distance to form a weighted Euclidean distance, which controls the weights of each dimensional feature vector. The traditional Euclidean distance d_{ij} and the Euclidean distance d_{ij}^* with the addition of weight w are expressed in Equations (32) to (33):

$$d_{ij} = \sqrt{|x_{i1} - x_{j1}|^2 + |x_{i2} - x_{j2}|^2 + \cdots + \left|x_{ik} - x_{jk}\right|^2} \tag{32}$$

$$d_{ij}^* = \sqrt{w_1|x_{i1} - x_{j1}|^2 + w_2|x_{i2} - x_{j2}|^2 + \cdots + w_n\left|x_{ik} - x_{jk}\right|^2} \tag{33}$$

After DBFCM adds weight w, the clustering center c_i does not change, but the objective function J_{DBFCM}, the affiliation u_{ij}, and the Euclidean distance from sample x_j to the clustering center c_i are all affected by the weighted Euclidean distance d_{ij}^*.

In summary, the objective function and affiliation of DBFCM are shown in Equations (32) to (33):

$$J_{DBFCM} = \sum_{i=1}^{K}\sum_{j=1}^{N} u_{ij}^m \|x_j - c_i\|_*^2 / v_j \tag{34}$$

$$u_{ij}^* = \frac{\frac{v_j}{\|x_j - c_i\|_*^{\frac{2}{m-1}}}}{\sum_{q=1}^{K} \frac{v_q}{\|x_j - c_i\|_*^{\frac{2}{m-1}}}} \tag{35}$$

3.2.3. Clustering Validity Quality Evaluation

The clustering criterion used in the clustering process is usually the sum of squared error (SSE). The analytical formula of SSE is shown in the Equation (36):

$$SSE(q) = \sum_{i=1}^{q}\sum_{x \in p_i} \|c_i - x\|_2^2 \tag{36}$$

In the formula, c_i is the i-th cluster center, and p_i represents the aggregation of data points in the i-th cluster. As the number of clusters q increases, the samples are classified more accurately, and the SSE decreases. Theoretically, the smaller the SSE, the better the clustering effect. As the value of q increases to a certain level, the rate of the decline in SSE slows down. As shown in Figure 8, when $q = 5$, it is the inflection point of the curve. The most suitable cluster number is 5.

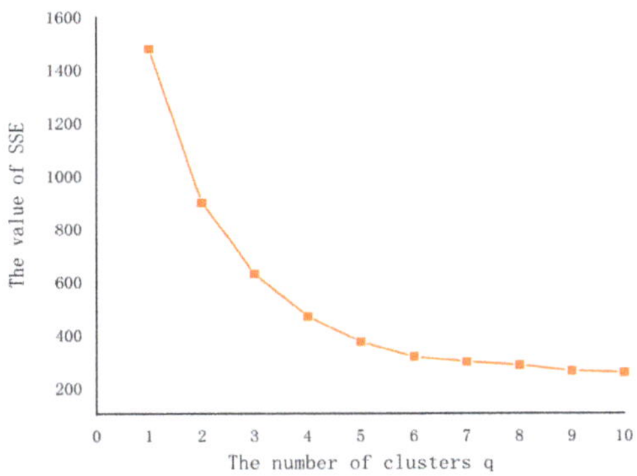

Figure 8. The gradient graph of *SSE* with increasing cluster number q from 1 to 10.

In this paper, two clustering validity indicators are used to assess the quality of clustering: the Calinski–Harabasz indicator (*CHI*) and the Davies–Bouldi indicator (*DBI*). The *CHI* is obtained by the ratio of the compactness to the separation [38]. Thus, the larger the index is, the more compact the result is. The *DBI* metric estimates intraclass closeness by the distance from the sample point within a class to the center of the class to which it belongs, and the distance between class centers indicates the interclass dispersion [39]. Thus, the smaller the index is, the better the effect. Equations (37) to (38) describe these metrics:

$$CHI = \frac{Tr(B_k)}{Tr(W_k)} \times \frac{n-k}{k-1} \tag{37}$$

$$DBI = \frac{1}{k}\sum_{i=1}^{k} \max_{j=1\ldots k, i \neq j} \frac{s_i + s_j}{\|c_i - c_j\|_2^2} \tag{38}$$

3.3. The Proposed CNN-IVIA-BLSTM Forecasting Model

Using the CNN-IVIA-BLSTM model to forecast the power load data:

Step 1. Feature fusion extraction. CNN is used to extract the data feature.

Step 2. Forecasting. IVIA optimizes the parameters related to the BLSTM network, then uses the optimized BLSTM to predict.

Step 3. Power load forecast evaluation. The load data forecasting effect analyzed using RMSE, MAE, and MAPE.

3.3.1. CNN: Convolutional Neural Network

A CNN can automatically extract potential features between massive loads of continuous and discontinuous data to build a compressed complete feature vector for the top fully connected layer [40]. It is a hierarchical neural feedforward network and consists of a series of network layers with different functions. It mainly comprises the input layer, implied layer, and output layer. Among these, the implied layer takes the convolutional layer as the core, and the main function of the convolution layer is to extract the feature fusion of the input data as an arithmetic of the convolution layer. In addition, the CNN includes pooling layers (to reduce the number of parameters of the neural network) and the Dropout layer (to prevent data overfitting). The 1D convolutional kernel is two-dimensional and has a length and width; however, there is a sliding window in the width or height direction, and multiplication is transformed into addition. In this paper, a 1D CNN is used, with a convolutional kernel size of 3×3, a sliding window of 10, and a step size of 1. The overall structure of the convolutional network is shown in Figure 9. A convolutional block is a combination of M convolutional layers and b convergence layers. It can be stacked with N consecutive convolutional blocks followed by K fully connected layers.

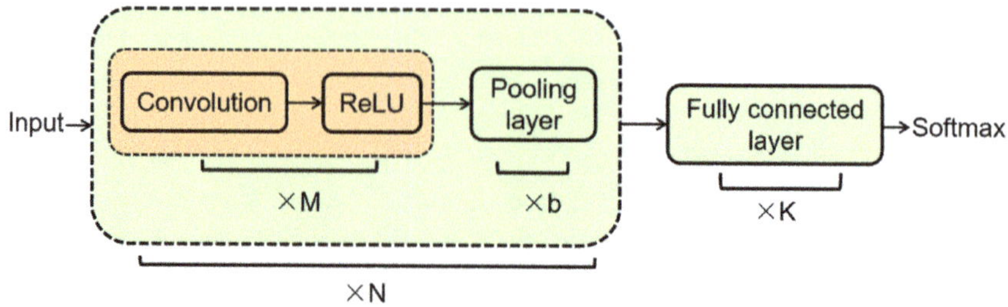

Figure 9. Overall structure of a CNN.

3.3.2. BLSTM Bidirectional Long Short-Term Memory Network

BLSTM can remember a sequence well, can solve the dependency problem of longer time spans, and has a strong advantage in improving time series correlation. Based on the LSTM (long short-term memory) network, the forward and reverse network structures form a closed loop of information, which can better verify and correct the process error information while maintaining the bidirectional data information, which has stronger robustness [41]. The LSTM network structure is shown in Figure 10, and the relevant calculation equations are shown in (39)–(44).

$$Input\ gate:\quad i_t = \sigma(W_i x_t + U_i h_{t-1} + b_i) \tag{39}$$

$$Forget\ gate:\quad f_t = \sigma\left(W_f x_t + U_f h_{t-1} + b_f\right) \tag{40}$$

$$Output\ gate:\quad o_t = \sigma(W_0 x_t + U_o h_{t-1} + b_o) \tag{41}$$

$$New\ memory\ unit:\ c_{t'} = \tanh(W_c x_t + U_c h_{t-1} + b_c) \tag{42}$$

$$Final\ memory\ unit:\ c_t = f_t \otimes c_{t-1} + i_t \otimes c_{t'} \tag{43}$$

$$Output:\quad h_t = o_t \otimes \tanh(c_t) \tag{44}$$

where W_i, W_f, and W_o represent input weight vectors; U_i, U_f, and U_o represent upper output weight vectors; and b_i, b_f, b_o, and b_c are bias vectors. Sigmoid is generally selected as the excitation function for σ, which mainly plays a role of gating. It has an output between 0 and 1, which matches the physical definition of gating and is very close to 1 or 0 when the input is large or small, thus ensuring that the gate is open or closed. The tanh is an option to generate the new memory unit c_t' due to a faster convergence rate with an output between -1 and 1, which coincides with the center of the feature distribution being 0 in most scenarios. The related formulas are given in Equations (45) and (46):

$$sigmoid(x) = \frac{1}{1+e^{-x}} \tag{45}$$

$$\tanh(x) = \frac{e^x - e^{-x}}{e^x + e^{-x}} \tag{46}$$

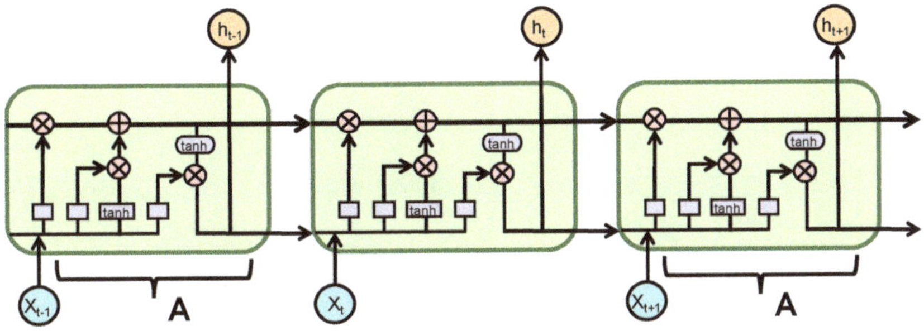

Figure 10. LSTM network structure diagram.

LSTM can only be passed in one direction, and the unit computation of the bidirectional neural network is connected with the unidirectional one. BLSTM consists of an input layer, forward LSTM layer, reverse LSTM layer, and output layer. However, the hidden layer of the bidirectional neural network must save two values, i.e., *A*, which is involved in the forward calculation, and *A′*, which is involved in the reverse calculation. The final output value depends on the sum. The BLSTM network structure is shown in Figure 11.

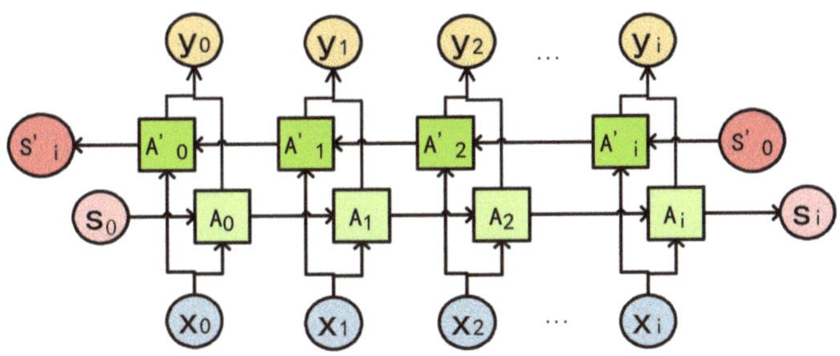

Figure 11. BLSTM network structure diagram.

However, a common problem is that BLSTM networks have many parameters, and without proper optimization, the model may be overfitted or slow to train, resulting in inefficiency. Too few nodes in the hidden layer will cause the model to not have the necessary learning ability and information processing capability, and too many nodes will increase the complexity of the network structure and make the learning process easily fall into local minima, which makes the network slow. When the learning rate is too high, the cost function is not easy to reduce to the lowest point, it is not easy to converge at the lowest point, and the convergence effect is poor. When too much training is performed, the gradient descent process may cross the nadir, which causes the training rate to be too low. In the case of reasonable tuning parameters, the more layers and neurons there are, the higher the accuracy rate; however, this can also lead to the overfitting phenomenon, and a regularization process can be used to solve the overfitting problem. In this paper, the number of implicit layer neurons, training times, learning rates, and regularization coefficients in BLSTM are optimized using the IVIA to improve the performance of the sequence modelling task.

3.3.3. Hybrid Forecasting Model

The CNN-IVIA-BLSTM model is a combined prediction model consisting of a CNN and BLSTM. BLSTM is a commonly used model architecture of deep learning models for sequence modelling tasks, and it has achieved good performance in many tasks. The CNN is first used to extract the feature vectors consisting of load influencing factors, which can be considered a "feature extractor", then the extracted feature vectors are used for load prediction with the BLSTM model optimized by the IVIA. The hybrid prediction model structure is shown in Figure 12, which takes full advantage of both the CNN and BLSTM network to ensure the accuracy of the power load.

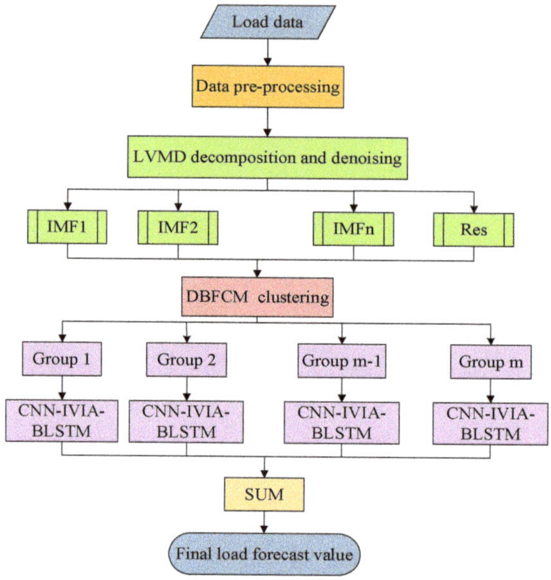

Figure 12. Hybrid forecasting model structure diagram.

The forecasting steps are as follows:

1. Selected information as model input.
2. LVMD decomposes and denoises the original sequence, and DBFCM performs clustering.
3. The IVIA population size N, the maximum number of iterations M, and the initial search range of the parameters (the number of neurons in the hidden layer H, training number E, learning rate η, and regularization factor $L2$) are set. The root mean square error (y_{RMSE}) is used as the objective function in the optimization algorithm, and finally, the model of the influenza virus immunization algorithm coupled with the bidirectional long and short-term memory network is developed.
4. The 1D CNN reads the load sequence with a sliding time window of 10 and a step size of 1 for feature extraction.
5. m prediction models are obtained by inputting the CNN-IVIA-BLSTM prediction models for each component separately.
6. Finally, the predicted values of the m prediction models are combined to obtain the predicted values of the load.

3.3.4. Power Load Forecast Evaluation Indicator

Three evaluation indexes are set as y_{MAPE} (mean absolute percentage error, MAPE), y_{RMSE} (root mean square error, RMSE), and y_{MAE} (mean absolute error, MAE). The equations are shown in (47) to (49):

$$y_{MAPE} = \frac{1}{n}\sum_{i=1}^{n} \left| \frac{X_{act}(i) - X_{pred}(i)}{X_{act}(i)} \right| \tag{47}$$

$$y_{RMSE} = \sqrt{\frac{\sum_{i=1}^{n} (X_{act}(i) - X_{pred}(i))^2}{n}} \tag{48}$$

$$y_{MAE} = \frac{1}{n}\sum_{i=1}^{n}\left|X_{act}(i) - X_{pred}(i)\right| \quad (49)$$

In the above equation, n is the total number of predictions, $X_{act}(i)$ is the real value of the load at moment i, and $X_{pred}(i)$ is the predicted value of the load at moment i.

4. Case Analysis

In this paper, historical power load data from 2016 to 2018 in H city were used for prediction with a sampling interval of 15 min. LSTM, BLSTM, CNN-BLSTM, LVMD-DBFCM-CNN-BLSTM, and the proposed model in this paper are selected to compare the prediction results.

4.1. Analysis of Data Processing Results

4.1.1. Data Denoising and Decomposition

The original load data are decomposed and denoised using VMD, as shown in Figure 13, and the original load data sequence is decomposed into 8 IMF components. From the figure, we can see that the IMF7–IMF8 values are small and contain more noisy data, which are removed from the original data, then the first six IMF components are used to reconstruct the data to achieve data denoising.

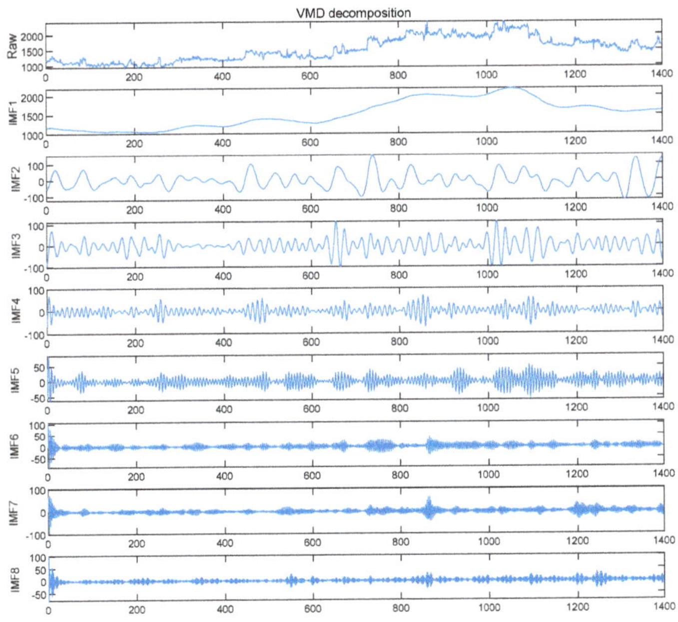

Figure 13. Result of VMD decomposition.

The second step of LVMD is to redecompose the denoised data with LMD. The denoised data are decomposed into a total of six PF components, and the decomposition results are shown in Figure 14. The values of PF1 are larger and maintain the same trend as the original data, containing the main valid information of the data. The values of PF2–PF6 are smaller, among which the periodic changes of PF2–PF4 are more obvious and PF6 is monotonically increasing, which is convenient for the prediction of each component. The randomness of the F5 sequence volatility is strong and the prediction accuracy of the F5 component is not high when making predictions.

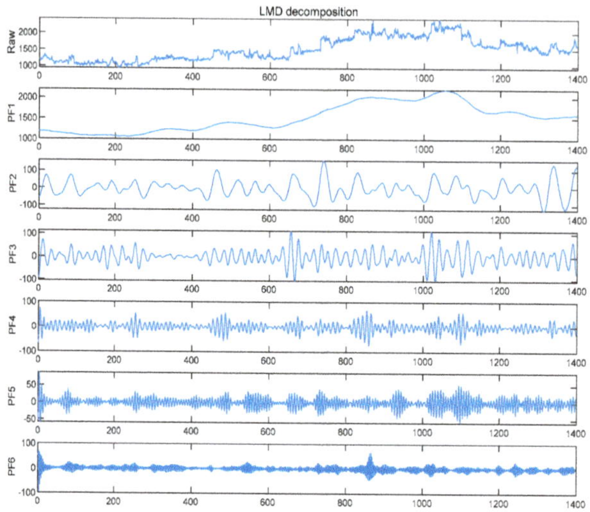

Figure 14. Result of LMD decomposition.

4.1.2. Analysis of the Clustering Results

The daily load curve after LVMD-DBFCM clustering is shown in Figure 15.

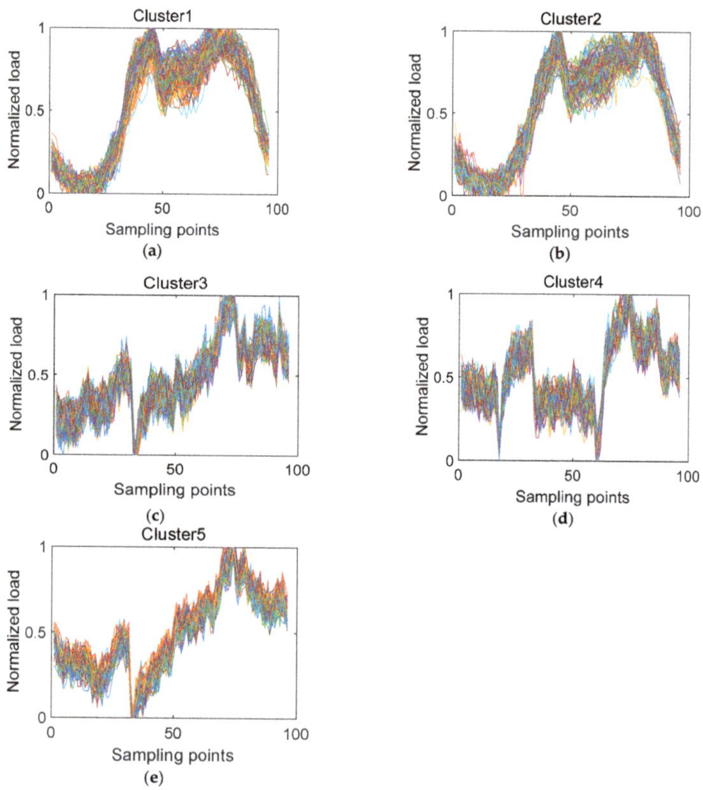

Figure 15. (**a**–**e**) Daily load curve after LVMD-DBFCM clustering when the number of clusters is determined to be 5.

The clustering algorithm proposed in this paper is compared with k-means [42], FCM [43], and DBFCM. Their comparison is displayed in Table 3 and Figure 16. From the results, it can be seen that the maximum value of the CHI metric is 1142.509 and the minimum value of the DBI metric is 1.083. The clustering validity of the method proposed in this paper is better than that of the other three methods.

Table 3. The comparative table displaying the clustering validity indicators, CHI and DBI, of the K-means, FCM, DBFCM, and LVMD-DBFCM methods.

Methods	CHI	DBI
K-means	721.201	1.549
FCM	814.014	1.401
DBFCM	1024.743	1.216
LVMD-DBFCM	**1142.509**	**1.083**

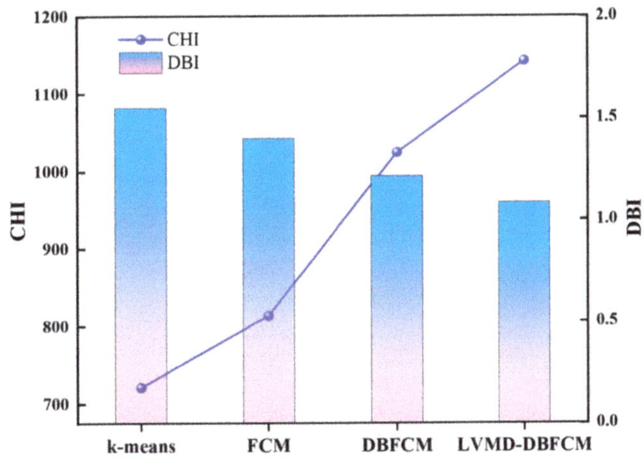

Figure 16. Comparative graph displaying the clustering validity indicators, CHI and DBI, of the K-means, FCM, DBFCM, and LVMD-DBFCM methods.

4.2. Analysis of the Prediction Results

The comparison between the prediction model proposed in this paper and the four models LSTM, BLSTM, CNN-BLSTM, and LVMD-DBFCM+CNN-BLSTM using three load evaluation indices, RMSE, MAE, and MAPE, is shown in Table 4 and Figure 17. From the results, we can see that the prediction models proposed in this paper have better prediction results than the other four models in all evaluation indices.

Table 4. The comparative table displaying the clustering validity indicators, CHI and DBI, of the K-means, FCM, DBFCM, and LVMD-DBFCM methods.

Models	RMSE	MAE	MAPE
LSTM	101.4817	80.0980	5.7087%
BLSTM	83.1498	61.4826	4.4460%
CNN-BLSTM	79.3340	58.0245	4.1782%
LVMD-DBFCM+CNN-BLSTM	57.7316	42.3669	2.9946%
Proposed	**31.9942**	**23.3691**	**1.6421%**

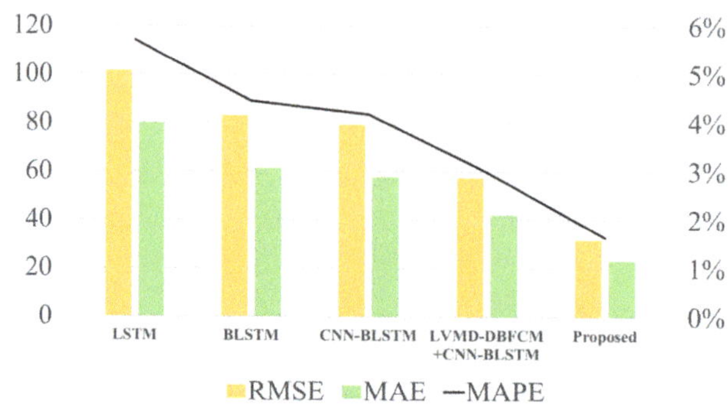

Figure 17. Comparison of the RMSE, MAE, and MAPE results for 5 models' predictions.

Figure 18 shows the prediction graphs of the five prediction models. From the figure, it can be clearly seen that the pink line, representing the model proposed in this paper, fits so closely to the target curve represented by the black line, which can better reflect the trend of the target, indicating that the model proposed in this paper has the better prediction effect than addressed models. Figure 19 shows the prediction error comparison of the 5 models, from which it can be clearly seen that the model proposed in this paper, represented by yellow, has the lowest error.

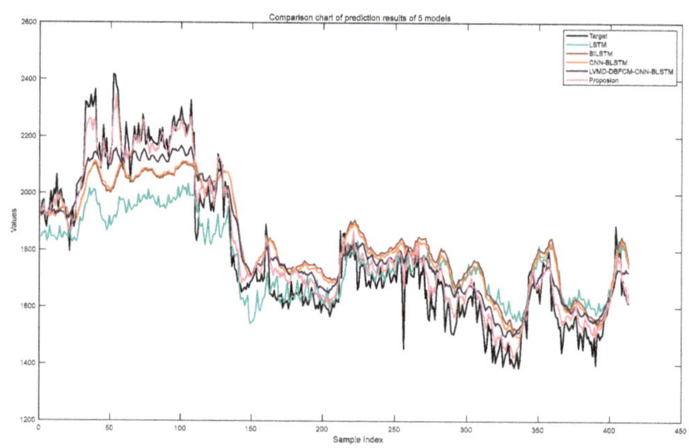

Figure 18. Comparison chart of the prediction results of 5 models.

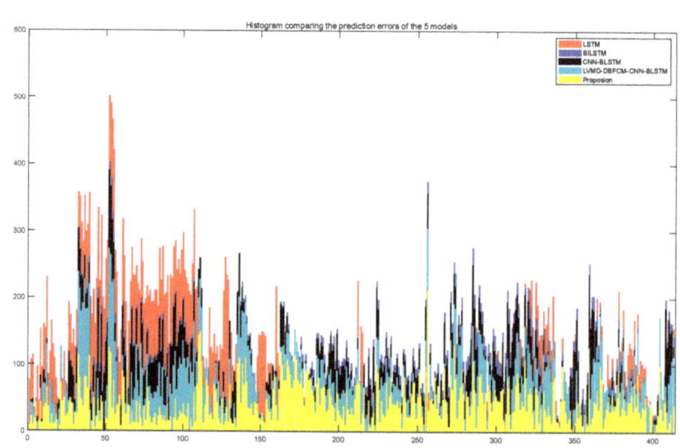

Figure 19. Comparison of prediction errors of the 5 models.

5. Conclusions

In this paper, a new model based on LVMD-DBFCM load curve clustering and a CNN-IVIA-BLSTM hybrid model for power load forecasting is proposed. This comprehensive technique takes historical load data and influencing factors (meteorology, economy, and data type) into account, where historical load data, meteorological factors, and economic factors are normalized, and the data types are uniquely heat coded.

The novel LVMD-DBFCM algorithm improves the continuity and stability of the data, and the values of the CHI and DBI quality assessment indicators are 1142.509 and 1.083, respectively, both of which reflect the good validity of the clustering method used in this paper. In the new CNN-IVIA-BLSTM model, a CNN is used for feature extraction, BLSTM is used for load forecasting, and the IVIA is used to optimize the relevant parameters in the BLSTM network. The results of the three electric load forecasting evaluation metrics of the hybrid forecasting model show that the RMSE is 31.9942, the MAE is 23.3691, and the MAPE is 1.6421%. The prediction effect of the electric load fits well with the target, and the prediction error is minimized.

Author Contributions: Conceptualization, L.H., J.W., Z.G. and T.Z.; Data curation, L.H., J.W., Z.G. and T.Z.; Formal analysis, L.H., J.W. and T.Z.; Investigation, L.H; Methodology, L.H. and J.W; Software, J.W. and Z.G.; Project administration, L.H.; Resources, L.H.; Supervision, L.H.; Validation, L.H.; Visualization, L.H.; Writing—original draft, L.H., J.W., Z.G. and T.Z.; Writing—review & editing, J.W. and Z.G. All authors have read and agreed to the published version of the manuscript.

Funding: This research received no external funding.

Institutional Review Board Statement: Not applicable.

Informed Consent Statement: Not applicable.

Data Availability Statement: Not applicable.

Conflicts of Interest: The authors declare no conflict of interest.

References

1. Huang, Y. Medium-Long Term Power Load Forecasting of a Region. Master's Thesis, Xihua University, Chengdu, China, 2015.
2. Han, F.J.; Wang, X.H.; Qiao, J. Review on Artificial Intelligence based Load Forecasting Research for the New-type Power System. *Proc. CSEE* **2023**, 1–24.
3. Zhang, Y.; Wang, A.H.; Zhang, H. Overview of smart grid development in China. *Power Syst. Prot. Control* **2021**, *49*, 180–187.
4. Gao, D.D.; Gao, S.T. Review on medium-long term power load forecasting study. *Sci. Technol. Innov. Her.* **2014**, *11*, 25.
5. Shang, C.; Gao, J.; Liu, H.; Liu, F. Short-term load forecasting based on PSO-KFCM daily load curve clustering and CNN-LSTM model. *IEEE Access* **2021**, *9*, 50344–50357. [CrossRef]
6. Huang, J.B. Based on trend extrapolation method analysis of city complexes load development characteristicsitle of the chapter. *Rural Electrif.* **2019**, *7*, 39–42.
7. Song, F.; Liu, J.; Zhang, T. The Grey Forecasting Model for the Medium-and Long-Term Load Forecasting. *J. Phys. Conf. Ser.* **2020**, *1654*, 012104. [CrossRef]
8. Fan, G.F.; Qing, S.; Wang, H. TSupport Vector Regression Model Based on Empirical Mode Decomposition and Auto Regression for Electric Load Forecasting. *Energies* **2013**, *6*, 1887–1901. [CrossRef]
9. Qiu, X.P. Neural Network and Deep Learning. *J. Chin. Inf. Process.* **2020**, *34*, 4.
10. Jarquin, C.S.S.; Gandelli, A.; Grimaccia, F.; Mussetta, M. Short-Term Probabilistic Load Forecasting in University Buildings by Means of Artificial Neural Networks. *Forecasting* **2023**, *5*, 390–404. [CrossRef]
11. Liu, J.H. Power load forecasting in Shanghai based on CNN-LSTM combined model. *Proc. SPIE* **2022**, *12254*, 125–132.
12. Li, J.W.; Luong, T.; Dan, J. When are tree structures necessary for deep learning of representations. In Proceedings of the 2015 Conference on Empirical Methods in Natural Language Processing, Lisbon, Portugal, 17–21 September 2015; Association for Computational Linguistics: Stroudsburg, PA, USA, 2015.
13. Gu, H.C.; Mou, P.; Li, J.W. Modeling and application of ethylene cracking furnace based on cross-iterative BLSTM network. *CIESC J.* **2019**, *70*, 548–555.
14. Xu, Y.; Xiang, Y.F.; Ma, T.X. Short-term Power Load Forecasting Method Based on EMD-CNN-LSTM Hybrid Model. *J. North China Electr. Power Univ.* **2022**, *49*, 81–89.
15. Gao, W.C. Urban Gas Load Forecasting Based on EWT-CNN-LSTM Model. Master's Thesis, Shanghai Normal University, Shanghai, China, 2021.

16. Sun, G.L.; Li, B.J.; Xu, D.M.; Li, Y.P. Monthly Runoff Prediction Model Based on VMD-SSA-LSTM. *Water Resour. Power* **2022**, *40*, 18–21.
17. Yang, Y. Research and Application of Medium and Long Term Load Forecasting Technology. Master's Thesis, Shenyang Institute of Computing Technology, Chinese Academy of Science, Shenyang, China, 2021.
18. Mojtaba, G.; Faraji, D.I.; Ebrahim, A.; Abolfazl, R. A novel and effective optimization algorithm for global optimization and its engineering applications: Turbulent Flow of Water-based Optimization (TFWO). *Eng. Appl. Artif. Intell.* **2020**, 92. [CrossRef]
19. Abdolkarim, M.B.; Mahmoud, D.N.; Adel, A. Golden eagle optimizer: A nature-inspired metaheuristic algorithm. *Comput. Ind. Eng.* **2021**, *152*, 107050.
20. Mohamed, A.A.; Hassan, S.A.; Hemeida, A.M.; Alkhalaf, S. Parasitism-Predation algorithm (PPA): A novel approach for feature selection. *Ain Shams Eng. J.* **2019**, *11*, 293–308. [CrossRef]
21. Dhiman, G.; Garg, M.; Nagar, A.; Kumar, V. A novel algorithm for global optimization: Rat Swarm Optimizer. *J. Ambient Intell. Humaniz. Comput.* **2020**, *12*, 8457–8482. [CrossRef]
22. Mirjalili, S.; Mirjalili, S.M.; Lewis, A. Grey Wolf Optimizer. *Adv. Eng. Softw.* **2014**, *69*, 46–61. [CrossRef]
23. Ramya, S.; Rajesh, N.B.; Viswanathan, B. Particle Swarm Optimization (PSO) based optimum Distributed Generation (DG) location and sizing for Voltage Stability and Loadability Enhancement in Radial Distribution System. *Int. Rev. Autom. Control IREACO* **2014**, *7*, 288–293.
24. Jhila, N.; Modarres, K.F.; Akiko, Y. A whale optimization algorithm (WOA) approach for clustering. *Cogent Math. Stat.* **2018**, *5*, 1483565.
25. Karol, S.; Ewelina, H.; Katarzyna, L.; Katarzyna, K. Spread of Influenza Viruses in Poland and Neighboring Countries in Seasonal Terms. *Pathogens* **2021**, *10*, 316.
26. Oxford, J.S. Special Article: What Is the True Nature of Epidemic Influenza Virus and How Do New Epidemic Viruses Spread. *Epidemiol. Infect.* **1987**, *99*, 1–3. [CrossRef] [PubMed]
27. Zhu, Z.W. Study on the Impact of Weather Factors on Characteristics of Electric Load. Ph.D. Thesis, Zhejiang University, Hangzhou, China, 2008.
28. Xu, F.; Weng, G.Q. Research on Load Forecasting Based on CNN-LSTM Hybrid Deep Learning Model. In Proceedings of the 2022 IEEE 5th International Conference on Electronics Technology (ICET), Chengdu, China, 13–16 May 2022; pp. 1332–1336.
29. Hu, Y.C. Research on Power Load Pattern Recognition Method Based on Improved k-means Clustering Algorithm. Master's Thesis, University of Electronic Science and Technology of China, Chengdu, China, 2018.
30. Zhang, H.; Zhang, Y.; Xu, Z. Thermal Load Forecasting of an Ultra-short-term Integrated Energy System Based on VMD-CNN-LSTM. In Proceedings of the 2022 International Conference on Big Data, Information and Computer Network (BDICN), Sanya, China, 20–22 January 2022; pp. 264–269.
31. Dragomiretskiy, K.; Zosso, D. Variational mode decomposition. *IEEE Trans. Signal Process.* **2014**, *62*, 531–544. [CrossRef]
32. Yang, S.S.; Zhou, H.; Zhao, H.Y. Fault diagnosis method for the bearing of reciprocating compressor based on LMD multiscale entropy and SVM. *J. Mech. Transm.* **2015**, *39*, 119–123.
33. Wu, D.S.; Yang, Q.; Zhang, J.Y. Ensemble fault diagnosis method based on VMD-LMD-CNN. *Bearing* **2020**, *10*, 57–63.
34. Li, P.S.; Li, X.R.; Chen, H.H. The characteristics classification and synthesis of power load based on fuzzy clustering. *Proc. CSEE* **2005**, *25*, 73–78.
35. Zhou, K.L. Theoretical and Applied Research on Fuzzy C-means Clustering and Its Cluster Validation. Ph.D. Thesis, Hefei University of Technology, Hefei, China, 2014.
36. Dong, T. Research on Electrical Load Pattern Recognition and Load Forecasting Based on Deep Learning. Master's Thesis, Jilin University, Changchun, China, 2022.
37. Ma, Z.B.; Xu, S.A.; Zhu, S.B. Power Load Classification Based on Feature Weighted Fuzzy Clustering. *ELECTRIC POWER* **2022**, *55*, 25–32.
38. Tomasini, C.; Emmendorfer, L.; Borges, E.N. A methodology for selecting the most suitable cluster validation internal indices. In Proceedings of the 31st Annual ACM Symposium on Applied Computing (SAC '16), Pisa, Italy, 4–8 April 2016; pp. 901–903.
39. Hao, X.H.; Li, Y.L.; Gu, Q. Power load data clustering algorithm based on DTW histogram. *Transducer Microsyst. Technol.* **2020**, *39*, 140–142.
40. Liu, Y.H.; Zhao, Q. Ultra-short-term power load forecasting based on cluster empirical mode decomposition of CNN-LSTM. *Power Syst. Technol.* **2021**, *45*, 4444–4451.
41. Zhu, H.S. Design and Implementation of CNN-BLSTM Speech Separation Algorithm Fused With Self-attention Mechanism. Master's Thesis, Hebei University of Science and Technology, Shijiazhuang, China, 2021.
42. Wang, J.D.; Gu, Z.C.; Ge, L.J. Load Clustering Characteristic Analysis of the Distribution Network Based on the Combined Improved Firefly Algorithm and K-means Algorithm. *J. Tianjin Univ. Sci. Technol.* **2023**, *56*, 137–147.
43. Pan, X.G. Research of Fuzzy Clustering Algorithm on Complicated Data and Feature Weight Learning Techniques. Ph.D. Thesis, Jiangnan University, Wuxi, China, 2022.

mathematics

Article

Hermite–Hadamard-Type Inequalities for Coordinated Convex Functions Using Fuzzy Integrals

Muhammad Amer Latif

Department of Basic Sciences, Deanship of Preparatory Year, King Faisal University, Hofuf 31982, Al-Hasa, Saudi Arabia; m_amer_latif@hotmail.com or mlatif@kfu.edu.sa

Abstract: In this paper, some estimates of third and fourth inequalities in Hermite–Hadamard-type inequalities for coordinated convex functions are proved using the non-additivity of the integrals and Fubini's theorem for fuzzy integrals. That is, the results are obtained in the fuzzy context and using the Lebesgue measure. Several examples are provided on how to evaluate these estimates in order to illustrate the obtained results.

Keywords: fuzzy measure; Sugeno integral; Hermite–Hadamard's inequality; Fubini's theorem

MSC: 26A15; 26A51

Citation: Amer Latif, M. Hermite–Hadamard-Type Inequalities for Coordinated Convex Functions Using Fuzzy Integrals. *Mathematics* **2023**, *11*, 2432. https://doi.org/10.3390/math11112432

Academic Editor: Janusz Brzdęk

Received: 4 April 2023
Revised: 13 May 2023
Accepted: 22 May 2023
Published: 24 May 2023

1. Introduction

Sugeno investigated the theory of fuzzy measures and fuzzy integrals as a tool for modeling non-deterministic issues. The fuzzy integral (also known as the Sugeno integral) has profoundly amazing mathematical characteristics that have been noted by several authors. Sugeno pioneered the study of the theory of nonadditive measures and integrals, frequently referred to as fuzzy measures and fuzzy integrals [1]. Ralescu and Adams [2] investigated numerous equivalent definitions of fuzzy integral, whereas Wang and Klir [3–5] offered an outline of fuzzy measure theory and generalized fuzzy measure theory, respectively. Fuzzy measures and the Sugeno integral have also been successfully employed in a variety of domains, including decision-making [6] and artificial intelligence [7]. Integral inequalities are useful tools in a variety of theoretical and practical applications. The investigation of inequalities for the Sugeno integral was started by Román-Flores et al. in [8–13], and was subsequently expanded upon by Ouyang et al. in [14–16].

Many researchers studied celebrated inequalities using the Sugeno integral; for example, Hu [17] proved the following Chebyshev-type inequalities for a Sugeno-like integral by using a binary operation called g-seminorm.

Theorem 1 ([17]). *Let $H : B^n \to B$ be a left continuous and non-decreasing n-place function. Let $u : B \to B$ be any strictly monotone increasing bijection, and suppose that $f_1, \ldots, f_n : X \to B$ is any comonotone system. If the g-seminorm G satisfies*

$$G(\varphi(H(x_1,\ldots,x_n)),c) \geq \left\{ \begin{array}{c} H(G(\varphi(x_1),c),\varphi(x_2),\ldots,\varphi(x_n)) \\ H(\varphi(x_1),G(\varphi(x_2),c),\ldots,\varphi(x_n)) \\ \cdots \\ H(\varphi(x_1),\varphi(x_2),\ldots,G(\varphi(x_n),c)) \end{array} \right\}$$

for all $(x_1,\ldots,x_n) \in B$, then for any $A \subseteq \Sigma$ we have

$$\int_{GA} \varphi(H(f_1,\ldots,f_n))d\mu \geq H\left(\int_{GA} \varphi(f_1)d\mu,\ldots,\int_{GA} \varphi(f_n)d\mu\right).$$

Caballero and Sadarangani [18] developed a Cauchy–Schwarz-type inequality for the Sugeno integral.

Theorem 2 ([18]). *Let (X, Σ, μ) be a fuzzy measure space, $f, g \in \mathcal{F}_+(X)$, ($\mathcal{F}_+(X)$ is the set of all non-negative measurable functions with respect to Σ) and f and g comonotone functions, $A \in \Sigma$ with $\int_A f \cdot g d\mu \leq 1$. Then,*

$$\left(\int_A f \cdot g d\mu\right)^2 \leq \left(\int_A f^2 d\mu \vee \int_A g^2 d\mu\right).$$

Agahi et al. [19] shown a generalization of the Stolarsky inequality for a Sugeno integral as follows:

Theorem 3 ([19]). *Let $f : [0,1] \to [0,1]$ be a nonincreasing function, $([0,1], \mathcal{B}([0,1]), m)$ a fuzzy measure space, and define $h : [0,1] \to [0,1]$ by $h(a) = m([0,m])$ for $a \in [0,1]$. Let β, γ be automorphisms on $[0,1]$ (i.e., $\beta, \gamma : [0,1] \to [0,1]$ are increasing bijections) and $\alpha = \left(\beta^{-1} \bigstar \alpha^{-1}\right)^{-1}$ is a continuous aggregation function that is jointly strictly increasing and bounded from above by* min*, and which is dominated by h, i.e., for all x, y from $[0,1]$, it holds*

$$h(x \bigstar y) \geq h(x) \bigstar h(y),$$

then

$$(S) \int_0^1 f(\alpha) dm \geq (S) \int_0^1 f(\beta) dm \bigstar (S) \int_0^1 f(\gamma) dm,$$

where $f(\alpha)$ means the composite function defined on $[0,1]$ and given by $f(\alpha)(x) = f(\alpha(x))$.

We also refer the reader to Ouyang et al. [16], which contains the study of the Minkowski type for the Sugeno integral on abstract spaces.

Theorem 4 ([16]). *Let μ be an arbitrary fuzzy measure on $[0,a]$ and $f, g : [0,a] \to \mathbb{R}$ be two real-valued measurable functions such that $(S) \int_0^a f d\mu \leq 1$ and $(S) \int_0^a g d\mu \leq 1$. If f, g are both non-decreasing, then the inequality*

$$(S) \int_0^a f g d\mu \geq \left((S) \int_0^a f d\mu\right)\left((S) \int_0^a g d\mu\right)$$

holds.

Caballero and Sadarangani [20] have shown that the classical Hermite–Hadamard inequalities [21,22] do not hold true for fuzzy integrals in general and established some Hermite–Hadamard-type inequalities for the Sugeno integral with peculiar examples to validate their results.

The main results from [20] are stated in the following theorems.

Theorem 5 ([20]). *Let $f : [a,b] \to [0, \infty)$ be a convex function and μ the Lebesgue measure on $\mathbb{R}$.*

(a) If $f(a) < f(b)$, then

$$\int_a^b f d\mu \leq \min\left\{\frac{(b-a) f(b)}{f(b) - f(a) + b - a}, b - a\right\}.$$

(b) If $f(a) = f(b)$, then

$$\int_a^b f d\mu \leq \min\{f(a), b - a\}.$$

(c) *If $f(a) > f(b)$, then*

$$\int_a^b f d\mu \leq \min\left\{\frac{(b-a)f(a)}{f(a)-f(b)+b-a}, b-a\right\}.$$

Recently, the integral inequalities for Sugeno integrals using different kinds of convexities and other results on several other types of inequalities based on Sugeno integrals are a thought-provoking topic to many authors in the field of fuzzy integrals, see for instance [20,23–37] and the references cited therein.

The integral inequalities for coordinated convex functions in the context of Sugeno integrals have not been investigated in any study so far. Hence, motivated by the ongoing research about the integral inequalities for the Sugeno integrals involving the different kinds of convex functions, it is the main objective of this paper to obtain Hermite–Hadamard-type inequalities for Sugeno double integrals by using coordinated convex functions.

2. Main Results

In order to proceed to our results, we first give some basic notations and properties of Sugeno integral.

Definition 1 ([20]). *Suppose that Σ is σ-algebra of subsets of $\mathbb{R}$ and that $\mu : \Sigma \to [0, \infty)$ is non-negative extended real valued set function, then μ is said to be fuzzy measure if and only if:*

1. $\mu(\emptyset) = 0$,
2. *$E, F \in \Sigma$ and $E \subseteq F$ imply that $\mu(E) \leq \mu(F)$ (monotonicity),*
3. *$\{E_n\} \subseteq \Sigma$, $E_1 \subseteq E_2 \subseteq \ldots$, imply $\lim_{n\to\infty} \mu(E_n) = \mu\left(\bigcup_{n=1}^{\infty} E_n\right)$ (continuity from below),*
4. *$\{E_n\} \subseteq \Sigma$, $E_1 \subseteq E_2 \subseteq \ldots$, $\mu(E_1) < \infty$, imply $\lim_{n\to\infty} \mu(E_n) = \mu\left(\bigcap_{n=1}^{\infty} E_n\right)$ (continuity from above).*

If f is a non-negative real-valued function defined on $\mathbb{R}$, we will denote by $L_\alpha f = \{x \in \mathbb{R} : f(x) \geq \alpha\} = \{f \geq \alpha\}$ the α-level of f, for $\alpha > 0$ and $L_0 f = \overline{\{x \in \mathbb{R} : f(x) > 0\}}$ =supp$f$, the support of f. It may be noted that if $\alpha \leq \beta$, then $\{f \leq \alpha\} \subseteq \{f \leq \beta\}$. If μ is fuzzy measure on $(\mathbb{R}, \Sigma)$, by $\mathcal{F}^\mu(\mathbb{R})$, we mean all μ-measurable functions from $\mathbb{R}$ to $[0, \infty)$.

Definition 2 ([20]). *Suppose that μ is a fuzzy measure on $(\mathbb{R}, \Sigma)$. If $f \in \mathcal{F}^\mu(\mathbb{R})$ and $A \in \Sigma$ then the Sugeno integral (or fuzzy integral) of f on A with respect to the fuzzy measure μ is defined as:*

$$\int_A f d\mu = \bigvee_{\alpha \geq 0} [\alpha \wedge \mu(A \cap \{f \geq \alpha\})],$$

where $\vee$ and $\wedge$ denote the supremum and infimum on $[0, \infty)$, respectively. In particular, if $A = X$, then

$$\int_X f d\mu = \int f d\mu = \bigvee_{\alpha \geq 0} [\alpha \wedge \mu(\{f \geq \alpha\})].$$

The following properties of the Sugeno integral are well known and can be found in [5].

Proposition 1 ([5]). *If μ is a fuzzy measure on $(\mathbb{R}, \Sigma)$, $A \in \Sigma$ and $f, g \in \mathcal{F}^\mu(\mathbb{R})$, then*

1. $\int_A f d\mu \leq \mu(A)$.
2. $\int_A k\, d\mu = k \wedge \mu(A)$.
3. *If $g \leq f$ on A, then $\int_A g d\mu \leq \int_A f d\mu$.*
4. $\mu(A \cap \{f \geq \alpha\}) \geq \alpha \Rightarrow \int_A f d\mu \geq \alpha$.
5. $\mu(A \cap \{f \geq \alpha\}) \leq \alpha \Rightarrow \int_A f d\mu \leq \alpha$.
6. *$\int_A f d\mu < \alpha$ if and only if there exists $\gamma < \alpha$ such that $\mu(A \cap \{f \geq \gamma\}) < \alpha$.*

7. $\int_A f d\mu > \alpha$ *if and only if there exists* $\gamma > \alpha$ *such that* $\mu(A \cap \{f \geq \gamma\}) > \alpha$.

Definition 3 ([20]). *If* μ *is a fuzzy measure on* $(\mathbb{R}, \Sigma)$, $A \in \Sigma$ *and* $f \in \mathcal{F}^{\mu}(\mathbb{R})$, *then the survival function* F *associated to* f *on* A *is defined by*

$$F(\alpha) = \mu(A \cap \{f \geq \alpha\}),$$

where $\alpha \geq 0$.

Remark 1 ([20]). *Consider the survival function* F *associated to* f *on* A, *that is,* $F(\alpha) = \mu(A \cap \{f \geq \alpha\})$. *If* $\mu(A \cap \{f \geq \alpha\}) \geq \alpha$ *and* $\mu(A \cap \{f \geq \alpha\}) \leq \alpha$, *from (4) and (5) of Proposition 1, we obtain*

$$F(\alpha) = \alpha \Rightarrow \int_A f d\mu = \alpha.$$

The above equation implies that any fuzzy integral can be calculated by solving the equation $F(\alpha) = \alpha$.

Theorem 6 ([38]). *Let* $(X, \mathcal{X}, \mu)$ *and* $(Y, \mathcal{Y}, v)$ *be two fuzzy measure spaces and* f *be a* $\mathcal{X} \times \mathcal{Y}$-*measurable function. Then, there exists a fuzzy measure* m *on* $X \times Y$ *such that*

$$\int_{\mathcal{Y}} \left(\int_{\mathcal{X}} f(x,y) d\mu \right) dv = \int_{\mathcal{X}} \left(\int_{\mathcal{Y}} f(x,y) dv \right) d\mu = \int_{\mathcal{X} \times \mathcal{Y}} f(x,y) dm$$

Considering the characteristic function of $A \times B \in \mathcal{X} \times \mathcal{Y}$, *we have* $m = \mu \wedge v$.

Remark 2 ([13]). *In the sequel,* μ *will denote the Lebesgue measure on* $\mathbb{R}$ *and* $\mu \times \mu$ *will denote the Lebesgue measure on* $\mathbb{R} \times \mathbb{R}$. *We recall that if* A, B *are two* μ-*measurable subsets of* $\mathbb{R}$, *then*

$$(\mu \times \mu)(A \times B) = \mu(A)\mu(B).$$

For more details on Sugeno integral we refer the interested readers to [1,5].

The classical Hermite–Hadamard inequalities provide estimates of the mean value of a convex function $f : [a, b] \to \mathbb{R}$

$$f\left(\frac{a+b}{2}\right) \leq \frac{1}{b-a} \int_a^b f(x) dx \leq \frac{f(a) + f(b)}{2}. \tag{1}$$

The inequalities (1) were first discovered by C. Hermite [22] in 1893 and J. Hadamard proved it independently again in [21]. The inequalities (1) have numerous generalizations and extensions, we the refer the reader to [20,35,37,39–47] and the references cited therein.

We recall some definitions and Hermite–Hadamard-type integral inequalities for coordinated convex functions on $[a, b] \times [c, d]$ in $\mathbb{R}^2$.

Definition 4 ([39]). *Let* $\Delta =: [a, b] \times [c, d]$ *in* $\mathbb{R}^2$ *with* $a < b$ *and* $c < d$ *be a bidimensional interval. A mapping* $f : \Delta \to \mathbb{R}$ *is said to be convex on* Δ *if the inequality*

$$f(\lambda x + (1 - \lambda)z, \lambda y + (1 - \lambda)w) \leq \lambda f(x, y) + (1 - \lambda) f(z, w)$$

holds for all $(x, y), (z, w) \in \Delta$ *and* $\lambda \in [0, 1]$.

Dragomir [39] modified Definition 4 of convex functions on Δ, known as coordinated convex functions as follows.

Definition 5 ([39]). *A function* $f : \Delta \to \mathbb{R}$ *is said to be convex on the coordinates on* Δ *if the partial mappings* $f_y : [a, b] \to \mathbb{R}$, $f_y(u) = f(u, y)$ *and* $f_x : [c, d] \to \mathbb{R}$, $f_x(v) = f(x, v)$ *are convex where defined for all* $x \in [a, b]$, $y \in [c, d]$.

Remark 3 ([41]). *It is clear that if a function $f : \Delta \to \mathbb{R}$ is convex on the coordinates on Δ, then*

$$\begin{aligned} & f(tx + (1-t)z, sy + (1-s)w) \\ & \le tsf(x,y) + t(1-s)f(x,w) + s(1-t)f(z,y) + (1-t)(1-s)f(z,w), \end{aligned}$$

holds for all $(t,s) \in [0,1] \times [0,1]$ and $x, z \in [a,b], y, w \in [c,d]$.

The following Hermite–Hadamard-type for coordinated convex functions on the rectangle from the plane $\mathbb{R}^2$ were proved in ([39], Theorem 1, page 778):

Theorem 7 ([39]). *If $f : \Delta \to \mathbb{R}$ is coordinated convex on Δ, then*

$$\begin{aligned} & f\left(\frac{a+b}{2}, \frac{c+d}{2}\right) \\ & \le \frac{1}{2}\left[\frac{1}{b-a}\int_a^b f\left(x, \frac{c+d}{2}\right)dx + \frac{1}{d-c}\int_c^d f\left(\frac{a+b}{2}, y\right)dy\right] \\ & \le \frac{1}{(b-a)(d-c)}\int_a^b\int_c^d f(x,y)dydx \\ & \le \frac{1}{4}\left[\frac{1}{b-a}\int_a^b [f(x,c) + f(x,d)]dx + \frac{1}{d-c}\int_c^d [f(a,y) + f(b,y)]dy\right] \\ & \le \frac{f(a,c) + f(a,d) + f(b,c) + f(b,d)}{4}. \end{aligned} \quad (2)$$

The above inequalities are sharp.

We will see that this inequality does not hold true for fuzzy integrals in general.

Example 1. *Take $X \times Y = [0,1] \times [0,1]$ and let μ be the Lebesgue measure on X and Y. Then, $\mu \times \mu$ will denote the Lebesgue measure on $X \times Y$. Let $f(x,y) : X \times Y \to [0,\infty)$ be defined as $f(x,y) = \frac{1}{4}xy$, then the function is convex on the coordinates on $X \times Y$. Now, we calculate the Sugeno integral $\int_{[0,1]^2} \frac{1}{4}xyd(\mu \times \mu)$. By using the Fubini theorem for fuzzy integrals, we observe that*

$$\int_{[0,1]^2} \frac{1}{4}xyd(\mu \times \mu) = \int_{[0,1]} \left(\int_{[0,1]} \frac{1}{4}xyd\mu\right)d\mu$$

Let F be the survival function associated to $f(x,y) = \frac{1}{4}xy$ on $[0,1]$, then

$$\alpha = F(\alpha) = \mu\left([0,1] \cap \left\{\frac{1}{4}xy \ge \alpha\right\}\right) = \mu\left([0,1] \cap \left\{y \ge \frac{4\alpha}{x}\right\}\right) = 1 - \frac{4\alpha}{x}.$$

Thus,

$$\alpha = \frac{x}{x+4}$$

According to Remark 1, we obtain

$$\int_{[0,1]} \frac{1}{4}xyd\mu = \frac{x}{x+4}.$$

Hence

$$\int_{[0,1]^2} \frac{1}{4}xyd(\mu \times \mu) = \int_{[0,1]} \frac{x}{x+4}d\mu = 3 - 2\sqrt{2} \approx 0.1716.$$

$$\frac{1}{4}\left[\int_{[0,1]} \frac{1}{4}xd\mu + \int_{[0,1]} \frac{1}{4}yd\mu\right] = \frac{1}{4}\left[\frac{1}{5} + \frac{1}{5}\right] = 0.1.$$

Finally,

$$\frac{f(0,0)+f(0,1)+f(1,0)+f(1,1)}{4}=0.0625.$$

We can see that the third and the fourth inequalities in (2) are not satisfied in the fuzzy context.

Example 2. *Take $X\times Y=[0,1]\times[0,1]$ and let μ be the Lebesgue measure on X and Y. Then, $\mu\times\mu$ will denote the Lebesgue measure on $X\times Y$. Let $f(x,y):X\times Y\to[0,\infty)$ be defined as $f(x,y)=4xy$, then the function is convex on the coordinates on $X\times Y$. Now, we calculate the Sugeno integral $\int_{[0,1]^2}4xyd(\mu\times\mu)$. By using the Fubini theorem for fuzzy integrals and Remark 1, we observe that $f\left(\frac{1}{2},\frac{1}{2}\right)=1$ and*

$$\int_{[0,1]^2}4xyd(\mu\times\mu)=\int_{[0,1]}\left(\int_{[0,1]}4xyd\mu\right)d\mu.$$

Suppose that F is the survival function associated to $f(x,y)=4xy$ on $[0,1]$. Then,

$$\alpha=F(\alpha)=\mu([0,1]\cap\{4xy\geq\alpha\})=\mu\left([0,1]\cap\left\{y\geq\frac{\alpha}{4x}\right\}\right)=1-\frac{\alpha}{4x}.$$

Thus,

$$\alpha=\frac{4x}{4x+1}$$

According to Remark 1, we obtain

$$\int_{[0,1]}4xyd\mu=\frac{4x}{4x+1}.$$

Hence,

$$\int_{[0,1]^2}4xyd(\mu\times\mu)=\int_{[0,1]}\frac{4x}{4x+1}d\mu=\frac{9-\sqrt{17}}{8}\approx 0.6096.$$

Lastly,

$$\frac{1}{2}\left[\int_{[0,1]}f\left(x,\frac{1}{2}\right)d\mu+\int_{[0,1]}f\left(\frac{1}{2},y\right)d\mu\right]=\frac{1}{2}\left[\int_{[0,1]}2xd\mu+\int_{[0,1]}2yd\mu\right]=\frac{1}{3}\approx 0.3333,$$

which shows that the first and the second inequalities in (2) are also not satisfied in the fuzzy framework.

Now we prove estimates for the third and the fourth inequalities in (2) but for the Sugeno integral. In order to obtain our main results, we will use a non-additivity assumption of integrals together with the Lebesgue measure, that is, we use the fuzzy context to prove the results.

Our first result gives estimates of the integrals involved in the fourth inequality in (2) for fuzzy integrals over the interval $[0,1]$.

Theorem 8. *Let $g:[0,1]\times[0,1]\to[0,\infty)$ be a convex function on the coordinates on $[0,1]\times[0,1]$. Let $\mu\times\mu$ be the Lebesgue measure on $[0,1]\times[0,1]$.*

1. *If $g(1,1)+g(1,0)>g(0,1)+g(0,0)$, then*

$$\int_{[0,1]}[g(x,0)+g(x,1)]d\mu\leq\min\left\{1,\frac{g(1,1)+g(1,0)}{1+g(1,1)+g(1,0)-g(0,1)-g(0,0)}\right\}. \quad (3)$$

2. *If* $g(1,1)+g(0,1)>g(1,0)+g(0,0)$, *then*

$$\int_{[0,1]}[g(0,y)+g(1,y)]d\mu \leq \min\left\{1, \frac{g(1,1)+g(0,1)}{1+g(1,1)+g(0,1)-g(1,0)-g(0,0)}\right\}. \quad (4)$$

Proof. By the coordinated convexity of g on $[0,1]\times[0,1]$ that

$$\begin{aligned} g(x,0)+g(x,1) &= g((1-x)\cdot 0+x\cdot 1,1)+g((1-x)\cdot 0+x\cdot 1,0) \\ &\leq (1-x)g(0,1)+xg(1,1)+(1-x)g(0,0)+xg(1,0) \\ &= g(0,1)+g(0,0)+x[g(1,1)+g(1,0)-g(0,1)-g(0,0)] \end{aligned}$$

and hence by 3. of Proposition 1, we obtain

$$\begin{aligned} &\int_{[0,1]}[g(x,0)+g(x,1)]d\mu \\ &\leq \int_{[0,1]}\{g(0,1)+g(0,0)+x[g(1,1)+g(1,0)-g(0,1)-g(0,0)]\}d\mu \\ &= \int_{[0,1]}h_1(x)d\mu. \end{aligned}$$

If we consider the survival function F together with the assumption $g(1,1)+g(1,0)>g(0,1)+g(0,0)$. Then, according to Remark 1, we obtain

$$\begin{aligned} \alpha = \mu\left([0,1]\cap\left\{\begin{array}{c} g(0,1)+g(0,0) \\ +x[g(1,1)+g(1,0)-g(0,1)-g(0,0)]\geq\alpha \end{array}\right\}\right) \\ = 1-\frac{\alpha-g(0,1)-g(0,0)}{g(1,1)+g(1,0)-g(0,1)-g(0,0)}. \quad (5) \end{aligned}$$

The solution of the Equation (5) is $\alpha=\frac{g(1,1)+g(1,0)}{1+g(1,1)+g(1,0)-g(0,1)-g(0,0)}$.
Applying 1. of Proposition 1, we obtain

$$\int_{[0,1]}[g(x,0)+g(x,1)]d\mu \leq \int_{[0,1]}h_1(x)d\mu \leq \mu([0,1])=1. \quad (6)$$

The solution $\alpha=\frac{g(1,1)+g(1,0)}{1+g(1,1)+g(1,0)-g(0,1)-g(0,0)}$ of the Equation (5) together with (6) prove the inequality (3).
Since g is convex on the coordinates on $[0,1]\times[0,1]$, we find that

$$\begin{aligned} g(0,y)+g(1,y) &= g(0,(1-y)\cdot 0+y\cdot 1)+g(1,(1-y)\cdot 0+y\cdot 1) \\ &\leq (1-y)g(0,0)+yg(0,1)+(1-y)g(1,0)+yg(1,1) \\ &= g(1,0)+g(0,0)+y[g(1,1)+g(0,1)-g(1,0)-g(0,0)] \end{aligned}$$

and hence by 3. of Proposition 1, we obtain

$$\begin{aligned} &\int_{[0,1]}[g(0,y)+g(1,y)]d\mu \\ &\leq \int_{[0,1]}\{g(1,0)+g(0,0)+y[g(1,1)+g(0,1)-g(1,0)-g(0,0)]\}d\mu \\ &= \int_{[0,1]}h_2(y)d\mu. \end{aligned}$$

Suppose F is the survival function and $g(1,1)+g(0,1) > g(1,0)+g(0,0)$, then according to Remark 1, we obtain

$$\beta = \mu\left([0,1] \cap \left\{ \begin{array}{c} g(1,0)+g(0,0) \\ +y[g(1,1)+g(0,1)-g(1,0)-g(0,0)] \geq \beta \end{array} \right\}\right) = 1 - \frac{\beta - g(1,0) - g(0,0)}{g(1,1)+g(0,1)-g(1,0)-g(0,0)}. \quad (7)$$

The solution of the Equation (7) is $\beta = \frac{g(1,1)+g(0,1)}{1+g(1,1)+g(0,1)-g(1,0)-g(0,0)}$.
Applying 1. of Proposition 1, we obtain

$$\int_{[0,1]} [g(0,y)+g(1,y)]d\mu \leq \int_{[0,1]} h_2(y)d\mu \leq \mu([0,1]) = 1. \quad (8)$$

The solution $\beta = \frac{g(1,1)+g(0,1)}{1+g(1,1)+g(0,1)-g(1,0)-g(0,0)}$ of the Equation (7) together with (8) prove the inequality (4). □

Remark 4. *If* $g(1,1)+g(1,0) = g(0,1)+g(0,0)$, *then*

$$\int_{[0,1]} [g(x,0)+g(x,1)]d\mu \leq 1 \wedge [g(1,1)+g(1,0)]. \quad (9)$$

If $g(1,1)+g(0,1) = g(1,0)+g(0,0)$, *then*

$$\int_{[0,1]} [g(0,y)+g(1,y)]d\mu \leq 1 \wedge [g(1,1)+g(0,1)]. \quad (10)$$

The second result provides an estimate of the first integral of the third inequality in (2) for fuzzy integrals over the interval $[0,1]$.

Theorem 9. *Let* $g : [0,1] \times [0,1] \to [0,\infty)$ *be a convex function on the coordinates on* $[0,1] \times [0,1]$ *such that* $g(0,1) > g(0,0)$ *and* $g(0,0)+g(1,1) > g(1,0)+g(0,1)$. *Let* $\mu \times \mu$ *be the Lebesgue measure on* $[0,1] \times [0,1]$, *then*

$$\int_{[0,1]^2} g(x,y)d(\mu \times \mu) \leq \min\{1,\alpha\}, \quad (11)$$

where α *is a positive solution of the equation*

$$(g(0,0)-g(1,0)-g(0,1)+g(1,1))\alpha^2 + (g(1,0)+g(0,1)-2g(1,1)-1)\alpha + g(1,1) = 0. \quad (12)$$

Proof. Since g is a convex function on the coordinates on $[0,1] \times [0,1]$. Therefore, for $(x,y) \in [0,1] \times [0,1]$, we obtain

$$\begin{aligned} g(x,y) &= g((1-x)\cdot 0 + x\cdot 1, (1-y)\cdot 0 + y \cdot 1) \\ &\leq (1-x)(1-y)g(0,0) + (1-x)yg(0,1) \\ &\quad + x(1-y)g(1,0) + xyg(1,1) = h(x,y). \end{aligned}$$

Suppose F is the survival function with respect to the variable x together with $g(0,1) > g(0,0)$ and $g(0,0)+g(1,1) > g(1,0)+g(0,1)$. By 3. of Proposition 1 and by using the Fubini theorem for fuzzy integrals and Remark 1, we have

$$\int_{[0,1]^2} g(x,y)d(\mu\times\mu) \leq \int_{[0,1]^2}[(1-x)(1-y)g(0,0)+(1-x)yg(0,1)$$
$$+x(1-y)g(1,0)+xyg(1,1)]d(\mu\times\mu) = \int_{[0,1]}\left(\int_{[0,1]} h(x,y)d\mu\right)d\mu.$$
$$= \int_{[0,1]} \frac{g(0,1)+[g(1,1)-g(0,1)]x}{1+g(0,1)-g(0,0)+[g(0,0)+g(1,1)-g(1,0)-g(0,1)]x}d\mu = \alpha, \quad (13)$$

where α is a positive solution of the equation

$$(g(0,0)-g(1,0)-g(0,1)+g(1,1))\alpha^2$$
$$+(g(1,0)+g(0,1)-2g(1,1)-1)\alpha+g(1,1)=0. \quad (14)$$

However, according to 1. of Proposition 1, we obtain

$$\int_{[0,1]^2} g(x,y)d(\mu\times\mu) \leq \mu\times\mu([0,1]\times[0,1]) = \mu([0,1])\mu([0,1]) = 1. \quad (15)$$

A positive solution of (14) and (15), we obtain (11). □

Remark 5. *If $g(0,0)+g(1,1)=g(1,0)+g(0,1)$, $g(1,1)=g(0,1)$ and $g(0,1)>g(0,0)$ in Theorem 9, then from (13) we obtain*

$$\int_{[0,1]^2} g(x,y)d(\mu\times\mu) \leq 1 \wedge \frac{g(0,1)}{1+g(0,1)-g(0,0)}. \quad (16)$$

Another estimate of the first integral of the third inequality in (2) for fuzzy integrals over the interval $[0,1]$ can be determined as given in the following remark.

Remark 6. *Since g is a convex function on the coordinates on $[0,1]\times[0,1]$, we obtain*

$$g(x,y) = g((1-x)\cdot 0 + x\cdot 1, y) \leq (1-x)g(0,y)+xg(1,y) \leq g(0,y)+g(1,y)$$

and

$$g(x,y) = g(x,(1-y)\cdot 0 + y\cdot 1) \leq (1-y)g(x,0)+yg(x,1) \leq g(x,0)+g(x,1).$$

Hence, by 1., 3. of Proposition 1 and Fubini theorem for fuzzy integrals, we obtain

$$\int_{[0,1]^2} g(x,y)d(\mu\times\mu) \leq \int_{[0,1]}[g(0,y)+g(1,y)]d\mu \quad (17)$$

and

$$\int_{[0,1]^2} g(x,y)d(\mu\times\mu) \leq \int_{[0,1]}[g(x,0)+g(x,1)]d\mu. \quad (18)$$

Thus, from (3), (4), (17) and (18), we find that

$$\int_{[0,1]^2} g(x,y)d(\mu\times\mu) \leq \min\left\{1, \frac{g(1,1)+g(1,0)}{1+g(1,1)+g(1,0)-g(0,1)-g(0,0)}, \right.$$
$$\left. \frac{g(1,1)+g(0,1)}{1+g(1,1)+g(0,1)-g(1,0)-g(0,0)}\right\}. \quad (19)$$

It is clear that from (11) and (19) that the inequality

$$\int_{[0,1]^2} g(x,y)d(\mu\times\mu) \le \min\left\{1, \alpha, \frac{g(1,1)+g(1,0)}{1+g(1,1)+g(1,0)-g(0,1)-g(0,0)}, \frac{g(1,1)+g(0,1)}{1+g(1,1)+g(0,1)-g(1,0)-g(0,0)}\right\} \quad (20)$$

holds, where α is a positive solution of the Equation (12).

The following example illustrates how to obtain an estimate for the first integral of the third inequality in (2) using fuzzy integrals.

Example 3. *Take $X=[0,1]$, $Y=[0,1]$ and let μ be the Lebesgue measure on X and Y. Suppose that $g:[0,1]\times[0,1]\to[0,\infty)$ is defined as $g(x,y)=x^2e^y$ and $\mu\times\mu$ is the Lebesgue measure on $X\times Y$.*
Since $g(1,1)+g(1,0)>g(0,1)+g(0,0)$, hence by 1. of Theorem 8, we have

$$\int_{[0,1]}(1+e)x^2d\mu \le \min\left\{1, \frac{e+1}{2+e}\right\} = \frac{e+1}{2+e} \approx 0.78806.$$

We observe that $g(1,1)+g(0,1)>g(1,0)+g(0,0)$, hence by 2. of Theorem 8, we obtain

$$\int_{[0,1]} e^y d\mu \le 1.$$

Finally, (20) gives

$$\int_{[0,1]^2} x^2e^y d(\mu\times\mu) \le \min\left\{1, \alpha, \frac{e+1}{2+e}\right\},$$

where α is a positive root of the equation

$$(e-1)\alpha^2 - 2e\alpha + e = 0.$$

The solution of this equation is

$$\alpha_1 = \frac{e+e^{\frac{1}{2}}}{e-1} \approx 2.5415$$

and

$$\alpha_2 = \frac{e-e^{\frac{1}{2}}}{e-1} \approx 0.62246.$$

Thus,

$$\int_{[0,1]^2} x^2e^y d(\mu\times\mu) \le \frac{e-e^{\frac{1}{2}}}{e-1} \approx 0.62246.$$

The next result gives different estimates of the integral involved in the fourth inequality in (2) for fuzzy integrals over the interval $[0,1]$.

Theorem 10. *Let $g:[0,1]\times[0,1]\to[0,\infty)$ be a convex function on the coordinates on $[0,1]\times[0,1]$. Let $\mu\times\mu$ be the Lebesgue measure on $[0,1]\times[0,1]$.*

1. *If $g(0,0)+g(0,1)<g(1,0)+g(1,1)$, then*

$$\int_{[0,1]}[g(x,0)+g(x,1)]d\mu \le \min\left\{1, \frac{g(0,0)+g(0,1)}{1+g(0,0)+g(0,1)-g(1,0)-g(1,1)}\right\}. \quad (21)$$

2. *If* $g(1,0)+g(0,0)<g(1,1)+g(0,1)$, *then*

$$\int_{[0,1]}[g(0,y)+g(1,y)]d\mu \leq \min\left\{1, \frac{g(0,0)+g(1,0)}{1+g(1,0)+g(0,0)-g(1,1)-g(0,1)}\right\}. \quad (22)$$

Proof. By the coordinated convexity of g on $[0,1]\times[0,1]$, we find that

$$\begin{aligned} g(x,0)+g(x,1) &= g((1-x)\cdot 0+x\cdot 1,1)+g((1-x)\cdot 0+x\cdot 1,0) \\ &\leq (1-x)g(0,1)+xg(1,1)+(1-x)g(0,0)+xg(1,0) \\ &= g(0,1)+g(0,0)+x[g(1,1)+g(1,0)-g(0,1)-g(0,0)] \end{aligned}$$

and hence by 3. of Proposition 1, we obtain

$$\begin{aligned} &\int_{[0,1]}[g(x,0)+g(x,1)]d\mu \\ &\leq \int_{[0,1]}\{g(0,1)+g(0,0)+x[g(1,1)+g(1,0)-g(0,1)-g(0,0)]\}d\mu \\ &= \int_{[0,1]}h_1(x)d\mu. \end{aligned}$$

If we consider the survival function F together with the condition $g(0,0)+g(0,1)<g(1,0)+g(1,1)$, then according to Remark 1, we obtain

$$\begin{aligned} \alpha = \mu\left([0,1]\cap\left\{\begin{array}{c} g(0,1)+g(0,0) \\ +x[g(1,1)+g(1,0)-g(0,1)-g(0,0)]\geq\alpha \end{array}\right\}\right) \\ = \frac{\alpha-g(0,1)-g(0,0)}{g(1,1)+g(1,0)-g(0,1)-g(0,0)}. \quad (23) \end{aligned}$$

The solution of the Equation (23) is $\alpha=\frac{g(0,0)+g(0,1)}{1+g(0,0)+g(0,1)-g(1,0)-g(1,1)}$. Applying 1. of Proposition 1, we obtain

$$\int_{[0,1]}[g(x,0)+g(x,1)]d\mu \leq \int_{[0,1]}h_1(x)d\mu \leq \mu([0,1])=1. \quad (24)$$

The solution of Equation (23) together with (24) give us the required inequality (21). Since g is convex on the coordinates on $[0,1]\times[0,1]$, we find that

$$\begin{aligned} g(0,y)+g(1,y) &= g(0,(1-y)\cdot 0+y\cdot 1)+g(1,(1-y)\cdot 0+y\cdot 1) \\ &\leq (1-y)g(0,0)+yg(0,1)+(1-y)g(1,0)+yg(1,1) \\ &= g(1,0)+g(0,0)+y[g(1,1)+g(0,1)-g(1,0)-g(0,0)] \end{aligned}$$

and hence by 3. of Proposition 1, we obtain

$$\begin{aligned} &\int_{[0,1]}[g(0,y)+g(1,y)]d\mu \\ &\leq \int_{[0,1]}\{g(1,0)+g(0,0)+y[g(1,1)+g(0,1)-g(1,0)-g(0,0)]\}d\mu \\ &= \int_{[0,1]}h_2(y)d\mu. \end{aligned}$$

Suppose F is the survival function and $g(1,0)+g(0,0)<g(1,1)+g(0,1)$, then according to Remark 1, we obtain

$$\beta=\mu\left([0,1]\cap\left\{\begin{array}{c} g(1,0)+g(0,0) \\ +y[g(1,1)+g(0,1)-g(1,0)-g(0,0)]\geq\beta \end{array}\right\}\right)$$
$$=\frac{\beta-g(1,0)-g(0,0)}{g(1,1)+g(0,1)-g(1,0)-g(0,0)}. \tag{25}$$

The solution of Equation (25) is $\beta=\frac{g(0,0)+g(1,0)}{1+g(1,0)+g(0,0)-g(1,1)-g(0,1)}$.
Applying 1. of Proposition 1, we obtain

$$\int_{[0,1]}[g(0,y)+g(1,y)]d\mu\leq\int_{[0,1]}h_2(y)d\mu\leq\mu([0,1])=1. \tag{26}$$

The solutions of Equation (25) and inequality (26) give us inequality (22). □

Remark 7. *If* $g(1,1)+g(1,0)=g(0,1)+g(0,0)$ *in Theorem 8, then* $g(0,0)+g(0,1)=g(1,0)+g(1,1)$

$$\int_{[0,1]}[g(x,0)+g(x,1)]d\mu\leq 1\wedge[g(0,1)+g(0,0)] \tag{27}$$

and if $g(1,1)+g(0,1)=g(1,0)+g(0,0)$, *then*

$$\int_{[0,1]}[g(0,y)+g(1,y)]d\mu\leq 1\wedge[g(1,0)+g(0,0)]. \tag{28}$$

We can obtain some different estimates of the integral involved in the fourth inequality in (2) for fuzzy integrals over the interval $[0,1]$ if we replace "$>$" with "$<$" in the assumptions of Theorem 10.

Theorem 11. *Let* $g:[0,1]\times[0,1]\to[0,\infty)$ *be a convex function on the coordinates on* $[0,1]\times[0,1]$ *such that* $g(0,0)<g(0,1)$ *and* $g(0,0)+g(1,1)<g(1,0)+g(0,1)$. *Let* $\mu\times\mu$ *be the Lebesgue measure on* $[0,1]\times[0,1]$, *then*

$$\int_{[0,1]^2}g(x,y)d(\mu\times\mu)\leq\min\{1,\alpha\}, \tag{29}$$

where α *is a positive solution of the equation*

$$(g(0,0)-g(1,0)-g(0,1)+g(1,1))\alpha^2$$
$$+(g(1,0)+g(0,1)-2g(0,0)-1)\alpha+g(0,0)=0. \tag{30}$$

Proof. Since g is a convex function on the coordinates on $[0,1]\times[0,1]$. Hence, for $(x,y)\in[0,1]\times[0,1]$, we have the inequality

$$g(x,y)=g((1-x)\cdot 0+x\cdot 1,(1-y)\cdot 0+y\cdot 1)$$
$$\leq(1-x)(1-y)g(0,0)+(1-x)yg(0,1)$$
$$+x(1-y)g(1,0)+xyg(1,1)=h(x,y).$$

Suppose F is the survival function with respect to the variable x together with $g(0,0)<g(0,1)$ and G is the survival function with respect to the variable y together with $g(1,0)+$

$g(0,1) < g(0,0) + g(1,1)$. By 3. of Proposition 1 and by using the Fubini theorem for fuzzy integrals and Remark 1, we have

$$\begin{aligned} \int_{[0,1]^2} g(x,y)d(\mu\times\mu) &\le \int_{[0,1]^2} [(1-x)(1-y)g(0,0) + (1-x)yg(0,1) \\ &\quad + x(1-y)g(1,0) + xyg(1,1)]d(\mu\times\mu) = \int_{[0,1]}\left(\int_{[0,1]} h(x,y)d\mu\right)d\mu. \\ &= \int_{[0,1]} \frac{g(0,0) + [g(1,0) - g(0,0)]x}{1 + g(0,0) - g(0,1) + [g(1,0) + g(0,1) - g(0,0) - g(1,1)]x} d\mu = \alpha, \end{aligned} \tag{31}$$

where α is a positive solution of the equation

$$\begin{aligned} &(g(0,0) - g(1,0) - g(0,1) + g(1,1))\alpha^2 \\ &\qquad + (g(1,0) + g(0,1) - 2g(0,0) - 1)\alpha + g(0,0) = 0. \end{aligned} \tag{32}$$

However, according to 1. of Proposition 1, we obtain

$$\int_{[0,1]^2} g(x,y)d(\mu\times\mu) \le \mu\times\mu([0,1]\times[0,1]) = \mu([0,1])\mu([0,1]) = 1. \tag{33}$$

A positive solution of (32) and inequality (33) give us the desired inequality (29). □

Remark 8. *If $g(0,0) + g(1,1) = g(1,0) + g(0,1)$, $g(0,0) = g(1,0)$ and $g(0,0) < g(0,1)$ in Theorem 11, then from (31) we obtain*

$$\int_{[0,1]^2} g(x,y)d(\mu\times\mu) \le 1 \wedge \frac{g(0,0)}{1 + g(0,0) - g(0,1)}. \tag{34}$$

A complementary estimate of the first integral of the third inequality in (2) for fuzzy integrals over the interval $[0,1]$ is given in the remark below.

Remark 9. *Since g is a convex function on the coordinates on $[0,1]\times[0,1]$, we obtain*

$$g(x,y) = g((1-x)\cdot 0 + x\cdot 1, y) \le (1-x)g(0,y) + xg(1,y) \le g(0,y) + g(1,y)$$

and

$$g(x,y) = g(x,(1-y)\cdot 0 + y\cdot 1) \le (1-y)g(x,0) + yg(x,1) \le g(x,0) + g(x,1).$$

Hence by 1., 3. of Proposition 1 and Fubini theorem for fuzzy integrals, we obtain

$$\int_{[0,1]^2} g(x,y)d(\mu\times\mu) \le \int_{[0,1]} [g(0,y) + g(1,y)]d\mu \tag{35}$$

and

$$\int_{[0,1]^2} g(x,y)d(\mu\times\mu) \le \int_{[0,1]} [g(x,0) + g(x,1)]d\mu. \tag{36}$$

Thus, from (21), (22), (35) and (36), we obtain

$$\begin{aligned} \int_{[0,1]^2} g(x,y)d(\mu\times\mu) \le \min\Bigg\{1, &\frac{g(0,0) + g(0,1)}{1 + g(0,0) + g(0,1) - g(1,0) - g(1,1)}, \\ &\frac{g(0,0) + g(1,0)}{1 + g(1,0) + g(0,0) - g(1,1) - g(0,1)}\Bigg\}. \end{aligned} \tag{37}$$

It is clear from (29) and (37) that the inequality

$$\int_{[0,1]^2} g(x,y)d(\mu\times\mu) \le \min\left\{1, \alpha, \frac{g(0,0)+g(0,1)}{1+g(0,0)+g(0,1)-g(1,0)-g(1,1)}, \frac{g(0,0)+g(1,0)}{1+g(1,0)+g(0,0)-g(1,1)-g(0,1)}\right\} \quad (38)$$

holds, where α is a positive solution of the Equation (30).

The following example illustrates the method for the calculation of the estimates of the first integral of the third inequality in (2).

Example 4. *Take $X=[0,1]$, $Y=[0,1]$ and let μ be the Lebesgue measure on X and Y. Suppose that $g:[0,1]\times[0,1]\to[0,\infty)$ is defined as $g(x,y)=(1+x-\ln(x+1))e^{-y}$ and $\mu\times\mu$ be the Lebesgue measure on $X\times Y$.*
Since $g(0,0)+g(0,1)<g(1,0)+g(1,1)$, hence by 1. of Theorem 9, we have

$$\int_{[0,1]}\left(1+e^{-1}\right)(1+x-\ln(x+1))d\mu \le \frac{1+e^{-1}}{(1+e^{-1})\ln 2-e^{-1}} \approx 2.3573.$$

We observe that $g(1,0)+g(0,0)<g(1,1)+g(0,1)$, hence by 2. of Theorem 9, we obtain

$$\int_{[0,1]}(3-\ln 2)e^{-y}d\mu \le \frac{3-\ln 2}{(e^{-1}-1)\ln 2-3e^{-1}+4} \approx 0.93843.$$

Finally, (30) gives

$$\left(1-e^{-1}\right)(\ln 2-1)\alpha^2+\left(e^{-1}-\ln 2-1\right)\alpha+1=0.$$

The solutions of this equation are

$$\alpha_1 = \frac{e^{-1}\left(e-\sqrt{5e^2-6e+e^2\ln^2 2+2e\ln 2-2e^2\ln 2+1}-1+e\ln 2\right)}{2(1-e^{-1})(\ln 2-1)} \approx 0.68574$$

and

$$\alpha_2 = \frac{e^{-1}\left(e+\sqrt{5e^2-6e+e^2\ln^2 2+2e\ln 2-2e^2\ln 2+1}-1+e\ln 2\right)}{2(1-e^{-1})(\ln 2-1)} \approx -7.5181$$

Thus,

$$\int_{[0,1]^2}[x+1-\ln(1+x)]e^{-y}d(\mu\times\mu) \le \frac{e^{-1}\left(e-\sqrt{5e^2-6e+e^2\ln^2 2+2e\ln 2-2e^2\ln 2+1}-1+e\ln 2\right)}{2(1-e^{-1})(\ln 2-1)} \approx 0.68574.$$

In the sequel, we will prove the general case of Theorems 8–11 and examples related to these theorems.

Theorem 12. *Let $g:[a,b]\times[c,d]\to[0,\infty)$ be a convex function on the coordinates on $[a,b]\times[c,d]$. Let $\mu\times\mu$ be the Lebesgue measure on $[a,b]\times[c,d]$.*

1. *If* $g(b,d)+g(b,c)>g(a,d)+g(a,c)$, *then*

$$\int_{[a,b]}[g(x,c)+g(x,d)]d\mu \le \min\left\{b-a,\frac{(b-a)(g(b,d)+g(b,c))}{b-a+g(b,d)+g(b,c)-g(a,d)-g(a,c)}\right\}. \quad (39)$$

2. *If* $g(a,d)+g(b,d)>g(a,c)+g(b,c)$, *then*

$$\int_{[c,d]}[g(a,y)+g(b,y)]d\mu \le \min\left\{d-c,\frac{(d-c)(g(a,d)+g(b,d))}{d-c+g(a,d)+g(b,d)-g(a,c)-g(b,c)}\right\}. \quad (40)$$

Proof. By the coordinated convexity of g on $[a,b]\times[c,d]$, we find that

$$\begin{aligned}
&g(x,c)+g(x,d)\\
&=g\left(\left(1-\frac{x-a}{b-a}\right)\cdot a+\frac{x-a}{b-a}\cdot b,d\right)+g\left(\left(1-\frac{x-a}{b-a}\right)\cdot a+\frac{x-a}{b-a}\cdot b,c\right)\\
&\le\left(\frac{b-x}{b-a}\right)g(a,d)+\left(\frac{x-a}{b-a}\right)g(b,d)+\left(\frac{b-x}{b-a}\right)g(a,c)+\left(\frac{x-a}{b-a}\right)g(b,c).
\end{aligned}$$

and hence by 3. of Proposition 1, we obtain

$$\begin{aligned}
&\int_{[a,b]}[g(x,c)+g(x,d)]d\mu\\
&\le\int_{[a,b]}\left\{\left(\frac{b-x}{b-a}\right)g(a,d)+\left(\frac{x-a}{b-a}\right)g(b,d)\right.\\
&\left.+\left(\frac{b-x}{b-a}\right)g(a,c)+\left(\frac{x-a}{b-a}\right)g(b,c)\right\}d\mu=\int_{[a,b]}h_1(x)d\mu.
\end{aligned}$$

If we consider the survival function F and $g(b,d)+g(b,c)>g(a,d)+g(a,c)$, then according to Remark 1, we obtain

$$\begin{aligned}
\alpha&=\mu\left([a,b]\cap\left\{\begin{array}{c}\left(\frac{b-x}{b-a}\right)g(a,d)+\left(\frac{x-a}{b-a}\right)g(b,d)\\+\left(\frac{b-x}{b-a}\right)g(a,c)+\left(\frac{x-a}{b-a}\right)g(b,c)\ge\alpha\end{array}\right\}\right)\\
&=b-\frac{\alpha(b-a)+ag(b,d)-bg(a,d)+ag(b,c)-bg(a,c)}{g(a,d)+g(b,c)-g(b,d)-g(a,c)}. \quad (41)
\end{aligned}$$

The solution of Equation (41) is $\alpha=\frac{(b-a)(g(b,d)+g(b,c))}{b-a+g(b,d)+g(b,c)-g(a,d)-g(a,c)}$.
Applying 1. of Proposition 1, we obtain

$$\int_{[a,b]}[g(x,c)+g(x,d)]d\mu\le\int_{[a,b]}h_1(x)d\mu\le\mu([a,b])=b-a. \quad (42)$$

The solutions of Equations (41) and (42) give us (39).
Since g is convex on the coordinates on $[a,b]\times[c,d]$, we find that

$$\begin{aligned}
&g(a,y)+g(b,y)\\
&=g\left(a,\left(1-\frac{y-c}{d-c}\right)\cdot c+\left(\frac{y-c}{d-c}\right)\cdot d\right)+g\left(b,\left(1-\frac{y-c}{d-c}\right)\cdot c+\frac{y-c}{d-c}\cdot d\right)\\
&\le\left(\frac{d-y}{d-c}\right)g(a,c)+\left(\frac{y-c}{d-c}\right)g(a,d)+\left(\frac{d-y}{d-c}\right)g(b,c)+\left(\frac{y-c}{d-c}\right)g(b,d).
\end{aligned}$$

and hence by 3. of Proposition 1, we obtain

$$\int_{[c,d]}[g(a,y)+g(b,y)]d\mu \leq \int_{[c,d]}\left\{\left(\frac{d-y}{d-c}\right)g(a,c)+\left(\frac{y-c}{d-c}\right)g(a,d)\right.$$
$$\left.+\left(\frac{d-y}{d-c}\right)g(b,c)+\left(\frac{y-c}{d-c}\right)g(b,d)\right\}d\mu=\int_{[c,d]}h_2(y)d\mu.$$

Suppose F is the survival function and $g(a,d)+g(b,d) > g(a,c)+g(b,c)$, then according to Remark 1, we obtain

$$\beta=\mu\left([c,d]\cap\left\{\begin{array}{c}\left(\frac{d-y}{d-c}\right)g(a,c)+\left(\frac{y-c}{d-c}\right)g(a,d)\\ +\left(\frac{d-y}{d-c}\right)g(b,c)+\left(\frac{y-c}{d-c}\right)g(b,d)\geq\beta\end{array}\right\}\right)$$
$$=d-\frac{(d-c)\beta+cg(a,d)+cg(b,d)-dg(a,c)-dg(b,c)}{g(a,d)+g(b,d)-g(a,c)-g(b,c)}. \quad (43)$$

The solution of Equation (43) is $\beta=\frac{(d-c)(g(a,d)+g(b,d))}{d-c+g(a,d)+g(b,d)-g(a,c)-g(b,c)}$. Applying 1. of Proposition 1, we obtain

$$\int_{[c,d]}[g(c,y)+g(d,y)]d\mu \leq \int_{[c,d]}h_2(y)d\mu \leq \mu([c,d])=d-c. \quad (44)$$

The solution of Equation (43) and the inequality (44) give us the inequality (40). □

Remark 10. *If $g(b,d)+g(b,c)=g(a,d)+g(a,c)$ in Theorem 12, then*

$$\int_{[a,b]}[g(x,c)+g(x,d)]d\mu \leq (b-a)\wedge(g(b,d)+g(b,c)) \quad (45)$$

and if $g(a,d)+g(b,d)=g(a,c)+g(b,c)$, then

$$\int_{[c,d]}[g(a,y)+g(b,y)]d\mu \leq (d-c)\wedge(g(a,d)+g(b,d)). \quad (46)$$

Theorem 13. *Let $g:[a,b]\times[c,d]\to[0,\infty)$ be a convex function on the coordinates on $[a,b]\times[c,d]$ such that $bg(a,d)+ag(b,c) > bg(a,c)+ag(b,d)$ and $g(b,d)+g(a,c) > g(a,d)+g(b,c)$. Let $\mu\times\mu$ be the Lebesgue measure on $[a,b]\times[c,d]$, then*

$$\int_{[a,b]\times[c,d]}g(x,y)d(\mu\times\mu)\leq\min\{(b-a)(d-c),\alpha\}, \quad (47)$$

where α is a positive solution of the equation

$$(g(b,d)-g(a,d)-g(b,c)+g(a,c))\alpha^2+(bc+ad-ac-bd$$
$$+(d-c)g(a,d)+(a-b+c-d)g(b,d)+(b-a)g(b,c))\alpha$$
$$+(ac-bc-ad+bd)g(b,d)=0. \quad (48)$$

Proof. Since g is a convex function on the coordinates on $[a,b] \times [c,d]$. Therefore, for $(x,y) \in [a,b] \times [c,d]$, we obtain

$$\begin{aligned} g(x,y) &= g\left(\left(1-\frac{x-a}{b-a}\right)\cdot a + \left(\frac{x-a}{b-a}\right)\cdot b, \left(1-\frac{y-c}{d-c}\right)\cdot c + \left(\frac{y-c}{d-c}\right)\cdot d\right) \\ &\leq \left(\frac{b-x}{b-a}\right)\left(\frac{d-y}{d-c}\right)g(a,c) + \left(\frac{b-x}{b-a}\right)\left(\frac{y-c}{d-c}\right)g(a,d) \\ &\quad + \left(\frac{x-a}{b-a}\right)\left(\frac{d-y}{d-c}\right)g(b,c) + \left(\frac{x-a}{b-a}\right)\left(\frac{y-c}{d-c}\right)g(b,d) = h(x,y). \end{aligned}$$

Suppose F is the survival function with respect to the variable x together with $bg(a,d) + ag(b,c) > bg(a,c) + ag(b,d)$ and $g(b,d) + g(a,c) > g(a,d) + g(b,c)$. By 3. of Proposition 1 and by using the Fubini theorem for fuzzy integrals and Remark 1, we have

$$\begin{aligned} \int_{[a,b]\times[c,d]} g(x,y)d(\mu\times\mu) &\leq \int_{[a,b]\times[c,d]} \left[\left(\frac{b-x}{b-a}\right)\left(\frac{d-y}{d-c}\right)g(a,c)\right. \\ &\quad + \left(\frac{x-a}{b-a}\right)\left(\frac{d-y}{d-c}\right)g(b,c) + \left(\frac{b-x}{b-a}\right)\left(\frac{y-c}{d-c}\right)g(a,d) \\ &\quad \left. + \left(\frac{x-a}{b-a}\right)\left(\frac{y-c}{d-c}\right)g(b,d)\right]d(\mu\times\mu) = \int_{[a,b]}\left(\int_{[c,d]} h(x,y)d\mu\right)d\mu. \\ &= \int_{[a,b]} \frac{(d-c)(bg(a,d) - ag(b,d) + (g(b,d) - g(a,d))x)}{\begin{array}{c}(b-a)(d-c) + b(g(a,d) - g(a,c)) + a(g(b,c) - g(b,d)) \\ +(g(b,d) + g(a,c) - g(a,d) - g(b,c))x\end{array}} d\mu = \alpha, \end{aligned} \tag{49}$$

where α is a positive solution of the equation

$$\begin{aligned} &(g(b,d) - g(a,d) - g(b,c) + g(a,c))\alpha^2 + (bc + ad - ac - bd \\ &\quad +(d-c)g(a,d) + (a-b+c-d)g(b,d) + (b-a)g(b,c))\alpha \\ &\quad + (ac - bc - ad + bd)g(b,d) = 0. \end{aligned} \tag{50}$$

However, according to 1. of Proposition 1, we obtain

$$\begin{aligned} \int_{[a,b]\times[c,d]} g(x,y)d(\mu\times\mu) &\leq \mu\times\mu([a,b]\times[c,d]) \\ &= \mu([a,b])\mu([c,d]) = (b-a)(d-c). \end{aligned} \tag{51}$$

A positive solution of (50) and the inequality (51) give us the desired inequality (47). □

Remark 11. *If in Theorem 13 the conditions $bg(a,d) + ag(b,c) > bg(a,c) + ag(b,d)$, $g(b,d) + g(a,c) = g(a,d) + g(b,c)$ and $g(b,d) = g(a,d)$ hold, then*

$$\begin{aligned} &\int_{[a,b]\times[c,d]} g(x,y)d(\mu\times\mu) \\ &\quad \leq \frac{(b-a)(d-c)(bg(a,d) - ag(b,d))}{(b-a)(d-c) + b(g(a,d) - g(a,c)) + a(g(b,c) - g(b,d))}. \end{aligned} \tag{52}$$

Remark 12. *Since g is a convex function on the coordinates on $[a,b] \times [c,d]$, we obtain*

$$\begin{aligned} g(x,y) &= g\left(\left(1-\frac{x-a}{b-a}\right)\cdot a + \left(\frac{x-a}{b-a}\right)\cdot b, \frac{y-c}{d-c}\right) \\ &\leq \left(1-\frac{x-a}{b-a}\right)g(a,y) + \left(\frac{x-a}{b-a}\right)g(b,y) \leq g(a,y) + g(b,y) \end{aligned}$$

and

$$\begin{aligned} g(x,y) &= g\left(x, \left(1-\frac{y-c}{d-c}\right)\cdot c + y\cdot d\right) \\ &\leq \left(1-\frac{y-c}{d-c}\right)g(x,c) + \frac{y-c}{d-c}g(x,d) \leq g(x,c)+g(x,d). \end{aligned}$$

Hence, by 1., and 3. of Proposition 1 and the Fubini theorem for fuzzy integrals, we obtain

$$\int_{[a,b]\times[c,d]} g(x,y)d(\mu\times\mu) \leq (b-a) \wedge \int_{[c,d]} [g(a,y)+g(b,y)]d\mu \tag{53}$$

and

$$\int_{[a,b]\times[c,d]} g(x,y)d(\mu\times\mu) \leq (d-c) \wedge \int_{[a,b]} [g(x,c)+g(x,d)]d\mu. \tag{54}$$

Thus, from (39), (40), (53) and (54), we observe that the inequality

$$\begin{aligned} \int_{[a,b]\times[c,d]} g(x,y)d(\mu\times\mu) \leq \min\Big\{&b-a, d-c, \\ &\frac{(b-a)(g(b,d)+g(b,c))}{b-a+g(b,d)+g(b,c)-g(a,d)-g(a,c)}, \frac{(d-c)(g(a,d)+g(b,d))}{d-c+g(a,d)+g(b,d)-g(a,c)-g(b,c)}\Big\} \end{aligned} \tag{55}$$

holds.

It is clear from (54) and (55) that the inequality

$$\begin{aligned} \int_{[a,b]\times[c,d]} g(x,y)d(\mu\times\mu) \leq \min\{&(b-a)(d-c), b-a, d-c, \alpha, \\ &\frac{(b-a)(g(b,d)+g(b,c))}{b-a+g(b,d)+g(b,c)-g(a,d)-g(a,c)}, \frac{(d-c)(g(a,d)+g(b,d))}{d-c+g(a,d)+g(b,d)-g(a,c)-g(b,c)} \end{aligned} \tag{56}$$

holds, where α is a positive solution of the Equation (48).

Theorem 14. *Let $g : [a,b]\times[c,d] \to [0,\infty)$ be a convex function on the coordinates on $[a,b] \times [c,d]$. Let $\mu\times\mu$ be the Lebesgue measure on $[a,b]\times[c,d]$.*

1. *If $g(b,d)+g(b,c) < g(a,d)+g(a,c)$, then*

$$\begin{aligned} &\int_{[a,b]} [g(x,c)+g(x,d)]d\mu \\ &\qquad \leq \min\left\{b-a, \frac{(b-a)(g(a,d)+bg(a,c))}{b-a+g(a,d)+g(a,c)-g(b,c)-g(b,d)}\right\}. \end{aligned} \tag{57}$$

2. *If $g(a,d)+g(b,d) < g(a,c)+g(b,c)$, then*

$$\begin{aligned} &\int_{[c,d]} [g(a,y)+g(b,y)]d\mu \\ &\qquad \leq \min\left\{d-c, \frac{(d-c)(g(a,c)+g(b,c))}{d-c+g(a,c)+g(b,c)-g(a,d)-g(b,d)}\right\}. \end{aligned} \tag{58}$$

Proof. By the coordinated convexity of g on $[a,b]\times[c,d]$ that

$$\begin{aligned}
&g(x,c)+g(x,d)\\
&=g\left(\left(1-\frac{x-a}{b-a}\right)\cdot a+\frac{x-a}{b-a}\cdot b,d\right)+g\left(\left(1-\frac{x-a}{b-a}\right)\cdot a+\frac{x-a}{b-a}\cdot b,c\right)\\
&\leq\left(\frac{b-x}{b-a}\right)g(a,d)+\left(\frac{x-a}{b-a}\right)g(b,d)+\left(\frac{b-x}{b-a}\right)g(a,c)+\left(\frac{x-a}{b-a}\right)g(b,c).
\end{aligned}$$

and hence by 3. of Proposition 1, we obtain

$$\begin{aligned}
&\int_{[a,b]}[g(x,c)+g(x,d)]d\mu\\
&\qquad\leq\int_{[a,b]}\left\{\left(\frac{b-x}{b-a}\right)g(a,d)+\left(\frac{x-a}{b-a}\right)g(b,d)\right.\\
&\qquad\qquad\left.+\left(\frac{b-x}{b-a}\right)g(a,c)+\left(\frac{x-a}{b-a}\right)g(b,c)\right\}d\mu=\int_{[a,b]}h_1(x)d\mu.
\end{aligned}$$

If we consider the survival function F and $g(b,d)+g(b,c)<g(a,d)+g(a,c)$, then according to Remark 1, we obtain that

$$\begin{aligned}
\alpha&=\mu\left([a,b]\cap\left\{\begin{array}{c}\left(\frac{b-x}{b-a}\right)g(a,d)+\left(\frac{x-a}{b-a}\right)g(b,d)\\+\left(\frac{b-x}{b-a}\right)g(a,c)+\left(\frac{x-a}{b-a}\right)g(b,c)\geq\alpha\end{array}\right\}\right)\\
&=\frac{(b-a)\alpha+ag(b,d)-bg(a,d)-bg(a,c)+ag(b,c)}{g(a,d)+g(a,c)-g(b,d)-g(b,c)}-a.
\end{aligned}\tag{59}$$

The solution of Equation (59) is $\alpha=\frac{(b-a)(g(a,d)+g(a,c))}{b-a+g(a,d)+g(a,c)-g(b,c)-g(b,d)}$. Applying 1. of Proposition 1, we obtain

$$\int_{[a,b]}[g(x,c)+g(x,d)]d\mu\leq\int_{[a,b]}h_1(x)d\mu\leq\mu([a,b])=b-a.\tag{60}$$

The solution of Equation (59) and the inequality (60) give the inequality (57). Since g is convex on the coordinates on $[a,b]\times[c,d]$, we find that

$$\begin{aligned}
&g(a,y)+g(b,y)\\
&=g\left(a,\left(1-\frac{y-c}{d-c}\right)\cdot c+\left(\frac{y-c}{d-c}\right)\cdot d\right)+g\left(b,\left(1-\frac{y-c}{d-c}\right)\cdot c+\frac{y-c}{d-c}\cdot d\right)\\
&\leq\left(\frac{d-y}{d-c}\right)g(a,c)+\left(\frac{y-c}{d-c}\right)g(a,d)+\left(\frac{d-y}{d-c}\right)g(b,c)+\left(\frac{y-c}{d-c}\right)g(b,d).
\end{aligned}$$

and hence by 3. of Proposition 1, we obtain

$$\begin{aligned}
&\int_{[c,d]}[g(a,y)+g(b,y)]d\mu\\
&\qquad\leq\int_{[c,d]}\left\{\left(\frac{d-y}{d-c}\right)g(a,c)+\left(\frac{y-c}{d-c}\right)g(a,d)\right.\\
&\qquad\qquad\left.+\left(\frac{d-y}{d-c}\right)g(b,c)+\left(\frac{y-c}{d-c}\right)g(b,d)\right\}d\mu=\int_{[c,d]}h_2(y)d\mu.
\end{aligned}$$

Suppose F is the survival function and if $g(a,d)+g(b,d) < g(a,c)+g(b,c)$, then according to Remark 1, we obtain

$$\beta = \mu\left([c,d] \cap \left\{ \begin{array}{c} \left(\frac{d-y}{d-c}\right)g(a,c) + \left(\frac{y-c}{d-c}\right)g(a,d) \\ +\left(\frac{d-y}{d-c}\right)g(b,c) + \left(\frac{y-c}{d-c}\right)g(b,d) \geq \beta \end{array} \right\}\right)$$
$$= \frac{\beta(d-c) + cg(a,d) + cg(b,d) - dg(a,c) - dg(b,c)}{g(a,d)+g(b,d)-g(a,c)-g(b,c)} - c. \quad (61)$$

The solution of Equation (61) is $\beta = \frac{(d-c)(g(a,c)+g(b,c))}{d-c+g(a,c)+g(b,c)-g(a,d)-g(b,d)}$.
Applying 1. of Proposition 1, we obtain

$$\int_{[c,d]} [g(c,y)+g(d,y)]d\mu \leq \int_{[c,d]} h_2(y)d\mu \leq \mu([c,d]) = d-c. \quad (62)$$

The solution of Equation (61) and inequality (62) yield the inequality (58). □

Remark 13. *If $g(b,d)+g(b,c) = g(a,d)+g(a,c)$ in Theorem 14, then*

$$\int_{[a,b]} [g(x,c)+g(x,d)]d\mu \leq (b-a) \wedge (g(a,d)+bg(a,c)) \quad (63)$$

and if $g(a,d)+g(b,d) = g(a,c)+g(b,c)$, then

$$\int_{[c,d]} [g(a,y)+g(b,y)]d\mu \leq (d-c) \wedge (g(a,c)+g(b,c)). \quad (64)$$

Theorem 15. *Let $g : [a,b] \times [c,d] \to [0,\infty)$ be a convex function on the coordinates on $[a,b] \times [c,d]$ such that $bg(a,c)+ag(b,d) < bg(a,d)+ag(b,c)$ and $g(a,d)+g(b,c) < g(b,d)+g(a,c)$. Let $\mu \times \mu$ be the Lebesgue measure on $[a,b] \times [c,d]$, then*

$$\int_{[a,b]\times[c,d]} g(x,y)d(\mu\times\mu) \leq \min\{(b-a)(d-c), \alpha\}, \quad (65)$$

where α is a positive solution of the equation

$$(g(a,d)-g(b,d)-g(a,c)+g(b,c))\alpha^2 + (ac+bd-bc-ad$$
$$+(c-d)g(b,c) + (b-a+d-c)g(a,c) + (a-b)g(a,d))\alpha$$
$$+ (bc-ac-bd+ad)g(a,c) = 0. \quad (66)$$

Proof. Since g is a convex function on the coordinates on $[a,b] \times [c,d]$. Therefore, for $(x,y) \in [a,b] \times [c,d]$, we obtain

$$g(x,y) = g\left(\left(1-\frac{x-a}{b-a}\right)\cdot a + \left(\frac{x-a}{b-a}\right)\cdot b, \left(1-\frac{y-c}{d-c}\right)\cdot c + \frac{y-c}{d-c}\cdot d\right)$$
$$\leq \left(\frac{b-x}{b-a}\right)\left(\frac{d-y}{d-c}\right)g(a,c) + \left(\frac{b-x}{b-a}\right)\left(\frac{y-c}{d-c}\right)g(a,d)$$
$$+ \left(\frac{x-a}{b-a}\right)\left(\frac{d-y}{d-c}\right)g(b,c) + \left(\frac{x-a}{b-a}\right)\left(\frac{y-c}{d-c}\right)g(b,d) = h(x,y).$$

Suppose F is the survival function with respect to the variable x together with the assumptions $bg(a,c)+ag(b,d) < bg(a,d)+ag(b,c)$ and $g(a,d)+g(b,c) < g(b,d)+g(a,c)$.

By 3. of Proposition 1 and by using the Fubini theorem for fuzzy integrals and Remark 1, we have

$$\begin{aligned}\int_{[a,b]\times[c,d]} g(x,y)d(\mu\times\mu) \leq \int_{[a,b]\times[c,d]}\Bigg[\left(\frac{b-x}{b-a}\right)\left(\frac{d-y}{d-c}\right)g(a,c)\\ +\left(\frac{b-x}{b-a}\right)\left(\frac{y-c}{d-c}\right)g(a,d)+\left(\frac{x-a}{b-a}\right)\left(\frac{d-y}{d-c}\right)g(b,c)\\ +\left(\frac{x-a}{b-a}\right)\left(\frac{y-c}{d-c}\right)g(b,d)\Bigg]d(\mu\times\mu)=\int_{[a,b]}\left(\int_{[c,d]}h(x,y)d\mu\right)d\mu.\\ =\int_{[a,b]}\frac{(d-c)(bg(a,c)-ag(b,c)+(g(b,c)-g(a,c))x)}{\begin{array}{c}(b-a)(d-c)+b(g(a,c)-g(a,d))+a(g(b,d)\\ -g(b,c))+(g(a,d)+g(b,c)-g(b,d)-g(a,c))x\end{array}}d\mu=\alpha,\end{aligned} \tag{67}$$

where α is a positive solution of the equation

$$\begin{aligned}(g(a,d)-g(b,d)-g(a,c)+g(b,c))\alpha^2+(ac+bd-bc-ad\\ +(c-d)g(b,c)+(b-a+d-c)g(a,c)+(a-b)g(a,d))\alpha\\ +(bc-ac-bd+ad)g(a,c)=0.\end{aligned} \tag{68}$$

However, according to 1. of Proposition 1, we obtain

$$\begin{aligned}\int_{[a,b]\times[c,d]} g(x,y)d(\mu\times\mu) \leq \mu\times\mu([a,b]\times[c,d])\\ =\mu([a,b])\mu([c,d])=(b-a)(d-c).\end{aligned} \tag{69}$$

A positive solution of (68) and the inequality (69) prove the desired inequality (66). □

Remark 14. *Suppose that $bg(a,c)+ag(b,d) < bg(a,d)+ag(b,c)$, $g(b,c)=g(a,c)$ and $g(a,d)+g(b,c)=g(b,d)+g(a,c)$ in Theorem 15, then*

$$\begin{aligned}\int_{[a,b]\times[c,d]} g(x,y)d(\mu\times\mu)\\ \leq \frac{(b-a)(d-c)(bg(a,c)-ag(b,c))}{(b-a)(d-c)+b(g(a,c)-g(a,d))+a(g(b,d)-g(b,c))}.\end{aligned} \tag{70}$$

Remark 15. *Since g is a convex function on the coordinates on $[a,b]\times[c,d]$, we obtain*

$$\begin{aligned}g(x,y)&=g\left(\left(1-\frac{x-a}{b-a}\right)\cdot a+\left(\frac{x-a}{b-a}\right)\cdot b,\frac{y-c}{d-c}\right)\\ &\leq\left(1-\frac{x-a}{b-a}\right)g(a,y)+\left(\frac{x-a}{b-a}\right)g(b,y)\leq g(a,y)+g(b,y)\end{aligned}$$

and

$$\begin{aligned}g(x,y)&=g\left(x,\left(1-\frac{y-c}{d-c}\right)\cdot c+y\cdot d\right)\\ &\leq\left(1-\frac{y-c}{d-c}\right)g(x,c)+\left(\frac{y-c}{d-c}\right)g(x,d)\leq g(x,c)+g(x,d).\end{aligned}$$

Hence, by 1. and 3. of Proposition 1 and the Fubini theorem for fuzzy integrals, we obtain

$$\int_{[a,b]\times[c,d]} g(x,y)d(\mu\times\mu)\leq(b-a)\wedge\int_{[c,d]}[g(a,y)+g(b,y)]d\mu \tag{71}$$

and

$$\int_{[a,b]\times[c,d]} g(x,y)d(\mu\times\mu) \le (d-c)\wedge\int_{[a,b]}[g(x,c)+g(x,d)]d\mu. \tag{72}$$

Thus, from (57), (58), (71) and (72), we find that

$$\int_{[a,b]\times[c,d]} g(x,y)d(\mu\times\mu) \le \min\left\{b-a, d-c, \frac{(b-a)(g(b,d)+g(b,c))}{b-a+g(b,d)+g(b,c)-g(a,d)-g(a,c)}, \frac{(d-c)(g(a,d)+g(b,d))}{d-c+g(a,d)+g(b,d)-g(a,c)-g(b,c)}\right\}. \tag{73}$$

It is clear from (72) and (73), that the inequality

$$\int_{[a,b]\times[c,d]} g(x,y)d(\mu\times\mu) \le \min\left\{(b-a)(d-c), b-a, d-c, \alpha, \frac{(b-a)(g(b,d)+g(b,c))}{b-a+g(b,d)+g(b,c)-g(a,d)-g(a,c)}, \frac{(d-c)(g(a,d)+g(b,d))}{d-c+g(a,d)+g(b,d)-g(a,c)-g(b,c)}\right\}, \tag{74}$$

holds, where α is a positive solution of the Equation (66).

Example 5. *Take $X=[0,1]$, $Y=[0,1]$ and let μ be the Lebesgue measure on X and Y. Suppose that $g:[0,1]\times[0,1]\to[0,\infty)$ is defined as $g(x,y)=(x-\arctan x)e^y$ and $\mu\times\mu$ be the Lebesgue measure on $X\times Y$.*
Since $g(b,d)+g(b,c)>g(a,d)+g(a,c)$, hence by 1. of Theorem 12, we have

$$\int_{[0,1]}(1+e)(x-\arctan x)d\mu \le \min\left\{1, \frac{4e-\pi-\pi e+4}{4e-\pi-\pi e+8}\right\}.$$

We observe that $g(a,d)+g(b,d)>g(a,c)+g(b,c)$, hence by 2. of Theorem 12, we obtain

$$\int_{[0,1]}\left(1-\frac{\pi}{4}\right)e^y d\mu \le \min\left\{1, \frac{(4-\pi)e}{\pi+4e-\pi e}\right\}.$$

Finally, (56) gives

$$\int_{[0,1]^2}(x-\arctan x)e^y d(\mu\times\mu) \le \min\left\{1, \alpha, \frac{4e-\pi-\pi e+4}{4e-\pi-\pi e+8}, \frac{(4-\pi)e}{\pi+4e-\pi e}\right\},$$

where α is a positive root of the equation

$$(4-\pi)(e-1)\alpha^2-(2(4-\pi)e+\pi)\alpha+(4-\pi)e=0.$$

The solution of this equation is

$$\alpha_1=\frac{\sqrt{64e-16\pi e+\pi^2}-8e-\pi+2\pi e}{2(\pi-4)(e-1)}\approx 0.31793$$

and

$$\alpha_2=-\frac{\pi+8e+\sqrt{64e-16\pi e+\pi^2}-2\pi e}{2(\pi-4)(e-1)}\approx 4.9759.$$

Thus,

$$\int_{[0,1]^2}(x - \arctan x)e^y d(\mu \times \mu)$$
$$\leq \frac{\sqrt{64e - 16\pi e + \pi^2} - 8e - \pi + 2\pi e}{2(\pi - 4)(e - 1)} \approx 0.31793.$$

Example 6. *Take $X = [0,1]$, $Y = [1,2]$ and let μ be the Lebesgue measure on X and Y. Suppose that $g : [0,1] \times [1,2] \to [0,\infty)$ is defined as $g(x,y) = (x^2 - 3x + \frac{5}{2})e^{-\sqrt{y}}$ and $\mu \times \mu$ be the Lebesgue measure on $X \times Y$.*
Since $g(b,d) + g(b,c) < g(a,d) + g(a,c)$, hence by 1. of Theorem 14, we have, then

$$\int_{[0,1]}\left(e^{-1} + e^{-\sqrt{2}}\right)\left(x^2 - 3x + \frac{5}{2}\right)d\mu \leq \frac{5\left(e^{-\sqrt{2}} + e^{-1}\right)}{2\left(1 + 2e^{-\sqrt{2}} + 2e^{-1}\right)}.$$

We observe that $g(a,d) + g(b,d) < g(a,c) + g(b,c)$, hence by 2. of Theorem 14, we obtain

$$\int_{[1,2]} 2e^{-\sqrt{y}}d\mu \leq \frac{2e^{-1}}{1 + 2e^{-1} - 2e^{-\sqrt{2}}}.$$

Finally, (74) gives

$$\int_{[0,1]\times[1,2]}\left(x^2 - 3x + \frac{5}{2}\right)e^{-\sqrt{y}}d(\mu \times \mu)$$
$$\leq \min\left\{1, \alpha, \frac{5\left(e^{-\sqrt{2}} + e^{-1}\right)}{2\left(1 + 2e^{-\sqrt{2}} + 2e^{-1}\right)}, \frac{e^{-1}}{1 + 2e^{-1} - 2e^{-\sqrt{2}}}\right\},$$

where α is a positive root of the equation

$$4\left(e^{-\sqrt{2}} - e^{-1}\right)\alpha^2 + \left(2 + 9e^{-1} - 5e^{-\sqrt{2}}\right)\alpha - 5e^{-1} = 0.$$

The solution of this equation is

$$\alpha_1 \approx 0.476\,85$$

and

$$\alpha_2 \approx 7.729\,4.$$

Thus,

$$\int_{[0,1]\times[1,2]}\left(x^2 - 3x + \frac{5}{2}\right)e^{-\sqrt{y}}d(\mu \times \mu) \leq \frac{e^{-1}}{1 + 2e^{-1} - 2e^{-\sqrt{2}}}.$$

3. Conclusions

In this study, we discussed the theory of fuzzy integrals, also known as Sugeno integrals, and their properties. Since integral inequalities are useful tools in several theoretical and applied fields, this is why mathematicians have developed analogues of a number of integral inequalities such as Cauchy–Schwarz-type inequality, Stolarsky inequality and Minkowski inequality using fuzzy integrals. It has been shown that classical Hermite–Hadamard inequalities do not hold true for fuzzy integrals in general. In this research, we prove that the Hermite–Hadamard-type integral inequalities (2) established for coordinated convex functions do not hold for the Sugeno integral. We also prove some estimates of some inequalities involved in (2) using the Sugeno integral and Fubini theorem of fuzzy integrals and support our study by providing peculiar examples. This study could have the potential to encourage the researchers already working in this field to further explore the topic of mathematical inequalities in the fuzzy context of functions of two or more variables.

The research conducted in this paper can also be extended for other generalizations of coordinated convex functions and not limited to the Hermite–Hadamard-type inequalities for coordinated convex functions, but also for coordinated Jensen's type inequalities.

Funding: This work is supported by the Deanship of Scientific Research, King Faisal University under the Ambitious Researcher Track with Grant Number GRANT2633.

Data Availability Statement: No data have been used in the manuscript.

Acknowledgments: The author is very thankful to all the anonymous referees for their very useful and constructive comments in order to improve the paper.

Conflicts of Interest: The author declares no conflicts of interest.

References

1. Román-Flores, H.; Flores-Franulič, A.; Bassanezi, R.; Rojas-Medar, M. On the level-continuity of fuzzy integrals. *Fuzzy Set. Syst.* **1996**, *80*, 339–344. [CrossRef]
2. Ralescu, D.; Adams, G. The fuzzy integral. *J. Math. Anal. Appl.* **1980**, *75*, 562–570. [CrossRef]
3. Ouyang, Y.; Mesiar, R.; Agahi, H. An inequality related to Minkowski type for Sugeno integrals. *Inf. Sci.* **2010**, *180*, 2793–2801. [CrossRef]
4. Struk, P. Extremal fuzzy integrals. *Soft Comput.* **2006**, *10*, 502–505. [CrossRef]
5. Wang, Z.; Klir, G. *Generalized Measure Theory*; Springer: New York, NY, USA, 2009.
6. Ying, M. Linguistic quantifiers modeled by Sugeno integrals. *Artif. Intell.* **2006**, *170*, 581–606. [CrossRef]
7. Xu, Q.; Ouyang, Y. A note on a Carlson-type inequality for the Sugeno integral. *Appl. Math. Lett.* **2012**, *25*, 619–623. [CrossRef]
8. Flores-Franulič, A.; Román-Flores, H.A Chebyshev type inequality for fuzzy integrals. *Appl. Math. Comput.* **2007**, *190*, 1178–1184. [CrossRef]
9. Román-Flores, H.; Flores- Franulič, A.; Chalco-Cano, Y. The fuzzy integral for monotone functions. *Appl. Math. Comput.* **2007**, *185*, 492–498. [CrossRef]
10. Román-Flores, H.; Flores-Franulič, A.; Chalco-Cano, Y. A convolution type inequality for fuzzy integrals. *Appl. Math. Comput.* **2008**, *195*, 94–99. [CrossRef]
11. Román-Flores, H.; Chalco-Cano, Y. Sugeno integral and geometric inequalities. *Int. J. Uncertain. Fuzziness-Knowl.-Based Syst.* **2007**, *15*, 1–11. [CrossRef]
12. Román-Flores, H.; Flores-Franulič, A.; Chalco-Cano, Y. A Jensen type inequality for fuzzy integrals. *Inf. Sci.* **2007**, *177*, 3192–3201. [CrossRef]
13. Román-Flores, H.; Flores-Franulič, A.; Chalco-Cano, Y. A two-dimensional Hardy type inequality for fuzzy integrals. *Int. J. Uncertain. Fuzziness Knowl.-Based Syst.* **2013**, *21*, 165–173. [CrossRef]
14. Li, D.-Q.; Cheng, Y.-H.; Wang, X.-S.; Qiao, X. Berwald-type inequalities for Sugeno integral with respect to $(\alpha, m, r)_g$-concave functions. *J. Inequal. Appl.* **2016**, *25*, 17. [CrossRef]
15. Ouyang, Y.; Fang, J. Sugeno integral of monotone functions based on Lebesgue measure. *Comput. Math. Appl.* **2008**, *56*, 367–374. [CrossRef]
16. Ouyang, Y.; Fang, J.; Wang, L. Fuzzy Chebyshev type inequality. *Int. J. Approx. Reason.* **2008**, *48*, 829–835. [CrossRef]
17. Hu, Y. Chebyshev type inequalities for general fuzzy integrals. *Inf. Sci.* **2014**, *278*, 822–825. [CrossRef]
18. Caballero, J.; Sadarangani, K. A Cauchy-Schwarz type inerquality for fuzzy integrals. *Nonlinear Anal.* **2010**, *73*, 3329–3335. [CrossRef]
19. Agahi, H.; Mesiar, R., Ouyang, Y. Pap, E.; Štrboja, M. On Stolarsky inequality for Sugeno and Choquet integrals. *Inf. Sci.* **2014**, *266*, 134–139. [CrossRef]
20. Caballero, J.; Sadarangani, K. Hermite-Hadamard inequality for fuzzy integrals. *Appl. Math. Comput.* **2009**, *215*, 2134–2138. [CrossRef]
21. Hadamard, J. Étude sur les propietés des functions entiéres et en particulies d'une function considerée par Riemann. *J. Math. Pure. Appl.* **1893**, *58*, 175–215.
22. Hermite, C. Sur deux limites d'une intégrale définie. *Mathesis* **1883**, *3*, 82.
23. Abbaszadeh, S.; Eshaghi, M.; de la Sen, M. The Sugeno fuzzy integral of log-convex functions. *J. Inequal. Appl.* **2015**, *362*, 12. [CrossRef]
24. Abbaszadeh, S; Eshaghi, M. A Hadamard-type inequality for fuzzy integrals based on r-convex functions. *Soft Comput.* **2016**, *20*, 3117–3124. [CrossRef]
25. Abbaszadeh, S; Eshaghi, M. Nonlinear integrals and Hadamard-type inequalities. *Soft Comput.* **2018**, *22*, 2843–2849. [CrossRef]
26. Agahi, H.; Mohammadpour, A.; Mesiar, R.; Vaezpour, S. M. Liapunov-type inequality for universal integral. *Int. J. Intell. Syst.* **2012**, *27*, 908–925. [CrossRef]
27. Agahi, H.; Mesiar, R.; Ouyang, Y. Further development of Chebyshev type inequalities for Sugeno integrals and T-(S-)evaluators. *Kybernetika* **2010**, *46*, 83–95.

28. Babakhani, A.; Agahi, H.; Mesiar, R. A ($\star$,$*$)-based Minkowski's inequality for Sugeno fractional integral of order $\alpha > 0$. *Fract. Calc. Appl. Anal.* **2015**, *18*, 862–874. [CrossRef]
29. Boczek, M.; Kaluszka, M. On the Minkowski-Hölder type inequalities for generalized Sugeno integrals with an application. *Kybernetika* **2016**, *52*, 329–347. [CrossRef]
30. Caballero, J.; Sadarangani, K. Sandor's inequality for Sugeno integrals. *Appl. Math. Comput.* **2011**, *218*, 1617–1622. [CrossRef]
31. Caballero, J.; Sadarangani, K. A Markov-type inequality for seminormed fuzzy integrals. *Appl. Math. Comput.* **2013**, *219*, 10746–10752. [CrossRef]
32. Caballero, J.; Sadarangani, K. Chebyshev inequality for Sugeno integrals. *Fuzzy Sets Syst.* **2010**, *161*, 1480–1487. [CrossRef]
33. Hong, D.H. Berwald and Favard type inequalities for fuzzy integrals. *Int. J. Uncertain. Fuzziness Knowl.-Based Syst.* **2016**, *24*, 47–58. [CrossRef]
34. Kaluszka, M.; Okolewski, A.; Boczek, M. On Chebyshev type inequalities for generalized Sugeno integrals. *Fuzzy Sets Syst.* **2014**, *244*, 51–62. [CrossRef]
35. Liao, J.; Wub, S.; Dua, T. -S. The Sugeno integral with respect to α-preinvex functions. *Fuzzy Sets Syst.* **2020**, *379*, 102–114. [CrossRef]
36. Li, D.-Q.; Cheng, Y.-H.; Wang, X.-S. Sandor type inequalities for Sugeno integral with respect to general (α, m, r)-convex functions. *J. Funct. Spaces* **2015**, *460–520*, 13. [CrossRef]
37. Li, D.-Q.; Song, X.-Q.; Yue, T. Hermite-Hadamard type inequality for Sugeno integrals. *Appl. Math. Comput.* **2014**, *237*, 632–638. [CrossRef]
38. Narukawa, Y.; Torra, V. Multidimensional generalized fuzzy integral. *Fuzzy Sets Syst.* **2009**, *160*, 802–815. [CrossRef]
39. Dragomir, S.S. On Hadamard's inequality for convex functions on the coordinates in a rectangle from the plane. *Taiwan. J. Math.* **2001**, *4*, 775–788.
40. Hwang, D.Y.; Tseng, K.L.; Yang, G.S. Some Hadamard's inequalities for coordinated convex functions in a rectangle from the plane. *Taiwan. J. Math.* **2007**, *11*, 63–73. [CrossRef]
41. Latif, M.A.; Alomari, M. Hadamard-type inequalities for product two convex functions on the coordinates. *Int. Math. Forum* **2009**, *4*, 2327–2338.
42. Latif, M.A.; Alomari, M. On the Hadamard-type inequalities for h-convex functions on the coordinates. *Int. J. Math. Anal.* **2009**, *3*, 1645–1656.
43. Latif, M.A.; Dragomir, S.S. On Some New inequalities for differentiable coordinated convex functions. *J. Inequal. Appl.* **2012**, *2012*, *28*. [CrossRef]
44. Özdemir, M.E.; Latif M.A.; Akdemir, A.O. On some Hadamard-type inequalities for product of two s-convex functions on the coordinates. *J. Inequal. Appl.* **2012**, *2012*, 21. [CrossRef]
45. Budak, H.; Tunç, T.; Sarilaya, M. Z. Fractional Hermite-Hadamard-type inequalities for interval-valued functions. *Proc. Amer. Math. Soc.* **2020**, *148*, 705–718. [CrossRef]
46. Khan, M.B.; Treant, S.; Budak, H. Generalized p-convex fuzzy-interval-valued functions and inequalities based upon the fuzzy-order relation. *Fractal Fract.* **2022**, *6*, 63. [CrossRef]
47. Wang, Z.; Klir, G. *Fuzzy Measure Theory*; Plenum: New York, NY, USA, 1992.

mathematics MDPI

Article

Improved and Provably Secure ECC-Based Two-Factor Remote Authentication Scheme with Session Key Agreement

Fairuz Shohaimay [1,2] and Eddie Shahril Ismail [1,*]

[1] Department of Mathematical Sciences, Faculty of Science and Technology, Universiti Kebangsaan Malaysia (UKM), Bangi 43600, Selangor, Malaysia
[2] Mathematical Sciences Studies, College of Computing, Informatics and Media, Universiti Teknologi MARA (UiTM) Pahang Branch, Raub Campus, Raub 27600, Pahang, Malaysia
* Correspondence: esbi@ukm.edu.my

Abstract: The remote authentication scheme is a cryptographic protocol incorporated by user–server applications to prevent unauthorized access and security attacks. Recently, a two-factor authentication scheme using hard problems in elliptic curve cryptography (ECC)—the elliptic curve discrete logarithm problem (ECDLP), elliptic curve computational Diffie–Hellman problem (ECCDHP), and elliptic curve factorization problem (ECFP)—was developed, but was unable to address several infeasibility issues while incurring high communication costs. Moreover, previous schemes were shown to be vulnerable to privileged insider attacks. Therefore, this research proposes an improved ECC-based authentication scheme with a session key agreement to rectify the infeasible computations and provide a mechanism for the password change/update phase. The formal security analysis proves that the scheme is provably secure under the random oracle model (ROM) and achieves mutual authentication using BAN logic. Based on the performance analysis, the proposed scheme resists the privileged insider attack and attains all of the security goals while keeping the computational costs lower than other schemes based on the three hard problems. Therefore, the findings suggest the potential applicability of the three hard problems in designing identification and authentication schemes in distributed computer networks.

Keywords: elliptic curve cryptography; key agreement; provable security; password authentication; smart card

MSC: 94A60

Citation: Shohaimay, F.; Ismail, E.S. Improved and Provably Secure ECC-Based Two-Factor Remote Authentication Scheme with Session Key Agreement. *Mathematics* **2023**, *11*, 5. https://doi.org/10.3390/math11010005

Academic Editor: Antanas Cenys

Received: 1 October 2022
Revised: 5 December 2022
Accepted: 13 December 2022
Published: 20 December 2022

1. Introduction

Currently, more Internet users depend on user–server-based applications for e-commerce, banking services, and operational networks because of their convenience and efficiency. These applications allow users to obtain numerous services remotely at any time, anywhere. This type of communication between the user and server usually involves data transmission and financial transactions over a public channel, such as the Internet. Unfortunately, sharing sensitive information over the public channel is insecure, exposing both parties to greater security risks and attacks. Therefore, a remote authentication scheme is imperative for verifying legal users and defending against unauthorized usage.

The first password authentication scheme by Lamport [1] is called a single-factor-based scheme because the user only needs to present a password to be verified by the server. However, studies have shown that single-factor-based schemes are subjected to security pitfalls. Since then, remote authentication schemes were designed based on two or more factors to increase the systems' security. For example, in addition to a password, the user is required to possess a registered smart card or a mobile device. For a multi-factor scheme, the user may also need to present a biometric trait such as a fingerprint.

Many two-factor smart-card-based remote authentication schemes have been proposed. Figure 1 depicts the general architecture of the two-factor authentication scheme, which consists of multiple users and a single server. In this system, the remote user must register with a valid identity and secret password with the server. Next, the server issues a legal smart card to the first-time registered user to access the required services. The smart card is employed to store the registered user's secret credentials for future login requests and perform cryptographic computations during the authentication process.

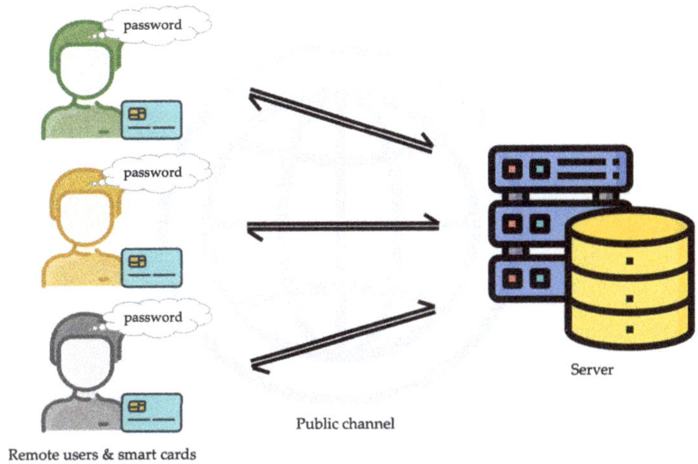

Figure 1. Architecture of two-factor remote authentication with multiple users and a single server.

Like other cryptographic schemes and protocols, the two-factor-based authentication schemes rely on the security primitives of one-way hash functions (e.g., SHA-2 [2]) and number-theoretic computational hard problems in public-key cryptography. For instance, the works by [3–5] were developed based on only one-way hash functions. Other schemes were built based on the intractability of hard problems, including the integer factorization problem in RSA [6], discrete logarithm problem [7], and elliptic curve discrete logarithm problem [8,9] in elliptic curve cryptography (ECC). ECC-based schemes are more prevalent than RSA-based schemes due to the smaller key size requirement [10].

In 2008, Juang et al. [11] first proposed an ECC-based remote password authentication scheme with a session key agreement for the client–server environment. The security of their scheme depended on two hard problems in ECC: the elliptic curve discrete logarithm problem (ECDLP) and elliptic curve computational Diffie–Hellman problem (ECCDHP). They claimed that their scheme preserved all of the security merits of the scheme by Fan et al. [12] and reduced the computational cost. Subsequently, Sun et al. [13] and Li et al. [14] found weaknesses in the design of the scheme by Juang et al. [11] in terms of the password-change phase and session key distribution, the inefficiency of using two secret keys, and user anonymity. Hence, both suggested enhancements to fix the design flaws.

Several improvements by [15–17] were presented to overcome the problems in the Sun et al. [13] scheme to resist offline password-guessing attacks, denial-of-service attacks, smart card loss attacks, and key compromise impersonation attacks. Later, Liu and Ma [18] found that the scheme by Sun et al. [13] still lacked user untraceability and resolved the issue with an improved efficiency. Then, the scheme by Li et al. [14] was found to suffer from desynchronization attacks [17,19]. Hence, Tsai et al. [19] and Byun [20] proposed new schemes with formal security model proofs to strengthen the scheme's security. Both schemes maintained the security of their schemes based on the hard problems of ECDLP and ECCDHP.

Wang et al. [21] proposed a two-factor authentication scheme for a ubiquitous computing environment based on ECDLP. Their scheme was shown to provide better security features and a lower computational cost. Subsequently, Wu et al. [22] showed that the scheme by Wang et al. [21] could not resist offline dictionary attacks, known session key attacks, denial-of-service attacks, and impersonation attacks using a compromised smart card. Meanwhile, Chang et al. [23] pointed out that the scheme by Wang et al. [21] did not satisfy mutual authentication and could not be incorporated using a multipurpose smart card. Hence, both [22,23] proposed new improved schemes to overcome these weaknesses.

In another study, Wang [24] found some security flaws in the design of the single-factor scheme by Islam and Biswas [25]. They developed a new scheme using smart cards with the security foundation of ECDLP, ECCDHP, and a one-way hash function. They claimed that their scheme offered resistance to impersonation attacks and improved the computational efficiency by removing the expensive bilinear pairing operation. Later, Odelu et al. [26] presented further improvements to resist the offline password attack and provide user anonymity. They proved that their scheme could withstand various security attacks and showed that it was provably secure under the random oracle model (ROM).

Other recent works have also proposed ECC-based two-factor schemes with added security features. Madhusudhan et al. [27] suggested a new scheme based on ECC and a fuzzy verifier for quick password verification. They showed that their scheme could resist replay attacks and provide security of the secret key, user untraceability, and perfect forward secrecy. Finally, Kumari et al. [28] designed a novel scheme that provides resistance against offline password-guessing attacks, lost smart card attacks, replay attacks, impersonation attacks, desynchronization attacks, and insider attacks.

1.1. Motivations and Contributions

In 2014, Qu and Tan [29] proposed a two-factor scheme based on the security of a collision-resistant one-way hash function, ECDLP, ECCDHP, and the elliptic curve factorization problem (ECFP). Later, Huang et al. [30] suggested security enhancements to overcome the offline password-guessing attack and user impersonation attack. However, in 2016, Maitra et al. [31] showed that the scheme by Huang et al. [30] was vulnerable to a new forgery attack. They also pointed out that the scheme by [30] could not be implemented in real-world problems because of some computational infeasibility issues. Later, both Chaudhry et al. [32] and Mehmood et al. [33] also suggested improvements to repel the user impersonation attack.

Although Maitra et al. [31] suggested security enhancements to the scheme by Huang et al. [30], their scheme exacted a higher computational cost compared to schemes by [29,30,32,33]. Even though the schemes designed by Chaudhry et al. [32] and Mehmood et al. [33] improved the efficiency of the scheme by Huang et al. [30], their schemes overlooked the computational infeasibility issues. In addition, their schemes did not provide a mechanism for the password change phase and were unable to withstand the privileged insider attack. Moreover, previously improved schemes by Maitra et al. [31] and Mehmood et al. [33] did not maintain all three hard problems in ECC: ECDLP, ECCDHP, and ECFP.

Therefore, this study proposes a new ECC-based two-factor remote authentication scheme based on Chaudhry et al. [32] to resolve these shortcomings. The scheme retains all of the security attributes of the scheme by Maitra et al. [31], including user traceability and efficient local password changeability. In addition, the proposed scheme is proven to withstand offline password-guessing attacks, replay attacks, privileged insider attacks, stolen-verifier attacks, and key-compromise impersonation attacks. Based on the formal security analysis, the proposed scheme is provably secure under ROM against adversary threats. Furthermore, the analysis showed that the proposed scheme is more efficient than the scheme by Maitra et al. [31].

1.2. Structure of the Article

Section 2 briefly describes the security fundamentals, adversary model, a review of Chaudhry et al. [32], and the drawbacks that are considered when developing the proposed scheme. Next, Section 3 explains the new proposed scheme. The formal security proof, informal security analysis, and formal verification using BAN logic are presented in Section 4. The proposed scheme is compared with other chosen schemes in the performance analysis according to the security and efficiency aspects given in Section 5. Then, Section 6 discusses the potential applications of the proposed scheme and future research considerations. Finally, Section 7 presents the conclusion.

2. Preliminaries

This section provides a brief overview of the mathematical concepts, formal definitions, adversary model, security goals, and BAN logic that served as the foundation in the design of the proposed scheme. Table 1 shows the notations and descriptions used in this paper.

Table 1. Notations and descriptions.

Notation	Description	Notation	Description
S	Server	P	Base point on G_p of order n such that $[n]P = \mathcal{O}$ and n is the smallest integer > 0
U_i	User i	$\mathbb{Z}_n^*$	Multiplicative group mod n
$\mathcal{A}$	Adversary	s	Secret key, random integer such that $s \in \mathbb{Z}_n^*$
ID_i	U_i's identity	P_{pub}	Public key, $P_{pub} = [s]P \in G_p$
pw_i	U_i's password	$h(\cdot)$	One-way hash function, $h : \{0,1\}^* \rightarrow \mathbb{Z}_n^*$
SC_i	U_i's smart card	$\Longrightarrow$	Secure channel
p	k-bit prime number, k is at least 512 bits	$\longrightarrow$	Public channel
$E(\mathbb{F}_p)$	The set of points on an elliptic curve over a finite field $\mathbb{F}_p$	$\|$	String concatenation operation
G_p	Additive cyclic subgroup of $E(\mathbb{F}_p)$, where $G_p = E(\mathbb{F}_p) \cup \{\mathcal{O}\}$	$\oplus$	Bitwise XOR operation
$\mathcal{O}$	The point at infinity that is an identity element of $E(\mathbb{F}_p)$		

2.1. Hash Function

A cryptographic one-way function $h : \{0,1\}^* \rightarrow \{0,1\}^l$ has the following properties:

- The function h takes an arbitrary length input $x \in \{0,1\}^*$ and returns a fixed l-bit length message digest $y \in \{0,1\}^l$.
- The function h is one-way; it is trivial to compute $y = h(x)$, but computationally infeasible to find the inverse $x = h^{-1}(y)$.
- The function h is collision-resistant; it is computationally infeasible to find two inputs $x_1 \neq x_2$ such that $h(x_1) = h(x_2)$.

Examples of secure hash algorithms, such as the SHA-2 family of hash functions [2], can be adopted in the proposed scheme.

Definition 1. *An adversary $\mathcal{A}$'s advantage in finding a collision is the probability of $\mathcal{A}$ selecting the pair (x_1, x_2) at random within polynomial time t_1 so that $x_1 \neq x_2$ and $h(x_1) = h(x_2)$, defined formally as*

$$Adv_{\mathcal{A}}^{\mathcal{H}ash}(t_1) = \Pr[(x_1, x_2) \Leftarrow_R \mathcal{A} : x_1 \neq x_2 \wedge h(x_1) = h(x_2)]. \quad (1)$$

If $Adv_{\mathcal{A}}^{\mathcal{H}ash}(t_1) \leq \varepsilon_1$, for any sufficiently small negligible function $\varepsilon_1 > 0$, the one-way hash function $h(\cdot)$ is collision-resistant.

2.2. Elliptic Curve over Finite Fields

The elliptic curve over a finite field $\mathbb{F}_p$ is defined as $E_p(a,b) : y^2 = x^3 + ax + b \pmod{p}$, where p is prime and satisfies the condition $4a^3 + 27b^2 \neq 0 \pmod{p}$. If point $P \in E(\mathbb{F}_p)$ and

$k \in \mathbb{Z}_p^*$, then the elliptic point multiplication operation $[k]P$ is the repeated point addition k times on point P.

$$[k]P = \overbrace{P + P + \cdots + P}^{k \text{ times}}$$

The security of the elliptic curve cryptosystem is based on the following computational hard problems.

Definition 2. *Given two points P, $Q(= [s]P) \in E(\mathbb{F}_p)$, the elliptic curve discrete logarithm problem (ECDLP) is to find the integer $s \in \mathbb{Z}_p^*$. The advantage of an adversary $\mathcal{A}$ in solving the ECDLP within execution time t_2 is defined as*

$$Adv_{\mathcal{A}}^{\mathcal{E}CDLP}(t_2) = \Pr[s \in \mathbb{Z}_p^* : P, Q = [s]P \in E(\mathbb{F}_p)]. \tag{2}$$

For any probabilistic polynomial time-bounded algorithm $\mathcal{A}$ and for any sufficiently small negligible function $\varepsilon_2 > 0$, if $Adv_{\mathcal{A}}^{\mathcal{E}CDLP}(t_2) \leq \varepsilon_2$, then the ECDLP is intractable.

Definition 3. *Given three points P, $Q(= [s]P)$, $R(= [t]P) \in E(\mathbb{F}_p)$, the elliptic curve computational Diffie–Hellman problem (ECCDHP) is to find the point $[s \cdot t]P \in E(\mathbb{F}_p)$ where $s, t \in \mathbb{Z}_p^*$. The advantage of an adversary $\mathcal{A}$ in solving the ECCDHP within execution time t_3 is defined as*

$$Adv_{\mathcal{A}}^{\mathcal{E}CCDHP}(t_3) = \Pr[[s \cdot t]P \in E(\mathbb{F}_p) : P, Q = [s]P, R = [t]P \in E(\mathbb{F}_p) \wedge s, t \in \mathbb{Z}_p^*]. \tag{3}$$

For any probabilistic polynomial time-bounded algorithm $\mathcal{A}$ and for any sufficiently small negligible function $\varepsilon_3 > 0$, if $Adv_{\mathcal{A}}^{\mathcal{E}CCDHP}(t_3) \leq \varepsilon_3$, then the ECCDHP is intractable.

Definition 4. *Given two points P, $Q(= [s + t]P) \in E(\mathbb{F}_p)$, the elliptic curve factorization problem (ECFP) is to find two points $[s]P, [t]P \in E(\mathbb{F}_p)$, where $s, t \in \mathbb{Z}_p^*$. The advantage of an adversary $\mathcal{A}$ in solving the ECFP within execution time t_4 is defined as*

$$Adv_{\mathcal{A}}^{\mathcal{E}CFP}(t_4) = \Pr[[s]P, [t]P \in E(\mathbb{F}_p) : P, Q = [s]P + [t]P \in E(\mathbb{F}_p) \wedge s, t \in \mathbb{Z}_p^*]. \tag{4}$$

For any probabilistic polynomial time-bounded algorithm $\mathcal{A}$ and for any sufficiently small negligible function $\varepsilon_4 > 0$, if $Adv_{\mathcal{A}}^{\mathcal{E}CFP}(t_4) \leq \varepsilon_4$, then the ECFP is intractable.

2.3. Adversary Model

The adversary model by Dolev and Yao [34] was considered for communications over an insecure public channel, and the following assumptions were made.

- *A1*: An adversary $\mathcal{A}$ can trap, delete, or alter the messages transmitted over the public channel.
- *A2*: An adversary $\mathcal{A}$ can retrieve the information stored in the smart card using power monitoring techniques as explained in [35,36].
- *A3*: An adversary $\mathcal{A}$ can guess the identity or password using a dictionary attack. However, A cannot guess both the identity and password simultaneously within polynomial time [37].
- *A4*: An adversary $\mathcal{A}$ can be a non-registered user who tries to attack the authentication system [31].
- *A5*: The server is considered a trusted authority, and the adversary $\mathcal{A}$, as a privileged insider, cannot extract the server's secret key s.

2.4. Security Goals

The following goals are defined for an ideal authentication scheme, as listed in [31,38].

- Mutual authentication: Both the server and the user can authenticate each other. No adversary can impersonate a legal user or server.

- Session key agreement: A session key should be created as the final step in the mutual authentication phase. Afterward, the communication between both parties can be encrypted using the shared session key.
- Forward secrecy: Even if the long-term private keys are compromised, the previous session keys cannot be used by any adversary to forge other session keys.
- User anonymity: A user's identity should not be transmitted explicitly over an insecure channel. This ensures that the user's sensitive information is protected from an adversary $\mathcal{A}$, even with the knowledge of login information or access to the server.
- User traceability: The server should be able to trace the sender of the login request message to avoid the denial-of-service attack. A database of registered users should be maintained by the server.
- Local password verification: A smart card can verify the user identity and password in the login phase before generating the login request message. This way, the smart card can reduce computational overhead by avoiding unnecessary calculations.
- Local password changeability: Users can update/change their passwords independently without the server's assistance. The smart card must be able to detect unauthorized password update requests through the wrong input of the user identity and old password.

2.5. BAN Logic

Burrows–Abadi–Needham (BAN) logic [39] is a set of rules based on belief modal logic for analyzing authentication protocols. The notations used in BAN logic and their descriptions are provided in Table 2. Table 3 lists the BAN logic rules, descriptions, and symbolic forms that are used in proving the mutual authentication property of the proposed scheme.

Table 2. BAN logic notations and descriptions.

Notation	Description
$P \mid\equiv X$	P believes X
$P \lhd X$	P sees X
$P \mid\sim X$	P once said X
$\#(X)$	Message X is fresh
$\langle X \rangle_Y$	Formula X is combined with secret Y
$(X)_Y$	Formula X hashed with secret Y
$P \overset{K}{\leftrightarrow} Q$	P and Q communicate with a shared secret key K
$P \overset{X}{\rightleftharpoons} Q$	Only P and Q share the formula X which is a secret

Table 3. BAN logic rules, descriptions, and symbolic forms.

Rule	Description	Symbolic Form
Message-meaning rule	If P sees $\langle X \rangle_K$ and P believes secret K is shared with Q, then P believes Q once said X.	$\frac{P \lhd \langle X \rangle_K,\, P \mid\equiv (P \overset{K}{\leftrightarrow} Q)}{P \mid\equiv Q \mid\sim X}$
Freshness-conjuncatenation rule	If P believes X is fresh, then P believes (X, Y) is fresh.	$\frac{P \mid\equiv \#(X)}{P \mid\equiv \#(X, Y)}$
Nonce-verification rule	If P believes X is fresh and P believes Q once said X, then P believes Q believes X.	$\frac{P \mid\equiv \#(X),\, P \mid\equiv Q \mid\sim X}{P \mid\equiv Q \mid\equiv X}$
Belief rule	If P believes X and P believes Y, then P believes (X, Y).	$\frac{P \mid\equiv X,\, P \mid\equiv Y}{P \mid\equiv (X, Y)}$

Table 3. *Cont.*

Rule	Description	Symbolic Form
Session-key rule	If P believes Q believes a necessary parameter X of the session key K and P believes X is fresh, then P believes session key K is shared with Q.	$\frac{P \mid\equiv Q \mid\equiv X, P \mid\equiv \#(X)}{P \mid\equiv (P \overset{K}{\leftrightarrow} Q)}$
Session-key verification rule [40]	If P believes that X is fresh and P believes Q believes X and P believes session key K is shared with Q, then P believes Q believes session key K is shared between P and Q.	$\frac{P \mid\equiv \#(X), P \mid\equiv Q \mid\equiv X, P \mid\equiv (P \overset{K}{\leftrightarrow} Q)}{P \mid\equiv Q \mid\equiv (P \overset{K}{\leftrightarrow} Q)}$

2.6. Review of the Scheme by Chaudhry et al.

In this section, a brief description of the scheme by Chaudhry et al. [32] is presented. The authentication scheme by Chaudhry et al. [32] is an improvement of the scheme proposed by Huang et al. [30]. Their scheme consists of four phases: (1) system initialization, (2) user registration, (3) user login, and (4) mutual authentication. Figure 2 summarizes the authentication scheme by Chaudhry et al. [32]. Each of the phases is reviewed as follows.

User Registration Phase		
User U_i		**Server S**
Choose ID_i, pw_i, and b_i $hpw_i = h_1(ID_i \| pw_i \| b_i)$	$\{ID_i, hpw_i\}$ $\Longrightarrow$	$CID_i = h_1(ID_i \oplus s)$ $AID_i = [CID_i + hpw_i]P$ $BID_i = h_2(h_1(ID_i) \cdot hpw_i)$ Store $SC_i \leftarrow \{AID_i, BID_i\}$
Store $SC_i \leftarrow \{b_i\}$ $SC_i = \{AID_i, BID_i, b_i\}$	SC_i $\Longleftarrow$	
User Login and Mutual Authentication Phase		
User U_i and Smart Card SC_i		**Server S**
Enter ID_i and pw_i $hpw_i = h_1(ID_i \| pw_i \| b_i)$ $BID_i' = h_2(h_1(ID_i) \cdot hpw_i)$ Check $BID_i' \overset{?}{=} BID_i$ Select r_i $R_i = [r_i]P$ $M_i = [r_i]P_{pub}$ $TID_i = AID_i - [hpw_i]P$ $DID_i = ID_i \oplus M_i$ $EID_i = h_3(h_4(TID_i \| M_i) \| R_i \| M_i)$	$\{DID_i, EID_i, R_i\}$ $\longrightarrow$	$M_i' = [s]R_i$ $ID_i' = DID_i \oplus M_i'$ $CID_i' = h_1(ID_i' \oplus s)$ $TID_i' = [CID_i']P$ $EID_i' = h_3(h_4(TID_i' \| M_i') \| R_i \| M_i')$ Check $EID_i' \overset{?}{=} EID_i$ Select r_s $R_s = [r_s]R_i$ $Z_s = R_s \oplus M_i'$ $H_s = h_3(EID_i' \| R_s \| TID_i')$
$R_s' = Z_s \oplus M_i$ $H_s' = h_3(EID_i \| R_s' \| TID_i)$ Check $H_s' \overset{?}{=} H_s$ $H_i = h_2(M_i \| R_s')$	$\{Z_s, H_s\}$ $\longleftarrow$	
	$\{H_i\}$ $\longrightarrow$	$H_i' = h_2(M_i' \| R_s)$ Check $H_i' \overset{?}{=} H_i$
	$Sk = h_5(M_i \| R_s \| R_i \| TID_i)$ $\longleftrightarrow$	

Figure 2. Scheme by Chaudhry et al. [32] based on ECC.

(1) System initialization phase

The Server S selects an elliptic curve $E_p(a,b)$ over $\mathbb{F}_p$, where p is k-bit prime, and a base point P of order n from G_p of $E_p(a,b)$, where n is a large number for security purposes. Then, S computes the secret key and public key pair (s, P_{pub}) such that $P_{pub} = [s]P$, where s is a random integer $s \in \mathbb{Z}_n^*$. The Server S also chooses five distinct one-way hash functions $h_i : \{0,1\}^* \rightarrow \mathbb{Z}_p^*$, where $i = 1, 2, \ldots, 5$. Finally, the Server S publishes $\{E_p(a,b), P, P_{pub}, h(\cdot)\}$ and keeps s secret.

(2) User registration phase

In this phase, the user U_i chooses an identity ID_i, a password pw_i, and a random integer b_i. Then, the user U_i computes $hpw_i = h_1(ID_i \| pw_i \| b_i)$ and sends $\{ID_i, hpw_i\}$ to S through a secure channel. Next, the Server S computes $CID_i = h_1(ID_i \oplus s)$, $AID_i = [CID_i + hpw_i]P$, and $BID_i = h_2(h_1(ID_i) \cdot hpw_i)$. The Server S stores $\{AID_i, BID_i\}$ into the smart card SC_i and issues the card securely to U_i. Once the user U_i receives the smart card SC_i, the user will update the value b_i into SC_i. Hence, the smart card $SC_i = \{AID_i, BID_i, b_i\}$.

(3) User login phase

In the login phase, the registered user U_i inserts the smart card SC_i into a remote terminal and enters the identity and password, $\{ID_i, pw_i\}$. Next, the smart card SC_i computes $hpw_i = h_1(ID_i \| pw_i \| b_i)$ and $BID_i' = h_2(h_1(ID_i) \cdot hpw_i)$, and checks if the equation $BID_i' = BID_i$ holds. Otherwise, the login phase is aborted. Then, the smart card SC_i selects a random integer r_i and computes $R_i = [r_i]P$, $M_i = [r_i]P_{pub}$, $TID_i = AID_i - [hpw_i]P$, $DID_i = ID_i \oplus M_i$, and $EID_i = h_3(h_4(TID_i \| M_i) \| R_i \| M_i)$. The smart card SC_i submits the login request message $\{DID_i, EID_i, R_i\}$ to S through a public channel.

(4) Mutual authentication phase

Once the Server S receives the login request message, it computes $M_i' = [s]R_i$, $ID_i' = DID_i \oplus M_i'$, $CID_i' = h_1(ID_i' \oplus s)$, $TID_i' = [CID_i']P$, and $EID_i' = h_3(h_4(TID_i' \| M_i')R_i \| M_i')$, and checks if the equation $EID_i' = EID_i$ holds. If the equation does not hold, the login request is rejected. Otherwise, the Server S generates a random integer r_s and computes $R_s = [r_s]R_i$, $Z_s = R_s \oplus M_i'$, and $H_s = h_3(EID_i' \| R_s \| TID_i)$, and sends the response message $= \{Z_s, H_s\}$ to U_i through the public channel.

After receiving the response message, the user U_i computes $R_s' = Z_s \oplus M_i$, and $H_s' = h_3(EID_i \| R_s' \| TID_i)$, and checks if $H_s' = H_s$ holds. If the equation does not hold, the user U_i disconnects from S. Otherwise, the user U_i computes $H_i = h_2(M_i \| R_s')$ and sends the message $= \{H_i\}$ to S. Next, the Server S computes $H_i' = h_2(M_i' \| R_s)$ and checks if $H_i' = H_i$. If the equation holds, the user U_i and the Server S achieve mutual authentication and agree on the session key $Sk = h_5(M_i \| R_s \| R_i \| TID_i)$. Otherwise, the session is terminated.

2.7. Drawbacks of Scheme by Chaudhry et al.

This section highlights the security drawbacks of the scheme by Chaudhry et al. [32].

(1) Computational infeasibility

During the mutual authentication phase, once the Server S has verified the equation $EID_i' = EID_i$, it then computes the value $Z_s = R_s \oplus M_i' = [r_s]R_i \oplus [s]R_i$. Then, the user retrieves the value of R_S' as $R_s' = Z_s \oplus M_i = ([r_s]R_i \oplus [s]R_i) \oplus [r_i]P_{pub}$. However, the XOR operation is undefined on the elliptic curve since it is not a closed operation under the elliptic curve group. The undefined XOR operation on two elliptic curve points was highlighted by Maitra et al. [31] as a drawback of the scheme by Huang et al. [30]. However, Chaudhry et al. [32] did not address the issue in the modification of their scheme. Hence, their scheme maintained the infeasible computations of the scheme by Huang et al. [30].

(2) Weakness to privileged insider attack

Consider an adversary $\mathcal{A}$ being a privileged insider who can monitor data transmission over a secure channel. In the registration phase of the scheme by Chaudhry et al. [32], the user U_i submits $\{ID_i, hpw_i\}$ to the Server through a secure channel. Hence, $\mathcal{A}$ has access to ID_i and hpw_i. If $\mathcal{A}$ possesses a lost/stolen smart card SC_i, then it is possible for $\mathcal{A}$ to launch an offline password-guessing attack. For example, assume that $\mathcal{A}$ has the values ID_i, hpw_i, and $\{AID_i, BID_i, b_i\}$ retrieved from SC_i by Assumption $A2$. Then, $\mathcal{A}$ can obtain the correct password pw_i by checking the equation $hpw_i = h(ID_i \| pw_i \| b_i) = h(ID_i \| \widetilde{pw}_i \| b_i)$, where $\widetilde{pw}_i$ is the guessed password. Therefore, the scheme by Chaudhry et al. [32] cannot resist the privileged insider attack.

(3) Unable to trace user

After receiving the login request message $\{DID_i, EID_i, R_i\}$, the Server computes all of the values M_i', ID_i', CID_i', and TID_i' straight away before verifying the value EID_i'. Based on discussions in [31], the Server was shown to be vulnerable to forgery attacks because it is unable to check if the login request comes from a registered user. Maitra et al. [31] also highlighted that the user untraceability feature is undesirable since the Server cannot provide user-specific services. In the scheme by Chaudhry et al. [32], the Server did not save any information about the registered users; therefore, it cannot trace the sender of the login request message.

(4) No mechanism for password change/update

In the scheme by Chaudhry et al. [32], they rectified the computation of AID_i during the user registration phase to overcome the user impersonation attack. Specifically, the value AID_i was computed as $AID_i = [CID_i + hpw_i]P = [h_1(ID_i \oplus s) + h_1(ID_i \| pw_i \| b_i)]P$. Note that the value of AID_i is stored in the memory of the smart card SC_i and its value depends on the password pw_i and random integer b_i. Consequently, the corresponding computation for the new value of AID_i^* should also be rectified when a user changes/updates a new password pw_i^* and new random integer b_i^*. However, the password change/update phase was not discussed. Therefore, their scheme did not provide a mechanism for the password change/update.

3. Proposed Scheme

This section presents the proposed ECC-based two-factor remote authentication scheme. Following the scheme by [32], the Server acts as a trusted authority that is responsible for preparing the global parameters and public and secret keys, as well as issuing smart cards to newly registered users. The proposed scheme also incorporates timestamps to verify the freshness of transmitted messages, similar to the design by [31]. Generally, the scheme consists of five phases: (1) system initialization, (2) user registration, (3) user login, (4) mutual authentication, and (5) password change/update. Figure 3 presents an overview of the proposed scheme.

User Registration Phase		
User U_i		**Server** S
Choose ID_i and pw_i		
Generate $b_i \in_R \mathbb{Z}_n^*$		
$hpw_i = h(pw_i \| b_i)$	$\{ID_i, hpw_i\}$ $\Longrightarrow$	
		$CID_i = h(ID_i \oplus s)$
		Check CID_i in Database
		Store Database $\leftarrow \{CID_i\}$
		$AID_i = [CID_i + hpw_i]P$
		$BID_i = h(h(ID_i) \oplus hpw_i)$
	SC_i $\Longleftarrow$	Store $SC_i \leftarrow \{AID_i, BID_i\}$
$\widehat{b}_i = h(ID_i \| pw_i) \oplus b_i$		
Store $SC_i \leftarrow \{\widehat{b}_i\}$		
$SC_i = \{AID_i, BID_i, \widehat{b}_i\}$		

User Login and Mutual Authentication Phase		
User U_i **and Smart Card** SC_i		**Server** S
Enter ID_i' and pw_i'		
$b_i' = \widehat{b}_i \oplus h(ID_i' \| pw_i')$		
$hpw_i' = h(pw_i' \| b_i')$		
$BID_i' = h(h(ID_i') \oplus hpw_i')$		
Check $BID_i' \stackrel{?}{=} BID_i$		
Select $r_i \in_R \mathbb{Z}_n^*$		
$R_i = [r_i]P = (x_{Ri}, y_{Ri})$		
$M_i = [r_i]P_{pub} = (x_{Mi}, y_{Mi})$		
$TID_i = AID_i - [hpw_i]P = (x_{Ti}, y_{Ti})$		
$DID_i = ID_i \oplus y_{Mi}$		
$EID_i = h(x_{Ti} \| x_{Mi} \| T_{i1})$	$\{DID_i, EID_i, R_i, T_{i1}\}$ $\longrightarrow$	
		Check $(T_{s1} - T_{i1}) \leq \Delta T$
		$M_i' = [s]R_i = (x_{Mi}', y_{Mi}')$
		$ID_i' = DID_i \oplus y_{Mi}'$
		Check $CID_i' = h(ID_i' \oplus s)$ in Database
		$TID_i' = [CID_i']P = (x_{Ti}', y_{Ti}')$
		$EID_i' = h(x_{Ti}' \| x_{Mi}' \| T_{i1})$
		Check $EID_i' \stackrel{?}{=} EID_i$
		Select $r_s \in_R \mathbb{Z}_n^*$
		$R_s = [r_s]R_i = (x_{Rs}, y_{Rs})$
		$Z_s = R_s + M_i'$
	$\{Z_s, H_s, T_{s1}\}$ $\longleftarrow$	$H_s = h(EID_i' \| x_{Rs} \| T_{s1} \| x_{Ti}')$
Check $(T_{i2} - T_{s1}) \leq \Delta T$		
$R_s' = Z_s - M_i = (x_{Rs}', y_{Rs}')$		
$H_s' = h(EID_i \| x_{Rs}' \| T_{s1} \| x_{Ti})$		
Check $H_s' \stackrel{?}{=} H_s$		
$H_i = h(x_{Mi} \| x_{Rs})$	$\{H_i, T_{i2}\}$ $\longrightarrow$	
		Check $(T_{s2} - T_{i2}) \leq \Delta T$
		$H_i' = h(x_{Mi}' \| x_{Rs})$
		Check $H_i' \stackrel{?}{=} H_i$
$Sk_{us} = h(y_{Ri} \| y_{Rs}' \| y_{Mi} \| y_{Ti} \| T_{i2} \| T_{s1})$	$\longleftrightarrow$	$Sk_{su} = h(y_{Ri} \| y_{Rs} \| y_{Mi}' \| y_{Ti}' \| T_{i2} \| T_{s1})$

Password-Change/Update Phase		
User U_i		**Smart card** SC_i
Enter ID_i' and pw_i'	$\{ID_i', pw_i'\}$ $\Longrightarrow$	
		$b_i' = \widehat{b}_i \oplus h(ID_i' \| pw_i')$
		$hpw_i' = h(pw_i' \| b_i')$
		$BID_i' = h(h(ID_i') \oplus hpw_i')$
		Check $BID_i' \stackrel{?}{=} BID_i$
Enter new password pw_i^*	$\{pw_i^*\}$ $\Longrightarrow$	
		Generate $b_i^* \in_R \mathbb{Z}_n^*$
		$\widehat{b}_i^* = h(ID_i' \| pw_i^*) \oplus b_i^*$
		$hpw_i^* = h(pw_i^* \| b_i^*)$
		$AID_i^* = AID_i + [hpw_i^* - hpw_i']P$
		$BID_i^* = h(h(ID_i') \oplus hpw_i^*)$
		Update $SC_i = \{AID_i^*, BID_i^*, \widehat{b}_i^*\}$

Figure 3. The proposed ECC-based remote user password authentication scheme.

3.1. System Initialization Phase

1. The Server S selects an elliptic curve $E_p(a,b)$ over $\mathbb{F}_p$, where p is k-bit prime and a base point P of order n from G_p of $E_p(a,b)$, where n is a large number for security purposes.
2. The Server S computes the secret key and public key pair (s, P_{pub}) such that $P_{pub} = [s]P$, where s is a random integer $s \in \mathbb{Z}_n^*$.
3. The Server S chooses a cryptographic one-way hash function $h : \{0,1\}^* \rightarrow \mathbb{Z}_n^*$.
4. The Server S publishes $\{E_p(a,b), P, P_{pub}, h(\cdot)\}$ and keeps s secret.

3.2. User Registration Phase

A new user must register with the Server S before requesting access to the services. The registration phase is detailed as follows:

1. The user U_i chooses an identity ID_i and password pw_i, and generates a random integer $b_i \in \mathbb{Z}_n^*$. Then, the user U_i computes $hpw_i = h(pw_i \| b_i)$ and sends $\{ID_i, hpw_i\}$ to S through a secure channel.
2. The Server S computes $CID_i = h(ID_i \oplus s)$ and checks the availability of CID_i. If the value CID_i is in the database of registered users, the user U_i will be asked to input a new ID_i. Otherwise, the Server stores CID_i into the database. Following the approach taken by [31], this step is added to allow S to trace the user during the login phase.
3. The Server S computes $AID_i = [CID_i + hpw_i]P$, $BID_i = h(h(ID_i) \oplus hpw_i)$, stores $\{AID_i, BID_i\}$ into the smart card SC_i, and issues the card securely to U_i.
4. Once the user U_i receives the smart card SC_i, the user computes $\widehat{b_i} = h(ID_i \| pw_i) \oplus b_i$ and stores the value $\widehat{b_i}$ into SC_i. Hence, the smart card $SC_i = \{AID_i, BID_i, \widehat{b_i}\}$.

3.3. User Login Phase

In the login phase, a user U_i submits a login request message to the Server S for access to the system. First, the user U_i inserts the smart card SC_i into a remote terminal and enters the identity and password, $\{ID_i', pw_i'\}$. The SC_i executes the following steps.

1. The smart card SC_i computes $b_i' = \widehat{b_i} \oplus h(ID_i' \| pw_i')$, $hpw_i' = h(pw_i' \| b_i')$, and $BID_i' = h(h(ID_i') \oplus hpw_i')$, and checks if $BID_i' = BID_i$ holds. If the equation holds, then U_i has entered the correct identity and password, $ID_i' = ID_i$ and $pw_i' = pw_i$, respectively. Otherwise, the login phase is aborted.
2. The smart card SC_i selects a random integer $r_i \in \mathbb{Z}_n^*$ and computes $R_i = [r_i]P = (x_{Ri}, y_{Ri}) \in G_p$, where x_{Ri} and y_{Ri} are the x-component and y-component of the point R_i, respectively.
3. The smart card SC_i computes $M_i = [r_i]P_{pub} = (x_{Mi}, y_{Mi}) \in G_p$, $TID_i = AID_i - [hpw_i]P = (x_{Ti}, y_{Ti}) \in G_p$, $DID_i = ID_i \oplus y_{Mi}$, and $EID_i = h(x_{Ti} \| x_{Mi} \| T_{i1})$, where T_{i1} is the timestamp of U_i's login request submission.
4. The smart card SC_i submits the login request message $= \{DID_i, EID_i, R_i, T_{i1}\}$ to S through a public channel.

3.4. Mutual Authentication Phase

Once the Server S receives the login request message at time T_{s1}, it proceeds with the following steps.

1. The Server S checks if $(T_{s1} - T_{i1}) \leq \Delta T$, where ΔT is the allowed time transmission delay. If the time difference does not hold, the login request is rejected.
2. The Server S computes $M_i' = [s]R_i = (x_{Mi}', y_{Mi}') \in G_p$ in order to retrieve the identity $ID_i' = DID_i \oplus y_{Mi}'$ and $CID_i' = h(ID_i' \oplus s)$. Then, the Server S checks the validity of ID_i' by searching the value of CID_i' in the registered users' database. If CID_i' is not in the database, the login request is rejected.

3. The Server S computes $TID_i' = [CID_i']P = (x'_{Ti}, y'_{Ti}) \in G_p$ and $EID_i' = h(x'_{Ti} \| x'_{Mi} \| T_{i1})$, and checks if $EID_i' = EID_i$ holds. If the equation does not hold, the login request is rejected.
4. The Server S generates a random integer $r_s \in \mathbb{Z}_n^*$, computes $R_s = [r_s]R_i = (x_{Rs}, y_{Rs}) \in G_p$, $Z_s = R_s + M_i'$, and $H_s = h(EID_i' \| x_{Rs} \| T_{s1} \| x'_{Ti})$, and sends the response message $= \{Z_s, H_s, T_{s1}\}$ to U_i through the public channel.
5. Once the user U_i receives the response message at time T_{i2}, the user U_i checks if $(T_{i2} - T_{s1}) \leq \Delta T$. If the time difference does not hold, the user U_i disconnects from the Server S.
6. The user U_i computes $R_s' = Z_s - M_i = (x'_{Rs}, y'_{Rs}) \in G_p$, and $H_s' = h(EID_i \| x'_{Rs} \| T_{s1} \| x_{Ti})$, and checks if $H_s' = H_s$ holds. If the equation does not hold, the user U_i disconnects from S.
7. The user U_i computes $H_i = h(x_{Mi} \| x_{Rs})$ and sends the message $= \{H_i, T_{i2}\}$ to S.
8. The Server S checks if $(T_{s2} - T_{i2}) \leq \Delta T$. If the time difference does not hold, the session is terminated.
9. The Server S computes $H_i' = h(x'_{Mi} \| x_{Rs})$ and checks if $H_i' = H_i$. If it holds, the user U_i and the Server S achieve mutual authentication and agree on the session key $Sk_{us} = h(y_{Ri} \| y'_{Rs} \| y_{Mi} \| y_{Ti} \| T_{i2} \| T_{s1}) = h(y_{Ri} \| y_{Rs} \| y'_{Mi} \| y'_{Ti} \| T_{i2} \| T_{s1}) = Sk_{su}$. Otherwise, the session is terminated.

3.5. Password Change/Update Phase

The user U_i can change or update the password during this phase by initially inserting the smart card SC_i into a remote terminal with the identity and password $\{ID_i', pw_i'\}$. Then, the smart card performs the following steps.

1. The smart card SC_i computes $b_i' = \widehat{b}_i \oplus h(ID_i' \| pw_i')$, $hpw_i' = h(pw_i' \| b_i')$, and $BID_i' = h(h(ID_i') \oplus hpw_i')$, and checks if $BID_i' = BID_i$. If the equation is true, the smart card SC_i asks the user U_i to submit a new password pw_i^*. Otherwise, the request is rejected.
2. Once the user U_i enters the new password pw_i^*, the smart card SC_i generates a new random integer $b_i^* \in \mathbb{Z}_n^*$ and computes the new values $\widehat{b}_i^* = h(ID_i \| pw_i^*) \oplus b_i^*$, $hpw_i^* = h(pw_i^* \| b_i^*)$, $AID_i^* = AID_i + [hpw_i^* - hpw_i]P$, and $BID_i^* = h(h(ID_i) \oplus hpw_i^*)$.
3. Finally, the smart card SC_i updates the values as $SC_i = \{AID_i^*, BID_i^*, \widehat{b}_i^*\}$.

3.6. Proof of Correctness

The propositions and proof of correctness are presented below for the sake of completeness.

Proposition 1. *If the user U_i enters the identity and password $\{ID_i, pw_i\}$ correctly, and the user login phase and Steps 1-2 of the mutual authentication phase run smoothly, then the Server S will obtain the correct $TID_i' = TID_i$, which is shown as follows.*

$$
\begin{aligned}
TID_i &= AID_i - [hpw_i]P \\
&= [CID_i + hpw_i]P - [hpw_i]P \\
&= [CID_i]P + [hpw_i]P - [hpw_i]P \\
&= [CID_i]P \\
&= [CID_i']P, \quad CID_i = CID_i' \text{ in Database} \\
&= TID_i'
\end{aligned}
$$

Proposition 2. *Assume that the user U_i receives the response message from the Server S and passes the timestamp check in Step 5 of the mutual authentication phase. The equation in Step 6 will retrieve the correct R'_s value as follows.*

$$\begin{aligned} R'_s &= Z_s - M_i \\ &= (R_s + M'_i) - [r_i]P_{pub} \\ &= (R_s + [s]R_i) - [r_i]P_{pub} \\ &= R_s + [s][r_i]P - [r_i][s]P \\ &= R_s \end{aligned}$$

Proposition 3. *If the user U_i enters the correct identity and password $\{ID_i, pw_i\}$, and the equation $BID'_i = BID_i$ holds, the smart card SC_i can compute the new value AID^*_i without the knowledge of CID_i, which is shown as follows.*

$$\begin{aligned} AID^*_i &= AID_i + [hpw^*_i - hpw_i]P \\ &= [CID_i + hpw_i]P + [hpw^*_i - hpw_i]P \\ &= [CID_i]P + [hpw_i]P + [hpw^*_i]P - [hpw_i]P \\ &= [CID_i + hpw^*_i]P \end{aligned}$$

4. Security Analysis of the Proposed Scheme

This section analyzes the security aspect of the proposed scheme. First, the formal security proof is presented based on the ROM using the proof by contradiction technique, which is similar to [26,32,33]. Next, the attainment of security goals is discussed. Then, the proposed scheme is shown to withstand several identified security attacks. Finally, the formal verification of the scheme using BAN logic is provided to prove the mutual authentication property.

4.1. Formal Security Analysis

The formal proof demonstrates that the proposed scheme is provably secure against an adversary $\mathcal{A}$ from obtaining the identity ID_i, secret key s, and shared session key $Sk_{us}(= Sk_{su})$. In this approach, a mathematical proof is presented to show that the security of the proposed scheme is reduced to the ability of the adversary to break four computationally intractable problems: the collision-resistant one-way hash function, ECDLP, ECCDHP, and ECFP.

The formal proof begins by assuming the adversary $\mathcal{A}$ knows the values for the parameters $\{AID_i, BID_i, \widehat{b_i}\}$ stored in the smart card, and the messages $\{DID_i, EID_i, R_i, T_{i1}\}$, $\{Z_s, H_s, T_{s1}\}$, and $\{H_i, T_{i2}\}$ transmitted in the public channel, as described in the adversary model in Section 2.3. In addition, the adversary $\mathcal{A}$ is assumed to have access to the following oracles.

- $\mathcal{O}_{\mathcal{H}ash}$: Given the input $h(x)$, the oracle yields the output x.
- $\mathcal{O}_{\mathcal{E}CDLP}$: Given the input P and $Q = [a]P$, the oracle yields the output a.
- $\mathcal{O}_{\mathcal{E}CCDHP}$: Given the input P, $Q = [a]P$, and $R = [b]P$, the oracle yields the output $[a \cdot b]P$.
- $\mathcal{O}_{\mathcal{E}CFP}$: Given the input P and $Q = [a]P + [b]P = [a + b]P$, the oracle yields the output $[a]P$ and $[b]P$.

Theorem 1. *Assuming that the cryptographic one-way hash function $h(\cdot)$ acts like a true random oracle, and ECDLP, ECCDHP, and ECFP are computationally intractable problems, then the proposed ECC-based authentication scheme is provably secure against an adversary $\mathcal{A}$ for deriving the identity ID_i, secret key s, and session key $Sk_{us}(= Sk_{su})$.*

Proof. Suppose an adversary $\mathcal{A}$ is constructed to derive the identity ID_i, secret key s, and session key $Sk_{us}(= Sk_{su})$ by running the algorithm $ALG^{Oracle}_{\mathcal{A},ECPAS}$, as shown in Algorithm 1 for the proposed ECC-based scheme. Based on Assumptions *A1* and *A2* in Section 2.3, the adversary $\mathcal{A}$ can obtain the transmitted messages $\{DID_i, EID_i, R_i, T_{i1}\}$, $\{Z_s, H_s, T_{s1}\}$, and $\{H_i, T_{i2}\}$, and the parameters $\{AID_i, BID_i, \widehat{b_i}\}$ stored in the smart card. Then, the success probability of $ALG^{Oracle}_{\mathcal{A},ECPAS}$ is given as $Succ^{Oracle}_{\mathcal{A},ECPAS} = 2\Pr[Adv^{Oracle}_{\mathcal{A},ECPAS} = 1] - 1$. The advantage for the $ALG^{Oracle}_{\mathcal{A},ECPAS}$ is the maximum of the success probability taken over all $\mathcal{A}$ with execution time t, $Adv^{Oracle}_{\mathcal{A},ECPAS}(t, q_1, q_2, q_3, q_4) = max_{\mathcal{A}}\{Succ^{Oracle}_{\mathcal{A},ECPAS}\}$, where q_1, q_2, q_3, and q_4 denote the number of queries made to oracles $\mathcal{O}_{\mathcal{H}ash}$, $\mathcal{O}_{\mathcal{E}CDLP}$, $\mathcal{O}_{\mathcal{E}CCDHP}$, and $\mathcal{O}_{\mathcal{E}CFP}$, respectively.

Algorithm 1 $ALG^{Oracle}_{\mathcal{A},ECPAS}$ for deriving identity ID_i, secret key s, and session key $Sk_{us}(= Sk_{su})$.

Eavesdrop the login message $\{DID_i, (EID)_i, R_i, T_{i1}\}$
Call $\mathcal{O}_{\mathcal{H}ash}$ on input $EID_i = h(x_{Ti}\|x_{Mi}\|T_{i1})$ to obtain $(x^\star_{Ti}\|x^\star_{Mi}\|T^\star_{i1}) \leftarrow \mathcal{O}_{\mathcal{H}ash}(EID_i)$
Call $\mathcal{O}_{\mathcal{E}CCDHP}$ on input P_{pub}, R_i, and P to obtain $M_i^{\star\star}$ as $(M_i^{\star\star} = (x^{\star\star}_{Mi}, y^{\star\star}_{Mi})) \leftarrow \mathcal{O}_{\mathcal{E}CCDHP}(P_{pub}, R_i, P)$
if $x^\star_{Mi} = x^{\star\star}_{Mi}$ **then**
  Call $\mathcal{O}_{\mathcal{E}CFP}$ on input AID_i and P to obtain $TID_i^{\star\star}$ and $([hpw_i]P)^\star$ as $(TID_i^{\star\star} = (x^{\star\star}_{Ti}, y^{\star\star}_{Ti}), ([hpw_i]P)^\star) \leftarrow \mathcal{O}_{\mathcal{E}CFP}(AID_i, P)$
  if $x^\star_{Ti} = x^{\star\star}_{Ti}$ **then**
    Compute $ID_i^\star = DID_i \oplus y^\star_{Mi}$
    Compute $EID_i^\star = h(x^\star_{Ti}\|x^\star_{Mi}\|T_{i1})$
    if $EID_i^\star = EID_i$ **then**
      Accept $ID_i^\star$ as the correct user's identity
      Call $\mathcal{O}_{\mathcal{E}CDLP}$ on input $TID_i^\star$ and P to obtain $CID_i^\star$ as $(CID_i^\star) \leftarrow \mathcal{O}_{\mathcal{E}CDLP}(TID_i^\star, P)$
      Call $\mathcal{O}_{\mathcal{H}ash}$ on input $CID_i^\star = h(ID_i \oplus s)$ to obtain $(ID_i \oplus s)^\star$ as $(ID_i \oplus s)^\star \leftarrow \mathcal{O}_{\mathcal{H}ash}(CID_i^\star)$
      Compute $s^\star = (ID_i \oplus s)^\star \oplus ID_i^\star$
      Eavesdrop the message $\{Z_s, H_s, T_{s1}\}$
      Compute $R_s^\star = Z_s - M_i^\star = (x^\star_{Rs}, y^\star_{Rs})$
      Compute $H_s^\star = h(EID_i^\star\|x^\star_{Rs}\|T_{s1}\|x^\star_{Ti})$
      if $H_s^\star = H_s$ **then**
        Accept $s^\star$ as the correct secret key
        Eavesdrop the message $\{H_i, T_{i2}\}$
        Compute $H_i^\star = h(x^\star_{Mi}\|x^\star_{Rs})$
        if $H_i^\star = H_i$ **then**
          Compute $Sk_{us} = h(y_{Ri}\|y^\star_{Rs}\|y^\star_{Mi}\|y^\star_{Ti}\|T_{i2}\|T_{s1}) = Sk_{su}$ as the correct shared session key
          return 1 (Success)
        else
          return 0 (Fail)
        end if
      else
        return 0 (Fail)
      end if
    else
      return 0 (Fail)
    end if
  else
    return 0 (Fail)
  end if
else
  return 0 (Fail)
end if

Based on algorithm $ALG^{Oracle}_{\mathcal{A},ECPAS}$, suppose the adversary $\mathcal{A}$ can compute the inverse of a cryptographic one-way hash functions, and solve ECDLP, ECCDHP, and ECFP by using the oracles $\mathcal{O}_{\mathcal{H}ash}$, $\mathcal{O}_{\mathcal{E}CDLP}$, $\mathcal{O}_{\mathcal{E}CCDHP}$, and $\mathcal{O}_{\mathcal{E}CFP}$. Then, the adversary $\mathcal{A}$ wins the game and successfully obtains ID_i, s, and $Sk_{us}(= Sk_{su})$. However, according to Definitions 1–4, the advantages $Adv^{\mathcal{H}ash}_{\mathcal{A}}(t_1) \leq \varepsilon_1$, $Adv^{\mathcal{E}CDLP}_{\mathcal{A}}(t_2) \leq \varepsilon_2$, $Adv^{\mathcal{E}CCDHP}_{\mathcal{A}}(t_3) \leq \varepsilon_3$, and $Adv^{\mathcal{E}CFP}_{\mathcal{A}}(t_4) \leq \varepsilon_4$, for any sufficiently small negligible functions $\varepsilon_1, \varepsilon_2, \varepsilon_3, \varepsilon_4 > 0$. Hence, it must be that $Adv^{Oracle}_{\mathcal{A},ECPAS}(t, q_1, q_2, q_3, q_4) \leq \varepsilon$ for any sufficiently small $\varepsilon > 0$. Therefore, the theorem is proven. □

4.2. Attainment of Security Goals

This section analyzes the proposed scheme's attainment of security goals as explained in Section 2.4.

(1) Mutual authentication

The proposed scheme includes mutual authentication steps for verifying the legality of the user and the Server. The Server authenticates the user by checking the value $CID_i = h(ID_i' \oplus s)$ in the registered users' database. Next, the user authenticates the Server by checking the value of $H_s' = h(EID_i \| x_{Rs}' \| T_{s1} \| x_{Ti})$. Although an adversary may obtain the value of EID_i, Z_s and AID_i by Assumptions *A1* and *A2*, the adversary needs to compute the values of $R_s' = Z_s - M_i$ and $TID_i = AID_i - [hpw_i]P$, which are not transmitted in the public channel. Furthermore, R_s' and TID_i are secured by the ECDLP and ECFP. Therefore, the proposed scheme provides mutual authentication.

(2) Session key agreement

After completing the mutual authentication steps, both the user and Server compute a shared session key $Sk_{us} = h(y_{Ri} \| y_{Rs}' \| y_{Mi} \| y_{Ti} \| T_{i2} \| T_{s1}) = h(y_{Ri} \| y_{Rs} \| y_{Mi}' \| y_{Ti}' \| T_{i2} \| T_{s1}) = Sk_{su}$. Since the adversary does not know R_s, M_i, and TID_i, the session key cannot be computed directly due to the cryptographic one-way hash function. Hence, the shared session key is protected in the proposed scheme.

(3) Forward secrecy

In the proposed scheme, the session keys are computed using the values $R_i = [r_i]P$ and $R_s = [r_s]P$, which are calculated based on random numbers r_i and r_s. Even if an adversary obtains the secret key s, the adversary still cannot obtain any information from the previous session keys. Thus, the proposed scheme provides forward secrecy.

(4) User anonymity

According to Assumption *A2*, an adversary may extract all of the values $\{AID_i, BID_i, \widehat{b}_i\}$ in the smart card. The ID_i is contained in the parameters AID_i and BID_i. However, the adversary needs to invert a one-way hash output, which is impossible in polynomial time, as shown in Theorem 1. As a result, the proposed scheme provides user anonymity.

(5) User traceability

Following Maitra et al. [31], the server should be able to trace the sender of the login request message by confirming that the sender is indeed a user registered in the database. The proposed scheme still maintains user anonymity because the user's ID_i is hidden and secured by the secret key s in the parameter $CID_i = h(ID_i \oplus s)$. Therefore, the proposed scheme allows the Server to trace the user.

(6) Local password verification

The proposed scheme provides wrong password input detection by the smart card during the login phase by checking the value $BID_i' = h(h(ID_i') \oplus hpw_i') = BID_i$. The incorrect combination of ID_i' and pw_i' will be detected before preparing the login request message. Hence, the proposed scheme provides local password verification.

(7) Local password changeability

The password change/update phase permits the user to modify the password without contacting the Server. Since the smart card can verify the password and identity locally through a remote terminal, it can compute and update the parameters $\{AID_i^*, BID_i^*, \widehat{b}_i^*\}$. Therefore, the proposed scheme provides efficient local password changeability.

4.3. Resistance to Security Attacks

This section presents the proposed scheme's ability to withstand several security attacks.

(1) Offline password-guessing attack

Suppose that an adversary $\mathcal{A}$ obtains a lost/stolen smart card SC_i and retrieves $\{AID_i, BID_i, \widehat{b_i}\}$. The adversary $\mathcal{A}$ must guess the user U_i's identity ID_i and password pw_i to compute b_i, hpw_i, and $BID_i' = BID_i = h(h(ID_i) \oplus hpw_i)$. However, according to Assumption *A3*, it is impossible to guess both ID_i and pw_i within polynomial time. Hence, the proposed scheme can withstand the offline password-guessing attack.

(2) Replay attack

By Assumption *A1*, an adversary $\mathcal{A}$ can intercept all of the messages transmitted through the public channel. Since the messages are generated using the random numbers (r_i, r_s) and timestamps (T_{i1}, T_{i2}, T_{s1}), the Server S will notice the repeated message submissions. Hence, it is impossible for $\mathcal{A}$ to replay intercepted messages. Therefore, the proposed scheme can resist replay attacks.

(3) Privileged insider attack

In this attack, suppose a privileged insider $\mathcal{A}$ as an active adversary who obtains the identity ID_i by monitoring data transmitted over a secure channel during the registration phase. In addition, assume that $\mathcal{A}$ extracts the values AID_i, BID_i, and $\widehat{b_i}$ from a lost/stolen smart card SC_i, as in Assumption *A2*. In the proposed scheme, $\mathcal{A}$ cannot launch the password-guessing attack because the password pw_i is secured by $hpw_i = h(pw_i \| b_i)$. The adversary $\mathcal{A}$ can try to retrieve the random number b_i from $\widehat{b_i} = h(ID_i \| pw_i) \oplus b_i$. However, $\mathcal{A}$ has to guess both pw_i and b_i simultaneously within polynomial time, which contradicts Assumption *A3*. Thus, the proposed scheme can withstand the privileged insider attack.

(4) Stolen-verifier attack

If an adversary $\mathcal{A}$ gains access to the database of registered users, the adversary $\mathcal{A}$ can try to extract the ID_i of a legal user U_i. However, the Server's database stores the value ID_i in $CID_i = h(ID_i \oplus s)$ secured by the collision-resistant one-way hash function. In addition, $\mathcal{A}$ also cannot obtain the secret key s since it is protected by ECDLP. It is impossible for $\mathcal{A}$ to retrieve ID_i. Therefore, the proposed scheme can resist stolen-verifier attacks.

(5) Key-compromised impersonation attack

Assume an adversary $\mathcal{A}$ obtains a compromised or stolen secret key s. Then, the adversary $\mathcal{A}$ can try to impersonate a legal user U_i to cheat the Server S. Still, the $\mathcal{A}$ must first pass the verification check $BID_i' = BID_i$. Furthermore, the $\mathcal{A}$ cannot create the login message $\{DID_i, EID_i, R_i, T_{i1}\}$ because it is not possible to compute hpw_i. Thus, the proposed scheme can withstand the key-compromised impersonation attack.

4.4. Formal Verification Using BAN Logic

This section provides the verification of the mutual authentication property for the proposed scheme using BAN logic [39]. The BAN logic analysis consists of four main steps: (1) defining the verification goals, (2) transforming the proposed scheme to its idealized form, (3) expressing the initial state assumptions, and (4) proving the security goals by using the BAN logic rules as in Table 3.

(1) Verification goals

First, the BAN logic goals for the proposed scheme are defined and listed as follows.

- Goal 1: $U_i \mid\equiv (U_i \stackrel{Sk}{\longleftrightarrow} S)$
- Goal 2: $U_i \mid\equiv S \mid\equiv (U_i \stackrel{Sk}{\longleftrightarrow} S)$
- Goal 3: $S \mid\equiv (U_i \stackrel{Sk}{\longleftrightarrow} S)$
- Goal 4: $S \mid\equiv U_i \mid\equiv (U_i \stackrel{Sk}{\longleftrightarrow} S)$

(2) Idealization of the proposed scheme

Next, the proposed scheme is transformed into the idealized form as follows.

- Message 1: $U_i \rightarrow S : \langle ID_i \rangle_{M_i}, (M_i, T_{i1})_{TID_i}, R_i, T_{i1}$
- Message 2: $S \rightarrow U_i : \langle R_s \rangle_{M_i}, (EID_i, R_s, T_{s1})_{TID_i}, T_{s1}$
- Message 3: $U_i \rightarrow S : (R_s)_{M_i}, T_{i2}$

(3) Initial state assumptions

The assumptions made on the initial state of the proposed scheme are listed below.

- A1: $S \mid\equiv (U_i \overset{M_i}{\rightleftharpoons} S)$;
- A2: $U_i \mid\equiv (U_i \overset{TID_i}{\rightleftharpoons} S)$;
- A3: $S \mid\equiv (U_i \overset{TID_i}{\rightleftharpoons} S)$;
- A4: $U_i \mid\equiv \#(T_{s1})$;
- A5: $S \mid\equiv \#(T_{i1}, T_{i2})$.

(4) Proof using BAN logic

The security proof analysis is presented based on the goals, initial state assumptions, and BAN logic rules.

- Step 1: From Message 1, $S \lhd (\langle ID_i \rangle_{M_i}, (M_i, T_{i1})_{TID_i}, R_i, T_{i1})$.
- Step 2: According to Step 1, A3, and applying the message-meaning rule, the statement $S \mid\equiv U_i \mid\sim (M_i, R_i, T_{i1})$ is deduced.
- Step 3: By the freshness-conjuncatenation rule and A5 yields, $S \mid\equiv \#(M_i, R_i, T_{i1})$.
- Step 4: From Step 2, Step 3, and the nonce-verification rule, then $S \mid\equiv U_i \mid\equiv (M_i, R_i, T_{i1})$.
- Step 5: From Message 3, $S \lhd ((R_s)_{M_i}, T_{i2})$.
- Step 6: Applying the message-meaning rule to Step 5 and A1, then $S \mid\equiv U_i \mid\sim (R_s, T_{i2})$.
- Step 7: By the freshness-conjuncatenation rule and A5 yields, $S \mid \#(R_s, T_{i2})$.
- Step 8: From Steps 6 and 7 using the nonce-verification rule, then $S \mid\equiv U_i \mid\equiv (R_s, T_{i2})$.
- Step 9: By the belief rule, Step 4, and Step 8, $S \mid\equiv U_i \mid\equiv (M_i, R_i, R_s, T_{i1}, T_{i2})$.
- Step 10: From Step 9, A5, and the session key rule, then $S \mid\equiv (U_i \overset{Sk}{\leftrightarrow} S)$ (**Goal 3**).
- Step 11: From A5, Step 9, Step 10, and the session-key verification rule, then $S \mid\equiv U_i \mid\equiv (U_i \overset{Sk}{\leftrightarrow} S)$ (**Goal 4**).
- Step 12: From Message 2, $U_i \lhd (\langle R_s \rangle_{M_i}, (EID_i, R_s, T_{s1})_{TID_i}, T_{s1})$.
- Step 13: Applying the message-meaning rule, from Step 12 and A2, then the statement $U_i \mid\equiv S \mid\sim (EID_i, R_s, T_{s1})$ is obtained.
- Step 14: By the freshness conjuncatenation rule and A4 yields, $U_i \mid\equiv \#(EID_i, R_s, T_{s1})$.
- Step 15: According to Step 13, Step 14, and applying the nonce verification rule, then $U_i \mid\equiv S \mid\equiv (EID_i, R_s, T_{s1})$.
- Step 16: By the session-key rule, Step 14, and Step 15, then $U_i \mid\equiv (U_i \overset{Sk}{\leftrightarrow} S)$ (**Goal 1**).
- Step 17: Finally, from A4, Step 15, Step 16, and the session-key verification rule, $U_i \mid\equiv S \mid\equiv (U_i \overset{Sk}{\leftrightarrow} S)$ (**Goal 2**).

Based on BAN logic analysis, all of the defined goals are achieved. Therefore, the proposed scheme is demonstrated to provide mutual authentication using the shared session key between U_i and S.

5. Performance Analysis

This section explains the performance of the proposed scheme compared to similar schemes and improvements by [29–33]. Since this study focuses on the schemes that have been improved based on Qu and Tan [29], the compared schemes are chosen based on the underlying security of three hard problems in ECC (i.e., ECDLP, ECCDHP, and ECFP) in the general user–server application. Based on the literature search, to the best of the authors' knowledge, only the works by Huang et al. [30] and Chaudhry et al. [32] fit this scope. The schemes by Maitra et al. [31] and Mehmood et al. [33] are also included in the performance comparison since they proposed enhancements based on Huang et al. [30].

Table 4 summarizes the security goals attainment and resistance to security attacks of every scheme based on the discussions in Section 4. The proposed scheme has been shown to achieve all of the security goals as given in Maitra et al. [31], which are formal security proof, mutual authentication, session key agreement, forward secrecy, user anonymity, user traceability, local password verification, and local password changeability. The proposed scheme has also been shown to withstand replay attacks, offline password-guessing attacks, privileged insider attacks, stolen-verifier attacks, insider attacks, and key-compromised impersonation attacks. Overall, the proposed scheme and Maitra et al. [31] outperformed other considered schemes in terms of security goals attainment. The proposed scheme performs better than all considered schemes based on the resistance to security attacks.

Table 4. Attainment of security goals and resistance to security attacks of the proposed scheme and other similar schemes.

	Schemes					
	Proposed	Qu and Tan [29]	Huang et al. [30]	Maitra et al. [31]	Chaudhry et al. [32]	Mehmood et al. [33]
Attainment of security goals						
Formal security proof	✓	✗	✗	✓	✓	✓
Mutual authentication	✓	✓	✓	✓	✓	✓
Session key agreement	✓	✓	✓	✓	✓	✓
Forward secrecy	✓	✓	✓	✓	✓	✓
User anonymity	✓	✓	✓	✓	✓	✓
User traceability	✓	✗	✗	✓	✗	✗
Local password verification	✓	✓	✓	✓	✓	✓
Local password changeability	✓	✓	✓	✓	✗	✗
Resistance to security attacks						
Replay attack	✓	✓	✓	✓	✓	✓
Offline password-guessing attack	✓	✗	✓	✓	✓	✓
Privileged insider attack	✓	✓	✗	✓	✗	✗
Stolen-verifier attack	✓	✓	✓	✓	✓	✗
Key-compromised impersonation attack	✓	✗	✗	✗	✗	✗

(✓): Yes, (✗): Not discussed.

For the computational cost analysis, the approximate running time is based on the performance evaluation by Kilinc and Yanik [41] using the PBC Library [42]. The running times of arithmetic and cryptographic operations were measured using the experimental platform, which is the Ubuntu 12.04.1 LTS 32bit operating system with Intel Pentium Dual CPU E2200 2.20 GHz processor and 2048 MB of RAM. Based on their findings, the order of the time complexity for the elliptic curve point multiplication operation (T_{em}), elliptic curve point addition operation (T_{ea}), symmetric encryption/decryption operation (T_{sym}), and hash operation (T_h) is stated as $T_{em} \gg T_{ea} > T_{sym} > T_h$. The estimated running times for T_{em}, T_{ea}, T_{sym}, and T_h are 2.226 ms, 0.0288 ms, 0.0046 ms, and 0.00023 ms, respectively. The modular multiplication/division operation (T_m) and the bitwise XOR ($\oplus$) operation recorded negligible running times and are hence ignored.

The computational cost is the total time complexity of operations executed in the user registration, user login, and mutual authentication phases. As shown in Table 5, the proposed scheme requires the computational cost of 7 T_{em} + 3 T_{ea} + 18 T_h and a running time of approximately 15.710 ms. In terms of the number of T_{em} operations executed, the proposed scheme maintains 7 T_{em} operations as in Huang et al. [30] and

Chaudhry et al. [32], which is four T_{em} operations less than Maitra et al. [31]. The running times for Qu and Tan [29], Huang et al. [30], Maitra et al. [31], Chaudhry et al. [32], and Mehmood et al. [33] are approximately 20.215 ms, 15.767 ms, 24.521 ms, 15.708 ms, and 9.003 ms, respectively. As seen in Figure 4a, the proposed scheme requires only a 0.02 ms higher running time than Chaudhry et al. [32]. This slight increase in running time is insignificant given that the proposed scheme is more secure than Chaudhry et al. [32] based on Table 4. Furthermore, the proposed scheme's running time is 8.811 ms less than that of Maitra et al. [31], which is noteworthy considering that both schemes attain the same security goals.

Table 5. Computational cost for executed operations in the proposed scheme and other similar schemes.

Schemes	Computational Cost	Running Time
Proposed	$7\ T_{em} + 3\ T_{ea} + 18\ T_h$	$\approx$ 15.710 ms
Qu and Tan [29]	$9\ T_{em} + 5\ T_{ea} + 16\ T_h$	$\approx$ 20.215 ms
Huang et al. [30]	$7\ T_{em} + 5\ T_{ea} + T_m + 18\ T_h$	$\approx$ 15.767 ms
Maitra et al. [31]	$11\ T_{em} + 2\ T_m + 15\ T_h$	$\approx$ 24.521 ms
Chaudhry et al. [32]	$7\ T_{em} + 3\ T_{ea} + 2\ T_m + 17\ T_h$	$\approx$ 15.708 ms
Mehmood et al. [33]	$4\ T_{em} + 2\ T_{ea} + 3\ T_m + 2\ T_{sym} + 14\ T_h$	$\approx$ 9.003 ms

T_{em}: Elliptic curve multiplication operation, T_{ea}: Elliptic curve point addition operation, T_m: Modular multiplication/division operation, T_{sym}: Symmetric encryption/decryption operation, T_h: Hash operation.

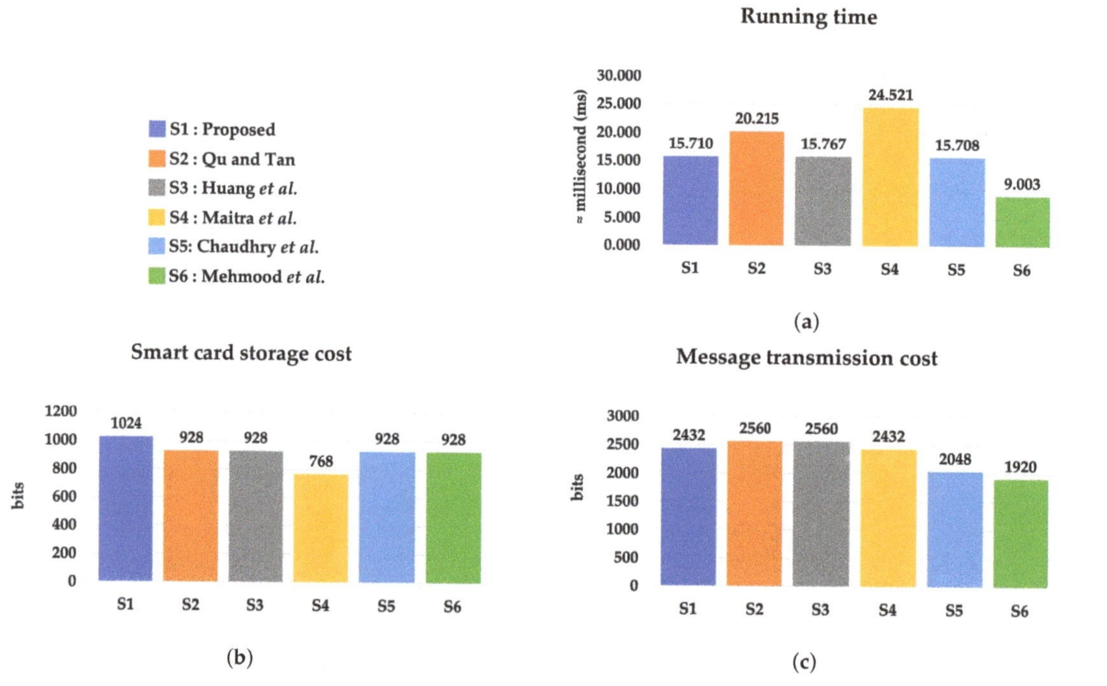

Figure 4. Comparisons of ECC-based schemes' performance in terms of (**a**) running time; (**b**) smart card storage cost; (**c**) message transmission cost.

For the smart card storage and message transmission costs analysis, the following assumptions are made. The sizes for the identity ID_i, password pw_i, and random numbers $\{b_i, r_i, r_s\}$ are 160 bits each. The hash function outputs $\{BID_i, \widehat{b}_i, EID_i, H_s, H_i\}$ are 256 bits, assuming the use of the SHA-256 [2] algorithm. The elliptic curve points $\{AID_i, R_i, Z_s\}$ are 512 bits each, whereas the x/y-coordinate is 256 bits. The timestamps $\{T_{i1}, T_{s1}, T_{i2}\}$ are 128 bits.

In the proposed scheme, the parameters $\{AID_i, BID_i, \widehat{b_i}\}$ are stored in the smart card SC_i. The storage cost required for the smart card is $512 + 256 + 256 = 1024$ bits, which is the highest among other schemes as shown in Figure 4b. The proposed scheme's storage cost incurs 96 more bits than schemes by [29,30,32,33] since the parameter $\widehat{b_i}$ is stored as a hash output to mask the random number b_i. Furthermore, the proposed scheme requires a 296-bit higher storage cost than the scheme by Maitra et al. [31] because the parameter AID_i is stored as an elliptic curve point instead of a hash output. Nevertheless, the proposed scheme's higher storage cost is justified given that the proposed scheme provides better security features than other schemes.

The message transmission cost is the total bit size of the messages $\{DID_i, EID_i, R_i, T_i\}$, $\{Z_s, H_s, T_{s1}\}$, and $\{H_i, T_{i2}\}$, which are exchanged during the user login phase and mutual authentication phase. For the proposed scheme, the transmission cost is $(4 \times 256) + (2 \times 512) + (3 \times 128) = 2432$ bits, which is comparable to that of Maitra et al. [31] and 128 bits lower than [29,30]. However, the proposed scheme's transmission cost is 384 bits and 512 bits higher than Chaudhry et al. [32] and Mehmood et al. [33], respectively. Note that the proposed scheme and Maitra et al. [31] require clock synchronization, unlike other schemes. Hence, the transmission of timestamps during the login and authentication phases explains the message transmission cost being higher than [32,33], as shown in Figure 4c. Even with timestamps, the proposed scheme and Maitra et al. [31] managed to keep their transmission cost lower than [29,30].

Overall, the computational cost and running time of the proposed scheme are lower than [29–31]. In terms of the message transmission cost, the proposed scheme performs the same as Maitra et al. [31]. As the proposed scheme maintains all of the hard problems (ECDLP, ECCDHP, and ECFP) of Qu and Tan [29] and attains all of the security goals of Maitra et al. [31] as shown in Table 5, the higher smart card storage cost is an acceptable trade-off. In conclusion, the proposed scheme is better than all considered schemes.

6. Applications

In the future, it is suggested to investigate the applicability of adopting the three hard problems, i.e., ECDLP, ECCDHP, and ECFP, in developing user/client identification and authentication cryptographic schemes in distributed computer networks [43–45]. The integration of distributed computer networks with physical and social systems has evolved tremendously to many applications in cyber–physical systems and cyber–physical social systems. These systems connect many low-powered devices, such as smart mobile applications and wireless sensor nodes, that are deployed in unsupervised environments. The communication and data sharing between the physical components and cyber components demand attention toward security requirements and privacy issues [46,47]. ECC is favored in many public-key-based cryptographic schemes due to its efficiency; hence, it is important to study the feasibility of implementing three hard problems (ECDLP, ECCDHP, and ECFP) in designing secure and efficient schemes.

7. Conclusions

This study highlighted several drawbacks of the scheme by Chaudry et al. The aim of this study was to propose an ECC-based two-factor remote authentication scheme with a session key agreement based on Chaudhry et al.'s scheme to solve these drawbacks. The proposed scheme is provably secure under the ROM using the formal definitions of ECDLP, ECCDHP, and ECFP. Based on the security and performance analyses with other previous schemes, the proposed scheme offers better security attributes and is more efficient in terms of the computational cost and running time. Future work is suggested to build better identification and authentication schemes based on the same hard problems (ECDLP, ECCDHP, and ECFP) for applications in cyber–physical systems.

Author Contributions: Conceptualization, E.S.I.; methodology, F.S.; validation, E.S.I.; formal analysis, F.S.; writing—original draft preparation, F.S.; writing—review and editing, E.S.I.; visualization, F.S.; supervision, E.S.I.; funding acquisition, E.S.I. All authors have read and agreed to the published version of the manuscript.

Funding: This research was funded by UKM grant number GUP-2020-029.

Data Availability Statement: The data presented in this study is contained within the article.

Acknowledgments: The authors deeply appreciate all of the comments and suggestions from the anonymous reviewers and the editor for improving the paper. The authors would like to thank Universiti Kebangsaan Malaysia, Universiti Teknologi MARA Malaysia, and the Ministry of Higher Education, Malaysia, for providing the facilities and financial support to conduct this research. Sincere thanks to Alena Lee Sanusi and Aziana Ismail for taking the time to proofread our paper.

Conflicts of Interest: The authors declare no conflict of interest.

References

1. Lamport, L. Password authentication with insecure communication. *Commun. ACM* **1981**, *24*, 770–772. [CrossRef]
2. NIST. *FIPS 180-4 Secure Hash Standard (SHS)*; Technical Report; National Institute of Standard and Technology: Gaithersburg, MD, USA, 2015.
3. Wang, X.M.; Zhang, W.F.; Zhang, J.S.; Khan, M.K. Cryptanalysis and improvement on two efficient remote user authentication scheme using smart cards. *Comput. Stand. Interfaces* **2007**, *29*, 507–512. [CrossRef]
4. Chaudhry, S.A.; Farash, M.S.; Naqvi, H.; Kumari, S.; Khan, M.K. An enhanced privacy preserving remote user authentication scheme with provable security. *Secur. Commun. Netw.* **2015**, *8*, 3782–3795. [CrossRef]
5. Madhusudhan, R.; Hegde, M. Cryptanalysis and improvement of remote user authentication scheme using smart card. In Proceedings of the 2016 International Conference on Computer and Communication Engineering (ICCCE), Kuala Lumpur, Malaysia, 26–27 July 2016; IEEE: Piscataway, NJ, USA, 2016; pp. 84–89.
6. Rivest, R.L.; Shamir, A.; Adleman, L. A method for obtaining digital signatures and public-key cryptosystems. *Commun. ACM* **1978**, *21*, 120–126. [CrossRef]
7. Diffie, W.; Hellman, M. New directions in cryptography. *IEEE Trans. Inf. Theory* **1976**, *22*, 644–654. [CrossRef]
8. Miller, V.S. Use of elliptic curves in cryptography. In Proceedings of the Conference on the Theory and Application of Cryptographic Techniques, Linz, Austria, 9–11 April 1985; Springer: Berlin/Heidelberg, Germany, 1985; pp. 417–426.
9. Koblitz, N. Elliptic curve cryptosystems. *Math. Comput.* **1987**, *48*, 203–209. [CrossRef]
10. Gura, N.; Patel, A.; Wander, A.; Eberle, H.; Shantz, S.C. Comparing elliptic curve cryptography and RSA on 8-bit CPUs. In Proceedings of the International Workshop on Cryptographic Hardware and Embedded Systems, Cambridge, MA, USA, 11–13 August 2004; Springer: Berlin/Heidelberg, Germany, 2004; pp. 119–132.
11. Juang, W.S.; Chen, S.T.; Liaw, H.T. Robust and efficient password-authenticated key agreement using smart cards. *IEEE Trans. Ind. Electron.* **2008**, *55*, 2551–2556. [CrossRef]
12. Fan, C.I.; Chan, Y.C.; Zhang, Z.K. Robust remote authentication scheme with smart cards. *Comput. Secur.* **2005**, *24*, 619–628. [CrossRef]
13. Sun, D.Z.; Huai, J.P.; Sun, J.Z.; Li, J.X.; Zhang, J.W.; Feng, Z.Y. Improvements of Juang et al.'s password-authenticated key agreement scheme using smart cards. *IEEE Trans. Ind. Electron.* **2009**, *56*, 2284–2291.
14. Li, X.; Qiu, W.; Zheng, D.; Chen, K.; Li, J. Anonymity enhancement on robust and efficient password-authenticated key agreement using smart cards. *IEEE Trans. Ind. Electron.* **2010**, *57*, 793–800.
15. He, D.; Chen, J.; Hu, J. Further improvement of Juang et al.'s password-authenticated key agreement scheme using smart cards. *Kuwait J. Sci. Eng.* **2011**, *38*, 55–68.
16. Li, X.; Zhang, Y. A simple and robust anonymous two-factor authenticated key exchange protocol. *Secur. Commun. Netw.* **2013**, *6*, 711–722. [CrossRef]
17. Jiang, Q.; Ma, J.; Li, G.; Yang, L. Robust two-factor authentication and key agreement preserving user privacy. *Int. J. Netw. Secur.* **2014**, *16*, 229–240.
18. Liu, C.; Ma, C.G. An efficient and provable secure PAKE scheme with robust anonymity. In Proceedings of the International Conference on Information Computing and Applications, Chengde, China, 14–16 September 2012; Springer: Berlin/Heidelberg, Germany, 2012; pp. 722–729.
19. Tsai, J.L.; Lo, N.W.; Wu, T.C. Novel anonymous authentication scheme using smart cards. *IEEE Trans. Ind. Inform.* **2012**, *9*, 2004–2013. [CrossRef]
20. Byun, J.W. On the secure design of hash-based authenticator in the smartcard authentication system. *Wirel. Pers. Commun.* **2019**, *109*, 2329–2352. [CrossRef]
21. Wang, R.C.; Juang, W.S.; Lei, C.L. Robust authentication and key agreement scheme preserving the privacy of secret key. *Comput. Commun.* **2011**, *34*, 274–280. [CrossRef]

22. Wu, S.; Zhu, Y.; Pu, Q. Robust smart-cards-based user authentication scheme with user anonymity. *Secur. Commun. Netw.* **2012**, *5*, 236–248. [CrossRef]
23. Chang, C.C.; Lin, I.C.; Wu, C.C. A multipurpose key agreement scheme in ubiquitous computing environments. *Mob. Inf. Syst.* **2015**, *2015*, 934716. [CrossRef]
24. Wang, L. Analysis and enhancement of a password authentication and update scheme based on elliptic curve cryptography. *J. Appl. Math.* **2014**, *2014*, 247836. [CrossRef]
25. Islam, S.H.; Biswas, G. Design of improved password authentication and update scheme based on elliptic curve cryptography. *Math. Comput. Model.* **2013**, *57*, 2703–2717. [CrossRef]
26. Odelu, V.; Das, A.K.; Goswami, A. An efficient ECC-based privacy-preserving client authentication protocol with key agreement using smart card. *J. Inf. Secur. Appl.* **2015**, *21*, 1–19. [CrossRef]
27. Madhusudhan, R.; Hegde, M.; Memon, I. A secure and enhanced elliptic curve cryptography-based dynamic authentication scheme using smart card. *Int. J. Commun. Syst.* **2018**, *31*, e3701.
28. Kumari, A.; Jangirala, S.; Abbasi, M.Y.; Kumar, V.; Alam, M. ESEAP: ECC based secure and efficient mutual authentication protocol using smart card. *J. Inf. Secur. Appl.* **2020**, *51*, 102443. [CrossRef]
29. Qu, J.; Tan, X.L. Two-factor user authentication with key agreement scheme based on elliptic curve cryptosystem. *J. Electr. Comput. Eng.* **2014**, *2014*, 16. [CrossRef]
30. Huang, B.; Khan, M.K.; Wu, L.; Muhaya, F.T.B.; He, D. An efficient remote user authentication with key agreement scheme using elliptic curve cryptography. *Wirel. Pers. Commun.* **2015**, *85*, 225–240. [CrossRef]
31. Maitra, T.; Obaidat, M.S.; Islam, S.H.; Giri, D.; Amin, R. Security analysis and design of an efficient ECC-based two-factor password authentication scheme. *Secur. Commun. Networks* **2016**, *9*, 4166–4181. [CrossRef]
32. Chaudhry, S.A.; Naqvi, H.; Mahmood, K.; Ahmad, H.F.; Khan, M.K. An improved remote user authentication scheme using elliptic curve cryptography. *Wirel. Pers. Commun.* **2017**, *96*, 5355–5373. [CrossRef]
33. Mehmood, Z.; Chen, G.; Li, J.; Albeshri, A. An untraceable ECC-based remote user authentication scheme. *KSII Trans. Internet Inf. Syst. (TIIS)* **2017**, *11*, 1742–1760.
34. Dolev, D.; Yao, A. On the security of public key protocols. *IEEE Trans. Inf. Theory* **1983**, *29*, 198–208. [CrossRef]
35. Kocher, P.; Jaffe, J.; Jun, B. Differential power analysis. In Proceedings of the Annual International Cryptology Conference, Santa Barbara, CA, USA, 15–19 August 1999; Springer: Berlin/Heidelberg, Germany, 1999; pp. 388–397.
36. Messerges, T.S.; Dabbish, E.A.; Sloan, R.H. Examining smart-card security under the threat of power analysis attacks. *IEEE Trans. Comput.* **2002**, *51*, 541–552. [CrossRef]
37. Sood, S.K.; Sarje, A.K.; Singh, K. Cryptanalysis of password authentication schemes: Current status and key issues. In Proceedings of the 2009 International Conference on Methods and Models in Computer Science (ICM2CS), New Delhi, India, 14–15 December 2009; IEEE: Piscataway, NJ, USA, 2009; pp. 1–7.
38. Wu, F.; Xu, L.; Kumari, S.; Li, X.; Alelaiwi, A. A new authenticated key agreement scheme based on smart cards providing user anonymity with formal proof. *Secur. Commun. Netw.* **2015**, *8*, 3847–3863. [CrossRef]
39. Burrows, M.; Abadi, M.; Needham, R.M. A logic of authentication. *Proc. R. Soc. Lond. Math. Phys. Sci.* **1989**, *426*, 233–271.
40. Sowjanya, K.; Dasgupta, M.; Ray, S. An elliptic curve cryptography based enhanced anonymous authentication protocol for wearable health monitoring systems. *Int. J. Inf. Secur.* **2020**, *19*, 129–146. [CrossRef]
41. Kilinc, H.H.; Yanik, T. A survey of SIP authentication and key agreement schemes. *IEEE Commun. Surv. Tutorials* **2013**, *16*, 1005–1023. [CrossRef]
42. Lynn, B. The Pairing-Based Cryptography (PBC) Library. Available online: https://crypto.stanford.edu/pbc/ (accessed on 30 September 2022).
43. Tsai, J.L. Weaknesses and improvement of Hsu-Chuang's user identification scheme. *Inf. Technol. Control* **2010**, *39*, 48–50.
44. Chang, C.C.; Lee, C.Y. A secure single sign-on mechanism for distributed computer networks. *IEEE Trans. Ind. Electron.* **2011**, *59*, 629–637. [CrossRef]
45. Chen, Y.C.; Liu, C.L.; Horng, G. Cryptanalysis of some user identification schemes for distributed computer networks. *Int. J. Commun. Syst.* **2014**, *27*, 2909–2917. [CrossRef]
46. Ghaffar, Z.; Ahmed, S.; Mahmood, K.; Islam, S.H.; Hassan, M.M.; Fortino, G. An improved authentication scheme for remote data access and sharing over cloud storage in cyber-physical-social-systems. *IEEE Access* **2020**, *8*, 47144–47160. [CrossRef]
47. Lu, Y.; Wang, D.; Obaidat, M.S.; Vijayakumar, P. Edge-assisted intelligent device authentication in cyber-physical systems. *IEEE Internet Things J.* **2022**, 1–14.. [CrossRef]

mathematics

MDPI

Article

Population-Based Meta-Heuristic Algorithms for Integrated Batch Manufacturing and Delivery Scheduling Problem

Yong-Jae Kim and Byung-Soo Kim *

Department of Industrial and Management Engineering, Incheon National University, 119, Academy-ro, Yeonsu-gu, Incheon 22012, Korea; yongjae@inu.ac.kr
* Correspondence: bskim@inu.ac.kr

Abstract: This paper addresses an integrated scheduling problem of batch manufacturing and delivery processes with a single batch machine and direct-shipping trucks. In the manufacturing process, some jobs in the same family are simultaneously processed as a production batch in a single machine. The batch production time depends only on the family type assigned to the production batch and it is dynamically adjusted by batch deterioration and rate-modifying activities. Each job after the batch manufacturing is reassigned to delivery batches. In the delivery process, each delivery batch is directly shipped to the corresponding customer. The delivery time of delivery batches is determined by the distance between the manufacturing site and customer location. The total volume of jobs in each production or delivery batch must not exceed the machine or truck capacity. The objective function is to minimize the total tardiness of jobs delivered to customers with different due dates. To solve the problem, a mixed-integer linear programming model to find the optimal solution for small problem instances is formulated and meta-heuristic algorithms to find effective solutions for large problem instances are presented. Sensitivity analyses are conducted to find the effect of problem parameters on the manufacturing and delivery time.

Keywords: scheduling; supply chain management; meta-heuristic algorithms; mixed-integer linear programming; batch production; batch delivery

MSC: 90B06

Citation: Kim, Y.-J.; Kim, B.-S. Population-Based Meta-Heuristic Algorithms for Integrated Batch Manufacturing and Delivery Scheduling Problem. *Mathematics* **2022**, *10*, 4127. https://doi.org/10.3390/math10214127

Academic Editor: Ioannis G. Tsoulos

Received: 11 October 2022
Accepted: 2 November 2022
Published: 4 November 2022

1. Introduction

Recently, many studies have been conducted on individual manufacturing and delivery problems, both of which are an important part of supply chain management (SCM). The methodologies for an integrated scheduling problem (ISP) generally provide better performance to improve the efficiency of the entire supply chain than individual manufacturing and delivery problems [1]. The study on ISPs is difficult even if ISPs provide better performance because of the complexity of the supply chain and the conflict of stakeholders in the supply chain. Nevertheless, ISPs are required for many sectors of industry such as ceramics, food, port cargo handling, and freight logistics [2]. In this study, we confine our study to ISPs regarding the manufacturing and delivery process. We apply the batch loading and scheduling problem (BLSP) in the manufacturing process [3] and the direct-shipping problem in the delivery process [1,4].

In the manufacturing process, jobs can be processed simultaneously on a batch processing machine, and a set of jobs that are processed simultaneously is called a production batch. The volumes of jobs are different. The total volume in a production batch must not exceed the machine capacity. Jobs with different families must not be assigned to the same production batch. The batch production time depends only on the family type assigned to the production batch. Furthermore, we consider deterioration and rate-modifying activities. In collaborative works between operators and machines, such as machining, assembling, and maintenance, the batch production time can increase due to operator fatigue or machine failure, where the increased portion of this batch production time is called deterioration. The recovery process from the deteriorated state to the original state is called rate-modifying activity (RMA) [5]. In this study, the batch production time increases in proportion to the interval between the recent RMA and the start time of the batch because we assume that the deterioration effect occurs linearly.

In the delivery process, each job after batch manufacturing is reassigned to delivery batches. The delivery batches are directly shipped by fixed numbers of homogeneous trucks. In this study, we consider that the delivery batch is independent of the production batch. The total volume in a delivery batch must not exceed the truck capacity. Jobs from different customers must not be assigned to the same delivery batch. The truck can deliver only one delivery batch at a time. The truck leaving the factory returns immediately after shipping the delivery batch to the customer. The delivery time including return time depends only on the customer of that particular delivery batch. The objective function is minimizing the total tardiness of jobs delivered to customers with different due dates.

Figure 1 describes a Gantt chart example for the presented ISP. The number of jobs, families, and customers are 5, 2, and 2, respectively. Jobs 1 and 2 belong to Family 1 and Jobs 3, 4, and 5 belong to Family 2. Jobs 1, 3, and 5 are requested from Customer 1, and Jobs 2 and 4 are requested from Customer 2. The production time for the family and the delivery time for the customer are (50, 100) and (229, 161), respectively. The due date and volume of jobs are (264, 235, 401, 477, 459) and (5, 10, 7, 14, 10), respectively. The machine capacity, truck capacity, deterioration rate, and RMA processing time are 20, 20, 0.3, and 20, respectively. The jobs are assigned to production batches (B^M) while keeping the constraints on the family compatibility and machine capacity. The production batches are sequenced with RMAs inserted between them. The production time of Batch 2 increases in proportion to the interval between the start time of Batches 2 and 1 due to deterioration. The original production time of Batch 2 is 100. The interval between the start time of Batches 2 and 1 is 50. The deterioration rate is 20. Thus, the production time of Batch 2 is 115 (=100 + 50 $\times$ 0.3). Assuming the RMA was performed before Batch 3 is processed, the deterioration for Batch 3 is restored and the batch production time is not increased. The manufacturing completion time of jobs is (50, 50, 165, 285, 165). Jobs that have been processed are assigned to delivery batches (B^D) while keeping the constraints on the customer compatibility and the truck capacity. Truck 1 transports Batch 2 at time 279, but the manufacturing completion time of Job 4 is 285. Therefore, the waiting time is 6 (= 285 – 279) between Batches 2 and 3 in Truck 1. The completion time of jobs in the batches for the corresponding customer is (279, 211, 440, 446, 440). By comparing the due dates of each job, the total tardiness of each job is 54 (=15 + 0 + 39 + 0 + 0).

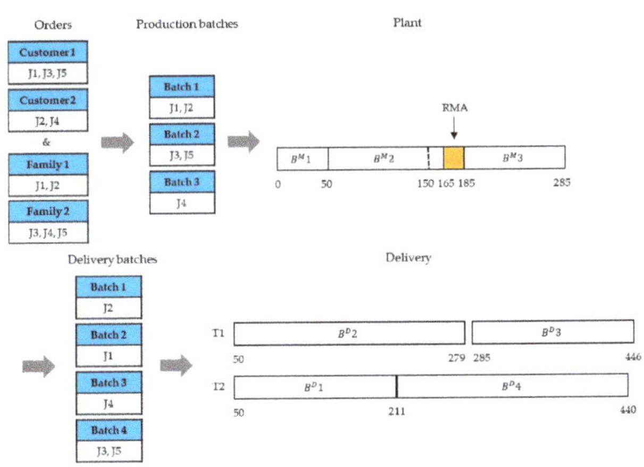

Figure 1. A Gantt chart example of schedules for the presented ISP.

2. Literature Review

In this section, we survey studies on ISPs, including batch processing. For ISPs including deterioration or RMA, we focus on their scheme.

For ISPs with a direct-shipping method, Liu [6] dealt with a two-stage delivery problem. The first stage of delivery is to deliver jobs from the warehouse to the batching machine by crane. The second stage of delivery is to deliver the processed jobs to the customer by only one vehicle. He proposed genetic algorithms to minimize the sum of makespan and the total setup cost. Jia et al. [2] studied a problem with parallel batch processing machines with different capacities. They proposed several heuristic algorithms for minimizing the total weighted delivery time of jobs. Selvarajah and Steiner [7] assumed that only items with the same customer and product belong to one batch. They presented a polynomial algorithm for minimizing the sum of total inventory holding cost and the batch delivery cost. Gao et al. [8] studied a problem with limited vehicle capacity. The jobs are batched without breaking the vehicle capacity constraints before being processed. They presented polynomial-time algorithms for two special cases with the same production time and delivery time of order, respectively. Furthermore, they provided a heuristic to solve a general problem. Cheng et al. [9], Cheng et al. [10], and Jia et al. [11] assumed that the vehicle capacity is an integer multiple of the machine capacity. Cheng et al. [9] and Cheng et al. [10] assumed that the batches are packaged in the same size of boxes or pallets and propose each $O(nlogn)$ time algorithm for identical and arbitrary job sizes. Jia et al. [11] dealt with a problem with parallel batch machines. They present two hybrid meta-heuristic algorithms based on ant colony optimization and a deterministic heuristic for minimizing total weighted delivery time. In addition, they proposed a lower bound for evaluating the presented algorithms. Li et al. [12] studied a problem with unbounded parallel-batch and job families. They defined the family as the customer who requested the job. They assumed that jobs with the same family have identical sizes in a vehicle, and jobs with different families are not delivered together. They showed that the problem is NP-hard and proposed a heuristic algorithm for minimizing completion time. Li et al. [13] studied a problem with both machine and vehicle capacity. Jobs have different sizes and the total volume of jobs in each batch does not exceed the machine capacity. Likewise, the total volume of jobs in the delivery batch does not exceed the vehicle capacity. They proposed a polynomial algorithm for identical job sizes and heuristics for different job sizes to maximize the total profit. Zhang et al. [14] dealt with a problem including the order-picking process. The orders are batched without breaking the capacity constraint of picking devices. They proposed an on-line algorithm for minimizing the makespan and total delivery cost. Nogueira et al. [15] and Feng and Xu [16] studied ISPs with parallel batching machines. Nogueira [15] assumed that job size and production time are generic. They presented a mathematical formulation model and several heuristic algorithms to maximize the total profits. Feng and Xu [16] developed a 0–1 mixed-integer programming (MIP) model. Jia et al. [17] further considered parallel non-identical batch machines. He et al. [18] proposed an enhanced branch-and-price algorithm for integrated 3D printing with JIT delivery systems. Li et al. [19] developed a MIP formulation and proposed a column generation-based approach for an ISP with dual delivery modes.

For ISPs with vehicle routing problems (VRP), Karaoğlan and Kesen [20] dealt with the problem of distributing products with a limited shelf life to customers in a vehicle. They proposed a branch-and-cut (B&C) algorithm to minimize lead time. Low et al. [21] and Low et al. [22] studied the problem of delivering the product to the customer after processing it in the distribution center. They provided adaptive genetic algorithms (AGAs) to minimize total cost, including delivery cost, vehicle cost, and penalty cost. Li [23] considered the bi-objective problem minimizing both customer waiting time and vehicle delivery cost.

For ISPs with deterioration or RMA, Kong et al. [24] considered the integrated problem of CCHR and delivery scheduling in steel production. They assumed that the rolling time is linearly proportional to the starting time of slabs. Liu et al. [25] dealt with the integrated problem with parallel batching machines and deteriorating jobs. The production time of a job increases non-linearly concerning the starting time, and the production time of a batch is assumed to be the maximum value of jobs belonging to that batch. Yin et al. [26] studied batch delivery scheduling on a single machine with RMA. They assumed that processing a job after RMA would reduce the original production time by modifying rate times.

Table 1 shows the classification for studies on ISPs with batch processing. The studies are categorized according to compatibility in a production batch, vehicle number, deterioration, RMA, and shipping method.

Table 1. Classification for Studies on ISP with batch processing.

	Compatibility in Production Batches			Vehicles Number	Deterioration	RMA	Shipping Method	
	Incompatible Product	Incompatible Family	Incompatible Customer				Direct-Shipping	VRP
Liu [6]				1			✓	
Jia et al. [2]				Limited			✓	
Selvarajah and Steiner [7]	✓		✓	1			✓	
Gao et al. [8]				1			✓	
Cheng et al. [9]				1			✓	
Cheng et al. [10]				1			✓	
Jia et al. [11]				Limited			✓	
Li et al. [12]			✓	1			✓	
Li et al. [13]				Limited			✓	
Zhang et al. [14]				1			✓	
Nogueira et al. [15]				Limited			✓	
Feng and Xu [16]				Unlimited			✓	
Jia et al. [17]				Limited			✓	
He et al. [18]			✓	Limited			✓	
Li et al. [19]			✓	Limited			✓	
Karaoğlan and Kesen [20]				1				✓
Low et al. [21]				Unlimited				✓
Low et al. [22]				Limited				✓
Li et al. [23]				Limited				✓
Kong et al. [24]				Unlimited	✓		✓	
Liu et al. [25]				1	✓		✓	
Yin et al. [26]				Unlimited		✓	✓	
This study		✓		Limited	✓	✓	✓	

To the best of our knowledge, an ISP simultaneously considering the family compatibility, batch deterioration with multiple RMAs, and direct-shipping method has received very limited attention; however, several ISP scheduling problems with batch manufacturing and delivery processes are often dealt with (See Table 1).

3. Mixed-Integer Linear Programming Model

In this section, the proposed mixed-integer linear programming (MILP) model is formulated; the notation of the formulation that follows is shown below:

Indices	
i, j	jobs
f	families
k, l	production batches
m, n	delivery batches
u	buckets
t	trucks
c	customers
Parameters	
J	set of jobs
F	set of families
B^M	set of production batches
U	set of buckets
B^D	set of delivery batches
T	set of trucks
C	set of customers
p_f	production time of family $f \in F$
F_j^J	family of job $j \in J$
F_k^B	family of production batches $k \in B^M$
h_c	delivery time for customer $c \in C$
R_{jc}^C	1 if job $j \in J$ is required by customer $c \in C$; 0 otherwise
v_j	volume of job $j \in J$
d_j	due time of job $j \in J$

DR	deterioration rate
V^M	machine capacity
Q	RMA processing time
V^T	truck capacity
M	a large number
Continuous variables	
x_k	production starting time of production batch k
I_k	time interval between starting time of production batch k and completion time of the most recent RMA before production batch k
c_u	completion time of bucket u
r_m	shipping starting time of delivery batch m
τ_j	tardiness of job j
Binary variables	
$y_{ik}^{B^M}$	1 if production batch k assigns job i; 0 otherwise
y_{ku}^{U}	1 if bucket u assigns production batch k; 0 otherwise
z_{klu}^{U}	1 if production batch k immediately precedes production batch l at bucket u; 0 otherwise
$y_{im}^{B^D}$	1 if delivery batch m assigns job i; 0 otherwise
y_{mt}^{T}	1 if truck t assigns delivery batch m; 0 otherwise
y_{mc}^{C}	1 if customer c assigns delivery batch m; 0 otherwise
z_{mnt}^{T}	1 if delivery batch m immediately precedes delivery batch n in truck t; 0 otherwise

The MILP formulation using the above notation is as follows:

$$Minimize\ z = \sum_{i \in J} \tau_i \tag{1}$$

Subject to

$$\sum_{\substack{k \in B^M \\ F_k^B = F_i^J}} y_{ik}^{B^M} = 1 \qquad \forall i \in J \tag{2}$$

$$\sum_{i \in J} v_i \cdot y_{ik}^{B^M} \le V^M \qquad \forall k \in B^M \tag{3}$$

$$\sum_{i \in J} y_{ik}^{B^M} \le M \cdot \sum_{u \in U} y_{ku}^{U} \qquad \forall k \in B^M \tag{4}$$

$$\sum_{u \in U} y_{ku}^{U} \le 1 \qquad \forall k \in B^M \tag{5}$$

$$\sum_{l \in B^M} z_{lku}^{U} = y_{ku}^{U} \qquad \forall k \in B^M; u \in U \tag{6}$$

$$\sum_{\substack{l \in B^M \\ l \ne k}} z_{klu}^{U} \le y_{ku}^{U} \qquad \forall k \in B^M; u \in U \tag{7}$$

$$\sum_{k \in B^M} z_{kku}^{U} \le 1 \qquad \forall u \in U \tag{8}$$

$$I_k \cdot (1 + DR) + p_{F_k^B} \le I_l + M \cdot \left(1 - \sum_{u \in U} z_{klu}^{U}\right) \qquad \forall k, l \in B^M; k \ne l \tag{9}$$

$$I_k \cdot (1 + DR) + p_{F_k^B} \le c_u + M \cdot \left(1 - y_{ku}^U\right) \qquad \forall k \in B^M; u \in U \qquad (10)$$

$$I_k + \sum_{\substack{v \in U \\ v < u}} c_v + Q \cdot \sum_{\substack{w \in U \\ w < u}} \sum_{l \in B^M} z_{llw}^U \le x_k + M \cdot \left(1 - y_{ku}^U\right) \qquad \forall k \in B^M; u \in U \qquad (11)$$

$$\sum_{m \in B^D} y_{im}^{B^D} = 1 \qquad \forall i \in J \qquad (12)$$

$$\sum_{i \in J} v_i \cdot y_{im}^{B^D} \le V^T \qquad \forall m \in B^D \qquad (13)$$

$$y_{im}^{B^D} + y_{jm}^{B^D} \le 1 + \sum_{c \in C} R_{ic}^C \cdot R_{jc}^C \qquad \forall m \in B^D; i, j \in J; i < j \qquad (14)$$

$$r_m \ge x_k + I_k \cdot DR + p_{F_k^B} - M \cdot \left(2 - y_{im}^{B^D} - y_{ik}^{B^M}\right) \qquad \forall i \in J; m \in B^D; k \in B^M \qquad (15)$$

$$\sum_{c \in C} y_{mc}^C \le 1 \qquad \forall m \in B^D \qquad (16)$$

$$y_{mc}^C \ge R_{ic}^C \cdot y_{im}^{B^D} \qquad \forall i \in J; m \in B^D; c \in C \qquad (17)$$

$$r_m + \sum_{c \in C} h_c \cdot y_{mc}^C \le r_n + M \cdot \left(1 - \sum_{t \in T} z_{mnt}^T\right) \qquad \forall m, n \in B^D; m \ne n \qquad (18)$$

$$\sum_{t \in T} y_{mt}^T = 1 \qquad \forall m \in B^D \qquad (19)$$

$$\sum_{n \in B^D} z_{nmt}^T = y_{mt}^T \qquad \forall m \in B^D; t \in T \qquad (20)$$

$$\sum_{\substack{n \in B^D \\ n \ne m}} z_{mnt}^T \le y_{mt}^T \qquad \forall m \in B^D; t \in T \qquad (21)$$

$$\sum_{m \in B^D} z_{mmt}^T \le 1 \qquad \forall t \in T \qquad (22)$$

$$r_m + \sum_{c \in C} h_c \cdot y^C_{mc} - d_i \leq \tau_i + M \cdot \left(1 - y^{B^D}_{im}\right) \quad \forall i \in J; m \in B^D \tag{23}$$

$$x_k, I_k, c_u, r_m, \tau_i \geq 0 \quad \forall k \in B^M; u \in U;\ m \in B^D; t \in T; c \in C \tag{24}$$

$$y^{B^M}_{ik}, y^U_{ku}, y^{B^D}_{im}, y^T_{mt}, y^C_{mc} = 0 \text{ or } 1 \quad \forall i \in J; k \in B^M; u \in U; m \in B^D; t \in T; c \in C \tag{25}$$

$$z^U_{klu}, z^T_{mnt} = 0 \text{ or } 1 \quad \forall k, l \in B^M; u \in U; m, n \in B^D; t \in T \tag{26}$$

Constraint (2) denotes a restriction wherein each job must be assigned to one of the production batches. Constraint (3) confirms that the total volumes of jobs in each production batch must not exceed machine capacity. Constraints (4) and (5) guarantee that non-empty production batches are assigned to one bucket. The bucket is defined as a set of batches processed between RMAs [27]. Constraints (6–8) ensure that production batches assigned to the same bucket are processed once in a specific sequence. $z^U_{kku} = 1$ means that production batch k is in the first position in each bucket.

Constraint (9) determines the precedence relation of production batches within the same bucket and calculates the interval between their starting time and the completion time of the recent RMA. Constraint (10) calculates the completion time of buckets. Constraint (11) calculates the starting time of production batches. Constraint (12) guarantees that each job is assigned to one delivery batch. Constraint (13) confirms that the total volumes of jobs in each delivery batch must not exceed truck capacity. Constraint (14) ensures that jobs in the same delivery batch are shipped to the same customer. Constraint (15) guarantees that the shipping starting time of delivery batches is larger than the completion time for all jobs in that delivery batch. The completion time of each job is defined as the completion time of production batches to which that job is assigned. Constraint (16) denotes a restriction wherein each delivery batch is shipped to at most one customer. Constraint (17) enforces a customer–delivery batch relationship through job–customer and job–delivery batch relationships. Constraint (18) determines the precedence relation of delivery batches within a truck and calculates the shipping starting time of each delivery batch. Constraint (19) denotes a restriction wherein delivery batch must be assigned to one truck. Constraints (20)–(22) guarantee that delivery batches assigned to the same truck are shipped once in a specific sequence. $z^T_{mmt} = 1$ means that delivery batch m is in the first position in each truck. Constraint (23) calculates the tardiness of jobs.

4. Meta-Heuristic Algorithms

An ISP is generally an NP-hard problem, and since the proposed problem is an ISP with batch processing, it is NP-hard. Therefore, other efficient algorithms that can solve large problem instances quickly are required instead of the proposed MILP model. In many scheduling problem papers, the problem is effectively and efficiently solved through meta-heuristic algorithms [28–30]. Due to this reason, three meta-heuristic algorithms, namely particle swarm optimization (PSO), the imperialist competitive algorithm (ICA), and the genetic algorithm (GA), are presented. The three meta-heuristic algorithms have the same decoding process.

4.1. Solution Representation and Decoding Method

The solution is divided into two parts: manufacturing and delivery. Thus, there are two one-dimensional arrays; one represents batching and scheduling for the manufacturing process, and the other represents truck assignment and scheduling for the delivery process. Figures 2 and 3 show an illustrative example of the decoding process for an encoded manufacturing and delivery solution using the meta-heuristic algorithms proposed in Sections 4.2–4.4. In all the presented meta-heuristic algorithms, the two one-dimensional encoded arrays are formed independently.

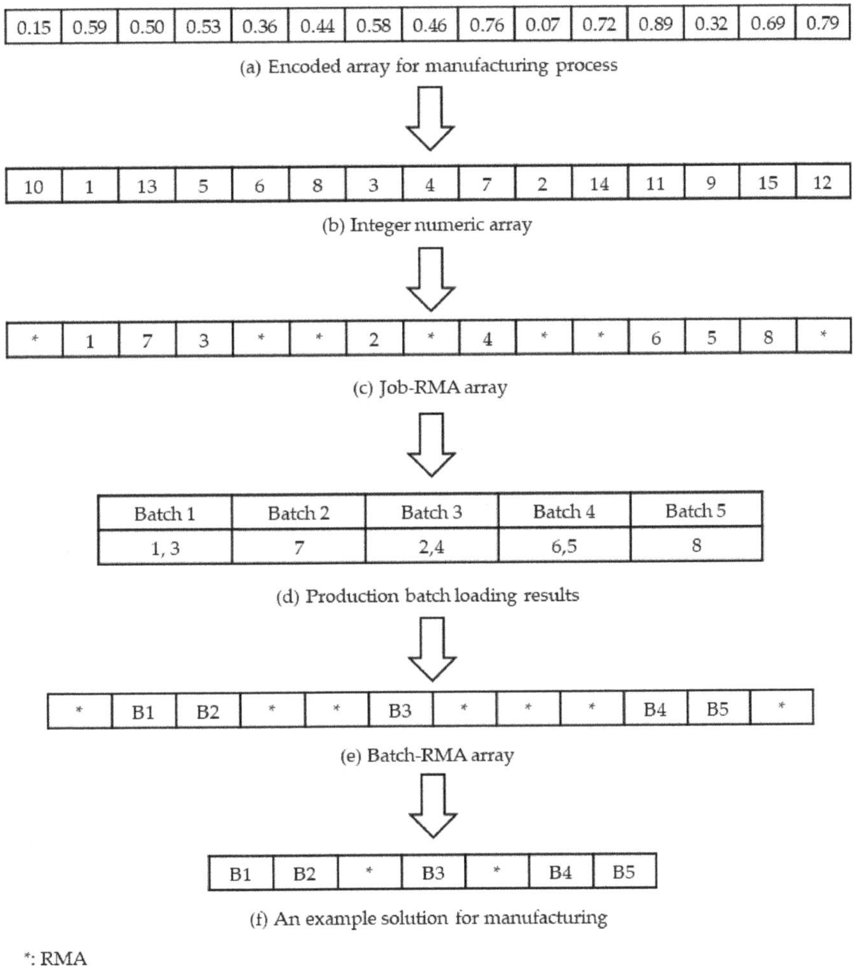

Figure 2. An example of decoding procedure for manufacturing.

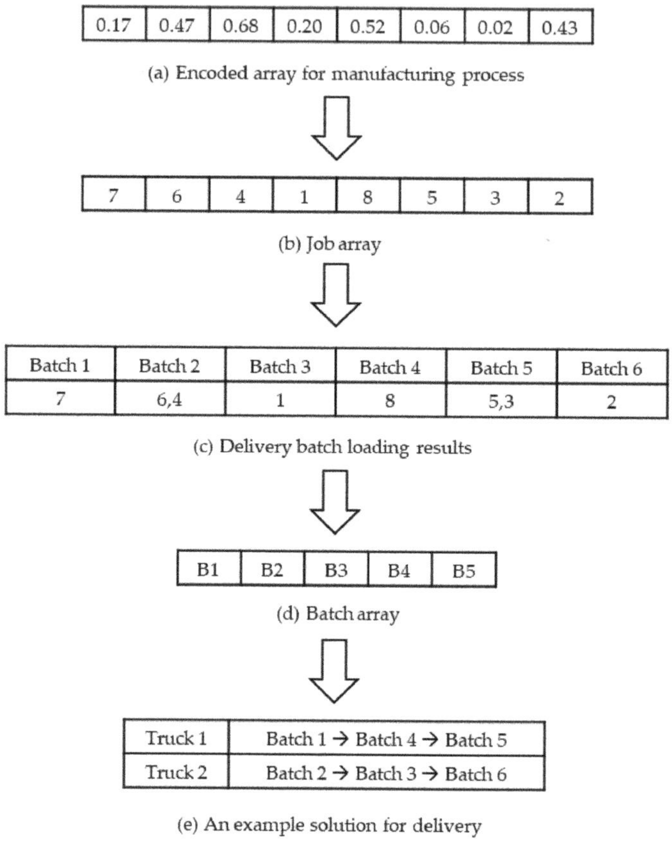

Figure 3. An example of decoding procedure for delivery.

In Figure 2, Figure 2a is converted to Figure 2b by the random-key method [31]. The main idea of random keys is that real numbers in the range [0, 1] represent the sequence of integers. In Figure 2a, the smallest number, 0.07, is in the 10th position. So, the first number in Figure 2b is 10. In Figure 2a, the smallest number after 0.07, 0.15, exists in the first position. So, the second number in Figure 2b is 1. In the same way, Figure 2a is converted to Figure 2b. Each element in Figure 2b represents a job or RMA. Suppose that the number of jobs is n. Then, the maximum number of RMAs is $n - 1$, assigned to the position between the jobs. Thus, $2n - 1$ elements are required for Figure 2c. Since the number of jobs is 8 in Figure 2b, the number of elements becomes 15 (= 8 + 7). Odd numbers are converted to (original number $+ 1)/2$, indicating the job index. All even numbers are converted to RMAs. Figure 2c contains information about the job index and RMA, which is converted into the manufacturing solution in several steps. Suppose that Jobs 1, 2, 3, and 4 belong to Family 1, and Jobs 5, 6, 7, and 8 belong to Family 2. The volume of jobs is (10, 9, 3, 7, 8, 6, 15, 11). The machine capacity is 20. The orders of jobs in Families 1 and 2 are (1,3,2,4) and (7,6,5,8) from Figure 2c, respectively. For each family, jobs belonging to the family are assigned to batches in the corresponding order of Figure 2c and satisfy the machine capacity constraint. According to Figure 2d, Figure 2c is converted to Figure 2e. The position of batches in Figure 2e is the same as the position of each job index located at the front of Figure 2c among jobs belonging to the batch. Finally, the first and last RMAs are removed and consecutive RMAs are considered as one.

In Figure 3, Figure 3a is converted to a job array using the random-key method. Each element in Figure 3b represents a job. Since the number of jobs is eight, the number of elements in Figure 3b is eight. Jobs are assigned to delivery batches in the order of Figure 3b while simultaneously satisfying the truck capacity constraint and the customer compatibility constraint. If jobs requested from different customers between two jobs requested from the same customer exist in the job array, they must not be assigned to the same delivery batch. Suppose that Jobs 1, 3, 5, and 7 are requested from Customer 1, and Jobs 2, 4, 6, and 8 are requested from Customer 2. The volume of each job is (8, 7, 5, 13, 6, 6, 9, 10). The truck capacity is 20. The manufacturing completion time of each job is (200, 350, 300, 65, 220, 70, 55, 160). The delivery time for the customer is (100, 150). The orders of jobs for Customers 1 and 2 are (7,1,5,3) and (6,4,8,2) from Figure 3b, respectively. Jobs 7 and 1 must not be assigned to the same batch. Although the total volume of Jobs 7 and 1 does not exceed the truck capacity, there are Jobs 6 and 4 between them. According to Figure 3c, Figure 3b is converted to Figure 3d. The manufacturing completion time of delivery batches is equal to the maximum value of manufacturing completion times in each job assigned in the delivery batch. Thus, the manufacturing completion time of delivery batches is (55, 70, 200, 160, 300, 350). Batches are assigned to a truck with the smallest value of differences between the manufacturing completion time of delivery batches and the available time of trucks according to their order in Figure 3d. If multiple trucks are assigned to a batch, the delivery batch is arbitrarily assigned to one of these trucks.

An encoded solution of three meta-heuristic algorithms is introduced using the decoding process presented in Figures 2 and 3. PSO, ICA, and GA are presented in Sections 4.2–4.4.

4.2. Particle Swarm Optimization (PSO)

The position and velocity consist of two one-dimensional arrays representing the manufacturing and delivery process, respectively. The position and velocity are independently initialized by the uniform distribution of real numbers between 0 and 1 ($U(0,1)$). After that, the best solution for specific particles (P_i) and the global best solution (P_g) are updated. The velocity (v_i) and position (X_i) of each particle are updated using Equations (27) and (28) based on P_i and P_g, respectively.

$$v_i \leftarrow v_i + c_1 \cdot U(0,1) \cdot (P_i - X_i) + c_2 \cdot U(0,1) \cdot (P_g - X_i), \tag{27}$$

$$X_t^i \leftarrow X_{t-1}^i + v_t^i. \tag{28}$$

The PSO procedure is shown in Algorithm 1.

Algorithm 1: The PSO procedure

Input iteration (*Iter*), population size (S_p), and acceleration weight (c_1) and (c_2).
Randomly generate initial positions and velocities through $U(0,1)$.
While ($g \leq Iter$)
  $g \leftarrow g + 1$
  For (i = 1 to S_p)
    If ($X_i < P_i$)
      $P_i \leftarrow X_i$
    End if
    If ($X_i < P_g$)
      $P_g \leftarrow X_i$
    End if
  End for
  For (i = 1 to S_p)
    $v_i \leftarrow v_i + c_1 \cdot U(0,1) \cdot (P_i - X_i) + c_2 \cdot U(0,1) \cdot (P_g - X_i)$
    $X_i \leftarrow X_i + v_i$
  End for
End while

4.3. Imperialist Competitive Algorithm (ICA)

The countries consist of two one-dimensional arrays representing the manufacturing and delivery process, respectively. The countries are independently initialized following the distribution $U(0,1)$. Afterward, N_{imp} powerful countries become imperialist. Any country that is not imperialist becomes a colony. Colonies are assigned to imperialist countries, and many colonies are assigned to powerful imperialist countries. To measure the power of imperialist countries, the normalized objective function value of nth imperialist country is calculated as follows:

$$f_n = z_{max} - z_n, \tag{29}$$

where z_n and z_{max} are objective function values for nth imperialist country and maximum objective function values for all the imperialist countries, respectively. The power of nth imperialist country is defined as follows:

$$p_n = \frac{f_n}{\sum_{k=1}^{N_{imp}} f_k} \tag{30}$$

The initial number of colonies of the nth imperialist country (NC_n) is calculated as follows:

$$NC_n = round(p_n \cdot N_{col}) \tag{31}$$

Colonies (X_c) move toward the direction of their imperialist $\left(X_I^c\right)$. The degree of approach is determined by β and a random number from the distribution $U(0,1)$.

$$X_c \leftarrow U(0,1) \cdot \beta \cdot (X_I^c - X_c) \tag{32}$$

Each element of countries probabilistically reset the value to $U(0,1)$. This probability is called the revolution rate and is set in parameter calibration. After performing the moving and revolution process, the imperialist countries of each empire are updated. Among all the countries including the existing imperialist countries, the country with the smallest objective function value becomes the new imperialist country. Afterward, the weakest colony in the weakest empire is taken away by other empires. This is called imperialistic competition. The total power of an empire is a measure of imperialistic competition. It is determined by the imperialist and colony power of each empire. The total objective function value of the nth empire (Tf_n) is calculated as follows:

$$Tf_n = z_n + \xi \cdot mean(z_c), \tag{33}$$

where ξ is the weight for the colony power and z_c is the objective function value for colonies belonging to empire n. Based on the Tf_n, the normalized total objective function value of the nth empire (NTf_n) is calculated as follows:

$$NTf_n = Tf_n - \max(Tf_n), \tag{34}$$

where $\max(Tf_n)$ is the maximum total objective function value for all empires. The possession probability ($pemp_n$) is calculated as follows:

$$pemp_n = \left| \frac{NTf_n}{\sum_{k=1}^{N_{imp}} NTf_k} \right| \tag{35}$$

The ICA procedure is shown in Algorithm 2.

Algorithm 2: The ICA procedure

Input iteration (*Iter*), population size (S_p), the number of imperialist countries (N_{imp}), revolution rate (p_r), assimilation constant (β), and coefficient of colonies' power (ξ).
*i*th country and nth imperialist country denoted by X^i and X_n^I, respectively.

Generate initial countries and determine the imperialist countries and colonies.
While ($g \leq Iter$)
  $g \leftarrow g + 1$
  For (i = 1 to S_p)
    Move to the colony toward its imperialist.
    If ($U(0,1) < p_r$)
      Conduct revolution.
    End if
    If $f(X^i) \leq f(X_n^I)$
      $X_n^I \leftarrow X^i$
    End if
  End for
Calculate the total cost of empires.
Conduct imperialistic competition.
End while

4.4. Genetic Algorithm (GA)

The chromosomes consist of two one-dimensional arrays representing the manufacturing and delivery process, respectively. The chromosomes are independently initialized by $U(0,1)$. The one-cut point crossover and uniform mutation are used as genetic operators. Crossover and mutation also proceed independent of two chromosomes. The uniform mutation operator is to replace the numeric value of a gene with a random number that follows $U(0,1)$, The roulette wheel selection is used as a selection method. The fitness function for chromosome i (F_i) used in the roulette wheel method is as follows:

$$F_i = z_{max} - z_i, \tag{36}$$

where z_i and z_{max} are objective function values for the *i*th chromosome and maximum objective function values for all the chromosomes, respectively. The objective function value for the ith chromosome is calculated as the aggregate solution of the ith manufacturing and delivery chromosomes. The GA procedure is shown in Algorithm 3.

Algorithm 3: The GA procedure

Input generation size (S_g), population size (S_p), crossover rate (p_c), and mutation rate (p_m).
Randomly generate initial population through $U(0,1)$.
$g \leftarrow 1$
While ($g \leq S_g$)
  For (i = 1 to S_p)
    If ($U(0,1) < p_c$)
      Perform the crossover operator for two different randomly selected chromosomes.
    End if
  End for
  For (i = 1 to S_p)
    For (n = 1 to N)
      If ($U(0,1) < p_m$)
        Perform the mutation operation.
      End if
    End for
  End for
Perform the roulette wheel selection.
$g \leftarrow g + 1$
End while

5. Computational Results

Problem instances for evaluating the performance of the proposed meta-heuristic algorithms are divided into large and small problem instances. In the experiment of small problem instances, the performances of PSO, ICA, and GA are validated by comparing them with the performance of the MILP model. The MILP is solved by CPLEX solver 12.7 using IBM ILOG CPLEX Optimization Studio. In the experiment of large problem instances, the performance of GA is relatively measured by comparing with performances of PSO and ICA. All meta-heuristic algorithms are implemented in C# and all computational experiments are performed by PCs with 3.60 GHz Intel Core i7-7700 CPUs.

5.1. Calibration of the Algorithm Parameters

Calibrating meta-heuristic algorithm parameters can significantly affect the performance of algorithms. The Taguchi method was used to find the best parameter combinations for PSO, ICA, and GA. The algorithm parameters are set at five levels. Tables 2 and 3 show values for each level and an orthogonal array $L_{25}(5^4)$ of GA parameters, respectively.

Table 2. The value of each level of GA parameters.

Parameters	Levels				
	1	2	3	4	5
G_s	200	400	600	800	1000
P_s	20	40	60	80	100
p_c	0.1	0.3	0.5	0.7	0.9
p_m	0.001	0.002	0.003	0.004	0.004

Table 3. Orthogonal arrays for GA parameters.

Run	G_s	P_s	p_c	p_m
1	$G_s(1)$	$P_s(1)$	$p_c(1)$	$p_m(1)$
2	$G_s(1)$	$P_s(2)$	$p_c(2)$	$p_m(2)$
3	$G_s(1)$	$P_s(3)$	$p_c(3)$	$p_m(3)$
4	$G_s(1)$	$P_s(4)$	$p_c(4)$	$p_m(4)$
5	$G_s(1)$	$P_s(5)$	$p_c(5)$	$p_m(5)$
6	$G_s(2)$	$P_s(1)$	$p_c(2)$	$p_m(3)$
7	$G_s(2)$	$P_s(2)$	$p_c(3)$	$p_m(4)$
8	$G_s(2)$	$P_s(3)$	$p_c(4)$	$p_m(5)$
9	$G_s(2)$	$P_s(4)$	$p_c(5)$	$p_m(1)$
10	$G_s(2)$	$P_s(5)$	$p_c(1)$	$p_m(2)$
11	$G_s(3)$	$P_s(1)$	$p_c(3)$	$p_m(5)$
12	$G_s(3)$	$P_s(2)$	$p_c(4)$	$p_m(1)$
13	$G_s(3)$	$P_s(3)$	$p_c(5)$	$p_m(2)$
14	$G_s(3)$	$P_s(4)$	$p_c(1)$	$p_m(3)$
15	$G_s(3)$	$P_s(5)$	$p_c(2)$	$p_m(4)$
16	$G_s(4)$	$P_s(1)$	$p_c(4)$	$p_m(2)$
17	$G_s(4)$	$P_s(2)$	$p_c(5)$	$p_m(3)$
18	$G_s(4)$	$P_s(3)$	$p_c(1)$	$p_m(4)$
19	$G_s(4)$	$P_s(4)$	$p_c(2)$	$p_m(5)$
20	$G_s(4)$	$P_s(5)$	$p_c(3)$	$p_m(1)$
21	$G_s(5)$	$P_s(1)$	$p_c(5)$	$p_m(4)$
22	$G_s(5)$	$P_s(2)$	$p_c(1)$	$p_m(5)$
23	$G_s(5)$	$P_s(3)$	$p_c(2)$	$p_m(1)$
24	$G_s(5)$	$P_s(4)$	$p_c(3)$	$p_m(2)$
25	$G_s(5)$	$P_s(5)$	$p_c(4)$	$p_m(3)$

For each run, six problem instances are randomly generated and repeated five times in each instance. The number of combinations for the algorithm parameters is 25, set by the Taguchi method. The smaller-the-better approach is used because the objective function is minimizing the total tardiness. Since various instances are used, the relative deviation index (*RDI*) is used instead of the objective function for the S/N ratio. *RDI* and the S/N ratio are represented by Equations (37) and (38), respectively.

$$RDI = \frac{Obj_{sol} - Best}{Worst - Best} \tag{37}$$

$$\text{S/N ratio}_k = -10 \cdot \log\left(\sum_{i=1}^{6}\sum_{j=1}^{5} RDI_{ijk}^2\right) for\ k = 1, 2, \ldots, 25 \quad (38)$$

where *Best* and *Worst* are the objective function values from the best and worst of the three algorithms (PSO, ICA, and GA) for each problem instance, respectively.

Figure 4a shows the mean plots of the S/N ratio for GA parameters. To find out parameters that significantly affect the difference among the S/N ratios, an analysis of variance (ANOVA) for the S/N ratio is tested. Table 4 shows the results of ANOVA for the S/N ratio. A parameter with the smallest sum square (SS) p_m is considered an error [32].

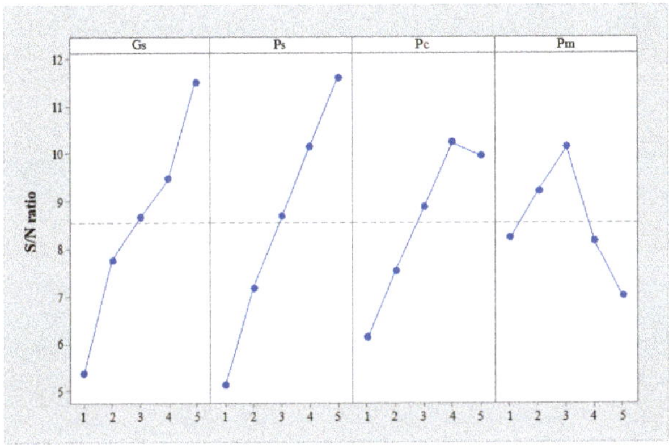

(**a**)

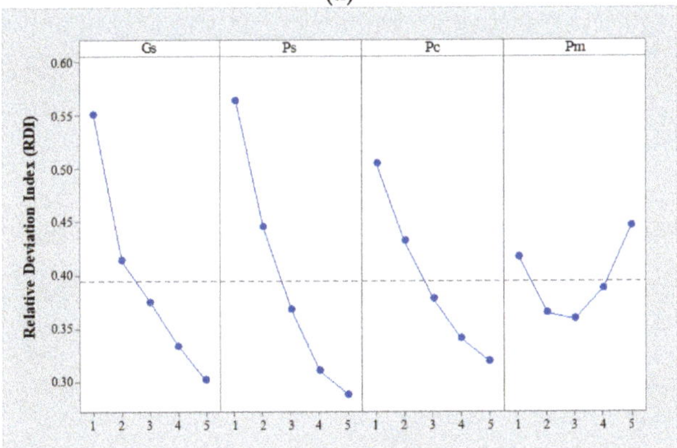

(**b**)

Figure 4. The mean S/N ratio and RDI plot for each GA parameter. (**a**) S/N ratio, (**b**) RDI.

Table 4. ANOVA result for S/N ratio of GA parameters.

Parameters	SS	df	V	F_0	*p*-Value
G_s	4.0983	4	1.0246	3.6219	**0.1202**
P_s	5.1190	4	1.2797	4.5239	**0.0865**
p_c	2.3853	4	0.5963	2.1080	0.2439
p_m(error)	1.1315	4	0.2829	-	-
Total	12.7342	16	-	-	-

The significance level is set to 15%, and G_s and P_s with a p-value less than 15% are judged to be significant. G_s and P_s are set to $G_s(5)$ and $P_s(5)$, respectively. P_c and P_m, which are parameters for which the difference in the S/N ratio between levels is not significant, are selected as the level with the smallest RDI. Figure 4b shows the mean RDI ratio plot for each GA parameter. P_c and P_m are set to $G_s(5)$ and $P_s(3)$, respectively. Parameter calibration for PSO and ICA is also executed in the same way as for GA. For PSO, the best G_s, P_s, c_1, and c_2 are 1000, 100, 0.1, and 0.8, respectively. For the ICA, the best G_s, P_s, N_{imp}, p_r, β, and CPW are 1000, 60, 3, 0.05, 2.5, and 0.25, respectively.

5.2. Setting of the Problem Parameters

Two problem instance groups are generated based on the number of $|J|$, $|T|$, $|C|$, and the expected ratio of tardy jobs (δ) that determine the complexity of problems. The planning horizons (PH) of small and large problem instances are one day (= 8 h = 480 min) and five days (= 8×5 h = 2400 min), respectively. The expected lead time of the last job ($E[Lead_{max}]$) for each instance should be approximately equal to PH. The expected lead time of the last job is calculated as:

$$E[Lead_{max}] = \frac{|J| \times E[v]}{V^M} \times E[p] \times DC + \frac{|J| \times E[v]}{V^M} \times RFC \times Q + E[h] \tag{39}$$

where DC and RFC are the deterioration coefficient and RMA frequency coefficient, respectively.

The first term is the expected total batch production time including the deterioration. The second term is the expected value of total RMA processing time. The third term is the expected delivery time of the last job. Through preliminary experiments, DC and RFC are set to 1.25 and 0.4. The generating conditions of each instance are summarized in Table 5. d_j is generated from the discrete distribution of $U[(1-0.75) \times \mu,\ (1+0.75) \times \mu]$, where $\mu = (1-\delta) \times PH$. p_f is generated from the range given in the table. h_c and v_j are generated from the discrete distributions of $U\ [2 \times p_{min},\ 2 \times p_{max}]$ and $U[5,\ 10]$, respectively. V_M, V_T, DR, and Q are set to 50, 20, 0.3, and $(2 \times E[p])$, respectively. For the small problem instances, $|T|$, $|C|$, and $|F|$ are fixed as 1 and 2. For the large problem instances, $|T|$ and $|C|$ are fixed as 10, 15, and 20, and $|F|$ is generated from $U[5,\ 10]$.

Table 5. Problem parameter setting.

Group	PH	$\|J\|$	p_f	h_c	v_j	V^M	V^T	DR	Q	d_j	
										τ = 0.6	τ = 0.3
Small problem instances	480	5	[65,100]	[130,200]	[5,10]	20	20	0.3	165	[48,336]	NA
		6	[55,90]	[110,180]	[5,10]	20	20	0.3	145		
Large problem instances	2400	200	[30,45]	[60,90]	[5,10]	50	20	0.3	75	[240,1680]	[420,2940]
		250	[25,35]	[50,70]	[5,10]	50	20	0.3	60		
		300	[20,30]	[40,60]	[5,10]	50	20	0.3	50		

For example, if $|J| = 200$,

$$E[Lead_{max}] = \frac{200 \times 7.5}{50} \times 37.5 \times 1.25 + \frac{200 \times 7.5}{50} \times 0.4 \times 75 + 75 = 2381.25 \cong 2400$$

5.3. Experimental Results in the Small Problem Instances

For small problem instances, to validate the performances of PSO, ICA, and GA, these are compared to the optimal solution. The performance of meta-heuristic algorithms is represented by the objective function value (Obj_{sol}) and the CPU time. All meta-heuristic algorithms are tested with 30 replications for each instance. Table 6 shows the performance of the MILP model, PSO, ICA, and GA for instances with $\tau = 0.6$. If the MILP model is not able to find the optimal solution within 2 h, CPU time and $Opt.$ are expressed as 7200.00++ and NA, respectively.

Table 6. Computational results for small problem instances ($\tau = 0.6$).

$\|J\|$	$\|T\|$	$\|C\|$	$\|F\|$	CPLEX		PSO		ICA		GA	
				Opt.	Time	Obj_{sol}	Time	Obj_{sol}	Time	Obj_{sol}	Time
5	1	1	1	584.00	1.05	584.00	0.26	584.00	0.11	584.00	0.27
			2	844.00	2.03	844.00	0.26	844.83	0.11	846.66	0.28
		2	1	484.00	0.77	484.00	0.25	484.00	0.11	484.00	0.27
			2	980.00	0.34	980.00	0.25	980.00	0.11	980.00	0.28
	2	1	1	675.80	4.57	675.80	0.24	675.80	0.11	675.80	0.26
			2	949.40	25.03	949.40	0.24	949.40	0.11	949.40	0.26
		2	1	365.00	27.30	365.00	0.26	365.00	0.11	365.00	0.28
			2	437.00	11.60	437.00	0.25	437.00	0.11	437.00	0.28
6	1	1	1	633.00	4.13	633.00	0.31	639.72	0.14	633.00	0.33
			2	770.00	11.83	770.00	0.31	771.47	0.14	770.00	0.33
		2	1	629.00	1.62	629.00	0.32	629.43	0.14	629.00	0.34
			2	821.00	2.67	821.00	0.30	821.00	0.13	821.00	0.32
	2	1	1	NA	7200.00++	1031.33	0.32	1031.33	0.14	1031.33	0.33
			2	671.03	104.65	671.03	0.31	671.03	0.14	671.03	0.34
		2	1	NA	7200.00++	886.98	0.35	887.83	0.14	886.98	0.34
			2	NA	7200.00++	608.70	0.35	618.07	0.15	599.32	0.36
	Average					710.64	0.29	711.87	0.13	710.22	0.30

The sample means of Obj_{sol}s for PSO, ICA, and GA are 710.64, 711.87, and 710.22, respectively. The sample means of CPU times for PSO, ICA, and GA are 0.29, 0.13, and 0.30, respectively. PSO, ICA, and GA all found near-optimal solutions.

5.4. Experimental Results in the Large Problem Instances

For large problem instances, the performance of PSO, ICA, and GA is measured by comparison with each other. The performance of algorithms is represented by the RDI for large problem instances and the CPU time.

All the algorithms are tested with 30 replications for each instance. Table 7 shows the performance of GA, ICA, and PSO for instances. At $\delta = 0.6$, the sample means of RDIs for PSO, ICA, and GA for each instance are 0.91, 0.42, and 0.08, respectively. The sample means of CPU times for PSO, ICA, and GA for each instance are 95.10, 56.42, and 95.54, respectively. At $\delta = 0.3$, the sample means of RDIs for PSO, ICA, and GA are 0.93, 0.58, and 0.08, respectively. The sample means of CPU times for PSO, ICA, and GA are 95.62, 56.50, and 95.70, respectively. The ranking of meta-heuristic algorithms from the best to the worst performance is GA, ICA, and PSO. PSO and GA with the same population size show similar CPU time, and ICA with a relatively smaller population size than PSO and GA shows less CPU time. We execute an additional experiment for ICA with extended CPU time. However, no significant improvement in the RDI of ICA is shown. Therefore, GA shows the best RDI among PSO, ICA, and GA under similar CPU time in large problem instances.

For analysis reasons regarding the performance differences for PSO, ICA, and GA, a convergence test is performed. An instance with $|J| = 300, |T| = 20, |C| = 20$ is used for the test and repeated 10 times. Figure 5 shows the convergence graph for PSO, ICA, and GA. The objective function values of the initial solution are similar for all three algorithms, but PSO and ICA converge faster to a value with a higher objective function than GA. Therefore, GA shows better performance in terms of objective function than PSO and ICA.

To verify the significant difference in RDI between algorithms, the Tukey HSD test was performed. Figure 6 shows the mean plots and Tukey HSD intervals ($\alpha = 0.05$) for all instances in Table 7. Figure 6 shows that the confidence intervals between all the algorithms do not overlap. In the other words, the difference in RDI between PSO, ICA, and GA is statistically significant.

Figure 7 shows the mean plots and Tukey HSD intervals ($\alpha = 0.05$) for $|J|$, $|T|$, $|C|$, and δ groups. For $|J|$, $|T|$, $|C|$, and δ groups, the performance is ranked in the order of

GA, ICA, and PSO, and the difference in RDI between PSO, ICA, and GA is statistically significant. In particular, GA provides the best RDI and robustness among all algorithms.

Table 7. Computational results for large problem instances.

$\|J\|$	$\|T\|$	$\|C\|$	$\tau = 0.6$						$\tau = 0.3$					
			PSO		ICA		GA		PSO		ICA		GA	
			RDI	Time	RDI	Time	RDI	Time	RDI	Time	RDI	Time	RDI	Time
200	10	10	0.90	61.01	0.43	35.95	0.09	61.41	0.90	61.44	0.60	36.22	0.08	61.52
		15	0.91	61.32	0.43	36.30	0.06	61.30	0.91	61.78	0.56	36.19	0.07	61.63
		20	0.89	61.58	0.44	36.37	0.05	61.88	0.93	61.87	0.60	36.15	0.06	61.87
	15	10	0.89	61.66	0.46	36.51	0.12	62.42	0.92	62.08	0.58	36.75	0.07	62.57
		15	0.90	62.15	0.41	37.11	0.07	62.59	0.93	62.44	0.61	36.95	0.06	62.66
		20	0.87	62.23	0.46	37.01	0.08	62.36	0.91	62.60	0.65	36.97	0.08	63.47
	20	10	0.90	62.35	0.45	36.84	0.07	62.79	0.92	62.48	0.58	36.94	0.08	62.55
		15	0.93	62.59	0.44	37.01	0.11	63.25	0.93	63.07	0.65	37.06	0.08	62.69
		20	0.91	62.91	0.50	37.28	0.10	63.29	0.93	63.05	0.64	37.24	0.11	62.78
250	10	10	0.92	92.04	0.48	54.58	0.09	92.22	0.92	92.22	0.56	54.31	0.09	91.90
		15	0.91	92.29	0.46	54.80	0.05	92.04	0.91	92.84	0.59	54.65	0.08	92.39
		20	0.93	92.56	0.44	54.71	0.06	92.71	0.94	93.15	0.54	54.82	0.09	92.60
	15	10	0.92	92.71	0.38	55.46	0.10	93.34	0.95	93.12	0.59	55.23	0.09	93.55
		15	0.89	92.86	0.43	54.99	0.08	93.40	0.92	93.69	0.61	55.57	0.11	93.55
		20	0.93	94.02	0.42	55.56	0.10	93.64	0.94	94.12	0.62	55.64	0.10	93.10
	20	10	0.88	92.97	0.39	55.52	0.11	93.88	0.91	93.72	0.55	55.59	0.04	93.50
		15	0.93	93.95	0.43	56.12	0.11	94.18	0.93	94.67	0.57	56.05	0.09	94.78
		20	0.88	94.15	0.42	56.08	0.05	94.42	0.93	94.51	0.58	56.00	0.07	94.28
300	10	10	0.92	128.67	0.42	76.30	0.07	129.12	0.94	129.57	0.54	76.79	0.07	129.18
		15	0.90	129.55	0.44	76.98	0.06	129.14	0.89	130.33	0.47	76.68	0.05	130.27
		20	0.94	130.14	0.39	77.45	0.10	130.51	0.93	130.49	0.53	76.80	0.07	130.07
	15	10	0.91	129.47	0.39	77.02	0.08	129.98	0.94	130.03	0.58	77.43	0.07	130.23
		15	0.95	130.62	0.37	78.09	0.10	132.10	0.93	131.17	0.59	77.20	0.08	130.83
		20	0.92	130.78	0.38	78.12	0.06	131.46	0.93	131.81	0.55	77.85	0.08	131.70
	20	10	0.92	130.49	0.39	77.74	0.08	131.14	0.95	131.34	0.54	78.01	0.07	132.78
		15	0.91	131.13	0.35	78.68	0.09	132.44	0.93	131.89	0.57	77.93	0.11	133.99
		20	0.92	131.58	0.36	78.08	0.10	132.49	0.94	132.35	0.58	78.51	0.08	133.41
Average			0.91	95.10	0.42	56.42	0.08	95.54	0.93	95.62	0.58	56.50	0.08	95.70

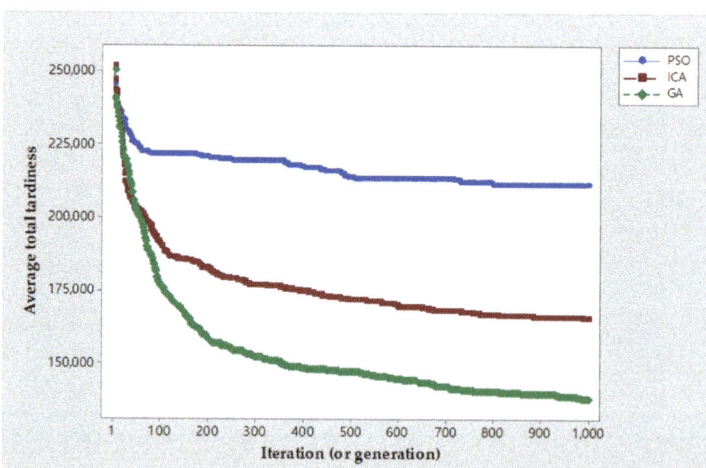

Figure 5. The convergence graph for PSO, ICA, and GA.

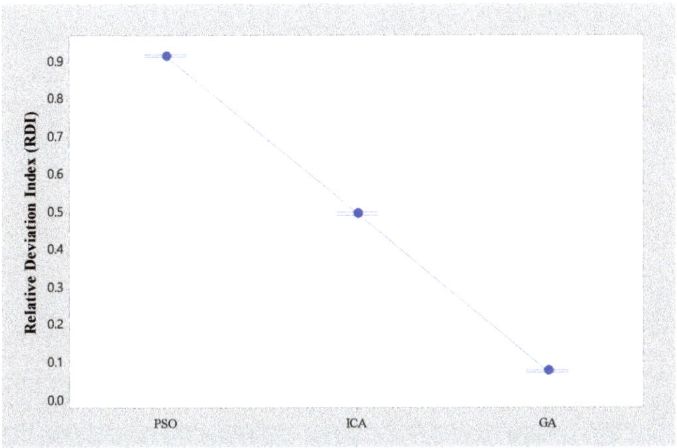

Figure 6. The mean plots and Tukey HSD intervals ($\alpha = 0.05$) for PSO, ICA, and GA.

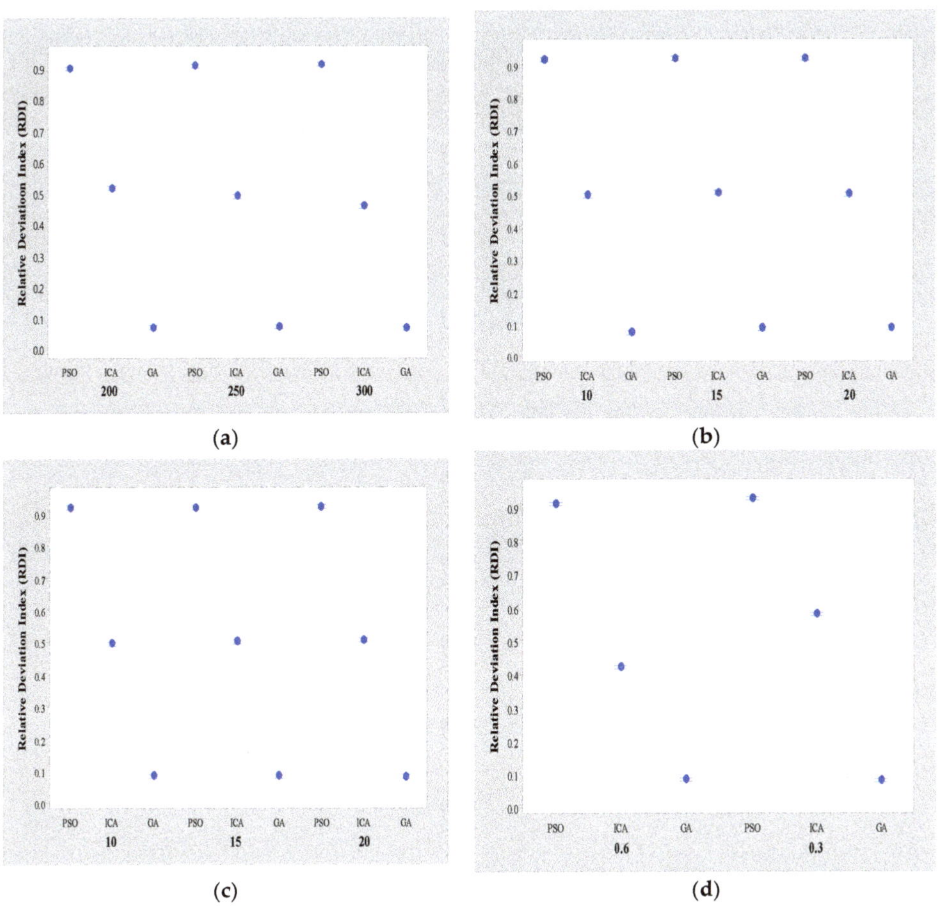

Figure 7. The mean plots and Tukey HSD intervals ($\alpha = 0.05$) for $|J|$, $|T|$, $|C|$, and δ groups. (**a**) $|J|$ group. (**b**) $|T|$ group. (**c**) $|C|$ group. (**d**) δ group.

6. Sensitivity Analysis

To reduce total tardiness for the ISP, scheduling problems of the manufacturing and delivery process are important. It is difficult to find whether manufacturing or delivery scheduling impacts the proposed total tardiness. To find this impact, the total manufacturing completion time (t^M) and the total delivery time (t^T) are presented as a performance measure. t^M and t^T are defined as follows:

$$t^M = \sum_{j=1}^{|J|} c^j \tag{40}$$

$$t^T = \sum_{j=1}^{|J|} \left(Lead^j - c^j \right) \tag{41}$$

where c^j and $Lead^j$ are the manufacturing completion time and lead time of job j, respectively.

Several problem parameters affect the scheduling of ISPs. Each problem parameter related to the manufacturing and delivery process affects t^M and t^T, respectively. For example, obviously, t^M decreases when V^M increases, or p_f decreases and t^T decreases when $|T|$ or V^T increases or h_c decreases. However, the effects of parameters related to the manufacturing process on t^T and the effects of parameters related to the delivery process on t^M are not obvious. To find these effects, an additional experiment is conducted by only using GA with the best performance shown.

Graphs (a), (b), and (c) in Figure 8 show the change in t^M according to the change in the parameters $|T|$, V^T, and h_c, respectively. In graphs (a), (b), and (c), t^M decreases when $|T|$ or V^T increases or h_c decreases. Meanwhile, the graphs (d) and (e) in Figure 8 show the change in t^T according to the change in the parameters p_f and V^M. According to the graphs (d) and (e), t^t decreases when V^M increases or p_f decreases. In summary, the parameters related to the delivery affect t^M, and the parameters related to manufacturing affect t^T. This is because if the time on one side decreases, the flexibility of decision making on the other side increases. One of the reasons is that the difference in manufacturing completion time between jobs decreases and various decisions in the delivery process become possible as p_f decreases.

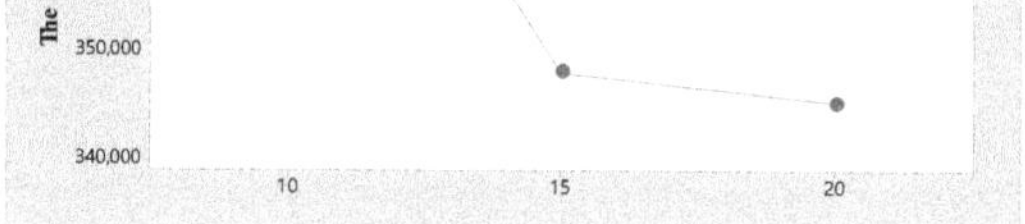

(a)

Figure 8. *Cont.*

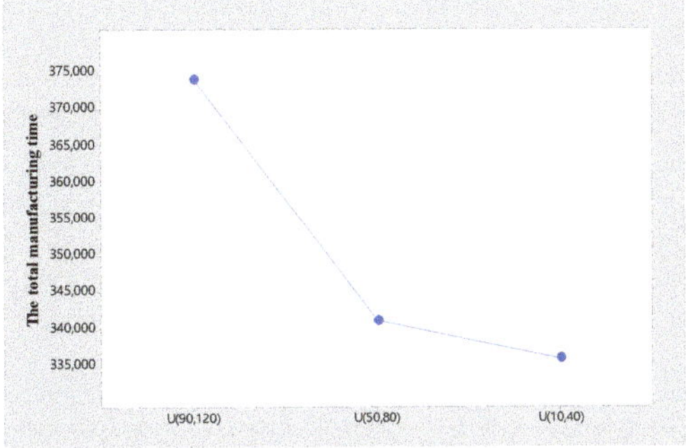

(b)

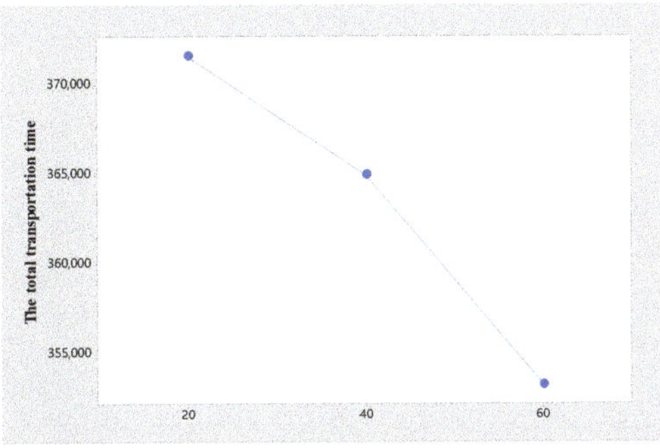

(c)

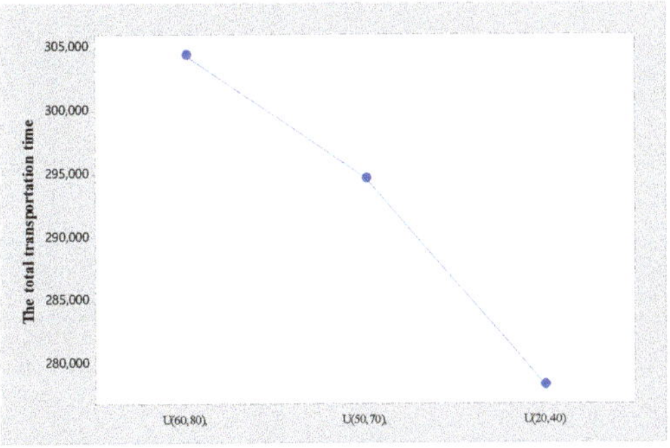

(d)

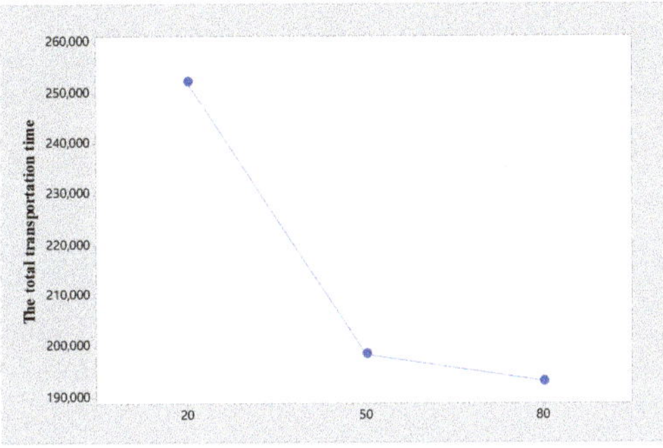

(e)

Figure 8. Total manufacturing and delivery time under different $|T|$, h_c, V^T, p_f, and V^M. (**a**) Total manufacturing time under different $|T|$. (**b**) Total manufacturing time under different h_c. (**c**) Total manufacturing time under different V^T. (**d**) Total delivery time under different p_f. (**e**) Total delivery time under different V^M.

7. Conclusions

In this research, the ISP with a batching machine, time-dependent batch deterioration, and RMAs is considered. A MILP model was formulated to solve small problem instances. Meta-heuristic algorithms were proposed to solve the large problem instances. The solution structure of meta-heuristics consists of two one-dimensional arrays for manufacturing and delivery. For small problem instances, we found the optimal solution using the developed MILP model. Additionally, we verified the performance of meta-heuristic algorithms by showing the near-optimal solution and comparing it with the MILP model in small problem instances. Three meta-heuristic algorithms, GA, ICA, and PSO, are proposed and relatively compared by using the relative deviation index (RDI) in large problem instances. The ranking of meta-heuristic algorithms from the best to the worst performance was GA, ICA, and PSO. Sensitivity analysis was conducted for GA with the best performance shown. We found that as the time for either manufacturing or delivery was reduced, the time for the other also decreased in this analysis.

However, this study has several limitations. For example, the real enterprise data considered for solving real industry problems are not used, and the objective function is simply set to total tardiness. The objective function can be modeled as a cost including setup, inventory, and tardiness costs. As for further work to extend this study, the problem could apply VRP to our delivery method. In addition, the problem can be extended to optimization problems resolving the conflict of stakeholders between manufacturing and third-party logistics (3PL). Finally, matheuristic and simheuristic approaches can be considered.

Author Contributions: Conceptualization, B.-S.K.; methodology, Y.-J.K.; software, Y.-J.K.; validation, B.-S.K.; formal analysis, Y.-J.K.; investigation, B.-S.K. and Y.-J.K.; resources, B.-S.K.; data curation, Y.-J.K.; writing—original draft preparation, Y.-J.K.; writing—review and editing, B.-S.K.; visualization, Y.-J.K.; supervision, B.-S.K. All authors have read and agreed to the published version of the manuscript.

Funding: This work was supported by the Incheon National University Research Grant in 2021.

Data Availability Statement: Not applicable.

Conflicts of Interest: The authors declare no conflict of interest.

References

1. Joo, C.M.; Kim, B.S. Rule-based meta-heuristics for integrated scheduling of unrelated parallel machines, batches, and heterogeneous delivery trucks. *Appl. Soft Comput. J.* **2017**, *53*, 457–476. [CrossRef]
2. Jia, Z.; Huo, S.; Li, K.; Chen, H. Integrated scheduling on parallel batch processing machines with non-identical capacities. *Eng. Optim.* **2019**, *52*, 715–730. [CrossRef]
3. Dobson, G.; Nambimadom, R.S. The Batch Loading and Scheduling Problem. *Oper. Res.* **2001**, *49*, 52–65. [CrossRef]
4. Joo, C.M.; Kim, B.S. Variable Neighborhood Search Algorithms for an Integrated Manufacturing and Batch Delivery Scheduling Minimizing Total Tardiness. *Appl. Sci.* **2019**, *9*, 4702. [CrossRef]
5. Lee, C.-Y.; Leon, V.J. Machine scheduling with a rate-modifying activity. *Eur. J. Oper. Res.* **2001**, *128*, 119–128. [CrossRef]
6. Liu, C.H. Using genetic algorithms for the coordinated scheduling problem of a batching machine and two-stage transportation. *Appl. Math. Comput.* **2011**, *217*, 10095–10104. [CrossRef]
7. Selvarajah, E.; Steiner, G. Batch scheduling in a two-level supply chain-a focus on the supplier. *Eur. J. Oper. Res.* **2006**, *173*, 226–240. [CrossRef]
8. Gao, S.; Qi, L.; Lei, L. Integrated batch production and distribution scheduling with limited vehicle capacity. *Int. J. Prod. Econ.* **2015**, *160*, 13–25. [CrossRef]
9. Cheng, B.-Y.; Leung, J.Y.-T.; Li, K.; Yang, S.-L. Single batch machine scheduling with deliveries. *Nav. Res. Logist.* **2015**, *62*, 470–482. [CrossRef]
10. Cheng, B.Y.; Leung, J.Y.T.; Li, K. Integrated scheduling on a batch machine to minimize production, inventory and distribution costs. *Eur. J. Oper. Res.* **2017**, *258*, 104–112. [CrossRef]
11. Jia, Z.; Zhuo, X.; Leung, J.Y.T.; Li, K. Integrated production and transportation on parallel batch machines to minimize total weighted delivery time. *Comput. Oper. Res.* **2019**, *102*, 39–51. [CrossRef]
12. Li, S.; Yuan, J.; Fan, B. Unbounded parallel-batch scheduling with family jobs and delivery coordination. *Inf. Process. Lett.* **2011**, *111*, 575–582. [CrossRef]
13. Li, K.; Jia, Z.H.; Leung, J.Y.T. Integrated production and delivery on parallel batching machines. *Eur. J. Oper. Res.* **2015**, *247*, 755–763. [CrossRef]
14. Zhang, J.; Wang, X.; Huang, K. On-line scheduling of order picking and delivery with multiple zones and limited vehicle capacity. *Omega* **2018**, *79*, 104–115. [CrossRef]
15. Nogueira, T.H.; Bettoni, A.B.; de Oliveira Mendes, G.T.; dos Santos, A.G.; Ravetti, M.G. Problem on the integration between production and delivery with parallel batching machines of generic job sizes and processing times. *Comput. Ind. Eng.* **2020**, *146*, 106573. [CrossRef]
16. Feng, X.; Xu, Z. Integrated Production and Transportation Scheduling on Parallel Batch-Processing Machines. *IEEE Access* **2019**, *7*, 148393–148400. [CrossRef]
17. Jia, Z.; Cui, Y.; Li, K. An ant colony-based algorithm for integrated scheduling on batch machines with non-identical capacities. *Appl. Intell.* **2022**, *52*, 1752–1769. [CrossRef]
18. He, P.; Li, K.; Kumar, P.N.R. An enhanced branch-and-price algorithm for the integrated production and transportation scheduling problem. *Int. J. Prod. Res.* **2021**, *60*, 1874–1889. [CrossRef]
19. Li, K.; He, P.; Ram Kumar, P.N. A column generation based approach for an integrated production and transportation scheduling problem with dual delivery modes. *Int. J. Prod. Res.* 2022. [CrossRef]
20. Karaoğlan, İ.; Kesen, S.E. The coordinated production and transportation scheduling problem with a time-sensitive product: A branch-and-cut algorithm. *Int. J. Prod. Res.* **2017**, *55*, 536–557. [CrossRef]
21. Low, C.; Chang, C.M.; Li, R.K.; Huang, C.L. Coordination of production scheduling and delivery problems with heterogeneous fleet. *Int. J. Prod. Econ.* **2014**, *153*, 139–148. [CrossRef]
22. Low, C.; Chang, C.M.; Gao, B.Y. Integration of production scheduling and delivery in two echelon supply chain. *Int. J. Syst. Sci. Oper. Logist.* **2017**, *4*, 122–134. [CrossRef]
23. Li, K.; Zhou, C.; Leung, J.Y.T.; Ma, Y. Integrated production and delivery with single machine and multiple vehicles. *Expert Syst. Appl.* **2016**, *57*, 12–20. [CrossRef]
24. Kong, L.; Li, H.; Luo, H.; Lieyun, D.; Luo, X.; Skitmore, M. Optimal single-machine batch scheduling for the manufacture, transportation and JIT assembly of precast construction with changeover costs within due dates. *Autom. Constr.* **2017**, *81*, 34–43. [CrossRef]

25. Liu, X.; Lu, S.; Pei, J.; Pardalos, P.M. A hybrid VNS-HS algorithm for a supply chain scheduling problem with deteriorating jobs. *Int. J. Prod. Res.* **2018**, *56*, 5758–5775. [CrossRef]
26. Yin, Y.; Cheng, T.C.E.; Wu, C.C.; Cheng, S.R. Single-machine batch delivery scheduling and common due-date assignment with a rate-modifying activity. *Int. J. Prod. Res.* **2014**, *52*, 5583–5596. [CrossRef]
27. Joo, C.M.; Kim, B.S. Genetic algorithms for single machine scheduling with time-dependent deterioration and rate-modifying activities. *Expert Syst. Appl.* **2013**, *40*, 3036–3043. [CrossRef]
28. Vélez-Gallego, M.C.; Damodaran, P.; Rodríguez, M. Makespan minimization on a single batch processing machine with unequal job ready times. *Int. J. Ind. Eng. Theory Appl. Pract.* **2011**, *18*, 536–546.
29. Wang, H.M.; Chou, F. Der Solving the parallel batch-processing machines with different release times, job sizes, and capacity limits by metaheuristics. *Expert Syst. Appl.* **2010**, *37*, 1510–1521. [CrossRef]
30. Marandi, F.; Fatemi Ghomi, S.M.T. Integrated multi-factory production and distribution scheduling applying vehicle routing approach. *Int. J. Prod. Res.* **2019**, *57*, 722–748. [CrossRef]
31. Bean, J.C. Genetic Algorithms and Random Keys for Sequencing and Optimization. *Inf. J. Comput.* **1994**, *6*, 154–160. [CrossRef]
32. Mosadegh, H.; Zandieh, M.; Ghomi, S.M.T.F. Simultaneous solving of balancing and sequencing problems with station-dependent assembly times for mixed-model assembly lines. *Appl. Soft Comput. J.* **2012**, *12*, 1359–1370. [CrossRef]

mathematics

MDPI

Article

cpd: An R Package for Complex Pearson Distributions

María José Olmo-Jiménez *,†, Silverio Vílchez-López † and José Rodríguez-Avi †

Department of Statistics and Operations Research, University of Jaén, 23071 Jaén, Spain
* Correspondence: mjolmo@ujaen.es; Tel.: +34-953-211909
† These authors contributed equally to this work.

Abstract: The complex Pearson (CP) distributions are a family of probability models for count data generated by the Gaussian hypergeometric function with complex arguments. The complex triparametric Pearson (CTP) distribution and its biparametric versions, the complex biparametric Pearson (CBP) and the extended biparametric Waring (EBW) distributions, belong to this family. They all have explicit expressions of the probability mass function (pmf), probability generating function and moments, so they are easy to handle from a computational point of view. Moreover, the CTP and EBW distributions can model over- and underdispersed count data, whereas the CBP can only handle overdispersed data, but unlike other well-known overdispersed distributions, the overdispersion is not due to an excess of zeros but other low values of the variable. Finally, the EBW distribution allows the variance to be split into three uniquely identifiable components: randomness, liability and proneness. These properties make the CP distributions of interest in the modeling of a great variety of data. For this reason, and for trying to spread their use, we have implemented an R package called cpd that contains the pmf, distribution function, quantile function and random generation for these distributions. In addition, the package contains fitting functions according to the maximum likelihood. This package is available from the Comprehensive R Archive Network (CRAN). In this work, we describe all the functions included in the cpd package, and we illustrate their usage with several examples. Moreover, the release of a plugin in order to use the package from the interface R Commander tries to contribute to the spreading of these models among non-advanced users.

Keywords: count data models; overdispersion; underdispersion; R package

MSC: 60-04

Citation: Olmo-Jiménez, M.J.; Vílchez-López, S.; Rodríguez-Avi, J. cpd: An R Package for Complex Pearson Distributions. *Mathematics* **2022**, *10*, 4101. https://doi.org/10.3390/math10214101

Academic Editor: Manuel Franco

Received: 21 September 2022
Accepted: 31 October 2022
Published: 3 November 2022

1. Introduction

The use of discrete distributions to model count data is widely illustrated in the literature. The first model, which describes the pure random case for an infinite range, is the Poisson distribution. This is a uniparametric model which assumes that data have equidispersion; that is to say, the variance is equal to the mean. Nevertheless, in real studies, data often exhibit overdispersion (i.e., the variance is greater than the mean) and less often exhibit underdispersion (i.e., the variance is less than the mean). For these situations, a great variety of models has been developed, with many of them obtained from the Poisson distribution. Among them, we find well-known models, such as the negative binomial (NB) [1], univariate generalized Waring (UGW) [2,3], generalized Poisson (GP) [4], zero-inflated [5] or hurdle models [6], as well as many other new models (see, for instance, [7–10]). One of these new models is the complex triparametric Pearson distribution with parameters a, b and γ, which are denoted by $CTP(a, b, \gamma)$. This distribution belongs to the family of discrete distributions generated by the Gaussian hypergeometric function when the two first parameters are complex conjugated numbers (i.e., ${}_2F_1(a + ib, a - ib; \gamma; 1)$, where i is the imaginary unit), and it has been widely studied in [11,12]. Two particular cases with two parameters have also been developed: the complex biparametric Pearson (CBP) distribution [13,14] and the extended bivariate Waring (EBW) distribution [15,16]. It is interesting to take into account the fact that the CTP and EBW models can handle both over- and underdispersion, whereas the CBP model can only handle overdispersion. Nevertheless, the overdispersion of these three models, unlike other well-known overdispersed models, is not due to an excess of zeros but other low values of the variable. Specifically, the CBP model is useful when there is overdispersion, and the probability of zero is similar to that of a Poisson distribution [14]. In addition to the fact that the CTP and EBW models can be underdispersed, another advantage is that they do not have computational problems since there are explicit expresions of the pmf, pgf and moments, as they do occur in other models such as the CMP, HP or GP models [12]. All these properties make the CP distributions of interest in the modeling of a great variety of data.

For these reasons, it is essential to facilitate their use, which is accomplished through their implementation in different statistical software. In this sense, R, the free software environment for statistical computing and graphics [17], not only allows for using the most common distributions to compute the probabilities and quantiles or generate random numbers but also model the data. Thus, for example, the stats package contains functions for handling many standard univariate probability distributions, and the extraDistr package adds more univariate and multivariate distributions to the list. In the MASS package, the maximum likelihood modeling of several models is available via the fitdistr functions, and the fitdistrplus package implements several methods for fitting univariate parametric distributions. In addition, there are also specific built-in functions related to these aspects in other R packages. To sum up, this allows us to propose different models for a given data set, estimating the corresponding parameters and, in addition, comparing them to choose the more adequate one. However, all of these packages include the CP distributions, making their use inaccessible to most researchers.

In trying to solve this problem, we implemented an R package called cpd for the CP distributions. Thus, in this work, we present and describe a package which allows for obtaining the pmf, distribution function, quantile function and random number generation for the three distributions. Moreover, this package offers the possibility of estimating the parameters by the maximum likelihood method and also provides several goodness-of-fit tests and graphics as well as additional fit criteria. The package is available from the Comprehensive R Archive Network (CRAN) at https://CRAN.R-project.org/package=cpd (accessed on 1 September 2022). In addition, we implemented a plugin for the R Commander GUI that allows non-advanced R users to work with these models without using R code. The plugin is also a package called RcmdrPlugin.cpd available from CRAN at https://CRAN.R-project.org/package=RcmdrPlugin.cpd (accessed on 1 September 2022).

The remainder of this paper is organized as follows. Section 2 reviews the definitions and properties of the CTP, CBP and EBW distributions. In Section 3, the functions of the cpd package are detailed, including several examples to illustrate their use. In the final section, this paper concludes with a summary of the main characteristics of the package implemented.

2. Complex Pearson Distributions

2.1. Brief Description and Properties

The complex triparametric Pearson (CTP) distribution was first developed in [11]. It is a triparametric discrete distribution of an infinite range generated by the Gaussian hypergeometric function ${}_2F_1$ with complex parameters, so it belongs to the Gaussian hypergeometric distributions (GHD) family [1]. Specifically, X follows a $CTP(a, b, \gamma)$ distribution when its pmf has the following expression:

$$f(x) = f_0 \, \frac{(a+ib)_x (a-ib)_x}{(\gamma)_x} \frac{1}{x!}, \quad x = 0, 1, \ldots \tag{1}$$

where i is the imaginary unit, $a, b \in \mathbb{R}$ and $\gamma > max\{0, 2a\}$. $(\alpha)_r$ is the Pochhammer symbol (also known as a rising factorial) defined as $\Gamma(\alpha + r)/\Gamma(\alpha)$, with $\Gamma(\cdot)$ as the gamma function and f_0 as the normalizing constant given by

$$_2F_1(a+ib, a-ib; \gamma; 1)^{-1} = \frac{\Gamma(\gamma - a - ib)\Gamma(\gamma - a + ib)}{\Gamma(\gamma)\Gamma(\gamma - 2a)}.$$

An alternative expression of Equation (1) in terms of the gamma function is

$$f(x) = C \cdot \frac{\Gamma(a+ib+x)\Gamma(a-ib+x)}{\Gamma(\gamma+x)} \frac{1}{x!}, \quad x = 0, 1, \ldots \tag{2}$$

where C is the normalizing constant

$$C = \frac{\Gamma(\gamma - a - ib)\Gamma(\gamma - a + ib)}{\Gamma(\gamma - 2a)\Gamma(a+ib)\Gamma(a-ib)}.$$

The probability generating function (pgf) is given by

$$G(t) = \frac{\Gamma(\gamma - a - ib)\Gamma(\gamma - a + ib)}{\Gamma(\gamma)\Gamma(\gamma - 2a)} \, {}_2F_1(a+ib, a-ib; \gamma; t), \quad t \in \mathbb{R}. \tag{3}$$

Aside from its pmf, the model also has explicit expressions of the mean μ and the variance σ^2:

$$\mu = \frac{a^2 + b^2}{\gamma - 2a - 1}, \quad \sigma^2 = \mu \frac{\mu + \gamma - 1}{\gamma - 2a - 2},$$

which exist if $\gamma > 2a + 1$ and $\gamma > 2a + 2$, respectively.

This is unimodal with the mode in $\left[\frac{(a-1)^2+b^2}{\gamma-2a+1}\right]$ when $\frac{(a-1)^2+b^2}{\gamma-2a+1} \notin \mathbb{Z}$, where $[\cdot]$ is the integer part; otherwise, the distribution has two consecutive modes in the values:

$$\frac{(a-1)^2 + b^2}{\gamma - 2a + 1} - 1 \text{ and } \frac{(a-1)^2 + b^2}{\gamma - 2a + 1}.$$

Then, if $a^2 + b^2 < \gamma$, then there is only one mode in zero. As a consequence, the pmf is *J*-shaped or bell-shaped. Moreover, the CTP is skewed to the right since its third central moment is always positive. For further details about the model, see [11,12].

One of the main properties of the CTP distribution is that it can be underdispersed, equidispersed or overdispersed. In particular, if $a \geq 0$, then the CTP is always overdispersed. Thus, the model has a great versatility in the modeling of count data, especially when the overdispersion of the data is due to a higher frequency of non-zero values. In addition, the fact is that having explicit expressions of the pmf, mean and variance prevents the computational problems of other well-known models that cope with over- and underdispersion, such as the Conway–Maxwell–Poisson [18] (CMP) and the hyper-Poisson [19] (HP) models.

The CTP model is a generalization of the complex biparametric Pearson ($CBP(b, \gamma)$) distribution, since the latter appears when $a = 0$ [13]. This model has the advantage of having one less parameter, but it is always overdispersed. It can be compared to an NB distribution, except for the fact that the probability of zero is less than that provided by an NB model and similar to that of a Poisson distribution.

The EBW distribution has two parameterizations: $EBW(\alpha, \rho)$ with $\alpha, \rho = \gamma - 2\alpha > 0$ and $EBW(\alpha, \gamma)$ with $\alpha < 0$ and $\gamma > 0$. This model can also be seen as a particular case of the $CTP(a, b, \gamma)$ distribution when $b = 0$ and as a particular case of the $UGW(a, k, \rho)$ distribution when $a = k = \alpha > 0$. In fact, given a UGW distribution, there exists an EBW distribution that is very close to the former with the benefit of having one less parameter. In addition, the *EBW* distribution allows the variance to be split into three uniquely identifiable components: randomness, liability and proneness (see more details in [16]), solving the existing indeterminacy in the UGW model. Specifically, if $X \sim EBW(\alpha, \rho)$, then these components are

$$\sigma^2 = \frac{\alpha^2}{\rho - 1} + \frac{\alpha^2(\alpha+1)}{(\rho-1)(\rho-2)} + \frac{\alpha^3(\alpha+\rho-1)}{(\rho-1)^2(\rho-2)}. \tag{4}$$

2.2. Maximum Likelihood Estimation

Under the i.i.d. sample assumption, the parameters a, b and γ are by default estimated by maximizing the log-likelihood function, defined as

$$\ln L_{x_1,\ldots,x_n}(a,b,\gamma) = \sum_{j=1}^{n} \ln f(x_j|a,b,\gamma) = 2Re[\ln\Gamma(\gamma - a + ib)] - 2Re[\Gamma(a+ib)] \tag{5}$$
$$+ 2Re[\Gamma(a+ib+x)] - \ln\Gamma(\gamma - 2a) - \ln\Gamma(\gamma + x) - \ln x!$$

with $x_1, \ldots, x_n$, where n is the number of observations of the variable $X \sim CTP(a,b,\gamma)$. It should be noticed that $\Gamma(z) = \overline{\Gamma(\overline{z})}$, where z and $\overline{z}$ are conjugate complex numbers.

The log-likelihood for the CBP distribution is obtained when $a = 0$, and that for the EBW distribution is obtained when $b = 0$.

3. Using the cpd Package

3.1. Overview

The cpd package provides the functions to compute the probability mass function, distribution function, quantile function and random generation for the complex triparametric Pearson (CTP), complex biparametric Pearson (CBP) and extended biparametric Waring (EBW) distributions. In addition, the package contains maximum-likelihood fitting functions for these models.

The source code is available from the Comprehensive R Archive Network (CRAN) repository (https://CRAN.R-project.org/package=cpd, accessed on 1 September 2022), with all the information about its functions and parameters in the package's help file. It can be installed and loaded by typing the following commands in R:

```
R> install.packages(''cpd'')
R> library(cpd)
```

The package is open-source, so it is also available from GitHub (https://github.com/ujaen-statistics/cpd, accessed on 1 September 2022), where updates and comments can be submitted.

Specifically, the cpd package allows for computing the probability mass, distribution and quantile funtions of a CBP distribution through the following respective code:

```
dcbp(x, b, gamma)
pcbp(q, b, gamma, lower.tail = TRUE)
qcbp(p, b, gamma, lower.tail = TRUE)
```

where x is a vector of the non-negative integer values, q is a vector of the quantiles, p is a vector of the probabilities and b and gamma are the parameters of the distribution. In the pcbp and qcbp functions, the argument lower.tail has to be specified to consider $P(X \leq x)$ (if it is TRUE) or $P(X \geq x)$ (if it is FALSE). It is also possible to generate n random numbers from a CBP distribution with parameters b and gamma using the rcbp function, whose sentence is rcbp(n, b, gamma).

For the CTP distribution, the probability mass, distribution and quantile functions, as well as the random generation, are analogous:

```
dctp(x, a, b, gamma)
```

```
pctp(q, a, b, gamma, lower.tail = TRUE)
qctp(p, a, b, gamma, lower.tail = TRUE)
rctp(n, b, gamma)
```

In the case of the EBW distribution, there are two possible parameterizations depending on the sign of its first parameter α. Thus, if $\alpha < 0$, then the usual parametrization is $EBW(\alpha, \gamma)$ with $\gamma > 0$, whereas if $\alpha > 0$, then the usual parametrization is $EBW(\alpha, \rho)$ with $\rho = \gamma - 2\alpha > 0$. (This constraint guarantees the existence of the probability distribution.) Then, the corresponding probability mass, distribution and quantile functions, together with the random generation, take into account these two parameterizations:

```
debw(x, alpha, gamma, rho)
pebw(q, alpha, gamma, rho, lower.tail = TRUE)
qebw(p, alpha, gamma, rho, lower.tail = TRUE)
rebw(n, alpha, gamma, rho)
```

Moreover, the cpd package provides functions for fitting the CTP, CBP and EBW distributions to count data by the maximum likelihood method. These functions are fitctp, fitcbp and fitebw, respectively. Thus, the usage for the fitcbp function is

```
fitcbp(x, bstart = NULL, gammastart = NULL, method = ''L-BFGS-B'',
+ control = list(), ...),
```

for the CTP distribution

```
fitctp(x, astart = NULL, bstart = NULL, gammastart = NULL,
+ method = ''L-BFGS-B'', control = list(), ...)
```

as well as for the EBW distribution

```
fitebw(x, alphastart = NULL, gammastart = NULL, rhostart = NULL,
+ method = ''L-BFGS-B'', control = list(), ...)
```

These fitting functions estimate the distribution parameters by maximizing the log-likelihood function given in Equation (5) using the optim function of the stats package with "L-BFGS-B" [20] as the default fitting method, which allows box constraints. Other alternative methods of the optim function include "Nelder–Mead", "BFGS", "CG" and "SANN" (see the function help information for more details). Nonlinear minimization using a Newton-type algorithm is also possible (see the nlm function of the stats package). If the fitting method is not "L-BFGS-B", then the parameters have to be reparameterized as $\alpha = e^{\alpha_0}$ and $\rho = e^{\rho_0}$ or $\alpha = -e^{\alpha_0}$ and $\gamma = e^{\gamma_0}$ with α_0, ρ_0 and $\gamma_0 \in \mathbb{R}$, in order to satisfy the corresponding constraints in each model. In this case, the standard errors provided by the fitting function are for the estimates of α_0 and ρ_0 or γ_0. The starting values for the optimization process are the estimates obtained by the method of the moments, unless the user introduces other values. These estimates are obtained by solving the system of equations:

$$\begin{pmatrix} m'_1 & -m'_1 & -1 \\ m'_2 & -m'_2 - m'^2_1 & -m'_1 + 1 \\ m'_3 & -m'_3 - 2m'_2 - m'_1 & -m'_2 - 2m'_1 - 1 \end{pmatrix} \begin{pmatrix} \theta_1 \\ \theta_2 \\ \theta_3 \end{pmatrix} = \begin{pmatrix} 0 \\ m_2 \\ 2m_3 + m_2 \end{pmatrix} \quad (6)$$

where m'_r is the rth sample raw moment, $\theta_1 = \widehat{\gamma} - 1$, $\theta_2 = 2\widehat{a}$ and $\theta_3 = \widehat{a}^2 + \widehat{b}^2$.

In the case of the EBW distribution, the method of moments could provide two sets of starting values. In such a case, the optimization process would be carried out twice (one with each set of starting values), and the solution with less AIC will be shown. These fitting functions return S3 objects of the classes fitCBP, fitCTP and fitEBW for which the print, summary and plot methods are provided.

The summary of an object of the classes fitCBP, fitCTP and fitEBW provides the ML parameter estimates, their standard errors and the statistic and p values of the Wald test to check if the parameters are significant. This summary also shows the loglikelihood, AIC and BIC values, as well as the results for the χ^2 goodness of fit test and the Kolmogorov–Smirnov test for discrete variables [21,22]. Finally, the correlation matrix of the parameter estimates also appears.

In addition, when the ML estimate of α is positive, the function varcomp—applied to an object of the class fitEBW—allows us to obtain the decomposition of the variance in the fitted EBW model (see the components in Equation (4)). This fact is useful to know the origin of the data variability.

The plot of an object of class fitCBP, fitCTP or fitEBW provides, by default, the observed and theoretical frequencies against the values of the variable, the CDF plot of both the empirical distribution and the fitted distribution or a PP plot representing the empirical distribution function evaluated at each data point (y axis) against the fitted distribution function (x axis).

3.2. Examples

3.2.1. Probability Mass, Distribution, Quantile and Random Generation Functions

We illustrate the use of the probability mass, distribution and quantile functions and how to generate random numbers from the CP distributions.

First, we consider $X \sim CBP(3, 2.5)$, and we compute $P(X = 0), P(X = 1)$ and $P(X = 2)$:

```
R> library(cpd)
R> cpd::dcbp(c(0, 1, 2), 3, 2.5)
[1] 0.02985882 0.10749176 0.15355965
```

The following sentences allow for computing $P(X \leq 3), P(X \leq 5)$ and $P(X > 2)$:

```
R> cpd::pcbp(c(3, 5), 3, 2.5)
[1] 0.4387825 0.6528353
R> cpd::pcbp(c(2), 3, 2.5, lower.tail = FALSE)
[1] 0.7090898
```

To obtain the quartiles of X and the 95th percentile, the sentence and the R output are

```
R> cpd::qcbp(c(0.25, 0.5, 0.75, 0.95), 3, 2.5)
[1]  2  4  7 17
```

Finally, to generate 300 numbers from X, we type

```
R> set.seed(123)
R> x <- cpd::rcbp(300, 3, 2.5)
R> table(x)
x
 0  1  2  3  4  5  6  7  8  9 10 11 12 13 14 15 16 17 18 19 20 21 22 26
 5 28 54 48 35 27 24 14 14  6  4  9  6  4  1  2  1  2  4  1  2  2  1  1
28 29 30 45
 1  1  2  1
```

Figure 1a shows the bar plot of the random generated data which is obtained with the code

```
R> barplot(table(x), xlab = ‘‘values’’, ylab = ‘‘frequencies’’)
```

We can observe that the mode of these data is at a value of two, and they exhibit overdispersion ($\overline{x} = 5.676667 < s^2 = 32.21952$).

Next we consider $Y \sim CTP(-1.5, 2, 2)$, which is underdispersed, and we compute $P(X = 0), P(X = 1), P(X = 2), P(X = 3)$:

```
R> cpd::dctp(c(0:3),-1.5, 2, 2)
[1] 0.1331089 0.4159654 0.2946422 0.1043524
```

Now, the cumulative probabilities $P(X \leq 1), P(X \leq 3)$ and $P(X > 1)$ are

```
R> cpd::pctp(c(1,3), −1.5, 2, 2)
[1] 0.5490744 0.9480690
R> cpd::pctp(1, −1.5, 2, 2, lower.tail = FALSE)
[1] 0.4509256
```

The quartiles and 95th percentile are obtained as follows:

```
R> cpd::qctp(c(0.25, 0.5, 0.75, 0.95), −1.5, 2, 2)
[1] 1 1 2 4
```

To generate 500 random values from Y, we use the code

```
R> set.seed(123)
R> y <- cpd::rctp(500, −1.5, 2, 2)
```

The frequency table is

```
R> table(y)
y
  0   1   2   3   4   5   6   7  10
 57 227 142  41  21   8   2   1   1
```

Additionally, Figure 1b shows a bar plot of the random generated data. The mode of these data is at a value of one, and they exhibit underdispersion ($\overline{y} = 1.574 > s^2 = 1.367259$).

Finally, to conclude this section, let us consider $X_1 \sim EBW(\alpha = 2, \rho = 5)$ and $X_2 \sim EBW(\alpha = -1.2, \gamma = 0.75)$, with the first being overdispersed and the second being underdispersed, and let us compute $P(X_i = 0), P(X_i = 1), P(X_i = 2), P(X_i = 3)$ and $P(X_i = 4), i = 1, 2$:

```
R> cpd::debw(c(0:4), 2, rho = 5)
[1] 0.53571429 0.23809524 0.10714286 0.05194805 0.02705628
R> cpd::debw(c(0:4), −1.2, gamma = 0.75)
[1] 0.3396452713 0.6521189210 0.0074527877 0.0005781556 0.0001248816
```

The cumulative probabilities $P(X_i \leq 2)$, $P(X_i \leq 4)$ and $P(X_i > 2), i = 1, 2$ are obtained as follows:

```
R> cpd::pebw(c(2,4), 2, rho = 5)
[1] 0.8809524 0.9599567
R> cpd::pebw(2, 2, rho = 5, lower.tail = FALSE)
[1] 0.1190476
R> cpd::pebw(c(2,4), −1.2, gamma = 0.75)
[1] 0.9981288 0.9998974
R> cpd::pebw(3, −1.2, gamma = 0.75, lower.tail = FALSE)
[1] 0.0002048643
```

The corresponding quartiles and 99th percentile are given by

```
R> cpd::qebw(c(0.25, 0.5, 0.75, 0.99), 2, rho = 5)
[1] 0 0 1 8
R> cpd::qebw(c(0.25, 0.5, 0.75, 0.99), −1.2, gamma = 0.75)
[1] 0 1 1 1
```

To generate 1000 random values of X_1 and X_2, we type the code

```
R> set.seed(123)
R> x1 <- cpd::rebw(1000, 2, rho = 5)
R> x2 <- cpd::rebw(1000, −1.2, gamma = 0.75)
```

The respective frequency tables are

```
R> table(x1)
x1
  0   1   2   3   4   5   6   7   8   9  10  11  18
542 236 105  51  25  16   8   6   5   3   1   1   1
R> table(x2)
x2
  0   1   2   3
332 657  10   1
```

In addition, the bar plots of these two datasets may be seen in Figure 1c,d. The modes of these data were zero and one, respectively. The first dataset was overdispersed ($\overline{x}_1 = 0.975 < s_1^2 = 2.657032$), and the second one was underdispersed ($\overline{x}_2 = 0.68 > s_2^2 = 0.2438438$).

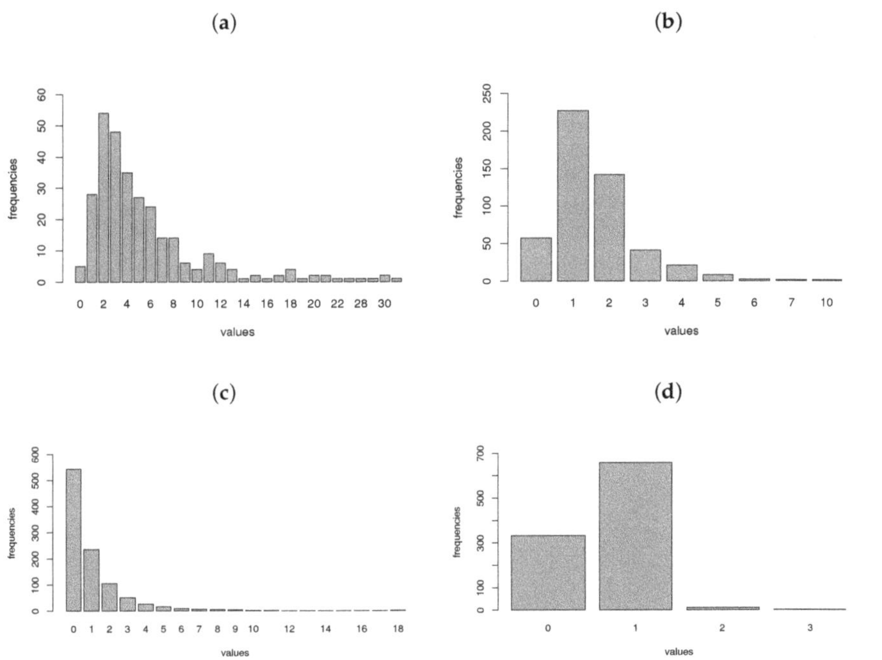

Figure 1. Bar plots of the random generated data from (**a**) $CBP(3, 2.5)$, (**b**) $CTP(-1.5, 2, 2)$, (**c**) $EBW(2, 5)$ and (**d**) $EBW(-1.2, 0.75)$.

3.2.2. Fitting Functions

To illustrate the use of the fitting functions, we provide three examples: two overdispersed and one underdispersed.

The first data set refers to the number of fire outbreaks by municipality in the region of Andalusia (Spain). Data were obtained from the Nature Databank of the Ministry of the Environment (Spain) and counting the number of fire outbreaks from 2001 to 2014. A fire outbreak was defined as a wildfire whose total area was less than one hectare. Moreover, municipalities whose forest land was zero were removed from the data. A description of these data appears in Table 1, which contains the mean, variance, quartiles, minimum and maximum of the data. It is clear that these data exhibit overdispersion since $s^2 > \overline{x}$.

Table 1. Descriptive summary of data.

	$\bar{x}$	s^2	Q_1	Q_2	Q_3	Min	Max
Fire outbreaks	3.528	28.964	1	2	4.75	0	56
Schools	1.431	10.3807	0	1	1	0	48
Syllables-1	1.889	0.910	1	2	2	0	4

We fit a CBP model to the data, considering as the initial values the estimates by the method of moments:

```
R> fireoutbreaks.cbp <- cpd::fitcbp(fireoutbreaks)
```

The output shows the ML estimates and their standard errors in parentheses:

```
     b              gamma
  1.486206        1.494944
(0.08849089)   (0.12183708)
```

Using the summary method, the output is more complete. The argument grouping = TRUE is the setting for grouping in classes with an expected frequency greater than or equal to five in the χ^2 goodness of fit test, since the default value is FALSE:

```
R> summary(fireoutbreaks.cbp, grouping = TRUE)

Parameters:
         Estimate    Std. Error    z-value       Pr(>|z|)
b        1.486206    0.08849089    16.79502   2.654186 × 10^-63
gamma    1.494944    0.12183708    12.27003~1.312062 × 10^-34

Loglikelihood: −1637.21    AIC: 3278.43    BIC: 3287.5

Goodness-of-fit tests:
Chi-2: 60.05902 (p-value: 5.11525627536094 × 10^-7)
Kolmogorov-Smirnov: 0.04732388 (p-value: 0.033)

Correlation Matrix:
              b     gamma
b     1.0000000 0.9296264
gamma 0.9296264 1.0000000
```

The AIC for the CBP fit is lower than the AIC related to an usual NB fit for these data:

```
R> library(MASS)
R> fireoutbreaks.nb<-MASS::fitdistr(fireoutbreaks, ''negative binomial'')
R> fireoutbreaks.nb
     size           mu
 0.80061445   3.52752644
(0.05537946) (0.16624492)
R> AIC(fireoutbreaks.nb)
[1] 3292.707
```

Next we model the data using the CTP distribution. However, the method of moments does not provide any estimates, so we introduce starting values for these parameters:

```
R> fireoutbreaks.ctp <- cpd::fitctp(fireoutbreaks)
Error in fitctp(fireoutbreaks) :
    The method of moments does not provide any estimates. Enter
    initial values for the~parameters.

R> fireoutbreaks.ctp<-cpd::fitctp(fireoutbreaks,astart=0,bstart= 1,
+ gammastart = 2.1)
R> summary(fireoutbreaks.ctp, grouping = TRUE)

Parameters:
       Estimate   Std. Error    z-value     Pr(>|z|)
a      1.880214    0.8151401   2.306615   0.021076319
b      1.579993    0.5102531   3.096489   0.001958269
gamma  6.441561    2.1065484   3.057874~0.002229130

Loglikelihood: −1623.73   AIC: 3253.45   BIC: 3260.53

Goodness-of-fit tests:
Chi-2: 23.21167 (p-value: 0.0569114065884523)
Kolmogorov-Smirnov: 0.01796262 (p-value: 0.746)

Correlation Matrix:
              a          b        gamma
a       1.0000000 −0.9502083 0.9952855
b      −0.9502083  1.0000000 −0.9193242
gamma   0.9952855 −0.9193242  1.0000000
```

Once again, the fitted model was improved compared with the previous ones.

Finally, we carry out an EBW fit:

```
R> fireoutbreaks.ebw <- cpd::fitctp(fireoutbreaks)
R> summary(fireoutbreaks.ebw, grouping = TRUE)

Parameters:
        Estimate    Std. Error    z-value       Pr(>|z|)
alpha   2.749528    0.1621672    16.954903    1.770541 × 10^-64
rho     3.139183    0.3171861     9.896977    4.290494 × 10^-23
gamma   8.638240    0.6293458    13.725746~7.119262 × 10^-43

Loglikelihood: −1624.12   AIC: 3252.25   BIC: 3261.32

Goodness-of-fit tests:
Chi-2: 24.05791 (p-value: 0.0641165517298056)
Kolmogorov-Smirnov: 0.01778027 (p-value: 0.773)

Correlation Matrix:
          alpha       rho
alpha 1.0000000 0.9247998
rho   0.9247998 1.0000000
```

This is the most accurate of the four fits using the AIC. Moreover, the goodness-of-fit tests show that the EBW distribution is a reasonable model for the fire outbreak data. Figure 2 includes the observed and expected frequencies, CDFs and PP plots for this fit obtained with the following sentences:

```
R> plot(fireoutbreaks.ebw)
R> plot(fireoutbreaks.ebw, plty = ‘‘CDF’’)
R> plot(fireoutbreaks.ebw, plty = ‘‘PP’’)
```

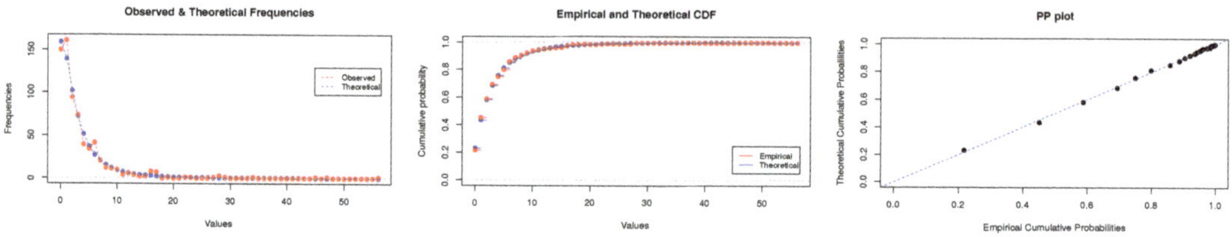

Figure 2. Observed and expected frequencies, CDFs and PP plots for the EBW fit of fire outbreak data.

As $\widehat{\alpha} = 2.7495 > 0$, the absolute value and the ratio of the variance components of the EBW fit can be obtained by typing the command

```
R> cpd::varcomp(fireoutbreaks.ebw)

                 Value     Ratio
Randomness   3.534015  0.1019654
Liability   11.631922  0.3356107
Proneness   19.493035  0.5624239
```

The results indicate that 10.1965% of the variability in fire outbreaks was due to ramdomness, 33.5611% was due to liability (which refers to the general and external conditions of the municipality), and 56.2424% was due to proneness (related to the specific and internal characteristics of the municipality).

The second data set refers to the number of compulsory secondary schools by municipality in Andalusia (Spain) in the academic year 2020–2021. Data have been obtained through the Multiterritorial Information System of Andalusia (SIMA). The main descriptive statistics for these data appear in Table 1. These data reveal severe overdispersion caused by a value of one, as can be seen from the table of frequencies:

```
R> table(cs_schools)

  0   1   2   3   4   5   6   7   8   9  10  11  12  13  14  18  20
201 437  67  31  17   6   6   2   3   2   2   1   2   1   1   1~1

 26  28  47  48
  1   1   1   1
```

First, we fit an NB model:

```
R> schools.nb <- MASS::fitdistr(cs_schools, ‘‘negative binomial’’)
R> schools.nb

    size          mu
1.19336385   1.43057377
(0.10212557) (0.06330097)

R> AIC(schools.nb)

2584.456
```

As expected, the results for the CTP fit (with initial values $a = 0, b = 1$ and $\gamma = 0.5$) improved remarkably with respect to the previous ones:

```
R> schools.ctp <- cpd::fitctp(schools, astart = 0, bstart = 1,
+ gammastart = 0.5)
R> summary(schools.ctp)

Parameters:
        Estimate    Std. Error    z-value      Pr(>|z|)
a      −0.5141782   0.03519091   −14.611108   2.386050 × 10^-48
b       0.4165253   0.09994774     4.167431   3.080518 × 10^-5
gamma   0.2020829   0.05620251     3.595620~3.236198 × 10^-4

Loglikelihood: −1058.91   AIC: 2123.81   BIC: 2131.14

Goodness-of-fit tests:
Chi-2: 40.81333 (p-value: 0.649840570375861
Kolmogorov-Smirnov: 0.007734697 (p-value: 0.932)

Correlation Matrix:
               a          b         gamma
a       1.0000000 −0.8066776 −0.7935876
b      −0.8066776  1.0000000  0.9637505
gamma  −0.7935876  0.9637505  1.0000000
```

The goodness-of-fit tests also support the adequacy of the model fitted. Figure 3 includes the observed and expected frequencies, CDFs and PP plots for the CTP fit obtained with the code sentences

```
R> plot(schools.ctp)
R> plot(schools.ctp, plty = ''CDF'')
R> plot(schools.ctp, plty = ''PP'')
```

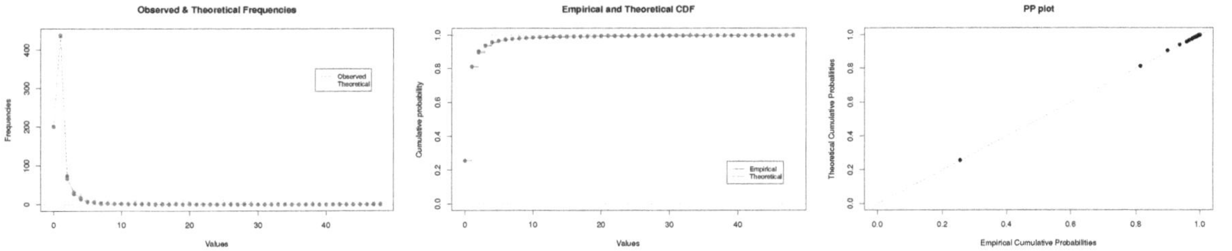

Figure 3. Observed and expected frequencies, CDFs and PP plots for the EBW fit of compulsory secondary school data.

In this example, it makes no sense to fit a CBP or a EBW model, since the parameters *a* and *b* in the CTP are statistically significant.

The third data set contained lengths of words (numbers of syllables) in a Slovak poem [23]. A description of these data appears in Table 1, where we considered the response variable $X - 1$, as though the data were generated by adding one to the distribution. These types of data related to word length often exhibit underdispersion [24].

As the CBP model is always overdispersed, it made no sense to fit it to these data, so we fitted a CTP model, considering as the initial values the estimates by the method of moments. In this example, the argument grouping is missing, so it is set to FALSE by default, since there are not enough degrees of freedom to group the classes with an expected frequency greater than or equal to five. Thus, the code is

```
R> syllables.ctp <- cpd::fitctp(syllables)
R> summary(syllables.ctp)
```

and the output

```
Parameters:
         Estimate       Std. Error       z-value        Pr(>|z|)
a      −5.7319721810    0.8127761     −7.0523387550 1.759352 × 10^-12
b       0.0008637403    6.0264402      0.0001433251 9.998856 × 10^-1
gamma  6.9298746437     3.3472112      2.0703428298~3.842025 × 10^-2

Loglikelihood: −159.33    AIC: 324.67    BIC: 328.19

Goodness-of-fit tests:
Chi-2: 0.6018625 (p-value: 0.437868271837147)
Kolmogorov-Smirnov: 0.01181905 (p-value: 0.994)

Correlation Matrix:
                  a            b          gamma
a        1.000000000  0.002883042 −0.970588862
b        0.002883042  1.000000000 −0.001203176
gamma  −0.970588862 −0.001203176  1.000000000
```

Now, we model these data using EBW distribution by typing

```
R> syllables.ebw <- cpd::fitebw(syllables)
R> summary(syllables.ebw)

Parameters:
         Estimate     Std. Error     z-value         Pr(>|z|)
alpha   −5.731882      0.812740     −7.052541     1.756792 × 10^-12
gamma    6.929558      3.347026     2.070363~3.841840 × 10^-2

Loglikelihood: −159.33    AIC: 322.67    BIC: 328.19

Goodness-of-fit tests:
Chi-2: 0.6019236 (p-value: 0.74010605961245)
Kolmogorov-Smirnov: 0.01181588 (p-value: 0.994)

Correlation Matrix:
            alpha        gamma
alpha  1.0000000 −0.9705887
gamma −0.9705887  1.0000000
```

Let us notice that the EBW fit was the best one according to the AIC and goodness-of-fit tests.

In addition, we used the sentences

```
R> plot(syllables.ebw)
R> plot(syllables.ebw, plty = ‘‘CDF’’)
R> plot(syllables.ebw, plty = ‘‘PP’’)
```

to obtain the plots in Figure 4.

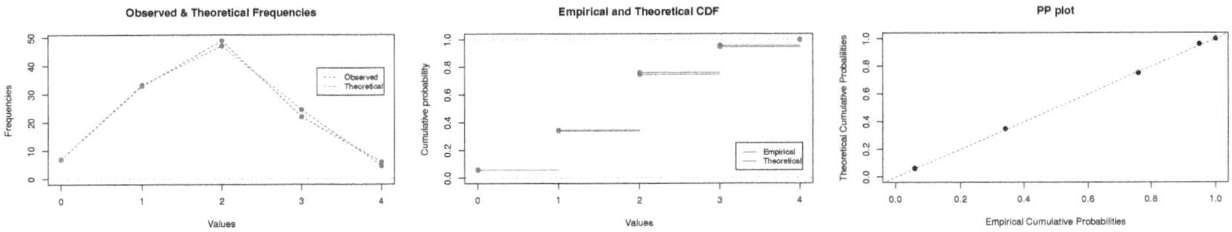

Figure 4. Observed and expected frequencies, CDFs and PP plots for the CTP fit of Slovak poem data.

4. Conclusions

The cpd package has been designed for computing probabilities and quantiles as well as for generating random numbers from the CBP, CTP and EBW distributions. These functions have also been included in a plugin for the GUI R Commander with the aim of facilitating their use by non-advanced R users. In addition, the package contains fitting functions to obtain the ML estimates of their parameters. In this way, we give more visibility to these models, which allows for modeling overdispersed data in which the overdispersion is not due to a value of zero but to low values of the variable $(1, 2, \ldots)$ and also underdispersed data. Thus, the probability of zero in the CBP is lower than in the corresponding Poisson with the same mean, so the CBP can be seen as an adequate model for overdispersed data without too many zeros. Regarding the CTP and EBW models, they do not have the computational problems of other well-known models for both over- and underdispered data such as the GP, the CMP or the HP.

Author Contributions: Data curation, J.R.-A.; Formal analysis, M.J.O.-J. and S.V.-L.; Investigation, M.J.O.-J., S.V.-L. and J.R.-A.; Methodology, M.J.O.-J., S.V.-L. and J.R.-A.; Software, M.J.O.-J. and S.V.-L.; Supervision, M.J.O.-J., S.V.-L. and J.R.-A.; Writing—original draft, M.J.O.-J. and J.R.-A.; Writing—review & editing, M.J.O.-J., S.V.-L. and J.R.-A. All authors have read and agreed to the published version of the manuscript.

Funding: This research received no funding.

Institutional Review Board Statement: Not applicable.

Informed Consent Statement: Not applicable.

Data Availability Statement: Data in Section 3.2.2 have been obtained from https://www.miteco.gob.es/es/biodiversidad/servicios/banco-datos-naturaleza/ (accessed on 1 September 2022) (Example 1) and https://www.juntadeandalucia.es/institutodeestadisticaycartografia/sima/index2-en.htm (accessed on 1 September 2022) (Example 2).

Conflicts of Interest: The authors declare no conflict of interest.

References

1. Johnson, N.L.; Kemp, A.W.; Kotz, S. *Univariate Discrete Distributions*, 3rd ed.; Wiley: New York, NY, USA, 2005.
2. Irwin, J.O. The generalized Waring distribution. Part I. *J. R. Stat. Soc. Ser. A* **1975**, *138*, 18–31. [CrossRef]
3. Rodríguez-Avi, J.; Conde-Sánchez, A.; Sáez-Castillo, A.J.; Olmo-Jiménez, M.J. A new generalization of the Waring distribution. *Comput. Stat. Data Anal.* **2007**, *51*, 6138–6150. [CrossRef]
4. Joe, H.; Zhu, R. Generalized Poisson Distribution: The Property of Mixture of Poisson and Comparison with Negative Binomial Distribution. *Biom. J.* **2005**, *45*, 219–229. [CrossRef] [PubMed]
5. Vieira, A.M.C.; Hinde, J.P.; Demetrio, C.G.B. Zero-inflated proportion data models applied to a biological control assay. *J. Appl. Stat.* **2000**, *27*, 373–389. [CrossRef]
6. Conceição, K.S.; Louzada, F.; Andrade, M.G.; Helou, E.S. Zero-modified power series distribution and its Hurdle distribution version. *J. Stat. Comput. Simul.* **2017**, *87*, 1842–1862. [CrossRef]
7. Sáez-Castillo, A.J.; Conde-Sánchez, A. Detecting over- and under-dispersion in zero inflated data with the hyper-Poisson regression model. *Stat. Pap.* **2017**, *58*, 19–33. [CrossRef]
8. da Silva, W.B.; Ribeiro, A.M.T.; Conceição, K.S.; Andrade, M.G.; Neto, F.L. On Zero-Modified Poisson-Sujatha Distribution to Model Overdispersed Count Data. *Austrian J. Stat.* **2018**, *47*, 1–19. [CrossRef]

9. Bonat, W.H.; Jørgensen, B.; Kokonendji, C.C.; Hinde, J.; Demétrio, C.G.B. Extended Poisson–Tweedie: Properties and regression models for count data. *Stat. Model.* **2018**, *18*, 24–49. [CrossRef]
10. Satheesh Kumar, C.; Harisankar, S. On some aspects of a general class of Yule distribution and its applications. *Commun. Stat.-Theory Methods* **2019**, *49*, 1–11. [CrossRef]
11. Rodríguez-Avi, J.; Conde-Sánchez, A.; Sáez-Castillo, A.J.; Olmo-Jiménez, M.J. A triparametric discrete distribution with complex parameters. *Stat. Pap.* **2004**, *45*, 81–95. [CrossRef]
12. Olmo-Jiménez, M.J.; Rodríguez-Avi, J.; Cueva-López, V. A review of the CTP distribution: A comparison with other over- and underdispersed count data models. *J. Stat. Comput. Simul.* **2018**, *88*, 2684–2706. [CrossRef]
13. Rodríguez-Avi, J.; Conde-Sánchez, A.; Sáez-Castillo, A.J. A new class of discrete distributions with complex parameters. *Stat. Pap.* **2003**, *44*, 67–88. [CrossRef]
14. Rodríguez-Avi, J.; Olmo-Jiménez, M.J. A regression model for overdispersed data without too many zeros. *Stat. Pap.* **2017**, *58*, 749–773. [CrossRef]
15. Cueva-López, V.; Olmo-Jiménez, M.J.; Rodríguez-Avi, J. EM algorithm for an extension of the Waring distribution. *Comput. Math. Methods* **2019**, *1*, e1046. [CrossRef]
16. Cueva-López, V.; Olmo-Jiménez, M.J.; Rodríguez-Avi, J. An over- and underdispersed biparametric extension of the Waring distribution. *Mathematics* **2021**, *9*, 170. [CrossRef]
17. R Core Team. *R: A Language and Environment for Statistical Computing*; R Foundation for Statistical Computing: Vienna, Austria, 2019.
18. Sellers, K.F.; Borle, S.; Shmueli, G. The COM-Poisson model for count data: A survey of methods and applications. *Appl. Stoch. Model. Bus. Ind.* **2012**, *28*, 104–116. [CrossRef]
19. Sáez-Castillo, A.J.; Conde-Sánchez, A. A hyper-Poisson regression model for overdispersed and underdispersed count data. *Comput. Stat. Data Anal.* **2013**, *61*, 148–157. [CrossRef]
20. Byrd, R.H.; Lu, P.; Nocedal, J.; Zhu, C. A limited memory algorithm for bound constrained optimization. *SIAM J. Sci. Comput.* **1995**, *16*, 1190–1208. [CrossRef]
21. Conover, W.J. A Kolmogorov goodness-of-fit test for discontinuous distributions. *J. Am. Stat. Assoc.* **1972**, *67*, 591–596. [CrossRef]
22. Gleser, L.J. Exact power of goodness-of-fit tests of Kolmogorov type for discontinuous distributions. *J. Am. Stat. Assoc.* **1985**, *80*, 954–958. [CrossRef]
23. Wimmer, G.; Kohler, R.; Grotjahn, R.; Altmann, G. Toward a theory of word length distributions. *J. Quant. Ling.* **1994**, *1*, 98–106. [CrossRef]
24. DjurasErnst, G.; Stadlober, S. *Text and Language: Structures Function Interrelations Quantitative Perspectives*; Chapter Modeling Word Length Frequencies by the Singh-Poisson Distribution; Praesens Verlag: Wien, Austria, 2010; pp. 37–46.

mathematics

MDPI

Article

A Complementary Dual of Single-Valued Neutrosophic Entropy with Application to MAGDM

Sonam Sharma and Surender Singh *

School of Mathematics, Shri Mata Vaishno Devi University, Katra 182320, India
* Correspondence: surender1976@gmail.com or surender.singh@smvdu.ac.in

Abstract: A single-valued neutrosophic set (SVNS) is a subcategory of neutrosophic set that is used to represent uncertainty and fuzziness in three tiers, namely truthfulness, indeterminacy, and falsity. The measure of entropy of a SVNS plays an important role to determine the ambiguity in a variety of situations. The knowledge measure is a dual form of entropy and is helpful in certain counterintuitive situations. In this paper, we introduce a knowledge measure for the SVNS and contrast the same with existing measures. The comparative study reveals that the proposed knowledge measure is more effective in modeling the structured linguistic variables. We provide the relations of the proposed knowledge measure with single valued neutrosophic similarity and distance measures. We also investigate the application of the proposed measure in multi-attribute group decision making (MAGDM). The proposed MAGDM model is helpful when the decision makers in the group have varied background and the hiring organization is unable to assign the level of importance or weight to a decision-maker.

Keywords: single valued neutrosophic set; knowledge measure; MAGDM; correlation coefficient; neutrosophic similarity

MSC: 03E72; 90B50; 03E75

Citation: Sharma, S.; Singh, S. A Complementary Dual of Single-Valued Neutrosophic Entropy with Application to MAGDM. *Mathematics* **2022**, *10*, 3726. https://doi.org/10.3390/math10203726

Academic Editor: Gia Sirbiladze

Received: 8 September 2022
Accepted: 2 October 2022
Published: 11 October 2022

1. Introduction

The first successful attempt to model the imprecision or ambiguity of human reasoning in a mathematical framework was put forward by Zadeh [1]. The quantitative representation of the linguistic knowledge of the human observations or cognition was investigated under the notion of fuzzy theory. Atanassov [2] further extended the theory and introduced 'Intuitionistic Fuzzy Set' in which an element of the universal set belongs or does not belong to a set to a certain extent. In both these concepts, we get a set single-valued and 2-tuple quantitative representation of the vagueness associated with an element of the universal set. However, there are some complex situations in real life where these two representations are not sufficient to handle such situations. Smarandache [3] brings out the notion of a neutrosophic set (NS) from a philosophical point of view to investigate the indeterminate or inconsistent information that commonly occurs in real-life circumstances. The neutrosophic set is based on three tiers—belongingness, indeterminacy, and non-belongingness. In short, the neutrosophic set is beyond the fuzzy set and intuitionistic fuzzy set. Most of the attributes in a complex situation where decision-makers use linguistic variables can be easily expressed with the help of a neutrosophic value. Wang et al. [4] introduced a subclass of neutrosophic set and termed it as a single-valued neutrosophic set. Various operations (union, intersection, complement) have been studied. Since the advent of the neutrosophic set, major work especially in the field of multi-attribute decision-making and pattern-recognition played a vital role. Ma et al. [5] studied Archimedean t-norm or t-conorom using intuitionistic fuzzy aggregation operator for multi-criteria decision making. Symmetric intuitionistic fuzzy weighted mean operators concerning extensive weighted

Archimedean t-norm and t-conorom put forward for dealing with membership and non-membership information. Wu et al. [6] discussed the entropy, similarity measure, and cross-entropy of information measure in SVNSs. The given information measure was used to handle MADM problems to check the effectiveness. Similarity measures and entropy of single-valued neutrosophic sets was proposed by Qin and Wang [7]. Smarandache [8] proposed a neutrosophic hedge algebra. Different operations in neutrosophic hedge algebra were also studied to aggregate the neutrosophic linguistic value. Hanafy [9] proposed a correlation coefficient formula for neutrosophic data. Singh et al. [10] studied the correlation coefficient in an intuitionistic fuzzy set. In the application part, a generalized correlation coefficient was used to solve MADM. Normalized correlation efficiency was considered as the weight of decision-makers. Biswas et al. [11] proposed a new technique concerning the TOPSIS method in the single-valued neutrosophic environment. Jin et al. [12] proposed an information measure for SVN entropy and similarity measure based on sine and cosine function. Comparative analysis was studied to check the effectiveness and rationality of the given method. Knowledge measure in a fuzzy set was given by Singh et al. [13] to check the effectiveness of the proposed method. In a hesitant fuzzy set, knowledge measure was computed by Lalotra and Singh [14]. The knowledge concerning the attributes in context with some available alternatives can be represented in various frameworks. A neutrosophic theory equips us with a kind of representation of the knowledge base that removes certain pitfalls of the fuzzy and intuitionistic fuzzy representation. The neutrosophic entropy provides a valuation of the uncertainty or ambiguity entailed in a given neutrosophic set. In the MADM problems, the neutrosophic entropy is utilized to compute the objective weights of the attribute. Sometimes, the entropy measures suffer from certain counterintuitive situations and render inappropriate results. The counterintuitive situations arise when an entropy/knowledge measure cannot distinguish two different neutrosophic sets. Moreover, the problems of multi-attribute group decision making if the higher organization is not aware of the expertise and knowledge base of decision-makers. Then how to assign the weightage to the decision experts. These two reasons motivated us to derive an alternative entropy-like measure for objective weight computation and to propose some mechanism for the weight assigned to the decision-makers. The main contribution of this paper is as follows:

- We propose an entropy-like measure in the neutrosophic settings and termed it a single-valued neutrosophic knowledge measure.
- We also discuss certain properties of the neutrosophic knowledge measure and establish its connection with the single-valued neutrosophic similarity and dissimilarity measure.
- An algorithm of MAGDM is proposed and implemented with the help of a numerical example.
- Comparative analysis to check the effectiveness of the proposed knowledge measure has also been presented.

The remaining part of the paper is organized is as follows: Section 2 presents the fundamental concepts regarding this paper. In Section 3, we propose a knowledge measure in the single-valued neutrosophic environment. Section 4 presents the relation of single valued neutrosophic similarity and distance measure with the single-valued neutrosophic knowledge measure. In Section 5, we consider an algorithm for the MAGDM problem. Section 6 deals with the comparative studies. Finally, concluding remarks are given in Section 7.

2. Preliminaries

In this section, we present some definitions and concepts concerning single-valued neutrosophic sets.

Definition 1. **([1]).** *Let* $Y = \{y_1, y_2, y_3 \ldots\ldots y_n\}$ *be the universal set then a fuzzy set in* Y *is defined as*

$$B = \{(y, T_B(y)) : y \in Y\}.$$

where $T_B(y) : Y \to [0, 1]$ *determines the truth membership of* y *in* B*. The value of* $T_B(y)$ *gives the degree of belongingness of* y *in* B*.*

Definition 2. **([2]).** *Intuitionistic fuzzy set* B *on a universal set* Y *is defined*

$$B = \{(y, T_B(y), F_B(y)) : y \in Y\}.$$

where $T_B(y) : Y \to [0, 1]$ *and* $F_B(y) : Y \to [0, 1]$ *determines the degree of membership and degree of non—membership respectively with the condition* $0 \le T_B(y) + F_B(y) \le 1$ *and the value of* $T_B(y)$ *and* $F_B(y)$ *gives the value of membership and non-membership of* y *in* B*, respectively.*

Definition 3. **([3,4]).** *A single-valued neutrosophic set on a universal set* B *is defined as*

$$B = \{(y, T_B(y), I_B(y), F_B(y)) : y \in Y\},$$

where $T_B(y) : Y \to [0, 1]$, $I_B(y) : Y \to [0, 1]$ *and* $F_B(y) : Y \to [0, 1]$ *assigns the degree of membership, degree of indeterminacy and degree of non—membership respectively with the condition* $0 \le T_B(y) + I_B(y) + F_B(y) \le 3$. $T_B(y)$, $I_B(y)$, *and* $F_B(y)$ *gives the degree of truth membership, degree of indeterminacy, and degree of false membership respectively in [0, 1].*

Remark 1: *Further in this paper, for a particular element* $x \in B$, *the 3-tuple* $(T_B(x), I_B(x), F_B(x))$ *will be termed as a single–valued neutrosophic element (SVNE) or single-valued neutrosophic value (SVNV).*

Operations on single-valued neutrosophic values (SVNVs) [4]:

Let $B = \{(y, T_B(y), I_B(y), F_B(y)) : y \in Y\}$ and $C = \{(y, T_C(y), I_C(y), F_C(y)) : y \in Y\}$ be two SVNVs then, we have the following operations.

Union: $B \cup C = \begin{pmatrix} \max(T_B(y), T_C(y)), \max(I_B(y), I_C(y)), \\ min\ (F_B(y), F_C(y)) : y \in Y \end{pmatrix}$.

Intersection: $B \cap C = \begin{pmatrix} \min(T_B(y), T_C(y)), \min(I_B(y), I_C(y)), \\ max\ (F_B(y), F_C(y))\ : y \in Y \end{pmatrix}$.

Complement: $B^C = (1 - T_B(y), 1 - I_B(y), 1 - F_B(y))$.

Definition 4. *Let N(Y) be the set of all single-valued neutrosophic elements on a universal set Y. Let* $B = (T_B(y), I_B(y), F_B(y))$ *and* $C = (T_C(y), I_C(y), F_C(y))$ *be two member of N(Y) s.t* $B \subseteq C$ *iff* $T_B(y) \le T_C(y)$, $I_B(y) \le I_C(y)$, $F_B(y) \ge F_C(y)$.

Then $\subseteq$ is a partially ordered set also, if

$$B \vee C = Sup\ \{B, C\} = \{\max(T_B(y), T_C(y)), \max(I_B(y), I_C(y)), \min(F_B(y), F_C(y))\}$$

and $B \wedge C = Inf\ \{B, C\} = \{\min\ (T_B(y), T_C(y)), \min\ (I_B(y), I_C(y)), \max\ (F_B(y), F_C(y))\}$ then $B \vee C$ and $B \wedge C$ are $Sup\ \{B, C\}$ and $Inf\ \{B, C\}$, respectively. Therefore, N(Y) is a lattice.

This lattice is used to describe the valuation of the single-valued neutrosophic knowledge measure.

Definition 5. **([9]).** *The correlation coefficient* ρ *between two neutrosophic sets B and C is defined as*

$$\rho(B, C) = \frac{\alpha\ (B,\ C)}{(L_B \cdot L_C)^{1/2}}$$

where α $(B,\ C)$ is a correlation measure between B and C given as follows.

$$\alpha(B,\ C) = \sum_{i=1}^{n}[T_B(y_i)\cdot T_C(y_i) + I_B(y_i)\cdot I_C(y_i) + \mathrm{F_B(y_i)\cdot F_C(y_i)}]$$

and $L_B = \sum_{i=1}^{n}\left[T^2{}_B(y_i) + I^2{}_B(y_i) + F^2{}_B(y_i)\right]$,

$$L_C = \sum_{i=1}^{n}\left[T^2{}_C(y_i) + I^2{}_C(y_i) + F^2{}_C(y_i)\right].$$

Now we define correlation efficiency and normalized correlation efficiency for the determination of objective weights of decision makers in the MAGDM problem.

Definition 6. *The correlation efficiency of a SVNV B_k with respect to SVNVs B_m, m, k = 1, 2, 3, ..., n is defined as*

$$\gamma_{NS}(B_k) = \frac{\sum_{m=1}^{n}\rho(B_k,\ B_m)}{n-1}$$

Definition 7. *Normalized correlation efficiency of $B_n$is defined as*

$$N_\gamma(B_k) = \frac{\gamma_{NS}(B_k)}{\sum_{m=1}^{n}\gamma_{NS}(B_m)}$$

Definition 8. ([12]). *The similarity measure S between two neutrosophic sets B and C is a function $S : B \times C \to [0,1]$ which satisfies the given condition:*

NSM1: *$0 \leq S$ (B, C) ≤ 1;*
NSM2: *S (B, C) = 1 if B = C;*
NSM3: *S (B, C) = S (C, B);*
NSM4: *S (A, C) $\leq$ S (A, B); S (A, C) $\leq$ S (B, C), if $A \subseteq B \subseteq C$.*

Definition 9. ([6]). *An entropy E on a single-valued neutrosophic element, $\psi = (\psi_1,\ \psi_2,\ \psi_3)$ is a function $E : N \to [0,1]$ which satisfies the following condition:*

NSE1: *E_{NS} (ψ) = 0 if ψ is a crisp set i.e., $\psi = (1,\ 0,\ 0)$ or $(0,\ 0, 1)$;*
NSE2: *E_{NS} (ψ)= 1 if (ψ_1, ψ_2, ψ_3) = (0.5, 0.5, 0.5);*
NSE3: *E_{NS} (θ) $\geq$ E_{NS} (ψ) if θ is more uncertain;*
NSE4: *E (ψ) = E (ψ^C), where $\psi^C = (1-\psi_1,\ 1-\psi_2,\ 1-\psi_3)$.*

Remark 2: *Let $\psi = (\psi_1,\ \psi_2,\ \psi_3)$ and $\theta = (\theta_1,\ \theta_2,\ \theta_3)$ be two single-valued neutrosophic elements in N(Y) then ψ_i and θ_i ($i = 1,\ 2,\ 3$) independently assumes their values in [0, 1].*

In fuzzy theory, a fuzzy set A^* is said to be a sharpened (less uncertain) version of a fuzzy set A with membership function $\mu_A : Y \to [0,1]$ if

$\mu_A{}^*(x) \leq \mu_A(x)$ for $\mu_A(x) \leq \frac{1}{2}$ i.e., $\mu_A(x) - \mu_A^c(x) \leq 0$

and $\mu_A{}^*(x) \geq \mu_A(x)$ for $\mu_A(x) \geq \frac{1}{2}$ i.e., $\mu_A(x) - \mu_A^c(x) \geq 0$.

Based on similar logic, a single-valued neutrosophic elements θ is more uncertain than ψ if $\psi_t \leq \theta_t$ for $\theta_t - \theta_t^c \leq 0$ and $\psi_t \geq \theta_t$ for $\theta_t - \theta_t^c \geq 0$ where $t = 1,\ 2,\ 3$.

3. A Knowledge Measure on Single-Valued Neutrosophic Set

We provide the following axiomatic framework for defining a knowledge measure of SVNV. Let *N(Y)* be the set of all single-valued neutrosophic values, then a knowledge measure on a single-valued neutrosophic value/element $\psi = (\psi_1,\ \psi_2,\ \psi_3)$ is a function *K*: *N(Y)*$\to$ [0, 1] that satisfies the following conditions:

NSK1: *K* (ψ) = 1 if and only if $\psi_t = 0$ or $\psi_t = 1$; *t* = 1, 2, 3;

NSK2: $K(\psi) = 0$ if and only if $(\psi_1, \psi_2, \psi_3) = (0.5, 0.5, 0.5)$;
NSK3: $K(\psi) = K(\psi^c)$;
NSK4: $K(\psi) \geq K(\theta)$ if $\theta = (\theta_1, \theta_2, \theta_3)$ is more uncertain than ψ i.e., $\theta_t \geq \psi_t$ when $\theta_t - \theta_t^c \leq 0$; $t = 1, 2, 3$ or $\theta_t \leq \psi_t$ when $\theta_t - \theta_t^c \geq 0$.

We propose a knowledge measure for SVNV ψ as

$$\mathrm{K}(\psi) = -\frac{1}{3}\sum_{t=1}^{3}\left[\cos\left(\frac{(\psi_t - {\psi_t}^c)}{2}\right)\pi - 1\right]. \tag{1}$$

The following theorem establishes the validity of the proposed knowledge measure.

Theorem 1. *$K(\psi)$ given in Equation (1) is a valid knowledge measure for SVNV ψ.*

Proof. For this, it is sufficient to show that $K(\psi)$ satisfies the axiomatic requirements NSK1—NSK4.
NSK1: We have $\psi = (\psi_1, \psi_2, \psi_3)$. Let us suppose that $\psi_t = 0$ or $\psi_t = 1$ then $\psi_t - \psi_t^c = 1$ or $\psi_t - \psi_t^c = -1$ for $t = 1, 2, 3$.

Using (1), we have $K(\psi) = 1$.

On the other hand, we assume that $K(\psi) = 1$

then, $\psi_t - \psi_t^c = (\psi_t - (1 - \psi_t) = 2\psi_t - 1$.

$\Rightarrow \psi_t - \psi_t^c \,\epsilon\, [-1, 1]$.

Therefore, every term in the summation of (1) is positive. As $K(\psi) = 1$, then every term in this summation should be equal to one, i.e.,

$$-\frac{1}{3}\left[\cos\left(\frac{(\psi_t - {\psi_t}^c)}{2}\right)\pi - 1\right] = 1. \tag{2}$$

and Equation (2) holds if and only if $\psi_t - \psi_t^c = -1$ or $\psi_t - \psi_t^c = 1$ for $t = 1, 2, 3$.

Hence, $\mathrm{K}(\psi) = 1$ if and only if $\psi_t = 0$ or $\psi_t = 1$ for $t = 1, 2, 3$.

NSK2: If $(\psi_1, \psi_2, \psi_3) = (0.5, 0.5, 0.5)$, we have $\psi_t - \psi_t^c = 0$. Then, from Equation (1), $K(\psi) = 0$. On the other hand, from the above analysis, we have $\psi_t - \psi_t^c \,\epsilon\, [-1, 1]$, it is obvious that $0 \leq K(\psi) \leq 1$.

If $K(\psi) = 0$ then $\psi_t - \psi_t^c = 0$ for t = 1, 2, 3.

It follows that $\psi_t = 0.5$, $t = 1, 2, 3$ i.e., $(\psi_1, \psi_2, \psi_3) = (0.5, 0.5, 0.5)$.

NSK3: Since $\psi^c = (1 - \psi_1, 1 - \psi_2, 1 - \psi_3)$ then $(\psi^c)^c = \psi$. Thus,

$$\begin{aligned}\mathrm{K}(\psi^c) &= -\tfrac{1}{3}\left[\cos\left(\tfrac{(\psi_t^c - ({\psi_t}^c)^c)}{2}\right)\pi - 1\right]\\ &= -\tfrac{1}{3}\left[\cos\left(\tfrac{(\psi_t^c - \psi_t)}{2}\right)\pi - 1\right]\\ &= -\tfrac{1}{3}\left[\cos\left(\tfrac{(\psi_t - \psi_t^c)}{2}\right)\pi - 1\right].\end{aligned}$$

Therefore, $K(\psi^c) = K(\psi)$.

NSK4: Assume that θ is more uncertain than ψ. Therefore, in view of Remark 2, we have two cases: $\psi_t \leq \theta_t$ for $\theta_t - \theta_t^c \leq 0$ and $\psi_t \geq \theta_t$, when $\theta_t - \theta_t^c \geq 0$; $t = 1, 2, 3$.

Case 1: Let $\psi_t \leq \theta_t$ for $\theta_t - \theta_t^c \leq 0$; $t = 1, 2, 3$.

Since $\theta_t - \theta_t^c \leq 0 \Rightarrow \theta_t \leq \theta_t^c$.

$$\text{Moreover, } \psi_t \leq \theta_t \tag{3}$$

$$\Rightarrow \theta_t^c \leq \psi_t^c \tag{4}$$

Using (3) and (4), we have

$$\begin{aligned}&\psi_t - {\psi_t}^c \leq \theta_t - {\theta_t}^c\\ &\Rightarrow \psi_t - {\psi_t}^c \leq \theta_t - {\theta_t}^c \leq 0.\end{aligned} \tag{A}$$

Further, $\psi_t - \psi_t^c \in [-1, 0] \; \forall \; t = 1, 2, 3.$

$$\therefore -1 \leq \psi_t - \psi_t^c \tag{B}$$

Using (A) and (B), we have

$$-1 \leq \psi_t - \psi_t{}^c \leq \theta_t - \theta_t{}^c \leq 0. \tag{C}$$

Note that, the generating function $f(x) = -\left(\cos\left(\frac{\pi x}{2}\right) - 1\right)$ of the knowledge measure defined in Equation (1) is a decreasing function of x in $[-1, 0]$.

Therefore, in view of Equations (1) and (C), we have

$$K(\psi) \geq K(\theta).$$

Case II: Let $\psi_t \geq \theta_t$ for $\theta_t - \theta_t^c \geq 0; t = 1, 2, 3.$

On the same lines as that of Case I, we can obtain

$$K(\theta) \geq K(\psi).$$

This shows that whenever θ is more uncertain than ψ, we have
$K(\theta) \geq K(\psi)$ and hence the proof of axiom **NSK4**. □

Theorem 2. *Let $K(\psi)$ and $K(\theta)$ be a knowledge measure of single-valued neutrosophic element $\psi = (\psi_1, \psi_2, \psi_3)$ and $\theta = (\theta_1, \theta_2, \theta_3)$ in the lattice N(Y) (Definition 4) then*

$$K(\psi \cup \theta) + K(\psi \cap \theta) = K(\psi) + K(\theta).$$

Proof. Here, we have two cases:

Case 1: when $\psi \supseteq \theta$. Then, from Equation (1), we have

$$\begin{aligned} K(\psi \cup \theta) &= -\frac{1}{3}\sum_{t=1}^{3}\left[\cos\left(\frac{(\psi_t - \psi_t^c) \cup (\theta_t - \theta_t^c)}{2}\right)\pi - 1\right] \\ &= -\frac{1}{3}\sum_{t=1}^{3}\left[\cos\left(\frac{max.((\psi_t - \psi_t{}^c), (\theta_t - \theta_t{}^c))}{2}\right)\pi - 1\right] \\ &= -\frac{1}{3}\sum_{t=1}^{3}\left[\cos\left(\frac{(\theta_t - \theta_t^c)}{2}\right)\pi - 1\right]. \\ \Rightarrow K(\psi \cup \theta) &= K(\theta). \end{aligned} \tag{5}$$

$$\begin{aligned} K(\psi \cap \theta) &= -\frac{1}{3}\sum_{t=1}^{3}\left[\cos\left(\frac{(\psi_t - \psi_t^c) \cap (\theta_t - \theta_t^c)}{2}\right)\pi - 1\right] \\ &= -\frac{1}{3}\sum_{t=1}^{3}\left[\cos\left(\frac{min.((\psi_t - \psi_t^c), (\theta_t - \theta_t^c))}{2}\right)\pi - 1\right] \\ &= -\frac{1}{3}\sum_{t=1}^{3}\left[\cos\left(\frac{(\psi_t - \psi_t^c)}{2}\right)\pi - 1\right]. \\ &\Rightarrow K(\psi \cap \theta) = K(\psi). \end{aligned} \tag{6}$$

From (5) and (6) we get,

$$K(\psi \cup \theta) + K(\psi \cap \theta) = K(\psi) + K(\theta).$$

Case 2: when $\psi \subseteq \theta$, then Equation (1) gives the similar results as that of Case I, i.e.,

$$K(\psi \cup \theta) + K(\psi \cap \theta) = K(\psi) + K(\theta).$$

□

In the next section, we establish the connections between similarity/distance measure for single-valued neutrosophic sets and the single-valued neutrosophic knowledge measure.

4. Single-Valued Neutrosophic Similarity Measure and Distance Measure

4.1. Single-Valued Neutrosophic Similarity Measure

Suppose $\psi = (\psi_1, \psi_2, \psi_3)$ and $\theta = (\theta_1, \theta_2, \theta_3)$ are two single-valued neutrosophic sets. The similarity $S(\psi, \theta)$ between ψ and θ should satisfy the following requirements [15].

NS1: $S(\psi,\theta) = 0$ if and only if $\psi_t - \theta_t = 1$ or $\psi_t - \theta_t. = -1$, $t = 1, 2, 3$;
NS2: $S(\psi,\theta) = 1$ if and only if $(\psi_1, \psi_2, \psi_3) = (\theta_1, \theta_2, \theta_3)$; $t = 1, 2, 3$;
NS3: $S(\psi, \theta) = S(\theta, \psi)$;
NS4: $S(\psi, \varphi) \leq S(\psi, \theta)$, $S(\psi, \varphi) \leq S(\theta, \varphi)$ if $\psi_t \leq \theta_t \leq \varphi_t$ or $\psi_t \geq \theta_t \geq \varphi_t$, $t = 1, 2, 3$.

Theorem 3. *Let $\psi = (\psi_1, \psi_2, \psi_3)$ be a SVNE, then $1 - S(\psi, \psi^c)$ is a single—valued neutrosophic knowledge measure i.e.,*

$$K(\psi) = 1 - S(\psi, \psi^c) \text{ is a knowledge measure.}$$

Proof. It is sufficient to show that $1 - S(\psi, \psi^c)$ satisfies the requirement NSK1–NSK4.

(a) $K(\psi) = 1$ if and only if $1 - S(\psi, \psi^c) = 1$ which implies $S(\psi, \psi^c) = 0$ if and only if $\psi_t, -\psi_t{}^c = 1$ or $\psi_t - \psi_t{}^c = -1$ i.e., $K(\psi) = 1$ if and only if $\psi_t - \psi_t{}^c = 1$ or $\psi_t, -\psi_t{}^c = -1$.
(b) $K((\psi) = 0$ which implies that $1 - S(\psi, \psi^c) = 0$ if and only if $S(\psi, \psi^c) = 1$ and $S(\psi, \psi^c) = 1$ if and only $\psi_t = \psi_t{}^c$ i.e., $K(\psi) = 0$ if and only if $(\psi_1, \psi_2, \psi_3) = (\psi_1^c, \psi_2^c, \psi_3^c)$ i.e., $K((\psi) = 0$ if and only if $(\psi_1, \psi_2, \psi_3) = (0.5, 0.5, 0.5)$.
(c) $K(\psi^c) = 1 - S(\psi^c, (\psi^c)^c)$ which implies $K(\psi^c) = 1 - S(\psi^c, \psi) = K(\psi)$.
(d) Let $\psi = (\psi_1, \psi_2, \psi_3)$ *and* $\theta = (\theta_1, \theta_2, \theta_3)$ be two SVNEs. Suppose that $\psi_t \leq \theta_t$ when $\theta_t - \theta_t{}^c \leq 0$ then $\theta_t \leq 1 - \theta_t \leq 0$ i.e., $\theta_t \leq 1 - \theta_t$ and we have

$$\psi_t \leq \theta_t \leq 1 - \theta_t \leq 1 - \psi_t$$

i.e., $\psi_t \leq \theta_t \leq \theta_t{}^c \leq \psi_t{}^c$.
Therefore, by definition of the similarity measure of SVNE (NS_4), it is deduced that
$S(\psi, \psi^c) \leq S(\theta, \psi^c) \leq S(\theta, \theta^c)$
or $1 - S(\psi, \psi^c) \geq 1 - S(\theta, \psi^c) \geq 1 - S(\theta, \theta^c)$
or $1 - S(\psi, \psi^c) \geq 1 - S(\theta, \theta^c)$.
which implies, $K(\psi) \geq K(\theta)$.
Similarly, $K(\psi) \geq K(\theta)$ when $\theta_t - \theta_t{}^c \geq 0$. □

4.2. Single-Valued Neutrosophic Distance Measure

If ψ and θ are two SVNVs then the distance measure between ψ and θ should satisfy the given conditions [16]:

NSD1: $d(\psi, \theta) = d(\theta, \psi)$;
NSD2: $d(\psi, \theta) = 1$ if and only if $\psi_t = 0$ or $\psi_t = 1$ for $t = 1, 2, 3$;
NSD3: $d(\psi, \theta) = 0$ if and only if $(\psi_1, \psi_2, \psi_3) = (\theta_1, \theta_2, \theta_3)$;
NSD4: $d(\psi, \varphi) \geq d(\psi, \theta)$; $d(\psi, \varphi) \geq d(\theta, \varphi)$ if $\psi_t \leq \theta_t \leq \varphi_t$ or $\psi_t \geq \theta_t \geq \varphi_t$.

Theorem 4. *Let $\psi = \psi_1, \psi_2, \psi_3$ be a SVNV, then $d(\psi, \psi^c)$ is a single-valued neutrosophic knowledge measure i.e., $K(\psi) = d(\psi, \psi^c)$ is a single-valued neutrosophic knowledge measure.*

Proof. It is sufficient to show that $d(\psi, \psi^c)$ satisfies the requirement NSD1–NSD4.

NSD1: As we know $K((\psi) = 1$ which implies $d(\psi, \psi^c) = 1$ if and only if $\psi_t - \psi_t{}^c = 1$ or $\psi_t. - \psi_t{}^c = -1$ i.e., $\psi_t - (1 - \psi_t) = 1$ or $\psi_t - (1 - \psi_t) = -1$ and this equation holds if and only $\psi_t = 0$ or $\psi_t = 1$.

NSD 2: $K(\psi) = 0$ which implies that $d(\psi, \psi^c) = 0$ if and only if $(\psi_1, \psi_2, \psi_3) = (\psi_1{}^c, \psi_2{}^c, \psi_3{}^c)$ or $(\psi_1, \psi_2, \psi_3) = (1 - \psi_1, 1 - \psi_2, 1 - \psi_3)$ or $\psi_t = 1 - \psi_t$ which implies $\psi_t = 0.5$ for $t = 1, 2, 3$ i.e., $(\psi_1, \psi_2, \psi_3) = (0.5, 0.5, 0.5)$.

NSD 3: $K(\psi^c) = d(\psi^c, \psi^c)^c)$ which implies $d(\psi^c, \psi,) = K(\psi^c) = K(\psi)$.

NSD 4: Let $\psi = (\psi_1, \psi_2, \psi_3)$ *and* $\theta = (\theta_1, \theta_2, \theta_3)$ be two SVNEs. Suppose that $\psi_t \leq \theta_t$ when $\theta_t - \theta_t{}^c \leq 0$ then $\theta_t - (1 - \theta_t) \leq 0$ or $\theta_t \leq 1 - \theta_t$ and we have, $\psi_t \leq \theta_t \leq 1 - \theta_t \leq 1 - \psi_t$ or $\psi_t \leq \theta_t \leq \theta_t{}^c \leq \psi_t{}^c$. Therefore, by NSD4
$d(\psi, \psi^c) \geq d(\theta, \psi^c) \geq d(\theta, \theta^c)$
i.e., $d(\psi, v^c) \geq d(\theta, \theta^c)$.
$\Rightarrow K(\psi) \geq K(\theta)$.

Consequently, $K(\psi) \geq K(\theta)$ for $\theta_t - \theta_t{}^c \geq 0$. □

In the next section, we investigate the application of the proposed knowledge measure in MAGDM problem.

5. The MAGDM Problem

Suppose we are given m attributes of each of the n available alternatives and a person or an organization seeks the best alternative with the help of k decision-makers. Let $C_1, C_2, C_3, \ldots, C_m$ attributes be the attributes pertaining to the alternatives $A_1, A_2, A_3 \ldots, A_n$ and $D_1, D_2, D_3 \ldots, D_k$ be the decision-makers. Then, each decision-maker gives neutrosophic ratings based on satisfaction level to the attribute in context of the available alternative. Objective weights of attributes are computed using entropy/knowledge measure.

This problem can be considered as multiple attribute group decision-making (MAGDM). The theory of neutrosophy provides an alternative and an efficient tool to design decision-making models with vague information. Such a mechanism is more practical and sophisticated for obtaining the reasonable and appropriate solution of MAGDM problem.

A framework based on cognitive logic to solve an MAGDM problem requires the following information.

(1) Decision matrices/table based on the neutrosophic knowledge-base of each decision maker.
(2) A unified decision table aggregating the opinion of the decision makers with different knowledge and background. The procedure of opinion aggregation essentially needs to consider the level of expertise of each of the decision-makers. Therefore, some level of importance or weight should be assigned to each decision expert. The weight computed in this manner may be considered as the level of expertise. Now question arises how to compute this weight. In such a scenario, the objective weights of decision-makers can be obtained using some mathematical procedure connecting the information base of the decision-makers. The correlation coefficient among the neutrosophic knowledge base of experts gives the linear association or degree of agreement in the opinion of the experts. The normalized correlation efficiency computes the relative agreement level of each of the expert at normalized scale. Thus, normalized correlation efficiency can be perceived as the weight to the expertise of a decision-maker.
(3) The weights of the decision-makers are utilized to obtain the collective decision matrix. The fusion of decision matrices also requires a suitable aggregation operator. In the present scenario, we use a single-valued neutrosophic weighted averaging operator.
(4) Finally, the rating of alternatives can be obtained.

On the basis of this discussion, we developed the flowchart given in Figure 1 and the following algorithm.

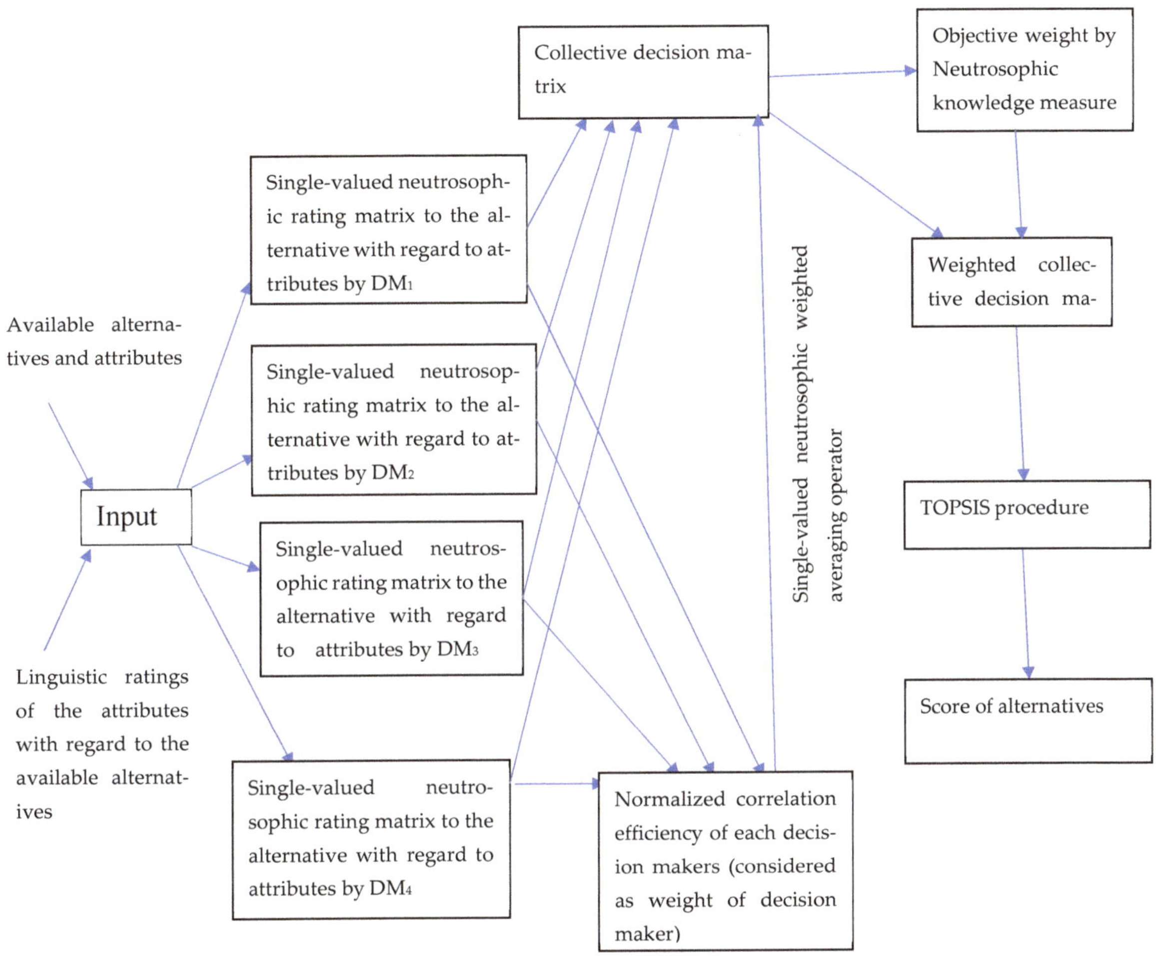

Figure 1. Flowchart for algorithm following in the MADM problem.

5.1. Algorithm for MADM Problem in Neutrosophic Set

Step 1: There may be lots of attributes in a decision-making problem. Among them, only some of the attributes are appropriate and technically sound. Therefore, appropriate attributes are identified with the help of the domain experts.

Step 2: Different alternatives may be good in different attributes. On the basis of their performance level, some ratings are given to each alternative with regard to each attribute by decision-makers and these scores are given in the form of linguistic terms. The alternatives with neutrosophic ratings of attributes are shown in the following decision matrix D:

$$\begin{array}{c} \\ A_1 \\ A_2 \\ \vdots \\ A_n \end{array} \begin{array}{c} \quad C_1 \qquad\qquad C_2 \ldots \quad C_n \\ \begin{bmatrix} (T_{11}, I_{11}, F_{11}) & (T_{12}, I_{12}, F_{12})\ldots & (T_{1m}, I_{1m}, F_{1m}) \\ (T_{21}, I_{21}, F_{21}) & (T_{22}, I_{22}, F_{22})\ldots & (T_{2m}, I_{2m}, F_{2m}) \\ (T_{n1}, I_{n1}, F_{n1}) & (T_{n2}, I_{n2}, F_{n2})\ldots & (T_{nm}, I_{nm}, F_{nm}) \end{bmatrix} \end{array}$$

In the given matrix $d_{ij} = (T_{ij}, I_{ij}, F_{ij})$ represent degree of truthness, indeterminacy, and falsity respectively.

Step 3: In the next step, we construct decision matrices for k decision-maker with the help of linguistic term. It is necessary to find out the weights of each decision-makers because each decision-maker has their own significance. For this, the linguistic terms for each decision-maker are rated with the help of neutrosophic number $(T_{ij},\ I_{ij},\ F_{ij})$. Using the correlation coefficient formula given in Definition 5 between each decision-maker helps us to find the correlation measure and correlation coefficient between linguistic opinions of the decision-makers. The correlation efficiency of each decision-maker can be considered as a more realistic weight of the decision-maker as it computes the objective and subjective assessments.
Step 4: Correlation efficiency concerning the intuitionistic fuzzy sets was computed by Singh et al. [10]. Analogously, we compute the correlation efficiency γ and normalized correlation efficiency Nγ in the neutrosophic environment in Definition 6 and Definition 7. The normalized correlation efficiency of each decision maker was considered as the weights of the decision-maker.
Step 5: With the help of the decision-maker's assessment, construct the aggregated single-valued neutrosophic decision matrix was constructed with the help of SVNWA (single-valued neutrosophic weighted averaging) operator (Biswas et al. [11])

$$(d_{ij})_{m\times n}=\left\{\begin{array}{c}\left(1-\prod_{k=1}^{n}(1-T_{ij}{}^{p})^{\lambda k}\right),\\ \prod_{k=1}^{n}(\ I_{ij}{}^{p})^{\lambda k},\ \prod_{k=1}^{n}(F_{ij}{}^{p})^{\lambda k}\end{array}\right\}. \tag{7}$$

where p is no. of decision-makers and λ_k is the weight of a decision-maker.
Step 6: The knowledge measure of the selected attribute is calculated using Equation (1). From the knowledge measure of attributes, we can find the weight of an attribute as follows.

$$w_i=\frac{K(C_i)}{\sum K(C_i)},\ \text{where } i=1,\ 2,\ 3,\ldots,n. \tag{8}$$

Step 7: Aggregation-weighted neutrosophic decision matrix is constructed for each alternative with respect to each attribute with the help of the given formula:

$$(T_{ij},I_{ij},F_{ij})\to(w_iT_{ij},w_iI_{ij},w_iF_{ij}). \tag{9}$$

Step 8: Obtain relative neutrosophic positive ideal (RNPIS) and relative neutrosophic negative ideal solution (RNNIS) for each attribute from aggregated single neutrosophic decision matrix as follows.

$$Q^{i+}=\{max.(T_{ij}),\ min.(I_{ij}),\ min.(F_{ij})\}. \tag{10}$$

$$Q^{i-}=\{min.(T_{ij}),\ max.(I_{ij}),\ max.(F_{ij})\}. \tag{11}$$

Step 9: The distance measure of each alternative from RNPIS (Q^{i+}) and RNNIS (Q^{i-}) is determined with the help of the given formula:

$$d^{i+}=\sqrt{\frac{1}{3n}\sum_{i=1}^{n}\begin{array}{c}(T_{ij}{}^{wi}(x_i)-T_i^{w+}(x_i))^2+\\(I_{ij}{}^{wi}(x_i)-I_i^{w+}(x_i))^2+\\(F_{ij}{}^{wi}(x_i)-F_i^{w+}(x_i))^2.\end{array}} \tag{12}$$

$$d^{i-}=\sqrt{\frac{1}{3n}\sum_{i=1}^{n}\begin{array}{c}(T_{ij}{}^{wi}(x_i)-T_i^{w-}(x_i))^2+\\(I_{ij}{}^{wi}(x_i)-I_i^{w-}(x_i))^2+\\(F_{ij}{}^{wi}(x_i)-F_i^{w-}(x_i))^2.\end{array}} \tag{13}$$

where T_i^{w+} *and* T_i^{w-} are the max. and min. value of truth membership, I_i^{w+} *and* I_i^{w-} are the max. and min. value of indeterminacy, and F_i^{w+}*and* F_i^{w-} are the max. and min. value of falsity membership.

Step 10: Finally, the relative closeness coefficient to the neutrosophic ideal solution is obtained which is defined as follows:

$$C_i^* = \frac{d^{i-}}{d^{i+} + d^{i-}}. \quad (14)$$

The larger relative closeness coefficient value depicts the most suitable and appropriate alternative.

5.2. Numerical Example Based on MADM (Multiple Attribute Decision-Making)

Suppose there are four decision-makers in a selection committee and they want to select the most suitable and deserving candidate for the managerial position in an organization. Let there be four candidates (alternatives) A_1, A_2, A_3, A_4 for the post. The candidates have been selected based on four attributes C_1, C_2, C_3, C_4 where C_1: Hardworker, C_2 : Leadership, C_3: Domain knowledge, and C_4: Visionary. Based on these attribute decision-makers DM_1, DM_2, DM_3, and DM_4 select the most suitable candidate. Now, we implement the proposed algorithm with the help of the numerical example:

Step 1: Each attribute attains its own significance. Some are very major and some are not. Similarly, each decision-maker has their own importance according to their background knowledge base, power, and position in an organization. The importance of attributes is expressed by linguistic term and these terms are rated as single-valued neutrosophic numbers as shown below in Table 1:

Table 1. Ranking of attributes and decision makers with linguistic terms.

Linguistic Term	SVNNs
Extremely good	(1.0, 0.0, 0.0)
Very good	(0.95, 0.15, 0.05)
Good	(0.75, 0.25, 0.10)
Medium	(0.50, 0.40, 0.30)
Bad	(0.20, 0.60, 0.60)
Very bad	(0.10, 0.80, 0.95)

Step 2: The linguistic term taken from Table 2 can be expressed as single-valued neutrosophic number using Table 1 for rating the opinion of each decision-maker. We present these ratings in the form of four decision matrices which subsequently helps to assess the objective weights of the decision-makers. The correlation measures of the neutrosophic values were calculated between each possible pair of decision-makers as shown in Table 3. In Table 4, the correlation coefficient between each pair of decision-makers is obtained with the help of the formula given in Definition 5. Further, we determine the correlation efficiency and normalized correlation efficiency as shown in Tables 5 and 6 by using the formula given in Definition 6 and Definition 7. We consider the normalize correlation efficiency of each decision-maker as the weights of decision-makers.

Table 2. Linguistic rating for four attributes for the four available alternatives by decision-makers.

Alternative	Decision-Maker	C_1	C_2	C_3	C_4
	DM_1	G	G	G	G
A_1	DM_2	VG	G	A	G
	DM_3	G	VG	VG	A
	DM_4	VG	G	A	VG
	DM_1	VG	G	A	A
A_2	DM_2	VG	A	G	G
	DM_3	VG	VG	G	G
	DM_4	VG	A	G	G
	DM_1	VG	VG	VG	VG
A_3	DM_2	G	G	G	G
	DM_3	A	A	A	A
	DM_4	G	VG	VG	G
	DM_1	G	G	G	G
A_4	DM_2	A	A	G	G
	DM_3	VG	G	G	G
	DM_4	G	G	G	G

For decision-maker DM_1, DM_2, DM_3, DM_4 linguistic term are given as below:
For decision-maker DM_1

$$\begin{array}{c} A_1 \\ A_2 \\ A_3 \\ A_4 \end{array} \begin{pmatrix} <G & G & G & G> \\ <VG & G & A & A> \\ <VG & VG & VG & VG> \\ <G & G & G & G> \end{pmatrix}$$

For decision-maker DM_2

$$\begin{array}{c} A_1 \\ A_2 \\ A_3 \\ A_4 \end{array} \begin{pmatrix} <VG & G & A & G> \\ <VG & A & G & A> \\ <G & G & G & G> \\ <A & G & G & G> \end{pmatrix}$$

For decision-maker DM_3

$$\begin{array}{c} A_1 \\ A_2 \\ A_3 \\ A_4 \end{array} \begin{pmatrix} <G & VG & VG & A> \\ <VG & VG & G & G> \\ <A & A & A & A> \\ <VG & G & G & G> \end{pmatrix}$$

For decision-maker DM_4

$$\begin{array}{c} A_1 \\ A_2 \\ A_3 \\ A_4 \end{array} \begin{pmatrix} <VG & G & A & VG> \\ <VG & A & G & G> \\ <G & VG & VG & G> \\ <G & G & G & G> \end{pmatrix}$$

For decision-maker DM_1, DM_2, DM_3, DM_4 single-valued neutrosophic values corresponding to linguistic terms are

$$\begin{pmatrix} <(0.75,0.25,0.10) & (0.75,0.25,0.10) & (0.75,0.25,0.10) & (0.75,0.25,0.10)> \\ <(0.95,0.15,0.05) & (0.75,0.25,0.10) & (0.50,0.40,0.30) & (0.50,0.40,0.30)> \\ <(0.95,0.15,0.05) & (0.95,0.15,0.05) & (0.95,0.15,0.05) & (0.95,0.15,0.05)> \\ <(0.75,0.25,0.10) & (0.75,0.25,0.10) & (0.75,0.25,0.10) & (0.75,0.25,0.10)> \end{pmatrix}$$

For decision maker DM_2, the single-valued neutrosophic value corresponding to linguistic terms is

$$\begin{pmatrix} <(0.95,0.15,0.05) & (0.75,0.25,0.10) & (0.50,0.40,0.30) & (0.75,0.25,0.10)> \\ <(0.95,0.15,0.05) & (0.50,0.40,0.30) & (0.75,0.25,0.10) & (0.50,0.40,0.30)> \\ <(0.75,0.25,0.10) & (0.75,0.25,0.10) & (0.75,0.25,0.10) & (0.75,0.25,0.10)> \\ <(0.50,0.40,0.30) & (0.50,0.40,0.30) & (0.75,0.25,0.10) & (0.75,0.25,0.10)> \end{pmatrix}$$

For decision maker DM_3, the single-valued neutrosophic value corresponding to linguistic terms is

$$\begin{pmatrix} <(0.75,0.25,0.10) & (0.95,0.15,0.05) & (0.95,0.15,0.05) & (0.50,0.40,0.30)> \\ <(0.95,0.15,0.05) & (0.95,0.15,0.05) & (0.75,0.25,0.10) & (0.75,0.25,0.10)> \\ <(0.50,0.40,0.30) & (0.50,0.40,0.30) & (0.50,0.40,0.30) & (0.50,0.40,0.30)> \\ <(0.95,0.15,0.05) & (0.75,0.25,0.10) & (0.75,0.25,0.10) & (0.75,0.25,0.10)> \end{pmatrix}$$

For decision maker DM_4, the single-valued neutrosophic value corresponding to linguistic terms is

$$\begin{pmatrix} <(0.95,0.15,0.05) & (0.75,0.25,0.10) & (0.50,0.40,0.30) & (0.95,0.15,0.05)> \\ <(0.95,0.15,0.05) & (0.50,0.40,0.30) & (0.75,0.25,0.10) & (0.75,0.25,0.10)> \\ <(0.50,0.40,0.30) & (0.95,0.15,0.05) & (0.95,0.15,0.05) & (0.75,0.25,0.10)> \\ <(0.75,0.25,0.10) & (0.75,0.25,0.10) & (0.75,0.25,0.10) & (0.75,0.25,0.10)> \end{pmatrix}$$

Table 3. Correlation measures of neutrosophic pair for each decision-maker.

	DM_1	DM_2	DM_3	DM_4
DM_1	11.336	10.262	10.193	10.984
DM_2	10.262	10.194	9.65	10.562
DM_3	10.193	9.65	10.939	10.118
DM_4	10.984	10.562	10.118	11.341

Table 4. Correlation coefficient of single-valued neutrosophic sets.

	DM_1	DM_2	DM_3	DM_4
DM_1	1	0.954	0.915	0.968
DM_2	0.954	1	0.914	0.982
DM_3	0.915	0.914	1	0.908
DM_4	0.968	0.982	0.908	1

Table 5. Correlation efficiency of decision-makers.

Decision-Maker	Correlation Efficiency
DM_1	0.945
DM_2	0.95
DM_3	0.912
DM_4	0.952

Table 6. Normalized correlation efficiency of decision-makers.

Decision-Maker	Normalized Correlation Efficiency
DM_1	0.251
DM_2	0.252
DM_3	0.242
DM_4	0.253

Where 0.251, 0.252, 0.242, and 0.253 are weights of decision-makers. Take $\lambda_1 = 0.251$, $\lambda_2 = 0.252$, $\lambda_3 = 0.242$, and $\lambda_4 = 0.253$.

Step 3: We construct aggregated neutrosophic decision matrix as given in Table 7 with the help of SVNWA (single-valued neutrosophic weighted averaging aggregation operator).

Table 7. Aggregated neutrosophic decision matrix.

	C_1	C_2	C_3	C_4
A_1	(0.889, 0.192, 0.070)	(0.830, 0.221, 0.084)	(0.759, 0.280, 0.147)	(0.803, 0.246, 0.109)
A_2	(0.949, 0.149, 0.050)	(0.759, 0.280, 0.147)	(0.702, 0.281, 0.131)	(0.702, 0.281, 0.131)
A_3	(0.802, 0.246, 0.109)	(0.868, 0.216, 0.092)	(0.764, 0.278, 0.144)	(0.803, 0.246, 0.109)
A_4	(0.798, 0.248, 0.111)	(0.702, 0.281, 0.132)	(0.750, 0.250, 0.100)	(0.750, 0.250, 0.100)

The weights of the attributes are determined with the help of Equation (8), i.e., $w_1 = 0.313$, $w_2 = 0.246$, $w_3 = 0.207$, and $w_4 = 0.232$.

Step 4: We construct an aggregated weighted neutrosophic decision matrix using Equation (9) and present in Table 8.

Table 8. Aggregated weighted neutrosophic decision matrix.

	C_1	C_2	C_3	C_4
A_1	(0.278, 0.060, 0.021)	(0.204, 0.054, 0.020)	(0.157, 0.057, 0.030)	(0.186, 0.057, 0.025)
A_2	(0.297, 0.046, 0.015)	(0.186, 0.068, 0.036)	(0.145, 0.058, 0.027)	(0.162, 0.065, 0.030)
A_3	(0.251, 0.076, 0.034)	(0.213, 0.053, 0.022)	(0.158, 0.057, 0.029)	(0.186, 0.057, 0.025)
A_4	(0.249, 0.077, 0.034)	(0.172, 0.069, 0.032)	(0.155, 0.051, 0.020)	(0.174, 0.058, 0.023)

Step 5: The neutrosophic relative positive ideal solution and relative negative solution is obtained from aggregated weighted neutrosophic decision matrix given in Equations (10) and (11).

Neutrosophic relative positive ideal solution (Q^{i+}):

$$Q^{i+} = \left\{ \begin{array}{lll} < 0.297, & 0.046, & 0.015 >, \\ < 0.213, & 0.053, & 0.020 >, \\ < 0.158, & 0.051, & 0.020 >, \\ < 0.186, & 0.057, & 0.023 >, \end{array} \right\}.$$

Neutrosophic relative positive ideal solution (Q^{i-}):

$$Q^{i-} = \left\{ \begin{array}{lll} < 0.249, & 0.077, & 0.034 >, \\ < 0.172, & 0.069, & 0.036 >, \\ < 0.145, & 0.058, & 0.030 >, \\ < 0.162, & 0.065, & 0.030 >, \end{array} \right\}.$$

Step 6: The distance measures ($d^{i+} and\ d^{i-}$) of each alternative from RNPIS and RNNIS are determined with the help of normalized Euclidean distance measure given in Equations (12) and (13). Finally, we obtain the relative closeness coefficient ($\mathbf{C_i}^*$) with the help of formula given in Equation (14).

The largest relative closeness value indicates the most suitable and appropriate alternative. From Table 9, it can be seen that the largest value of relative closeness coefficient is corresponding to A_1. Hence, A_1 is the most suitable alternative.

Table 9. Closeness coefficients of the alternatives.

Alternatives	d^{i+}	d^{i-}	C_i^*
A_1	0.0079	0.0173	0.6865
A_2	0.0134	0.0178	0.5705
A_3	0.0170	0.0157	0.4801
A_4	0.0220	0.0065	0.2280

Using the VIKOR method (Kamal et al. [17]) to the data of the numerical problem considered in this section, the ranking results are $A_1 > A_3 > A_4 > A_2$. We observe that the best alternative due to the VIKOR Method and proposed method remains the same. Therefore, the proposed method is consistent with VIKOR method for finding the best alternative.

6. Comparative Study

In this section, we investigate the effectiveness of the proposed knowledge measure in MCDM problems and comparative study of proposed similarity measure.

6.1. *Effectiveness of the Proposed Knowledge Measure against Different Existing Entropies*

To check the usefulness of our proposed knowledge measure in MCDM problems, we consider the numerical example given in Section 5.2. We calculate the weight of the attribute with the help of the proposed single-valued neutrosophic knowledge measure. We use a same numerical problem and calculate the weight of the attributes by using different existing entropies as shown in Table 10.

Table 10. Ranking result using existing entropies and the proposed knowledge measure.

Existing Entropies	Ranking
$EY_1 = \frac{1}{3(\sqrt{2}-1)} \sum_{t=1}^{3} \left(\left(\sqrt{2} \cos \frac{\alpha_t - \alpha_t{}^c}{4} \times \pi \right) - 1 \right)$ (Wu et al. [6])	$A_1 > A_3 > A_2 > A_4$
$EY_2 = \frac{1}{3(\sqrt{2})-1} \sum_{t=1}^{3} \left(\frac{\sin \pi(\alpha_t - \alpha_t{}^c + 1)}{4} \right) + \left(\frac{\cos \pi(\alpha_t - \alpha_t{}^c + 1)}{4} \right) - 1$ (Jin **et al.** [12])	$A_1 > A_3 > A_2 > A_4$
$EY_3 = 1 - \frac{1}{n} \sum_{t=1}^{m} (\alpha_t + \gamma_t) \lvert 2\beta_t - 1 \rvert$ (Elshabshery and Fattouh [18])	$A_1 > A_3 > A_2 > A_4$
$EY_4 = \frac{1}{n} \sum \left(1 - \frac{1}{b-a} \int_a^b \lvert \alpha_t - \gamma_t \rvert \lvert \beta_t - \beta_t{}^c \rvert \right) dx$ (Aydogdu [15])	$A_1 > A_3 > A_2 > A_4$
K (Proposed Knowledge measure)	$A_1 > A_2 > A_3 > A_4$

From Table 10, we observe that the best alternative using the existing single valued entropy measures and our proposed knowledge measure remains same, i.e., A_1. However, overall ranking of the alternatives is different due to our proposed measure. Moreover, as pointed out in the Section 5, the ranking of alternatives using VIKOR method is $A_1 > A_3 > A_2 > A_4$ which is consistent with the ranking due to single-valued neutrosophic entropy measures EY_i (i = 1, 2, 3, 4), since the VIKOR method is suitable for the problems with conflicting criteria as it gives the compromise-type solutions. Thus, in view of these facts, we can say that the existing entropy measures are more suitable to MAGDM problems with conflicting criteria and our proposed measure is suitable to the problems with non-conflicting criteria.

6.2. Comparison between the Existing Entropies of the Neutrosophic Sets and the Proposed Knowledge Measure of Neutrosophic Sets on the Basis of Linguistic Hedges

In this subsection, we compare the existing entropies of single-valued neutrosophic sets and proposed knowledge measure on the basis of linguistic hedges.

Zadeh [19] introduced the notion of the linguistic hedge in two parts: 1. Very, much more, or less which are used in direct situations; 2. hedges essentially, practically, techniques are used in complex situations. Zadeh gave different types of operations, two of which are concentration and dilation. The concentration is defined as

$$\mathrm{CON}(A) = A^2.$$

and dilation is the somehow opposite of concentration and is defined as

$$\mathrm{DIL}\ (A) = A^{0.5}.$$

Singh et al. [13] also investigated the superiority of the fuzzy knowledge measure using structured linguistic framework.

The entropy for a fuzzy set A on the basis of mathematical operation should follow the order as

$$E\left(A^{\frac{1}{2}}\right) > E(A) > E\left(A^2\right) > E\left(A^3\right). \tag{15}$$

Because of this, knowledge measure of fuzzy set, should follow the order

$$K\left(A^{\frac{1}{2}}\right) < K(A) < K\left(A^2\right) < K\left(A^3\right). \tag{16}$$

Neutrosophy is a broader sense of capturing the vagueness, so, the inequality order given in Equations (15) and (16) must also be followed by an entropy and a knowledge measure, respectively, in the single-valued neutrosophic framework. To analyze the practicality of the proposed measure, we consider the following empirical scenario.

Let $NS(A) = (x_1, (0.1, 0.2, 0))\ (x_2, (0.6, 0.1, 0.3)), (x_3, (0.6, 0.3, 0.1)), (x_4, (0, 0.2, 0.2))$ be a SVNS in which x_1, x_2, x_3, x_4 assumes three values, each of which can be regarded as membership value, indeterminacy, and non-membership value. Here, we consider A as linguistic hedge good, $A^{1/2}$ as average, A^2 as very good, and A^3 as extremely good. The modifier for the NS(A) is given as

$$K(A^n) = (x,\ (T_A(x))^n, (I_A(x))^n, (F_A(x))^n\ )$$

$$EY_1 = \frac{1}{3\left(\sqrt{2}-1\right)} \sum_{t=1}^{3}\left(\left(\sqrt{2}\cos\frac{\alpha_t - {\alpha_t}^c}{4} \times \pi\right) - 1\right).$$

(Wu et al. [6])

$$EY_2 = \frac{1}{3\left(\sqrt{2}\right) - 1} \sum_{t=1}^{3}\left(\frac{\sin\pi(\alpha_t - {\alpha_t}^c + 1)}{4}\right) + \left(\frac{\cos\pi(\alpha_t - {\alpha_t}^c + 1)}{4}\right) - 1.$$

(Jin et al. [12])

$$EY_3 = 1 - \frac{1}{n}\sum_{t=1}^{m}(\alpha_t + \gamma_t\ )|2\,\beta_t - 1\,|.$$

(Elshabshery and Fattouh [18])

$$EY_4 = \frac{1}{\mathrm{n}}\sum\left(1 - \frac{1}{\mathrm{b}-\mathrm{a}}\int_{\mathrm{a}}^{\mathrm{b}} |\alpha_{\mathrm{t}} - \gamma_{\mathrm{t}}||\beta_{\mathrm{t}-} - {\beta_{\mathrm{t}}}^{\mathrm{c}}|\right)dx.$$

(Aydogdu [15])

$$EY_5 = \begin{cases} 1 - \frac{|2\beta_t - 1|}{2} & if \quad \alpha_t = \gamma_t = 0.5 \\ \frac{2-|2\alpha_t - 1| - |2\gamma_t - 1|}{4} & otherwise. \end{cases}$$

(Qin and Wang [7])

$$E_{Y6} = 1 - \frac{1}{n}\sum_{t=1}^{m}(\alpha_t + \gamma_t)|\beta_{t-} - \beta_t{}^c|$$

(Majumdar and Samanta [20])

$$E_{Y7} = 1 - \frac{1}{n}\sum_{t=1}^{m}\frac{|\alpha_t - 0.5| + |\gamma_t - 0.5| + |\beta_{t-} - 0.5| + |\beta_t{}^c - 0.5|}{2}.$$

(Thao and Smarandache [21])

$$E_{Y8} = \frac{1}{n}\sum_{t=1}^{m}\frac{2 - |\alpha_t - \gamma_t| - \beta_{t-} - \beta_t{}^c|}{2 + |\alpha_t - \gamma_t| + \beta_{t-} - \beta_t{}^c|}$$

(Aydogdu and Sahin [22])

The comparative results for the following entropies of SVNS are given in Table 11.

Table 11. Computed entropies/knowledge measures of SVNSs.

NS	EY_1	EY_2	EY_3	EY_4	EY_5	E_{Y6}	E_{Y7}	E_{Y8}	$K(A)$
$A^{1/2}$	0.948	2.966	0.884	2.965	0.247	0.833	0.663	0.581	0.270
A	0.845	2.171	0.705	2.169	0.187	0.825	0.387	0.394	0.482
A^2	0.850	1.009	0.802	1.006	0.108	0.802	0.153	0.301	0.762
A^3	0.899	0.547	0.885	0.544	0.058	0.885	0.069	0.299	0.873

Now, from Table 11, we observed that

$$EY_1(A^{1/2}) > EY_1(A) < EY_1(A^2) < EY_1(A^3)$$
$$EY_2(A^{1/2}) > EY_2(A) > EY_2(A^2) > EY_2\ (A^3)$$
$$EY_3(A^{1/2}) > EY_3(A) < EY_3(A^2) < EY_3\ (A^3)$$
$$EY_4(A^{1/2}) > EY_4(A) > EY_4(A^2) > EY_4\ (A^3)$$
$$EY_5(A^{1/2}) > EY_5(A) > EY_5(A^2) > EY_5\ (A^3)$$
$$EY_6(A^{1/2}) > EY_6(A) > EY_6(A^2) < EY_6\ (A^3)$$
$$EY_7(A^{1/2}) > EY_7(A) > EY_7(A^2) > EY_7\ (A^3)$$
$$EY_8(A^{1/2}) > EY_8(A) > EY_8(A^2) > EY_8\ (A^3)$$
$$K(A^{1/2}) < K(A) < K(A^2) < K(A^3).$$

From the above result, it has been observed that entropies EY_2, EY_4, EY_7, and EY_8 follow the pattern given in Equation (15). The knowledge measure of the neutrosophic set is also shown above, which satisfies the order given in Equation (16). Now, we consider another example of a NS.

$$NS(A) = ((x_1,\ (0.1,\ 0.3,\ 0.7)), (x_2,\ (0,\ 0.1,\ 0.9)),\ (x_3,\ (0.8,\ 0.2,\ 0.3)),\ (x_4, (0.8,\ 0.1,\ 0.4))$$

The entropies and knowledge measure table of the given NSs is shown below

From the Tables 11 and 12, it was observed that our proposed knowledge measure produces theoretical valid results, i.e., $K(A^{1/2}) < K(A) < K(A^2) < K(A^3)$, while SVN entropies produce unreasonable results in different instances. Therefore, the performance of our knowledge measure is better than the conventional entropy measures in the neutrosophic settings.

Table 12. Computed entropies/knowledge measures of SVNSs.

NS	EY_1	EY_4	EY_5	EY_8	$K(A)$
$A^{1/2}$	0.660	0.660	0.195	0.471	0.358
A	0.697	0.561	0.2	0.405	0.468
A^2	0.455	0.365	0.207	0.143	0.613
A^3	0.430	0.345	0.2102	0.148	0.663

7. Conclusions

In this paper, we have proposed a knowledge measure in the single-valued neutrosophic framework. The single-valued neutrosophic knowledge measure has been found to have a general relationship with the similarity and distance measures. Comparative studies demonstrated that the given knowledge measure is more effective and suitable than the existing entropy measure while dealing with the linguistic hedges and in MADM problems. We have also developed a mechanism to handle a MAGDM problem incorporating the proposed single-valued knowledge measure, an existing correlation measure, and an aggregation operator. This algorithm is found novel in sense that it identifies the level of expertise of each decision maker in the group even if the hiring organization has no information about their domain knowledge. The investigation of a problem related to MAGDM using the proposed method improves upon the existing methods in two ways. It offers independent choice of truthiness, falsity, and indeterminacy to the decision-makers for creating numerical data from the vague knowledge base and derives the weightage to decision-makers from the model itself. However, the current study demonstrates the practical applications using artificially generated data; the creation of real single-valued neutrosophic database seems to make this study more pragmatic. Our future studies will focus on some other areas of applications such as pattern recognition, image processing, etc., where the single-valued neutrosophic information seems to play a significant role.

Author Contributions: Conceptualization, S.S. (Sonam Sharma) and S.S. (Surender Singh); Formal analysis, S.S. (Sonam Sharma); Investigation, S.S. (Sonam Sharma); Methodology, S.S. (Surender Singh); Project administration, S.S. (Surender Singh); Resources, S.S. (Sonam Sharma) and S.S. (Surender Singh); Supervision, S.S. (Surender Singh); Visualization, S.S. (Sonam Sharma); Writing—original draft, S.S. (Sonam Sharma); Writing—review & editing, S.S. (Surender Singh). All authors have read and agreed to the published version of the manuscript.

Funding: This research received no external funding.

Institutional Review Board Statement: Not applicable.

Informed Consent Statement: Not applicable.

Data Availability Statement: Not applicable.

Acknowledgments: Authors are highly thankful to anonymous referee and the editor for their critical and constructive suggestions to bring this article in the present form.

Conflicts of Interest: Authors declare that they have no conflict of interest.

References

1. Zadeh, L.A. Fuzzy sets. *Inf. Control* **1965**, *8*, 338–353. [CrossRef]
2. Atanassov, K. Intuitionistic fuzzy sets. *Fuzzy Sets Syst.* **1986**, *20*, 87–96. [CrossRef]
3. Smarandache, F. Neutrosophic Set-A Generalization of the Intuitionistic Fuzzy Set. *Int. J. Pure Appl. Math.* **2005**, *24*, 287–297.

4. Wang, H.; Smarandache, F.; Zhang, Y.; Sunderraman, R. Single valued neutrosophic sets. In *Multispace & Multistructure-Neutrosophic Transdisciplinarity (100 Collected Papers of Sciences)*; North-European Scientific Publishers: Hanko, Finland, 2010; pp. 410–413.
5. MA, M.Z.; Yang, W. Symmetric Intuitionistic Fuzzy Weighted Mean Operators Based on Weighted Archimedian t-Norm and t-Conorms for Multi-Criteria Decision Making. *Informatica* **2020**, 3, 89–112. [CrossRef]
6. Wu, H.; Yuan, Y.; Wei, L.; Pei, L. On entropy, similarity measure and cross entropy of single–valued neutrosophic sets and their application in multi–attribute decision making. *Soft Comput.* **2018**, *22*, 7367–7376. [CrossRef]
7. Qin, K.; Wang, L. New similarity and entropy measures of single–valued neutrosophic sets with applications in multi–attribute decision making. *Soft Comput.* **2020**, *24*, 16165–16176. [CrossRef]
8. Smarandache, F. Neutrosophic Hedge Algebras. *Broad Res. Artif. Intell. Neurosci.* **2019**, *10*, 117–123.
9. Hanafy, I.; Salama, A.; Mahfouz, K. Correlation of Neutrosophic Data. *Int. J. Eng. Sci.* **2012**, *1*, 39–43.
10. Singh, S.; Sharma, S.; Lalotra, S. Generalized Correlation Coefficients of Intuitionistic Fuzzy Sets with Application to MAGDM and Clustering Analysis. *Int. J. Fuzzy Syst.* **2020**, *22*, 1582–1595. [CrossRef]
11. Biswas, P.; Pramanik, S.; Giri, B. TOPSIS method for multi–attribute group decision–making under single–valued neutrosophic environment. *Neural Comput. Appl.* **2016**, *27*, 727–737. [CrossRef]
12. Jin, F.; Ni, Z.; Zhu, X.; Chen, H.; Langari, R.; Mao, X.; Yuan, H. Single-valued neutrosophic entropy and similarity measures to solve supplier selection problems. *J. Intell. Fuzzy Syst.* **2018**, *35*, 6513–6523. [CrossRef]
13. Singh, S.; Lalotra, S.; Sharma, S. Dual concepts in fuzzy theory: Entropy and knowledge measure. *Int. J. Intell. Syst.* **2019**, *34*, 1034–1059. [CrossRef]
14. Lalotra, S.; Singh, S. Knowledge measure of hesitant fuzzy set and its application in multi–attribute decision–making. *Comp. Appl. Math.* **2020**, *39*, 1–31. [CrossRef]
15. Aydodgu, A. On Similarity and Entropy of Single Valued Neutrosophic Sets. *ICSRS* **2015**, *29*, 67–74.
16. Chai, J.S.; Selvachandran, G.; Smarandache, F.; Gerogiannis, V.C.; Son, L.H.; Bui, Q.; Vo, B. New similarity measures for single-valued neutrosophic sets with applications in pattern recognition and medical diagnosis problems. *Complex Intell. Syst.* **2021**, *7*, 703–723. [CrossRef]
17. Kamal, N.L.A.M.; Abdullah, L.; Yee, F.M.; Abdullah, I.; Vafaei, N. Single Valued Neutrosophic VIKOR and Its Application to Wastewater Treatment Selection. *Neutrosophic Sets Syst.* **2021**, *47*, 251–272.
18. Elshabshery, A.; Fattouh, M. On some Information Measures of Single–Valued Neutrosophic Sets and their Applications in MCDM Problems. *Int. J. Eng. Res. Technol.* **2021**, *10*, 406–415.
19. Zadeh, L.A. A Fuzzy–Set–Theoretic Interpretation of Linguistic Hedges. *J. Cybern.* **1972**, *2*, 4–34. [CrossRef]
20. Majumdar, P.; Samanta, S.K. On Similarity and entropy of neutrosophic sets. *J. Intell. Fuzzy Syst.* **2014**, *26*, 1245–1252. [CrossRef]
21. Thao, N.X.; Smarandache, F. Apply new entropy-based similarity measures of single valued neutrosophic sets to select supplier material. *J. Intell. Fuzzy Syst.* **2020**, *39*, 1005–1019. [CrossRef]
22. Aydodgu, A.; Sahin, R. New entropy measures based on neutrosophic set and their applications to multi -criteria decision making. *Suleyman Demirel Univ. Fen Bilim. Enst. Derg.* **2019**, *23*, 40–45.

applied sciences

MDPI

Article

Projection Pursuit Multivariate Sampling of Parameter Uncertainty

Oktay Erten *, Fábio P. L. Pereira and Clayton V. Deutsch

Centre for Computational Geostatistics, 6-247 Donadeo Innovation Centre for Engineering, 9211-116 Street, University of Alberta, Edmonton, AB T6G 1H9, Canada
* Correspondence: oktay.erten@outlook.com

Abstract: The efficiency of sampling is a critical concern in Monte Carlo analysis, which is frequently used to assess the effect of the uncertainty of the input variables on the uncertainty of the model outputs. The projection pursuit multivariate transform is proposed as an easily applicable tool for improving the efficiency and quality of a sampling design in Monte Carlo analysis. The superiority of the projection pursuit multivariate transform, as a sampling technique, is demonstrated in two synthetic case studies, where the random variables are considered to be uncorrelated and correlated in low (bivariate) and high (five-variate) dimensional sampling spaces. Five sampling techniques including Monte Carlo simulation, classic Latin hypercube sampling, maximin Latin hypercube sampling, Latin hypercube sampling with multidimensional uniformity, and projection pursuit multivariate transform are employed in the simulation studies, considering cases where the sample sizes (n) are small (i.e., $10 \leq n \leq 100$), medium (i.e., $100 < n \leq 1000$), and large (i.e., $1000 < n \leq 10{,}000$). The results of the case studies show that the projection pursuit multivariate transform appears to yield the fewest sampling errors and the best sampling space coverage (or multidimensional uniformity), and that a significant amount of computer effort could be saved by using this technique.

Keywords: Monte Carlo analysis; Latin hypercube sampling; projection pursuit multivariate transform; multidimensional uniformity

Citation: Erten, O.; Pereira, F.P.L.; Deutsch, C.V. Projection Pursuit Multivariate Sampling of Parameter Uncertainty. *Appl. Sci.* **2022**, *12*, 9668. https://doi.org/10.3390/app12199668

Academic Editor: Arcangelo Castiglione

Received: 30 July 2022
Accepted: 22 September 2022
Published: 26 September 2022

1. Introduction

Mathematical models are frequently used in many disciplines (i.e., natural sciences, social sciences, engineering) in order to realistically estimate the physical processes in question. To construct such models (or outputs), in most cases, one must use a number of input variables. For example, to calculate the original oil in place (OOIP) for a reservoir, five input variables including the thickness of the deposit, deposit area, net oil to gross volume, net porosity, and water saturation should be considered [1]. However, due to the physical and financial constraints related to the sampling scheme, there are generally a limited number of measurements (or observations) of the input variables available for modeling. Therefore, it is imperative that the effect of uncertainty associated with the input variables of the model output be taken into account [2].

There are many sampling techniques that are used to assess the uncertainty associated with the parameters of the models. For example, Monte Carlo simulation (MCS), which relies on a repeated random sampling and statistical analysis, is generally used for this purpose [3,4]. In MCS, the population is assumed to be independent and identically distributed, and the realizations of a sample are randomly chosen from the population with an equal probability. A pseudo-random number generator, which satisfies a series of statistical tests for randomness [5,6], is used to generate a sequence of independent numbers (or random variates) from a uniform distribution $U(0,1)$ [7]. A major drawback of MCS is that the realizations that are chosen completely at random tend to form clusters, which leaves gaps that are not investigated in the sampling space. If one takes a large sample of realizations, the accuracy of the predicted model output increases and the sampling errors become negligible. However, this is only achievable provided that the run time of MCS is reasonably short. If the MCS run is computationally expensive, then the sampling techniques with better sampling efficiency such as Latin hypercube sampling (LHS) (or stratified random sampling) [8,9] and its variants should be employed.

The original and most simple form of LHS is referred to as classic Latin hypercube sampling (CLHS). In CLHS, the population is divided into a number of non-overlapping strata and MCS is used to generate a realization from each stratum. Because the population is stratified, the heterogeneity in each stratum becomes less, which results in realizations that are more uniform and representative. The stratification is maximized when the number of strata n is equal to the sample size (n), i.e., $[0,1/n],[1/n,2/n],\ldots,[(n-1)/n,1]$ [10]. In CLHS, as mentioned earlier, the population is marginally stratified, that is, it accounts for only the univariate uniformity of the realizations and does not enforce any multivariate uniformity. To improve the space-filling properties of CLHS designs, many studies in the literature make use of mainly two performance criteria: (1) minimizing the pairwise correlation between the realizations, and (2) maximizing the minimum distance between the realizations [11–16]. Considering the correlated (or dependent) random variables, several studies [17–20] propose various methodologies to generate a sample whose correlation matrix is approximately equal to the given (or target) correlation matrix, that is, the joint distribution is reproduced.

The two important variants of LHS that can be used to generate realizations that have improved space-filling characteristics are (1) maximin Latin hypercube sampling (maximin LHS) [21] and (2) Latin hypercube sampling with multidimensional uniformity (LHSMDU) [1]. In the former approach, the aim is to generate a sample that maximizes the minimum Euclidean distance between any pair of realizations, which is achieved by generating a large number (i.e., thousands) of sampling designs and choosing the one that has a maximized distance between any pair of realizations. Due to the maximization of the distance between any pair, the realizations of the sample tend to spread out across the sampling space, resulting in a better multidimensional uniformity. The latter approach expands the univariate uniformity obtained by CLHS to the multivariate context. The algorithm first generates more realizations than are required, and the realizations that are close to each other (or redundant realizations) are sequentially eliminated in the multidimensional space. The post-processing of the realizations is then carried out to enforce the uniformity in the high-dimensional space.

Considering both maximin LHS and LHSMDU, it is important to note two things: (1) the realizations generated by both techniques are still based on CLHS, and (2) both techniques initially require a large number of sampling designs to be generated, which significantly increases the central processing unit (CPU) run time. A better approach for enforcing the sparsity in the sampling designs, however, would be to use projection pursuit [22,23] iterations. The idea of projection pursuit is that the projected data is expected to have a univariate Gaussian distribution if the original data is multivariate Gaussian. The original data is, therefore, generally first transformed to normal scores and then sphered so that the projection index only measures the deviation of the distribution of any projected data from the standard Gaussian distribution $N(0,1)$. The projection pursuit multivariate transform (PPMT) proposed by [24] makes use of this idea and applies a normal score transformation along a projection vector in an iterative fashion so that the original data will be eventually transformed to the uncorrelated and multivariate Gaussian scores. In fact, this amounts to saying that the multidimensional uniformity of the sampling design can be ensured through this technique.

We demonstrate the applicability of PPMT as an efficient sampling technique in two synthetic case studies using the low (bivariate) and high (five-variate) dimensional sampling spaces. The results of the simulation studies indicate that considering the various sample sizes, PPMT yields much fewer sampling errors and exhibits better space-filling characteristics than the other sampling techniques.

2. Sampling Techniques

2.1. Monte Carlo Simulation

MCS (or simple random sampling) is a technique through which a sample of the population is constructed using a random sequence of numbers. The deterministic parameters of the population can then be estimated from each sample [25]. The inverse transform method (or inversion sampling) is used to generate a realization through MCS [4]. Let Z be a random variable whose cumulative distribution function (CDF), which is monotonically non-decreasing, is denoted by $\{F(z),\ a \le z \le b\}$, and the inverse CDF (or a quantile function) of Z is defined by $F^{-1}(u) = \inf\{z \in [a,b] : F(z) \ge u\},\ 0 < u < 1$. Considering that $U \sim U(0,1)$ is also a random variable that has a standard uniform distribution, then $z = F^{-1}(u)$, which can also be observed from $\Pr(Z \le z) = \Pr(F^{-1}(u) \le z) = \Pr(u \le F(z)) = F(z)$.

For example, consider that $n = 5$ is the required number of realizations and $k = 2$ is the number of independent Gaussian random variables, $X_1 \perp X_2 \sim N(0,1)$, that is, the sampling space is two-dimensional and orthogonal. Independent random numbers from a uniform distribution $U(0,1)$ (i.e., the pairs exhibit a uniform distribution in the unit square, and similarly, the triplets also have a uniform distribution in the unit cube) are generated, that is, $\mathbf{p}_i = [p_{1i}\ p_{2i} \ldots p_{ni}]^T,\ i = 1,\ldots,k = 2$ and $n = 5$. These numbers are then established in a matrix $\mathbf{P}$ where $(p_{ij}) \in \mathbb{R}^{5x2}$:

$$\mathbf{P} = \begin{bmatrix} 0.06 & 0.09 \\ 0.93 & 0.51 \\ 0.82 & 0.66 \\ 0.99 & 0.40 \\ 0.14 & 0.76 \end{bmatrix}$$

Each element of the matrix $\mathbf{P}$ is mapped according to a target CDF, which yields the independent realizations in the Gaussian unit $\mathbf{x}_i = [x_{1i}\ x_{2i} \ldots x_{ni}]^T,\ i = 1,\ldots,k = 2$ and $n = 5$. For instance, considering a probability of $p_{42} = 0.40$ associated with the variable X_2, the corresponding realization can be obtained as $x_{42} = F^{-1}(0.40) = -0.24$, where $F^{-1}(\cdot)$ denotes the inverse of the Gaussian CDF for the variable X_2. The MCS sampling design indicating the realizations given in matrix $\mathbf{P}$ and mapping these probabilities according to the given CDF are presented in Figure 1a,b, respectively.

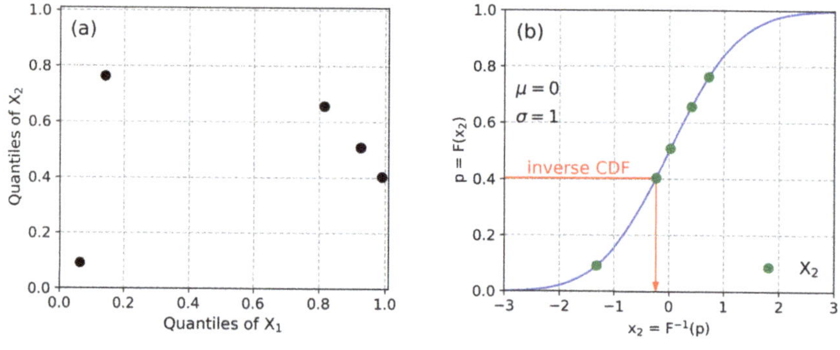

Figure 1. (**a**) A sampling design with $n = 5$ realizations generated by MCS in a two-dimensional sampling space where the random variables are independent, (**b**) Mapping the quantiles according to the standard Gaussian distribution using the inverse transform method.

The matrix $\mathbf{X}$, $(x_{ij}) \in \mathbb{R}^{5x2}$ whose elements are the realizations of X_1 and X_2 is then defined by

$$\mathbf{X} = \begin{bmatrix} -1.54 & -1.32 \\ 1.44 & 0.03 \\ 0.90 & 0.41 \\ 2.31 & -0.24 \\ -1.08 & 0.72 \end{bmatrix}$$

In the case where X_1 and X_2 are correlated according to a given target correlation matrix $\mathbf{C}$ where $(c_{ij}) \in \mathbb{R}^{2x2}$, the linear dependency between X_1 and X_2, can be added via Cholesky decompsition [26,27] of $\mathbf{C}$, that is,

$$\mathbf{C} = \mathbf{L} \cdot \mathbf{L}^T, \quad (1)$$

where $\mathbf{L}$ is the lower triangular matrix and $\mathbf{L}^T$ (where the superscript T denotes transposition) is the upper triangular matrix. The correlated realizations of X_1 and X_2 are computed by multiplying the matrix $\mathbf{L}$ by the matrix $\mathbf{X}^T$:

$$\mathbf{X}^* = \mathbf{L} \cdot \mathbf{X}^T, \quad (2)$$

The resulting matrix $\mathbf{X}^*$ where $(x^*_{ij}) \in \mathbb{R}^{2x5}$ contains the realizations of X_1 and X_2 that have a correlation matrix, which is close to the target correlation matrix $\mathbf{C}$. The corresponding dependent quantiles (or probabilities) $0 \leq p_{ij} \leq 1;\ i = 1, \ldots, n = 5$ and $j = 1, \ldots,$ $k = 2$ can be drawn from the standard Gaussian CDF as $p_{ij} = F(x^*_{ij})$.

2.2. Latin Hypercube Sampling

CLHS, which was proposed by McKay et al. [8], partitions each CDF of the sample of size (n) from k variables into n contiguous intervals. An independent random number from the uniform distribution $p_i \in [0,1], i = 1, \ldots, n$ is then selected from each interval, resulting in n random numbers for each of the k variables. The aforementioned n random numbers are then randomly combined without replacements to generate the ordered quantiles.

For example, consider that the sampling space is two-dimensional $(k = 2)$ and the required number of realizations is five $(n = 5)$. The elements of the following matrix $\mathbf{P}$ consist of the random numbers $[0,1]$ selected from each interval of the CDFs of two random variables. The matrix $\mathbf{R}$ contains the random permutations.

$$\mathbf{P} = \begin{bmatrix} 0.53 & 0.53 \\ 0.51 & 0.71 \\ 0.26 & 0.75 \\ 0.88 & 0.15 \\ 0.67 & 0.64 \end{bmatrix} \quad \mathbf{R} = \begin{bmatrix} 4 & 5 \\ 2 & 4 \\ 3 & 1 \\ 5 & 2 \\ 1 & 3 \end{bmatrix}$$

The ordered quantiles are then generated by

$$\mathbf{H} = \frac{1}{n}(\mathbf{R} - \mathbf{P}), \quad (3)$$

The pairwise elements of the following matrix $\mathbf{H}$ indicate a stratigraphic sampling design, that is, one realization from each row and each column is generated from the sampling space, as shown in Figure 2a.

$$\mathbf{H} = \begin{bmatrix} 0.69 & 0.89 \\ 0.30 & 0.66 \\ 0.55 & 0.05 \\ 0.82 & 0.37 \\ 0.06 & 0.47 \end{bmatrix}$$

Each element of the matrix **H** can then be mapped according to a target CDF as $x_{ij} = F^{-1}(h_{ij});\ i = 1, \ldots, n = 5$ and $j = 1, \ldots, k = 2$, as shown in Figure 2b.

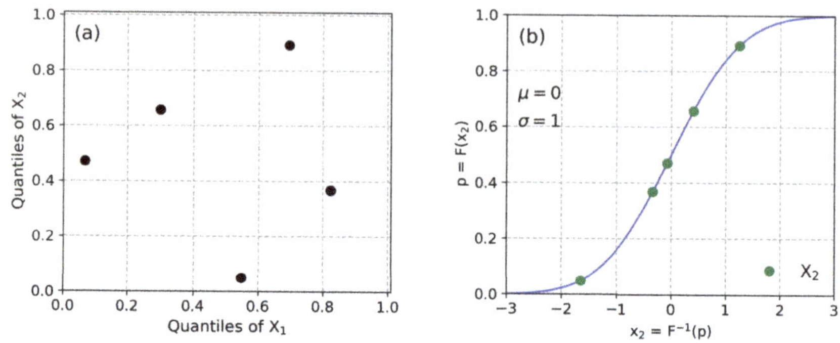

Figure 2. (**a**) A sampling design with $n = 5$ realizations generated by CLHS in a two-dimensional sampling space where the random variables are independent, (**b**) Mapping the quantiles according to the standard Gaussian distribution using the inverse transform method.

In the case of correlated random variables, the linear dependency can be added to the realizations using the Cholesky decomposition (Equation (1)).

As for maximin LHS, which is based on the distance-based criterion proposed by Johnson et al. [21], the realizations are generated by CLHS so that the minimum Euclidean distance between all of the realizations is maximized. Let w be the variable indicating the minimum distance between all of the realizations. The optimization problem can then be defined by

$$\begin{aligned} &\text{maximize } w \\ &\text{subject to } w \le \left\|\mathbf{h}_i - \mathbf{h}_j\right\|,\ (i,j) \in \mathcal{J} \\ &\qquad\qquad \mathbf{h}_i \in \mathcal{F},\ i = 1, \ldots, n, \end{aligned} \tag{4}$$

where $\mathcal{J} = \{(i,j) \mid 1 \le i < j \le n\}$, $\mathbf{h} = (\mathbf{h}_1^T, \ldots, \mathbf{h}_n^T)$ and $\mathbf{h}_i$ is the vector of coordinates for realization i in $\mathbb{R}^d$ with $d = 2$. Considering a large number of iterations, the realizations tend to be separated from each other, allowing for a more uniform space coverage. The algorithm steps of the maximin LHS are given as follows:

1. Set the initial value of the minimum Euclidean distance to zero, $w_{initial} = 0$.
2. Generate a sampling design $D_l, l = 1, \ldots, L$ using CLHS.
3. Calculate the minimum Euclidean distance w_l from the CLHS design D_l generated in step 2.
4. If $w_l > w_{initial}$, with $l = 1, \ldots, L$, set the new initial minimum Euclidean distance value as w_l, that is, $w_{initial} = w_l$.
5. Return to step 2 and repeat the steps L times.
6. End.

The algorithm steps given above are used to generate a maximin LHS design which takes into account a multidimensional uniformity through the maximization of the minimum Euclidean distance calculated from the predefined number of CLHS designs. In step 1, the initial value of the minimum Euclidean distance is equal to zero. In step 2, a CLHS design is generated. In step 3, the minimum Euclidean distance is calculated from that CLHS design, and step 4 checks if the minimum Euclidean distance calculated from the CLHS design is greater than the initial value of the Euclidean distance, which is zero. If so, the calculated minimum Euclidean distance is set as the new initial value. In step 5, the iteration is carried out a predefined number of times, and in step 6, the algorithm completes all of the iterations. A maximin LHS design where the minimum Euclidean distance is maximized considering 1000 iterations is shown in Figure 3a.

The LHSMDU algorithm proposed by Deutsch and Deutsch [1] combines CLHS with a realization elimination algorithm [28] in order to increase the multidimensional uniformity of the sampling matrix. Consider that $n = 5$ represents the required number of realizations and $k = 2$ is the number of random variables that are uncorrelated (or orthogonal to one another). A sampling design is generated by LHSMDU using the following steps for the algorithms:

1. Generate $k \cdot (m \cdot n)$ random numbers from a uniform distribution $U(0,1)$, where m is an integer greater than one and the common value of m is 5 (readers are referred to Section 3 in [1] on how an appropriate m value is selected.).
2. For each realization $i = 1, \ldots, (m \cdot n)$, calculate the Euclidean distance to other realizations and average the two smallest calculated distances.
3. For the realization i, save the average distance and return step 2 until all of the average distances are calculated for all of the realizations $i = 1, \ldots, (m \cdot n)$.
4. Remove the realization $(m \cdot n) = (m \cdot n) - 1$ for which the smallest Euclidean distance is calculated in step 2.
5. Return to step 2 and repeat the steps until the remaining number of realizations is equal to the number of realizations n that is selected initially, that is, $(m \cdot n) = n$.
6. For variable j, $j = 1, \ldots, k$, rank the n realizations and use these rankings as random permutations (or a stratum).
7. Generate random numbers $U(0,1)$ for the n number of strata.
8. Sample the CDF of the variable j using the random numbers generated in step 7.
9. Increment j, $(j = j + 1)$ and return to step 6 until the ranking and sampling are carried out for all k variables.
10. End.

In the case of correlated random variables, the linear dependency can be added to the realizations using the Cholesky decomposition (Equation (1)). A sampling design generated by LHSMDU with the m value equal to 5 is shown in Figure 3b.

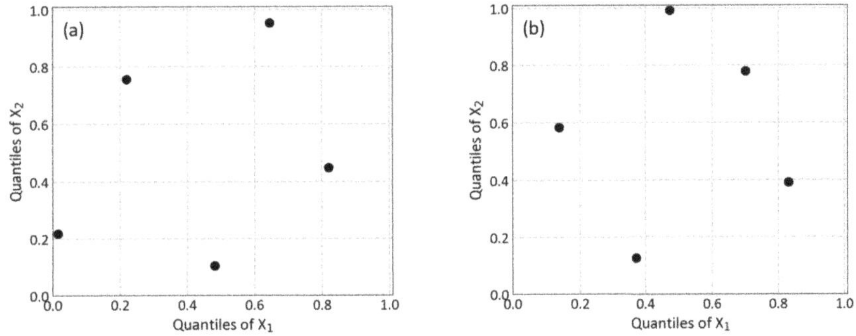

Figure 3. A sampling design with $n = 5$ realizations generated by (**a**) maximin LHS and (**b**) LHSMDU in a two-dimensional sampling space where the random variables are independent.

2.3. Projection Pursuit Multivariate Transform

Considering a two-dimensional sampling space ($\mathbb{R}^d$ where $d = 2$), a pattern (or a structure), such as clusters, outliers, and skewness, can be instantaneously discovered by simply observing the scatterplot. However, it is not possible to detect the aforementioned patterns when the sampling space is greater than three ($\mathbb{R}^d$ where $d > 3$). Projection pursuit, which was first proposed by [22] and first implemented by [23], can be used to detect these structures in the datasets defined in a high-dimensional sampling space. It makes use of a projection index $I(\mathbf{u}, \mathbf{v})$ that measures the degree of 'interestingness' of the data projected onto the plane spanned by the orthogonal vectors $(\mathbf{u}, \mathbf{v})$. The plane that maximizes the projection index is determined by numerical optimization. The dataset is generally transformed to normal scores and sphered in advance, that is, the transformed data has a mean of zero, a variance of one, and an identity correlation matrix. The projection index then measures the discrepancy between the distribution of the projected data and the standard Gaussian distribution $N(0,1)$.

For example, if the data is multivariate Gaussian, all of the projections are expected to be Gaussian and no 'interestingness' will be found. Also, as proved by [29], most projections of the multivariate data appear to be approximately Gaussian under appropriate conditions. If one considers that $I(\mathbf{u},\mathbf{v}) = 0$ is the perfectly Gaussian case, any projected data that has a non-Gaussian distribution will increase the value of $I(\mathbf{u},\mathbf{v})$, which indicates the 'interestingness' (or non-Gaussianity). One can also consider many other projection indices that measure the deviation from the standard Gaussian distribution [30–32]. We use the Fortran program called `PPMT.EXE`, which was proposed by [24] and is publicly available through the link http://www.ccgalberta.com/resources/select-software/, accessed on 20 September 2022, in order to demonstrate the generation of a sampling design by the PPMT technique. Let us consider again that the sampling space is two-dimensional, that is, $k = 2$ represents the number of random variables, and $n = 5$ is the number of realizations. The steps of the PPMT procedure for generating the required number of realizations $(k \cdot n)$ are summarized as follows:

1. Generate $(k \cdot n)$ random numbers from a uniform distribution $U(0,1)$ using MCS and establish these random numbers in a (5×2) matrix $\mathbf{P}$.
2. Transform the elements of matrix $\mathbf{P}$ to the standard Gaussian values, that is, $\mathbf{Y} = G^{-1}[\mathrm{F}(\mathbf{P})]$, where $G^{-1}[\mathrm{F}(\cdot)]$ is the normal score transform.
3. Compute the variance–covariance matrix of $\mathbf{Y}$, that is, $\boldsymbol{\Sigma} = (1/n)[\mathbf{Y}\mathbf{Y}^T]$.
4. Diagonalize $\boldsymbol{\Sigma}$, that is, $\boldsymbol{\Sigma} = \mathbf{Q}_1 \wedge_1 \mathbf{Q}_1^T$, where $\mathbf{Q}_1$ denotes an orthogonal matrix of the eigenvectors and $\wedge_1$ denotes the diagonal matrix of the eigenvalues.
5. Sphere the elements of matrix $\mathbf{Y}$; that is, $\mathbf{Y}' = \mathbf{S}^{-1/2}\mathbf{Y}$, where $\mathbf{S}^{-1/2} = \mathbf{Q}_1 \wedge_1^{-1/2} \mathbf{Q}_1^T$.
6. Project $\mathbf{Y}'$ onto k-dimensional unit length vector $\boldsymbol{\theta}$, that is, $\mathbf{p} = \boldsymbol{\theta}\mathbf{Y}'$.
7. Determine $\boldsymbol{\theta}$ maximizing the projection index $I(\boldsymbol{\theta})$ that measures the univariate non-Gaussianity.
8. Transform $\mathbf{Y}'$ to the standard Gaussian values $\hat{\mathbf{Y}}$ so that the projection $\hat{\mathbf{p}} = \boldsymbol{\theta}\hat{\mathbf{Y}}$ is univariate Gaussian. The steps for Gaussian transformation along a projection vector of $\mathbf{Y}'$ can be found in Barnett et al. [33].
9. Return to step 7 until the projection index $I(\boldsymbol{\theta})$ reaches convergence. The stopping criteria for the optimization can be found in [24].
10. Establish the final PPMT scores as a matrix $\hat{\mathbf{Y}}$ where $(\hat{y}_{ij}) \in \mathbb{R}^{5x2}$.
11. Draw the probabilities from the standard Gaussian distribution and establish them in a matrix $\mathbf{D} = F^{-1}[G(\hat{\mathbf{Y}})]$, where $(d_{ij}) \in \mathbb{R}^{5x2}$, where $\mathbf{D}$ indicates a PPMT sampling design.
12. End.

The steps given above generate a sampling design through the projection pursuit iterations. In step 1, the random realizations are generated using MCS. In steps 2–4, the realizations are transformed into the standard Gaussian values and its variance–covariance matrix is diagonalized. In step 5, the normalized realizations are sphered. The projection pursuit iterations are carried out in steps 6–9. In step 10, the final PPMT scores are generated, and in step 11, the probabilities $p_i \in [0,1]$, $i = 1,\ldots,n$ are drawn from the standard Gaussian distribution. The Cholesky decomposition (Equation (1)) can be used to impose the linear correlations (or target correlation matrix) among the independent variables. A flowchart indicating the steps of the aforementioned algorithm is presented in Figure 4.

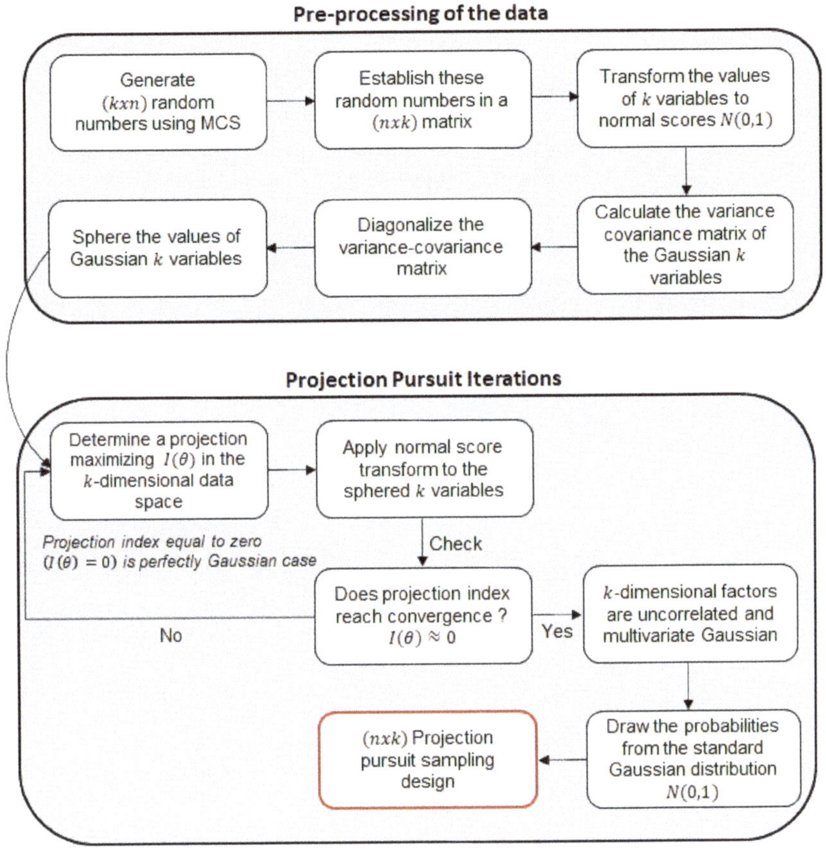

Figure 4. A flowchart indicating the steps for generating a sampling design using PPMT.

Considering the two-dimensional sampling space ($k = 2$) and ($n = 5$) realizations, a sampling design generated by PPMT mapping the probabilities according to the given CDF are shown in Figure 5a,b, respectively.

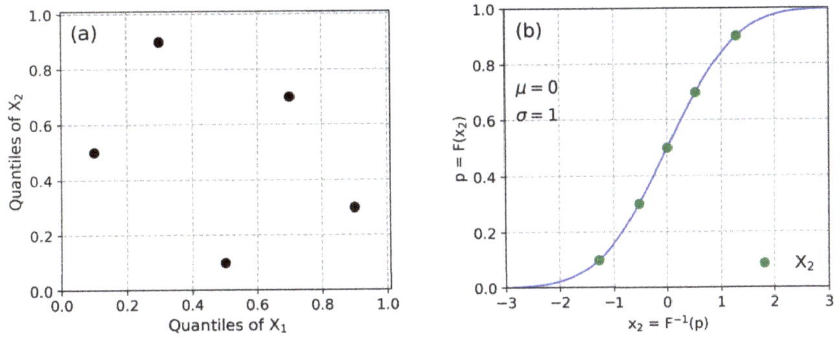

Figure 5. (**a**) A sampling design with $n = 5$ realizations generated by PPMT in a two-dimensional sampling space where the random variables are independent, (**b**) Mapping the quantiles according to the standard Gaussian distribution using the inverse transform method.

3. Case Studies

3.1. Synthetic Bivariate Case

In the first case study, we consider that the sampling space is two-dimensional ($k = 2$) and that 20 realizations ($n = 20$) of each random variable are generated using MCS, CLHS, maximin LHS, LHSMDU, and PPMT, considering the cases where the random variables are uncorrelated and correlated. It is noted that the values of the realizations that are generated are only in the range of $[0, 1]$, that is, $[0, 1]$ bounds are interpreted as the probability. To stabilize the distance measures between all of the realizations, 500 sets of the sample of $n = 20$ are generated using each sampling technique. To assess the quality of the sampling designs yielded by each technique, we use the Wraparound L2 (WL2) statistics [34] that measure the discrepancy between the number of design realizations per subvolume in comparison to the same number of uniformly distributed realizations across the sampling space.

$$WL2 = -\left(\frac{4}{3}\right)^p + \frac{1}{n^2}\sum_{i=1}^{n}\sum_{j=1}^{n}\prod_{k=1}^{p}\left(\frac{3}{2} - \left|z_i^k - z_j^k\right|\left(1 - \left|z_i^k - z_j^k\right|\right)\right), \tag{5}$$

where z_i^k and z_j^k are the elements of the vectors $\mathbf{z}^1, \ldots, \mathbf{z}^n$; $i, j = 1, \ldots, n$ denotes the number of realizations, and $k = 1, \ldots, p$ denotes the number of random variables.

We first consider the case where the two random variables are uncorrelated. The realizations are straightforwardly generated by MCS and CLHS. Considering maximin LHS and LHSMDU, the additional parameters required by the procedures include the number of iterations in maximin LHS and m value in LHSMDU, as explained in Section 2.2. We select the number of iterations as 1000 in the maximin LHS procedure and consider the m value to be equal to 5 in the LHSMDU procedure. As for PPMT, the entire procedure, as explained in Section 2.3, consists of generating realizations by MCS, mapping these realizations according to a standard Gaussian CDF, using these realizations to generate PPMT scores, and back-transforming the PPMT scores to the uniform distribution. Figure 6 shows four sets of sample with $n = 20$ realizations generated by MCS, CLHS, maximin LHS, LHSMDU, and PPMT.

The contours shown in Figure 6 are the probability contours of the multivariate Gaussian distribution. Because the random variables are independent, their covariance matrix is an identity matrix. Therefore, the probability contours, as shown in Figure 6, represent a circle shape.

We now consider the case where the random variables are positively correlated according to the following covariance matrix $\mathbf{C}$:

$$\mathbf{C} = \begin{bmatrix} 1 & 0.85 \\ 0.85 & 1 \end{bmatrix}$$

As can be seen from the elements of the matrix $\mathbf{C}$, the variances of both random variables are one and the strength of the linear relation between the random variables, as determined by the covariance (or correlation coefficient), is 0.85. The target correlation is imposed to the realizations of the independent random variables through the Cholesky decomposition (Equation (1)). Figure 7 shows the four sets of samples with $n = 20$ correlated realizations generated by each sampling technique.

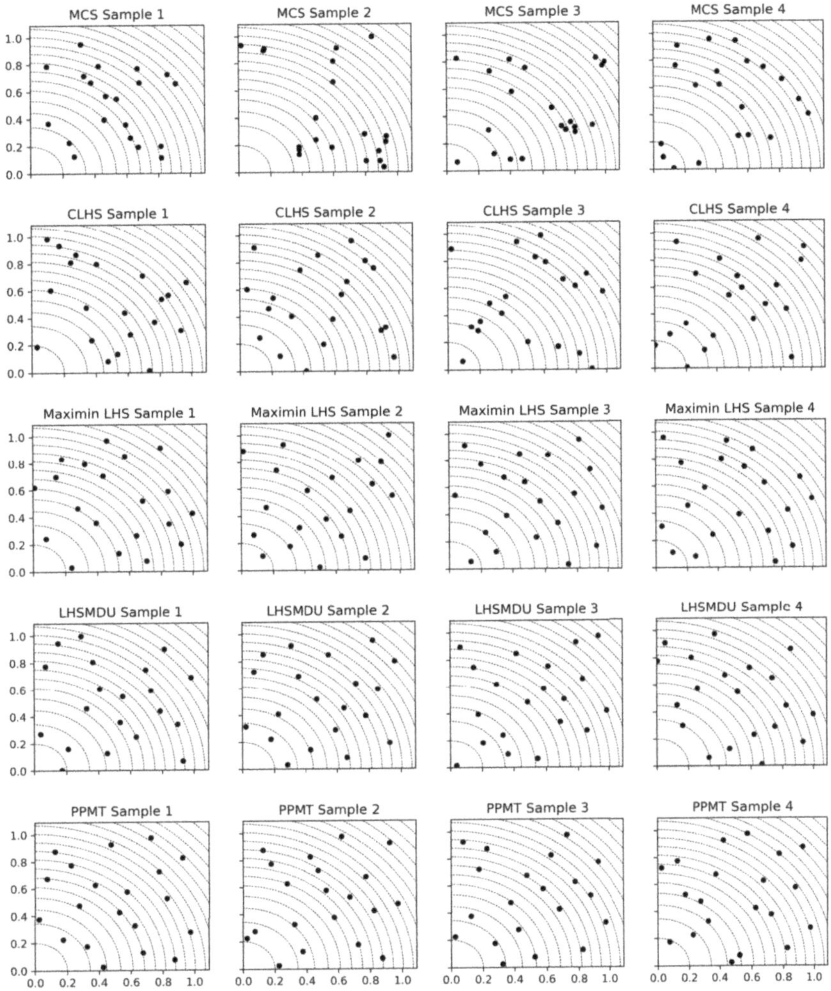

Figure 6. Four sets of samples with $n = 20$ realizations generated by each sampling technique in a two-dimensional sampling space where the random variables are independent.

The contours shown in Figure 7 represent the probability contours of the multivariate Gaussian distribution. Considering the target correlation matrix **C**, the probability contours exhibit an elliptical shape.

3.2. Synthetic Five-Variate Case

To further investigate the efficiency of each sampling technique, we present another case study where the sampling space is considered to be five-dimensional. We use the petroleum reservoir's OOIP as a variable to be sampled. The formula for calculating the OOIP is given as follows:

$$\mathrm{OOIP} = \mathrm{CAT} \cdot \mathrm{NTG} \cdot \phi_{net}(1 - \mathrm{S}_w), \tag{6}$$

where C is the constant that accounts for units and is considered to be one; A represents the area of the deposit; T is the thickness of the deposit; NTG is the net oil to gross volume; ϕ_{net} is the net porosity and S_w is the water saturation. As given in Equation (6), the value of OOIP is calculated as a function of five variables. To assess the quality of the sampling designs generated by each technique, the underlying (or truth) distribution of each OOIP variable is simulated using 10 million realizations generated by MCS. The selected distributions and their parameters for each variable are contained in Table 1.

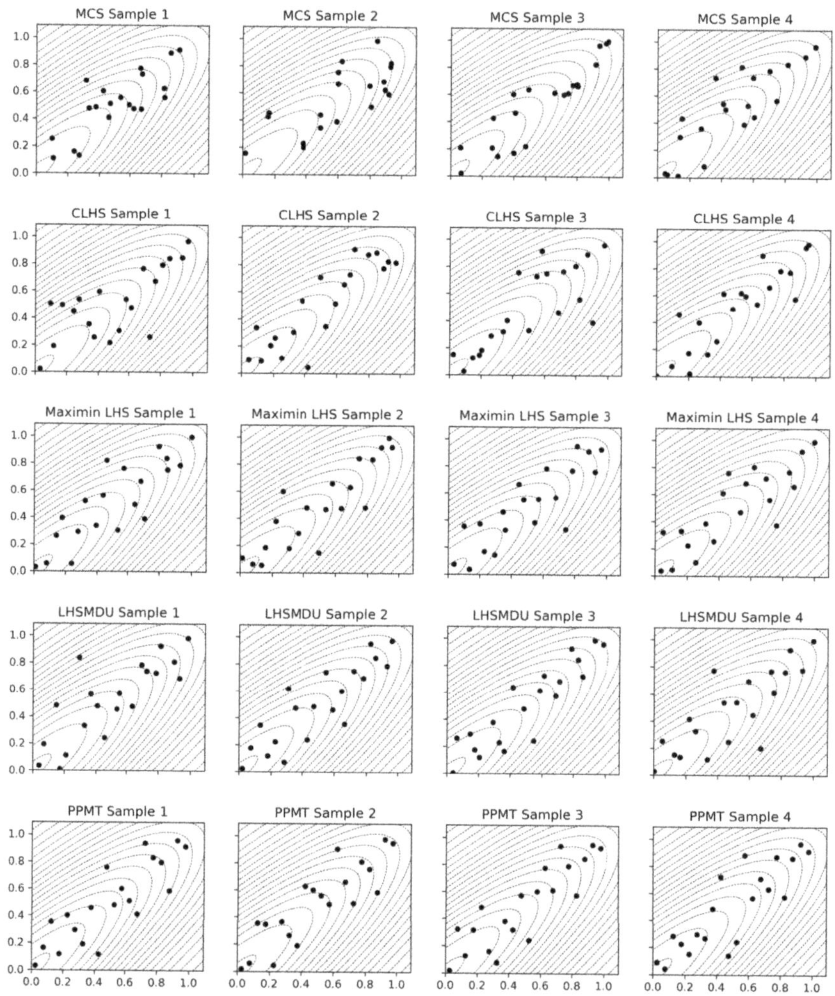

Figure 7. Four sets of samples with $n = 20$ realizations generated by each sampling technique in a two-dimensional sampling space where the random variables are positively correlated.

Table 1. The parametric distributions and their parameters used for each OOIP variable.

Variable	Distribution	Parameters
A	Triangular	$a = 2, b = 4, c = 6$
T	Gaussian	$m = 10, \sigma = 1$
NTG	Uniform	$a = 0.6, b = 0.8$
ϕ_{net}	Triangular	$a = 0.15, b = 0.25, c = 0.35$
S_w	Triangular	$a = 0.15, b = 0.2, c = 0.3$

To make the results comparable to the ones shown in the study presented by Deutsch and Deutsch [1], we use the same distribution types and their parameters given in that paper. In the first part of this case study, we consider that all of the OOIP variables are uncorrelated. Given the nature of sampling, it is clear that if one takes a larger sample of realizations, the accuracy of the parameter estimation will increase. Therefore, we consider several scenarios where the sample sizes (n) are relatively small ($10 \leq n \leq 100$), medium ($100 < n \leq 1000$), and large ($1000 < \text{n} \leq 10{,}000$). For each case, 100 sets of the predetermined sample size are generated using MCS, CLHS, maximin LHS, LHSMDU, and PPMT, that is, each of the OOIP variables is sampled from their underlying CDFs and the empirical OOIP CDF is then constructed for every sample size and every sampling technique. The efficiency of the sampling designs generated by each technique is assessed based on a criterion that is similar to the Kolmogorov–Smirnov D statistics [35], which is given as follows:

$$e = \max\left|F_{ref}^{-1}(p) - F_{emp}^{-1}(p)\right|, \tag{7}$$

where e is the error value indicating the maximum discrepancy between the empirical CDF generated by the particular sampling technique and the underlying CDF; $F_{ref}^{-1}(p)$; $p = 0.1, 0.2, \ldots, 0.9$ are the quantile values read from the underlying CDF; and $F_{emp}^{-1}(p)$; $p = 0.1, 0.2, \ldots, 0.9$ are the quantile values read from the empirical CDF. Figure 8 shows the underlying OOIP CDF and the randomly selected empirical CDFs of the OOIP generated by each sampling technique.

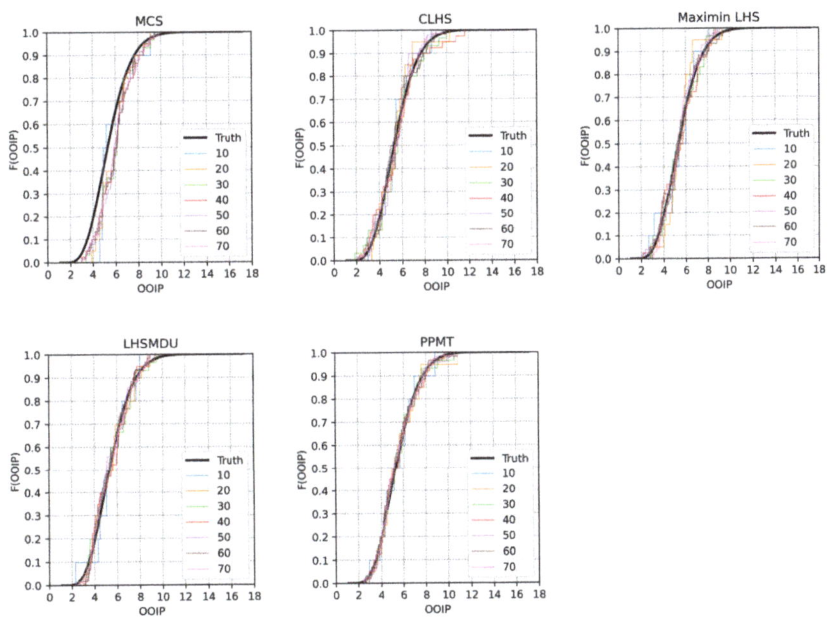

Figure 8. The underlying CDF of the OOIP in the case where the OOIP variables are uncorrelated and the empirical CDFs are generated by each sampling technique considering the small sample sizes.

It can be seen in Figure 8 that only the empirical CDFs of the OOIP generated based on the small sample sizes are shown; this is because the greatest discrepancy between the sampling techniques should be observed in cases where the sample size is small. As the sample size increases, one should expect that the average e values calculated for each sampling technique tend to get close to one another, which can be observed in Figure 9.

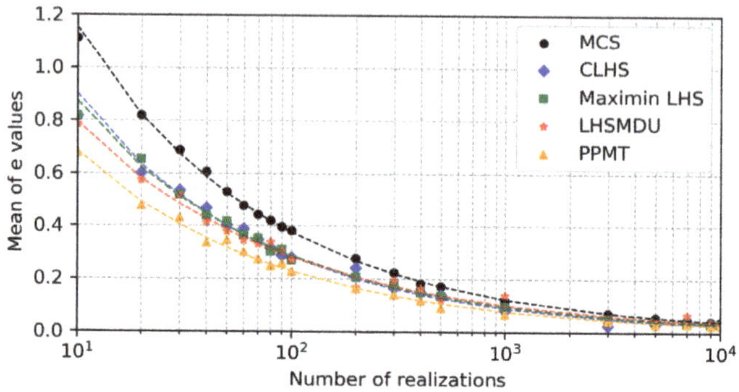

Figure 9. The averages of the *e* values versus the number of realizations generated by each sampling technique considering that the OOIP variables are uncorrelated.

The power law functions can be seen in Figure 9, which are in the form of the following equation:

$$e_{mean} = a \cdot L^b, \quad (8)$$

(where L denotes the number of realizations) are fitted to the pairwise points of the average e values and the number of realizations. It is noted that the logarithmic scale is used for the x-axis shown in Figure 9. The coefficients (a and b) of the power law functions, which are contained in Table 2, are calculated through the ordinary least squares regression.

Table 2. The coefficients of the power law functions in the case where the OOIP variables are uncorrelated.

Coefficients	MCS	CLHS	Maximin LHS	LHSMDU	PPMT
a	1.267	1.056	0.980	0.803	0.717
b	−0.488	−0.504	−0.484	−0.449	−0.477

It can be seen in Table 2 that the value of the coefficient b is approximately equal to −0.5 for all of the sampling techniques, the largest value of the coefficient a is obtained from the model fitted to the MCS case, and the smallest value of the coefficient a is computed from the model fitted to the PPMT case.

In the second part of this case study, we consider that the OOIP variables are somewhat correlated with one another through the following target correlation matrix **C**:

$$\mathbf{C} = \begin{bmatrix} 1 & 0 & 0 & 0 & 0 \\ 0 & 1 & 0.3 & 0.25 & -0.4 \\ 0 & 0.3 & 1 & 0.4 & -0.5 \\ 0 & 0.25 & 0.4 & 1 & -0.6 \\ 0 & -0.4 & -0.5 & -0.6 & 1 \end{bmatrix}$$

It is clear from the elements of the matrix **C** that the first variable (thickness, T) does not have any linear correlations with any of the remaining OOIP variables. The rest of the variables appear to be either positively or negatively correlated with one another. The Cholesky decomposition of the target correlation matrix **C** is carried out, and the independent realizations of the OOIP variables are correlated. Figure 10 shows the underlying OOIP CDF and the randomly selected empirical CDFs of the OOIP generated by each sampling technique.

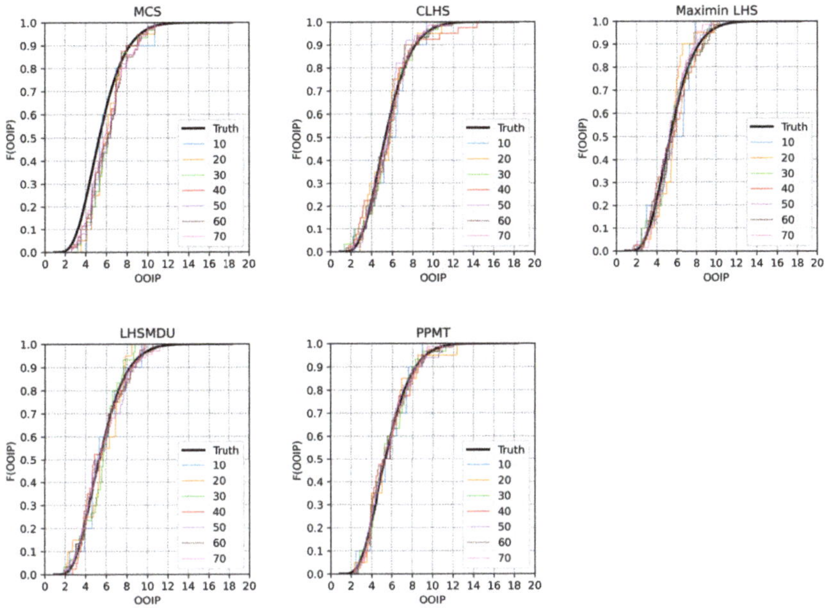

Figure 10. The underlying CDF of the OOIP in the case where the OOIP variables are correlated and the empirical CDFs are generated by each sampling technique considering the small sample sizes.

The quality of the sampling designs generated by each technique is again assessed according to the criterion given in Equation (7). Figure 11 shows the plot of the averages of the *e* values versus the number of realizations along with the fitted power law functions.

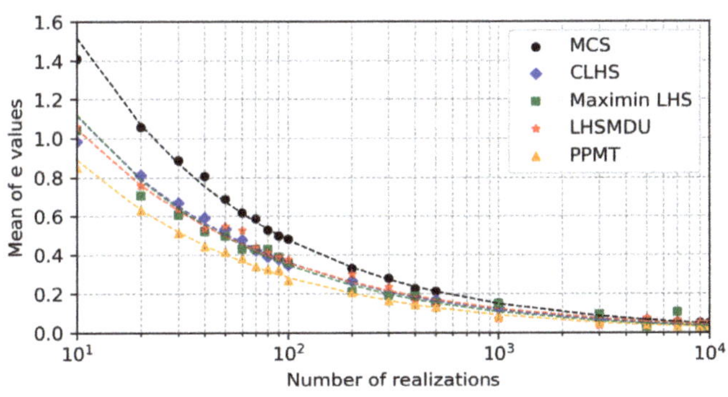

Figure 11. The averages of the *e* values versus the number of realizations generated by each sampling technique considering that the OOIP variables are correlated.

The coefficients of the fitted models for each sampling technique are given in Table 3.

Table 3. The coefficients of the power law functions in the case where the OOIP variables are correlated.

Coefficients	MCS	CLHS	Maximin LHS	LHSMDU	PPMT
a	1.577	1.297	1.244	1.118	1.028
b	−0.504	−0.492	−0.513	−0.466	−0.497

One can see in Table 3 that the values of the coefficient b are approximately equal to -0.5 for all sampling techniques, and the values of the coefficient a appear to be systematically higher than those estimated for the uncorrelated case shown in Table 2.

3.3. Quality Assessments of Sampling Designs

The quality of each design generated by the sampling techniques is visually inspected and numerically assessed according to the magnitude of the values of each criterion given in Equations (5) and (7). Considering the first case study where the sampling space is two-dimensional, one can see in Figures 6 and 7 that given the realizations generated by MCS, a large area of the sampling space is not investigated. This is mainly because MCS generates realizations completely at random; therefore, it is expected that a cluster of realizations is formed in the sampling space. The sampling designs generated by CLHS appear to be better than the ones generated by MCS. This is due to the fact that CLHS enforces the univariate uniformity of the realizations, that is, the realizations are drawn from each stratum, which guarantees that there is only one realization from each row and each column of the sampling space. However, no multidimensional uniformity is considered in CLHS. As for maximin LHS, LHSMDU, and PPMT through which one can take into account the multidimensional uniformity, the sampling designs shown in Figures 6 and 7 clearly indicate significant improvements in terms of the space-filling properties of the realizations.

In addition to the visual inspection of the designs, we also numerically assess the quality of each sampling design using $WL2$ statistics (Equation (5)). The box plots indicating the distribution of the aforementioned statistic considering the cases where the random variables are uncorrelated and correlated are presented in Figure 12.

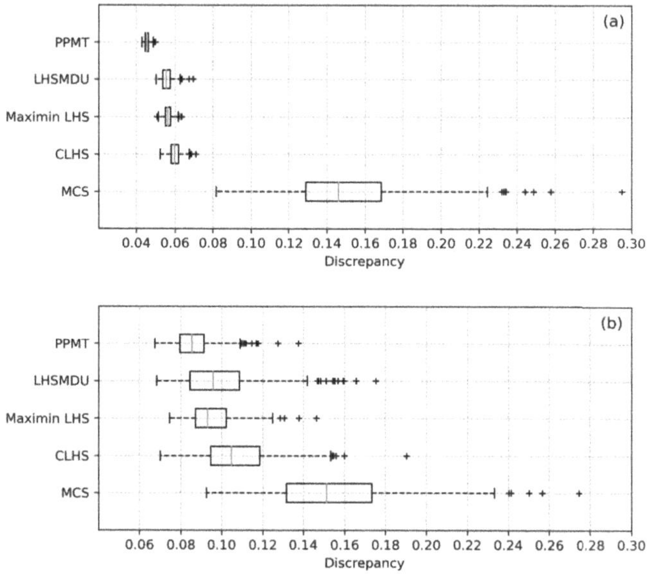

Figure 12. The box plots of the $WL2$ statistics: (**a**) uncorrelated case, and (**b**) correlated case.

It can clearly be seen in Figure 12a,b that MCS yields the largest discrepancy values according to the $WL2$ statistic, and PPMT, on the other hand, appears to outperform CLHS, maximin LHS, and LHSMDU. This is mainly due to the fact that the sampling designs generated by PPMT have no restrictions due to the Latin hypercube design. The median value of the discrepancy generated by MCS is approximately twice as much as the those generated by CLHS and its variants. The medians of the discrepancy values generated by maximin LHS and LHSMDU are rather close and slightly lower than that generated by CLHS.

Considering the second case study, one can see in Figures 8 and 10 that the empirical CDFs of the samples of varying sizes generated by PPMT appear to be closer to the underlying OOIP CDF than those generated by the other sampling techniques. It is again clear in the same figures that the most noticeable discrepancy between the empirical CDFs and the underlying OOIP CDF is in the MCS case. Similarly, Figures 9 and 11 clearly show that the fewest errors indicating the discrepancy between each empirical OOIP CDF and the underlying OOIP CDF are generated by PPMT. Considering the case where the random variables are correlated, the magnitudes of the error values appear to be greater for all of the sampling techniques than the uncorrelated case; however, PPMT outperforms the other sampling techniques in the correlated case as well.

In addition to the aforementioned comparisons, one should also know how many realizations should be generated by MCS to ensure a specified statistical accuracy. This can be calculated using the error values generated by each sampling technique in the power law function (Equation (8)) that is fitted to the MCS error values versus the number of realizations. Table 4 contains the approximate number of realizations that should be generated by MCS to ensure the same statistical accuracy achieved by the other sampling techniques.

Table 4. The equivalent number of realizations to be generated by MCS ensuring the specified sampling accuracy.

	MCS Equivalent Number of Realizations							
Reals #	**CLHS**		**Maximin LHS**		**LHSMDU**		**PPMT**	
10	20	**24**	22	**25**	26	**30**	30	**32**
100	195	**200**	199	**204**	202	**208**	281	**312**
1000	1631	**1657**	1690	**1703**	1699	**1715**	3398	**3401**
10,000	15,948	**16,144**	17,055	**17,899**	17,064	**17,956**	21,034	**21,945**

The approximate number of realizations that should be generated by MCS is calculated using the coefficients (Tables 2 and 3) of the power law functions fitted to the MCS case considering the cases where the random variables are deemed to be uncorrelated and correlated. For example, considering the uncorrelated case, the estimated coefficients ($a = 1.267$ and $b \approx -0.5$) of the power law function fitted to the MCS case are used along with the error value yielded by each sampling technique for each sample size. In other words, if we want to generate 10 realizations using CLHS, the corresponding error value is used in the power law function with the coefficients estimated for the MCS case and the equivalent number of realizations to be generated by MCS is calculated. The columns where the number of realizations are given in bold in Table 4 represent the case where the random variables are correlated. For example, considering 100 realizations, in order to maintain the same statistical accuracy observed in the PPMT case, one should generate 281 and 312 realizations using MCS in the uncorrelated and correlated cases, respectively.

4. Conclusions

The main objective of this paper was to introduce PPMT as an efficient and easily applicable tool for the assessment of parameter uncertainty of the models defined in the multidimensional sampling spaces. The study considered four other sampling techniques for comparison including MCS, which is a general technique for random sampling from a given distribution, CLHS, which is a stratified random sampling technique, and two variants, the (maximin LHS and LHSMDU) of the LHS technique. Two synthetic case studies where various sample sizes (ranging from $n = 10$ to $n =$ 10,000) were used considering two- and five-dimensional sampling spaces were conducted in order to assess the sampling performance of PPMT.

In the first case study, the sampling space was considered to be two-dimensional and $n = 20$ realizations of two random variables were generated 500 times. The visual and numerical results shown in Figures 6, 7, and 12 clearly indicated that in comparison to the other sampling techniques, PPMT appeared to be the best technique that enforces the multidimensional uniformity among the realizations defined in a two-dimensional sampling space. When the two random variables were correlated according to the target correlation matrix, PPMT again outperformed the other sampling techniques by yielding the fewest discrepancy values.

As for the second case study where the sampling space was considered to be five-dimensional, the results of the simulation study presented in Figures 8–11 clearly indicated that PPMT yielded the fewest sampling errors in comparison to all other sampling techniques in question and generated realizations whose empirical CDFs were rather close to the underlying CDF of the variable of interest (OOIP). The other important outcome of the simulation study, which is worth mentioning, is that the CPU capacity (or run time) required for the PPMT procedure was a lot less than those required for maximin LHS and LHSMDU. Therefore, PPMT can be considered an easily applicable technique that can be used for random sampling in Monte Carlo analysis. The Python code demonstrating the implementation of the PPMT as a sampling technique in the case of correlated random variables is available through https://github.com/Oktay-Erten/ppmt_parameter_uncertainty, accessed on 20 September 2022.

Author Contributions: Conceptualization, C.V.D.; methodology, C.V.D. and O.E.; software, O.E. and F.P.L.P.; writing—original draft preparation, O.E.; writing—review and editing, O.E. and F.P.L.P.; supervision, C.V.D. All authors have read and agreed to the published version of the manuscript.

Funding: This research received no external funding.

Institutional Review Board Statement: Not applicable.

Informed Consent Statement: Not applicable.

Data Availability Statement: Not applicable.

Acknowledgments: The authors thank the industrial sponsors of the Centre for Computational Geostatistics (CCG) for providing the resources to prepare this manuscript.

Conflicts of Interest: The authors declare no conflicts of interest.

Sample Availability: Not applicable.

Abbreviations

CDF	cumulative distribution function
CLHS	classic Latin hypercube sampling
CPU	central processing unit
LHS	Latin hypercube sampling
LHSMDU	Latin hypercube sampling with multidimensional uniformity
Maximin LHS	maximin Latin hypercube sampling
MCS	Monte Carlo simulation
OOIP	original oil in place
PPMT	projection pursuit multivariate transform
WL2	Wraparound L2

References

1. Deutsch, J.L.; Deutsch, C.V. Latin hypercube sampling with multidimensional uniformity. *J. Stat. Plan. Inference* **2012**, *142*, 763–772. [CrossRef]
2. Erten, O.; Deutsch, C. Bootstrap. In *Encyclopedia of Mathematical Geosciences*; Daya Sagar, B., Cheng, Q., McKinley, J., Agterberg, F., Eds.; Springer International Publishing: Cham, Switzerland, 2021; pp. 1–5.
3. James, F. Monte Carlo theory and practice. *Rep. Prog. Phys.* **1980**, *43*, 1145. [CrossRef]
4. Fishman, G. *Monte Carlo: Concepts, Algorithms, and Applications*; Springer Science & Business Media: Berlin, Germany, 2013.

5. Law, A.M.; Kelton, W.D.; Kelton, W.D. *Simulation Modeling and Analysis*; McGraw-Hill: New York, NY, USA, 2000; Volume 3.
6. Ortiz, J.C.; Deutsch, C.V. Testing pseudo-random number generators. In *Third Annual Report of the Centre for Computational Geostatistics*; University of Alberta: Edmonton, AB, Canada, 2001.
7. Khodadadian, A.; Taghizadeh, L.; Heitzinger, C. Optimal multilevel randomized quasi-Monte-Carlo method for the stochastic drift–diffusion-Poisson system. *Comput. Methods Appl. Mech. Eng.* **2018**, *329*, 480–497. [CrossRef]
8. McKay, M.D.; Beckman, R.J.; Conover, W.J. A comparison of three methods for selecting values of input variables in the analysis of output from a computer code. *Technometrics* **1979**, *21*, 239–245.
9. McKay, M.D. Latin hypercube sampling as a tool in uncertainty analysis of computer models. In Proceedings of the 24th Conference on Winter Simulation, Arlington, VA, USA, 13–16 December 1992; pp. 557–564.
10. Pebesma, E.J.; Heuvelink, G.B.M. Latin hypercube sampling of Gaussian random fields. *Technometrics* **1999**, *41*, 303–312. [CrossRef]
11. Jin, R.; Chen, W.; Sudjianto, A. An efficient algorithm for constructing optimal design of computer experiments. *J. Stat. Plan. Inference* **2005**, *134*, 268–287. [CrossRef]
12. Tang, B. Orthogonal Array-Based Latin Hypercubes. *J. Am. Stat. Assoc.* **1993**, *88*, 1392–1397. [CrossRef]
13. Tang, B. Selecting Latin hypercubes using correlation criteria. *Stat. Sin.* **1998**, *8*, 965–977.
14. Ye, K.Q.; Li, W.; Sudjianto, A. Algorithmic construction of optimal symmetric Latin hypercube designs. *J. Stat. Plan. Inference* **2000**, *90*, 145–159. [CrossRef]
15. Morris, M.D.; Mitchell, T.J. Exploratory designs for computational experiments. *J. Stat. Plan. Inference* **1995**, *43*, 381–402. [CrossRef]
16. Park, J.S. Optimal Latin-hypercube designs for computer experiments. *J. Stat. Plan. Inference* **1994**, *39*, 95–111. [CrossRef]
17. Iman, R.L.; Conover, W.J. A distribution-free approach to inducing rank correlation among input variables. *Commun. Stat.-Simul. Comput.* **1982**, *11*, 311–334. [CrossRef]
18. Olsson, A.M.J.; Sandberg, G.E. Latin hypercube sampling for stochastic finite element analysis. *J. Eng. Mech.* **2002**, *128*, 121–125. [CrossRef]
19. Owen, A.B. Controlling correlations in Latin hypercube samples. *J. Am. Stat. Assoc.* **1994**, *89*, 1517–1522. [CrossRef]
20. Stein, M. Large sample properties of simulations using Latin hypercube sampling. *Technometrics* **1987**, *29*, 143–151. [CrossRef]
21. Johnson, M.E.; Moore, L.M.; Ylvisaker, D. Minimax and maximin distance designs. *J. Stat. Plan. Inference* **1990**, *26*, 131–148. [CrossRef]
22. Kruskal, J.B. Toward a practical method which helps uncover the structure of a set of multivariate observations by finding the linear transformation which optimizes a new "index of condensation". In *Statistical Computation*; Elsevier: Amsterdam, The Netherlands, 1969; pp. 427–440.
23. Friedman, J.H.; Tukey, J.W. A projection pursuit algorithm for exploratory data analysis. *IEEE Trans. Comput.* **1974**, *100*, 881–890. [CrossRef]
24. Barnett, R.M.; Manchuk, J.G.; Deutsch, C.V. Projection Pursuit Multivariate Transform. *Math. Geosci.* **2014**, *46*, 337–359. [CrossRef]
25. Halton, J.H. A retrospective and prospective survey of the Monte Carlo method. *Siam Rev.* **1970**, *12*, 1–63. [CrossRef]
26. Watkins, D.S. *Fundamentals of Matrix Computations*; John Wiley & Sons: Hoboken, NJ, USA, 2004; Volume 64.
27. Deutsch, C.V.; Journel, A.G. *GSLIB Geostatistical Software Library and User's Guide*, 2nd ed.; Oxford University Press: New York, NY, USA, 1997; p. 369.
28. Deutsch, C.; Begg, S. The use of ranking to reduce the required number of realizations. Centre for Computational Geostatistics (CCG) Annual Report. 2001. Available online: http://www.ccgalberta.com/ccgresources/report03/2001-115_value_of_ranking.pdf (accessed on 20 September 2022).
29. Diaconis, P.; Freedman, D. Asymptotics of graphical projection pursuit. *Ann. Stat.* **1984**, *12*, 793–815. [CrossRef]
30. Cook, D.; Buja, A.; Cabrera, J.; Hurley, C. Grand tour and projection pursuit. *J. Comput. Graph. Stat.* **1995**, *4*, 155–172.
31. Hall, P. On polynomial-based projection indices for exploratory projection pursuit. *Ann. Stat.* **1989**, *17*, 589–605. [CrossRef]
32. Klinke, S. Exploratory Projection Pursuit: The Multivariate and Discrete Case. 1995. Available online: https://citeseerx.ist.psu.edu/viewdoc/download?doi=10.1.1.45.4509&rep=rep1&type=pdf (accessed on 20 September 2022).
33. Barnett, R.M.; Manchuk, J.G.; Deutsch, C.V. The Projection-Pursuit Multivariate Transform for Improved Continuous Variable Modeling. *SPE J.* **2016**, *21*, 2010–2026. [CrossRef]
34. Hickernell, F.J. Lattice Rules: How Well Do They Measure Up? In *Random and Quasi-Random Point Sets*; Hellekalek, P., Larcher, G., Eds.; Springer: New York, NY, USA, 1998; pp. 109–166.
35. Johnson, R.A.; Wichern, D.W. (Eds.) *Applied Multivariate Statistical Analysis*; Prentice-Hall, Inc.: Upper Saddle River, NJ, USA, 1988.

mathematics

Article

Visibility Adaptation in Ant Colony Optimization for Solving Traveling Salesman Problem

Abu Saleh Bin Shahadat [1], M. A. H. Akhand [1,*] and Md Abdus Samad Kamal [2,*]

[1] Department of Computer Science and Engineering, Khulna University of Engineering & Technology, Khulna 9203, Bangladesh; shahadat1407007@stud.kuet.ac.bd
[2] Graduate School of Science and Technology, Gunma University, Kiryu 376-8515, Japan
* Correspondence: akhand@cse.kuet.ac.bd (M.A.H.A.); maskamal@gunma-u.ac.jp (M.A.S.K.)

Abstract: Ant Colony Optimization (ACO) is a practical and well-studied bio-inspired algorithm to generate feasible solutions for combinatorial optimization problems such as the Traveling Salesman Problem (TSP). ACO is inspired by the foraging behavior of ants, where an ant selects the next city to visit according to the pheromone on the trail and the visibility heuristic (inverse of distance). ACO assigns higher heuristic desirability to the nearest city without considering the issue of returning to the initial city or starting point once all the cities are visited. This study proposes an improved ACO-based method, called ACO with Adaptive Visibility (ACOAV), which intelligently adopts a generalized formula of the visibility heuristic associated with the final destination city. ACOAV uses a new distance metric that includes proximity and eventual destination to select the next city. Including the destination in the metric reduces the tour cost because such adaptation helps to avoid using longer links while returning to the starting city. In addition, partial updates of individual solutions and 3-Opt local search operations are incorporated in the proposed ACOAV. ACOAV is evaluated on a suite of 35 benchmark TSP instances and rigorously compared with ACO. ACOAV generates better solutions for TSPs than ACO, while taking less computational time; such twofold achievements indicate the proficiency of the individual adoption techniques in ACOAV, especially in AV and partial solution update. The performance of ACOAV is also compared with the other ten state-of-the-art bio-inspired methods, including several ACO-based methods. From these evaluations, ACOAV is found as the best one for 29 TSP instances out of 35 instances; among those, optimal solutions have been achieved in 22 instances. Moreover, statistical tests comparing the performance revealed the significance of the proposed ACOAV over the considered bio-inspired methods.

Keywords: ant colony optimization; adaptive visibility; traveling salesman problem; partial solution update; 3-opt local search

MSC: 68Q07; 68R05; 68T20; 68W50

Citation: Shahadat, A.S.B.; Akhand, M.A.H.; Kamal, M.A.S. Visibility Adaptation in Ant Colony Optimization for Solving Traveling Salesman Problem. *Mathematics* **2022**, *10*, 2448. https://doi.org/10.3390/math10142448

Academic Editor: Ioannis G. Tsoulos

Received: 8 June 2022
Accepted: 11 July 2022
Published: 13 July 2022

1. Introduction

The Traveling Salesman Problem (TSP) is one of the challenging combinatorial optimization problems that attracted the bio-inspired research community. TSP falls in the class of combinatorial optimization problems with various practical applications [1]. In TSP, a traveler visits a set of cities and finally returns to the initial city, making a closed-loop tour [2]. A TSP instance contains a list of cities having geographical coordinates (i.e., x and y) of individual ones. The distance between two cities is generally the Euclidean distance, and tour cost is the total distance to be passed by the traveler for visiting all the cities [3,4]. The problem is to complete the tour with minimum cost. Due to the importance of optimal route generation issues in many real-life applications (e.g., vehicle routing problem, task scheduling, and logistics), TSP is the most studied optimization task.

Various techniques and methods have been investigated to solve TSP in the last few decades; among those, bio-inspired swarm intelligence (SI) based methods are the most popular ones. In SI, Ant Colony Optimization (ACO) is the pioneer method to solve TSP [5,6]. Later, Artificial Bee Colony (ABC) [7], Particle Swarm Optimization (PSO) [8], Cuckoo Search (CS) [9], and other SIs are used to solve TSP. An SI algorithm starts with an initial population, and then knowledge from different agents (i.e., members) is utilized to build a better solution. Recently, several ACO-based improved methods with modifications in individual ACO components have been investigated to solve TSP [10–13]. In [11], a parallel cooperative ACO (PACO) is proposed by splitting the ants into several groups to form multiple independent ant colonies. Each colony shares its best solution with the others to generate the overall best solution. The heterogeneous adaptive ACO (HAACO) [10] initializes the control parameters and employs the elitism method to replace the worst ant with the child of the best ant. In [13], a dynamic evaporation strategy of pheromone is adopted to enhance ACO, called DEACO, which clusters the cities and intelligently chooses the first city to find the shortest path. In addition, several hybrid methods by integrating ACO with other SI methods have also been proposed recently [14–16].

This study investigates an efficient bio-inspired ACO method, incorporating a new heuristic function and solution updating strategy with the conventional ACO for solving TSPs. In ACO, an ant selects the next city to visit according to the pheromone trail and the visibility heuristic [17]. ACO assigns a higher visibility heuristic to the nearest city without considering returning to the initial city or destination. Few studies identified that choosing the nearest city is not always a good decision [18,19]. A common scenario of ACO is that it got stuck to local optima, and its convergence speed is low [20]. Considering the above-mentioned issues regarding conventional ACO, a new ACO-based method is proposed in this study, adopting transition rules emphasizing distance to the returning city. Notably, recent ACO-based studies [10–12,14], focused on various issues without considering such an important feature that the ants have to return to the initial city. In the proposed ACO with adaptive visibility (ACOAV) method, partial updates of individual solutions and the 3-Opt algorithm are adopted to achieve better performance. The proposed method is tested on a set of benchmark TSPs, and it outperformed ACO and some other recent methods. The research contributions of adaptive visibility formulation and partial updating of individual solutions made the proposed ACOAV an effective TSP solving method.

The remaining paper is organized into four sections. Section 2 presents TSP basics and reviews recent methods to solve it. Section 3 describes the proposed method, including the formulation of the visibility heuristic function. Experimental results and performance comparison are provided in Section 4. Section 5 discusses major contributions and main achievements briefly. Finally, Section 6 gives a brief conclusion of the present study and directions for further work.

2. TSP and Recent Methods to Solve It

This section describes the traveling salesman problem (TSP) with its mathematical formulation and its importance. Next, a brief review of recent prominent TSP solving methods, including ACO-based techniques and other bio-inspired methods, is provided.

2.1. TSP and Its Importance

TSP is the problem of visiting a set of cities and returning to the initial city with the minimum cost. TSP can be represented by a complete weighted graph, $G = (V, E)$; vertex set $V = \{v_1, v_2, v_3, \ldots, v_N.\}$, here vertex v_i represents the ith city; edge set $E = \{(v_i, v_j) \mid 1 \leq i, j \leq N \text{ and } i \neq j\}$, where N is the number of vertices, and each edge represents the direct link between the cities v_i and v_j. Distance between two cities s and l is calculated using the following Euclidean distance formula.

$$d_{sl} = \sqrt{(x_s - x_l)^2 + (y_s - y_l)^2} \tag{1}$$

Suppose, $C = \{c_1, c_2, c_3, \ldots, c_N, c_1\}$ is a complete solution for a TSP of N cities, where $c_i \in V, 1 \leq i \leq N$. C is the order the traveler visits cities. Visiting each city exactly once and returning to the starting city makes a Hamiltonian loop in the graph. The sum of the distance of the edges used by the traveler is the total cost of the tour:

$$f(c) = \sum_{i=1}^{N-1} d_{c_i c_{i+1}} + d_{c_N c_1}. \tag{2}$$

TSP aims to complete the tour with minimum cost (i.e., the lowest value of Equation (2)). Therefore, the main task of TSP is to choose the city order to visit that gives a minimum tour length. In other words, TSP belongs to the class of combinatorial optimization problems to select N appropriate edges from $N(N-1)/2$ edges to complete the tour optimally with the minimum cost [2].

The significance of TSP is that it represents many practical applications [1]; a method for TSP may also be applied for solving other optimization problems [2]. It has practical applications in the traveling purchaser problem, vehicle routing problem [21], cellular manufacturing, frequency assignment problem, circuit wiring, clustering, job-shop scheduling, etc. [1,22]. The TSP model is also applicable in different cases such as X-ray crystallography, overhauling gas turbine engines, warehouse order-picking problems, data analysis in psychology, and wallpaper cutting [22]. TSP is intensively studied in theoretical computer science, operations research, and engineering [13,22]. TSPs are NP-hard and cannot be solved in polynomial time [2]. However, several metaheuristic methods, including SI methods, have been developed to handle TSP effectively.

2.2. *Solving TSP with ACO and Its Updated Models*

Ant colony optimization (ACO) is a pioneer SI method to solve TSP. ACO is developed based on the natural ants foraging for food interacting through deposited pheromone on the paths. In ACO, an ant selects the order of the cities according to a probability function considering pheromone level and distances between individual cities. Two control parameters α and β in ACO regulate the pheromone and visibility, respectively, to select an edge. A brief description of ACO is available in the coming section before presenting the proposed model. However, at a glance, the intensity of pheromone increases on the edges that are visited mostly by the ants, and finally, tours of individual ants converge to a single tour path. Though ACO discovers good solutions rapidly, it can easily be trapped into local optima [13]. Several studies investigated to increase searching diversity and overcome other limitations of ACO [23]. Some recently proposed modified ACO on solving TSP are discussed below.

The heterogeneous adaptive ACO (HAACO) method [10] adaptively initializes the control parameters (i.e., α and β) of the conventional ACO, resulting in homogenous behavior for early iterations and heterogeneous behavior at later iterations exhibited by ants. Ants give a higher preference towards the heuristic value (e.g., β = 10) over the pheromone trail (e.g., α = 0) at the initial phase. HAACO ensures retaining the fittest individual in the population through an elitism mechanism where the best ant undergoes mutation, and the offspring replaces the worst ant. Both uniform and Gaussian mutation operators are investigated and Gaussian mutation showed better performance than uniform mutation. It also considered the 3-Opt local search method to improve individual solutions. These mechanisms explore the search space and prevent converging to local optima.

The dynamic pheromone evaporation (DPE) strategy is adopted with ACO in [13] to propose DEACO. Dynamic pheromone effectiveness and dynamic pheromone convergence speed parameters are introduced for DPE; thus, the evaporation rate changes dynamically over iterations. Such a dynamic evaporation rate prevents premature convergence and reinforces search space discovery. DEACO considers a new objective function to determine the worth of the solutions, which can also be used for heuristic value calculation when the tour is incomplete. It also chooses the first city to find the better solution using a clustering method.

The parallel cooperative ACO (PACO) [11] operates multiple ant colonies in parallel and shares the local best and global best tours with other colonies to determine the globally optimum tour. PACO fragments the population of ants into subpopulations to mitigate the premature stagnation problem of the conventional single colony in the ACO algorithm. Every colony exchanges information in a master-slave paradigm, and the greedy information swap shows better performance over the other strategies. A best-fit mitigation interval is chosen through experiments to specify how often colonies exchange information. Each colony employs the 3-Opt algorithm after a fixed number of iterations to improve itself. This process stops on a pre-specified termination condition and returns the global best tour.

In the hybrid PSO-ACO method [14], PSO is used to optimize the control parameters. Initially, a small amount of pheromone is deposited on every edge, considering the number of ants and the number of cities. Ants complete their first tours considering only the visibility heuristic value. PSO considers ACO parameters to encode particles, and parameter values of individual ants are used to initialize the particles. After PSO operations, optimal parameters are sent to ACO for further actions, i.e., update pheromone. After the termination of hybrid operations of ACO and PSO, the 3-Opt algorithm is used to update the individual solutions. Some other recent hybrid methods integrating ACO with other metaheuristic methods are Slime Mold-Ant Colony Fusion Algorithm [24], ACO with Levy Flight [15], Density Peaks Clustering and ACO with K-Opt algorithm [16], Coordinating PSO, ACO, and K-Opt [25], ACO-based Memetic Algorithm with local search [26], and ACO with Immigrants Schemes [23].

2.3. Solving TSP with Other Prominent Bio-Inspired Methods

Similar to ACO, several bio-inspired algorithms are found efficient in solving TSP. Individual bio-inspired optimization algorithms are developed mimicking actions, interactions, or survivals of natural or living organisms. Among those, group behavior of swarm or natural organisms (such as bees, lions, spider monkeys, fish, cuckoos, etc.) is well studied to develop different Swarm Intelligent (SI) algorithms for different optimization tasks, including TSP [7,9,27–29]. This section briefly describes several prominent bio-inspired algorithms for solving TSP.

The Artificial Bee Colony (ABC) algorithm [30] is a well-known method for continuous optimization (e.g., function optimization), and Swap Sequence-based ABC (SSABC) [31] incorporates different types of swap operations and a 3-Opt algorithm to tackle TSP efficiently. The SSABC considered a group of eight different rules, and one is chosen based on the proposed probability mechanism to perform a swap operation for interaction among individual bees. There are three different functional bee types (i.e., employed, onlooker, and scout bees). Bees use the roulette wheel technique to select a rule and update the solution. 3-Opt algorithm is applied on stagnant solution i.e., not improved for a number of iterations.

Lion Swarm Optimization (LSO) is inspired by lion hunting behavior and its discrete version, called discrete LSO (DLSO) [32], is a recent algorithm for TSP. In DSLO, each lion represents an individual solution by discrete coding of TSP, and it uses a proposed order crossover mechanism for movement. DLSO divides the swarm into three categories (i.e., lion king, lioness, and lion cubs) and operates in parallel. A ring topology is used in DLSO to transfer relevant information. A complete 2-Opt (C2-Opt) local search is incorporated in DLSO to improve individual solutions.

Spider Monkey Optimization (SMO) is based on the Fission-Fusion style movement of spider monkeys during their food searching, and the Discrete SMO (DSMO) [33] is the modified one for TSP. Every spider monkey is assigned an initial random solution, and monkeys do different types of interactions to upgrade the solution quality. The monkey population is split into subgroups, and the monkey with the best solutions in the group is known as the local leader, and the monkey containing the overall best solution is the global leader. In six stages, spider monkeys continue to search for better food sources, i.e., TSP solutions.

Discrete Symbiotic Organisms Search (DSOS) [34] algorithm extends the continuous SOS algorithm for TSP. The SOS algorithm is a population-based metaheuristic inspired by the symbiotic interactions among different organisms in nature. In the initial phase, random organisms (i.e., solutions) are created in the ecosystem, and these organisms are adapted through different symbiosis strategies to increase fitness and chances of survival. Symbiosis is performed in three phases: mutualism, commensalism, and parasitism phases. A transformation methodology is proposed to generate a neighbor state by randomly swapping, inserting, and inverting the order of two cities.

A hybrid algorithm combining genetic algorithm (GA), multiagent reinforcement learning (MARL) heuristic, and nearest insertion into the convex hull (NICH) local search is investigated for TSP in [35]. The primary tour is constructed using the MARL algorithm, and some of the best solutions are given to the GA. GA improves the solutions by using varieties of crossover and mutation operators. Then the tours further improved using 2-Opt and NICH local search algorithms. Creating a convex hull for a partial tour, NICH manipulates the cities of the partial tour and the rest of the tour by the nearest insertion method.

Some other recent bio-inspired algorithms for TSP are Velocity Tentative PSO (VTPSO) [8], Discrete Cat Swarm Optimization (DCSO) [36], Discrete Grey Wolf Optimizer (DGWO) [37], Discrete Cuckoo Search (DCS) algorithm [9], ABC algorithm with variable degree of perturbation [38], and Whale Optimization Algorithm (WOA) [39].

3. ACO with Adaptive Visibility (ACOAV) for TSP

This section first discusses conventional ACO [5,17] and draws attention to its visibility heuristic to update it for solving TSP more efficiently. Next, the proposed ACOAV with a new solution updating strategy is explained in detail.

3.1. Review of Conventional ACO

An ant colony consists of a group of ants where each ant uses a metaheuristic-based approach to build a solution for TSP individually. Ants maintain indirect interaction among themselves through the deposited pheromone on the accessed routes [40]. An ant selects the order of the cities according to a probability function based on the pheromone on the path and visibility heuristic value. All the ants continue this process until converging to a single path. The rules of transition (ants' movement) and pheromone update are the key points in ACO.

In ACO, an ant on city s moves to city l according to the following probability function

$$P_{sl} = \frac{{\tau_{sl}}^{\alpha} \times \eta_{sl}^{\beta}}{\sum_{m \in CitiesToOrder} \tau_{sm}^{\alpha} \times \eta_{sm}^{\beta}}, \tag{3}$$

where $CitiesToOrder$ is the list of the remaining cities to visit, τ_{sl} is the pheromone value over the (s, l) edge, and n_{sl} is visibility heuristic value between the nodes; parameters α and β are empirical variables for controlling importance toward pheromone trail and visibility heuristic values, respectively. The visibility between the nodes is the inverse of the distance.

$$n_{sl} = \frac{1}{d_{sl}}, \tag{4}$$

where d_{sl} is the distance between city s and city l.

At every iteration, the pheromone trail is updated, considering individual solutions developed by the ants. Pheromone is deposited on the edges that the ant traveled. A part of the pheromone is ao considered to evaporate at a constant rate. The well-known pheromone trail updating formula [17] over the (s, l) edge is

$$\tau_{sl} = (1-\rho)\ \tau_{sl} + \sum_{i=1}^{PopSize} \Delta{\tau_{sl}}^{i}, \tag{5}$$

where ρ ($0 < \rho < 1$) is the pheromone trail evaporation rate, $\Delta\tau_{sl}{}^{i}$ is the amount of the deposited pheromone by ant i, which is calculated as of Equation (6).

$$\Delta\tau_{sl}{}^{i} = \begin{cases} \frac{1}{L^i}, & \text{if edge } (s,\ l) \text{ is used by ant } i \\ 0, & \text{otherwise.} \end{cases} \tag{6}$$

Here, L^i is the tour cost of ant i using Equation (2).

In ACO, the visibility heuristic (n_{sl}) used in Equation (3), simply inversing the distance (d_{sl}) using Equation (4), has limitations in solving TSP. ACO assigns higher heuristic desirability to the nearest city without considering the issue of returning to the initial city or the origin of the tour. Once the last city is visited, it completes the TSP tour returning to the initial city. Pheromone updates on the most visited paths converge individual solutions into a single solution. However, an initial fixed amount of pheromone on all the paths motivates the individual ants to choose the nearest city, which might be different (divert) from the optimal tour. Thus, a common scenario of ACO is that it got stuck to local optima, and its convergence speed is low [20]. Therefore, a different visibility heuristic considering the target city might be more realistic to solve TSP by generating a better solution, which is the main motivation of this study.

3.2. *Adaptive Visibility Integration to ACO for TSP*

This section proposes a new visibility heuristic to alter the visibility heuristic in conventional ACO (i.e., Equation (4)). A partial solution updating strategy with the new heuristic is explained for better solution development. The 3-Opt algorithm is also explained briefly, which is employed to overcome the stagnation problem. The proposed method is a population-based method, and initialization is the first common issue in it, similarly to conventional ACO and other bio-inspired algorithms. The major steps of the proposed method are explained first, and then the complete process is summarized in an algorithm.

3.2.1. Population Initialization

ACOAV starts with an initial population similarly to conventional ACO. A small amount of pheromone is deposited on each link at this initialization stage. Values of other parameters (e.g., evaporation rate) are also assigned in this stage. Each ant is initialized by a random tour taking cities to visit randomly, each city exactly once, and returns to the starting city. Ants are assigned constant relative importance towards pheromone and visibility heuristic value. These initial tours are updated over the iterations.

3.2.2. Adaptive Visibility (AV) Heuristic and Formulation

The motivation for developing a new AV technique is easily understandable for a case of optimal path selection from the start/source to the destination. Suppose there are four cities (C1, C2, C3, and C4) in Figure 1, and an ant is in city C1 and its destination C4 visiting C2, C3. Distances between the cities for path options are assigned with city indexes. As an example, d_{12} (=d_{21}) is the distance between C1 and C2. The tour cost will be different on the visiting priority between C2 and C3. There are two tour options: C1-C2-C3-C4 having the tour cost ($d_{12} + d_{23} + d_{34}$) and C1-C3-C2-C4 having tour costs ($d_{13} + d_{23} + d_{24}$), which are shown in Figure 1b,c, respectively. Their tour cost difference is

$$\Delta\text{TC} = (d_{12} + d_{23} + d_{34}) - (d_{13} + d_{23} + d_{24}) = (d_{12} - d_{24}) - (d_{13} - d_{34}). \tag{7}$$

The above formulation can be generalized based on the distances of the intended city, l from the current city, s and destination city, e by the following equation of adaptive distance (ad)

$$\text{adaptive distance, } ad_{sle} = d_{sl} - d_{le} \tag{8}$$

Equation (7) can be rewritten using the adaptive distance of Equation (8) as

$$\Delta TC = ad_{124} - ad_{134}, \tag{9}$$

where ad_{124}, ad_{134} are the adaptive distances of C2 and C3, considering C1 as the current city and C4 as the destination city. According to Equation (9), $\Delta TC < 0$, i.e., $ad_{124} < ad_{134}$ means the tour choosing C2 first (i.e., C1-C2-C3-C4) is better than the second one choosing C3 first (i.e., C1-C3-C2-C4). On the other hand, the second tour (choosing C3 first) is better than the first one if $\Delta TC > 0$, i.e., $ad_{124} > ad_{134}$. It concludes that choosing the city first with a lesser adaptive distance may lead to a shorter path. It is evident for producing the optimal tour that the next city should be selected considering an overall path distance regardless of the nearest issue. Thus, Equation (8) depicts an interesting hypothesis for choosing the next city, emphasizing the distance between the intended city and the destination city: prioritizing the city with a long distance from the destination city.

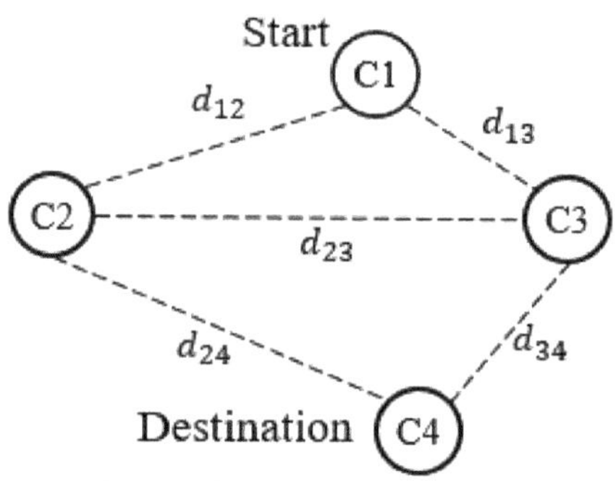

(**a**) Distances between the cities for path options are assigned with city indexes.

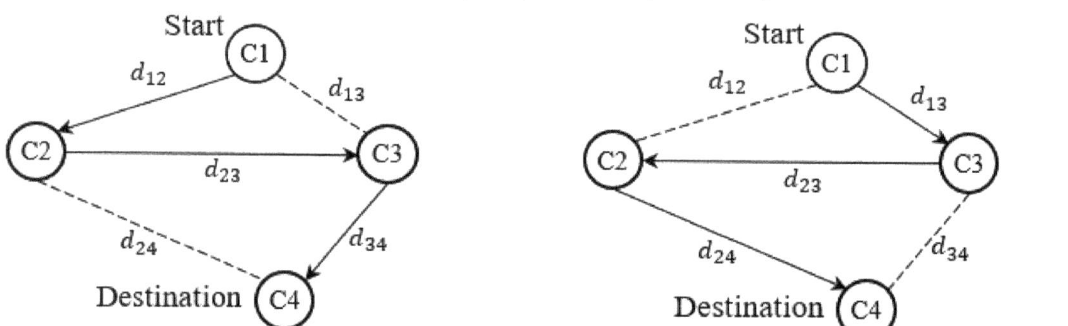

(**b**) Tour option C1-C2-C3-C4 having the tour cost ($d_{12} + d_{23} + d_{34}$) (**c**) Tour option C1-C3-C2-C4 having tour cost ($d_{13} + d_{23} + d_{24}$)

Figure 1. Adaptive visibility demonstration for optimal path selection from start city (C1) to destination city (C4) visiting C2 and C3. Tour option C1-C2-C3-C4 will be optimal when $(d_{12} - d_{24}) < (d_{13} - d_{34})$; otherwise, tour option C1-C3-C2-C4 will be optimal having $(d_{12} - d_{24}) > (d_{12} - d_{34})$.

Figure 2 illustrates the proposed AV technique in solving a sample small-sized TSP problem compared with conventional ACO. Suppose there are five cities (C1, C2, C3, C4, and C5), and the distances between the cities are mentioned in the graph in Figure 2a. An ant in city C1 will visit C2, C3, C4, C5 and return to city C1. The tour cost will be different depending on the visiting priority between the cities. If the ant chooses the nearest city first, the path will be C1-C2-C5-C4-C3-C1; the total route distance is 2000 = (100 + 300 + 500 + 400 + 700) with ACO visibility without considering the destination (Figure 2b). Considering C1 is both the current and destination city at the beginning and with AV (Equation (8)), the adaptive distances for all the cities are zero, and we are taking C2 as the next city same as ACO, to understand further effects of AV. After traveling from C1, C2 becomes the current city, and the destination city C1 remains the same for all the stages. Now the adaptive distances for cities C3, C4, and C5 are −200, 0, and 100, respectively. So, AV selects city C3 to visit next for its less adaptive distance (Equation (8)) and C3 becomes the current city. At the next stage, the adaptive distances for the remaining cities C4 and C5 are −200 and 500, respectively. Therefore, C4 is the next city to be visited, and finally, C5 will be visited as the only unvisited one. Thus, the AV technique chooses cities C3, C4, and C5 successively, and the final tour path becomes C1-C2-C3-C4-C5-C1. Figure 2c shows the tour path with AV having the tour cost of 1700 (=100 + 500 + 400 + 500 + 200), which holds a lower cost than the previous path.

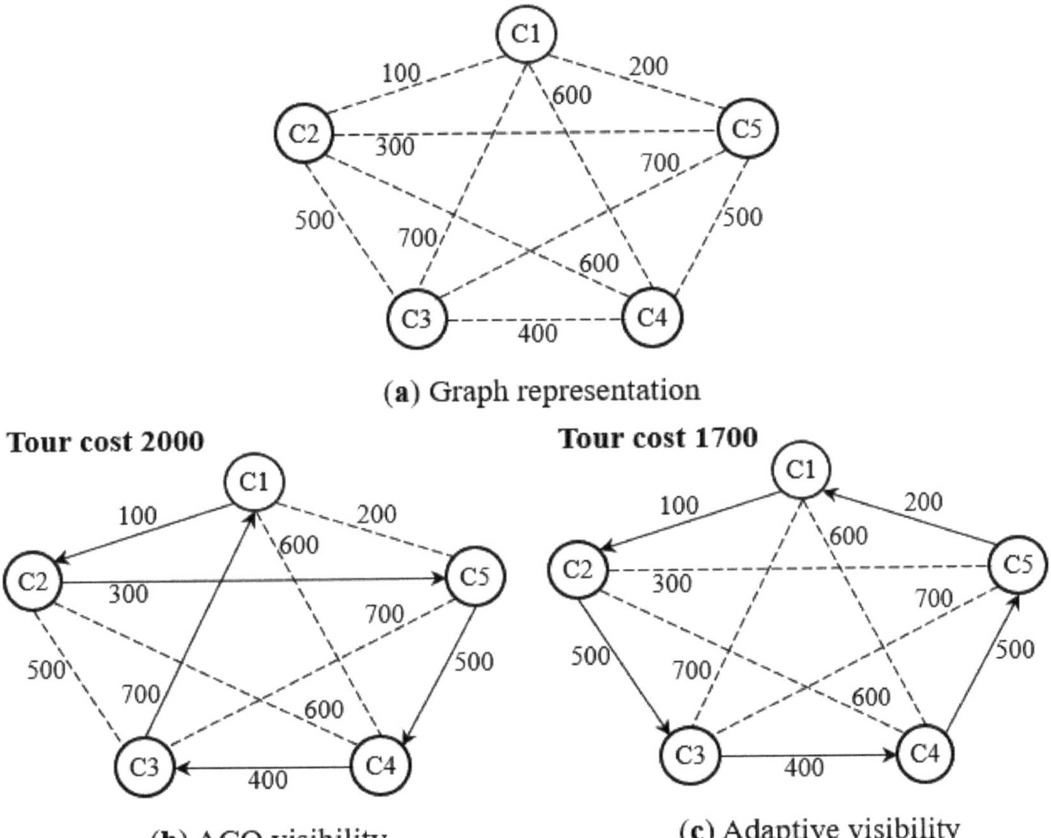

Figure 2. Demonstration of TSP solving with the proposed adaptive visibility compared with ACO visibility. Tour cost emphasizing nearest city for ACO is 2000 and tour cost with proposed adaptive visibility is 1700.

Based on the above demonstration in Figure 2, a generalized formula is developed for adaptive visibility. The AV formulation considers both the distances to the current city (s) and destination city (e) from the intended city (l). Based on distances d_{sl} and d_{le}, the weighted adaptive distance f_{sle} based on Equation (8) is

$$f_{sle} = w1 \times d_{sl} - w2 \times d_{le} - \min_{k \in CitiesToOrder}(w1 \times d_{sk} - w2 \times d_{ke}) + 1, \quad (10)$$

where $w1$ and $w2$ are weighting factors, denoting the respective importance of the distances on the scale from 0 to 1. Increasing $w1$ prioritizes the cities close to the current city, while increasing $w2$ prioritizes the cities far from the destination city. The first portion of Equation (10) is the simple extended version of Equation (8) employing weights $w1$ and $w2$, and it may produce a negative value that is not appropriate to place as a distance factor in calculating transition probability. The last portion of the equation $(-\min_{k \in CitiesToOrder}(w1 \times d_{sk} - w2 \times d_{ke}) + 1)$ is added to map the values into a positive form while maintaining rationality and a 1 for the lowest value for any case, even though the first portion gives a negative value. Suppose, from the current location, three values of the first portion for their intended cities (i.e., *CitiesToOrder*) are −200, 0, and 100. Thus, $\min_{k \in CitiesToOrder}(w1 \times d_{sk} - w2 \times d_{ke})$ returns the minimum value among the three values, and it is −200. Therefore, outcomes of Equation (10) for the three intended cases are 1, 201, and 301, those are appropriate as well as distinguishable for distance measure. Finally, the adaptive visibility (AV) value n_{sle} is calculated by inversing f_{sle} as

$$n_{sle} = \frac{1}{f_{sle}}. \tag{11}$$

Notably, the proposed weighted adaptive distance (Equation (10)) is a generalized case for distance calculation; for $w1 = 1$ and $w2 = 0$, AV is the same as ACO. In the proposed ACO with AV (ACOAV), transition probability, P_{sle} is calculated using the same probability function of ACO (Equation (3)) by putting deposited pheromone and adaptive visibility.

$$P_{sle} = \frac{{\tau_{sl}}^{\alpha} \times \eta_{sle}^{\beta}}{\sum_{m \in CitiesToOrder} \tau_{sm}^{\alpha} \times \eta_{sme}^{\beta}} \tag{12}$$

Finally, the ant chooses the next city to visit that gets the highest probability value from Equation (12) as

$$nextcity = argmax_{l:l \in CitiesToOrder} P_{sle}. \tag{13}$$

3.2.3. Partial Solution Update with AV

In each iteration, an ant in ACO updates its complete solution following Equation (13). ACOAV may also update the complete solution of individual ants, considering the starting city as the destination city. However, a portion of the tour is also possible to update; such a partial update is identified as an effective approach. In the partial update, the transition rule is applied for a selected portion of a tour by choosing two random numbers, $r1$ and $r2$ ($1 \leq r1, r2 \leq NoOfCities$), where cities at indexes $r1$ and $r2$ are considered as start and destination cities, respectively. For the transition rule with AV, an ant considers the start city as the current city *CitiesToOrder* contains the cities between $r1$ and $r2$ indexes. The current city is updated every time after visiting a city, *CitiesToOrder* is reduced by one until the ant reaches the destination city. In the case of $r1 > r2$, updating is performed considering a solution circularly, i.e., the update starts from $r1$, continues up to the end, and next continues from the starting to $r2$. The updated tour portion is deployed in the ant's solution to ensure improvement with the partial update if an improvement is observed in the selected portion through the AV-based operation. For such a partial update, ACOAV keeps a copy of the last tour of the previous iteration to update a random portion of it. In contrast, conventional ACO does not use the last tour since it generates a complete tour [18].

3.2.4. 3-Opt Algorithm Adaptation

The K-Opt algorithm is one of the best local search algorithms and has been considered in several recent TSP studies [10,11,14]. The 3-Opt algorithm, a special case of the K-Opt algorithm where K = 3 [41], has been adopted in the proposed ACOAV. It removes three edges of a tour and makes three sub-tours. These sub-tours are reversed and positionally changed to generate a new tour. Three sub-tours can be reconnected in eight different combinations. Suppose a complete tour is divided into three segments A, B, C. A-B-C is the original tour, and the other 3-Opt movements are A-B^r-C, A-B-C^r, A-B^r-C^r, A-C-B, A-C^r-B,

A-C-B^r, A-C^r-B^r, where B^r and C^r are the reversed tours of B and C segments. The tour is updated according to the combination, which has the lowest cost [12]. Embedding the 3-Opt algorithm helps to overcome stagnation problems and increase searching capability.

3.3. ACOAV Algorithm

Figure 3 shows the flowchart for the proposed algorithm with major operational steps already explained in the previous section. Algorithm 1 shows the complete pseudocode of the proposed ACOAV for solving TSP as described above. In the algorithm, solution update (Line 6) is the most crucial step and is described separately in Algorithm 2. Specifically, the notable difference between ACOAV and ACO is the implication of the proposed AV in the partial solution update mode. The 3-OptOperation (Line 8) on the individual solution is also an additional operation with respect to ACO.

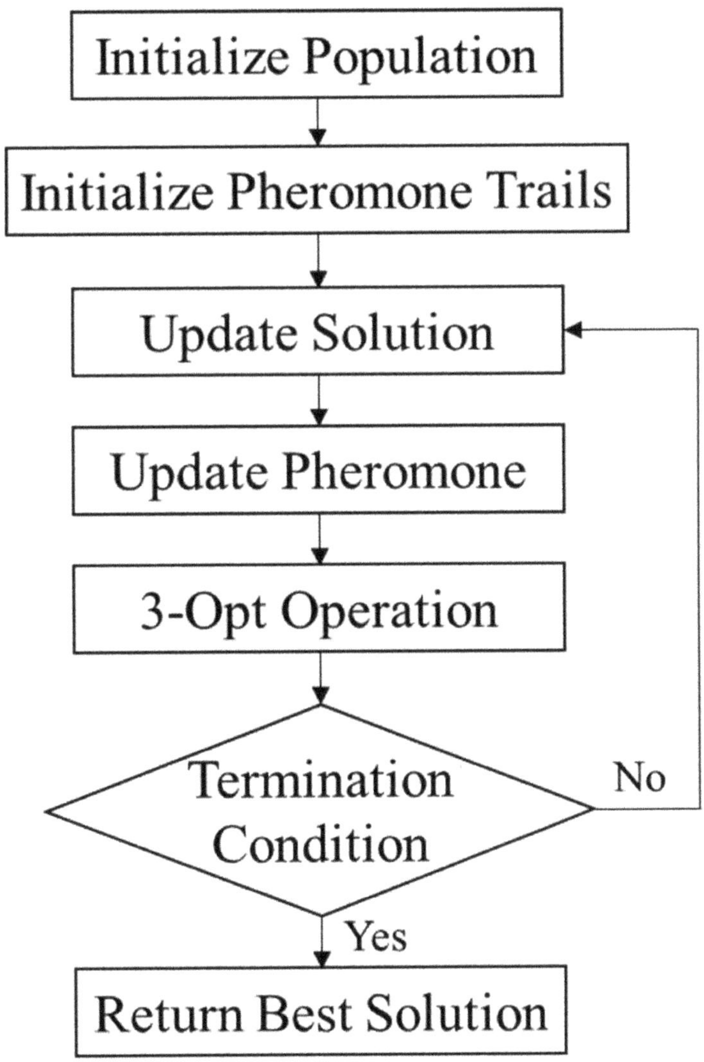

Figure 3. The flowchart of the proposed method.

Algorithm 1 ACOAV

```
t ← 1
Sol ← InitializePopulation(PopSize)
InitializePheromoneTrails(τ0)
While (t < MaxIt) do
    for i:1 to PopSize
        Sol^i ← UpdateSolution(Sol^i) // Algorithm 2
        UpdatePheromone(Sol^i) // Equation (5)
        Sol^i ← 3-OptOperation(Sol^i)
    end for
    t ← t +1
end while
BestSol ← FindBest (Sol)
End
```

Algorithm 2 UpdateSolution(Sol^i)//Partial Solution Update

```
TempSol ← Sol^i
r1 ← RandInt(1, NoOfCities), r2 ← RandInt(1, NoOfCities)
r ← r1 + 1
CitiesToOrder ← Ø
while (r != r2) do
    CitiesToOrder ← CitiesToOrder ∪ {TempSol.city[r]}
    if (r < NoOfCities)
        r ← r + 1 // Increase the index number by one
    else
        r ← 1//Reset the index number to start from the first visited city
    end if
end while
r ← r1 + 1
s ← TempSol.city[r1]
e ←TempSol.city[r2]
while (CitiesToOrder ≠ Ø or Null)
    TempSol.city[r] ← argmax_{l:l∈CitiesToOrder} P_sle //Equation (13)
    s ← TempSol.city[r]
    CitiesToOrder ← CitiesToOrder − {s}
    if (r < NoOfCities)
        r ← r + 1
    else
        r ← 1
    end if
end while
if (TempSol.Cost < Sol^i.Cost)
    Sol^i ← TempSol
end if
return Sol^i
```

4. Experimental Studies

This section describes the implementation details and verification of the effectiveness of ACOAV by comparing its performance with conventional ACO and several state-of-the-art methods, including the latest ACO-based methods.

4.1. Experimental Setup

For the proposed ACOAV, every edge is assigned an initial pheromone value ($\tau_0 = 1$), and control parameters are set as $\alpha = 1$, $\beta = 4$, $\rho = 0.1$, $w1 = 1$ and $w2 = 0.5$. The parameter values were set based on several trial runs. The algorithm is run for a maximum iteration (*MaxIt*) 1000, and results are taken from 20 independent trials. A suite of 35 TSP instances is taken from TSPLIB, which is a well-known TSP benchmark repository [3]. The size of the selected problems varies from 51 (i.e., eil51) to 2392 (i.e., pr2392), which brings a diverse test set. The tour length (TL) obtained from the solutions is compared to evaluate the quality of the solution. Note that a lower value of the TL indicates a better quality of the solution. Moreover, the error rate is calculated by comparing the tour length with the known optimal solution according to Equation (14).

$$\text{Error rate} = \frac{\text{Achieved Best Solution TL} - \text{Optimal Solution TL}}{\text{Optimal Solution TL}} \times 100\%. \tag{14}$$

A PC equipped with Intel(R) Core (TM) i5-4210U CPU (1.70–2.40 GHz) with 8.00 GB RAM and Windows 10 OS is used to conduct the experiments and analysis. The algorithm is implemented using C++ language in Visual Studio 2019 program.

4.2. Experimental Results and Performance Comparison

Performance comparisons of the ACOAV algorithm are presented on the benchmark TSP instances in two steps. At first, performance is compared with conventional ACO, and then the same is compared with several latest ACO-based methods and prominent bio-inspired methods. It is already mentioned that ACOAV transforms to conventional ACO when $w1 = 1$ and $w2 = 0$ and considers only full solution updates.

Figure 4 presents tour costs improvement through iterations for conventional ACO and proposed ACOAV on four selected problems: berlin52, eil76, kroA200, and pr439. In the experiments, ACOAV without 3-Opt on full solution update (FSU) and partial solution update (PSU) is also included to realize the effect of AV and PSU in solving TSPs. At a glance, ACOAV (FSU) is the implication of AV on conventional ACO, and ACOAV (PSU) is the implication of AV in PSU mode on conventional ACO. It is shown from Figure 4 that AV employment (on FSU or PSU) improves conventional ACO and the proposed ACOAV with 3-Opt significantly better than ACO for all four problems. As an example, for the berlin52 problem with 500 iterations, ACO and ACOAV achieved tour costs of 7989 and 7542, respectively. For the same problem, ACOAV (FSU) and ACOAV (PSU) show tour costs of 7777 and 7664, respectively. These outcomes reflect the performance improvement with PSU instead of FSU. The notion of performance for other problems is also similar.

The graphs in Figure 4 give an outline of the performance improvement over iterations. When comparing the graphs between ACO and ACOAV (FSU), it shows that the introduction of the proposed AV leads to faster convergence with lower tour costs than conventional ACO, for all four problems. An interesting observation is that FSU shows better performance than PSU at early iterations, and PSU leads to a better solution than FSU with iterations increasing. Such a phenomenon is understandable, as partial updates usually require several iterations to develop a complete solution sequence. A small improvement due to the 3-Opt operation in the proposed ACOAV is also visualized while comparing with ACOAV (PSU) for the problems. More importantly, the significant performance of the proposed method is realized by the graphs that proposed ACOAV and ACOAV (PSU or FSU) without 3-Opt converge faster and generate better quality solutions than ACO.

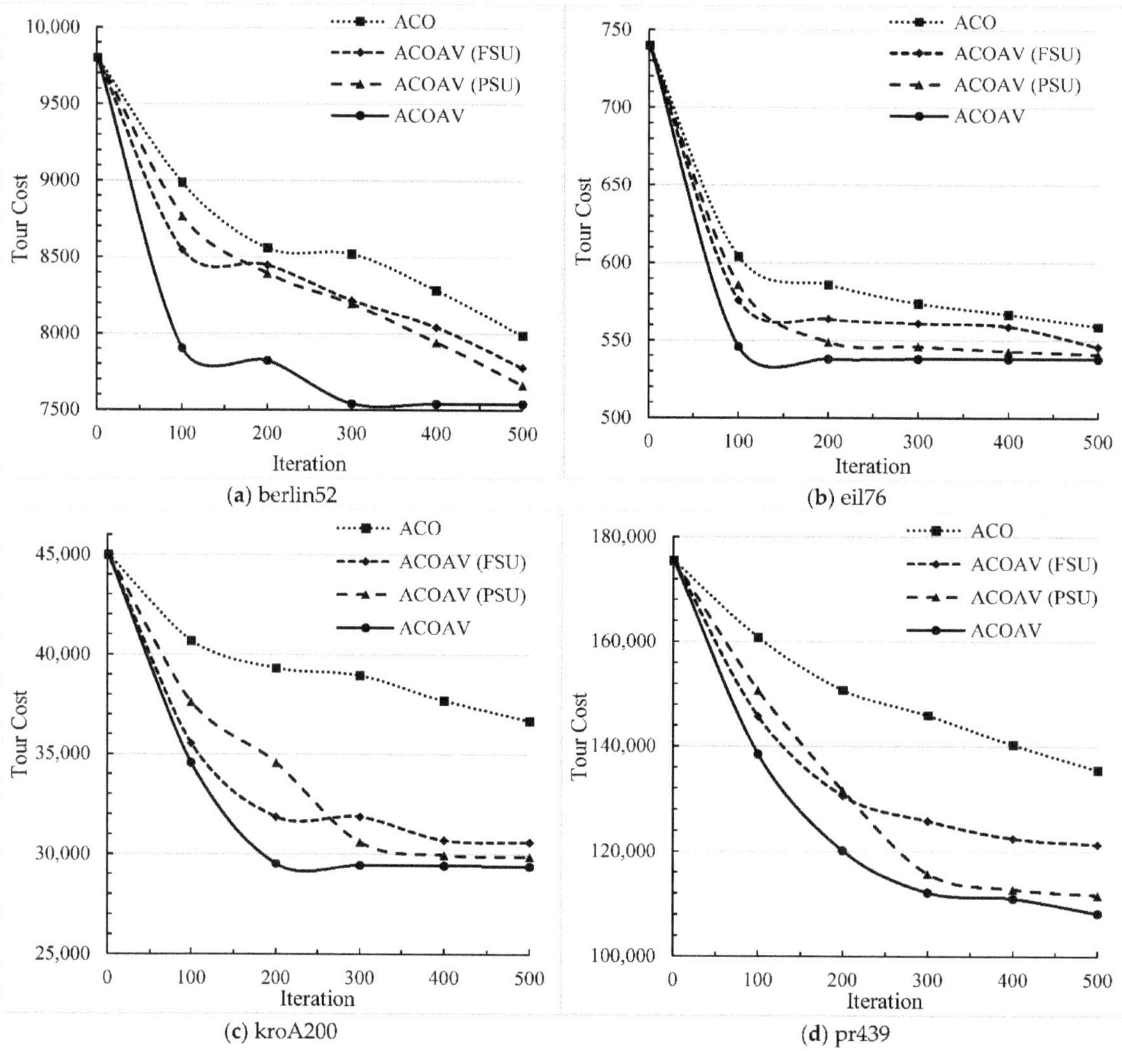

Figure 4. Tour cost improvement over iterations for conventional ACO, with different methods with adaptive visibility in different modes. ACOAV (FSU) is ACO + AV with full solution update mode, ACOAV (PSU) is ACO + AV with partial solution update mode, and ACOAV is the proposed model as ACOAV (PSU) + 3-Opt.

It might be interesting to observe the cost-effectiveness of the new features, i.e., AV with PSU and 3-Opt operation, which is introduced in ACOAV. Figure 5 shows the training time of the experiments on the four selected problems for the achieved tour costs presented in Figure 4. It is observed that ACOAV (FSU) took a little longer time than ACO as AV employment holds additional operations using Equation (10). As an example, for 500 iterations on the pr439 problem, ACO and ACOAV (FSU) took 647 and 763 s, respectively. Most interestingly, ACOAV (PSU) took much less time than ACO since updating a portion (in PSU mode) takes less time than updating the full solution as in ACO. For the same pr439 problem, ACOAV (PSU) took 254 s for 500 iterations, which is less than half of ACO. Finally, the 3-Opt operation in the proposed ACOAV is a computational overhead on ACOAV (PSU), and ACOAV took 366 s for 500 iterations on the pr439 problem. At a glance, the proposed ACOAV is more cost-effective than conventional ACO for any problem, and its efficiency for the larger problem is remarkable. For 500 iterations on the eil76 problem, ACO and ACOAV took 21 and 9 s, respectively. On the other hand, for 500 iterations on the kroA200 problem, ACO and ACOAV took 135 and 69 s, respectively. Such analysis though experimental outcomes, reveals the cost-effectiveness of individual components in the proposed ACOAV, especially AV and PSU.

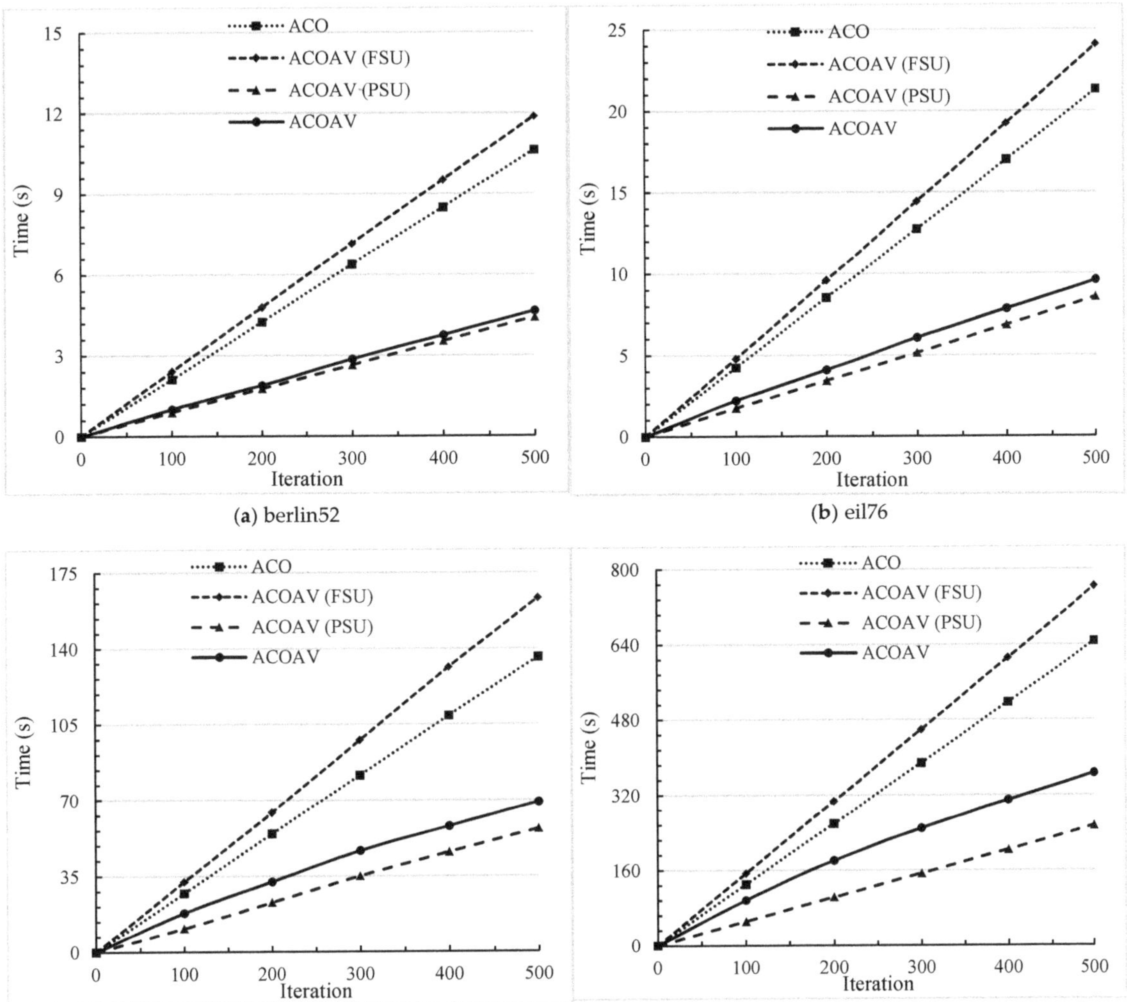

(**a**) berlin52 (**b**) eil76 (**c**) kroA200 (**d**) pr439

Figure 5. Process time (in the second) comparison over iterations for conventional ACO, with different methods with adaptive visibility in different modes. ACOAV (FSU) is ACO + AV with full solution update mode, ACOAV (PSU) is ACO + AV with partial solution update mode, and ACOAV is the proposed model as ACOAV (PSU) + 3-Opt.

Table 1 illustrates the detailed performance comparison among ACO, ACOAV (FSU), ACOAV (PSU), and proposed ACOAV. For a fair comparison, all four methods were run for the same number of iterations and trials. Comparison is performed based on achieved best and average tour costs among 20 independent runs. The table also contains the optimal tour length of individual problems, and the ER is presented following Equation (14). The comparison between ACO and ACOAV (FSU) illustrates that AV adaptation significantly improves the solution quality for TSP instances; ACOAV (FSU) is better than ACO for 30 TSP cases on best tour length, and it also achieved an average best TL for 33 TSP cases out of 35. FSU has a lesser standard deviation because of the first converging. Results comparison between ACOAV (FSU) and ACOAV (PSU) shows the effect of PSU in solving TSP instances. For all TSP instances, ACOAV (PSU) outperformed ACOAV (FSU) for all the cases based on the best and average TLs. The table shows that employing 3-Opt on ACOAV (PSU) leads ACOAV towards the optimal tour. The achieved optimal count and best count (among four methods) are given at the bottom of the table for overall performance comparison. ACO and ACOAV (FSU) cannot reach optimal or best tours for both best and average TLs. ACOAV (PSU) generates the optimal and best tour for only eil51 (the smallest problem). Whereas ACOAV generates the overall best solution for all TSP instances, and for 22 cases of TSP instances, it achieved optimal tour. Moreover, ACOAV has a lesser standard deviation in general which indicates better global optimization ability.

Table 2 compares the proposed ACOAV with several state-of-the-art bio-inspired methods based on the achieved best tour length and error rate respecting the optimal tour length. Among the considered methods, PSO-ACO [14], PACO [11], DEACO [13], and HAACO [10] are the recent ACO-based methods. At the same time, the other five methods are the prominent bio-inspired ones for TSP. Results presented for ACOAV are collected from Table 1, and outcomes of the compared methods are collected from the respective articles. Several case outcomes are not available for a particular method, which are marked in '-'. The optimal and best tour count is placed at the end of the table to understand the performance of the methods. Among the methods, DSMO [33] and DLSO [32] show the worst performance and cannot generate any optimal or overall best tour. PSO-ACO and PACO could generate the optimal and best tours for 4 and 6 TSP instances, respectively. SSABC [31] and HAACO show good performance, but tour cost for larger TSP instances is not available. Both of the algorithms generate optimal and best tours for 7 TSP instances. DSOS [34] shows optimal and best tours for 6 and 7 TSP instances. Among the compared methods, GA-MARL + NICH-LS [35] and DEACO are comparatively better than others and show optimal tours for 15 and 14 TSP instances, respectively. On the other hand, the proposed ACOAV shows the optimal solution for 22 cases. Among the ten methods in the table, ACOAV generates the overall best solution for 30 TSP instances while GA-MARL + NICH-LS and DEACO for 16 instances, respectively. At a glance, ACOAV reaches optimal tours for up to 200 city TSP problems and several larger problems, such as fl417 and pcb442.

Table 3 compares ACOAV with other state-of-the-art bio-inspired methods in terms of the average tour cost. The table also contains the optimal tour length of individual problems. Among the 35 TSP instances, DSMO, DLSO, and PSO-ACO failed to get the best tour for any TSP instances. SSABC, PACO, and HAACO generate the best tour for a small-sized TSP instance, berlin52. Among the compared methods, DEACO showed the highest number of the best tours in 16 cases and achieved the average TLs equal to the optimal TLs with zero standard deviation (SD) (i.e., all individual solutions reached the same optimal TLs) in 13 cases. On the other hand, ACOAV showed optimal values for 19 cases with zero SD for small-sized TSP instances. In the case of larger TSP instances, ACOAV also has lower SD values than other methods, which indicates better stability of the solutions generated by ACOAV. The comparison claims that the proposed ACOAV performs the best as it can show the best average tour costs for 31 TSP cases. Finally, the result compared in Tables 2 and 3 revealed ACOAV as a promising method for solving TSP instances.

Table 1. Performance Comparison among ACO, ACOAV (FSU), ACOAV (PSU), and ACOAV on 20 trials.

Sl.	TSP Instance	Optimal Tour Length	Best Tour Length (TL) and Error Rate (ER) Comparison								Average Tour Length (TL) and Standard Deviation (SD) Comparison							
			ACO		ACOAV (FSU)		ACOAV (PSU)		ACOAV		ACO		ACOAV (FSU)		ACOAV (PSU)		ACOAV	
			Best TL	ER (%)	Best TL	ER (%)	Best TL	ER (%)	Best TL	ER (%)	Avg. TL	SD	Avg. TL	SD	Avg. TL	SD	Avg. TL	SD
1	eil51	426	435	2.11	436	2.35	426	**0**	**426**	**0**	441.3	3.86	436.3	0.9	426	0	**426**	**0**
2	berlin52	7542	7689	1.95	7677	1.79	7569	0.36	**7542**	**0**	7795.8	81.61	7677	**0**	7659.25	20.7	**7542**	**0**
3	st70	675	696	3.11	710	5.19	683	1.19	**675**	**0**	712.75	6.11	716	3.46	683.9	0.3	**675**	**0**
4	eil76	538	548	1.86	543	0.93	541	0.56	**538**	**0**	559	4.25	556.85	2.48	542.05	1.36	**538**	**0**
5	pr76	108,159	119,176	10.19	115,930	7.18	110,465	2.13	**108,159**	**0**	126,121	2784.97	115,930	**0**	110,928	201.09	**108,159**	**0**
6	rat99	1211	1264	4.38	1274	5.2	1244	2.73	**1211**	**0**	1297.4	16.9	1285.1	9.41	1244.5	0.97	**1211**	**0**
7	kroA100	21,282	24,210	13.76	22,788	7.08	21,521	1.12	**21,282**	**0**	24,904.4	334.39	22,788	**0**	21,585.6	14.82	**21,282**	**0**
8	kroB100	22,141	25,191	13.78	23,852	7.73	22,416	1.24	**22,141**	**0**	25,847.7	298.76	23,852	**0**	22,488.5	38.92	**22,141**	**0**
9	rd100	7910	8547	8.05	8556	8.17	7943	0.42	**7910**	**0**	8880.95	148.21	8556	**0**	7974.65	22.54	**7910**	**0**
10	eil101	629	682	8.43	652	3.66	636	1.11	**629**	**0**	692.95	7.05	653.65	1.31	637.2	0.51	**629**	**0**
11	lin105	14,379	15,714	9.28	14,803	2.95	14,549	1.18	**14,379**	**0**	16310	197.1	14,803	**0**	14,601.6	16.56	**14,379**	**0**
12	pr107	44,303	48,512	9.5	50,356	13.66	44,566	0.59	**44,303**	**0**	49,129.8	359.13	50,356	**0**	44,897.9	137.2	**44,303**	**0**
13	pr124	59,030	66,702	13	62,977	6.69	59,990	1.63	**59,030**	**0**	68,970.2	910.83	62,977	**0**	60,049.8	22.11	**59,030**	**0**
14	ch130	6110	6859	12.26	6581	7.71	6204	1.54	**6110**	**0**	7012.3	77.03	6581	**0**	6234.8	13.83	**6110**	**0**
15	ch150	6528	7236	10.85	6941	6.33	6621	1.42	**6528**	**0**	7421.5	100.03	6962.45	29.34	6630.2	5.47	**6528**	**0**
16	kroA150	26,524	31,857	20.11	29,330	10.58	27,092	2.14	**26,524**	**0**	33,233.9	503.85	29,330	**0**	27,168.5	37.65	**26,524**	**0**
17	kroB150	26,130	31,262	19.64	28,386	8.63	26,537	1.56	**26,130**	**0**	33011.1	504.53	28,386	**0**	26,736.5	79.45	**26,130**	**0**
18	rat195	2323	2524	8.65	2428	4.52	2362	1.68	**2326**	**0.13**	2582.2	32.64	2430.9	4.09	2371.4	4.19	**2330.2**	**1.6**
19	d198	15,780	18,503	17.26	16,961	7.48	15,997	1.38	**15,780**	**0**	19,023.9	202.59	17,136.5	88.53	16,111.5	47.25	**15,780**	**0**
20	kroA200	29,368	36,628	24.72	31,866	8.51	29,725	1.22	**29,368**	**0**	38,404.1	639.51	31,886.7	49.28	29,830.5	56.59	**29,368**	**0**
21	kroB200	29,437	37,305	26.73	32,304	9.74	29,857	1.43	**29,438**	**0.003**	38,516.3	605.58	32,304	**0**	30,155.8	112.8	**29,439.5**	0.67
22	tsp225	3861	4592	18.93	4208	8.99	3997	3.52	**3923**	**1.61**	4733.4	61.17	4208	**0**	4047.15	15.41	**3956.9**	11.62
23	pr226	80,369	96,539	20.12	86,109	7.14	80,993	0.78	**80,369**	**0**	102,238	1712.21	86,127.4	**9.2**	81,289.2	71.16	**80,369.6**	**1.43**
24	a280	2579	3089	19.78	2778	7.72	2656	2.99	**2581**	**0.08**	3241.75	46.44	2816.8	17.1	2666.8	8.53	**2589.4**	**3.83**

Table 1. *Cont.*

Sl.	TSP Instance	Optimal Tour Length	Best Tour Length (TL) and Error Rate (ER) Comparison								Average Tour Length (TL) and Standard Deviation (SD) Comparison							
			ACO		ACOAV (FSU)		ACOAV (PSU)		ACOAV		ACO		ACOAV (FSU)		ACOAV (PSU)		ACOAV	
			Best TL	ER (%)	Best TL	ER (%)	Best TL	ER (%)	Best TL	ER (%)	Avg. TL	SD	Avg. TL	SD	Avg. TL	SD	Avg. TL	SD
25	pr299	48,191	63,759	32.3	54,470	13.03	50,159	4.08	**48,215**	**0.05**	67,450.6	1132.05	54,470.3	**0.48**	50,458.7	146.52	**48,519.3**	70.16
26	lin318	42,029	57,586	37.01	44,892	6.81	43,079	2.5	**42,203**	**0.41**	58,481.8	514.22	44,892	**0**	43243.2	76.5	**42,220.9**	9.99
27	rd400	15,281	20,664	35.23	16,899	10.59	15,973	4.53	**15,467**	**1.22**	21,210.2	199.2	16,899	**0**	16,055.5	42.85	**15,576**	44.23
28	fl417	11,861	14,370	21.15	12,948	9.16	11,921	0.51	**11,861**	**0**	14,725.4	149.15	12,948	**0**	11,937	8.23	**11,861.2**	0.4
29	pr439	107,217	135,080	25.99	121,360	13.19	111,024	3.55	**107,613**	**0.37**	139,745	1847.98	121,360	**0**	111,490	251.24	**107,965**	182.42
30	pcb442	50,778	72,682	43.14	56,991	12.24	53,833	6.02	**50,778**	**0**	75,440.8	1069.75	56,991.8	**3.49**	54,150.7	171.18	**50,945.2**	54.98
31	rat575	6773	9012	33.06	7344	8.43	7118	5.09	**6935**	**2.39**	9213.35	85.31	7345.85	**3.6**	7141.95	11.87	**6972.25**	15.73
32	rat783	8806	12,286	39.52	9712	10.29	9407	6.82	**9050**	**2.77**	12,554.5	119.38	9712.8	**0.98**	9451.45	21.13	**9101.1**	20.77
33	pr1002	259,045	371,087	43.25	300,757	16.1	284,985	10.01	**266,155**	**2.74**	374,401	2224.55	300,972	**214.5**	288,991	1397.71	**268,111**	678.67
34	fl1400	20,127	29,486	46.5	22,673	12.65	21,157	5.12	**20,215**	**0.44**	30,313.1	395.21	22,799.7	128.91	21,271.4	92.34	**20,226.7**	**5.39**
35	pr2392	378,032	581,878	53.92	439,849	16.35	428,523	13.36	**392,461**	**3.82**	588,396	3977.43	440,033	**140.07**	431,504	1610.47	**397,871**	1805.49
	Optimal/Best Count		0/0		0/0		1/1		**22/35**		0/0		0/0		1/1		**19/35**	
	Win-Draw-Loss over ACO		-		30-0-5		35-0-0		**35-0-0**		-		33-0-2		35-0-0		**35-0-0**	
	Win-Draw-Loss over ACOAV (FSU)		5-0-30		-		35-0-0		**35-0-0**		2-0-33		-		35-0-0		**35-0-0**	
	Win-Draw-Loss over ACOAV (PSU)		0-0-35		0-0-35		-		**34-1-0**		0-0-35		0-0-35		-		**34-1-0**	

Table 2. Best Tour Length (TL) and Error Rate (ER) Comparison of ACOAV with Existing State-of-the-Art Bio-inspired Methods.

Sl.	TSP Instance	Optimal Tour Length	GA-MARL + NICH-LS [35]		DSOS [34]		SSABC [31]		DSMO [33]		DLSO [32]		PSO-ACO [14]		PACO [11]		DEACO [13]		HAACO [10]		Proposed ACOAV	
			Best TL	ER (%)	Best TL	ER (%)	Best TL	ER (%)	Best TL	ER (%)	Best TL	ER (%)	Best TL	ER (%)	Best TL	ER (%)	Best TL	ER (%)	Best TL	ER (%)	Best TL	ER (%)
1	eil51	426	**426**	**0**	427	0.23	427	0.23	428.86	0.67	428.87	0.67	**426**	**0**	426	0	**426**	**0**	**426**	**0**	**426**	**0**
2	berlin52	7542	**7542**	**0**	**7542**	**0**	**7542**	**0**	7544.37	0.03	7544.37	0.03	**7542**	**0**	**7542**	**0**	**7542**	**0**	**7542**	**0**	**7542**	**0**
3	st70	675	**675**	**0**	**675**	**0**	**675**	**0**	677.11	0.31	677.11	0.31	676	0.15	676	0.15	**675**	**0**	**675**	**0**	**675**	**0**
4	eil76	538	**538**	**0**	542	0.74	**538**	**0**	558.68	3.84	–	–	**538**	**0**	**538**	**0**	541.6	0.67	**538**	**0**	**538**	**0**
5	pr76	108,159	**108,159**	**0**	**108,159**	**0**			108,159.4	0.0004	108,159.43	0.01	–	–	–	–	–	–	–	–	**108,159**	**0**
6	rat99	1211	**1211**	**0**	1224	1.07	**1211**	**0**	1225.56	1.2	–	–	1224	1.07	1213	0.17	**1211**	**0**	**1211**	**0**	**1211**	**0**
7	kroA100	21,282	**21,282**	**0**	**21,282**	**0**	**21,282**	**0**	21,298.21	0.08	21,285.44	0.02	21,301	0.09	**21,282**	**0**	**21,282**	**0**	**21,282**	**0**	**21,282**	**0**
8	kroB100	22,141	**22,141**	**0**	**22,141**	**0**	–	–	22,308	0.75	22,142.07	0.01	–	–	–	–	22,141	0	–	–	**22,141**	**0**

Table 2. *Cont.*

Sl.	TSP Instance	Optimal Tour Length	GA-MARL + NICH-LS [35]		DSOS [34]		SSABC [31]		DSMO [33]		DLSO [32]		PSO-ACO [14]		PACO [11]		DEACO [13]		HAACO [10]		Proposed ACOAV	
			Best TL	ER (%)	Best TL	ER (%)	Best TL	ER (%)	Best TL	ER (%)	Best TL	ER (%)	Best TL	ER (%)	Best TL	ER (%)	Best TL	ER (%)	Best TL	ER (%)	Best TL	ER (%)
9	rd100	7910	–	–	–	–	–	–	8041.3	1.66	–	–	–	–	–	–	7910	0	–	–	**7910**	**0**
10	eil101	629	**629**	**0**	640	1.75	**629**	**0**	648.66	3.13	642.53	2.15	631	0.32	**629**	**0**	**629**	**0**	630	0.16	**629**	**0**
11	lin105	14,379	**14,379**	**0**	14,381	0.01	**14,379**	**0**	14383	0.03	14,383.0	0.03	**14,379**	**0**	**14,379**	**0**	**14,379**	**0**	**14,379**	**0**	**14,379**	**0**
12	pr107	44,303	**44,303**	**0**	44,314	0.02	–	–	44,385.86	0.19	–	–	–	–	–	–	–	–	–	–	**44,303**	**0**
13	pr124	59,030	**59,030**	**0**	**59,030**	**0**	–	–	60,285.21	2.13	–	–	–	–	–	–	59,074	0.07	–	–	**59,030**	**0**
14	ch130	6110	6132	0.36	–	–	–	–	–	–	6158.08	0.79	–	–	–	–	**6110**	**0**	–	–	**6110**	**0**
15	ch150	6528	**6528**	**0**	6542	0.21	–	–	–	–	6530.90	0.04	6538	0.15	6570	0.64	**6528**	**0**	6566	0.58	**6528**	**0**
16	kroA150	26,524	26,579	0.21	–	–	–	–	27,591.44	4.02	–	–	–	–	–	–	26,572	0.18	–	–	**26,524**	**0**
17	kroB150	26,130	**26,130**	**0**	–	–	–	–	26,601.94	1.81	–	–	–	–	–	–	**26,130**	**0**	–	–	**26,130**	**0**
18	rat195	2323	–	–	–	–	–	–	2372.89	2.15	–	–	–	–	–	–	2340	0.73	–	–	**2326**	**0.13**
19	d198	15,780	–	–	–	–	–	–	15,978.13	1.26	15,808.93	0.18	–	–	–	–	–	–	–	–	**15,780**	**0**
20	kroA200	29,368	29,435	0.23	29,477	0.37	29450	0.28	30,481.35	3.79	29,519.83	0.52	29,468	0.34	29,533	0.56	**29,368**	**0**	29,483	0.39	**29,368**	**0**
21	kroB200	29,437	–	–		–	–	–	30,716.5	4.35	29,652.94	0.73	–	–	–	–	–	–	–	–	**29,438**	**0.003**
22	tsp225	3861	**3865**	**0.1**	3877	0.41	–	–	4013.68	3.95	3929.51	1.77	–	–	–	–	–	–	–	–	3923	1.61
23	pr226	80,369	**80,369**	**0**	80,407	0.05	–	–	83,587.98	4.01	–	–	–	–	–	–	–	–	–	–	**80,369**	**0**
24	a280	2579	2595	0.62		–	–	–	–	–	2609.54	1.18	–	–	–	–	–	–	–	–	**2581**	**0.08**
25	pr299	48,191	48,637	0.93	49162	2.01	–	–	50,579.82	4.96	–	–	–	–	–	–	48,455	0.55	–	–	**48,215**	**0.05**
26	lin318	42,029	42,255	0.54	**42,201**	**0.4**	–	–	44,118.66	4.97	42,744.96	1.7	–	–	–	–	–	–	–	–	42,203	0.41
27	rd400	15,281	–	–	–	–	–	–	–	–	–	–	–	–	15,578	1.94	**15,323**	**0.27**	15,603	2.11	15,467	1.22
28	fl417	11,861	–	–	–	–	–	–	12,218.98	3.02	–	–	–	–	11,972	0.94	11,866	0.04	11,960	0.83	11,861	0
29	pr439	107,217	107,833	0.57	–	–	–	–	112,105.2	4.56	–	–	–	–	108,482	1.18	–	–	108,730	1.41	**107,613**	**0.37**
30	pcb442	50,778	–	–	51,418	1.26	–	–	–	–	52,330.24	3.06	–	–	51,962	2.33	50,964.5	0.37	51,780	1.97	**50,778**	0
31	rat575	6773	–	–	7073	4.43	–	–	–	–	–	–	–	–	7003	3.4	**6773**	**0**	–	–	6935	2.39
32	rat783	8806	–	–	9045	2.71	–	–	–	–	–	–	–	–	9111	3.46	**8916.0**	**1.25**	–	–	9050	2.77
33	pr1002	259,045	266,886	3.03	272,381	5.15	–	–	–	–	273,696.03	5.66	–	–	–	–	–	–	–	–	**266,155**	**2.74**
34	fl1400	20,127	20,304	0.88	–	–	–	–	–	–	–	–	–	–	–	–	–	–	–	–	**20,215**	**0.44**
35	pr2392	378,032	397,314	5.1	419,246	10.9	–	–	–	–	–	–	–	–	–	–	–	–	–	–	**392,461**	**3.82**
Optimal/Best Count			15/16		6/7		7/7		0/0		0/0		4/4		6/6		14/16		7/7		**22/30**	

Table 3. Average Tour Length (TL) and Standard Deviation (SD) Comparison of ACOAV with Existing State-of-the-Art Bio-inspired Methods.

Sl.	TSP Instance	Optimal Tour Length	GA-MARL + NICH-LS [35]		DSOS [34]		SSABC [31]		DSMO [33]		DLSO [32]		PSO-ACO [14]		PACO [11]		DEACO [13]		HAACO [10]		Proposed ACOAV	
			Avg. TL	SD	Avg. TL	SD	Avg. TL	SD	Avg. TL	SD	Avg. TL	SD	Avg. TL	SD	Avg. TL	SD	Avg. TL	SD	Avg. TL	SD	Avg. TL	SD
1	eil51	426	427.4	–	427.90	1.20	427.01	0.46	436.96	4.73	429.7	1.61	426.45	0.61	426.35	0.49	**426**	**0**	427.5	–	**426**	**0**
2	berlin52	7542	7550.7	–	7542.60	0	**7542**	**0**	7633.6	85.4	7544.37	0	7543.2	2.37	**7542**	**0**	**7542**	**0**	**7542**	–	**7542**	**0**
3	st70	675	679.43	–	679.2	2.8	675.77	1.17	702.64	15.04	678.78	3.38	678.2	1.47	677.85	0.99	**675**	**0**	676.5	–	**675**	**0**
4	eil76	538	545.3	–	542,547.4	3.9	538.15	0.60	572.7	7.56	–	–	538.3	0.47	539.85	1.09	541.6	0.6	542	–	**538**	**0**
5	pr76	108,159	109,556.57	–	–	–			111,299.3	2050.48	108,572.35	341.96	–	–	–	–	–	–	–	–	**108,159**	**0**
6	rat99	1211	1223.3	–	1228.37	14.32	1211.50	0.67	1291.93	21.07	–	–	1227.4	1.98	1217.1	4.01	1211.7	–	1214.1	–	**1211**	**0**
7	kroA100	21,282	21,354.4	–	21,409.50	149.15	21,287.19	8.10	22,024.27	508.89	21,370.09	44.66	21,445.1	78.24	21,326.8	33.72	**21,282**	**0**	21,364.2	–	**21,282**	**0**
8	kroB100	22,141	22,283.4	–	22,339.20	230.18	–	–	23,022.37	277.32	22,270.58	95.52	–	–	–	–	**22,141**	**0**	–	–	**22,141**	**0**
9	rd100	7910	–	–			–	–	8377.76	209.4	–	–	–	–	–	–	**7910**	**0**			**7910**	**0**
10	eil101	629	642.6	–	650.60	4.57	630.59	2.37	674.4	10.97	649.05	–	632.7	2.12	630.55	2.63	**629**	**0**	632.5	–	**629**	**0**
11	lin105	14,379	14,385.63	–	14,431.73		14,379.10	1.30	15,114	500.76	14,433.33	34.23	14,379.15	0.48	14,393	19.76	**14,379**	**0**	14,411.8	–	**14,379**	**0**
12	pr107	44,303	44,424.73	–	44,445.10	181.35	–	–	45,666.99	1300.43	–	–	–	–	–	–	–	–		–	**44,303**	**0**
13	pr124	59,030	59,208.83	–	59,030	264.08	–	–	62,443.49	1644.93	–	–	–	–	–	–	–	–		–	**59,030**	**0**
14	ch130	6110	6204.17	–	–	–	–	–	–	–	6201.98	30.96	–	–	–	–	**6110**	**0**	–	–	6110	**0**
15	ch150	6528	6547.67	–	6552.58		–	–	–	–	6597.83	38.83	6563.95	27.58	6601.4	15.01	**6528**	**0**	6578.8	–	6528	**0**
16	kroA150	26,524	26,891.83	–	–		–	–	28,354.09	524.91	–	–	–	–	–	–	**26,524**	**0**	–	–	**26,524**	**0**
17	kroB150	26,130	26,477.33	–	–		–	–	27,576.16	625.26	–	–	–	–	–	–	**26,130**	**0**	–	–	**26,130**	**0**
18	rat195	2323	–	–	–		–	–	2488.55	50.48	–	–	–	–	–	–	–	–	–	–	**2330.2**	**1.6**
19	d198	15,780	–	–	–		–	–	16,270.47	171.2	15,896.48	35.21	–	–	–	–	–	–	–	–	**15,780**	**0**
20	kroA200	29,368	29,621	–	29,651.23		29,469	20.03	31,828.64	652.32	29,766.27	118.37	29,646.05	114.71	29,644.5	53.43	**29,368**	**0**	29,633.2	–	**29,368**	**0**
21	kroB200	29,437	–	–	–	–	–	–	31,781.62	487.39	29,994.08	226.62	–	–	–	–	29,440	5.1	–	–	**29,439.5**	**0.67**
22	tsp225	3861	**3925.33**	–	–	–	–	–	4162.79	66.08	3977.53	21.05	–	–	–	–	–	–	–	–	3956.9	**11.62**
23	pr226	80,369	80,638.6	–	–	–	–	–	85,935.69	2105.13	–	–	–	–	–	–	–	–	–	–	**80,369.6**	**1.43**
24	a280	2579	2655.47	–	–	–	–	–	–	–	2650.49	33.83	–	–	–	–	–	–	–	–	**2589.4**	**3.83**
25	pr299	48,191	49,200.57	–	50,335.20	905.42	–	–	51,747.99	863.32	–	–	–	–	–	–	–	–	–	–	**48,519.3**	**70.16**
26	lin318	42,029	42,996.63	–	42,972.42	2037.43	–	–	45,460.25	660.47	43,172.51	235.18	–	–	–	–	42,225	47	–	–	**42,220.9**	**9.99**
27	rd400	15,281	–	–	–	–	–	–	–	–	–	–	–	–	15,613.9	–	**15,385**	–	15,644.2	–	15,576	**44.23**
28	fl417	11,861	–	–	–	–	–	–	12,950.77	360.99	–	–	–	–	11,987.4	–	11,875	–	11,979.5	–	**11,861.2**	**0.4**
29	pr439	107,217	109,577.87	–	–	–	–	–	116,379.2	2462.82	–	–	–	–	108,702	–	–	–	108,950.6	–	**107,965**	**182.42**
30	pcb442	50,778	–	–	–	–	–	–	–	–	52,841.26	230.75	–	–	52,202.4	–	50,965	–	52,179.8	–	**50,945.2**	**54.98**
31	rat575	6773	–	–	7117.32	171.65	–	–	–	–	–	–	–	–	7012.4	–	**6804.0**	**10.3**	–	–	6972.25	15.73
32	rat783	8806	–	–	9102.67	37.28	–	–	–	–	–	–	–	–	9127.3	–	**8935.9**	**12.44**	–	–	9101.1	20.77
33	pr1002	259,045	269,845.97	–	278,381.51	4328.62	–	–	–	–	275,825	1189.80	–	–	–	–	–	–	–	–	**268,111**	**678.67**
34	fl1400	20,127	20,444.33	–	–	–	–	–	–	–	–	–	–	–	–	–	20,229	31.15	–	–	**20,226.7**	**5.39**
35	pr2392	378,032	400,171.73	–	425,431.78	4352.75	–	–	–	–	–	–	–	–	–	–	–	–	–	–	**397,871**	**1805.49**
Optimal/Best Count			0/1		1/1		1/1		0/0		0/0		0/0		1/1		13/16		1/1		**19/31**	

4.3. Statistical Analysis of Presented Results

Statistical tests are used to detect the significance of performance differences among a group of methods. Two well-known tests were conducted in this study. At first, the Friedman test is conducted to compare the identicalness of the algorithms. Then Post-hoc test is performed to compare the performance of the algorithms pairwise.

4.3.1. Friedman Test

The Friedman test [42] is a nonparametric statistical test that can detect if there exists any significant difference among several samples or not. The following statements are the two assumptions for this test:

- Observations are mutually independent. That is, the results within one row do not affect the results of other rows.
- For each row, results can be ranked based on their performance.

H_0 and H_1 are two hypotheses for the Friedman test. The null hypothesis, H_0 states that results came from the algorithms are similar. The alternative hypothesis, H_1 states that results are different for at least one of the algorithms.

In Table 4, a rank is given to every algorithm based on its average cost for each TSP instance from Table 3. In the case of ties in performance, an average rank is assigned to the algorithms [43,44]. Here, we considered seven popular TSP instances for the Friedman test because the tour costs (i.e., lengths) for other TSP instances are not available. Here, the number of algorithms, $k = 10$, and the number of TSP instances, $n = 7$. The Friedman test statistics, T_F is

$$T_F = \frac{12}{nk(k+1)} \sum_{j=1}^{k} R_{.j}^2 - 3n(k+1)$$
$$= \frac{12}{7 \times 10 \times (10+1)} (45^2 + 55^2 + 24^2 + 70^2 + 57^2 + 43^2 + 30^2 + 12^2 + 37^2 + 12^2) - 3 \times 7 \times (10+1) \quad (15)$$
$$= 52.34$$

Table 4. Ranking of the Friedman Test among proposed ACOAV and Existing State-of-the-Art Bio-inspired Methods.

n	Method	GA-MARL + NICH-LS [35]	DSOS [34]	SSABC [31]	DSMO [33]	DLSO [32]	PSO-ACO [14]	PACO [11]	DEACO [13]	HAACO [10]	Proposed ACOAV
	Rank(R)	R_{i1}	R_{i2}	R_{i3}	R_{i4}	R_{i5}	R_{i6}	R_{i7}	R_{i8}	R_{i9}	R_{i10}
1	eil51	6	8	5	10	9	4	3	1.5	7	1.5
2	berlin52	9	6	3	10	8	7	3	3	3	3
3	st70	9	8	3	10	7	6	5	1.5	4	1.5
4	kroA100	5	8	3	10	7	9	4	1.5	6	1.5
5	eil101	7	9	4	10	8	6	3	1.5	5	1.5
6	lin105	5	8	3	10	9	4	6	1.5	7	1.5
7	kroA200	4	8	3	10	9	7	6	1.5	5	1.5
$R_{.j} = \sum_{i=1}^{n} R_{ij}$ for $i = 1,2,3\ldots,n$		$R_{.1} = 45$	$R_{.2} = 55$	$R_{.3} = 24$	$R_{.4} = 70$	$R_{.5} = 57$	$R_{.6} = 43$	$R_{.7} = 30$	$R_{.8} = 12$	$R_{.9} = 37$	$R_{.10} = 12$
Average rank ($\frac{R_{.j}}{n}$)		6.43	7.86	3.43	10	8.14	6.14	4.29	**1.71**	5.29	**1.71**

However, the test statistic T_F is a Chi-square-based approximation that is poor and conservative when the number of algorithms increases. Therefore, the test statistic T_{ID} proposed by Iman and Davenport [42] is also measured.

$$T_{ID} = \frac{(n-1) \times T_F}{n(k-1) - T_F} = \frac{(7-1) \times 52.34}{7 \times (9-1) - 52.34} = 29.46 \quad (16)$$

The null hypothesis H_0 is rejected if $T_{ID} > F_{k-1;\,(k-1)(n-1);\,1-\alpha}$ where $F_{k-1;\,(k-1)(n-1);\,1-\alpha}$ is the critical value of F distribution with significance level α and degrees of freedom $df1 = k - 1$, $df2 = (k - 1) \times (n - 1)$.

From the F distribution table at $\alpha = 0.01$, the critical value of $F_{k-1;\,(k-1)(n-1);\,1-\alpha} = F_{9;\,54;\,0.99} \simeq 2.75$.

Since $T_{ID} > F_{9;\,54;\,0.99}$, according to the Friedman test, the performances of some algorithms are significantly different from others. As of Table 4, ACOAV and DEACO hold the same average rank value 1.71. Therefore, the proposed ACOAV is significantly better than all the methods except DEACO.

4.3.2. Post-Hoc Test

The post-hoc test can be performed to find out the comparative difference between the algorithms when Friedman's test rejects the null hypothesis. Fisher's LSD test [42] is considered in this study from several post-hoc test methods. When $R_{.i}$ and $R_{.j}$ are the summation of ranks for two algorithms, then the algorithms are considered significantly different if

$$|R_{.i} - R_{.j}| > t_{(n-1)(K-1);1-\frac{\alpha}{2}} \sqrt{2\frac{n\sum_{i=1}^{n}\sum_{j=1}^{k} R_{ij}^2 - \sum_{j=1}^{k} R_{.j}^2}{(n-1)(k-1)}}, \tag{17}$$

where $t_{(n-1)(K-1);\,1-\frac{\alpha}{2}}$ is the critical value from t distribution table with significance level α and degrees of freedom $(n-1)(k-1)$. Again, the number of algorithms, $k = 10$, and the number of TSP instances, $n = 7$. Thus,

$$t_{(n-1)(K-1);1-\frac{\alpha}{2}} \times \sqrt{2 \times \frac{n\sum_{i=1}^{n}\sum_{j=1}^{k} R_{ij}^2 - \sum_{j=1}^{k} R_{.j}^2}{(n-1)(k-1)}} = 2.67 \times \sqrt{2 \times \frac{7 \times 2682 - 18181}{(7-1)\times(10-1)}} = 12.51$$

Table 5 contains the absolute difference in the summation of ranks of other algorithms with ACOAV. Notably, the statistical tests were performed for only seven TSP instances as outcomes for those problems are available for all the compared methods. The post-hoc test concludes that ACOAV significantly outperforms seven methods (showing a test value greater than the critical value 12.51) and is competitive with DEACO/SSABC. At a glance, ACOAV outperformed all the compared methods considering results presented in Tables 2 and 3; and, the outperformance is significant with respect to most of the methods as on the basis of statistical test measures are shown in Tables 4 and 5.

Table 5. Post-hoc Test of proposed ACOAV with Existing State-of-the-Art Bio-inspired Methods.

$\lvert R_{.i} - R_{.j}\rvert$	GA-MARL + NICH-LS [35] ($R_{.1}$)	DSOS [34] ($R_{.2}$)	SSABC [31] ($R_{.3}$)	DSMO [33] ($R_{.4}$)	DLSO [32] ($R_{.5}$)	PSO-ACO [14] ($R_{.6}$)	PACO [11] ($R_{.7}$)	DEACO [13] ($R_{.8}$)	HAACO [10] ($R_{.9}$)
ACOAV ($R_{.10}$)	33	43	12	58	45	31	18	0	25

5. Discussion

For a particular TSP problem, visibility values among different cities remain fixed as distance based on coordinates are constant. Therefore, the nearest city gets the highest visibility heuristic desirability in the conventional ACO regardless of the destination city or return to the originating city. Such a constraint does not reflect the issue of destination or returning to the starting city of TSP solution development. Adaptive visibility (AV), the different better approach, is the main contribution of the study to cop the destination city issue in solving TSP.

Through the concept and step-by-step processes to formulate AV, the development of the generalized formula of Equation (10) is an interesting technical contribution to devising an efficient ACO-based method for TSP, called ACOAV. Equation (12) only modifies the heuristic desirability and applies to ACO instead of the distance between cities, while other factors remain unchanged. Conventional ACO feature is also achieved from AV for setting $w1 = 1$ and $w2 = 0$ in Equation (10). Although Equation (10) seems computationally heavy, the overall computational proficiency of an algorithm depends on its adaptation technique. Another innovative idea of partial solution update (PSU) is adopted in the proposed ACOAV to minimize the computational cost. In the PSU mode, only a portion of a TSP solution is considered to update, i.e., reorder the nodes of the selected portion for a better TSP solution. The proficiency of AV and PSU is rigorously investigated with experimental outcomes on TSP instances.

The proposed ACOAV is an effective updated model of ACO and efficient concerning the recent ACO-based methods. Several ACO-based methods adapt the control parameters using different techniques. HAACO [10] uses different parameter settings for initial phases with heterogeneous ants. DEACO [13] uses a dynamic pheromone evaporation strategy, and PSO-ACO [14] uses PSO to adjust the control parameters of ACO. On the other hand, PACO runs multiple ant colonies in parallel to increase search diversity. In short, recent ACO-based methods only focused on increasing the searching diversity of the ants. Nevertheless, none of these methods focused on improving the visibility heuristic that plays an indispensable role in ant movements. Ants perform intelligent movement using AV, and that makes ACOAV superior to other existing ACO-based methods. Moreover, other methods employ complex strategies, such as multiple ant colonies or other algorithms, to improve ACO performance. On the other hand, ACOAV skillfully updates the visibility heuristic, which enhances the performance significantly.

6. Conclusions

In this study, ant colony optimization with adaptive visibility (ACOAV) is proposed as a better method than the conventional ACO to solve TSP. The visibility heuristic value in conventional ACO is given by the simple inverse of the distance between nodes without considering the necessity of returning to the tour originating city at the last step. In contrast, the proposed ACOAV intelligently managed the issue of returning to the originating city (the final destination) by introducing a generalized AV formula. In addition, partial updates of individual solutions are adopted in the proposed ACOAV to accelerate convergence. The proposed ACOAV is tested on several benchmark TSP instances, and rigorously compared with ACO. At first, the new functionalities introduced in ACOAV are evaluated individually to justify their incorporation. Specifically, it is found from the simulation that the AV enables ACOAV to have a better-converging speed showing relatively better results with fewer iterations. Furthermore, it is found that partial updates effectively push the solution towards the best tour eventually. In such a way, both functionalities (i.e., AV and partial update) play an essential role indirectly in achieving faster convergence towards the best tour, and therefore, they eventually make the ACOAV the best solution for TSP. It is remarkable that ACOAV significantly outperformed ACO in twofold achievements, i.e., yielding better TSP solutions and requiring less computational time. The performance of the proposed ACOAV is also compared with ten state-of-the-art methods, including several recent ACO-based methods. The proposed ACOAV can be stated as the best method for generating optimal solutions for several cases. Furthermore, the statistical test also justifies the significance of the performance of ACOAV over recent bio-inspired methods.

Different future research scopes have emerged from the present study. Proposed ACOAV is shown to achieve optimal tours for the small-sized TSPs, especially for the problem size of fewer than 200 cities. Further experiments while varying parameter values might generate optimal tours for larger TSP instances. The present study considered a 3-Opt local search; the effects of other local search algorithms might improve the performance of ACOAV. An exciting but challenging future research topic might be to develop a more efficient and robust visibility function for ACO to solve TSP and other combinatorial optimization problems, such as job shop scheduling, the vehicle routing problem, etc.

Author Contributions: Conceptualization, M.A.H.A.; Data curation, A.S.B.S.; Methodology, A.S.B.S. and M.A.H.A.; Software, A.S.B.S.; Supervision, M.A.H.A.; Visualization, A.S.B.S., M.A.H.A. and M.A.S.K.; Writing—original draft, A.S.B.S. and M.A.H.A.; Writing—review & editing, M.A.H.A. and M.A.S.K. All authors have read and agreed to the published version of the manuscript.

Funding: This research received no external funding.

Institutional Review Board Statement: Not applicable.

Informed Consent Statement: Not applicable.

Data Availability Statement: Not applicable.

Conflicts of Interest: The authors declare no conflict of interest.

References

1. Lenstra, J.K.; Kan, A.H.G.R. Some Simple Applications of the Travelling Salesman Problem. *J. Oper. Res. Soc.* **1975**, *26*, 717–733. [CrossRef]
2. Hoffman, K.L.; Padberg, M.; Rinaldi, G. *Encyclopedia of Operations Research and Management Science*; Springer: Boston, MA, USA, 2013.
3. Reinelt, G. Tsplib 95. *Interdiszip. Zent. Für Wiss. Rechn. (IWR) Heidelb.* **1995**, *338*, 1–16.
4. Reinelt, G. TSPLIB—A Traveling Salesman Problem Library. *ORSA J. Comput.* **1991**, *3*, 376–384. [CrossRef]
5. Dorigo, M.; Gambardella, L. Ant colony system: A cooperative learning approach to the traveling salesman problem. *IEEE Trans. Evol. Comput.* **1997**, *1*, 53–66. [CrossRef]
6. Selvi, V.; Umarani, R. Comparative Analysis of Ant Colony and Particle Swarm Optimization Techniques. *Int. J. Comput. Appl.* **2010**, *5*, 1–6. [CrossRef]
7. Li, W.H.; Yang, Y.; Liao, H.Q.; Li, J.L.; Zheng, X.P. Artificial Bee Colony Algorithm for Traveling Salesman Problem. *Adv. Mater. Res.* **2011**, *314–316*, 2191–2196. [CrossRef]
8. Akhand, M.A.H.; Akter, S.; Rashid, M.A.; Yaakob, S.B. Velocity tentative PSO: An optimal velocity implementation based particle swarm optimization to solve traveling salesman problem. *IAENG Int. J. Comput. Sci.* **2015**, *42*, 221–232.
9. Ouaarab, A.; Ahiod, B.; Yang, X.-S. Discrete cuckoo search algorithm for the travelling salesman problem. *Neural Comput. Appl.* **2013**, *24*, 1659–1669. [CrossRef]
10. Tuani, A.F.; Keedwell, E.; Collett, M. Heterogenous Adaptive Ant Colony Optimization with 3-opt local search for the Travelling Salesman Problem. *Appl. Soft Comput.* **2020**, *97*, 106720. [CrossRef]
11. Gülcü, Ş.; Mahi, M.; Baykan, Ö.K.; Kodaz, H. A parallel cooperative hybrid method based on ant colony optimization and 3-Opt algorithm for solving traveling salesman problem. *Soft Comput.* **2018**, *22*, 1669–1685. [CrossRef]
12. Dahan, F.; El Hindi, K.; Mathkour, H.; AlSalman, H. Dynamic Flying Ant Colony Optimization (DFACO) for Solving the Traveling Salesman Problem. *Sensors* **2019**, *19*, 1837. [CrossRef] [PubMed]
13. Ebadinezhad, S. DEACO: Adopting dynamic evaporation strategy to enhance ACO algorithm for the traveling salesman problem. *Eng. Appl. Artif. Intell.* **2020**, *92*, 103649. [CrossRef]
14. Mahi, M.; Baykan, Ö.K.; Kodaz, H. A new hybrid method based on Particle Swarm Optimization, Ant Colony Optimization and 3-Opt algorithms for Traveling Salesman Problem. *Appl. Soft Comput.* **2015**, *30*, 484–490. [CrossRef]
15. Liu, Y.; Cao, B. A Novel Ant Colony Optimization Algorithm with Levy Flight. *IEEE Access* **2020**, *8*, 67205–67213. [CrossRef]
16. Liao, E.; Liu, C. A Hierarchical Algorithm Based on Density Peaks Clustering and Ant Colony Optimization for Traveling Salesman Problem. *IEEE Access* **2018**, *6*, 38921–38933. [CrossRef]
17. Dorigo, M.; Maniezzo, V.; Colorni, A. Ant system: Optimization by a colony of cooperating agents. *IEEE Trans. Syst. Man Cybern. Part B Cybern.* **1996**, *26*, 29–41. [CrossRef]
18. Li, B.; Wang, L.; Song, W. Ant Colony Optimization for the Traveling Salesman Problem Based on Ants with Memory. In Proceedings of the 2008 Fourth International Conference on Natural Computation, Jinan, China, 18–20 October 2008; Volume 7, pp. 496–501. [CrossRef]
19. Guntsch, M.; Middendorf, M. A Population Based Approach for ACO. *Lect. Notes Comput. Sci. Incl. Subser. Lect. Notes Artif. Intell. Lect. Notes Bioinform.* **2002**, *2279*, 72–81. [CrossRef]
20. Dorigo, M.; Gambardella, L.M. A study of some properties of Ant-Q. In *Parallel Problem Solving from Nature—PPSN IV. PPSN 1996*; Lecture Notes in Computer Science; Springer: Berlin/Heidelberg, Germany, 1996. [CrossRef]
21. Peya, Z.J.; Akhand, M.A.H.; Sultana, T.; Rahman, M.M.H. Distance based Sweep Nearest Algorithm to Solve Capacitated Vehicle Routing Problem. *Int. J. Adv. Comput. Sci. Appl.* **2019**, *10*, 259–264. [CrossRef]
22. Punnen, A.P. The Traveling Salesman Problem: Applications, Formulations and Variations. In *The Traveling Salesman Problem and Its Variations*; Springer: Boston, MA, USA, 2007; Volume 12, pp. 1–28. [CrossRef]
23. Mavrovouniotis, M.; Yang, S. Ant colony optimization with immigrants schemes for the dynamic travelling salesman problem with traffic factors. *Appl. Soft Comput.* **2013**, *13*, 4023–4037. [CrossRef]
24. Liu, M.; Li, Y.; Li, A.; Huo, Q.; Zhang, N.; Qu, N.; Zhu, M.; Chen, L. A Slime Mold-Ant Colony Fusion Algorithm for Solving Traveling Salesman Problem. *IEEE Access* **2020**, *8*, 202508–202521. [CrossRef]
25. Khan, I.; Maiti, M.K.; Maiti, M. Coordinating Particle Swarm Optimization, Ant Colony Optimization and K-Opt Algorithm for Traveling Salesman Problem. *Commun. Comput. Inf. Sci.* **2017**, *655*, 103–119. [CrossRef]
26. Mavrovouniotis, M.; Muller, F.M.; Yang, S. Ant Colony Optimization with Local Search for Dynamic Traveling Salesman Problems. *IEEE Trans. Cybern.* **2016**, *47*, 1743–1756. [CrossRef] [PubMed]
27. Yazdani, M.; Jolai, F. Lion Optimization Algorithm (LOA): A nature-inspired metaheuristic algorithm. *J. Comput. Des. Eng.* **2015**, *3*, 24–36. [CrossRef]

28. Sharma, H.; Hazrati, G.; Bansal, J.C. Spider Monkey Optimization Algorithm. In *Studies in Computational Intelligence*; Springer International Publishing: Berlin/Heidelberg, Germany, 2019; Volume 779, pp. 43–59.
29. Ahmad, Y.; Ullah, M.; Khan, R.; Shafi, B.; Khan, A.; Zareei, M.; Aldosary, A.; Mohamed, E.M. SiFSO: Fish Swarm Optimization-Based Technique for Efficient Community Detection in Complex Networks. *Complexity* **2020**, *2020*, 6695032. [CrossRef]
30. Li, L.; Cheng, Y.; Tan, L.; Niu, B. A Discrete Artificial Bee Colony Algorithm for TSP Problem. *Int. J. Adv. Comput. Technol.* **2012**, *4*, 566–573.
31. Khan, I.; Maiti, M.K. A swap sequence based Artificial Bee Colony algorithm for Traveling Salesman Problem. *Swarm Evol. Comput.* **2019**, *44*, 428–438. [CrossRef]
32. Daoqing, Z.; Mingyan, J. Parallel discrete lion swarm optimization algorithm for solving traveling salesman problem. *J. Syst. Eng. Electron.* **2020**, *31*, 751–760. [CrossRef]
33. Akhand, M.; Ayon, S.I.; Shahriyar, S.; Siddique, N.; Adeli, H. Discrete Spider Monkey Optimization for Travelling Salesman Problem. *Appl. Soft Comput.* **2019**, *86*, 105887. [CrossRef]
34. Ezugwu, A.E.-S.; Adewumi, A.O. Discrete symbiotic organisms search algorithm for travelling salesman problem. *Expert Syst. Appl.* **2017**, *87*, 70–78. [CrossRef]
35. Alipour, M.M.; Razavi, S.N.; Derakhshi, M.R.F.; Balafar, M.A. A hybrid algorithm using a genetic algorithm and multiagent reinforcement learning heuristic to solve the traveling salesman problem. *Neural Comput. Appl.* **2017**, *30*, 2935–2951. [CrossRef]
36. Bouzidi, A.; Riffi, M.E. Discrete Cat Swarm Optimization to Resolve the Traveling Salesman Problem. *Int. J. Adv. Res. Comput. Sci. Softw. Eng.* **2013**, *3*, 13–18.
37. Panwar, K.; Deep, K. Discrete Grey Wolf Optimizer for symmetric travelling salesman problem. *Appl. Soft Comput.* **2021**, *105*, 107298. [CrossRef]
38. Pandiri, V.; Singh, A. An artificial bee colony algorithm with variable degree of perturbation for the generalized covering traveling salesman problem. *Appl. Soft Comput.* **2019**, *78*, 481–495. [CrossRef]
39. Zhang, J.; Hong, L.; Liu, Q. An Improved Whale Optimization Algorithm for the Traveling Salesman Problem. *Symmetry* **2020**, *13*, 48. [CrossRef]
40. Colorni, A.; Dorigo, M.; Maniezzo, V. Distributed Optimization by ant colonies. *Proc. First Eur. Conf. Artif. Life* **1991**, *142*, 134–142.
41. Helsgaun, K. General k-opt submoves for the Lin–Kernighan TSP heuristic. *Math. Program. Comput.* **2009**, *1*, 119–163. [CrossRef]
42. Pereira, D.G.; Afonso, A.; Medeiros, F.M. Overview of Friedman's Test and Post-hoc Analysis. *Commun. Stat. Simul. Comput.* **2014**, *44*, 2636–2653. [CrossRef]
43. Riffenburgh, R.H. Tests on Ranked Data. In *Statistics in Medicine*; Elsevier: Amsterdam, The Netherlands, 2012; pp. 221–248.
44. Derrac, J.; García, S.; Molina, D.; Herrera, F. A practical tutorial on the use of nonparametric statistical tests as a methodology for comparing evolutionary and swarm intelligence algorithms. *Swarm Evol. Comput.* **2011**, *1*, 3–18. [CrossRef]

mathematics

MDPI

Article

Swarm-Intelligence Optimization Method for Dynamic Optimization Problem

Rui Liu [1], Yuanbin Mo [2,3,*], Yanyue Lu [4], Yucheng Lyu [3], Yuedong Zhang [1] and Haidong Guo [3]

[1] School of Electronic Information, Guangxi Minzu University, Nanning 530006, China; liurui777@stu.gxun.edu.cn (R.L.); zhangyuedong@stu.gxun.edu.cn (Y.Z.)
[2] Guangxi Key Laboratory of Hybrid Computation and IC Design Analysis, Guangxi Minzu University, Nanning 530006, China
[3] Institute of Artificial Intelligence, Guangxi Minzu University, Nanning 530006, China; ychlyu@stu.gxun.edu.cn (Y.L.); haidongguo@stu.gxun.edu.cn (H.G.)
[4] School of Chemistry and Chemical Engineering, Guangxi Minzu University, Nanning 530006, China; 20070041@gxun.edu.cn
* Correspondence: moyuanbin2020@gxun.edu.cn

Abstract: In recent years, the vigorous rise in computational intelligence has opened up new research ideas for solving chemical dynamic optimization problems, making the application of swarm-intelligence optimization techniques more and more widespread. However, the potential for algorithms with different performances still needs to be further investigated in this context. On this premise, this paper puts forward a universal swarm-intelligence dynamic optimization framework, which transforms the infinite-dimensional dynamic optimization problem into the finite-dimensional nonlinear programming problem through control variable parameterization. In order to improve the efficiency and accuracy of dynamic optimization, an improved version of the multi-strategy enhanced sparrow search algorithm is proposed from the application side, including good-point set initialization, hybrid algorithm strategy, Lévy flight mechanism, and Student's t-distribution model. The resulting augmented algorithm is theoretically tested on ten benchmark functions, and compared with the whale optimization algorithm, marine predators algorithm, harris hawks optimization, social group optimization, and the basic sparrow search algorithm, statistical results verify that the improved algorithm has advantages in most tests. Finally, the six algorithms are further applied to three typical dynamic optimization problems under a universal swarm-intelligence dynamic optimization framework. The proposed algorithm achieves optimal results and has higher accuracy than methods in other references.

Keywords: dynamic optimization; swarm intelligence; control variable parameterization; nonlinear programming problem; sparrow search algorithm

MSC: 49M37; 68T20

Citation: Liu, R.; Mo, Y.; Lu, Y.; Lyu, Y.; Zhang, Y.; Guo, H. Swarm-Intelligence Optimization Method for Dynamic Optimization Problem. *Mathematics* **2022**, *10*, 1803. https://doi.org/10.3390/math10111803

Academic Editor: Srinivas R. Chakravarthy

Received: 27 March 2022
Accepted: 20 May 2022
Published: 25 May 2022

1. Introduction

Dynamic optimization, also known as optimal control, a core part of industrial process design, directly affects the approval of multiple performance indicators such as the overall output, material loss, and efficiency improvement of the control system. It has long been an important means to maximize the value in process control of the chemical industry [1,2]. Affected by the upgrading of industry and the expansion of system scale, the established mathematical model is often full of high-dimensional, strongly nonlinear, and other complex characteristics that are difficult to deal with. Therefore, how to achieve an effective solution to this kind of dynamic optimization problem is not only a challenging but also an urgent and practical research topic. With the continuous development and deepening of optimization technology, the swarm-intelligence optimization technique, as an emerging branch, is becoming an attractive alternative to solve dynamic optimization problems [3], which is increasingly favored by academia and industry.

The earliest method applied to dynamic optimization is the indirect method [4], which has rigorous and accurate results. However, the mathematical process is often complex and difficult to implement, and there is no analytical solution [5], so it is very limited in practical application. Different from the indirect method, the direct method [6] discretizes the variables of the dynamic optimization problem into a form that can be solved by numerical methods. Among them, the CVP method [7] only discretizes the control variables. It has higher efficiency when solving the system and has become the mainstream direct method. Furthermore, the CVP method provides an effective time-domain discretization strategy, which transforms the dynamic optimization problem into a finite-dimensional NLP problem, so that the swarm-intelligence algorithm, a practical parameter optimization technique [8,9], can be used. At present, swarm-intelligence optimization techniques have attracted extensive attention in the application of various optimization problems [10,11], and have the advantages of low dependence on prior knowledge, high robustness based on population search, and no need to calculate the gradient information of the objective function.

In recent years, scholars have used swarm-intelligence algorithms to solve dynamic optimization problems, and proposed solutions including particle swarm optimization (PSO) [12], beetle antennae search (BAS) [13], ant colony optimization (ACO) [14], seagull optimization algorithm (SOA) [15], sailfish optimizer (SFO) [16], and cultural algorithm (CA) [17]. In this context, these successful application cases confirm the effectiveness of swarm-intelligence algorithms for dynamic optimization. However, in the current literature description, a universal framework of swarm-intelligence algorithms for dynamic optimization is generally ignored, which is not conducive to the further research of various algorithms with different performances, thus limiting the long-term development of swarm-intelligence dynamic optimization methods. Therefore, it is very necessary to establish a universal framework of the swarm-intelligence dynamic optimization method, which is a core topic to be solved in this paper. Furthermore, the efficiency and accuracy of solving specific problems in the existing research still need to be improved, which often requires an approach with better performance. Therefore, this paper introduces an improved version of the sparrow search algorithm (SSA) applied to dynamic optimization problems, and uses other well-known swarm-intelligence algorithms, including whale optimization algorithm (WOA) [18], marine predators algorithm (MPA) [19], harris hawks optimization (HHO) [20], and social group optimization (SGO) [21] compared under a universal swarm-intelligence dynamic optimization framework. The boosted abilities of the proposed algorithm for typical dynamic optimization problems is successfully verified.

The SSA was chosen as the base for augmentation as it has been validated as having a better optimization performance and solving ability [22–26] compared to PSO, grey wolf optimizer (GWO) [27], gravitational search algorithm (GSA) [28], and sine cosine algorithm (SCA) [29]. It has been successfully applied in various domains, including UAV track planning [30], density peak clustering [31], BP neural network optimization [32], robot path planning [33], and micro-grid operation [34], showing great potential. However, it is also established that the basic SSA suffers from insufficient search scope, weak resistance to local extremum, and a slow convergence rate, which needs to be further enhanced. Hybridization is a popular algorithm design approach [35], by integrating the advantages of different algorithms, a hybrid algorithm with better performance can be constructed. In this paper, SGO is introduced into the SSA optimization framework. On this basis, the good-point set, inertia weight factor, and Lévy flight are used to modify the details, and the structure of the optimization algorithm is modified by using the Student's t-distribution model. Then, a cooperative-mutation hybrid-swarm-intelligence algorithm (CM-HSSA) is proposed to solve the dynamic optimization problem.

The main objective behind the universal swarm-intelligence dynamic optimization framework proposed in this study is to further improve the SSA from the application side, and increase the efficiency and accuracy of solving specific dynamic optimization problems. Furthermore, since the potential for algorithms with different performances under a universal swarm-intelligence dynamic optimization framework has not been investigated enough, five other well-known swarm-intelligence techniques were also implemented and tested for three typical cases. In a nutshell, the significant characteristics of our paper are listed as:

- A universal swarm-intelligence dynamic optimization method is summarized and proposed, which lays a theoretical foundation for subsequent research on using the swarm-intelligence technique to solve dynamic optimization problems.
- A novel modified SSA is implemented from the application side and utilized to improve the efficiency and accuracy of typical dynamic optimization problems.
- Other well-known swarm-intelligence techniques for dynamic optimization are further investigated under a universal optimization framework.

The rest of the paper is organized in the following manner. Section 2 describes the fundamental methods used (the CVP method and a universal swarm-intelligence dynamic optimization method). Section 3 introduces the modified version of the algorithm as well as the original one and tests other algorithms on benchmark functions. Section 4 deals with the optimization of typical dynamic optimization problems with six algorithms under a universal swarm-intelligence dynamic optimization framework. Finally, Section 5 summarizes some conclusions and prospects for future work.

2. Preliminaries

In this section, firstly, the standard mathematical model of the dynamic optimization problem is introduced. Secondly, the basic principle of the CVP strategy is introduced, and then a universal swarm-intelligence dynamic optimization framework is summarized and proposed. In particular, the general implementation scheme and flow chart of this method are given.

2.1. Dynamic Optimization Problem Description

Generally, dynamic optimization problems are common in the control systems of industrial processes and widely exist in the chemical industry. The research object is mainly aimed at dynamic time-varying systems [36]. The established mathematical model is often described in the form of a differential–algebraic equation (DAE), which contains constraints and an objective function. Therefore, the essence of solving the dynamic optimization problem is to apply the control effect to the variables in the model and then select the appropriate optimization scheme to make the performance index in the process reach the best state. The mathematical model of a typical dynamic optimization problem can be described as follows:

$$\begin{aligned} \min J &= \Phi[x(t_f)] + \int_{t_0}^{t_f} L[x(t), u(t), t]dt \\ s.t. &\begin{cases} \frac{dx}{dt} = f[x(t), u(t), t] \\ x(t_0) = x_0 \\ u_{lb} \leq u(t) \leq u_{ub} \\ t \in [t_0, t_f] \end{cases} \end{aligned} \tag{1}$$

where J is the objective function, also known as the performance index, which is composed of the final value term $\Phi[x(t_f)]$ at the process termination time t_f and the integral term $\int_{t_0}^{t_f} L[x(t), u(t), t]dt$ existing on the whole time period $[t_0, t_f]$, $u(t) = [u_1(t), u_2(t), \cdots, u_m(t)]^T$ is the m-dimensional control variable, and constrained by the upper boundary u_{ub} and the lower boundary u_{lb}, $x(t) = [x_1(t), x_2(t), \cdots, x_n(t)]^T$ is the n-dimensional state variable. Therefore, Equation (1) can be briefly described as looking for the control variable $u(t)$ that makes the target J obtain the optimal value under the condition of the initial state $x(t_0) = x_0$, and the value of $u(t)$ should meet the requirements of the feasible region.

2.2. CVP Strategy

As the mainstream numerical calculation method in the direct solution, the principle of the CVP method is to use the basic function with finite parameters to approach the control effect. Specifically, the strategy first discretizes the time domain ($[t_0, t_f]$) into NE sub-interval ($[t_{k-1}, t_k]$ $(k = 1, 2, \cdots, NE)$), that is, $t_0 \le t_1 \le \cdots \le t_{N-1} \le t_{NE} = t_f$, and further uses the basis function to approximate the components on each sub-interval, then $u(t)$ can be expressed as the cumulative sum of each component on the whole $[t_0, t_f]$:

$$u(t) = \sum_{k=1}^{NE} \sigma_j^k(t)\ j = 1, 2, \cdots, m\ k = 1, 2, \cdots, NE \tag{2}$$

where $\sigma_j^k(t)$ is the linear combination of the basic function of the known structure of each component ($u_j(t)$) in the time interval ($[t_{k-1}, t_k]$), which is determined by limited parameters. The mathematical model of the optimization problem transformed by the CVP method can be described as:

$$\begin{gathered} \min \widetilde{J} = \varphi[\sigma(t)] \\ s.t. u_{lb} \le \sum_{k=1}^{NE} \sigma^k(t) \le u_{ub} \end{gathered} \tag{3}$$

where $\sigma(t) = \left[\sigma^1(t), \sigma^2(t), \cdots, \sigma^{NE}(t)\right]^T$ is the parameter vector to be optimized. Therefore, the CVP method provides an effective transformation method, and an infinite-dimensional dynamic optimization problem is transformed into a finite-dimensional static optimization problem with a finite number of parameters.

2.3. Swarm-Intelligence Dynamic Optimization Method Based on CVP Strategy

For dynamic optimization problems, after CVP processing, the control variables, state variables, objective functions, and constraints of the system are all determined by the parameter vector, thus forming the NLP problem which can be solved by the swarm-intelligent optimization algorithm. Depending on the type of basis function, the approximation effect is also different. As the most important type of basis function, the piecewise constant approximation strategy is the most reasonable choice from theoretical analysis to practical calculation, and has the characteristics of simplicity and effectiveness. Figure 1 shows the control curve approximated by piecewise constant when NE = 7.

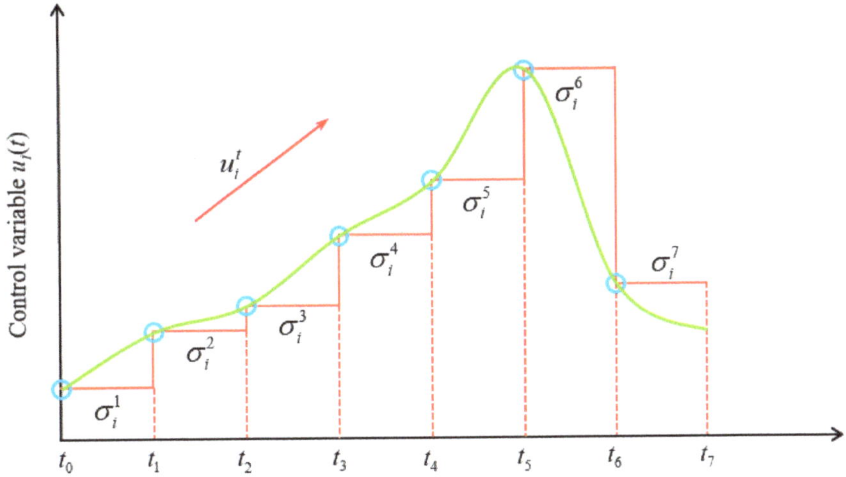

Figure 1. Piecewise constant approximation of CVP method.

In particular, different from the traditional deterministic optimization algorithm based on gradient, the swarm-intelligence optimization algorithm established by randomness generally does not need to calculate the gradient information about the objective function, so the relevant gradient calculation process is not included in the solution structure. The calculation steps of the swarm-intelligence dynamic optimization method based on the CVP strategy are as follows. Figure 2 shows the calculation framework of this method.

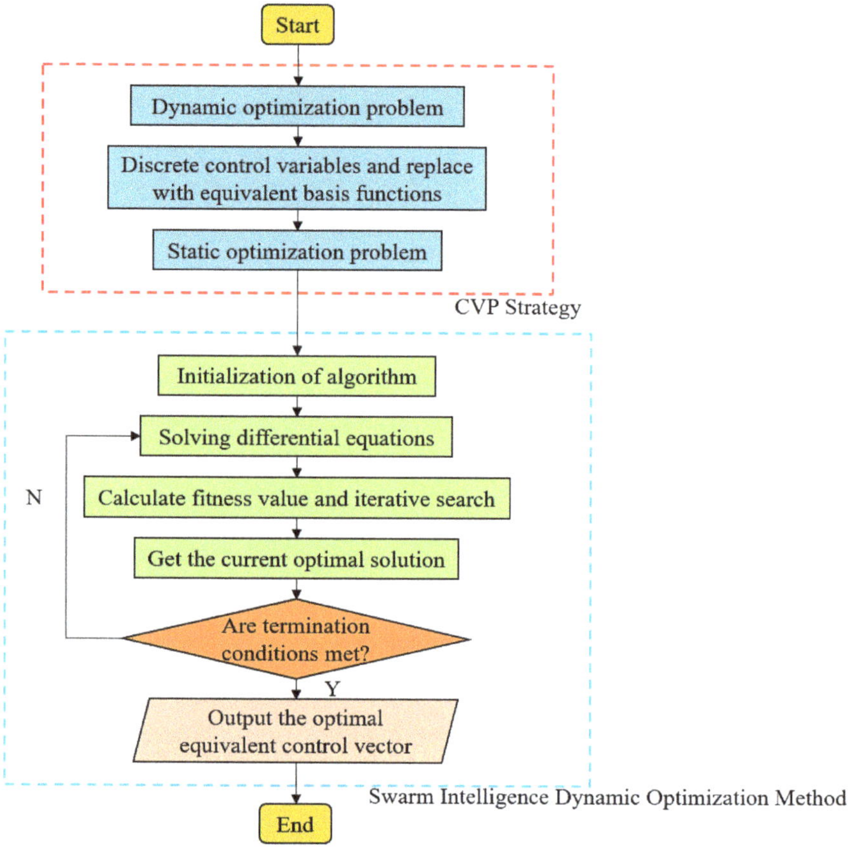

Figure 2. Swarm-Intelligence Dynamic Optimization Method Based on CVP Strategy.

(1) Through the CVP strategy, $u(t)$ is transformed into $\sigma(t)$, and the dynamic optimization problem shown in Equation (1) is transformed into the static optimization problem form shown in Equation (3).
(2) Set relevant parameters, such as population size, the maximum number of iterations, and algorithm parameters.
(3) Initialize the population.
(4) Evaluate and sort the fitness values of individuals in the population and record the current optimal value.
(5) According to the evolution strategy of the algorithm, a new population is generated.
(6) Compare the fitness value of the new solution and replace it if it is better than the current value.
(7) Determine whether the current condition meets the stop criterion; if so, terminate the algorithm and output the optimal solution. Otherwise, return to (4) and continue to execute, and set $t = t + 1$.

3. Mathematical Models and Algorithms

This beginning of the section introduces the basic implementation of the SSA algorithm, followed by a discussion about the known and observed flaws and drawbacks of the original version, and a detailed description of the proposed modified algorithm that is devised to specifically overcome these flaws of the original algorithm is provided. In the end, ten groups of benchmark functions are used to test the performance of the proposed algorithm, WOA, MPA, HHO, SSA, and SGO.

3.1. Sparrow Search Algorithm

The mathematical model of SSA mainly refers to the foraging habits of sparrows, idealizes the individual behavior in the population, formulates the corresponding iterative rules, and divides the individual into two roles of producers and scroungers in each generation according to the fitness value. In addition, SSA also designed an early warning process, which is to randomly select some individuals in the population called scouters, and update their locations in each iteration. Finally, it searches for the global optimal solution through a certain number of iterations.

The locations of producers are updated as follows:

$$X_i^{t+1} = \begin{cases} X_i^t \cdot \exp\left(-\frac{i}{\alpha \cdot Iter_{\max}}\right), & R_2 < ST \\ X_i^t + Q \cdot L, & R_2 \geq ST \end{cases} \tag{4}$$

where t represents the current iteration, $Iter_{\max}$ is the maximum number of iterations, α is a random number in the range of (0, 1], Q is a random number subject to normal distribution, L is a $1 \times D$ matrix with each element value of 1, $R_2 (R_2 \in [0,1])$ and $ST(ST \in [0.5,1])$ represent the alarm value and safety threshold, respectively. It can be seen from Equation (4) that their values determine the update mode of producers

The locations of scroungers are updated as follows:

$$X_i^{t+1} = \begin{cases} Q \cdot \exp\left(\frac{X_{worst}^t - X_i^t}{i^2}\right), & i > n/2 \\ X_p^t + \left|X_i^t - X_p^t\right| \cdot A^+ \cdot L, & otherwise \end{cases} \tag{5}$$

where n represents the number of sparrows, X_p is the best foraging location occupied by the current producers, X_{worst}^t is the current worst foraging location, is a A matrix with element values of 1 or -1, and $A^+ = A^T(AA^T)^{-1}$. Q and L are the same as in Equation (4).

The locations of scouters are updated as follows:

$$X_i^{t+1} = \begin{cases} X_{best}^t + \beta \cdot \left|X_i^t - X_{best}^t\right|, & f_i > f_g \\ X_i^t + K \cdot \left(\frac{\left|X_i^t - X_{worst}^t\right|}{(f_i - f_w) + \varepsilon}\right), & f_i = f_g \end{cases} \tag{6}$$

where f_i is the individual fitness value of scouters, f_g represents the global optimal fitness value, X_{best}^t is the global optimal foraging location, β is K are step control parameters, and ε is a minimal constant to avoid the denominator being zero.

The flowchart of SSA is shown in Figure 3.

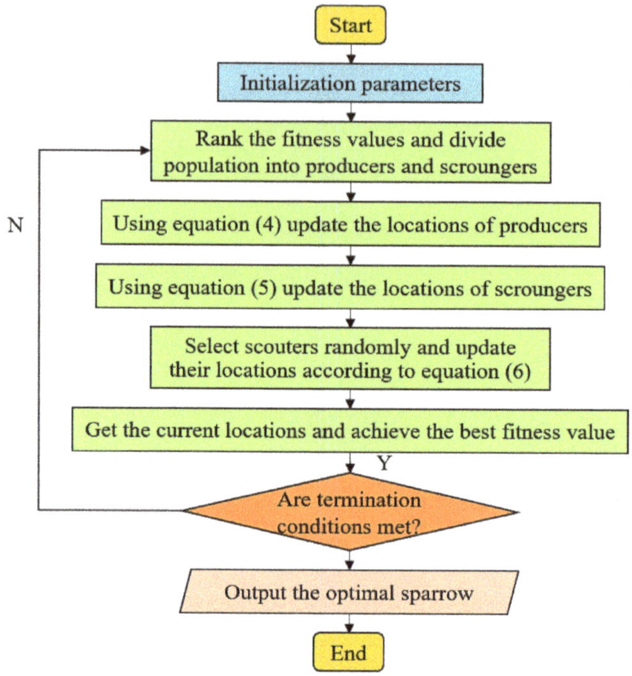

Figure 3. The flowchart of SSA.

3.2. *Multi-Strategy Improved Hybrid Swarm-Intelligence Optimization Algorithm*

3.2.1. Good-Point Set Theory

Previous studies have shown that the impact of the distribution of the initial population on swarm-intelligence algorithms cannot be ignored [9]. To improve the uniformity of the initial population search in solution space, scholars mostly use a chaotic map strategy to solve this problem. At present, the commonly used chaotic map models include circle map [37], tent map [38], piecewise map [39], cat map [40], logistic map [41], and Gauss map [42]. However, although this initialization method based on chaotic mapping has achieved some results, it still has considerable randomness, so it cannot effectively ensure the search breadth of the initial population.

To solve the above problems, this paper applies the good-point set theory [43] to the initial population stage. Its construction principle is: set G_s be the unit cube in s-dimensional Euclidean space, and if $r \in G_s$, the shape is as follows:

$$P_n(k) = \left\{ \left(\left\{ r_1^{(n)} \cdot k \right\}, \left\{ r_2^{(n)} \cdot k \right\}, \cdots, \left\{ r_s^{(n)} \cdot k \right\} \right), 1 \le k \le n \right\} \tag{7}$$

If the deviation $\varphi(n) = C(r, \varepsilon) n^{-1+\varepsilon}$ is satisfied, where $C(r, \varepsilon)$ is the constant only related to r and $\varepsilon(\varepsilon > 0)$, then $P_n(k)$ is the good-point set and r is the good point. $\left\{ r_s^{(n)} \cdot k \right\}$ indicates the decimal part, n is the number of samples, and we set $r = \{2\cos(2\pi k/p), 1 \le k \le s\}$ and p as the minimum prime number satisfying $(p-3) \ge s$. Mapping the good points of G_s to the search space is:

$$X_{i,j} = \left\{ r_j^{(i)} \cdot k \right\} \cdot (ub_j - lb_j) + lb_j \tag{8}$$

with the same number of points, a consistent distribution effect can be obtained each time using the good-point set to initialize the population. Because the construction of the good-point set is independent of the dimension of the sample, it plays a better role in solving high-dimensional problems. Through calculation and analysis, the deviation of the good-point set is $O(n^{-1+\varepsilon})$, while the deviation of the random method is $O(n^{-1/2}(\log\log n)^{1/2})$. Compared with the random method, the deviation of the good-point set is reduced to the square-root level. Therefore, the good-point set theory provides a stable and effective uniform point selection strategy for population initialization.

To intuitively compare the two initialization methods, the population distribution generated by the random method and good-point set method when $N = 100$ on $[0, 1]$ is provided in Figure 4. In addition, we further compared the six commonly used chaotic maps mentioned above with the good-point set method. Considering the randomness of chaotic maps, we carried out 10 experiments, and the average value distributions of each method when $N = 100$ are shown in Figure 5.

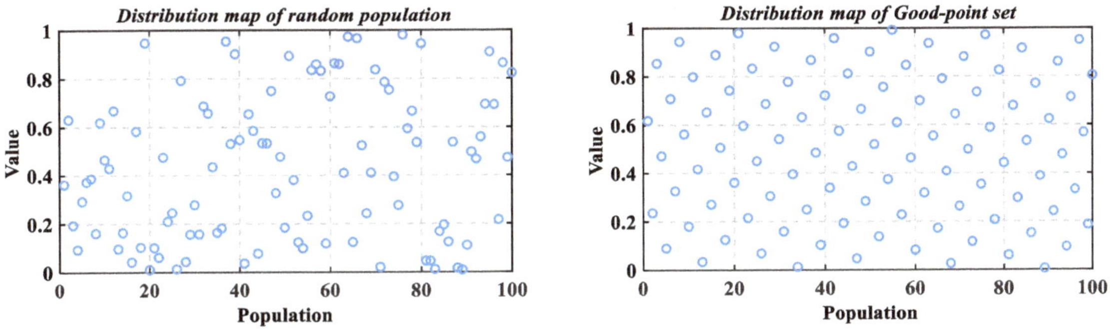

Figure 4. Comparison of two different initialization strategies.

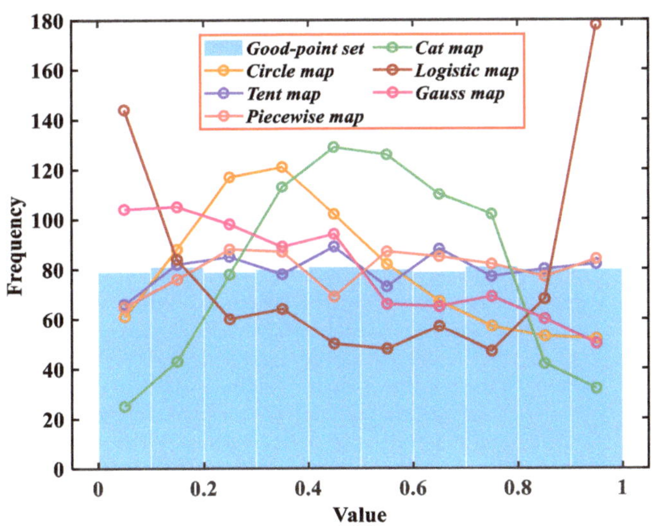

Figure 5. Average value distributions of seven different initialization strategies.

3.2.2. Hybrid Algorithm Strategy

In SSA, the producers represent the sparrows with better fitness in the population, and have the function of guiding other individuals to move to the best foraging location. Therefore, the location update process of the producers will closely affect the optimization ability of SSA. According to Equation (4), the producers have two ways of updating their locations. When $R_2 \geq ST$, the individuals will move randomly near the current locations according to the normal distribution; when $R_2 < ST$, the update of the locations is affected by Equation (9):

$$f(x) = \exp(-\frac{x}{\alpha \cdot Iter_{\max}}) \tag{9}$$

when $Iter_{\max} = 1000$, $N = 1000$, $D = 1$ and $\alpha = 1$, the value distribution of the producers is shown in Figure 6. It can be seen that the search scope of producers shows an obvious reduction trend with iterations, and finally decreases to less than half of the initial range, all concentrated in the range of 0 to 0.4. The reduction of the search range is bound to reduce the population diversity in the optimization process, resulting in a search blind area, which increases the risk of SSA falling into the local extremum in the later stages of the iteration.

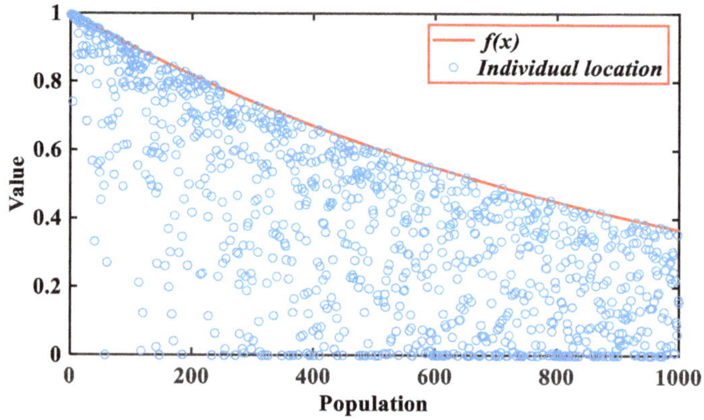

Figure 6. Producers' location update trends ($R_2 < ST$).

To improve the deficiency of the location update strategy of the producers, we decided to introduce the improvement phase of SGO to replace Equation (4). The mathematical model of the improvement phase of SGO is as follows:

$$X_i^{t+1} = c \cdot X_i^t + r \cdot (X_{best}^t - X_i^t) \tag{10}$$

where t represents the current number of iterations, X_i^t and X_{best}^t are the current individual location and the global optimal individual location, respectively, $c \in (0,1)$ is the self-reflection parameter, which is 0.2 in the original reference [15], and r is a random number satisfying uniform distribution from $[0,1]$. It can be seen that the location update will be guided by the current optimal individual, which is conducive to improving the global exploration ability and the convergence rate of the algorithm in the initial stage. However, as the self-reflection parameter, the constant attribute of c leads to an invariance dependence on the location information with iterations. We change c into an inertia weight factor whose value changes dynamically [44], as shown in Equation (11):

$$c^t = \frac{c_s(c_s - c_e)(Iter_{\max} - t)}{Iter_{\max}} \tag{11}$$

where c^t represents the inertia weight factor, c_s and c_e represent the adjustment parameters, $c_s = 0.9$. and $c_s = 0.4$ are set, respectively. Therefore, by introducing c^t, the adaptive regulation of the participation degree of its location information is achieved. The decreasing characteristic of c^t makes the algorithm maintain a good global exploration ability at the early stage of iteration and helps the algorithm have a more effective local development ability at the later stage of iteration. Figure 7 describes the changing trend of c^t with iterations. The new producers' update strategy is shown in Equation (12):

$$X_i^{t+1} = c^t \cdot X_i^t + r \cdot (X_{best}^t - X_i^t) \tag{12}$$

where c^t is the inertia weight factor, X_i^{t+1}, X_i^t and X_{best}^t are the next-generation location, current location, and current optimal location of the producer, respectively.

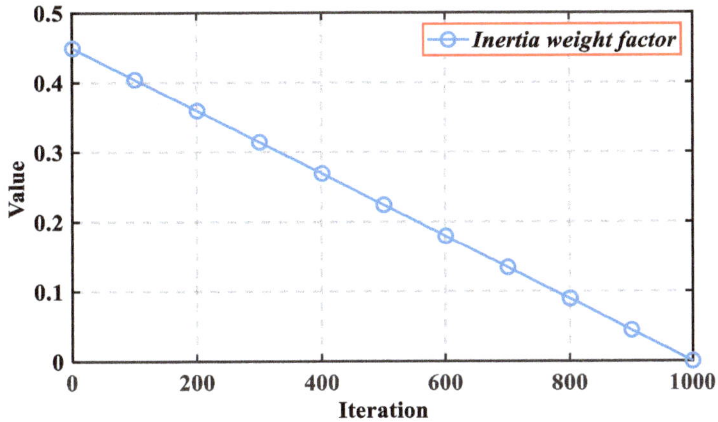

Figure 7. The changing trend of the inertia weight factor.

3.2.3. Stagnation Disturbance Strategy Based on Lévy Flight

When the iteration proceeds to a certain extent, affected by the local extremum, the update range of the producers will become smaller or move only near the current region. At this time, more and more producers will change into scroungers, which indicates that there is no solution in the nearby region, causing the algorithm to stagnate.

Lévy flight is a random walk. Studies have shown that the movement patterns of many animals can be described by it [45]. Since the generation of its step is affected by the heavy-tailed distribution, there will be a jump performance with a large span during the random walk. Therefore, Lévy flight is applied to the update of individuals as a disturbance, which will enable the search of the algorithm to enter a broader area and improve the ability of global exploration. Furthermore, to further illustrate that Lévy flight can adapt to larger-scale search, Brownian motion trajectory and Lévy flight trajectory simulated by the Mantegna method [46] are revealed in Figure 8.

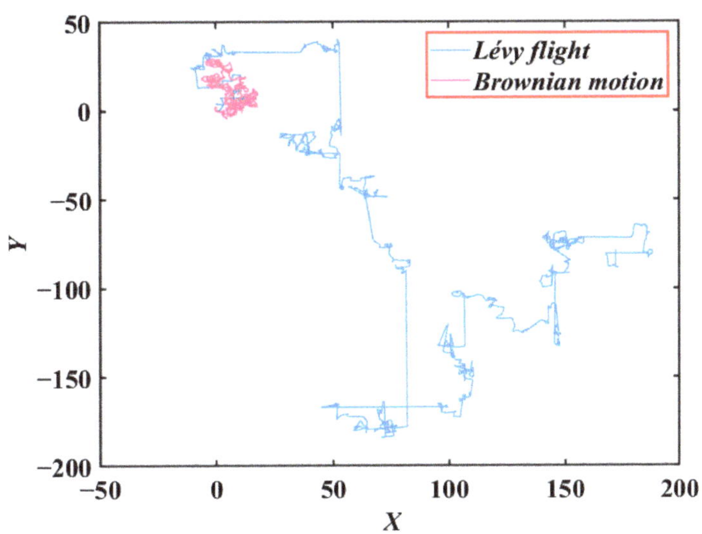

Figure 8. Lévy flight and Brownian motion.

The calculation method of Lévy flight in this paper is as follows [47]:

$$Levy(x) = 0.01 \times \frac{\theta_1 \cdot \sigma}{|\theta_2|^{\frac{1}{\beta}}} \tag{13}$$

where θ_1 and θ_2 are parameters subject to normal distribution, β is a constant, which is taken as 1.5 in this paper, and σ is calculated as follows:

$$\sigma = \left[\frac{\Gamma(1+\beta) \cdot \sin(\pi \cdot \beta/2)}{\Gamma((1+\beta)/2) \cdot \beta \cdot 2^{(\beta-1)/2}} \right]^{\frac{1}{\beta}} \tag{14}$$

where $\Gamma(x)$ is the *gamma* function and $\Gamma(x) = (x-1)!$ and x belongs to the set of natural numbers.

The new scroungers' update strategy is shown in Equation (16):

$$X_i^{t+1} = \begin{cases} X_i^t + X_i^t \cdot Levy(d), & i > n/2 \\ X_p^t + \left| X_i^t - X_p^t \right| \cdot A^+ \cdot L, & otherwise \end{cases} \tag{15}$$

where X_i^{t+1}, X_i^t and X_p^t are the next-generation location, current location, and current optimal location of the scrounger, respectively, and d represents the dimension of the location vector.

3.2.4. Early Warning Process Based on Student's *t*-Distribution Mutation Factor

According to Equation (6), the update of the early warning process is related to the fitness value of the individual. When $f_i > f_g$, the individual will move towards the current optimal location, When $f_i = f_g$, the individual will move randomly in the area near itself, and the distance is related to the current worst location and the worst fitness value. Therefore, the early warning process of SSA is essentially the furthest disturbance to the population location after the iteration of producers and scroungers.

Student's *t*-distribution is an important distribution type. Its curved shape is related to the change in degrees of freedom n. When $n = 1$, *t*-distribution is Cauchy distribution; when $n \to \infty$, *t*-distribution is Gaussian distribution, that is, Cauchy distribution and Gaussian distribution are two special cases of Student's *t*-distribution.

In this paper, the degrees of freedom for *t*-distribution are taken from the current iteration, and a mutation factor based on the Student's *t*-distribution that changes with iterations can be obtained. This is applied to scouters in the early warning process as a random disturbance. The mathematical model of the new early warning process is shown in Equation (16).

$$X_i^{t+1} = X_p^t + X_p^t \cdot trnd(t) \tag{16}$$

where $trnd(t)$ is the *t*-distribution mutation factor with the current iteration as the degree of freedom, and X_i^{t+1} and X_p^t are the next-generation location and current optimal location of the scouter, respectively. Moreover, the mutation factor combines the advantages of Cauchy distribution and Gaussian distribution and generates different disturbance ranges through changing degrees of freedom, which can effectively balance the global exploration ability and local development ability of the algorithm. The improved algorithm based on the *t*-distribution mutation factor is defined as the collaborative-mutation hybrid sparrow search algorithm (CM-HSSA). The pseudo-code of CM-HSSA is shown in Algorithm 1:

Algorithm 1: The framework of CM-HSSA

Input: *Max_Iter*: the maximum iteration; N: the population size; PD: the proportion of producers; SD: the proportion of early warning sparrows; c_s, c_e: the inertia weight adjustment parameters.
Output: X_{best}: the optimal individual location; f_g: the fitness value of the optimal individual.
/* Initialization*/

for i = 1 to N do
  for j = 1 to d do
    Initialize the location of N sparrows using equation (8);
  end for
end for

/*Iterative search*/

Calculate the fitness value and record the current optimal individual;
for (t < *Max_iter*)
  for i = 1 to PD*N do
    Update the location of producers according to equation (12);
  end for
  for i = PD*N + 1 to N do
    Update the location of scroungers according to equation (15);
  end for
  for i = 1 to SD*N do
    Update the location of early warning sparrows according to equation (16);
  end for
  Evaluate the fitness value of the new location and update if it is better;
end for

/*Algorithm terminated*/

Return X_{best}, f_g

3.3. Benchmark Function Experiments

This section provides ten groups of classical benchmark functions to test the optimization performance of six algorithms, including four unimodal functions with only one global optimal value and five multimodal functions with multiple local extremums. F_1–F_8 are 30 dimensions and F_9–F_{10} are 2 dimensions. Table 1 gives the relevant information on these benchmark functions. The range represents the search scope, Opt represents the theoretical optimal value and D represents the dimension of the problem. Among them, F_1–F_4 can test the convergence rate, accuracy, and local development ability, while F_5–F_{10} can test the anti-local extremum ability and global exploration ability.

3.3.1. Parameter Settings

To verify the significance of the improvement, we compared the optimization effects of WOA, MPA, HHO, SSA, SGO, and CM-HSSA on benchmark functions. To ensure the objectivity of the experiments, the population is set to 30 and the maximum iteration is 100. The specific parameter settings of each algorithm are as follows. For WOA, the logarithmic spiral shape parameter $b = 1$. For MPA, the fish aggregating device $FADs = 0.2$. For HHO, the prey energy factor E is a random number between $(-1, 1)$. For SSA, the safety threshold $ST = 0.8$, the proportion of producers $PD = 0.2$, and the proportion of scouters $SD = 0.1$. For SGO, the self-reflection parameter $c = 0.2$. For CM-HSSA, the proportion of producers $PD = 0.2$, the proportion of scouters $SD = 0.1$, and weight adjustment parameters $c_s = 0.9$ and $c_e = 0.4$. It is worth noting that the above parameters are taken from the original references. The values of these artificially set parameters are obtained based on experience, which can maximize the optimization performance of the algorithms.

Table 1. Information on benchmark functions.

Benchmark Function	Formula	Range	Opt
Sphere Model	$F_1(x) = \sum_{i=1}^{n} x_i{}^2$	[−100, 100]	0
Schwefel's problem 2.22	$F_2(x) = \sum_{i=1}^{n} \lvert x_i \rvert + \prod_{i=1}^{n} \lvert x_i \rvert$	[−10, 10]	0
Schwefel's problem 1.2	$F_3(x) = \sum_{i=1}^{n} \left(\sum_{j=1}^{i} x_j \right)^2$	[−100, 100]	0
Schwefel's problem 2.21	$F_4(x) = \max_i \{ \lvert x_i \rvert, 1 \leq i \leq n \}$	[−100, 100]	0
Generalized Schwefel's problem 2.26	$F_5(x) = \sum_{i=1}^{n} -x_i \sin \sqrt{\lvert x_i \rvert}$	[−500, 500]	−4.18.9829D
Generalized Rastrigin's Function	$F_6(x) = \sum_{i=1}^{n} [x_i^2 - 10\cos(2\pi x_i) + 10]$	[−5.12, 5.12]	0
Ackley's Function	$F_7(x) = -20\exp\left(-0.2\sqrt{\frac{1}{n}\sum_{i=1}^{n} x_i{}^2} \right) - \exp\left(\frac{1}{n}\sum_{i=1}^{n} \cos(2\pi x_i) \right) + 20 + e$	[−32, 32]	0
Generalized Griewank Function	$F_8(x) = \frac{1}{4000}\sum_{i=1}^{n} x_i^2 - \prod_{i=1}^{n} \cos \frac{x_i}{\sqrt{i}} + 1$	[−600, 600]	0
Branin Function	$F_9(x) = \left(x_2 - \frac{5.1}{4\pi^2} x^2 + \frac{5}{\pi} x_1 - 6 \right)^2 + 10\left(1 - \frac{1}{8\pi}\right)\cos x_1 + 10$	[−5, 5]	0.398
Goldstein–Price Function	$F_{10}(x) = [1 + (x_1 + x_2 + 1)^2(19 - 14x_1 + 3x_1^2 - 14x_2 + 6x_1x_2 + 3x_2^2)] \times [30 + (2x_1 - 3x_2)^2(18 - 32x_1 + 12x_1^2 + 48x_2 - 36x_1x_2 + 27x_2^2)]$	[−2, 2]	3

3.3.2. Statistical Result Comparison

To ensure the fairness of the experiments, each algorithm runs 20 times independently, and the mean value, standard deviation, and average calculation time are recorded. Among them, the mean value reflects the optimization accuracy, the standard deviation reflects the robustness, and TIC/TOC is used to calculate the running time of each algorithm. Through the experimental results listed in Table 2, we can see the different optimization performances of the algorithms.

Table 2. Experimental results of six algorithms.

Function	Result	WOA	MPA	HHO	SSA	SGO	CM-HSSA
F_1	Mean	3.5706×10^{-10}	1.9437	1.3052×10^{-20}	2.9595×10^{-33}	4.3773×10^{-135}	0
	Std.	7.1180×10^{-10}	1.0336	5.8268×10^{-20}	1.3235×10^{-33}	2.9367×10^{-136}	0
	TIC/TOC	0.075297	0.260806	0.122167	0.098561	0.121907	0.102957
F_2	Mean	9.7066×10^{-9}	9.6357×10^{-2}	4.3809×10^{-13}	2.0208×10^{-21}	1.5103×10^{-68}	0
	Std.	2.1815×10^{-8}	3.3799×10^{-2}	1.0991×10^{-12}	8.9306×10^{-21}	1.6423×10^{-69}	0
	TIC/TOC	0.060907	0.197253	0.121605	0.090216	0.127647	0.096940
F_3	Mean	9.8067×10^{4}	2.1566×10^{2}	1.8472×10^{-13}	4.1637×10^{-33}	2.2632×10^{-135}	0
	Std.	2.8622×10^{4}	1.8756×10^{2}	8.2479×10^{-13}	1.8621×10^{-32}	9.9703×10^{-136}	0
	TIC/TOC	0.098011	0.306055	0.232791	0.116329	0.246341	0.148943
F_4	Mean	5.6341×10^{1}	4.5856×10^{-1}	3.6428×10^{-13}	3.9747×10^{-21}	1.0950×10^{-68}	0
	Std.	2.8796×10^{1}	1.0924×10^{-1}	6.1075×10^{-13}	1.7745×10^{-20}	5.3688×10^{-70}	0
	TIC/TOC	0.060370	0.203315	0.113611	0.106351	0.126110	0.117772
F_5	Mean	-8.6688×10^{3}	-7.2695×10^{3}	-1.2356×10^{4}	-6.2868×10^{3}	-6.9435×10^{3}	-1.06×10^{4}
	Std.	1.0522×10^{3}	4.7419×10^{2}	7.9240×10^{2}	1.6650×10^{3}	6.5873×10^{2}	7.8299×10^{2}
	TIC/TOC	0.067170	0.251228	0.157786	0.079425	0.111361	0.098025

Table 2. *Cont.*

Function	Result	WOA	MPA	HHO	SSA	SGO	CM-HSSA
	Mean	1.2998×10^{-8}	8.6017	0	0	0	0
F_6	Std.	5.3053×10^{-8}	7.2342	0	0	0	0
	TIC/TOC	0.060027	0.217248	0.177083	0.076583	0.130681	0.106733
	Mean	3.7669×10^{-7}	8.9223×10^{-2}	7.3576×10^{-12}	1.0658×10^{-15}	8.8818×10^{-16}	8.8818×10^{-16}
F_7	Std.	5.5884×10^{-7}	2.8663×10^{-2}	2.2196×10^{-12}	7.9441×10^{-16}	0	0
	TIC/TOC	0.071849	0.177886	0.127309	0.079305	0.122893	0.098612
	Mean	9.1942×10^{-1}	2.7695×10^{-1}	0	0	0	0
F_8	Std.	2.8307×10^{-1}	1.4355×10^{-1}	0	0	0	0
	TIC/TOC	0.073628	0.198219	0.159379	0.077422	0.131028	0.115157
	Mean	4.0011×10^{-1}	3.9789×10^{-1}	3.9853×10^{-1}	3.9789×10^{-1}	3.9789×10^{-1}	3.9789×10^{-1}
F_9	Std.	3.7678×10^{-3}	5.0943×10^{-11}	1.1419×10^{-3}	6.3089×10^{-7}	2.3781×10^{-8}	0
	TIC/TOC	0.051275	0.173310	0.137616	0.098129	0.096289	0.083473
	Mean	8.4292	3.0201	3.0231	3.0023	3.0001	3.0000
F_{10}	Std.	1.1139×10^{1}	1.8833×10^{-10}	1.4497×10^{-4}	2.7527×10^{-7}	1.9782×10^{-7}	2.2017×10^{-15}
	TIC/TOC	0.056070	0.183948	0.155738	0.071500	0.107347	0.081171

The simulation software used in experiments was MATLAB R2018b. It is worth mentioning that iterations are generally positively correlated with the accuracy, while the maximum iteration set in this paper is 100, which can better reflect the optimization performance of the algorithms in short iterations. According to Table 2, CM-HSSA can obtain the stable optimal convergence accuracy for unimodal functions $F_1 - F_4$, and the optimization performance is better than other algorithms. For multimodal functions, HHO has the highest accuracy on F_5, followed by CM-HSSA. The mean value of the two algorithms has reached the level of -1×10^4, which is higher than other algorithms. For F_6 and F_8, CM-HSSA, HHO, SSA, and SGO can obtain the best optimization accuracy. Although their convergence behavior is different, they all successfully find the global optimal solution in the limited iteration. For F_7, F_9, and F_{10}, CM-HSSA has the highest accuracy and the smallest standard deviation, which is better than other algorithms, indicating that CM-HSSA has stronger local development ability and the ability to jump out of the local extremum. In terms of calculation time, WOA is the fastest, CM-HSSA is close to that of SSA, faster than HHO and SGO, and MPA takes the longest time. To improve the visualization of results and the significance of CM-HSSA, we selected the boxplot and Wilcoxon test [48] to further analyze the data in Table 3.

Table 3. The *p*-value test results over benchmark functions.

Function	CM-HSSA vs. WOA	CM-HSSA vs. MPA	CM-HSSA vs. HHO	CM-HSSA vs. SSA	CM-HSSA vs. SGO
F_1	8.0065×10^{-9}	8.0065×10^{-9}	8.0065×10^{-9}	8.0065×10^{-9}	8.0065×10^{-9}
F_2	8.0065×10^{-9}	8.0065×10^{-9}	8.0065×10^{-9}	8.0065×10^{-9}	8.0065×10^{-9}
F_3	8.0065×10^{-9}	8.0065×10^{-9}	8.0065×10^{-9}	2.992×10^{-8}	8.0065×10^{-9}
F_4	8.0065×10^{-9}	8.0065×10^{-9}	8.0065×10^{-9}	8.0065×10^{-9}	8.0065×10^{-9}
F_5	2.6609×10^{-6}	6.7004×10^{-8}	6.1833×10^{-4}	6.8341×10^{-7}	6.7004×10^{-8}
F_6	2.9868×10^{-8}	8.0065×10^{-9}	N/A	N/A	N/A
F_7	8.0065×10^{-9}	8.0065×10^{-9}	1.0433×10^{-7}	3.4211×10^{-4}	N/A
F_8	8.0065×10^{-9}	8.0065×10^{-9}	N/A	N/A	N/A
F_9	1.1597×10^{-4}	6.7956×10^{-8}	1.0581×10^{-4}	8.0065×10^{-9}	5.0209×10^{-5}
F_{10}	8.0065×10^{-9}	8.0065×10^{-9}	2.1025×10^{-7}	4.0137×10^{-8}	1.9299×10^{-3}

The boxplot in Figure 9 shows the characteristic information of the results of six algorithms, including maximum, minimum, and median. Table 3 shows the difference between the results of CM-HSSA and other algorithms through p-value comparison with the Wilcoxon test. When the p-value is less than 5%, there is an obvious difference between the two algorithms; otherwise, it means that the difference is not obvious, and N/A means that the two algorithms have the same performance and cannot be compared. According to the data recorded in Table 3, in most tests (42/50), the p-value is less than 5%, indicating that the optimization performance of CM-HSSA is significantly different from that of other algorithms, and the optimization ability is much higher than that of SSA. To further analyze the differences in convergence modes of each algorithm, Figures 10–19 show the convergence trajectories of the six algorithms, and plot y-coordinates using a base-10 logarithmic scale on the y-axis.

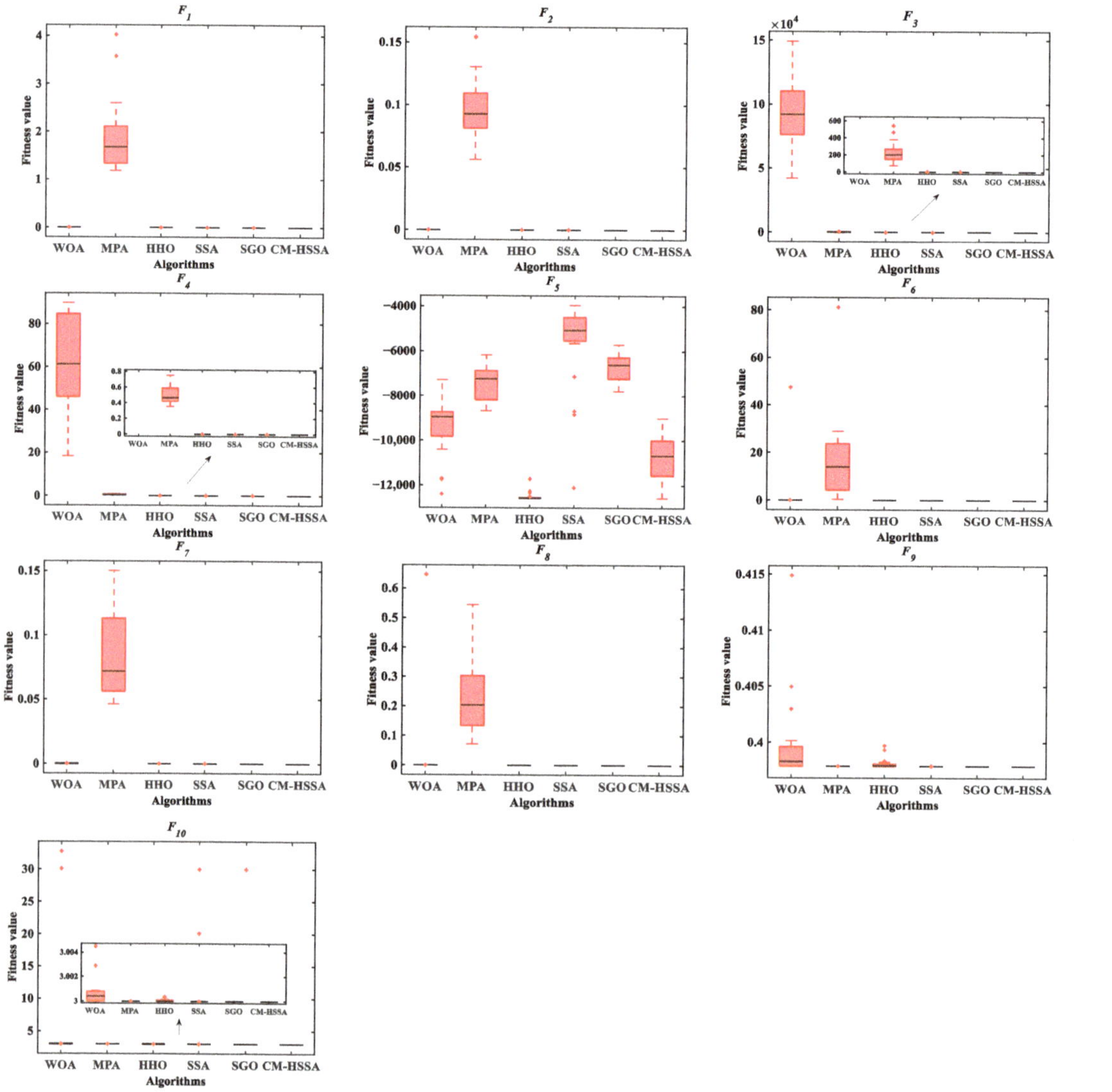

Figure 9. Boxplot of six algorithms for benchmark functions.

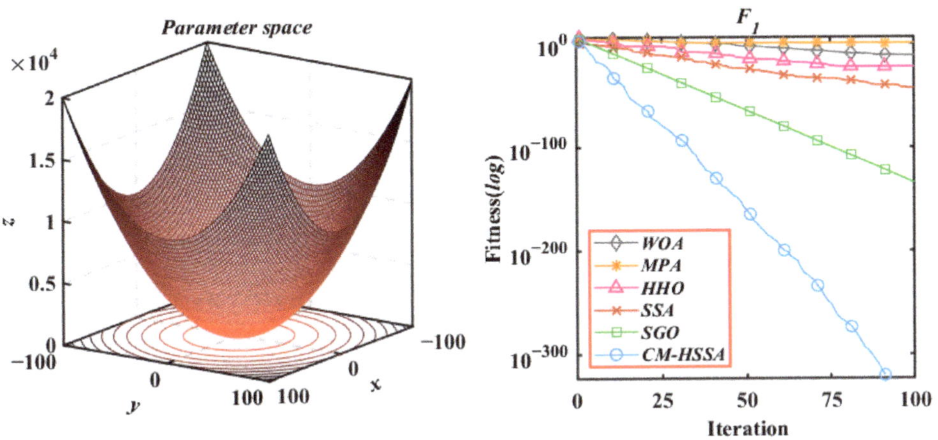

Figure 10. Sphere Model.

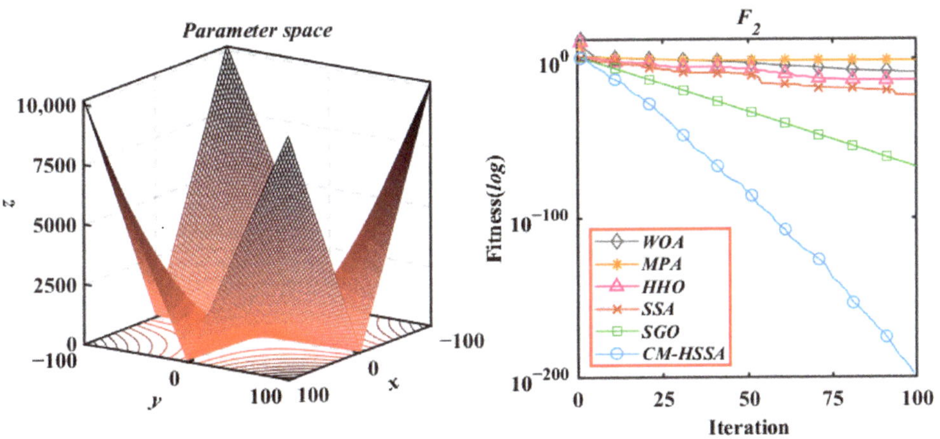

Figure 11. Schwefel's problem 2.22.

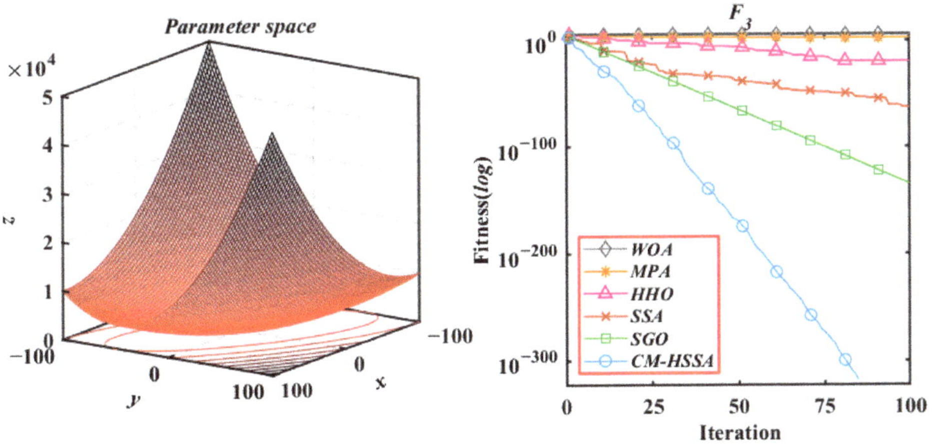

Figure 12. Schwefel's problem 1.2.

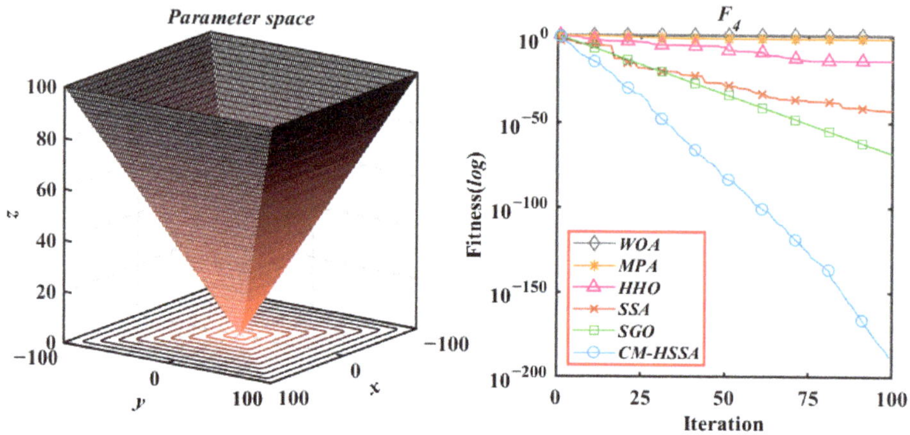

Figure 13. Schwefel's problem 2.21.

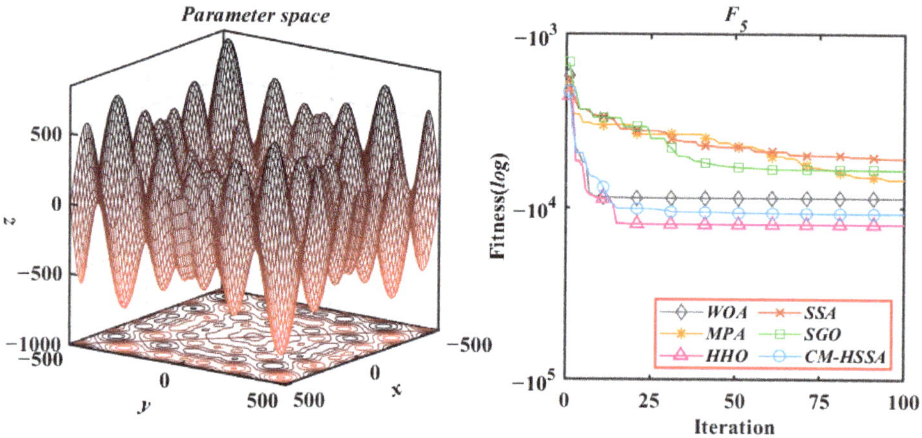

Figure 14. Generalized Schwefel's problem 2.26.

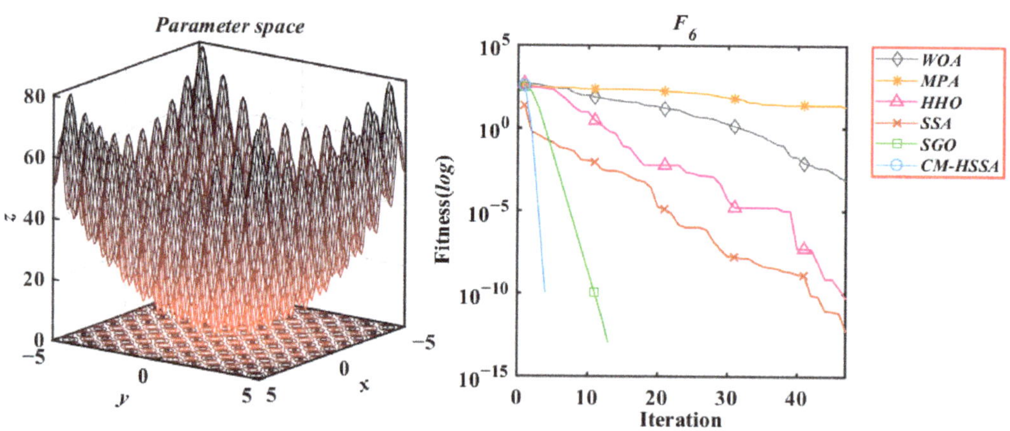

Figure 15. Generalized Rastrigin's Function.

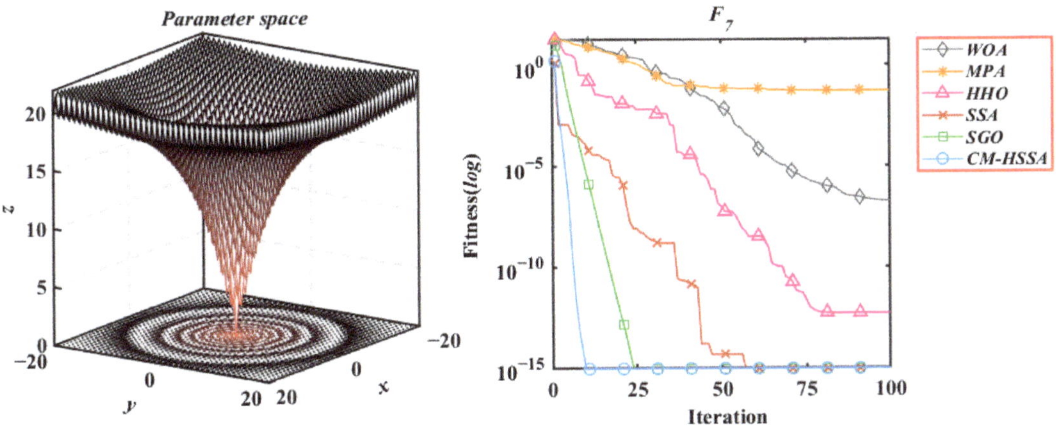

Figure 16. Ackley's Function.

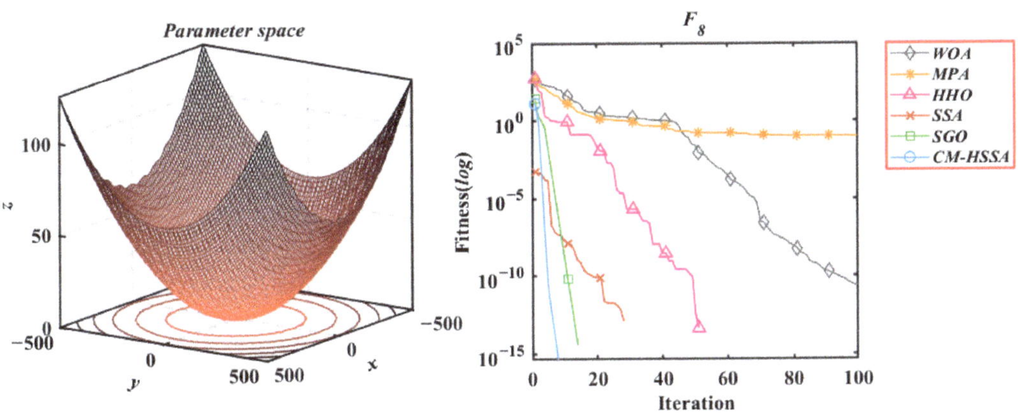

Figure 17. Generalized Griewank Function.

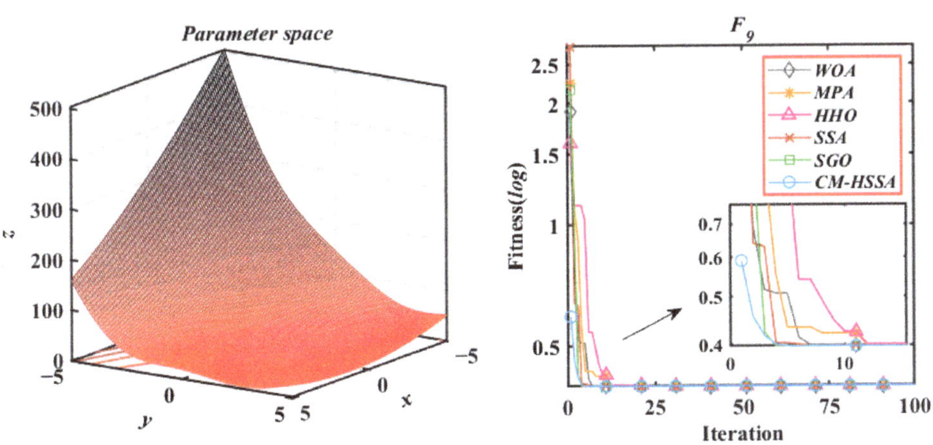

Figure 18. Branin Function.

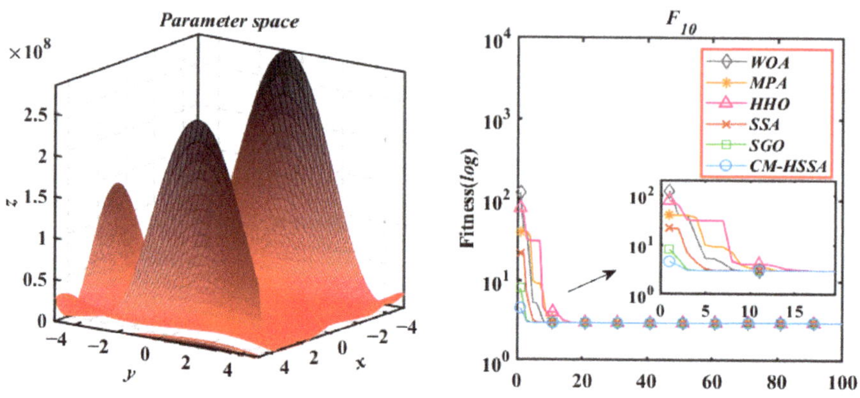

Figure 19. Goldstein−Price Function.

According to the iterative trajectories of the above six algorithms, the convergence behavior of the algorithms can be summarized into the following three types. The first type is that the convergence rate is significantly accelerated with iterations, which is mainly reflected in $f_1 \sim f_4$ of CM-HSSA. It shows that the adaptive mechanism of CM-HSSA effectively finds a meaningful search space in the initial iteration, and finds the global optimal solution more quickly. The second convergence behavior is to converge to the optimal only at the end of the iteration, which is mainly reflected in the optimization of other algorithms except for CM-HSSA. Compared with the first convergence behavior, the convergence rate of this type is significantly slower. The third type of convergence behavior is to accelerate the convergence from the initial iteration, which is reflected in the optimization of all multimodal functions of the four algorithms, and this ability of CM-HSSA is more obvious. For $f_6 \sim f_8$, based on the good-point set population distribution, CM-HSSA only needs 10 iterations to search for the optimal solution, which has a faster convergence rate compared with other algorithms. When CM-HSSA determines the search direction, it can quickly converge to the optimal accuracy, which is also reflected in the optimization of $f_9 \sim f_{10}$.

In summary, through the performance test of the benchmark functions, it is preliminarily verified that the improved strategy is effective. Compared with other algorithms, the results of CM-HSSA have significant advantages in most tests (42/50), improved the convergence rate and accuracy of the original SSA, and also obtain a stable enhancement in robustness. In the next section, the performance of six algorithms for dynamic optimization problems is further investigated under a universal swarm-intelligence dynamic optimization framework.

4. Case Studies in Dynamic Optimization

In this section, three typical dynamic optimization problems are selected as the research targets. A universal swarm-intelligence dynamic optimization framework is used to further analyze the performance of WOA, MPA, HHO, SSA, SGO, and the proposed CM-HSSA for dynamic optimization problems, and the results are compared with existing references. Specifically, the piecewise constant based on the equal division method is used to approximate the control variable, and an infinite-dimensional dynamic optimization problem is transformed into a finite-dimensional static optimization problem, which can be solved by six algorithms. To calculate the values of state variables and objective functions, the fourth-order Runge–Kutta method is used to solve the initial value problem of differential equations in each interval to obtain high-precision numerical solutions. In addition, the three cases are calculated by segments $NE = 100$. All algorithms set the population to 200 and the maximum iteration to 1000. The specific parameters of each algorithm are the same as those in 3.3.1. Each case is tested 20 times independently and the mean value, standard deviation, and calculation time (s) of the results are recorded.

4.1. Problem 1: Batch Reactor Consecutive Reaction

The batch reactor consecutive reaction is a classic dynamic optimization problem that has been widely cited as a research object. For a batch reactor with a constructive chemical reaction, temperature control plays a key role in the formation of products. In the initial stage, it is necessary to provide a higher temperature to meet the conditions of reaction start-up. With the progress of the reaction, the temperature needs to be continuously reduced to ensure the maximum concentration of the target product. Therefore, the optimization goal of this problem is to determine an optimal temperature control trajectory to optimize the concentration of target product *B* generated by reactant *A* within 1 h of reaction. The mathematical model of batch reactor constructive reaction problem is described as follows [49]:

$$\max J(t_f) = C_B(t_f)$$
$$s.t.\begin{cases} \frac{dC_A}{dt} = -k_1 C_A^2 \\ \frac{dC_B}{dt} = k_1 C_A^2 - k_2 C_B \\ t_f = 1 \\ k_1 = 4 \times 10^3 \times e^{-2500/T} \\ k_2 = 6.2 \times 10^5 \times e^{-5000/T} \\ 298 \le T \le 398,\ C_A(0) = 1,\ C_B(0) = 0 \end{cases} \tag{17}$$

where C_A is the reactant concentration, C_B is the target product concentration, T is the reaction temperature, and t_f is the reaction termination time. Figure 20 shows the iterative trajectories of six algorithms to solve problem 1 when NE = 100. Table 4 records the mean value (mol/L), standard deviation, and mean calculation time (t/s) in 20 experiments. From the experimental results, we can see the difference between CM-HSSA and other algorithms.

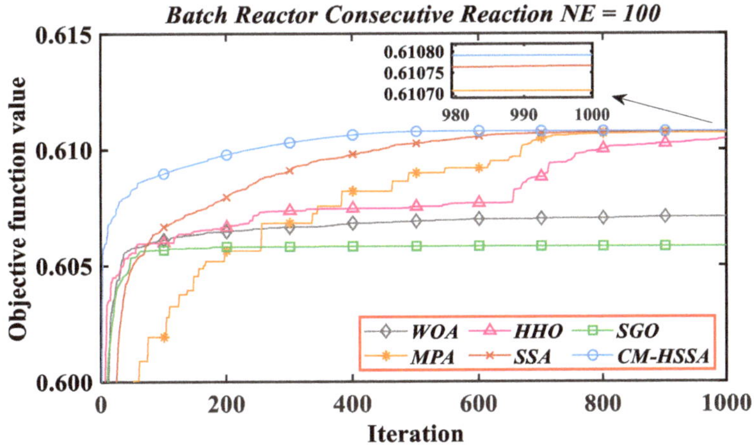

Figure 20. Iterative trajectories of six algorithms for Problem 1.

Table 4. Comparison of optimization results for Problem 1.

Method	Mean	Std.	TIC/TOC
WOA	0.60718532	5.9120×10^{-4}	359.5657
MPA	0.61070726	7.8608×10^{-4}	344.6441
HHO	0.61047035	1.9521×10^{-3}	1092.0483
SSA	0.61077333	2.4912×10^{-7}	351.2811
SGO	0.60584429	9.7315×10^{-4}	767.8168
CM-HSSA	0.61079200	2.9799×10^{-7}	347.2429

By comparing the results in Table 4, CM-HSSA has the highest accuracy, and the small standard deviation shows that the result is stable. Ranking of other algorithms: SSA > MPA > HHO > WOA > SGO. In terms of calculation time, the difference between WOA, MPA, SSA, and CM-HSSA is within 15 s, while SGO and HHO are much longer. According to the literature [3], 99.95% of the highest average accuracy of the six algorithms is defined as a satisfactory solution, and their performance of the convergence rate is evaluated according to the iterations of reaching the satisfactory solution. For problem 1, the satisfactory solution is 0.6104866. WOA, HHO, and SGO failed to reach the satisfactory solution. CM-HSSA takes 346 iterations to reach the satisfactory solution, while MPA and SSA take 706 iterations and 573 iterations. That is to say, compared with MPA and SSA, iterations are reduced by 50.99% and 39.61%, respectively, with CM-HSSA. Figure 21 shows the optimal control trajectory and optimal state variable trajectory of CM-HSSA solving problem 1. To further illustrate the advantages of the obtained results, the data in different references are recorded and compared with CM-HSSA, as shown in Table 5.

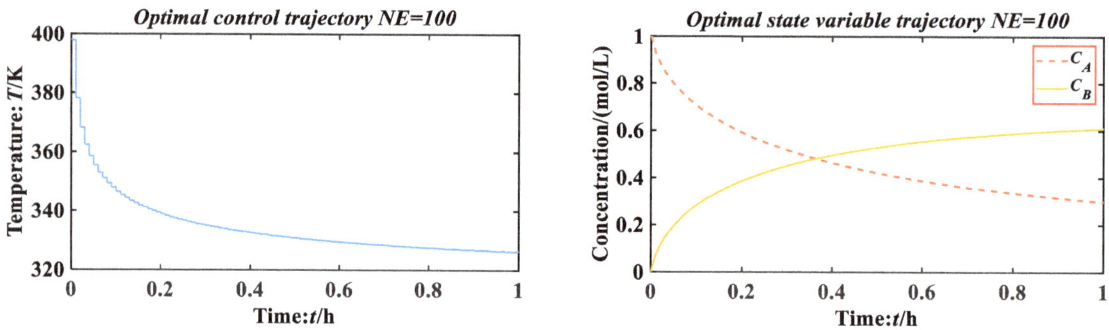

Figure 21. CM-HSSA for Problem 1.

Table 5. Comparison of optimization results for Problem 1.

Method	*NE*	*J*/(mol/L)
OC [50]	-	0.61
SQP [51]	80	0.610775
IDP [52]	80	0.610775
PSO-CVP [12]	-	0.6105359
	10	0.6101
IKEA [53]	20	0.610426
	100	0.610781–0.610789
HIGA [54]	10	0.61007
	20	0.61046
	10	0.6101
IKBCA [17]	20	0.610454
	100	0.610779–0.610787
	10	0.610558922
EBSO [13]	20	0.61064758
	80	0.61078114
MSFO [16]	50	0.610771–0.610785
ISOA [15]	30	0.61059223
CVP-PSO [3]	-	0.6107847
CVP-APSO [3]	-	0.6107850
This work (CM-HSSA)	100	0.61079200

According to Table 5: Renfro et al. [50] obtained a result of 0.61 using the orthogonal collocation (OC) method, Logsdon et al. [51] obtained 0.610775 by using the SQP strategy, while the iterative dynamic programming (IDP) method used in reference [52] obtained the same result, Shi et al. [12] used PSO to solve the problem under the CVP framework and obtained 0.6105359, Peng et al. [53] obtained 0.610781 to 0.610789 by using the proposed

IKEA, and the accuracy is slightly better than HIGA [54], which is generally consistent with the results of IKBCK [17]. The EBSO proposed by Lyu et al. [13] is better than the three algorithms mentioned above when the number of segments is small, but the accuracy improved by the algorithm is not obvious through the increased segments, The MSFO used by Zhang et al. [16] best obtained 0.610785. The ISOA proposed by Xu et al. [15] obtained 0.61059223 in the case of equal division of 30 segments. The results are poor compared with EBSO [13], and the accuracy is limited due to the small number of segments. In this paper, CM-HSSA is used to solve problem 1, and 0.61079200 is obtained in the case of equal division of 100 segments. By comparing the above references, it can be seen that the result obtained by CM-HSSA is the best, reaching the level of 0.61079, which is slightly better than the best value of 0.6107850 in reference [3], which shows that CM-HSSA is feasible and effective to solve the batch reactor consecutive reaction problem.

4.2. Problem 2: Catalyst Mixing Reaction in Tubular Reactor

The problem of the catalyst mixing reaction in the tubular reactor was first proposed by Gunn et al. [55] in 1965. This problem can be briefly described as: in a tubular reactor with a certain length, the two catalysts A and B are mixed to produce the target product C. Therefore, the dynamic optimization involved in this problem is to optimize the output of target product C in a fixed-length tubular reactor by regulating the catalyst concentration in the mixture. It is worth noting that the reaction process takes place in an isothermal tubular reactor by default. The mathematical model of catalyst mixing reaction in the tubular reactor is as follows:

$$\max J(z_f) = 1 - x_A(z_f) - x_B(z_f)$$
$$s.t.\begin{cases} \frac{dx_A}{dz} = -u(z)[10 \times x_B(z) - x_A(z)] \\ \frac{dx_B}{dz} = u(z)[10 \times x_B(z) - x_A(z)] - [1 - u(z)] \times x_B(z) \\ z_f = 12 \\ 0 \le u(z) \le 1,\ x_A(0) = 1,\ x_B(0) = 0 \end{cases} \tag{18}$$

where x_A and x_B are the mole fractions of A and B in the mixture, z_f is the length of the tubular reactor, and $u(z)$ is the mixing fraction of catalyst A. Figure 22 shows the iterative trajectories of six algorithms to solve problem 2 when NE = 100. Table 6 records the mean value (mol/L), standard deviation, and mean calculation time (t/s) in 20 experiments.

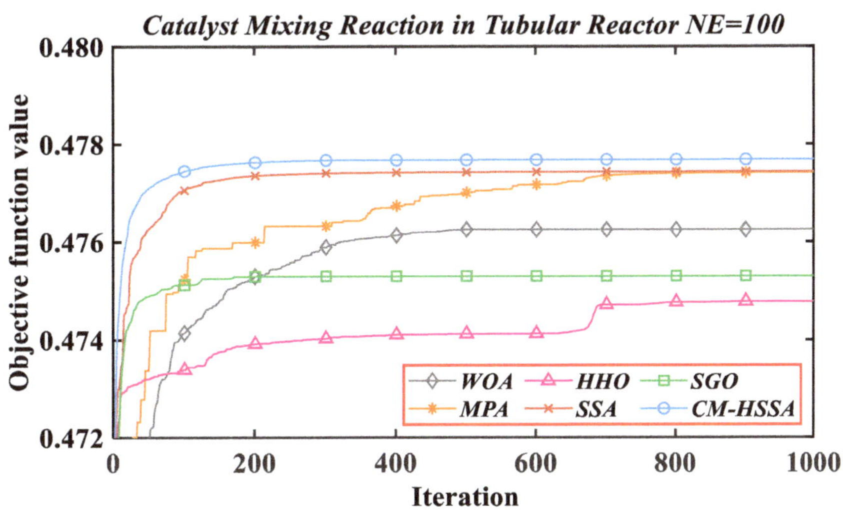

Figure 22. Iterative trajectories of six algorithms for Problem 2.

Table 6. Comparison of optimization results for Problem 2.

Method	Mean	Std.	TIC/TOC
WOA	0.47625678	5.8233×10^{-4}	426.7259
MPA	0.47742011	6.4257×10^{-4}	1034.2457
HHO	0.47478338	1.7305×10^{-3}	1213.1210
SSA	0.47744034	9.2403×10^{-4}	503.7366
SGO	0.47530289	3.8821×10^{-4}	1161.0388
CM-HSSA	0.47770179	2.7368×10^{-5}	457.1058

It can be seen from the results in Table 6 that CM-HSSA has the most stable and highest solution accuracy. In terms of optimization accuracy, ranking of other algorithms is as follows: SSA > MPA > WOA > SGO > HHO. In terms of calculation time, the difference between WOA, SSA, and CM-HSSA is within 78 s, while MPA, SGO, and HHO are much longer. In terms of convergence rate, only CM-HSSA reaches a satisfactory solution and takes 103 iterations. MPA and SSA are close to the satisfactory solution of 0.47746294, but they have not achieved this value with 1000 iterations. Figure 23 shows the optimal control trajectory and optimal state variable trajectory of CM-HSSA solving problem 2. To further illustrate the advantages of the obtained results, the data in different references are recorded and compared with CM-HSSA, as shown in Table 7.

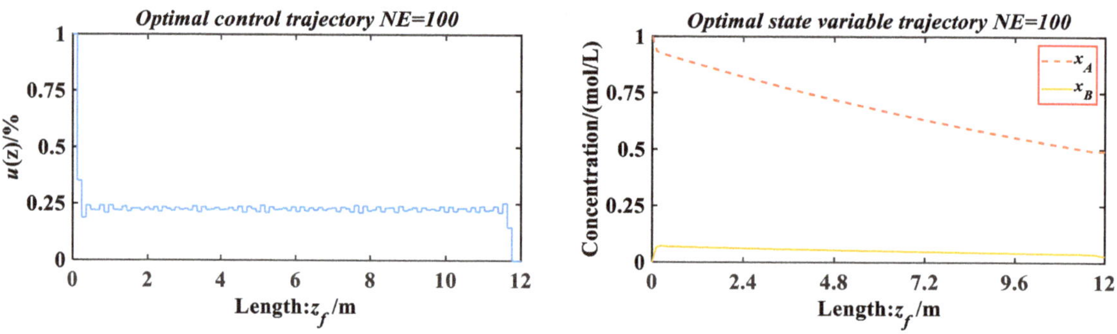

Figure 23. CM-HSSA for Problem 2.

Table 7. Comparison of optimization results for Problem 2.

Methods	*NE*	*J*/(mol/L)
IDP [52]	20	0.47527
	40	0.47695
ACO [14]	-	0.47615
IKEA [53]	10	0.475
	20	0.4757
	100	0.47761–0.47768
IKBCA [17]	20	0.4753
	100	0.47768–0.47770
EBSO [13]	10	0.47502183
	20	0.47627191
	40	0.47697288
MSFO [16]	20	0.47562
	70	0.477544–0.47760
ISOA [15]	40	0.47721
This work (CM-HSSA)	100	0.47770179

It can be seen from Table 7 that the highest accuracy of IDP [52] is 0.47695, which is slightly better than the result solved by Rajesh et al. [14] using ACO. The result obtained by MSFO [16] is 0.477544–0.47760, which is inferior to the accuracy achieved by IKEA [53] and

IKBCA [17]. ISOA [15] achieved 0.47721, and its accuracy is better than that of EBSO [13] in the same segmentation. When the segment is 100, CM-HSSA obtains a result of 0.47770179 for solving the catalyst mixing reaction in a tubular reactor compared with other references; only the proposed algorithm can stably reach the level of 0.4777, while the accuracy of different methods hovers in the range of 0.475–0.476. Therefore, CM-HSSA has a better solution effect than other methods, which further proves the effectiveness of the algorithm proposed in this paper.

4.3. Problem 3: Parallel Reactions in Tubular Reactor

The parallel reaction problem in the tubular reactor is a dynamic optimization problem with saturation characteristics of control variables [56], and it has been cited by many researchers. In the tubular reactor, there is a side reaction process ($A \rightarrow C$) parallel to the main reaction ($A \rightarrow B$), so the optimization goal of this problem is to maximize the target product B of the main reaction at the end by determining an optimal control trajectory. Similarly, all reactions of this problem occur in an isothermal tubular reactor by default. The mathematical model of parallel reactions in the tubular reactor is as follows:

$$\max J(t_f) = x_B(t_f)$$
$$s.t. \begin{cases} \frac{dx_A}{dt} = -[u(t) + 0.5u^2(t)]x_A(t) \\ \frac{dx_B}{dt} = u(t)x_A(t) \\ t_f = 1 \\ 0 \le u(t) \le 5,\ x_A(0) = 1,\ x_B(0) = 0 \end{cases} \tag{19}$$

where x_A is the concentration of reactant A, x_B is the concentration of target product B, t_f is the reaction termination time, and $u(t)$ is the saturation of the control variable. Figure 24 shows the iterative trajectories of six algorithms to solve problem 3 when NE = 100. Table 8 records the mean value (mol/L), standard deviation, and mean calculation time (t/s) in 20 experiments.

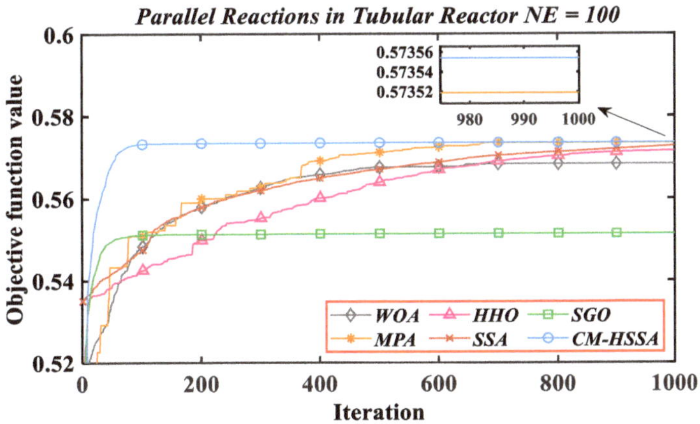

Figure 24. Iterative trajectories of six algorithms for Problem 3.

Table 8. Comparison of optimization results for Problem 3.

Method	Mean	Std.	TIC/TOC
WOA	0.56836465	1.4475×10^{-3}	364.3671
MPA	0.57349880	1.0338×10^{-3}	381.3942
HHO	0.57152795	2.9227×10^{-3}	1062.2493
SSA	0.57269740	8.7921×10^{-3}	349.0882
SGO	0.55138595	3.5277×10^{-4}	685.3328
CM-HSSA	0.57355371	4.2218×10^{-6}	376.5377

By comparing the results in Table 8, CM-HSSA also obtains the highest accuracy and the smallest standard deviation among the six algorithms, and the ranking of other algorithms is as follows: MPA > SSA > HHO > WOA > SGO. In terms of calculation time, except for SGO and HHO, the difference between the other four algorithms is within 33 s. In terms of convergence rate, CM-HSSA takes only 98 iterations to achieve a satisfactory solution of 0.57326693, while MPA takes 671 iterations, so the iterations are reduced by 85.39% with CM-HSSA. Other algorithms failed to reach this value. Through comprehensive comparison, CM-HSSA has more advantages and wider applicability in optimization performance and calculation efficiency. Figure 25 shows the optimal control trajectory and optimal state variable trajectory of CM-HSSA solving problem 3. To further confirm the superiority of the results obtained, the data in different references are recorded and compared with CM-HSSA, as shown in Table 9.

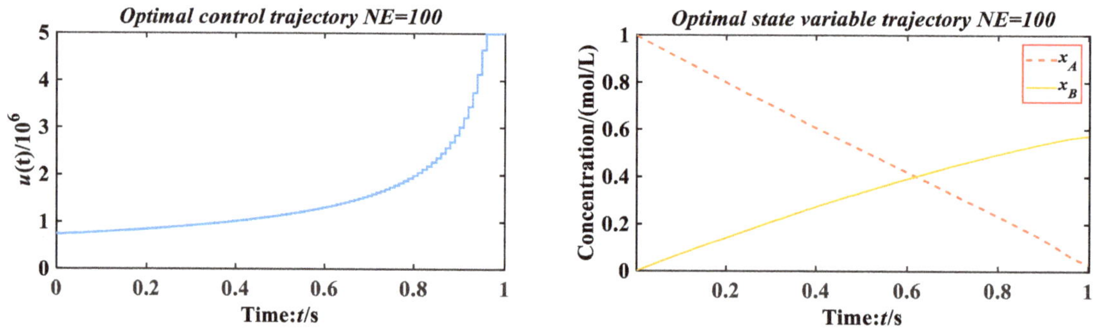

Figure 25. CM-HSSA for Problem 3.

Table 9. Comparison of optimization results for Problem 3.

Methods	*NE*	*J*/(mol/L)
IDP [52]	20	0.57330
	40	0.57348
	80	0.57353
CVP [57]	-	0.56910
CVI [57]	-	0.57322
ACO [14]	-	0.57284
CP-PSO [3]	-	0.573543
CP-APSO [3]	-	0.573544
ISOA [15]	40	0.573073
This work (CM-HSSA)	100	0.57355371

According to Table 9, the IDP [52] divided into 80 segments has the best result of 0.57353. Biegler [57] proposed combining successive quadratic programming and orthogonal collocation, and obtained 0.56910 and 0.57322 based on CVP and control vector iteration (CVI), respectively. ACO [14] solved the problem and obtained 0.57284, which is lower than other methods. Zhou et al. [3] proposed a dynamic optimization control parameter solution and obtained 0.573544 using APSO, which is improved compared with PSO. Xu et al. [15] divided it into 40 segments and obtained the best result of 0.573073. Compared with these references, the solution of CM-HSSA has reached the highest accuracy level of the current optimization for this problem, and is slightly better than the result of 0.573544 obtained in reference [3], which further verifies the ability of CM-HSSA to solve dynamic optimization problems.

5. Conclusions

This manuscript introduced a novel SSA algorithm named CM-HSSA that further enhanced both the exploration and exploitation abilities of the original method. Through the benchmark function experiments, compared with WOA, MPA, HHO, SSA, and SGO, the statistical results verify that CM-HSSA has more advantages in stability, accuracy, and convergence rate. Under a universal swarm-intelligence dynamic optimization framework, the above six algorithms are used to solve three typical chemical dynamic optimization problems, and the simulation results further validate the applicability of CM-HSSA to solve the dynamic optimization problems. Compared with different methods in the literature, CM-HSSA also achieved the best results.

For dynamic optimization problems, in addition to the performance of the optimization algorithm affecting the final results, the segmentation of the time domain and the selection of the approximation method for control variables will also have different effects on the solution. In future work, we will conduct more research on these two aspects, and use high-performance optimization algorithms to solve complex dynamic optimization problems with strong nonlinearity.

Author Contributions: Conceptualization, R.L. and Y.M.; methodology, R.L.; software, Y.L. (Yucheng Lyu); validation, Y.M. and Y.L. (Yanyue Lu); formal analysis, H.G.; investigation, Y.Z.; writing—original draft preparation, R.L.; writing—review and editing, R.L. and Y.M.; visualization, R.L.; supervision, Y.M. All authors have read and agreed to the published version of the manuscript.

Funding: This work was supported by the National Natural Science Foundation of China (21466008, 21968008); the Project of the Natural Science Foundation of Guangxi Province (2019GXNSFAA185017); the Scientific Research Project of Guangxi University for Nationalities (2021MDKJ004); the Innovation Project of Guangxi University for Nationalities Graduate Education (gxun-chxs2021064).

Institutional Review Board Statement: Not applicable.

Informed Consent Statement: Not applicable.

Data Availability Statement: Not applicable.

Acknowledgments: The authors acknowledge support by the National Natural Science Foundation of China (21466008); the Project of the Natural Science Foundation of Guangxi Province (2019GXNSFAA185017); the Scientific Research Project of Guangxi University for Nationalities (2021MDKJ004); the Innovation Project of Guangxi University for Nationalities Graduate Education (gxun-chxs2021064); the National Natural Science Foundation of China (21968008).

Conflicts of Interest: The authors declare no potential conflict of interest.

References

1. Zhang, B.; Chen, D.; Wu, X. Graded optimization strategy and its application to chemical dynamic optimization with fixed boundary. *CIESC J.* **2005**, *56*, 1276–1280.
2. Mo, Y.; Chen, D. Chaos particle swarm optimization algorithm and its application in biochemical process dynamic optimization. *CIESC J.* **2006**, *57*, 2123–2127.
3. Zhou, Y.; Liu, X. Control parameterization-based adaptive particle swarm approach for solving chemical dynamic optimization problems. *Chem. Eng. Technol.* **2014**, *37*, 692–702. [CrossRef]
4. Sun, Y.; Zhang, M.; Liang, X. Improved Gauss pseudospectral method for solving a nonlinear optimal control problem with complex constraints. *Acta Autom. Sin.* **2013**, *39*, 672–678. [CrossRef]
5. Chachuat, B.; Mitsos, A.; Barton, P.I. Optimal design and steady-state operation of micro power generation employing fuel cells. *Chem. Eng. Sci.* **2005**, *60*, 4535–4556. [CrossRef]
6. Peng, H.; Gao, Q.; Wu, Z.; Zhong, W. A mixed variable variational method for optimal control problems with applications in aerospace control. *Acta Autom. Sin.* **2011**, *37*, 1248–1255.
7. Pollard, G.P.; Sargent, R.W.H. Off line computation of optimum controls for a plate distillation column. *Automatica* **1970**, *6*, 59–76. [CrossRef]
8. Schranz, M.; Di Caro, G.A.; Schmickl, T.; Elmenreich, W. Swarm intelligence and cyber-physical systems: Concepts, challenges and future trends. *Swarm Evol. Comput.* **2021**, *60*, 100762. [CrossRef]
9. Hussain, K.; Mohd, M.N.; Cheng, S. Metaheuristic research: A comprehensive survey. *Artif. Intell. Rev.* **2019**, *52*, 2191–2233. [CrossRef]
10. Bacanin, N.; Stoean, R.; Zivkovic, M.; Petrovic, A.; Rashid, T.A.; Bezdan, T. Performance of a novel chaotic firefly algorithm with enhanced exploration for tackling global optimization problems: Application for dropout regularization. *Mathematics* **2021**, *9*, 2705. [CrossRef]

11. Malakar, S.; Ghosh, M.; Bhowmik, S.; Sarkar, R.; Nasipuri, M. A GA based hierarchical feature selection approach for handwritten word recognition. *Neural Comput. Appl.* **2020**, *32*, 2533–2552. [CrossRef]
12. Shi, B.; Yin, Y.; Liu, F. Optimal control strategies combined with PSO and control vector parameterization for batchwise chemical process. *CIESC J.* **2019**, *70*, 979–986.
13. Lyu, Y.; Mo, Y.; Lu, Y.; Liu, R. Enhanced Beetle Antennae Algorithm for Chemical Dynamic Optimization Problems' Non-Fixed Points Discrete Solution. *Processes* **2022**, *10*, 148. [CrossRef]
14. Asgari, S.A.; Pishvaie, M.R. Dynamic Optimization in Chemical Processes Using Region Reduction Strategy and Control Vector Parameterization with an Ant Colony Optimization Algorithm. *Chem. Eng. Technol. Ind. Chem. Plant Equip. Process. Eng. Biotechnol.* **2008**, *31*, 507–512. [CrossRef]
15. Xu, L.; Mo, Y.; Lu, Y. Improved Seagull Optimization Algorithm Combined with an Unequal Division Method to Solve Dynamic Optimization Problems. *Processes* **2021**, *9*, 1037. [CrossRef]
16. Zhang, Y.; Mo, Y. Dynamic optimization of chemical processes based on modified sailfish optimizer combined with an equal division method. *Processes* **2021**, *9*, 1806. [CrossRef]
17. Liu, Z.; Du, W.L.; Qi, R. Dynamic optimization in chemical processes using improved knowledge-based cultural algorithm. *CIESC J.* **2010**, *61*, 2889–2895.
18. Mirjalili, S.; Lewis, A. The whale optimization algorithm. *Adv. Eng. Softw.* **2016**, *95*, 51–67. [CrossRef]
19. Faramarzi, A.; Heidarinejad, M.; Mirjalili, S. Marine Predators Algorithm: A nature-inspired metaheuristic. *Expert Syst. Appl.* **2020**, *152*, 113377. [CrossRef]
20. Heidari, A.A.; Mirjalili, S.; Faris, H. Harris hawks optimization: Algorithm and applications. *Future Gener. Comput. Syst.* **2019**, *97*, 849–872. [CrossRef]
21. Satapathy, S.; Naik, A. Social group optimization (SGO): A new population evolutionary optimization technique. *Complex Intell. Syst.* **2016**, *2*, 173–203. [CrossRef]
22. Fathy, A.; Alanazi, T.M.; Rezk, H.; Yousri, D. Optimal energy management of micro-grid using sparrow search algorithm. *Energy Rep.* **2022**, *8*, 758–773. [CrossRef]
23. Zhang, Z.; Han, Y. Discrete sparrow search algorithm for symmetric traveling salesman problem. *Appl. Soft Comput.* **2022**, *118*, 108469. [CrossRef]
24. Yan, S.; Yang, P.; Zhu, D. Improved Sparrow Search Algorithm Based on Iterative Local Search. *Comput. Intell. Neurosci.* **2021**, *2021*, 6860503. [CrossRef]
25. Xiong, Q.; Zhang, X.; He, S. A Fractional-Order Chaotic Sparrow Search Algorithm for Enhancement of Long Distance Iris Image. *Mathematics* **2021**, *9*, 2790. [CrossRef]
26. Zhang, C.; Ding, S. A stochastic configuration network based on chaotic sparrow search algorithm. *Knowl. Based Syst.* **2021**, *220*, 106924. [CrossRef]
27. Mirjalili, S.; Mirjalili, S.M.; Lewis, A. Grey wolf optimizer. *Adv. Eng. Softw.* **2014**, *69*, 46–61. [CrossRef]
28. Rashedi, E.; Nezamabadi-Pour, H.; Saryazdi, S. GSA: A gravitational search algorithm. *Inf. Sci.* **2009**, *179*, 2232–2248. [CrossRef]
29. Mirjalili, S. SCA: A sine cosine algorithm for solving optimization problems. *Knowl. Based Syst.* **2016**, *96*, 120–133. [CrossRef]
30. Liu, G.; Shu, C.; Liang, Z. A modified sparrow search algorithm with application in 3d route planning for UAV. *Sensors* **2021**, *21*, 1224. [CrossRef]
31. Ma, J.; Hao, Z.; Sun, W. Enhancing sparrow search algorithm via multi-strategies for continuous optimization problems. *Inf. Process. Manag.* **2022**, *59*, 102854. [CrossRef]
32. Shi, L.; Ding, X.; Li, M. Research on the capability maturity evaluation of intelligent manufacturing based on firefly algorithm, sparrow search algorithm, and BP neural network. *Complexity* **2021**, *2021*, 5554215. [CrossRef]
33. Ouyang, C.; Qiu, Y.; Zhu, D. Adaptive spiral flying sparrow search algorithm. *Sci. Program.* **2021**, *2021*, 6505253. [CrossRef]
34. Nguyen, T.T.; Ngo, T.G.; Dao, T.K. Microgrid Operations Planning Based on Improving the Flying Sparrow Search Algorithm. *Symmetry* **2022**, *14*, 168. [CrossRef]
35. Chen, X.; Tianfield, H.; Mei, C. Biogeography-based learning particle swarm optimization. *Soft Comput.* **2017**, *21*, 7519–7541. [CrossRef]
36. Luyben, W.L. Optimum product recovery in chemical process design. *Ind. Eng. Chem. Res.* **2014**, *53*, 16044–16050. [CrossRef]
37. Tavazoei, M.S.; Haeri, M. Comparison of different one-dimensional maps as chaotic search pattern in chaos optimization algorithms. *Appl. Math. Comput.* **2007**, *187*, 1076–1085. [CrossRef]
38. Li, Y.; Han, M.; Guo, Q. Modified whale optimization algorithm based on tent chaotic mapping and its application in structural optimization. *KSCE J. Civ. Eng.* **2020**, *24*, 3703–3713. [CrossRef]
39. Varol Altay, E.; Alatas, B. Bird swarm algorithms with chaotic mapping. *Artif. Intell. Rev.* **2020**, *53*, 1373–1414. [CrossRef]
40. Han, X.H.; Xiong, X.; Duan, F. A new method for image segmentation based on BP neural network and gravitational search algorithm enhanced by cat chaotic mapping. *Appl. Intell.* **2015**, *43*, 855–873. [CrossRef]
41. Sayed, G.I.; Tharwat, A.; Hassanien, A.E. Chaotic dragonfly algorithm: An improved metaheuristic algorithm for feature selection. *Appl. Intell.* **2019**, *49*, 188–205. [CrossRef]
42. Koyuncu, H. GM-CPSO: A new viewpoint to chaotic particle swarm optimization via Gauss map. *Neural Process. Lett.* **2020**, *52*, 241–266. [CrossRef]

43. Yuan, J.; Liu, Z.; Lian, Y. Global Optimization of UAV Area Coverage Path Planning Based on Good Point Set and Genetic Algorithm. *Aerospace* **2022**, *9*, 86. [CrossRef]
44. Tang, Y.; Wang, Z.; Fang, J. Feedback learning particle swarm optimization. *Appl. Soft Comput.* **2011**, *11*, 4713–4725. [CrossRef]
45. Brockmann, D.; Hufnagel, L.; Geisel, T. The scaling laws of human travel. *Nature* **2006**, *439*, 462–465. [CrossRef]
46. Mantegna, R.N. Fast, accurate algorithm for numerical simulation of Levy stable stochastic processes. *Phys. Rev. E* **1994**, *49*, 4677. [CrossRef]
47. Mirjalili, S. Dragonfly algorithm: A new meta-heuristic optimization technique for solving single-objective, discrete, and multi-objective problems. *Neural Comput. Appl.* **2016**, *27*, 1053–1073. [CrossRef]
48. Derrac, J.; García, S.; Molina, D. A practical tutorial on the use of nonparametric statistical tests as a methodology for comparing evolutionary and swarm intelligence algorithms. *Swarm Evol. Comput.* **2011**, *1*, 3–18. [CrossRef]
49. Rajesh, J.; Gupta, K.; Kusumakar, H.S. Dynamic optimization of chemical processes using ant colony framework. *Comput. Chem.* **2001**, *25*, 583–595. [CrossRef]
50. Renfro, J.G.; Morshedi, A.M.; Asbjornsen, O.A. Simultaneous optimization and solution of systems described by differential/algebraic equations. *Comput. Chem. Eng.* **1987**, *11*, 503–517. [CrossRef]
51. Logsdon, J.S.; Biegler, L.T. A relaxed reduced space SQP strategy for dynamic optimization problems. *Comput. Chem. Eng.* **1993**, *17*, 367–372. [CrossRef]
52. Dadebo, S.A.; McAuley, K.B. Dynamic optimization of constrained chemical engineering problems using dynamic programming. *Comput. Chem. Eng.* **1995**, *19*, 513–525. [CrossRef]
53. Peng, X.; Qi, R.; Du, W.; Qian, F. An improved knowledge evolution algorithm and its application to chemical process dynamic optimization. *CIESC J.* **2012**, *63*, 841–850.
54. Sun, F.; Du, W.; Qi, R. A hybrid improved genetic algorithm and its application in dynamic optimization problems of chemical processes. *Chin. J. Chem. Eng.* **2013**, *21*, 144–154. [CrossRef]
55. Gunn, D.J.; Thomas, W.J. Mass transport and chemical reaction in multifunctional catalyst systems. *Chem. Eng. Sci.* **1965**, *20*, 89–100. [CrossRef]
56. Liu, X.; Chen, L.; Hu, Y. Solution of chemical dynamic optimization using the simultaneous strategies. *Chin. J. Chem. Eng.* **2013**, *21*, 55–63. [CrossRef]
57. Biegler, L.T. Solution of dynamic optimization problems by successive quadratic programming and orthogonal collocation. *Comput. Chem. Eng.* **1984**, *8*, 243–247. [CrossRef]

applied sciences

MDPI

Article

Harvesting the Aggregate Computing Power of Commodity Computers for Supercomputing Applications

Dereje Regassa [1], Heonyoung Yeom [1] and Yongseok Son [2,*]

[1] Department of Computer Science and Engineering, Seoul National University, Seoul 08826, Korea; dereje@snu.ac.kr (D.R.); yeom@snu.ac.kr (H.Y.)
[2] Department of Computer Science and Engineering, Chung-Ang University, Seoul 06974, Korea
* Correspondence: sysganda@cau.ac.kr

Abstract: Distributed supercomputing is becoming common in different companies and academia. Most of the parallel computing researchers focused on harnessing the power of commodity processors and even internet computers to aggregate their computation powers to solve computationally complex problems. Using flexible commodity cluster computers for supercomputing workloads over a dedicated supercomputer and expensive high-performance computing (HPC) infrastructure is cost-effective. Its scalable nature can make it better employed to the available organizational resources, which can benefit researchers who aim to conduct numerous repetitive calculations on small to large volumes of data to obtain valid results in a reasonable time. In this paper, we design and implement an HPC-based supercomputing facility from commodity computers at an organizational level to provide two separate implementations for cluster-based supercomputing using Hadoop and Spark-based HPC clusters, primarily for data-intensive jobs and Torque-based clusters for Multiple Instruction Multiple Data (MIMD) workloads. The performance of these clusters is measured through extensive experimentation. With the implementation of the message passing interface, the performance of the Spark and Torque clusters is increased by 16.6% for repetitive applications and by 73.68% for computation-intensive applications with a speedup of 1.79 and 2.47 respectively on the HPDA cluster. We conclude that the specific application or job could be chosen to run based on the computation parameters on the implemented clusters.

Keywords: HPC; shared memory; optimization; commodity hardware; big data

Citation: Dereje, R.; Yeom, H.; Yongseok, S. Harvesting the Aggregate Computing Power of Commodity Computers for Supercomputing Applications. *Appl. Sci.* **2022**, *12*, 5113. https://doi.org/10.3390/app12105113

Academic Editor: Agostino Forestiero

Received: 28 April 2022
Accepted: 16 May 2022
Published: 19 May 2022

1. Introduction

High-performance computing (HPC) is defined in terms of distributed, parallel computing infrastructure with high-speed interconnecting networks and high-speed network interfaces, including switches and routers specially designed to provide an aggregate performance of many-core and multicore systems, computing clusters, in a cloud or the form of a grid. Whereas the grid spans a large area and uses WAN protocols, clusters are suitable for performing scientific computations in a small area. Cluster-based supercomputing can be classified into different types based on the hardware, inter-networking standards, devices (e.g., switches), number of sockets, number of cores in a rack, and so on. Recently, applications of HPC in scientific research have greatly increased as an alternative to dedicated supercomputers [1–3], because they are costly, difficult to program, and require sophisticated expertise. Supercomputers are generally stand-alone, and scalability is a problem in supercomputer-based data centers [4]. The HPC cluster-based supercomputing platforms have been successfully implemented in various universities [1,5]. A list of open-source cluster management software [6] and some of the commercially available schedulers are available in various domains.

The university environment, where the prime goal is the transition from isolated, stand-alone computers to an HPC-based cluster computing environment, is substantially different from individual and production-constrained computational processes in many aspects. The study team expects that the transition from stand-alone laboratory experiments to computationally intensive industrial-grade experiments must be performed so that researchers can understand the establishment of the cluster environment from scratch, the node configuration, and the data network, rather than any specific hardware requirements. These seamlessly provide access to the working cluster using commodity computers connected by a high-speed network [1].

The expectations and utility of the cluster environment vary across cluster users. For example, a cluster user sometimes does not even need to know that the job is completed with the help of a backbone a supercomputing facility. Some users may need to know only the details of the available memory and the number of necessary cores. In contrast, expert users may need to know almost everything, including protocols, scheduling policies, privileges, priorities, and everything directly or indirectly relevant to optimizing application performance.

The approach of using commodity computers to form a cheap cluster is found in different universities that aim to upgrade their stand-alone lab sessions into powerful, scalable, and large-scale experiments, such as computational climate models, computational fluid dynamics, numerical weather prediction models, and many more.

In this study, we focus on obtaining expertise in commodity clusters prior to any purchases through surveying a primitive set of applications intended for demonstration. Once most of the users understand the platform, the underlying hardware can be upgraded to the actual HPC standards, which benefits data rates and computation power.

Unlike in industry, university research clusters are flexible and less demanding in terms of failure after effects. In the absence of a real-world cost of a crash or failure of an application running on the commodity cluster, laboratory experiments can tolerate the breakdown of experiments due to a node or power failure, exhaustive use of available memory, system shutdown without notice, and other issues. We identified different challenges while introducing cluster computing in a university compared to a production-grade industry setup. These challenges influence many design decisions, not only the cluster implementation but also the culture of students and instructors. The users move from a single-threaded programming paradigm to a large-scale, threaded, distributed, and parallel form of programs, a fundamental element of success in the cluster-based supercomputing facility. Training and workshops may need to be organized to introduce the available resources to the staff and students at the university to provide basic libraries to support their experiments. Hence, we develop a cost-effective distributed supercomputing platform from cheap computers around us which can be used by various researchers for different applications.

The contribution of this work is as follows:

- We analyze the existing HPC infrastructure design options, review existing cluster solutions, and identify significant challenges related to the hardware, software, and networks in the implementation of cluster computing.
- We review the existing cluster solutions implemented at different university level and advanced research laboratories.
- We implement a cluster from 30 commodity computers to aggregate Gigaflops of computing power and use an efficient scheduler to run jobs on the cluster infrastructure.
- perform scientific experiments on a newly implemented cluster with distributed optimization and machine learning for a showcase.
- compare the practical performance of the proposed system under different configurations, considering various algorithms and loads.

The rest of this paper is organized as follows. Section 2 describes the background and motivation. Section 3 presents the design and implementation of the proposed idea using a cluster of commodity computers. Section 4 shows the experimental evaluation and its results. Section 5 discusses the related work and Section 6 concludes the paper with recommendations for the use of the output on a similar scale.

2. Background and Motivation

Since its inception in 1991, HPCC provided vital enabling technologies and disseminated progress in the field to extend the lead in the field of HPC. Later, more agencies joined in a coordinated effort to develop HPC. Most development and progress work in HPC was made available on the HPCC website as annual reports. This initiative led to significant improvements in HPC research.

The HPC cluster is used in implementations and experiments to compute the simulated Gaussian density function to solve a density estimation problem based on shared memory and the parallel execution of the kernel density estimation algorithm on a graphic processing unit (GPU)-enabled system [4,7], the OpenMP library [8]. The study results reveal significant improvement in the time taken by the HPC cluster in estimating the density function. Although GPU-based computing provides easy access to higher teraflops with fewer machines, this setup is only suitable for parallel jobs, whereas fine-grain parallelism, synchronization of threads, and programming GPU-enabled systems are complex tasks. Therefore, several studies have been performed on the Intel Xeon processor-based HPC cluster. These clusters have the advantages of being easy to program, supporting SIMD and MIMD models, and providing low-cost alternatives to a few teraflops with the ease of scalability.

The floating-point operations (FLOPs) [9–11] are used in scientific computing community to measure the processing power of individual computers, different types of HPC, Grid, and supercomputing facilities. China has made significant progress in developing HPC systems in recent years. Since the development of the supercomputer by their national defense university, the nation's first win of the Gordon Bell Prize at Supercomputing 2016 (SC16) also represents an accomplishment in HPC applications. As a result, various academic institutions, private companies, and research groups have collaborate and contributed to the development of HPC on their premises [3,10,12–14]. Major applications of HPC are in data storage and analysis, data mining, simulation and modeling, scientific calculations, bioinformatics, big data challenges, and complex visualizations.

However, the purchase of a supercomputer for academic and collaborative research may cost an investment of millions of dollars and therefore most universities avoid directly purchasing the supercomputers. But to provide a platform for the students and researchers to work on the most challenging scientific problems related to different fields such as computational fluid dynamics, medical imaging, graphics, higher dimensional visualization, big data, parallel machine learning, and data mining multidisciplinary optimization. Many universities have implemented low-cost supercomputing using clusters of Intel processors. Commodity computers fail to process complex jobs with very large data and computations as memory requirements and millions of times complex function evaluations. To solve bigger problems, scientists have invented cost-effective solutions to aggregate individual gigaflops of individual computers by clustering and facilitating the high-speed I/O and communication to provide a platform that performs the large-scale experiments. In this study, most reviewed research ideas are taken from HPC applications in scientific computing because the objective is to provide an HPC-based supercomputing infrastructure for various scientific computing applications.

3. Design and Implementation

This study started with specific design goals and objectives restricted to the installation of a pilot HPC cluster for future research projects. This work brings the ideas of alternative options to develop a supercomputing environment from available commodity computers for those researchers who need a supercomputing platform for their research yet can not buy the HPC. We followed the standard methodology for information technology/information and communication technology projects during the implementation of DeepStack and high-performance-data analytics (HPDA) clusters. This study used a hybrid methodology involving qualitative, quantitative, and empirical research aspects at various stages. The two stages of this study are identified as (i) the design and implementation of the HPC clusters as a platform and (ii) the empirical performance measure of various parameters of the newly implemented platform. The first objective relies on qualitative research aspects aimed at the platform design, whereas the second stage focuses on quantitative and empirical performance measures under the careful design of relevant experiments.

As the industry estimates that we are creating Quintilian bytes of data every year which makes data processing, transmission, and storage difficult it has created entirely new sets of challenges and has forced us to find new ways to handle Big data effectively [15]. From document analysis of the existing HPC infrastructure [16–19], we choose the HPC environments that are expected to perform CPU-bound operations with a heavy asymmetric load on compute nodes. Therefore, we choose the Deep Stack cluster and HDPA-based HPC clusters as the computing power of the nodes as well as message passing infrastructure between the processes plays an important role in the performance of application programs.

In our design-oriented approach to the installation of the HPC cluster platform, the initially necessary data collection from various sources and experts has been done, including standards, benchmarks, heuristics, networks, topology, workload patterns, optimization methods, node configuration profiles, and detailed description of protocol selection procedures. The analyzed data were used for planning, topology design, network installation, and the actual design and implementation phase for the HPC cluster platform. After the platform was successfully set up, the empirical methodology was used, and the design of experiments, performance analysis, and data analysis tasks was completed.

The following methodologies are followed during the design and implementation of this research:

- **Data Collection**—different data sources were used to collect necessary information regarding hardware, software details and design issues, and availability of cluster management systems. Data was collected from various websites related to university-level HPC cluster systems from Stanford University, FSU, MIT, and Yale. Expert interviews were performed at different levels of professionals in the fields of HPC to get specific requirements for the design and implementation of the proposed platforms.
- **Data pre-processing**—Once the data were collected and encoded in a standard format, the pre-processing of data was performed including checks for consistency, validity, missing values, errors, and updates.
- **Planning and design of HPC platforms**—Based on the input of various experts and available HPC installations in the industry, these design parameters for the proposed systems were set. This included selection of appropriate network topology, protocols, hardware, and software components to enable the HPC server and compute nodes each.
- **Implementation**—The actual implementation of two separate types of HPC clusters was initiated within separate initiatives of data science and intelligent systems research groups. The implementation procedure was divided into various phases as per the life cycle of an ICT/IT project. On each HPC server and compute-node internetworking, protocol configuration, installation, and configuration of the servers and client software were completed sequentially.
- **Verification and Validation** —the recent installation against the requirements and objectives of the proposed research are verified. The correctness of installation and availability of HPC cluster and system-level services of the new HPC system was checked.
- **Design of experiments**—as the platform was implemented on available hardware, we have designed structured workloads for various experiments and tested the performance of the underlying platform under different settings. Since it is experimental research, we have designed two groups of experiments—the control group and the experiment groups. In the control group experiment, we have used a standard algorithms like the FIFO scheduling algorithm with default parameters for the installed platform that will be used as a base for comparison, in the experiment group we have an experimental setup with varying platform parameters like the number of

cores, memory, virtual memory, default wall-times, network topology and allocated bandwidth. These experiments were performed on introductory level workloads and the optimum performance profile of the underlying platform was deducted.
- **Data Analysis**—after the execution of planned stages and experiments on real time data, data analysis was performed to understand the performance of the system, and interpretation and implications of various parameters changes in newly installed HPC clusters.

3.1. DeepStack Cluster

To save the readers from confusion regarding general torque specifications and our specific implementation we have given a unique name for our implementation of torque cluster as the DeepStack HPC cluster. This cluster includes three 5-node of identical Intel core i-7 clusters with approximately 15 TB storage and 120 GB of RAM in the experimental phase with specifications given in Table 1. However, to get a better result, we have projections of 30-core permanently connected compute nodes under DeepStack and 100 Ad-hoc compute nodes during runtime but not guaranteed. It is possible to extend the capabilities of DeepStack as a source of the highest computing power and a full-fledged implementation of the scientific libraries for various scientific jobs. Using Torque with this setup enabled us in achieving our objectives of running Torque with MPICH2 and OpenMPI for large-scale, distributed memory jobs over available processors from configured compute nodes, and support various scientific libraries.

Table 1. DeepStack Cluster Specifications.

Parameters	Used Values
Master Machine	CPU 2.25 Ghz RAM—8 GB HDD 1 TB OS—CentOS 7
Computing Nodes	CPU 2.25 Ghz RAM—8 GB HDD 1 TB OS—CentOS 7
Libraries	netcdf/intel/3.6.3 netcdf/intel/4.1.1 Stata/11, R/intel/2.9.2, matlab/2017/b
Software Packages	Cent OS Enterprise Server 7, Torque 6.1.1,OpenMPI MPICH
Language Support	C++, Java, Fortran

In this study, each Torque subcluster is implemented from scratch and involves installation and configuring various server-side dependencies and libraries. The DeepStack cluster architecture is given in Figure 1.

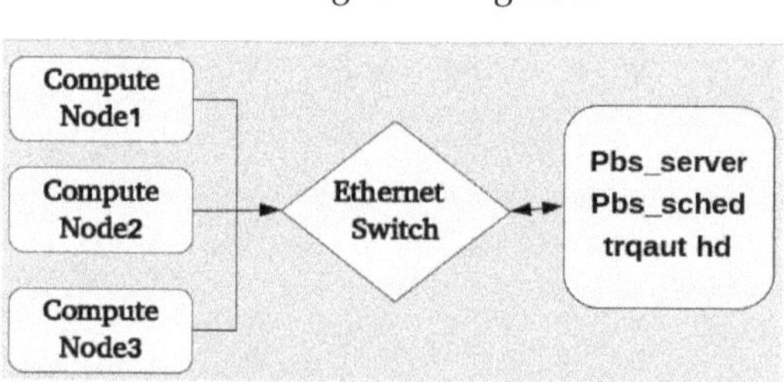

Figure 1. DeepStack cluster Architecture.

3.1.1. Resource Manager Components

Each torque resource manager designated as a server computer contains three components:

1. Torque Server module (PBS_server)—In this architecture, each Torque server is given the same hostname and server name per the recommendation of Adaptive Computing Pvt. Ltd., Naples, FL USA, which developed Torque as open-source cluster management software. Each Torque server is an instance of PBS Server. Each Torque server comes with a default queue called the batch. Certain configuration steps should be done in sequence to ensure the server installs with its dependencies.
2. Scheduler module (PBS_sched)—Torque provides a very basic built-in scheduler capable of scheduling jobs in first-in-first-out order in job-exclusive and shared mode on compute nodes. If this cluster is used for scientific simulations, it must be configured with advanced schedulers, such as MOAB and MUAI, which are not free like the default scheduler. The default scheduler is enough for the start cluster computing. Each server instance is configured with the default scheduler.
3. Authentication server module (TRQAUTHD)—as described as trquat hd in Figure 1 —Torque provides an authentication server(TRQAUTHD) whose job is to allow only pre-configured clients to submit jobs. An instance of TRQAUTHD is required for the PBS_server along with the PBS_sched demon. By default, the server computer is allowed to submit jobs on the server, but the purpose of TRQAUTHD is to register the clients with a server so that jobs can be submitted by the external IP address registered using TRQAUTHD.

3.1.2. Torque Compute Node Components

The Torque compute node contains the PBS Multi operation machine (MoM) modules that participate in job execution in the processor exclusive and processor shared modes. If a compute node is enabled to submit jobs for execution, the TRQAUTHD client-side module must be configured with compute node.

1. The PBS MoM module communicates with the PBS server and participates in the actual execution of jobs. It provides various resources for job execution, such as processors, memory, and wall-time (the actual time that a clock on the wall or a stopwatch in hand measures from the start of a process to the end of the process). Each MoM receives its jobs in a queue and executes one job at a time in basic settings. Each job acquires a lock over the MoM node and proceeds to completion before the processor is freed for other jobs on the server in the waiting queues. During the implantation, various communication patterns were noticed between the MoM node and PBS server, which will be presented as case studies to help researchers troubleshoot various issues in job executions, debugging, and performance enhancement.
2. Authentication Module (TRQAUTHD)—each MoM node is allowed to submit jobs at a server and must have an authentication module (TRQAUTHD) installed and configured. For MoM nodes, it is an optional package.

3.1.3. Torque Client Node

The Torque client modules were developed at the time of installation of the Torque server and can be copied to clients with the necessary scripts for their registry in the system services. Torque clients are designated nodes that can submit jobs to the PBS server. Each client can be registered with the server by its hostname and IP address. A change in the IP and hostname of a client crash the job submission system, and the whole processes of registering and service invocation must be repeated to bring the system to the operating mode.

Each set of sub-clusters contains one server, an instance of the scheduler, and one authentication daemon at the server. For failover capability, identical servers are installed in similar clusters. Currently, five Compute nodes are attached to every server and one client is also configured to submit the jobs. The complete architecture of the implemented cluster is given in Figure 1.

3.2. High-Performance Data Analytics Cluster

The low-cost, distributed, and data-intensive cluster, known as the HPDA cluster, has been set up in the designated laboratory. Distributed computing is just a distributed system where multiple machines perform specific work simultaneously. While performing the work, machines communicate with each other by passing messages between them. Distributed computing is practical when fast processing (computation) is required on big data. Apache Hadoop [20] and Apache Spark [21] are well-known examples of big data processing systems. The HPDA cluster comprises Apache Hadoop and Apache Spark. The cluster specifications and diagram are illustrated in Table 2 and Figure 2 respectively.

Table 2. HPDA Cluster Specifications.

Parameters	Used Values
Number of Nodes: Cores in Use: Availed Memory: Availed HDFS:	10 (1 Master, 9 Slaves) 72 24.9 GB 6.4 TB
Master Machine	CPU 2.25 Ghz RAM—4 GB HDD 1 TB OS—Ubuntu 16.04 LTS
Ethernet Switch	16 Port 10/100 Mbp
Software Tools	Hadoop 2.7, Spark 2.1.0, Scala 2.12.1, OpenJDK 1.8 Hive 2.0.0, HBase 1.1.4 Flume 1.6.0, Anaconda3 4.3.0 R 3.3.2, RStudio 1.0.136
Language Support	Scala, Java, Python, R, & MapReduce

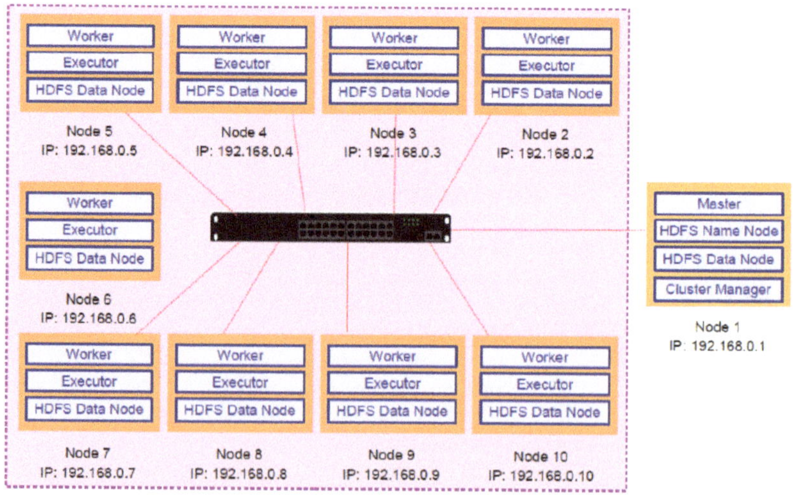

Figure 2. HPDA cluster diagram.

3.2.1. Apache Hadoop Setup

The Apache Hadoop software is an open-source framework built for reliable, scalable distributed computing tasks with huge data sets over a cluster of multiple computers. The features used for this implementation are (i) the large data set distribution across clusters of computers using a simple programming model (ii) became the de facto standard for storing, processing, and analyzing hundreds of terabytes and petabytes of data and (iii) is cheaper to use in comparison to other traditional proprietary technologies and can handle all type of data.

Generally, a Hadoop system comprises a computer acting as the master node and multiple computers acting as slave nodes, as shown in Figure 2. Hadoop has two modules, in total, including the HDFS and MapReduce Framework. The HDFS usually only has one NameNode, which manages the directory tree and metadata of related files for the HDFS. It could also own a secondary NameNode that can be employed to backup mirror files, combine logs and mirror files periodically and send them back to NameNode. In general, NameNode and Secondary NameNode are deployed on the master node. In addition, the DataNode of HDFS is responsible for storing data and sending processed data back to NameNode and is usually deployed on the slave node.

3.2.2. Apache Spark Setup

Apache Spark is installed on top of Hadoop. Spark [21] is a fault-tolerant and distributed data analytics tool capable of implementing large-scale data-intensive applications on commodity hardware. Hadoop and other technologies have already popularized acyclic data flow techniques for building data-intensive applications on commodity clusters. However, these are unsuitable for applications that reuse a working dataset for multiple parallel operations. Some of these applications are iterative machine learning algorithms and interactive data analysis tools. Spark addresses these problems, and is also scalable and fault-tolerant. To accommodate these goals, Spark introduces data storage and processing abstraction called RDDs.

Resilient Distributed Dataset is a collection that has been distributed all over the Spark cluster [22]. RDDs' main purpose is to support higher-level, parallel operations on data in a straightforward manner. Spark can run tasks up to 100 times faster, when it utilizes the in-memory computations and 10 times faster when it uses disk than traditional map-reduce tasks. Spark performs well in these cases, where Hadoop users have reported deficiency with MapReduce. The features like the optimized parameter for iterative jobs using gradient descent and the interactive analytics interfaces used to run queries on large data sets using Hadoop. The behavior of the First Come First Serve (FCFS) scheduling algorithm on a single processor and the gain in speedup with an increasing number of processors used for our experiment is designed as seen in Table 3.

Table 3. Experimental Setup.

Environment	Setup
Jobs	Arrival time and CPU burst time
Granularity level	Process
Type	Fully parallel
Expected Speedup	Linear
Scheduling Algorithm	FCFS
Computing Nodes	Exclusive
Process level parameters	Individual process, virtual memory
Overall performance parameters	Speedup in latency
No. of queues	One
Language Support	Scala, Java, Python, R, & MapReduce

4. Experimental Evaluation and Results

In this section, we present the experimental design and results to determine a more efficient option. While experimentation, careful selection of the cluster type, installation, and collection of necessary data help to get a better result. Two clusters are selected for the experiment that could represent different HPC applications. Hence, DeepStack and HPDA clusters are selected in this study for compute-bound and memory-bound jobs.

4.1. DeepStack Cluster

The benchmark experiments on the HPC cluster have passed through the test and debugging level programs to test the availability of the cluster server compute nodes and then data with intermediate complexity including Markov chains, Monte-Carlo simulation, distributed PageRank algorithm, HITS algorithm, Distributed Gradient descent. After successful tests of these stages benchmarking programs are used. Benchmark programs include the problems based on highly optimized libraries for core scientific research with competitive baseline figures in the form of linear algebra, genetics, and simulation studies using LPACK, LINPACK, BLAS, etc.

Based on the theoretical basis, the experiment is tested for speedups. Three types of speedups were used in the experiment, and comparable latency and throughput can be achieved using our cluster architecture:

1. linear speedup(fully parallel case experimented),
2. sublinear speedup(sequential parallel mixed case),
3. superliner speedup(cached case—future work)

Items (1) and (2) can be explained by Ahmdal's diminishing returns, whereas (3) is explained by the principle of locality of data (i.e., distributed caching architecture is used). To perform the experiments and compare the results based on the theoretical basis, the sample workload taken for experimentation is seen in Table 4. The latency (L) of architecture is the time taken per unit of workload and is given by the following formula:

$$L = \frac{W}{T} \quad (1)$$

where, T is the total time taken by the workload on this architecture, and W denotes the total workload in the number of instructions/jobs. From the relative performance of two systems processing the same problem as speedup, we measured the speedup/performance improvements from speedup latency and speedup throughput perspectives. Hence, the throughput is the execution rate of a task as follows: density ρ, the number of processors. In addition, Q is inversely proportional to the latency of the architecture:

$$Q = \frac{\rho \cdot AW}{T} = \frac{\rho \cdot A}{L} \quad (2)$$

where A—is the number of processors, ρ—execution density(the number of stages in an instruction pipeline, W—denotes the total workload executed, and T represents the total time taken.

Regarding the speedup in latency, by making the architecture parallel, we intend to speed up the system latency. The following formula is used to compute the speed up in latency between architectures 1 and 2:

$$S_{latency} = \frac{L_1}{L_2} = \frac{T_1 W_2}{T_2 W_1} \quad (3)$$

where S denotes the speedup in latency from architecture 1 to 2. In addition, L_1 is the latency on architecture 1, and L_2 is the latency on architecture 2.

The following formula defines the speedup in throughput:

$$S_{throughput} = \frac{Q_2}{Q_1} = \frac{\rho 2 A_2 T_1 W_2}{\rho 1 A_1 T_2 W_1} = \frac{\rho 2 A_2}{\rho 1 A_1} S_{latency} \quad (4)$$

where $S_{throughput}$ represents the speedup in throughput of Architecture 2 with respect to Architecture 1. Moreover, Q_1 denotes the throughput of architecture 1, and Q_2 is the throughput of architecture 2.

Regarding Ahmdal's law, if some part of the program cannot be parallelized because of dependencies, then the actual speedup cannot be linearly scaled in proportion to the number of newly added processors. In addition, the speedup is inversely proportional to the amount of sequentially in the program with the following formula:

$$S_{latency}(S) = \frac{1}{(1-p) + (p/s)} \quad (5)$$

where $S_{latency}$ is the theoretical speedup of the execution of the whole task, s is the speedup of part of the task that benefits from improved system resources, p is the proportion of execution time that the part benefiting from improved resources originally occupied.

Table 4. Sample Workload.

SNo.	Job Id	Arrival Time	CPU Burst Time
1	P1	4	200
2	P2	10	500
3	P3	6	400
4	P4	2	300
5	P5	5	200
6	P6	3	100
7	P7	12	250
8	P8	14	320
9	P9	20	250
10	P10	22	150

The performance of the DeepStack cluster increased as the compute nodes were added. Theoretical improvement in performance should scale linearly for fully parallelizable tasks, but in a real implementation, a 10% loss in speedup on two-node and a 15% loss in speedup occurred for the three-node cluster. This loss was entirely due to the latency of the network (Ethernet) cable and switch performance. Adding more nodes in a cluster increases the speedup in latency and TAT, decreases the average waiting time of individual processes, and increases communication cost.

In linear speedup, per Ahmdal's law on a fixed parallelizability level in a program, if we increase the number of processors, Torque exhibits an approximately linear increase in speedup as we add new processors.

For the sublinear speed up, if the program has an s%(<100%) sequential part, it exhibits sublinear speedup on Torque. The exact penalty can be computed by Ahmdal's equation and implementation of the program. During experimentation, single compute node, FCFS scheduling performance is used as a benchmark which has execution order as given in Table 5 below and the average baseline performance is seen in Table 6. Note: AT: Arrival time, BT: Burst time, CT: completion time, TAT: Turn around time, and WT: waiting time.

Table 5. Torque job execution order on a single node.

P4	P6	P1	P5	P3	P2	P7	P8	P9	P10

Table 6. Torque job execution order on a single node.

Process	AT	BT	Start	CT	TAT	WT
P4	2	300	2	302	300	0
P6	3	100	302	402	399	299
P1	4	200	402	602	598	398
P5	5	200	602	802	797	597
P3	6	400	802	1202	1196	796
P2	10	500	1202	1702	1692	1192
P7	12	250	1702	1952	1940	1690
P8	14	320	1952	2272	2258	1938
P9	20	250	2272	2522	2502	2252
P10	22	150	2522	2672	2650	2500
			Avg. Baseline	**1443**	**1433.2**	**1166.2**

The cluster-level performance of DeepStack with a 2-node cluster and DeepStack with a 3-node cluster and the expected speedup with the actual gain from the setup is also seen in Table 7 and Table 8, respectively.

Table 7. Torque job execution order on 2-node cluster.

	DeepStack	Speedup	Expected Speedup
Avg. CT	804.315	1443/804.315 = 1.79	2
Avg. TAT	791.818	1.8	2
Avg. WT	525.81	2.12	NA
Avg. Latency	8.28	NA	NA

Table 8. Torque job execution order on 3-node cluster.

	DeepStack	Speedup
Avg. CT	583.75	1443/583.75 = 2.47
Avg. TAT	573.8	2.49
Avg. WT	301.5	3.87
Avg. Latency	8.25	NA

As per Amdahl's law in this case the speedup in CT and TAT should be twice as of the single processor benchmark but due to internal latencies, it is decreased by 10% for the 2-node cluster and by 15% for the 3-node cluster as seen on Table 9. This is because the performance of clusters declines as described by Amdahl's law. This happens because, there is a delay from locking of processes by a scheduling algorithm used, high internal bandwidth consumption, and poor performance of Ethernet cable capacity. The use of Gigabits per second Ethernet platform among Pbs_servers and compute nodes helps to get better performance.

Table 9. Performance Comparison of DeepStack Cluster with Single Node cluster.

Metric	1 Node	2 Node	Speedup	3 Node	Speedup
Avg. CT	1443	804.315	1.79	583.75	2.47
Avg. TAT	1443.2	791.818	1.8	573.8	2.49
Avg. WT	1166.2	525.81	2.12	301.5	3.87
Avg. Latency	0	8.28	NA	8.25	NA

4.2. HPDA Cluster

As part of this study, the evaluation was conducted on the HPDA cluster, running a WordCount program that counts the number of words specified in a given large text file and calculates *pi*. The following are benchmark applications on the HPDA cluster:

- iterative Jobs,
- interactive Analytics,
- distributed Machine Learning,
- streaming Analytics.
- distributed Graphs Processing.

Figure 3 depicts the processing time on a single node and multinode clusters.

The execution time depends on the network communication (I/O), the number of search operations of words, size of the input file. The whole process of MapReduce processes and builds a cost function that explicitly models the relationship between the volume of input data, available system resources (map and reduce slots), and complexity of the reduce function for the target MapReduce job.

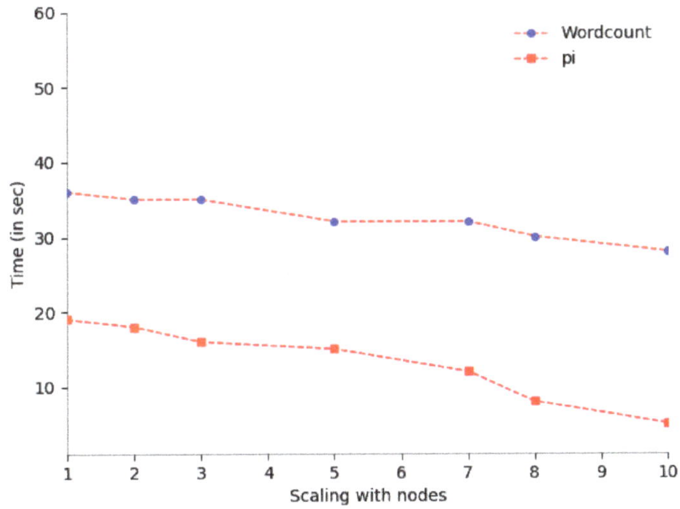

Figure 3. HPDA performance Comparison.

The job execution order is compared with a single compute node for base-line comparison. Single Compute Node, FCFS scheduling performance is used as a benchmark for this experiment and the performance result is seen in Table 9.

As per Amdahl's law, in this case, the speedup in compute time (CT) and total time (TAT) should be twice as of the single processor benchmark. But due to Ethernet latencies it is decreased by 15%. Similarly, we have increased the number of compute nodes in the experiment and computed the performance parameters in each setup. Since workload and other conditions are fixed we can use speedup as a measure of performance of our HPC cluster.

From the Hadoop cluster setup of 10 nodes, we were able to aggregate a distributed file system of 6.45 terabytes (TB) from 1 master and 9 live workers running. From this cluster setup, we get 24.9 GB of memory which is enough to run medium size data-intensive applications or jobs. On this cluster, the authors run a WordCount application of 1.23 GB on HDFS and observed when the file is replicated to the worker nodes as this is based on the size of the input file.

Similarly, the researchers were also able to run the finding *pi* application on the HPDA cluster and result that the submitted WordCount application used 72 cores and successfully processed in 5 s only as seen in Table 10. We have run the above-mentioned applications i.e., WordCount and *pi* on a single machine that has 4 cores and 4 GB RAM. Then, the WordCount application successfully processed in 36 s, and *pi* application processed in 19 s which is been shown in Table 10.

Table 10. Performance Comparison of HPDA Cluster with Single Node Cluster.

Cluster Type/App	WordCount App	*pi* App
Single Node Cluster (4 GB RAM, 1 TB HDD)	36 s	19 s
Multi-Node Cluster (24.9 GB RAM, 6.4 TB HDD)	30 s	5 s

It has been observed from Table 10, that the execution time depends on the network communication (I/O), the number of search operations of words, size of the input file. The whole process of MapReduce processing and building up a cost function that explicitly models the relationship between the amount of input data, the available system resources (Map and Reduce slots), and the complexity of the Reduce function for the target MapReduce job.

4.3. Result Analysis

OpenMP, MPI, and MapReduce are the most widely recognized parallel or distributed programming frameworks. The performance study of three parallel programming frameworks was done [23,24]. The comparative studies have been conducted for two problem sets the all-pairs-shortest-path problem and a joining problem for large data sets. OpenMP [21] is the defacto standard model for shared memory systems, MPI [25] is the defacto standard for distributed memory systems, and MapReduce [24,26] is recognized as the defacto standard framework intended for big data processing. For each problem, the parallel programs have been developed in terms of the three models, and their performance has been observed. The experiment results concluded that if a problem is small enough to be accommodated and the computing resources such as cores and memory are sufficient, OpenMP is a good choice. When the data size is moderate and the problem is computation-intensive, MPI can be considered the framework. When the data size is large and the tasks do not require iterative processing, MapReduce can be an excellent framework. OpenMP is the easiest to use because there is no special attention needed to be paid. After all, it just needs to place some directives in the sequential code. MapReduce is relatively easy to use once we can abstract an application into Map and Reduce steps. The programmers do not have to consider workload partitioning and synchronization. MapReduce programs, however, take a considerable time for the problems requiring many iterations, like the all-pairs shortest-path problem. MPI allows more flexible control structures than MapReduce; hence, MPI is a good choice when a program is needed to be executed in a parallel and distributed manner with complicated coordination among processes.

As per the research questions, we have analyzed the role of commodity hardware in cluster establishment and showed that at minimum an HPC cluster can be configured with an Ethernet switch and cabling inter-connectivity among available commodity computers. We have shown the optimum configuration of the HPC cluster system for various types of workloads and the choice of software and libraries is highly influenced by the type of parallelisms available in application-level programs. For shared-memory programming, we have identified that Torque was a good choice, for distributed memory scatter-gather pattern it was Hadoop based HPC cluster which showed better performance and for distributed in-memory computing it was a spark cluster. Also with the help of the MPI library, Torque supports distributed memory programming paradigm. Lastly, we have analyzed the performance of various clusters under variable load conditions, and we have designed rather simple experiments which are limited to computing the latency profile of various clusters implemented.

5. Related Work

In addition to advances in hardware and communication infrastructure, the HPC community has witnessed a growing set of cluster management solutions, including Torque, Apache Hadoop, Open-Mosiac, Rocks, OSCAR, OpenPBS, Alchemy, and HTcondor [3,10,21,23,27–30]. Several open-source libraries exist, such as OpenCV, CUDA, OpenMP, OpenMPI, MPICH, and many implementations of the message passing (MPI) programming paradigm are on almost all reviewed cluster platforms [21,31]. The libraries for individual research domains, corresponding benchmarks, and quantitative performance evaluations of benchmark problems are also available for most of the HPC implementations [16,32].

In this race of cost-effective HPC implementations, various universities have developed a dedicated HPC clusters. Data centers provide researchers access to HPC resources for cutting-edge research [3,13,17–19] where others have tried to introduce the use of low cost HPC cluster from inexpensive hardware. However, these may not be exhaustively tested for different applications that will affect the performance of the cluster machines [33]. The authors' team believes this represents the importance of infrastructure implementation in academic sectors with relatively fewer costs while getting a nearly equal performing HPC infrastructure for their research activities . The University of Columbia implemented a 167-node cluster with 2672 cores on Dual Intel E5-2650v2 Processors(2.6 GHz) with a Torque/Moab job scheduler in 2009, which was upgraded to support research projects in various application areas [17]. Similarly, Yale University provides an HPC-based computing environment with an excellent publication records. The Stanford HPC center provides a million core compute nodes [18], and the FSU HPC cluster center has more than 10,000 cores and 201,449 Gflops, with a 3 million job capacity [3,10,19,34–36]. Looking to this achievements one can raise a question that, if having HPC can help to solve many research questions, how can I develop HPC by myself?. This question should be answered by someone that knows how. Therefore, we wanted to contribute to answer this question through designing HPC from a cheap computers around us so that others will do the same to solve their own problems.

There is no HPC platform that can be used for varying research activities at the campus level that can be used by the researchers for research that satisfies their needs. This is a bottleneck for researchers in science, engineering, and biological and genetic simulation research [6]. In addition to these domains, core computer science research also requires a specialized HPC centers with commodity computers available in laboratories to research new domains, such as big data, distributed machine learning, the internet of things, and cloud computing [2,3,37].

The parallel or distributed programming frameworks OpenMP, MPI, and MapReduce are the most widely recognized, and the performance of these three parallel programming frameworks with comparative studies has been assessed for two problems sets: the all-pairs-shortest-path problem and the join-problem for large data sets [23]. OpenMP is the defacto standard model for shared memory systems, and MPI [25] is the defacto standard for distributed memory systems. Finally, MapReduce [26] is recognized as the defacto standard framework for big data processing.

For each problem, parallel programs have been developed regarding the three models, and their performance has been observed. The experimental results indicated that if a problem is small enough to accommodate sufficient computing resources, such as cores and memory, OpenMP is a good choice. When the data size is moderate and the problem is computationally intensive, MPI can be considered for the framework. When the data size is large and the tasks do not require iterative processing, MapReduce can be an excellent framework.

OpenMP is the easiest to use because no special attention is needed, as it just requires to place some directives in the sequential code. MapReduce is relatively easy to use once we abstract an application into the map and reduce steps, and programmers do not have to consider workload partitioning and synchronization. MapReduce programs, however, take a considerable time for problems requiring numerous iterations, such as the all-pairs shortest-path problems. Moreover, MPI allows more flexible control structures than MapReduce; hence, MPI is a good choice when a program must be executed in a parallel and distributed manner with complicated coordination among processes.

The language-independent MPI is a communications protocol for parallel computing where point-to-point and collective communication are supported [26]. However, the standard does not currently support fault tolerance [25] because it primarily addresses HPC problems. Another MPI drawback is that it is unsuitable for the small-grain level of parallelism, for example, to exploit the parallelism of multicore platforms for shared memory multiprocessing. In contrast, OpenMP is an Application Programming Interface (API) that supports multi-platform shared memory multiprocessing programming on most processor architectures and operating systems [21]. OpenMP is becoming the standard for shared memory parallel programming due to its high performance; however, it is unsuitable for distributed memory systems. The idea of extending this API to cope with this issue is now a growing field of research [38]. OpenMP's user-friendly interface allows it to easily parallelize complex algorithms, unlike MPI, because the code must be heavily re-engineered to obtain relevant performance improvements.

Spark is a state-of-the-art framework for HPC designed to efficiently deal with iterative computational procedures that recursively perform operations on the same data [39,40], such as supervised machine learning algorithms. Spark is based on the concept of maintaining data in memory rather than on disk, as done by other well known approaches such as Apache Mahout, which require data reloading and incur considerable latencies. Experiments have shown that Spark outperforms conventional MapReduce jobs in terms of speed by up to two orders of magnitude [20,40,41]. The core data units in Spark are called resilient distributed datasets (RDDs). They are a distributed, immutable, and fault-tolerant memory abstraction that collects an element set in which an operation set can be applied to produce other RDDs (transformations) or return values (actions). The RDDs can reside in memory, on the disk, or a combination of these. However, they are only computed on actions following a lazy evaluation strategy to perform minimal computation and prevent unnecessary memory usage. The RDDs are not cached in memory by default, therefore, when data are reused, a persist method is needed to avoid re-computation.

Various cluster management options are available for running Spark. The options range from the simple Spark integrated Stand-alone Scheduler to other widespread cluster managers, such as Apache Mesos and Hadoop YARN [30,42,43]. To get benefited from this reach features, this study deploys Spark in a Hadoop cluster. Apache Hadoop is an open-source software platform for distributed big data processing over commodity cluster architectures [30,44]. It has three main elements: (a) a MapReduce programming model that separates data processing into mapping to perform data operations locally, shuffling to redistribute network data and reduce data summarization; (b) a Hadoop distributed file system (HDFS) with high-throughput data access; and (c) a cluster manager (YARN) handling available computing resources and job scheduling. Nevertheless, Spark on Hadoop may be preferred [30,43] because it also

- offers a distributed file system with failure and data replication management.
- allows the addition of new nodes at run time, and
- provides a set of tools for data analysis and management that is easy to use, deploy and maintain.

The authors identified specific gaps in the HPC platform that can easily developed from commodity computers around us for research activities. To our knowledge, there is no HPC developed from cheap computers to serve a fully functional HPC infrastructure the can be used for any research activities ranging from simple simulation to a very complicated computation that requires high computing power.There are attempts to design HPC from low cost machines and Raspberry Pi [33,45] but the first approach didn't test the environment for range of job types where the second mainly compared the energy consumption of their approach with the commodity servers. Authors from [46] also applies similar concept yet they designed the cluster for odd-even sorting which cannot represent all ranges of applications. Therefore, to address this technical gaps and come up with comprehensive solution, we designed a cost-effective HPC cluster from commodity computers around us. The novelty of this approach is, we put together the best existing programming language independent tools that any one can easily access and state-of-the-art frameworks to deal with it. Hence we implemented, tested, and compared our scheme that helps interested researcher to develop and use HPC from commodity computers.

6. Conclusions

The study was conducted with the available commodity hardware and open-source tools. The study team developed two types of clusters. Depending on the type and complexity of the computation, individual clusters can be chosen to perform jobs. The DeepStack cluster supports second-generation or low-level programming languages, such as C, C++, and Fortran. The HPDA cluster supports the latest programming or high-level programming languages such as Java, Scala, Python, and R.

Many pieces of research indicate that the DeepStack cluster is a good choice if a problem is small enough to be accommodated, the computing resources, such as cores and memory, are sufficient, and the data size is moderate. When the data size is large and the task requires high-speed iterative processing, then the HPDA cluster is a good choice.

From the results, we have analyzed the role of commodity hardware in cluster establishment. We have identified that at minimum an HPC cluster can be configured with an Ethernet switch and cabling interconnectivity among available desktop computers with the help of suitable cluster management software. But, during the experiment, we found that the rate of failure of commodity computers is more as compared to the specially designed computers, also there is less than 50 Mbps bit rate due to various types of delays between compute-node and server-nodes, which is quite less than the rate at which connected hard disk can supply data to a processor using SCSI or PCI.

We also found the optimum configuration of the HPC cluster system for various types of workloads. The choice of software and libraries is highly influenced by the type of parallelisms available in application-level programs. For shared-memory programming, we have identified that Torque was a good choice, for distributed memory scatter-gather pattern it was a Hadoop-based HPC cluster which showed better performance and for distributed in-memory computing, it was a spark cluster. Also with the help of the MPI library Torque supports distributed memory programming paradigm.

Lastly, we have analyzed the performance of various clusters under variable load conditions, but as per the scope of this research, we have designed rather simple experiments which are limited to computing the latency profile of various clusters implemented. More sophisticated experiments involving the performance gain at compiler level optimizations, use of libraries, programming paradigms, and different types of workloads shall be studied in the future.

Author Contributions: Conceptualization, D.R. and Y.S.; methodology, D.R. and H.Y.; investigation, Y.S. and H.Y.; validation, H.Y.; visualization, D.R. and Y.S.; writing—original draft preparation, D.R.; writing—review and editing, D.R., H.Y. and Y.S. All authors have read and agreed to the published version of the manuscript.

Funding: This work was supported by the National Research Foundation of Korea (NRF) grant funded by the Korean government (MSIT) (No.2021R1C1C1010861). This work was supported in part by the Korea Institute for Advancement of Technology (KIAT) grant funded by Korea government (MOTIE) (P0012724, The Competency Development Program for Industry Specialist) (Corresponding Author: Yongseok Son).

Institutional Review Board Statement: Not applicable.

Informed Consent Statement: Not applicable.

Data Availability Statement: For more, we put the link to the installation manual and configuration steps at the following https://github.com/Derejereg/HPC/blob/main/Installation_manual.pdf (accessed on 27 April 2022) using GitHub.

Conflicts of Interest: The authors declare no conflict of interest.

References

1. Plaza, A.; Chang, C.I.; Plaza, J.; Valencia, D. Commodity cluster and hardware-based massively parallel implementations of hyperspectral imaging algorithms. In Proceedings of the Algorithms and Technologies for Multispectral, Hyperspectral, and Ultraspectral Imagery XII, Kissimmee, FL, USA, 17–20 April 2006; Volume 6233, p. 623316.
2. Yeo, C.S.; Buyya, R.; Pourreza, H.; Eskicioglu, R.; Graham, P.; Sommers, F. Cluster computing: High-performance, high-availability, and high-throughput processing on a network of computers. In *Handbook of Nature-Inspired and Innovative Computing*; Springer: Berlin/Heidelberg, Germany, 2006; pp. 521–551.
3. Es-Sabery, F.; Hair, A. Big data solutions proposed for cluster computing systems challenges: A survey. In Proceedings of the 3rd International Conference on Networking, Information Systems & Security, Marrakech, Morocco, 31 March–2 April 2020; pp. 1–7.
4. Hager, G.; Wellein, G. *Introduction to High Performance Computing for Scientists and Engineers*; CRC Press: Boca Raton, FL, USA, 2010.
5. García-Risueño, P.; Ibáñez, P.E. A review of High Performance Computing foundations for scientists. *Int. J. Mod. Phys. C* **2012**, *23*, 1230001. [CrossRef]
6. Makino, J.; Taiji, M. *Scientific Simulations with Special-Purpose Computers—The GRAPE Systems*; Wiley-VCH: Weinheim, Germany, 1998; ISBN 0-471-96946-X.
7. Brightwell, R.; Fisk, L.A.; Greenberg, D.S.; Hudson, T.; Levenhagen, M.; Maccabe, A.B.; Riesen, R. Massively parallel computing using commodity components. *Parallel Comput.* **2000**, *26*, 243–266. [CrossRef]
8. Jin, H.; Jespersen, D.; Mehrotra, P.; Biswas, R.; Huang, L.; Chapman, B. High performance computing using MPI and OpenMP on multi-core parallel systems. *Parallel Comput.* **2011**, *37*, 562–575. [CrossRef]
9. Lee, J.; Kang, H.; Yu, S.; Kim, C.; Yea, S.J. Whole cancer genome analysis using an I/O aware job scheduler on high performance computing resource. In Proceedings of the 2014 IEEE International Conference on Bioinformatics and Biomedicine (BIBM), Belfast, UK, 2–5 November 2014; pp. 10–11.
10. Balaprakash, P.; Dongarra, J.; Gamblin, T.; Hall, M.; Hollingsworth, J.K.; Norris, B.; Vuduc, R. Autotuning in high-performance computing applications. *Proc. IEEE* **2018**, *106*, 2068–2083. [CrossRef]
11. Hellwagner, H.; Karl, W.; Leberecht, M. Enabling a PC Cluster for High-Performance Computing. *SPEEDUP J.* **1997**, 11. Available online: https://citeseerx.ist.psu.edu/viewdoc/download?doi=10.1.1.454.6607&rep=rep1&type=pdf (accessed on 27 April 2022).
12. Lee, J.; Abuzaghleh, O. Implementing an affordable high performance computing platform for teaching-oriented computer science curriculum. In Proceedings of the 2011 ASEE Annual Conference & Exposition, Vancouver, BC, Canada, 26–29 June 2011; pp. 22–816.
13. Sterling, T.L. *Beowulf Cluster Computing with Linux*; MIT Press: Cambridge, MA, USA, 2002.
14. LexisNexis. HPCC Systems. Available online: https://hpccsystems.com/ (accessed on 7 October 2021).
15. Cisco. *Cisco Annual Internet Report*; Cisco Public (2018–2023); Cisco Systems, Inc.: San Jose, CA USA, 2022.
16. Harney, J.; Lim, S.H.; Sukumar, S.; Stansberry, D.; Xenopoulos, P. On-demand data analytics in HPC environments at leadership computing facilities: Challenges and experiences. In Proceedings of the 2016 IEEE International Conference on Big Data (Big Data), Washington, DC, USA, 5–8 December 2016; pp. 2087–2096.
17. XSEDE. HPC at University of Columbia. Available online: https://www.xsede.org/ (accessed on 12 September 2021).
18. Stanford. HPC at Stanford University. Available online: https://hpcc.stanford.edu/ (accessed on 19 September 2021).
19. FSU. FSU HPC Cluster. Available online: https://rcc.fsu.edu/services/hpc (accessed on 26 August 2021).
20. Xin, R.S.; Rosen, J.; Zaharia, M.; Franklin, M.J.; Shenker, S.; Stoica, I. Shark: SQL and rich analytics at scale. In Proceedings of the 2013 ACM SIGMOD International Conference on Management of Data, New York, NY, USA, 22–27 June 2013; pp. 13–24.
21. OpenMP Architecture Review Board. *Openmp Application Program Interface*; Techical Report; OpenMP Architecture Review Board: Beaverton, OR, USA, 2013.
22. Olston, C.; Reed, B.; Srivastava, U.; Kumar, R.; Tomkins, A. Pig latin: A not-so-foreign language for data processing. In Proceedings of the 2008 ACM SIGMOD International Conference on Management of Data, Vancouver, BC, Canada, 10–12 June 2008; pp. 1099–1110.
23. Kang, S.J.; Lee, S.Y.; Lee, K.M. Performance comparison of OpenMP, MPI, and MapReduce in practical problems. *Adv. Multimed.* **2015**, *2015*, 7. [CrossRef]
24. Hashem, I.A.T.; Anuar, N.B.; Marjani, M.; Ahmed, E.; Chiroma, H.; Firdaus, A.; Abdullah, M.T.; Alotaibi, F.; Ali, W.K.M.; Yaqoob, I.; et al. MapReduce scheduling algorithms: A review. *J. Supercomput.* **2020**, *76*, 4915–4945. [CrossRef]

25. Snir, M.; Gropp, W.; Otto, S.; Huss-Lederman, S.; Dongarra, J.; Walker, D. *MPI—The Complete Reference: The MPI Core*; MIT Press: Cambridge, MA, USA, 1998; Volume 1.
26. Dean, J.; Ghemawat, S. MapReduce: Simplified data processing on large clusters. *Commun. ACM* **2008**, *51*, 107–113. [CrossRef]
27. Sloan, J.D. *High Performance Linux Clusters with OSCAR, Rocks, OpenMosix, and MPI: A Comprehensive Getting-Started Guide*; O'Reilly Media, Inc.: Sebastopol, CA USA, 2004; ISBN 9780596005702.
28. Staples, G. TORQUE resource manager. In Proceedings of the 2006 ACM/IEEE conference on Supercomputing, Tampa, FL, USA, 11–17 November 2006; p. 8. [CrossRef]
29. Apache. Apache Hive. Available online: https://hive.apache.org/ (accessed on 13 August 2021).
30. Venkatraman, S.; Subramanyam, R. Data Aware Distributed Storage (DAS) for Performance Improvement across a Hadoop Commodity Cluster. In *Advances in Decision Sciences, Image Processing, Security and Computer Vision*; Springer: Cham, Switzerland, 2019; Volume 3, p. 350.
31. Bell, G.; Gray, J. What's next in high-performance computing? *Commun. ACM* **2002**, *45*, 91–95. [CrossRef]
32. Hoffman, A.R. *Supercomputers: Directions in Technology and Applications*; Number 04, e-Book; National Academy of Science and National Research Board—National Academies Press: Washington DC, USA, 1989.
33. Kumar, D.; Memon, S.; Thebo, L.A. Design, implementation & performance analysis of low cost high performance computing (HPC) clusters. In Proceedings of the 2018 12th International Conference on Signal Processing and Communication Systems (ICSPCS), Cairns, Australia, 17–19 December 2018; pp. 1–6.
34. Hmes, V.; Kureshi, I. Developing high performance computing resources for teaching cluster and grid computing courses. *Procedia Comput. Sci.* **2015**, *51*, 1714–1723.
35. Apon, A.; Buyya, R.; Jin, H.; Mache, J. Cluster computing in the classroom: Topics, guidelines, and experiences. In Proceedings of the First IEEE/ACM International Symposium on Cluster Computing and the Grid, Brisbane, QLD, Australia, 15–18 May 2001; pp. 476–483.
36. Staples, G. Torque Resource Manager. 2013. Available online: https://www.hpcadvisorycouncil.com/pdf/RADIOSS_Analysis_and_Profiling_AMD.pdf (accessed on 3 July 2021).
37. Dongarra, J.; Sterling, T.; Simon, H.; Strohmaier, E. High-performance computing: Clusters, constellations, MPPs, and future directions. *Comput. Sci. Eng.* **2005**, *7*, 51–59. [CrossRef]
38. Basumallik, A.; Min, S.J.; Eigenmann, R. Programming distributed memory sytems using OpenMP. In Proceedings of the 2007 IEEE International Parallel and Distributed Processing Symposium, Long Beach, CA, USA, 26–30 March 2007; pp. 1–8.
39. Zaharia, M.; Chowdhury, M.; Franklin, M.J.; Shenker, S.; Stoica, I. Spark: Cluster computing with working sets. *HotCloud* **2010**, *10*, 95.
40. Yu, Z.; Bei, Z.; Qian, X. Datasize-aware high dimensional configurations auto-tuning of in-memory cluster computing. In Proceedings of the Twenty-Third International Conference on Architectural Support for Programming Languages and Operating Systems, Williamsburg, VA, USA, 24–28 March 2018; pp. 564–577.
41. Zaharia, M.; Chowdhury, M.; Das, T.; Dave, A.; Ma, J.; McCauly, M.; Franklin, M.J.; Shenker, S.; Stoica, I. Resilient distributed datasets: A fault-tolerant abstraction for in-memory cluster computing. In Proceedings of the 9th USENIX Symposium on Networked Systems Design and Implementation (NSDI 12), San Jose, CA, USA, 25–27 April 2012; pp. 15–28.
42. Karau, H.; Konwinski, A.; Wendell, P.; Zaharia, M. *Learning Spark: Lightning-Fast Big Data Analysis*; O'Reilly Media, Inc.: Sebastopol, CA, USA, 2015; ISBN 978-1-44-935862-4.
43. Reyes-Ortiz, J.L.; Oneto, L.; Anguita, D. Big data analytics in the cloud: Spark on hadoop vs mpi/openmp on beowulf. *Procedia Comput. Sci.* **2015**, *53*, 121–130. [CrossRef]
44. White, T. *Hadoop: The Definitive Guide*; O'Reilly Media, Inc.: Sebastopol, CA USA, 2015; ISBN 9781491901632.
45. Bourhnane, S.; Abid, M.R.; Zine-Dine, K.; Elkamoun, N.; Benhaddou, D. High-Performance Computing: A Cost Effective and Energy Efficient Approach. *Adv. Sci. Technol. Eng. Syst. J.* **2020**, *5*, 1598–1608. [CrossRef]
46. Myint, K.N.; Aung, W.T.; Zaw, M.H. Research and Analysis of Parallel Performance with MPI Odd-Even Sorting Algorithm on Super Cheap Computing Cluster. Ph.D. Thesis, University of Computer Studies, Yangon, Myanmar, 2019.

applied sciences

MDPI

Article

Suitability Evaluation of the Lining Form Based on Combination Weighting–Set Pair Analysis

Chen Xing [1], Leihua Yao [1,*], Yingdong Wang [2] and Zijuan Hu [1]

[1] School of Engineering and Technology, China University of Geosciences (Beijing), Beijing 100083, China; xingc@cugb.edu.cn (C.X.); m18370956200@163.com (Z.H.)
[2] Beijing Engineering Corporation Limited, Beijing 100024, China; timothyboy@163.com
* Correspondence: leihuayao@163.com

Abstract: Aiming at the many uncertain factors in the suitability evaluation of reinforced concrete lining of high-pressure pipelines, the set pair analysis (SPA) theory is used to establish the suitability evaluation model. By summarizing the key influencing factors of typical lining design criteria, five suitability evaluation indices are determined from three criteria, i.e., the minimum overburden criterion, the minimum principal stress criterion, and the hydraulic fracturing criterion. In order to fully consider the subjective and objective factors, the combination ordered weighted averaging (C-OWA) operator and the criteria importance through intercriteria correlation (CRITIC)-entropy weighting model (EWM) were used to construct a combination weighting method, and the weight coefficients of each index were comprehensively determined. Based on the SPA theory and calculation rules, combined with the lining suitability grading criteria, the five-element connection degree of each index and the comprehensive connection degree of each working point were calculated. In this study, the model is applied to the suitability evaluation of reinforced concrete lining at each drilling point of the high-pressure pipeline of a pumped storage power station (PSPS) in Shanxi Province. The results show that the proposed model consisting of subjective weight and objective weight can effectively avoid the error caused by a single weight method, which improves the evaluation sensitivity and rationality.

Keywords: high-pressure pipeline; suitability evaluation; reinforced concrete lining; C-OWA operator; CRITIC-EWM; combination weighting–SPA

Citation: Xing, C.; Yao, L.; Wang, Y.; Hu, Z. Suitability Evaluation of the Lining Form Based on Combination Weighting–Set Pair Analysis. *Appl. Sci.* **2022**, *12*, 4896. https://doi.org/10.3390/app12104896

Academic Editors: Nuno Lau, Rui Araújo, António Pedro Aguiar, Rodrigo Ventura and João Fabro

Received: 6 April 2022
Accepted: 10 May 2022
Published: 12 May 2022

1. Introduction

The pumped storage power station (PSPS) is a special power supply with many functions, such as peak shaving, valley filling, phase modulation, and emergency standby [1,2]. It not only has a flexible storage capacity which could support the deployment of wind and solar energy, but also helps to ensure the secure and stable operation of power grid [3,4]. In 2016, the signing of the "Paris Agreement" resulted in a quick development of green and low-carbon energy [5]. With rapid development of new energy, the construction of PSPS has once again entered the vision of decision makers in some major countries. The International Renewable Energy Agency (IRENA) proposed that by 2030, the global installed capacity of energy storage will increase by 42% to 68% by 2017, and the increase in installed capacity of pumped storage will be in a range from about 40% to 50%. With commitment from the Chinese government, the peak of carbon dioxide emission and carbon neutrality will be achieved before 2030 and 2060, respectively, and it is estimated that the installed capacity of PSPS will reach about 120 million kW by 2030, continuing to maintain the world's leading level. The opportunity of rapid development for PSPS could be expected both globally and domestically. Therefore, it is very urgent to speed up the construction of PSPS [6].

The key part of PSPS is the water conveyance system that connects the upper and lower reservoirs, which is similar to the blood vessels of the human body. In the design scheme, the water level difference and internal water pressure of the pipeline are generally high. Therefore, the lining must be able to have the ability to deal with problems caused by high water level difference and internal water pressure [7]. At present, the common forms of high-pressure pipelines are steel plate lining and reinforced concrete lining. The former has high price and complicated construction; the latter is economical and simple in manufacturing processes [8]. When designing and selecting the lining forms, it is necessary to consider the influence of a variety of uncertain factors to prevent unpredictable losses. From an economic and reasonable point of view, it is very important to choose an appropriate high-pressure pipeline lining form.

In order to select the appropriate lining form and avoid hydraulic fracturing under the effect of high water head, some design guidelines are commonly used for high-pressure pipelines, such as the minimum overburden criterion, the minimum principal stresses criterion, and the hydraulic fracturing criterion [9–11]. Zhang [10] discussed the interrelationship between these three guidelines and the application conditions, and confirmed that the engineering design for high-pressure pipelines should be focused on the reconnaissance and utilization of the surrounding rock geotechnical conditions and crustal stress state as well as the permeability of rock mass under high water pressure, etc. Schleiss [12] discussed the influence of seepage pressure and stress distribution on lining cracking based on design criteria for reinforced concrete-lined tunnels. Kang Bian et al. [11] used the minimum overburden criterion and the minimum principal stress criterion to eliminate the possibility of cracking caused by improper lining design in the investigation of the causes of high-pressure pipe linings cracking in Huizhou PSPS. In addition, many scholars have used various mathematical models to simulate the hydraulics interactions in high-pressure pipelines in order to explore the key factors affecting the stability of reinforced concrete linings. Chen et al. [13] used a three-dimensional finite element method (3D-FEM) to effectively reflect the main influencing factors (such as the property of surrounding rocks, the internal water pressure, and the interaction and initial gap between surrounding rocks and lining) of the lining stress characteristics. Dadashi et al. [7] used the ABAQUS finite element program based on a direct-coupled method to analyze the hydraulic interaction of high-pressure pipelines and found that the reinforcement distribution and the lining thickness are the key factors affecting the lining stability. Zhou et al. [8] and Su [9] found that in situ stress, rock permeability, and the rock mass deformation modulus are all important parameters for the design of reinforced concrete linings by analyzing the hydraulics interactions in high-pressure hydraulic pipelines.

At present, the methods of selecting the lining form are mostly based on a quantitative perspective. In actual projects, there are also some cases in which the experts directly select the lining form qualitatively according to their engineering experiences. However, whether from a qualitative or quantitative point of view, few scholars have considered the interaction between deterministic and uncertain factors in the suitability evaluation of lining forms. The subjective and objective combination weighting–SPA can be well dealt with in this regard. In this study, a combination weighting–SPA evaluation model of a five-element connection degree was established based on the reinforced concrete lining form. In this model, C-OWA operator was used to calculate the subjective weights of evaluation indices, and the objective weight was determined by the CRITIC method and EWM. Based on the above model, the suitability evaluation indices of the lining form were discussed and selected.

2. The Suitability Level of Lining Form

The established suitability evaluation model in this study can not only utilize the results of current drilling exploration, reduce the workload, and improve the efficiency of design scheme, but also considers the decision of expert evaluation. The five evaluation indices are related to the minimum overburden thickness, the minimum in situ stress, the hydraulic fracturing pressure, the hydraulic gradient, and the surrounding rock classification, which can be recorded as A, B, C, D, and E, respectively. The suitability of the reinforced concrete lining form is divided into five evaluation levels: extremely suitable (I), more suitable (II), basically suitable (III), relatively unsuitable (IV), and extremely unsuitable (V). The relationship between indices and evaluation level of the lining is shown in Table 1.

Table 1. Relationship between Indices and Evaluation Level of the Lining.

Evaluation Level	Extremely Suitable (I)	More Suitable (II)	Basically Suitable (III)	Relatively Unsuitable (IV)	Extremely Unsuitable (V)
A	>1.5	1.3~1.5	1.2~1.3	1.0~1.2	<1.0
B	>1.3	1.2~1.3	1.1~1.2	1.0~1.1	<1.0
C	>1.3	1.2~1.3	1.1~1.2	1.0~1.1	<1.0
D	<3	3~6	6~10	10~15	>15
E	100~50	50~30	30~10	10~3	<3

(a) Evaluation index A is selected with reference to the minimum overburden criterion. It is calculated by Equation (1). This index determines the suitability level of the safety degree of reinforced concrete linings by the ratio of the minimum overburden thickness of rock mass at the design location of high-pressure pipelines to the hydrostatic pressure. In the Norwegian criterion, the safety degree value is in the range of 1.05–1.1. However, the rock weathering in China is relatively deep. Professor Gu Zhaoqi's suggestion is mostly adopted in the practical engineering applications and the evaluation level is more suitable when the index value is in the range of 1.3–1.5 [14].

$$A \geq \frac{r_R \cos \alpha C_{RM}}{\gamma_w h_s} \tag{1}$$

where C_{RM} is the minimum overburden thickness of rock mass (m); h_s is the hydrostatic pressure head at the selected position of the high-pressure pipeline (m); γ_w is the weight of water, which is taken as 10 kN/m^3; γ_R is the weight of rock mass (kN/m^3); and α is the slope angle of the mountain body at the selected position of the high-pressure pipeline.

(b) Evaluation index B is selected according to the hydraulic fracturing criterion, which is used mainly to check the permeability of rock mass and fracture and determine whether they meet the requirements of permeability stability. The index is calculated by comparing the normal stress σ_n on the joint or fracture surface with the hydrostatic pressure of high-pressure pipelines, which could help to determine the suitability level of the safety degree of reinforced concrete linings. When Equation (2) is satisfied, tensile stress does not occur on the joint or fracture surface of the surrounding rock under water pressure, and hydraulic fracturing does not occur on the surrounding rock. Generally, the index should have a certain safety margin. According to the engineering experience, it is more suitable for the value range of 1.2–1.3 [15].

$$B \geq \frac{\sigma_n}{\gamma_w h_s} \tag{2}$$

(c) Evaluation index C is selected according to the minimum principal stress criterion. This criterion is actually an engineering generalization of the hydraulic fracturing criterion. In engineering applications, the normal stress σ_n is difficult to obtain on the joints or fracture surfaces of high-pressure pipelines, and therefore it is replaced by the minimum principal stress σ_{min}. The index uses the ratio of minimum principal stress σ_{min} to the hydrostatic pressure in high-pressure pipelines to determine the suitability level of safety degree, taking the same range of values as the evaluation indicator B. It is obtained by Equation (3).

$$C \geq \frac{\sigma_{min}}{\gamma_w h_s} \tag{3}$$

(d) Evaluation index D is based on the hydraulic gradient of high-pressure pipelines to evaluate the suitability of reinforced concrete lining. The hydraulic gradient is closely related to the hydraulic fracturing criterion and is an important parameter for determining the permeability stability of the surrounding rock. According to the material and fracture development in the borehole fault fracture zone of high-pressure pipelines, there is a time-varying phenomenon along the fault fracture zone, and it is relatively strong. Combined with similar engineering experience, the high-pressure hydraulic gradient is generally not greater than 10~15 under I~III class hard surrounding rock conditions, and the hydraulic gradient in fault fracture zone and influence zone is recommended to be 3.0~6.0. Therefore, the hydraulic gradient value range is 6~10, which is basically suitable.

(e) Evaluation index E evaluates the suitability of reinforced concrete linings according to rock mass classification of high-pressure pipelines. This index is also an important impact parameter for the above criterions. In drilling exploration, the rock quality designation (RQD) is usually selected to determine the rock mass classification. However, many researchers pointed out that the RQD cannot directly reflect the rock structure type and rock block composition [16–18]. As shown in Figure 1, when the distance (intercept) between joints is 10 cm or less, the RQD is 0%, whereas when the distance is 11 cm or more, the RQD is 100%. Compared with RQD, the rock mass block index (RBI) is able to reflect the size of the rock mass [19]. For a rock mass with RQD equal to 90%, although its block composition varies among 10–30, >50, >100 cm, etc., the RBI values corresponding to each case are different. RBI is more accurate than RQD in reflecting the structural characteristics of the rock mass [20]. Therefore, RBI is an improvement and supplement to RQD, and the formula is shown in Equation (4).

$$\text{RBI} = 3C_{r3} + 10C_{r10} + 30C_{r30} + 50C_{r50} + 100C_{r100} \tag{4}$$

where, Cr_3, Cr_{10}, Cr_{30}, $Cr_{50,}$ and Cr_{100} are the core acquisition rates with lengths of 3~10, 10~30, 30~50, 50~100, and >100 cm, respectively. Furthermore, 3, 10, 30, 50, and 100 are all coefficients.

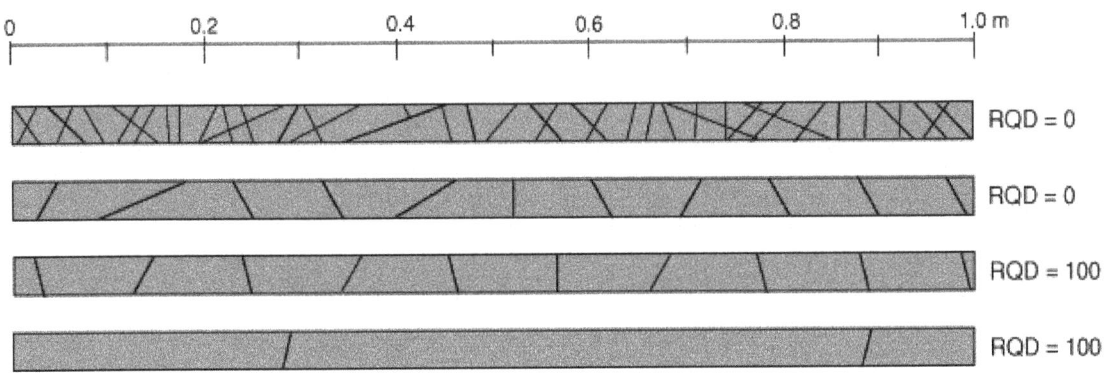

Figure 1. Examples of minimum and maximum values of RQD for various joint densities along drill cores (Reprinted with permission from Ref. [17]. Copyright 2022, Elsevier).

3. The Proposed Methodology

Figure 2 presents the flowchart of the proposed methodology for the suitable level of reinforced concrete lining. The main steps of the methodology are shown below.

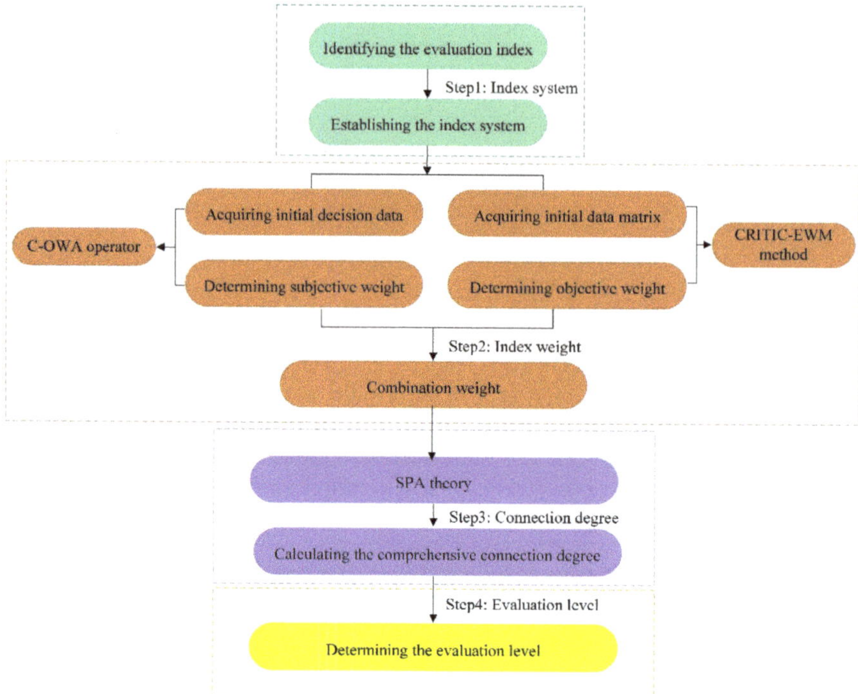

Figure 2. Flowchart of the proposed methodology.

3.1. C-OWA Operator

A commonly used aggregation operator is the ordered weighted averaging (OWA) operator [21]. The C-OWA operator is an improved weight calculation method based on OWA operator. It eliminates the subjectivity and extreme value of expert weight assigned by the orderly weighted average operation, making weight more objective and reasonable [22,23]. Steps are as follows:

(1) Firstly, n experts were invited to score the importance of indices at the same level (using a 10-point system) to form the initial decision data $A = (x_{1j}, x_{2j}, \ldots, x_{mj})$. Then, the initial decision data are arranged in descending order to acquire new decision data $B = (y_{0j}, y_{2j}, \ldots, y_{(m-1)j})$.

(2) The weighted vector u_i of the decision data B is calculated by Equation (5).

$$u_i = \frac{C_{m-1}^{i}}{\sum\limits_{k=0}^{m-1} C_{m-1}^{k}} = \frac{C_{m-1}^{i}}{2^{m-1}} \tag{5}$$

where C_{m-1}^{i} represents the combination number of i data selected from $m-1$ data, $i \in [0, m-1]$.

(3) The absolute weight of the assessment index P_j is obtained by weighting the decision data B with the weighted vector u_i. The equation is as follows:

$$P_j = \sum_{i=0}^{m-1} u_i y_{ij} \tag{6}$$

where j represents the jth index.

(4) According to the absolute weight, the relative weight of the assessment index λ_j is calculated by Equation (7).

$$\lambda_j = \frac{P_j}{\sum\limits_{j=1}^{n} P_j} \tag{7}$$

3.2. CRITIC-EWM Method

3.2.1. CRITIC Method

The CRITIC method was proposed by Diakoulaki et al. in 1995, and the basic idea is to use two parameters (i.e., correlation coefficient and standard deviation) to determine the objective weight of the index [24]. The standard deviation indicates the contrast intensity of the index, and the larger the standard deviation is, the greater the value difference of each scheme [25]. Meanwhile, the correlation coefficient is a quantitative index of conflict, and the conflict decreases with the increase in the correlation coefficient [26]. It is mainly divided into five steps:

(1) The initial indicator data matrix X is defined as follows:

$$X = \left(x_{ij}\right)_{m\times n} \tag{8}$$

where x_{ij} are the raw data of the jth index corresponding to the ith object.

(2) Initial data normalization

In order to eliminate the impact of different dimensions on the evaluation results, it is necessary to standardize each index.

If the index value used is larger, the forward treatment is generally adopted by Equation (9).

$$y_{ij} = \frac{x_{ij} - \min(x_{ij})}{\max(x_{ij}) - \min(x_{ij})} \tag{9}$$

If the index value is smaller, the reverse treatment is generally used by Equation (10).

$$y_{ij} = \frac{\max(x_{ij}) - x_{ij}}{\max(x_{ij}) - \min(x_{ij})} \tag{10}$$

(3) The standard deviation of each index is calculated by Equation (11).

$$\sigma_j = \sqrt{\frac{1}{m-1}\sum_{i=1}^{m}(y_{ij} - \overline{y}_j)^2} \tag{11}$$

where σ_j is the standard deviation of the index x_j; $\overline{y}_j$ is the average value of the jth index.

(4) Correlation coefficient of the indices is calculated by Equation (12).

$$r_{ij} = \frac{\sum\limits_{i=1}^{n}(y_i - \overline{y}_i)(y_j - \overline{y}_j)}{\sqrt{\sum\limits_{i=1}^{n}(y_i - \overline{y}_i)^2\sum\limits_{j=1}^{n}(y_j - \overline{y}_j)^2}} \tag{12}$$

where r_{ij} is the correlation coefficient between index x_i and x_j.

(5) The objective weight of the indices is calculated by Equations (13) and (14).

$$B_j = \sigma_j\sum_{i=1}^{n}(1 - r_{ij}) \tag{13}$$

$$\varphi_j = \frac{B_j}{\sum\limits_{j=1}^{n} B_j} \tag{14}$$

where B_j is the composite information of the jth index.

3.2.2. Improved EWM

EWM is an objective weighting method for characterizing the dispersion of information on the indices [27,28]. In order to extend the scope of application of EWM and make it more convenient to use [29], the improved EWM to determine the objective weights of evaluation indices was adopted. Steps are as follows:

(1) Based on the CRITIC method, n indices and m samples are set in the evaluation, and the measured value of the jth index in the ith sample is recorded by Equation (15). The entropy value R_j of the jth index is defined by Equation (16).

$$r_{ij} = \frac{y_{ij} + 1}{\sum\limits_{i=1}^{m} (y_{ij} + 1)} \tag{15}$$

$$R_j = \frac{-\sum\limits_{i=1}^{m} r_{ij} \ln r_{ij}}{\ln m} \tag{16}$$

(2) The improved entropy weight of the jth index is calculated by Equation (17).

$$e_j = \frac{1 - R_j}{\sum\limits_{j=1}^{n} (1 - R_j)} \tag{17}$$

3.2.3. Determining the Objective Weight

The CRITIC method can eliminate the influence of some indices with strong correlation and reduce the information overlap between indices. However, this method ignores the dispersion between data, and the improved EWM can effectively make up for this deficiency. The integration of these two methods can fully take into account the dispersion, correlation, and comparison intensity of the data, and more effectively reflect the differences and correlation of the objective data of the evaluation indices [30]. The specific formula of combined objective weight ω_j is calculated by Equation (18).

$$\omega_j = (\varphi_j + e_j)/2 \tag{18}$$

3.3. Combination Weighting

The combination weighting method integrates expert experience and objective data. It makes the weighting of the evaluation indices reach the unity of subjective and objective, and improves the rationality of evaluation results. The multiplicative synthesis method [31] is used to determine the comprehensive weight. The calculation equation is:

$$\omega'_j = \frac{\lambda_j \omega_j}{\sum\limits_{j=1}^{n} \lambda_j \omega_j} \tag{19}$$

where ω'_j represents the combination weight of the jth index and λ_j and ω_j are the subjective and objective weights of the jth index, respectively.

3.4. SPA

The SPA method is an improved uncertainty theory which is an integrated certainty–uncertainty system [32–34]. It illustrates unsureness in three angles, namely "identity", "discrepancy", and "contrary", and the connection degree is one of its main features [35,36]. These two objects are a system of interconnection, restriction, and interaction with each other, forming a pair of certainty and uncertainty sets *H (A, B)*. The degree of connectedness of the two objects is determined as follows:

$$\mu = \frac{S}{K} + \frac{F}{K}i + \frac{P}{K}j = a + bi + cj \tag{20}$$

where μ is the connection degree between sets A and B, K stands for the total number of elements; S elements show identical properties, F elements demonstrate discrepant properties, and P elements display contrary properties; a, b, and c represent the degree of "identity", "discrepancy", and "contrary", respectively, and $a + b + c = 1$; i is the uncertainty coefficient in $[-1, 1]$; and j is the contrary coefficient, $j = -1$.

The evaluation of a high-pressure pipe lining form is a complex work affected by a variety of uncertain factors. In order to accurately express the certain and uncertain relationships between the indices, the discrepancy in the ternary connection degree of its core elements was further subdivided to obtain a five-element connection degree expression in this study. The equation is as follows:

$$\begin{cases} u = a + b_1 i_1 + b_2 i_2 + b_3 i_3 + cj = a + bu + cv + dw + el \\ a + b + c + d + e = 1 \\ a, b, c, d, e \in 0, 1 \\ u, v, w \ \in -1,\ 1,\ l = -1 \end{cases} \tag{21}$$

where, a, b, c, d, e are the connection degrees of the evaluation levels I, II, III, IV, and V, respectively, and u, v, w, and l are connection component coefficients.

Combining the comprehensive weight of each index, the comprehensive five-element connection degree of the lining suitability level is calculated, and the calculation formula is obtained by Equation (22).

$$\mu = \sum_{j=1}^{n} a_j \omega'_j + \sum_{j=1}^{n} b_j \omega'_j u + \sum_{j=1}^{n} c_j \omega'_j v + \sum_{j=1}^{n} d_j \omega'_j w + \sum_{j=1}^{n} e_j \omega'_j l \tag{22}$$

4. Case Study

4.1. Project Overview

A PSPS in Shanxi Province has a proposed installed capacity of 1500 MW. The water conveyance system is arranged in the mountain beam between the upper and lower reservoirs, consisting of the diversion system and the tail water system, both of which adopt the arrangement of one pipe and two machines. Two high-pressure main pipes are arranged in parallel, with horizontal spacing of 28.04 m. The lengths of 1# and 2# high-pressure main pipes are 1338.93 m and 1370.11 m, respectively. The high-pressure main pipe is arranged by a buried three inclined shaft. The middle-up flat section and the middle-down flat section are set up with elevations of 1830 m and 1580 m, respectively.

The lithology of the project area is mainly biotite plagiogneiss and diopside-bearing biotite plagioclase amphibolite. The gneiss is a massive structure, and the primary gneissosity is mainly streamlined. It is affected by the late tectonic movement, and faults are more developed in the rock mass. The amphibolite is a massive structure and xenolith form. There is no obvious lithologic boundary between the two lithologies, but mainly in the form of a fracture and melting contact relationship.

Currently, it is in the peak period of new energy policy expansion and energy storage/storage power station construction demand. Due to the requirements of the construction period and design plan of the PSPS and the restriction of using hole-detecting explosives, there is no hole-detecting data for high-pressure pipelines. Therefore, the drilling data of the high-pressure pipeline (Table 2) were used to establish a combination weighting–SPA evaluation model and to analyze the suitability range of reinforced concrete linings. The survey and drilling schematic of the hydropower station is shown in Figure 3.

Table 2. High-pressure pipeline section analysis data.

Drilling Number	*A*	*B*	*C*	*D*	*E*
1	2.43	4.79	7.07	5.89	33.93
2	2.04	2.93	4.96	8.20	69.12
3	1.82	2.79	3.67	8.20	86.40
4	1.80	2.44	4.52	9.38	15.98
5	1.66	0.76	1.71	13.09	46.77
6	1.59	0.37	2.99	17.19	74.11
7	1.60	1.34	1.76	25.32	6.82

Figure 3. Survey and drilling schematic of hydropower station. (**a**): topography of the project area; (**b**): drilling of the high-pressure pipeline; (**c**): hole-detecting of the powerhouse; (**d**): photograph of rock cores.

γ_R = 26.5 kN/m^3. The in situ stress of the borehole is measured by hydraulic fracturing method. σ_{min} is derived from the law of linear relationship between the evolution of the in situ stress of the borehole with depth [37,38].

4.2. Calculation of Index Weight

The C-OWA operator is used for group decision-making weighting and subjective weights are calculated for evaluation indices. Eight experts from the field of geological engineering and hydraulic engineering, as listed in Table 3, were invited to score the importance of each evaluation index on a 10-point scale. The expert's scores on the importance of each index were presented in Table 4, in which the higher score means higher importance.

Table 3. Expert Information.

No.	Professional Field	Position	Educational Level
Expert 1	Geological Engineering	Senior Engineer	Master
Expert 2	Geological Engineering	Senior Engineer	Master
Expert 3	Geological Engineering	Professor	Doctor
Expert 4	Geological Engineering	Professor	Doctor
Expert 5	Hydraulic Engineering	Senior Engineer	Master
Expert 6	Hydraulic Engineering	Senior Engineer	Master
Expert 7	Hydraulic Engineering	Professor	Doctor
Expert 8	Hydraulic Engineering	Professor	Doctor

Table 4. Index importance scores.

Index	Expert 1	Expert 2	Expert 3	Expert 4	Expert 5	Expert 6	Expert 7	Expert 8
A	3	2.5	2	1	1.5	1	1.5	1.2
B	1	2	1.5	2	1.5	2	1	1.5
C	2	2.5	2.5	2.5	2.5	3	2.5	3
D	2.5	2.5	2	2.5	1.5	2	2.5	1.5
E	1.5	1	2	2	3	2.5	2.5	3

Firstly, the subjective weights of the indices λ_i are calculated by Equations (5)–(7):
λ_j = (0.1546, 0.1560, 0.2492, 0.2188, 0.2215);
Then, the combined objective weight of the indices ω_j is calculated by Equations (8)–(18):
ω_j = (0.3343, 0.1612, 0.0924, 0.2471, 0.1650);
Finally, the combination weight of the indices ω'_j is calculated by Equation (19):
ω'_j = (0.2714, 0.1320, 0.1209, 0.2838, 0.1918).

4.3. Level Determination

According to the equipartition principle, $u = 0.5$, $v = 0$, and $w = -0.5$ are taken to be located in the three quarter positions of the interval $[-1, 1]$, and the value of the five-element connection degree of the comprehensive evaluation is calculated to determine the suitability level of the reinforced concrete lining form. The corresponding relationship between the five-element connection degree and the suitability level of the reinforced concrete lining form is shown in Table 5.

Table 5. The Relationship between the Five-element Connection Degree and the Suitability Level.

Suitability Level	Extremely Suitable (I)	More Suitable (II)	Basically Suitable (III)	Relatively Unsuitable (IV)	Extremely Unsuitable (V)
μ	[0.6, 1]	[0.2, 0.6)	[−0.2, 0.2)	[−0.6, −0.2)	[−1, −0.6)

4.4. Suitability Evaluation

According to Equation (22) and combined with Tables 2 and 5, the suitability evaluation of seven groups of drilling samples for the reinforced concrete lining is carried out. The five-element connection degree of working points is shown in Table 6, and the results of each evaluation method are shown in Table 7. According to the evaluation results, all the high-pressure pipelines of the PSPS are basically suitable for reinforced concrete linings. From No. 1 to No. 7, with the increasing underground depth, the anisotropy of the high-pressure pipeline section gradually increases and the suitability of the reinforced concrete lining form gradually decreases, which is mainly due to the influence of in situ stress and surrounding rock classification.

Table 6. Five-element Connection Degree of Comprehensive Evaluation of Each Working Point.

Working Point	1	2	3	4	5	6	7
a	0.5243	0.7162	0.7162	0.5243	0.5222	0.5842	0.5243
b	0.3051	0.0000	0.0000	0.0000	0.0620	0.0000	0.0000
c	0.1706	0.2710	0.2710	0.3315	0.0000	0.0000	0.0045
d	0.0000	0.0129	0.0129	0.1442	0.2168	0.0000	0.1873
e	0.0000	0.0000	0.0000	0.0000	0.1990	0.4158	0.2838
μ	0.6769	0.7097	0.7097	0.4522	0.2457	0.1683	0.1468

Table 7. Comparison of Evaluation Results.

Working Point	1	2	3	4	5	6	7
Expert subjective evaluation method	I	I	I	II	II	III	IV
Combined objective weighting–SPA	I	I	I	II	II	III	II
Combination weighting–SPA	I	I	I	II	II	III	III

By comparison, the results of the three methods are basically the same, except that the results of the three evaluation methods for No. 7 are slightly different (Table 7). The expert evaluation method shows that the rating level of the site is relatively unsuitable. The reason for this is that the experts believe that the surrounding rock classification has declined and the hydraulic gradient value is too large, which makes the use of reinforced concrete lining more unfavorable. However, the involved experts might have different knowledge backgrounds and insights, and consequently the determination of the weight coefficient varies from person to person. The evaluation level calculated by the objective combination weighting–SPA method is more suitable. Objective weight combination weighting is based on the measured data, using certain mathematical methods to objectively assign the weight of the evaluation index. The method lacks subjective control and is often prone to be inconsistent with the actual situation. In summary, for the evaluation process of the suitability of reinforced concrete lining form, using only one of the methods is not comprehensive enough, which might result in over- or under-estimation of the results. Therefore, the combination weighting method integrates the advantages of the subjective and objective weighting methods, combines expert knowledge while fully considering the objectivity of the sample data, minimizes the loss of information, and improves the evaluation reliability.

5. Discussion

(1) Based on the example of the lining form of high-pressure pipelines in PSPS and combining the C-OWA operator, CRITIC method, improved EWM theory, and SPA theory, a new suitability evaluation method is proposed. Compared with the traditional evaluation methods, the advantages of this method are summarized as follows: (1) The combination weighting method consisting of subjective weight and objective weight can effectively avoid the error caused by a single weight method and make the prediction result closer to the actual situation. (2) The case study shows that the multivariate connection degree model is reliable and can be accepted for evaluating the reinforced concrete lining forms suitability and conveniently for practical applications. In addition, the connection degree indicates the interconnection and mutual influence of each index, and can represent the certainty and uncertainty degree from a whole perspective view, and the identical–discrepancy–contrary relationships between the sample and the classification grade. (3) In the selection of evaluation indices, the RQD index, which conventionally characterizes the structural features of rock mass, is improved and supplemented. The RBI index is used to replace RQD, so that the results can more accurately reflect the size of the rock mass, which has obvious engineering practical significance for the quantitative analysis of the variation characteristics of physical and mechanical parameters of rock mass.

(2) The five evaluation indices were selected by the combination weighting–SPA model for the selection of lining forms of diversion tunnels, according to the three design criteria commonly used in hydraulic tunnels. Therefore, this model has certain representativeness and applicability. For the high-pressure pipelines to be evaluated, a similar suitability rating model for lining forms can be established by selecting evaluation indices with similar property characteristics. However, for different hydraulic tunnels, there may be different geological characteristics and risk environments. Therefore, the evaluation indices can be increased or decreased appropriately to meet the suitability evaluation of reinforced concrete lining forms for different high-pressure pipelines.

(3) There are still shortcomings in this study. The hydraulic tunnel design criterion in the proposed model is relatively conservative and basic. For complex cases, how to integrate weights into the evaluation model for the selection of lining forms and how to consider the effects of the index on lining stability remain to be clarified, and further efforts need to be made on accurately selecting an appropriate lining form to prevent accidents in the future. Therefore, a more scientific, rational, and universal classification standard needs to be realized through further research.

6. Conclusions

The relevant factors affecting the lining form were summarized by analyzing the three design criteria and constructs a relatively complete evaluation system for the suitability of reinforced concrete linings. The subjective–objective combination weighting–SPA model not only takes into account the influence of subjective judgment of experts, but also reflects the difference and relevance of objective data of evaluation indices, which improves the evaluation sensitivity and makes the evaluation results more in line with reality. The consistency of the evaluation results with those of other methods verifies the feasibility of the proposed method. This approach optimizes the evaluation process of the lining form and makes the results more comprehensive, objective, and rational. It has exceptional application value in the construction of hydraulic tunnels and can provide reference value and reference significance for the selection of lining forms of other PSPS.

Author Contributions: Data curation, Y.W.; Formal analysis, C.X.; Methodology, C.X.; Supervision, L.Y.; Visualization, Z.H.; Writing—original draft, C.X. All authors have read and agreed to the published version of the manuscript.

Funding: This research received no external funding.

Acknowledgments: We sincerely thank the anonymous reviewers for their time and effort devoted to improving the manuscript.

Conflicts of Interest: The authors declare that they have no known competing financial interest or personal relationships that could have appeared to influence the work reported in this paper.

Abbreviations

The following abbreviations are used in this manuscript:

PSPS	Pumped storage power station
OWA	Ordered weighted averaging
C-OWA	Combination ordered eeighted averaging
CRITIC	Criteria importance through intercriteria correlation
EWM	Entropy eeighting model
SPA	Set pair analysis
IRENA	International Renewable Energy Agency
3D-FEM	Three-dimensional finite element method
RQD	Rock quality designation
RBI	Rock mass block index

References

1. Perez-Diaz, J.I.; Chazarra, M.; Garcia-Gonzalez, J.; Cavazzini, G.; Stoppato, A. Trends and challenges in the operation of pumped-storage hydropower plants. *Renew. Sustain. Energy Rev.* **2015**, *44*, 767–784. [CrossRef]
2. Zhang, S.F.; Andrews-Speed, P.; Perera, P. The evolving policy regime for pumped storage hydroelectricity in China: A key support for low-carbon energy. *Appl. Energy* **2015**, *150*, 15–24. [CrossRef]
3. Barbour, E.; Wilson, I.A.G.; Radcliffe, J.; Ding, Y.L.; Li, Y.L. A review of pumped hydro energy storage development in significant international electricity markets. *Renew. Sustain. Energy Rev.* **2016**, *61*, 421–432. [CrossRef]
4. Xue, F.F.; Xu, C.; Shen, W.Z.; Li, L.M. Ventilation in pumped storage power stations: Influence of dehumidifiers in an underground tunnel. *Appl. Therm. Eng.* **2020**, *172*, 115162. [CrossRef]
5. Gao, Y.; Gao, X.; Zhang, X. The 2 °C Global Temperature Target and the Evolution of the Long-Term Goal of Addressing Climate Change-From the United Nations Framework Convention on Climate Change to the Paris Agreement. *Engineering* **2017**, *3*, 272–278. [CrossRef]
6. Bao, Y.D.; Chen, J.P.; Sun, X.H.; Han, X.D.; Li, Y.C.; Zhang, Y.W.; Gu, F.F.; Wang, J.Q. Debris flow prediction and prevention in reservoir area based on finite volume type shallow-water model: A case study of pumped-storage hydroelectric power station site in Yi County, Hebei, China. *Environ. Earth Sci.* **2019**, *78*, 577. [CrossRef]
7. Dadashi, E.; Noorzad, A.; Shahriar, K.; Goshtasbi, K. Hydro-mechanical interaction analysis of reinforced concrete lining in pressure tunnels. *Tunn. Undergr. Space Technol.* **2017**, *69*, 125–132. [CrossRef]
8. Zhou, Y.F.; Su, K.; Wu, H.G. Hydro-mechanical interaction analysis of high pressure hydraulic tunnel. *Tunn. Undergr. Space Technol.* **2015**, *47*, 28–34. [CrossRef]
9. Su, K.; Wu, H.G. Analysis of Hydro-Mechanical Interaction to Pervious Pressure Tunnel. In Proceedings of the Asia-Pacific Power and Energy Engineering Conference, Chengdu, China, 28–31 March 2010.
10. Zhang, C.S. Design criteria and engineering application of concrete-lined high pressure power tunnels. *J. Hydroelectr. Eng.* **2009**, *28*, 80–84. (In Chinese)
11. Bian, K.; Liu, J.; Xiao, M.; Liu, Z.P. Cause investigation and verification of lining cracking of bifurcation tunnel at Huizhou Pumped Storage Power Station. *Tunn. Undergr. Space Technol.* **2016**, *54*, 123–134. [CrossRef]
12. Schleiss, A.J. Design of reinforced concrete linings of pressure tunnels and shafts. *Hydropower Dams* **1997**, *4*, 88–94.
13. Chen, J.T.; Yang, Y.; Ye, C.; Yang, Y.; Xiao, M. Three-Dimensional Numerical Analysis of Compound Lining in Complex Underground Surge-Shaft Structure. *Math. Probl. Eng.* **2015**, *2015*, 387379. [CrossRef]
14. Gu, Z.Q. *Experiences in Norwegian Hydropower Engineering*; Tapir Publishers: Trondhe, Norway, 1985.
15. Hu, Y.J.; Fang, J.P.; Huang, D.J.; Feng, S.N. Advances in Pressure Tunnel Design and Structure Computation. *Water Power* **2011**, *37*, 15–18. (In Chinese)
16. Du, S.G.; Xu, S.F.; Yang, S.F. Application of rock quality designation(RQD) to engineering classification of rocks. *J. Eng. Geol.* **2000**, *8*, 351–356. (In Chinese)
17. Palmstrom, A. Measurements of and correlations between block size and rock quality designation (RQD). *Tunn. Undergr. Space Technol.* **2005**, *20*, 362–377. [CrossRef]
18. Azimian, A. A New Method for Improving the RQD Determination of Rock Core in Borehole. *Rock Mech. Rock Eng.* **2016**, *49*, 1559–1566. [CrossRef]
19. Hu, X.W.; Zhong, P.L.; Ren, Z.G. Rock mass block index and its engineering practice significance. *J. Hydraul. Eng.* **2002**, *33*, 80–83. (In Chinese)
20. Huang, R.Q.; Huo, J.J. Quantitative analysis of rock mass block index for dam foundation of jinping I hydropower station. *Chin. J. Rock Mech. Eng.* **2011**, *30*, 449–453. (In Chinese)
21. Yager, R.R. On ordered weighted averaging aggregation operators in multicriteria decision-making. *Ieee Trans. Syst. Man Cybern.* **1988**, *18*, 183–190. [CrossRef]
22. Li, X.H.; Zhang, L.Y.; Zhang, R.R.; Yang, M.; Li, H. A semi-quantitative methodology for risk assessment of university chemical laboratory. *J. Loss Prev. Process Ind.* **2021**, *72*, 104553. [CrossRef]
23. Zhao, J.X.; Meng, W.; Sun, F. Construction Risk Assessment of Metro Elevated Station Based on C-OWA Operator and Improved Extenics. *IOP Conf. Ser. Earth Environ. Sci.* **2020**, *525*, 012012. [CrossRef]
24. Diakoulaki, D.; Mavrotas, G.; Papayannakis, L. determining objective weights in multiple criteria problems—The critic method. *Comput. Oper. Res.* **1995**, *22*, 763–770. [CrossRef]
25. Wang, X.G.; Zhou, Z.; Sun, L.C.; Xie, G.H.; Lou, Q.H. Research on the evaluation index system of "new energy cloud" operation mode based on CRITIC weighting method and AHP method. *IOP Conf. Ser. Earth Environ. Sci.* **2021**, *831*, 012017. [CrossRef]
26. Shi, H.T.; Li, Y.F.; Jiang, Z.N.; Zhang, J. Comprehensive power quality evaluation method of microgrid with dynamic weighting based on CRITIC. *Meas. Control* **2021**, *54*, 1097–1104. [CrossRef]
27. Wang, Z.; Xing, X.G.; Yan, F. An abnormal phenomenon in entropy weight method in the dynamic evaluation of water quality index. *Ecol. Indic.* **2021**, *131*, 108137.
28. Zou, Z.H.; Sun, J.N.; Ren, G.P. Study and Application on the Entropy method for Determination of Weight of evaluating indicators in Fuzzy Synthetic Evaluation for Water Quality Assessment. *Acta Sci. Circumstantiae* **2005**, *25*, 552–556. (In Chinese)
29. Ning, B.Q.; Liu, J.; Li, R.H.; Li, L.Y. A multi-attribute decision ranking method based on grey correlation analysis and relative entropy. *Math. Pract. Theory* **2018**, *48*, 2240086.

30. Song, D.M.; Liu, C.X.; Shen, C. Multiple Objective and Attribute Decision Making Based on the Subjective and Objective Weighting. *J. Shandong Univ.* **2015**, *45*, 1–9. (In Chinese)
31. Luo, Z.Z.; Jiang, H.F.; Fu, J.L.; Ding, G.F. Combination Weighting-based Comprehensive Evaluation for Discrete Workshop Production Plan. *J. Syst. Simul.* **2021**, *33*, 1856–1865. (In Chinese)
32. Zhao, K.Q.; Xuan, A.L. Set pair theory-a new method of non-define and its applications. *Syst. Eng.* **1996**, *1*, 18–23. (In Chinese)
33. Chen, W.; Zhang, G.H.; Jiao, Y.Y.; Wang, H. Unascertained Measure-Set Pair Analysis Model of Collapse Risk Evaluation in Mountain Tunnels and Its Engineering Application. *Ksce J. Civ. Eng.* **2021**, *25*, 451–467. [CrossRef]
34. Miao, T.S.; Lu, W.X.; Luo, J.N.; Guo, J.Y. Application of set pair analysis and uncertainty analysis in groundwater pollution assessment and prediction: A case study of a typical molybdenum mining area in central Jilin province, China. *Environ. Earth Sci.* **2019**, *78*, 323. [CrossRef]
35. Giao, N.T.; Nhien, H.T.H.; Anh, P.K.; Van Ni, D. Classification of water quality in low-lying area in Vietnamese Mekong delta using set pair analysis method and Vietnamese water quality index. *Environ. Monit. Assess.* **2021**, *193*, 319. [CrossRef] [PubMed]
36. Su, F.M.; Li, P.Y.; He, X.D.; Elumalai, V. Set Pair Analysis in Earth and Environmental Sciences: Development, Challenges, and Future Prospects. *Expo. Health* **2020**, *12*, 343–354. [CrossRef]
37. Worotnicki, G.; Denham, D. The state stress in the upper part of the Earth's crust in Australia according to measurements in tunnels and mines and from seismic observation. In *Investigation of Stress in Rock—Advances in Stress Measurement*; Institution of Engineers: Sydney, Australia, 1976; pp. 71–82.
38. Brown, E.T.; Hoek, E. Technical note trends in relationships between measured in-situ stress and depth. *Int. J. Rock Mech. Min. Sci. Geomech. Abstr.* **1978**, *15*, 211–215. [CrossRef]

mathematics

MDPI

Article

Adaptive Differential Evolution Algorithm Based on Fitness Landscape Characteristic

Liming Zheng and Shiqi Luo *

Department of Electronic Engineering, School of Information Science and Technology, Jinan University, Guangzhou 510632, China; lmzheng@jnu.edu.cn
* Correspondence: shiqi1996@stu2019.jnu.edu.cn

Abstract: Differential evolution (DE) is a simple, effective, and robust algorithm, which has demonstrated excellent performance in dealing with global optimization problems. However, different search strategies are designed for different fitness landscape conditions to find the optimal solution, and there is not a single strategy that can be suitable for all fitness landscapes. As a result, developing a strategy to adaptively steer population evolution based on fitness landscape is critical. Motivated by this fact, in this paper, a novel adaptive DE based on fitness landscape (FL-ADE) is proposed, which utilizes the local fitness landscape characteristics in each generation population to (1) adjust the population size adaptively; (2) generate DE/current-to-pcbest mutation strategy. The adaptive mechanism is based on local fitness landscape characteristics of the population and enables to decrease or increase the population size during the search. Due to the adaptive adjustment of population size for different fitness landscapes and evolutionary processes, computational resources can be rationally assigned at different evolutionary stages to satisfy diverse requirements of different fitness landscapes. Besides, the DE/current-to-pcbest mutation strategy, which randomly chooses one of the top $p\%$ individuals from the archive *cbest* of local optimal individuals to be the pcbest, is also an adaptive strategy based on fitness landscape characteristic. Using the individuals that are approximated as local optimums increases the algorithm's ability to explore complex multimodal functions and avoids stagnation due to the use of individuals with good fitness values. Experiments are conducted on CEC2014 benchmark test suit to demonstrate the performance of the proposed FL-ADE algorithm, and the results show that the proposed FL-ADE algorithm performs better than the other seven highly performing state-of-art DE variants, even the winner of the CEC2014 and CEC2017. In addition, the effectiveness of the adaptive population mechanism and DE/current-to-pcbest mutation strategy based on landscape fitness proposed in this paper are respectively verified.

Keywords: fitness landscape; differential evolution; population size adaptation; mutation strategy

MSC: 68W50

Citation: Zheng, L.; Luo, S. Adaptive Differential Evolution Algorithm Based on Fitness Landscape Characteristic. *Mathematics* **2022**, *10*, 1511. https://doi.org/10.3390/math10091511

Academic Editor: Fabio Caraffini

Received: 18 March 2022
Accepted: 26 April 2022
Published: 1 May 2022

1. Introduction

Optimization problems are universal in nature, because most problems in real life can be expressed by optimization models, involving multiple standards and goals. Researchers are concentrating their efforts on broad algorithms that may be used for a variety of issues. The development of such general purpose algorithms, which can be categorized as meta-heuristics, has accelerated over the last few decades. Differential evolution (DE) is a basic but powerful algorithm that belongs to the evolutionary algorithms category (EA). DE is a population-based algorithm of metaheuristics, conceptualized by Storn and Price [1]. DE has become a popular choice among researchers for tackling optimization issues in a variety of disciplines over the last two decades [2], such as vehicle routing problems [3], power engineering problems [4], wireless sensors [5], the medical field [6], and chemical engineering [7].

Similar to other evolutionary algorithms (EAs), such as genetic algorithm (GA) [8], memetic algorithm (MA) [9], and estimation of distribution algorithm (EDA) [10], DE also contains three evolutionary operators, namely, mutation, crossover, and selection. The most significant phase in DE is mutation, which is also a crucial step in the production of new individuals by adding a weighted difference vector between two individuals in a population to the third. The essence of mutation is a search behavior, which can prevent premature convergence by increasing population diversity. Over the past decades, many mutation strategies, such as DE/rand-to-best/1, DE/lbest/1 [11], DE/current-to-*p*best/1 [12], and ranking-based mutation [13], have been developed for DE. Shen and He [14] presented a mixed mutation strategy. The algorithm collects local fitness landscape characteristics based on each individual's fitness changes over the distance between each person and the best individual and then uses the Gaussian Cauchy mutation to choose variants of the probability distribution of an optimal mixing strategy. Generally, different strategies display distinct characteristics and are suitable for different fitness landscape characteristics of the evolutionary process.

DE is made up of three main parameters: (1) population size N, which specifies the number of selection operations performed in each generation, (2) scaling factor F, which limits the differential mutation operation, and (3) crossover rate CR, which specifies how many parameters in the target vector altered during the crossover operation [15]. Three parameters suggested respectively by DE inventors Storn and Price [16] are: (1) $N \in$ [5D,10D], D is the dimension of the problem, and $N = 100$ is recommended as the initial value; (2) $F \in [0.4,1]$, and $F = 0.5$ is sufficient for obtaining an optimal solution; (3) for unimodal separable functions, $CR \in [0, 1]$ and $CR = 0.1$ are appropriate initial values, while $CR = 0.9$ is a good initial value for multimodal and nonseparable functions. Besides, other corresponding DE control parameters are also suggested in references [17,18]. Compared to fixed control parameter values, the adaptive control parameter mechanism performs better. Different adaptive or self-adaptive mechanisms (such as FADE [19], ADE [20], jDE [21], JADE [12], SHADE [22], SinDE [23]) are introduced to dynamically update control parameters F and CR without a user's prior knowledge of the relationship between parameter settings and the features of an optimization problem. Apparently, good adaptive parameter design can improve the convergence performance of the algorithm. Since 2013, Tanabe and Fukunaga's SHADE [22] has become a common denominator for the top DE variations [24]. To assist the selection of future control parameter values, this method leveraged a historical recollection of previous control parameter choices.

Although the number of parameters of the differential evolution is small, different hyperparameters and differential evolution variants, such as different mutation operators and F and CR parameters, may introduce structural biases. Structural bias is a form of bias in which artifacts in an algorithm lead to a preference for specific regions in the search space, independent of the objective function [25,26]. What's more, previous work [26,27] on structural bias has pointed out that the strategy for dealing with infeasible solutions can have a huge impact on the presence and strength of structural bias. The choice of the strategy for dealing with infeasible solutions is of high importance, in particular for highly multidimensional problems, as it is more likely to generate infeasible solutions [28]. After analyzing the structural bias of a large number of DE algorithms, Diederick et al. [25] concluded that it is safest to use the dismiss strategy because this strategy for dealing with infeasible solutions shows almost no structural bias, which is consistent with the intersection of the operator being independent of population size. The proposed FL-ADE algorithm inherits the dismiss strategy for dealing with infeasible solutions of LSHADE [29].

However, most of the proposed adaptive DE variants modify the values of F and CR, change the DE strategy, but use a fixed population size. Only a few papers deal with the adaptation of the population size even though the population size is also a crucial parameter of DE, which affects the search efficiency [30]. It can be said that the most popular method of population size control is applied in L-SHADE variant [29]. It is a simple and effective deterministic population size that linearly decreases with the number of evaluations proposed by Tanabe and Fukunaga. The efficiency of the L-SHADE algorithm was confirmed by its first position in the competition of CEC2014 [31]. The same "L" approach was also used in another DE variant typically derived from the L-SHADE algorithm. However, there are few studies on the size of adaptive population. Poláková et al. [32] proposed an adaptive mechanism of population size according to current population diversity. Zhan et al. proposed adaptive distributed differential evolution ADDE [33], which describes an adaptive population parameter setting based on best solution improvement (BSI). This article proposes a new mechanism for population size adaptation based on the fitness landscape.

In recent years, there have been some DE variants based on the fitness landscape. Based on fitness landscape research, Huang et al. devised a self-feedback strategy differential evolution (SFSDE) method [34]. By extracting the features of the local fitness landscape, Li et al. [35] proposed a new self-feedback DE algorithm (SFDE) that selected the optimal mutation strategy and calculated the probability distributions of single-modality and multimodality. Tan et al. recently proposed a DE with an adaptive mutation operator based on fitness landscape (FLDE) [36], which analyzed the fitness landscape features of 45 benchmark training functions before training the relationship between three mutation strategies and fitness landscape features offline using random forest (RF). Finally, throughout the evolutionary process, the trained RF is used to anticipate which mutation technique should be employed to perform mutation operator for each problem. The suggested FLDE algorithm is very competitive with the other five well-known DE algorithms, according to the experimental results. Although there are some DE variants based on the fitness landscape, they all focus on selecting mutation strategies.

The fitness landscape can be used to assess the complexity of an optimization issue in general. However, even though fitness landscape can be used to define the optimization problem and study the evolutionary process, it is rarely used to create algorithms [36]. An adaptable DE based on fitness landscape features (FL-ADE) is proposed in this work. The following are the paper's major contributions:

1. A novel adaptive mechanism of the population size based on the fitness landscape enables the reduction of the population size when exploration is needed and the increase in the population size when exploitation is needed. Most importantly, all these changes in population size are adaptive by extracting the local fitness landscape characteristics and do not require the introduction of any additional parameters.
2. A new mutation strategy: DE/current-to-pcbest, which utilizes the individuals of the approximate local optimum, increases the capability of exploration in multimodal fitness landscape and avoids falling into local optimal due to the use of good function values.

As we can see, FL-ADE takes advantage of the approximate local optimal individuals for the mutation strategy DE/current-to-pcbest to search for different potential regions according to the fitness landscape of different functions. In addition, the fitness landscape characteristics of each generation are calculated, so as to adjust the population size adaptively. The novelty of FL-ADE is that there is no need to use different strategies and set the corresponding population size in advance according to the characteristics of the function. Regardless of any kind of problem, unimodal, multimodal, composite, and hybrid functions, FL-ADE can adaptively focus on different search strategies. On unimodal functions, FL-ADE makes more use of good fitness values, while maintaining a relatively small population size and quickly converging to the global optimum. In contrast, in the face of multimodal functions and complex functions, FL-ADE will use more potential approximate local optimal values, while maintaining a large population size in the early stage to maintain the diversity of the population. In the later stage or when it almost converges to the optimal region, it reduces the population size to increase the iteration speed and quickly converge. The experimental results on CEC2014 benchmark test suite compared with the seven powerful state-of-art DE, even the winner of the CEC2014 and CEC2017.

The rest of the paper is laid out as follows: The classic DE and LSHADE's adaptive parameter process, as well as fitness landscape characteristics, are briefly described in Section 2. The planned FL-ADE is then shown in Section 3. Section 4 contains the experimental results as well as a discussion. The Section 5 gives the conclusions and makes a prospect for the future research.

2. Material Method

In this section, the classic DE is introduced. In addition, the adaptive parameter mechanism and linear population size reduction mechanism of LSHADE is described. Finally, the fitness landscape characteristics are introduced.

2.1. Differential Evolution

The standard DE algorithm consists of four parts.

(1) *Initialization:* The initial population is randomly generated within a given boundary domain as:

$$x_{i,j}^{G} = x_j^{min} + rand(0,1) \times \left(x_j^{max} - x_j^{min}\right) \tag{1}$$

where $i = 1, 2, \ldots, N$ and $j = 1, 2, \ldots, D$. Herein, N represents the population size, D is the problem dimension, *rand*(0,1) is a set of random numbers uniformly distributed in the interval of (0, 1), and x_j^{max} and x_j^{min} denote the upper and lower boundaries of the jth dimension, respectively.

(2) *Mutation operator:* At each generation, a mutation vector v_i is generated based on the difference between two individuals. Here, we list some classic mutation strategies as follows: *DE/rand/1:*

$$v_i = x_{r1} + F \times (x_{r2} - x_{r3}) \tag{2}$$

DE/best/1:

$$v_i = x_{best} + F \times (x_{r1} - x_{r2}) \tag{3}$$

DE/current-to-best/1:

$$v_i = x_i + F \times (x_{best} - x_i) + F \times (x_{r1} - x_{r2}) \tag{4}$$

DE/current-to-best/1:

$$v_i = x_i + F \times \left(x_{best}^{p} - x_i\right) + F \times (x_{r1} - \widetilde{x}_{r2}) \tag{5}$$

where $v_i = [v_{i,1}, v_{i,2}, \ldots v_{i,D}]$, $r1, r2, r3 \in [1, 2, \ldots, N]$, $i \neq r1 \neq r2 \neq r3$ and F is the scaling factor. x_{best} is the best individual, which has the best fitness value in the current population. x_{best}^{p} is randomly chosen from the top $100 \times N \times p\%$ individuals in the current population with $p \subset (0,1)$. $\widetilde{x}_{r2}$ is randomly chosen from the union of P and A, where P is the set of the current population and A is the set of archived inferior solutions [12].

(3) *Crossover Operator:* Trial vector u_i is formed by the individuals x_i and v_i, where $u_i = [u_{i,1}, u_{i,2}, \ldots, u_{i,D}]$. In general, there are two classic crossover operators, namely, binomial crossover and exponential crossover. In this paper, the binomial cross is adopted. In the binomial crossover, each dimension of u_i is separately determined to come from v_i and x_i by the parameter of crossover rate CR as:

$$u_{i,j} = \begin{cases} v_{i,j},\ if\ rand(0,1) \leq CR_i\ or\ j = j_{rand} \\ x_{i,j},\ otherwise \end{cases} \tag{6}$$

where *rand*(0,1) is a random number between 0 and 1, while the *jrand* is a random index in [1, 2, ..., *D*] to ensure that at least one dimension of u_i comes from v_i.

(4) *Selection Operator:* The selection operation procedure is to compare the objective values of target vector x_i and trial vector u_i for the minimization problem by using Equation (7), which means that the better one will be selected for the next generation.

$$x_i^{G+1} = \begin{cases} u_i^G, \; if \; f(u_i^G) < f(x_i^G) \\ x_i^G, \; otherwise \end{cases} \tag{7}$$

The DE repeats the above mutation, crossover, and selection operators until it satisfies the terminal conditions.

2.2. LSHADE

The SHADE [22] algorithm is a better variant of JADE because it employs a different parameter adaptation mechanism based on success history-based adaptation, which is a useful technique for parameter adaptation based on the historical memory of successful parameter sets throughout iteration. Success history-based parameter adaptation employs a historical memory M_{CR}, M_F, which retains a set of CR and F values that have worked successfully in the past and produces new CR and F pairs by selecting them at random. By introducing a linear population size reduction method as well as changes to parameters such as H, $|A|$, and CR, where H is the number of entries into the successfully historical memory pool and $|A|$ is the size of external archive, the LSHADE [29] algorithm enhanced the overall optimization performance of SHADE. All of these adjustments helped LSHADE win the first place in the CEC2014 competition. Recently proposed strong DE variants, such as iLSHADE [37], jSO [38], PaDE [39], and LSHADE_cnEpSin [40], are all variants of LSHADE. LSHADE mainly includes the adaptive parameter mechanism based on success history, as well as a linear population size reduction mechanism (LPSR). This paper also uses LSHADE's adaptive parameter mechanism based on success history, and the proposed adaptive population size mechanism FL-APS is also combined with LPSR, so this chapter will introduce LSHADE. For the linear population size reduction mechanism, Equation (8) depicts the detailed change of N during evolution for the population size reduction mechanism:

$$N_{G+1}^{Linear} = round\left[\left(\frac{N^{min} - N^{init}}{FESmax}\right) \times FES\right] + N^{init} \tag{8}$$

where the N^{init} is the initial population size, N^{min} is the minimum population size, FES_max denotes the maximum number of function evaluations, FES denotes the current number of function evaluations, and $round[\cdot]$ means "round to the nearest integer".

This paper uses the parameter adaptation mechanism of LSHADE. In each generation, in Equation (7), CR and F values that succeed in generating a trial vector u_i^G which is better than the parent individual x_i^G are recorded as S_{CR}, S_F, and at the end of the generation, the memory of M_{CR}, M_F is updated using Algorithm 1. The position in memory to update is determined by the index $k(1 \leq k \leq H)$. The k-th element in the memory is updated in generation G. k is set to 1 at the start of the search and is incremented whenever a new element is added to the history. If $k > H$, k is set to 1. S_{CR} and S_F denote the success sets of control parameters F and CR, and "success" means that a certain individual employing these control parameters produces an offspring with better fitness value. The memory is not updated when all individuals in generation G fail to develop a trial vector that is better than the parent, i.e., $S_{CR} = S_F = \emptyset$.

As M_{CR} is updated, if $M_{CR,k,G=\perp}$ (where $\perp$ denotes a special, "terminal value") or $\max(S_{CR}) = 0$ (i.e., all elements of S_{CR} are 0), $M_{CR,k,G+1}$ is set to $\perp$. The weighted Lehmer mean $mean_{WL}$ (S) is computed using the formula below, and as with [41], the amount of fitness improvement Δf_k is used in order to influence the parameter adaptation (S refers to either S_{CR} or S_F).

$$\begin{cases} mean_{WL}(S) = \frac{\sum_{k=1}^{|S|} w_k \times S_k^2}{\sum_{k=1}^{|S|} w_k \times S_k} \\ w_k = \frac{\Delta f_k}{\sum_{l=1}^{|S_{CR}|} \Delta f_l} \\ \Delta f_k = \left| f(u_k^G) - f(x_k^G) \right| \end{cases} \tag{9}$$

Algorithm 1: Memory update algorithm in LSHADE

```
Input: the success set S_CR and S_F
Output: the historical memory M_CR and M_F
If S_CR ≠ ∅ and S_F ≠ ∅ then
    If M_CR,k,G = ⊥ or max(S_CR) = 0 then
        M_CR,k,G+1 = ⊥;
    Else
        M_CR,k,G+1 = mean_WL(S_CR);
    End If
    M_F,k,G+1 = mean_WL(S_F);
    k++;
    If k > H, then k = 1; End If
Else
    M_CR,k,G+1 = M_CR,k,G;
    M_F,k,G+1 = M_F,k,G;
End If
```

2.3. Fitness Landscape Characteristics

2.3.1. Definition of Fitness Landscape

Fitness landscape, as defined by Wright, is a static model of the problem that has been shown to be effective for examining evolution methods [42]. Consider a continuous function $f(x)$, where $x \in R^n$, where n is the dimension of problem. A fitness landscape in a continuous space is represented by the triple (R^n, f, d), where $d(x,y)$ is the Euclidean distance between two points x and y. It is simple to express the qualities of a fitness landscape in three dimensions using terminology such as ridges, valleys, and basins.

The number of optima in a fitness landscape is used as a statistic (including both local and global optima). Rugged describes a fitness landscape with several local optima, which implies that the terrain is irregular. The difficulty of a fitness landscape is directly proportional to the number of optima. The greater the number of optima, the more difficult the fitness landscape. Counting the exact number of optima or calculating a statistical estimate both take a lengthy time to compute. Instead, we are looking for a simpler method with a reduced calculation cost.

Complexity of an optimization problem can be judged by fitness landscape. However, complex fitness landscapes generally contain different local fitness landscapes, such as ridges, valleys, and basins. For a population $X = [x_1, x_2, \ldots, x_N]$ a local fitness landscape is a part of a fitness landscape.

2.3.2. Local Fitness Landscape

The fitness landscape that corresponds to the evolutionary algorithm's optimization issue is typically quite complicated, including discontinuity, nonlinearity, and nondifferentiability, and it might be unimodal, multimodal, or high dimensional [43].

Shen et al. [14] introduced a simple approach to count the number of optima in the landscape to make computations easier. To begin, sort other people in the population based on their distances from the best person in the population, then examine how their fitness changes over time. The local fitness landscape is similar to a unimodal landscape if fitness increases with distance; otherwise, it is similar to a multimodal environment. The following is a description of how it works:

For a population $P = x_1, x_2, \ldots, x_N$, where each $x_i = (x_{i,1}, x_{i,2}, \ldots, x_{i,D})$ is a solution on R^D.

1. Find the optimal solution of the population and denote it as x^*. Then figure out the distance between each x_i, $i = 1, 2, \ldots, N$ and the optimal solution x^* with the Equation (10):

$$d(i) = \sum_{j=1}^{D} \left| x_{i,j} - x_j^* \right| \tag{10}$$

2. Sort the individuals based on value $d(i)$ calculated above from smallest to largest and denote as $k_1, k_2, \ldots, k_{N-1}$ in order.
3. Set $\theta = 0$ initially. Then, the value of θ will be increased by 1, if $f\left(x_{k_{m+1}}\right) < f(x_{k_m})$ $(m = 1, 2, \ldots, N-1)$. θ is the parameter value for calculating the local fitness landscape feature.
4. Normalize θ by dividing the population size:

$$\frac{\theta}{N-1} \tag{11}$$

where N is the population size. Intuitively, the ruggedness of a fitness landscape is proportional to the number of optima. The normalized θ is used to measure the overall ruggedness of the fitness landscape observation.

3. The Proposed FL-ADE Algorithm

The description of the novel FL-ADE algorithm is divided into three parts in this section: the first part describes the method for extracting fitness landscape characteristics; the second part describes the novel adaptive population size mechanism; and the last part implements a novel mutation strategy "DE/current-to-pcbest" as well as the overall procedure of FL-ADE.

3.1. Extraction of Fitness Landscape Characteristics

For a population $P = x_1, x_2, \ldots, x_N$, where each $x_i = (x_{i,1}, x_{i,2}, \ldots, x_{i,D})$ is a solution on R^D.

(1) Find the optimal solution of the population and denote it as x^*. Then figure out the distance between each x_i, $i = 1, 2, \ldots, N$ and the optimal solution x^* with the Equation (10) (the same as step 1 in Section 2.3.2).

(2) Sort the individuals based on value $d(i)$ calculated above from smallest to largest and denote as $k_1, k_2, \ldots, k_{N-1}$ in order (the same as step 2 in Section 2.3.2).

(3) Set $c = 0$ initially. Then, the value of c will be increased by 1, if $f(x_{k_m}) < f\left(x_{k_{m-1}}\right)$ and $f(x_{k_m}) < f\left(x_{k_{m+1}}\right)$ $(m = 2, 3, \ldots, N-1)$. Finally, c is taken as the number of optimal values for calculating the local fitness landscape feature. It should be emphasized that x_{k_m} is only the optimal value estimated from the sample to reflect the fitness landscape attributes, which is not the true optimum. Moreover, the x_{k_m} is put into the archive *cbest*.

(4) Normalizing c by dividing the population size:

$$\varphi = \frac{c}{N} \times \alpha \tag{12}$$

where φ is the local fitness landscape's simplified observation feature value, which is considered as a normalization of the number of optimal values observed in the fitness landscape, $\varphi \in [0, 1]$, $\alpha \in [0, 1]$. When φ is close to 0, it is closer to the unimodal local fitness landscape; in contrast, it is a multimodal local fitness landscape when φ is close to 1 [44]. The pseudocode to calculate the local fitness landscape characteristic φ and obtain the archive of *cbest* is given in Algorithm 2.

Algorithm 2: calculate φ and get the archive of *cbest*

Input: population $P = [x_1, x_2, \ldots, x_N]$
Output: the local fitness landscape characteristic c and the archive *cbest*
1. Find the optimal solution of the population x_{best};
2. **For** $i = 1$ to N
3. $d(i) = \sum_{j=1}^{D} |x_{i,j} - x_{best,j}|$
4. **End For**
5. Sort the individuals based on value $d(i)$ calculated above from the smallest to the largest and denote as $k_1, k_2, \ldots, k_u$ in order;
6. **For** $m = 2$ **to** $N - 1$ **do**
7. **If** $f(x_{k_m}) < f(x_{k_{m-1}})$ **and** $f(x_{k_m}) < f(x_{k_{m+1}})$ **then**
8. $c++$;
9. $cbest(c) = x_{k_m}$;
10. $cbest(c+1) = x_{best}$;
11. **End If**
12. **End For**
13. $\varphi = \frac{c}{N} \times \alpha$

3.2. DE/Current-to-Pcbest

DE/rand/1 is the first mutation strategy developed for DE [1]. It has proven to be the most successful and widely used strategy. However, DE/rand/1 has poor convergence because it does not use the best individuals to guide the evolution of the population. Zhang proposed DE/current-to-pbest/1 strategy by using one of the top 100p% individuals as x^{p}_{best} to guide the current individuals. However, the DE/current-to-$p$best/1 strategy only focuses on the individuals with good fitness, making it easy to fall into the local optimal.

In order to explore more potential areas while maintaining good convergence, DE/current-to-pcbest/1 strategy is proposed, which utilizes the individuals of the approximate local optimum, increases the capability of exploration in multimodal fitness landscape and the diversity of the population, and enhances the ability to jump out of local optimal. It also can accelerate the convergence in unimodal landscape fitness. That is, for an individual x_i, the DE/current-to-pcbest/1 strategy is as:

$$v_i = x_i + F_i \times \left(x^{p}_{cbest} - x_i\right) + F_i \times (x_{r1} - \tilde{x}_{r2}) \quad (13)$$

where x^{p}_{cbest} is randomly chosen as one of the top 100p% individuals in archive *cbest* with $p \in (0,1]$. The pseudocode to calculate the local fitness landscape characteristic φ and obtain the archive of *cbest* is given in Algorithm 2. Note that the individuals in the population are sorted based on their fitness value before the mutation operation.

For each dimension j, if the mutant vector element $v_{i,j}$ is outside the search range boundaries $[x_j^{min}, x_j^{max}]$, this is an infeasible solution. As mentioned in the introduction, the dismiss strategy for dealing with infeasible solutions shows almost no structural bias, so this FL-ADE is applied the same correction dismiss strategy for dealing with infeasible solutions performed in LSHADE [29]:

$$v^{G}_{i,j} = \begin{cases} (x_j^{min} + x^{G}_{i,j})/2 \ \ if \ v^{G}_{i,j} < x_j^{min} \\ (x_j^{max} + x^{G}_{i,j})/2 \ \ if \ v^{G}_{i,j} > x_j^{max} \end{cases} \quad (14)$$

3.3. Adaptive Population Size Mechanism Based on Fitness Landscape (FL-APS)

In this part, a novel adaptive population size scheme based on fitness landscape is proposed to adjust the appropriate population size dynamically. As we all know, the LSHADE proposed linear population size reduction (LPSR) scheme has proven to be an outstanding scheme for population size adaptation. However, when the number of evaluations increases, the LPSR mechanism simply declines linearly and does not adjust to the different landscape of objective functions. That is why we proposed the adaptive population size scheme based on fitness landscape (FL-APS). The detail of proposed mechanism is given in Equation (15):

$$N_{G+1}^{FL} = round\left[\left(N^{init} - N^{min}\right) \times \varphi + N^{min}\right] \tag{15}$$

where N^{min} and N^{init} denote the minimum and initial value of population size, and $N^{min} = 4$ because DE/current-to-pcbest mutation operator requires at least four individuals. φ is the value of local fitness characteristics; when φ is close to 0, it is close to the unimodal local fitness landscape, N is close to N^{min}, and convergence is accelerated by a smaller value of N; when φ is close to 1, it is close to the multi-modal local fitness landscape, N is close to N^{init}, and N is large to maintain the population diversity. Therefore, the population size N can be adjusted adaptively by extracting fitness landscape characteristics each generation in the iterative process.

After introducing a FES_t into the FL-APS, a balance between population size and generation number could be achieved, and the FL-APS can be altered to Equation (16):

$$N_{G+1} = \begin{cases} round\left[(N_{G+1}^{FL} + N_{G+1}^{Linear})/2\right],\ if\ FES < FES_t \\ \min(N_{G+1}^{FL}, N_{G+1}^{Linear}),\ otherwise \end{cases} \tag{16}$$

where N_{G+1}^{Linear} is the LPSR mechanism calculated by Equation (8). At the early stage of iteration, the relationship between population size and the number of fitness evaluation is balanced by calculating the average value of N_{G+1}^{FL} and N_{G+1}^{Linear}. It can not only maintain a relatively large population size for exploration but also avoid wasting too much computing resources in the early stage. At the later stage of iteration, the minimum values of N_{G+1}^{FL} and N_{G+1}^{Linear} are selected to accelerate exploitation and lead to rapid convergence. To better illustrate the advantages of FL-APS, Figure 1 shows the reduction curve of FLADE population size on f_6 benchmark function in CEC2014; the dimension number is set to D = 30, initial population size is set to $N^{init} = 25\log(D)\sqrt{D}$, the minimum population size is set to N^{min} = 4, and the maximum number of function evaluations is set to FES_max = 10,000 D. The FL-APS with the $FES_t = 1/2 * FES\_max$ is the default setting for the adaptation of population size. A more in-depth of the parameters of FL-APS will be discussed in Section 4.4. We can see from Figure 1 that the population size of FL-APS decreased more slowly than LPSR with the reduction of FES at the beginning of the evolution, which enables the population to maintain enough diversity to explore more areas in the early stage. In the last stage of the iteration, the population size decreased rapidly to accelerate the convergence as the fitness landscape characteristics φ decreased.

Figure 2 shows the number of generations of FL-APS and LSHADE on different types of CEC2014 functions. As can be seen from Figure 2, FL-APS can have more generations in the unimodal function to accelerate convergence, maintain a large population size in the multimodal function and hybrid function to maintain diversity, and adjust population size adaptively in the composition function according to the complexity of landscape fitness.

3.4. Complete Procedure of the Proposed FL-ADE

The parameters CR and F of FL-ADE are consistent with the parameter mechanism of LSHADE, the mutation strategy uses the proposed DE/current-to-pcbest/1, and the population size mechanism is FL-APS. The pseudocode of the proposed FL-ADE is presented in Algorithm 3. To better introduce the proposed FL-ADE, the specific operation is as follows:

Step 1—*Initialization*: The initialization of FL-ADE is the same as in classic DE. Randomly initialize a population of N individuals $P^G = (x_{1,G}, \ldots, x_{N,G})$ with $x_{i,G} = \left(x_{i,G}^1, x_{i,G}^2, \ldots, x_{i,G}^D\right)$, and each individual is uniformly distributed in the range $[X_{min}, X_{max}]$ with $i = (1, 2, \ldots, N)$ according to Equation (1). Set up the maximum generation number FE = 10,000 × D, the generation index G = 1, the initial population size $N^{init} = 25\log(D)\sqrt{D}$, the minimum population size N^{min} = 4, and the percentage of top individuals in archive *cbest* $pc = 0.03$. Set the other parameters to be the same as in LSHADE: $H = 6$, set all values in M_{CR}, M_F to 0.5, and archive $A = \emptyset$, as Line 1–Line 4 in Algorithm 3.

Step 2—For each generation G, all individuals are re-indexed in ascending order of their distance with the best individual $x_{best,G}$. Then calculate the local fitness landscape feature value φ and the approximate local optimal individual archive *cbest* (Algorithm 2). Generating parameters F and CR with successful parameter memory M_{CR} and M_F is same as in LSHADE, where $randn_i(\mu, \sigma^2)$, $randc_i(\mu, \sigma^2)$ are values selected randomly from normal and Cauchy distribution with mean μ and variance σ^2 as in Lines 6–14 in Algorithm 3.

Step 3—*Mutation*: Randomly choose one of the top p% individuals from the archive *cbest* as x^p_{cbest}. Generate mutant vector $v_{i,G}$ via the DE/current-to-pcbest/1 according to Equation (13); then deal with infeasible solutions according to Equation (14), as stated in Lines 15–17 in Algorithm 3.

Step 4—*Crossover*: Use the binomial crossover of the classical DE to generate the trial vector $u_{i,G}$ according to Equation (6).

Step 5—*Selection*: As in classic DE, after comparing each target individual with the $x_{i,G}$ and $u_{i,G}$, the individual with better fitness value will enter the next generation, and the parameters CR and F are stored in the successful parameter archive S_{CR} and S_F, as in Lines 18–28 in Algorithm 3.

Step 6: Update memories M_{CR} and M_F according to Algorithm 1. Adaptively adjust population size according to Equation (16). Repeat step 3 to step 6 until the number of evaluations is greater than or equal to FES_{max}, as in Lines 29–31 in Algorithm 3.

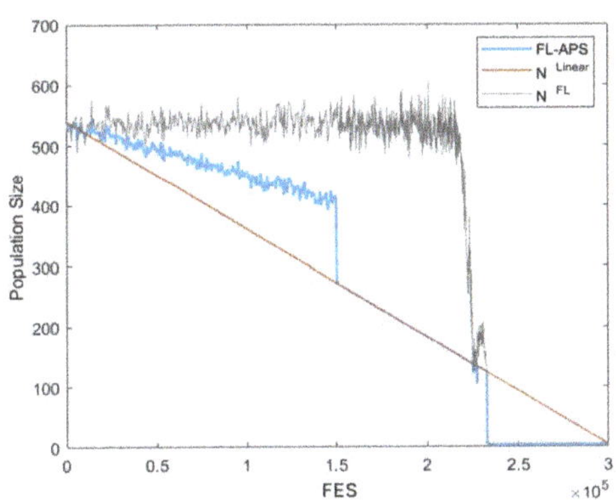

Figure 1. Illustration of the population size of FL-APS on f_6.

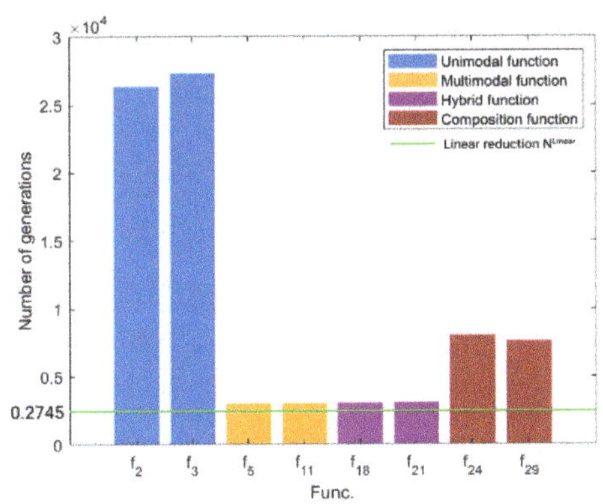

Figure 2. Illustration of the number of generations between the FL-ADE and LSHADE on different types of benchmark functions.

Algorithm 3: FL-ADE algorithm

Input: Bound constraints $[X_{min}, X_{max}]$, the fixed maximum number of function evaluations FES_{max}, benchmark functions $f(X)$;
Output: Best fitness value $f(x_{best})$, best individual X_{best}
// Initialization phase
1. $G = 1$, $N_G = N^{init}$, Archive $A = \varnothing$;
2. Initialize population $P_G = (x_{1,G}, \ldots, x_{N,G})$ randomly;
3. Evaluate P_G, $FES = N_G$;;
4. Set all values in M_{CR}, M_F to 0.5;
// Main loop
5. While $FES < FES_{max}$ do
6. $S_{CR} = \varnothing, S_F = \varnothing$;
7. For $i = 1$ to N do
8. r_i = Select from $[1, H]$ randomly;
9. If $M_{CR,r_I} = \perp$, $CR_{i,G} = 0$. Otherwise
10. $CR_{i,G} = randn_i(M_{CR}, ri,\ 0.1)$;
11. $F_{i,G} = randc_i(M_F, ri,\ 0.1)$;
12. End If
13. End For
// Adaptively mixed mutation strategy
14. Calculate φ and get the archive of *cbest* (Algorithm 2);
15. Dealing with infeasible solutions according to Equation (14);
16. Randomly choose one of the top $p\%$ individuals from the archive *cbest* as x^p_{cbest};
17. Generate trial vector $u_{i,G}$ according to DE/current-to-*pcbest*/1/bin in Equation (13);
18. For $i = 1$ to N do
19. If $f(u_{i,G}) \leq f(x_{i,G})$ then
20. $x_{i,G+1} = u_{i,G}$;
21. Else
22. $x_{i,G+1} = x_{i,G}$;
23. End If
24. If $f(u_{i,G}) < f(x_{i,G})$ then
25. $x_{i,G} \rightarrow A$;
26. $CR_{i,G} \rightarrow S_{CR}$, $F_{i,G} \rightarrow S_F$;
27. End If
28. End For
29. If necessary, delete randomly selected individuals from the archive such that the archive size is $|A|$.
30. Update memories M_{CR} and M_F (Algorithm 1);
// FL-APS
31. Adaptively adjust population size according to Equation (16);
32. $G{+}{+}$;
33. End While

4. Experiment Analysis of FL-ADE Algorithm

In this section, we demonstrate the performance of FL-ADE through experimental results and discuss the influence of its components.

4.1. Experiment Environment

FL-ADE was tested on 30 benchmark test functions developed for IEEE CEC2014 listed in Table 1. For these benchmarks, they can be categorized into four groups:

(1) Unimodal functions $f_1 - f_3$.
(2) Simple multimodal functions $f_4 - f_{16}$.
(3) Hybrid functions $f_{17} - f_{22}$.
(4) Composition functions $f_{23} - f_{30}$.

Table 1. The benchmark functions of IEEE CEC2014.

Type	Func.	Functions	Search Range	$f(x^*)$
Unimodal functions	f_1	Rotated high conditioned elliptic function	$[-100, 100]^D$	100
	f_2	Rotated Bent Cigar function	$[-100, 100]^D$	200
	f_3	Rotated Discus function	$[-100, 100]^D$	300
Simple multimodal functions	f_4	Shifted and rotated Rosenbrock's function	$[-100, 100]^D$	400
	f_5	Shifted and rotated Ackley's function	$[-100, 100]^D$	500
	f_6	Shifted and rotated Weierstrass function	$[-100, 100]^D$	600
	f_7	Shifted and rotated Griewank's function	$[-100, 100]^D$	700
	f_8	Shifted Rastrigin's function	$[-100, 100]^D$	800
	f_9	Shifted and rotated Rastrigin's function	$[-100, 100]^D$	900
	f_{10}	Shifted Schwefel's function	$[-100, 100]^D$	1000
	f_{11}	Shifted and rotated Schwefel's function	$[-100, 100]^D$	1100
	f_{12}	Shifted and rotated Katsuura function	$[-100, 100]^D$	1200
	f_{13}	Shifted and rotated HappyCat function	$[-100, 100]^D$	1300
	f_{14}	Shifted and rotated HGBat function	$[-100, 100]^D$	1400
	f_{15}	Shifted and rotated expanded Griewank's plus Rosenbrock's function	$[-100, 100]^D$	1500
	f_{16}	Shifted and rotated expanded Scaffer's function	$[-100, 100]^D$	1600
Hybrid functions	f_{17}	Hybrid function 1 (N = 3)	$[-100, 100]^D$	1700
	f_{18}	Hybrid function 2 (N = 3)	$[-100, 100]^D$	1800
	f_{19}	Hybrid function 3 (N = 4)	$[-100, 100]^D$	1900
	f_{20}	Hybrid function 4 (N = 4)	$[-100, 100]^D$	2000
	f_{21}	Hybrid function 5 (N = 5)	$[-100, 100]^D$	2100
	f_{22}	Hybrid function 6 (N = 5)	$[-100, 100]^D$	2200
Composition functions	f_{23}	Composition function 1 (N = 5)	$[-100, 100]^D$	2300
	f_{24}	Composition function 2 (N = 3)	$[-100, 100]^D$	2400
	f_{25}	Composition function 3 (N = 3)	$[-100, 100]^D$	2500
	f_{26}	Composition function 4 (N = 5)	$[-100, 100]^D$	2600
	f_{27}	Composition function 5 (N = 5)	$[-100, 100]^D$	2700
	f_{28}	Composition function 6 (N = 5)	$[-100, 100]^D$	2800
	f_{29}	Composition function 7 (N = 3)	$[-100, 100]^D$	2900
	f_{30}	Composition function 8 (N = 3)	$[-100, 100]^D$	3000

The performance of an algorithm is estimated by the fitness error $\Delta f = f - f^*$ (f was the best result obtained by the corresponding algorithm on a particular function, and f^* is the global optimal of the function). If Δf is smaller than $eps = 10 \times 10^{-8}$ it is considered as 0.

To demonstrate the efficiency of the proposed FL-ADE, seven high-performing state-of-art DE variants that appeared from 2011 to 2019 are selected in this research. The first competitor is EPSDE [45], which employed an ensemble of mutation strategies and control parameters. MPEDE [46] is the second algorithm to compare, which simultaneously consists of three mutation strategies. The third competitor is CoBiDE [47], in which the covariance matrix learning and the bimodal distribution parameters setting are incorporated into the DE framework. The fourth peer algorithm SHADE [22] uses a different parameter adaptation mechanism based on the success history to improve JADE [12]. The fifth contender is LSHADE [29], which enhances SHADE's overall optimization performance by implementing a linear population size reduction technique, which helped LSHADE win the CEC2014 championship. The sixth competitor is the second place of the CEC2017 competition LSHADE_cnEpSin [40], which is an improved version of the first place of the CEC2016 competition LSHADE_EpSin [48]. The last comparison algorithm is PaDE [39], which proposes a novel control parameter adaptation schemes and a novel parabolic population size reduction scheme.

4.2. Parameter Settings of the Contrasted Algorithms

As previously stated, this paper compares seven DE variants: EPSDE, MPEDE, SHADE, LSHADE, LSHADE cnEpSin, PaDE, and the newly proposed FL-ADE algorithm. All these algorithms adopted the recommended parameter settings of the cited references in our experiments as summarized in Table 2.

Table 2. Recommended parameter settings of all these contrasted algorithms.

No.	Algorithms	Parameters Initial Settings
1	EPSDE [45]	$F = [0.4, 0.9],\ CR = [0.1, 0.9],\ N = 50;$
2	MPEDE [46]	$F \sim C(\mu F,\ 0.1),\ \mu F = 0.5,\ Cr \sim N(\mu Cr,\ 0.1),\ \mu Cr = 0.5,\ N = 250,\ c = 0.1,\ p = 0.05,\ \lambda_1 = \lambda_2 = \lambda_3 = 0.2,\ ng = 20;$
3	CoBiDE [47]	$N = 60,\ pb = 0.4,\ ps = 0.5;$
4	SHADE [22]	$F \sim C(\mu F,\ 0.1),\ \mu F = 0.5,\ Cr \sim N(\mu Cr,\ 0.1),\ \mu Cr = 0.5,\ N = 100,\ p = 0.2,\ H = 100;$
5	LSHADE [29]	$F\&Cr$ same as SHADE, $N = 18D \sim 4,\ r^{arc} = 2.6,\ p = 0.11,\ H = 6;$
6	LSHADE_cnEpSin [40]	$\mu F = 0.5,\ \mu Cr = 0.5,\ \mu Freq = 0.5,\ ps = 0.5,\ pc = 0.4,\ \mathrm{H} = 5;\ N = 18D \sim 4,\ r^{arc} = 1.4,\ p = 0.11;$
7	PaDE [39]	$\mu F = 0.8,\ \mu Cr = 0.6,\ F\&Cr$ same as LSHADE, $k = 4,\ p = 0.11,\ N = 25log(D)\sqrt{D} \sim 4,\ r^{arc} = 1.6,\ T_0 = 70,\ r^d = 0.04,\ H = 4;$
8	FL-ADE	$F\&Cr\&H\&r^{arc}$ same as LSHADE, $N = 25log(D)\sqrt{D} \sim 4,\ pc = 0.3,\ FES_t = (1/2) * FES_{max}.$

In EPSDE, the population size $N = 50$ is maintained constant, the pool of crossover rate CR values is taken in the range 0.1–0.9 in steps of 0.1, and the pool of scale factor F values is taken in the range 0.4–0.9 in steps of 0.1.

In MPEDE, the population size is set to $N = 250$, and the two newly introduced ones, namely the ratio λ_1 (as $\lambda_1 = \lambda_2 = \lambda_3$) between indicator population and whole population, are set to $\lambda_1 = \lambda_2 = \lambda_3 = 0.2$, and generation gap ng for determining the recent best performing mutation strategy periodically is set to $ng = 20$. The scale factor F obeys semifixed Cauchy distribution, $F \sim C(\mu F,\ 0.1)$, and μF is the location parameter with its initial value is equal to 0.5. The crossover rate CR obeys semifixed Normal distribution, $Cr \sim N(\mu Cr,\ 0.1)$, μCr denotes the mean value, μCr denotes the mean value, and the initial value of μCr is set to $\mu Cr = 0.5$. The ratio of top superior individuals $p = 0.05$, and balance parameter $c = 0.1$.

In CoBiDE the population size $N = 60$, $pb = 0.4$ denotes the probability to execute DE according to the covariance matrix learning, and $ps = 0.5$ denotes the proportion of the individuals chosen from the current population to calculate the covariance matrix.

In SHADE, the control parameter settings of F and CR are the same as MPEDE, the population size $N = 100$, and the ratio of top superior individuals $p = 0.2$. Moreover, historical success values of F and Cr are recorded in a H-entry pool in SHADE, where $H = 100$.

LSHADE employs the same initial values and distributions of control parameters F and Cr as SHADE, and a linear population size reduction scheme is also employed in LSHADE with the initial population size equaling to $N = 18 \times D$, and the minimum population size is equal to 4. Furthermore, the parameter $r^{arc} = 2.6$, defining the factor of external archive size; the parameter $H = 6$, defining entry number in the memory pool; and the parameter $p = 0.11$, denoting the ratio of top superior individuals, are also different from the ones in SHADE.

In LSHADE_cnEpSin, the crossover operator is performed based on the covariance matrix learning with Euclidean neighborhood with a probability $pc = 0.4$, and $ps = 0.5$ is the proportion of individuals that are used to generate the covariance matrix; $ufreq$ represents the mean frequency of the sinusoidal function is set to 0.5. The initial values of all μF, μCR are both set to 0.5, the initial population size equaling to $N = 18 \times D$, the minimum population size is equal to 4, the factor of external archive size $r^{arc} = 1.4$, the entry number in the memory pool $H = 5$, and the ratio of top superior individuals $p = 0.11$.

In PaDE, control parameters F and Cr obey Cauchy distribution $C(\mu F, 0.1)$ and Normal distribution $N(\mu Cr, 0.1)$, respectively, and the initial values of μF and μCr are set to $\mu F = 0.8$ and $\mu Cr = 0.6$. The initial population size is set to $N = 25\log(D)\sqrt{D}$. All individuals in the PaDE algorithm are categorized into k groups, and k is set to a constant value, $k = 4$. Moreover, a time stamp scheme is employed in the external archive, and the timestamp threshold is set to $T0 = 70$. Parameters r^{arc} and p are set tuned values under the time stamp scheme, $r^{arc} = 1.6$ and $p = 0.11$, and the default setting of decay rate r^d is a fixed constant value $r^d = 0.04$.

In the proposed FL-ADE algorithm, the parameter $F\&Cr\&H\&r^{arc}$ is consistent with LSHADE, the initial population size $N = 25\log(D)D_{\surd}$ and the minimum population size equaling to 4. The ratio of top superior individuals in archive *cbest* is set to $pc = 0.3$; the threshold $FES_t = (1/2) * FES_{max}$.

4.3. Comparison with State-of-the-Art DE Algorithms

In this section, the proposed FL-ADE was compared with the seven of DE variants: EPSDE [45], MPEDE [46], CoBiDE [47], SHADE [22], LSHADE [29], LSHADE_cnEpSin [40], and PaDE [39]. The experiments are conducted on $f_1 - f_{30}$ benchmarks of CEC2014 on 10-D, 30D, and 50D optimization, respectively. The allowed maximum number of function evaluations (FES_{max}) of each run was set to 10,000 $\times$ D (D is the dimension of the problem) based on the guideline provided in the special session of CEC2014 [31], and 51 runs were conducted on each benchmark.

Figures 3–5 summarize the comparison results between FL-ADE and other DE algorithms, while the detailed comparison results are shown in Tables 3–5. Tables 3–5 present the mean(std) (mean value and standard deviation) of fitness error for 51 runs of 10D, 30D, and 50D optimization, respectively. To have statistically sound conclusions, Wilcoxon rank-sum test is employed to show the differences between two algorithms on a single problem. The mean error and standard deviation (in bracket) of the function error values are provided in the tables. Three symbols "+, −, =" indicate that FL-ADE is significantly better than, significantly worse than, and almost the same as the corresponding competitor algorithm, respectively. The best result for each problem is shown in boldface. All of these are measured under Wilcoxon's signed rand test with a significance level $\alpha = 0.05$.

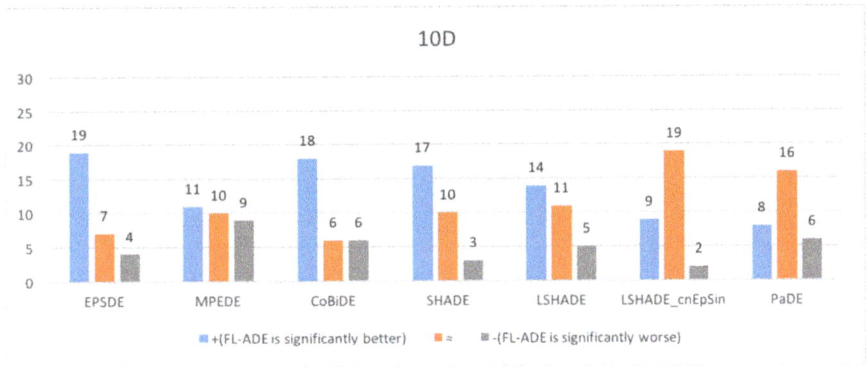

Figure 3. Summarized results between FL-ADE and state-of-art DE on 10D problems.

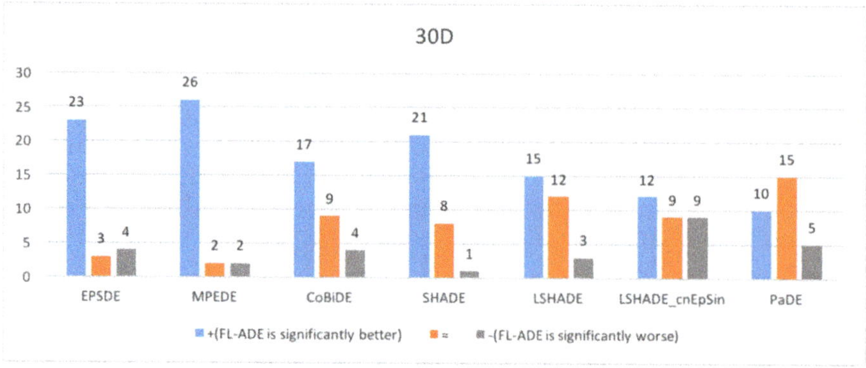

Figure 4. Summarized results between FL-ADE and state-of-art DE on 30D problems.

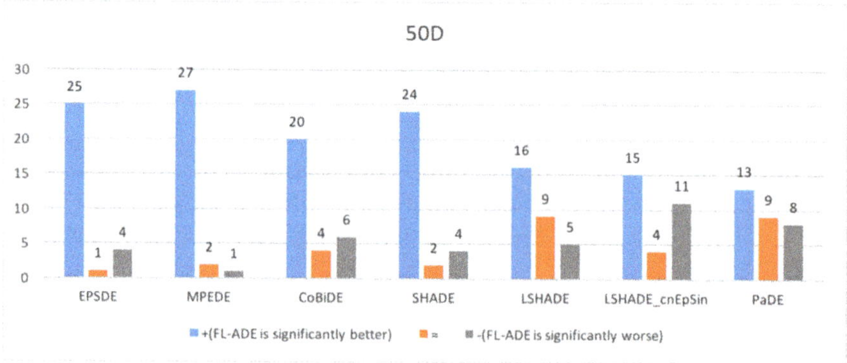

Figure 5. Summarized results between FL-ADE and state-of-art DE on 50D problems.

Table 3. Comparison results between EPSDE, MPEDE, CoBiDE, SHADE, LSHADE, LSHADE-cnEpSin, PaDE, and FL-ADE on 10D functions.

D = 10	EPSDE [45]	MPEDE [46]	CoBiDE [47]	SHADE [22]	LSHADE [29]	LASHDE_cnEpSin [40]	PaDE [39]	FL-ADE
f_1	**0.00E+00** **(0.00E+00) =**	**0.00E+00** **(0.00E+00) =**	**0.00E+00** **(0.00E+00) =**	**0.00E+00** **(0.00E+00) =**	**0.00E+00** **(0.00E+00) =**	**0.00E+00** **(0.00E+00) =**	**0.00E+00** **(0.00E+00) =**	**0.00E+00** **(0.00E+00)**
f_2	**0.00E+00** **(0.00E+00) =**	**0.00E+00** **(0.00E+00) =**	**0.00E+00** **(0.00E+00) =**	**0.00E+00** **(0.00E+00) =**	**0.00E+00** **(0.00E+00) =**	**0.00E+00** **(0.00E+00) =**	**0.00E+00** **(0.00E+00) =**	**0.00E+00** **(0.00E+00)**
f_3	**0.00E+00** **(0.00E+00) =**	**0.00E+00** **(0.00E+00) =**	**0.00E+00** **(0.00E+00) =**	**0.00E+00** **(0.00E+00) =**	**0.00E+00** **(0.00E+00) =**	**0.00E+00** **(0.00E+00) =**	**0.00E+00** **(0.00E+00) =**	**0.00E+00** **(0.00E+00)**
f_4	**7.82E−02** **(5.58E−01)**−	1.46E+01 (1.71E+01)−	7.75E+00 (1.36E+01)−	2.28E+01 (1.65E+01)−	2.33E+01 (1.64E+01) −	1.77E+01 (1.06E+01)−	2.59E+01 (1.53E+01)−	3.34E+01 (6.82E+00)
f_5	2.00E+01 (1.43E−02)+	**1.12E+01** **(9.28E+00) =**	1.96E+01 (2.34E+00)+	1.66E+01 (6.38E+00)+	1.64E+01 (6.88E+00) =	1.36E+01 (8.89E+00) =	1.25E+01 (8.73E+00) =	1.37E+01 (8.98E+00)
f_6	2.86E+00 (5.95E−01)+	**0.00E+00** **(0.00E+00) =**	4.82E−06 (1.47E−05)+	2.75E−04 (1.19E−03)+	1.44E−01 (2.13E−01)+	**0.00E+00** **(0.00E+00) =**	**0.00E+00** **(0.00E+00) =**	**0.00E+00** **(0.00E+00)**
f_7	1.46E−02 (9.01E−03)+	**0.00E+00** **(0.00E+00)**−	2.83E−02 (2.03E−02)+	2.53E−03 (2.33E−03)+	3.38E−04 (1.71E−03) =	1.45E−04 (1.04E−03) =	1.66E−03 (4.47E−03) =	1.25E−03 (5.19E−03)
f_8	**0.00E+00** **(0.00E+00) =**	**0.00E+00** **(0.00E+00) =**	**0.00E+00** **(0.00E+00) =**	**0.00E+00** **(0.00E+00) =**	**0.00E+00** **(0.00E+00) =**	**0.00E+00** **(0.00E+00) =**	**0.00E+00** **(0.00E+00) =**	**0.00E+00** **(0.00E+00)**
f_9	4.13E+00 (1.07E+00)+	1.71E+00 (7.75E−01)+	4.04E+00 (1.52E+00)	3.22E+00 (8.70E−01)+	3.66E+00 (1.05E+00)+	1.85E+00 (6.90E−01)+	2.09E+00 (9.60E−01)+	**4.68E−01** **(8.53E−01)**
f_{10}	3.18E−02 (4.02E−02) =	**0.00E+00** **(0.00E+00)**−	1.02E−02 (3.08E−02)+	1.22E−03 (8.75E−03)−	1.22E−03 (8.75E−03)−	1.22E−03 (8.75E−03)−	4.90E−03 (1.70E−02)−	1.71E−02 (3.08E−02)
f_{11}	3.49E+02 (1.07E+02)+	4.97E+01 2.29E+01)+	7.75E+01 (8.29E+01)+	1.15E+02 (8.84E+01)+	5.13E+01 (4.75E+01)+	2.33E+01 (3.11E+01) =	2.37E+01 (3.03E+01) =	**1.49E+01** **(1.31E+01)**
f_{12}	3.24E−01 (5.99E−02)+	1.99E−01 (4.09E−02)+	3.41E−01 (1.76E−01)+	1.81E−01 (3.00E−02)+	8.41E−02 (2.41E−02)+	**6.78E−02** **(1.74E−02) =**	9.57E−02 (2.92E−02)+	6.91E−02 (2.26E−02)
f_{13}	1.26E−01 (2.42E−02)+	7.17E−02 (1.88E−02)+	1.11E−01 (4.81E−02)+	1.06E−01 (1.72E−02)+	5.64E−02 (1.62E−02)+	4.65E−02 (1.12E−02)+	4.84E−02 (1.59E−02)+	**4.04E−02** **(1.37E−02)**

Table 3. *Cont.*

D = 10	EPSDE [45]	MPEDE [46]	CoBiDE [47]	SHADE [22]	LSHADE [29]	LASHDE_cnEpSin [40]	PaDE [39]	FL-ADE
f_{14}	1.31E−01 (4.19E−02)+	1.21E−01 (3.16E−02)+	1.35E−01 (3.70E−02)+	1.20E−01 (3.61E−02)+	**6.91E−02 (1.80E−02)**−	8.46E−02 (3.48E−02) =	7.97E−02 (2.80E−02) =	8.14E−02 (2.70E−02)
f_{15}	7.37E−01 (1.18E−01)+	4.58E−01 (9.13E−02)+	6.18E−01 (2.37E−01)+	5.21E−01 (9.76E−02)+	3.90E−01 (9.10E−02) =	3.81E−01 (6.82E−02) =	3.87E−01 (8.09E−02) =	**3.73E−01 (7.06E−02)**
f_{16}	2.57E+00 (2.32E−01)+	1.29E+00 (2.45E−01)+	2.01E+00 (5.10E−01)+	1.71E+00 (3.19E−01)+	1.55E+00 (2.80E−01)+	1.03E+00 (2.92E−01)+	1.13E+00 (3.00E−01)+	**8.18E−01 (2.64E−01)**
f_{17}	5.47E+01 (7.23E+01)+	1.05E+00 (2.36E+00) =	**3.82E−01 (3.84E−01)** =	1.16E+00 (2.01E+00) =	1.50E+00 (1.10E+00)+	2.58E+01 (4.01E+01)+	1.59E+00 (1.84E+00)+	7.43E−01 (7.87E−01)
f_{18}	1.26E+00 (8.96E−01)+	**3.83E−02 (1.41E−01)**−	5.86E−02 (1.95E−01)−	1.21E−01 (1.29E−01) =	2.16E−01 (1.79E−01) =	3.46E−01 (4.68E−01) =	8.78E−02 (9.66E−02)−	1.76E−01 (1.81E−01)
f_{19}	1.43E+00 (2.27E−01)+	7.90E−02 (2.65E−02)+	2.82E−01 (1.17E−01)+	1.67E−01 (7.76E−02)+	1.59E−01 (7.47E−02)+	3.58E−01 (4.43E−01)+	**5.09E−02 (2.47E−02)** =	5.61E−02 (2.96E−02)
f_{20}	1.62E−01 (8.75E−02) =	5.72E−02 (3.52E−02)−	**3.09E−02 (5.17E−02)**−	2.61E−01 (7.59E−02)+	1.64E−01 (7.96E−02)+	2.70E−01 (2.00E−01) =	1.22E−01 (9.60E−02) =	1.69E−01 (1.57E−01)
f_{21}	2.39E+00 (4.42E+00)+	**7.37E−02 (8.89E−02)**−	1.05E−01 (1.58E−01)−	2.11E−01 (1.96E−01) =	2.01E−01 (1.63E−01) =	1.52E+00 (4.01E+00)+	2.51E−01 (2.43E−01) =	3.22E−01 (2.58E−01)
f_{22}	2.01E+01 (1.11E+00)+	9.14E−02 (2.06E−02)+	6.89E−01 (7.69E−01)+	1.10E−01 (3.56E−02) =	1.49E−01 (6.33E−02)+	1.60E+00 (4.91E+00)+	**7.67E−02 (2.56E−02)**−	1.05E−01 (4.25E−02)
f_{23}	**3.29E+02 (2.87E−13)** =	**3.29E+02 (2.87E−13)+**	**3.29E+02 (2.87E−13)** =	**3.29E+02 (2.87E−13)** =	**3.29E+02 (2.87E−13)** =	**3.29E+02 2.87E−13)** =	**3.29E+02 (2.87E−13)** =	**3.29E+02 (2.87E−13)**
f_{24}	1.12E+02 (1.72E+00)+	**1.03E+02 (3.52E+00)**−	1.11E+02 (2.85E+00)+	1.09E+02 (1.97E+00)+	1.10E+02 (1.55E+00)+	1.07E+02 (1.79E+00) =	1.07E+02 (3.06E+00) =	1.07E+02 (2.11E+00)
f_{25}	1.91E+02 (2.54E+01)+	1.14E+02 (7.52E+00)+	1.27E+02 (3.00E+01)+	1.19E+02 (4.53E+00)+	1.30E+02 (3.43E+01)+	1.24E+02 (2.04E+01)+	1.13E+02 (1.21E+01)+	**1.10E+02 (1.34E+01)**
f_{26}	1.00E+02 (2.66E−02)+	1.00E+02 (2.22E−02)+	1.00E+02 (4.49E−02)+	1.00E+02 (2.00E−02)+	1.00E+02 (1.32E−02)+	**1.00E+02 (1.22E−02)** =	**1.00E+02 (1.95E−02)** =	**1.00E+02 (1.78E−02)**
f_{27}	3.40E+02 (1.32E+02)+	9.46E+00 (5.58E+01)+	7.64E+01 (1.53E+02)+	7.48E+01 (1.42E+02)+	2.32E+01 (8.75E+01)+	1.40E+02 (1.71E+02)+	3.85E+01 (1.14E+02)+	**4.81E+01 (1.20E+02)**

Table 3. *Cont.*

D = 10	EPSDE [45]	MPEDE [46]	CoBiDE [47]	SHADE [22]	LSHADE [29]	LASHDE_cnEpSin [40]	PaDE [39]	FL-ADE
f_{28}	**3.06E+02 (1.17E−02)**−	3.67E+02 (1.42E+01)−	3.60E+02 (1.41E+01)−	3.85E+02 (4.16E+01)−	3.67E+02 (5.52E+00)−	3.80E+02 (2.95E+01) =	3.90E+02 (5.04E+01)−	3.80E+02 (3.08E+01)
f_{29}	**2.02E+02 (4.50E−01)**−	2.22E+02 (7.97E−02)−	2.18E+02 (1.77E+01)−	2.20E+02 (1.27E+01) =	2.22E+02 (5.26E−01)−	2.22E+02 (5.65E−01) =	2.22E+02 (1.93E−01)−	2.22E+02 (5.61E−01)
f_{30}	**2.33E+02 (5.35E+00)**−	4.64E+02 (1.24E+01)=	4.63E+02 (5.24E+00)+	4.70E+02 (1.72E+01)+	4.63E+02 (5.22E+00)=	4.70E+02 (1.83E+01)=	4.66E+02 (1.14E+01)+	4.66E+02 (1.18E+01)
+/=/−	19/7/4	11/10/9	18/6/6	17/10/3	14/11/5	9/19/2	8/16/6	−/−/−

Table 4. Comparison results between EPSDE, MPEDE, CoBiDE, SHADE, LSHADE, LSHADE-cnEpSin, PaDE, and FL-ADE on 30D functions.

D = 30	EPSDE [45]	MPEDE [46]	CoBiDE [47]	SHADE [22]	LSHADE [29]	LASHDE_cnEpSin [40]	PaDE [39]	FL-ADE
f_1	1.10E+04 (2.17E+04)+	3.82E+01 (1.89E+02)+	1.43E+04 (1.24E+04)+	1.60E+03 (2.29E+03)+	**0.00E+00 (0.00E+00)** =	**0.00E+00 (0.00E+00)** =	**0.00E+00 (0.00E+00)** =	**0.00E+00 (0.00E+00)**
f_2	**0.00E+00 (0.00E+00)** =	**0.00E+00 (0.00E+00)**+	**0.00E+00 (0.00E+00)** =	**0.00E+00 (0.00E+00)** =	**0.00E+00 (0.00E+00)** =	**0.00E+00 (0.00E+00)** =	**0.00E+00 (0.00E+00)** =	**0.00E+00 (0.00E+00)**
f_3	**0.00E+00 (0.00E+00)** =	**0.00E+00 (0.00E+00)**+	**0.00E+00 (0.00E+00)** =	**0.00E+00 (0.00E+00)** =	**0.00E+00 (0.00E+00)** =	**0.00E+00 (0.00E+00)** =	**0.00E+00 (0.00E+00)** =	**0.00E+00 (0.00E+00)**
f_4	3.57E+00 (2.44E+00)+	3.16E−04 (1.29E−03)+	9.70E−06 (3.46E−05)+	**0.00E+00 (0.00E+00)** =	**0.00E+00 (0.00E+00)** =	**0.00E+00 (0.00E+00)** =	**0.00E+00 (0.00E+00)** =	**0.00E+00 (0.00E+00)**
f_5	2.03E+01 (3.82E−02)+	2.04E+01 (3.65E−02)+	2.03E+01 (2.63E−01) =	2.02E+01 (2.55E−02)+	2.01E+01 (2.83E−02)+	2.01E+01 (3.27E−02)+	2.02E+01 (4.42E−02)+	**2.01E+01 (3.07E−02)**
f_6	1.89E+01 (1.69E+00)+	2.08E+00 (1.63E+00)+	1.18E+00 (1.11E+00)+	6.02E+00 (3.88E+00)+	8.61E+00 (2.90E+00)+	2.63E−06 (7.26E−06)−	**0.00E+00 (0.00E+00)**−	1.97E−01 (5.06E−01)

Table 4. *Cont.*

D = 30	EPSDE [45]	MPEDE [46]	CoBiDE [47]	SHADE [22]	LSHADE [29]	LASHDE_cnEpSin [40]	PaDE [39]	FL-ADE
f_7	1.16E−03 (4.30E−03)+	1.06E−03 (4.12E−03)+	**0.00E+00 (0.00E+00) =**	3.87E−04 (1.93E−03) =	**0.00E+00 (0.00E+00) =**	**0.00E+00 (0.00E+00) =**	**0.00E+00 (0.00E+00) =**	**0.00E+00 (0.00E+00)**
f_8	**0.00E+00 (0.00E+00) =**	**0.00E+00 (0.00E+00)+**	**0.00E+00 (0.00E+00) =**	**0.00E+00 (0.00E+00) =**	**0.00E+00 (0.00E+00) =**	**0.00E+00 (0.00E+00) =**	**0.00E+00 (0.00E+00) =**	**0.00E+00 (0.00E+00)**
f_9	4.27E+01 (6.94E+00)+	3.15E+01 (8.16E+00)+	3.97E+01 (1.11E+01)+	2.24E+01 (3.57E+00)+	1.16E+01 (2.06E+00)+	1.34E+01 (2.47E+00)+	8.30E+00 (1.57E+00)+	**5.21E+00 (1.97E+00)**
f_{10}	1.25E+00 (7.65E+00)+	2.24E+01 (3.69E+00)+	5.78E+01 (1.31E+01)+	4.49E−03 (8.65E−03) =	2.48E−03 (6.78E−03)+	3.66E−03 (9.15E−03)+	1.22E−03 (4.95E−03) =	**4.08E−03 (1.02E−02)**
f_{11}	3.54E+03 (4.02E+02)+	2.94E+03 (3.02E+02)+	1.69E+03 (3.79E+02)+	1.80E+03 (1.81E+02)+	1.34E+03 (1.92E+02)+	1.26E+03 (2.07E+02)+	1.21E+03 (1.82E+02) =	**1.17E+03 (1.88E+02)**
f_{12}	5.03E−01 (6.24E−02)+	5.84E−01 (7.76E−02)+	**2.56E−01 (3.26E−01)−**	2.42E−01 (2.93E−02)+	1.77E−01 (2.81E−02)+	1.92E−01 (2.76E−02)+	1.83E−01 (2.77E−02)+	1.49E−01 (2.21E−02)
f_{13}	2.56E−01 (3.74E−02)+	2.03E−01 (3.01E−02)+	2.50E−01 (5.49E−02)+	2.23E−01 (3.30E−02)+	1.46E−01 (2.21E−02)+	1.35E−01 (1.59E−02)+	1.08E−01 (1.54E−02)+	**9.53E−02 (1.26E−02)**
f_{14}	2.96E−01 (8.06E−02)+	2.06E−01 (2.80E−02) =	2.36E−01 (3.68E−02)+	2.28E−01 (2.99E−02)+	2.14E−01 (2.65E−02) =	**1.87E−01 (2.45E−02)−**	2.23E−01 (2.17E−02) =	2.15E−01 (2.86E−02)
f_{15}	5.45E+00 (7.70E−01)+	4.83E+00 (9.71E−01)+	3.08E+00 (7.71E−01)+	3.10E+00 (4.66E−01)+	2.57E+00 (5.19E−01)+	2.31E+00 (2.57E−01)+	2.17E+00 (2.47E−01) =	**2.11E+00 (2.74E−01)**
f_{16}	1.11E+01 (3.22E−01)+	1.05E+01 (2.87E−01)+	9.84E+00 (7.78E−01)+	9.67E+00 (3.59E−01)+	9.11E+00 (3.53E−01)+	8.43E+00 (4.71E−01) =	8.48E+00 (3.98E−01) =	**8.40E+00 (4.01E−01)**
f_{17}	5.19E+04 (5.95E+04)+	7.51E+02 (3.08E+02)+	2.78E+02 (1.94E+02) =	9.90E+02 (3.13E+02)+	1.77E+02 (7.68E+01)−	**1.26E+02 (8.76E+01)−**	1.66E+02 (9.89E+01)−	2.62E+02 (1.09E+02)
f_{18}	2.55E+02 (4.50E+02)+	5.91E+01 (1.02E+01)+	1.14E+01 (4.38E+00)+	6.70E+01 (2.63E+01)+	4.40E+00 (1.76E+00) =	6.97E+00 (2.48E+00)+	5.31E+00 (2.42E+00) =	**4.38E+00 (2.15E+00)**
f_{19}	1.29E+01 (9.99E−01)+	4.20E+00 (6.85E−01)+	2.87E+00 (4.65E−01)−	4.58E+00 (7.68E−01)+	3.40E+00 (6.01E−01)−	**2.75E+00 (6.37E−01)−**	3.43E+00 (5.76E−01)−	3.88E+00 (7.01E−01)

Table 4. *Cont.*

D = 30	EPSDE [45]	MPEDE [46]	CoBiDE [47]	SHADE [22]	LSHADE [29]	LASHDE_cnEpSin [40]	PaDE [39]	FL-ADE
f_{20}	4.86E+01 (4.87E+01)+	1.29E+01 (6.46E+00)+	7.55E+00 (3.20E+00)+	1.25E+01 (6.18E+00)+	3.95E+00 (7.84E−01)+	**2.03E+00** **(9.54E−01)**−	2.66E+00 (1.11E+00) =	2.62E+00 (1.19E+00)
f_{21}	1.24E+04 (1.79E+04)+	1.91E+02 (8.70E+01)+	1.18E+02 (1.05E+02) =	2.43E+02 (1.28E+02)+	9.82E+01 (7.02E+01) =	8.04E+01 (8.75E+01) =	7.82E+01 (7.80E+01) =	**1.01E+02** **(8.46E+01)**
f_{22}	2.24E+02 (9.80E+01)+	1.21E+02 (5.40E+01)+	1.21E+02 (8.90E+01)+	1.46E+02 (5.79E+01)+	2.92E+01 (8.71E+00)+	6.16E+01 (5.54E+01)+	6.84E+01 (5.34E+01)+	**2.97E+01** **(2.48E+01)**
f_{23}	**3.14E+02** **(9.56E−13)**−	3.15E+02 (4.02E−13) =	3.15E+02 (4.02E−13) =	3.15E+02 (4.02E−13) =	3.15E+02 (4.02E−13) =	3.15E+02 (1.25E−01)−	3.15E+02 (3.18E−13) =	3.15E+02 (4.02E−13)
f_{24}	2.29E+02 (6.01E+00)+	2.24E+02 (2.02E+00)+	2.23E+02 (1.10E+00)+	2.25E+02 (3.15E+00)+	2.23E+02 (1.32E+00)+	2.08E+02 (1.04E+01)−	2.24E+02 (1.18E+00)+	**2.01E+02** **(5.15E+00)**
f_{25}	2.00E+02 (6.29E−01)+	**2.01E+02** **(2.21E+00)**−	2.03E+02 (3.91E−01)+	2.08E+02 (2.54E+00)+	2.03E+02 (3.20E−02)+	2.03E+02 (3.81E−02)+	2.03E+02 (1.43E−01)+	2.00E+02 (5.09E−01)
f_{26}	1.00E+02 (4.83E−02)+	1.00E+02 (3.19E−02)+	1.00E+02 (4.72E−02)+	1.00E+02 (3.24E−02)+	1.00E+02 (1.82E−02)+	1.00E+02 (1.51E−02)+	1.00E+02 (1.40E−02)+	**1.00E+02** **(1.33E−02)**
f_{27}	8.38E+02 (1.04E+02)+	3.62E+02 (4.20E+01)+	3.78E+02 (4.21E+01)+	3.38E+02 (4.39E+01)+	3.00E+02 (4.94E−05) =	3.00E+02 (1.85E−13)−	**3.00E+02** **(1.77E−13)**−	3.00E+02 (1.83E+00)
f_{28}	**3.97E+02** **(1.37E+01)**−	8.70E+02 (3.20E+01)+	8.18E+02 (2.83E+01)−	8.11E+02 (3.65E+01)−	8.26E+02 (2.03E+01)−	8.31E+02 (1.42E+01)−	8.60E+02 (1.41E+01)+	8.44E+02 (1.76E+01)
f_{29}	**2.14E+02** **(1.40E+00)**−	7.00E+02 (9.77E+01)+	5.99E+02 (2.26E+02) =	7.25E+02 (1.15E+01)+	7.16E+02 (2.99E+00) =	7.20E+02 (5.28E+00)+	6.94E+02 (1.13E+02)+	7.17E+02 (4.49E+00)
f_{30}	**5.77E+02** **(1.44E+02)**−	6.61E+02 (1.73E+02)−	7.07E+02 (2.16E+02)−	1.26E+03 (3.80E+02)=	1.79E+03 (7.80E+02)+	1.06E+03 (2.88E+02)=	5.99E+02 (2.27E+02)−	1.18E+03 (4.30E+02)
+/=/−	23/3/4	26/2/2	17/9/4	21/8/1	15/12/3	12/9/9	10/15/5	-/-/-

Table 5. Comparison results between EPSDE, MPEDE, CoBiDE, SHADE, LSHADE, LSHADE-cnEpSin, PaDE, and FL-ADE on 50D functions.

D = 50	EPSDE [45]	MPEDE [46]	CoBiDE [47]	SHADE [22]	LSHADE [29]	LASHDE_cnEpSin [40]	PaDE [39]	FL-ADE
f_1	2.28E+06 (5.96E+06)+	2.80E+05 (1.23E+05)+	3.05E+05 (1.27E+05)+	2.19E+04 (1.01E+04)+	1.37E+03 (1.42E+03) =	**0.00E+00** **(0.00E+00)−**	4.11E+02 (5.63E+02)−	2.18E+03 (2.24E+03)
f_2	1.38E−08 (1.90E−08)+	6.23E+01 (3.17E+02)+	1.72E−01 (4.28E−01)+	**0.00E+00** **(0.00E+00) =**	**0.00E+00** **(0.00E+00) =**	**0.00E+00** **(0.00E+00) =**	**0.00E+00** **(0.00E+00) =**	**0.00E+00** **(0.00E+00)**
f_3	2.20E−04 (1.06E−03)+	3.82E+00 (6.59E+00)+	5.81E−03 (8.13E−03)+	9.37E−08 (2.76E−07)+	**0.00E+00** **(0.00E+00) =**	**0.00E+00** **(0.00E+00) =**	**0.00E+00** **(0.00E+00) =**	**0.00E+00** **(0.00E+00)**
f_4	3.21E+01 (2.04E+01) =	5.84E+01 (3.42E+01) =	5.04E+01 (4.04E+01) =	**1.09E+01** **(2.95E+01)−**	5.84E+01 (4.54E+01) =	4.66E+01 (3.75E+01)−	1.18E+01 (3.18E+01)−	5.59E+01 (4.89E+01)
f_5	2.06E+01 (6.49E−02)+	2.06E+01 (3.52E−02)+	2.02E+01 (3.37E−01)−	2.03E+01 (2.45E−02)+	2.03E+01 (2.89E−02)+	2.03E+01 (3.52E−02)+	2.03E+01 (6.28E−02)+	**2.02E+01** **(3.03E−02)**
f_6	4.57E+01 (2.46E+00)+	1.11E+01 (2.85E+00)+	4.30E+00 (2.36E+00)+	8.53E+00 (4.90E+00)+	4.27E+00 (8.13E+00) =	**1.05E−02** **(7.26E−02)−**	9.10E−02 (2.78E−01)−	9.89E−01 (1.08E+00)
f_7	7.63E−03 (9.81E−03)+	4.11E−03 (5.66E−03)+	**0.00E+00** **(0.00E+00) =**	3.24E−03 (4.53E−03)+	**0.00E+00** **(0.00E+00) =**	**0.00E+00** **(0.00E+00) =**	**0.00E+00** **(0.00E+00) =**	**1.45E−04** **(1.04E−03)**
f_8	3.90E−02 (1.95E−01)+	1.05E+01 (1.73E+00)+	**5.07E−09** **(2.98E−08) =**	**0.00E+00** **(0.00E+00) =**	1.96E−07 (2.08E−07)+	3.36E−08 (2.54E−08)+	**0.00E+00** **(0.00E+00) =**	**0.00E+00** **(0.00E+00)**
f_9	1.48E+02 (1.86E+01)+	8.74E+01 (2.67E+01)+	8.35E+01 (2.22E+01)+	4.96E+01 (7.18E+00)+	1.98E+01 (2.68E+00)+	2.79E+01 (7.30E+00)+	1.59E+01 (2.14E+00)+	**1.49E+01** **(5.40E+00)**
f_{10}	7.05E+02 (6.65E+02)+	4.52E+02 (4.86E+01)+	2.70E+02 (6.45E+01)+	**5.88E−03** **(8.04E−03)−**	7.55E−01 (4.37E−01)+	1.25E+00 (5.40E−01)+	1.11E−02 (1.19E−02)−	9.32E−03 (8.60E−03)
f_{11}	8.98E+03 (4.95E+02)+	7.57E+03 (3.03E+02)+	4.04E+03 (7.13E+02)+	3.93E+03 (4.04E+02)+	3.60E+03 (3.78E+02)+	3.26E+03 (3.38E+02)+	**3.18E+03** **(3.28E+02) =**	**3.10E+03** **(3.09E+02)**
f_{12}	8.61E−01 (1.38E−01)+	8.86E−01 (8.62E−02)+	**8.33E−02** **(8.03E−02)−**	2.39E−01 (2.99E−02)+	2.26E−01 (2.68E−02)+	2.60E−01 (3.67E−02)+	2.19E−01 (3.96E−02)+	1.88E−01 (2.81E−02)
f_{13}	3.73E−01 (5.19E−02)+	3.77E−01 (4.02E−02)+	3.50E−01 (5.53E−02)+	3.25E−01 (4.23E−02)+	1.98E−01 (2.30E−02)+	2.11E−01 (2.21E−02)+	1.71E−01 (2.24E−02)+	**1.52E−01** **(1.87E−02)**

Table 5. *Cont.*

D = 50	EPSDE [45]	MPEDE [46]	CoBiDE [47]	SHADE [22]	LSHADE [29]	LASHDE_cnEpSin [40]	PaDE [39]	FL-ADE
f_{14}	3.31E−01 (6.28E−02)+	2.93E−01 (2.53E−02)−	2.66E−01 (3.07E−02)−	2.92E−01 (2.86E−02)−	2.68E−01 (2.14E−02)−	**1.86E−01 (2.29E−02)**−	2.83E−01 (2.60E−02)−	3.09E−01 (2.39E−02)
f_{15}	1.82E+01 (2.18E+00)+	1.36E+01 (3.35E+00)+	5.98E+00 (1.29E+00)+	8.35E+00 (1.11E+00)+	6.11E+00 (9.60E−01)+	5.67E+00 (4.90E−01)+	5.02E+00 (4.84E−01) =	**5.28E+00 (8.87E−01)**
f_{16}	2.07E+01 (4.32E−01)+	1.97E+01 (3.61E−01)+	1.84E+01 (1.06E+00)+	1.81E+01 (4.20E−01)+	1.76E+01 (3.89E−01)+	1.69E+01 (3.94E−01)+	1.66E+01 (4.20E−01) =	**1.66E+01 (5.08E−01)**
f_{17}	2.19E+05 (1.31E+05)+	2.21E+03 (4.81E+02)+	1.07E+04 (7.97E+03)+	2.51E+03 (6.77E+02)+	1.05E+03 (3.42E+02)−	**3.61E+02 (1.60E+02)**−	1.64E+03 (4.27E+02) =	1.69E+03 (4.08E+02)
f_{18}	2.70E+03 (3.86E+03)+	1.05E+02 (1.93E+01)+	1.04E+02 (1.38E+02)−	1.38E+02 (3.24E+01)+	7.56E+01 (2.16E+01)−	**1.98E+01 (5.08E+00)**−	1.05E+02 (1.30E+01)+	9.42E+01 (1.37E+01)
f_{19}	2.47E+01 (1.17E+00)+	1.17E+01 (1.68E+00)+	**6.82E+00 (1.10E+00)**−	1.67E+01 (8.00E+00)+	1.05E+01 (9.70E−01)+	9.83E+00 (1.23E+00)+	7.99E+00 (2.04E+00) =	7.78E+00 (1.69E+00)
f_{20}	3.48E+02 (3.26E+02)+	1.22E+02 (6.22E+01)+	3.18E+01 (1.18E+01)+	2.01E+02 (5.32E+01)+	1.43E+01 (3.74E+00)+	**6.16E+00 (1.92E+00)**−	1.50E+01 (6.42E+00)+	9.57E+00 (2.71E+00)
f_{21}	6.48E+04 (5.11E+04)+	1.00E+03 (3.06E+02)+	2.50E+03 (2.54E+03)+	1.26E+03 (3.57E+02)+	4.36E+02 (1.07E+02) =	**3.12E+02 (9.82E+01)**−	5.67E+02 (1.70E+02)+	5.01E+02 (1.84E+02)
f_{22}	7.58E+02 (1.57E+02)+	5.37E+02 (1.67E+02)+	5.12E+02 (1.69E+02)+	4.51E+02 (1.46E+02)+	1.78E+02 (9.55E+01)+	**8.20E+01 (5.88E+01)** =	1.40E+02 (5.59E+01)+	**9.22E+01 (6.00E+01)**
f_{23}	**3.37E+02 (2.50E−12)**−	3.44E+02 (0.00E+00) =	3.44E+02 (4.11E−13)−	3.44E+02 (5.23E−13)−	3.44E+02 (4.86E−13)−	3.42E+02 (5.79E−01)−	3.44E+02 (4.89E−13)−	3.44E+02 (4.64E−13)
f_{24}	2.72E+02 (6.14E+00)+	2.71E+02 (6.89E+00)+	2.67E+02 (3.22E+00)+	2.75E+02 (2.19E+00)+	2.75E+02 (8.75E−01)+	2.68E+02 (1.50E+00)+	2.75E+02 (8.36E−01)+	**2.30E+02 (3.01E+01)**
f_{25}	2.02E+02 (3.96E+00)+	2.00E+02 (0.00E+00)+	2.07E+02 (1.13E+00)+	2.21E+02 (4.08E+00)+	2.05E+02 (2.47E−01)+	2.05E+02 (1.48E−01)+	2.06E+02 (4.57E−01)+	**2.00E+02 (2.46E−13)**
f_{26}	1.00E+02 (5.83E−02)+	1.24E+02 (4.27E+01)+	1.06E+02 (2.37E+01)+	1.04E+02 (1.95E+01)+	1.00E+02 (2.04E−02)+	1.00E+02 (2.87E−02)+	1.10E+02 (3.00E+01)+	**1.04E+02 (1.96E+01)**
f_{27}	1.57E+03 (4.19E+01)+	5.52E+02 (6.74E+01)+	4.18E+02 (6.37E+01)+	5.18E+02 (6.07E+01)+	3.30E+02 (3.01E+01)−	**3.11E+02 (1.96E+01)**−	3.11E+02 (2.33E+01)−	3.59E+02 (3.87E+01)

Table 5. *Cont.*

D = 50	EPSDE [45]	MPEDE [46]	CoBiDE [47]	SHADE [22]	LSHADE [29]	LASHDE_cnEpSin [40]	PaDE [39]	FL-ADE
f_{28}	**3.92E+02 (1.45E+01)**−	1.20E+03 (4.86E+01)+	1.16E+03 (4.64E+01)+	1.17E+03 (4.87E+01)+	1.13E+03 (3.45E+01) =	1.15E+03 (2.86E+01)+	1.26E+03 (7.44E+01)+	1.13E+03 (3.01E+01)
f_{29}	**2.25E+02 (2.41E+00)**−	8.20E+02 (5.19E+01)+	1.11E+03 (2.12E+02)+	8.76E+02 (6.92E+01)+	7.24E+05 (5.16E+06)+	8.13E+02 (3.82E+01)+	6.03E+02 (1.18E+02)−	7.97E+02 (4.89E+01)
f_{30}	**1.04E+03 (2.31E+02)**−	9.54E+03 (8.16E+02)+	8.83E+03 (4.05E+02)=	9.96E+03 (7.85E+02)+	8.76E+03 (5.19E+02)=	8.27E+03 (4.84E+02)−	9.59E+03 (6.36E+02)+	8.88E+03 (5.31E+02)
+/=/−	25/1/4	27/2/1	20/4/6	24/2/4	16/9/5	15/4/11	13/9/8	-/-/-

Figure 3 summarizes the comparison results between FL-ADE and other DE algorithms on 10D, while the detailed comparison results are shown in Table 3. Table 3 shows that all contrasting methods perform equally well on benchmarks f_1, f_2, f_3, f_8, and f_{23} and that all eight algorithms can discover the global optima during 51 runs on these four benchmarks. Furthermore, when compared to the EPSDE algorithm, the proposed FL-ADE algorithm achieves 26 better or similar results out of 30 benchmarks; it also achieves 21 better or similar results when compared to the MPEDE algorithm, 24 better or similar results when compared to the CoBiDE algorithm, 27 better or similar results when compared to the SHADE algorithm, 25 better or similar results when compared to the LSHADE algorithm, 28 better or similar results when compared to the LSHADE_cnEpSin algorithm, and 24 better or similar results when compared to the PaDE algorithm. Furthermore, when compared to the seven contrasting methods, the suggested FL-ADE algorithm performs the best on benchmarks f_9, f_{13}, f_{16}, f_{25}, f_{27}. It is worth noticing that the suggested FL-ADE outperforms the competition on nine of the sixteen multimodal functions f_4–f_{16}. This is due to the adaptive mixed approach's current-to-pcbest method, which gives the proposed FL-ADE archive good multimodal performance. In a nutshell, the new proposed FL-ADE achieves a superior overall performance on 10D optimization when compared to the CEC2014 benchmark functions.

Figure 4 summarizes the comparison results between FL-ADE and other DE algorithms on 30D, while the detailed comparison results are shown in Table 4. Table 4 on 30D optimization reveals that on benchmarks f_2, f_3, and f_8, all contrasting algorithms can locate the global optima. Furthermore, when compared to the EPSDE algorithm, the proposed FL-ADE algorithm achieves 26 better or similar results out of 30 benchmarks, 28 better or similar results when compared to the MPEDE algorithm, 26 better or similar results when compared to the CoBiDE algorithm, 29 better or similar results when compared to the SHADE algorithm, 27 better or similar results when compared to LSHADE algorithm, 21 better or similar results when compared to the LSHADE_cnEpSin algorithm, and 25 better or similar results when compared to the PaDE algorithm. Importantly, FL-ADE works almost as well at 30D optimization, as it does at 10D optimization for multimodal functions, achieving 10 of the best out of 16. Furthermore, when compared to the other seven contrasting methods, the suggested FL-ADE algorithm performs the best on benchmarks f_9, f_{13}, f_{15}, f_{22}, and f_{26}. As a results, the new proposed FL-ADE is still competitive on 30D optimization under CEC14 benchmark functions with the other seven high-performance DE variant algorithms.

Figure 5 summarizes the comparison results between FL-ADE and other DE algorithms on 50D, while the detailed comparison results are shown in Table 5. There is no test function that that allows each algorithm to find the global optimal, as shown in Table 5 on 50D optimization. In comparison to the EPSDE algorithm, the novel proposed FL-ADE algorithm achieves 26 better or similar results out of 30 benchmarks, 29 better or similar results in comparison to the MPEDE algorithm, 24 better or similar results in comparison to the CoBiDE algorithm, 26 better or similar results in comparison to SHADE, 25 better or similar results in comparison to the LSHADE algorithms, 19 better or similar results in comparison to the LSHADE-cnEpSin algorithm, and 22 better or similar results when compared to the PaDE algorithm. To summarize, the new suggested FL-ADE on 50D optimization under the CEC2014 test suite is still competitive with the contrasting state-of-the-art DE variations.

For a comprehensive comparison, twelve 30D functions are chosen from IEEE CEC2014 benchmark test suit, including one unimodal function f_1, six multimodal functions f_9, f_{11}, f_{12}, f_{13}, f_{15}, f_{16}, three hybrid functions f_{20}, f_{21}, f_{22} and two composition function f_{24}, f_{26} to show the convergence performance of FL-ADE.

(1) From Figure 6a, we can see that although FL-ADE can find the global optimal on the unimodal function f_1, the convergence speed is not as fast as PaDE and LSHADE-cnEpSin. This is because FL-ADE pays more attention to exploration in the early stage. It can be seen that after FES_t, FL-ADE quickly converges to the global optimum.

(2) From Figure 6b–f, we can see that FL-ADE can always find better solutions than other algorithms on the multimodal function $f_9, f_{11}, f_{13}, f_{15}, f_{16}$. This is because the mutation strategy DE/current-to-pcbest gives FL-ADE a considerable advantage in multimodal functions.
(3) Figure 6g,h, show that FL-ADE has good convergence on complex functions and can also find better solutions than other algorithms.
(4) From Figure 6i–l, we can see that FL-ADE has excellent convergence speed and accuracy on composition function f_{24}, f_{25}. Furthermore, this is because the fitness landscape of composition function is very complex, and FL-ADE can quickly locate the optimal area based on the feedback of fitness landscape characteristics.

4.4. The Effectiveness of FL-APS

In this part we mainly discuss the new proposed FL-APS and the former proposed linear population. As we know, the CEC competitions employed the fixed maximum number of function evaluations, FES_{max} = 10,000 × D. Therefore, a larger population size means that fewer generations are available during the whole evolution. A larger population size in each generation during the evolution will improve the perception ability of the algorithm to the fitness landscape characteristic, while a smaller population size will fall into some local optima. However, larger population size will lead to the number of fitness evaluations reduction and poor convergence. Therefore, we need to balance the population size during each generation and the total generations available during the evolution; thus, the FES_t in Equation (16) was introduced.

Linear population size reduction (LPSR) scheme proposed in LSHADE was proven to be an excellent reduction scheme for many optimization problems. However, LPSR simply decreases the population size with the increase in the number of fitness evaluations, and the quick reduction of population size at the beginning of the evolution usually leads to a bad perception of the landscape of some objective functions; therefore, we proposed the adaptive population size based on fitness landscape characteristic (FL-APS). The FL-APS can adaptively increase or decrease population size based on fitness landscape characteristic so as to reasonably allocate fixed number of function evaluation. The optimization performance comparisons under CEC2014 benchmark test suits on 10D, 30D, and 50D optimization of the FL-ADE with linear population size reduction scheme (FL-ADE_linear) and the default FL-ADE with FL-APS in Table 6.

We can see from Table 6 that the FL-ADE with FL-APS obtains 26 better or similar performances in comparison with FL-ADE_linear on 10D, 26 better or similar performances in comparison with FL-ADE_linear on 30D, and 23 better or similar performances in comparison with FL-ADE_linear on 50D. Figure 7 summarizes the results between default FL-ADE and FL-ADE_linear on CEC2014 benchmark on 10D, 30D, and 50D. To summarize, the novel adaptive population size based on fitness landscape characteristic is meaningful and effective.

4.5. The Effectiveness of DE/Current-to-Pcbest

To demonstrate the effectiveness of DE/current-to-pcbest mutation strategy, DE/best/1, DE/best/2, DE/rant-to-best/1, DE/current-to-best/1 and DE/current-to-pbest/1 are employed in this experiment. The same parameter setting of crossover rate $CR = 0.9$, scaling factor $F = 0.5$, and the population $N = 100$ are used for all mutation strategies. The experimental results are presented in Table 7 according to the Wilcoxon rank-sum test and the Friedman test; DE/current-to-pcbest/1 gets the first ranking among the six mutation strategies. The result of Wilcoxon rank-sum is summarized in Figure 8; obviously, the DE with strategy DE/current-to-pcbest/1 is better than the DE with other strategies.

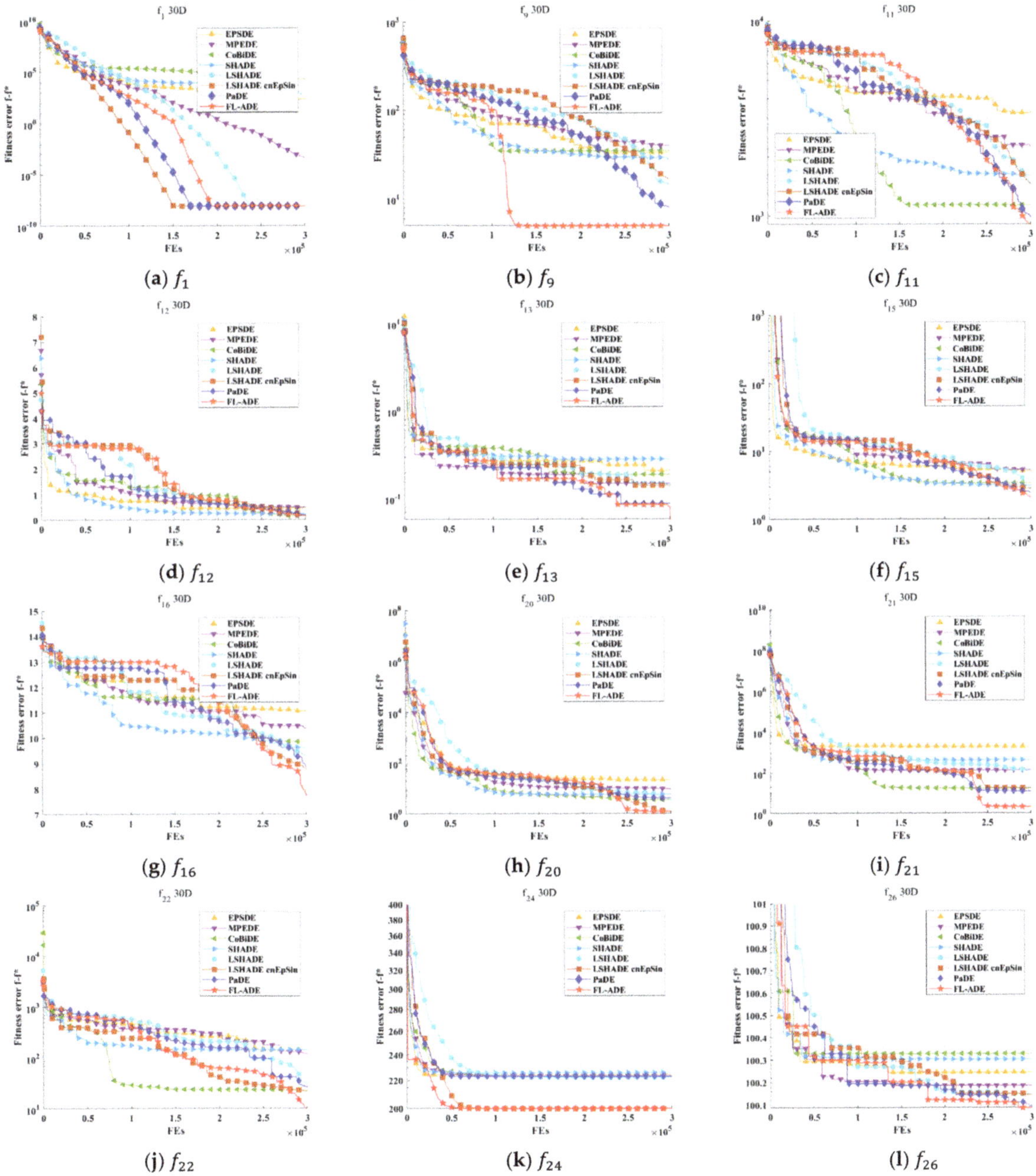

(a) f_1 **(b)** f_9 **(c)** f_{11}

(d) f_{12} **(e)** f_{13} **(f)** f_{15}

(g) f_{16} **(h)** f_{20} **(i)** f_{21}

(j) f_{22} **(k)** f_{24} **(l)** f_{26}

Figure 6. Convergence curves of FL-ADE and other state-of-the-art DE variants on f_1, f_9, f_{11}, f_{12}, f_{13}, f_{15}, f_{16}, $f_{20} f_{21}$, f_{22}, f_{24}, f_{26}, when D = 30.

Table 6. Comparison results of FL-ADE_linear and the default FL-ADE on CEC2014 benchmark on 10D, 30D, and 50D.

	10D			30D			50D		
Func.	FL-ADE_Linear		FL−ADE	FL-ADE_Linear		FL−ADE	FL-ADE_Linear		FL-ADE
f_1	**0.00E+00 ± (0.00E+00)**	=	**0.00E+00 ± (0.00E+00)**	**0.00E+00 ± (0.00E+00)**	=	**0.00E+00 ± (0.00E+00)**	**1.69E+03 ± (1.79E+03)**	=	**2.18E+03 ± (2.24E+03)**
f_2	**0.00E+00 ± (0.00E+00)**	=	**0.00E+00 ± (0.00E+00)**	**0.00E+00 ± (0.00E+00)**	=	**0.00E+00 ± (0.00E+00)**	**0.00E+00 ± (0.00E+00)**	=	**0.00E+00 ± (0.00E+00)**
f_3	**0.00E+00 ± (0.00E+00)**	=	**0.00E+00 ± (0.00E+00)**	**0.00E+00 ± (0.00E+00)**	=	**0.00E+00 ± (0.00E+00)**	**0.00E+00 ± (0.00E+00)**	=	**0.00E+00 ± (0.00E+00)**
f_4	**3.01E+01 ± (3.01E+01)**	−	3.34E+01 ± (6.82E+00)	**0.00E+00 ± (0.00E+00)**	=	**0.00E+00 ± (0.00E+00)**	**4.09E+01 ± (4.83E+01)**	=	**5.59E+01 ± (4.89E+01)**
f_5	**1.45E+01 ± (1.45E+01)**	=	**1.37E+01 ± (8.98E+00)**	**2.01E+01 ± (2.27E−02)**	=	**2.01E+01 ± (3.07E−02)**	**2.02E+01 ± (3.92E−02)**	=	**2.02E+01 ± (3.03E−02)**
f_6	**0.00E+00 ± (0.00E+00)**	=	**0.00E+00 ± (0.00E+00)**	**0.00E+00 ± (0.00E+00)**	−	1.97E−01 ± (5.06E−01)	**6.49E−01 ± (8.40E−01)**	−	9.89E−01 ± (1.08E+00)
f_7	**1.30E−03 ± (1.30E−03)**	=	**1.25E−03 ± (5.19E−03)**	**0.00E+00 ± (0.00E+00)**	=	**0.00E+00 ± (0.00E+00)**	**0.00E+00 ± (0.00E+00)**	=	**1.45E−04 ± (1.04E−03)**
f_8	**0.00E+00 ± (0.00E+00)**	=	**0.00E+00 ± (0.00E+00)**	**0.00E+00 ± (0.00E+00)**	=	**0.00E+00 ± (0.00E+00)**	**0.00E+00 ± (0.00E+00)**	=	**0.00E+00 ± (0.00E+00)**
f_9	2.34E+00 ± (2.34E+00)	+	**4.68E−01 ± (8.53E−01)**	6.67E+00 ± (1.74E+00)	+	**5.21E+00 ± (1.97E+00)**	**1.15E+01 ± (1.71E+00)**	−	1.49E+01 ± (5.40E+00)
f_{10}	**6.12E−03 ± (6.12E−03)**	−	1.71E−02 ± (3.08E−02)	**3.67E−03 ± (9.03E−03)**	=	**4.08E−03 ± (1.02E−02)**	**8.36E−03 ± (8.50E−03)**	−	9.32E−03 ± (8.60E−03)
f_{11}	3.74E+01 ± (3.74E+01)	+	**1.49E+01 ± (1.31E+01)**	**1.16E+03 ± (2.30E+02)**	=	**1.17E+03 ± (1.88E+02)**	**3.09E+03 ± (2.61E+02)**	=	**3.10E+03 ± (3.09E+02)**
f_{12}	**7.01E−02 ± (7.01E−02)**	=	**6.91E−02 ± (2.26E−02)**	**1.48E−01 ± (2.54E−02)**	=	**1.49E−01 ± (2.21E−02)**	**1.85E−01 ± (1.88E−02)**	=	**1.88E−01 ± (2.81E−02)**
f_{13}	5.16E−02 ± (5.16E−02)	+	**4.04E−02 ± (1.37E−02)**	1.19E−01 ± (1.19E−02)	+	**9.53E−02 ± (1.26E−02)**	1.71E−01 ± (1.88E−02)	+	**1.52E−01 ± (1.87E−02)**
f_{14}	**7.42E−02 ± (7.42E−02)**	=	**8.14E−02 ± (2.70E−02)**	2.42E−01 ± (2.64E−02)	+	**2.15E−01 ± (2.86E−02)**	**3.08E−01 ± (2.14E−02)**	=	**3.09E−01 ± (2.39E−02)**
f_{15}	**3.64E−01 ± (3.64E−01)**	=	**3.73E−01 ± (7.06E−02)**	**2.11E+00 ± (2.45E−01)**	=	**2.11E+00 ± (2.74E−01)**	**4.81E+00 ± (4.35E−01)**	−	5.28E+00 ± (8.87E−01)
f_{16}	1.14E+00 ± (1.14E+00)	+	**8.18E−01 ± (2.64E−01)**	**8.47E+00 ± (5.08E−01)**	=	**8.40E+00 ± (4.01E−01)**	**1.68E+01 ± (3.48E−01)**	=	**1.66E+01 ± (5.08E−01)**
f_{17}	**9.00E−01 ± (9.00E−01)**	=	**7.43E−01 ± (7.87E−01)**	**2.01E+02 ± (1.15E+02)**	−	2.62E+02 ± (1.09E+02)	**1.51E+03 ± (4.67E+02)**	−	1.69E+03 ± (4.08E+02)
f_{18}	**1.54E−01 ± (1.54E−01)**	=	**1.76E−01 ± (1.81E−01)**	5.94E+00 ± (2.60E+00)	+	**4.38E+00 ± (2.15E+00)**	1.01E+02 ± (1.73E+01)	+	**9.42E+01 ± (1.37E+01)**
f_{19}	6.96E−02 ± (6.96E−02)	+	**5.61E−02 ± (2.96E−02)**	**3.76E+00 ± (6.73E−01)**	=	**3.88E+00 ± (7.01E−01)**	**8.05E+00 ± (1.98E+00)**	=	**7.78E+00 ± (1.69E+00)**
f_{20}	**1.30E−01 ± (1.30E−01)**	=	**1.69E−01 ± (1.57E−01)**	**2.78E+00 ± (1.22E+00)**	=	**2.62E+00 ± (1.19E+00)**	1.70E+01 ± (7.22E+00)	+	**9.57E+00 ± (2.71E+00)**
f_{21}	**3.11E−01 ± (3.11E−01)**	=	**3.22E−01 ± (2.58E−01)**	**8.36E+01 ± (7.54E+01)**	=	**1.01E+02 ± (8.46E+01)**	**5.42E+02 ± (1.47E+02)**	=	**5.01E+02 ± (1.84E+02)**
f_{22}	**8.87E−02 ± (8.87E−02)**	−	1.05E−01 ± (4.25E−02)	**2.81E+01 ± (1.86E+01)**	=	**2.97E+01 ± (2.48E+01)**	1.21E+02 ± (7.48E+01)	+	**9.22E+01 ± (6.00E+01)**
f_{23}	3.29E+02 ± (3.29E+02)	=	3.29E+02 ± (2.87E−13)	3.15E+02 ± (3.73E−13)	=	3.15E+02 ± (4.02E−13)	3.44E+02 ± (4.78E−13)	−	3.44E+02 ± (4.64E−13)
f_{24}	**1.08E+02 ± (1.08E+02)**	+	**1.07E+02 ± (2.11E+00)**	**2.24E+02 ± (1.13E+00)**	+	**2.01E+02 ± (5.15E+00)**	**2.75E+02 ± (6.35E−01)**	+	**2.30E+02 ± (3.01E+01)**
f_{25}	1.36E+02 ± (1.36E+02)	+	**1.10E+02 ± (1.34E+01)**	2.03E+02 ± (6.15E−02)	+	**2.00E+02 ± (5.09E−01)**	2.05E+02 ± (4.64E−01)	+	**2.00E+02 ± (2.46E−13)**
f_{26}	**1.00E+02 ± (1.00E+02)**	=	**1.00E+02 ± (1.78E−02)**	1.00E+02 ± (1.50E−02)	+	**1.00E+02 ± (1.33E−02)**	1.00E+02 ± (2.01E−02)	+	**1.04E+02 ± (1.96E+01)**
f_{27}	6.58E+01 ± (6.58E+01)	+	**4.81E+01 ± (1.20E+02)**	**3.00E+02 ± (1.58E−13)**	−	3.00E+02 ± (1.83E+00)	**3.34E+02 ± (3.15E+01)**	−	3.59E+02 ± (3.87E+01)
f_{28}	**3.77E+02 ± (3.77E+02)**	−	3.80E+02 ± (3.08E+01)	**8.33E+02 ± (1.80E+01)**	−	8.44E+02 ± (1.76E+01)	**1.11E+03 ± (3.01E+01)**	=	**1.13E+03 ± (3.01E+01)**
f_{29}	**2.22E+02 ± (2.22E+02)**	=	**2.22E+02 ± (5.61E−01)**	**7.16E+02 ± (3.11E+00)**	=	**7.17E+02 ± (4.49E+00)**	8.04E+02 ± (3.94E+01)	+	**7.97E+02 ± (4.89E+01)**
f_{30}	**4.64E+02 ± (4.64E+02)**	=	**4.66E+02 ± (1.18E+01)**	**1.14E+03 ± (4.12E+02)**	=	**1.18E+03 ± (4.30E+02)**	**8.94E+03 ± (5.58E+02)**	=	**8.88E+03 ± (5.31E+02)**
+/=/−	8/18/4			7/19/4			8/15/7		

Table 7. Comparison results of DE with different strategies on 30D.

D = 30	DE/best/1		DE/best/2		DE/rand-to-best/1		DE/current-to-best/1		DE/current-to-pbest/1		DE/current-to-pcbest/1
f_1	6.41E+07 ± (4.78E+07)	+	**6.91E+04 ± (1.26E+05)**	−	1.51E+07 ± (1.28E+07)	+	2.10E+07 ± (1.67E+07)	+	3.37E+06 ± (2.55E+06)	+	2.29E+06 ± (1.74E+06)
f_2	1.02E+10 ± (4.37E+09)	+	**0.00E+00 ± (0.00E+00)**	−	1.32E+09 ± (9.53E+08)	+	3.47E+09 ± (1.77E+09)	+	5.10E+08 ± (6.49E+08)	=	2.80E+08 ± (2.71E+08)
f_3	1.07E+04 ± (1.01E+04)	+	**0.00E+00 ± (0.00E+00)**	−	4.88E+03 ± (4.24E+03)	+	3.22E+03 ± (2.85E+03)	+	1.14E+03 ± (9.87E+02)	=	7.69E+02 ± (5.98E+02)
f_4	7.99E+02 ± (4.45E+02)	+	**1.70E+01 ± (2.75E+01)**	−	2.76E+02 ± (1.12E+02)	+	3.42E+02 ± (1.15E+02)	+	1.65E+02 ± (4.98E+01)	=	1.49E+02 ± (3.98E+01)
f_5	**2.04E+01 ± (1.34E−01)**	−	2.09E+01 ± (8.77E−02)	−	2.09E+01 ± (6.26E−02)	=	2.09E+01 ± (4.85E−02)	=	2.09E+01 ± (5.75E−02)	=	2.09E+01 ± (5.25E−02)
f_6	2.14E+01 ± (3.19E+00)	+	9.06E+00 ± (4.47E+00)	+	9.47E+00 ± (2.62E+00)	+	1.19E+01 ± (2.67E+00)	+	4.56E+00 ± (1.67E+00)	+	**2.76E+00 ± (9.83E−01)**
f_7	9.26E+01 ± (3.84E+01)	+	**8.78E−03 ± (1.28E−02)**	−	1.56E+01 ± (9.73E+00)	+	2.90E+01 ± (1.46E+01)	+	5.75E+00 ± (4.14E+00)	+	3.35E+00 ± (1.94E+00)
f_8	1.24E+02 ± (2.16E+01)	+	1.74E+02 ± (1.65E+01)	+	5.19E+01 ± (1.59E+01)	+	6.03E+01 ± (1.80E+01)	+	**2.72E+01 ± (1.38E+01)**	=	**4.67E+01 ± (4.50E+01)**
f_9	1.46E+02 ± (2.97E+01)	+	2.00E+02 ± (2.20E+01)	+	5.76E+01 ± (1.45E+01)	−	6.78E+01 ± (1.79E+01)	−	**4.35E+01 ± (3.59E+01)**	−	1.11E+02 ± (5.22E+01)
f_{10}	3.08E+03 ± (7.27E+02)	−	3.12E+03 ± (1.78E+03)	−	1.27E+03 ± (3.66E+02)	−	**1.59E+03 ± (1.06E+03)**	−	3.53E+03 ± (1.98E+03)	−	4.68E+03 ± (1.32E+03)
f_{11}	3.74E+03 ± (7.24E+02)	−	6.33E+03 ± (1.27E+03)	+	**2.25E+03 ± (1.55E+03)**	−	5.23E+03 ± (1.79E+03)	−	6.39E+03 ± (4.72E+02)	=	6.31E+03 ± (3.42E+02)
f_{12}	**2.82E−01 ± (1.23E−01)**	−	2.41E+00 ± (2.61E−01)	=	2.40E+00 ± (2.68E−01)	=	2.38E+00 ± (2.62E−01)	=	2.38E+00 ± (3.35E−01)	=	2.41E+00 ± (2.64E−01)
f_{13}	2.38E+00 ± (8.52E−01)	+	4.45E−01 ± (8.54E−02)	+	4.25E−01 ± (2.28E−01)	+	5.84E−01 ± (4.25E−01)	+	**2.79E−01 ± (6.86E−02)**	=	**2.69E−01 ± (5.94E−02)**
f_{14}	3.19E+01 ± (1.47E+01)	+	5.13E−01 ± (2.71E−01)	+	1.94E+00 ± (3.63E+00)	+	1.06E+01 ± (9.36E+00)	+	**2.41E−01 ± (5.23E−02)**	=	**2.33E−01 ± (4.08E−02)**
f_{15}	1.91E+03 ± (3.22E+03)	+	1.69E+01 ± (3.37E+00)	+	1.83E+01 ± (2.19E+01)	+	4.37E+01 ± (9.76E+01)	+	**8.19E+00 ± (4.19E+00)**	=	**9.52E+00 ± (3.05E+00)**
f_{16}	1.15E+01 ± (7.23E−01)	+	**1.20E+01 ± (3.74E−01)**	+	1.08E+01 ± (4.48E−01)	=	1.11E+01 ± (3.41E−01)	+	1.07E+01 ± (4.03E−01)	=	1.09E+01 ± (3.23E−01)
f_{17}	2.07E+05 ± (4.20E+05)	+	1.04E+03 ± (5.36E+02)	−	6.43E+04 ± (1.06E+05)	+	4.76E+04 ± (1.03E+05)	+	5.11E+03 ± (6.01E+03)	+	**4.38E+03 ± (6.11E+03)**
f_{18}	3.53E+05 ± (2.45E+06)	+	6.88E+02 ± (3.59E+03)	+	2.29E+02 ± (7.44E+01)	+	2.03E+02 ± (6.96E+01)	+	1.20E+02 ± (4.53E+01)	+	**1.02E+02 ± (3.43E+01)**
f_{19}	5.29E+01 ± (3.08E+01)	+	8.29E+00 ± (2.07E+00)	+	2.45E+01 ± (2.16E+01)	+	2.13E+01 ± (1.95E+01)	+	8.63E+00 ± (2.34E+00)	+	**6.93E+00 ± (1.54E+00)**
f_{20}	5.78E+02 ± (6.88E+02)	+	9.27E+01 ± (5.39E+01)	=	3.10E+02 ± (1.63E+02)	+	2.38E+02 ± (1.22E+02)	+	1.27E+02 ± (9.04E+01)	+	**9.69E+01 ± (1.08E+02)**
f_{21}	4.07E+04 ± (1.01E+05)	+	**5.41E+02 ± (3.59E+02)**	=	9.86E+03 ± (1.91E+04)	+	4.64E+03 ± (6.25E+03)	+	9.15E+02 ± (6.14E+02)	+	6.57E+02 ± (4.60E+02)
f_{22}	5.86E+02 ± (2.00E+02)	+	3.32E+02 ± (1.59E+02)	+	2.73E+02 ± (1.24E+02)	+	2.78E+02 ± (1.01E+02)	+	**1.91E+02 ± (8.33E+01)**	=	**1.60E+02 ± (9.26E+01)**
f_{23}	3.77E+02 ± (2.92E+01)	+	**3.15E+02 ± (4.55E−13)**	−	3.30E+02 ± (8.04E+00)	+	3.36E+02 ± (9.35E+00)	+	3.19E+02 ± (2.08E+00)	+	3.18E+02 ± (1.71E+00)
f_{24}	2.70E+02 ± (9.70E+00)	+	2.33E+02 ± (6.73E+00)	+	2.43E+02 ± (5.39E+00)	+	2.47E+02 ± (5.63E+00)	+	2.33E+02 ± (6.40E+00)	+	**2.29E+02 ± (5.69E+00)**
f_{25}	2.17E+02 ± (6.69E+00)	+	**2.04E+02 ± (1.61E+00)**	−	2.11E+02 ± (2.41E+00)	+	2.11E+02 ± (3.42E+00)	+	2.07E+02 ± (1.41E+00)	+	2.06E+02 ± (1.46E+00)
f_{26}	1.28E+02 ± (5.33E+01)	+	1.13E+02 ± (5.14E+01)	+	1.29E+02 ± (4.49E+01)	+	1.17E+02 ± (3.68E+01)	+	**1.14E+02 ± (3.47E+01)**	=	**1.04E+02 ± (1.96E+01)**
f_{27}	8.70E+02 ± (2.26E+02)	+	5.79E+02 ± (1.15E+02)	+	5.68E+02 ± (1.38E+02)	+	6.27E+02 ± (1.58E+02)	+	4.98E+02 ± (6.74E+01)	+	**4.28E+02 ± (4.69E+01)**
f_{28}	1.98E+03 ± (5.00E+02)	+	**1.12E+03 ± (2.62E+02)**	=	1.47E+03 ± (3.29E+02)	+	1.57E+03 ± (3.26E+02)	+	1.10E+03 ± (2.36E+02)	+	**1.02E+03 ± (1.61E+02)**
f_{29}	8.28E+06 ± (1.58E+07)	+	**1.44E+06 ± (4.04E+06)**	−	2.99E+06 ± (8.01E+06)	+	4.39E+06 ± (9.84E+06)	+	1.73E+06 ± (4.99E+06)	=	1.52E+06 ± (5.92E+06)
f_{30}	7.07E+04 ± (6.36E+04)	+	**3.28E+03 ± (1.72E+03)**	=	2.45E+04 ± (2.41E+04)	+	3.28E+04 ± (6.06E+04)	+	4.87E+03 ± (2.66E+03)	+	**3.68E+03 ± (4.19E+03)**
+/=/−	26/0/4		14/5/11		24/3/3		25/2/3		14/14/2		-/-/-
Rank	5.30		2.77		3.77		4.33		2.63		2.20

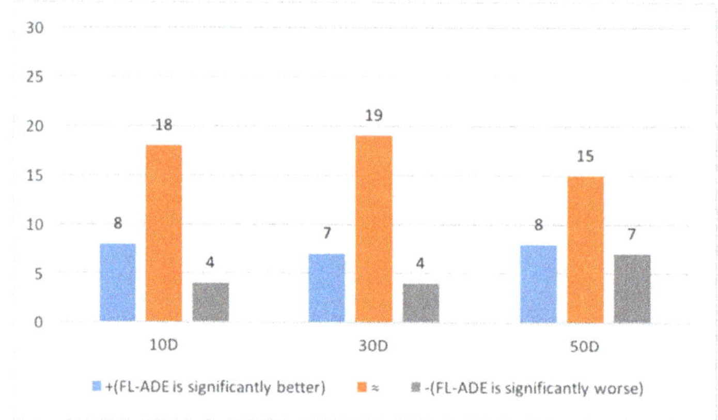

Figure 7. Summarized results between default FL-ADE and FL-ADE_linear on CEC2014 benchmark on 10D, 30D and 50D.

4.6. The Sensitivities of Parameters

4.6.1. The Value of Parameter N^{init} and FES_t

To investigate the impact of N^{init} and FES_t in the proposed population scheme FL-APS, FL-ADE with five different N^{init}, i.e., $25\log(D)\sqrt{D}$, $10D$, $12D$, $18D$, $20D$, and three FES_t, i.e., $(1/3) * FES_{max}$, $(1/2) * FES_{max}$, $(2/3) * FES_{max}$, were conducted on the 30D benchmark functions. The results obtained by Friedman test are listed in Table 8. From the result of Friedman test, it can be observed that $N = 25\log(D)\sqrt{D}$ and $FES_t = (1/2) * FES_{max}$ achieve a better average ranking, so the values of these two parameters are adopted in this paper.

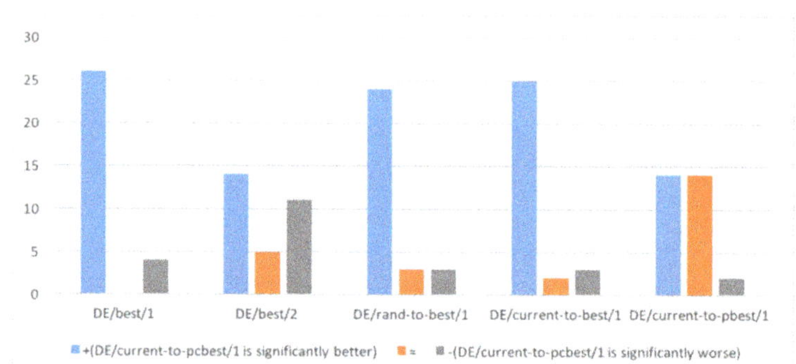

Figure 8. Summarized results between of DE with different strategies on 30D.

Table 8. Result of Friedman test for parameter N^{init} and FES_t.

30D	25log D$\sqrt{}$D	10D	12D	18D	20D	Average
1/3	8.0	9.35	8.42	7.43	9.12	8.46
1/2	6.93	8.68	7.78	6.43	7.98	**7.56**
2/3	6.83	9.68	7.82	7.97	7.57	7.97
average	**7.25**	9.24	8.01	7.28	8.22	

4.6.2. The Value of Parameter *pc*

The value of *pc* directly determines the performance of the strategy; thus, the impact of *pc* needs to be investigated. For this purpose, FL-ADE configured with three different *pc* was conducted on the 10D, 30Ds and 50D benchmark functions. The results obtained by Friedman test are listed in Table 9, and $pc = 0.3$ saw the best average ranking, so the default FL-ADE parameters use these parameter settings.

Table 9. Result of Friedman test for parameter p^c.

	pc = 0.1	pc = 0.2	pc = 0.3	pc = 0.4	pc = 0.5
10D Ranking	3.68	2.88	**2.42**	2.85	3.17
30D Ranking	3.43	3.00	**2.83**	2.87	2.87
50D Ranking	3.28	3.28	**2.48**	2.97	2.98
Average Ranking	3.46	3.05	**2.58**	2.90	3.01

4.7. Algorithm Complexity

All the experiments in this paper were performed using MATLAB R2018b, running on a Hewlett-Packard PC from Guangzhou China with Intel Core i7−9700 (3.00 GHz) CPU and 8 GB of RAM on a Windows 10 system. Table 10 shows the complexity of the FL-ADE algorithm for testing the problem of 10, 30, and 50 dimensions. The variable T0 is the computing time to run the code in Algorithm 4 [49]. T_1 is the computing time to execute f_{18} for 200,000 evaluations of a certain dimension. T_2 is obtained by operating 200,000 evaluations with the tested algorithm for f_{18} of the same dimension, whose average value for five times is represented by $\hat{T}_2$, and the runtime complexity for the tested algorithm can be reflected by $\left(\hat{T}_2 - T_1\right)/T_0$.

The results in Table 10 show that the proposed FL-ADE inherits the linear complexity of classical DE; this is a real plus when comparing it to powerful recent metaheuristics algorithm. This is because the FL-ADE does not increase the number of evaluations, and the increased calculation is simple, mainly calculating the local fitness landscape characteristics φ and the population size of each generation.

Algorithm 4: The code for calculationg the time T_0

Input: $T_{max} = 1,000,000.$
1. tic
2. for $i = 1 \ : \ T_{max}$
3. $x = 0.55 + (double)\ i;\ x = x + x;\ x = x/2;\ x = x + x;$
4. $x = sqrt(x);\ x = \log(x);\ x = exp(x);\ x = x/(x+2);$
5. end
6. toc
Output: the time T_0

Table 10. Algorithm complexity of FL-ADE.

	T_0	T_1	$\hat{T}_2$	$\left(\hat{T}_2 - T_1\right)/T_0$
$D = 10$		0.063057	1.9117746	29.25278648
$D = 30$	0.063198	0.186583	1.9586996	28.04070698
$D = 50$		0.433899	2.3953128	31.03601063

5. Conclusions

Different search strategies are designed for different fitness landscape conditions to find the optimal solution, and there is no single strategy that can be suitable for all fitness landscapes. In this paper, a novel adaptive DE based on fitness landscape (FL-ADE) was proposed, which utilizes the local fitness landscape characteristics in each generation population to perform the FL-APS mechanism and the DE/current-to-pcbest mutation strategy. The FL-APS mechanism makes the population size change adaptively based on the local fitness landscape characteristics at each generation. The DE/current-to-pcbest mutation strategy, which randomly chooses one of the locally optimal individuals to be the pcbest, is designed to enhance the exploration capability for multimodal fitness landscape. Based on the FL-APS, computational resources can be rationally assigned at different evolutionary stages and satisfy diverse requirements of different fitness landscapes. The experimental results, using the Wilcoxon rank-sum test, confirm that the FL-ADE algorithm is highly competitive when compared to seven high-performance state-of-art DE variants, i.e., EPSDE, MPEDE, CoBiDE, SHADE, LSHADE, LSHADE-cnEpSin, and PaDE, based on 30 benchmark functions from CEC2014. The results also show that the fitness landscape information can effectively guide the adaptive evolution of the population. In future research, we plan to use fitness landscape information to adjust other parameters of DE adaptively.

Author Contributions: Conceptualization, L.Z.; methodology, L.Z.; software, S.L.; validation, L.Z. and S.L.; formal analysis, L.Z.; investigation, S.L.; resources, L.Z.; data curation, S.L.; writing—original draft preparation, S.L.; writing—review and editing, S.L.; visualization, L.Z.; supervision, S.L.; project administration, L.Z.; funding acquisition, L.Z. All authors have read and agreed to the published version of the manuscript.

Funding: The work described in this paper was supported by the Applied Science and Technology Research and Development Special Fund Project of Guangdong Province (No. 2016B010126004) and the National Special Project Number for International Cooperation (No. 2015DFR11050).

Institutional Review Board Statement: Not applicable.

Informed Consent Statement: Not applicable.

Data Availability Statement: The study did not report any data.

Acknowledgments: Thanks for the experimental environment provided by Jinan University and and the financial support given by the fund.

Conflicts of Interest: The authors declare no conflict of interest.

References

1. Storn, R.; Price, K. Differential evolution–a simple and efficient heuristic for global optimization over continuous spaces. *J. Glob. Optim.* **1997**, *11*, 341–359. [CrossRef]
2. Pant, M.; Zaheer, H.; Garcia-Hernandez, L.; Abraham, A. Differential Evolution: A review of more than two decades of research. *Eng. Appl. Artif. Intell.* **2020**, *90*, 103479. [CrossRef]
3. Sethanan, K.; Pitakaso, R. Differential evolution algorithms for scheduling raw milk transportation. *Comput. Electron. Agric.* **2016**, *121*, 245–259. [CrossRef]
4. Hultmann Ayala, H.V.; Coelho, L.d.S.; Mariani, V.C.; Askarzadeh, A. An improved free search differential evolution algorithm: A case study on parameters identification of one diode equivalent circuit of a solar cell module. *Energy* **2015**, *93*, 1515–1522. [CrossRef]
5. Prauzek, M.; Krömer, P.; Rodway, J.; Musilek, P. Differential evolution of fuzzy controller for environmentally-powered wireless sensors. *Appl. Soft Comput.* **2016**, *48*, 193–206. [CrossRef]
6. Koutny, T. Using meta-differential evolution to enhance a calculation of a continuous blood glucose level. *Comput. Methods Programs Biomed.* **2016**, *133*, 45–54. [CrossRef]
7. Chen, X.; Du, W.; Qian, F. Solving chemical dynamic optimization problems with ranking-based differential evolution algorithms. *Chin. J. Chem. Eng.* **2016**, *24*, 1600–1608. [CrossRef]
8. Michalewicz, Z.; Hartley, S.J. Genetic algorithms+ data structures= evolution programs. *Math. Intell.* **1996**, *18*, 71.
9. Moscato, P.; Norman, M.G. A "Memetic" Approach for the Traveling Salesman Problem Implementation of a Computational Ecology for Combinatorial Optimization on Message-Passing Systems. *Parallel Comput. Transput. Appl.* **1992**, *1*, 177–186.
10. Mühlenbein, H.; Paass, G. From recombination of genes to the estimation of distributions I. Binary parameters. In *International Conference on Parallel Problem Solving from Nature*; Springer: Berlin/Heidelberg, Germany, 1996; pp. 178–187.
11. Yu, W.; Shen, M.; Chen, W.; Zhan, Z.; Gong, Y.; Lin, Y.; Liu, O.; Zhang, J. Differential Evolution With Two-Level Parameter Adaptation. *IEEE Trans. Cybern.* **2014**, *44*, 1080–1099. [CrossRef]
12. Zhang, J.; Sanderson, A.C. JADE: Adaptive Differential Evolution With Optional External Archive. *IEEE Trans. Evol. Comput.* **2009**, *13*, 945–958. [CrossRef]
13. Gong, W.; Cai, Z. Differential Evolution With Ranking-Based Mutation Operators. *IEEE Trans. Cybern.* **2013**, *43*, 2066–2081. [CrossRef]

14. Shen, L.; He, J. A mixed strategy for Evolutionary Programming based on local fitness landscape. In Proceedings of the IEEE Congress on Evolutionary Computation, Barcelona, Spain, 18–23 July 2010; pp. 1–8.
15. Das, S.; Suganthan, P.N. Differential Evolution: A Survey of the State-of-the-Art. *IEEE Trans. Evol. Comput.* **2011**, *15*, 4–31. [CrossRef]
16. Price, K.; Storn, R.M.; Lampinen, J.A. *Differential Evolution: A Practical Approach to Global Optimization*; Springer Science & Business Media: Berlin/Heidelberg, Germany, 2006.
17. Mezura-Montes, E.; Velázquez-Reyes, J.; Coello, C.A.C. A comparative study of differential evolution variants for global optimization. In Proceedings of the 8th Annual Conference on Genetic and Evolutionary Computation, Seattle, WA, USA, 8–12 July 2006; pp. 485–492.
18. Gämperle, R.; Müller, S.D.; Koumoutsakos, P. A parameter study for differential evolution. *Adv. Intell. Syst. Fuzzy Syst. Evol. Comput.* **2002**, *10*, 293–298.
19. Liu, J.; Lampinen, J. A Fuzzy Adaptive Differential Evolution Algorithm. *Soft Comput.* **2005**, *9*, 448–462. [CrossRef]
20. Zaharie, D. Parameter adaptation in differential evolution by controlling the population diversity. In Proceedings of the International Workshop on Symbolic and Numeric Algorithms for Scientific Computing, Timisoara, Romania, 25–29 September 2005; pp. 385–397.
21. Brest, J.; Greiner, S.; Boskovic, B.; Mernik, M.; Zumer, V. Self-Adapting Control Parameters in Differential Evolution: A Comparative Study on Numerical Benchmark Problems. *IEEE Trans. Evol. Comput.* **2006**, *10*, 646–657. [CrossRef]
22. Tanabe, R.; Fukunaga, A. Success-history based parameter adaptation for Differential Evolution. In Proceedings of the 2013 IEEE Congress on Evolutionary Computation, Cancun, Mexico, 20–23 June 2013; pp. 71–78.
23. Draa, A.; Bouzoubia, S.; Boukhalfa, I. A sinusoidal differential evolution algorithm for numerical optimisation. *Appl. Soft Comput.* **2015**, *27*, 99–126. [CrossRef]
24. Al-Dabbagh, R.D.; Neri, F.; Idris, N.; Baba, M.S. Algorithmic design issues in adaptive differential evolution schemes: Review and taxonomy. *Swarm Evol. Comput.* **2018**, *43*, 284–311. [CrossRef]
25. Vermetten, D.; van Stein, B.; Kononova, A.V.; Caraffini, F. Analysis of Structural Bias in Differential Evolution Configurations. In *Differential Evolution: From Theory to Practice*; Kumar, B.V., Oliva, D., Suganthan, P.N., Eds.; Springer: Singapore, 2022; pp. 1–22.
26. Stein, B.V.; Caraffini, F.; Kononova, A.V. Emergence of structural bias in differential evolution. In Proceedings of the Genetic and Evolutionary Computation Conference Companion, Lille, France, 10–14 July 2021; pp. 1234–1242.
27. Kononova, A.V.; Caraffini, F.; Wang, H.; Bäck, T. *Can Compact Optimisation Algorithms Be Structurally Biased?* Springer International Publishing: Cham, Switzerland, 2020; pp. 229–242.
28. Kononova, A.V.; Caraffini, F.; Bäck, T. Differential evolution outside the box. *Inf. Sci.* **2021**, *581*, 587–604. [CrossRef]
29. Tanabe, R.; Fukunaga, A.S. Improving the search performance of SHADE using linear population size reduction. In Proceedings of the 2014 IEEE Congress on Evolutionary Computation (CEC), Beijing, China, 6–11 July 2014; pp. 1658–1665.
30. Bujok, P. An Evaluative Study of Adaptive Control of Population Size in Differential Evolution. In *Artificial Intelligence and Soft Computing*; Rutkowski, L., Scherer, R., Korytkowski, M., Pedrycz, W., Tadeusiewicz, R., Zurada, J.M., Eds.; Springer International Publishing: Cham, Switzerland, 2019; pp. 421–431.
31. Liang, J.; Qu, B.; Suganthan, P.; Hernández-Díaz, A.G. *Ranking Results of CEC14 Special Session and Competition on Real-Parameter Single Objective Optimization*; Technical Report; Zhengzhou University: Zhengzhou, China; Nanyang Technological University: Singapore, 2014.
32. Poláková, R.; Tvrdík, J.; Bujok, P. Differential evolution with adaptive mechanism of population size according to current population diversity. *Swarm Evol. Comput.* **2019**, *50*, 100519. [CrossRef]
33. Zhan, Z.H.; Wang, Z.J.; Jin, H.; Zhang, J. Adaptive Distributed Differential Evolution. *IEEE Trans. Cybern.* **2020**, *50*, 4633–4647. [CrossRef]
34. Huang, Y.; Li, W.; Ouyang, C.; Chen, Y. A self-feedback strategy differential evolution with fitness landscape analysis. *Soft Comput.* **2018**, *22*, 7773–7785. [CrossRef]
35. Li, W.; Li, S.; Chen, Z.; Zhong, L.; Ouyang, C. Self-feedback differential evolution adapting to fitness landscape characteristics. *Soft Comput.* **2019**, *23*, 1151–1163. [CrossRef]
36. Tan, Z.; Li, K.; Wang, Y. Differential evolution with adaptive mutation strategy based on fitness landscape analysis. *Inf. Sci.* **2021**, *549*, 142–163. [CrossRef]
37. Brest, J.; Maučec, M.S.; Bošković, B. iL-SHADE: Improved L-SHADE algorithm for single objective real-parameter optimization. In Proceedings of the 2016 IEEE Congress on Evolutionary Computation (CEC), Vancouver, BC, Canada, 24–29 July 2016; pp. 1188–1195.
38. Brest, J.; Maučec, M.S.; Bošković, B. Single objective real-parameter optimization: Algorithm jSO. In Proceedings of the 2017 IEEE Congress on Evolutionary Computation (CEC), Donostia-San Sebastián, Spain, 5–8 June 2017; pp. 1311–1318.
39. Meng, Z.; Pan, J.-S.; Tseng, K.-K. PaDE: An enhanced Differential Evolution algorithm with novel control parameter adaptation schemes for numerical optimization. *Knowl. -Based Syst.* **2019**, *168*, 80–99. [CrossRef]
40. Awad, N.H.; Ali, M.Z.; Suganthan, P.N. Ensemble sinusoidal differential covariance matrix adaptation with Euclidean neighborhood for solving CEC2017 benchmark problems. In Proceedings of the 2017 IEEE Congress on Evolutionary Computation (CEC), Donostia-San Sebastián, Spain, 5–8 June 2017; pp. 372–379.

41. Fei, P.; Tang, K.; Guoliang, C.; Yao, X. Multi-start JADE with knowledge transfer for numerical optimization. In Proceedings of the 2009 IEEE Congress on Evolutionary Computation, Trondheim, Norway, 18–21 May 2009; pp. 1889–1895.
42. Wright, S. The Roles of Mutation, Inbreeding, Crossbreeding, and Selection in Evolution. In Proceedings of the Sixth International Congress of Genetics, Ithaca, NY, USA, 24–31 August 1932; Volume 1.
43. Wang, X.; Sheng, M.; Ye, K.; Lin, J.; Mao, J.; Chen, S.; Sheng, W. A multilevel sampling strategy based memetic differential evolution for multimodal optimization. *Neurocomputing* **2019**, *334*, 79–88. [CrossRef]
44. Li, W.; Meng, X.; Huang, Y. Fitness distance correlation and mixed search strategy for differential evolution. *Neurocomputing* **2020**, *458*, 514–525. [CrossRef]
45. Mallipeddi, R.; Suganthan, P.N.; Pan, Q.K.; Tasgetiren, M.F. Differential evolution algorithm with ensemble of parameters and mutation strategies. *Appl. Soft Comput.* **2011**, *11*, 1679–1696. [CrossRef]
46. Wu, G.; Mallipeddi, R.; Suganthan, P.N.; Wang, R.; Chen, H. Differential evolution with multi-population based ensemble of mutation strategies. *Inf. Sci.* **2016**, *329*, 329–345. [CrossRef]
47. Wang, Y.; Li, H.-X.; Huang, T.; Li, L. Differential evolution based on covariance matrix learning and bimodal distribution parameter setting. *Appl. Soft Comput.* **2014**, *18*, 232–247. [CrossRef]
48. Awad, N.H.; Ali, M.Z.; Suganthan, P.N.; Reynolds, R.G. An ensemble sinusoidal parameter adaptation incorporated with L-SHADE for solving CEC2014 benchmark problems. In Proceedings of the 2016 IEEE Congress on Evolutionary Computation (CEC), Vancouver, BC, Canada, 24–29 July 2016; pp. 2958–2965.
49. Liang, J.J.; Qu, B.Y.; Suganthan, P.N. *Problem Definitions and Evaluation Criteria for the CEC 2014 Special Session and Competition on Single Objective Real-Parameter Numerical Optimization*; Technical Report; Computational Intelligence Laboratory, Zhengzhou University: Zhengzhou, China; Nanyang Technological University: Singapore, 2013; Volume 635.

applied sciences

MDPI

Article

Deriving Situation-Adaptive Policy for Container Stacking in an Automated Container Terminal

Taekwang Kim [1] and Kwang Ryel Ryu [2,*]

[1] Department of Information Convergence Engineering, Pusan National University, Busan 46241, Korea; kimt86@pusan.ac.kr
[2] School of Computer Science and Engineering, Pusan National University, Busan 46241, Korea
* Correspondence: krryu@pusan.ac.kr

Abstract: Determining where to stack the containers at the storage yard of a container terminal is an important problem because that decision critically affects the efficiency of container handling in the yard and, eventually, the efficiency of the vessel operations, which is considered the most important for the productivity of the whole terminal. One limitation of the stacking policies previously proposed is that they are static in nature. Although good locations for stacking may change as the workload of vessel operation changes, the previous policies are insensitive to such changes. Failure to recommend good locations leads to elongated operations of yard cranes and thus makes it hard for them to keep up with the workload of vessel operation. In this paper, we propose a method for deriving a dynamic policy that can adapt to the workload of vessel operation that changes over time. Our method derives two boundary policies: one for very high workload and the other for very low. Then, a policy appropriate for any intermediate workload can be synthesized from the two boundary policies through interpolation. Simulation experiments showed that the proposed policy significantly reduced overall container handling time compared to the previous static policy. When measured in terms of the time the transportation vehicles wait for container handling services, the improvement was approximately 19%.

Keywords: container terminal; storage yard; container stacking; situation-adaptive policy; optimization; genetic algorithm

Citation: Kim, T.; Ryu, K.R. Deriving Situation-Adaptive Policy for Container Stacking in an Automated Container Terminal. *Appl. Sci.* **2022**, *12*, 3892. https://doi.org/10.3390/app12083892

Academic Editors: Nuno Lau, Rui Araújo, António Pedro Aguiar, Rodrigo Ventura and João Fabro

Received: 1 March 2022
Accepted: 11 April 2022
Published: 12 April 2022

1. Introduction

One of the most important operational goals of a container terminal is to minimize the vessel turnaround time by maximizing the efficiency of *vessel operations* with regard to loading or unloading containers onto or from the vessels. The *outbound* containers brought in by the trucks from inland are stored in the yard until they are loaded onto the vessels. On the other hand, the *inbound* containers unloaded from the vessels dwell in the storage yard until they are claimed by the external trucks for inland transportation. As buffer storage for both inbound and outbound containers, the operational efficiency in the storage yard critically affects the overall productivity of the terminal. A very important factor that affects the operational efficiency in the storage yard is the determination of the stacking locations of the containers that arrive at the yard. If, for example, a container just unloaded from a vessel is stacked on top of another to be loaded soon, the upper one has to be relocated to a different stack when retrieving the lower one. Such *rehandling* of containers should be minimized in order to maximize the operational efficiency in the yard. When the containers are loaded onto a vessel, they follow a predetermined sequence. The loading sequence is determined at the planning stage, taking into account the vessel stability, ports of destination, and the efficiency of operation at the storage yard. Still, rehandling is the major cause of loading delay because the retrieval schedules are usually unknown at the time the containers arrive at the yard, and thus, they can be stacked at the wrong locations. In this paper, we deal with the problem of determining good locations for stacking not only the containers that newly arrive at the yard but also those that are relocated within the yard.

Perhaps the simplest approach to determining the stacking locations in the yard is using a heuristic rule based on a simple criterion such as preferring the nearest location or preferring the stack-top of containers of the same category [1–3]. More recent and advanced methods use multiple criteria to evaluate candidate stacking locations from various perspectives [4–6]. These methods use *scoring functions* to calculate the score of a candidate location by a weighted sum of the results of evaluations based on various criteria. Since the scores and the resulting best stacking location depend on the weight combination, or *weight vector*, used in the scoring function, the weight vector can be regarded as a *stacking policy*. To find a good weight vector, i.e., a good stacking policy, References [4,6] use genetic algorithms (GA) in which each candidate policy is evaluated through simulations of applying the policy to various scenarios of operations in the yard and measuring the resulting performance. However, the policy found in this way is the one whose *average* performance in various scenarios is the best. In certain situations, there may be other policies that work better than the average policy.

What we propose in this paper is a method for deriving a stacking policy, which can be adapted to changing situations by having its weight vectors adjusted. As an indicator of the situation, we use the workload of vessel operation because good stacking locations are very much dependent on it. Our method is based on the idea that a new policy may be synthesized if we are given two boundary policies: one for a very low workload and the other for a very high workload. We assume that a good policy for any intermediate situation can be derived by taking an interpolation of the weight vectors of the two boundary policies. We use a GA to search for not only the two boundary policies but also the two numeric values that quantify the two fuzzy terms 'very low' and 'very high'. Our GA can be seen as a reinforcement learning algorithm conducting a search in the policy space [7] instead of learning a value function based on rewards from the environment. Experimental results show that our method performs better by dynamically adapting the stacking policy to varying situations than the previous methods that use a static best on-average policy.

The rest of the paper is organized as follows. Section 2 gives a detailed description of the operations in the storage yard of an automated container terminal. Section 3 reviews the related works, and Section 4 describes the stacking policy based on the scoring functions. Section 5 explains how we derive a situation-adaptive stacking policy by using a GA. Section 6 reports the results of experiments, and Section 7 discusses how we can extend our method to the cases with multiple situation indicators. Finally, Section 8 gives some concluding remarks.

2. Operations in the Storage Yard of an Automated Container Terminal

The descriptions in this section are mostly based on the material given in [4,6,8]. As can be seen in Figure 1, the automated container terminal can be largely divided into four regions: quay, apron, storage yard, and hinterland. The quay is where the vessels berth and a number of quay cranes (QC) load/unload containers onto/from the vessels. The apron is the area for the automated guided vehicles (AGV) to deliver containers between the quay and the storage yard. The hinterland is where the external trucks (ET) bring in containers to the storage yard or bring out the containers picked up from the storage yard. The storage yard consists of dozens of rectangular blocks that are laid out in the perpendicular direction to the quay. Each block consists of hundreds of container stacks several tiers high, where the stacks are arranged in dozens of bays in the perpendicular direction and in several rows in the horizontal direction. Since a bay is of the length of a 20 ft container, a stack of 40 ft container spans two consecutive bays. For safety reasons, the containers cannot be stacked together if their sizes are not the same. Each block is equipped with two automated stacking cranes (ASC) to handle the containers. Since the two ASCs are of the same size, one cannot move across the other and thus may interfere with each other at close ranges. The container transfer to and from an AGV is made by the seaside

ASC at a seaside handover point (HP) located at the seaside end of the block, while the transfer to and from an ET is made by the landside ASC at a landside HP located at the opposite end.

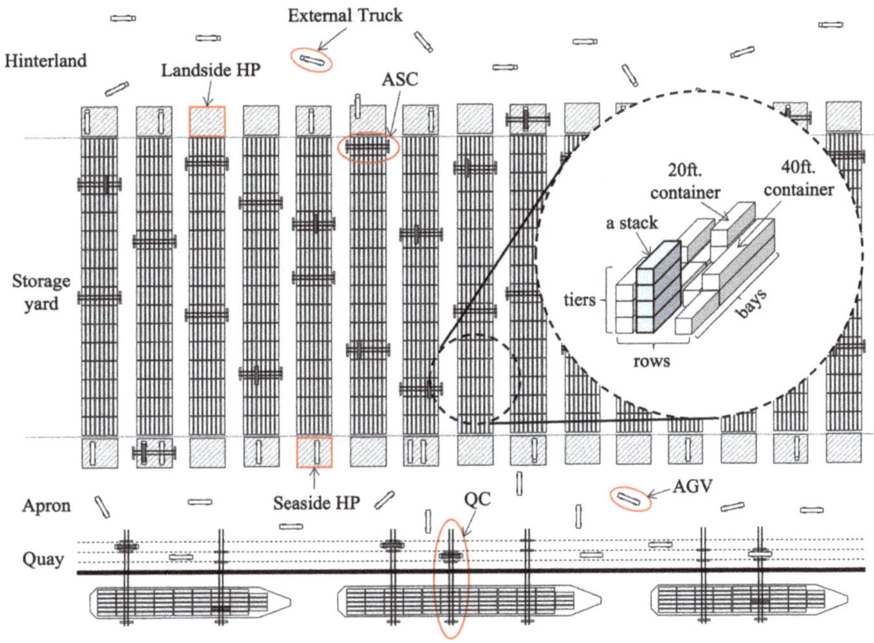

Figure 1. Layout of an automated container terminal.

The containers in the storage blocks are categorized into three groups by their directions of the flow of logistics: inbound, outbound, and *transshipment* containers. The inbound and outbound containers are already mentioned in the previous section. The transshipment containers are those that are unloaded from certain vessels and stored in the yard, but unlike the inbound containers, they are to be loaded onto other vessels for further sea transportation. A container stored in a block is often relocated to some other place within the block. If container X is placed on top of container Y in a stack but Y has to be taken out earlier than X, then X has to be moved to another location before Y can be retrieved. This relocation, or rehandling, is hard to avoid because the retrieval schedule of containers is usually unknown at the time the containers arrive at the yard and are piled up. Thus, we need a good policy that can select good stacking locations for the incoming containers so that both the container handling time and the possibility of rehandling are minimized. Note that a stacking policy should also be able to recommend good stacking locations for the rehandled containers to minimize further rehandling. Rehandling is considered the biggest cause of delay of container handling in the storage yard.

There arises another difficulty with container handling when the two ASCs of a block cannot move across each other. To avoid collision, an ASC sometimes has to stop and wait until the other one finishes its job and backs up. This *interference* deteriorates the throughput of the ASCs. Interferences are more likely to occur as the travel distance of an ASC becomes longer because it gets close to the other ASC with a higher probability. Unfortunately, the ASCs often cannot avoid long-distance travel in such a block layout as that shown in Figure 1. The outbound containers brought in by the ETs enter the block through the landside HPs and are usually stored at locations near the landside end. However, they eventually go out of the block through the seaside HPs when they are loaded onto their target vessels. On the other hand, the inbound containers unloaded from the vessels enter the block through the seaside HPs and then later exit through the landside HPs. These long-distance movements of containers in opposite directions easily lead to ASC interferences. One way of minimizing such interference is to have the containers tactically relayed through cooperation between the two ASCs. As an example, an inbound container stacked near the seaside end of the block can be moved first to a certain intermediate location by the seaside ASC and then to its final destination HP at the landside by the landside ASC. The first movement of this relay operation is called *repositioning*. Note that the stacking locations for the containers to be repositioned should also be determined by the stacking policy.

Given a stacking policy, container handling in a storage block is typically performed in the following way. Whenever an ASC finishes its current job, it selects the most urgent job from the job queue that contains all the jobs requested for the next horizon of length, say thirty minutes. The jobs in the queue include those to be undertaken according to the loading/unloading schedules and those requested from the ETs that have already arrived at the landside HPs but have not been serviced yet. The ETs expected to arrive at the block during the next horizon are not counted because their arrival time is highly unpredictable. If the current ASC is the landside ASC, the most urgent job would be the ET job with the longest waiting time. If the current ASC is the seaside ASC, the most urgent job would be either a loading or an unloading job with the earliest deadline according to their schedules. When the job selected is a stacking job, the stacking policy examines all the available slots in the block and recommends the best one as the stacking location. An available slot can be found at the top of every stack unless it has already reached the allowed maximum tier. However, the stack should not belong to the bay where the other ASC is currently working. When the job selected is a retrieval job, no reference to the stacking policy is necessary unless there are other containers above the target container. If there are some containers above, they all must be relocated one after another to the locations recommended by the stacking policy. When the retrieval job selected requires a travel distance longer than a given threshold, the target container should be repositioned to the location recommended by the stacking policy before it can be sent to its destination HP by the other ASC.

The efficiency of ASC operation in a block can be assessed by measuring the AGV delay and the ET waiting time, where the former is counted far more important than the latter. A good stacking policy makes the ASC operation efficient by reducing the interference, rehandling, and the overall container handling time, which makes it possible for the ASCs to provide the AGVs and ETs with faster services. However, it takes quite a long period of observation to see how good a stacking policy is. Whether or not the stacking locations recommended are good can be seen better when the containers are retrieved out of the block than when they come into the block. Since the average dwell time of containers in container terminals is often longer than a week, the AGV delay and ET waiting time should be measured for a long period during which time enough containers are retrieved, under the condition that the containers keep coming in and going out fairly constantly during that period of time.

3. Related Works

There are some previous works on container stacking that deal with the problem of allocating the storage spaces for incoming containers. Reference [9] showed how to organize the storage area to minimize the number of container handling moves given a fixed amount of space, based on simple models that capture the relationship between the handling moves and the amount of available space. Reference [10] developed a space allocation method for inbound containers so as to minimize the expected number of rehandles. References [11,12] used a mixed-integer programming model together with some heuristics to allocate storage space for outbound and transshipment containers, respectively. Reference [13] applied a constraint satisfaction technique to allocate spaces to outbound containers. Reference [14] developed a space allocation method that can cope with the uncertainties in loading/unloading times of vessels. Reference [15] used a genetic algorithm to optimize the space allocation problem to avoid bottlenecks in storage yard operations and to minimize vessel service time. All these methods allocate a bulk of storage locations for reservation prior to the arrival of containers. If the arrival plan changes, the storage space must be reallocated. In contrast, the methods discussed below designate specific storage locations for each individual container at the time of its arrival.

The majority of the previous works on stacking locations have used rules or heuristics. Reference [16] considered the configuration of the container stack and the weight distribution of containers in the yard-bay to derive a decision tree model that determines the storage location of each outbound container. The decision tree can be deemed as another representation of a set of rules. Reference [1] proposed stacking rules that recommend the containers belonging to the same category be stacked together. Containers of the same category are of the same weight class and size, have the same destination port, and are loaded onto the same vessel. Reference [2] proposed rules that consider not only the category but also the height of the stacking position. Reference [3] suggested a heuristic rule for determining the locations of relocated containers to minimize the number of relocations during the retrieval process. Reference [17] conducted simulation studies to investigate the effect of using information about container departure times and the tradeoff between stacking farther away versus stacking close to the HPs. It used simple stacking rules that are designed to work in perpendicularly laid out storage blocks in an automated container terminal. Reference [18] presented what they call a hybrid sequence stacking method that determines the stacking locations of outbound containers considering the container weights. Proposing an ideal configuration of a yard bay to avoid rehandling, this method tries to stack the incoming containers so that their positions are as close to the ideal configuration as possible.

Some previous works tried to adopt more AI (artificial intelligence) techniques based upon rules or heuristics. Reference [19,20] derived stacking policies for outbound containers considering the uncertainties in their weights with the purpose of minimizing rehandling. Their policy consists of three precedence rules, each for a container weight group, where the rules are optimized by a simulated annealing algorithm. Reference [21] proposed a heuristic method to stack outbound containers. This method evaluates each candidate stacking location for an incoming container through a simulation and selects the best one. In the simulation, after stacking the container at the candidate location, the remaining containers arriving in a random sequence are stacked following a heuristic priority rule and the resulting performance is measured. Reference [22] used simple rules that determine the stacking positions based on the stack height and the estimated time of retrieval. The rules adopt fuzzy logic to represent their conditions to deal with a high degree of uncertainty in the arrival of containers at the yard. Reference [23] proposed a multi-agent system for container stacking, in which the stack agent recommends a stacking position by consulting the knowledge base composed of if-then rules that check various conditions such as the container types, the configuration of the storage space, and the occurrence of exceptional events. Each stacking decision is evaluated by the evaluation agent that rejects the decision when unacceptable. If a decision turns out to be unacceptable, a learning mechanism is activated to add a new rule to the knowledge base so that the rules responsible for the wrong decision can be disabled. This learning mechanism makes the proposed system adaptive to changes, while the adaptation is mainly focused on the disturbances and unexpected events. Reference [24] investigated the impact of container stacking methods regarding how they deal with uncertainties in container terminals and reduce container handling costs. The stacking methods studied, however, determine only the best yard-bays but not the specific stacking slots.

Compared to the works discussed above, References [4,6] are much more closely related to our work. The stacking policy in these researches employs scoring functions that evaluate a candidate stacking location from various perspectives using different criteria, where the score of a location is calculated as the weighted sum of the scores for those criteria. The policy uses different scoring functions for different container types because each container type requires its own evaluation criteria for stacking. However, all those scoring functions with different weight vectors are together treated as a single policy. This policy is optimized by a search using a GA, where a candidate policy is evaluated by simulating the operations at a block under the policy for a certain period of time and measuring the resulting performance. For this simulation, Reference [4] provided a pool of operation scenarios of various kinds for a more accurate policy evaluation. To evaluate a policy, it is applied to a randomly selected subset of those scenarios, and the resulting performances were averaged. Therefore, the policy thus optimized can be said to be the best on average. Given a certain situation, there may be some other policy that works better than the on-average best policy. Another limitation with such policy is that the policy cannot change as the operational environment changes. As an effort to overcome this problem, Reference [5] proposed an online search algorithm that dynamically adjusts and optimizes a stacking policy by continuously generating variants of stacking policies and evaluating them while they are actually being applied for determining the stacking positions. However, this online search cannot keep up with the rapid changes in a situation. When the situation changes, the performance of the current policy begins to deteriorate, at which time that of a variant of the current policy may show a better performance. If this happens, the current policy would be switched to that variant, but only after experiencing some deterioration. In general, we cannot expect such a good variant to appear quickly at the right moment. Therefore, we can say the online search is not really reactive to changes, but it just gradually adapts to changes. Another drawback is that the online search cannot find really good policies because of its limited explorative capacity. Since all the variants must be simulated and tested online, it is hard to generate and test a number of variants under a real-time constraint, which deteriorates the search performance.

The stacking policy derived by the method proposed in this paper is a significant improvement on the policy proposed in [4,6]. While [4,6] look for a policy whose average performance in various situations is the best, our method derives a policy that can quickly adapt to changing situations. When the situation changes, our policy immediately reacts to the change and provides a newly customized action. This improvement was possible because the policy that we deal with was based on scoring functions and the weight vectors used in those functions were easily adjustable. Most previous works reviewed above in the second and the third paragraphs of this section use policies based on if-then rules. Those rules are carefully crafted by the designers rather than being optimized by any algorithms. They are hard to be automatically modified upon situation changes. While the method proposed in [5] looks closest to ours in that the policy can adapt to changing situations, its adaptation is slow or gradual rather than immediate or reactive. The different characteristics of the related works discussed so far are compared and summarized in Table 1. The works reviewed in the first paragraph of this section are not included in the table because they have little relevance to our work.

Table 1. Comparison of the proposed method with the related works that determine the stacking locations for individual containers.

Characteristics	Related References and Their Characteristics				
	[1–3,16–18,21–24]	[19,20]	[4,6]	[5]	Proposed Method
Rule-based	Yes	Yes	No	No	No
Score-based	No	No	Yes	Yes	Yes
Optimized	No	Yes	Yes	Yes	Yes
Situation adaptive	No	No	No	Yes	Yes
Reactive to changes	No	No	No	No	Yes

4. Stacking Policy Based on Scoring Functions

This section describes the stacking policies proposed by [4,6]. The stacking location of an incoming container is determined in two stages. First, the container is assigned to a block in the storage yard taking into account various operational conditions in all of the blocks. Second, a specific location within the assigned block is selected from among the candidate locations based on several criteria such as the distance to the destination, the stack height, the likelihood of rehandling, and so on. The stacking policies described in this paper are used in the second stage to determine a specific stacking location within the designated block.

To determine a stacking location within a block, all the available slots in the block are evaluated by using a scoring function and then the one with the best score is chosen. Note that the slot determination is required not only for the containers newly coming into the block but also for those that are rehandled or repositioned within the block. Furthermore, a good target slot can be different depending on whether the container is an inbound, an outbound, or a transshipment container. Table 2 shows that the stacking policy uses different scoring functions for different container types. The score $s_i(x)$ of a slot x for the ith container type is calculated by the weighted sum given below:

$$s_i(x) = \sum_i w_{i,j} C_{i,j}(x) \quad (1)$$

where $C_{i,j}(x)$ is the evaluation value of slot x according to the jth criterion for the ith container type and $w_{i,j}$ is the weight for $C_{i,j}$. The stacking policy of Table 2 consists of seven scoring functions, each of which employs a different subset of eight criteria. Notice that the decision by the policy may change as the values of the weights of the criteria change.

Table 2. Scoring function for each of the seven container types.

Container Type		Scoring Function
Incoming	Inbound	$s_1 = w_1 D_{to} + w_2 D_{from} + w_3 H + w_4 E + w_5 S$
	Outbound	$s_2 = w_6 D_{to} + w_7 D_{from} + w_8 H + w_9 E + w_{10} S + w_{11} G + w_{12} P$
	Tranship	$s_3 = w_{13} D_{to} + w_{14} H + w_{15} E + w_{16} S + w_{17} G$
Rehandle	Inbound	$s_4 = w_{18} D_{to} + w_{19} D_{from} + w_{20} H + w_{21} E + w_{22} S$
	Outbound and Tranship	$s_5 = w_{23} D_{to} + w_{24} D_{from} + w_{25} H + w_{26} E + w_{27} S + w_{28} G$
Reposition	Inbound	$s_6 = w_{29} D_{to} + w_{30} D_{from} + w_{31} H + w_{32} E + w_{33} S + w_{34} T$
	Outbound and Tranship	$s_7 = w_{35} D_{to} + w_{36} D_{from} + w_{37} H + w_{38} E + w_{39} S + w_{40} G + w_{41} T$

The criterion D_{to} is the distance to the candidate stacking location from the current location of the target container, and D_{from} is the distance from the candidate location to the outgoing HP of the container. These two criteria are calculated differently for different types of containers, as illustrated in Figure 2. D_{to} and D_{from} give the same value if the target container is a transshipment container. H is the height of the stack underneath the candidate location. The higher the stack, the greater the likelihood of rehandling. E is an indicator of whether or not the candidate location is an empty ground. The empty grounds have to be saved as much as possible in preparation for the possible shortage of stacking locations. S is the amount of reduction in empty ground slots available for container stacking. For safety reasons, containers of different sizes cannot be stacked together. A 40 ft container occupies two adjacent stacks, as shown in Figure 1. Therefore, placing a 20 ft container on one of the two consecutive empty ground slots not only uses one ground slot for a 20 ft container but also reduces the availability of ground slots for 40 ft containers.

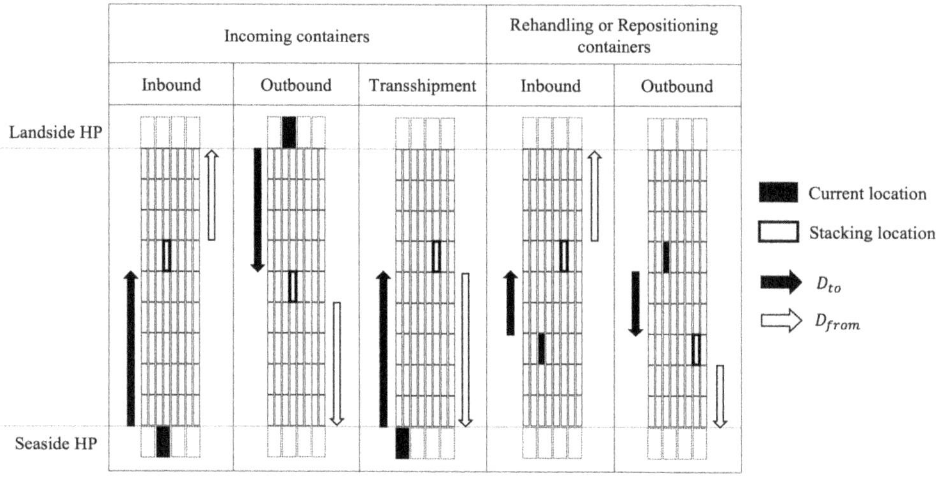

Figure 2. D_{to} and D_{from} depending on the type of container.

T indicates whether or not the container right underneath the candidate location has been temporarily repositioned to that slot. T is used only in the scoring functions for the containers to be repositioned. Since a repositioned container will soon be moved to an outgoing HP, rehandling is quite likely to occur if there is some other container on top of it. However, if the container on top of it is also a repositioned container, rehandling can be avoided by simply moving the upper one to its own outgoing HP before moving the lower one. Criterion T encourages the repositioned containers to be kept together in the same stacks. This is desirable to save stacking slots available for the containers newly coming in or for those rehandled. G is the estimated likelihood of the occurrence of rehandling when a container to be loaded to a vessel is stacked on top of others. If the container was just stacked and all the containers underneath belong to the same category, no rehandling occurs during loading. Otherwise, the underneath containers belonging to different categories can cause rehandlings. As mentioned before, the containers belonging to the same category are to be loaded onto the same vessel, have the same port of destination, are of the same size, and are of the same weight class.

P appears only in the scoring function s_2, which is specialized for the outbound containers brought in by the ETs. It is the preference value of a candidate location depending on which region of the block the location belongs to. It is preferable that the outbound containers to be loaded sooner are stacked closer to the seaside. Figure 3 illustrates the distributions of the preferences over different regions in a block for the outbound containers of different loading times. The block is divided into five regions, and there are three different urgency levels for loading. The preferences are distributed differently for each urgency level, resulting in fifteen preference values each for a region and an urgency level. In the previous work [6], these preference values as well as all the weight values in Table 2 were determined by running a GA-based search algorithm.

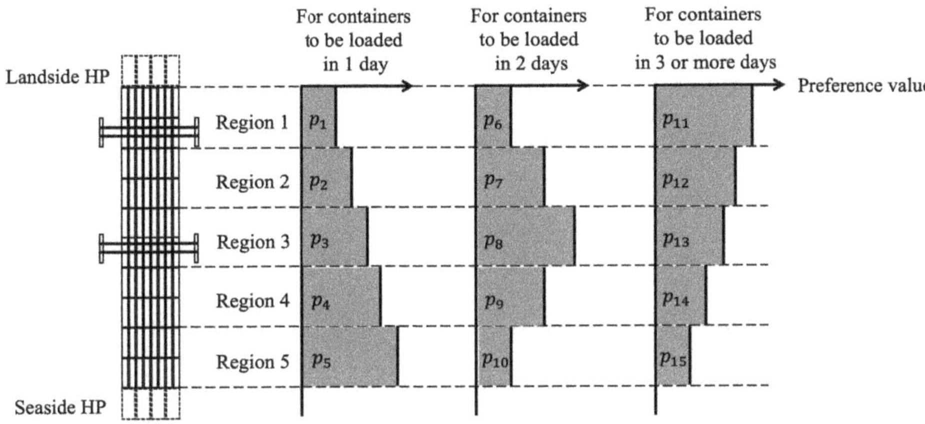

Figure 3. An example of preference distributions over the stacking regions within a block for the outbound containers with different loading times.

Table 3 summarizes the eight criteria explained above. To calculate the weighted sum of the respective subsets of these criteria, the policy of Table 2 uses 41 weights. The evaluation values of all the criteria are normalized to [0, 1] and the weight values are constrained to [−1, 1]. Note that a weight can be negative if the value of the corresponding criterion affects the policy adversely. As the decision made by the policy depends on the weight combination or the weight vector, the weight vector is considered as the policy. When the policy is optimized by using a GA, each candidate policy is evaluated by applying it to a variety of scenarios of operations in a block and averaging the resulting performances. In this way, the optimization algorithm derives a policy that works the best on average. Given a certain situation, however, a different policy might perform better than this best on-average policy. In the next section, we explain how we derive a policy that can be dynamically adapted to changing situations.

Table 3. Evaluation criteria used in the scoring functions.

Criterion	Description
D_{to}	Distance to the stacking location from the container pick-up position
D_{from}	Distance from the stacking location to a departure HP
H	Height of the stack at the candidate location
E	Indicator for an empty ground
S	Reduction in empty space availability
P	Regional preference within a block for the incoming outbound containers
G	Likelihood of rehandling
T	Indicator of having a repositioned container underneath

5. Proposed Method

In this research, we consider the current workload of vessel operation as the only important indicator of the current operational situation in a storage block. The workload of vessel operation from the standpoint of a block is the workload of its seaside ASC that handles the containers to be loaded to or unloaded from the vessels. Since any delay by the seaside ASC leads to a delay of the vessel operation at the quay, its efficient operation is critical. It may not be desirable, for example, that the seaside ASC spends too much time in container stacking when the seaside workload is high. The containers arriving at the seaside HPs are better stacked at locations not far from those HPs in order not to have other vessel operations delayed. However, a desired amount of such adjustment of travel distance for stacking cannot be made in any obvious way. It is difficult to invent a formula relating the amount of adjustment with the seaside workload that is continuously changing over time. One intuitive approach to dealing with continuously changing situations would be to divide the situations into a finite number of representative situations and then to derive a specialized policy for each representative situation. Note that a coarse division would not be effective enough because each constituent policy would suffer from the same problem of showing only the best on-average performance although to a lower degree. On the other hand, a very fine division requires too many policies that all must be derived through computationally expensive optimization search.

Another approach one may think of would be to turn the criteria used in the scoring functions of the policy into functions of the seaside workload. While most criteria are clearly independent of the seaside workload, the regional preference P seems to be dependent on it. When the seaside workload is heavy, preferring the seaside regions is not desirable because the chances of interference with the seaside ASC get high. In fact, the preference as a function of the seaside workload seems necessary more for the inbound than the outbound containers because the seaside ASC may not want to travel a long distance for container stacking when its load is heavy. Furthermore, there are some weights whose desirable values seem to depend on the seaside workload, although the corresponding criteria are not. Some of the examples are the weights for D_{to} and D_{from}. The value of criterion D_{to} for an inbound container should be considered more importantly (i.e., should be given a larger weight) to save the travel time as the seaside workload gets higher. However, we do not know how exactly the values of the criteria or weights should change as a function of the seaside workload. In our proposed method, therefore, we exclude P from the policy, and instead, we synthesize a new policy from two boundary policies whenever needed: one for a very low workload and the other for a very high workload. For the synthesis, we take an interpolation of the two boundary policies. When we use a GA to search for the two boundary policies, we simultaneously search for the two threshold values to quantify the two fuzzy terms 'very low' and 'very high'.

Let s represent the current workload that is measured by adding up the estimated processing times of all the vessel jobs scheduled to be done by the seaside ASC within the next horizon of length h seconds from the current point of time. A vessel job is either a loading or an unloading job. For a loading job, the seaside ASC makes an empty trip from its current location to the location of the target container, picks up the container, makes a loaded trip to a seaside HP, and puts the container down on top of an AGV waiting there. Among these actions, container pickup can take longer if it involves rehandlings. For an unloading job, the seaside ASC undertakes an empty trip from its current location to the seaside HP where the AGV bringing the target container is parked, picks up the container from the AGV, undertakes a loaded trip to the designated stacking location, and puts the container down at that location. Since the loading and unloading schedules for each vessel are predetermined at the planning stage well before the real operation starts, the workload of vessel operations within a horizon can be easily estimated.

We use θ_l and θ_h to denote the threshold values for the extreme or boundary workloads; the workload is said to be very low if $s \leq \theta_l$ and very high if $s \geq \theta_h$. The seaside ASC is said to be overloaded if $s > h$, as the time taken to finish the works planned for the horizon exceeds the length of the horizon. Let π_l and π_h be the policies specialized for the situations of very low workload and very high workload, respectively. Then, the policy π_s for workload s with $\theta_l < s < \theta_h$ can be synthesized from π_l and π_h by deriving new weight values to be used in the scoring functions of π_s through interpolations between the corresponding weights in π_l and π_h. The ith weight $w_{s,i}$ to be used in policy π_s is calculated as

$$w_{s,i} = \frac{(\theta_h - s)w_{l,i} + (s - \theta_l)w_{h,i}}{\theta_h - \theta_l} \tag{2}$$

where $w_{l,i}$ and $w_{h,i}$ are the ith weights in π_l and π_h, respectively. As s gets closer to θ_h, $w_{s,i}$ is influenced more by $w_{h,i}$ than $w_{l,i}$, or the other way around. Note that the score $\pi_s(x)$ for a candidate slot x by the synthesized policy π_s can be directly calculated from the scores $\pi_l(x)$ and $\pi_h(x)$ as

$$\pi_s(x) = \frac{(\theta_h - s)\pi_l(x) + (s - \theta_l)\pi_h(x)}{\theta_h - \theta_l} \tag{3}$$

without actually deriving the individual weights constituting π_s because the values θ_h, θ_l, and s in Equation (2) are independent of i.

We use a GA for our optimization, which is basically the same as that used in [4]. We optimize not only π_l and π_h but also the two threshold values θ_l and θ_h. Figure 4 shows the representation of the candidate solution adopted by our GA. Since each policy consists of 40 weights after dropping out the criterion P, there are 82 real values to be optimized in total, where v_l and v_h are constrained to be in [0, 2] and the weight values in [–1, 1]. During the evaluation, v_l and v_h are decoded to θ_l and θ_h, respectively, by having them multiplied to the length of horizon h. The reason for setting the upper bound of v_l and v_h to 2 is that the workload of vessel operation measured in time can exceed the length of the horizon when overloaded.

θ_l	θ_h	π_l				π_h			
v_l	v_h	$w_{l,1}$	$w_{l,2}$	. . .	$w_{l,40}$	$w_{h,1}$	$w_{h,2}$	. . .	$w_{h,40}$

Figure 4. Representation of candidate solution.

To evaluate a candidate policy during the search, the policy is applied through simulation to a set of scenarios randomly chosen from the provided pool and the resulting performances are averaged. The pool contains various scenarios of different difficulty levels; a scenario is difficult if the workloads of the ASCs are high. The length of a scenario is three weeks, which is long enough to measure the efficiency of the ASC operation because enough of the containers that arrived during this period are retrieved. During the first two weeks, the stacking yard, or block, is initialized starting from an empty yard without simulating the ASC's movements. Then, from the beginning of the third week, the efficiency of the ASC operation is measured with their movements simulated realistically, reflecting acceleration, deceleration, and interferences. More details on this crane simulation can be found in [25]. For an evaluation of the performance in the third week of a scenario, a candidate policy $\langle\theta_l, \theta_h, \pi_l, \pi_h\rangle$ is applied to the scenario. When a container has to be stacked, the workload s of vessel operation for the next horizon of length h seconds is estimated. If $s \leq \theta_l$ or $s \geq \theta_h$, then π_l or π_h becomes the policy to be used, respectively. Otherwise, a new policy π_s specialized for the workload s is synthesized through interpolation and then applied for stacking. Note that we need to synthesize only one of the seven scoring functions shown in Table 2 depending on the type of container to be stacked. This synthesis and application of a new policy are repeated every time a container is stacked. When the simulation of a scenario is over, the performance of the candidate policy is measured by the following objective function:

$$f(\pi) = W_1 \cdot D_{AGV}(\pi) + W_2 \cdot D_{ET}(\pi) \tag{4}$$

where π is the stacking policy under evaluation, $D_{AGV}(\pi)$ is the average (per container) AGV delay observed under π, $D_{ET}(\pi)$ is the average waiting time of ETs under π, and W_1 and W_2 are the respective weights for D_{AGV} and D_{ET}. W_1 is usually much larger than W_2 because the seaside operations are considered much more important than the landside operations. The final evaluation is obtained by averaging the objective values measured from all the scenarios.

6. Experimental Results

We used the algorithm named NTGA for policy optimization, which is the same one as that used in [4] to derive the static stacking policy described in Section 4. Our parameter setting of NTGA is shown in Table 4. Since NTGA requires a random subset of operation scenarios selected from a pool to evaluate each candidate policy, we generated 1000 scenarios to constitute a pool. Each scenario consists of container handling jobs to be completed for three weeks in a block that is 46 bays long, 8 rows wide, and 5 tiers high. The job requests are made by the AGVs and ETs that arrive at the seaside and landside HPs, respectively. They either bring in a container to be stored in the block or ask for a container to be picked up from the block. The average number of AGVs arriving per day is approximately from 220 to 300, and that of ETs is from 70 to 110. More requests are from the AGVs than ETs because there are transshipment containers whose proportion among the containers unloaded from the vessels is about 50% in our scenarios. The average daily workload of the ASCs increases with the number of requests, but the workload continuously changes within a day as the requests are not evenly distributed over time.

Table 4. Parameter setting of NTGA in our experiments.

Parameter	Setting Value
Population size	100
Mating pool size	2
Buffer pool size	50
Sampling size	2 (initially 4)
Crossover operator	Simulated binary crossover
Crossover probability	0.9
Mutation operator	Polynomial mutation
Mutation probability	0.01
Number of evaluations	100,000

Using this pool of scenarios, we derived both the static stacking policy of [4] and the dynamic stacking policy proposed in this paper. Then, the two policies were applied to 100 scenarios that were separately generated following the same distribution as that used for generating the scenarios of the above pool. The two weight values W_1 and W_2 of Equation (4) were empirically set to 50 and 1, respectively, not only when the policies were derived but also when they were tested. The results obtained by measuring the AGV delay and ET waiting time are shown in Table 5. We can see that the proposed policy outperforms the static policy in terms of both AGV delay and ET waiting time. The improvement is 22.3% for AGV delay and 16.3% for ET waiting time. We also confirmed that the proposed policy performs significantly better than the static policy by using a paired t-test with a confidence level of higher than 99.99%. The data for our experiments and the execution of our program can be found in [26].

Table 5. Performance comparison of the two policies.

	Static Policy	Dynamic Policy
AGV delay	80.0	62.2
ET waiting time	285.3	238.8

To obtain some hints about how the dynamic policy works, we investigated the behaviors of the two boundary policies, i.e., π_l for a very low workload and π_h for a very high workload. The two threshold values θ_l and θ_h (see Figure 4) found by our search algorithm for distinguishing the very low and very high workloads were 274 s and 1892 s, respectively. Recall that the length of our horizon is 1800 s; the workload of 274 s really looks very low, and that of 1892 s is over the capacity. If we want to quantify the overall stack preference of a boundary policy π_b for the ith container type (see Table 2 for the seven container types), we apply π_b to a scenario as if it is a static policy and pay special attention to the moments of stacking the ith type containers. Whenever we come across such a moment during the simulation, we not only apply π_b to stack the container at the best location as usual, but also calculate the scores for all of 46 × 8 slots as if they are all candidate locations and just save the scores separately. For the latter calculation, the constraint of a maximum possible tier of five is relaxed. Furthermore, we assume that we can stack 20 ft containers even on top of 40 ft ones. When the simulation is over, we obtain the stack preference by averaging the separately saved scores for every slot in the block.

Figure 5a compares the stack preferences of π_l and π_h for the incoming inbound containers, where the slots of better scores are indicated by a darker shade. When the seaside workload is very low, the best locations by π_l are distributed toward the landside end. This is quite reasonable because the inbound containers will eventually leave the block through the landside HPs. Since the seaside workload is low, the seaside ASC does not hesitate to travel a long distance to stack the containers at the landside end so that later retrieval by the ETs is expedited. However, we can see that the locations toward the seaside end of the block are considered not the worst but somewhat preferable by π_l. Note that π_l is used not only in the situations of very low workloads but also in the situations of intermediate workloads through interpolation with π_h. If we separately derived a static policy specialized for a very low workload, it might not prefer any seaside locations at all. On the other hand, when the seaside workload is very high, π_l prefers the locations closer to the seaside end than those farther away. When the seaside ASC is very busy, it should avoid long-distance travel as not to delay the services to the AGVs waiting at the seaside HPs. Figure 5b shows the overall stack preferences of π_l and π_h for the incoming outbound containers. We can see that the stack preferences are almost the opposite of what we have seen for the inbound containers. Figure 5c shows the stack preferences of the static policy for the inbound (shown in the upper part) and outbound (shown in the lower part) containers. It seems that the static policy generally prefers the locations closer to the departure HPs regardless of the seaside workload.

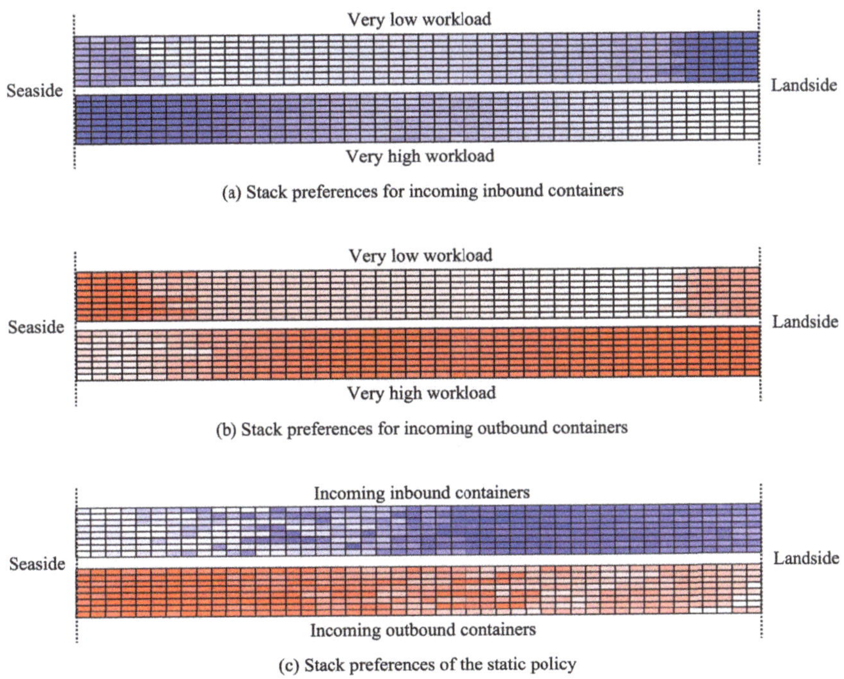

Figure 5. Comparison of the stack preferences.

7. Cases with Multiple Situation Indicators

Thus far, we have been assuming that the seaside workload is the only indicator we consider to represent the situation. Although rare practically in container terminals, we can imagine the cases in which the situation is represented by multiple indicators. Our method described in Section 5 can be extended to cover such cases by generalizing the interpolation to a weighted average of multiple relevant terms. Consider, for simplicity of explanation, the case with two indicators s_1 and s_2. We need two threshold values for each indicator to distinguish between very low and very high values, i.e., $\theta_{1,l}$ and $\theta_{1,h}$ for s_1, and $\theta_{2,l}$ and $\theta_{2,h}$ for s_2. This leads to four extreme or boundary situations $B_{l,l}$, $B_{l,h}$, $B_{h,l}$, and $B_{h,h}$, where $B_{l,l}$ is the set of situations with $s_1 \leq \theta_{1,l}$ and $s_2 \leq \theta_{2,l}$, $B_{l,h}$ with $s_1 \leq \theta_{1,l}$ and $s_2 \geq \theta_{2,h}$, $B_{h,l}$ with $s_1 \geq \theta_{1,h}$ and $s_2 \leq \theta_{2,l}$, and $B_{h,h}$ with $s_1 \geq \theta_{1,h}$ and $s_2 \geq \theta_{2,h}$. We use $\pi_{l,l}$, $\pi_{l,h}$, $\pi_{h,l}$, and $\pi_{h,h}$ to denote the policies specialized for the boundary situations $B_{l,l}$, $B_{l,h}$, $B_{h,l}$, and $B_{h,h}$, respectively. Given an intermediate situation s other than those boundary situations, the policy π_s for s can be synthesized from $\pi_{l,l}$, $\pi_{l,h}$, $\pi_{h,l}$, and $\pi_{h,h}$ by taking a weighted average after normalizing the indicator values. The normalization is necessary to compensate the different scales of different indicators.

Figure 6a represents the space of all situations on a two-dimensional plane formed by two coordinates, one for indicator s_1 and the other for indicator s_2. We can see how the areas of the four boundary situations $B_{l,l}$, $B_{l,h}$, $B_{h,l}$, and $B_{h,h}$ are located in relation to the threshold values of the two indicators. A and B in the figure are two situations other than the boundary situations. A is an intermediate situation whose indicator values do not go over any threshold. B is not quite an intermediate situation because one of its indicator s_2 takes a value below the lower threshold $\theta_{2,l}$. Figure 6b shows the situations in Figure 6a after a normalization, where each indicator value s_i is transformed to $(s_i - \theta_{i,l})/(\theta_{i,h} - \theta_{i,l})$. In Figure 6b, the areas of boundary situations are marked by the corresponding policies, and the distances to those areas from A and B are indicated by $d_{i,j}$'s. Let π_A denote the policy for situation A in the figure. Then, the score $\pi_A(x)$ for a candidate slot x in situation A can be calculated by a weighted average of the scores given by $\pi_{l,l}$, $\pi_{l,h}$, $\pi_{h,l}$, and $\pi_{h,h}$:

$$\pi_A(x) = \frac{\frac{1}{d_{l,l}(A)}\pi_{l,l}(x) + \frac{1}{d_{l,h}(A)}\pi_{l,h}(x) + \frac{1}{d_{h,l}(A)}\pi_{h,l}(x) + \frac{1}{d_{h,h}(A)}\pi_{h,h}(x)}{\frac{1}{d_{l,l}(A)} + \frac{1}{d_{l,h}(A)} + \frac{1}{d_{h,l}(A)} + \frac{1}{d_{h,h}(A)}} \quad (5)$$

where $d_{i,j}(A)$ is the distance from A to the area of $\pi_{i,j}$, and $1/d_{i,j}$ can be interpreted as the respective closeness. If π_B denotes the policy for situation B, the score $\pi_B(x)$ for a candidate slot x in situation B can be similarly calculated as

$$\pi_B(x) = \frac{\frac{1}{d_{-,l}(B)}\pi_{l,l}(x) + \frac{1}{d_{-,h}(B)}\pi_{h,l}(x)}{\frac{1}{d_{-,l}(B)} + \frac{1}{d_{-,h}(B)}} \quad (6)$$

where terms related to $\pi_{l,h}$ and $\pi_{h,h}$ are not included because s_2 is extremely low in B and thus they are irrelevant. Note that Equation (6) is equivalent to the linear interpolation we calculated in Equation (3).

The formulation given above can be easily extended to the cases with more than two situation indicators in principle. However, extensions to such cases would be practically infeasible because the number of boundary policies increases exponentially to 2^n, where n is the number of situation indicators. As we have seen in Figure 4, our chromosome for the search of the policy already consisted of 82 real-numbered genes when there was a single indicator. If there were two situation indicators, the number of genes should have increased to 164. This number doubles each time another indicator is added, resulting in a huge search space.

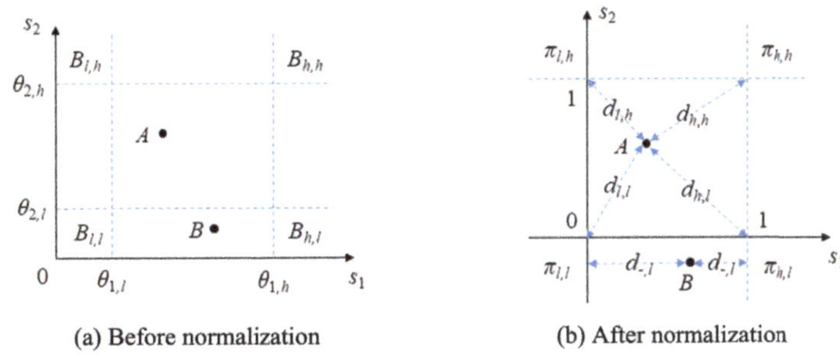

Figure 6. Situation space displayed on a two-dimensional plane formed by the two coordinates of the situation indicators s_1 and s_2.

8. Concluding Remarks

We derived a dynamic or situation-adaptive stacking policy and compared its performance with that of a static policy that is insensitive to the situation change. The result of simulation experiments tells us that the dynamic policy clearly outperforms the static policy. In the experiments, the policy was optimized using randomly generated operation scenarios and then tested using a different set of scenarios separately generated from the same distribution. This separation was an effort to guarantee the generality of the observed performance of the policy in similar situations. However, the policy needs to be newly optimized and tested for being applicable to a new container terminal whose distribution of the operations is different from that of our target container terminal.

The way we proposed to obtain a dynamic policy is to synthesize a new policy from two boundary policies: one for a very low and the other for a very high workload. For the synthesis, we recommended a linear interpolation of the two boundary policies. We used a GA-based search algorithm to optimize not only the two boundary policies but also the two threshold values for distinguishing the very low and very high workloads. The idea of using linear interpolation is based on the assumption that the values of the weight vectors in the policy are linearly related to the seaside workload, which is the only indicator we consider to represent the situation. Although this linearity assumption might not be really correct, our empirical study has shown that the method based on linear interpolation results in a significant improvement over the previous works. As emphasized toward the end of the literature review in Section 3, no previous work has ever tried to solve the kind of problem that we deal with in this paper. The linearity assumption, however, is not scalable to large problems because it needs boundary policies whose number increases exponentially with the number of situation indicators. Our future work, therefore, will be to drop the linearity assumption and use a multi-layer perceptron to represent a policy instead of the linear scoring functions so that we do not need to use boundary policies. Another viable approach would be to apply a reinforcement learning algorithm that learns value functions based on the rewards from the environment.

Author Contributions: Conceptualization, T.K. and K.R.R.; methodology, T.K. and K.R.R.; software, T.K.; validation, T.K. and K.R.R.; formal analysis, T.K and K.R.R.; investigation, T.K. and K.R.R.; resources, T.K.; data curation, T.K. and K.R.R.; writing—original draft preparation, T.K.; writing—review and editing, K.R.R.; visualization, T.K.; supervision, K.R.R.; project administration, K.R.R.; funding acquisition, K.R.R. All authors have read and agreed to the published version of the manuscript.

Funding: This work was supported by Institute of Information & communications Technology Planning & Evaluation(IITP) grant funded by the Korea government(MSIT) (No. 2020-0-01450, Artificial Intelligence Convergence Research Center [Pusan National University]).

Institutional Review Board Statement: Not applicable.

Informed Consent Statement: Not applicable.

Data Availability Statement: Not applicable.

Conflicts of Interest: The authors declare no conflict of interest.

References

1. Dekker, R.; Voogd, P.; Van Asperen, E.T. Advanced methods for container stacking. *Contain. Termin. Cargo Syst.* **2006**, *28*, 563–586. [CrossRef]
2. Duinkerken, M.B.; Evers, J.J.M.; Ottjes, J.A. A simulation model for integrating quay transport and stacking policies on automated container terminals. In Proceedings of the 15th European Simulation Multiconference, Prague, Czech, 6–9 June 2001; pp. 909–916.
3. Kim, K.H.; Hong, G.P. A heuristic rule for relocating blocks. *Comput. Oper. Res.* **2006**, *33*, 940–954. j.cor.2004.08.005. [CrossRef]
4. Jang, H.; Choe, R.; Ryu, K.R. Deriving a Robust policy for container stacking using a noise-tolerant genetic algorithm. In Proceedings of the 2012 ACM Research in Applied Computation Symposium, San Antonio, TX, USA, 23–26 October 2012; ACM Press: New York, NY, USA, 2012; p. 31. [CrossRef]
5. Park, T.; Choe, R.; Kim, Y.H.; Ryu, K.R. Dynamic adjustment of container stacking policy in an automated container terminal. *Int. J. Prod. Econ.* **2011**, *133*, 385–392. [CrossRef]
6. Park, T.; Sohn, M.; Ryu, K.R. Optimizing stacking policies using an MOEA for an automated container terminal. In Proceedings of the 40th International Conference on Computers and Industrial Engineering, Awaji, Japan, 25–28 July 2010; pp. 1–6. [CrossRef]
7. Sutton, R.S.; Barto, A.G. *Reinforcement Learning: An Introduction*, 2nd ed.; The MIT Press: Cambridge, MA, USA; London, UK, 2018.
8. Kim, J.; Hong, E.J.; Yang, Y.; Ryu, K.R. Noisy Optimization of Dispatching Policy for the Cranes at the Storage Yard in an Automated Container Terminal. *App. Sci.* **2021**, *11*, 6922. [CrossRef]
9. Taleb-Ibrahimi, M.; De Castilho, B.; Daganzo, C.F. Storage space versus handling work in container terminals. *Transport. Res. B Methodol.* **1993**, *27*, 13–32. [CrossRef]
10. Kim, K.H.; Kim, H.B. Segregating space allocation models for container inventories in port container terminals. *Int. J. Prod. Econ.* **1999**, *59*, 385–392. [CrossRef]
11. Kim, K.H.; Park, K.T. A note on a dynamic space-allocation method for outbound containers. *Eur. J. Oper. Res.* **2003**, *148*, 92–101. [CrossRef]
12. Lee, L.H.; Chew, E.P.; Tan, K.C.; Han, Y. An optimization model for storage yard management in transshipment hubs. *Contain. Termin. Cargo Syst.* **2006**, *28*, 539–561._6. [CrossRef]
13. Kim, K.H.; Lee, J.S. Satisfying constraints for locating export containers in port container terminals. In Proceedings of the 2006 International Conference on Computational Science and Its Applications—Volume Part III, Glasgow, UK, 8–11 May 2006; pp. 564–573. doi: 10.1007/11751595_60. [CrossRef]
14. Zhen, L. Storage allocation in transshipment hubs under uncertainties. *Int. J. Prod. Res.* **2014**, *52*, 72–88. 00207543.2013.828166. [CrossRef]
15. Said, G.A.E.N.A.; El-Horbaty, E.S.M. An intelligent optimization approach for storage space allocation at seaports: A case study. In Proceedings of the 2015 IEEE 7th International Conference on Intelligent Computing and Information Systems, Cairo, Egypt, 12–14 December 2015; pp. 66–72. [CrossRef]
16. Kim, K.H.; Park, Y.M.; Ryu, K.R. Deriving decision rules to locate export containers in container yards. *Eur. J. Oper. Res.* **2000**, *123*, 89–101. [CrossRef]
17. Borgman, B.; Asperen, E.van; Dekker, R. Online rules for container stacking. *Contain. Termin. Cargo Syst.* **2010**, *32*, 687–716. [CrossRef]
18. Chen, L.; Lu, Z. The storage location assignment problem for outbound containers in a maritime terminal. *Int. J. Prod. Econ.* **2012**, *135*, 73–80. [CrossRef]
19. Kang, J.; Ryu, K.R.; Kim, K.H. Determination of storage locations for incoming containers of uncertain weight. In Proceedings of the 19th international conference on Advances in Applied Artificial Intelligence: Industrial, Engineering and Other Applications of Applied Intelligent Systems, Annecy, France, 27–30 June 2006; pp. 1159–1168._123. [CrossRef]
20. Kang, J.; Ryu, K.R.; Kim, K.H. Deriving stacking strategies for export containers with uncertain weight information. *J. Intell. Manuf.* **2006**, *17*, 399–410. [CrossRef]
21. Zhang, C.; Wu, T.; Zhong, M.; Zheng, L.; Miao, L. Location assignment for outbound containers with adjusted weight proportion. *Comput. Oper. Res.* **2014**, *52*, 84–93. [CrossRef]
22. Ries, J.; Gonzalez-Ramírez, R.G.; Miranda, P. A fuzzy logic model for the container stacking problem at container terminals. *Lect. Notes Comput. Sci.* **2014**, *8760*, 93–111._7. [CrossRef]
23. Rekik, I.; Elkosantini, S. A multi agent system for the online container stacking in seaport terminals. *J. Comput. Sci.* **2019**, *35*, 12–24. [CrossRef]
24. Gunawardhana, J.A.; Perera, H.N.; Thibbotuwawa, A. Rule-based dynamic container stacking to optimize yard operations at port terminals. *Marit. Transp. Res.* **2021**, *2*, 100034. [CrossRef]

25. Park, T.; Choe, R.; Ok, S.M.; Ryu, K.R. Real-time scheduling for twin RMGs in an automated container yard. *Contain. Termin. Cargo Syst.* **2010**, *32*, 593–615. [CrossRef]
26. Kim, T.; Ryu, K.R. Experimental data and the executable to test the performance of situation-adaptive policy for container stacking in an automated container terminal. *Mendeley Data* **2022**, *V1*. [CrossRef]

applied sciences

MDPI

Article

IORand: A Procedural Videogame Level Generator Based on a Hybrid PCG Algorithm

Marco A. Moreno-Armendáriz [1], Hiram Calvo [1,*], José A. Torres-León [1] and Carlos A. Duchanoy [2]

1 Computational Cognitive Sciences Laboratory, Center for Computing Research, Instituto Politécnico Nacional, Mexico City 07738, Mexico; mam_armendariz@cic.ipn.mx (M.A.M.-A.); jtorresl2019@cic.ipn.mx (J.A.T.-L.)
2 Gus Chat, Av. Paseo de la Reforma 26-Piso 19, Mexico City 06600, Mexico; carlos.duchanoy@gus.chat
* Correspondence: hcalvo@cic.ipn.mx

Abstract: In this work we present the intelligent orchestrator of random generators (IORand), a hybrid procedural content generation (PCG) algorithm, driven by game experience, based on reinforcement learning and semi-random content generation methods. Our study includes a presentation of current PCG techniques and why a hybridization of approaches has become a new trend with promising results in the area. Moreover, the design of a new method for evaluating video game levels is presented, aimed at evaluating game experiences, based on graphs, which allows identifying the type of interaction that the player will have with the level. Then, the design of our hybrid PCG algorithm, IORand, whose reward function is based on the proposed level evaluation method, is presented. Finally, a study was conducted on the performance of our algorithm to generate levels of three different game experiences, from which we demonstrate the ability of IORand to satisfactorily and consistently solve the generation of levels that provide specific game experiences.

Keywords: procedural content generation; artificial intelligence; reinforcement learning; semi-random generation; hybrid algorithms

Citation: Moreno-Armendáriz, M.A.; Calvo, H.; Torres-León, J.A.; Duchanoy, C.A. IORand: A Procedural Videogame Level Generator Based on a Hybrid PCG Algorithm. *Appl. Sci.* **2022**, *12*, 3792. https://doi.org/10.3390/app12083792

Academic Editor: Federico Divina

Received: 9 March 2022
Accepted: 7 April 2022
Published: 9 April 2022

1. Introduction

Procedural content generation (PCG) is a discipline that focuses on generating content algorithmically, mainly practiced in video games. PCG is commonly used for creating game content as levels, maps, quests, characters, or even rules during the run time or as a design tool. It is motivated by the need to personalize the video game content for each player, increase the replayability, or reduce the production costs in the industry [1]. Nowadays, small independent companies (*indies*) can produce and distribute creative applications; however, they usually lack adequate numbers of employees to develop a substantial amount of content manually. These indie companies look at PCG as a viable alternative for producing content at an affordable price.

The high impact of this area has generated a lot of interest in the academic community and has led to a considerable number of books and surveys in the last two decades [2–8], each one proposing different kinds of taxonomies for categorizing all of the work in the area. There are three main taxonomies used for classifying the works: the taxonomy of content, the taxonomy of methods, and the taxonomy of roles. The taxonomy of content classifies works by the type of content generated as map-level, narrative, texture, music, faces among others [2,6]; the taxonomy of roles segments them by the algorithm role and applications in the design; and the taxonomy of methods divide the works by technology type as machine learning, deep learning, evolutionary algorithms, among others [5,6].

PCG utilizes numerous techniques, each one with strengths and weaknesses. Togelius in [7] proposes a classification of the PCG approaches according to the content generation process in three categories; a search-based algorithm, constructive, and a simple generate and test.

The search-based approach to PCG is generally performed using some search optimization algorithm to look for content with the desired qualities. The main components of the search approach are the optimization algorithm, the content representation, and the evaluation function. The optimization algorithm is the core of the search-based method, mainly a simple evolutionary algorithm [9–14], and in some cases, other stochastic algorithms. Recently, reinforcement learning has been explored as an alternative for PCG search-base, describing the generation task as a Markov decision process where a model is trained to select the best action to improve the content quality [15–20].

The constructive approach consists of a series of steps to generate content and does not evaluate the output to regenerate it. In this approach, the algorithm uses a set of rules to generate content to warrant a level with the desired qualities. The experience of the designers is used to dictate the rules and are commonly captured using grammar [21–24] for structural generation, such as a dungeon, or linguistical generation, such as a quest. Moreover, simple rules that react to local spatial information, such as cellular automata [25], have been used to generate structures, such as dungeons or caves. However, the difficulty of dictating a set of rules has led to the implementation of machine learning algorithms [26–30] for content generation. These algorithms are capable of inferring the rules from data of similar games or previously designed levels.

The generate and test approach generates content using an algorithm that usually does not guarantee a good result and then tests the performance of the generated content according to some criteria. If the content does not accomplish the performance, it is regenerated; this process can iterate as much as is needed. This paradigm is a middle point between the constructive and the search-based approaches, it considers that the constructive method does not warranty playable content. Commonly, this approach uses random [31,32] or semi-random algorithms, such as fractals [33].

To provide a comparison between approaches, it is necessary to understand the criteria by which a procedural generation algorithm is analyzed. In [8], Togelius proposes a taxonomy to measure the algorithms in three dimensions. The first dimension is the degree of determinism or randomness that the algorithm possesses and is measured by the amount of variations of outcomes in different runs of the same algorithm with the same parameters. The second is the degree of controllability the designer has over the algorithm to influence the generated outcome. The third and final dimension is the level of iterability or number of iterations the algorithm needs to perform to produce content. While these criteria provide insight into the main characteristics of the algorithm, they do not describe the performance of the content generated. In [34], the creativity of the algorithms are analyzed, and the authors propose two metrics, novelty, and playability, to measure how creative the generated content is. Novelty is defined as the variability from the reference levels and the output ones, measured by the different levels the algorithm produces. Playability is defined as the percentage of the generated content that is useful.

We gathered the algorithm comparisons and taxonomy classifications from different sources [2–8,34]; we present a side-by-side comparison of the numerous techniques used for PCG in Table 1. In this comparison, we use a scale that considers levels (high, moderate, fair, and low) to describe the performance of each algorithm metric.

Table 1. Comparison of techniques for PCG.

Technique	Generation Process	Determinism	Controllability	Interactivity	Novelty	Playability
Evol. Algorithms [9–14,35]	Search-based	Moderate	Moderate	High	Fair	High
Reinforcement learning [15–20]	Search-based	Moderate	Moderate	High	Fair	High
Cellular Automata [25]	Gen. and test	Fair	Low	Fair	Moderate	Fair
Grammars [21–24]	Constructive	High	High	Low	Low	High
Machine Learning Supervised Learning [26–30] Adversarial Learning [36–38]	Constructive	High	High	Low	Low	High
Random Generators [31,32]	Gen. and test	Low	Low	Moderate	High	Low
Rule-based [39]	Constructive	High	High	Low	Low	High

In this table, we can observe that different algorithms that share the same generation process behave similarly in each metric, and many approaches have strengths that oppose weaknesses of the others. This would suggest that hybrid approaches could effectively improve performance and minimize weaknesses. The idea of hybridizing PCG algorithms through composition was original proposed in [40] where they experiment with combining an evolutive algorithm with a constructive method; the evolutive algorithm uses the constructive approach as part of its inner working and tests the idea in a simple dungeon generator. In [41], a hybrid approach that combines grammar and cellular automata to generate dungeon layouts is presented. This algorithm generates levels with high playability and slightly improves the novelty of the output levels due to the cellular automata.

The hybridization of algorithms has shown a lot of potential. However, there are still many research questions to be addressed in regard to creating a PCG algorithm that is fully controllable, fast, and capable of generating novel and playable content. For this reason, we propose a novel hybrid algorithm that uses a reinforced learning approach to orchestrate a group of semi-random generators. The reinforced learning algorithm has the objective of providing controllability to search for a specific player experience and provide results with high playability. This core algorithm is complemented with a group of semi-random generators to improve the novelty of the generated content.

Our proposal includes the following contributions:

- We introduce new metrics for any game level: risk, obstruction, precision, reward, motivation, and distance. The motivation metric considers the potential reward available in an interaction and the precision required.
- Through the designed reward function, our hybrid algorithm of PCG generates a diversity of levels that meet any given gaming experience.

2. Problem Statement

As a case study, we developed a game called "Pingu run", a game of the platform genre. Pingu run was designed and implemented in Pygame for this purpose and is available in [42]. This game consists of moving a penguin called Pingu from an initial point to an endpoint by travelling through a series of platforms and avoiding enemies; a video of the gameplay is shown in [43]. You can collect different rewards, such as extra lives or bonuses, in the final score. This video game is made up of four different types of elements: platforms, bonuses, enemies, and markers. The platforms are those elements in which the player can stand or support to jump. We consider three types of platforms: static, mobile, and bouncing. The bonuses are classified in score, lives, and sub-missions. Four types of enemies are used: a ranged enemy (troll), a flying enemy (eagle), a melee enemy (bear), and a static enemy (lava). Finally, the markers are elements that signal essential points in the game. We have three kinds of markers: start point, mid-level, and endpoint. These elements and their uses in a level are shown in Figure 1.

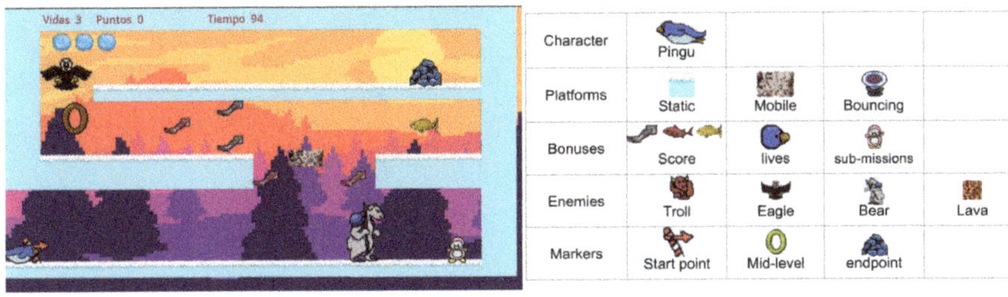

(**a**) (**b**)

Figure 1. Pingu Run video game. (**a**) Pingu run level example. (**b**) Game elements.

When building a level, the designer plans a series of challenges that he/she wants the player to face. With this, the designer seeks to convey emotions and narratives that define a specific game experience. An analytical link between the game features and the player experience has been proposed in [44] using the concept of affordance. Affordance refers to the opportunities or choices given by the environment to the player. The game design defines its possible uses and makes clear the actions the player may perform. For this reason, we consider it feasible to measure the experience indirectly from the characteristics of the game content.

To probe this hypothesis, we analyzed three different experiences denominated: simple, obstacles, and jumps. The simple experience is characterized by flat levels that allow the player to take risks to acquire the greatest amount of rewards without fear of being penalized for some error; they are generally easy to complete and are usually relaxing or restful stages for the player. The obstacle experience conveys to the player the need to dodge multiple enemies to navigate the level generally transmits emotion and requires concentration to complete it successfully, but the player is aware that an error is not definitive and can be corrected. Finally, the jumping experience provides greater emotion and tension in the player since it requires precision to go through the level with the risk that a mistake will penalize him/her with an absolute loss in the game. In Figure 2, a visual example of each experience is shown.

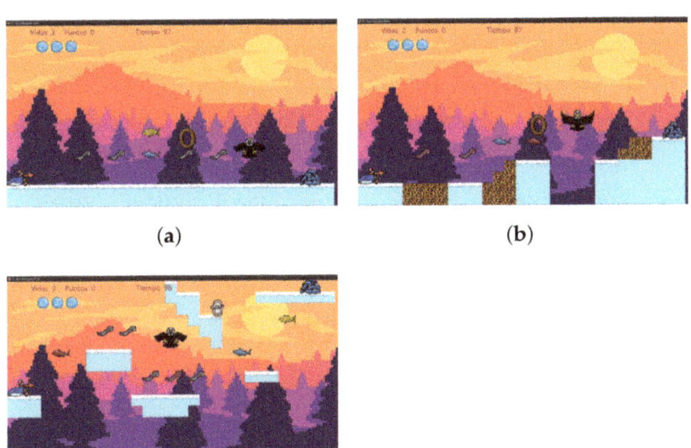

(**a**) (**b**)

(**c**)

Figure 2. Pingu Run experience examples. (**a**) Simple experience. (**b**) Obstacle experience. (**c**) Jump experience.

In the next section, we present our proposed level evaluation process that may serve as the basis of the quantitative measure of the game experience.

3. Proposed Level Evaluation

Content evaluation metrics are necessary to guide the search algorithms and to verify the generated content quality. However, it is difficult to define metrics that capture subjective factors, such as the player's experience or aesthetics. In [45], the authors propose to measure the reactions of the players through cameras during gameplay. Despite the acceptable results, this approach is impractical for content generation. In this work, we propose to measure the content characteristics and then search for affordance and its relation to the player experience. A graph-based description has proven useful to describe the game construction at a conceptual level, capturing the interactions between elements [22,46,47]. The proposed methodology splits the game elements into functional objects considering each platform, bonus, enemy, and marker as objects, and analyzes the interactions between them. In the first step, we selected the platform objects as the node-set (V), and each node was analyzed considering the risk (rk) and reward (rw). The risk represents the level of threat that may harm a player when standing in a node [48]. The reward is a measure of the level of bonification a player may have while standing in a node [49]. The second step consists of determining all possible transitions (E) between nodes using a reachability metric [50,51]. Each transition is described within six measures: risk (rk), reward (rw), obstruction (o), precision (p), motivation (m), and distance (d). While risk and reward are the same measures, with one difference they are computed for the threats and rewards a player may have or suffer during the transition. The obstruction measures the degree a platform may interfere during a transition [52], and the precision measures how exact a jump between platforms needs to be performed [50,51].

The reward metric is divided into two types, the *level reward* (lrw) which measures the points obtained for reaching the marker rings and completing the level, and the *bonus reward* (brw) that measures points earned for defeating enemies and for reaching bonus items (bonuses). This division is made in order to have a better description of the type of reward that the player can obtain on a platform or during a transition. Motivation is the measure of reward attenuated by accuracy, it represents the degree of interest the players will have in transitioning, given the points they can earn and the degree of skill such a transition would entail. Being a measure derived from the reward, it is also divided into *level motivation* (lm) and *bonus motivation* (bm). Distance is a measure of the length of a line or path between two points. For the purposes of this project, the Euclidean distance was used, which measures the length of the shortest line connecting the two points. Each level is represented by a set of nodes V, Equation (1), and a set of transitions, Equation (2). We describe each node v_i using its object identification number ID_{ei}, the risk and reward of the node (Equation (3)), and each transition e_j using the nodes it connects in order v_p, v_k, its risk, reward, obstruction, precision, motivation, and distance (Equation (4)).

$$V = \{v_1, v_2, v_3, ..., vi\} \tag{1}$$

$$E = \{e_1, e_2, e_3, ..., e_j\} \tag{2}$$

$$v_i = [ID_{vi}, rk_{vi}, lrw_{vi}, brw_{vi}] \tag{3}$$

$$e_j = [e_{v_p,v_k}, rk_{ej}, lrw_{ej}, brw_{ej}, o_{ej}, p_{ej}, lm_{ej}, bm_{ej}, d_{ej}] \tag{4}$$

The evaluation process of a game is exemplified in Figure 3. In it, the nodes or platforms are shown in green, the markers in blue, and with black lines, the transitions that allow the player to go through the level from the starting point to the final point are illustrated. The red circles mark the enemies that are used for risk calculation and the yellow circles mark the bonus used in the reward metrics.

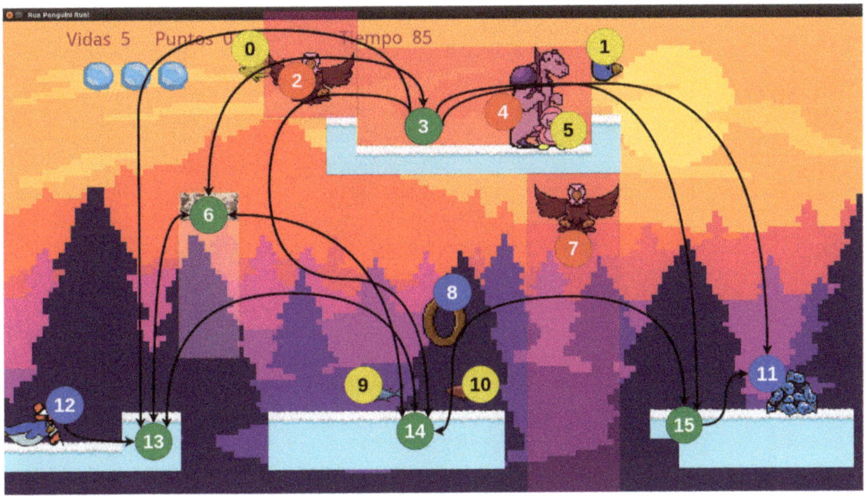

Figure 3. Node annotation of an example level.

The main objective of the metrics was to characterize the experience of the player while traveling in the game, which motivated us to analyze the route that the player would travel. However, a set of different routes that allow the player to travel from the start point to the endpoint can be found in the graph that describes a level. Each route is defined by a set of nodes and transitions and can represent a different experience for the player. Some routes may be more interesting for the players than others, and some routes may be more useful for the analysis. To decide which routes and metrics are more important for describing the player experience, we analyze all critical routes and weigh their importance. We define a critical route as an optimal route that shows the maximum or minimum values for one of the eight proposed measures (risk, level reward, bonus reward, obstruction, precision, level motivation, bonus motivation, and distance). To find the critical routes, we used the Dijkstra algorithm [53].

The implemented node selection method (relaxation function) in Dijkstra's algorithm is strictly "<" to find the paths of minimum sum, and ">" to find the paths of the maximum sum; therefore, to discern between two nodes that can form a critical path, the algorithm always keeps the first one it found.

When executing Dijkstra's algorithm, all possible critical paths in the level graph can be calculated; that is, those that have a maximum sum and a minimum sum in the measured characteristic, for example, one of these paths would be the "maximum risk" path, the one whose sum of risk, from the initial node to the end of the level, is the highest possible. However, it is necessary to select the route or routes that allow the game experiences to be identified. For our case study, the player must always travel interacting with all level markers (doors (rings) and goal); therefore, the route of interest is the one in which these markers are always present. Therefore, the "maximum level reward" route was selected, as it is the route in which the maximum sum—due to the points received for reaching the markers—is accumulated, i.e., they are always present in it.

4. Feature Selection

Once the critical path is established, it is necessary to define which measurements will be taken from each characteristic and which will be useful to identify the gaming experiences. For this purpose, a set of three example levels was created for each of the predefined game experiences, resulting in nine example levels. These levels were evaluated against the eight metrics and the path of interest (maximum level reward) was calculated. For each metric, three types of data were stored: the maximum value, the minimum value, and the rhythm (the latter is a measure of the number of peaks and valleys of the curve drawn by the measured values of said characteristics along the evaluated route), giving a total of 24 measures on the critical path, from each level. Then, a file containing the measurements (pace, maximum, and minimum value) taken from each level was formed, adding a class label, corresponding to the type of game experience to which each level belonged. Based on this dataset, the extra-trees classifier algorithm was used as a dimension reduction method, which in this case serves to determine the most convenient measurements of the 24 taken, to be used to identify game experiences. The extra-trees classifier is a classification algorithm that generates a set of decision trees, where each tree is built from the analysis of a random sample of the original dataset. Each tree calculates the Gini index of each characteristic to determine how much of a relationship they have with the class label; the lower the value of this index, the greater its relationship with the class label. The algorithm averages and normalizes the Gini indices of each generated tree, giving, as a result, the Gini importances, for which, the higher its value, the more relationship a characteristic has with the class label; the sum of all the importances is equal to 1. This measure is used in the extra-trees classifier algorithm as the decision-making mechanism.

To identify the measures of interest, this algorithm was run to learn how to classify the levels of the aforementioned file, the one that contains the measures of the critical path. For our case, the classification task was not relevant, but the importance of each characteristic, which were stored in a new file.

Finally, the top five values of importance were chosen to identify the gaming experiences, which are: rhythm of level reward, rhythm of level motivation, rhythm of risk, maximum value of bonus motivation, and rhythm of level reward. Additionally, it was decided to add the measure of "maximum distance value" since it is considered to be related to the visual harmony of the elements of the level, although its calculated importance was not high.

5. The IORand Algorithm

The PCG algorithm proposed in this work is based on the combination of two different approaches, first, the search-based approach, with the use of the deep Q-network (DQN) reinforcement learning (RL) algorithm, and second, by the random generators approach, with the semi-random content generation algorithm. The flow of the IORand algorithm is shown in Figure 4.

Our algorithm starts at block 1 by initializing the step counter (n) and generating an initial state (s), i.e., a slot to create a "Pingu run" level. This level is sent to the agent in block 2, as the current state of the environment in its first step. The agent decides which semi-random generator will be executed and with what type of platform (one of those defined in Figure 1b). This pair, type of platform, and generator, make up the action (a) of the agent. Depending on the semi-random generator that was chosen, one of the blocks from 3 to 7 is executed, with the type of platform specified. Each one of these blocks (3–7) is a semi-random content generation algorithm, which changes the environment according to an action that a human designer would do (create, insert, move, change, or remove platforms). After the selected generator calculates its effect on the environment, it is implemented in the level slice at block 8. At this point, one step of the algorithm is complete, so the counter is incremented. If "n" steps have been executed, where "n" is defined by the user, then the algorithm terminates and the resulting level slice (s) from performing the agent action is stored on a file (in JSON format) in block 10, otherwise this state is sent to block 9, where a calculation of the reward is made, to indicate to the agent how good his/her action was to achieve the desired gaming experience. Block 9 sends the reward (r) and the current state of the environment (s), which is also the resulting state (s') of its previous action. Both the reward (r) and the resulting state (s') are used only during the training phase; once the agent is trained, it is only necessary to send the current state of the environment (s).

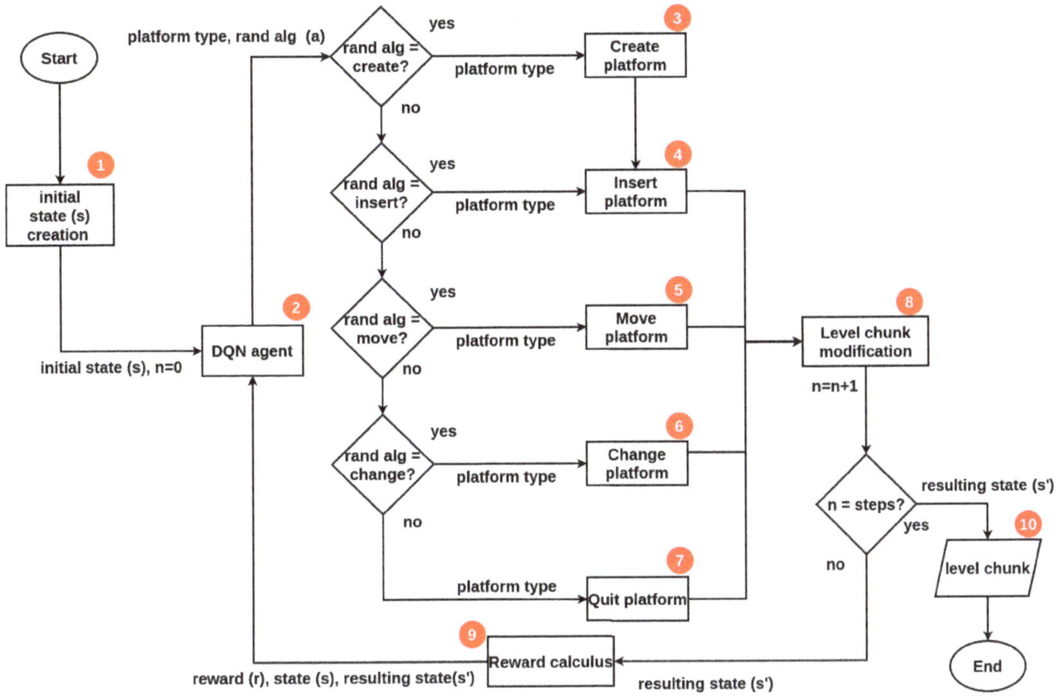

Figure 4. IORand flow chart diagram.

The idea of this hybrid algorithm is that the part based on reinforcement learning guides the content generation, such as an orchestrator, which seeks to maximize the value of an objective function, which in this case should reflect that the generated content provides a certain gaming experience to the player. Meanwhile, the part based on random generation provides a mechanism to increase the diversity of the results generated thanks to the RL algorithm.

Every RL algorithm is mainly made up of three parts—the agent, the environment, and the reward. The agent is the "intelligent" part of the algorithm; it is capable of interacting with the environment, modifying it depending on the actions it decides to execute. The goal of the agent is to learn a certain behavior in its environment. To do this, the environment feeds it back with information on task completion and the state of the environment itself. This information must be processed by the agent to adjust its decision-making process (policy), so that this adjustment serves to the decisions one makes, to help one accomplish the objective task. The environment is linked to the specific problem or task that the agent is expected to learn to solve; in our case, the environment is a space where "Pingu run" level slices can be built. The reward is a number that represents how well the agent performs in solving the specific task; this reward is calculated by analyzing the fulfillment of the objective task derived from the change in the environment due to the agent's actions in it. In our case, the reward should serve to indicate that a certain gaming experience is provided in the level slice of the environment.

Next, each part of the proposed algorithm is explained in greater depth. First, the design of the environment is presented. Next, the reward function used during training to identify gaming experiences is explained. Finally, the agent is introduced, explaining each of the DQN components in our implementation.

5.1. Environment

The environment is that in which the agent interacts and that contains the problem or task to be solved in its structure. The agent modifies the environment through actions causing a change in the state of the environment. To make its decisions, the agent receives an observation of the current state of the environment, which can be a complete or partial representation of the said state, depending on the type of training or learning that is sought in the agent.

Defining the environment for our case study means defining its states and observations. Since the problem to be solved is the generation of a space in two dimensions (level slice), the state and the observations are a matrix representation of the said space, in which each square of the matrix corresponds to an element of the game used to build the level. For the purposes of this project, these matrices have 16 rows and 29 columns, which are equivalent to the number of elements that can be seen on the screen.

These arrays can contain only the alphanumeric representations of the game elements to build the "Pingu run" levels.

For practical purposes, the initial state is one of these arrays built under the following conditions: the initial and final points are predefined and cannot be modified by the agent. A predefined number of floor platforms are placed in the space between the start and end platforms.

5.2. Reward

Definition of the reward. The reward function is the indicator of how close the agent is to meeting the goal, which is defined from an evaluation of the state to which the agent brings the environment due to its actions. Usually, the higher the reward, the better the agent's performance.

For our case study, the reward must evaluate the performance of the agent to generate a level that makes the player feel a specific gaming experience.

To achieve this, we propose the identification of game experiences based on the type of interaction that the player will have with the level along the critical path. For this, it is proposed to evaluate the characteristics of the level, measured from the use of the designed metrics, evaluating these measurements along the critical path of interest. The idea of this evaluation is to ensure that the rhythm and the range of values (maximum and minimum value) of the characteristics are close to those of a prototype level of a gaming experience. In other words, our evaluation of compliance with the gaming experience is based on the comparison of the measured/evaluated level with an ideal/prototype level, where the comparison is made through the rhythm and the range of values of the characteristics of interest.

Therefore, our reward function is divided into two parts, the "pacing accuracy" and the "value range accuracy"; they are defined as "accuracy" because the closer the measured values are to the target values. The greater the certainty, the evaluated level will provide the desired gaming experience. For practical purposes, we set the reward values in the range of [0, 1] , where '0' is the worst performance and '1' is the ideal performance.

Rhythm accuracy. For this part of the reward, a bell-shaped function was used, whose value was calculated based on Equation (5). For this function, 'x_i' is the rhythm of the i-th feature, 'σ' is the desired variance of the rhythm, which serves to widen or reduce the width of the bell, making the evaluation of the rhythm more relaxed (wider) or tighter (smaller), 'μ' is the target beat, indicating that the bell will have its center and, therefore, maximum value, at this point.

$$f(x_i) = \frac{1}{\mu\sqrt{2\pi}} e^{\frac{1}{2}\cdot(\frac{x-\mu}{\sigma})^2} \tag{5}$$

This function is normalized to fit its values to the range [0, 1] . Therefore, the first accuracy function for the reward is shown in Equation (6).

$$a_r = \frac{f(x_i)}{f(\mu)} \quad (6)$$

Value range accuracy. For this portion of the reward, the goal is to assess how close the range of measured values is to the range of target values. For this purpose, two different comparisons between the ranges are proposed. To define how equal two ranges are, it is proposed to compare the amplitude of the range and its center, the closer two ranges are in terms of their centers and their amplitudes, the more similar they are. Let γ be the maximum measured value, δ be the minimum measured value, λ be the maximum target value, τ be the minimum target value; then the first of these comparisons is shown in Equation (7), the amplitude similarity (S_a), which is given by 1 minus the normalized distance between the amplitudes.

$$S_a = 1 - \frac{|(\gamma - \delta) - (\lambda - \tau)|}{max(\gamma, \lambda)} \quad (7)$$

The second comparison is shown in Equation (8), the similarity of centers (S_c), which is given by 1 minus the normalized distance between centers.

$$S_c = 1 - \frac{[(\frac{1}{2}(\gamma - \delta) + \delta) - (\frac{1}{2}(\lambda - \tau) + \tau)]}{max(\gamma, \lambda)} \quad (8)$$

These equations are valid if and only if $max(\gamma, \lambda) \neq 0$, otherwise the Equations (9) and (10) should be used

$$S_a = 1 - |(\gamma - \delta) - (\lambda - \tau)| \quad (9)$$

$$S_c = 1 - [(\frac{1}{2}(\gamma - \delta) + \delta) - (\frac{1}{2}(\lambda - \tau) + \tau)] \quad (10)$$

Finally, the range of value accuracies is shown in Equation (11)

$$a_v = \frac{1}{2}S_a + \frac{1}{2}S_c \quad (11)$$

In addition to these two accuracies, a penalty factor p was added to the reward function, which acquires a value of 0.5 when the generated level is not playable (that is, the end point cannot be reached from the starting point), and a value of 1 when it is. Then, the reward function R for a feature is a weighted sum of both accuracies a_r and a_v, attenuated by the penalty factor p, as shown in Equation (12), where $\alpha + \beta = 1$. The weight values α and β allow one to define which accuracy is more important—that of the rhythm or that of the range of values.

$$R = p(\alpha \cdot a_r + \beta \cdot a_v) \quad (12)$$

Given a game experience, a reward must be calculated according to Equation (12) for each characteristic of interest and finally make a weighted sum of each of them. The reward of a gaming experience is defined in Equation (13), where w_i is the weight of the i-th feature and R_i is the reward function of that same feature, where $sum_i w_i = 1$.

$$R_{GX} = \sum_i w_i \cdot R_i \quad (13)$$

Prototype levels. To calculate the reward value of each of the defined game experiences (simple, jumps, obstacles), it is necessary to set their target values (μ, σ, λ, τ, α, β and w). The goal is that, when adjusting the reward, Equation (13), using the target values of a specific gaming experience, and this equation is used to evaluate a level of that same gaming experience, a value close to 1 is obtained, indicating that the said level meets the desired gaming experience, and also that when evaluating a level that provides another gaming experience, the calculated value is close to 0, indicating that the level does not meet the desired gaming experience. Therefore, it is necessary to find an adequate combination of target values for each game experience, in order to maximize the reward at the levels of that same experience and minimize the reward at the levels of the rest of the experiences.

To determine the target values for each game experience, we used our three example levels for each game experience defined (simple, obstacle, jump). We then applied our evaluation process on each of them, obtaining the graphs for each example level (see example in Figure 3). From those graphs, we analyzed the rhythm and value range through the route of interest (maximum level reward), of the selected important features (level reward, level motivation, risk, bonus motivation, bonus reward, and distance). After proposing various sets of target values, those shown in Table 2 were found. In general, the values of rhythm (μ), variation of rhythm (σ), maximum value (λ), and minimum value (τ) were defined based on the most frequent value or in the average of the measurements observed in the example levels. On the other hand, the weightings of the rhythm accuracy (α), value range (β), and weight (w) were proposed when considering the Gini importance of each measure. If one measure has a greater Gini importance than another, then its weight (w) should be greater in the reward function. To do so, we summed the Gini importances of each selected feature and calculated its proportion in that sum to obtain its weight (w). If the rhythm of a feature has a greater Gini importance than its maximum or minimum value, then the rhythm accuracy (α) must be greater than the value range accuracy weight (β), and vice versa. For this purpose, we experimented with various (α) and (β) values, considering the Gini importances and looking for a better reward evaluation for the example levels.

It should be clarified that, during the search for these reward functions, it was identified that the *bonus motivation*, in the case of the *simple* game experience, caused a misidentification of the game experiences, reducing the reward of the level corresponding to the desired experience; therefore, for that experience, the "bonus motivation" feature was removed from the reward function.

Based on these target values for the reward functions, the crafted levels obtained the evaluations shown in Table 3, which shows that each level receives a reward close to 1 when evaluated with the reward function corresponding to the game experience and an evaluation close to 0 for the functions corresponding to the other experiences, thus fulfilling the objective set for the reward function and the respective objective values for each game experience.

Table 2. Objective values for each metric in each game experience.

Game Experience	Metric	μ	σ	λ	τ	α	β	w
Simple	level reward	2	3	916	0	0.95	0.05	0.29
	level motivation	0	3	1406	0	0.75	0.25	0.26
	risk	2	0.001	491	0	0.65	0.35	0.22
	bonus reward	2	2	265	0	0.85	0.15	0.15
	distance	4	4	5	1	0	1	0.08
Obstacle	level reward	4	3	800	0	0.95	0.01	0.25
	level motivation	2	3	1384	−70	0.75	0.25	0.22
	risk	6	4	243	0	0.65	0.35	0.19
	bonus motivation	2	0.001	160	−184	0.35	0.65	0.13
	bonus reward	4	4	90	0	0.85	0.15	0.11
	distance	4	4	11	1	0	1	0.1

Table 2. *Cont.*

Game Experience	Metric	μ	σ	λ	τ	α	β	w
Jump	level reward	1	0.001	1000	0	0.95	0.05	0.25
	level motivation	1	0.001	1805	−56	0.75	0.25	0.22
	risk	5	2	155	0	0.65	0.35	0.19
	bonus motivation	1	2	530	−68	0.35	0.65	0.13
	bonus reward	5	3	350	0	0.85	0.15	0.11
	distance	4	4	8	1	0	1	0.1

Table 3. Calculated rewards for example levels using objective values from Table 2.

Level	Reward Function		
	Obstacle	Simple	Jump
obstacle 1	**0.748515**	0.216647	0.169015
obstacle 2	**0.741509**	0.33168	0.192525
obstacle 3	**0.695833**	0.253077	0.285230
Simple 1	0.083789	**0.843283**	0.08732
Simple 2	0.197562	**0.765775**	0.178005
Simple 3	0.204252	**0.829947**	0.183384
Jump 1	0.189115	0.255675	**0.975981**
Jump 2	0.158059	0.268668	**0.702896**
Jump 3	0.181520	0.261360	**0.970456**

5.3. Agent

The agent is the part of the RL algorithm that is responsible for solving the problem. To implement this module, there are a variety of algorithms that can be used; in this project, the deep Q-network (DQN) algorithm was chosen. In short, this algorithm uses an artificial neural network to regress the expected reward; that is, this ANN estimates how convenient it is to execute each possible action in a given state.

The DQN algorithm works as follows:

- The agent receives an observation of the environment. In our implementation, this translates to a 16×29 matrix with symbols representing game items or empty spaces in the level.
- If necessary, the observation is preprocessed to be propagated by the ANN; in our implementation, the symbols of the matrix are already in numerical format, so no preprocessing was necessary.
- This numerical matrix is propagated through the ANN and a vector whose components have values between 0 and 1, each associated with a possible action, is obtained as a network output. The closer this value is to 1, the greater the reward that the ANN has estimated for said action.
- The ϵ-greedy policy is applied, in which the agent chooses an action at random with probability ϵ and with the complementary probability, 1-ϵ, chooses the most convenient action according to the ANN estimation.
- The chosen action (a) is executed on the observed state of the environment (s), causing a transition to the resulting state (s') and the calculation of a reward (r).
- The agent receives, from the environment, the observation of the resulting state and the reward of its previous action.

- The agent stores the data of the previous transition, the observed state (s), the chosen action (a), the resulting state (s′), and the reward obtained (r).
- To adjust the weights and bias of the neural network, it is necessary to:
  - Store transitions (agent experience) in a replay memory.
  - Once there is a certain amount of data in memory, the transitions stored in it are randomly queried to train the agent's neural network. The adjustments of the weights and bias are made only in the neurons associated with the chosen action in each transition.
- With each training step, the value of ϵ is reduced by a percentage of its current value (our implementation multiplies it by 0.996).

To better understand the behavior of the agent, it is necessary to delve into the design of the neural network and the repetition memory used for its training, in addition to clearly defining the action space.

Replay memory. The replay memory acts as a training set for the agent's neural network based on the DQN algorithm. In our implementation, this memory was set to store 150,000 transitions. These transitions are stored in the form of a row, in which, if an additional element arrives that overflows the row, then the first stored element is eliminated and the new one is stored; preserving the most recent experience. This memory is randomly queried to see if the inputs to the neural network are uncorrelated.

Implemented artificial neural network architecture. For our implementation, a convolutional neural network (CNN) was used in conjunction with a vanilla network. The CNN fulfills the function of processing the level matrix, detecting spatial relationships in it and extracting features from the input data. The vanilla network fulfills the function of the regressor, processing the features extracted through the CNN and calculating from them, the expected reward for executing each of the actions on the observed state (network input). The architecture of the presented network is shown in Figure 5.

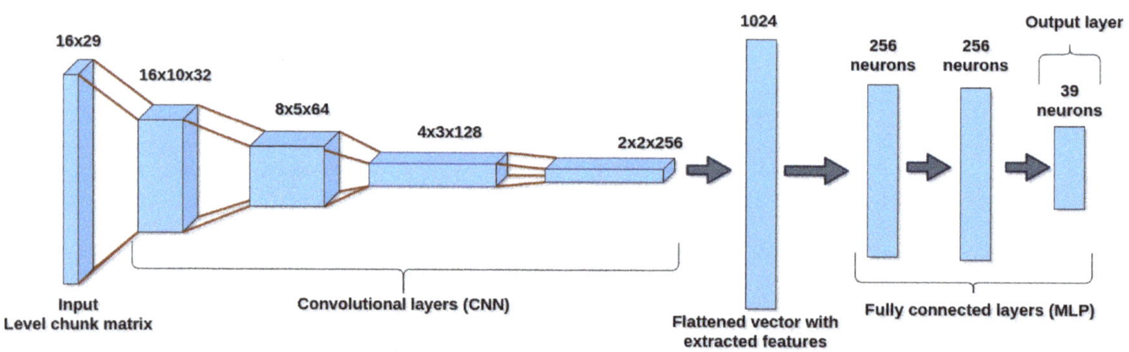

Figure 5. Deep neural network architecture.

A CNN was used because it was considered the best deep architecture option given the nature of the states of our environment, which contained a representation of the content displayed on the screen and, therefore, considered a type of a reduced image of the level. Therefore, a CNN was chosen, given its proven image processing capabilities. The output layer has 39 neurons, one for each possible action that the agent can perform. Next, it will be explained why the action space has that dimension.

Action space. The action space is everything that the agent can choose to do. The actions alter the environment causing a transition of states in it.

In our case, the agent can choose to run a semi-random content generator using a specific block type. In this way, the agent only chooses what modification will be made and on what type of content; however, it is not the agent who directly makes the modification.

The semi-random algorithms used seek to simulate actions that a human designer can perform on a level, for which the generators are proposed to: create and insert a block, insert a block, move a block, remove a block, change the type of a block, and change the address of a block.

In Table 4, the relationship between the type of point block and the generator algorithm that can be executed on it, is shown. This relationship between elements and generators to execute on them arise from an analysis of the actions that each generator implies. For example, the generator that "changes direction" modifies the label of the game element to indicate that its movement will be another; this action only makes sense for elements that can move in the game. By counting the number of 'X's in the table, the 39 possible actions are obtained, the product indicating—to each of the six semi-random generators—which of the 11 types of element it will use at a given moment.

Table 4. Available action vs. game element relationship.

Game Element	Semi-Random Content Generator					
	Create and Insert	Insert	Move	Quit	Change Type	Change Direction
Platform: static	X		X	X		
Enemy: lava	X		X	X		
Bonuses: score (except golden fish)		X	X	X	X	
Platform: mobile		X	X	X	X	X
Enemies: Bear and Troll		X	X	X	X	X
Enemy: Eagle		X	X	X		X
Bonus: sub-mission		X	X	X		
Marker: Mid-level (ring)		X	X	X		
Platform: bouncing		X	X	X		
Bonus: Score - golden fish		X	X	X		
Bonus: lives		X	X	X		

5.4. Semi-Random Content Generators

Each of the generators performs a specific action in the environment. Below is an explanation of the effect each generator produces.

Move. The agent detects platforms from the map array, then creates a list of platforms of the selected type. From this list, the generator randomly chooses an element and moves all of its blocks to an adjacent position. This action is only executed if there are enough empty spaces to place the selected element in the adjacent position.

Remove. The agent detects platforms from the map array, then creates a list of platforms of the selected type. From this list, the generator chooses an element at random and changes all of the blocks that compose it to empty spaces.

Change (type or address). The agent detects platforms from the map matrix, then creates a list of platforms of the selected type. Of these platforms, the generating algorithm chooses one at random. Depending on the type of platform, there is a list of possible values that it can take. The algorithm selects a new type at random, making sure it is different from the old one.

Insert. The algorithm determines the space required to place the selected platform on the level slice, so to place the platform, it picks random coordinates in the space between the starting and ending platforms, until it finds one that is surrounded by as much empty space as required; space was determined, inserting the label(s) that make up the platform.

Create and insert. In order to create platforms, the generator must be told the dimensions of the platform to be created, as well as the number of blocks that must be used to build it. This information is provided by the user before executing the PCG algorithm. To create the platforms, the algorithm creates an array with empty spaces of the specified dimensions and selects a random coordinate within that array. In this coordinate, the generating algorithm inserts the label corresponding to the type of platform selected. Then, in positions adjacent to the platform being created, adds the label again. This process is repeated until adding as many labels as the number of blocks defined or until the matrix is filled. Once the platform is created, the previously defined "insert" method is executed.

All generators are subject to a maximum number of platform restrictions. This limit is set by the user before running the algorithm.

6. Metrics to Evaluate PCG Algorithm Performance

To evaluate the performance of the proposed algorithm, the following metrics are used for three of the criteria mentioned in Table 1, playability, novelty, and iterativity, which will allow us to compare our proposal with the state-of-the-art, and add a measure of game experience and one of effort. Although the rest of the criteria (generation process, determinism, controllability) must also be evaluated, they acquire a value according to the design of the algorithm itself. The criteria of this evaluation and the way to measure them are as follows:

Playability. This measure indicates if the generated level is playable or not; that is, if the player will be able to finish the game. In this case, it means that a level is playable if the player can obtain the penguin from the start point to the end point of the level. It is a binary metric; this characteristic is fulfilled or not, so the possible values are {0, 1}.

Game experience. It refers to the fulfillment of the objective task; that is, the levels provide the desired gaming experiences. To measure this feature, we use the reward evaluation obtained at each generated level. Our reward returns values in the range [0, 1].

Novelty. This measurement requires the calculation of the degree of difference between two slices A and B, which is obtained by calculating the number of operations necessary to transform slice A into slice B. This degree of difference is normalized with respect to the maximum number of operations to transform one string into another; therefore, their values are in the range [0, 1]. The novelty of a slice is the average degree of difference between it and the rest of the slices produced by the algorithm in one run.

Effort. For this measure, the degree of difference is also used. The effort of a slice is given by the degree of difference between it and the initial slice from which it was produced, indicating the degree of changes necessary to reach the final level from the initial point of content generation.

Iterativity. It relates to the number of iterations needed by the algorithm to produce content.

To measure iterativeness, we establish four ranges given a different number of actions that an agent can perform to produce a level. For the experiments, the iterativity ranges were established, associated with the number of actions according to the following relationship:

- High—500 actions (steps);
- Moderate—300 shares;
- Fair—150 shares;
- Low—75 actions.

To adjust the rest of the measurements to these four ranges, the values associated with each of them are shown in Table 5. With the exception of gaming experience, the rest of the measures are equally distributed (in two, 0 and 1, or four ranges of the same density, between 0 and 1). An exception is made for that particular metric because the reward value equal to 0.72 was defined as the target gaming experience fulfillment threshold, taking into consideration the minimum reward values earned by the crafted levels in the set example data in Table 3.

Table 5. Value ranges for each PCG metric.

PCG Metric	Ranges			
	High	**Moderate**	**Fair**	**Low**
Playability	1	-	-	0
Game experience	[1, 0.72]	(0.72, 0.45]	(0.45, 0.225]	[0.225, 0]
Novelty	[1, 0.75]	(0.75, 0.5]	(0.5, 0.25]	(0.25, 0]
Effort	[1, 0.75]	(0.75, 0.5]	(0.5, 0.25]	(0.25, 0]

7. Experiments and Results

Each agent was trained for 2000 epochs, with 300 steps per epoch. After every five steps, the agent executes a learning step, adjusting its ANN, taking 64 random transition examples from its replay memory. As input, each agent received their target values, as shown in Table 2, for the reward calculation. On every training epoch, each agent started with its own initial state of the environment, and it was the same through the 2000 epochs. Once the three agents (one for each game experience) were trained, experiments were carried out with each agent for each of the iterative ranges presented. In each experiment, the agents generated 50 level slices and each was evaluated with the metrics of gaming experience, playability, novelty, and effort.

For this article, the best experimental results are presented, which were obtained with high iteration, i.e., with 500 actions. Compared to those presented, the rest of the experiments did not show any relevant improvement or data of interest.

In Table 6, the performance of each agent in the experiment of 500 actions is presented. This table shows the percentage of level slices whose evaluations fall into each of the ranges presented by metric. From this table, it stands out that the three agents produce a higher percentage of playable levels than non-playable ones. Moreover, regarding the achievement of the target gaming experience, most of the slices produced are in the moderate and low ranges, between 60% and 70% of the levels are in those ranges. For the novelty value, the agents maintained the highest percentages of the levels in a moderate range. In addition, it should be noted that the 'obstacle' agent had an outstanding performance since 100% of the slices it generated were in the range. moderate novelty. Finally, we can see that effort seems to be related to novelty, as their results are quite similar to the naked eye, suggesting that a slice with high novelty also requires high effort to produce.

In addition to the table, we present some graphs that allow us to evaluate in greater depth the performance of the agents. First, the graphs of the degree of performance by the PCG metric for each of the agents are presented. These plots are presented in Figure 6, which allows us to visually assess the distribution of novelty, effort, and gaming experience ratings. These graphs show that the agent who manages to obtain higher reward values is the 'jump' agent (in Figure 6a), and, in general, this is the agent that presents better behavior for this experiment, since, even when it does not reach the novelty values of the 'obstacle' agent, in Figure 6b, the reward values obtained by this agent exceed those obtained by the other two, in addition to maintaining the minimum values in this same metric above the 'simple' and 'obstacle' agents. Of these three agents, the least outstanding is the 'simple', from Figure 6b, since both the agent 'jump' and 'obstacle' obtain better results in both novelty and reward.

Table 6. Experimental results: 50 levels produced with High iterativity.

Agent	PCG Metric	Ranges			
		High	Moderate	Fair	Low
Simple	Playability	**64%**	-	-	36%
	Game experience	4%	**38%**	20%	**38%**
	Novelty	0%	**56%**	44%	0%
	Effort	0%	**66%**	34%	0%
Jump	Playability	**70%**	-	-	30%
	Game experience	14%	20%	24%	**42%**
	Novelty	0%	**60%**	40%	0%
	Effort	0%	**74%**	26%	0%
Obstacles	Playability	**76%**	-	-	24%
	Game experience	4%	**40%**	28%	28%
	Novelty	0%	**100%**	0%	0%
	Effort	4%	**90%**	6%	0%

In the plotted scatter graphics, each point represents a level slice produced by an agent in a reward vs. novelty space, allowing us to analyze the relationship between these two characteristics in the results of the proposed PCG algorithm. Four types of level slices are presented in these graphs. The first are the 'suitable levels' (adequate levels), which are slices with high evaluations in gaming experience, moderate novelty, and are playable; these levels are the ideal solutions of the PCG algorithm by successfully solving the target tasks (high reward and playable) and also doing it creatively (moderate novelty). The second ones are the *interesting levels* , which are playable slices with moderate novelty, although their rewards are not this in any particular range, these slices are interesting because they offer a real and applicable solution to the problem of procedural generation of video game levels, by producing creative slices that are indeed playable, even though they do not make it feel like the desired gaming experience. The third are the *plain levels*, slices that are playable, but do not meet any other attribute of interest; these slices represent good solutions for the objective problem; however, their low novelty and non-compliance with the gaming experience make them much less relevant. Finally, the *non-playable levels*, are those slices in which the player cannot reach the finish point; these are completely undesirable slices and cannot be used in real applications. In addition, the reward threshold is shown in these graphs with a dotted line (0.72) to facilitate the identification of the levels that obtain high rewards.

These plots are presented in Figures 7–9. These graphic reports allow interpreting the results from another perspective. The first interesting aspect is the absence of flat levels in the 'obstacle' agent in Figure 8, which, although it does produce 'non-playable levels', these are less compared to the two other agents. Taking a look at the 'jump' agent in Figure 7, we see a higher number of slices approaching or crossing the reward threshold, suggesting that this is the best agent to produce levels that deliver the desired game experience. Of the agents 'simple', in Figure 9, and 'jump' it can be stated that a moderate degree of novelty is not related to a high degree of reward, since 'interesting' levels are shown both at the upper and lower limits of the reward range. In contrast, a high reward always implies gameplay and, at least half the time, a moderate degree of novelty. The data and code necessary to reply these results are available at [54].

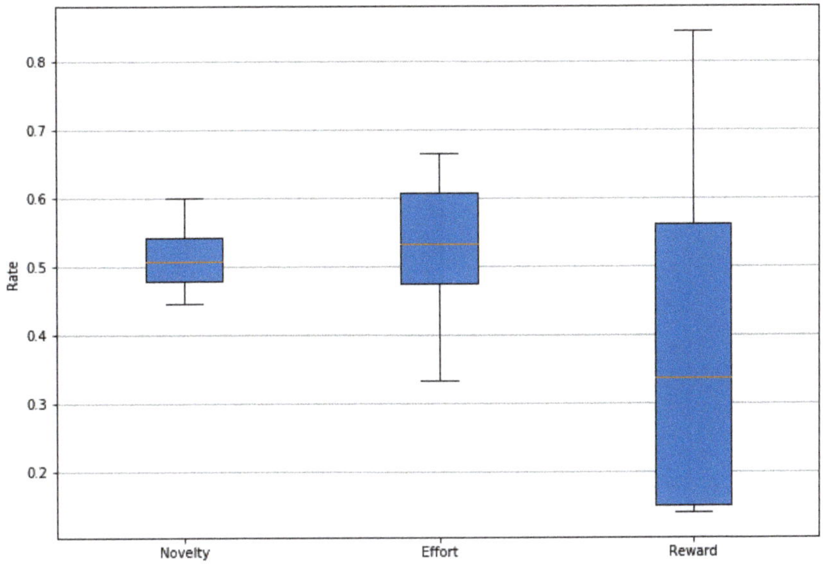

(**a**)

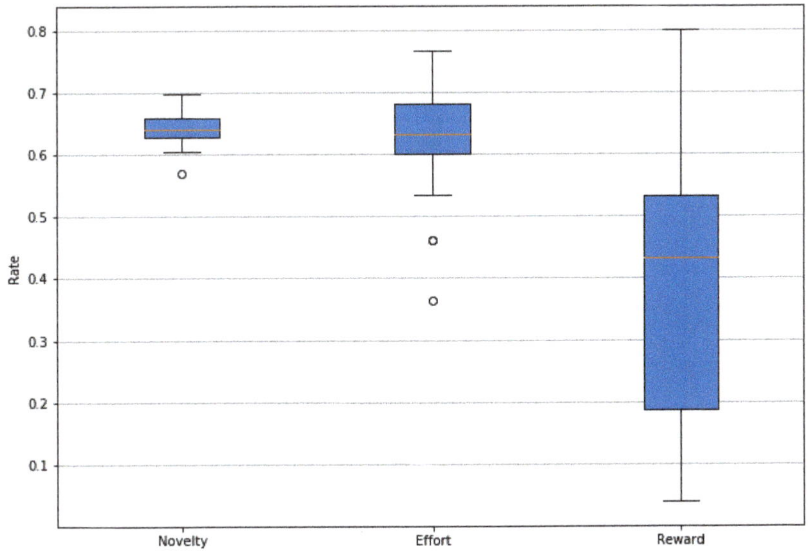

(**b**)

Figure 6. *Cont.*

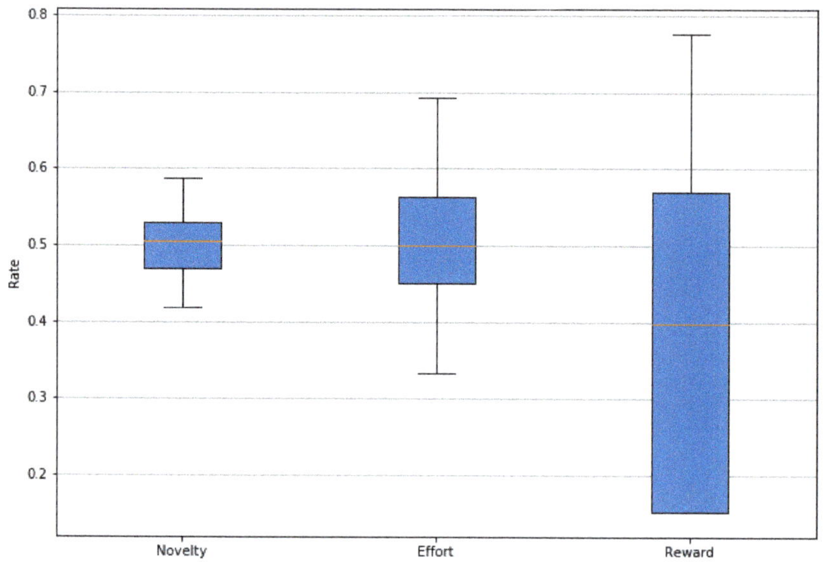

(c)

Figure 6. Trained agents' performance rate. (**a**) Jump agent. (**b**) Obstacle agent. (**c**) Simple experience.

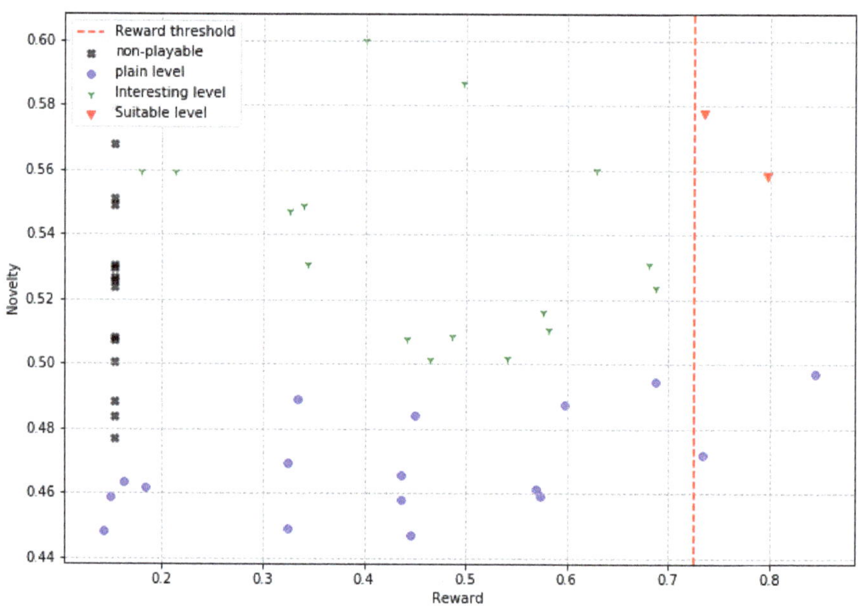

Figure 7. Jump levels produced.

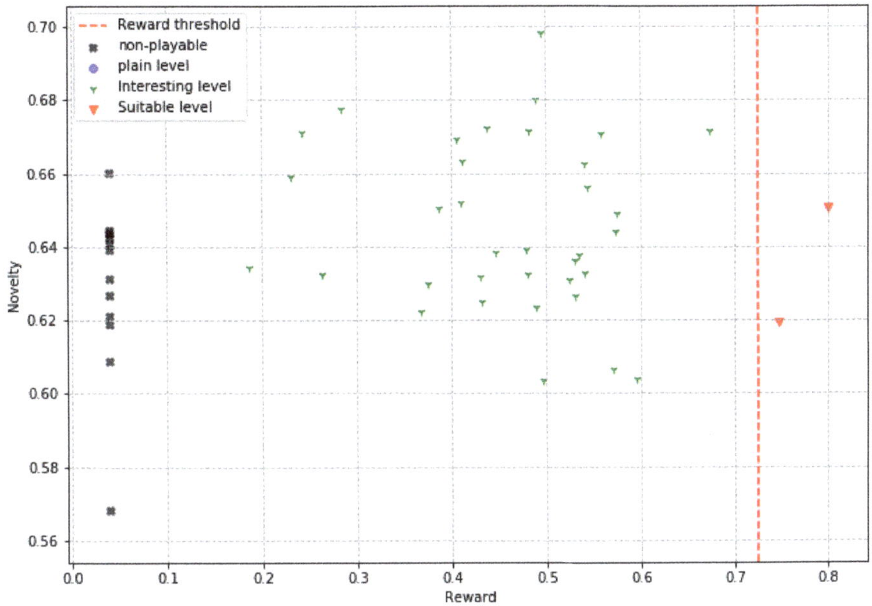

Figure 8. Obstacle levels produced.

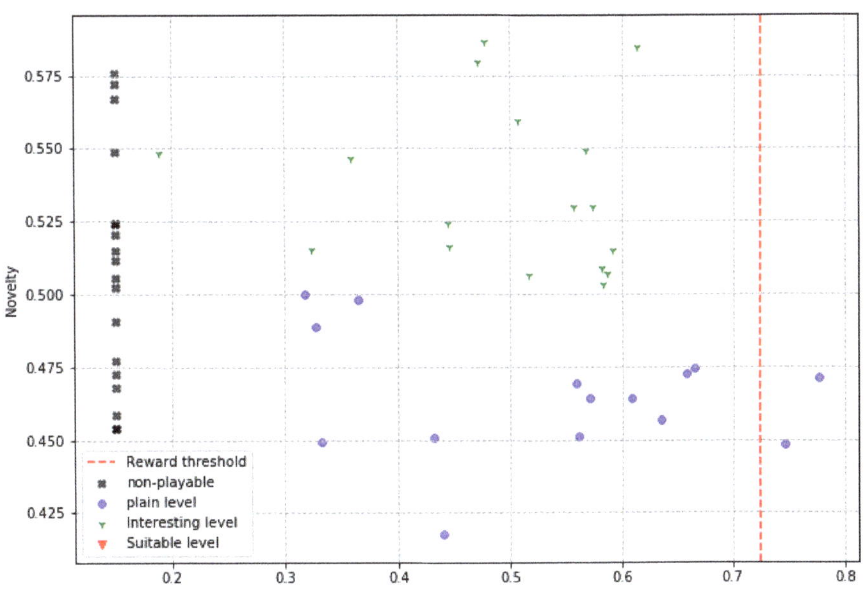

Figure 9. Simple levels produced.

8. Discussion

To compare our algorithm with the rest of the proposals of the state-of-the-art, Table 7 is presented. In it, the generation process of our algorithm is first reported, which is a hybrid based on the combination of the search-based and generate and test approaches. The degree of determinism of the algorithm is also mentioned, a characteristic that is inversely proportional to the degree of novelty, which, given the experimental results, acquires a value of "moderate"; therefore, the degree of determinism, being its opposite, acquires a value of "fair". On the other hand, the iterativity is obtained from the range in which the reported experiments are found; that is, our algorithm obtains a "high" value in iterativity, since it is the range in which its behavior is adequate. Regarding gameplay, our algorithm is in the "high" range since a higher percentage of levels produced are playable. That means that in contrast to the search based proposals, IORand generates content with a higher novelty measure (moderate instead of fair) but keeps the controllability in a moderate range as the rest of those same proposals. In comparison with the generate and test proposals, IORand manages to generate content with a high playability measure. Moreover, in contrast with the constructive proposals, IORand keeps a good relationship between the controllability, novelty, and playability, as there are no other algorithms that maintain moderate controllability, moderate novelty, and high playability. As a whole, this makes IORand a solution for the experience-driven PCG problem whose behavior is definitely different from the state-of-the-art proposals.

Table 7. IORand performance.

Technique	Generation Process	Determinism	Controllability	Iterativity	Novelty	Playability
IORand	Hybrid (Search-based + Generate and test)	Fair	Moderate	High	Moderate	High

The experimental results give us enough evidence to conclude that our PCG algorithm, together with the metrics and the designed reward function, are able to solve the generation of level slices that provide a desired gaming experience. This is affirmed since, once the agents have been trained, they are capable of generating novel, playable levels that provide specific gaming experiences. The usefulness of the proposed metrics, the reward function, and the hybrid algorithm to solve the game experience-oriented level generation problem have been shown, representing a contribution in the PCG area. Finally, IORand is a PCG algorithm that can be used in many kinds of problems and games. As it works by using building blocks guided by a reward function that indicates how good or bad its results are, it needs a new reward and new building blocks given any new problem. IORand is the result of studying experience-driven PCG solutions, but it can be seen as a generative design tool, more than just a PCG one. This algorithm creates content based on an optimization problem; it maximizes the similarity to our description of game experience. However, it could solve different problems given a specification on the available blocks to build a solution and a quantitative measure of its performance or objective as reward.

9. Conclusions

In this work, we presented a new hybrid PCG algorithm capable of solving the problem of a procedural generation of video game levels, focused on providing desired gaming experiences. In turn, an evaluation method based on a critical path analysis is presented in a graph (annotated with the designed metrics), to identify game experiences based on expected interactions with the game elements present in said paths.

By combining the hybrid algorithm of PCG with the evaluation method, we obtained agents capable of consistently solving the task of generating level slices that provide a certain gaming experience, while the slices are playable, and, additionally, it was possible to improve the degree of novelty for PCG algorithms based on RL reported in the state-of-the-art, obtaining results in a moderate range, while the state-of-the-art is reported with an acceptable range.

These results demonstrate that the hybridization of PCG algorithms is a feasible solution to find more and better answers to the unanswered questions in this area.

In future works of this proposal, there are other challenges that we will deal with. Here we list some of them; generalization capability of our agent; how much to change for other games or game experiences? Is there any test or fact to prove it? Finally, concerning the proposed similarity measure for game experiences, how do we use it for other games? Could it be parametrized correctly with other game metrics?

Author Contributions: Conceptualization, methodology, M.A.M.-A., J.A.T.-L., C.A.D. and H.C.; investigation and resources, M.A.M.-A., H.C., C.A.D. and J.A.T.-L.; software, visualization, and data curation, J.A.T.-L.; validation H.C. and M.A.M.-A.; formal analysis, M.A.M.-A.; writing—original draft preparation, J.A.T.-L., C.A.D.; writing—review and editing, H.C. and M.A.M.-A.; supervision, project administration and funding acquisition, M.A.M.-A. All authors have read and agreed to the published version of the manuscript.

Funding: This work was possible thanks to the support from the Mexican government (CONACyT) and Instituto Politecnico Nacional through SIP-IPN research grants SIP-2083, SIP-20220553, SIP-20221064, and SIP 20220533; IPN-COFAA and IPN-EDI.

Institutional Review Board Statement: Not applicable.

Informed Consent Statement: Not applicable.

Data Availability Statement: The authors are committed to providing access to all of the necessary information so that readers can fully reproduce the results presented in this work. Used datasets are publicly available.

Conflicts of Interest: The authors declare no conflict of interest.

References

1. Togelius, J.; Kastbjerg, E.; Schedl, D.; Yannakakis, G.N. What is procedural content generation. *Mario Borderline. Pcgames* **2010**, *11*, 3:1–3:6. .
2. De Kegel, B.; Haahr, M. Procedural puzzle generation: A survey. *IEEE Trans. Games* **2019**, *12*, 21–40. [CrossRef]
3. Risi, S.; Togelius, J. Increasing generality in machine learning through procedural content generation. *Nat. Mach. Intell.* **2020**, *2*, 428–436. [CrossRef]
4. Shaker, N.; Togelius, J.; Nelson, M.J. *Procedural Content Generation in Games*; Springer: Berlin/Heidelberg, Germany, 2016.
5. Summerville, A.; Snodgrass, S.; Guzdial, M.; Holmgård, C.; Hoover, A.K.; Isaksen, A.; Nealen, A.; Togelius, J. Procedural content generation via machine learning (PCGML). *IEEE Trans. Games* **2018**, *10*, 257–270. [CrossRef]
6. Liu, J.; Snodgrass, S.; Khalifa, A.; Risi, S.; Yannakakis, G.N.; Togelius, J. Deep learning for procedural content generation. *Neural Comput. Appl.* **2021**, 33, 19–37. [CrossRef]
7. Togelius, J.; Yannakakis, G.N.; Stanley, K.O.; Browne, C. Search-based procedural content generation: A taxonomy and survey. *IEEE Trans. Comput. Intell. Games* **2011**, *3*, 172–186. [CrossRef]
8. Yannakakis, G.N.; Togelius, J. *Artificial Intelligence and Games*; Springer: Berlin/Heidelberg, Germany, 2018. Available online: gameaibook.org (accessed on 8 April 2022).
9. Alvarez, A.; Dahlskog, S.; Font, J.; Togelius, J. Empowering quality diversity in dungeon design with interactive constrained map-elites. In Proceedings of the 2019 IEEE Conference on Games (CoG), London, UK, 20–23 August 2019; pp. 1–8.
10. Ashlock, D.; Lee, C.; McGuinness, C. Search-based procedural generation of maze-like levels. *IEEE Trans. Comput. Intell. Games* **2011**, *3*, 260–273. [CrossRef]
11. Frade, M.; de Vega, F.F.; Cotta, C. Evolution of artificial terrains for video games based on obstacles edge length. In Proceedings of the IEEE Congress on Evolutionary Computation, Barcelona, Spain, 18–23 July 2010; pp. 1–8.
12. Gravina, D.; Khalifa, A.; Liapis, A.; Togelius, J.; Yannakakis, G.N. Procedural content generation through quality diversity. In Proceedings of the 2019 IEEE Conference on Games (CoG), London, UK, 20–23 August 2019; pp. 1–8.
13. Togelius, J.; Preuss, M.; Beume, N.; Wessing, S.; Hagelbäck, J.; Yannakakis, G.N.; Grappiolo, C. Controllable procedural map generation via multiobjective evolution. *Genet. Program. Evolvable Mach.* **2013**, *14*, 245–277. [CrossRef]
14. Valtchanov, V.; Brown, J.A. Evolving dungeon crawler levels with relative placement. In Proceedings of the Fifth International C* Conference on Computer Science and Software Engineering; Montreal, QC, Canada, 27 June 2012 ; pp. 27–35.
15. Earle, S. Using Fractal Neural Networks to Play SimCity 1 and Conway's Game of Life at Variable Scales. *arXiv* **2020**, arXiv:2002.03896.
16. Chen, Z.; Amato, C.; Nguyen, T.H.D.; Cooper, S.; Sun, Y.; El-Nasr, M.S. Q-deckrec: A fast deck recommendation system for collectible card games. In Proceedings of the 2018 IEEE conference on Computational Intelligence and Games (CIG), Reno, NV, USA, 14–17 August 2018; pp. 1–8.
17. Khalifa, A.; Bontrager, P.; Earle, S.; Togelius, J. Pcgrl: Procedural content generation via reinforcement learning. In Proceedings of the AAAI Conference on Artificial Intelligence and Interactive Digital Entertainment, Worchester, MA, USA, 19–23 October 2020; Volume 16, pp. 95–101.

18. Guzdial, M.; Liao, N.; Chen, J.; Chen, S.Y.; Shah, S.; Shah, V.; Reno, J.; Smith, G.; Riedl, M.O. Friend, collaborator, student, manager: How design of an ai-driven game level editor affects creators. In Proceedings of the 2019 CHI Conference on Human Factors in Computing Systems, Scotland, UK, 4–9 May 2019; pp. 1–13.
19. Delarosa, O.; Dong, H.; Ruan, M.; Khalifa, A.; Togelius, J. Mixed-Initiative Level Design with RL Brush. *arXiv* **2020**, arXiv:2008.02778.
20. Mott, J.; Nandi, S.; Zeller, L. Controllable and coherent level generation: A two-pronged approach. In Proceedings of the 6th Experimental AI in Games Workshop at AIIDE 2019, Honolulu, HI, USA, 27 January 2019.
21. Doran, J.; Parberry, I. A prototype quest generator based on a structural analysis of quests from four MMORPGs. In Proceedings of the 2nd International Workshop on Procedural Content Generation in Games, New York, NY, USA, 28 June 2011; pp. 1–8.
22. Dormans, J. Adventures in level design: Generating missions and spaces for action adventure games. In Proceedings of the 2010 Workshop on Procedural Content Generation in Games, Monterey, CA, USA, 18 June 2010; pp. 1–8.
23. Dormans, J. Level design as model transformation: A strategy for automated content generation. In Proceedings of the 2nd International Workshop on Procedural Content Generation in Games, New York, NY, USA, 28 June 2011; pp. 1–8.
24. Dormans, J.; Bakkes, S. Generating missions and spaces for adaptable play experiences. *IEEE Trans. Comput. Intell. Games* **2011**, *3*, 216–228. [CrossRef]
25. Johnson, L.; Yannakakis, G.N.; Togelius, J. Cellular automata for real-time generation of infinite cave levels. In Proceedings of the 2010 Workshop on Procedural Content Generation in Games, New York, NY, USA, 18 June 2010; pp. 1–4.
26. Karavolos, D.; Liapis, A.; Yannakakis, G.N. Pairing character classes in a deathmatch shooter game via a deep-learning surrogate model. In Proceedings of the 13th International Conference on the Foundations of Digital Games, Malmö, Sweden, 7–10 August 2018; pp. 1–10.
27. Guzdial, M.; Reno, J.; Chen, J.; Smith, G.; Riedl, M. Explainable PCGML via game design patterns. *arXiv* **2018**, arXiv:1809.09419.
28. Guzdial, M.; Liao, N.; Riedl, M. Co-creative level design via machine learning. *arXiv* **2018**, arXiv:1809.09420.
29. Summerville, A.; Guzdial, M.; Mateas, M.; Riedl, M. Learning player tailored content from observation: Platformer level generation from video traces using lstms. In Proceedings of the AAAI Conference on Artificial Intelligence and Interactive Digital Entertainment, Burlingame, CA, USA, 8–9 October 2016; Volume 12.
30. Karavolos, D.; Liapis, A.; Yannakakis, G. Learning the patterns of balance in a multi-player shooter game. In Proceedings of the 12th International Conference on the Foundations of Digital Games, Hyannis, MA, USA, 14–17 August 2017; pp. 1–10.
31. Kamal, K.R.; Uddin, Y.S. Parametrically controlled terrain generation. In Proceedings of the 5th International Conference on Computer Graphics and Interactive Techniques in Australia and Southeast Asia, Perth, Australia, 1–4 December 2007; pp. 17–23.
32. Tobin, J.; Fong, R.; Ray, A.; Schneider, J.; Zaremba, W.; Abbeel, P. Domain randomization for transferring deep neural networks from simulation to the real world. In Proceedings of the 2017 IEEE/RSJ international conference on intelligent robots and systems (IROS), Vancouver, BC, Canada, 24–28 September 2017; pp. 23–30.
33. Belhadj, F. Terrain modeling: A constrained fractal model. In Proceedings of the 5th international Conference on Computer Graphics, Virtual Reality, Visualisation and Interaction in Africa, Grahamstown, South Africa, 29–31 October 2007 ; pp. 197–204.
34. Guzdial, M.J.; Riedl, M.O. *Combinatorial Creativity for Procedural Content Generation via Machine Learning*; AAAI Workshops: New Orleans, LA, USA, 2018; pp. 557–564.
35. Petrovas, A.; Bausys, R. Procedural Video Game Scene Generation by Genetic and Neutrosophic WASPAS Algorithms. *Appl. Sci.* **2022**, *12*, 772. [CrossRef]
36. Di Liello, L.; Ardino, P.; Gobbi, J.; Morettin, P.; Teso, S.; Passerini, A. Efficient Generation of Structured Objects with Constrained Adversarial Networks. *Adv. Neural Inf. Process. Syst.* **2020**, *33*, 14663–14674.
37. Fontaine, M.C.; Liu, R.; Togelius, J.; Hoover, A.K.; Nikolaidis, S. Illuminating mario scenes in the latent space of a generative adversarial network. *arXiv* **2020**, arXiv:2007.05674.
38. Awiszus, M.; Schubert, F.; Rosenhahn, B. TOAD-GAN: Coherent style level generation from a single example. In Proceedings of the AAAI Conference on Artificial Intelligence and Interactive Digital Entertainment, held virtually 19–23 October 2020; Volume 16, pp. 10–16.
39. On, C.K.; Foong, N.W.; Teo, J.; Ibrahim, A.A.A.; Guan, T.T. Rule-based procedural generation of item in role-playing game. *Int. J. Adv. Sci. Eng. Inf. Technol* **2017**, *7*, 1735.
40. Togelius, J.; Justinussen, T.; Hartzen, A. Compositional procedural content generation. In Proceedings of the third workshop on Procedural Content Generation in Games, Raleigh, NC, USA, 29 May–1 June 2012; pp. 1–4.
41. Gellel, A.; Sweetser, P. A Hybrid Approach to Procedural Generation of Roguelike Video Game Levels. In Proceedings of the International Conference on the Foundations of Digital Games, Bugibba, Malta, 15–18 September 2020; pp. 1–10.
42. Torres, J.A. Pingu Run Github Repository. Available online: https://github.com/JAlbertoTorres/Pingu-run (accessed on 23 February 2022).
43. Torres, J.A. Gameplay de Pingu Run. Available online: https://youtu.be/TZza1W5kSOI (accessed on 23 February 2022).
44. IJsselsteijn, W.A.; de Kort, Y.A.; Poels, K. *The Game Experience Questionnaire*; Eindhoven University of Technology: Eindhoven, The Netherlands, 2013.
45. Karpouzis, K.; Yannakakis, G.N.; Shaker, N.; Asteriadis, S. The platformer experience dataset. In Proceedings of the 2015 International Conference on Affective Computing and Intelligent Interaction (ACII), Xi'an, China, 21–24 September 2015; pp. 712–718.

46. Karavolos, D.; Bouwer, A.; Bidarra, R. Mixed-Initiative Design of Game Levels: Integrating Mission and Space into Level Generation. In Proceedings of the Foundations of Digital Games, Pacific Grove, CA, USA, 22–25 June 2015.
47. Forsyth, W. Globalized random procedural content for dungeon generation. *J. Comput. Sci. Coll.* **2016**, *32*, 192–201.
48. Aponte, M.V.; Levieux, G.; Natkin, S. Difficulty in videogames: An experimental validation of a formal definition. In Proceedings of the 8th International Conference on Advances in Computer Entertainment Technology, Lisbon, Portugal, 8–11 November 2011; pp. 1–8.
49. Summerville, A.; Mariño, J.R.; Snodgrass, S.; Ontañón, S.; Lelis, L.H. Understanding mario: An evaluation of design metrics for platformers. In Proceedings of the 12th International Conference on the Foundations of Digital Games, Hyannis, MA, USA, 14–17 August 2017; pp. 1–10.
50. Mourato, F.; Birra, F.; dos Santos, M.P. Difficulty in action based challenges: Success prediction, players' strategies and profiling. In Proceedings of the 11th Conference on Advances in Computer Entertainment Technology, Madeira, Portugal, 11–14 November 2014; pp. 1–10.
51. Mourato, F.; Santos, M.P.D. Measuring difficulty in platform videogames. In Proceedings of the 4a Conferencia Nacional em Interacao PessoaMquina, Grupo Portugues de Computaao Grfica/Eurographics, Aveiro, Portugal, 13–15 October 2010; pp. 173–180.
52. Dıaz, A.C. Procedural Generation Applied to a Video Game Level Design. Bachelor's Thesis, Universitat politécnica de Catalunyia, Facultat d'Informàtica de Barcelona, Barcelona, Spain, 2015; pp. 41–44.
53. Cormen, T.H.; Leiserson, C.E.; Rivest, R.L.; Stein, C. *Introduction to Algorithms*; MIT Press: Cambridge, MA, USA, 2009.
54. Pingu-Run. Pingu-Run Code. Available online: https://github.com/JAlbertoTorres/ChunksCreator (accessed on 8 March 2022).

mathematics

MDPI

Article

A Dimension Group-Based Comprehensive Elite Learning Swarm Optimizer for Large-Scale Optimization

Qiang Yang [1], Kai-Xuan Zhang [1], Xu-Dong Gao [1,*], Dong-Dong Xu [1], Zhen-Yu Lu [1], Sang-Woon Jeon [2] and Jun Zhang [2,3]

1 School of Artificial Intelligence, Nanjing University of Information Science and Technology, Nanjing 210044, China; qiang_yang@nuist.edu.cn (Q.Y.); kaixuan.zhang@nuist.edu.cn (K.-X.Z.); 001601@nuist.edu.cn (D.-D.X.); 001114@nuist.edu.cn (Z.-Y.L.)

2 Department of Electrical and Electronic Engineering, Hanyang University, Ansan 15588, Korea; sangwoonjeon@hanyang.ac.kr (S.-W.J.); junzhang@ieee.org (J.Z.)

3 Department of Computer Science and Information Engineering, Chaoyang University of Technology, Taichung 413310, Taiwan

* Correspondence: 003103@nuist.edu.cn

Abstract: High-dimensional optimization problems are more and more common in the era of big data and the Internet of things (IoT), which seriously challenge the optimization performance of existing optimizers. To solve these kinds of problems effectively, this paper devises a dimension group-based comprehensive elite learning swarm optimizer (DGCELSO) by integrating valuable evolutionary information in different elite particles in the swarm to guide the updating of inferior ones. Specifically, the swarm is first separated into two exclusive sets, namely the elite set (***ES***) containing the top best individuals, and the non-elite set (***NES***), consisting of the remaining individuals. Then, the dimensions of each particle in ***NES*** are randomly divided into several groups with equal sizes. Subsequently, each dimension group of each non-elite particle is guided by two different elites randomly selected from ***ES***. In this way, each non-elite particle in ***NES*** is comprehensively guided by multiple elite particles in ***ES***. Therefore, not only could high diversity be maintained, but fast convergence is also likely guaranteed. To alleviate the sensitivity of DGCELSO to the associated parameters, we further devise dynamic adjustment strategies to change the parameter settings during the evolution. With the above mechanisms, DGCELSO is expected to explore and exploit the solution space properly to find the optimum solutions for optimization problems. Extensive experiments conducted on two commonly used large-scale benchmark problem sets demonstrate that DGCELSO achieves highly competitive or even much better performance than several state-of-the-art large-scale optimizers.

Keywords: large-scale optimization; particle swarm optimization; dimension group-based comprehensive elite learning; high-dimensional problems; elite learning

MSC: 37N40; 46N10; 47N10

Citation: Yang, Q.; Zhang, K.-X.; Gao, X.-D.; Xu, D.-D.; Lu, Z.-Y.; Jeon, S.-W.; Zhang, J. A Dimension Group-Based Comprehensive Elite Learning Swarm Optimizer for Large-Scale Optimization. *Mathematics* **2022**, *10*, 1072. https://doi.org/10.3390/math10071072

Academic Editors: Ioannis G. Tsoulos, Hsien-Chung Wu and Frank Werner

Received: 4 February 2022
Accepted: 24 March 2022
Published: 26 March 2022

1. Introduction

Large-scale optimization problems, also called high-dimensional problems, are ubiquitous in daily life and industrial engineering in the era of big data and the Internet of Things (IoT), such as water distribution optimization problems [1], cyber-physical systems design problems [2], control of pollutant spreading on social networks [3], and offshore wind farm collector system planning problems [4]. As the dimensionality of optimization problems increases, most existing optimization methods encounter the degradation of optimization effectiveness, due to the "curse of dimensionality" [5,6].

Specifically, the increase of dimensionality results in the following challenges for existing optimization algorithms: (1) With the growth of dimensionality, the properties of optimization problems become much more complicated. In particular, in the high-dimensional environment, optimization problems usually are non-convex, non-differentiable, or even non-continuous [7–9]. This makes traditional gradient-based optimization algorithms become infeasible. (2) The solution space grows exponentially as the dimensionality increases [10–13]. This greatly challenges the optimization efficiency of most existing algorithms. (3) The landscape of optimization problems becomes more complex in a high-dimensional space. On the one hand, some unimodal problems may become multimodal with the increase of dimensionality; on the other hand, in some multimodal problems, not only does the number of local optimal regions increase rapidly, but also the local regions become much wider and flatter [11,12,14]. This likely leads to premature convergence and stagnation of existing optimization techniques.

As a kind of metaheuristic algorithm, particle swarm optimization (PSO) maintains a population of particles, each of which represents a feasible solution to optimization problems, to search the solution space for the global optimum solutions [15–17]. By means of its great merits, such as strong global search ability, independence in the mathematic properties of optimization problems, and inherent parallelism [17], PSO has witnessed rapid development and excellent success in solving complex optimization problems [18–22] since it was proposed in 1995 [15]. As a result, PSO has been widely employed to solve real-world optimization problems in daily life and industrial engineering [1,23].

However, most existing PSOs are initially designed for low-dimensional optimization problems. Confronted with large-scale optimization problems, their effectiveness usually deteriorates due to the previously mentioned challenges [24–26]. To improve the optimization effectiveness of PSO in tackling high-dimensional problems, researchers have been devoted to designing novel and effective evolution mechanisms for PSO. Broadly speaking, existing large-scale PSOs can be divided into two categories [27], namely cooperative coevolutionary large-scale PSOs [6,28,29] and holistic large-scale PSOs [24,26,30–32].

Cooperative coevolutionary PSOs (CCPSOs) [6,28,29,33] adopt the divide-and-conquer technique to decompose one large-scale optimization problem into several exclusive smaller sub-problems and then optimize these sub-problems individually by traditional PSOs designed for low-dimensional problems to find the optimal solution to the large-scale optimization problem. Since the decomposed subproblems are separately optimized, the key component of CCPSOs is the decomposition strategy [6,28]. Ideally, a good decomposition strategy should place interacted variables into the same sub-problem, so that they can be optimized together. However, without prior knowledge, it is considerably difficult to decompose a large-scale problem accurately. As a result, current research on CCPSOs lies in developing novel decomposition strategies to divide the large-scale optimization problem as accurately as possible. Hence, many effective decomposition strategies [6,34–38] have been put forward.

However, CCPSOs heavily rely on the quality of the decomposition strategies. According to the no free lunch theorem, there is no decomposition strategy suitable for all large-scale problems. Therefore, some researchers attempt to design large-scale PSOs from another perspective, namely the holistic large-scale PSOs [5,26,30,39].

In contrast to CCPSOs, holistic large-scale PSOs [5,26,30,39,40] still optimize all variables simultaneously such as traditional PSOs. Since the learning strategy in updating the velocity of particles plays the most important role in PSO [15,16,18], the key to improving the effectiveness of PSO in coping with large-scale optimization is to devise effective learning strategies for particles, which should not only help particles explore the solution space efficiently to locate promising areas fast, but also aid particles to exploit the promising areas effectively to obtain high-quality solutions. Along this line, researchers have developed many remarkable learning strategies for PSO to solve high-dimensional problems, such as the competitive learning scheme [26], the social learning strategy [30], the two-phase learning method [1], and the level-based learning approach [25]. Recently, some researchers even have attempted to develop novel coding schemes for PSO to improve its optimization performance in solving large-scale optimization problems [41].

Although the above-mentioned large-scale PSOs have presented excellent optimization performance in solving some large-scale optimization problems, they still encounter limitations, such as premature convergence and stagnation into local areas, in solving complicated high-dimensional problems, especially those with overlapping correlated variables or fully non-separable variables. Therefore, the optimization performance of PSOs in tackling large-scale optimization still deserves improvement, which still remains an open and hot topic to study in the evolutionary computation community.

In nature, individuals with better fitness usually preserve more valuable evolutionary information than those with worse fitness, to guide the evolution of one species [42]. Moreover, in general, different individuals usually preserve different useful genes. Inspired by these observations, in this paper, we propose a dimension group-based comprehensive elite learning swarm optimizer (DGCELSO) by integrating useful genes embedded in different elite individuals to guide the update of particles to search the large-scale solution space effectively and efficiently. Specifically, the main components of the proposed DGCELSO are summarized as follows:

(1) A dimension group-based comprehensive elite learning scheme is proposed to guide the update of inferior particles by learning from multiple superior ones. Instead of learning from only at most two exemplars in existing holistic large-scale PSOs [24–26,30], the devised learning strategy first randomly divides the dimensions of each inferior particle into several equally sized groups and then employs different superior particles to guide the update of different dimension groups. Moreover, unlike existing elite strategies that only use one elite to direct the evolution of an individual [43,44], it employs a random dimension group-based recombination techniques to try to integrate valuable evolutionary information in multiple elites to guide the update of each non-elite particle. In this way, the learning diversity of particles could be largely promoted, which is beneficial for particles to avoid falling into local traps. Moreover, it is also possible that useful evolutionary information embedded in different superior particles could be integrated to direct the learning of inferior particles, which may be profitable for particles to approach promising areas quickly.

(2) Dynamic adjustment strategies for the control parameters involved in the proposed learning strategy are further designed to cooperate with the learning strategy to help PSO search the large-scale solution space properly. With these dynamic strategies, the developed DGCELSO could appropriately compromise the intensification and diversification of the search process at the swarm level and the particle level.

To verify the effectiveness of the proposed DGCELSO, extensive experiments are conducted to compare DGCELSO with several state-of-the-art large-scale optimizers on the widely used CEC'2010 [7] and CEC'2013 [8] large-scale benchmark optimization problem sets. Meanwhile, deep investigations on DGCELSO are also conducted to discover what contributes to its good performance.

The rest of this paper is organized as follows. Section 2 introduces the classical PSO and large-scale PSO variants. Then, the proposed DGCELSO is elucidated in detail in Section 3. Section 4 conducts extensive experiments to verify the effectiveness of the proposed DGCELSO. Finally, Section 5 concludes this paper.

2. Related Work

In this paper, a D-dimensional single-objective minimization optimization problem is considered, which is defined as follows:

$$\min f(x),\ x \in R^D \tag{1}$$

where x consisting of D variables is a feasible solution to the optimization problem, and D is the dimension size. In this paper, we directly use the function value as the fitness value of one particle.

2.1. Canonical PSO

In the canonical PSO [15,16], each particle is represented by two vectors, namely the position vector $\boldsymbol{x}$ and the velocity vector $\boldsymbol{v}$. During the evolution, in the canonical PSO [15,16], each particle is guided by its historically personal best position and the historically best position of the whole swarm. Specifically, each particle is updated as follows:

$$v_i^d \leftarrow wv_i^d + c_1 r_1(\boldsymbol{pbest}_i^d - x_i^d) + c_2 r_2(\boldsymbol{gbest}^d - x_i^d) \quad (2)$$

$$x_i^d \leftarrow x_i^d + v_i^d \quad (3)$$

where v_i^d is the dth dimension of the velocity of the ith particle, x_i^d is the dth dimension of the position of the ith particle, $\boldsymbol{pbest}_i^d$ is the dth dimension of the historically personal best position found by the ith particle, and $\boldsymbol{gbest}^d$ is the dth dimension of the historically global best position found by the whole swarm. As for the parameters, c_1 and c_2 are two acceleration coefficients, while r_1 and r_2 are two real random numbers uniformly generated within [0, 1]. w represents the inertia weight.

As shown in Equation (2), in the canonical PSO, each particle is cognitively directed by its ***pbest*** (the second part in the right hand of Equation (2) and socially guided by ***gbest*** of the whole swarm (the third part in the right hand of Equation (2). Due to the greedy attraction of ***gbest***, the swarm in the canonical PSOs usually becomes trapped in local areas when tackling multimodal problems [18,45]. Therefore, to improve the effectiveness of PSO in searching multimodal space with many local areas, researchers developed many novel learning strategies to guide the learning of particles, such as the comprehensive learning strategy [46], the genetic learning strategy [47], the scatter learning strategy [18], and the orthogonal learning strategy [48], etc.

Though a lot of novel learning strategies have helped PSO achieve very promising performance in solving multimodal problems, most of them are particularly designed for low-dimensional optimization problems. Encountered with large-scale optimization problems, most existing PSOs lose their effectiveness due to the "curse of dimensionality" and the aforementioned challenges in high-dimensional problems.

2.2. Large-Scale PSO

To solve the previously mentioned challenges of large-scale optimization, researchers devoted extensive attention to designing novel PSOs. As a result, numerous large-scale PSO variants have sprung up [1,26]. In a broad sense, existing large-scale PSOs can be classified into the following two categories.

2.2.1. Cooperative Coevolutionary Large-Scale PSO (CCPSO)

Cooperative coevolutionary PSOs (CCPSOs) [6,29,49] mainly use the divide-and-conquer technique to separate all variables of one high-dimensional problem into several exclusive groups, and then optimize each group of variables independently to obtain the optimal solution to the high-dimensional problem. Bergh and Engelbrecht put forward the earliest CCPSO [49]. In this algorithm, all variables in a large-scale optimization problem are randomly divided into K groups with each containing D/K variables (where D is the dimension size). Then the canonical PSO described in Section 2.1 is employed to optimize each group of variables. Nevertheless, the performance of this algorithm heavily relies on the setting of the number of groups (namely K). To alleviate this issue, in [29], an improved CCPSO, named CCPSO2, was proposed by first predefining a set of group numbers and then randomly selecting a group number in each iteration to separate variables into groups. In the above two algorithms, the correlations between variables are not taken into account explicitly. Hence, their optimization effectiveness degrades dramatically in solving problems with many interacted variables [11,12].

To alleviate the above issue, researchers have attempted to design effective variable grouping strategies to separate variables into groups by detecting the correlations between variables [6,35–37]. In the literature, the most representative grouping strategy is the differential grouping (DG) method [6], which uses the differential function values to detect the correlation between any two variables by exerting the same disturbance on the two variables. Based on the detected correlations between variables, DG could separate variables into groups satisfactorily. However, this method has two drawbacks. (1) It cannot detect the indirect interaction between variables [36], and (2) it consumes a lot of fitness evaluations ($O(D^2)$, D is the number of variables) in the variable decomposition stage [35,37].

To fill the first gap, Sun et al. devised an extended DG (XDG) [36], and Mei et al. brought up a global DG (GDG) [50] to detect both the direct and indirect interactions between variables. To alleviate the second predicament, a fast DG, named DG2 [35], and a recursive DG (RDG) [37] were put forward to reduce the consumption of fitness evaluations in the variable grouping stage. To further improve the detection efficiency of RDG, an efficient recursive differential grouping (ERDG) [51] was devised to reduce the used fitness evaluations in the decomposition stage, and to alleviate the sensitivity of RDG to parameters, an improved version, named RDG2, was developed [52] by adaptively adjusting the setting of parameters. In [53], Ma et al. proposed a merged differential grouping method based on subset-subset interaction and binary search by first identifying separable variables and non-separable variables, and putting all separable variables into the same subset, while dividing the non-separable variables into multiple subsets via a binary-tree-based iterative merging method. To further promote the variable grouping accuracy, Liu et al. proposed a deep grouping method by considering both the variable interaction and the essentialness of the variable to decompose one high-dimensional problem [54]. Instead of decomposing a large-scale optimization problem into fixed variable groups, Zhang et al. developed a dynamic grouping strategy to dynamically separate variables into groups during the evolution [55]. Specifically, the proposed algorithm first evaluates the contribution of variables based on the historical information and then constructs dynamic variable groups for the next generation based on the evaluated contribution and the detected interaction information.

By means of their promising performance in solving large-scale optimization problems, cooperative coevolutionary algorithms have been widely applied to solve various industrial engineering problems. For instance, Neshat et al. [56] proposed a novel multi-swarm cooperative co-evolution algorithm with the multi verse optimizer algorithm, the equilibrium optimization method, and the moth flame optimization approach, to optimize the layout of offshore wave energy converters. To tackle distributed flowshop group scheduling problems, Pan et al. [57] proposed a cooperative co-evolutionary algorithm with a collaboration model and a re-initialization scheme to tackle them. In [58], a hybrid cooperative co-evolution algorithm with a symmetric local search plus Nelder–Mead was devised to optimize the positions and the power-take-off settings of wave energy converters. In [59], Liang et al. developed a cooperative coevolutionary multi-objective evolutionary algorithm to tackle the transit network design and frequency setting problem.

Although the above-mentioned cooperative coevolutionary algorithms including CCPSOs achieved good performance in dealing with certain kinds of high-dimensional problems and have been applied to solve real-world problems, they are still confronted with limitations in tackling complicated high-dimensional problems. On the one hand, according to the theorem of No Free Lunch, there is no universal grouping method that could accurately separate variables into groups for all types of large-scale optimization problems; on the other hand, faced with high-dimensional problems with overlapping variable correlations, most existing variable grouping strategies would separate all these variables into the same group, leading to a very large variable group. Under this situation, traditional PSOs designed for low-dimensional problems used in CCPSO still cannot effectively optimize such a large group of variables. As a result, some researchers have attempted to design large-scale PSOs from another perspective to be elucidated next.

2.2.2. Holistic Large-Scale PSO

Unlike CCPSOs, holistic large-scale PSOs [18,26] still consider all variables as a whole and optimize them simultaneously like in traditional low-dimensional PSOs [16]. To solve the previously mentioned challenges of large-scale optimization, the key to holistic large-scale PSOs is to devise effective and efficient learning strategies for particles to largely promote the swarm diversity so that particles could explore the exponentially increased solution space efficiently and exploit the promising areas extensively to obtain high-quality solutions.

In [60], a dynamic multi-swarm PSO along with the Quasi-Newton local search method (DMS-L-PSO) was proposed to optimize large-scale optimization problems by dynamically separating particles into smaller sub-swarms in each generation. Taking inspiration from the competitive learning scheme in human society, Cheng and Jin proposed a competitive swarm optimizer (CSO) [26]. Specifically, this optimizer first separates particles into exclusive pairs and then lets each pair of particles compete with each other. After the competition, the winner is not updated and thus directly enters the next generation, while the loser is updated by learning from the winner. Likewise, inspired by the social learning strategy in animals, a social learning PSO (SLPSO) [61] was devised to let each particle probabilistically learn from those which are better than itself. By extending the pairwise competition mechanism in CSO to a tri-competitive strategy, Mohapatra et al. [62] developed a modified CSO (MCSO) to accelerate the convergence speed of the swarm to tackle high-dimensional problems. Taking inspiration from the comprehensive learning strategy designed for low-dimensional problems [46] and the competitive learning approach in CSO [26], Yang et al. designed a segment-based predominant learning swarm optimizer (SPLSO) [30] to cope with large-scale optimization. Specifically, this optimizer first uses the pairwise competition mechanism in CSO to divide particles into two groups, namely the relatively good particles and the relatively poor particles. Then, it further randomly separates the dimensions of each relatively poor particle into a certain number of exclusive segments, and subsequently randomly selects a relatively good particle to direct the update of each segment of the inferior particle.

Unlike the above large-scale PSOs [26,30,62], which let the updated particle learn from only one superior, Yang et al. devised a level-based learning swarm optimizer (LLSO) [25] by taking inspiration from the teaching theory in pedagogy. Specifically, this optimizer first separates particles into different levels and then lets each particle in lower levels learn from two random superior exemplars selected from higher levels. Inspired by the cooperative learning behavior in human society, Lan et al. put forward a two-phase learning swarm optimizer (TPLSO) [24]. This optimizer separates the learning of each particle into the mass learning phase and the elite learning phase. In the former learning phase, the tri-competitive mechanism is employed to update particles, while in the elite learning phase, the elite particles are picked out to learn from each other to further exploit promising areas to refine the found solutions. Similarly, Wang et al. proposed a multiple strategy learning particle swarm optimization (MSL-PSO) [40], in which different learning strategies are used to update particles in different evolution stages. In the first stage, each particle learns from those with better fitness and the mean position of the swarm to probe promising positions. Then, all the best probed positions are sorted based on their fitness and the top best ones are used to update particles in the second stage. In [41], Jian et al. developed a novel region encoding scheme to extend the solution representation from a single point to a region, and a novel adaptive region search strategy to keep the search diversity. These two schemes are then embedded into SLPSO to tackle large-scale optimization problems.

To find a good compromise between exploration and exploitation, Li et al. devised a learning structure to decouple exploration and exploitation for PSO in [63] to solve large-scale optimization. In particular, an exploration learning strategy was devised to direct particles to sparse areas based on a local sparseness degree measurement, and then an adaptive exploitation learning strategy was developed to let particles exploit the found promising areas. Deng et al. [39] devised a ranking-based biased learning swarm optimizer

(RBLSO) based on the principle that the fitness difference between learners and exemplars should be maximized. In particular, in this algorithm, a ranking paired learning (RPL) scheme was designed to let the worse particles learn peer-to-peer from the better ones, and at the same time, a biased center learning (BCL) strategy was devised to let each particle learn from the weighted mean position of the whole swarm. Lan et al. [64] proposed a hierarchical sorting swarm optimizer (HSSO) to tackle large-scale optimization. Specifically, this optimizer first divides particles into a good swarm and a bad swarm with equal sizes based on their fitness. Then, particles in the bad group are updated by learning from those in the good one. Subsequently, the good swarm is taken as a new swarm to execute the above swarm division and particle updating operations until there is only one particle in the good swarm. Kong et al. [65] devised an adaptive multi-swarm particle swarm optimizer to cope with high-dimensional problems. Specifically, it first adaptively divides particles into several sub-swarms and then employs the competition mechanism to select exemplars for particle updating. Huang et al. [66] put forward a convergence speed controller to cooperate with PSO to deal with large-scale optimization. Specifically, this controller is triggered periodically to produce an early warning to PSO before it falls into premature convergence.

Though most existing large-scale PSOs have presented their success in solving certain kinds of high-dimensional problems, their effectiveness still degrades in solving complicated high-dimensional problems [11,12,27,67], especially on those with many wide and flat local areas. Therefore, promoting the effectiveness and efficiency of PSO in solving large-scale optimization still deserves extensive attention and thus this research direction is still an active and hot topic in the evolutionary computation community.

3. Dimension Group-Based Comprehensive Elite Learning Swarm Optimizer

In nature, during the evolution of one species, those elite individuals with better adaptability to the environment usually preserve more valuable evolutionary information, such as genes, to direct the evolution of the species [42]. Moreover, different individuals may preserve different useful genes. Likewise, during the evolution of the swarm in PSO, different particles may contain useful variable values that may be close to the true global optimal solutions. Therefore, a natural idea is to integrate those useful values embedded in different particles to guide the evolution of the swarm. To this end, this paper proposes a dimension group-based comprehensive elite learning swarm optimizer (DGCELSO) to tackle large-scale optimization. The detailed components of this optimizer are elucidated as follows.

3.1. Dimension Group-Based Comprehensive Elite Learning

Given that NP particles are maintained in the swarm, the proposed DGCEL strategy first partitions the swarm into two exclusive sets, namely the elite set, denoted by ***ES***, and the non-elite set, denoted by ***NES***. Specifically, ***ES*** contains the best es particles in the swarm, while ***NES*** consists of the rest $nes = (NP - es)$ particles. Since the size of ***ES***, namely es, is related to NP, we set $es = \lceil tp * NP \rceil$ (where tp is the ratio of the elite particles in ***ES*** out of the whole swarm), for the convenience of parameter fine-tuning.

Since elite particles usually preserve more valuable evolutionary information than the non-elite ones, in this paper, we first develop an elite learning strategy (EL). Specifically, we let the elite particles in ***ES*** directly enter the next generation, while only updating the non-elite particles in ***NES***. Moreover, the elite particles in ***ES*** are employed to guide the learning of non-elite particles in ***NES***.

With respect to the elite particles, during the evolution, though they may be far from the global optimal area, they usually contain valuable genes that are very close to the true global optimal solution. To integrate the useful evolutionary information embedded in different elites, we propose a dimension group-based comprehensive learning strategy (DGCL). Specifically, during the update of each non-elite particle, the whole dimensions of this particle are first randomly shuffled and then are partitioned into NDG dimension groups (where NDG denotes the number of dimension groups), with each group containing D/NDG dimensions. In this way, the dimensions of each non-elite particle are randomly divided into NDG groups, namely $\boldsymbol{DG} = [DG^1, DG^2, \ldots, DG^{NDG}]$.

Here, it should be mentioned that for each non-elite particle, the dimensions are randomly shuffled, and thus it is likely that the division of dimension groups is different for different non-elite particles. In addition, if $D\%NDG$ is not zero, then the remaining dimensions are equally allocated to the first ($D\%NDG$) groups, i.e., each of the first ($D\%NDG$) groups contains ($D/NDG + 1$) dimensions.

Subsequently, unlike most existing large-scale PSOs [25,26,30] which use the same exemplars to update all dimensions of one inferior particle, the proposed DGCL uses one exemplar to update each dimension group of each non-elite particle, and thus one non-elite particle could learn from different exemplars.

Incorporating the proposed EL into the DGCL, the DGCEL is developed by using the elite particles in $\boldsymbol{ES}$ to direct the update of each dimension group of a non-elite particle. Specifically, each non-elite particle is updated as follows:

$$V_{NES_j}^{DG_i} \leftarrow r_1 V_{NES_j}^{DG_i} + r_2(X_{ES_{r1}}^{DG_i} - X_{NES_j}^{DG_i}) + \phi r_3(X_{ES_{r2}}^{DG_i} - X_{NES_j}^{DG_i}) \tag{4}$$

$$X_{NES_j}^{DG_i} \leftarrow V_{NES_j}^{DG_i} + X_{NES_j}^{DG_i} \tag{5}$$

where NES_j represents the jth non-elite particle in $\boldsymbol{NES}$; DG_i denotes the ith dimension group of the jth non-elite particle; $X_{NES_j}^{NG_i}$ and $V_{NES_j}^{DG_i}$ are the ith dimension group of the position and velocity of the jth particle in $\boldsymbol{NES}$, respectively; ES_{r1} and ES_{r2} are two different elite particles randomly selected from $\boldsymbol{ES}$; r_1, r_2, and r_3 are three random real parameters uniformly sampled within [0, 1]; $\phi \in [0, 1]$ is a control parameter in charge of the influence of the second elite particle.

As for the update of each non-elite particle in $\boldsymbol{NES}$, as shown in Equation (4), the following details should be paid careful attention:

(1) As previously mentioned, for each non-elite particle, the dimensions are randomly shuffled. As a result, the partition of dimension groups is different for different non-elite particles.
(2) For each dimension group $\boldsymbol{DG_i}$, two different elite particles $\boldsymbol{X_{ES_{r1}}}$ and $\boldsymbol{X_{ES_{r2}}}$ are first randomly selected from $\boldsymbol{ES}$. Then, the better one between these two elites (suppose it is $\boldsymbol{X_{ES_{r1}}}$) acts as the first exemplar in Equation (4), while the worse one (suppose it is $\boldsymbol{X_{ES_{r2}}}$) acts as the second exemplar to guide the update of the dimension group of the non-elite particle.
(3) The two elite particles guiding the update of each dimension group are both randomly selected. Therefore, they are likely to be different for different dimension groups.

As a whole, a complete flowchart of the proposed DGCEL is shown in Figure 1. Taking deep analysis on Equation (4) and Figure 1, we find that the proposed DGCEL strategy brings the following advantages to PSO:

(1) Instead of using historical evolutionary information, such as the historically global best position (***gbest***), the personal best positions (***pbest***), and the neighborhood best position (***nbest***), in traditional PSOs [18,47], the devised DGCEL employs the elite particles in the current swarm to direct the learning of the non-elite particles. In contrast to the historical information, which may remain unchanged for many generations, particles in the swarm are usually updated generation by generation. Therefore, in the proposed DGCEL, the selected two guiding exemplars are not only likely different for different particles but also probably different for the same particle in different generations. This is very beneficial for the promotion of swarm diversity.
(2) Instead of updating each particle with the same exemplars for all dimensions in most existing large-scale PSOs [5,24–26,30], the proposed DGCEL updates non-elite particles at the dimension group level. Therefore, for different dimension groups, the

two guiding exemplars are likely different. In this way, not only could one non-elite particle learn from multiple different elite ones, but also the useful genes hidden in different elites could be incorporated to direct the evolution of the swarm. As a result, not only the learning diversity of particles could be improved, but also the learning efficiency of particles could be promoted.

(3) In DGCEL, each dimension group of a non-elite particle is guided by two randomly selected elite particles in ***ES***. With the guidance of multiple elites, each non-elite particle is expected to approach promising areas quickly. In addition, since the elite particles in ***ES*** are not updated and directly enter the next generation, the useful evolutionary information in the current swarm is protected from being destroyed by uncertain updates. Therefore, the elites in ***ES*** become better and better as the evolution iterates, and at last, it is expected that these elites converge to the optimal areas.

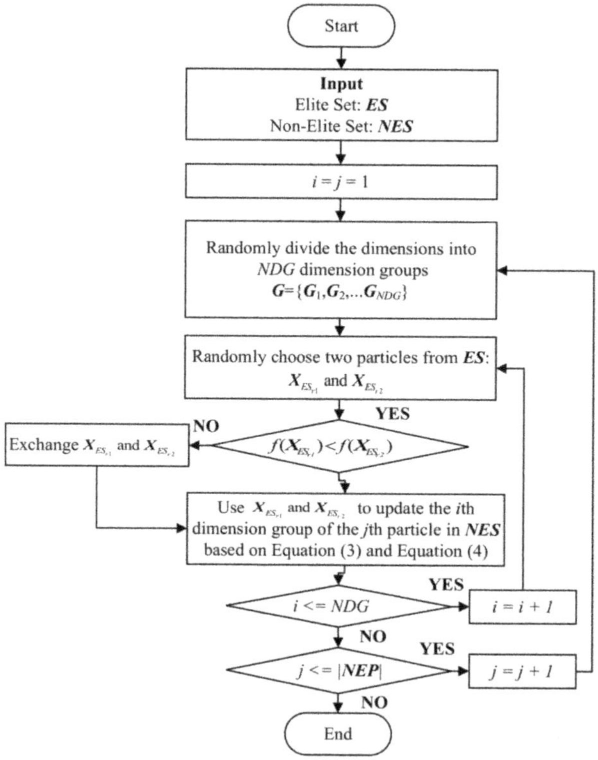

Figure 1. Flowchart of the proposed DGCEL strategy.

Remark

To the best of our knowledge, there are four existing PSOs that are very similar to the proposed DGCELSO. They are CLPSO [46], OLPSO [48], GLPSO [47], and SPLSO [30]. The first three were originally designed for low-dimensional problems, while the last one was initially devised for large-scale optimization. Compared with these existing PSOs, the developed DGCELSO distinguishes from them in the following ways:

(1) In contrast to the three low-dimensional PSOs [46–48], the proposed DGCELSO uses the elite particles in the swarm to comprehensively guide the learning of the non-elite particles at the dimension group level. First, the three low-dimensional PSOs all use the personal best positions (***pbests***) of particles to construct only one guiding exemplar for each updated particle, whereas DGCELSO leverages the elite

particles in the current swarm to construct two different guiding exemplars for each non-elite particle. Second, the three low-dimensional PSOs construct the guiding exemplar dimension by dimension. Nevertheless, DGCELSO constructs the two guiding exemplars group by group. With these two differences, DGCELSO is expected to construct more promising guiding exemplars for the updated particles, and thus the learning effectiveness and efficiency of particles could be largely promoted to explore the large-scale solution space.

(2) In contrast to the large-scale PSO, namely SPLSO [30], DGCELSO uses two different elite particles to direct the update of each dimension group of each non-elite particle. First, the partition of the swarm in DGCELSO is very different from the one in SPLSO. In DGCELSO, the swarm is divided into two exclusive sets according to the fitness of particles, with the best *es* particles entering ***ES*** and the rest entering ***NES***. However, in SPLSO, particles in the swarm are paired together and each paired two particles compete with each other, with the winner entering the relatively good set and the loser entering the relatively poor set. Second, for each non-elite particle, DGCELSO adopts two random elites in ***ES*** to guide the update of each dimension group, whereas in SPLSO, each dimension group of a loser is updated by only one random relatively good particle with the other exemplar being the mean position of the relatively good set, which is shared by all updated particles. Therefore, it is expected that the learning effectiveness and efficiency of particles in DGCELSO are higher than in SPLSO. Hence, DGCELSO is expected to explore and exploit the large-scale solution space more appropriately than SPLSO.

3.2. Adaptive Strategies for Control Parameters

Taking deep investigation on the proposed DGCELSO, we find that except for the swarm size *NP*, it has three control parameters, namely the ratio of elite particles out of the whole swarm *tp*, the number of dimension groups *NDG*, and the control parameter ϕ in Equation (4). The swarm size *NP* is a common parameter for all evolutionary algorithms, which is usually problem-dependent and thus remains fine-tuned. As for ϕ, it subtly controls the influence of the second guiding exemplar in the velocity update. We also leave it to be fine-tuned in the experiment as *NP*. For the other two control parameters, we devise the following dynamic adjustment schemes to alleviate the sensitivity of DGCELSO to them.

3.2.1. Dynamic Adjustment for *tp*

With respect to the ratio of elite particles out of the whole swarm *tp*, it determines the size of the elite set ***ES***. When *tp* is large, on the one hand, a large number of particles are preserved and enter the next generation directly; on the other hand, the learning of non-elite particles is diversified due to a large number of candidate exemplars, namely the elite particles. In this situation, the swarm biases to explore the solution space. In contrast, when *tp* is small, only a small number of elites are preserved. In this case, the learning of non-elite particles is concentrated to exploit the promising areas where the elites locate. Therefore, the swarm biases to exploit the solution space. However, it should be mentioned that such a bias is not at the serious sacrifice of swarm diversity because the guiding exemplars are both randomly selected for each dimension group of each non-elite particle.

Based on the above consideration, it seems rational not to keep *tp* fixed during the evolution. To this end, we devise a dynamic adjustment strategy for *tp* as follows:

$$tp = 0.4 - 0.2 \times \frac{fes}{Fes_{max}} \tag{6}$$

where *fes* represents the number of fitness evaluations used so far, and Fes_{max} is the maximum number of fitness evaluations.

From Equation (6), it is found that *tp* is linearly decreased from 0.4 to 0.2. Therefore, at the early stage, *tp* is high, while at the late stage, *tp* is small. As a result, as the evolution proceeds, the swarm gradually tends to exploit the solution space. This just matches the expectation that the swarm should explore the solution fully in the early stages to find promising areas while exploiting the found promising areas in the late stage to obtain high-quality solutions. The effectiveness of this dynamic adjustment scheme will be verified in the experiments in Section 4.3.

3.2.2. Dynamic Adjustment for NDG

In terms of the number of dimension groups *NDG*, it directly affects the learning of non-elite particles. A large *NDG* leads to a large number of elite particles that might participate in the learning of non-elite particles. This might be useful when the useful genes are scattered in very diversified dimensions. In this situation, with a large *NDG*, the chance of integrating the useful genes together to direct the learning of non-elite particles could be promoted. By contrast, when the useful genes are scattered in centered dimensions, a small *NDG* is preferred. However, without prior knowledge of the positions of useful genes embedded in the elite particles, it is difficult to give a proper setting of *NDG*.

To alleviate the above concern, we devise the following dynamic adjustment of *NDG* for each non-elite particle based on the Cauchy distribution:

$$NDG_{NES_j} \sim Cauchy(60, 10) \tag{7}$$

$$NDG_{NES_j} = floor(NDG_{NES_j}/10) * 10 + \begin{cases} 0 & if \mathrm{mod}(NDG_{NES_j}, 10) < 5 \\ 10 & otherwise \end{cases} \tag{8}$$

where NDG_{NES_j} denotes the setting of *NDG* for the *j*th particle in ***NES***, *Cauchy* (60, 10) is a Cauchy distribution with the position parameter 60 and scaling parameter 10. *floor*(x) is a function that returns the largest integer smaller than x. *mod*(x,y) is a function that returns the remainder when x/y.

In Equations (7) and (8), two details deserve careful attention. First, the Cauchy distribution is used here because it can generate values around the position parameter with a long fat tail. With this distribution, the generated *NDG*s for different non-elite particles are likely diversified. Second, with Equation (8), we keep the setting of *NDG* for each non-elite particle at multiple times of 10. This setting is adopted here for promoting the difference between two different values of *NDG* to improve the learning diversity of non-elite particles and for the convenience of computation.

From Equations (7) and (8), it is found that different non-elite particles likely preserve different *NDG*s. On the one hand, the learning diversity of non-elite particles could be further improved. On the other hand, the chance of integrating useful genes embedded in different elite particles is likely promoted with different settings of *NDG*. The effectiveness of this dynamic adjustment scheme for *NDG* will be verified in the experiments in Section 4.3.

3.3. *Overall Procedure of DGCELSO*

By integrating the above components, DGCELSO is developed with the overall procedure outlined in Algorithm 1 and the complete flowchart shown in Figure 2. Specifically, after the swarm is initialized and evaluated (Line 1), the algorithm goes to the main iteration loop (Lines 2~17). First, the swarm is partitioned into the elite set (***ES***) and the non-elite set (***NES***) as shown in Lines 3 and 4. Then, each particle in ***NES*** is updated as shown in Lines 5~16. During the update of one non-elite particle, the dimensions of this particle are first separated into several dimension groups (Lines 6 and 7). Then, for each dimension group of the non-elite particle, two different elite particles are randomly selected from ***ES*** (Line 9), and then the dimension group is updated by learning from these two elites (Line 13). The above process iterates until the termination condition is met. At the end of the algorithm, the best solution in the swarm is output (Line 18).

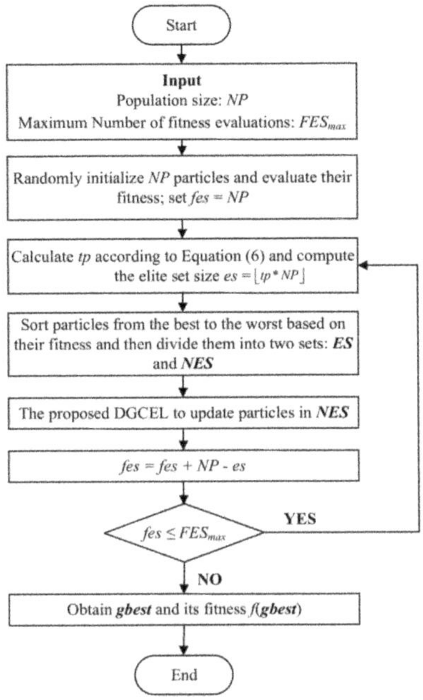

Figure 2. Flowchart of the proposed DGCELSO.

With respect to the computational complexity in time, from Algorithm 1, it is found that in each generation, it takes $O(NP\log_2 NP)$ to sort the swarm and $O(NP)$ to partition the swarm into two sets in Line 4; then, it takes $O(NP*D)$ to shuffle the dimensions and $O(NP*D)$ to partition the shuffled dimensions into groups for all non-elite particles (Line 7); at last, it takes $O(NP*D)$ to update all non-elite particles (Lines 8~14). To sum up, the time complexity of DGCELSO is $O(NP*D)$ based on the consideration that the swarm size is usually much smaller than the dimension size in large-scale optimization.

Algorithm 1: The Pseudocode of DGCELSO.

Input: Population size NP, Maximum number of fitness evaluations FES_{max}, Control parameter ϕ;
1: Initialize NP particles randomly and calculate their fitness; $fes = NP$;
2: **While** ($fes \leq FES_{max}$) **do**
3: Calculate tp according to Equation (6) and obtain the elite set size $es = \lfloor tp * NP \rfloor$;
4: Sort particles based on their fitness and divide them into two sets, namely ***ES*** and ***NES***;
5: **For** each non-elite particle $\boldsymbol{NES_j}$ in ***NES*** **do**
6: Generate NDG_{NES_j} based on Equation (7);
7: Random shuffle the dimensions and then split the dimensions into NDG_{NES_j} groups;
8: **For** each dimension group $\boldsymbol{DG_i}$ **do**
9: Randomly select two different elite particles from ***ES***: $\boldsymbol{X_{ESr1}}$ and $\boldsymbol{X_{ESr2}}$;
10: **If** ($f(\boldsymbol{X_{ESr2}}) < f(\boldsymbol{X_{ESr1}})$) **then**
11: Swap $ESr1$ and $ESr2$;
12: **End If**
13: Update the dimension group of $\boldsymbol{NES_j}$ according to Equations (3) and (4);
14: **End For**
15: Calculate the fitness of the updated $\boldsymbol{NES_j}$, and fes ++;
16: **End For**
17: **End While**
18: Obtain the best solution in the swarm ***gbest*** and its fitness $f(\boldsymbol{gbest})$

Output: $f(\boldsymbol{gbest})$ and ***gbest***

Regarding the computational complexity in space occupation, in Algorithm 1, we can see that except for $O(NP*D)$ to store the positions of all particles and $O(NP*D)$ to store the velocities of all particles, it only takes extra $O(NP)$ to store the index of particles in the two sets, and $O(D)$ to store the dimension groups. Comprehensively, DGCELSO only takes $O(NP*D)$ space.

Based on the above time and space complexity analysis, it is found that the proposed DGCELSO remains as efficient as the classical PSO, which also takes $O(NP*D)$ time in each generation and $O(NP*D)$ space.

4. Experimental Section

To verify the effectiveness of the proposed DGCELSO, extensive experiments are conducted on two sets of large-scale optimization problems, namely the CEC'2010 [7] and the CEC'2013 [8] large-scale benchmark sets in this section. The CEC'2010 set contains 20 high-dimensional problems with 1000 dimensions, while the CEC'2013 set consists of 15 problems with 1000 dimensions as well. In particular, the CEC'2013 set is an extension of the CEC'2010 set by introducing more complicated features, such as overlapping interactions among variables and imbalance contribution of variables. Therefore, compared with the CEC'2010 problems, the CEC'2013 problems are more complicated and more difficult to optimize. For more detailed information on the two benchmark large-scale problem sets, readers are referred to [7,8].

In this section, we first investigate the settings of two key parameters (namely the swarm size NP and the control parameter ϕ) for DGCELSO in Section 4.1. Then, extensive experiments are conducted on the two benchmark sets to compare DGCELSO with several state-of-the-art large-scale optimizers in Section 4.2. At last, a deep investigation into the proposed DGCELSO is performed to observe what contributes to the good performance of DGCELSO.

In the experiments, unless otherwise stated, the maximum number of fitness evaluations is set as $3000 \times D$, where D is the dimension size. In this paper, the dimension size of all optimization problems is 1000, and thus the total number of fitness evaluations is 3×10^6. To make fair and comprehensive comparisons, the median, the mean, and the standard deviation (Std) values over 30 independent runs are used to evaluate the performance of all algorithms. Moreover, to tell the statistical significance, the Wilcoxon rank-sum test at the significance level of "$\alpha = 0.05$" was conducted to compare two different algorithms. Furthermore, to obtain the overall ranks of different algorithms on one whole benchmark set, the Friedman test at the significance level of "$\alpha = 0.05$" was conducted on each benchmark set.

Lastly, it is worth noting that we use the C programming language and Code Blocks software to implement the proposed DGCELO. Moreover, all experiments were run on a PC with 8 Intel Core i7-10700 2.90-GHz CPUs, 8-GB memory, and the 64-bit Ubuntu 12.04 LTS system.

4.1. Parameter Setting

Due to the proposed two dynamic adjustment strategies of the associated parameters in DGCELSO, there are only two parameters, namely the swarm size NP and the control parameter ϕ that need fine-tuning. Therefore, to investigate the optimal setting of the two parameters for DGCELSO in solving 1000-D large-scale optimization problems, we conduct experiments by varying NP from 100 to 600 and ϕ ranging from 0.1 to 0.9 for DGCELSO on the CEC'2010 benchmark set. Table 1 shows the mean fitness values obtained by DGCELSO with different settings of NP and ϕ on the CEC'2010 set. In this table, the best results are highlighted in bold, and the average rank of each configuration is also presented, which was obtained using the Friedman test at the significance level of "$\alpha = 0.05$".

Table 1. Comparison results among DGCELSO with different settings of NP and ϕ on the 1000-D CEC'2010 problems.

F	$NP = 100$									$NP = 200$								
	$\phi = 0.1$	$\phi = 0.2$	$\phi = 0.3$	$\phi = 0.4$	$\phi = 0.5$	$\phi = 0.6$	$\phi = 0.7$	$\phi = 0.8$	$\phi = 0.9$	$\phi = 0.1$	$\phi = 0.2$	$\phi = 0.3$	$\phi = 0.4$	$\phi = 0.5$	$\phi = 0.6$	$\phi = 0.7$	$\phi = 0.8$	$\phi = 0.9$
F_1	3.31×10^{2}	5.18×10^{7}	8.23×10^{7}	1.34×10^{7}	1.12×10^{3}	2.92×10^{-23}	9.51×10^{-20}	5.27×10^{5}	1.01×10^{8}	5.12×10^{-26}	6.22×10^{-29}	5.73×10^{-27}	0.00×10^{0}	1.10×10^{-26}	2.11×10^{-22}	9.04×10^{2}	5.08×10^{7}	1.22×10^{9}
F_2	2.93×10^{3}	3.58×10^{3}	3.64×10^{3}	3.22×10^{3}	2.50×10^{3}	1.62×10^{3}	1.12×10^{3}	9.12×10^{3}	1.13×10^{4}	1.16×10^{3}	1.61×10^{3}	1.69×10^{3}	1.40×10^{3}	8.84×10^{2}	2.95×10^{3}	1.07×10^{4}	1.14×10^{4}	1.19×10^{4}
F_3	5.74×10^{0}	1.13×10^{1}	1.14×10^{1}	8.23×10^{0}	3.24×10^{0}	2.18×10^{-1}	6.43×10^{-14}	3.89×10^{-1}	1.36×10^{1}	3.47×10^{-14}	2.90×10^{-2}	1.19×10^{-1}	3.42×10^{-14}	3.81×10^{-14}	4.88×10^{-14}	3.63×10^{-1}	1.27×10^{1}	1.71×10^{1}
F_4	4.77×10^{11}	5.57×10^{12}	5.93×10^{12}	2.88×10^{12}	1.66×10^{11}	1.14×10^{11}	1.53×10^{11}	2.90×10^{11}	7.08×10^{11}	1.74×10^{11}	1.96×10^{11}	4.21×10^{11}	1.28×10^{11}	1.25×10^{11}	1.67×10^{11}	2.48×10^{11}	6.04×10^{11}	2.14×10^{13}
F_5	2.96×10^{7}	3.16×10^{7}	3.01×10^{7}	3.54×10^{7}	1.30×10^{8}	2.75×10^{8}	2.86×10^{8}	2.96×10^{8}	3.05×10^{8}	2.81×10^{8}	2.36×10^{8}	2.24×10^{8}	2.55×10^{8}	2.77×10^{8}	2.84×10^{8}	2.91×10^{8}	3.04×10^{8}	3.09×10^{8}
F_6	1.99×10^{1}	2.02×10^{1}	2.02×10^{1}	2.01×10^{1}	1.99×10^{1}	2.01×10^{1}	2.15×10^{1}	2.15×10^{1}	2.03×10^{1}	1.94×10^{1}	1.97×10^{1}	1.97×10^{1}	1.98×10^{1}	1.96×10^{1}	4.00×10^{-9}	3.82×10^{-1}	1.34×10^{1}	1.78×10^{1}
F_7	2.94×10^{6}	9.82×10^{8}	1.28×10^{9}	3.37×10^{8}	7.90×10^{5}	7.40×10^{5}	1.17×10^{5}	8.40×10^{4}	7.80×10^{5}	3.34×10^{-6}	3.70×10^{4}	1.75×10^{6}	1.40×10^{3}	8.73×10^{-6}	3.14×10^{-1}	2.51×10^{4}	5.02×10^{5}	1.76×10^{7}
F_8	3.39×10^{7}	4.89×10^{7}	4.72×10^{7}	4.42×10^{7}	1.67×10^{5}	6.68×10^{4}	1.58×10^{7}	4.17×10^{7}	4.89×10^{7}	3.33×10^{5}	5.47×10^{6}	2.47×10^{7}	1.86×10^{3}	3.95×10^{3}	1.73×10^{7}	3.97×10^{7}	4.52×10^{7}	4.62×10^{7}
F_9	8.72×10^{7}	1.03×10^{9}	1.17×10^{9}	6.34×10^{8}	2.80×10^{7}	1.75×10^{7}	4.08×10^{7}	4.70×10^{8}	1.36×10^{10}	1.97×10^{7}	3.44×10^{7}	6.48×10^{7}	1.98×10^{7}	1.47×10^{7}	4.03×10^{7}	3.65×10^{9}	2.29×10^{10}	4.11×10^{10}
F_{10}	3.14×10^{3}	3.85×10^{3}	3.97×10^{3}	3.43×10^{3}	2.65×10^{3}	1.69×10^{3}	2.66×10^{3}	1.09×10^{4}	1.16×10^{4}	1.20×10^{3}	1.76×10^{3}	1.82×10^{3}	1.48×10^{3}	9.59×10^{2}	1.01×10^{4}	1.08×10^{4}	1.14×10^{4}	1.20×10^{4}
F_{11}	7.08×10^{1}	9.71×10^{1}	9.38×10^{1}	8.60×10^{1}	5.22×10^{1}	3.03×10^{1}	2.47×10^{1}	2.53×10^{1}	6.32×10^{1}	1.57×10^{1}	2.00×10^{1}	2.03×10^{1}	2.00×10^{1}	1.09×10^{1}	1.85×10^{-13}	1.38×10^{0}	5.04×10^{1}	1.41×10^{2}
F_{12}	9.54×10^{4}	1.03×10^{6}	1.13×10^{6}	6.89×10^{5}	4.82×10^{3}	8.18×10^{2}	6.14×10^{4}	5.07×10^{6}	6.64×10^{6}	2.35×10^{3}	1.71×10^{4}	7.25×10^{4}	1.72×10^{3}	1.92×10^{3}	2.36×10^{6}	5.14×10^{6}	6.61×10^{6}	7.97×10^{6}
F_{13}	5.89×10^{3}	3.83×10^{6}	4.40×10^{6}	9.77×10^{5}	5.29×10^{3}	3.06×10^{3}	2.35×10^{3}	6.58×10^{4}	1.36×10^{8}	6.55×10^{2}	8.02×10^{2}	1.02×10^{3}	5.58×10^{2}	4.97×10^{2}	5.48×10^{2}	2.96×10^{3}	3.95×10^{7}	9.38×10^{9}
F_{14}	2.68×10^{8}	2.11×10^{9}	2.30×10^{9}	1.45×10^{9}	8.31×10^{7}	4.56×10^{7}	1.39×10^{8}	3.47×10^{9}	3.23×10^{10}	5.82×10^{7}	1.12×10^{8}	2.20×10^{8}	6.07×10^{7}	4.61×10^{7}	2.11×10^{8}	2.02×10^{10}	5.18×10^{10}	7.58×10^{10}
F_{15}	3.33×10^{3}	4.07×10^{3}	4.13×10^{3}	3.53×10^{3}	2.81×10^{3}	1.11×10^{4}	1.09×10^{4}	1.12×10^{4}	1.17×10^{4}	1.07×10^{4}	3.71×10^{3}	3.33×10^{3}	1.07×10^{4}	1.05×10^{4}	1.05×10^{4}	1.08×10^{4}	1.14×10^{4}	1.21×10^{4}
F_{16}	1.89×10^{2}	2.56×10^{2}	2.56×10^{2}	2.22×10^{2}	1.47×10^{2}	8.35×10^{1}	5.10×10^{1}	6.61×10^{1}	2.58×10^{2}	6.10×10^{-1}	1.60×10^{1}	2.71×10^{1}	6.75×10^{0}	3.42×10^{-2}	2.93×10^{-13}	1.35×10^{1}	2.52×10^{2}	3.39×10^{2}
F_{17}	3.02×10^{5}	1.61×10^{6}	1.71×10^{6}	1.27×10^{6}	3.52×10^{4}	1.06×10^{4}	2.08×10^{6}	9.92×10^{6}	1.40×10^{7}	4.93×10^{4}	9.98×10^{4}	2.77×10^{5}	2.24×10^{4}	1.18×10^{5}	6.85×10^{6}	1.08×10^{7}	1.47×10^{7}	1.80×10^{7}
F_{18}	1.70×10^{4}	7.87×10^{8}	1.16×10^{9}	3.75×10^{7}	2.67×10^{3}	1.71×10^{3}	2.72×10^{3}	6.17×10^{6}	4.49×10^{10}	1.95×10^{3}	2.54×10^{3}	3.90×10^{3}	1.66×10^{3}	1.30×10^{3}	1.59×10^{3}	1.71×10^{7}	2.95×10^{10}	1.40×10^{11}
F_{19}	2.34×10^{6}	4.66×10^{6}	4.74×10^{6}	3.92×10^{6}	1.72×10^{6}	6.52×10^{6}	1.48×10^{7}	2.01×10^{7}	2.49×10^{7}	9.10×10^{6}	2.46×10^{6}	2.41×10^{6}	5.98×10^{6}	1.09×10^{7}	1.60×10^{7}	2.09×10^{7}	2.58×10^{7}	3.04×10^{7}
F_{20}	8.41×10^{3}	1.01×10^{9}	1.39×10^{9}	4.32×10^{7}	2.93×10^{3}	1.29×10^{3}	1.25×10^{3}	7.61×10^{6}	4.93×10^{10}	1.41×10^{3}	2.13×10^{3}	2.72×10^{3}	1.51×10^{3}	1.10×10^{3}	1.02×10^{3}	2.27×10^{7}	3.25×10^{10}	1.47×10^{11}
Rank	3.75	6.35	7.05	5.30	2.80	2.45	3.25	5.75	8.30	3.25	4.25	5.25	3.15	2.45	3.95	6.25	7.70	8.75
F	$NP = 300$									$NP = 400$								
	$\phi = 0.1$	$\phi = 0.2$	$\phi = 0.3$	$\phi = 0.4$	$\phi = 0.5$	$\phi = 0.6$	$\phi = 0.7$	$\phi = 0.8$	$\phi = 0.9$	$\phi = 0.1$	$\phi = 0.2$	$\phi = 0.3$	$\phi = 0.4$	$\phi = 0.5$	$\phi = 0.6$	$\phi = 0.7$	$\phi = 0.8$	$\phi = 0.9$
F_1	9.78×10^{-27}	0.00×10^{0}	0.00×10^{0}	0.00×10^{0}	0.00×10^{0}	7.97×10^{-10}	5.83×10^{5}	1.86×10^{8}	2.23×10^{9}	6.50×10^{-24}	0.00×10^{0}	0.00×10^{0}	0.00×10^{0}	8.36×10^{-24}	4.18×10^{-3}	3.39×10^{6}	3.30×10^{8}	3.01×10^{9}
F_2	6.93×10^{2}	1.06×10^{3}	1.15×10^{3}	8.88×10^{2}	5.90×10^{2}	1.04×10^{4}	1.09×10^{4}	1.15×10^{4}	1.21×10^{4}	5.75×10^{2}	8.12×10^{2}	8.78×10^{2}	6.57×10^{2}	9.82×10^{3}	1.05×10^{4}	1.10×10^{4}	1.16×10^{4}	1.22×10^{4}
F_3	3.36×10^{-14}	3.05×10^{-14}	3.15×10^{-14}	3.18×10^{-14}	3.88×10^{-14}	1.15×10^{-7}	5.14×10^{0}	1.47×10^{1}	1.77×10^{1}	3.51×10^{-14}	2.99×10^{-14}	2.98×10^{-14}	3.15×10^{-14}	3.98×10^{-14}	3.29×10^{-4}	7.64×10^{0}	1.54×10^{1}	1.79×10^{1}
F_4	2.22×10^{11}	2.13×10^{11}	2.00×10^{11}	1.60×10^{11}	1.57×10^{11}	2.06×10^{11}	3.80×10^{11}	1.31×10^{12}	6.16×10^{13}	2.88×10^{11}	2.60×10^{11}	2.27×10^{11}	1.96×10^{11}	1.82×10^{11}	2.48×10^{11}	5.19×10^{11}	3.12×10^{12}	1.09×10^{14}
F_5	2.83×10^{8}	2.81×10^{8}	2.76×10^{8}	2.80×10^{8}	2.82×10^{8}	2.86×10^{8}	2.93×10^{8}	3.02×10^{8}	3.18×10^{8}	2.82×10^{8}	2.82×10^{8}	2.81×10^{8}	2.78×10^{8}	2.83×10^{8}	2.89×10^{8}	2.94×10^{8}	3.06×10^{8}	3.16×10^{8}
F_6	4.00×10^{-9}	6.18×10^{0}	1.86×10^{1}	4.00×10^{-9}	4.00×10^{-9}	2.08×10^{-7}	5.36×10^{0}	1.54×10^{1}	1.84×10^{1}	4.00×10^{-9}	3.88×10^{-9}	4.00×10^{-9}	4.00×10^{-9}	4.00×10^{-9}	4.72×10^{-4}	8.08×10^{0}	1.61×10^{1}	1.86×10^{1}
F_7	3.43×10^{-3}	1.20×10^{-3}	3.63×10^{-2}	2.15×10^{-5}	2.54×10^{-3}	3.60×10^{2}	1.83×10^{5}	9.66×10^{5}	4.68×10^{8}	8.32×10^{-1}	3.50×10^{-2}	3.16×10^{-2}	7.06×10^{-3}	1.03×10^{0}	7.18×10^{3}	3.40×10^{5}	5.23×10^{6}	1.55×10^{9}
F_8	1.37×10^{7}	3.51×10^{5}	3.72×10^{4}	4.36×10^{3}	9.82×10^{5}	3.01×10^{7}	4.30×10^{7}	4.58×10^{7}	4.65×10^{7}	2.23×10^{7}	1.00×10^{7}	3.36×10^{6}	6.67×10^{5}	1.33×10^{7}	3.52×10^{7}	4.41×10^{7}	4.61×10^{7}	4.67×10^{7}
F_9	2.47×10^{7}	2.28×10^{7}	2.49×10^{7}	1.77×10^{7}	2.14×10^{7}	1.69×10^{8}	1.40×10^{10}	3.19×10^{10}	5.08×10^{10}	3.12×10^{7}	2.42×10^{7}	2.41×10^{7}	2.01×10^{7}	3.01×10^{7}	1.36×10^{9}	1.92×10^{10}	3.70×10^{10}	5.55×10^{10}
F_{10}	8.49×10^{2}	1.13×10^{3}	1.22×10^{3}	9.23×10^{2}	9.75×10^{3}	1.05×10^{4}	1.09×10^{4}	1.15×10^{4}	1.22×10^{4}	9.74×10^{3}	8.59×10^{2}	9.25×10^{2}	1.11×10^{3}	1.02×10^{4}	1.05×10^{4}	1.10×10^{4}	1.16×10^{4}	1.22×10^{4}
F_{11}	1.25×10^{-13}	2.23×10^{-1}	6.68×10^{0}	1.10×10^{-13}	1.17×10^{-13}	3.45×10^{-7}	8.82×10^{0}	7.59×10^{1}	1.59×10^{2}	1.30×10^{-13}	1.11×10^{-13}	1.05×10^{-13}	1.06×10^{-13}	1.31×10^{-13}	7.60×10^{-4}	1.51×10^{1}	9.53×10^{1}	1.67×10^{2}
F_{12}	2.50×10^{4}	4.39×10^{3}	5.55×10^{3}	2.55×10^{3}	8.74×10^{4}	4.13×10^{6}	5.75×10^{6}	7.14×10^{6}	8.45×10^{6}	1.71×10^{5}	9.83×10^{3}	7.41×10^{3}	1.18×10^{4}	2.03×10^{6}	4.62×10^{6}	6.16×10^{6}	7.49×10^{6}	8.77×10^{6}
F_{13}	5.69×10^{2}	5.35×10^{2}	5.91×10^{2}	5.15×10^{2}	4.69×10^{2}	4.85×10^{2}	1.08×10^{5}	5.63×10^{8}	1.67×10^{10}	5.31×10^{2}	5.36×10^{2}	5.46×10^{2}	4.93×10^{2}	4.50×10^{2}	4.80×10^{2}	3.72×10^{5}	1.46×10^{9}	2.23×10^{10}
F_{14}	7.79×10^{7}	6.96×10^{7}	7.62×10^{7}	5.17×10^{7}	7.69×10^{7}	5.73×10^{9}	3.85×10^{10}	6.50×10^{10}	8.96×10^{10}	1.14×10^{8}	7.11×10^{7}	7.20×10^{7}	6.01×10^{7}	1.43×10^{8}	1.68×10^{10}	4.63×10^{10}	7.22×10^{10}	9.64×10^{10}
F_{15}	1.04×10^{4}	1.05×10^{4}	1.05×10^{4}	1.04×10^{4}	1.03×10^{4}	1.06×10^{4}	1.10×10^{4}	1.16×10^{4}	1.22×10^{4}	1.04×10^{4}	1.03×10^{4}	1.04×10^{4}	1.03×10^{4}	1.03×10^{4}	1.06×10^{4}	1.11×10^{4}	1.17×10^{4}	1.23×10^{4}
F_{16}	2.15×10^{-13}	5.86×10^{-2}	9.78×10^{-2}	1.55×10^{-13}	2.01×10^{-13}	2.36×10^{-6}	1.02×10^{2}	2.92×10^{2}	3.52×10^{2}	2.39×10^{-13}	1.61×10^{-13}	1.53×10^{-13}	1.60×10^{-13}	2.44×10^{-13}	7.14×10^{-3}	1.52×10^{2}	3.07×10^{2}	3.57×10^{2}
F_{17}	1.53×10^{6}	5.71×10^{4}	5.65×10^{4}	6.57×10^{4}	4.44×10^{6}	8.75×10^{6}	1.28×10^{7}	1.65×10^{7}	1.98×10^{7}	4.81×10^{6}	1.72×10^{5}	1.03×10^{5}	7.07×10^{5}	6.20×10^{6}	1.02×10^{7}	1.37×10^{7}	1.72×10^{7}	2.06×10^{7}
F_{18}	1.56×10^{3}	1.65×10^{3}	1.56×10^{3}	1.31×10^{3}	1.13×10^{3}	1.26×10^{3}	1.57×10^{9}	5.42×10^{10}	1.76×10^{11}	1.31×10^{3}	1.31×10^{3}	1.33×10^{3}	1.25×10^{3}	1.12×10^{3}	3.88×10^{3}	4.92×10^{9}	6.87×10^{10}	1.94×10^{11}
F_{19}	1.33×10^{7}	8.92×10^{6}	8.14×10^{6}	1.02×10^{7}	1.43×10^{7}	1.88×10^{7}	2.25×10^{7}	2.75×10^{7}	3.21×10^{7}	1.45×10^{7}	1.09×10^{7}	1.02×10^{7}	1.21×10^{7}	1.53×10^{7}	1.94×10^{7}	2.47×10^{7}	2.87×10^{7}	3.36×10^{7}
F_{20}	1.12×10^{3}	1.32×10^{3}	1.40×10^{3}	1.08×10^{3}	9.79×10^{2}	1.00×10^{3}	1.88×10^{9}	5.74×10^{10}	1.82×10^{11}	9.89×10^{2}	1.15×10^{3}	1.10×10^{3}	9.85×10^{2}	9.82×10^{2}	1.72×10^{3}	5.69×10^{9}	7.40×10^{10}	2.03×10^{11}
Rank	3.80	3.48	4.03	2.03	2.88	5.00	6.90	7.95	8.95	3.90	2.90	2.50	2.05	3.95	5.70	7.00	8.00	9.00

Table 1. *Cont.*

F	NP = 500 $\phi = 0.1$	$\phi = 0.2$	$\phi = 0.3$	$\phi = 0.4$	$\phi = 0.5$	$\phi = 0.6$	$\phi = 0.7$	$\phi = 0.8$	$\phi = 0.9$
F_1	8.02×10^{-22}	0.00×10^{0}	0.00×10^{0}	0.00×10^{0}	6.91×10^{-21}	4.51×10^{0}	8.52×10^{6}	4.72×10^{8}	3.65×10^{9}
F_2	8.73×10^{3}	6.81×10^{2}	7.34×10^{2}	5.74×10^{2}	1.01×10^{4}	1.06×10^{4}	1.11×10^{4}	1.16×10^{4}	1.23×10^{4}
F_3	3.92×10^{-14}	2.96×10^{-14}	2.90×10^{-14}	3.12×10^{-14}	8.55×10^{-14}	1.24×10^{-2}	9.15×10^{0}	1.58×10^{1}	1.80×10^{1}
F_4	3.35×10^{11}	3.11×10^{11}	2.86×10^{11}	2.49×10^{11}	2.16×10^{11}	3.21×10^{11}	6.31×10^{11}	1.05×10^{13}	1.22×10^{14}
F_5	2.80×10^{8}	2.76×10^{8}	2.78×10^{8}	2.77×10^{8}	2.77×10^{8}	2.90×10^{8}	2.95×10^{8}	3.06×10^{8}	3.20×10^{8}
F_6	4.00×10^{-9}	4.00×10^{-9}	4.00×10^{-9}	4.00×10^{-9}	4.00×10^{-9}	1.69×10^{-2}	9.72×10^{0}	1.66×10^{1}	1.88×10^{1}
F_7	2.61×10^{1}	1.21×10^{0}	7.81×10^{-1}	4.75×10^{-1}	3.43×10^{1}	3.08×10^{4}	4.80×10^{5}	3.01×10^{7}	2.65×10^{9}
F_8	2.74×10^{7}	1.73×10^{7}	1.16×10^{7}	9.50×10^{6}	2.05×10^{7}	3.79×10^{7}	4.47×10^{7}	4.63×10^{7}	4.68×10^{7}
F_9	3.94×10^{7}	2.71×10^{7}	2.62×10^{7}	2.33×10^{7}	4.11×10^{7}	4.17×10^{9}	2.23×10^{10}	4.01×10^{10}	5.87×10^{10}
F_{10}	1.00×10^{4}	1.35×10^{3}	7.94×10^{2}	9.73×10^{3}	1.02×10^{4}	1.06×10^{4}	1.11×10^{4}	1.17×10^{4}	1.23×10^{4}
F_{11}	1.67×10^{-13}	1.11×10^{-13}	1.04×10^{-13}	1.10×10^{-13}	5.51×10^{-13}	2.59×10^{-2}	1.94×10^{1}	1.08×10^{2}	1.72×10^{2}
F_{12}	1.54×10^{6}	2.34×10^{4}	1.47×10^{4}	4.12×10^{4}	3.02×10^{6}	4.87×10^{6}	6.37×10^{6}	7.75×10^{6}	9.01×10^{6}
F_{13}	5.27×10^{2}	4.92×10^{2}	4.67×10^{2}	4.65×10^{2}	4.69×10^{2}	4.80×10^{2}	1.08×10^{6}	2.45×10^{9}	2.62×10^{10}
F_{14}	1.66×10^{8}	8.14×10^{7}	7.67×10^{7}	7.21×10^{7}	3.28×10^{8}	2.36×10^{10}	5.14×10^{10}	7.78×10^{10}	1.30×10^{11}
F_{15}	1.03×10^{4}	1.03×10^{4}	1.03×10^{4}	1.03×10^{4}	1.03×10^{4}	1.07×10^{4}	1.11×10^{4}	1.17×10^{4}	1.24×10^{4}
F_{16}	2.83×10^{-13}	1.65×10^{-13}	1.55×10^{-13}	1.66×10^{-13}	1.34×10^{-12}	2.74×10^{-1}	1.83×10^{2}	3.16×10^{2}	3.61×10^{2}
F_{17}	5.98×10^{6}	7.91×10^{5}	2.80×10^{5}	3.07×10^{6}	7.05×10^{6}	1.07×10^{7}	1.43×10^{7}	1.76×10^{7}	2.10×10^{7}
F_{18}	1.18×10^{3}	1.26×10^{3}	1.18×10^{3}	1.13×10^{3}	1.00×10^{3}	1.10×10^{5}	8.30×10^{9}	7.98×10^{10}	2.08×10^{11}
F_{19}	1.55×10^{7}	1.23×10^{7}	1.15×10^{7}	1.31×10^{7}	1.64×10^{7}	2.02×10^{7}	2.50×10^{7}	2.92×10^{7}	3.32×10^{7}
F_{20}	9.94×10^{2}	1.02×10^{3}	1.06×10^{3}	9.70×10^{2}	9.78×10^{2}	1.10×10	9.17×10^{9}	8.52×10^{10}	2.20×10^{11}
Rank	4.25	2.75	2.00	2.00	4.15	5.85	7.00	8.00	9.00

F	NP = 600 $\phi = 0.1$	$\phi = 0.2$	$\phi = 0.3$	$\phi = 0.4$	$\phi = 0.5$	$\phi = 0.6$	$\phi = 0.7$	$\phi = 0.8$	$\phi = 0.9$
F_1	2.72×10^{-19}	2.86×10^{-26}	0.00×10^{0}	5.33×10^{-26}	9.43×10^{-16}	1.90×10^{2}	1.57×10^{7}	6.02×10^{8}	4.18×10^{9}
F_2	9.87×10^{3}	6.03×10^{2}	6.59×10^{2}	4.15×10^{3}	1.02×10^{4}	1.06×10^{4}	1.11×10^{4}	1.17×10^{4}	1.23×10^{4}
F_3	8.00×10^{-13}	2.97×10^{-14}	2.90×10^{-14}	3.16×10^{-14}	6.07×10^{-11}	1.06×10^{-1}	1.02×10^{1}	1.61×10^{1}	1.82×10^{1}
F_4	4.16×10^{11}	3.83×10^{11}	3.30×10^{11}	2.84×10^{11}	2.67×10^{11}	3.79×10^{11}	7.59×10^{11}	1.73×10^{13}	1.39×10^{14}
F_5	2.80×10^{8}	2.79×10^{8}	2.76×10^{8}	2.79×10^{8}	2.84×10^{8}	2.87×10^{8}	2.99×10^{8}	3.03×10^{8}	3.18×10^{8}
F_6	4.09×10^{-9}	3.88×10^{-9}	3.88×10^{-9}	4.00×10^{-9}	6.07×10^{-9}	1.45×10^{-1}	1.08×10^{1}	1.68×10^{1}	1.89×10^{1}
F_7	2.63×10^{2}	1.79×10^{1}	1.21×10^{1}	8.88×10^{0}	3.52×10^{2}	7.34×10^{4}	6.46×10^{5}	1.15×10^{8}	3.95×10^{9}
F_8	3.08×10^{7}	2.23×10^{7}	1.74×10^{7}	1.59×10^{7}	2.52×10^{7}	3.96×10^{7}	4.50×10^{7}	4.64×10^{7}	4.69×10^{7}
F_9	4.84×10^{7}	3.01×10^{7}	2.85×10^{7}	2.62×10^{7}	5.78×10^{7}	6.49×10^{9}	2.49×10^{10}	4.26×10^{10}	6.14×10^{10}
F_{10}	1.02×10^{4}	9.45×10^{3}	6.43×10^{3}	9.99×10^{3}	1.03×10^{4}	1.06×10^{4}	1.11×10^{4}	1.17×10^{4}	1.24×10^{4}
F_{11}	6.69×10^{-12}	1.12×10^{-13}	1.05×10^{-13}	1.13×10^{-13}	4.00×10^{-10}	1.90×10^{-1}	2.66×10^{1}	1.17×10^{2}	1.75×10^{2}
F_{12}	2.65×10^{6}	4.99×10^{4}	2.82×10^{4}	1.18×10^{5}	3.37×10^{6}	5.07×10^{6}	6.50×10^{6}	7.83×10^{6}	9.11×10^{6}
F_{13}	4.82×10^{2}	5.44×10^{2}	5.20×10^{2}	4.56×10^{2}	4.42×10^{2}	5.63×10^{2}	3.15×10^{6}	3.37×10^{9}	2.90×10^{10}
F_{14}	2.61×10^{8}	9.02×10^{7}	8.54×10^{7}	8.71×10^{7}	8.67×10^{8}	2.77×10^{10}	5.44×10^{10}	7.81×10^{10}	1.03×10^{11}
F_{15}	1.03×10^{4}	1.03×10^{4}	1.03×10^{4}	1.03×10^{4}	1.03×10^{4}	1.07×10^{4}	1.12×10^{4}	1.17×10^{4}	1.24×10^{4}
F_{16}	1.54×10^{-11}	1.74×10^{-13}	1.58×10^{-13}	1.80×10^{-13}	1.18×10^{-9}	2.41×10^{0}	2.05×10^{2}	3.22×10^{2}	3.63×10^{2}
F_{17}	6.68×10^{6}	2.56×10^{6}	1.14×10^{6}	4.24×10^{6}	7.51×10^{6}	1.11×10^{7}	1.44×10^{7}	1.81×10^{7}	2.12×10^{7}
F_{18}	1.08×10^{3}	1.22×10^{3}	1.22×10^{3}	1.05×10^{3}	9.59×10^{2}	1.41×10^{6}	1.12×10^{10}	8.67×10^{10}	2.19×10^{11}
F_{19}	1.62×10^{7}	1.31×10^{7}	1.23×10^{7}	1.40×10^{7}	1.72×10^{7}	2.06×10^{7}	2.48×10^{7}	2.97×10^{7}	3.49×10^{7}
F_{20}	9.91×10^{2}	9.85×10^{2}	9.94×10^{2}	9.77×10^{2}	9.86×10^{2}	1.53×10^{6}	1.28×10^{10}	9.33×10^{10}	2.30×10^{11}
Rank	4.10	2.65	1.85	2.30	4.20	5.90	7.00	8.00	9.00

From this table, we obtain the following findings. (1) From the perspective of the Friedman test, when NP is fixed, the setting of parameter ϕ is neither too small nor too large, and the optimal setting is usually within [0.3, 0.6]. Specifically, when NP is 100 and 200, the optimal ϕ is 0.6 and 0.5 respectively. When NP is within [300, 500], the optimal ϕ is consistently 0.4. When NP is 600, the optimal ϕ is 0.3. (2) More specifically, we find that when NP is small, such as 100, the optimal ϕ is usually large. This is because a small NP could not afford enough diversity for DGCELPSO to explore the solution space. Therefore, to improve the diversity, ϕ should be large to enhance the influence of the second guiding exemplar in Equation (4), which is in charge of preventing the updated particle from being greedily attracted by the first guiding exemplar. On the contrary, when NP is large, such as 600, a small ϕ is preferred. This is because a large NP offers too high diversity for DGCELPSO to slow down its convergence. Consequently, to let particles fully exploit the found promising areas, ϕ should be small to decrease the influence of the second guiding exemplar in Equation (4). (3) Taking comprehensive comparisons among all settings of NP along with the associated optimal settings of ϕ, we find that DGCELSO with NP = 300 and ϕ = 0.4 achieves the best overall performance.

Based on the above observation, NP = 300 and ϕ = 0.4 are adopted for DGCELSO in the experiments related to 1000-D optimization problems.

4.2. Comparisons with State-of-the-Art Methods

To comprehensively verify the effectiveness of the devised DGCELSO, this section conducts extensive comparison experiments to compare DGCELSO with several state-of-the-art large-scale algorithms. Specifically, nine popular and latest large-scale methods are selected, namely TPLSO [24], SPLSO (The source code can be downloaded from https://gitee.com/mmmyq/SPLSO, accessed on 1 January 2022) [30], LLSO (The source code can be downloaded from https://gitee.com/mmmyq/LLSO, accessed on 1 January 2022) [25], CSO (The source code can be downloaded from http://www.soft-computing.de/CSO_Matlab_New.zip, accessed on 1 January 2022) [26], SLPSO (The source code can be downloaded from http://www.soft-computing.de/SL_PSO_Matlab.zip, accessed on 1 January 2022) [61], DECC-GDG (The source code can be downloaded from https://ww2.mathworks.cn/matlabcentral/mlc-downloads/downloads/submissions/45783/versions/1/download/zip/CC-GDG-CMAES.zip, accessed on 1 January 2022) [50], DECC-DG2 (The source code can be downloaded from https://bitbucket.org/mno/differential-grouping2/src/master/, accessed on 1 January 2022) [35], DECC-RDG (The source code can be downloaded from https://www.researchgate.net/profile/Yuan-Sun-18/publications, accessed on 1 January 2022) [37], and DECC-RDG2 (The source code can be downloaded from https://www.researchgate.net/profile/Yuan-Sun-18/publications, accessed on 1 January 2022) [52]. The former five large-scale optimizers are state-of-the-art holistic large-scale PSO variants, while the latter four algorithms are state-of-the-art cooperative coevolutionary evolutionary algorithms. Compared with these nine different state-of-the-art large-scale optimizers, the effectiveness of DGCELSO is expected to be demonstrated.

Tables 2 and 4 display the comparison results between DGCELSO and the nine compared algorithms on the 1000-D CEC'2010 and the 1000-D CEC'2013 large-scale benchmark sets, respectively. In these two tables, the symbols, "+", "−", and "=" above the p-values obtained from the Wilcoxon rank test denote that the proposed DGCELSO is significantly superior to, significantly inferior to, and equivalent to the associated compared algorithms on the related functions, respectively. "*w/t/l*" in the second to last rows of the two tables count the numbers of functions where DGCELSO performs significantly better, equivalently, and significantly worse than the associated compared methods. Actually, they are the numbers of "+", "=" and "−", respectively. In the last rows of the two tables, the averaged ranks of all algorithms obtained from the Friedman test are presented as well.

Table 2. Fitness comparison between DECELSO and the compared algorithms on the 1000-D CEC'2010 problems with 3×10^6 fitness evaluations.

F	Quality	DGCELSO	TPLSO	SPLSO	LLSO	CSO	SLPSO	DECC-GDG	DECC-DG2	DECC-RDG	DECC-RDG2
F_1	Median	0.00×10^{0}	1.98×10^{-18}	7.70×10^{-20}	2.97×10^{-22}	4.64×10^{-12}	7.65×10^{-18}	6.53×10^{0}	1.95×10^{-1}	2.60×10^{-3}	1.05×10^{-3}
	Mean	0.00×10^{0}	1.93×10^{-18}	7.73×10^{-20}	3.13×10^{-22}	4.75×10^{-12}	7.73×10^{-18}	6.54×10^{0}	7.34×10^{-1}	6.42×10^{0}	8.08×10^{-3}
	Std	0.00×10^{0}	3.04×10^{-19}	6.95×10^{-21}	6.93×10^{-23}	7.77×10^{-13}	8.84×10^{-19}	9.35×10^{-1}	1.61×10^{0}	3.41×10^{1}	3.28×10^{-2}
	p-value	-	$4.32 \times 10^{-8+}$	$4.32 \times 10^{-8+}$	$4.32 \times 10^{-8+}$	$4.32 \times 10^{-8+}$	$4.32 \times 10^{-8+}$	$4.32 \times 10^{-8+}$	$4.32 \times 10^{-8+}$	$4.32 \times 10^{-8+}$	$4.32 \times 10^{-8+}$
F_2	Median	8.85×10^{2}	1.13×10^{3}	4.45×10^{2}	9.71×10^{2}	7.52×10^{3}	1.94×10^{3}	1.40×10^{3}	3.00×10^{3}	2.98×10^{3}	2.99×10^{3}
	Mean	8.88×10^{2}	1.11×10^{3}	4.45×10^{2}	9.78×10^{2}	7.48×10^{3}	1.93×10^{3}	1.40×10^{3}	3.00×10^{3}	2.98×10^{3}	3.00×10^{3}
	Std	4.13×10^{1}	8.28×10^{1}	1.63×10^{1}	5.17×10^{1}	2.60×10^{2}	8.05×10^{1}	2.67×10^{1}	1.34×10^{2}	1.16×10^{2}	1.35×10^{2}
	p-value	-	$4.32 \times 10^{-8+}$	$4.32 \times 10^{-8-}$	$4.32 \times 10^{-8+}$	$4.32 \times 10^{-8+}$	$4.32 \times 10^{-8+}$	$4.32 \times 10^{-8+}$	$4.32 \times 10-8^{+}$	$4.32 \times 10-8^{+}$	$4.32 \times 10-8^{+}$
F_3	Median	3.24×10^{-14}	1.44×10^{0}	2.56×10^{-13}	2.89×10^{-14}	2.56×10^{-9}	1.88×10^{0}	1.12×10^{1}	1.08×10^{1}	1.12×10^{1}	1.11×10^{1}
	Mean	3.18×10^{-14}	1.45×10^{0}	2.52×10^{-13}	2.76×10^{-14}	2.57×10^{-9}	1.84×10^{0}	1.11×10^{1}	1.09×10^{1}	1.11×10^{1}	1.10×10^{1}
	Std	1.32×10^{-15}	1.34×10^{-1}	1.86×10^{-14}	2.16×10^{-15}	1.82×10^{-10}	2.62×10^{-1}	5.69×10^{-1}	6.40×10^{-1}	6.46×10^{-1}	6.88×10^{-1}
	p-value	-	$4.32 \times 10^{-8+}$	$4.32 \times 10^{-8+}$	$4.32 \times 10^{-8-}$	$4.32 \times 10^{-8+}$	$4.32 \times 10^{-8+}$	$4.32 \times 10^{-8+}$	$4.32 \times 10^{-8+}$	$4.32 \times 10^{-8+}$	$4.32 \times 10^{-8+}$
F_4	Median	1.58×10^{11}	2.77×10^{11}	4.36×10^{11}	4.48×10^{11}	6.92×10^{11}	2.68×10^{11}	1.37×10^{14}	1.44×10^{12}	1.39×10^{12}	1.37×10^{12}
	Mean	1.60×10^{11}	2.89×10^{11}	4.30×10^{11}	4.54×10^{11}	6.87×10^{11}	2.83×10^{11}	1.38×10^{14}	1.69×10^{12}	1.49×10^{12}	1.44×10^{12}
	Std	3.72×10^{10}	9.22×10^{10}	8.17×10^{10}	1.29×10^{11}	1.76×10^{11}	8.77×10^{10}	2.68×10^{13}	6.16×10^{11}	6.33×10^{11}	5.35×10^{11}
	p-value	-	$1.00 \times 10^{0=}$	$3.49 \times 10^{-3+}$	$4.32 \times 10^{-8+}$	$1.02 \times 10^{-3+}$	$3.19 \times 10^{-7+}$	$4.32 \times 10^{-8+}$	$4.32 \times 10^{-8+}$	$4.32 \times 10^{-8+}$	$3.19 \times 10^{-7+}$
F_5	Median	2.82×10^{8}	1.63×10^{7}	5.97×10^{6}	1.09×10^{7}	2.00×10^{6}	2.89×10^{7}	3.84×10^{8}	1.72×10^{8}	1.75×10^{8}	1.72×10^{8}
	Mean	2.80×10^{8}	1.59×10^{7}	6.30×10^{6}	1.16×10^{7}	2.46×10^{6}	3.04×10^{7}	3.82×10^{8}	1.75×10^{8}	1.71×10^{8}	1.73×10^{8}
	Std	9.11×106	4.51×10^{6}	1.73×10^{6}	2.93×10^{6}	1.33×10^{6}	8.42×10^{6}	1.54×10^{7}	1.84×10^{7}	1.84×10^{7}	1.50×10^{7}
	p-value	-	$4.32 \times 10^{-8-}$	$4.32 \times 10^{-8-}$	$4.32 \times 10^{-8-}$	$4.32 \times 10^{-8-}$	$4.32 \times 10^{-8-}$	$4.32 \times 10^{-8+}$	$4.32 \times 10^{-8-}$	$4.32 \times 10^{-8-}$	$4.32 \times 10^{-8-}$
F_6	Median	4.00×10^{-9}	2.08×10^{0}	1.00×10^{-8}	4.00×10^{-9}	8.18×10^{-7}	2.14×10^{1}	3.51×10^{5}	8.81×10^{0}	1.07×10^{1}	1.06×10^{1}
	Mean	4.00×10^{-9}	2.20×10^{0}	9.44×10^{-9}	4.00×10^{-9}	8.16×10^{-7}	1.95×10^{1}	3.58×10^{5}	8.90×10^{0}	1.05×10^{1}	1.05×10^{1}
	Std	3.73×10^{-15}	3.74×10^{-1}	1.18×10^{-9}	8.27×10^{-25}	2.57×10^{-8}	4.13×10^{0}	4.27×10^{4}	6.50×10^{-1}	7.02×10^{-1}	6.84×10^{-1}
	p-value	-	$4.32 \times 10^{-8+}$	$4.32 \times 10^{-8+}$	$4.32 \times 10^{-8-}$	$4.32 \times 10^{-8+}$	$4.32 \times 10^{-8+}$	$4.32 \times 10^{-8+}$	$4.32 \times 10^{-8+}$	$4.32 \times 10^{-8+}$	$4.32 \times 10^{-8+}$
F_7	Median	1.89×10^{-5}	9.21×10^{2}	4.51×10^{2}	6.58×10^{0}	2.13×10^{4}	6.26×10^{4}	2.98×10^{10}	1.80×10^{3}	4.86×10^{1}	5.18×10^{1}
	Mean	2.15×10^{-5}	5.86×10^{3}	4.76×10^{2}	2.31×10^{1}	2.13×10^{4}	6.49×10^{4}	3.10×10^{10}	1.98×10^{3}	6.40×10^{1}	5.87×10^{1}
	Std	1.55×10^{-5}	1.03×10^{4}	1.29×10^{2}	7.45×10^{1}	4.53×10^{3}	3.81×10^{4}	4.19×10^{9}	9.49×10^{2}	4.67×10^{1}	3.71×10^{1}
	p-value	-	$2.07 \times 10^{-6+}$	$4.32 \times 10^{-8+}$	$4.32 \times 10^{-8+}$	$4.32 \times 10^{-8+}$	$4.32 \times 10^{-8+}$	$4.32 \times 10^{-8+}$	$4.32 \times 10^{-8+}$	$4.32 \times 10^{-8+}$	$4.32 \times 10^{-8+}$

Table 2. *Cont.*

F	Quality	DGCELSO	TPLSO	SPLSO	LLSO	CSO	SLPSO	DECC-GDG	DECC-DG2	DECC-RDG	DECC-RDG2
F_8	Median	4.28×10^{3}	4.78×10^{5}	3.11×10^{7}	2.33×10^{7}	3.86×10^{7}	7.51×10^{6}	6.78×10^{8}	6.05×10^{2}	6.57×10^{-1}	3.68×10^{-1}
	Mean	4.36×10^{3}	4.98×10^{5}	3.11×10^{7}	2.33×10^{7}	3.87×10^{7}	7.57×10^{6}	8.05×10^{8}	2.71×10^{5}	6.65×10^{5}	7.43×10^{-1}
	Std	4.17×10^{2}	1.43×10^{5}	9.43×10^{4}	2.96×10^{5}	8.47×10^{4}	2.44×10^{6}	4.70×10^{8}	9.94×10^{5}	1.49×10^{6}	1.24×10^{0}
	p-value	-	4.32×10^{-8}+	4.32×10^{-8}+	4.32×10^{-8}+	4.32×10^{-8}+	4.32×10^{-8}+	4.32×10^{-8}+	4.32×10^{-8}+	4.32×10^{-8}+	4.32×10^{-8}+
F_9	Median	1.76×10^{7}	4.25×10^{7}	4.57×10^{7}	4.64×10^{7}	6.65×10^{7}	3.31×10^{7}	7.45×10^{8}	2.15×10^{8}	1.76×10^{8}	1.77×10^{8}
	Mean	1.77×10^{7}	4.32×10^{7}	4.59×10^{7}	4.48×10^{7}	6.68×10^{7}	3.35×10^{7}	7.43×10^{8}	2.18×10^{8}	1.73×10^{8}	1.77×10^{8}
	Std	1.69×10^{6}	4.10×10^{6}	2.99×10^{6}	4.16×10^{6}	4.38×10^{6}	3.63×10^{6}	3.71×10^{7}	1.73×10^{7}	1.22×10^{7}	1.66×10^{7}
	p-value	-	4.32×10^{-8}+	1.00×10^{0}=	4.32×10^{-8}+	4.32×10^{-8}+	4.32×10^{-8}+	4.32×10^{-8}+	4.32×10^{-8}+	4.32×10^{-8}+	4.32×10^{-8}+
F_{10}	Median	9.18×10^{2}	9.67×10^{2}	7.99×10^{3}	8.87×10^{2}	9.58×10^{3}	2.59×10^{3}	4.16×10^{3}	6.73×10^{3}	6.32×10^{3}	6.27×10^{3}
	Mean	9.23×10^{2}	9.84×10^{2}	7.99×10^{3}	8.88×10^{2}	9.58×10^{3}	2.79×10^{3}	4.15×10^{3}	6.72×10^{3}	6.32×10^{3}	6.27×10^{3}
	Std	3.82×10^{1}	8.52×10^{1}	1.25×10^{2}	3.50×10^{1}	6.49×10^{1}	1.28×10^{3}	5.70×10^{1}	9.30×10^{1}	1.12×10^{2}	1.09×10^{2}
	p-value	-	1.06×10^{-2}+	4.32×10^{-8}+	4.32×10^{-8}−	4.32×10^{-8}+	4.32×10^{-8}+	4.32×10^{-8}+	4.32×10^{-8}+	4.32×10^{-8}+	4.32×10^{-8}+
F_{11}	Median	1.11×10^{-13}	3.48×10^{0}	3.02×10^{-12}	2.90×10^{0}	3.98×10^{-8}	2.37×10^{1}	5.58×10^{0}	5.39×10^{0}	4.76×10^{0}	4.86×10^{0}
	Mean	1.10×10^{-13}	3.50×10^{0}	3.05×10^{-12}	5.51×10^{0}	3.98×10^{-8}	2.42×10^{1}	5.53×10^{0}	5.59×10^{0}	4.75×10^{0}	4.86×10^{0}
	Std	2.36×10^{-15}	1.30×10^{0}	2.84×10^{-13}	5.43×10^{0}	3.19×10^{-9}	3.03×10^{0}	5.49×10^{-1}	6.12×10^{-1}	4.79×10^{-1}	3.88×10^{-1}
	p-value	-	4.32×10^{-8}+	4.32×10^{-8}+	4.32×10^{-8}+	4.32×10^{-8}+	4.32×10^{-8}+	4.32×10^{-8}+	4.32×10^{-8}+	4.32×10^{-8}+	4.32×10^{-8}+
F_{12}	Median	2.55×10^{3}	1.23×10^{4}	9.39×10^{4}	1.24×10^{4}	4.25×10^{5}	1.30×10^{4}	2.87×10^{5}	3.99×10^{4}	2.22×10^{4}	2.21×10^{4}
	Mean	2.55×10^{3}	1.23×10^{4}	9.53×10^{4}	1.23×10^{4}	4.37×10^{5}	1.54×10^{4}	2.87×10^{5}	3.94×10^{4}	2.21×10^{4}	2.19×10^{4}
	Std	2.13×10^{2}	1.30×10^{3}	6.64×10^{3}	1.32×10^{3}	6.49×10^{4}	7.06×10^{3}	1.10×10^{4}	2.17×10^{3}	1.28×10^{3}	1.45×10^{3}
	p-value	-	4.32×10^{-8}+	4.32×10^{-8}+	4.32×10^{-8}+	4.32×10^{-8}+	3.19×10^{-7}+	4.32×10^{-8}+	4.32×10^{-8}+	4.32×10^{-8}+	4.32×10^{-8}+
F_{13}	Median	4.64×10^{2}	7.29×10^{2}	4.50×10^{2}	7.82×10^{2}	4.68×10^{2}	8.87×10^{2}	1.39×10^{3}	1.65×10^{3}	8.25×10^{2}	8.17×10^{2}
	Mean	5.15×10^{2}	7.54×10^{2}	5.48×10^{2}	7.91×10^{2}	5.53×10^{2}	9.81×10^{2}	1.42×10^{3}	1.77×10^{3}	8.24×10^{2}	8.40×10^{2}
	Std	1.49×10^{2}	1.07×10^{2}	1.66×10^{2}	2.37×10^{2}	1.75×10^{2}	3.86×10^{2}	3.40×10^{2}	5.06×10^{2}	1.35×10^{2}	1.98×10^{2}
	p-value	-	5.90×10^{-5}+	3.49×10^{-3}+	4.32×10^{-8}+	2.73×10^{-1}=	2.85×10^{-2}+	3.49×10^{-3}+	5.90×10^{-5}+	1.18×10^{-5}+	2.07×10^{-6}+
F_{14}	Median	5.10×10^{7}	1.29×10^{8}	1.61×10^{8}	1.23×10^{8}	2.46×10^{8}	8.61×10^{7}	8.59×10^{8}	8.71×10^{8}	7.19×10^{8}	7.18×10^{8}
	Mean	5.17×10^{7}	1.32×10^{8}	1.60×10^{8}	1.22×10^{8}	2.46×10^{8}	8.55×10^{7}	8.64×10^{8}	8.60×10^{8}	7.23×10^{8}	7.25×10^{8}
	Std	2.76×10^{6}	9.33×10^{6}	8.42×10^{6}	6.41×10^{6}	1.29×10^{7}	7.57×10^{6}	3.30×10^{7}	4.17×10^{7}	3.65×10^{7}	3.44×10^{7}
	p-value	-	4.32×10^{-8}+	2.07×10^{-6}+	4.32×10^{-8}+	4.32×10^{-8}+	4.32×10^{-8}+	4.32×10^{-8}+	4.32×10^{-8}+	4.32×10^{-8}+	4.32×10^{-8}+

Table 2. *Cont.*

F	Quality	DGCELSO	TPLSO	SPLSO	LLSO	CSO	SLPSO	DECC-GDG	DECC-DG2	DECC-RDG	DECC-RDG2
F_{15}	Median	1.04×10^{4}	1.04×10^{4}	9.92×10^{3}	8.30×10^{2}	1.01×10^{4}	1.12×10^{4}	6.75×10^{3}	6.73×10^{3}	6.55×10^{3}	6.56×10^{3}
	Mean	1.04×10^{4}	8.88×10^{3}	9.91×10^{3}	8.97×10^{2}	1.01×10^{4}	1.12×10^{4}	6.76×10^{3}	6.73×10^{3}	6.55×10^{3}	6.55×10^{3}
	Std	6.65×10^{1}	3.41×10^{3}	6.31×10^{1}	3.47×10^{2}	6.48×10^{1}	1.19×10^{2}	8.82×10^{1}	7.27×10^{1}	8.86×10^{1}	8.39×10^{1}
	p-value	-	$1.44 \times 10^{-1=}$	$4.32 \times 10^{-8-}$	$4.32 \times 10^{-8-}$	$1.06 \times 10^{-2-}$	$4.32 \times 10^{-8+}$	$4.32 \times 10^{-8-}$	$4.32 \times 10^{-8-}$	$4.32 \times 10^{-8-}$	$4.32 \times 10^{-8-}$
F_{16}	Median	1.55×10^{-13}	1.78×10^{1}	4.66×10^{-12}	4.40×10^{0}	5.64×10^{-8}	2.12×10^{1}	3.98×10^{-4}	3.89×10^{-4}	1.92×10^{-5}	1.88×10^{-5}
	Mean	1.55×10^{-13}	1.89×10^{1}	4.68×10^{-12}	4.33×10^{0}	5.68×10^{-8}	2.36×10^{1}	3.97×10^{-4}	3.90×10^{-4}	1.93×10^{-5}	1.89×10^{-5}
	Std	2.66×10^{-15}	7.46×10^{0}	4.41×10^{-13}	2.50×10^{0}	6.21×10^{-9}	1.11×10^{1}	1.44×10^{-5}	1.33×10^{-5}	8.87×10^{-7}	8.30×10^{-7}
	p-value	-	$4.32 \times 10^{-8+}$	$4.32 \times 10^{-8+}$	$4.32 \times 10^{-8+}$	$4.32 \times 10^{-8+}$	$4.32 \times 10^{-8+}$	$4.32 \times 10^{-8+}$	$4.32 \times 10^{-8+}$	$4.32 \times 10^{-8+}$	$4.32 \times 10^{-8+}$
F_{17}	Median	6.70×10^{4}	9.65×10^{4}	6.90×10^{5}	9.17×10^{4}	2.19×10^{6}	8.64×10^{4}	2.64×10^{5}	2.64×10^{5}	1.99×10^{5}	1.97×10^{5}
	Mean	6.57×10^{4}	9.83×10^{4}	6.84×10^{5}	9.12×10^{4}	2.21×10^{6}	8.74×10^{4}	2.65×10^{5}	2.63×10^{5}	1.98×10^{5}	1.98×10^{5}
	Std	7.55×10^{3}	9.90×10^{3}	3.57×10^{4}	5.43×10^{3}	2.07×10^{5}	1.39×10^{4}	7.79×10^{3}	7.33×10^{3}	8.75×10^{3}	9.45×10^{3}
	p-value	-	$4.32 \times 10^{-8+}$	$7.15 \times 10^{-2+}$	$4.32 \times 10^{-8+}$	$4.32 \times 10^{-8+}$	$4.32 \times 10^{-8+}$	$4.32 \times 10^{-8+}$	$4.32 \times 10^{-8+}$	$4.32 \times 10^{-8+}$	$4.32 \times 10^{-8+}$
F_{18}	Median	1.25×10^{3}	2.29×10^{3}	1.25×10^{3}	2.49×10^{3}	1.38×10^{3}	2.95×10^{3}	1.15×10^{3}	1.14×10^{3}	1.08×10^{3}	1.11×10^{3}
	Mean	1.31×10^{3}	2.36×10^{3}	1.35×10^{3}	2.51×10^{3}	1.64×10^{3}	2.92×10^{3}	1.16×10^{3}	1.13×10^{3}	1.07×10^{3}	1.10×10^{3}
	Std	2.94×10^{2}	4.19×10^{2}	3.81×10^{2}	7.42×10^{2}	8.13×10^{2}	8.08×10^{2}	1.31×10^{2}	1.29×10^{2}	1.08×10^{2}	1.02×10^{2}
	p-value	-	$5.90 \times 10^{-5+}$	$3.49 \times 10^{-3+}$	$2.61 \times 10^{-4+}$	$2.73 \times 10^{-1=}$	$2.85 \times 10^{-2+}$	$3.19 \times 10^{-7-}$	$3.19 \times 10^{-7-}$	$3.19 \times 10^{-7-}$	$4.32 \times 10^{-8-}$
F_{19}	Median	1.02×10^{7}	3.94×10^{6}	8.19×10^{6}	1.85×10^{6}	9.78×10^{6}	5.20×10^{6}	2.11×10^{6}	2.09×10^{6}	1.96×10^{6}	1.93×10^{6}
	Mean	1.02×10^{7}	3.89×10^{6}	8.20×10^{6}	1.82×10^{6}	9.86×10^{6}	5.23×10^{6}	2.12×10^{6}	2.10×10^{6}	1.95×10^{6}	1.92×10^{6}
	Std	7.69×10^{5}	2.64×10^{5}	4.61×10^{5}	9.22×10^{4}	5.07×10^{5}	9.15×10^{5}	8.77×10^{4}	9.92×10^{4}	7.80×10^{4}	1.05×10^{5}
	p-value	-	$4.32 \times 10^{-8-}$	$4.32 \times 10^{-8-}$	$4.32 \times 10^{-8-}$	$4.32 \times 10^{-8-}$	$4.32 \times 10^{-8-}$	$4.32 \times 10^{-8-}$	$4.32 \times 10^{-8-}$	$4.32 \times 10^{-8-}$	$4.32 \times 10^{-8-}$
F_{20}	Median	1.06×10^{3}	2.04×10^{3}	9.79×10^{2}	1.88×10^{3}	9.87×10^{2}	1.73×10^{3}	5.43×10^{3}	5.33×10^{3}	4.32×10^{3}	4.25×10^{3}
	Mean	1.08×10^{3}	2.08×10^{3}	1.06×10^{3}	1.92×10^{3}	1.07×10^{3}	1.73×10^{3}	5.45×10^{3}	5.46×10^{3}	4.28×10^{3}	4.34×10^{3}
	Std	7.30×10^{1}	2.00×10^{2}	1.75×10^{2}	3.00×10^{2}	1.70×10^{2}	1.53×10^{2}	3.32×10^{2}	3.37×10^{2}	2.29×10^{2}	3.20×10^{2}
	p-value	-	$4.32 \times 10^{-8+}$	$5.90 \times 10^{-5-}$	$4.32 \times 10^{-8+}$	$5.90 \times 10^{-5-}$	$1.18 \times 10^{-5+}$	$4.32 \times 10^{-8+}$	$4.32 \times 10^{-8+}$	$4.32 \times 10^{-8+}$	$4.32 \times 10^{-8+}$
	w/t/l		16/2/2	14/1/5	15/0/5	15/2/3	18/0/2	17/0/3	16/0/4	16/0/4	16/0/4
Rank		2.75	4.80	4.65	3.70	6.25	6.05	8.20	7.10	5.85	5.65

Table 3. Fitness comparison between DECELSO and the compared algorithms on the 1000-D CEC'2013 problems with 3×10^6 fitness evaluations.

F	Quality	DGCELSO	TPLSO	SPLSO	LLSO	CSO	SLPSO	DECC-GDG	DECC-DG2	DECC-RDG	DECC-RDG2
F_1	Median	0.00×10^{0}	3.21×10^{-18}	1.17×10^{-19}	4.02×10^{-22}	7.92×10^{-12}	1.03×10^{-17}	7.06×10^{0}	3.46×10^{0}	2.04×10^{-2}	2.96×10^{-2}
	Mean	0.00×10^{0}	3.81×10^{-18}	1.18×10^{-19}	4.28×10^{-22}	7.88×10^{-12}	1.65×10^{-17}	7.43×10^{0}	6.31×10^{0}	3.51×10^{-2}	1.08×10^{-1}
	Std	0.00×10^{0}	1.57×10^{-18}	1.04×10^{-20}	1.29×10^{-22}	1.19×10^{-12}	3.25×10^{-17}	9.38×10^{-1}	7.78×10^{0}	3.88×10^{-2}	2.08×10^{-1}
	p-value	-	$4.32 \times 10^{-8+}$	$4.32 \times 10^{-8+}$	$9.63 \times 10^{-7+}$	$4.32 \times 10^{-8+}$	$4.32 \times 10^{-8+}$	$4.32 \times 10^{-8+}$	$4.32 \times 10^{-8+}$	$4.32 \times 10^{-8+}$	$4.32 \times 10^{-8+}$
F_2	Median	8.61×10^{2}	1.30×10^{3}	9.64×10^{2}	1.14×10^{3}	8.58×10^{3}	2.09×10^{3}	1.43×10^{3}	7.81×10^{3}	7.81×10^{3}	7.69×10^{3}
	Mean	8.77×10^{2}	1.34×10^{3}	1.06×10^{3}	1.14×10^{3}	8.58×10^{3}	2.10×10^{3}	1.43×10^{3}	7.88×10^{3}	7.74×10^{3}	7.74×10^{3}
	Std	4.28×10^{1}	1.75×10^{2}	4.38×10^{2}	5.00×10^{1}	1.76×10^{2}	1.61×10^{2}	2.43×10^{1}	4.07×10^{2}	3.47×10^{2}	3.56×10^{2}
	p-value	-	$4.32 \times 10^{-8+}$	$4.32 \times 10^{-8+}$	$4.32 \times 10^{-8+}$	$4.32 \times 10^{-8+}$	$4.32 \times 10^{-8+}$	$4.32 \times 10^{-8+}$	$4.32 \times 10^{-8+}$	$4.32 \times 10^{-8+}$	$4.32 \times 10^{-8+}$
F_3	Median	2.16×10^{1}	2.22×10^{1}	2.16×10^{1}	2.16×10^{1}	2.16×10^{1}	2.16×10^{1}	2.15×10^{1}	2.15×10^{1}	2.14×10^{1}	2.15×10^{1}
	Mean	2.16×10^{1}	2.31×10^{1}	2.16×10^{1}	2.16×10^{1}	2.16×10^{1}	2.16×10^{1}	2.15×10^{1}	2.15×10^{1}	2.14×10^{1}	2.15×10^{1}
	Std	6.26×10^{-3}	1.72×10^{0}	7.11×10^{-15}	7.11×10^{-15}	7.11×10^{-15}	2.37×10^{-1}	3.00×10^{-2}	4.23×10^{-2}	4.82×10^{-2}	4.90×10^{-2}
	p-value	-	$3.49 \times 10^{-3+}$	$2.61 \times 10^{-4-}$	$2.61 \times 10^{-4-}$	$2.07 \times 10^{-6-}$	$7.15 \times 10^{-1=}$	$4.65 \times 10^{-1=}$	$4.65 \times 10^{-1=}$	$7.15 \times 10^{-1=}$	$7.15 \times 10^{-1=}$
F_4	Median	2.55×10^{9}	4.23×10^{9}	9.14×10^{9}	6.40×10^{9}	1.22×10^{10}	4.28×10^{9}	4.15×10^{11}	8.12×10^{10}	7.45×10^{10}	6.10×10^{10}
	Mean	2.52×10^{9}	4.27×10^{9}	9.41×10^{9}	6.55×10^{9}	1.35×10^{10}	4.33×10^{9}	4.20×10^{11}	7.79×10^{10}	7.16×10^{10}	6.78×10^{10}
	Std	6.55×10^{8}	1.03×10^{9}	1.86×10^{9}	1.40×10^{9}	3.12×10^{9}	9.91×10^{8}	7.75×10^{10}	2.19×10^{10}	1.92×10^{10}	2.32×10^{10}
	p-value	-	$1.44 \times 10^{-1=}$	$1.02 \times 10^{-3+}$	$1.06 \times 10^{-2+}$	$4.32 \times 10^{-8+}$	$4.32 \times 10^{-8+}$	$4.32 \times 10^{-8+}$	$4.32 \times 10^{-8+}$	$4.32 \times 10^{-8+}$	$4.32 \times 10^{-8+}$
F_5	Median	7.83×10^{5}	6.80×10^{5}	6.43×10^{5}	6.51×10^{5}	5.90×10^{5}	8.89×10^{5}	8.62×10^{6}	6.10×10^{6}	5.81×10^{6}	5.72×10^{6}
	Mean	7.91×10^{5}	6.79×10^{5}	6.30×10^{5}	6.56×10^{5}	5.97×10^{5}	8.90×10^{5}	8.66×10^{6}	6.06×10^{6}	5.72×10^{6}	5.67×10^{6}
	Std	1.03×10^{5}	1.10×10^{5}	1.00×10^{5}	1.01×10^{5}	1.03×10^{5}	1.31×10^{5}	2.80×10^{5}	2.40×10^{5}	4.24×10^{5}	3.61×10^{5}
	p-value	-	$1.06 \times 10^{-2-}$	$1.18 \times 10^{-5-}$	$3.49 \times 10^{-3-}$	$2.61 \times 10^{-4-}$	$2.85 \times 10^{-2+}$	$4.32 \times 10^{-8+}$	$4.32 \times 10^{-8+}$	$4.32 \times 10^{-8+}$	$4.32 \times 10^{-8+}$
F_6	Median	1.06×10^{6}	1.17×10^{6}	1.06×10^{6}	1.06×10^{6}	1.06×10^{6}	1.06×10^{6}	1.06×10^{6}	1.06×10^{6}	1.06×10^{6}	1.06×10^{6}
	Mean	1.06×10^{6}	1.22×10^{6}	1.06×10^{6}	1.06×10^{6}	1.06×10^{6}	1.06×10^{6}	1.06×10^{6}	1.06×10^{6}	1.06×10^{6}	1.06×10^{6}
	Std	1.27×10^{3}	1.61×10^{5}	0.00×10^{0}	0.00×10^{0}	0.00×10^{0}	0.00×10^{0}	3.00×10^{3}	0.00×10^{0}	0.00×10^{0}	0.00×10^{0}
	p-value	-	$5.90 \times 10^{-5+}$	$4.65 \times 10^{-1=}$	$1.18 \times 10^{-5-}$	$2.73 \times 10^{-1=}$	$2.61 \times 10^{-4-}$	$4.65 \times 10^{-1=}$	$1.44 \times 10^{-1=}$	$2.73 \times 10^{-1=}$	$2.85 \times 10^{-2-}$
F_7	Median	7.93×10^{4}	1.22 × 106	5.42 × 106	1.70 × 106	5.45 × 106	1.47 × 106	7.45 × 108	7.36 × 107	2.84 × 108	8.36 × 107
	Mean	9.71×10^{4}	1.24 × 106	5.50 × 106	1.87 × 106	5.81 × 106	1.58 × 106	7.67 × 108	7.79 × 107	3.65 × 108	8.25 × 107
	Std	5.54×10^{4}	5.05 × 105	2.23 × 106	1.08 × 106	3.04 × 106	7.53 × 105	1.32 × 108	2.73 × 107	2.63 × 108	2.06 × 107
	p-value	-	$1.18 \times 10-5^{+}$	$1.18 \times 10-5^{+}$	$4.32 \times 10-8^{+}$	$2.61 \times 10-4^{+}$	$3.19 \times 10-7^{+}$	$4.32 \times 10-8^{+}$	$4.32 \times 10-8^{+}$	$4.32 \times 10-8^{+}$	$4.32 \times 10-8^{+}$
F_8	Median	5.58×10^{13}	7.07×10^{13}	1.56×10^{14}	1.37×10^{14}	2.43×10^{14}	9.65×10^{13}	1.70×10^{16}	9.35×10^{15}	6.96×10^{15}	5.83×10^{15}
	Mean	6.15×10^{13}	7.28×10^{13}	1.55×10^{14}	1.36×10^{14}	2.46×10^{14}	1.09×10^{14}	1.65×10^{16}	9.32×10^{15}	6.95×10^{15}	6.38×10^{15}
	Std	2.08×10^{13}	4.02×10^{13}	2.92×10^{13}	3.39×10^{13}	8.71×10^{13}	5.44×10^{13}	4.49×10^{15}	2.71×10^{15}	1.64×10^{15}	1.99×10^{15}
	p-value	-	$5.90 \times 10^{-5+}$	$1.44 \times 10^{-1=}$	$2.85 \times 10^{-2-}$	$1.18 \times 10^{-5-}$	$3.49 \times 10^{-3-}$	$4.32 \times 10^{-8-}$	$4.32 \times 10^{-8-}$	$4.32 \times 10^{-8-}$	$4.32 \times 10^{-8-}$

Table 4. Fitness comparison between DECELSO and the compared algorithms on the 1000-D CEC'2013 problems with 3×10^{6} fitness evaluations.

F	Quality	DGCELSO	TPLSO	SPLSO	LLSO	CSO	SLPSO	DECC-GDG	DECC-DG2	DECC-RDG	DECC-RDG2
F_9	Median	4.67×10^{7}	4.52×10^{7}	7.23×10^{7}	1.11×10^{8}	5.94×10^{7}	8.05×10^{7}	5.62×10^{8}	5.55×10^{8}	5.40×10^{8}	5.32×10^{8}
	Mean	4.47×10^{7}	4.28×10^{7}	8.08×10^{7}	1.29×10^{8}	6.08×10^{7}	7.99×10^{7}	5.61×10^{8}	5.59×10^{8}	5.38×10^{8}	5.31×10^{8}
	Std	1.37×10^{7}	7.49×10^{6}	2.21×10^{7}	8.85×10^{7}	1.29×10^{7}	1.18×10^{7}	3.24×10^{7}	2.93×10^{7}	3.03×10^{7}	2.33×10^{7}
	p-value	-	$4.65 \times 10^{-1=}$	$3.19 \times 10^{-7+}$	$4.32 \times 10^{-8+}$	$2.61 \times 10^{-4+}$	$3.19 \times 10^{-7+}$	$4.32 \times 10^{-8+}$	$4.32 \times 10^{-8+}$	$4.32 \times 10^{-8+}$	$4.32 \times 10^{-8+}$
F_{10}	Median	9.40×10^{7}	9.44×10^{7}	9.40×10^{7}	9.41×10^{7}	9.41×10^{7}	9.37×10^{7}	9.46×10^{7}	9.46×10^{7}	9.46×10^{7}	9.45×10^{7}
	Mean	9.40×10^{7}	9.52×10^{7}	9.39×10^{7}	9.41×10^{7}	9.40×10^{7}	9.27×10^{7}	9.46×10^{7}	9.46×10^{7}	9.46×10^{7}	9.45×10^{7}
	Std	2.95×10^{5}	1.70×10^{6}	2.18×10^{5}	2.23×10^{5}	2.14×10^{5}	1.99×10^{6}	2.57×10^{5}	2.51×10^{5}	1.98×10^{5}	2.78×10^{5}
	p-value	-	$1.02 \times 10^{-3+}$	$6.79 \times 10^{-2=}$	$1.18 \times 10^{-5+}$	$2.07 \times 10^{-6+}$	$1.02 \times 10^{-3-}$	$3.19 \times 10^{-7+}$	$3.19 \times 10^{-7+}$	$4.32 \times 10^{-8+}$	$2.07 \times 10^{-6+}$
F_{11}	Median	6.44×10^{7}	1.88×10^{8}	9.22×10^{11}	9.23×10^{11}	9.26×10^{11}	9.38×10^{11}	6.80×10^{8}	1.99×10^{10}	5.75×10^{8}	1.33×10^{10}
	Mean	7.14×10^{7}	1.83×10^{8}	9.27×10^{11}	9.28×10^{11}	9.29×10^{11}	9.34×10^{11}	6.84×10^{8}	2.52×10^{10}	5.68×10^{8}	1.49×10^{10}
	Std	2.45×10^{7}	5.62×10^{7}	9.35×10^{9}	9.68×10^{9}	9.63×10^{9}	8.96×10^{9}	1.09×10^{8}	1.38×10^{10}	9.23×10^{7}	7.57×10^{9}
	p-value	-	$3.49 \times 10^{-3+}$	$4.32 \times 10^{-8+}$	$4.32 \times 10^{-8+}$	$4.32 \times 10^{-8+}$	$4.32 \times 10^{-8+}$	$2.61 \times 10^{-4+}$	$4.32 \times 10^{-8+}$	$3.49 \times 10^{-3+}$	$4.32 \times 10^{-8+}$
F_{12}	Median	1.12×10^{3}	2.19×10^{3}	1.03×10^{3}	1.80×10^{3}	1.04×10^{3}	1.76×10^{3}	5.54×10^{3}	5.42×10^{3}	4.28×10^{3}	4.25×10^{3}
	Mean	1.14×10^{3}	2.13×10^{3}	1.05×10^{3}	1.82×10^{3}	1.08×10^{3}	1.77×10^{3}	5.51×10^{3}	5.59×10^{3}	4.34×10^{3}	4.30×10^{3}
	Std	9.96×10^{1}	2.72×10^{2}	5.45×10^{1}	1.52×10^{2}	7.45×10^{1}	1.69×10^{2}	3.67×10^{2}	7.64×10^{2}	3.24×10^{2}	2.48×10^{2}
	p-value	-	$4.32 \times 10^{-8+}$	$2.85 \times 10^{-2-}$	$4.32 \times 10^{-8+}$	$1.00 \times 10^{0=}$	$4.32 \times 10^{-8+}$	$4.32 \times 10^{-8+}$	$4.32 \times 10^{-8+}$	$4.32 \times 10^{-8+}$	$4.32 \times 10^{-8+}$
F_{13}	Median	4.89×10^{7}	2.01×10^{8}	1.20×10^{9}	2.98×10^{8}	7.08×10^{8}	4.01×10^{8}	1.56×10^{9}	1.43×10^{9}	2.87×10^{9}	7.08×10^{8}
	Mean	6.40×10^{7}	2.21×10^{8}	1.20×10^{9}	3.42×10^{8}	7.48×10^{8}	5.20×10^{8}	1.50×10^{9}	1.47×10^{9}	2.98×10^{9}	7.17×10^{8}
	Std	5.35×10^{7}	1.24×10^{8}	4.91×10^{8}	1.42×10^{8}	2.85×10^{8}	4.85×10^{8}	3.35×10^{8}	3.46×10^{8}	7.23×10^{8}	1.57×10^{8}
	p-value	-	$3.19 \times 10^{-7+}$	$1.00 \times 10^{0=}$	$4.32 \times 10^{-8+}$	$1.02 \times 10^{-3+}$	$4.32 \times 10^{-8+}$	$1.44 \times 10^{-1=}$	$2.73 \times 10^{-1=}$	$4.32 \times 10^{-8+}$	$3.49 \times 10^{-3+}$
F_{14}	Median	1.77×10^{7}	5.86×10^{7}	5.19×10^{9}	8.06×10^{7}	2.90×10^{9}	1.51×10^{8}	4.45×10^{9}	4.54×10^{9}	2.23×10^{9}	2.50×10^{9}
	Mean	1.78×10^{7}	6.05×10^{7}	8.31×10^{9}	1.59×10^{8}	3.67×10^{9}	2.51×10^{8}	5.28×10^{9}	4.58×10^{9}	2.78×10^{9}	3.33×10^{9}
	Std	2.62×10^{6}	1.34×10^{7}	6.56×10^{9}	2.27×10^{8}	3.32×10^{9}	2.25×10^{8}	3.84×10^{9}	1.83×10^{9}	1.85×10^{9}	2.09×10^{9}
	p-value	-	$4.32 \times 10^{-8+}$	$4.32 \times 10^{-8+}$	$4.32 \times 10^{-8+}$	$2.07 \times 10^{-6+}$	$3.19 \times 10^{-7+}$	$2.85 \times 10^{-2+}$	$6.79 \times 10^{-2=}$	$1.00 \times 10^{0=}$	$6.79 \times 10^{-2=}$
F_{15}	Median	3.53×10^{7}	1.29×10^{7}	4.13×10^{7}	4.58×10^{6}	7.60×10^{7}	5.99×10^{7}	8.60×10^{6}	8.82×10^{6}	7.75×10^{6}	8.04×10^{6}
	Mean	3.54×10^{7}	1.26×10^{7}	4.13×10^{7}	4.59×10^{6}	7.61×10^{7}	6.03×10^{7}	8.98×10^{6}	8.95×10^{6}	7.96×10^{6}	8.07×10^{6}
	Std	7.60×10^{6}	1.36×10^{6}	3.05×10^{6}	3.22×10^{5}	6.14×10^{6}	6.54×10^{6}	8.90×10^{5}	9.38×10^{5}	9.30×10^{5}	9.78×10^{5}
	p-value	-	$4.32 \times 10^{-8-}$	$4.32 \times 10^{-8+}$	$4.32 \times 10^{-8+}$	$1.18 \times 10^{-5+}$	$4.65 \times 10^{-1=}$	$4.32 \times 10^{-8-}$	$4.32 \times 10^{-8-}$	$4.32 \times 10^{-8-}$	$4.32 \times 10^{-8-}$
	w/t/l		11/2/2	8/4/3	12/0/3	11/2/2	11/2/2	11/3/1	10/4/1	11/3/1	11/2/2
Rank		2.73	4.47	4.87	4.13	5.80	5.00	8.13	7.40	6.47	6.00

In Table 2, the comparison results on the CEC'2010 set are summarized as follows. (1) From the perspective of the Friedman test, as shown in the last row, it is found that the proposed DGCELSO has the lowest rank value, which is much smaller than those of the compared algorithms. This means that DGCELSO achieves the best overall performance and shows great superiority to the compared algorithms. (2) With respect to the Wilcoxon rank-sum test, as shown in the second last row, it is observed that DGCELSO performs significantly better than the compared algorithms on at least 14 problems. In particular, competed with the four cooperative coevolutionary evolutionary algorithms, DGCELSO presents significant superiority to them on at least 16 problems and only shows inferiority in at most four problems. In comparison with the five holistic large-scale PSO variants, DGCELSO is significantly superior to SLPSO on 18 problems, achieves much better performance than TPLSO on 16 problems, outperforms both LLSO and CSO on 15 problems, and beats SPLSO down on 14 problems. The superiority of DGCELSO to the five holistic large-scale PSOs demonstrates the effectiveness of the proposed DGCEL strategy.

In Table 4, we summarize the comparison results on the CEC'2013 set as follows. (1) From the perspective of the Friedman test, as shown in the last row, it is found that the rank value of the proposed DGCELSO is still the lowest among the ten algorithms, and such a rank is still much smaller than those of the nine compared algorithms. This demonstrates that DGCELSO still achieves the best overall performance on the complicated CEC'2013 benchmark set and shows great dominance to the compared algorithms. (2) With respect to the Wilcoxon rank-sum test, as shown in the second to last row, it is observed that except for SPLSO, DGCELSO shows significantly better performance than the other eight compared algorithms on at least 10 problems and shows inferiority on at most three problems. Competed with SPLSO, DGCELSO beats it on eight problems and is defeated on only three problems. The superiority of DGCELSO to the compared algorithms on the CEC'2013 benchmark set demonstrates that it is promising for complicated large-scale optimization problems.

The above experiments demonstrated the effectiveness of the proposed DGCELSO. To further demonstrate its efficiency in solving large-scale optimization problems, we conduct experiments on the two large-scale benchmark sets to investigate the convergence speed of the proposed DGCELSO in comparison with the nine compared methods. In this experiment, the maximum number of fitness evaluations is set as 5×10^6. Figures 3 and 4 show the convergence comparison results on the CEC'2010 and the CEC'2013 benchmark sets, respectively.

In Figure 3, on the CEC'2010 benchmark set, the following findings can be obtained. (1) At first glance, it is found that the proposed DGCELSO obviously obtains faster convergence along with better solutions than all the nine compared algorithms on nine problems (F_1, F_4, F_7, F_9, F_{11}, F_{12}, F_{14}, F_{16}, and F_{17}). On F_3, F_{13}, F_{18}, and F_{20}, DGCELSO achieves very similar performance with some compared algorithms in terms of the solution quality but obtains much faster convergence than the associated compared algorithms. (2) More specifically, we find that DGCELSO obviously shows much better performance in both convergence speed and solution quality than the five holistic large-scale PSO variants, namely TPLSO, SPLSO, LLSO, CSO, and SLPSO on 17, 16, 15, 16, and 17, respectively. In the competition with the four cooperative coevolutionary evolutionary algorithms, namely DECC-DG, DECC-GD2, DECC-RDG, and DECC-RDG2, DGCELSO shows clear superiority in both convergence speed and solution quality on 17, 17, 17, and 15 problems, respectively.

From Figure 4, similar observations on the CEC'2013 benchmark set can be attained. (1) At first glance, it is found that the proposed DGCELSO obtains faster convergence along with better solutions than all the nine compared algorithms on six problems (F_1, F_4, F_7, F_{11}, F_{13}, and F_{14}). On F_8, F_9, and F_{12}, DGCELSO shows superiority in both convergence speed and solution quality to eight compared algorithms and is inferior to only one compared algorithm. (2) More specifically, we find that DGCELSO performs better with faster convergence speed and higher solution quality than TPLSO, SPLSO, LLSO, CSO, and SLPSO on 11, 11, 9, 12, and 10 problems, respectively. In competition with DECC-DG, DECC-GD2,

DECC-RDG, and DECC-RDG2, DGCELSO presents great dominance to them on 11, 9, 11, and 12 problems, respectively.

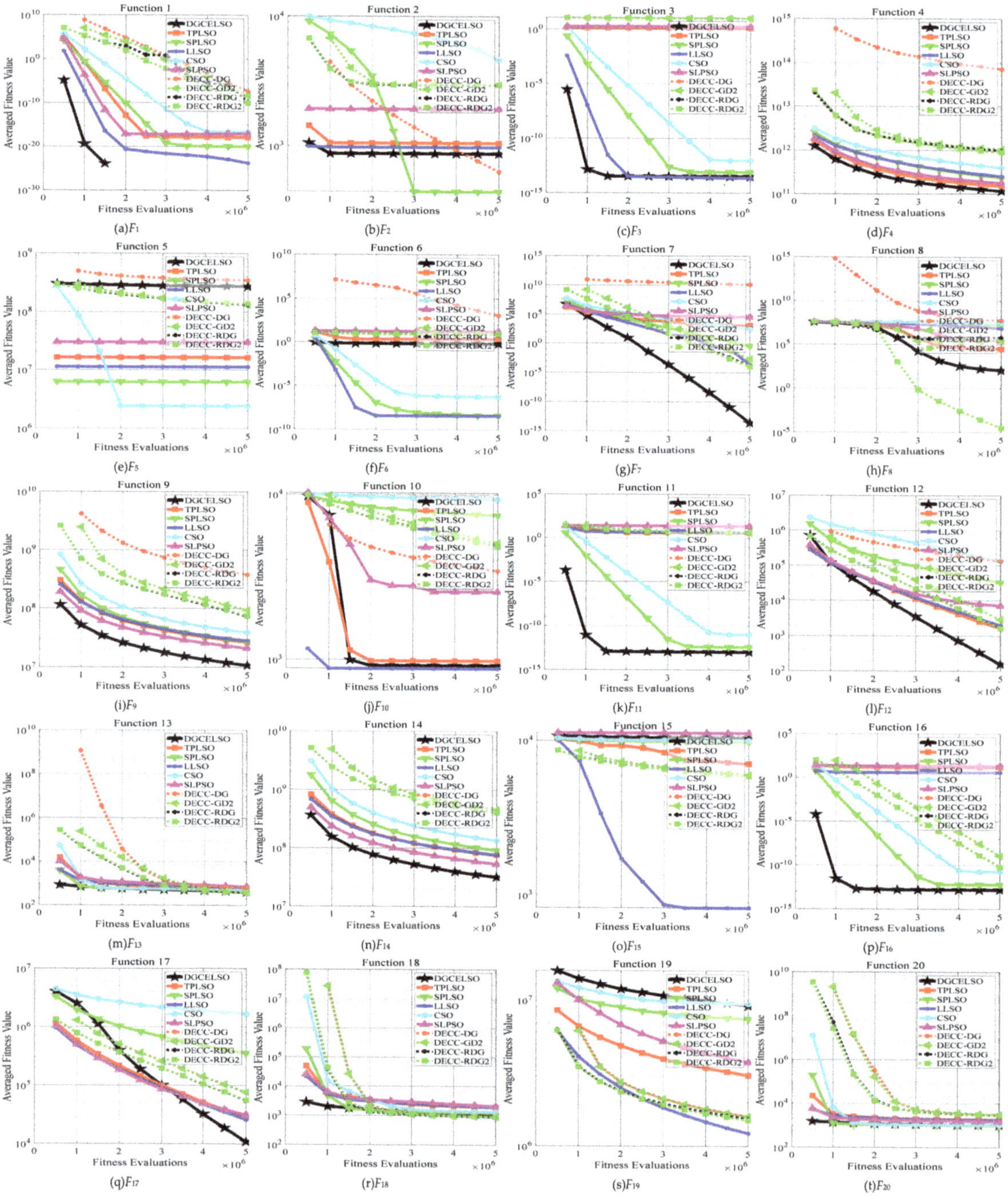

Figure 3. Convergence behavior comparison between DGCELSO and the compared algorithms on each 1000-D CEC'2010 benchmark problem.

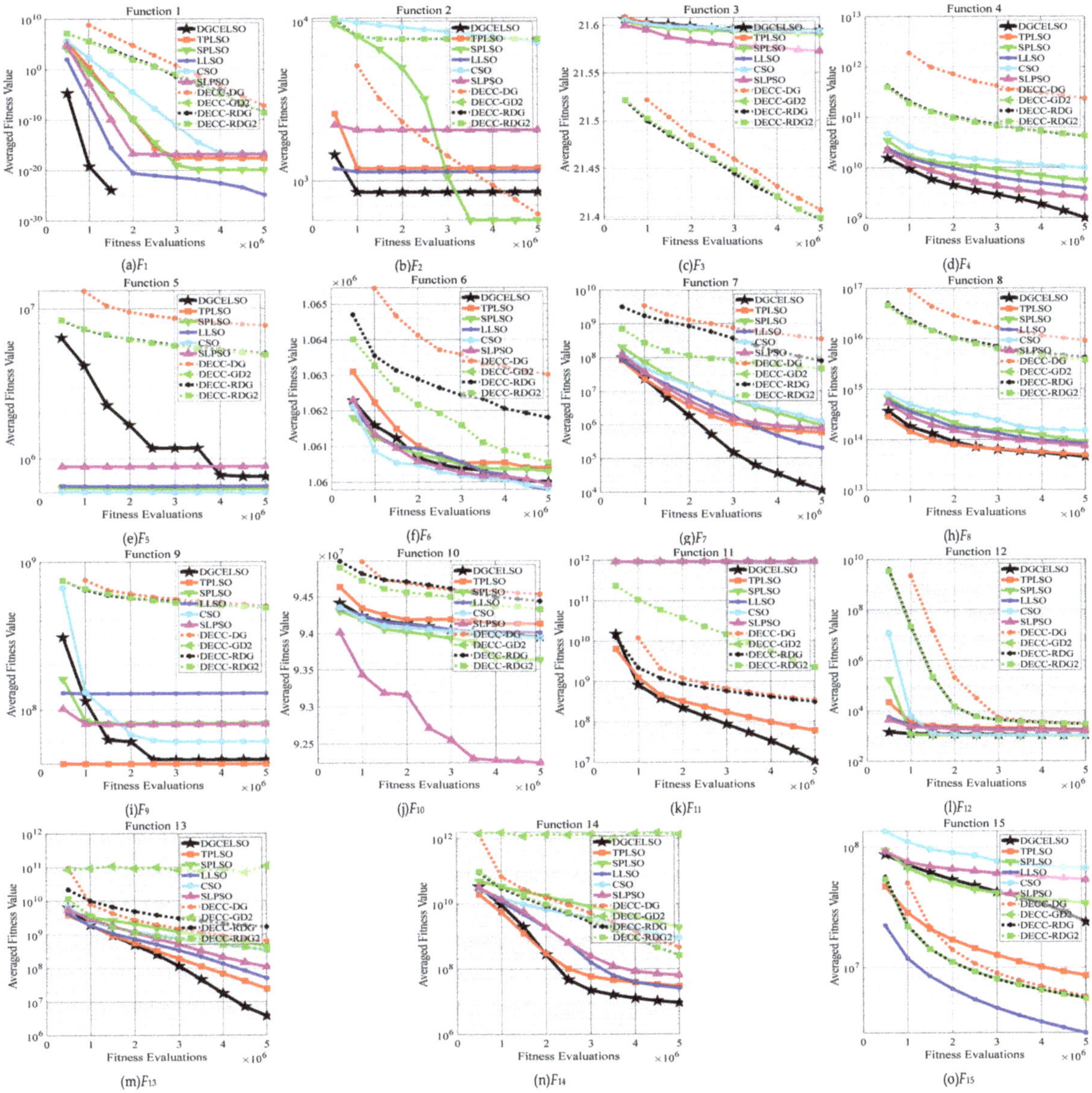

Figure 4. Convergence behavior comparison between DGCELSO and the compared algorithms on each 1000-D CEC'2013 benchmark problem.

To sum up, compared with these state-of-the-art large-scale algorithms, DGCELSO performs much better in both convergence speed and solution quality. The superiority of DGCELSO mainly benefits from the proposed DGCEL strategy, which could implicitly assemble useful information embedded in elite particles to guide the evolution of the swarm. In particular, the superiority of DGCELSO to the five holistic large-scale PSOs, which also adopt elite particles in the current swarm to direct the evolution of the swarm, demonstrates that the assembly of evolutionary information in elites is effective. Such assembly not only improves the learning diversity of particles due to the random selection of guiding exemplars from the elites but also promotes the learning effectiveness of particles because each updated particle could learn from multiple different elites with the help of the dimension group-based learning. As a result, DGCELSO could compromise search intensification and diversification well to explore and exploit the large-scale solution appropriately to locate satisfactory solutions.

4.3. Deep Investigation on DGCELSO

In this section, we conduct extensive experiments on the 1000-D CEC'2010 benchmark set to verify the effectiveness of the main components in the proposed DGCELSO.

4.3.1. Effectiveness of the Proposed DGCEL

First, we conduct experiments to investigate the effectiveness of the proposed DGCEL strategy. To this end, we first incorporate the segment-based predominance learning strategy (SPL) in SPLSO, which is the most similar work to the proposed DGCELSO, to replace the DGCEL strategy, leading to a new variant of DGCELSO, which we denote as "DGCELSO-SPL". In addition, we also develop two extreme cases of DGCELSO, where the number of dimension groups (*NDG*) is set as 1 and 1000, respectively. The former, which we denote as "DGCELSO-1", con all dimensions as a group, and thus can be considered a DGCELSO without the dimension group-based comprehensive learning, while the latter, which we denote as "DGCELSO-1000", considers each dimension as a group. This can be considered a DGCELSO by introducing the comprehensive learning strategy in CLPSO [46] to replace the dimension group-based comprehensive learning in DGCELSO. Then, we conduct experiments on the CEC'2010 benchmark set to compare the above four versions of DGCELSO. Table 5 shows the comparison results among the four versions of DGCELSO. In this table, the best results are highlighted in bold.

From Table 5, the following observations can be attained. (1) From the perspective of the Friedman test, it is found that the rank value of DGCELSO is the smallest among the four versions of DGCELSO. This demonstrates that DGCELSO achieves the best overall performance. (2) Comparing DGCELSO with DGCELSO-SPL, DGCELSO shows great superiority. This demonstrates that the proposed DGCEL strategy is much better than SPL. It should be mentioned that, like DGCEL, SPL also lets each particle learn from multiple elites in the swarm, based on the dimension group. The differences between DGCEL and SPL lie in two aspects. On the one hand, SPL lets particles learn from relatively better elites which are determined by the competition between randomly paired two particles, while DGCEL lets particles learn from absolutely better elites which are the top $tp*NP$ best particles in the swarm. On the other hand, the second exemplar in the velocity update in SPL is the mean position of the whole swarm, which is shared by all updated particles, while the second exemplar in DGCEL is also randomly selected from the elite particles. With the observed superiority of DGCEL to SPL, it is demonstrated that the exemplar selection in DGCEL is better than that in SPL. (3) Competed with DGCELSO-1 and DGCELSO-1000, DGCELSO presents great superiority. This superiority demonstrates the effectiveness of the proposed dimension group-based comprehensive learning strategy. Instead of learning from only two exemplars in DGCELSO-1, which consider all dimensions as a group, and learning from multiple exemplars dimension by dimension in DGCELSO-1000, which considers each dimension as a group, DGCELSO lets each updated particle learn from multiple exemplars based on dimension group. In this way, the potentially useful information embedded in different exemplars is more likely to be assembled in DGCELSO than in DGCELSO-1 and DGCELSO-1000.

Based on the above observations, it is found that the proposed DGCEL strategy is effective and plays a crucial role in helping DGCELSO achieve promising performance.

4.3.2. Effectiveness of the Proposed Dynamic Adjustment Schemes for Parameters

In this subsection, we conduct experiments to verify the effectiveness of the proposed dynamic adjustment schemes for the two control parameters, namely the elite ratio tp and the number of dimension groups *NDG*.

Table 5. Comparison results among different versions of DGCELSO on the 1000-D CEC'2010 problems.

F	DGCELSO	DGCELSO-1	DGCELSO-1000	DGCELSO-SPL
F_1	0.00×10^{0}	3.85×10^{-26}	0.00×10^{0}	1.81×10^{3}
F_2	8.88×10^{2}	1.98×10^{3}	8.70×10^{2}	1.54×10^{3}
F_3	3.18×10^{-14}	1.08×10^{0}	3.16×10^{-14}	1.97×10^{-2}
F_4	1.60×10^{11}	2.15×10^{11}	1.56×10^{11}	9.44×10^{11}
F_5	2.80×10^{8}	6.93×10^{7}	2.79×10^{8}	1.07×10^{7}
F_6	4.00×10^{-9}	1.96×10^{1}	4.00×10^{-9}	3.74×10^{-1}
F_7	2.15×10^{-5}	4.01×10^{3}	2.17×10^{-5}	6.15×10^{6}
F_8	4.36×10^{3}	6.84×10^{5}	4.26×10^{3}	3.27×10^{7}
F_9	1.77×10^{7}	3.28×10^{7}	1.77×10^{7}	1.05×10^{8}
F_{10}	9.23×10^{2}	2.02×10^{3}	9.34×10^{2}	3.63×10^{3}
F_{11}	1.10×10^{-13}	2.08×10^{1}	1.10×10^{-13}	6.66×10^{-1}
F_{12}	2.55×10^{3}	4.60×10^{3}	2.63×10^{3}	1.99×10^{5}
F_{13}	5.15×10^{2}	7.69×10^{2}	4.87×10^{2}	1.42×10^{3}
F_{14}	5.17×10^{7}	9.78×10^{7}	5.13×10^{7}	3.42×10^{8}
F_{15}	1.04×10^{4}	2.04×10^{3}	1.05×10^{4}	1.00×10^{4}
F_{16}	1.55×10^{-13}	2.92×10^{1}	2.93×10^{-2}	5.72×10^{-1}
F_{17}	6.57×10^{4}	4.30×10^{4}	7.12×10^{4}	7.10×10^{5}
F_{18}	1.31×10^{3}	2.30×10^{3}	1.33×10^{3}	2.38×10^{4}
F_{19}	1.02×10^{7}	1.33×10^{6}	1.06×10^{7}	6.52×10^{6}
F_{20}	1.08×10^{3}	1.98×10^{3}	1.08×10^{3}	2.11×10^{4}
Rank	1.80	2.90	1.90	3.40

First, we conduct experiments to investigate the effectiveness of the proposed dynamic scheme for *tp*. To this end, we first set *tp* as different fixed values from 0.1 to 0.9. Then, we compare the DGCELSO with the dynamic scheme with these DGCELSOs with different fixed *tp* values. Table 6 shows the comparison results between the DGCELSO with the dynamic scheme and the ones with different values of *tp* on the CEC'2010 benchmark set. In this table, the best results are highlighted in bold.

From Table 6, the following findings can be obtained. (1) From the perspective of the Friedman test, it is found that DGCELSO with the dynamic *tp* ranks first among all versions of DGCELSO with different settings of *tp*. This demonstrates that DGCELSO with the dynamic *tp* achieves the best overall performance. (2) More specifically, we find that DGCELSO with the dynamic strategy obtains the best results on 4 problems and its results on the other problems are very close to the best ones obtained by the DGCELSO with the associated optimal settings of *tp*. These two observations demonstrate that the dynamic strategy for *tp* is helpful in achieving good performance for DGCELSO.

Then, we conduct experiments to verify the dynamic scheme for the number of dimension groups (*NDG*). To this end, we first set *NDG* as different fixed values from 20 to 100. Subsequently, we conduct experiments on the CEC'2010 set to compare the DGCELSO with the dynamic scheme for *NDG* and the ones with different fixed values of *NDG*. Table 7 shows the comparison results among the above versions of DGCELSO. In this table, the best results are highlighted in bold.

Table 6. Comparison results between DGCELSO with the dynamic strategy for *tp* and the ones with different fixed settings of *tp* on the 1000-D CEC'2010 problems.

F	*tp* = 0.1	*tp* = 0.2	*tp* = 0.3	*tp* = 0.4	*tp* = 0.5	*tp* = 0.6	*tp* = 0.7	*tp* = 0.8	*tp* = 0.9	*Dynamic*
F_1	9.55×10^{-3}	0.00×10^{0}	0.00×10^{0}	0.00×10^{0}	0.00×10^{0}	0.00×10^{0}	0.00×10^{0}	0.00×10^{0}	3.31×10^{-26}	0.00×10^{0}
F_2	2.33×10^{3}	1.38×10^{3}	1.05×10^{3}	8.21×10^{2}	6.71×10^{2}	1.03×10^{3}	9.26×10^{3}	9.83×10^{3}	1.00×10^{4}	8.88×10^{2}
F_3	1.41×10^{0}	3.30×10^{-14}	3.17×10^{-14}	3.14×10^{-14}	2.98×10^{-14}	2.96×10^{-14}	2.99×10^{-14}	2.93×10^{-14}	2.98×10^{-14}	3.18×10^{-14}
F_4	6.60×10^{11}	1.64×10^{11}	1.80×10^{11}	1.89×10^{11}	2.01×10^{11}	2.24×10^{11}	2.28×10^{11}	2.52×10^{11}	2.53×10^{11}	1.60×10^{11}
F_5	5.90×10^{7}	2.64×10^{8}	2.75×10^{8}	2.76×10^{8}	2.83×10^{8}	2.79×10^{8}	2.81×10^{8}	2.82×10^{8}	2.83×10^{8}	2.80×10^{8}
F_6	1.99×10^{1}	2.00×10^{1}	1.98×10^{1}	4.00×10^{-9}	4.00×10^{-9}	4.00×10^{-9}	4.00×10^{-9}	3.88×10^{-9}	4.00×10^{-9}	4.00×10^{-9}
F_7	1.15×10^{6}	9.66×10^{-8}	5.36×10^{-5}	8.46×10^{-3}	3.32×10^{-1}	2.98×10^{0}	1.56×10^{1}	8.23×10^{1}	3.67×10^{2}	2.15×10^{-5}
F_8	4.15×10^{7}	1.10×10^{3}	7.82×10^{3}	1.19×10^{5}	2.07×10^{6}	6.99×10^{6}	1.07×10^{7}	1.36×10^{7}	1.57×10^{7}	4.36×10^{3}
F_9	1.21×10^{8}	1.93×10^{7}	1.85×10^{7}	1.78×10^{7}	2.05×10^{7}	2.03×10^{7}	2.21×10^{7}	2.17×10^{7}	2.34×10^{7}	1.77×10^{7}
F_{10}	2.44×10^{3}	1.53×10^{3}	1.06×10^{3}	2.19×10^{3}	9.48×10^{3}	9.80×10^{3}	1.01×10^{4}	1.02×10^{4}	1.02×10^{4}	9.23×10^{2}
F_{11}	2.86×10^{1}	2.04×10^{1}	1.05×10^{1}	1.11×10^{-13}	1.09×10^{-13}	1.11×10^{-13}	1.11×10^{-13}	1.13×10^{-13}	1.15×10^{-13}	1.10×10^{-13}
F_{12}	1.68×10^{5}	1.12×10^{3}	2.48×10^{3}	7.33×10^{3}	2.81×10^{4}	1.16×10^{5}	5.74×10^{5}	1.53×10^{6}	2.08×10^{6}	2.55×10^{3}
F_{13}	1.68×10^{3}	4.27×10^{2}	5.13×10^{2}	4.12×10^{2}	6.18×10^{2}	4.30×10^{2}	4.50×10^{2}	4.88×10^{2}	5.19×10^{2}	5.15×10^{2}
F_{14}	3.74×10^{8}	5.88×10^{7}	5.39×10^{7}	5.68×10^{7}	5.82×10^{7}	6.54×10^{7}	6.97×10^{7}	8.12×10^{7}	8.92×10^{7}	5.17×10^{7}
F_{15}	2.66×10^{3}	1.08×10^{4}	1.05×10^{4}	1.04×10^{4}	1.04×10^{4}	1.04×10^{4}	1.04×10^{4}	1.04×10^{4}	1.04×10^{4}	1.04×10^{4}
F_{16}	7.57×10^{1}	5.55×10^{0}	1.62×10^{-13}	1.64×10^{-13}	1.68×10^{-13}	1.76×10^{-13}	1.79×10^{-13}	1.87×10^{-13}	1.94×10^{-13}	1.55×10^{-13}
F_{17}	5.02×10^{5}	2.01×10^{4}	5.70×10^{4}	1.86×10^{6}	3.51×10^{6}	4.29×10^{6}	4.92×10^{6}	5.16×10^{6}	5.46×10^{6}	6.57×10^{4}
F_{18}	4.17×10^{3}	1.45×10^{3}	1.35×10^{3}	1.45×10^{3}	1.09×10^{3}	1.43×10^{3}	1.16×10^{3}	1.12×10^{3}	1.16×10^{3}	1.31×10^{3}
F_{19}	2.18×10^{6}	6.26×10^{6}	1.05×10^{7}	1.20×10^{7}	1.32×10^{7}	1.39×10^{7}	1.42×10^{7}	1.50×10^{7}	1.53×10^{7}	1.02×10^{7}
F_{20}	3.09×10^{3}	1.30×10^{3}	1.19×10^{3}	1.10×10^{3}	1.06×10^{3}	1.03×10^{3}	1.02×10^{3}	9.94×10^{2}	9.86×10^{2}	1.08×10^{3}
Rank	7.75	4.98	4.53	4.23	4.85	5.35	5.9	6.28	7.73	3.43

Table 7. Comparison results between DGCELSO with the dynamic strategy for *NDG* and the ones with different fixed settings of *NDG* on the 1000-D CEC'2010 problems.

F	NDG = 20	NDG = 30	NDG = 40	NDG = 50	NDG = 60	NDG = 70	NDG = 80	NDG = 90	NDG = 100	Dynamic
F_1	0.00×10^0	0.00×10^0	0.00×10^0	0.00×10^0	0.00×10^0	0.00×10^0	0.00×10^0	0.00×10^0	0.00×10^0	0.00×10^0
F_2	8.46×10^2	8.56×10^2	8.49×10^2	8.41×10^2	8.53×10^2	8.48×10^2	8.52×10^2	8.51×10^2	8.44×10^2	8.88×10^2
F_3	3.18×10^{-14}	3.18×10^{-14}	3.19×10^{-14}	3.22×10^{-14}	3.21×10^{-14}	3.24×10^{-14}	3.19×10^{-14}	3.21×10^{-14}	3.19×10^{-14}	3.18×10^{-14}
F_4	1.69×10^{11}	1.69×10^{11}	1.68×10^{11}	1.65×10^{11}	1.54×10^{11}	1.56×10^{11}	1.65×10^{11}	1.58×10^{11}	1.65×10^{11}	1.60×10^{11}
F_5	2.78×10^8	2.79×10^8	2.80×10^8	2.78×10^8	2.78×10^8	2.78×10^8	2.79×10^8	2.78×10^8	2.79×10^8	2.80×10^8
F_6	4.00×10^{-9}	4.00×10^{-9}	4.00×10^{-9}	4.00×10^{-9}	4.00×10^{-9}	4.00×10^{-9}	4.00×10^{-9}	4.00×10^{-9}	3.88×10^{-9}	4.00×10^{-9}
F_7	2.36×10^{-5}	3.22×10^{-5}	2.64×10^{-5}	2.58×10^{-5}	2.18×10^{-5}	1.92×10^{-5}	2.22×10^{-5}	2.00×10^{-5}	2.93×10^{-5}	2.15×10^{-5}
F_8	5.49×10^3	5.20×10^3	5.15×10^3	5.12×10^3	5.14×10^3	4.98×10^3	5.07×10^3	5.01×10^3	5.07×10^3	4.36×10^3
F_9	1.80×10^7	1.78×10^7	1.73×10^7	1.81×10^7	1.74×10^7	1.76×10^7	1.78×10^7	1.74×10^7	1.82×10^7	1.77×10^7
F_{10}	8.97×10^2	8.94×10^2	8.94×10^2	8.94×10^2	8.92×10^2	8.92×10^2	9.02×10^2	9.14×10^2	8.89×10^2	9.23×10^2
F_{11}	1.11×10^{-13}	1.11×10^{-13}	1.11×10^{-13}	1.11×10^{-13}	1.11×10^{-13}	1.10×10^{-13}	1.10×10^{-13}	1.11×10^{-13}	1.11×10^{-13}	1.10×10^{-13}
F_{12}	3.13×10^3	3.13×10^3	3.24×10^3	3.18×10^3	3.24×10^3	3.21×10^3	3.11×10^3	3.14×10^3	3.28×10^3	2.55×10^3
F_{13}	4.48×10^2	5.03×10^2	5.13×10^2	4.83×10^2	5.30×10^2	5.09×10^2	4.51×10^2	4.64×10^2	4.82×10^2	5.15×10^2
F_{14}	5.26×10^7	5.26×10^7	5.24×10^7	5.16×10^7	5.16×10^7	5.07×10^7	5.23×10^7	5.18×10^7	5.24×10^7	5.17×10^7
F_{15}	1.04×10^4	1.04×10^4	1.04×10^4	1.04×10^4	1.04×10^4	1.04×10^4	1.04×10^4	1.04×10^4	1.04×10^4	1.04×10^4
F_{16}	1.59×10^{-13}	1.57×10^{-13}	1.58×10^{-13}	1.58×10^{-13}	3.85×10^{-2}	1.58×10^{-13}	2.93×10^{-2}	1.59×10^{-13}	1.59×10^{-13}	1.55×10^{-13}
F_{17}	1.19×10^5	1.20×10^5	1.19×10^5	1.20×10^5	1.28×10^5	1.22×10^5	1.28×10^5	1.34×10^5	1.35×10^5	6.57×10^4
F_{18}	1.19×10^3	1.21×10^3	1.28×10^3	1.31×10^3	1.24×10^3	1.18×10^3	1.31×10^3	1.31×10^3	1.28×10^3	1.31×10^3
F_{19}	1.10×10^7	1.09×10^7	1.12×10^7	1.10×10^7	1.12×10^7	1.13×10^7	1.12×10^7	1.12×10^7	1.11×10^7	1.02×10^7
F_{20}	1.12×10^3	1.11×10^3	1.09×10^3	1.11×10^3	1.08×10^3	1.11×10^3	1.09×10^3	1.08×10^3	1.10×10^3	1.08×10^3
Rank	6.05	6.05	6.18	5.33	5.78	4.70	5.50	5.18	6.00	4.25

From Table 7, we can obtain the following findings. (1) From the perspective of the Friedman test, it is found that the rank value of the DGCELSO with the dynamic scheme for *NDG* is the smallest among all versions of DGCELSO with different settings of *NDG*. This demonstrates that DGCELSO with the dynamic strategy achieves the best overall performance. (2) More specifically, we find that DGCELSO with the dynamic strategy obtains the best results on nine problems, while DGCELSO with fixed *NDG* obtains the best results on at most four problems. In particular, on the other 11 problems where DGCELSO with the dynamic strategy does not achieve the best results, its optimization results are very close to the best ones obtained by DGCELSO with the associated optimal *NDG*. These two observations verify the effectiveness of the dynamic strategy for *NDG*.

To sum up, the above comparative experiments demonstrated the effectiveness and efficiency of DGCELSO in solving large-scale optimization problems. In particular, the deep investigation experiments have validated that it is the proposed DGCEL strategy along with the two dynamic strategies that play a crucial role in helping DGCELSO achieve promising performance.

5. Conclusions

This paper proposed a dimension group-based comprehensive elite learning swarm optimizer (DGCELSO) to effectively solve large-scale optimization problems. Specifically, this optimizer first partitions the swarm into two exclusive sets, namely the elite set and the non-elite set. Then, the non-elite particles are updated by learning from the elite ones with the elite particles directly entering the next generation. During the update of each non-elite particle, the dimensions are separated into several dimension groups. Subsequently, for each dimension group, two elites are randomly selected from the elite set and then act as the guiding exemplars to direct the update of the dimension group. In this way, each non-elite particle could comprehensively learn from multiple elites. Moreover, not only are the guiding exemplars for different non-elite particles different, but the guiding exemplars for different dimension groups of the same non-elite particle are also likely to be different. As a result, not only could the learning diversity of particles be improved, but the learning efficiency of particles could also be promoted. To further aid the optimizer to explore and exploit the solution space properly, we designed two dynamic adjustment strategies for the associated control parameters in the proposed DGCELSO.

Experiments conducted on the 1000-D CEC'2010 and CEC'2013 large-scale benchmark sets verified the effectiveness of the proposed DGCELSO by comparing it with nine state-of-the-art large-scale methods. Experimental results demonstrate that DGCELSO achieves highly competitive or even much better performance than the compared methods in terms of both the solution quality and the convergence speed.

Author Contributions: Q.Y.: Conceptualization, supervision, methodology, formal analysis, and writing—original draft preparation. K.-X.Z.: Implementation, formal analysis, and writing—original draft preparation. X.-D.G.: Methodology, and writing—review, and editing. D.-D.X.: Writing—review and editing. Z.-Y.L.: Writing—review and editing, and funding acquisition. S.-W.J.: Writing—review and editing. J.Z.: Conceptualization and writing—review and editing. All authors have read and agreed to the published version of the manuscript.

Funding: This work was supported in part by the National Natural Science Foundation of China under Grant 62006124 and U20B2061, in part by the Natural Science Foundation of Jiangsu Province under Project BK20200811, in part by the Natural Science Foundation of the Jiangsu Higher Education Institutions of China under Grant 20KJB520006, in part by the National Research Foundation of Korea (NRF-2021H1D3A2A01082705), and in part by the Startup Foundation for Introducing Talent of NUIST.

Institutional Review Board Statement: Not applicable.

Informed Consent Statement: Not applicable.

Data Availability Statement: Not applicable.

Conflicts of Interest: The authors declare no conflict of interest.

References

1. Jia, Y.H.; Mei, Y.; Zhang, M. A Two-Stage Swarm Optimizer with Local Search for Water Distribution Network Optimization. *IEEE Trans. Cybern.* **2021**. [CrossRef]
2. Cao, K.; Cui, Y.; Liu, Z.; Tan, W.; Weng, J. Edge Intelligent Joint Optimization for Lifetime and Latency in Large-Scale Cyber-Physical Systems. *IEEE Internet Things J.* **2021**. [CrossRef]
3. Chen, W.N.; Tan, D.Z.; Yang, Q.; Gu, T.; Zhang, J. Ant Colony Optimization for the Control of Pollutant Spreading on Social Networks. *IEEE Trans. Cybern.* **2020**, *50*, 4053–4065. [CrossRef]
4. Zuo, T.; Zhang, Y.; Meng, K.; Tong, Z.; Dong, Z.Y.; Fu, Y. A Two-Layer Hybrid Optimization Approach for Large-Scale Offshore Wind Farm Collector System Planning. *IEEE Trans. Ind. Inform.* **2021**, *17*, 7433–7444. [CrossRef]
5. Yang, Q.; Chen, W.N.; Gu, T.; Jin, H.; Mao, W.; Zhang, J. An Adaptive Stochastic Dominant Learning Swarm Optimizer for High-Dimensional Optimization. *IEEE Trans. Cybern.* **2020**, *52*, 1960–1976. [CrossRef]
6. Omidvar, M.N.; Li, X.; Mei, Y.; Yao, X. Cooperative Co-Evolution with Differential Grouping for Large Scale Optimization. *IEEE Trans. Evol. Comput.* **2014**, *18*, 378–393. [CrossRef]
7. Tang, K.; Li, X.; Suganthan, P.; Yang, Z.; Weise, T. *Benchmark Functions for the CEC 2010 Special Session and Competition on Large-Scale Global Optimization*; Nature Inspired Computation and Applications Laboratory, University of Science and Technology of China: Hefei, China, 2009.
8. Li, X.; Tang, K.; Omidvar, M.N.; Yang, Z.; Qin, K.; China, H. *Benchmark Functions for the CEC 2013 Special Session and Competition on Large-Scale Global Optimization*; Technical Report; Evolutionary Computation and Machine Learning Group, RMIT University: Melbourne, Australia, 2013.
9. Yang, Q.; Li, Y.; Gao, X.-D.; Ma, Y.-Y.; Lu, Z.-Y.; Jeon, S.-W.; Zhang, J. An Adaptive Covariance Scaling Estimation of Distribution Algorithm. *Mathematics* **2021**, *9*, 3207. [CrossRef]
10. Yang, Q.; Chen, W.N.; Gu, T.; Zhang, H.; Yuan, H.; Kwong, S.; Zhang, J. A Distributed Swarm Optimizer with Adaptive Communication for Large-Scale Optimization. *IEEE Trans. Cybern.* **2020**, *50*, 3393–3408. [CrossRef]
11. Omidvar, M.N.; Li, X.; Yao, X. A Review of Population-Based Metaheuristics for Large-Scale Black-Box Global Optimization: Part A. *IEEE Trans. Evol. Comput.* 2021. *in press*. Available online: https://ieeexplore.ieee.org/document/9627116 (accessed on 1 January 2022).
12. Omidvar, M.N.; Li, X.; Yao, X. A Review of Population-Based Metaheuristics for Large-Scale Black-Box Global Optimization: Part B. *IEEE Trans. Evol. Comput.* 2021. *in press*. Available online: https://ieeexplore.ieee.org/document/9627138 (accessed on 1 January 2022).
13. Yang, Q.; Xie, H.; Chen, W.; Zhang, J. Multiple Parents Guided Differential Evolution for Large Scale Optimization. In Proceedings of the IEEE Congress on Evolutionary Computation, Vancouver, BC, Canada, 24–29 July 2016; pp. 3549–3556.
14. Yang, Q.; Chen, W.N.; Li, Y.; Chen, C.L.P.; Xu, X.M.; Zhang, J. Multimodal Estimation of Distribution Algorithms. *IEEE Trans. Cybern.* **2017**, *47*, 636–650. [CrossRef]
15. Eberhart, R.; Kennedy, J. A New Optimizer Using Particle Swarm Theory. In Proceedings of the International Symposium on Micro Machine and Human Science, Nagoya, Japan, 4–6 October 1995; pp. 39–43.
16. Shi, Y.; Eberhart, R. A Modified Particle Swarm Optimizer. In Proceedings of the IEEE International Conference on Evolutionary Computation Proceedings: IEEE World Congress on Computational Intelligence, Anchorage, AK, USA, 4–9 May 1998; pp. 69–73.
17. Tang, J.; Liu, G.; Pan, Q. A Review on Representative Swarm Intelligence Algorithms for Solving Optimization Problems: Applications and Trends. *IEEE/CAA J. Autom. Sin.* **2021**, *8*, 1627–1643. [CrossRef]
18. Ren, Z.; Zhang, A.; Wen, C.; Feng, Z. A Scatter Learning Particle Swarm Optimization Algorithm for Multimodal Problems. *IEEE Trans. Cybern.* **2014**, *44*, 1127–1140. [CrossRef]
19. Zhang, J.; Lu, Y.; Che, L.; Zhou, M. Moving-Distance-Minimized PSO for Mobile Robot Swarm. *IEEE Trans. Cybern.* **2021**. [CrossRef]
20. Villalón, C.L.C.; Dorigo, M.; Stützle, T. PSO-X: A Component-Based Framework for the Automatic Design of Particle Swarm Optimization Algorithms. *IEEE Trans. Evol. Comput.* **2021**. [CrossRef]
21. Ding, W.; Lin, C.T.; Cao, Z. Deep Neuro-Cognitive Co-Evolution for Fuzzy Attribute Reduction by Quantum Leaping PSO with Nearest-Neighbor Memeplexes. *IEEE Trans. Cybern.* **2019**, *49*, 2744–2757. [CrossRef]
22. Yang, Q.; Hua, L.; Gao, X.; Xu, D.; Lu, Z.; Jeon, S.-W.; Zhang, J. Stochastic Cognitive Dominance Leading Particle Swarm Optimization for Multimodal Problems. *Mathematics* **2022**, *10*, 761. [CrossRef]
23. Bonavolontà, F.; Noia, L.P.D.; Liccardo, A.; Tessitore, S.; Lauria, D. A PSO-MMA Method for the Parameters Estimation of Interarea Oscillations in Electrical Grids. *IEEE Trans. Instrum. Meas.* **2020**, *69*, 8853–8865. [CrossRef]
24. Lan, R.; Zhu, Y.; Lu, H.; Liu, Z.; Luo, X. A Two-Phase Learning-Based Swarm Optimizer for Large-Scale Optimization. *IEEE Trans. Cybern.* **2020**, *51*, 6284–6293. [CrossRef]
25. Yang, Q.; Chen, W.; Deng, J.D.; Li, Y.; Gu, T.; Zhang, J. A Level-Based Learning Swarm Optimizer for Large-Scale Optimization. *IEEE Trans. Evol. Comput.* **2018**, *22*, 578–594. [CrossRef]
26. Cheng, R.; Jin, Y. A Competitive Swarm Optimizer for Large Scale Optimization. *IEEE Trans. Cybern.* **2015**, *45*, 191–204. [CrossRef]

27. Mahdavi, S.; Shiri, M.E.; Rahnamayan, S. Metaheuristics in Large-Scale Global Continues Optimization: A Survey. *Inf. Sci.* **2015**, *295*, 407–428. [CrossRef]
28. Ma, X.; Li, X.; Zhang, Q.; Tang, K.; Liang, Z.; Xie, W.; Zhu, Z. A Survey on Cooperative Co-Evolutionary Algorithms. *IEEE Trans. Evol. Comput.* **2018**, *23*, 421–441. [CrossRef]
29. Li, X.; Yao, X. Cooperatively Coevolving Particle Swarms for Large Scale Optimization. *IEEE Trans. Evol. Comput.* **2011**, *16*, 210–224.
30. Yang, Q.; Chen, W.; Gu, T.; Zhang, H.; Deng, J.D.; Li, Y.; Zhang, J. Segment-Based Predominant Learning Swarm Optimizer for Large-Scale Optimization. *IEEE Trans. Cybern.* **2017**, *47*, 2896–2910. [CrossRef]
31. Xie, H.Y.; Yang, Q.; Hu, X.M.; Chen, W.N. Cross-Generation Elites Guided Particle Swarm Optimization for Large Scale Optimization. In Proceedings of the IEEE Symposium Series on Computational Intelligence, Athens, Greece, 6–9 December 2016; pp. 1–8.
32. Song, G.W.; Yang, Q.; Gao, X.D.; Ma, Y.Y.; Lu, Z.Y.; Zhang, J. An Adaptive Level-Based Learning Swarm Optimizer for Large-Scale Optimization. In Proceedings of the IEEE International Conference on Systems, Man, and Cybernetics, Melbourne, Australia, 17–20 October 2021; pp. 152–159.
33. Potter, M.A.; De Jong, K.A. A Cooperative Co-Evolutionary Approach to Function Optimization. In Proceedings of the International Conference on Parallel Problem Solving from Nature, Berlin, Germany, 22–26 September 1994; pp. 249–257.
34. Yang, Q.; Chen, W.N.; Zhang, J. Evolution Consistency Based Decomposition for Cooperative Coevolution. *IEEE Access* **2018**, *6*, 51084–51097. [CrossRef]
35. Omidvar, M.N.; Yang, M.; Mei, Y.; Li, X.; Yao, X. DG2: A Faster and More Accurate Differential Grouping for Large-Scale Black-Box Optimization. *IEEE Trans. Evol. Comput.* **2017**, *21*, 929–942. [CrossRef]
36. Sun, Y.; Kirley, M.; Halgamuge, S.K. Extended Differential Grouping for Large Scale Global Optimization with Direct and Indirect Variable Interactions. In Proceedings of the Annual Conference on Genetic and Evolutionary Computation, Madrid, Spain, 11–15 July 2015; pp. 313–320.
37. Sun, Y.; Kirley, M.; Halgamuge, S.K. A Recursive Decomposition Method for Large Scale Continuous Optimization. *IEEE Trans. Evol. Comput.* **2017**, *22*, 647–661. [CrossRef]
38. Song, A.; Chen, W.N.; Gong, Y.J.; Luo, X.; Zhang, J. A Divide-and-Conquer Evolutionary Algorithm for Large-Scale Virtual Network Embedding. *IEEE Trans. Evol. Comput.* **2020**, *24*, 566–580. [CrossRef]
39. Deng, H.; Peng, L.; Zhang, H.; Yang, B.; Chen, Z. Ranking-Based Biased Learning Swarm Optimizer for Large-Scale Optimization. *Inf. Sci.* **2019**, *493*, 120–137. [CrossRef]
40. Wang, H.; Liang, M.; Sun, C.; Zhang, G.; Xie, L. Multiple-Strategy Learning Particle Swarm Optimization for Large-Scale Optimization Problems. *Complex Intell. Syst.* **2021**, *7*, 1–16. [CrossRef]
41. Jian, J.R.; Chen, Z.G.; Zhan, Z.H.; Zhang, J. Region Encoding Helps Evolutionary Computation Evolve Faster: A New Solution Encoding Scheme in Particle Swarm for Large-Scale Optimization. *IEEE Trans. Evol. Comput.* **2021**, *25*, 779–793. [CrossRef]
42. Kampourakis, K. *Understanding Evolution*; Cambridge University Press: Cambridge, UK, 2014.
43. Ju, X.; Liu, F. Wind Farm Layout Optimization Using Self-Informed Genetic Algorithm with Information Guided Exploitation. *Appl. Energy* **2019**, *248*, 429–445. [CrossRef]
44. Ju, X.; Liu, F.; Wang, L.; Lee, W.-J. Wind Farm Layout Optimization Based on Support Vector Regression Guided Genetic Algorithm with Consideration of Participation among Landowners. *Energy Convers. Manag.* **2019**, *196*, 1267–1281. [CrossRef]
45. Xia, X.; Gui, L.; Yu, F.; Wu, H.; Wei, B.; Zhang, Y.L.; Zhan, Z.H. Triple Archives Particle Swarm Optimization. *IEEE Trans. Cybern.* **2020**, *50*, 4862–4875. [CrossRef]
46. Liang, J.J.; Qin, A.K.; Suganthan, P.N.; Baskar, S. Comprehensive Learning Particle Swarm Optimizer for Global Optimization of Multimodal Functions. *IEEE Trans. Evol. Comput.* **2006**, *10*, 281–295. [CrossRef]
47. Gong, Y.; Li, J.; Zhou, Y.; Li, Y.; Chung, H.S.; Shi, Y.; Zhang, J. Genetic Learning Particle Swarm Optimization. *IEEE Trans. Cybern.* **2016**, *46*, 2277–2290. [CrossRef]
48. Zhan, Z.; Zhang, J.; Li, Y.; Shi, Y. Orthogonal Learning Particle Swarm Optimization. *IEEE Trans. Evol. Comput.* **2011**, *15*, 832–847. [CrossRef]
49. Van den Bergh, F.; Engelbrecht, A.P. A Cooperative Approach to Particle Swarm Optimization. *IEEE Trans. Evol. Comput.* **2004**, *8*, 225–239. [CrossRef]
50. Mei, Y.; Omidvar, M.N.; Li, X.; Yao, X. A Competitive Divide-and-Conquer Algorithm for Unconstrained Large-Scale Black-Box Optimization. *ACM Trans. Math. Softw.* **2016**, *42*, 1–24. [CrossRef]
51. Yang, M.; Zhou, A.; Li, C.; Yao, X. An Efficient Recursive Differential Grouping for Large-Scale Continuous Problems. *IEEE Trans. Evol. Comput.* **2021**, *25*, 159–171. [CrossRef]
52. Sun, Y.; Omidvar, M.N.; Kirley, M.; Li, X. Adaptive Threshold Parameter Estimation with Recursive Differential Grouping for Problem Decomposition. In Proceedings of the Genetic and Evolutionary Computation Conference, Kyoto, Japan, 15–19 July 2018.
53. Ma, X.; Huang, Z.; Li, X.; Wang, L.; Qi, Y.; Zhu, Z. Merged Differential Grouping for Large-scale Global Optimization. *IEEE Trans. Evol. Comput.* **2022**, *in press*. [CrossRef]
54. Liu, H.; Wang, Y.; Fan, N. A Hybrid Deep Grouping Algorithm for Large Scale Global Optimization. *IEEE Trans. Evol. Comput.* **2020**, *24*, 1112–1124. [CrossRef]

55. Zhang, X.Y.; Gong, Y.J.; Lin, Y.; Zhang, J.; Kwong, S.; Zhang, J. Dynamic Cooperative Coevolution for Large Scale Optimization. *IEEE Trans. Evol. Comput.* **2019**, *23*, 935–948. [CrossRef]
56. Neshat, M.; Mirjalili, S.; Sergiienko, N.Y.; Esmaeilzadeh, S.; Amini, E.; Heydari, A.; Garcia, D.A. Layout Optimisation of Offshore Wave Energy Converters Using a Novel Multi-swarm Cooperative Algorithm with Backtracking Strategy: A Case Study from Coasts of Australia. *Energy* **2022**, *239*, 122463. [CrossRef]
57. Pan, Q.K.; Gao, L.; Wang, L. An Effective Cooperative Co-Evolutionary Algorithm for Distributed Flowshop Group Scheduling Problems. *IEEE Trans. Cybern.* **2020**. [CrossRef]
58. Neshat, M.; Alexander, B.; Wagner, M. A Hybrid Cooperative Co-Evolution Algorithm Framework for Optimising Power Take off and Placements of Wave Energy Converters. *Inf. Sci.* **2020**, *534*, 218–244. [CrossRef]
59. Liang, M.; Wang, W.; Dong, C.; Zhao, D. A Cooperative Coevolutionary Optimization Design of Urban Transit Network and Operating Frequencies. *Expert Syst. Appl.* **2020**, *160*, 113736. [CrossRef]
60. Zhao, S.-Z.; Liang, J.J.; Suganthan, P.N.; Tasgetiren, M.F. Dynamic Multi-Swarm Particle Swarm Optimizer with Local Search for Large Scale Global Optimization. In Proceedings of the IEEE Congress on Evolutionary Computation, Hong Kong, China, 1–6 June 2008; pp. 3845–3852.
61. Cheng, R. A Social Learning Particle Swarm Optimization Algorithm for Scalable Optimization. *Inf. Sci.* **2015**, *291*, 43–60. [CrossRef]
62. Mohapatra, P.; Das, K.N.; Roy, S. A Modified Competitive Swarm Optimizer for Large Scale Optimization Problems. *Appl. Soft Comput.* **2017**, *59*, 340–362. [CrossRef]
63. Li, D.; Guo, W.; Lerch, A.; Li, Y.; Wang, L.; Wu, Q. An Adaptive Particle Swarm Optimizer with Decoupled Exploration and Exploitation for Large Scale Optimization. *Swarm Evol. Comput.* **2021**, *60*, 100789. [CrossRef]
64. Lan, R.; Zhang, L.; Tang, Z.; Liu, Z.; Luo, X. A Hierarchical Sorting Swarm Optimizer for Large-Scale Optimization. *IEEE Access* **2019**, *7*, 40625–40635. [CrossRef]
65. Kong, F.; Jiang, J.; Huang, Y. An Adaptive Multi-Swarm Competition Particle Swarm Optimizer for Large-Scale Optimization. *Mathematics* **2019**, *7*, 521. [CrossRef]
66. Huang, H.; Lv, L.; Ye, S.; Hao, Z. Particle Swarm Optimization with Convergence Speed Controller for Large-Scale Numerical Optimization. *Soft Comput.* **2019**, *23*, 4421–4437. [CrossRef]
67. LaTorre, A.; Muelas, S.; Peña, J.-M. A Comprehensive Comparison of Large Scale Global Optimizers. *Inf. Sci.* **2015**, *316*, 517–549. [CrossRef]

Σ mathematics

Article

Stochastic Triad Topology Based Particle Swarm Optimization for Global Numerical Optimization

Qiang Yang [1], Yu-Wei Bian [1], Xu-Dong Gao [1,*], Dong-Dong Xu [1], Zhen-Yu Lu [1], Sang-Woon Jeon [2] and Jun Zhang [2,3]

1 School of Artificial Intelligence, Nanjing University of Information Science and Technology, Nanjing 210044, China; qiang_yang@nuist.edu.cn (Q.Y.); 201983460001@nuist.edu.cn (Y.-W.B.); 001601@nuist.edu.cn (D.-D.X.); 001114@nuist.edu.cn (Z.-Y.L.)
2 Department of Electrical and Electronic Engineering, Hanyang University, Ansan 15588, Korea; sangwoonjeon@hanyang.ac.kr (S.-W.J.); junzhang@ieee.org (J.Z.)
3 Department of Computer Science and Information Engineering, Chaoyang University of Technology, Taichung 413310, Taiwan
* Correspondence: 003103@nuist.edu.cn

Abstract: Particle swarm optimization (PSO) has exhibited well-known feasibility in problem optimization. However, its optimization performance still encounters challenges when confronted with complicated optimization problems with many local areas. In PSO, the interaction among particles and utilization of the communication information play crucial roles in improving the learning effectiveness and learning diversity of particles. To promote the communication effectiveness among particles, this paper proposes a stochastic triad topology to allow each particle to communicate with two random ones in the swarm via their personal best positions. Then, unlike existing studies that employ the personal best positions of the updated particle and the neighboring best position of the topology to direct its update, this paper adopts the best one and the mean position of the three personal best positions in the associated triad topology as the two guiding exemplars to direct the update of each particle. To further promote the interaction diversity among particles, an archive is maintained to store the obsolete personal best positions of particles and is then used to interact with particles in the triad topology. To enhance the chance of escaping from local regions, a random restart strategy is probabilistically triggered to introduce initialized solutions to the archive. To alleviate sensitivity to parameters, dynamic adjustment strategies are designed to dynamically adjust the associated parameter settings during the evolution. Integrating the above mechanism, a stochastic triad topology-based PSO (STTPSO) is developed to effectively search complex solution space. With the above techniques, the learning diversity and learning effectiveness of particles are largely promoted and thus the developed STTPSO is expected to explore and exploit the solution space appropriately to find high-quality solutions. Extensive experiments conducted on the commonly used CEC 2017 benchmark problem set with different dimension sizes substantiate that the proposed STTPSO achieves highly competitive or even much better performance than state-of-the-art and representative PSO variants.

Keywords: particle swarm optimization; stochastic triad topology; guiding exemplar; multimodal problems; global optimization

MSC: 68-04; 65-04

Citation: Yang, Q.; Bian, Y.-W.; Gao, X.-D.; Xu, D.-D.; Lu, Z.-Y.; Jeon, S.-W.; Zhang, J. Stochastic Triad Topology Based Particle Swarm Optimization for Global Numerical Optimization. *Mathematics* **2022**, *10*, 1032. https://doi.org/10.3390/math10071032

Academic Editor: Janez Žerovnik

Received: 21 February 2022
Accepted: 21 March 2022
Published: 24 March 2022

1. Introduction

Particle swarm optimization (PSO) has witnessed tremendous success in solving optimization problems, especially non-differentiable ones [1–5], since its advent in 1995 [6,7]. Specifically, it maintains a swarm of particles, each of which represents a feasible solution, to iteratively search the solution space to find the global optima. Due to its good interpretability and great convenience of implementation [8–10], PSO has been widely applied to solve various real-world problems in daily life and industrial engineering, such as supply chain network design [11], control of pollutant spreading on social networks [12], and industrial power load forecasting [13].

In the classical PSO [6,7], a fully connected topology with all particles is utilized to select guiding exemplars for particles to update, leading to the global best position (usually denoted as ***gbest***) discovered by the whole swarm being shared by all particles. As a result, the learning diversity of particles is limited and thus the swarm falls into local areas when dealing with multimodal problems. To alleviate this limitation, researchers have paid extensive attention to designing novel learning strategies [14–19] to improve the optimization effectiveness of PSO.

In essence, the key to the learning strategies in PSO lies in the selection of guiding exemplars to direct the update of particles [17,20]. Broadly speaking, existing exemplar selection mechanisms can be classified into two categories, namely topology-based exemplar selection methods [15,16,21–23], and exemplar construction approaches [17–19,24,25].

Topology-based exemplar selection methods have been widely utilized in the research of PSO. In most cases, these methods aim to determine a less greedy exemplar to replace the social exemplar, namely ***gbest***, in the classical PSO [6,7]. Based on different topologies, an abundance of remarkable PSO variants have been developed [26–28], such as ring topology structure [26], pyramid topology structure [27], Von Neumann topology structure [29], random topology [22], etc. Different topologies usually preserve different characteristics and merits. Therefore, a natural idea is to hybridize them to ensemble the merits of different topologies to improve the optimization performance of PSO. Along this line, many PSO variants [28,30–33] have been developed based on different methods of hybridization. In addition, to alleviate the shortcoming of static topologies where each particle can only interact with fixed peers, some researchers further proposed dynamic topologies [34–36] to dynamically change the topologies (the topology type or the topology size) based on the evolution state of the swarm.

Although topology-based PSO variants have shown to be highly promising in solving optimization performance, the guiding exemplars selected by different topologies to direct the update of particles are all the historical promising positions found by particles. Therefore, the learning effectiveness of particles is limited by the historical positions [14,18,37]. Once all the historical positions converge to local areas, it is difficult for the swarm to jump out of the local basin. To alleviate this issue, researchers have attempted to develop novel PSOs from another perspective, namely constructing new guiding exemplars for particles to learn from [14,38].

Different from topology-based exemplar selection methods, exemplar construction methods generally build new guiding exemplars for particles by combining dimensions of historical positions. In general, it is highly possible that the built exemplars are not visited by particles. The most representative PSO variant in this direction is the comprehensive learning PSO (CLPSO) [17], which constructs a new guiding exemplar dimension by dimension from the personal best positions of all particles. Inspired from this method of constructing new exemplars, researchers have developed many other construction approaches, such as orthogonal learning PSO (OLPSO) [37], and genetic learning PSO (GLPSO) [18].

Although most existing PSO variants have shown very promising performances in simple optimization problems, such as unimodal problems and simple multimodal problems, their performance is confronted with great challenges or even deteriorates dramatically when dealing with complicated optimization problems, such as multimodal problems with many wide and flat local areas [39,40], and composition problems with many interacting variables. However, in the era of big data and Internet of Things (IoT) [41], optimization problems become increasingly complicated, which are ubiquitous in every field in daily life and industrial engineering [11,13]. As a consequence, the optimization ability of PSO to solve complicated optimization problems warrants urgent and careful research, rendering ongoing research into PSO an important frontier in evolutionary community [42,43].

To improve the optimization effectiveness of PSO in solving complicated problems, this paper proposes a stochastic triad topology-based PSO (STTPSO). Specifically, during the evolution, for each particle, two personal best positions are first randomly selected from those of the rest particles. Then, the personal best position of the particle to be updated and the two random best positions form a triad topology. Based on this topology, the best one in the topology and the mean position of the topology are employed as the guiding exemplars to direct the update of the particle. In this way, each particle likely preserves different guiding exemplars and thus the learning diversity of particles can be largely improved, which is beneficial for the swarm to escape from local areas.

Overall, the main components of the proposed STTPSO are summarized as follows:

(1) A stochastic triad topology is employed to connect the personal best position of each particle and two different personal best positions randomly selected from those of the rest particles to select guiding exemplars for particles to update. Different from existing studies [22,37], which only utilize the topologies to determine the best position to replace the social exemplar, namely ***gbest***, in the classical PSO (with another guiding exemplar as the personal best position of the particle), the proposed STTPSO utilizes the stochastic triad topology to select the best one and computes the mean position of the triad best positions as the two guiding exemplars to direct the update of each particle. Since the topology is stochastic, it is likely that different particles preserve different guiding exemplars. As a result, the learning diversity of particles can be largely promoted, and thus the probability of the swarm escaping from local areas can be promoted.
(2) An archive is maintained to store the obsolete personal best positions and then is combined with the personal best positions of all particles in the current generation to form the triad topologies for particles. In this way, valuable historical information can be utilized to direct the update of particles, which is helpful for improving swarm diversity.
(3) A random restart strategy is designed by randomly initializing a solution with a small probability. However, instead of employing this restart strategy on the swarm, we utilize it on the archive. That is to say, a randomly initialized solution is inserted into the archive with a small probability. In this way, the swarm diversity can be promoted without significant sacrifice of convergence speed.
(4) A dynamic strategy for the acceleration coefficients is devised to alleviate the sensitivity of STTPSO. Instead of utilizing fixed values for the two acceleration coefficients, this paper randomly samples the two acceleration coefficients based on the Gaussian distribution with the mean value set as the classical setting of the two coefficients and a small deviation. With this dynamic strategy, different particles can have different settings, and thus the learning diversity can be further promoted.

The above four components collaborate cohesively to help STTPSO explore and exploit the solution appropriately to locate the optima of optimization problems. In order to verify the effectiveness of the proposed STTPSO, extensive experiments are conducted on the widely used CEC 2017 benchmark problem set [44] with three different dimension sizes by comparing STTPSO with several state-of-the-art and popular PSO variants. In addition, deep investigations on the devised STTPSO are also conducted to observe the influence of each component in STTPSO.

The rest of this paper is organized as follows. Closely related studies are reviewed briefly in Section 2. In Section 3, the proposed STTPSO is elucidated in detail. Then, extensive comparison experiments and analysis of the associated results are conducted and discussed in Section 4. Finally, the conclusion of this paper is provided in Section 5.

2. Related Works

2.1. Basic PSO

PSO is a heuristic search algorithm and was first proposed in 1995 by Kennedy and Eberhart [6,7]. Specifically, PSO maintains a population of particles to search the solution space, and each particle is represented by a position vector $\boldsymbol{x}_i$ and a velocity vector $\boldsymbol{v}_i$, where the position vector represents a feasible solution to the problem, while the velocity vector represents the moving direction of the particle. Moreover, each particle memorizes its own historical best position $\boldsymbol{pbest}_i$, and the global best position $\boldsymbol{gbest}$ of the whole population is also memorized during the evolution. Then, each particle is updated by cognitively learning from its own experience, namely its personal best position $\boldsymbol{pbest}_i$, and socially learning from the experience of the whole swarm, namely the global best position $\boldsymbol{gbest}$. Specifically, each particle is updated as follows:

$$\boldsymbol{v}_i = w\boldsymbol{v}_i + c_1 r_1(\boldsymbol{pbest}_i - \boldsymbol{x}_i) + c_2 r_2(\boldsymbol{gbest} - \boldsymbol{x}_i) \tag{1}$$

$$\boldsymbol{x}_i = \boldsymbol{x}_i + \boldsymbol{v}_i \tag{2}$$

where w is the inertia weight, c_1 and c_2 are called acceleration coefficients, and r_1 and r_2 are two real random numbers uniformly sampled within [0, 1].

In Equation (1), the first part in the right hand is the "inertia part", which controls the memory of the velocity of each particle in the last generation. The second part is the "self-cognition" part, where each particle learns from its own experience. The third part is the social part, where each particle learns from the experience of the whole swarm.

As for the inertia weight w, in the literature [17,18,37,45], a linear decay method is widely utilized in PSO variants, which is presented below:

$$w = 0.9 - 0.5 \times \frac{fes}{FE_{max}} \tag{3}$$

In the literature [10,17,18,37,45–47], it is widely recognized that the learning strategy in Equation (1) plays the most important role in helping PSO achieve satisfactory performance. As a result, researchers have been devoted to designing novel effective learning strategies for PSO to improve its optimization ability.

2.2. Advanced Learning Strategies for PSO

To improve the optimization performance of PSO, many researchers have designed an ocean of effective learning strategies to improve the learning diversity and the learning effectiveness of particles [18,48–51]. As far as we are concerned, existing learning strategies for PSO could be classified into two main categories as shown in Table 1, namely topology-based learning strategies [22,26,52], and exemplar construction-based learning strategies [14,17,37,38,53].

Topology-based learning strategies [21,22,26,54,55], mainly utilize a specific topology to interact with particles to select guiding exemplars to update particles. In fact, the classical PSO [6,7] introduced above is a topology-based learning PSO, where the topology is the full topology connecting all particles. Such a full topology usually leads to an excessively greedy guiding exemplar (namely the global best position $\boldsymbol{gbest}$), which likely attracts particles into local areas. To alleviate this issue, many researchers have developed many kinds of local topologies to select less greedy guiding exemplars to direct the update of particles. For instance, in [26], the ring topology was utilized to organize particles into a ring, and then each particle interacts with its two neighbors to select one guiding exemplar to replace $\boldsymbol{gbest}$ in Equation (1). In [27], the pyramid topology with a three-dimensional wire-frame triangle was used to select the guiding exemplars for particles. In [29,55], the star topology was employed for particles to interact with others. In this topology, the central node shares information with other particles, and other particles also share information with the central node. Therefore, the communication is a two-way information exchange. In [29], the Von Neumann topology which is a two-dimensional lattice, was adopted to select guiding exemplars. Specifically, this topology connects the top, bottom, left and right neighbors of each point to form a neighborhood topology of size five. Such a topology can be regarded as a "two-dimensional" ring topology derived from a one-dimensional line into a two-dimensional plane.

Table 1. The rough classification of existing PSO variants.

Category			Methods	Characteristics
Topology-based Methods	Static Topology	Full Topology	PSO [6,7]	Each particle can only communicate with fixed peers. The learning diversity of particles is limited.
		Ring Topology	MRTPSO [26], GGL-PSOD [56]	
		Pyramid Topology	PMKPSO [27]	
		Star Topology	PSO-Star [29,55]	
		Von Neumann Topology	PSO-Von-Neumann [29]	
		Hybrid Topology	XPSO [23]	
	Dynamic Topology	Dynamic Topology	DNSPSO [28], DMSPSO [15], SPSO [22]	Each particle communicates with dynamic peers. The learning diversity of particles is high.
		Dynamic Size Topology	RPSO [16]	
Exemplar construction-based Methods	Random Construction		CLPSO_LS [14], CLPSO [17], HCLPSO [25], TCSPSO [19]	Randomly recombine dimensions of personal best positions. The exemplar construction efficiency is low, but it consumes no fitness evaluations in exemplar construction.
	Operator-based Construction		MPSOEG [24], GLPSO [18]	Recombine dimensions of personal best positions based evolutionary operators in other EAs. The exemplar construction efficiency is high, but it consumes many fitness evaluations in exemplar construction
	Orthogonal Recombination		OLPSO [37]	Recombine dimensions of personal best positions based on orthogonal experimental design. The exemplar construction efficiency is high, but it consumes a lot of fitness evaluation in exemplar construction

All topologies mentioned above are static topologies. In these topologies, each particle interacts with fixed peers during the evolution, and thus they bear limitations in improving the learning effectiveness of particles. To compensate for this shortcoming, researchers have attempted to develop dynamic topologies to select guiding exemplars for particles. Along this line, Liang and Suganthan [15] designed a random topology, which connects each particle with several randomly selected particles. Their experimental results demonstrated that the randomly constructed topological structure exhibits the best performance when its size is equal to three. In [22], each particle interacts with k particles randomly selected from the swarm. As for the setting of k, it is set between one and the population size. In particular, it can be the same for all particles and can also be different for different particles. In [16], the authors proposed adaptive adjustment of the size of the topology based on the evolution state. Specifically, in the early stage, a small size is maintained to preserve high search diversity, so that particles focus on exploring the solution space. Whereas, at the late stage, a large topology size is maintained to guarantee the convergence, so that particles focus on exploiting the solution space. In [23], the authors combined the global topology and the local topology to select guiding exemplars for particles, so that a good balance between exploration and exploitation could be maintained.

The aforementioned topology-based learning strategies mainly select guiding exemplars for particles based on existing personal best positions found by all particles. To further promote the learning effectiveness of particles, some researchers have attempted to construct novel exemplars, which might not be visited by particles, for particles, leading to exemplar construction-based learning strategies [14,17,19,24,37,38].

Different from topology-based learning strategies, exemplar construction-based learning strategies construct new exemplars that are not visited by particles based on the personal best positions of all particles. In this direction, the most representative algorithm is the comprehensive learning PSO (CLPSO) [17]. Specifically, this algorithm uses the binary tournament selection mechanism to select a learning object for each dimension of the current particle. From a macro perspective, CLPSO constructs a new position for each particle dimension by dimension that does not exist in the current population. Since CLPSO randomly recombines dimensions of different personal best positions, it ignores the correlation between variables and thus cannot effectively integrate useful evolutionary information together. To further improve the optimization performance of CLPSO, a heterogeneous CLPSO (HCLPSO) was proposed in [25]. This algorithm divides the population into two sub-populations, with one sub-population used for exploration, which is updated by the original CL strategy, and another sub-population used for exploitation, which is guided by the global best position.

Although the above introduced CLPSO variants have achieved good performance in solving multimodal problems, the construction of promising exemplars for particle is inefficient since the recombination of dimensions is totally random. To improve the exemplar construction effectiveness and efficiency, in [37], Zhan et al. proposed an orthogonal learning PSO (OLPSO) by using the orthogonal experimental design to seek useful dimension combinations of the historical positions found by particles. Specifically, OLPSO adopts an orthogonal matrix to evaluate the effectiveness of the dimension combinations and then combines the most useful dimension combinations to construct promising exemplars. Though the exemplar construction efficiency is improved, it consumes many fitness evaluations in the exemplar construction stage. To reduce fitness evaluation consumption in the exemplar construction, in [18], Gong et al. employed the genetic operators such as crossover, mutation, and selection, to construct guiding exemplars for particles, leading to a genetic learning PSO (GLPSO). With these operators, GLPSO is expected to generate diversified and high-quality exemplars for particles. To further improve its optimization effectiveness, in [56], a global GLPSO with a ring topology (GGL-PSOD) was devised, where the ring topology is adopted to generate diversified exemplars based on neighbor particles. Though the constructed exemplars are promising, GLPSO and its variants still consume many fitness evaluations in the exemplar construction. To further construct diversified but promising guiding exemplars for particles, in [19], terminal crossover and steering-based PSO with distribution (TCSPSO) was proposed by devising a new crossover mechanism to construct exemplars. Meanwhile, a global disturbance was designed to improve the population diversity to escape from local areas. In [24], a modified particle swarm optimization with effective guides (MPSOEG) was devised by generating two types of guiding exemplars with an optimal guide creation (OGC) module. In particular, a global exemplar is constructed by the OCG module to guide the swarm towards promising regions, whereas a local exemplar is constructed for each particle to escape from local areas.

Except for the abovementioned learning strategies, some researchers have even attempted utilizing multiple learning strategies to direct the evolution of the swarm in PSO. For instance, in [20], the concept of evolutionary game theory was introduced, and four classical learning strategies were taken as game strategies in the game theory. Then, the swarm adaptively selects the most suitable learning strategy based on the current evolution state. In [28], a dynamic-neighborhood-based switching PSO (DNSPSO) algorithm was proposed by adjusting the personal best position and the global best position based on a distance-based dynamic neighborhood and hybridizing the differential evolution algorithm to alleviate premature convergence.

Although many of the original limitations and shortcomings of PSO have been greatly improved since its introduction, its optimization performance in solving complex optimization problems with many interacting variables and a wide saddle-point region still encounters great challenges. Therefore, methods to improve the optimization performance of PSO in solving widely emerging complicated problems remains an open issue and deserves careful attention, which results in the research of PSO remaining a highly popular and frontier topic in the evolutionary computation community. To improve the learning effectiveness of particles in complicated environments, this paper proposes a stochastic triad topology-based PSO (STTPSO), which will be elucidated in the following section.

3. Stochastic Triad Topology-Based Particle Swarm Optimization

The most crucial aspect of PSO is the interaction among particles to select guiding exemplars to direct the update of particles [22,26,27,29]. Most existing topology-based PSO variants [21,22,55] only adopt the topologies to select one exemplar to replace the social exemplar (namely $\boldsymbol{gbest}$) in the classical PSO (shown in Equation (1)). Such utilization of topologies is limited since it only changes one exemplar in Equation (1), which results in limited improvement in the learning effectiveness and learning diversity of particles. To alleviate this predicament, this paper proposes a stochastic triad topology-based PSO (STTPSO), which utilizes a stochastic triad topology for each particle to select two novel guiding exemplars to replace the two ones in the classical PSO to promote the learning effectiveness and learning diversity of particles.

3.1. Stochastic Triad Topology

During the evolution, given that PS particles are maintained in the swarm, then for each particle x_i $(0 \le i \le PS)$, a stochastic triad topology is employed to connect the personal best position ($\boldsymbol{pbest_i}$) of this particle and two different personal best positions, which are randomly selected from those of other particles. Given that the two randomly selected personal best positions are $\boldsymbol{pbest_{r1}}$ and $\boldsymbol{pbest_{r2}}$, respectively, this paper utilizes the best one among the triad $\boldsymbol{pbest}$s, ($\boldsymbol{pbest_i}$, $\boldsymbol{pbest_{r1}}$, and $\boldsymbol{pbest_{r2}}$) and the mean position of these triad $\boldsymbol{pbest}$s as the two guiding exemplars to replace the two ones in Equation (1) to update each particle. Specifically, the velocity of each particle is updated as follows:

$$v_i = wv_i + c_1r_1(\boldsymbol{tpbest_i} - x_i) + c_2r_2(\overline{\boldsymbol{tmean_i}} - x_i) \tag{4}$$

where $\boldsymbol{tpbest_i}$ is the best one among the triad $\boldsymbol{pbest}$s, which is determined as follows:

$$\boldsymbol{tpbest_i} = \text{argmin}\{f(\boldsymbol{pbest_{r_1}}), f(\boldsymbol{pbest_{r_2}}), f(\boldsymbol{pbest_i})\} \tag{5}$$

$\overline{\boldsymbol{tmean_i}}$ represents the mean position of the triad $\boldsymbol{pbest}$s, which is calculated as follows:

$$\overline{\boldsymbol{tmean_i}} = \frac{1}{3}(\boldsymbol{pbest_{r_1}} + \boldsymbol{pbest_{r_2}} + \boldsymbol{pbest_i}) \tag{6}$$

Upon deep observation of Equation (4), we discover the following findings: (1) The first guiding exemplar ($\boldsymbol{tpbest_i}$) is likely different for different particles. This is because the triad topology of each particle is constructed by randomly selecting two different personal best positions ($\boldsymbol{pbest_{r1}}$ and $\boldsymbol{pbest_{r2}}$) from those of other particles along with the personal best position ($\boldsymbol{pbest_i}$) of this particle. Due to such randomness, the diversity of the first exemplar could be largely promoted. (2) The second guiding exemplar ($\overline{\boldsymbol{tmean_i}}$) is also likely different due to the random construction of the triad topology. Therefore, the diversity of the second exemplar is also promoted to a large extent. Along with high diversity of the first exemplar, we can see that the learning diversity of particles is high, which is of great benefit for particles to search the solution space dispersedly and thus is helpful for the swarm to escape from local areas. (3) The first exemplar is expectedly better than the personal best position of the particle to be updated. As a result, the learning effectiveness and efficiency of particles is expected to be promoted, which is beneficial for the swarm to rapidly converge to promising areas. (4) As for the second exemplar, it can be considered as a kind of distribution estimation of the triad $\boldsymbol{pbest}$s. Utilizing it as one guiding exemplar is also expected to direct the updated particle to promising areas fast. (5) However, compared with the first exemplar, the quality of the second exemplar is uncertain. In particular, we can consider that the first exemplar is responsible for convergence, while the second is in charge of swarm diversity. Therefore, in Equation (4), a promising balance between fast convergence and high diversity is implicitly maintained in the update of particles.

As for the triad topology structure, to guarantee the learning effectiveness of particles, instead of frequently changing the topology structure, we first keep the structure fixed for each particle. That is to say, the indexes ($\boldsymbol{pbest}_{r1}$ and $\boldsymbol{pbest}_{r2}$) of the randomly selected two personal best positions for each particle are not changed. Then, we observe the change of the personal best position of each particle. For particle $\boldsymbol{x}_i$ ($0 \leq i \leq PS$), if its personal best position $\boldsymbol{pbest}_i$ keeps unchanged for continuous $stop_{max}$ times, this indicates that the learning effectiveness of this particle under the triad structure degrades. In this situation, to improve the learning effectiveness of this particle, we randomly reselect two different personal best positions from those of other particles to rebuild the triad topology structure. In this way, the learning effectiveness and learning diversity of particles can be largely promoted.

In Section 4.3, investigative experiments are conducted to verify the effectiveness of the adaption strategy for the triad topology structure of each particle. Experimental results show that $stop_{max}$ = 30 helps STTPSO achieve the best overall performance, and thus in this paper, we set $stop_{max}$ = 30 for STTPSO.

Remark

In essence, the proposed stochastic triad topology belongs to a kind of random topology. In the literature, many studies [21,22,52,57] have adopted random topologies to select guiding exemplars for particles to learn from. Compared with these existing studies, this paper distinguishes itself from them in the following ways:

(1) Unlike existing studies that use the random topologies to determine only one guiding exemplar to replace the social exemplar ($\boldsymbol{gbest}$) in the classical PSO [6,7], the proposed STTPSO utilizes the stochastic triad topology for each particle to select the best one among the triad personal best positions and computes the mean position of these ***pbest***s as the two guiding exemplars to direct the update of this particle. In this way, due to the randomness of the triad topology, not only the diversity of the first exemplar is promoted largely, but also the diversity of the second exemplar is promoted to a large extent. Therefore, the learning diversity of particles is improved, which is beneficial for enhancing the chance of escaping from local areas for the swarm.

(2) Unlike existing studies that change the random topology structure every generation, this paper adaptively changes the triad topology structure based on the evolution state of each particle. In particular, we record stagnation times of each particle ($\boldsymbol{x}_i$), which is actually the number of continuous generations where the personal best position ($\boldsymbol{pbest}_i$) of the particle remains unchanged. When such a number exceeds a predefined threshold $stop_{max}$, the triad topology structure is reconstructed by randomly reselecting two different personal best positions from those of other particles. In this way, the triad topology structure of each particle is changed asynchronously, which guarantees the learning effectiveness of particles.

3.2. Dynamic Acceleration Coefficients

As for the parameters in Equation (4), with respect to the inertia weight w, we directly adopt the widely used dynamic strategy as shown in Equation (3) to dynamically adjust w during the evolution.

As for the acceleration coefficients c_1 and c_2, in the classical PSO, a large body of research has recommended to set them as $c_1 = c_2 = 1.49618$ [18]. Such a setting makes all particles share the same setting, which, as far as we are concerned, is not beneficial for improving the learning diversity of particles. Therefore, to alleviate this issue and to further enhance the learning diversity of particles, we first randomly samples two values v_1 and v_2 from the Gaussian distribution *Gaussian* (1.49618, 0.1) as follows:

$$\begin{cases} v_1 = Gaussian(1.49618,\ 0.1) \\ v_2 = Gaussian(1.49618,\ 0.1) \end{cases} \tag{7}$$

Then, we set c_1 and c_2 based on the sampled two values as follows:

$$\begin{cases} c_1 = max(v_1,\ v_2) \\ c_2 = min(v_1,\ v_2) \end{cases} \tag{8}$$

First, the Gaussian distribution *Gaussian* (1.49618, 0.1) with the mean value set as the classical setting of c_1 and c_2 and the standard deviation set as a small value allows the sampled two values to be close to the classical setting but with a small difference. In this way, the diversity of the settings of c_1 and c_2 is slightly increased, resulting in a slight improvement in the learning diversity of particles.

Second, between the two sampled values, the larger one is set to c_1, while the smaller one is set to c_2. This is because, as aforementioned in Equation (4), the first guiding exemplar (the best one among the triad ***pbest*s**) is expectedly better than the second guiding exemplar (the mean position of the triad ***pbest*s**), and thus we can consider that the first exemplar is responsible for the convergence, while the second exemplar takes charge of the diversity. Since the second exemplar is the mean position of the triad ***pbest*s**, which is expectedly different from the first exemplar, we set c_1 with the larger sampled value and c_2 with the smaller value to guarantee that the updated particle learns more from the first exemplar, so that it can approach promising areas fast without serious loss of diversity by learning slightly less from the second exemplar.

Lastly, experiments conducted in Section 4.3 will demonstrate the effectiveness of the proposed dynamic acceleration coefficient strategy.

3.3. *Historical Information Utilization*

In PSO, the obsolete historical information may also contain useful evolutionary information. As a consequence, many studies [2,58] have maintained an additional archive to store historical information to evolve the swarm. Inspired from this, this paper also maintains an archive of size $PS/2$ to store the obsolete personal best positions of particles.

Specifically, during the evolution, once a particle discovers a better position, its old personal best position is first inserted into the archive and then is replaced by the new better position. Once the archive is full, namely when its size exceeds $PS/2$, an obsolete personal best position is inserted into the archive by randomly replacing a solution in the archive.

During the update of particles, the archive along with the personal best positions of all particles are used to construct the triad topology structure of each particle. In particular, when the stagnation times of the personal best position of one particle exceeds $stop_{max}$, two different personal best positions are randomly selected from the archive and those of the other particles to rebuild the triad topology structure of this particle. In this way, the historical evolutionary information is employed to evolve the swarm.

Due to the utilization of historical information, the number of candidates used to build the triad topology of each particle is increased and thus the learning diversity of particles can be improved largely, which is beneficial for the swarm to escape from local areas. Experiments conducted in Section 4.3 will demonstrate the effectiveness of this additional archive.

3.4. *Random Restart Strategy*

To further enhance the chance of escaping from local areas for the swarm, this paper further proposes a random restart strategy to introduce initial solutions. Specifically, given a small restart probability *pm*, during the evolution, in each generation, when a uniformly sampled value within [0, 1] is smaller than pm, a feasible solution is randomly initialized within the range of variables. Then, instead of inserting this solution into the current swarm as noted in existing studies [59–63], this paper inserts this initialized solution into the archive. If the archive is full, it randomly replaces a solution in the archive.

In particular, such a restart strategy with the initialized solution inserted in the archive could not seriously break the convergence of the swarm but could improve the learning diversity of particles effectively. Most existing studies [60–63] only replace one particle in the current swarm with the initialized solution. Such a strategy usually leads to a very small improvement in the learning diversity and learning effectiveness of particles. This is because the personal best positions of all particles remain unchanged, leading to the learning effectiveness of most particles not improving. However, in our strategy, the randomly initialized solution is inserted into the archive, which is then used to build the triad topology structure of each particle. Therefore, we can see that once the initialized solution is selected to build the triad structure of one particle, at least the second exemplar (namely the mean position of the triad positions in the topology) is changed. Consequently, the learning diversity of particles can be effectively improved, which is beneficial for the swarm to escape from local areas. Experiments conducted in Section 4.3 will demonstrate the effectiveness of this restart strategy.

3.5. Overall Procedure

Integrating the above components together, the overall procedure of the developed STTPSO is shown in Algorithm 1. Specifically, as shown in Lines 1 to 4, the triad topology is constructed for each of PS particles after they are randomly initialized and evaluated. Moreover, the stagnation time of each particle is initialized as 0. Then, the algorithm goes into the main iteration (Lines 5~27). First, for each particle $\boldsymbol{x}_i$, the inertia weight w is computed (Line 7) and then the acceleration coefficients c_1 and c_2 are set based on Gaussian distribution (Lines 8–11). Subsequently, the particle is updated, and then the personal best position ($\boldsymbol{pbest}_i$) of this particle is updated with its stagnation time $stop_i$ updated as well (Lines 12–19). Once the stagnation time of particle $\boldsymbol{x}_i$ reaches the allowed maximum stagnation time $stop_{max}$, two different personal best positions are randomly selected from those of all particles and the archive to rebuild the triad topology structure (Lines 20~22). After all particles are updated, the random restart strategy is conditionally triggered to randomly insert an initialized solution into the archive (Lines 24~26). The above main iteration proceeds until the maximum number of fitness evaluations is exhausted and at the end of the program, the global best position is obtained as the final output.

From Algorithm 1, it can be observed that during each iteration, $O(PS)$ is needed to compute the parameters such as w, c_1 and c_2. Following this, $O(PS)$ is needed to obtain the best one among the triad $\boldsymbol{pbest}$s and $O(PS \times D)$ to calculate the mean position of the triad $\boldsymbol{pbest}$s for all particles. Then, $O(PS \times D)$ is used to update particles. During the update of the archive, $O(PS \times D)$ is needed in each generation. Overall, the time complexity of STTPSO is $O(PS \times D)$, which is the same as the classical PSO. Therefore, we can see that STTPSO remains as efficient as the classical PSO.

Algorithm 1: The pseudocode of STTPSO

Input: swarm size *PS*, maximum fitness evaluations *FEmax*, maximum stagnation times $stop_{max}$, restart probability *pm*;

Initialize *PS* particles randomly and calculate their fitness;
Set *fes* = *PS*, and set the archive empty;
Randomly select two different personal best positions ($\boldsymbol{pbest}_{r1}$ and $\boldsymbol{pbest}_{r2}$) from the personal best positions of other particles and the archive for each particle to form the associated triad topology;
Set the stagnation time $stop_i = 0$ ($1 \le i \le PS$) for each particle;
While ($fes \le FEmax$) **do**
  For i = 1:*PS* **do**
    Compute w according to Equation (3);
    Randomly sample c_1 and c_2 from Gaussian(1.49618,0.1);
    If $c_1 < c_2$ **then**
      Swap c_1 and c_2;
    End If
    Update $\boldsymbol{x}_i$ and $\boldsymbol{v}_i$ according to Equations (2) and (4);
    Calculate the fitness of the updated $\boldsymbol{x}_i$: $f(\boldsymbol{x}_i)$ and fes + = 1;
    If $f(\boldsymbol{x}_i) < f(\boldsymbol{pbest}_i)$ **then**
      Put $\boldsymbol{pbest}_i$ in the archive and set $stop_i = 0$;
      $\boldsymbol{pbest}_i = \boldsymbol{x}_i$;
    Else
      $stop_i$ += 1;
    End If
    If $stop_i >= stop_{max}$ **then**
      Reselect two different personal best positions ($\boldsymbol{pbest}_{r1}$ and $\boldsymbol{pbest}_{r2}$) from those of other particles and the archive for $\boldsymbol{x}_i$ to form the associated triad topology;
    End If
  End For
  If rand(0, 1) < *pm* **then**
    Randomly initialize a solution and store it into the archive;
  End If
End While
Obtain the global best solution ***gbest*** and its fitness $f(\boldsymbol{gbest})$;

Output: $f(\boldsymbol{gbest})$ **and** ***gbest***

4. Experiments

This section mainly verifies the effectiveness of the proposed STTPSO by extensive experiments conducted on the widely used CEC 2017 benchmark function set [44]. Specifically, this benchmark set contains 29 optimization problems with four categories, namely unimodal functions, simple multimodal functions, hybrid functions, and composition functions. Compared with the former two categories of optimization problems, the latter two kinds of optimization problems are more difficult to optimize.

4.1. Experimental Setup

Firstly, in order to verify the effectiveness of STTPSO effectively, we select seven most advanced PSO variants as the compared methods, namely DNSPSO [28], XPSO [23], DPLPSO [45], TCSPSO [19], GLPSO [18], HCLPSO [25] and CLPSO [17]. DNSPSO, XPSO and DPLPSO are state-of-the-art topology-based PSO variants, while TCSPSO, GLPSO, HCLPSO and CLPSO are state-of-the-art exemplar construction based PSO variants.

Secondly, in order to verify the optimization performance of STTPSO in a comprehensive way, we conduct comparative experiments on the CEC 2017 benchmark set with three dimension sizes, namely 30-*D*, 50-*D* and 100-*D* respectively. For the sake of fairness, the maximum number of function evaluation times is set as 10,000 × *D* for all algorithms.

Thirdly, for fair comparisons, except for the population size, we adopt the parameter settings for all key parameters in the compared PSO variants as recommended in the associated papers. As for the population size, due to its sensitivity to optimization problems, we fine-tune its settings for all compared PSO variants. After preliminary parameter fine-tuning experiments, Table 2 lists the specific parameter settings of all algorithms.

Table 2. Parameter settings of the proposed STTPSO and the compared algorithms.

Algorithm	*D*	Parameter Settings	
STTPSO	30	$PS = 300$	$AS = PS/2$; $w = 0.9\sim0.4$; $c\sim N(1.49618, 0.1)$; $pm = 0.01$; $stop_{max} = 30$
	50	$PS = 300$	
	100	$PS = 300$	
DNSPSO	30	$PS = 50$	$w = 0.4\sim0.9$; $k = 5$; $F = 0.5$; $CR = 0.9$;
	50	$PS = 50$	
	100	$PS = 60$	
XPSO	30	$PS = 100$	$\eta = 0.2$; $Stag_{max} = 5$; $p = 0.5$; $\sigma = 0.1$
	50	$PS = 150$	
	100	$PS = 150$	
TCSPSO	30	$PS = 50$	$w = 0.9\sim0.4$; $c_1 = c_2 = 2$
	50	$PS = 50$	
	100	$PS = 50$	
GLPSO	30	$PS = 40$	$w = 0.7298$; $c = 1.49618$; $pm = 0.1$; $sg = 7$
	50	$PS = 40$	
	100	$PS = 50$	
HCLPSO	30	$PS = 160$	$w = 0.99\sim0.2$; $c_1 = 2.5\sim0.5$; $c_2 = 0.5\sim2.5$; $c = 3\sim1.5$
	50	$PS = 180$	
	100	$PS = 180$	
DPLPSO	30	$PS = 40$	$c_1 = c_2 = 2$; $L = 50$
	50	$PS = 40$	
	100	$PS = 40$	
CLPSO	30	$PS = 40$	$Pc = 0.05\sim0.5$
	50	$PS = 40$	
	100	$PS = 40$	

Finally, in order to comprehensively evaluate the optimization performance of all algorithms, we independently execute each algorithm 30 times and use the median, the mean and the standard deviation to measure the optimization performance of each algorithm. To distinguish the statistical significance with respect to the performance difference between two algorithms, the Wilcoxon rank sum test at the significance level of 0.05 is conducted. Furthermore, to obtain the overall performance of each algorithm on the whole benchmark set, the Friedman test is conducted to obtain the overall ranks of all algorithms on the whole benchmark set.

4.2. Comparison with State-of-the-Art PSO Variants

In this section, we conduct extensive comparative experiments on the CEC 2017 benchmark set with the three dimension sizes to compare STTPSO with the seven state-of-art PSO variants. Tables 3–5, respectively, show the comparison results between the proposed STTPSO and the seven PSO variants on the 30-*D*, the 50-*D* and the 100-*D* CEC 2017 benchmark functions. In these tables, the symbols '+', '−' and '=' indicate that the proposed STTPSO is significantly superior to, significantly inferior to and roughly equivalent to the associated compared algorithm on the associated functions. As shown in the second to last row of each table, '$w/t/l$' count the number of functions where the proposed STTPSO achieves significantly better performance, obtains roughly equivalent performance, and exhibits significantly worse performance than the compared algorithms, respectively. In fact, they are the numbers of '+', '=' and '−'. In the last row of each table, the average rank of each algorithm obtained by the Friedman test is displayed. Moreover, the statistical comparison results between the proposed STTPSO and the seven state-of-the-art PSO variants on the CEC 2017 benchmark set with different dimension sizes in terms of "$w/t/l$" are summarized in Table 6.

As shown in Table 3, the comparison results on the 30-*D* CEC 2017 benchmark functions are summarized below:

(1) According to the Friedman test results as shown in the last row, STTPSO achieves the lowest rank among all eight algorithms and its rank value (1.86) is much smaller than those (at least 2.55) of the seven compared algorithms. This demonstrates that STTPSO achieves the best overall performance on the 30-*D* CEC 2017 benchmark functions, and presents significant superiority over the seven compared algorithms.

(2) The second last row of Table 2 shows that STTPSO is significantly superior to the compared algorithms on at least 21 problems except for XPSO, and only presents inferior performance on, at most, five problems. Compared with XPSO, STTPSO obtains significantly better performance on 18 problems, while only performing worse than XPSO on three problems.

(3) In terms of the comparison results on different types of optimization problems, STTPSO achieves highly competitive performance with all the compared algorithms on the two unimodal problems. In particular, it shows significant dominance to DNSPSO and DPLPSO both on the two problems. In terms of the six simple multimodal problems, except for DNSPSO, STTPSO shows significantly better performance than the other six compared algorithms on all these problems. Compared with DNSPSO, STTPSO presents significant superiority on five problems and shows inferiority on only one problem. Regarding the 10 hybrid problems, STTPSO shows much better performance than DPLPSO on all 10 problems. Compared with DNSPSO, TCSPSO, and HCLPSO, STTPSO obtains significantly better performance on seven, six, and seven problems, respectively, and only shows inferiority to them on, at most, two problems. In comparison with XPSO, GLPSO, and CLPSO, STTPSO achieves no worse performance on at least seven problems and displays inferiority to them on, at most, three problems. Concerning the 11 composition problems, STTPSO outperforms the seven compared algorithms on at least nine problems, and only shows inferiority on, at most, two problems. In particular, STTPSO significantly outperforms both TCSPSO and GLPSO on all these problems and obtains much better performance than both HCLPSO and DPLPSO on 10 problems with no inferiority to them on all the 11 problems. Overall, it is demonstrated that STTPSO shows promise in solving various kinds of problems and particularly obtains good performance on complicated problems, such as multimodal problems, hybrid problems, and composition problems.

Table 3. Comparison results between the proposed STTPSO and the 7 state-of-the-art and popular PSO variants on the 30-*D* CEC 2017 benchmark functions. The highlighted *p*-values means that the proposed STTPSO is significantly better than the associated compared algorithms on the corresponding problems.

f	Category	Quality	STTPSO	DNSPSO	XPSO	TCSPSO	GLPSO	HCLPSO	DPLPSO	CLPSO
f_1	Unimodal Functions	Median	1.19×10^{3}	1.95×10^{5}	2.26×10^{3}	3.20×10^{3}	2.30×10^{3}	5.49×10^{3}	2.64×10^{9}	1.52×10^{2}
		Mean	2.10×10^{3}	2.11×10^{5}	4.05×10^{3}	3.66×10^{3}	3.06×10^{3}	8.65×10^{3}	2.87×10^{9}	3.88×10^{2}
		Std	2.28×10^{3}	1.30×10^{5}	4.72×10^{3}	4.08×10^{3}	2.42×10^{3}	7.34×10^{3}	1.11×10^{9}	7.31×10^{2}
		p-value	-	$\mathbf{1.83 \times 10^{-6}+}$	$2.17 \times 10^{-1}=$	$1.95 \times 10^{-1}=$	$5.85 \times 10^{-2}=$	$\mathbf{8.55 \times 10^{-5}+}$	$\mathbf{1.83 \times 10^{-6}+}$	$7.84 \times 10^{-5}-$
f_3		Median	1.52×10^{4}	1.51×10^{5}	6.26×10^{-2}	9.94×10^{3}	1.14×10^{-13}	4.54×10^{1}	3.91×10^{4}	4.30×10^{4}
		Mean	1.53×10^{4}	1.54×10^{5}	7.91×10^{-1}	1.15×10^{4}	1.33×10^{-13}	6.87×10^{1}	3.87×10^{4}	4.41×10^{4}
		Std	4.39×10^{3}	3.37×10^{4}	2.00×10^{0}	3.68×10^{3}	5.36×10^{-14}	8.17×10^{1}	8.36×10^{3}	1.00×10^{4}
		p-value	-	$\mathbf{1.83 \times 10^{-6}+}$	$1.83 \times 10^{-6}-$	$2.51 \times 10^{-4}-$	$1.83 \times 10^{-6}-$	$1.83 \times 10^{-6}-$	$\mathbf{1.83 \times 10^{-6}+}$	$\mathbf{1.83 \times 10^{-6}+}$
$f_{1\text{–}3}$	$w/t/l$		-	2/0/0	0/1/1	0/1/1	0/1/1	1/0/1	2/0/0	1/0/1
f_4	Simple Multimodal Functions	Median	8.47×10^{1}	2.54×10^{1}	1.24×10^{2}	1.30×10^{2}	1.47×10^{2}	8.56×10^{1}	7.62×10^{2}	9.09×10^{1}
		Mean	8.48×10^{1}	2.56×10^{1}	1.20×10^{2}	1.32×10^{2}	1.48×10^{2}	8.73×10^{1}	7.99×10^{2}	9.12×10^{1}
		Std	3.74×10^{-1}	1.07×10^{0}	2.71×10^{1}	4.84×10^{1}	4.27×10^{1}	7.92×10^{0}	1.86×10^{2}	1.57×10^{0}
		p-value	-	$1.83 \times 10^{-6}-$	$\mathbf{3.56 \times 10^{-5}+}$	$\mathbf{3.89 \times 10^{-5}+}$	$\mathbf{4.97 \times 10^{-6}+}$	$\mathbf{7.84 \times 10^{-5}+}$	$\mathbf{1.83 \times 10^{-6}+}$	$\mathbf{1.82 \times 10^{-6}+}$
f_5		Median	4.98×10^{0}	2.04×10^{2}	4.18×10^{1}	8.56×10^{1}	6.77×10^{1}	6.45×10^{1}	2.00×10^{2}	7.66×10^{1}
		Mean	4.71×10^{0}	2.03×10^{2}	4.38×10^{1}	8.92×10^{1}	6.68×10^{1}	6.77×10^{1}	1.93×10^{2}	7.52×10^{1}
		Std	1.96×10^{0}	1.25×10^{1}	1.58×10^{1}	2.54×10^{1}	1.96×10^{1}	1.67×10^{1}	3.28×10^{1}	6.69×10^{0}
		p-value	-	$\mathbf{1.83 \times 10^{-6}+}$	$\mathbf{1.83 \times 10^{-6}+}$	$\mathbf{1.83 \times 10^{-6}+}$	$\mathbf{1.82 \times 10^{-6}+}$	$\mathbf{1.82 \times 10^{-6}+}$	$\mathbf{1.83 \times 10^{-6}+}$	$\mathbf{1.83 \times 10^{-6}+}$
f_6		Median	1.14×10^{-13}	1.87×10^{-1}	3.82×10^{-3}	8.01×10^{-1}	6.34×10^{-3}	1.62×10^{-4}	2.97×10^{1}	2.66×10^{-6}
		Mean	1.12×10^{-7}	1.91×10^{-1}	1.50×10^{-2}	1.04×10^{0}	9.69×10^{-3}	2.41×10^{-3}	2.98×10^{1}	2.86×10^{-6}
		Std	2.91×10^{-7}	5.92×10^{-2}	3.82×10^{-2}	1.15×10^{0}	8.77×10^{-3}	5.85×10^{-3}	5.82×10^{0}	1.90×10^{-6}
		p-value	-	$\mathbf{1.83 \times 10^{-6}+}$	$\mathbf{1.83 \times 10^{-6}+}$	$\mathbf{1.82 \times 10^{-6}+}$	$\mathbf{1.83 \times 10^{-6}+}$	$\mathbf{1.83 \times 10^{-6}+}$	$\mathbf{1.82 \times 10^{-6}+}$	$\mathbf{1.83 \times 10^{-6}+}$

Table 3. *Cont.*

f	Category	Quality	STTPSO	DNSPSO	XPSO	TCSPSO	GLPSO	HCLPSO	DPLPSO	CLPSO
f_7		Median	3.44×10^{1}	2.36×10^{2}	7.94×10^{1}	1.45×10^{2}	9.75×10^{1}	1.06×10^{2}	2.90×10^{2}	9.23×10^{1}
		Mean	3.46×10^{1}	2.33×10^{2}	8.17×10^{1}	1.42×10^{2}	9.86×10^{1}	1.01×10^{2}	2.88×10^{2}	9.05×10^{1}
		Std	1.12×10^{0}	1.69×10^{1}	1.81×10^{1}	2.87×10^{1}	1.53×10^{1}	1.86×10^{1}	2.45×10^{1}	7.88×10^{0}
		p-value	-	$\mathbf{1.83 \times 10^{-6}}$+	$\mathbf{1.83 \times 10^{-6}}$+	$\mathbf{1.83 \times 10^{-6}}$+	$\mathbf{1.83 \times 10^{-6}}$+	$\mathbf{1.83 \times 10^{-6}}$+	$\mathbf{1.83 \times 10^{-6}}$+	$\mathbf{1.83 \times 10^{-6}}$+
f_8		Median	3.98×10^{0}	2.01×10^{2}	3.83×10^{1}	9.55×10^{1}	5.97×10^{1}	5.66×10^{1}	1.94×10^{2}	8.09×10^{1}
		Mean	4.15×10^{0}	2.02×10^{2}	3.98×10^{1}	9.33×10^{1}	6.08×10^{1}	6.35×10^{1}	1.90×10^{2}	8.18×10^{1}
		Std	1.67×10^{0}	1.03×10^{1}	1.37×10^{1}	2.22×10^{1}	1.65×10^{1}	2.08×10^{1}	3.20×10^{1}	9.86×10^{0}
		p-value	-	$\mathbf{1.82 \times 10^{-6}}$+	$\mathbf{1.82 \times 10^{-6}}$+	$\mathbf{1.83 \times 10^{-6}}$+	$\mathbf{1.83 \times 10^{-6}}$+	$\mathbf{1.83 \times 10^{-6}}$+	$\mathbf{1.82 \times 10^{-6}}$+	$\mathbf{1.83 \times 10^{-6}}$+
f_9		Median	5.69×10^{-14}	1.50×10^{0}	1.45×10^{0}	3.01×10^{2}	5.98×10^{1}	4.90×10^{1}	1.27×10^{3}	6.58×10^{2}
		Mean	5.69×10^{-14}	2.09×10^{0}	2.73×10^{0}	3.85×10^{2}	7.13×10^{1}	8.90×10^{1}	1.50×10^{3}	6.76×10^{2}
		Std	5.69×10^{-14}	1.39×10^{0}	3.44×10^{0}	3.36×10^{2}	4.83×10^{1}	1.49×10^{2}	6.80×10^{2}	2.80×10^{2}
		p-value	-	$\mathbf{1.83 \times 10^{-6}}$+	$\mathbf{1.82 \times 10^{-6}}$+	$\mathbf{1.83 \times 10^{-6}}$+	$\mathbf{1.83 \times 10^{-6}}$+	$\mathbf{1.83 \times 10^{-6}}$+	$\mathbf{1.82 \times 10^{-6}}$+	$\mathbf{1.83 \times 10^{-6}}$+
$f_{4\text{–}9}$	*w/t/l*		-	5/0/1	6/0/0	6/0/0	6/0/0	6/0/0	6/0/0	6/0/0
f_{10}	Hybrid Functions	Median	1.81×10^{3}	6.21×10^{3}	2.80×10^{3}	2.98×10^{3}	3.26×10^{3}	2.87×10^{3}	6.39×10^{3}	3.00×10^{3}
		Mean	2.82×10^{3}	5.90×10^{3}	2.62×10^{3}	2.97×10^{3}	3.23×10^{3}	2.90×10^{3}	6.33×10^{3}	2.94×10^{3}
		Std	1.82×10^{3}	1.01×10^{3}	6.13×10^{2}	4.22×10^{2}	8.64×10^{2}	5.17×10^{2}	4.47×10^{2}	2.77×10^{2}
		p-value	-	$\mathbf{4.50 \times 10^{-6}}$+	5.37×10^{-1}=	7.89×10^{-1}=	2.41×10^{-1}=	8.69×10^{-1}=	$\mathbf{1.83 \times 10^{-6}}$+	8.53×10^{-1}=
f_{11}		Median	1.79×10^{1}	9.12×10^{1}	8.01×10^{1}	1.16×10^{2}	7.24×10^{1}	1.09×10^{2}	4.10×10^{2}	1.21×10^{2}
		Mean	2.79×10^{1}	9.19×10^{1}	8.63×10^{1}	1.18×10^{2}	7.82×10^{1}	1.08×10^{2}	4.23×10^{2}	1.16×10^{2}
		Std	2.33×10^{1}	8.54×10^{0}	4.60×10^{1}	4.27×10^{1}	3.83×10^{1}	4.49×10^{1}	1.19×10^{2}	1.72×10^{1}
		p-value	-	$\mathbf{1.83 \times 10^{-6}}$+	$\mathbf{9.77 \times 10^{-6}}$+	$\mathbf{4.50 \times 10^{-6}}$+	$\mathbf{1.68 \times 10^{-5}}$+	$\mathbf{3.03 \times 10^{-6}}$+	$\mathbf{1.83 \times 10^{-6}}$+	$\mathbf{2.02 \times 10^{-6}}$+
f_{12}		Median	5.38×10^{4}	5.52×10^{7}	2.64×10^{4}	1.86×10^{5}	1.13×10^{6}	2.39×10^{5}	1.55×10^{8}	1.74×10^{6}
		Mean	6.09×10^{4}	6.25×10^{7}	1.93×10^{5}	5.12×10^{5}	3.35×10^{6}	2.55×10^{5}	1.77×10^{8}	2.01×10^{6}
		Std	3.97×10^{4}	2.90×10^{7}	5.57×10^{5}	6.68×10^{5}	4.39×10^{6}	1.70×10^{5}	9.46×10^{7}	1.11×10^{6}
		p-value	-	$\mathbf{1.83 \times 10^{-6}}$+	2.17×10^{-1}=	$\mathbf{1.36 \times 10^{-5}}$+	$\mathbf{1.72 \times 10^{-5}}$+	$\mathbf{8.88 \times 10^{-6}}$+	$\mathbf{1.83 \times 10^{-6}}$+	$\mathbf{1.83 \times 10^{-6}}$+

Table 3. *Cont.*

f	Category	Quality	STTPSO	DNSPSO	XPSO	TCSPSO	GLPSO	HCLPSO	DPLPSO	CLPSO
f_{13}		Median	5.18×10^3	1.32×10^6	7.54×10^3	8.24×10^3	7.23×10^3	6.10×10^4	6.71×10^6	3.39×10^3
		Mean	1.10×10^4	1.52×10^6	9.83×10^3	3.28×10^5	1.19×10^4	3.80×10^4	2.47×10^7	3.40×10^3
		Std	1.11×10^4	7.16×10^5	1.02×10^4	1.14×10^6	1.45×10^4	2.64×10^4	7.42×10^7	1.42×10^3
		p-value	-	**1.83×10^{-6}+**	9.34×10^{-1}=	4.59×10^{-1}=	1.00×10^{0}=	**2.26×10^{-5}+**	**1.83×10^{-6}+**	1.14×10^{-2}–
f_{14}		Median	3.89×10^3	1.90×10^2	3.75×10^3	3.49×10^4	1.66×10^3	1.43×10^4	1.20×10^5	4.42×10^4
		Mean	6.63×10^3	1.93×10^2	5.36×10^3	5.20×10^4	3.38×10^4	1.55×10^4	1.66×10^5	4.93×10^4
		Std	6.41×10^3	2.22×10^1	4.38×10^3	7.90×10^4	7.81×10^4	9.99×10^3	2.35×10^5	3.41×10^4
		p-value	-	1.83×10^{-6}–	2.49×10^{-1}=	**5.93×10^{-4}+**	9.51×10^{-1}=	**2.51×10^{-4}+**	**4.08×10^{-6}+**	**4.50×10^{-6}+**
f_{15}		Median	3.91×10^3	4.24×10^4	1.61×10^3	1.08×10^4	5.46×10^3	8.96×10^3	1.28×10^4	4.08×10^2
		Mean	7.84×10^3	4.70×10^4	3.59×10^3	1.33×10^4	8.80×10^3	1.35×10^4	2.61×10^4	4.59×10^2
		Std	8.32×10^3	2.04×10^4	4.74×10^3	1.04×10^4	8.48×10^3	1.22×10^4	3.16×10^4	2.45×10^2
		p-value	-	**1.83×10^{-6}+**	8.04×10^{-2}=	**3.41×10^{-2}+**	7.89×10^{-1}=	**4.38×10^{-2}+**	**2.33×10^{-3}+**	3.34×10^{-6}–
f_{16}		Median	2.22×10^1	1.88×10^3	5.70×10^2	8.65×10^2	8.49×10^2	7.46×10^2	1.57×10^3	6.61×10^2
		Mean	5.93×10^1	1.87×10^3	5.33×10^2	8.54×10^2	8.21×10^2	7.11×10^2	1.52×10^3	6.24×10^2
		Std	6.79×10^1	1.61×10^2	1.96×10^2	2.56×10^2	2.22×10^2	2.08×10^2	2.55×10^2	1.64×10^2
		p-value	-	**1.83×10^{-6}+**	**1.83×10^{-6}+**	**1.83×10^{-6}+**	**1.83×10^{-6}+**	**1.83×10^{-6}+**	**1.83×10^{-6}+**	**1.83×10^{-6}+**
f_{17}		Median	4.48×10^1	8.55×10^2	1.64×10^2	3.18×10^2	2.02×10^2	3.12×10^2	4.36×10^2	1.96×10^2
		Mean	4.69×10^1	8.68×10^2	1.45×10^2	2.96×10^2	2.28×10^2	3.22×10^2	4.31×10^2	1.88×10^2
		Std	1.01×10^1	1.21×10^2	8.08×10^1	1.44×10^2	1.38×10^2	1.54×10^2	1.50×10^2	6.59×10^1
		p-value	-	**1.83×10^{-6}+**	**4.97×10^{-6}+**	**1.83×10^{-6}+**	**3.34×10^{-6}+**	**1.83×10^{-6}+**	**1.83×10^{-6}+**	**1.83×10^{-6}+**
f_{18}		Median	1.87×10^5	1.88×10^5	9.24×10^4	1.41×10^5	1.68×10^4	1.90×10^5	8.67×10^5	1.87×10^5
		Mean	2.64×10^5	2.16×10^5	1.46×10^5	2.78×10^5	1.23×10^5	2.05×10^5	1.03×10^6	2.46×10^5
		Std	2.51×10^5	8.08×10^4	1.27×10^5	2.93×10^5	4.87×10^5	1.47×10^5	8.56×10^5	1.55×10^5
		p-value	-	1.00×10^{0}=	5.58×10^{-2}=	9.18×10^{-1}=	1.20×10^{-4}–	9.34×10^{-1}=	6.60×10^{-5}+	9.18×10^{-1}=

Table 3. *Cont.*

f	Category	Quality	STTPSO	DNSPSO	XPSO	TCSPSO	GLPSO	HCLPSO	DPLPSO	CLPSO
f_{19}		Median	5.77×10^3	1.95×10^3	3.65×10^3	7.66×10^3	2.99×10^3	1.31×10^4	1.52×10^4	1.05×10^2
		Mean	1.10×10^4	2.25×10^3	4.56×10^3	1.49×10^4	7.89×10^3	1.63×10^4	3.47×10^4	1.35×10^2
		Std	1.32×10^4	1.04×10^3	4.83×10^3	1.58×10^4	1.05×10^4	1.76×10^4	7.74×10^4	8.36×10^1
		p-value	-	2.86×10^{-3}–	3.78×10^{-2}–	3.24×10^{-1}=	3.34×10^{-1}=	2.94×10^{-1}=	**1.80×10^{-2}+**	1.83×10^{-6}–
$f_{10\text{–}19}$	*w/t/l*		-	7/1/2	3/6/1	6/4/0	4/5/1	7/3/0	10/0/0	5/2/3
f	Category	Quality	STTPSO	DNSPSO	XPSO	TCSPSO	GLPSO	HCLPSO	DPLPSO	CLPSO
f_{20}	Composition Functions	Median	3.72×10^1	3.50×10^2	1.74×10^2	3.84×10^2	1.96×10^2	2.13×10^2	3.66×10^2	1.94×10^2
		Mean	4.65×10^1	3.87×10^2	1.84×10^2	3.70×10^2	2.16×10^2	2.09×10^2	4.00×10^2	1.89×10^2
		Std	3.35×10^1	1.20×10^2	6.75×10^1	1.39×10^2	1.01×10^2	1.04×10^2	1.28×10^2	6.57×10^1
		p-value	-	**1.83×10^{-6}+**	**2.48×10^{-6}+**	**1.83×10^{-6}+**	**3.69×10^{-6}+**	**5.48×10^{-6}+**	**1.83×10^{-6}+**	**3.69×10^{-6}+**
f_{21}		Median	2.12×10^2	4.04×10^2	2.35×10^2	2.81×10^2	2.66×10^2	2.75×10^2	4.02×10^2	2.88×10^2
		Mean	2.13×10^2	4.04×10^2	2.38×10^2	2.84×10^2	2.67×10^2	2.76×10^2	4.00×10^2	2.83×10^2
		Std	3.76×10^0	9.54×10^0	1.05×10^1	2.25×10^1	2.07×10^1	1.25×10^1	2.25×10^1	2.53×10^1
		p-value	-	**1.83×10^{-6}+**	**1.83×10^{-6}+**	**1.83×10^{-6}+**	**1.82×10^{-6}+**	**1.82×10^{-6}+**	**1.82×10^{-6}+**	**2.47×10^{-6}+**
f_{22}		Median	1.00×10^2	6.38×10^3	1.00×10^2	1.05×10^2	1.00×10^2	1.02×10^2	5.79×10^2	2.06×10^2
		Mean	1.00×10^2	6.29×10^3	4.30×10^2	1.70×10^3	1.02×10^2	8.37×10^2	5.92×10^2	8.25×10^2
		Std	0.00×10^0	7.48×10^2	9.91×10^2	1.75×10^3	3.08×10^0	1.49×10^3	1.44×10^2	1.18×10^3
		p-value	-	**1.82×10^{-6}+**	**3.82×10^{-3}+**	**1.43×10^{-4}+**	**1.63×10^{-3}+**	**4.67×10^{-4}+**	**1.83×10^{-6}+**	**1.83×10^{-6}+**
f_{23}		Median	3.85×10^2	5.83×10^2	3.99×10^2	4.44×10^2	4.26×10^2	4.53×10^2	6.77×10^2	4.46×10^2
		Mean	3.86×10^2	5.87×10^2	3.98×10^2	4.47×10^2	4.33×10^2	4.51×10^2	6.84×10^2	4.45×10^2
		Std	7.70×10^0	3.62×10^1	1.15×10^1	2.85×10^1	3.02×10^1	2.02×10^1	4.57×10^1	1.09×10^1
		p-value	-	**1.83×10^{-6}+**	**6.89×10^{-4}+**	**1.82×10^{-6}+**	**2.24×10^{-6}+**	**1.83×10^{-6}+**	**1.83×10^{-6}+**	**1.83×10^{-6}+**

Table 3. *Cont.*

f	Category	Quality	STTPSO	DNSPSO	XPSO	TCSPSO	GLPSO	HCLPSO	DPLPSO	CLPSO
f_{24}		Median	4.60×10^{2}	6.68×10^{2}	4.70×10^{2}	5.37×10^{2}	4.88×10^{2}	5.35×10^{2}	7.32×10^{2}	5.60×10^{2}
		Mean	4.61×10^{2}	6.82×10^{2}	4.73×10^{2}	5.38×10^{2}	4.99×10^{2}	5.39×10^{2}	7.37×10^{2}	5.60×10^{2}
		Std	8.66×10^{0}	4.48×10^{1}	2.40×10^{1}	5.08×10^{1}	3.90×10^{1}	2.34×10^{1}	3.79×10^{1}	1.66×10^{1}
		p-value	-	**1.83×10^{-6}+**	**1.28×10^{-2}+**	**2.24×10^{-6}+**	**2.06×10^{-5}+**	**1.83×10^{-6}+**	**1.82×10^{-6}+**	**1.83×10^{-6}+**
f_{25}		Median	3.87×10^{2}	3.79×10^{2}	3.91×10^{2}	4.14×10^{2}	4.09×10^{2}	3.89×10^{2}	6.00×10^{2}	3.89×10^{2}
		Mean	3.87×10^{2}	3.78×10^{2}	3.92×10^{2}	4.12×10^{2}	4.04×10^{2}	3.89×10^{2}	6.18×10^{2}	3.89×10^{2}
		Std	2.06×10^{-1}	1.04×10^{0}	4.86×10^{0}	1.56×10^{1}	1.24×10^{1}	8.23×10^{0}	8.24×10^{1}	5.62×10^{-1}
		p-value	-	1.77×10^{-6}−	**6.65×10^{-6}+**	**1.83×10^{-6}+**	**1.82×10^{-6}+**	**1.74×10^{-2}+**	**1.83×10^{-6}+**	**1.78×10^{-6}+**
f_{26}		Median	1.47×10^{3}	3.28×10^{3}	3.00×10^{2}	2.32×10^{3}	1.94×10^{3}	2.04×10^{3}	1.61×10^{3}	1.85×10^{3}
		Mean	1.49×10^{3}	3.45×10^{3}	6.97×10^{2}	2.23×10^{3}	1.92×10^{3}	1.82×10^{3}	1.95×10^{3}	1.58×10^{3}
		Std	1.09×10^{2}	4.89×10^{2}	6.06×10^{2}	6.97×10^{2}	4.77×10^{2}	6.36×10^{2}	1.02×10^{3}	4.70×10^{2}
		p-value	-	**1.83×10^{-6}+**	3.56×10^{-5}−	**1.97×10^{-4}+**	**1.67×10^{-4}+**	**1.52×10^{-2}+**	1.92×10^{-1}=	1.88×10^{-1}=
f_{27}		Median	5.13×10^{2}	5.00×10^{2}	5.36×10^{2}	5.61×10^{2}	5.48×10^{2}	5.14×10^{2}	8.09×10^{2}	5.11×10^{2}
		Mean	5.18×10^{2}	5.00×10^{2}	5.35×10^{2}	5.61×10^{2}	5.50×10^{2}	5.16×10^{2}	8.19×10^{2}	5.10×10^{2}
		Std	1.40×10^{1}	0.00×10^{0}	1.10×10^{1}	1.95×10^{1}	1.28×10^{1}	1.59×10^{1}	5.73×10^{1}	4.56×10^{0}
		p-value	-	2.24×10^{-6}−	**9.31×10^{-5}+**	**2.87×10^{-6}+**	**2.48×10^{-6}+**	8.05×10^{-1}=	**1.82×10^{-6}+**	1.70×10^{-2}−
f_{28}		Median	4.08×10^{2}	5.00×10^{2}	4.03×10^{2}	4.40×10^{2}	4.72×10^{2}	4.55×10^{2}	8.24×10^{2}	4.74×10^{2}
		Mean	3.79×10^{2}	5.00×10^{2}	3.86×10^{2}	4.51×10^{2}	4.51×10^{2}	4.48×10^{2}	8.80×10^{2}	4.83×10^{2}
		Std	5.79×10^{1}	0.00×10^{0}	6.85×10^{1}	5.21×10^{1}	7.00×10^{1}	3.48×10^{1}	1.61×10^{2}	2.88×10^{1}
		p-value	-	**1.81×10^{-6}+**	7.76×10^{-1}=	**3.73×10^{-4}+**	**1.72×10^{-3}+**	**1.54×10^{-4}+**	**1.83×10^{-6}+**	**2.02×10^{-6}+**
f_{29}		Median	4.84×10^{2}	1.66×10^{3}	5.66×10^{2}	8.62×10^{2}	7.74×10^{2}	6.92×10^{2}	1.34×10^{3}	6.58×10^{2}
		Mean	5.11×10^{2}	1.56×10^{3}	5.88×10^{2}	9.05×10^{2}	8.17×10^{2}	6.98×10^{2}	1.34×10^{3}	6.46×10^{2}
		Std	7.29×10^{1}	2.89×10^{2}	9.03×10^{1}	1.86×10^{2}	2.37×10^{2}	1.56×10^{2}	2.29×10^{2}	7.20×10^{1}
		p-value	-	**1.83×10^{-6}+**	**2.95×10^{-4}+**	**1.83×10^{-6}+**	**1.72×10^{-5}+**	**1.30×10^{-5}+**	**1.83×10^{-6}+**	**7.33×10^{-6}+**

Table 3. *Cont.*

f	Category	Quality	STTPSO	DNSPSO	XPSO	TCSPSO	GLPSO	HCLPSO	DPLPSO	CLPSO
f_{30}		Median	4.19×10^3	4.00×10^4	8.04×10^3	1.20×10^4	9.45×10^3	7.39×10^3	1.96×10^6	1.27×10^4
		Mean	5.03×10^3	6.20×10^4	9.78×10^3	1.80×10^4	2.08×10^4	8.59×10^3	2.55×10^6	1.37×10^4
		Std	2.02×10^3	5.12×10^4	6.72×10^3	1.77×10^4	2.96×10^4	4.52×10^3	1.97×10^6	3.99×10^3
		p-value	-	**1.83×10^{-6}+**	**$4.36 \times 10-4$+**	**4.08×10^{-6}+**	**1.88×10^{-5}+**	**1.54×10^{-3}+**	**1.83×10^{-6}+**	**2.02×10^{-6}+**
$f_{20\text{–}30}$	*w/t/l*		-	9/0/2	9/1/1	11/0/0	11/0/0	10/1/0	10/1/0	9/1/1
	w/t/l		-	23/1/5	18/8/3	23/5/1	21/6/2	24/4/1	28/1/0	21/3/5
	rank		**1.86**	6.00	2.55	5.48	4.00	4.41	7.45	4.24

Table 4. Comparison results between the proposed STTPSO and the 7 state-of-the-art and popular PSO variants on the 50-*D* CEC 2017 benchmark functions. The highlighted *p*-values means that the proposed STTPSO is significantly better than the associated compared algorithms on the corresponding problems.

f	Category	Quality	STTPSO	DNSPSO	XPSO	TCSPSO	GLPSO	HCLPSO	DPLPSO	CLPSO
f_1	Unimodal Functions	Median	2.89×10^3	5.13×10^3	8.51×10^2	5.47×10^3	3.11×10^3	9.83×10^3	1.92×10^{10}	2.06×10^3
		Mean	4.33×10^3	8.53×10^3	3.66×10^3	2.84×10^6	2.76×10^6	7.15×10^7	1.84×10^{10}	2.59×10^3
		Std	4.38×10^3	1.13×10^4	5.11×10^3	1.52×10^7	1.37×10^7	2.72×10^8	4.36×10^9	1.88×10^3
		p-value	-	2.02×10^{-1}=	4.97×10^{-1}=	2.02×10^{-1}=	7.69×10^{-2}=	**2.51×10^{-4}+**	**1.83×10^{-6}+**	2.67×10^{-1}=
f_3		Median	5.80×10^4	3.84×10^5	4.35×10^3	5.55×10^4	4.55E^{-13}	4.71×10^3	1.23×10^5	1.31×10^5
		Mean	5.85×10^4	3.82×10^5	4.72×10^3	5.83×10^4	3.92E^{-12}	5.11×10^3	1.22×10^5	1.31×10^5
		Std	8.60×10^3	6.15×10^4	1.64×10^3	9.39×10^3	1.49E^{-11}	2.41×10^3	1.62×10^4	2.22×10^4
		p-value	-	**1.83×10^{-6}+**	1.83×10^{-6}–	5.51×10^{-1}=	1.83×10^{-6}–	1.83×10^{-6}–	**1.83×10^{-6}+**	**1.83×10^{-6}+**
$f_{1\text{–}3}$	*w/t/l*		-	1/1/0	0/1/1	0/2/0	0/1/1	1/0/1	2/0/0	1/1/0
f_4	Simple Multimodal Functions	Median	1.75×10^2	4.57×10^1	2.46×10^2	2.88×10^2	3.22×10^2	1.61×10^2	3.81×10^3	1.90×10^2
		Mean	1.69×10^2	5.24×10^1	2.33×10^2	2.93×10^2	3.29×10^2	1.48×10^2	4.02×10^3	1.87×10^2
		Std	3.35×10^1	2.36×10^1	5.15×10^1	9.09×10^1	8.97×10^1	5.23×10^1	1.06×10^3	2.00×10^1
		p-value	-	2.24×10^{-6}–	**5.08×10^{-5}+**	**6.65×10^{-6}+**	**2.02×10^{-6}+**	9.99×10^{-2}=	**1.83×10^{-6}+**	**1.90×10^{-2}+**
f_5		Median	8.96×10^0	4.11×10^2	8.21×10^1	1.87×10^2	1.47×10^2	1.65×10^2	4.58×10^2	2.02×10^2
		Mean	9.09×10^0	4.12×10^2	8.33×10^1	1.91×10^2	1.42×10^2	1.66×10^2	4.46×10^2	1.98×10^2
		Std	2.70×10^0	1.88×10^1	1.50×10^1	3.82×10^1	3.83×10^1	3.23×10^1	4.64×10^1	1.55×10^1
		p-value	-	**1.82×10^{-6}+**	**1.83×10^{-6}+**	**1.83×10^{-6}+**	**1.83×10^{-6}+**	**1.82×10^{-6}+**	**1.82×10^{-6}+**	**1.83×10^{-6}+**

Table 4. *Cont.*

f	Category	Quality	STTPSO	DNSPSO	XPSO	TCSPSO	GLPSO	HCLPSO	DPLPSO	CLPSO
f_6		Median	1.22×10^{-6}	1.02×10^{-1}	5.59×10^{-2}	3.01×10^{0}	1.19×10^{-2}	1.85×10^{-3}	5.28×10^{1}	1.23×10^{-8}
		Mean	4.48×10^{-6}	1.16×10^{-1}	1.53×10^{-1}	3.93×10^{0}	2.00×10^{-2}	2.65×10^{-3}	5.18×10^{1}	2.45×10^{-3}
		Std	8.10×10^{-6}	4.66×10^{-2}	2.59×10^{-1}	3.68×10^{0}	2.08×10^{-2}	2.37×10^{-3}	4.60×10^{0}	1.32×10^{-2}
		p-value	-	**1.83×10^{-6}+**	**1.83×10^{-6}+**	**1.83×10^{-6}+**	**1.83×10^{-6}+**	**1.83×10^{-6}+**	**1.83×10^{-6}+**	**3.56×10^{-5}+**
f_7		Median	6.27×10^{1}	4.70×10^{2}	1.51×10^{2}	3.18×10^{2}	2.27×10^{2}	1.94×10^{2}	7.71×10^{2}	2.11×10^{2}
		Mean	6.33×10^{1}	4.70×10^{2}	1.53×10^{2}	3.35×10^{2}	2.36×10^{2}	2.02×10^{2}	7.70×10^{2}	2.10×10^{2}
		Std	3.22×10^{0}	1.85×10^{1}	2.57×10^{1}	6.19×10^{1}	3.89×10^{1}	3.06×10^{1}	6.79×10^{1}	1.44×10^{1}
		p-value	-	**1.83×10^{-6}+**	**1.83×10^{-6}+**	**1.83×10^{-6}+**	**1.83×10^{-6}+**	**1.83×10^{-6}+**	**1.83×10^{-6}+**	**1.83×10^{-6}+**
f_8		Median	8.96×10^{0}	3.99×10^{2}	8.81×10^{1}	1.97×10^{2}	1.36×10^{2}	1.56×10^{2}	4.60×10^{2}	1.96×10^{2}
		Mean	9.35×10^{0}	4.00×10^{2}	9.29×10^{1}	2.09×10^{2}	1.41×10^{2}	1.56×10^{2}	4.45×10^{2}	1.97×10^{2}
		Std	3.22×10^{0}	1.86×10^{1}	2.34×10^{1}	6.12×10^{1}	3.13×10^{1}	2.68×10^{1}	4.48×10^{1}	1.61×10^{1}
		p-value	-	**1.82×10^{-6}+**	**1.83×10^{-6}+**	**1.82×10^{-6}+**	**1.82×10^{-6}+**	**1.83×10^{-6}+**	**1.82×10^{-6}+**	**1.82×10^{-6}+**
f_9		Median	4.99×10^{-1}	1.72×10^{1}	1.55×10^{1}	2.85×10^{3}	4.51×10^{2}	1.25×10^{3}	1.08×10^{4}	3.95×10^{3}
		Mean	8.56×10^{-1}	3.56×10^{1}	4.97×10^{1}	3.34×10^{3}	5.98×10^{2}	1.26×10^{3}	1.10×10^{4}	4.22×10^{3}
		Std	1.05×10^{0}	5.56×10^{1}	8.96×10^{1}	1.87×10^{3}	5.23×10^{2}	5.35×10^{2}	1.90×10^{3}	1.20×10^{3}
		p-value	-	**3.34×10^{-6}+**	**1.83×10^{-6}+**	**1.83×10^{-6}+**	**1.83×10^{-6}+**	**1.83×10^{-6}+**	**1.83×10^{-6}+**	**1.83×10^{-6}+**
f_{4-9}	*w/t/l*		-	5/0/1	6/0/0	6/0/0	6/0/0	5/1/0	6/0/0	6/0/0
f_{10}		Median	3.60×10^{3}	1.16×10^{4}	5.17×10^{3}	5.57×10^{3}	5.59×10^{3}	5.58×10^{3}	1.23×10^{4}	6.66×10^{3}
		Mean	3.49×10^{3}	1.15×10^{4}	5.00×10^{3}	5.56×10^{3}	6.27×10^{3}	5.57×10^{3}	1.23×10^{4}	6.59×10^{3}
		Std	5.89×10^{2}	1.43×10^{3}	8.85×10^{2}	6.71×10^{2}	1.85×10^{3}	6.10×10^{2}	5.33×10^{2}	4.43×10^{2}
		p-value	-	**1.83×10^{-6}+**	**6.04×10^{-6}+**	**1.83×10^{-6}+**	**2.02×10^{-6}+**	**1.82×10^{-6}+**	**1.82×10^{-6}+**	**1.82×10^{-6}+**
f_{11}	Hybrid Functions	Median	6.11×10^{1}	2.04×10^{2}	1.58×10^{2}	2.15×10^{2}	2.90×10^{2}	2.08×10^{2}	2.34×10^{3}	1.93×10^{2}
		Mean	6.12×10^{1}	2.02×10^{2}	1.56×10^{2}	2.36×10^{2}	3.81×10^{2}	2.07×10^{2}	2.38×10^{3}	1.92×10^{2}
		Std	1.18×10^{1}	2.05×10^{1}	3.37×10^{1}	9.86×10^{1}	3.27×10^{2}	6.30×10^{1}	5.66×10^{2}	4.10×10^{1}
		p-value	-	**1.83×10^{-6}+**	**1.83×10^{-6}+**	**1.83×10^{-6}+**	**1.83×10^{-6}+**	**1.83×10^{-6}+**	**1.83×10^{-6}+**	**1.83×10^{-6}+**
f_{12}		Median	8.59×10^{5}	2.78×10^{7}	5.18×10^{5}	1.83×10^{6}	1.19×10^{6}	2.94×10^{6}	3.95×10^{9}	1.86×10^{7}
		Mean	8.89×10^{5}	3.23×10^{7}	9.85×10^{5}	8.82×10^{6}	5.72×10^{6}	3.61×10^{6}	4.04×10^{9}	1.96×10^{7}
		Std	5.26×10^{5}	1.68×10^{7}	1.13×10^{6}	2.40×10^{7}	1.24×10^{7}	2.48×10^{6}	1.72×10^{9}	9.49×10^{6}
		p-value	-	**1.83×10^{-6}+**	**7.11×10^{-1}=**	**2.97×10^{-5}+**	**1.52×10^{-2}+**	**2.24×10^{-6}+**	**1.83×10^{-6}+**	**1.83×10^{-6}+**

Table 4. *Cont.*

f	Category	Quality	STTPSO	DNSPSO	XPSO	TCSPSO	GLPSO	HCLPSO	DPLPSO	CLPSO
f_{13}		Median	1.88×10^4	2.62×10^6	2.30×10^3	3.81×10^3	3.32×10^3	2.15×10^4	3.11×10^8	1.11×10^4
		Mean	1.63×10^4	2.73×10^6	4.40×10^3	7.76×10^3	5.58×10^3	2.29×10^4	5.22×10^8	1.13×10^4
		Std	1.12×10^4	1.72×10^6	4.74×10^3	9.15×10^3	6.22×10^3	1.32×10^4	7.74×10^8	3.04×10^3
		p-value	-	$\mathbf{1.83 \times 10^{-6}}$+	1.30×10^{-4}–	1.44×10^{-2}–	1.54×10^{-4}–	9.99×10^{-2}=	$\mathbf{1.83 \times 10^{-6}}$+	4.60×10^{-2}–
f_{14}		Median	5.52×10^4	7.90×10^3	3.86×10^4	4.09×10^4	4.01×10^4	8.86×10^4	1.56×10^6	4.64×10^5
		Mean	7.09×10^4	8.09×10^3	3.97×10^4	2.30×10^5	1.43×10^5	1.22×10^5	2.21×10^6	5.29×10^5
		Std	5.64×10^4	3.51×10^3	2.64×10^4	5.29×10^5	2.20×10^5	1.15×10^5	2.01×10^6	2.65×10^5
		p-value	-	8.07×10^{-6}–	1.36×10^{-2}–	3.65×10^{-1}=	7.42×10^{-1}=	$\mathbf{4.38 \times 10^{-2}}$+	$\mathbf{2.02 \times 10^{-6}}$+	$\mathbf{1.83 \times 10^{-6}}$+
f_{15}		Median	1.18×10^4	4.55×10^5	2.69×10^3	7.12×10^3	3.89×10^3	1.81×10^4	2.93×10^6	8.15×10^2
		Mean	1.32×10^4	4.80×10^5	4.24×10^3	1.46×10^4	5.84×10^3	1.92×10^4	1.79×10^7	9.31×10^2
		Std	9.68×10^3	1.58×10^5	4.08×10^3	2.54×10^4	6.28×10^3	1.01×10^4	2.92×10^7	4.49×10^2
		p-value	-	$\mathbf{1.83 \times 10^{-6}}$+	4.71×10^{-4}–	2.76×10^{-1}=	1.07×10^{-3}–	6.41×10^{-2}=	$\mathbf{1.83 \times 10^{-6}}$+	6.65×10^{-6}–
f_{16}		Median	4.55×10^2	3.79×10^3	9.27×10^2	1.63×10^3	1.60×10^3	1.52×10^3	3.13×10^3	1.45×10^3
		Mean	4.23×10^2	3.78×10^3	9.56×10^2	1.70×10^3	1.57×10^3	1.53×10^3	3.06×10^3	1.41×10^3
		Std	1.90×10^2	2.33×10^2	3.27×10^2	4.35×10^2	4.06×10^2	3.57×10^2	5.41×10^2	2.00×10^2
		p-value	-	$\mathbf{1.83 \times 10^{-6}}$+	$\mathbf{6.04 \times 10^{-6}}$+	$\mathbf{1.83 \times 10^{-6}}$+	$\mathbf{1.83 \times 10^{-6}}$+	$\mathbf{1.83 \times 10^{-6}}$+	$\mathbf{1.83 \times 10^{-6}}$+	$\mathbf{1.83 \times 10^{-6}}$+
f_{17}		Median	2.45×10^2	2.41×10^3	8.83×10^2	1.15×10^3	1.03×10^3	1.30×10^3	1.85×10^3	1.06×10^3
		Mean	3.11×10^2	2.40×10^3	8.56×10^2	1.17×10^3	1.05×10^3	1.20×10^3	1.83×10^3	1.04×10^3
		Std	1.43×10^2	2.00×10^2	2.56×10^2	2.97×10^2	2.23×10^2	3.24×10^2	2.86×10^2	1.94×10^2
		p-value	-	1.82×10^{-6}+	3.69×10^{-6}+	1.83×10^{-6}+	1.83×10^{-6}+	1.83×10^{-6}+	1.83×10^{-6}+	2.02×10^{-6}+
f_{18}		Median	3.71×10^5	3.27×10^6	1.71×10^5	3.12×10^6	1.09×10^6	3.18×10^5	7.61×10^6	1.14×10^6
		Mean	4.28×10^5	3.47×10^6	3.94×10^5	5.58×10^6	3.42×10^6	4.33×10^5	1.04×10^7	1.31×10^6
		Std	2.34×10^5	1.61×10^6	4.88×10^5	5.71×10^6	5.02×10^6	3.40×10^5	9.88×10^6	7.63×10^5
		p-value	-	$\mathbf{1.83 \times 10^{-6}}$+	2.58×10^{-1}=	$\mathbf{4.50 \times 10^{-6}}$+	$\mathbf{1.90 \times 10^{-3}}$+	6.81×10^{-1}=	$\mathbf{1.83 \times 10^{-6}}$+	$\mathbf{2.24 \times 10^{-6}}$+
f_{19}		Median	1.60×10^3	2.67×10^4	9.19×10^3	1.30×10^4	1.42×10^4	1.03×10^4	1.44×10^6	3.36×10^2
		Mean	4.33×10^3	3.18×10^4	1.12×10^4	1.51×10^4	1.74×10^4	1.48×10^4	3.65×10^6	5.26×10^2
		Std	6.31×10^3	1.73×10^4	8.07×10^3	1.38×10^4	1.07×10^4	1.36×10^4	8.78×10^6	5.03×10^2
		p-value	-	$\mathbf{1.83 \times 10^{-6}}$+	$\mathbf{1.42 \times 10^{-4}}$+	$\mathbf{1.54 \times 10^{-3}}$+	$\mathbf{3.26 \times 10^{-5}}$+	$\mathbf{1.90 \times 10^{-3}}$+	$\mathbf{1.83 \times 10^{-6}}$+	1.10×10^{-4}–
$f_{10\text{–}19}$	*w/t/l*		-	9/0/1	5/2/3	7/2/1	7/1/2	7/3/0	10/0/0	7/0/3

Table 4. *Cont.*

f	Category	Quality	STTPSO	DNSPSO	XPSO	TCSPSO	GLPSO	HCLPSO	DPLPSO	CLPSO
f_{20}		Median	9.66×10^{1}	1.66×10^{3}	4.66×10^{2}	9.58×10^{2}	7.45×10^{2}	9.48×10^{2}	1.44×10^{3}	5.92×10^{2}
		Mean	1.76×10^{2}	1.59×10^{3}	4.70×10^{2}	9.01×10^{2}	7.43×10^{2}	8.97×10^{2}	1.38×10^{3}	6.14×10^{2}
		Std	1.38×10^{2}	3.96×10^{2}	2.19×10^{2}	2.96×10^{2}	2.66×10^{2}	2.33×10^{2}	2.58×10^{2}	1.30×10^{2}
		p-value	-	1.83×10^{-6}+	2.97×10^{-5}+	1.83×10^{-6}+	1.83×10^{-6}+	1.83×10^{-6}+	1.83×10^{-6}+	1.83×10^{-6}+
f_{21}		Median	2.22×10^{2}	6.01×10^{2}	2.79×10^{2}	3.77×10^{2}	3.37×10^{2}	3.86×10^{2}	6.59×10^{2}	4.21×10^{2}
		Mean	2.22×10^{2}	6.00×10^{2}	2.80×10^{2}	4.02×10^{2}	3.46×10^{2}	3.80×10^{2}	6.53×10^{2}	4.21×10^{2}
		Std	3.65×10^{0}	2.02×10^{1}	1.96×10^{1}	6.26×10^{1}	4.16×10^{1}	2.91×10^{1}	3.35×10^{1}	1.52×10^{1}
		p-value	-	**1.82×10^{-6}+**	**1.83×10^{-6}+**	**1.82×10^{-6}+**	**1.82×10^{-6}+**	**1.82×10^{-6}+**	**1.83×10^{-6}+**	**1.82×10^{-6}+**
f_{22}		Median	3.04×10^{3}	1.28×10^{4}	5.63×10^{3}	6.39×10^{3}	6.49×10^{3}	6.24×10^{3}	1.26×10^{4}	7.13×10^{3}
		Mean	3.10×10^{3}	1.27×10^{4}	4.80×10^{3}	6.14×10^{3}	5.88×10^{3}	5.76×10^{3}	1.13×10^{4}	7.14×10^{3}
		Std	1.34×10^{3}	9.78×10^{2}	2.20×10^{3}	1.52×10^{3}	3.61×10^{3}	1.63×10^{3}	3.48×10^{3}	2.75×10^{2}
		p-value	-	**1.83×10^{-6}+**	**1.01×10^{-2}+**	**1.18×10^{-5}+**	**1.33×10^{-3}+**	**8.55×10^{-5}+**	**2.48×10^{-6}+**	**2.02×10^{-6}+**
f_{23}	Composition Functions	Median	5.06×10^{2}	8.64×10^{2}	5.27×10^{2}	6.43×10^{2}	6.65×10^{2}	6.66×10^{2}	1.22×10^{3}	6.66×10^{2}
		Mean	5.09×10^{2}	8.88×10^{2}	5.24×10^{2}	6.50×10^{2}	6.81×10^{2}	6.65×10^{2}	1.22×10^{3}	6.66×10^{2}
		Std	1.71×10^{1}	6.87×10^{1}	3.11×10^{1}	5.85×10^{1}	8.50×10^{1}	3.99×10^{1}	6.75×10^{1}	1.90×10^{1}
		p-value	-	**1.83×10^{-6}+**	**4.38×10^{-2}+**	**1.82×10^{-6}+**	**1.83×10^{-6}+**	**1.83×10^{-6}+**	**1.83×10^{-6}+**	**1.83×10^{-6}+**
f_{24}		Median	5.81×10^{2}	1.06×10^{3}	5.90×10^{2}	7.06×10^{2}	7.38×10^{2}	7.27×10^{2}	1.31×10^{3}	8.04×10^{2}
		Mean	5.83×10^{2}	1.08×10^{3}	6.06×10^{2}	7.09×10^{2}	7.49×10^{2}	7.23×10^{2}	1.33×10^{3}	8.05×10^{2}
		Std	1.63×10^{1}	1.26×10^{2}	5.78×10^{1}	7.38×10^{1}	9.38×10^{1}	3.46×10^{1}	9.27×10^{1}	2.65×10^{1}
		p-value	-	**1.82×10^{-6}+**	**3.41×10^{-2}+**	**1.83×10^{-6}+**	**2.02×10^{-6}+**	**1.83×10^{-6}+**	**1.82×10^{-6}+**	**1.82×10^{-6}+**
f_{25}		Median	4.80×10^{2}	4.31×10^{2}	5.98×10^{2}	6.75×10^{2}	6.61×10^{2}	4.80×10^{2}	2.60×10^{3}	5.31×10^{2}
		Mean	5.06×10^{2}	4.41×10^{2}	5.97×10^{2}	6.76×10^{2}	6.66×10^{2}	5.01×10^{2}	2.77×10^{3}	5.30×10^{2}
		Std	3.55×10^{1}	2.14×10^{1}	2.43×10^{1}	6.58×10^{1}	7.00×10^{1}	3.54×10^{1}	6.68×10^{2}	6.29×10^{0}
		p-value	-	3.14×10^{-6}–	**1.83×10^{-6}+**	**1.82×10^{-6}+**	**2.02×10^{-6}+**	2.33×10^{-1}=	**1.82×10^{-6}+**	**5.91×10^{-4}+**
f_{26}		Median	2.23×10^{3}	6.70×10^{3}	9.43×10^{2}	3.98×10^{3}	2.96×10^{3}	3.70×10^{3}	7.22×10^{3}	3.60×10^{3}
		Mean	2.27×10^{3}	7.10×10^{3}	1.15×10^{3}	4.07×10^{3}	3.04×10^{3}	3.64×10^{3}	6.89×10^{3}	3.52×10^{3}
		Std	1.23×10^{2}	1.63×10^{3}	8.75×10^{2}	1.02×10^{3}	6.50×10^{2}	3.13×10^{2}	2.09×10^{3}	3.37×10^{2}
		p-value	-	**1.83×10^{-6}+**	1.42×10^{-5}–	**1.18×10^{-5}+**	**6.87×10^{-6}+**	**1.83×10^{-6}+**	**1.83×10^{-6}+**	**2.02×10^{-6}+**
f_{27}		Median	7.00×10^{2}	5.00×10^{2}	7.19×10^{2}	9.01×10^{2}	8.35×10^{2}	6.54×10^{2}	1.98×10^{3}	6.35×10^{2}
		Mean	6.94×10^{2}	5.00×10^{2}	7.35×10^{2}	9.06×10^{2}	8.39×10^{2}	6.89×10^{2}	1.99×10^{3}	6.33×10^{2}
		Std	5.44×10^{1}	0.00×10^{0}	8.89×10^{1}	9.24×10^{1}	8.57×10^{1}	1.06×10^{2}	1.76×10^{2}	2.76×10^{1}
		p-value	-	1.82×10^{-6}–	7.03×10^{-2}=	**1.83×10^{-6}+**	**7.69×10^{-6}+**	7.50×10^{-1}=	**1.83×10^{-6}+**	7.84×10^{-5}–

Table 4. *Cont.*

f	Category	Quality	STTPSO	DNSPSO	XPSO	TCSPSO	GLPSO	HCLPSO	DPLPSO	CLPSO
f_{28}		Median	5.08×10^2	5.00×10^2	5.38×10^2	6.68×10^2	7.00×10^2	4.92×10^2	2.91×10^3	1.71×10^3
		Mean	9.06×10^2	5.00×10^2	5.47×10^2	6.77×10^2	6.97×10^2	4.92×10^2	3.04×10^3	1.79×10^3
		Std	1.14×10^3	0.00×10^0	3.99×10^1	6.99×10^1	5.82×10^1	3.40×10^1	5.38×10^2	4.45×10^2
		p-value	-	6.21×10^{-2}=	7.35×10^{-2}=	1.52×10^{-2}−	1.52×10^{-2}−	9.56×10^{-2}=	**3.69×10^{-6}+**	**4.71×10^{-4}+**
f_{29}		Median	5.58×10^2	3.32×10^3	8.90×10^2	1.43×10^3	1.03×10^3	1.19×10^3	3.44×10^3	1.01×10^3
		Mean	5.90×10^2	3.27×10^3	8.66×10^2	1.43×10^3	1.08×10^3	1.16×10^3	3.52×10^3	1.02×10^3
		Std	1.75×10^2	2.69×10^2	1.87×10^2	2.49×10^2	2.75×10^2	3.49×10^2	5.07×10^2	1.49×10^2
		p-value	-	**1.83×10^{-6}+**	**2.07×10^{-5}+**	**1.83×10^{-6}+**	**7.33×10^{-6}+**	**2.02×10^{-6}+**	**1.83×10^{-6}+**	**2.48×10^{-6}+**
f_{30}		Median	8.17×10^5	1.79×10^6	1.93×10^6	2.00×10^6	2.00×10^6	1.16×10^6	1.38×10^8	7.30×10^5
		Mean	8.55×10^5	2.22×10^6	1.92×10^6	2.36×10^6	2.40×10^6	1.23×10^6	1.38×10^8	7.40×10^5
		Std	1.54×10^5	1.12×10^6	3.32×10^5	8.05×10^5	1.44×10^6	4.15×10^5	5.12×10^7	7.50×10^4
		p-value	-	**4.50×10^{-6}+**	**1.83×10^{-6}+**	**1.83×10^{-6}+**	**1.83×10^{-6}+**	**4.36×10^{-4}+**	**1.83×10^{-6}+**	1.15×10^{-3}−
$f_{20–30}$	*w/t/l*		-	8/1/2	8/2/1	10/0/1	10/0/1	8/3/0	11/0/0	9/0/2
	w/t/l		-	23/2/4	19/5/5	23/4/2	23/2/4	21/7/1	29/0/0	23/1/5
	rank		**2.17**	5.72	2.45	5.14	4.31	4.14	7.79	4.28

Table 5. Comparison results between the proposed STTPSO and the 7 state-of-the-art and popular PSO variants on the 100*D* CEC 2017 benchmark functions. The highlighted p-values means that the proposed STTPSO is significantly better than the associated compared algorithms on the corresponding problems.

f	Category	Quality	STTPSO	DNSPSO	XPSO	TCSPSO	GLPSO	HCLPSO	DPLPSO	CLPSO
f_1	Unimodal Functions	Median	2.53×10^3	3.31×10^3	3.50×10^3	2.62×10^3	6.15×10^3	1.37×10^4	1.10×10^{11}	1.63×10^9
		Mean	4.29×10^3	7.08×10^3	8.27×10^3	6.22×10^3	1.21×10^4	1.93×10^7	1.10×10^{11}	1.78×10^9
		Std	4.43×10^3	1.04×10^4	9.67×10^3	7.24×10^3	1.60×10^4	1.04×10^8	1.17×10^{10}	1.48×10^9
		p-value	-	7.89×10^{-1}=	2.41×10^{-1}=	4.34×10^{-1}=	**5.15×10^{-3}+**	**7.20×10^{-5}+**	**1.83×10^{-6}+**	**4.08×10^{-6}+**
f_3		Median	2.22×10^5	1.07×10^6	7.04×10^4	2.55×10^5	7.98×10^1	8.06×10^4	3.82×10^5	5.10×10^5
		Mean	2.23×10^5	1.04×10^6	6.94×10^4	2.53×10^5	3.93×10^3	8.23×10^4	3.73×10^5	5.12×10^5
		Std	2.17×10^4	1.25×10^5	9.80×10^3	3.05×10^4	9.85×10^3	2.40×10^4	4.19×10^4	3.83×10^4
		p-value	-	**1.83×10^{-6}+**	1.83×10^{-6}−	**1.74×10^{-4}+**	1.83×10^{-6}−	1.82×10^{-6}−	**1.83×10^{-6}+**	**1.82×10^{-6}+**
$f_{1–3}$	*w/t/l*		-	1/1/0	0/1/1	1/1/0	1/0/1	1/0/1	2/0/0	2/0/0

Table 5. *Cont.*

f	Category	Quality	STTPSO	DNSPSO	XPSO	TCSPSO	GLPSO	HCLPSO	DPLPSO	CLPSO
f_4	Simple Multimodal Functions	Median	2.16×10^2	1.99×10^2	4.80×10^2	6.28×10^2	8.16×10^2	2.45×10^2	2.20×10^4	3.17×10^2
		Mean	2.18×10^2	2.07×10^2	4.77×10^2	7.03×10^2	8.54×10^2	2.46×10^2	2.25×10^4	3.26×10^2
		Std	1.89×10^1	5.44×10^1	5.85×10^1	1.82×10^2	1.91×10^2	2.61×10^1	3.31×10^3	4.30×10^1
		p-value	-	2.41×10^{-1}=	**1.83×10^{-6}+**	**1.83×10^{-6}+**	**1.83×10^{-6}+**	**4.36×10^{-4}+**	**1.83×10^{-6}+**	**1.83×10^{-6}+**
f_5		Median	3.08×10^1	1.02×10^3	2.29×10^2	5.36×10^2	3.76×10^2	4.78×10^2	1.20×10^3	7.46×10^2
		Mean	2.96×10^1	1.03×10^3	2.28×10^2	5.55×10^2	3.85×10^2	4.96×10^2	1.20×10^3	7.46×10^2
		Std	4.78×10^0	4.29×10^1	4.81×10^1	1.09×10^2	5.70×10^1	7.09×10^1	4.04×10^1	4.13×10^1
		p-value	-	**1.82×10^{-6}+**	**1.83×10^{-6}+**	**1.83×10^{-6}+**	**1.83×10^{-6}+**	**1.83×10^{-6}+**	**1.82×10^{-6}+**	**1.83×10^{-6}+**
f_6		Median	1.13×10^{-3}	2.42×10^{-1}	3.95×10^0	1.85×10^1	4.94×10^{-2}	9.66×10^{-3}	7.63×10^1	1.07×10^{-2}
		Mean	1.90×10^{-3}	2.82×10^{-1}	4.43×10^0	1.78×10^1	5.62×10^{-2}	1.54×10^{-2}	7.59×10^1	2.65×10^{-2}
		Std	2.09×10^{-3}	1.54×10^{-1}	3.28×10^0	6.21×10^0	2.65×10^{-2}	1.99×10^{-2}	4.73×10^0	3.21×10^{-2}
		p-value	-	**1.83×10^{-6}+**	**1.83×10^{-6}+**	**1.83×10^{-6}+**	**1.83×10^{-6}+**	**1.07×10^{-5}+**	**1.83×10^{-6}+**	**5.09×10^{-4}+**
f_7		Median	1.80×10^2	1.14×10^3	4.41×10^2	1.22×10^3	7.93×10^2	6.74×10^2	2.78×10^3	7.07×10^2
		Mean	1.80×10^2	1.13×10^3	4.48×10^2	1.23×10^3	7.99×10^2	6.80×10^2	2.77×10^3	7.02×10^2
		Std	1.07×10^1	4.09×10^1	8.25×10^1	1.99×10^2	1.16×10^2	1.08×10^2	2.21×10^2	6.34×10^1
		p-value	-	1.83×10^{-6}+	1.83×10^{-6}+	1.83×10^{-6}+	1.82×10^{-6}+	1.83×10^{-6}+	1.83×10^{-6}+	1.83×10^{-6}+
f_8		Median	2.89×10^1	1.02×10^3	2.14×10^2	5.29×10^2	4.02×10^2	5.48×10^2	1.26×10^3	7.41×10^2
		Mean	2.98×10^1	1.02×10^3	2.15×10^2	5.54×10^2	4.08×10^2	5.46×10^2	1.26×10^3	7.48×10^2
		Std	5.21×10^0	3.27×10^1	4.03×10^1	8.00×10^1	6.80×10^1	8.92×10^1	4.77×10^1	3.22×10^1
		p-value	-	**1.82×10^{-6}+**	**1.83×10^{-6}+**	**1.83×10^{-6}+**	**1.83×10^{-6}+**	**1.83×10^{-6}+**	**1.83×10^{-6}+**	**1.83×10^{-6}+**
f_9		Median	2.53×10^1	8.78×10^2	4.97×10^2	1.38×10^4	8.82×10^3	7.70×10^3	5.07×10^4	2.24×10^4
		Mean	2.81×10^1	2.10×10^3	5.37×10^2	1.40×10^4	8.58×10^3	8.23×10^3	5.10×10^4	2.32×10^4
		Std	1.44×10^1	2.82×10^3	3.40×10^2	4.05×10^3	2.37×10^3	2.91×10^3	5.93×10^3	4.46×10^3
		p-value	-	**1.83×10^{-6}+**	**1.83×10^{-6}+**	**1.83×10^{-6}+**	**1.83×10^{-6}+**	**1.83×10^{-6}+**	**1.83×10^{-6}+**	**1.83×10^{-6}+**
$f_{4\text{–}9}$	$w/t/l$		-	5/1/0	6/0/0	6/0/0	6/0/0	6/0/0	6/0/0	6/0/0
f_{10}	Hybrid Functions	Median	8.46×10^3	3.03×10^4	1.23×10^4	1.37×10^4	3.05×10^4	1.35×10^4	2.90×10^4	2.17×10^4
		Mean	8.52×10^3	3.00×10^4	1.24×10^4	1.34×10^4	3.04×10^4	1.32×10^4	2.91×10^4	2.18×10^4
		Std	8.64×10^2	7.81×10^2	1.41×10^3	1.08×10^3	3.27×10^2	1.10×10^3	8.24×10^2	4.97×10^2
		p-value	-	**1.83×10^{-6}+**	**1.83×10^{-6}+**	**1.83×10^{-6}+**	**1.83×10^{-6}+**	**1.83×10^{-6}+**	**1.83×10^{-6}+**	**1.83×10^{-6}+**

Table 5. *Cont.*

f	Category	Quality	STTPSO	DNSPSO	XPSO	TCSPSO	GLPSO	HCLPSO	DPLPSO	CLPSO
f_{11}		Median	5.25×10^{2}	2.87×10^{4}	1.16×10^{3}	2.74×10^{3}	1.41×10^{4}	8.15×10^{2}	7.75×10^{4}	1.34×10^{3}
		Mean	5.22×10^{2}	2.90×10^{4}	1.20×10^{3}	3.42×10^{3}	1.61×10^{4}	7.93×10^{2}	7.82×10^{4}	1.34×10^{3}
		Std	1.17×10^{2}	7.16×10^{3}	2.46×10^{2}	1.92×10^{3}	7.38×10^{3}	2.03×10^{2}	9.38×10^{3}	1.60×10^{2}
		p-value	-	$\mathbf{1.83 \times 10^{-6}}$+	$\mathbf{1.83 \times 10^{-6}}$+	$\mathbf{1.83 \times 10^{-6}}$+	$\mathbf{1.83 \times 10^{-6}}$+	$\mathbf{3.89 \times 10^{-5}}$+	$\mathbf{1.83 \times 10^{-6}}$+	$\mathbf{1.83 \times 10^{-6}}$+
f_{12}		Median	7.69×10^{5}	1.49×10^{7}	1.15×10^{7}	5.19×10^{7}	5.53×10^{7}	1.55×10^{7}	3.01×10^{10}	7.81×10^{7}
		Mean	7.68×10^{5}	1.63×10^{7}	1.94×10^{7}	8.56×10^{7}	1.07×10^{8}	2.18×10^{7}	3.03×10^{10}	8.97×10^{7}
		Std	2.66×10^{5}	8.03×10^{6}	2.04×10^{7}	9.57×10^{7}	1.58×10^{8}	2.60×10^{7}	6.25×10^{9}	4.08×10^{7}
		p-value	-	$\mathbf{1.83 \times 10^{-6}}$+	$\mathbf{1.43 \times 10^{-5}}$+	$\mathbf{1.83 \times 10^{-6}}$+	$\mathbf{1.83 \times 10^{-6}}$+	$\mathbf{1.83 \times 10^{-6}}$+	$\mathbf{1.83 \times 10^{-6}}$+	$\mathbf{1.83 \times 10^{-6}}$+
f_{13}		Median	1.40×10^{3}	4.13×10^{3}	3.05×10^{3}	4.13×10^{3}	3.44×10^{3}	1.18×10^{4}	2.58×10^{9}	3.82×10^{4}
		Mean	4.47×10^{3}	9.58×10^{3}	4.62×10^{3}	6.81×10^{3}	1.87×10^{4}	1.86×10^{4}	2.78×10^{9}	4.13×10^{4}
		Std	5.03×10^{3}	1.16×10^{4}	4.04×10^{3}	5.71×10^{3}	7.41×10^{4}	1.27×10^{4}	1.14×10^{9}	1.59×10^{4}
		p-value	-	8.78×10^{-2}=	4.72×10^{-1}=	1.50×10^{-1}=	3.55×10^{-1}=	$\mathbf{2.26 \times 10^{-5}}$+	$\mathbf{1.83 \times 10^{-6}}$+	$\mathbf{1.83 \times 10^{-6}}$+
f_{14}		Median	1.92×10^{5}	2.02×10^{6}	2.23×10^{5}	9.21×10^{5}	8.66×10^{5}	4.17×10^{5}	1.20×10^{7}	4.85×10^{6}
		Mean	2.04×10^{5}	2.19×10^{6}	4.45×10^{5}	1.48×10^{6}	1.30×10^{6}	8.45×10^{5}	1.39×10^{7}	5.01×10^{6}
		Std	7.61×10^{4}	8.82×10^{5}	6.26×10^{5}	1.37×10^{6}	1.34×10^{6}	1.08×10^{6}	7.91×10^{6}	1.20×10^{6}
		p-value	-	$\mathbf{1.83 \times 10^{-6}}$+	8.78×10^{-2}=	$\mathbf{7.33 \times 10^{-6}}$+	$\mathbf{1.67 \times 10^{-4}}$+	$\mathbf{1.30 \times 10^{-5}}$+	$\mathbf{1.83 \times 10^{-6}}$+	$\mathbf{1.83 \times 10^{-6}}$+
f_{15}		Median	1.10×10^{3}	4.68×10^{4}	1.43×10^{3}	2.33×10^{3}	2.57×10^{3}	2.77×10^{4}	3.46×10^{8}	4.82×10^{3}
		Mean	3.87×10^{3}	1.16×10^{5}	2.18×10^{3}	4.88×10^{3}	3.92×10^{3}	1.85×10^{4}	3.92×10^{8}	5.32×10^{3}
		Std	5.76×10^{3}	1.61×10^{5}	1.96×10^{3}	5.47×10^{3}	4.10×10^{3}	1.10×10^{4}	2.20×10^{8}	2.70×10^{3}
		p-value	-	$\mathbf{1.83 \times 10^{-6}}$+	$5.65 \times 10_{-1}$=	3.76×10^{-1}=	6.51×10^{-1}=	$\mathbf{7.20 \times 10^{-5}}$+	$\mathbf{1.83 \times 10^{-6}}$+	$\mathbf{2.77 \times 10^{-2}}$+
f_{16}		Median	1.57×10^{3}	8.87×10^{3}	2.91×10^{3}	3.71×10^{3}	3.98×10^{3}	4.01×10^{3}	9.81×10^{3}	4.02×10^{3}
		Mean	1.58×10^{3}	8.88×10^{3}	2.84×10^{3}	3.82×10^{3}	4.06×10^{3}	4.16×10^{3}	9.62×10^{3}	4.07×10^{3}
		Std	4.82×10^{2}	3.59×10^{2}	4.77×10^{2}	6.74×10^{2}	8.46×10^{2}	6.93×10^{2}	5.78×10^{2}	3.37×10^{2}
		p-value	-	$\mathbf{1.83 \times 10^{-6}}$+	$\mathbf{4.08 \times 10^{-6}}$+	$\mathbf{1.83 \times 10^{-6}}$+	$\mathbf{2.02 \times 10^{-6}}$+	$\mathbf{1.83 \times 10^{-6}}$+	$\mathbf{1.83 \times 10^{-6}}$+	$\mathbf{1.83 \times 10^{-6}}$+
f_{17}		Median	1.16×10^{3}	5.97×10^{3}	2.52×10^{3}	3.23×10^{3}	2.77×10^{3}	3.72×10^{3}	6.77×10^{3}	3.23×10^{3}
		Mean	1.22×10^{3}	5.95×10^{3}	2.43×10^{3}	3.07×10^{3}	2.80×10^{3}	3.81×10^{3}	7.04×10^{3}	3.21×10^{3}
		Std	3.81×10^{2}	3.03×10^{2}	4.78×10^{2}	5.17×10^{2}	5.82×10^{2}	6.74×10^{2}	1.41×10^{3}	2.81×10^{2}
		p-value	-	$\mathbf{1.83 \times 10^{-6}}$+	$\mathbf{2.48 \times 10^{-6}}$+	$\mathbf{1.83 \times 10^{-6}}$+	$\mathbf{1.83 \times 10^{-6}}$+	$\mathbf{1.83 \times 10^{-6}}$+	$\mathbf{1.83 \times 10^{-6}}$+	$\mathbf{1.83 \times 10^{-6}}$+

Table 5. *Cont.*

f	Category	Quality	STTPSO	DNSPSO	XPSO	TCSPSO	GLPSO	HCLPSO	DPLPSO	CLPSO
f_{18}		Median	3.35×10^5	3.02×10^7	3.87×10^5	2.76×10^6	3.39×10^5	1.17×10^6	2.20×10^7	7.77×10^6
		Mean	3.70×10^5	3.11×10^7	4.91×10^5	3.38×10^6	5.25×10^5	1.63×10^6	2.45×10^7	7.50×10^6
		Std	1.72×10^5	1.18×10^7	3.58×10^5	2.09×10^6	4.73×10^5	1.25×10^6	1.25×10^7	2.41×10^6
		p-value	-	$\mathbf{1.83 \times 10^{-6}}$+	1.62×10^{-1}=	$\mathbf{1.83 \times 10^{-6}}$+	4.97×10^{-1}=	$\mathbf{2.74 \times 10^{-6}}$+	$\mathbf{1.83 \times 10^{-6}}$+	$\mathbf{1.83 \times 10^{-6}}$+
f_{19}		Median	5.57×10^3	4.75×10^3	3.13×10^3	2.30×10^3	4.20×10^9	1.28×10^4	6.56×10^8	1.79×10^3
		Mean	7.46×10^3	7.38×10^3	4.46×10^3	4.86×10^3	4.12×10^9	1.93×10^4	6.13×10^8	1.97×10^3
		Std	7.38×10^3	6.43×10^3	5.99×10^3	6.41×10^3	7.32×10^8	1.52×10^4	2.78×10^8	7.13×10^2
		p-value	-	9.67×10^{-1}=	1.33×10^{-1}=	8.04×10^{-2}=	$\mathbf{1.83 \times 10^{-6}}$+	$\mathbf{3.06 \times 10^{-3}}$+	$\mathbf{1.83 \times 10^{-6}}$+	2.18×10^{-3}–
f_{10-19}	$w/t/l$		-	8/2/0	5/5/0	7/3/0	7/3/0	10/0/0	10/0/0	9/0/1
f	Category	Quality	STTPSO	DNSPSO	XPSO	TCSPSO	GLPSO	HCLPSO	DPLPSO	CLPSO
f_{20}	Composition Functions	Median	7.56×10^2	5.57×10^3	2.10×10^3	2.88×10^3	5.26×10^3	2.86×10^3	4.69×10^3	2.29×10^3
		Mean	7.88×10^2	5.20×10^3	2.16×10^3	2.83×10^3	5.23×10^3	2.76×10^3	4.69×10^3	2.34×10^3
		Std	2.72×10^2	9.09×10^2	3.96×10^2	4.93×10^2	1.93×10^2	3.59×10^2	4.05×10^2	2.09×10^2
		p-value	-	$\mathbf{1.83 \times 10^{-6}}$+	$\mathbf{1.83 \times 10^{-6}}$+	$\mathbf{1.83 \times 10^{-6}}$+	$\mathbf{1.83 \times 10^{-6}}$+	$\mathbf{1.83 \times 10^{-6}}$+	$\mathbf{1.83 \times 10^{-6}}$+	$\mathbf{1.83 \times 10^{-6}}$+
f_{21}		Median	2.98×10^2	1.22×10^3	4.64×10^2	7.91×10^2	6.06×10^2	8.46×10^2	1.60×10^3	9.66×10^2
		Mean	2.96×10^2	1.22×10^3	4.66×10^2	7.87×10^2	6.26×10^2	8.42×10^2	1.60×10^3	9.62×10^2
		Std	9.91×10^0	3.32×10^1	5.19×10^1	8.43×10^1	7.42×10^1	6.64×10^1	8.73×10^1	3.02×10^1
		p-value	-	$\mathbf{1.83 \times 10^{-6}}$+	$\mathbf{1.83 \times 10^{-6}}$+	$\mathbf{1.83 \times 10^{-6}}$+	$\mathbf{1.82 \times 10^{-6}}$+	$\mathbf{1.83 \times 10^{-6}}$+	$\mathbf{1.82 \times 10^{-6}}$+	$\mathbf{1.83 \times 10^{-6}}$+
f_{22}		Median	8.16×10^3	3.08×10^4	1.36×10^4	1.46×10^4	1.80×10^4	1.41×10^4	3.08×10^4	2.25×10^4
		Mean	8.11×10^3	3.02×10^4	1.22×10^4	1.46×10^4	1.81×10^4	1.42×10^4	3.07×10^4	2.23×10^4
		Std	1.17×10^3	1.39×10^3	4.91×10^3	1.35×10^3	3.93×10^3	1.27×10^3	1.24×10^3	6.52×10^2
		p-value	-	$\mathbf{1.83 \times 10^{-6}}$+	$\mathbf{7.05 \times 10{-}3}$+	$\mathbf{1.83 \times 10^{-6}}$+	$\mathbf{1.83 \times 10^{-6}}$+	$\mathbf{1.83 \times 10^{-6}}$+	$\mathbf{1.83 \times 10^{-6}}$+	$\mathbf{1.83 \times 10^{-6}}$+
f_{23}		Median	7.10×10^2	1.84×10^3	8.13×10^2	1.04×10^3	1.11×10^3	8.83×10^2	2.82×10^3	9.01×10^2
		Mean	7.18×10^2	1.89×10^3	8.09×10^2	1.07×10^3	1.12×10^3	8.89×10^2	2.85×10^3	9.01×10^2
		Std	2.95×10^1	2.76×10^2	4.95×10^1	1.09×10^2	1.66×10^2	4.44×10^1	2.15×10^2	2.66×10^1
		p-value	-	$\mathbf{1.83 \times 10^{-6}}$+	$\mathbf{2.48 \times 10^{-6}}$+	$\mathbf{1.83 \times 10^{-6}}$+	$\mathbf{1.83 \times 10^{-6}}$+	$\mathbf{1.83 \times 10^{-6}}$+	$\mathbf{1.83 \times 10^{-6}}$+	$\mathbf{1.83 \times 10^{-6}}$+
f_{24}		Median	1.14×10^3	3.00×10^3	1.20×10^3	1.52×10^3	1.68×10^3	1.50×10^3	4.68×10^3	1.50×10^3
		Mean	1.14×10^3	3.14×10^3	1.25×10^3	1.55×10^3	1.64×10^3	1.51×10^3	4.69×10^3	1.49×10^3
		Std	5.71×10^1	6.83×10^2	1.17×10^2	1.46×10^2	2.19×10^2	6.83×10^1	4.23×10^2	2.50×10^1
		p-value	-	$\mathbf{1.83 \times 10^{-6}}$+	$\mathbf{1.67 \times 10{-}4}$+	$\mathbf{1.83 \times 10^{-6}}$+	$\mathbf{1.83 \times 10^{-6}}$+	$\mathbf{1.82 \times 10^{-6}}$+	$\mathbf{1.83 \times 10^{-6}}$+	$\mathbf{1.82 \times 10^{-6}}$+

Table 5. *Cont.*

f	Category	Quality	STTPSO	DNSPSO	XPSO	TCSPSO	GLPSO	HCLPSO	DPLPSO	CLPSO
f_{25}		Median	8.21×10^{2}	7.62×10^{2}	1.09×10^{3}	1.29×10^{3}	1.38×10^{3}	7.63×10^{2}	1.10×10^{4}	9.02×10^{2}
		Mean	7.97×10^{2}	7.67×10^{2}	1.10×10^{3}	1.35×10^{3}	1.37×10^{3}	7.80×10^{2}	1.10×10^{4}	9.08×10^{2}
		Std	5.15×10^{1}	5.19×10^{1}	7.56×10^{1}	2.93×10^{2}	2.10×10^{2}	6.35×10^{1}	1.35×10^{3}	4.80×10^{1}
		p-value	-	3.00×10^{-2}–	**1.83×10^{-6}+**	**1.83×10^{-6}+**	**1.83×10^{-6}+**	1.68×10^{-1}=	**1.82×10^{-6}+**	**1.83×10^{-6}+**
f_{26}		Median	6.53×10^{3}	2.88×10^{4}	5.34×10^{3}	1.06×10^{4}	8.54×10^{3}	1.13×10^{4}	2.89×10^{4}	1.09×10^{4}
		Mean	6.55×10^{3}	2.92×10^{4}	3.90×10^{3}	1.14×10^{4}	8.61×10^{3}	1.12×10^{4}	2.85×10^{4}	1.10×10^{4}
		Std	4.55×10^{2}	5.67×10^{3}	2.58×10^{3}	2.40×10^{3}	1.57×10^{3}	5.79×10^{2}	2.39×10^{3}	3.16×10^{2}
		p-value	-	**1.83×10^{-6}+**	4.97×10^{-6}–	**1.83×10^{-6}+**	**4.08×10^{-6}+**	**1.83×10^{-6}+**	**1.83×10^{-6}+**	**1.83×10^{-6}+**
f_{27}		Median	7.39×10^{2}	5.00×10^{2}	8.54×10^{2}	1.12×10^{3}	1.02×10^{3}	8.05×10^{2}	4.00×10^{3}	7.58×10^{2}
		Mean	7.53×10^{2}	5.00×10^{2}	8.75×10^{2}	1.12×10^{3}	1.01×10^{3}	8.17×10^{2}	4.05×10^{3}	7.59×10^{2}
		Std	4.28×10^{1}	0.00×10^{0}	7.68×10^{1}	1.75×10^{2}	8.73×10^{1}	8.19×10^{1}	4.27×10^{2}	2.21×10^{1}
		p-value	-	1.83×10^{-6}–	**3.34×10^{-6}+**	**1.83×10^{-6}+**	**1.82×10^{-6}+**	**1.15×10^{-3}+**	**1.83×10^{-6}+**	3.99×10^{-1}=
f_{28}		Median	5.85×10^{2}	5.00×10^{2}	8.26×10^{2}	1.33×10^{3}	1.27×10^{3}	5.85×10^{2}	1.43×10^{4}	1.28×10^{4}
		Mean	4.99×10^{3}	5.00×10^{2}	8.26×10^{2}	1.37×10^{3}	1.32×10^{3}	1.12×10^{3}	1.41×10^{4}	1.28×10^{4}
		Std	5.82×10^{3}	0.00×10^{0}	4.81×10^{1}	3.31×10^{2}	1.61×10^{2}	2.21×10^{3}	1.54×10^{3}	5.96×10^{1}
		p-value	-	1.82×10^{-6}–	3.88×10^{-1}=	3.88×10^{-1}=	3.88×10^{-1}=	2.33×10^{-1}=	**3.34×10^{-6}+**	**1.13×10^{-5}+**
f_{29}		Median	1.76×10^{3}	6.76×10^{3}	3.08×10^{3}	3.91×10^{3}	3.53×10^{3}	3.95×10^{3}	1.03×10^{4}	3.30×10^{3}
		Mean	1.82×10^{3}	6.79×10^{3}	3.10×10^{3}	3.91×10^{3}	3.71×10^{3}	3.94×10^{3}	1.05×10^{4}	3.30×10^{3}
		Std	3.67×10^{2}	3.36×10^{2}	5.11×10^{2}	5.41×10^{2}	7.04×10^{2}	6.27×10^{2}	1.15×10^{3}	2.71×10^{2}
		p-value	-	**1.83×10^{-6}+**	**1.83×10^{-6}+**	**1.83×10^{-6}+**	**1.83×10^{-6}+**	**1.82×10^{-6}+**	**1.83×10^{-6}+**	**1.82×10^{-6}+**
f_{30}		Median	4.41×10^{3}	7.97×10^{2}	2.59×10^{4}	1.04×10^{5}	2.24×10^{5}	1.13×10^{4}	2.38×109	5.76×10^{4}
		Mean	4.74×10^{3}	8.61×10^{2}	3.28×10^{4}	1.46×10^{5}	6.40×10^{5}	1.50×10^{4}	2.49×109	7.32×10^{4}
		Std	1.39×10^{3}	2.29×10^{2}	2.36×10^{4}	1.34×10^{5}	1.01×10^{6}	1.62×10^{4}	7.38×108	5.20×10^{4}
		p-value	-	1.83×10^{-6}–	**1.83×10^{-6}+**	**1.83×10^{-6}+**	**1.83×10^{-6}+**	**1.07×10^{-5}+**	**1.83×10^{-6}+**	**1.83×10^{-6}+**
$f_{20\text{–}30}$	*w/t/l*		-	7/0/4	9/1/1	10/1/0	10/1/0	9/2/0	11/0/0	10/1/0
	w/t/l		-	21/4/4	20/7/2	24/5/0	24/4/1	26/2/1	29/0/0	27/1/1
	rank		**1.52**	5.31	2.72	4.83	4.83	4.00	7.72	5.07

Table 6. Statistical comparison results between the proposed STTPSO and the 7 state-of-the-art and popular PSO variants on the CEC 2017 benchmark set with different dimensions in terms of "$w/t/l$".

Category	*D*	DNSPSO	XPSO	TCSPSO	GLPSO	HCLPSO	DPLPSO	CLPSO
Unimodal Functions	30	2/0/0	0/1/1	0/1/1	0/1/1	1/0/1	2/0/0	1/0/1
	50	1/1/0	0/1/1	0/2/0	0/1/1	1/0/1	2/0/0	1/1/0
	100	1/1/0	0/1/1	1/1/0	1/0/1	1/0/1	2/0/0	2/0/0
Simple Multimodal Functions	30	5/0/1	6/0/0	6/0/0	6/0/0	6/0/0	6/0/0	6/0/0
	50	5/0/1	6/0/0	6/0/0	6/0/0	5/1/0	6/0/0	6/0/0
	100	5/1/0	6/0/0	6/0/0	6/0/0	6/0/0	6/0/0	6/0/0
Hybrid Functions	30	7/1/2	3/6/1	6/4/0	4/5/1	7/3/0	10/0/0	5/2/3
	50	9/0/1	5/2/3	7/2/1	7/1/2	7/3/0	10/0/0	7/0/3
	100	8/2/0	5/5/0	7/3/0	7/3/0	10/0/0	10/0/0	9/0/1
Composition Functions	30	9/0/2	9/1/1	11/0/0	11/0/0	10/1/0	10/1/0	9/1/1
	50	8/1/2	8/2/1	10/0/1	10/0/1	8/3/0	11/0/0	9/0/2
	100	7/0/4	9/1/1	10/1/0	10/1/0	9/2/0	11/0/0	10/1/0
Whole Set	30	23/1/5	18/8/3	23/5/1	21/6/2	24/4/1	28/1/0	21/3/5
	50	23/2/4	19/5/5	23/4/2	23/2/4	21/7/1	29/0/0	23/1/5
	100	21/4/4	20/7/2	24/5/0	24/4/1	26/2/1	29/0/0	27/1/1

As shown in Table 4, the comparison results on the 50-*D* CEC 2017 benchmark problems are summarized below:

(1) According to the Friedman test results shown in the last row, STTPSO achieves the lowest rank. This indicates that STTPSO still achieves the best overall performance on the whole 50-*D* CEC 2017 benchmark set. In particular, except for XPSO, its rank value (2.17) is much smaller than those (at least 4.14) of the other six compared algorithms. This demonstrates that STTPSO displays significantly better overall performance than the six compared algorithms.
(2) From the perspective of the Wilcoxon rank sum test, as shown in the second to last row, STTPSO achieves significantly better performance than the seven compared algorithms on at least 19 problems and shows inferiority to them on, at most, five problems. In particular, compared with DNSPSO, TCSPSO, GLPSO, and CLPSO, STTPSO significantly dominates them all on 23 problems. In comparison with DPLPSO, STTPSO presents significant superiority on all the 29 problems.
(3) In terms of different types of optimization problems, STTPSO achieves highly competitive performance with the seven compared state-of-the-art PSO variants regarding the two unimodal problems. Particularly, STTPSO defeats DPLPSO concerning these two problems. On the six simple multimodal problems, STTPSO performs much better than the seven compared algorithms on at least five problems. In particular, STTPSO presents significant dominance to XPSO, TCSPSO, GLPSO, DPLPSO, and CLPSO on all the six problems. Regarding the 10 hybrid problems, except for XPSO, STTPSO is significantly superior to the seven compared algorithms on at least seven problems, and shows inferiority on, at most, three problems. In particular, STTPSO significantly outperforms DPLPSO on all the 10 problems and obtains significantly better performance than DNSPSO on nine problems. Concerning the 11 composition problems, STTPSO displays significantly better performance than the seven state-of-the-art PSO variants on at least eight problems, and performs worse than them on, at most, two problems. Particularly, STTPSO shows significant dominance to DPLPSO on all the 11 problems and obtains much better performance than both TCSPSO and GLPSO on 10 problems. Overall, it is still demonstrated that STTPSO is a promising approach for problem optimization and displays its sound optimization ability in solving complicated optimization problems, such as multimodal problems, hybrid problems, and composition problems.

As shown in Table 5, the comparison results on the 100-*D* CEC 2017 benchmark set are summarized below:

(1) According to the Friedman test results, STTPSO achieves the lowest rank among all algorithms. This verifies that STTPSO still obtains the best overall performance on the 100-*D* CEC 2017 benchmark set. In particular, its rank value (1.52) is much smaller than those (at least 2.72) of the seven compared algorithms. This further demonstrates that STTPSO displays significant dominance to the seven compared algorithms. Together with the observations on the 30-*D* and 50-*D* CEC 2017 benchmark set, we can see that STTPSO consistently performs the best on the CEC 2017 benchmark set with different dimension sizes among all eight algorithms, and consistently presents its significant superiority to the seven compared algorithms on the benchmark set with the three dimension sizes. Therefore, it is demonstrated that STTPSO preserves a good scalability to solve optimization problems.
(2) Regarding the Wilcoxon rank sum test, from the second to last row, it is observed that STTPSO achieves significantly better performance than the seven compared algorithms on at least 20 problems and shows inferiority to them on, at most, four problems. In particular, STTPSO outperforms DPLPSO significantly on all the 29 problems, and obtains much better performance than TCSPSO, GLPSO, HCLPSO, and CLPSO on 24, 24, 26, and 27 problems, respectively.
(3) With respect to the optimization performance on different types of optimization problems, STTPSO obtains highly competitive or even much better performance than the

seven compared algorithms on the two unimodal problems. Particularly, STTPSO shows significant dominance to DPLPSO and CLPSO on the two problems. As for the six simple multimodal problems, except for DNSPSO, STTPSO exhibits significant superiority to the other six compared algorithms on all these six problems. Competed with DNSPSO, STTPSO also shows much better performance on five problems. In terms of the 10 hybrid problems, except for XPSO, STTPSO is significantly superior to the other six compared algorithms on at least seven problems. Compared with XPSO, STTPSO illustrates significantly better performance on five problems and does not show inferiority on any of the problems. In particular, it is discovered that STTPSO is significantly better than HCLPSO and DPLPSO on all the 10 problems. Regarding the 11 composition problems, except for DNSPSO, STTPSO achieves much better performance than the other six compared algorithms on at least nine problems. Compared with DNSPSO, it still performs much better on seven problems. Particularly, STTPSO shows significant superiority to DPLPSO on all the 11 problems, and obtains much better performance than TCSPSO, GLPSO, and CLPSO on 10 problems and shows no inferiority to the three compared methods on these kinds of problems. Overall, it is demonstrated that STTPSO is still effective at solving optimization problems, especially complicated problems, such as multimodal problems, hybrid problems, and composition problems.

To summarize, as shown in Table 6, on the CEC 2017 benchmark set with different dimension sizes, we find that the proposed STTPSO not only shows highly competitive performance against the compared state-of-the-art PSO variants on simple optimization problems, such as unimodal problems, but also achieves much better performance on complicated optimization problems, such as multimodal problems, hybrid problems and composition problems. In particular, we find that the superiority of STTPSO to the compared state-of-the-art methods is far more conspicuous regarding complicated problems, such as hybrid problems and composition problems. On the other hand, it can be concluded that STTPSO preserves a good scalability to solve optimization problems, since it consistently achieves the best overall performance on the CEC 2017 set with the three dimension sizes. Moreover, it is found that as the dimensionality increases, the superiority of STTPSO to certain compared algorithms become much more evident.

The above extensive experiments have demonstrated the effectiveness of STTPSO in solving optimization problems. To further demonstrate its efficiency in tackling optimization problems, we conduct experiments on the 50D CEC 2017 benchmark set to form convergence comparisons between STTPSO and the seven compared algorithms. Figure 1 presents the comparison results on the 16 50D CEC 2017 problems of different categories.

From Figure 1, the following observations can be obtained. (1) At a first glance, STTPSO obtains much better performance in terms of both convergence speed and solution quality on 12 problems ($f_5, f_7, f_9, f_{11}, f_{12}, f_{16}, f_{17}, f_{20}, f_{21}, f_{23}, f_{24}$, and f_{29}). (2) On f_{19} and f_{26}, STTPSO shows clear dominance to six compared methods regarding both convergence speed and solution quality, and only presents inferiority to only one of the compared methods. (3) On f_1 and f_{14}, STTPSO displays conspicuously faster convergence speed and higher solution quality than five compared methods, and only presents slight inferiority to two compared methods. (4) Overall, it is demonstrated that STTPSO could solve optimization problems with both high effectiveness and efficiency.

The superiority of STTPSO mainly benefits from the proposed stochastic triad strategy along with the devised archive, the restart mechanism and the dynamic parameter adjustment strategy. These strategies cooperate cohesively to improve the learning diversity and learning effectiveness of particles, which help the swarm explore and exploit the solution space properly to find the optima of optimization problems. To investigate the influence of the four components, we will conduct a thorough investigation on STTPSO in the following subsection.

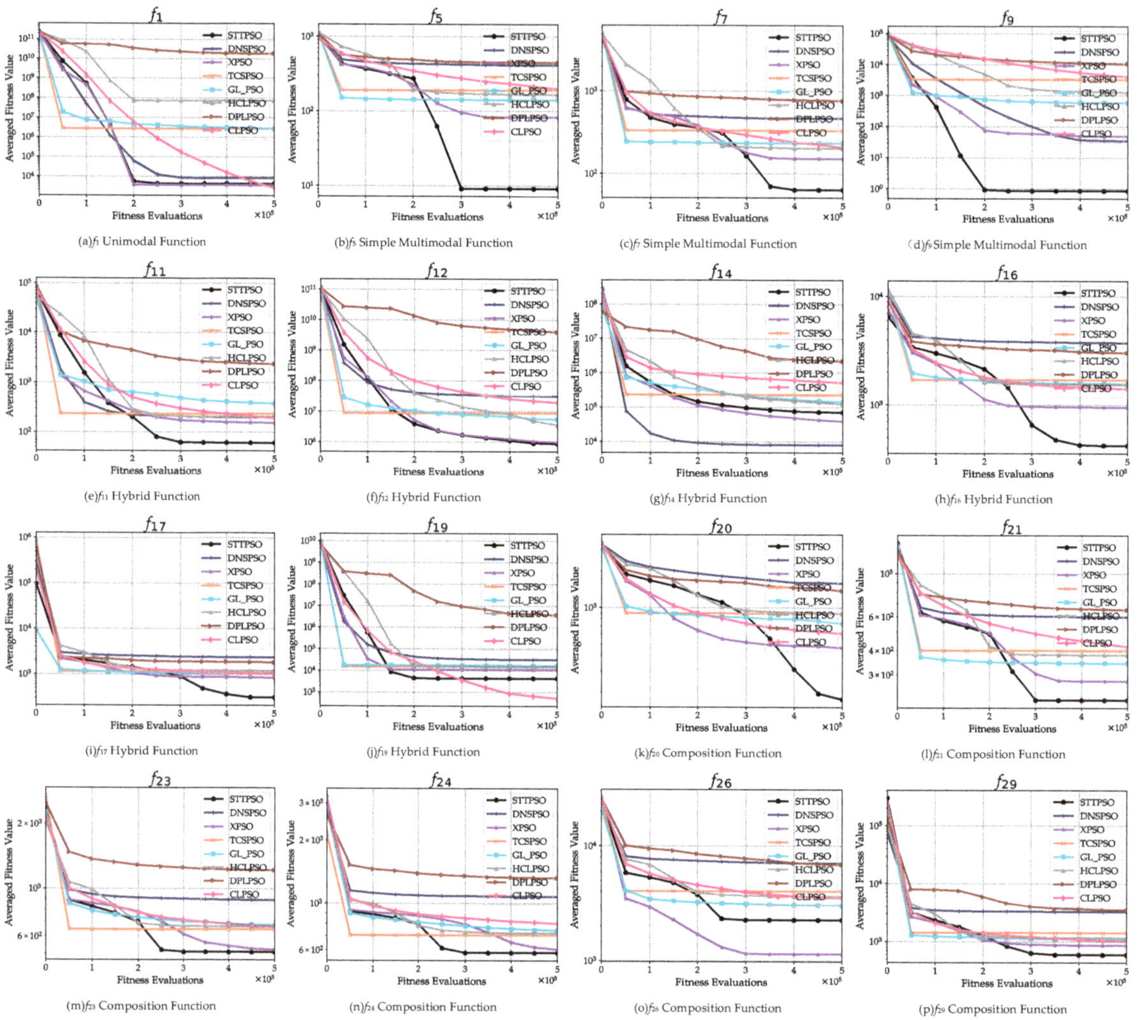

Figure 1. Convergence behavior comparisons between STTPSO and the seven compared algorithms on the 16 50*D* CEC 2017 benchmark problems.

4.3. Deep Investigation on STTPSO

In this section, we aim to verify the effectiveness of each component in STTPSO by conducting experiments on the 50-*D* CEC 2017 benchmark set.

4.3.1. Effectiveness of the Reformulation of the Stochastic Triad Topology

First, we conduct experiments to verify the effectiveness of the reformulation of the stochastic triad topology. In Section 3.1, we mentioned that in order to retain the learning effectiveness of particles, we let the triad topology structure remain fixed for each particle and then adjust it based on the evolution state of this particle. In particular, when the personal best position (***pbest***) of one particle keeps unchanged for $stop_{max}$ times, we then reformulate the triad topology by randomly selecting two personal positions from those of the current swarm and the archive. In this way, both the learning effectiveness and learning diversity of particles can be guaranteed.

To verify the effectiveness of this strategy, we fix $stop_{max}$ as different values as shown in Table 7. It should be mentioned that the larger the value of $stop_{max}$, the less frequently the triad topology structure changes. In particular, when $stop_{max}$ = 0, the triad structure is consequently changed in every generation. The comparison results among STTPSO with different settings of $stop_{max}$ is shown in Table 7.

From Table 7, it is found that (1) from the perspective of the Friedman test, $stop_{max}$ = 30 helps STTPSO achieve the best overall performance on the 50-*D* CEC 2017 benchmark set. In particular, we find that the rank value (2.14) of STTPSO with $stop_{max}$ = 30 is much smaller than those (at least 4.31) of STTPSO with other settings. This indicates that the superiority of $stop_{max}$ = 30 is much more significant than the other settings. Moreover, we also find that STTPSO with small $stop_{max}$, such as $stop_{max}$ = 0 and $stop_{max}$ = 5 achieve a much worse performance. This demonstrates an absence of beneficial effects regarding STTPSO frequently changing the triad structure. (2) Through meticulous observation, we find that STTPSO with $stop_{max}$ = 30 achieves the best performance on 22 problems. Concerning the other seven problems, its performance is extremely close to the STTPSO with the associated optimal settings of $stop_{max}$.

Based on the above observations, it is verified that the reformulation of the triad topology is very helpful for STTPSO to achieve promising performance. In particular, such reformulation should bear neither an excessively high frequency, nor an excessively low frequency.

4.3.2. Effectiveness of the Additional Archive and the Proposed Random Restart Strategy

Subsequently, we conduct experiments to verify the effectiveness of the additional archive and the random restart strategy. To this end, we first remove the additional archive from STTPSO, deriving a new version of STTPSO, which we name as STTPSO_WA. Then, we remove the restart strategy from STTPSO, deriving another version of STTPSO, which we denote as STTPSO_WR. Subsequently, we conduct experiments on the 50-*D* CEC 2017 benchmark set to compare the three versions of STTPSO. Table 8 displays the comparison results.

From Table 8, it is discovered that STTPSO with both the archive and the restart strategy achieves the best overall performance than the other two versions of STTPSO. In particular, we find that STTPSO_WR obtains the worst performance. This indicates that compared with the archive, the restart strategy is far more helpful. This is because compared with the archive, which stores the obsolete historical information, the restart strategy is more effective at improving the swarm diversity since it can introduce new solutions into the archive to promote the learning diversity of particles.

4.3.3. Effectiveness of the Dynamic Acceleration Coefficients

At last, we conduct experiments to verify the effectiveness of the devised dynamic acceleration coefficient strategy. In Section 3.2, instead of using fixed acceleration coefficients, the proposed STTPSO first randomly samples two different values based on the Gaussian distribution, and then the larger one between the two sampled values is utilized as c_1, while the smaller one is utilized as c_2. In this way, a promising balance between exploration and exploitation can be preserved. To validate this, we first denote the original strategy in this paper as "Dynamic". Then, we replace the settings of c_1 and c_2 with two other settings. The first is to directly utilize the sampled values as c_1 and c_2 without comparison, which we denote as "Dynamic2". The other is to utilize the smaller one between the two sampled values as c_1, and the larger one as c_2, which is a converse setting of the one used in this paper, which we denote as "Dynamic3". Lastly, as the baseline comparison, we adopt fixed settings for c_1 and c_2 by varying them from 1.0 to 2.0. Table 9 shows the comparison results between STTPSO with different settings of c_1 and c_2 on the 50-*D* CEC 2017 benchmark set.

Table 7. Comparison results between STTPSO with different settings of stagnation times on the 50*D* CEC 2017 functions. The best results are highlighted in bold.

f	$stop_{max}=0$	$stop_{max}=5$	$stop_{max}=10$	$stop_{max}=15$	$stop_{max}=20$	$stop_{max}=25$	$stop_{max}=30$	$stop_{max}=35$	$stop_{max}=40$
f_1	9.66×10^{6}	1.07×10^{4}	7.82×10^{3}	6.64×10^{3}	6.20×10^{3}	9.07×10^{3}	**4.33×10^{3}**	7.55×10^{3}	8.64×10^{3}
f_3	1.75×10^{5}	7.17×10^{4}	6.60×10^{4}	6.66×10^{4}	6.17×10^{4}	6.53×10^{4}	**5.85×10^{4}**	6.16×10^{4}	6.14×10^{4}
f_4	2.32×10^{2}	1.97×10^{2}	1.88×10^{2}	1.91×10^{2}	1.95×10^{2}	1.83×10^{2}	**1.69×10^{2}**	1.74×10^{2}	1.87×10^{2}
f_5	3.99×10^{2}	9.49×10^{1}	1.50×10^{1}	1.47×10^{1}	1.77×10^{1}	1.67×10^{1}	**9.09×10^{0}**	1.84×10^{1}	1.69×10^{1}
f_6	6.74×10^{0}	**8.56×10^{-7}**	1.51×10^{-6}	1.26×10^{-6}	1.78×10^{-6}	1.50×10^{-6}	4.48×10^{-6}	1.83×10^{-6}	2.53×10^{-6}
f_7	4.76×10^{2}	3.35×10^{2}	1.16×10^{2}	**6.12×10^{1}**	6.52×10^{1}	6.75×10^{1}	6.33×10^{1}	6.90×10^{1}	6.32×10^{1}
f_8	3.97×10^{2}	4.72×10^{1}	1.68×10^{1}	1.74×10^{1}	1.63×10^{1}	1.73×10^{1}	**9.35×10^{0}**	1.80×10^{1}	1.81×10^{1}
f_9	1.46×10^{2}	**6.68×10^{-1}**	1.62×10^{0}	1.39×10^{0}	1.92×10^{0}	1.93×10^{0}	8.56×10^{-1}	1.19×10^{0}	1.17×10^{0}
f_{10}	1.32×10^{4}	1.25×10^{4}	1.22×10^{4}	1.13×10^{4}	1.15×10^{4}	1.14×10^{4}	**3.49×10^{3}**	1.09×10^{4}	1.09×10^{4}
f_{11}	5.40×10^{2}	**4.67×10^{1}**	5.27×10^{1}	5.62×10^{1}	5.81×10^{1}	6.07×10^{1}	6.12×10^{1}	5.77×10^{1}	6.07×10^{1}
f_{12}	7.61×10^{7}	2.82×10^{6}	1.83×10^{6}	2.12×106	2.16×10^{6}	1.64×10^{6}	**8.89×10^{5}**	1.64×10^{6}	1.69×10^{6}
f_{13}	3.65×10^{4}	3.01×10^{4}	2.19×10^{4}	3.04×10^{4}	2.92×10^{4}	2.72×10^{4}	**1.63×10^{4}**	2.70×10^{4}	2.64×10^{4}
f_{14}	4.53×10^{5}	2.17×10^{5}	2.34×10^{5}	1.77×10^{5}	1.74×10^{5}	2.21×10^{5}	**7.09×10^{4}**	2.19×10^{5}	1.83×10^{5}
f_{15}	3.11×10^{4}	3.13×10^{4}	3.13×10^{4}	2.98×10^{4}	3.12×10^{4}	3.11×10^{4}	**1.32×10^{4}**	3.00×10^{4}	3.00×10^{4}
f_{16}	3.04×10^{3}	1.71×10^{3}	6.33×10^{2}	4.70×10^{2}	4.38×10^{2}	5.28×10^{2}	**4.23×10^{2}**	5.27×10^{2}	5.81×10^{2}
f_{17}	1.95×10^{3}	1.04×10^{3}	7.59×10^{2}	5.10×10^{2}	4.22×10^{2}	3.80×10^{2}	**3.11×10^{2}**	5.28×10^{2}	6.00×10^{2}
f_{18}	5.94×10^{6}	2.78×10^{6}	2.51×10^{6}	2.00×10^{6}	1.82×10^{6}	1.82×10^{6}	**4.28×10^{5}**	1.51×10^{6}	1.64×10^{6}
f_{19}	2.46×10^{3}	2.29×10^{3}	2.11×10^{3}	2.02×10^{3}	1.91×10^{3}	1.90×10^{3}	4.33×10^{3}	**1.69×10^{3}**	1.81×10^{3}
f_{20}	1.63×10^{3}	1.29×10^{3}	1.17×10^{3}	1.01×10^{3}	8.36×10^{2}	8.01×10^{2}	**1.76×10^{2}**	6.15×10^{2}	6.60×10^{2}
f_{21}	6.00×10^{2}	2.79×10^{2}	2.28×10^{2}	2.28×10^{2}	2.30×10^{2}	2.29×10^{2}	**2.22×10^{2}**	2.32×10^{2}	2.32×10^{2}
f_{22}	1.33×10^{4}	1.21×10^{4}	1.19×10^{4}	1.05×10^{4}	9.73×10^{3}	1.02×10^{4}	**3.10×10^{3}**	9.29×10^{3}	9.76×10^{3}
f_{23}	8.36×10^{2}	5.13×10^{2}	5.11×10^{2}	5.14×10^{2}	5.18×10^{2}	5.18×10^{2}	**5.09×10^{2}**	5.21×10^{2}	5.20×10^{2}
f_{24}	8.85×10^{2}	5.98×10^{2}	5.89×10^{2}	5.90×10^{2}	5.94×10^{2}	5.91×10^{2}	**5.83×10^{2}**	5.96×10^{2}	5.95×10^{2}
f_{25}	5.48×10^{2}	**4.81×10^{2}**	4.82×10^{2}	4.83×10^{2}	4.81×10^{2}	4.82×10^{2}	5.06×10^{2}	4.84×10^{2}	4.83×10^{2}
f_{26}	5.49×10^{3}	2.37×10^{3}	2.40×10^{3}	2.45×10^{3}	2.51×10^{3}	2.51×10^{3}	**2.27×10^{3}**	2.54×10^{3}	2.57×10^{3}
f_{27}	**6.92×10^{2}**	7.72×10^{2}	7.66×10^{2}	7.52×10^{2}	7.59×10^{2}	7.56×10^{2}	6.94×10^{2}	7.59×10^{2}	7.63×10^{2}
f_{28}	5.01×10^{3}	4.99×10^{3}	4.30×10^{3}	5.00×10^{3}	4.92×10^{3}	4.96×10^{3}	**9.06×10^{2}**	4.77×10^{3}	4.78×10^{3}
f_{29}	2.07×10^{3}	1.32×10^{3}	1.16×10^{3}	1.00×10^{3}	1.03×10^{3}	9.13×10^{2}	**5.90×10^{2}**	9.35×10^{2}	1.01×10^{3}
f_{30}	1.57×10^{6}	$1.52 \times 10_{6}$	1.45×10^{6}	1.36×10^{6}	1.37×10^{6}	1.34×10^{6}	**8.55×10^{5}**	1.15×10^{6}	1.24×10^{6}
rank	8.55	6.52	5.14	4.31	4.69	4.62	**2.14**	4.31	4.72

Table 8. Comparison results between STTPSO with and without the archive and the restart strategy on the 50-*D* CEC 2017 functions. The best results are highlighted in bold.

f	STTPSO	STTPSO_WR	STTPSO_WA
f_1	$\mathbf{4.33 \times 10^3}$	8.59×10^3	5.75×10^3
f_3	$\mathbf{5.85 \times 10^4}$	6.82×10^4	6.08×10^4
f_4	$\mathbf{1.69 \times 10^2}$	1.96×10^2	1.96×10^2
f_5	$\mathbf{9.09 \times 10^0}$	1.64×10^1	1.46×10^1
f_6	4.48×10^{-6}	$\mathbf{2.38 \times 10^{-6}}$	3.85×10^{-6}
f_7	6.33×10^1	9.71×10^1	$\mathbf{5.95 \times 10^1}$
f_8	$\mathbf{9.35 \times 10^0}$	1.56×10^1	1.45×10^1
f_9	$\mathbf{8.56 \times 10^{-1}}$	1.59×10^0	1.57×10^0
f_{10}	$\mathbf{3.49 \times 10^3}$	1.21×10^4	1.18×10^4
f_{11}	6.12×10^1	$\mathbf{5.07 \times 10^1}$	5.07×10^1
f_{12}	$\mathbf{8.89 \times 10^5}$	2.46×10^6	1.85×10^6
f_{13}	$\mathbf{1.63 \times 10^4}$	2.52×10^4	2.46×10^4
f_{14}	$\mathbf{7.09 \times 10^4}$	$2.17 \times 10_5$	1.88×10^5
f_{15}	$\mathbf{1.32 \times 10^4}$	3.13×10^4	2.70×10^4
f_{16}	$\mathbf{4.23 \times 10^2}$	6.87×10^2	6.41×10^2
f_{17}	$\mathbf{3.11 \times 10^2}$	$5.46 \times 10_2$	3.78×10^2
f_{18}	$\mathbf{4.28 \times 10^5}$	1.99×10^6	1.63×10^6
f_{19}	4.33×10^3	2.16×10^3	$\mathbf{2.03 \times 10^3}$
f_{20}	$\mathbf{1.76 \times 10^2}$	1.12×10^3	8.45×10^2
f_{21}	$\mathbf{2.22 \times 10^2}$	2.29×10^2	2.27×10^2
f_{22}	$\mathbf{3.10 \times 10^3}$	1.15×10^4	1.01×10^4
f_{23}	$\mathbf{5.09 \times 10^2}$	5.12×10^2	5.16×10^2
f_{24}	$\mathbf{5.83 \times 10^2}$	5.94×10^2	5.98×10^2
f_{25}	5.06×10^2	$\mathbf{4.81 \times 10^2}$	4.82×10^2
f_{26}	$\mathbf{2.27 \times 10^3}$	2.40×10^3	2.47×10^3
f_{27}	$\mathbf{6.94 \times 10^2}$	7.56×10^2	7.47×10^2
f_{28}	$\mathbf{9.06 \times 10^2}$	4.96×10^3	4.34×10^3
f_{29}	$\mathbf{5.90 \times 10^2}$	1.01×10^3	8.27×10^2
f_{30}	$\mathbf{8.55 \times 10^5}$	1.30×10^6	1.12×10^6
rank	**1.31**	2.62	2.07

From Table 9, it is observed that from the average rank obtained from the Friedman test, the proposed dynamic strategy for c_1 and c_2 helps STTPSO achieve the best overall performance among all setting versions of c_1 and c_2. This demonstrates that the proposed dynamic strategy is extremely effective for STTPSO to achieve good performance. In particular, compared with the fixed settings, the proposed dynamic strategy helps STTPSO achieve much better performance than all the fixed settings. This demonstrates the dynamic sampling of c_1 and c_2 is far more effective than fixed ones. In comparison with the other two dynamic strategies, STTPSO with the proposed dynamic strategy obtains significantly better performance than those with the other two dynamic strategies. This demonstrates utilization of the larger one between the sampled values as c_1 and the smaller one as c_2 is far more effective. Together, we can observe that the proposed dynamic acceleration coefficient strategy is helpful in order for STTPSO to achieve good performance.

Table 9. Comparison results between STTPSO with different acceleration coefficient settings on the 50*D* CEC 2017 functions. The best results are highlighted in bold.

f	Dynamic	Dynamic1	Dynamic2	$c_1 = 1.0$			$c_1 = 1.5$			$c_1 = 2.0$		
				$c_2 = 1.0$	$c_2 = 1.5$	$c_2 = 2.0$	$c_2 = 1.0$	$c_2 = 1.5$	$c_2 = 2.0$	$c_2 = 1.0$	$c_2 = 1.5$	$c_2 = 2.0$
f_1	4.33×10^3	7.70×10^3	8.72×10^3	$\mathbf{2.44 \times 10^3}$	4.08×10^3	7.78×10^3	6.41×10^3	7.75×10^3	1.74×10^4	1.37×10^4	1.93×10^4	5.31×10^7
f_3	5.85×10^4	6.84×10^4	6.43×10^4	6.42×10^4	7.20×10^4	7.39×10^4	$\mathbf{5.43 \times 10^4}$	6.76×10^4	9.44×10^4	6.18×10^4	9.54×10^4	1.46×10^5
f_4	1.69×10^2	1.91×10^2	1.90×10^2	$\mathbf{1.18 \times 10^2}$	1.75×10^2	1.94×10^2	1.91×10^2	1.84×10^2	1.93×10^2	1.80×10^2	1.93×10^2	2.33×10^2
f_5	9.09×10^0	1.68×10^1	1.82×10^1	9.78×10^0	$\mathbf{8.50 \times 10^0}$	1.08×10^1	1.72×10^1	1.71×10^1	2.90×10^2	2.89×10^1	2.94×10^2	3.90×10^2
f_6	4.48×10^{-6}	2.37×10^{-4}	2.39×10^{-6}	7.32×10^{-5}	5.44×10^{-6}	$\mathbf{1.27 \times 10^{-6}}$	8.07×10^{-4}	1.50×10^{-6}	3.54×10^{-4}	3.90×10^{-5}	3.99×10^{-3}	3.02×10^0
f_7	6.33×10^1	9.75×10^1	7.80×10^1	6.67×10^1	6.24×10^1	1.80×10^2	$\mathbf{6.07 \times 10^1}$	9.58×10^1	3.61×10^2	1.03×10^2	3.65×10^2	4.44×10^2
f_8	9.35×10^0	1.70×10^1	1.76×10^1	8.57×10^0	$\mathbf{8.18 \times 10^0}$	1.13×10^1	1.68×10^1	1.77×10^1	2.85×10^2	2.87×10^1	2.74×10^2	3.88×10^2
f_9	8.56×10^{-1}	1.74×10^0	1.60×10^0	6.85×10^{-1}	$\mathbf{3.14 \times 10^{-1}}$	8.52×10^{-1}	1.22×10^0	1.49×10^0	2.31×10^0	4.38×10^0	3.46×10^0	2.95×10^2
f_{10}	$\mathbf{3.49 \times 10^3}$	1.22×10^4	1.21×10^4	3.52×10^3	7.09×10^3	1.22×10^4	7.77×10^3	1.23×10^4	1.27×10^4	1.14×10^4	1.29×10^4	1.28×10^4
f_{11}	6.12×10^1	5.16×10^1	5.39×10^1	7.72×10^1	5.91×10^1	$\mathbf{4.65 \times 10^1}$	5.81×10^1	5.04×10^1	1.72×10^2	8.09×10^1	1.91×10^2	4.75×10^2
f_{12}	8.89×10^5	2.18×10^6	2.40×10^6	$\mathbf{7.14 \times 10^5}$	1.23×10^6	1.87×10^6	1.46×10^6	2.16×10^6	5.52×10^6	3.38×10^6	6.65×10^6	1.08×10^8
f_{13}	1.63×10^4	3.26×10^4	3.00×10^4	$\mathbf{9.33 \times 10^3}$	1.82×10^4	2.72×10^4	2.11×10^4	3.14×10^4	3.61×10^4	3.65×10^4	3.63×10^4	3.77×10^4
f_{14}	$\mathbf{7.09 \times 10^4}$	2.77×10^5	2.14×10^5	7.15×10^4	1.02×10^5	1.81×10^5	1.93×10^5	2.41×10^5	2.84×10^5	2.43×10^5	2.40×10^5	4.09×10^5
f_{15}	1.32×10^4	3.12×10^4	3.14×10^4	$\mathbf{3.71 \times 10^3}$	2.15×10^4	2.97×10^4	3.07×10^4	3.12×10^4	3.15×10^4	3.15×10^4	3.15×10^4	3.19×10^4
f_{16}	4.23×10^2	6.14×10^2	5.99×10^2	4.96×10^2	4.02×10^2	6.90×10^2	$\mathbf{3.92 \times 10^2}$	5.46×10^2	2.26×10^3	6.08×10^2	2.16×10^3	2.80×10^3
f_{17}	3.11×10^2	7.20×10^2	6.73×10^2	3.01×10^2	3.85×10^2	7.63×10^2	$\mathbf{2.39 \times 10^2}$	5.93×10^2	1.38×10^3	4.69×10^2	1.26×10^3	1.78×10^3
f_{18}	4.28×10^5	2.45×10^6	2.39×10^6	$\mathbf{3.80 \times 10^5}$	6.79×10^5	1.55×10^6	1.64×10^6	2.45×10^6	3.21×10^6	2.25×10^6	3.06×10^6	4.03×10^6
f_{19}	4.33×10^3	2.09×10^3	2.20×10^3	5.91×10^3	1.75×10^3	1.86×10^3	$\mathbf{1.70 \times 10^3}$	2.14×10^3	2.32×10^3	2.37×10^3	2.40×10^3	2.48×10^3
f_{20}	1.76×10^2	1.13×10^3	1.12×10^3	$\mathbf{1.73 \times 10^2}$	4.35×10^2	1.11×10^3	4.18×10^2	1.17×10^3	1.32×10^3	1.11×10^3	1.36×10^3	1.47×10^3
f_{21}	2.22×10^2	2.27×10^2	2.30×10^2	2.32×10^2	2.33×10^2	$\mathbf{2.22 \times 10^2}$	2.30×10^2	2.30×10^2	4.92×10^2	2.43×10^2	4.82×10^2	5.92×10^2
f_{22}	$\mathbf{3.10 \times 10^3}$	1.17×10^4	1.18×10^4	3.24×10^3	4.42×10^3	1.18×10^4	4.83×10^3	1.18×10^4	1.25×10^4	1.23×10^4	1.28×10^4	1.29×10^4
f_{23}	5.09×10^2	5.18×10^2	5.16×10^2	5.36×10^2	5.40×10^2	$\mathbf{5.04 \times 10^2}$	5.22×10^2	5.15×10^2	6.38×10^2	5.27×10^2	6.32×10^2	8.21×10^2
f_{24}	5.83×10^2	5.89×10^2	5.87×10^2	6.04×10^2	6.06×10^2	$\mathbf{5.73 \times 10^2}$	5.90×10^2	5.90×10^2	7.90×10^2	5.88×10^2	7.45×10^2	8.85×10^2
f_{25}	5.06×10^2	4.83×10^2	4.84×10^2	5.51×10^2	5.16×10^2	$\mathbf{4.80 \times 10^2}$	4.84×10^2	4.82×10^2	5.24×10^2	5.16×10^2	5.31×10^2	5.69×10^2
f_{26}	2.27×10^3	2.44×10^3	2.49×10^3	$\mathbf{2.15 \times 10^3}$	2.34×10^3	2.28×10^3	2.40×10^3	2.42×10^3	2.65×10^3	2.56×10^3	2.67×10^3	5.09×10^3
f_{27}	$\mathbf{6.94 \times 10^2}$	7.43×10^2	7.67×10^2	7.46×10^2	7.31×10^2	7.07×10^2	7.75×10^2	7.53×10^2	7.42×10^2	8.28×10^2	8.04×10^2	7.80×10^2
f_{28}	9.06×10^2	5.13×10^3	5.10×10^3	5.29×10^2	$\mathbf{5.27 \times 10^2}$	2.37×10^3	4.60×10^3	5.14×10^3	5.30×10^3	5.67×10^3	5.66×10^3	5.74×10^3
f_{29}	5.90×10^2	1.14×10^3	1.33×10^3	$\mathbf{5.62 \times 10^2}$	5.92×10^2	8.11×10^2	1.13×10^3	1.13×10^3	1.43×10^3	1.80×10^3	1.62×10^3	1.80×10^3
f_{30}	8.55×10^5	1.48×10^6	1.38×10^6	$\mathbf{8.27 \times 10^5}$	9.18×10^5	1.33×10^6	1.29×10^6	1.41×10^6	1.53×10^6	1.59×10^6	1.54×10^6	1.51×10^6
rank	**2.83**	6.45	6.21	3.66	3.79	4.66	4.55	5.97	9.72	8.10	10.38	11.69

5. Conclusions

This paper has devised a stochastic triad topology-based particle swarm optimization (STTPSO) algorithm to solve optimization problems. Specifically, in this optimizer, for each particle, a triad topology is utilized to connect the personal best position of this particle and two other personal best positions randomly selected from those of particles in the current swarm and an additional archive used to store obsolete historical best positions. Then, the best one in the topology and the mean position of the connected triad personal best positions are employed to update each particle. In addition, during the evolution, the triad topology structure of each particle is dynamically updated based on its evolution state. In this way, the learning diversity and learning effectiveness of particles could be largely promoted, so that the swarm could explore and exploit the solution space appropriately. To further improve the swarm diversity, a random restart strategy is proposed by randomly initializing a feasible solution and then inserting into the archive. To alleviate the sensitivity of STTPSO to the acceleration coefficients, a dynamic acceleration coefficient strategy is devised based on the Gaussian distribution. With the above mechanisms, the proposed STTPSO is expected to search the solution space with proper intensification and diversification to achieve promising performance.

Extensive comparative experiments conducted on the CEC 2017 benchmark set with three different dimension sizes have demonstrated the effectiveness of STTPSO. Compared with seven state-of-the-art PSO variants, the proposed STTPSO consistently achieves the best overall performance on the CEC 2017 set with the three dimension sizes. In particular, we find that STTPSO exhibits much better performance than the compared methods regarding complicated optimization problems, such as hybrid problems and composition problems. In addition, the experimental results verified that STTPSO preserves a good scalability to solve optimization problems. In depth investigation on the proposed STTPSO was also conducted to validate the effectiveness of each component in STTPSO. Experimental results demonstrated each component as being of great benefit for STTPSO to achieve good performance.

However, from Tables 3–5, we can see that the results obtained by STTPSO on certain problems remain far from the true optima. Therefore, its optimization performance still requires improvement. In this paper, we adjusted the parameters in STTPSO dynamically without considering the evolution state of particles and the difference between particles. As a result, in future, we will mainly focus on devising adaptive parameter adjustment strategies by considering both the difference between particles and the evolution state of particles to further promote the optimization ability of STTPSO.

Author Contributions: Q.Y.: Conceptualization, supervision, methodology, formal analysis, and writing—original draft preparation. Y.-W.B.: Implementation, formal analysis, and writing—original draft preparation. X.-D.G.: Methodology, and writing—review and editing. D.-D.X.: Writing—review and editing. Z.-Y.L.: Writing—review and editing, and funding acquisition. S.-W.J.: Writing—review and editing. J.Z.: Conceptualization and writing—review and editing. All authors have read and agreed to the published version of the manuscript.

Funding: This work was supported in part by the National Natural Science Foundation of China under Grant 62006124 and U20B2061, in part by the Natural Science Foundation of Jiangsu Province under Project BK20200811, in part by the Natural Science Foundation of the Jiangsu Higher Education Institutions of China under Grant 20KJB520006, in part by the National Research Foundation of Korea (NRF-2021H1D3A2A01082705), and in part by the Startup Foundation for Introducing Talent of NUIST.

Conflicts of Interest: The authors declare no conflict of interest.

References

1. Li, L.; Chang, L.; Gu, T.; Sheng, W.; Wang, W. On the Norm of Dominant Difference for Many-Objective Particle Swarm Optimization. *IEEE Trans. Cybern.* **2021**, *51*, 2055–2067. [CrossRef] [PubMed]
2. Xia, X.; Gui, L.; Yu, F.; Wu, H.; Wei, B.; Zhang, Y.L.; Zhan, Z.H. Triple Archives Particle Swarm Optimization. *IEEE Trans. Cybern.* **2020**, *50*, 4862–4875. [CrossRef] [PubMed]
3. Liu, W.; Wang, Z.; Yuan, Y.; Zeng, N.; Hone, K.; Liu, X. A Novel Sigmoid-Function-Based Adaptive Weighted Particle Swarm Optimizer. *IEEE Trans. Cybern.* **2021**, *51*, 1085–1093. [CrossRef] [PubMed]
4. Yang, Q.; Hua, L.; Gao, X.; Xu, D.; Lu, Z.; Jeon, S.-W.; Zhang, J. Stochastic Cognitive Dominance Leading Particle Swarm Optimization for Multimodal Problems. *Mathematics* **2022**, *10*, 761. [CrossRef]
5. Yang, Q.; Chen, W.-N.; Zhang, J. Probabilistic Multimodal Optimization. In *Metaheuristics for Finding Multiple Solutions*; Springer: Berlin/Heidelberg, Germany, 2021; pp. 191–228.
6. Eberhart, R.; Kennedy, J. A New Optimizer Using Particle Swarm Theory. In Proceedings of the International Symposium on Micro Machine and Human Science, Nagoya, Japan, 4–6 October 1995; pp. 39–43.
7. Shi, Y.; Eberhart, R.C. Empirical Study of Particle Swarm Optimization. In Proceedings of the Congress on Evolutionary Computation, Washington, DC, USA, 6–9 July 1999; pp. 1945–1950.
8. Tsekouras, G.E.; Tsimikas, J.; Kalloniatis, C.; Gritzalis, S. Interpretability Constraints for Fuzzy Modeling Implemented by Constrained Particle Swarm Optimization. *IEEE Trans. Fuzzy Syst.* **2018**, *26*, 2348–2361. [CrossRef]
9. Lin, C.; Chen, C.; Lin, C. Efficient Self-Evolving Evolutionary Learning for Neurofuzzy Inference Systems. *IEEE Trans. Fuzzy Syst.* **2008**, *16*, 1476–1490.
10. Yang, Q.; Chen, W.N.; Gu, T.; Jin, H.; Mao, W.; Zhang, J. An Adaptive Stochastic Dominant Learning Swarm Optimizer for High-Dimensional Optimization. *IEEE Trans. Cybern.* **2022**, *52*, 1960–1976. [CrossRef]
11. Zhang, X.; Du, K.J.; Zhan, Z.H.; Kwong, S.; Gu, T.L.; Zhang, J. Cooperative Coevolutionary Bare-Bones Particle Swarm Optimization with Function Independent Decomposition for Large-Scale Supply Chain Network Design with Uncertainties. *IEEE Trans. Cybern.* **2020**, *50*, 4454–4468. [CrossRef]
12. Chen, W.N.; Tan, D.Z.; Yang, Q.; Gu, T.; Zhang, J. Ant Colony Optimization for the Control of Pollutant Spreading on Social Networks. *IEEE Trans. Cybern.* **2020**, *50*, 4053–4065. [CrossRef]
13. Ge, Q.; Guo, C.; Jiang, H.; Lu, Z.; Yao, G.; Zhang, J.; Hua, Q. Industrial Power Load Forecasting Method Based on Reinforcement Learning and PSO-LSSVM. *IEEE Trans. Cybern.* **2022**, *52*, 1112–1124. [CrossRef]
14. Cao, Y.; Zhang, H.; Li, W.; Zhou, M.; Zhang, Y.; Chaovalitwongse, W.A. Comprehensive Learning Particle Swarm Optimization Algorithm With Local Search for Multimodal Functions. *IEEE Trans. Evol. Comput.* **2019**, *23*, 718–731. [CrossRef]
15. Liang, J.J.; Suganthan, P.N. Dynamic Multi-swarm Particle Swarm Optimizer. In Proceedings of the IEEE Swarm Intelligence Symposium, Pasadena, CA, USA, 8–10 June 2005; pp. 124–129.
16. Mirjalili, S.; Lewis, A. Obstacles and Difficulties for Robust Benchmark Problems: A Novel Penalty-based Robust Optimisation Method. *Inf. Sci.* **2016**, *328*, 485–509. [CrossRef]
17. Liang, J.J.; Qin, A.K.; Suganthan, P.N.; Baskar, S. Comprehensive Learning Particle Swarm Optimizer for Global Optimization of Multimodal Functions. *IEEE Trans. Evol. Comput.* **2006**, *10*, 281–295. [CrossRef]
18. Gong, Y.-J.; Li, J.-J.; Zhou, Y.; Li, Y.; Chung, H.S.-H.; Shi, Y.-H.; Zhang, J. Genetic Learning Particle Swarm Optimization. *IEEE Trans. Cybern.* **2015**, *46*, 2277–2290. [CrossRef] [PubMed]
19. Zhang, X.; Liu, H.; Zhang, T.; Wang, Q.; Wang, Y.; Tu, L. Terminal Crossover and Steering-based Particle Swarm Optimization Algorithm with Disturbance. *Appl. Soft Comput.* **2019**, *85*, 105841. [CrossRef]
20. Liu, Z.; Nishi, T. Strategy Dynamics Particle Swarm Optimizer. *Inf. Sci.* **2022**, *582*, 665–703. [CrossRef]
21. Kennedy, J.; Mendes, R. Population Structure and Particle Swarm Performance. In Proceedings of the Congress on Evolutionary Computation, Honolulu, HI, USA, 12–17 May 2002; pp. 1671–1676.
22. Clerc, M. Beyond Standard Particle Swarm Optimisation. *Int. J. Swarm Intell. Res.* **2010**, *1*. [CrossRef]
23. Xia, X.; Gui, L.; He, G.; Wei, B.; Zhang, Y.; Yu, F.; Wu, H.; Zhan, Z.-H. An Expanded Particle Swarm Optimization Based on Multi-exemplar and Forgetting Ability. *Inf. Sci.* **2020**, *508*, 105–120. [CrossRef]
24. Karim, A.A.; Isa, N.A.M.; Lim, W.H. Modified Particle Swarm Optimization with Effective Guides. *IEEE Access* **2020**, *8*, 188699–188725. [CrossRef]
25. Lynn, N.; Suganthan, P.N. Heterogeneous Comprehensive Learning Particle Swarm Optimization with Enhanced Exploration and Exploitation. *Swarm Evol. Comput.* **2015**, *24*, 11–24. [CrossRef]
26. Yue, C.; Qu, B.; Liang, J. A Multiobjective Particle Swarm Optimizer Using Ring Topology for Solving Multimodal Multiobjective Problems. *IEEE Trans. Evol. Comput.* **2018**, *22*, 805–817. [CrossRef]
27. Chakraborty, A.; Ray, K.S.; Dutta, S.; Bhattacharyya, S.; Kolya, A. Species Inspired PSO based Pyramid Match Kernel Model (PMK) for Moving Object Motion Tracking. In Proceedings of the Fourth International Conference on Research in Computational Intelligence and Communication Networks, Kolkata, India, 22–23 November 2018; pp. 152–157.
28. Nianyin, Z.; Zidong, W.; Weibo, L.; Hong, Z.; Kate, H.; Xiaohui, L. A Dynamic Neighborhood-based Switching Particle Swarm Optimization Algorithm. *IEEE Trans. Cybern.* **2020**, *8*, 701–717.
29. Vazquez, J.C.; Valdez, F. Fuzzy Logic for Dynamic Adaptation in PSO with Multiple Topologies. In Proceedings of the 2013 Joint IFSA World Congress and NAFIPS Annual Meeting, Edmonton, AB, Canada, 24–28 June 2013; pp. 1197–1202.

30. Liu, Q.; Wei, W.; Yuan, H.; Zhan, Z.-H.; Li, Y. Topology Selection for Particle Swarm Optimization. *Inf. Sci.* **2016**, *363*, 154–173. [CrossRef]
31. Lin, W.; Bo, Y.; Jeff, O. Particle Swarm Optimization Using Dynamic Tournament Yopology. *Appl. Soft Comput.* **2016**, *48*, 584–596.
32. Xia, X.; Gui, L.; Zhan, Z.-H. A Multi-swarm Particle Swarm Optimization Algorithm based on Dynamical Topology and Purposeful Detecting. *Appl. Soft Comput.* **2018**, *67*, 126–140. [CrossRef]
33. Zou, J.; Deng, Q.; Zheng, J.; Yang, S. A Close Neighbor Mobility Method Using Particle Swarm Optimizer for Solving Multimodal Optimization Problems. *Inf. Sci.* **2020**, *519*, 332–347. [CrossRef]
34. Parrott, D.; Xiaodong, L. Locating and Tracking Multiple Dynamic Optima by a Particle Swarm Model Using Speciation. *IEEE Trans. Evol. Comput.* **2006**, *10*, 440–458. [CrossRef]
35. Cervantes, A.; Galvan, I.M.; Isasi, P. AMPSO: A New Particle Swarm Method for Nearest Neighborhood Classification. *IEEE Trans. Syst. Man Cybern. Part B Cybern.* **2009**, *39*, 1082–1091. [CrossRef]
36. Janson, S.; Middendorf, M. A Hierarchical Particle Swarm Optimizer and Its Adaptive Variant. *IEEE Trans. Syst. Man Cybern. Part B Cybern.* **2005**, *35*, 1272–1282. [CrossRef]
37. Zhan, Z.; Zhang, J.; Li, Y.; Shi, Y. Orthogonal Learning Particle Swarm Optimization. *IEEE Trans. Evol. Comput.* **2011**, *15*, 832–847. [CrossRef]
38. Chen, S.; Hong, X.; Harris, C.J. Particle Swarm Optimization Aided Orthogonal Forward Regression for Unified Data Modeling. *IEEE Trans. Evol. Comput.* **2010**, *14*, 477–499. [CrossRef]
39. Yang, Q.; Chen, W.N.; Yu, Z.; Gu, T.; Li, Y.; Zhang, H.; Zhang, J. Adaptive Multimodal Continuous Ant Colony Optimization. *IEEE Trans. Evol. Comput.* **2017**, *21*, 191–205. [CrossRef]
40. Yang, Q.; Chen, W.N.; Li, Y.; Chen, C.L.P.; Xu, X.M.; Zhang, J. Multimodal Estimation of Distribution Algorithms. *IEEE Trans. Cybern.* **2017**, *47*, 636–650. [CrossRef] [PubMed]
41. Yang, Q.; Li, Y.; Gao, X.-D.; Ma, Y.-Y.; Lu, Z.-Y.; Jeon, S.-W.; Zhang, J. An Adaptive Covariance Scaling Estimation of Distribution Algorithm. *Mathematics* **2021**, *9*, 3207. [CrossRef]
42. Wei, F.F.; Chen, W.N.; Yang, Q.; Deng, J.; Zhang, J. A Classifier-Assisted Level-Based Learning Swarm Optimizer for Expensive Optimization. *IEEE Trans. Evol. Comput.* **2020**, *25*, 219–233. [CrossRef]
43. Yang, Q.; Chen, W.N.; Gu, T.; Zhang, H.; Yuan, H.; Kwong, S.; Zhang, J. A Distributed Swarm Optimizer with Adaptive Communication for Large-Scale Optimization. *IEEE Trans. Cybern.* **2020**, *50*, 3393–3408. [CrossRef]
44. Wu, G.; Mallipeddi, R.; Suganthan, P.N. *Problem Definitions and Evaluation Criteria for The CEC 2017 Competition on Constrained Real-Parameter Optimization*; Technical Report; National University of Defense Technology: Changsha, China; Kyungpook National University: Daegu, Korea; Nanyang Technological University: Singapore, 2017; pp. 1–16.
45. Shen, Y.; Wei, L.; Zeng, C.; Chen, J. Particle Swarm Optimization with Double Learning Patterns. *Comput. Intell. Neurosci.* **2016**, *2016*, 6510303. [CrossRef]
46. Xie, H.-Y.; Yang, Q.; Hu, X.-M.; Chen, W.N. Cross-generation Elites Guided Particle Swarm Optimization for large scale optimization. In Proceedings of the IEEE Symposium Series on Computational Intelligence, Athens, Greece, 6–9 December 2016; pp. 1–8.
47. Yang, Q.; Chen, W.N.; Gu, T.; Zhang, H.; Deng, J.D.; Li, Y.; Zhang, J. Segment-Based Predominant Learning Swarm Optimizer for Large-Scale Optimization. *IEEE Trans. Cybern.* **2017**, *47*, 2896–2910. [CrossRef]
48. Hesam, V.; Naser, S.H.; Mahsa, S. A Hybrid Generalized Reduced Gradient-based Particle Swarm Optimizer for Constrained Engineering Optimization Problems. *J. Comput. Civ. Eng.* **2021**, *5*, 86–119.
49. Riaan, B.; Engelbrecht, A.P.; van den Bergh, F. A Niching Particle Swarm Optimizer. In Proceedings of the Asia-Pacific Conference on Simulated Evolution and Learning, Orchid Country Club, Singapore, 18–22 November 2002; pp. 692–696.
50. Yousri, D.; Thanikanti, S.B.; Allam, D.; Ramachandaramurthy, V.K.; Eteiba, M.B. Fractional Chaotic Ensemble Particle Swarm Optimizer for Identifying the Single, Double, and Three Diode Photovoltaic Models' Parameters. *Energy* **2020**, *195*, 116979. [CrossRef]
51. Chen, X.; Tianfield, H.; Du, W. Bee-foraging Learning Particle SwarmOptimization. *Appl. Soft Comput.* **2021**, *102*, 107134. [CrossRef]
52. Zhan, Z.-H.; Shi, L.; Tan, K.C.; Zhang, J. A Survey on Evolutionary Computation for Complex Continuous optimization. *Artif. Intell. Rev.* **2021**, *55*, 59–110. [CrossRef]
53. Tao, X.; Li, X.; Chen, W.; Liang, T.; Qi, L. Self-Adaptive Two Roles Hybrid Learning Strategies-based Particle Swarm Optimization. *Inf. Sci.* **2021**, *578*. [CrossRef]
54. Xu, G.; Zhao, X.; Wu, T.; Li, R.; Li, X. An Elitist Learning Particle Swarm Optimization with Scaling Mutation and Ring Topology. *IEEE Access* **2018**, *6*, 78453–78470. [CrossRef]
55. Kennedy, J. Small Worlds and Mega-minds: Effects of Neighborhood Topology on Particle Swarm Performance. In Proceedings of the Congress on Evolutionary Computation, Washington, DC, USA, 6–9 July 1999; Volume 1933, pp. 1931–1938.
56. Lin, A.; Sun, W.; Yu, H.; Wu, G.; Tang, H. Global Genetic Learning Particle Swarm Optimization with Diversity Enhancement by Ring Topology. *Swarm Evol. Comput.* **2019**, *44*, 571–583. [CrossRef]
57. Turkey, M.; Poli, R. A Model for Analysing the Collective Dynamic Behaviour and Characterising the Exploitation of Population-Based Algorithms. *Evol. Comput.* **2014**, *22*, 159–188. [CrossRef]

58. Zhang, J.; Sanderson, A.C. JADE: Adaptive Differential Evolution with Optional External Archive. *IEEE Trans. Evol. Comput.* **2009**, *13*, 945–958. [CrossRef]
59. Djellali, H.; Ghoualmi, N. Improved Chaotic Initialization of Particle Swarm applied to Feature Selection. In Proceedings of the International Conference on Networking and Advanced Systems, Annaba, Algeria, 26–27 June 2019; pp. 1–5.
60. Watanabe, M.; Ihara, K.; Kato, S.; Sakuma, T. Initialization Effects for PSO Based Storage Assignment Optimization. In Proceedings of the Global Conference on Consumer Electronics, Kyoto, Japan, 12–15 October 2021; pp. 494–495.
61. Wang, C.J.; Fang, H.; Wang, C.; Daneshmand, M.; Wang, H. A Novel Initialization Method for Particle Swarm Optimization-based FCM in Big Biomedical Data. In Proceedings of the IEEE International Conference on Big Data, Santa Clara, CA, USA, 29 October–1 November 2015; pp. 2942–2944.
62. Farooq, M.U.; Ahmad, A.; Hameed, A. Opposition-based Initialization and A Modified Pattern for Lnertia Weight (IW) in PSO. In Proceedings of the IEEE International Conference on INnovations in Intelligent SysTems and Applications, Gdynia, Poland, 3–5 July 2017; pp. 96–101.
63. Guo, J.; Tang, S. An Improved Particle Swarm Optimization with Re-initialization Mechanism. In Proceedings of the International Conference on Intelligent Human-Machine Systems and Cybernetics, Hangzhou, China, 26–27 August 2009; pp. 437–441.

applied sciences

MDPI

Article

Dealing with Uncertainty in the MRCPSP/Max Using Discrete Differential Evolution and Entropy

Angela Hsiang-Ling Chen [1,*], Yun-Chia Liang [2,*] and José David Padilla [2]

1 Department of Industrial and Systems Engineering, Chung Yuan Christian University, Taoyuan 320, Taiwan
2 Department of Industrial Engineering and Management, Yuan Ze University, Taoyuan 320, Taiwan; s1028909@mail.yzu.edu.tw
* Correspondence: achen@cycu.edu.tw (A.H.-L.C.); ycliang@saturn.yzu.edu.tw (Y.-C.L.)

Abstract: In this paper, we investigate the characterization of MRCPSP/max under uncertainty conditions and emphasize managerial ability to recognize and handle positively disruptive events. This proposition is then demonstrated using the entropy approach to find disruptive events and response time intervals. The problem is solved using a resilient characteristic of the three-stage procedure gauged by schedule robustness and adaptivity; the resulting schedule absorbs the impact of an unexpected event without rescheduling during execution. The use of the differential evolution algorithm, known as DDE, in a discrete manner is proposed and evaluated against the best known optima (BKO). Our findings indicate the DDE is effective overall; moreover, compared against the BKO for every stage, the most significant difference is that the stability of the solutions provided by DDE under the three-stage framework proves to be sufficiently robust when practitioners add response times at certain range levels, in this case from 8% to 15%.

Keywords: resilience; uncertainty; MRCPSP/max; entropy; discrete differential evolution (DDE)

Citation: Chen, A.H.-L.; Liang, Y.-C.; Padilla, J.D. Dealing with Uncertainty in the MRCPSP/Max Using Discrete Differential Evolution and Entropy. *Appl. Sci.* **2022**, *12*, 3049. https://doi.org/10.3390/app12063049

Academic Editor: Valentino Santucci

Received: 10 December 2021
Accepted: 15 March 2022
Published: 16 March 2022

1. Introduction

The prevalence of uncertainty has exposed significant weakness and fragility in every business sector. The ubiquity and potential of uncertainty to impact the allocation and utilization of resources have motivated research into operation issues from various perspective, such as by stipulating scheduling policies [1], modeling uncertainty and causality in a project [2], improving decision-making [3], scheduling activities with stochastic durations [4–6] and during resource breakdowns [7], and evaluating resources shared under coalition conditions [8]. The aspects of resilience [9,10] and sustainability [11] have elevated the conventional concept of robustness, which often implies scheduling, with disrupted resource availability and dynamic resource demands [12]. However, common approaches generate initial project schedules that are static and deterministic and often involve the use of the critical path method (CPM) to build a baseline schedule. To ensure safety during the activity, a project manager implements a safety allowance, a project buffer augmented by a percentage of the initially estimated duration, which varies almost exclusively with the manager's experience and proficiency [13].

Recognizing the uncertainty inherent in project planning has induced many research efforts in project scheduling under conditions of uncertainty; see [14–17] for review articles. At the same time, countless efforts have been made to provide solution stability and quality to maintain a safety allowance, revealing a potential trade-off [13]. For example, in [7,18], the authors intentionally controlled the resource-interdependence and durations of the activities to mitigate the effect of time uncertainties. In [19–21], the authors maintained robustness by dealing with the activity starting times and duration tolerance levels. In addition, other effective objectives that represent robustness, such as weighted slack functions, path-based measures, slack variability measures, and combined cost (time) functions, were introduced and applied [22–25].

Nevertheless, as pointed out in [16], there has been relatively less research on robust optimization for the RCPSP in deterministic settings [17]. In our past work [26,27], we found very little research applying the entropy approach to the RCPSP. Only five studies regarding the project scheduling domain could be found; thus, we applied the entropy concept to handle uncertainties in the standard MRCPSP and successfully generated robust schedules with fewer elements required to be considered in the estimation. Furthermore, none of those five studies applied to MRCPSP/max. This realization motivated us to investigate the characterization of MRCPSP/max under uncertainty conditions and emphasized the managerial ability to recognize and handle positively disruptive events. Therefore, we attempted to construct resilient entropy schedules through our three-stage DDE approach. We hope to offer a means for the adaptive capacity of an organization to improve preparedness for dynamic environments and help managers to positively adjust projects to turbulence through the availability of resilient schedules.

The remainder of this paper is arranged as follows. Section 2 reviews uncertainty and resilience in relation to MRCPSP/max and introduces the models and solution concepts used in this paper. Section 3 presents the decision rules for mode assignment and activity lists and the discrete version of differential evolution (DDE) with enhancements in its implementation. Section 4 provides the experimental setup, computation, and analysis of robust makespans on benchmark sets. Lastly, conclusions are drawn in Section 5.

2. Coping with Uncertainty in Project Management

The dynamic behavior of real-world environments results in unanticipated conditions that may limit the implementation of ideal and non-restrictive schedules. Some view this inability to accurately predict (or control) project outcomes as being due to an aggregation of several risk factors, such as project magnitude and scope, the number of employees and suppliers, the amount of hardware and software, the set of work standards and skills, variations in design and engineering estimates, additional time required for rework and unreliable deliveries, and difficulties in assigning tasks or communicating. To enhance the managerial ability to cope with project uncertainties, we begin with a more specific discussion of RCPSP/max, followed by a resilient framework with entropy measures designed to handle uncertainty.

2.1. Problem Models

The classical RCPSP remains a generic model with simple constraints that guide the allocation of limited resources within a project, in which an activity executed in one specific way cannot start before its predecessor is completed. Later, the concept of modes representing various resource sets to be potentially utilized was introduced in [28], and this multimode characteristic extended the model to real industrial cases, which encompass the amount of man/machine resources available to complete a job in smart manufacturing, the skill levels, and the different labor contracts in the workforce required to provide services. Among many other extended models [17], the deterministic single-mode RCPSP/max [29] was modified, allowing minimal and maximal time lags between any two precedence-related activities. The objective of this problem is to assign each activity a start time, while satisfying all temporal and resource constraints within a minimum project makespan.

The multi-mode RCPSP/max (MRCPSP/max) problem consists of $n+2$ activities with the set $V = \{0, 1, \ldots n, n+1\}$, where activity i is to be executed in only one mode $\mu_i \in M$. Depending on the mode μ_i, each activity has a fixed duration or processing time d_{i,μ_i}, which is a non-negative real or integer number. In addition, dummy activities 0 and $n+1$ with $d_{0,\mu_0} = d_{n+1,\mu_{n+1}} = 0$ represent the beginning and the completion of the project, respectively. A start schedule S is an assignment of start times to all activities, i.e., a schedule vector $S = (S_i)$, where S_i represents the start time of activity i and S_0 is assumed to be 0. The end-time of activity i is denoted as C_i. As durations are deterministic and preemption is not allowed, we thus have C_i defined as in Equation (1):

$$S_i + d_{i,\mu_i} = C_i \quad \mu_i \in M,\ i \in V \tag{1}$$

In MRCPSP/max, schedules are subject to two constraints: temporal and resource constraints. Here, temporal constraints restrict the time lag between activities in an activity-on-node (AON) network $N = \langle V,\ E,\ \delta \rangle$, consisting of the node set V, the arc set E, and the arc weight function δ. Without considering the arc weight, the time lag depends on the mode μ_i of activity i and the mode μ_j of activity j $(j \neq i)$, and is either a minimum (maximum) time lag $l^{min}_{i,\mu_i,j,\mu_j}$ $\left(l^{max}_{i,\mu_i,j,\mu_j}\right)$ between the start times of two different activities i and j such that

$$S_i + l^{min}_{i,\mu_i,j,\mu_j} \leq S_j \leq S_i + l^{max}_{i,\mu_i,j,\mu_j} \quad \langle i,j \rangle \in E, \quad \mu_i \in M,\ i \in V \tag{2}$$

When both $l^{min}_{i,\mu_i,j,\mu_j} = 0$ and $l^{max}_{i,\mu_i,j,\mu_j} = 0$, activity j cannot be started before activity i begins. In this definition, time lags connect the start times of two related activities, known as start-to-start time lags. A scheduleS is time- feasible if all the time lag constraints are satisfied at the start times S_i $(i = 0, 1, \ldots, n+1)$. However, in this study, the arc weight denotes a user preference matrix, assigning the minimum and maximum time lags of $\delta^{min}_{i,\mu_i,j,\mu_j} = l^{min}_{i,\mu_i,j,\mu_j}$ and $\delta^{max}_{i,\mu_i,j,\mu_j} = l^{max}_{i,\mu_i,j,\mu_j}$ to each arc <*i*, *j*>. The inclusion of such time lags will lead to cycles in *N*; in a more realistic case, a project manager will consult with the customer about his/her specific requirements and hold a group discussion with team members for implementation. However, these interdependent activities may follow immediately or a few days later.

In terms of resource constraints, let $A(M,\ S,\ t)$ be the set of activities being processed at time instant t for schedule S, and $M = (\mu_i)$ be a mode vector used by activity i. The amount of non-renewable resources k used by activity i denotes $r^{\nu}_{i,\mu_i,k}$ $(i \in V,$ $\mu_i \in M, k \in R^{\nu})$ and renewable resources are denoted as $r^{\rho}_{i,\mu_i,k}$ $(i \in V,\ \mu_i \in M,\ k \in R^{\rho})$. Both are subject to non-renewable and renewable capacities, expressed as R^{ν}_k, and R^{ρ}_k, respectively. Since all non-dummy activities are executed in only one mode for a specific duration, depending on the resources consumed, a schedule S is resource-feasible if Equations (3) and (4) hold.

$$\sum\nolimits_{i \in A(M,S,t)} r^{\rho}_{i,\mu_i,k} \leq R^{\rho}_k \quad \mu_i \in M, \quad k \in R^{\rho}, \quad t \geq 0 \tag{3}$$

$$\sum\nolimits_{i \in A(M,S,t)} r^{\nu}_{i,\mu_i,k} \leq R^{\nu}_k \quad \mu_i \in M, \quad k \in R^{\nu}, \quad t \geq 0 \tag{4}$$

Furthermore, a schedule is called feasible if both time and resources are feasible. Thus, the objective of the deterministic MRCPSP/max scheduling problem is to find a feasible schedule so that the project makespan is defined as the start time of the final dummy activity S_{n+1}, and is minimized as in Equation (5).

$$Min\ S_{n+1} = \max_{i=1,\ldots,n} C_i \tag{5}$$

2.2. Understanding Uncertainty

Project characterization can directly influence a manager's potential response. As such, understanding project vulnerabilities is essential in the detection of uncertainty. Several researchers have proposed the information theory and entropy approach for this area [2,26,27,30–37]. For example, some authors in [26,27,34–36] applied the entropy model as a measure of duration uncertainty or a priority rule for scheduling all activities; others [37] focused on calibrating entropy measures to better estimate uncertainty in activity durations. In this study, we consider the implications of entropy presented in [30], where an appropriate measure of project uncertainty, Schedule Entropy U, borrowed from the theory of information in [38], is mathematically defined as in Equation (6):

$$U = -\sum_i p_i \ln p_i \tag{6}$$

where p_i indicates a discrete set of unfavorable activity probabilities, and the sum is extended over all the unfavorable sets of such a schedule. Since all activities durations have an estimated range between $d_{il} < d < d_{iu}$, in which "l" denotes the "lower bound" and "u" denotes the "upper bound", and these are uniformly distributed in the interval (d_{il}, d_{iu}), the determination of probabilities p_i can be expressed as in Equation (7):

$$p_i = \frac{\Delta_t}{d_{iu} - d_{il}} \tag{7}$$

Furthermore, the amount of time in which potentially disruptive events are within the estimated duration range, and hence still within the control of the project manager—and thus, the set of unfavorable events E_i for every activity—can be obtained from Equation (8),

$$E_i = EFT_{d_i} + (d_{iu} - d_i) - LFT_{d_i} = (d_{iu} - d_i) - l_i \tag{8}$$

where d_{iu} and d_i denote the longest and the most probable duration, respectively, and l_i refers to the float or slack of activity i, implying that it can be delayed without delaying subsequent activities and project completion dates. Additionally, the terms $EFT_{d_{iu}}$ and LFT_{d_i} refer to the earliest finish time with the longest duration and the latest finish time with the most probable duration, respectively.

The entropy U of the project schedule can be thought of as all the individual entropies U_i of activity i. Activity i can be potentially unfavorable or disruptive, if and only if it is critical to the CPM and its actual finish time is beyond the latest finish time. An entropy value for a single activity, i.e., U_i, can be determined as in Equation (9):

$$U_i = -\frac{E_i}{(d_{iu} - d_{il}) \ln\left(\frac{\Delta_t}{d_{iu} - d_{il}}\right)} \tag{9}$$

As such, Equation (10) shows the value for the total entropy in the project.

$$U = -\sum_i U_i \tag{10}$$

Based on these equations, it is understandable that an individual entropy value is subject to the estimation of the range for its duration, i.e., the more significant the difference in the interval (d_{il}, d_{iu}), the greater the manager's perception of the activity's uncertainty and therefore the higher its entropy value. In addition, as shown below, the schedule entropy is directly affected by the order in which activities are scheduled.

Figure 1 depicts the concept of unfavorable events E_i and the relevant time interval Δ_t. The solid gray bar symbolizes a scheduled activity with the most probable duration d_i, whereas the solid yellow bar shows its most prolonged duration d_{iu}. The slack l_i is the time interval between EFT_{d_i} and LFT_{d_i}. The time difference between LFT_{d_i} and $EFT_{d_{iu}}$ is denoted by E_i. The parameter Δ_t is determined by the decision-maker and is dependent on the nature of the project. Riskier projects require lower values for Δ_t, which acts as a checkpoint for the project manager to update the project's status and take control of its progress.

In this study, entropy is used in scheduling to handle disruptive activities, as introduced in [30]. The main purpose of this method is to determine disruptive events and response time intervals; this can be done through the use of available information such as activity durations and dependencies. In practice, the relevant time interval Δ_t refers to the period of detection and activation, recognized as event awareness from the managerial perspective. It is essential for managers to adjust positively to the impact of possible adverse events. A project manager sets up checkpoints to detect potential threats and keep the project scope and expected outcome as intact as possible. The higher the project stakes, or the more unstable the development environment, the greater the need for more frequent checkpoints. Since uncertain durations, resource requests, and capacities may likewise not be constant during a project's lifespan, Δ_t reflects how a manager perceives a disruptive event and takes the initiative if an incident affects the progress of the schedule. Once a decision is reached at each checkpoint, the options available for reaching the subsequent checkpoint decrease, thus reducing uncertainty.

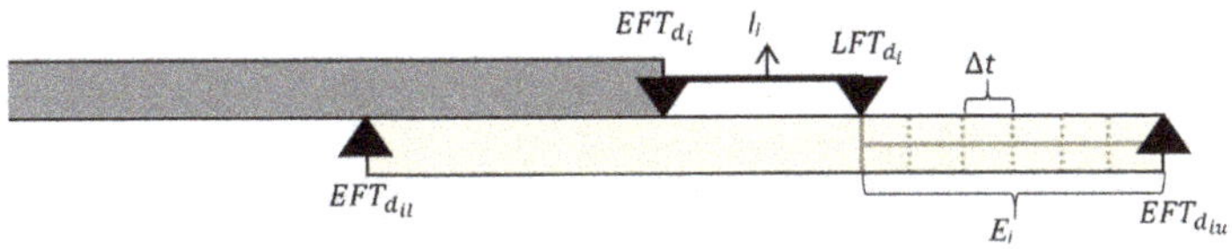

Figure 1. Graphical representation of E_i and Δ_t.

2.3. A Resilient Approach to Uncertainty

As discussed in the previous sections, project managers or decision-makers must assess different controls and operating conditions. A final project schedule embodying trade-offs among various aspects reflects a stable state, characterized by its makespan, cost, risk, and net present value. However, in the presence of disruptive events, the control of activity progress may be broken, resulting in cost overruns, makespan tardiness, performance degradation, or even project failure. Therefore, this section focuses on establishing a desirable algorithm characteristic called resilience. The authors in [39] defined resilience as a process of aligning a set of adaptive capacities to provide a positive course of functioning and adaptation after a disturbance. Given changing project constraints, variables, and structures, decision-makers must adapt their preferences or objectives to arrive at another stable state. In this sense, more than one stable condition exists for project scheduling problems. In [40,41], the term robustness is defined as "the ability (of a schedule) to cope with small-time increments in some activities resulting from uncontrollable factors." Thus, resilience is the magnitude of the disruptive event absorbed before a schedule degrades to the threshold, i.e., the minimum acceptable level of individual activity performance [9].

Researchers have worked on developing an algorithm that measures how the schedule robustness of scheduled activities deviates from the actual occurrences through a resilient scheduling algorithm. Lambrechts et al. [7] determined the expected increase in activity duration due to resource breakdown, proposing a buffer time to prevent schedule disruptions. In [42], the scenario-based proactive robustness optimization (SBPRO) method was developed using the critical-chain project management (CCPM) method. Moradi and Shadrokh [8] applied the CCPM and considered only renewable resources at the cost of recruiting additional resources. Balouka and Cohen [43] sought to minimize the worst-case project duration by deciding on activity modes, resource allocations, and a schedule baseline. In [44], the authors included multiple alternative execution modes and allowed the switching of possibilities between different modes for the same activity during scheduled construction projects. These studies focused on procedures designed to build a robust schedule through the use of time or resource buffers. Their procedures reflect the controllable flexibility needed to produce an incremental solution based on the subsequent revelation of contingent events.

Other researchers have developed optimization models for robustness measures (RMs). For instance, Chtourou and Haouari [45] proposed different slack-based models to predict a schedule's robustness in relation to the single-mode RCPSP. The authors in [9] addressed resilience in mean-variance models having two types of ratios, i.e., the average interval to activity duration and the free slack-to-activity duration. Finally, Milat et al. [10] improved resilience by maximizing free floats as the degree of perturbations absorbed rises. Their study depicted resilience through the alternative measure expressed in the objective function, which maximizes the highest proportion of resource-technology free floats for the activities early in the schedule.

In this study, after handling the issue of schedule uncertainty via entropy, we have aimed to achieve project resilience by extending the robustness model in [45] from the single-mode RCPSP to the multimode RCPSP/max, subjecting it to the constraints presented in [46]. This objective function indicates the relationships between each activity's precedence, resource usage, and slack in each available mode to maximize the robustness measures. The mathematical model designed to maximize the schedule's robustness is shown in Equation (11):

$$Max\ Z = \sum_{\mu_i}\sum_{i}\left(min\left(l_{i,\mu_i},\ \left(frac * d_{i\mu_i}\right)\right) * N_{succ_i} * \sum_{k} r^{\rho}_{i,\mu_i,k} \right) \tag{11}$$

where *frac* represents a threshold (%) of activity duration ($0 < frac < 1$), $d_{i\mu_i}$ refers to the duration of activity i executed in mode μ_i, N_{succ_i} denotes the number of immediate successors of activity i, and l_{i,μ_i} expresses the slack of activity i if executed in mode μ_i. As previously mentioned, the free slack is determined by $LFT_{d_i} - EFT_{d_i}$, or $LST_{d_i} - EST_{d_i}$.

3. Methods

In this study, we aimed to minimize the project makespan while maximizing its robustness for an optimal sequence of activities. The schedule-generating procedure begins by evaluating the benchmark instances' feasibility. An infeasibility is observed when a schedule with a mode combination consumes more non-renewable resources than the total amount available or its total completion time exceeds the required target. Otherwise, an instance is considered feasible, and, once selected, the process moves forward to the following three stages for baseline schedules.

First, Stage I produces a minimized makespan schedule using an optimization algorithm. Stage II uses Stage I's schedule as an input to generate an entropy-based upper-bound makespan schedule. Finally, Stage III generates maximized-robustness schedules with a makespan between Stage I and Stage II. The pseudo-code (Algorithm 1) for the execution of the method is as follows:

The idea is to enhance resilience in the schedule generation scheme. In this case, although the initial schedule (baseline) in Stage I seeks solely to minimize the makespan, this will leave no room for unexpected events that will almost certainly happen. On the other hand, in Stage II, the entropy-based schedule may render the project infeasible even before it begins. Thus, the resulting makespans from Stages I and II serve as lower- and upper-bound values. With this range of values, this schedule generation scheme absorbs the impact of unexpected events without rescheduling during execution. The progress of the makespan and robustness at each stage is conceptualized in Figure 2.

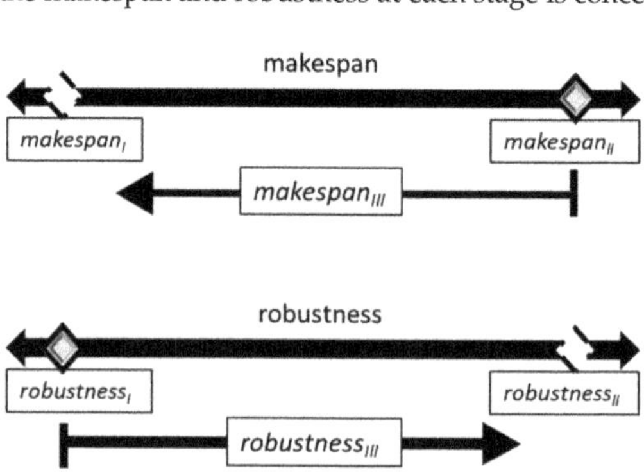

Figure 2. Graphical representation of the three-stage procedure in terms of makespan and robustness.

Algorithm 1: Repeat until all feasible instances are solved

```
Stage 1: Minimize Makespan (Target Makespan/makespan_I)
    Initialization Phase
    While i < population size (Np)
    Evaluate Mode Selection Rules (MSR)
    Evaluate Activity Priority Rules (APR)
    End
  Discrete Differential Evolution Algorithm
  End
  Stage 2: Determine Schedule's Entropy (Upper Bound Makespan/makespan_II)
    Initialization Phase
        While i < population size (Np)
              Evaluate activity risk and set checkpoint frequency
    Determine Unfavorable events
    Determine Event Entropies
    End
  Compute Schedule Entropy
  End
  Stage 3: Maximize Robustness (Robustness Measure/makespan_III)
    Initialization Phase
    If makespan > makespan_II, then
    Reject initial solution
    End if
    While i < population size (Np)
    Evaluate Mode Selection Rules (MSR)
    Evaluate Activity Priority Rules (APR)
    End
  Discrete Differential Evolution Algorithm
  End
  End
```

Part of the complexity involved when solving the MRCPSP/max relies upon selecting the execution modes and determining the order in which to execute activities. In this study, we consider the mode selection rules and activity priority rules used by Chen et al. [47] to determine the best execution mode for every activity and the order in which the activities are executed. The schedules are produced by means of a serial generation scheme (SGS) and improved by means of a discrete differential evolution algorithm.

3.1. Discrete Differential Evolution Algorithm

Differential evolution is an evolutionary-type, population-based algorithm to optimize functions over continuous solution spaces [48]. Characterized by simplicity, straightforwardness, and robustness, numerous applications have been developed to solve combinatorial optimization problems [49,50], such as the machine scheduling problem in production (MSPP) [51–60], the traveling salesman problem (TSP) [61–63], the linear ordering problem (LOP) [61,64], the multidimensional two-way number partitioning problem (MDTWNPP) [65], and the multidimensional knapsack problem (MKP) [66]. In [64,65], the authors addressed permutation-based optimization problems and proposed an algebraic structure and a binary operator that allowed the solutions to be directly expressed as permutations. The duality of geometric search operators was introduced in [61] for both continuous and combinatorial problems. Furthermore, in [67], angle modulation, a trigonometric base (i.e., a sin/cos function) technique, was developed to generate a bit string from continuous to binary problem spaces. A set-based encoding scheme that redefined all algorithmic operators for the discrete space was applied to TSP in [63]. In [66], the authors emphasized a selection operator based on the multiple probability estimation models and verified its usage in continuous and combinatorial problems.

Moreover, due to the interdependency between variables, the model's performance may be compromised in binary- or permutation-based problems. Nonetheless, the precedence relationship is fundamental to project scheduling, making the concept of decoding (i.e., converting continuous encoding vectors into permutations of activity lists) more complex than other approaches currently in use. Furthermore, when dealing with possible execution modes for activities in RCPSP, one needs to consider the task sequence and its appropriate mode almost in parallel. Only a few studies have examined either single- or multimode RCPSP [14,68–74], and even fewer have addressed the MRCPSP/max format, in which the encoding scheme is only visible from one viewpoint. Thus, in this paper, we propose the application of DDE to solve the MRCPSP/max. In our DDE method, the encoding of the DDE algorithm consists of two vectors, one representing the task sequence and another with the execution mode for each activity. Several potential resource conflicts and precedence constraints are taken into account in our design.

Conventional DE works in two phases: initialization and evolution. In the initialization phase, the population $S^g = \left\{X_i^g : i = 0, 1, \ldots N_p - 1\right\}$ at each generation g for the size of N_p contains candidate solutions (i.e., schedules). As shown in Equation (12), each solution consists of D-dimensional parameter vectors $X_i^g = \left\{x_{i,j}^g : j = 0, 1, \ldots D - 1\right\}$, generated as follows by a uniformly distributed random number $rand\ [0, 1]$.

$$X_i^g = X_{min}^{g-1} + rand\ [0, 1] \cdot \left(X_{max}^{g-1} - X_{min}^{g-1}\right) \quad (12)$$

The search space S^g is constrained by the maximum and minimum bounds $\left(X_{max}^{g-1},\ X_{min}^{g-1}\right)$. X_i^g is instantiated independently and further adjusted throughout the execution of the algorithm. The key is to generate a suitable number of trial parameter vectors to avoid stagnation and provide sufficient solution space for the next phase.

Mutation and crossover operators and population maintenance mechanisms begin their computing schemes in the evolution phase. The classical mutation and crossover operators generate new vectors, whereas the population maintenance mechanism determines which vector will survive the next generation. In this respect, a target vector X_i^g refers to a parent vector from the current generation g, whereas a mutant vector M_i^g obtained through the differential mutation operation, is called the donor vector. The offspring formed by recombining donor and target vectors are called trial vectors, denoted as T_i^g.

To show the discretization of the DE, i.e., the proposed DDE algorithm, consider a multimode project composed of six activities, with their precedence relationships shown in Table 1. The encoding of the DDE algorithm consists of two vectors, one representing the task sequence and another with the execution mode for each activity. For example, the first activity to be executed is activity 1, in mode 2, followed by activity 3 in mode 1, and so on. The mutation and crossover operations will not change the selected mode for each activity for this particular example, thus obviating the need to deal with infeasible solutions later on.

Table 1. The precedence relationships of the example with six activities.

Activity	Predecessor
0	-
1	0
2	0
3	1
4	2
5	3
6	4, 5
7	5, 6

3.2. Mutation

Unlike the genetic algorithm (GA), DDE's main search component to optimize solutions is a mutation, not a crossover. To mutate a solution, the DDE constructs the first population with N_p members and randomly selects three different feasible solutions, named X_0^{g-1}, X_1^{g-1}, X_2^{g-1}, where g denotes the generation number. Next, the mutant vector M_i^g is determined based on the scaled difference of any two of the three solution vectors, shown in Equation (13).

$$M_i^g = X_0^{g-1} + F{\cdot}rand_i^g\left(X_1^{g-1} - X_2^{g-1}\right) \quad (13)$$

where F refers to a scaling factor, a positive number that controls the directional hop length of two vectors. In this example, a value of 1.5 was selected arbitrarily, and the value of $rand_i^g$ falls between 0 and 1. Table 2 shows how mutation works in the proposed DDE. First, three feasible solutions, X_0^1, X_1^1, X_2^1, are randomly selected, and a vector with random numbers is provided, i.e., $rand_1^2$. As a numerical example, consider the second value (3.30), obtained by $3 + 1.5 \times 0.2(2 - 1) = 3.30$.

Table 2. An illustrative example of a mutant vector creation.

Sequence of Tasks	1	2	3	4	5	6
Solution X_0^1	1	3	5	2	4	6
Solution X_1^1	1	2	3	5	4	6
Solution X_2^1	2	1	4	3	5	6
$rand_1^2$	0.30	0.20	1.00	0.30	0.21	0.10
Mutated Vector M_1^2	0.55	3.30	3.50	2.90	3.69	6.00

3.3. Crossover

After mutation, the crossover operation is executed to enhance the population diversity. The mutant vector M_i^g exchanges its parameter with the target vector X_3^{g-1} selected from the current population. As a result, a trial vector, T_i^g, is determined using the following scheme in Equation (14):

$$T_i^g = \begin{cases} M_i^g \ \ if\left(rand_i^g \le C_r\right) \\ X_3^{g-1} \ \ Otherwise \end{cases} \quad (14)$$

The probability of crossover C_r acts as a control parameter of DDE, and its value ranges between 0 and 1. If $rand_i^g \le C_r$, the trial vector gets its value from the corresponding dimension of the newly generated mutant vector M_i^g. Otherwise, it is copied from the current vector X_3^{g-1}. Table 3 shows the target vector used for the crossover and the random numbers $rand_1^2$ generated to compare C_r; in our example, $C_r = 0.2$. The trial vector is then created, as shown in Table 3. For the value of $rand_1^2$ fewer than $C_r = 0.2$, elements of task 2 (i.e., activity 1) and task 6 (i.e., activity 6) from the mutant vector M_1^2 are copied, whereas tasks 1, 3, 4, and 5 from the target vector are used.

Table 3. An illustrative example of the target vector, the trial vector, and the decoded vector.

Sequence of Tasks	1	2	3	4	5	6
Target Vector X_3^1	2	1	3	5	4	6
Mutated Vector M_1^2	0.55	3.30	3.50	2.90	3.69	6.00
$rand_1^2$	0.40	0.14	0.90	0.85	1.00	0.02
Trial Vector T_1^2	2	3.30	3	5	4	6.00
Decoded Vector DT_1^2	1	3	2	5	4	6

Finally, to determine the task execution sequence, i.e., the permutation of activities, the values obtained in the trial vector are sorted, always satisfying the precedence constraints. For this example, the first task to be scheduled is selected between tasks 1 and 2. Since task 1 has a lower value than task 2 (2 < 3.3), task 1 is scheduled first. Next, the tasks with scheduling priority are tasks 2 and 3. Task 3 attains the second position because 3 < 3.3. Then, task 2 competes with task 5, which has a value of 4, for the third position. Task 2 wins the place, since 3.30 < 4, and so on until all the tasks are scheduled. For this particular decoded solution, the execution sequence of activities is 1-3-2-5-4-6, as illustrated in the last row of Table 3.

3.4. Selection

In this operation, the new solution DT_i^g is compared with the target vector X_i^g according to their fitness values. The vector with better fitness will survive into the next generation as in Equation (15).

$$X_i^g = \begin{cases} DT_i^g & if\ f\left(DT_i^g\right) \leq f\left(X_i^{g-1}\right) \\ X_i^{g-1} & Otherwise \end{cases} \tag{15}$$

Since the objective in this study is to minimize the makespan (at stage I) or maximize robustness (at stage III), if the new solution yields an equal or better objective value, it replaces the corresponding target vector in the next generation. Otherwise, the target vector is retained in the population. Once the population is updated, the evolution procedure is repeated until a predefined termination criterion is reached.

4. Results and Discussion

This section presents the results of the methodology introduced above in relation to the more complex MRCPSP/max. Experiments were conducted to evaluate the practicality and efficiency of solving the test instances generated in [75]. There are three benchmark sets with different activities (30, 50, and 100 activities); each set contains 270 instances, and every instance uses three renewable, three non-renewable, and three doubly constrained resources. Furthermore, except for the dummy activities (initial and final) with only one execution mode and with no duration and no resource consumption, every activity can include three, four, or five different execution modes. In the current study, the best-known optima (BKO) are not compared against the makespan, but against the artificial bee colony (ABC) results obtained in a previous study [47].

4.1. Parameter Settings

The parameters used in this study were set based on [47,68]. Sensitivity analyses were performed to select the best values for the relevant time interval, $\Delta_t = 1$, and *frac* = 0.25. For the algorithmic settings in Table 4, the population size (N_p) was 40, F (scaling factor) was 1.5, and C_r was 0.2 in the DDE. For the ABC in Table 5, the population size was 30, with an abandonment limit of 5, and MNC was 20.

Table 4. Parameter settings for the DDE algorithms.

Parameter	Setting
Population Size (N_p)	40
Scaling Factor (F)	1.5
Probability of Crossover (C_r)	0.2
Δ_t	1
frac	0.25

Table 5. Parameter settings for the ABC algorithms.

Parameter	Setting
Population Size (N_p)	30
Abandonment Limit	5
Maximum Number of Cycles (MNC)	20
Δ_t	1
frac	0.25

4.2. Computational Results

All experiments were carried out using an Intel i7 personal computer with 8 GB of RAM, and the problem was coded using MATLAB. The DDE iterates through the total problem instances available for each set and randomly selects a predetermined number of feasible instances. Table 6 shows the optimal solutions found and the average runtime for each benchmark when running each algorithm. For all algorithms, increasing the number of activities elevates the problem's complexity. Hence, the average run times increase, whereas the total numbers of optima decrease. Furthermore, the results indicate that the average runtime of DDE was slightly higher than that of ABC running all MRCPSP/max benchmarks. On average, the DDE algorithm takes 20.556 s to find a schedule with optimized robustness, whereas the takes 20.329 s. However, DDE obtains slightly higher numbers of optima than ABC. Thus, we can conclude that DDE is more effective than ABC in this regard.

Table 6. The average runtime per MRCPSP/Max benchmark set for each algorithm.

Benchmark Set	Optima Found (No.)		Average Run Time (s)	
	ABC	DDE	ABC	DDE
MM30	260	263	11.888	12.189
MM50	123	124	17.063	17.223
MM100	84	87	32.037	32.257

Furthermore, Table 7 presents the results obtained from evaluating all 270 instances of every benchmark set. The target, entropy-based, and resilient schedules in Stage I, II, and III are referred to as S1, S2, and S3, respectively. They were assessed based on two measures: the makespan (Avg. Dev.) and the robustness (Avg. RM.). Avg. Dev. refers to the average of all the deviations computed by $(M_s - \text{BKO})/\text{BKO}$. M_s denotes the schedule makespan at the current stage, whereas the best-known makespan (BKO) represents the reference makespan, i.e., the optimal solution when comparing the target schedule of Stage I and the upper-bound schedule of Stage III. Furthermore, Avg. RM. is a measure of average robustness. Finally, the algorithmic performance was evaluated for three benchmark sets (i.e., MM30, MM50, and MM100) against the ABC.

The three-stage procedure proved to be robust enough to produce results comparable to different optimization algorithms. These results are encouraging, given that practitioners can add anywhere between 8% to 15% of their original estimates as response time intervals (i.e., buffer times). Furthermore, both ABC and DDE algorithms performed better when considering that response time intervals used by practitioners rely primarily on intuition and experience. On the other hand, this three-stage procedure relies solely on information available to every project, including activity durations and precedence.

Meanwhile, Figures 3–5 show the results of the average deviation of every benchmark instance when compared against the BKO for every stage, divided according to the optimization algorithm applied. In addition, Figures 6–8 show the robustness measures obtained at each stage using the DDE and ABC algorithms on different sizes of benchmark instances, i.e., MM30, MM50, and MM100.

Table 7. Summary results for the MRCPSP/Max benchmark evaluations for each algorithm.

Stage	Measure	MM30		MM50		MM100	
		ABC	DDE	ABC	DDE	ABC	DDE
S1	Avg. Dev.	0.00176	0.00580	0.04571	0.03104	0.04424	0.04031
	Std. Dev.	0.00664	0.01952	0.03946	0.06002	0.03257	0.04090
	Avg. RM.	102.75556	116.62593	116.62593	117.31481	117.39630	115.85185
S2	Avg. Dev.	0.09690	0.09524	0.10132	0.09640	0.08497	0.07570
	Std. Dev.	0.05793	0.08394	0.05917	0.08180	0.04307	0.05348
	Avg. RM.	132.72593	131.86296	133.81481	136.22593	137.12963	134.46670
S3	Avg. Dev.	0.05041	0.02491	0.05387	0.03711	0.04373	0.04259
	Std. Dev.	0.04794	0.04856	0.05220	0.05615	0.04067	0.04371
	Avg. RM.	100.62593	123.80370	124.45926	125.99259	127.43704	124.52593

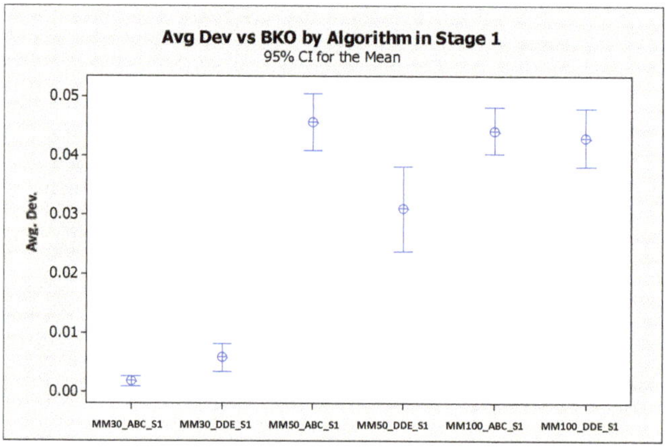

Figure 3. Avg dev vs. BKO CI for each algorithm in Stage 1 (MRCPSP/max).

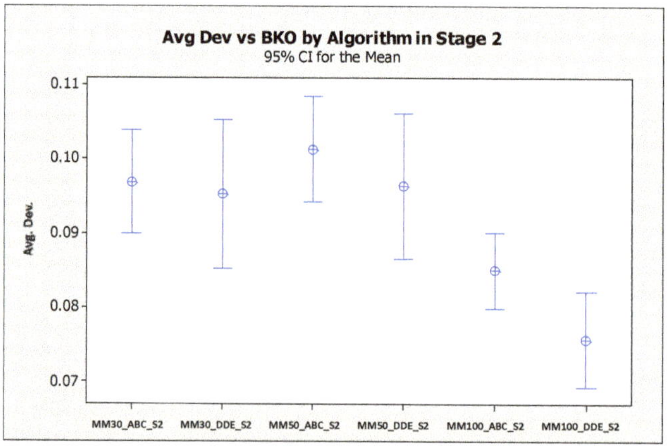

Figure 4. Avg dev vs. BKO CI for each algorithm in Stage 2 (MRCPSP/max).

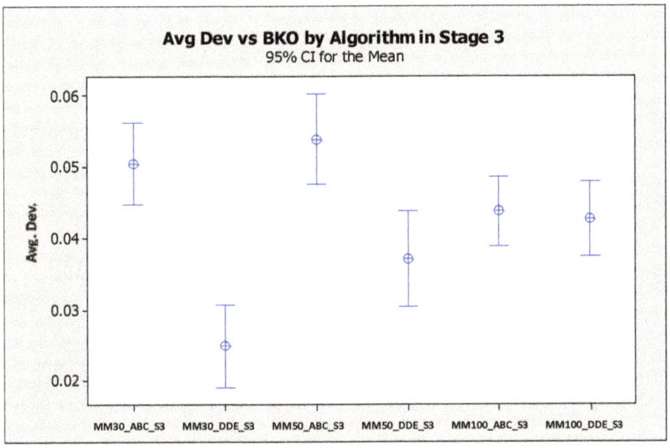

Figure 5. Avg dev vs. BKO CI for each algorithm in Stage 3 (MRCPSP/max).

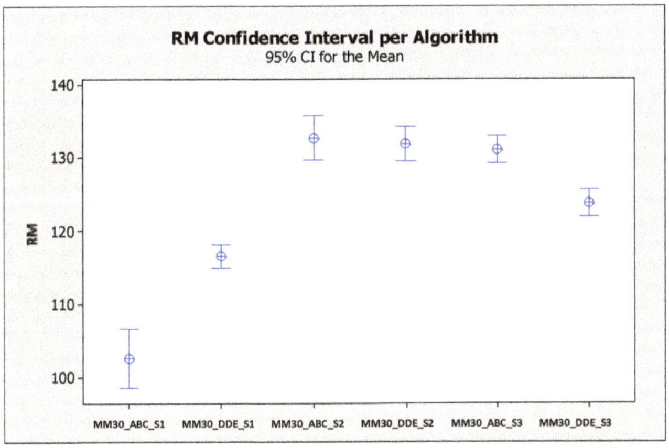

Figure 6. Robustness measure CI for each algorithm and stage for MM30 instances.

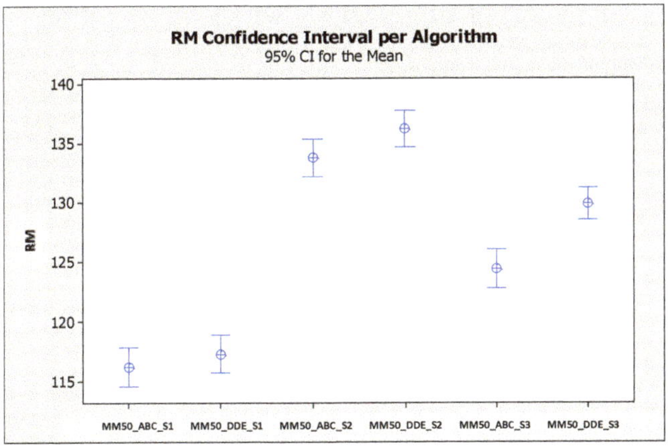

Figure 7. Robustness measure CI for each algorithm and stage for MM50 instances.

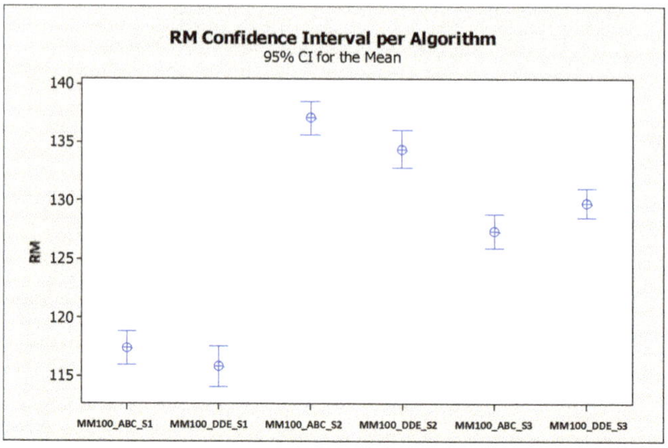

Figure 8. Robustness measure CI for each algorithm and stage for MM100 instances.

Compared to the previously obtained results (columns marked with ABC), the most significant difference observed for the obtained solutions is their stability. Though the standard deviation of both algorithms increases with complexity, this increase is lower and slower when using DDE. Furthermore, the methodology remains stable and yields schedules with robust makespans near the best-known optima. In summary, better implementations of optimization algorithms can further improve the performance of the proposed methodology.

5. Conclusions

Uncertainty greatly impacts the dynamic behavior of real-world environments. Adaptive capacity helps to improve preparedness in dynamic environments, and managers must respond effectively to changes in environmental conditions. Researchers and practitioners have sought to optimize schedules and quality in many studies, and improving schedules remains a pressing concern. In this paper, we tried to resolve the multi-mode resource-constrained project scheduling problem (MRCPSP/max), which is not a common domain for the DDE algorithm, and we specifically considered entropy, which helped to deal with uncertainty.

We focus on three main contributions in this paper. Initially, we explored the characterization of MRCPSP/max under uncertainty conditions and confirmed the need for managers to recognize and positively respond to disruptive events. Using entropy to determine disruptive events and response intervals in scheduling, we demonstrated this proposition. Then, we formulated the robustness attribute as a scheduling adaptability maximization problem and a three-stage schedule generation framework to enhance resilience by absorbing the impact of unexpected events, while rescheduling during execution. Our final contribution was a discrete framework for the differential evolution algorithm. In our application of DDE, the encoding of the DDE algorithm consisted of two vectors representing the task sequence and the execution modes for each activity. Several potential resource conflicts and priority constraints were considered in our design. The proposed DDE was evaluated by solving test instances of benchmark sets by comparing its performance to the best known optima (BKO) and the previous application based on the artificial bee colony (ABC) approach.

The findings indicated that, for all algorithms, the problem's complexity influences the number of optima found and the average run time. Overall, a more effective algorithm is the DDE algorithm, as it offers more optimal solutions and a higher number of them. Additionally, we were able to determine when practitioners need to add response time intervals at certain range levels, such as 8% to 15% in this case, to benefit from schedule robustness. Finally, compared to the BKO for every stage, the stability of the solutions provided by the DDE demonstrated its algorithmic advantage in terms of resilience. Unfortunately, the more realistic the academic model is, the more difficult it is to solve the problem; the MRCPSP/max is simply one of the very difficult problems.

Nonetheless, the encouraging computational results may lead to future implications along other lines. First of all, it is interesting to study which features make instances of MRCPSP/max difficult or easy to solve. In this sense, future studies may further enhance scheduling efficiency by examining various criteria for activity prioritization and mode selection. Another interesting line of research is investigating other potential encoding scheme frameworks in order to capture problem-specific aspects. Furthermore, efforts to examine other approaches to dealing with uncertainty in project scheduling and the verification of their performance using real-world data are also necessary.

Author Contributions: Conceptualization, A.H.-L.C., Y.-C.L. and J.D.P.; methodology, Y.-C.L. and J.D.P.; software, J.D.P.; validation, Y.-C.L. and J.D.P.; formal analysis, J.D.P.; writing—original draft preparation, A.H.-L.C., Y.-C.L. and J.D.P.; writing—review and editing, A.H.-L.C. and Y.-C.L.; visualization, Y.-C.L. and J.D.P.; supervision, A.H.-L.C.; project administration, A.H.-L.C.; funding acquisition, A.H.-L.C. All authors have read and agreed to the published version of the manuscript.

Funding: This research was funded by the Ministry of Science and Technology, Taiwan, ROC, grant numbers: MOST 103-2221-E-253-005, and MOST 104-2221-E-253-002.

Conflicts of Interest: The authors declare no conflict of interest.

References

1. Alipouri, Y.; Sebt, M.H.; Ardeshir, A.; Chan, W.T. Solving the FS-RCPSP with hyper-heuristics: A policy-driven approach. *J. Oper. Res. Soc.* **2019**, *70*, 403–419. [CrossRef]
2. Khodakarami, V.; Fenton, N.; Neil, M. Project Scheduling: Improved approach to incorporate uncertainty using Bayesian Networks. *Proj. Manag. J.* **2007**, *38*, 39–49. [CrossRef]
3. Rezakhani, P. Project scheduling and performance prediction: A fuzzy-Bayesian network approach. In *Engineering, Construction and Architectural Management*; Emerald Publishing Limited Publisher: Bingley, UK, 2021.
4. Ballestín, F. When it is worthwhile to work with the stochastic RCPSP? *J. Sched.* **2007**, *10*, 153–166. [CrossRef]
5. Klerides, E.; Hadjiconstantinou, E. A decomposition-based stochastic programming approach for the project scheduling problem under time/cost trade-off settings and uncertain durations. *Comput. Oper. Res.* **2010**, *37*, 2131–2140. [CrossRef]
6. Mokhtari, H.; Aghaie, A.; Rahimi, J.; Mozdgir, A. Project time–cost trade-off scheduling: A hybrid optimization approach. *Int. J. Adv. Manuf. Technol.* **2010**, *50*, 811–822. [CrossRef]
7. Lambrechts, O.; Demeulemeester, E.; Herroelen, W. Time slack-based techniques for robust project scheduling subject to resource uncertainty. *Ann. Oper. Res.* **2011**, *186*, 443–464. [CrossRef]
8. Moradi, H.; Shadrokh, S. A robust scheduling for the multimode project scheduling problem with a given deadline under uncertainty of activity duration. *Int. J. Prod. Res.* **2019**, *57*, 3138–3167. [CrossRef]
9. Xiong, J.; Chen, Y.; Zhou, Z. Resilience analysis for project scheduling with renewable resource constraints and uncertain activity durations. *J. Ind. Manag. Optim.* **2016**, *12*, 719.
10. Milat, M.; Knezic, S.; Sedlar, J. A new surrogate measure for a resilient approach to construction scheduling. *Procedia Comput. Sci.* **2021**, *181*, 468–476. [CrossRef]
11. Askarifard, M.; Abbasianjahromi, H.; Sepehri, M.; Zeighami, E. A robust multi-objective optimization model for project scheduling considering risk and sustainable development criteria. *Environ. Dev. Sustain.* **2021**, *23*, 11494–11524. [CrossRef]
12. Rahman, M.H.F.; Chakrabortty, R.K.; Ryan, M.J. Managing Uncertainty and Disruptions in Resource-Constrained Project Scheduling Problems: A Real-Time Reactive Approach. *IEEE Access* **2021**, *9*, 45562–45586. [CrossRef]
13. Van de Vonder, S.; Demeulemeester, E.; Herroelen, W.; Leus, R. The use of buffers in project management: The trade-off between stability and makespan. *Int. J. Prod. Econ.* **2005**, *97*, 227–240. [CrossRef]
14. Golab, A.; Gooya, E.; Falou, A.; Cabon, M. Review of conventional metaheuristic techniques for resource-constrained project scheduling problem. *J. Proj. Manag.* **2022**, *7*, 95–110. [CrossRef]
15. Habibi, F.; Barzinpour, F.; Sadjadi, S. Resource-constrained project scheduling problem: Review of past and recent developments. *J. Proj. Manag.* **2018**, *3*, 55–88. [CrossRef]
16. Hazır, Ö.; Ulusoy, G. A classification and review of approaches and methods for modeling uncertainty in projects. *Int. J. Prod. Econ.* **2020**, *223*, 107522. [CrossRef]
17. Hartmann, S.; Briskorn, D. An Updated Survey of Variants and Extensions of the Resource-Constrained Project Scheduling Problem. *Eur. J. Oper. Res.* **2022**, *297*, 1–14. [CrossRef]
18. Nansheng, P.; Qichen, M. Resource allocation in robust scheduling. *J. Oper. Res. Soc.* **2022**, 1–18. [CrossRef]

19. Bruni, M.E.; Pugliese, L.D.P.; Beraldi, P.; Guerriero, F. An adjustable robust optimization model for the resource-constrained project scheduling problem with uncertain activity durations. *Omega* **2017**, *71*, 66–84. [CrossRef]
20. Bold, M.; Goerigk, M. A compact reformulation of the two-stage robust resource-constrained project scheduling problem. *Comput. Oper. Res.* **2021**, *130*, 105232. [CrossRef]
21. Wang, Z.; Ng, T.S.; Pang, C.K. Minimizing activity exposures in project scheduling under uncertainty. *Expert Syst. Appl.* **2021**, *173*, 114635. [CrossRef]
22. Kreter, S.; Schutt, A.; Stuckey, P.J.; Zimmermann, J. Mixed-integer linear programming and constraint programming formulations for solving resource availability cost problems. *Eur. J. Oper. Res.* **2018**, *266*, 472–486. [CrossRef]
23. Li, H.; Xiong, L.; Liu, Y.; Li, H. An effective genetic algorithm for the resource levelling problem with generalised precedence relations. *Int. J. Prod. Res.* **2018**, *56*, 2054–2075. [CrossRef]
24. Mahalleh, M.K.K.; Ashjari, B.; Yousefi, F.; Saberi, M. A robust solution to resource-constraint project scheduling problem. *Int. J. Fuzzy Log. Intell. Syst.* **2017**, *17*, 221–227. [CrossRef]
25. Davari, M.; Demeulemeester, E. The proactive and reactive resource-constrained project scheduling problem. *J. Sched.* **2019**, *22*, 211–237. [CrossRef]
26. Chen, A.H.L.; Liang, Y.C.; Padilla, J.D. An entropy-based upper bound methodology for robust predictive multimode RCPSP schedules. *Entropy* **2014**, *16*, 5032–5067. [CrossRef]
27. Chen, A.H.L.; Liang, Y.C.; Padilla, J.D. Using discrete differential evolution and Entropy to solve the MRCPSP. In Proceedings of the 2017 IEEE Congress on Evolutionary Computation (CEC), Donostia, Spain, 5–8 June 2017; pp. 2437–2442.
28. Elmaghraby, S.E. *Activity Networks: Project Planning and Control by Network Models*; John Wiley & Sons: Hoboken, NJ, USA, 1977.
29. Bartusch, M.; Möhring, R.H.; Radermacher, F.J. Scheduling project networks with resource constraints and time windows. *Ann. Oper. Res.* **1988**, *16*, 199–240. [CrossRef]
30. Bushuyev, S.D.; Sochnev, S.V. Entropy measurement as a project control tool. *Int. J. Proj. Manag.* **1999**, *17*, 343–350. [CrossRef]
31. Asllani, A.; Ettkin, L. An entropy-based approach for measuring project uncertainty. *Acad. Inf. Manag. Sci. J.* **2007**, *10*, 31–45.
32. Song, H.; Wu, D.; Li, M.; Cai, C.; Li, J. An entropy based approach for software risk assessment: A perspective of trustworthiness enhancement. In Proceedings of the 2nd International Conference on Software Engineering and Data Mining, Chengdu, China, 23–25 June 2010; pp. 575–578.
33. Tseng, C.C.; Ko, P.W. Measuring schedule uncertainty for a stochastic resource-constrained project using scenario-based approach with utility-entropy decision model. *J. Ind. Prod. Eng.* **2016**, *33*, 558–567. [CrossRef]
34. Chenarani, A.; Druzhinin, E.A. A quantitative measure for evaluating project uncertainty under variation and risk effects. *Eng. Technol. Appl. Sci. Res.* **2017**, *7*, 2083–2088. [CrossRef]
35. Christodoulou, S.E. Entropy-based heuristic for resource-constrained project scheduling. *J. Comput. Civ. Eng.* **2017**, *31*, 04016068. [CrossRef]
36. Qiao, J.; Li, Y. Resource leveling using normalized Entropy and relative Entropy. *Autom. Constr.* **2018**, *87*, 263–272. [CrossRef]
37. Vanhoucke, M.; Batselier, J. A statistical method for estimating activity uncertainty parameters to improve project forecasting. *Entropy* **2019**, *21*, 952. [CrossRef]
38. Shannon, C.E. A mathematical theory of communication. *Bell Syst. Tech. J.* **1948**, *27*, 379–423. [CrossRef]
39. Norris, F.H.; Stevens, S.P.; Pfefferbaum, B.; Wyche, K.F.; Pfefferbaum, R.L. Community resilience as a metaphor, theory, set of capacities, and strategy for disaster readiness. *Am. J. Community Psychol.* **2008**, *41*, 127–150. [CrossRef]
40. Al-Fawzan, M.A.; Haouari, M. A bi-objective model for robust resource-constrained project scheduling. *Int. J. Prod. Econ.* **2005**, *96*, 175–187. [CrossRef]
41. Kobylański, P.; Kuchta, D. A note on the paper by MA Al-Fawzan and M. Haouari about a bi-objective problem for robust resource-constrained project scheduling. *Int. J. Prod. Econ.* **2007**, *107*, 496–501. [CrossRef]
42. Ma, G.; Gu, L.; Li, N. Scenario-based proactive robust optimization for critical-chain project scheduling. *J. Constr. Eng. Manag.* **2015**, *141*, 04015030. [CrossRef]
43. Balouka, N.; Cohen, I. A robust optimization approach for the multimode resource-constrained project scheduling problem. *Eur. J. Oper. Res.* **2019**, *291*, 457–470. [CrossRef]
44. Burgelman, J.; Vanhoucke, M. Project schedule performance under general mode implementation disruptions. *Eur. J. Oper. Res.* **2020**, *280*, 295–311. [CrossRef]
45. Chtourou, H.; Haouari, M. A two-stage-priority-rule-based algorithm for robust resource-constrained project scheduling. *Comput. Ind. Eng.* **2008**, *55*, 183–194. [CrossRef]
46. Talbot, F.B. Resource-constrained project scheduling with time-resource tradeoffs: The non-preemptive case. *Manag. Sci.* **1982**, *28*, 1197–1210. [CrossRef]
47. Chen, A.H.L.; Liang, Y.C.; Padilla, J.D. A practical and robust execution time-frame procedure for the multi-mode resource-constrained project scheduling problem with minimal and maximal time lags. *Algorithms* **2016**, *9*, 63. [CrossRef]
48. Storn, R.; Price, K. Differential evolution–a simple and efficient heuristic for global optimization over continuous spaces. *J. Glob. Optim.* **1997**, *11*, 341–359. [CrossRef]
49. Mohapatra, P.; Roy, S.; Das, K.N.; Dutta, S.; Raju, M.S.S. A review of evolutionary algorithms in solving large scale benchmark optimisation problems. *Int. J. Math. Oper. Res.* **2022**, *21*, 104–126. [CrossRef]

50. Pant, M.; Zaheer, H.; Garcia-Hernandez, L.; Abraham, A. Differential Evolution: A review of more than two decades of research. *Eng. Appl. Artif. Intell.* **2020**, *90*, 103479.
51. Han, Y.; Li, J.; Sang, H.; Liu, Y.; Gao, K.; Pan, Q. Discrete evolutionary multi-objective optimization for energy-efficient blocking flow shop scheduling with setup time. *Appl. Soft Comput.* **2020**, *93*, 106343. [CrossRef]
52. de Fátima Morais, M.; Ribeiro, M.H.D.M.; da Silva, R.G.; Mariani, V.C.; dos Santos Coelho, L. Discrete differential evolution metaheuristics for permutation flow shop scheduling problems. *Comput. Ind. Eng.* **2022**, *166*, 107956. [CrossRef]
53. Öztop, H.; Tasgetiren, M.F.; Kandiller, L.; Pan, Q.K. Metaheuristics with restart and learning mechanisms for the no-idle flowshop scheduling problem with makespan criterion. *Comput. Oper. Res.* **2022**, *138*, 105616. [CrossRef]
54. Yuan, Y.; Xu, H. Flexible job shop scheduling using hybrid differential evolution algorithms. *Comput. Ind. Eng.* **2013**, *65*, 246–260. [CrossRef]
55. Zhao, F.; Shao, Z.; Wang, J.; Zhang, C. A hybrid differential evolution and estimation of distribution algorithm based on neighbourhood search for job shop scheduling problems. *Int. J. Prod. Res.* **2016**, *54*, 1039–1060. [CrossRef]
56. Zhang, G.; Xing, K.; Cao, F. Discrete differential evolution algorithm for distributed blocking flow shop scheduling with makespan criterion. *Eng. Appl. Artif. Intell.* **2018**, *76*, 96–107. [CrossRef]
57. Wu, X.; Liu, X.; Zhao, N. An improved differential evolution algorithm for solving a distributed assembly flexible job shop scheduling problem. *Memetic Comput.* **2019**, *11*, 335–355. [CrossRef]
58. Ali, I.M.; Essam, D.; Kasmarik, K. A novel design of differential evolution for solving discrete traveling salesman problems. *Swarm Evol. Comput.* **2020**, *52*, 100607. [CrossRef]
59. Zhao, F.; Zhao, L.; Wang, L.; Song, H. An ensemble discrete differential evolution for the distributed blocking flow shop scheduling with minimizing makespan criterion. *Expert Syst. Appl.* **2020**, *160*, 113678. [CrossRef]
60. Yuan, S.; Li, T.; Wang, B. A discrete differential evolution algorithm for flow shop group scheduling problem with sequence-dependent setup and transportation times. *J. Intell. Manuf.* **2021**, *32*, 427–439. [CrossRef]
61. Moraglio, A.; Togelius, J.; Silva, S. Geometric differential evolution for combinatorial and programs spaces. *Evol. Comput.* **2013**, *21*, 591–624. [CrossRef] [PubMed]
62. Krömer, P.; Uher, V.; Snášel, V. Novel Random Key Encoding Schemes for the Differential Evolution of Permutation Problems. *IEEE Trans. Evol. Comput.* **2022**, *26*, 43–57. [CrossRef]
63. Liu, Y.; Chen, W.N.; Zhan, Z.H.; Lin, Y.; Gong, Y.J.; Zhang, J. A set-based discrete differential evolution algorithm. In Proceedings of the 2013 IEEE International Conference on Systems, Man, and Cybernetics, Manchester, UK, 13–16 October 2013; pp. 1347–1352.
64. Baioletti, M.; Milani, A.; Santucci, V. Variable neighborhood algebraic differential evolution: An application to the linear ordering problem with cumulative costs. *Inf. Sci.* **2020**, *507*, 37–52. [CrossRef]
65. Santucci, V.; Baioletti, M.; Di Bari, G. An improved memetic algebraic differential evolution for solving the multidimensional two-way number partitioning problem. *Expert Syst. Appl.* **2021**, *178*, 114938. [CrossRef]
66. Wang, L.; Fu, X.; Mao, Y.; Menhas, M.I.; Fei, M. A novel modified binary differential evolution algorithm and its applications. *Neurocomputing* **2012**, *98*, 55–75. [CrossRef]
67. Pampara, G.; Engelbrecht, A.P.; Franken, N. Binary differential evolution. In Proceedings of the 2006 IEEE International Conference on Evolutionary Computation, Vancouver, BC, Canada, 16–21 July 2006; pp. 1873–1879.
68. Damak, N.; Jarboui, B.; Siarry, P.; Loukil, T. Differential evolution for solving multimode resource-constrained project scheduling problems. *Comput. Oper. Res.* **2009**, *36*, 2653–2659. [CrossRef]
69. Kazemipoor, H.; Tavakkoli-Moghaddam, R.; Shahnazari-Shahrezaei, P.; Azaron, A. A differential evolution algorithm to solve multi-skilled project portfolio scheduling problems. *Int. J. Adv. Manuf. Technol.* **2013**, *64*, 1099–1111. [CrossRef]
70. Peng, W.; Huang, M. A critical chain project scheduling method based on a differential evolution algorithm. *Int. J. Prod. Res.* **2014**, *52*, 3940–3949. [CrossRef]
71. Zhang, L.; Luo, Y.; Zhang, Y. Hybrid particle swarm and differential evolution algorithm for solving multimode resource-constrained project scheduling problem. *J. Control Sci. Eng.* **2015**, *2015*, 48. [CrossRef]
72. Eshraghi, A. A new approach for solving resource constrained project scheduling problems using differential evolution algorithm. *Int. J. Ind. Eng. Comput.* **2016**, *7*, 205–216. [CrossRef]
73. Sallam, K.M.; Chakrabortty, R.K.; Ryan, M.J. A hybrid differential evolution with cuckoo search for solving resource-constrained project scheduling problems. In Proceedings of the 2019 IEEE International Conference on Industrial Engineering and Engineering Management (IEEM), Macao, China, 15–18 December 2019; pp. 1344–1348.
74. Quoc, H.D.; The, L.N.; Doan, C.N.; Thanh, T.P. New Effective Differential Evolution Algorithm for the Project Scheduling Problem. In Proceedings of the 2020 2nd International Conference on Computer Communication and the Internet (ICCCI), Nagoya, Japan, 26–29 June 2020; pp. 150–157.
75. Institute of Management and Economics, TU Clausthal. Multi Mode Project Duration Problem MRCPSP/Max. Available online: https://www.wiwi.tu-clausthal.de/en/ueber-uns/abteilungen/betriebswirtschaftslehre-insbesondere-produktion-und-logistik/research/research-areas/project-generator-progen/max-and-psp/max-library/multi-mode-project-duration-problem-mrcpsp/max (accessed on 7 October 2020).

mathematics

MDPI

Article

Techno-Economic and Environmental Analysis of Grid-Connected Electric Vehicle Charging Station Using AI-Based Algorithm

Mohd Bilal [1,*], Ibrahim Alsaidan [2], Muhannad Alaraj [2], Fahad M. Almasoudi [3] and Mohammad Rizwan [1]

1 Department of Electrical Engineering, Delhi Technological University, Delhi 110042, India; rizwan@dce.ac.in
2 Department of Electrical Engineering, College of Engineering, Qassim University, Buraydah 52571, Saudi Arabia; alsaidan@qu.edu.sa (I.A.); muhannad@qu.edu.sa (M.A.)
3 Department of Electrical Engineering, Faculty of Engineering, University of Tabuk, Tabuk 47913, Saudi Arabia; falmasoudi@ut.edu.sa
* Correspondence: bilal_2k17phdee05@dtu.ac.in

Abstract: The rapid growth of electric vehicles in India necessitates more power to energize such vehicles. Furthermore, the transport industry emits greenhouse gases, particularly SO_2, CO_2. The national grid has to supply an enormous amount of power on a daily basis due to the surplus power required to charge these electric vehicles. This paper presents the various hybrid energy system configurations to meet the power requirements of the electric vehicle charging station (EVCS) situated in the northwest region of Delhi, India. The three configurations are: (a) solar photovoltaic/diesel generator/battery-based EVCS, (b) solar photovoltaic/battery-based EVCS, and (c) grid-and-solar photovoltaic-based EVCS. The meta-heuristic techniques are implemented to analyze the technological, financial, and environmental feasibility of the three possible configurations. The optimization algorithm intends to reduce the total net present cost and levelized cost of energy while keeping the value of lack of power supply probability within limits. To confirm the solution quality obtained using modified salp swarm algorithm (MSSA), the popularly used HOMER software, salp swarm algorithm (SSA), and the gray wolf optimization are applied to the same problem, and their outcomes are equated to those attained by the MSSA. MSSA exhibits superior accuracy and robustness based on simulation outcomes. The MSSA performs much better in terms of computation time followed by the SSA and gray wolf optimization. MSSA results in reduced levelized cost of energy values in all three configurations, i.e., USD 0.482/kWh, USD 0.684/kWh, and USD 0.119/kWh in configurations 1, 2, and 3, respectively. Our findings will be useful for researchers in determining the best method for the sizing of energy system components.

Keywords: artificial intelligence; salp swarm algorithm; hybrid optimization of multiple energy resources; renewable energy; electric vehicle charging station

Citation: Bilal, M.; Alsaidan, I.; Alaraj, M.; Almasoudi, F.M.; Rizwan, M. Techno-Economic and Environmental Analysis of Grid-Connected Electric Vehicle Charging Station Using AI-Based Algorithm. *Mathematics* **2022**, *10*, 924. https://doi.org/10.3390/math10060924

Academic Editor: Alessandro Niccolai

Received: 2 February 2022
Accepted: 10 March 2022
Published: 14 March 2022

1. Introduction

Energy is a critical component of long-term development and poverty eradication. In India, electricity is accessible to 97.2 percent of the population. Natural gas and coal are the most common fuels used in the electricity sector. India has a total power generation capacity of 388.134 GW (including off-grid renewable energy), with renewable energy (RE) accounting for 21.26 percent [1]. The scarcity of natural gas and coal reserves could pose a serious threat to electricity production. To meet the country's rising electricity consumption, the Indian government has announced plans to install 275 GW of sustainable power by 2027. The government has a strategy to satisfy demand with 356.681 GW (Ministry of power, Delhi, India) of electricity generation from coal and nuclear power plants, yet these methods are not eco-friendly and are eroding our environment. RE-based power generation, which emits fewer greenhouse gases (GHGs) than other traditional power generation systems, is required in this framework. India, as a developing country, must provide quality and reliable energy for all of its citizens to achieve its long-term development goals. As per the information revealed by World Resources Institute, India is responsible for emitting 7.1 percent of worldwide emissions and has total emissions of 2.47 tCO_2e compared to a global average of 6.45 tCO_2e. According to World Bank data, per capita consumption of electricity in India has grown from 920 kWh in 2012–2013 to 1210 kWh in 2019–2020.

The transportation sector of India contributes significantly to GHGs emissions. CO_2 accounts for the majority of these GHGs emissions. Furthermore, the energy and agriculture industries are hastening CO_2 emissions. The rapid growth in the number of vehicles needed to serve the country's massive population is a worrisome symptom of pollution and fuel consumption. Aside from these circumstances, the use of electric vehicles (EVs) such as e-rickshaws is increasing regularly. These EVs emit fewer pollutants and no exhausts. The Ministry of Road Transport and Highways does not have any data on these EVs. Some of these EVs are certified, and data reveal that there are 517,322 registered EVs out of a total of 295,800,000 automobiles. However, the rapid expansion of these EVs necessitates a massive quantity of electricity every day from India's national grid. EVs are now charged in residential neighborhoods, with residential customers paying the electric bills. In such circumstances, charging those EVs does not generate any revenue for the electricity sector. Meanwhile, these EVs are creating a lot of stress on the grid network of India. A profitable strategy to generate electrical energy is needed to relieve the strain on the national system. However, to the best of the authors' knowledge, there are no endeavors in India to establish a hybrid-energy-based electric vehicle charging station (EVCS), which prompted us to conduct this study described here.

India has various sources of RE, which include solar photovoltaic (SPV), wind turbines (WT), biogas generators, and biomass. As a result, power might be generated using existing resources. The use of renewables can reduce harmful emissions while also lowering operating costs and maintaining cleanliness. In these systems, paying a bill every month is not an option. EV charging with RE resources enhances commercial profitability, reduces air pollution, and decreases noise in metropolitan regions.

Due to the enormous solar energy availability, it is the finest alternative for generating electrical energy to serve EVs. The solar-powered EVCS at Karna Lake Resort in Karnal is one of Bharat Heavy Electricals Limited's (BHEL) 20 such stations along the Delhi-Chandigarh expressway. It has chargers for all models of electric cars already on the road in the country, including AC001 (33.3 kW), DC001 (15 kW), and 72 kW (50 kW DC + 22 kW AC).

Solar irradiation is missing during rainy days and in foggy surroundings, and with fluctuations in wind energy, no energy is generated at that time. As a result, relying solely on solar and wind energy may compromise system stability and cause a project to fail. Hence, diesel generators (DGs), which can be a decent substitute for generating power even though solar and wind power are unavailable, are required.

1.1. Present Scenario of EVs and EVCS

In the early phases, companies put forward hybrid EV technology as the most reliable and eco-friendly technology. Even though this technology has made major developments, it still mixes traditional fuel practice with an electric engine, which is used when applications require minimal power capacity. Later, technological innovations and more effective battery storage led to the establishment of improved EV systems that completely rely on electrical energy for supply. These devices started to run using electrical energy provided by RE sources, such as SPV and WT. Some research was carried out for enhancing energy efficiency. For example, in Ref. [2], a complete investigation for the forecasting of effective charging profile for EVs by using multiple run-out of charging conditions is provided. Artificial neural networks to build algorithms that perform the optimized energy management of EVs are used [3]. The authors demonstrated how to use a smart nano grid system and plug-in electric automobiles to manage the optimized integration of hybrid SPV systems [4].

According to annual surveys, the number of EVs and charging stations has increased exponentially over the last decade. Battery-based EVs were first started in India in 2001. Now, these are running in nearly every part of the region. There are three varieties of

EVs comprising electric rickshaws, which carry between four and five passengers, auto-rickshaw for three passengers, and electric vans for delivering products. A fully charged electric rickshaw can go about 80–110 km and costs roughly one lakh seventy-five thousand rupees in the market, whereas auto-rickshaws and electric vans have a range of 50–80 km and cost around one lakh rupees. These were formerly imported from China, but today they are made in the country itself. An EV that is completely recharged can travel 80–110 km a day while consuming 8–12 kWh. The cost of power per unit is Rs. 6.90 on the commercial tariff, hence the cost incurred per km run is around Rs. 75.9. One of the e-rickshaw charging stations installed in Delhi is shown in Figure 1.

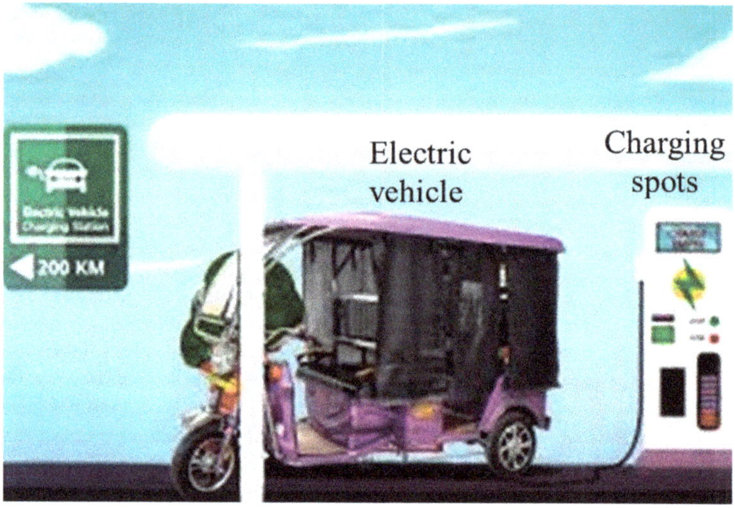

Figure 1. Electric vehicle charging station in Delhi.

The features of e-rickshaw in India are shown in Table 1. This mode of transportation is beneficial to the environment, emitting nearly no fumes and creating significantly fewer CO_2 emissions. It has brought a cultural shift to the transportation industry, becoming well known for its lower fares when related to other means of road transit, thus enhancing the satisfaction of low-income individuals, including vehicle drivers [5]. Some private ventures have launched electric-based Easy Bikes and Auto-rickshaw, which have gained traction in village areas and near metropolitan zones. Such EVs consume a significant amount of grid energy; however, there are a lot of measures underway to relieve the strain on the system. However, because power rickshaws are said to utilize a significant quantity of electricity from grid networks, the changes have yet to be widely implemented.

Table 1. Specification of EVs (e-rickshaw) in Delhi. Data from Ref. [6].

Particulars	E-Rickshaw
Dimension	2790 × 975 × 1730 mm
Carrying capacity	Four passengers + driver
Weight	210 kg (without battery)
Battery	4 Nos./48 V
Battery capacity	80 Ampere/hour
Approx. mileage	70 km per full charge
Charging time	7–8 h per day
Maximum speed	25 km/h
Power consumption	8–11 kWh

The authors performed a survey in Delhi's northwestern city for this study. Every day, more than 200 different varieties of electric rickshaws run here. There were just 20 manually operated rickshaws recorded. The majority of electric rickshaws are powered using electricity installed at home, which is not permitted by the Indian government. Additionally, power outages occur very often in the neighborhood. Therefore, charging EVs during that time is challenging.

It would be easier for those individuals to drive their battery-powered EVs under these conditions if services were created to charge EV batteries using other trustworthy energy sources in the surrounding part. EVs have had a seismic impact on road transport, particularly in rural and sub-urban regions, despite significant charging issues.

1.2. Literature Review

EVs adoption rate is rapidly increasing all over the world, promoting this energy usage industry as a potential area of research. This subsection focuses on the existing EV charging infrastructure, hybrid renewable power production technologies for EV charging, costs of electricity generation, and the reduction in GHG emissions by the charging system. The power and transport industries account for the majority of carbon footprints on the planet. RE can help to reduce harmful emissions in the electrical sector, while grid-powered EVs can greatly decrease pollution in the transportation sector. When RE sources and EVs are combined, they have a lot of potential to solve ecological and financial issues. EVs and RE sources have been discussed in terms of the operation of the system in several kinds of literature.

Optimal systems of various hybrid combinations, such as SPV/wind/diesel/battery systems, have been documented in several articles. On the other hand, SPV systems are less expensive and more applicable than WT [7]. Since the growth and deployment of hybrid RE systems have increased dramatically in the past few years [8], the proper design of these systems is critical for achieving a high degree of reliability at low prices. Under-sizing the system can result in operational limits and energy shortages, while over-sizing the system can lead to high expenses. Hence, sophisticated optimization methods are essential to tackle the optimal sizing problem. The components of the hybrid power system must be designed and integrated as part of the sizing technique to anticipate their efficiency based on meteorological conditions; this comprises predicting the energy output of renewable sources and the battery charge state.

A range of sizing strategies has been reported in the literature, which includes recursive approaches [9] and software applications, such as HOMER [10]. The use of traditional procedures based on recursive, quantitative, or statistical methods has fallen dramatically [11], whilst approaches utilizing optimization algorithms have attracted a lot of attention. Heuristic algorithms are quite efficient and have a lot of potential when it comes to handling optimization challenges. Several types of research have been performed to apply such algorithms to the challenge of sizing RE plants [12]. The practical limits of the EV market in China are investigated, and the influence of incentives on an auto manufacturer's optimal production and pricing decisions is analyzed [13]. A feasibility assessment for a solar-energy-based EVCS in Shenzhen, China, is evaluated using the HOMER software tool. This anticipated strategy addresses the grid power-related issues by incorporating SPV, as well as achieving the expected demand for EVs [14]. There is no other option for generating electricity on wet or foggy days. Another feasibility analysis based on solar-powered EVCS was undertaken in Bulgaria [15]. Many studies envision the incorporation of RE sources, such as a PV system, into charging stations, as the best solution for maximizing the financial and environmental advantages of EVs while also promoting the idea of a smart grid [16]. In Ref. [17], authors have described multiple aspects of the problem formulation, including the choice of design variables and the definition of a probability distribution function to improve efficacy and effectiveness. In Ref. [18], the authors investigated an operational strategy for EVs that employs sustainable RE sources, such as SPV, and gives an optimization method for charging park energy management.

The radial distribution network is affected by the uncertainty of the charging load of the EVCS and SPV unit. A study used a hybrid of genetic algorithm (GA)- and particle swarm optimization (PSO)-based optimum scheduling of EV load to minimize system losses, variations in voltage, cost of charging, and battery costs of EV [19]. The elastic demand was developed using either feedback from crowded travel and congested stations on route selections or the presumption that charging demand between origin and destination pairs follows a nonlinear inverse cost function without taking into account pre-generating pathways and charging configurations [20]. Authors envisioned a PSO for the optimal sizing of SPV and battery energy storage system (BESS) in a grid-connected EVCS, with a financial model as the goal [21]. In Ref. [22], authors suggested a hybrid solar and wind charging station that was deployed in Tangshan, China, and proposed an LPSP technique to evaluate the component capabilities. In Ref. [23], a cost-effective EV charging station is developed by comparing a standalone EV charging station to a grid-connected charging station while taking emissions into account. However, there was no cost evaluation for the hybrid system. The authors utilized linear programing to determine the optimum integration of CS, restricting the search area under the assumption that CSs can only be positioned in parking areas and service stations. The optimization technique considers the accessibility of CS within a reasonable driving range [24]. In Ref. [25], a charging station based on solar- and grid-connected stations for a worksite in the Netherlands is presented. Various charging possibilities for EVs were proposed. However, the work did not contain an economic analysis. The demand response systems and EV charging is investigated by integrating them into radial distribution network buses [26]. Particle swarm optimization is being used to resolve the issue of charging station placement [27]. A comprehensive design of an EVCS is proposed, in which EVCS are integrated with renewable sources of energy and a BESS to ease the burden on the utility grid. To optimize the design, a new technique is proposed, which is a hybrid of the crow search algorithm and PSO [28]. Another study was conducted at the University of Palermo in Italy to build a long-term recharging facility for EVs. It results in a low levelized cost of energy, implying that energy storage systems will require a significant amount of initial capital [29]. The GA is used to optimize the adoption of CS. The authors of this study specify some settlements in the urban environment. Settlements are intended to be the locations where EVs are parked for the majority of the time. GA defines CS positions by minimizing distances to settlements [30]. Based on local resource allocation, the author combined SPV and biogas-based resources with EV load in Bangladesh [5]. SPV energy is widely available because PV units can be easily installed on the rooftop. In Ref. [31], the author provides an optimization strategy for grid-connected SPV/BESS/EVCS to scale SPV, BESS, and establish the charging and discharging characteristics of BESS. This problem is implemented using the multi-agent PSO technique, which integrates the multi-agent system (MAS) and the PSO mechanism. The authors of Ref. [32] consider charging station implementation, taking into account the repercussions on travelers, cab drivers, electricity retail outlets, transport and distribution networks, and electricity consumers. A multi-objective optimization was suggested to handle the developed framework. A hybrid SPV and wind-based EVCS are designed and simulated in HOMER software [33]. An eco-friendly alternative for EV charging based on two RE sources, namely SPV and biogas, is presented. The HOMER package is used to assess the effectiveness and features of SPV and biogas-based EVCS. An SPV system, two biogas generators, and a bi-directional converter with BESS are all part of the proposed system. HOMER software analyzes the alteration of various costs, such as net present value, fixed cost, and energy cost for various SPV configurations [6]. The authors of Ref. [34] presented an enhanced version of chicken swarm optimization (CSO) to optimally locate the EVCS in an IEEE 33 node test distribution network. They first evaluated the network impact of projected EV charging load at the EVCSs in terms of voltage levels, average voltage deviation index, voltage stability index, and power loss, and then, using a feed-forward neural network, they investigated the solar power required to energize the EVCSs. In Ref. [35], the author presented a grid-integrated solar- and wind-based energy system to meet the power requirements of a shopping mall, as well as EVCS load. The components are sized to attain the lowest electricity cost while lessening the deficiency of power supply probability, using artificial colony bee and PSO. The optimum strategy and operation of a renewable-powered EVCS will be presented in this study, to lower cost and emissions. The proposed hybrid system comprises SPV, WT, biogas, batteries, and DG units [36]. In Ref. [37], the authors evaluate the technical viability of implementing a standalone fast charging station in the State of Qatar, which includes WT, SPV system, and a bio-generator as RE source, as well as multiple storage devices. The envisaged design is constructed, modeled, and simulated using the HOMER software to determine the best techno-economic configuration for fast charging 50 EVs on a daily basis. Space constraints, the deterministic nature of EV demand, and metrological conditions of the considered site are all taken into account. The goal of the research undertaken is to use HOMER pro software to deliver SPV power to EVs and feed excess power to the grid network. The author used GA to develop a modeling framework for an SPV/wind/battery for a deserted landmass [38]. A multi-objective PSO to size and analyze a PV/wind/hydroelectric power station with a pumped-storage connected energy system is suggested [39]. The authors investigated a probabilistic modeling approach of an EVCS with SPV, battery, and transformer, with an emphasis on charging demand and SPV power generation complexities. To begin, a detailed EV charging requirement model is developed by projecting the coupling between the EV dynamic charging pattern, charger specifications, and EV charging assignment model. The planning model is then developed to minimize the total cost of the charging station, with the uncertainties associated with EV charging being acknowledged through various limits [40]. A multi-objective moth flame optimization (MFO) that was used to optimize a hybrid microgrid system with SPV/wind/BESS/DG is designed [41]. In Ref. [42], the author focuses on a hybrid energy system that includes an SPV, WT, a biogas generator, and BESS for providing reliable electricity to a deserted island in Bangladesh called Saint Martin. The energy system's component is sized in accordance with energy cost and life cycle emissions under a certain level of reliability. The objectives are optimized by considering two well-known multi-objective optimization methods: non-dominated sorting GA-II and infeasibility-driven evolutionary algorithm. In Ref. [43], the authors presented the technical economic viability of solar- and wind-based EVCS at five different areas in the southern region of Tamil Nadu. The locations are selected such that wind and solar possibilities are of contrary nature in the required places. In Ref. [44], a comprehensive quantitative model is designed to identify the suitable SPV/WT/BESS sizes for distant locations. The suggested method is tested at Ras-Shaitan, a rural area in Egypt-Sinai. The goal of the optimization method is to accommodate power requirements by lowering the cost of energy under various power supply failure scenarios. To get the optimum objective value, the gray wolf optimizer is used to determine the SPV, WT, and BESS units.

Several heuristic methodologies for tackling the sizing issue of various system components have been suggested in the research publications. Because of their excellent ability to tackle difficult optimization issues, researchers are increasingly looking into these strategies. However, the efficiency of each optimization technique, when used for evaluating the optimal size, may vary in terms of precision and convergence rate. Although there is a range of sizing methodologies available, academics working on this problem are constantly looking for a robust and highly precise approach with a small implementation time. This is in order to develop and formulate energy systems that are both reliable and cost effective.

The control and power management of EVs in grid-connected systems are the primary focus of researchers. However, one of the important aspects that must be addressed is an economic analysis that takes into account the power exchange with the grid. The fast adoption of EVs poses both constraints and opportunities for the current electricity system. A small grid-connected SPV and DG-based hybrid system with EVs are presented in this article for a charging station in the northwest region of Delhi, India. The main objective is to formulate a statistical model of a solar and diesel generator-based hybrid system with EVs and a backup grid. Furthermore, the purpose of this research is to reduce power interchange with the grid. The salp swarm algorithm (SSA) and the modified salp swarm algorithm (MSSA) have never been introduced before for satisfying the EV load requirement using hybrid SPV and DG with utility grid as a backup. To our knowledge, no previous study has compared the performance of so many different sizing methodologies on a single energy system. Hence, the assessment described in this article will be a useful guide for researchers looking to select a technique for their sizing concern.

1.3. Contribution

A solar PV module, diesel generator, battery storage, and backup grid are all part of the hybrid energy system being investigated. The system is designed as a simple micro grid to test the algorithms' performance against a conventional framework. The system's total net present cost (TNPC) is minimized, while the lack of power supply probability (LPSP) is taken into account as a reliability restriction. In addition, the levelized cost of energy cost (LCOE) is used to evaluate the hybrid energy system. The optimization technique is based on real-time solar irradiation and ambient temperature data collected over one year.

In light of the preceding discussions, the following are the key objectives of this paper:

- Identify the optimal EVCS design using realistic inputs on physical, operating, and economic features of RE sources, such as SPV and DG, with the utility grid.
- To create a novel idea that combines a photovoltaic module with DG to provide electricity for charging plug-in electric vehicles.
- To assess the technological, financial, and ecological viability of the proposed model.
- Two novel methods, SSA and MSSA, are discussed and employed in the sizing problem of the system component for the first time. The efficacy of such methods is assessed and equated in terms of consistency, convergence cycle, and processing time.

The rest of this article is laid out as follows. The modeling of the various components of a hybrid energy system is performed in Section 2. The problem formulation is elaborated in Section 3. The proposed optimization methodology is discussed in Section 4. The results are thoroughly discussed in Section 5. Future challenges and opportunities are presented in Section 6. Finally, conclusions are stated in Section 7.

2. Hybrid System Components and Their Mathematical Modeling

The mathematical modeling of the hybrid energy system to meet the charging demand of EVs is described in this section. An SPV panel, DG, battery, inverter, and backup grid are all part of the hybrid energy system under consideration. To enhance load supply reliability, the battery is utilized to control the variation of RE generation. A graphical representation of the proposed SPV/DG/battery storage system is shown in Figure 2. The methodology for conducting this research work is shown in Figure 3.

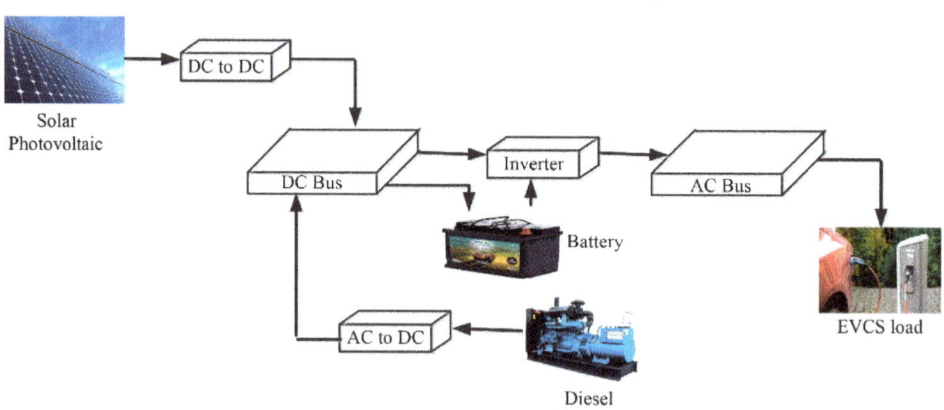

Figure 2. Schematic representation of the proposed hybrid energy system.

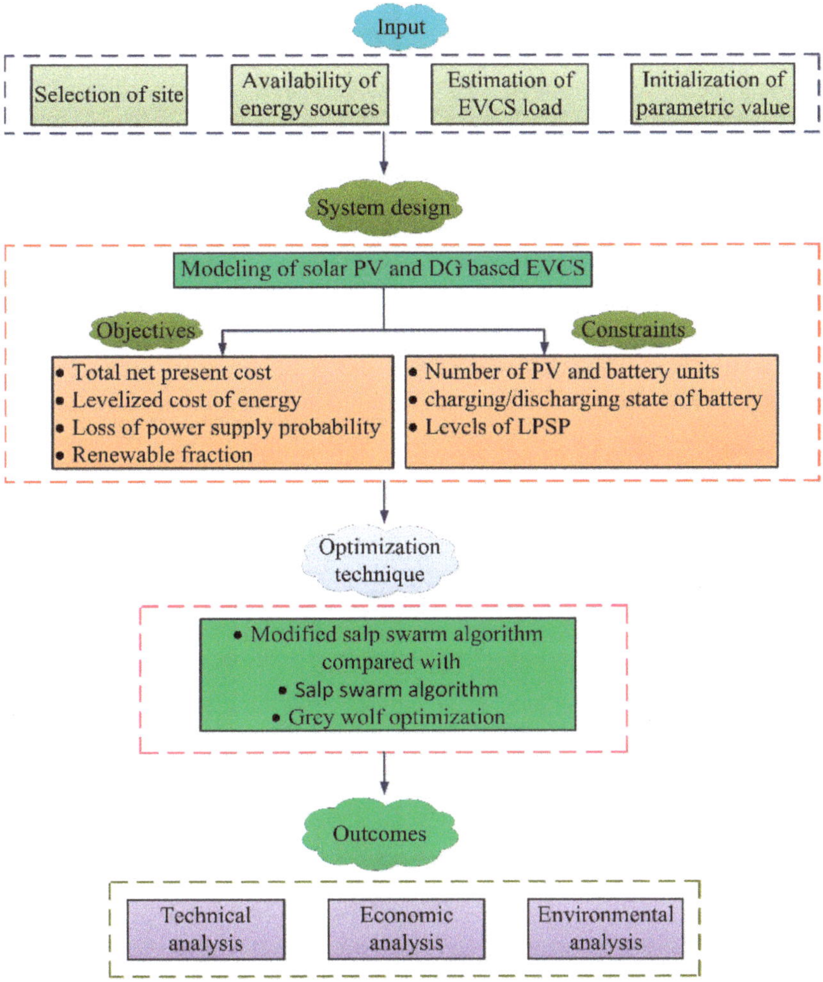

Figure 3. A schematic representation of the system analysis and design procedure.

2.1. SPV Panel Output

Solar radiation is the most prevalent component on the planet, as well as being the most robust RE source. It is comparatively simple to execute and can be made affordable in all remote areas of the country. Concerning the maximum output pattern of SPV, modules are widely used to estimate the quality of an SPV generation system to its peak power. SPV panels are used to generate electricity by collecting solar energy. Only a small portion of the solar radiation that penetrates the SPV panels is transformed to electricity, with the remainder being converted to heat. Several factors can influence the quality of the SPV power output. The power output obtained from SPV is dependent on the area of the panel, solar irradiance incident on the surface of the panel, temperature of the SPV cells, and geographical parameters, i.e., latitude and longitude of the panel location, using Equation (1):

$$P_{SPV\text{-}out}(t) = \eta_{SPV} A_{SPV} G_h(t)\{1 - 0.005(T_{cell} - 25)\} \quad (1)$$

where η_{SPV} is the efficiency of solar panels, A_{SPV} represents the surface area of the solar panel, $G_h(t)$ denotes the hourly solar irradiance falling on the surface of the solar panel, and T_{cell} is the temperature of the cell. The parameters taken for the selected PV panel are tabulated in Table 2.

The temperature of the cell and the fluctuation of power as a function of temperature are given by

$$T_{cell} = T_a + \frac{NOCT - 20}{0.8} G_h(t) \quad (2)$$

where T_a is the ambient temperature.

The rated maximum power output of solar panel can be expressed as

$$P_{max} = V_{max} I_{max} \quad (3)$$

where V_{max} denotes the maximum value of voltage, and I_{max} represents the maximum current, respectively.

The overall power generated using SPV modules can be estimated by

$$P^{total}_{SPV\text{-}out}(t) = N_{SPV} P_{SPV\text{-}out}(t) \quad (4)$$

where N_{SPV} is the number of the installed solar PV panels.

2.2. *Diesel Generator (DG)*

The quantity of energy storage necessary for isolated villages and rural enterprises can be lowered by using DGs, resulting in a profitable and reliable solution. In the event of battery degradation during peak loads, diesel can be used as a backup source of energy.

The DG's efficiency and hourly fuel usage should be accounted for when planning a hybrid system and may be represented using the equation below [34]:

$$u(t) = a \times P_{DG}(t) + b \times P_{rated\text{-}DG} \quad (5)$$

where u(t) is the fuel consumption in liters per hour, $P_{DG}(t)$ is produced power in kW, $P_{rated\text{-}DG}$ is the rated power, a (0.246) and b (0.08415) are constant parameters in liter per kW, which characterize the coefficients of fuel consumption.

The efficiency of DG can be computed by

$$\eta_{overall} = \eta_{brake\ thermal} \times \eta_{DG} \quad (6)$$

where $\eta_{overall}$ is the overall efficiency of DG, whereas $\eta_{brake\ thermal}$ represents the brake thermal efficiency of DG.

2.3. *Modeling of Battery Energy Storage System (BESS)*

Due to the erratic nature of SPV output, optimum battery sizing is essential to meet the load demand. The existing state of charge (SOC) is the most important decision variable for monitoring the charging/discharging states of the battery. Overcharging of the battery takes place when the hybrid model generates too much power or when the load demand is too low. When the battery's SOC reaches its maximum value, i.e., SOC (max), the control system gets involved and stops the charging mechanism, whereas when it acquires its minimum value, i.e., SOC (min), the control system disables the load to avoid the battery from being drained [45].

There are instances when the state of the battery varies depending on the power output and load demand. The battery performs an important role in the system's energy management by adjusting power demands and power supply. The battery is said to be charging when the generated power is higher than the required energy, and the charging state of the battery at time t is given by [45]

$$SOC_{bat}(t+1) = SOC_{bat}(t)(1-\sigma) + \left[\left(P_{SPV\text{-}out}(t) + \frac{P_{DG}(t)}{\eta_{bdinv}} \right) - P_{EVCS\text{-}dem}(t) \right] \times \eta_{bat} \quad (7)$$

On the other hand, when RE sources are unable to create enough power to meet demand, the battery is employed as a backup to feed the load system and operate in discharging state. The discharging state of the battery at time t is given by

$$SOC_{bat}(t+1) = SOC_{bat}(t)(1-\sigma) - \frac{\left[P_{EVCS\text{-}dem}(t) - \left(P_{SPV\text{-}out}(t) + \frac{P_{DG}(t)}{\eta_{bdinv}}\right)\right]}{\eta_{bat}} \quad (8)$$

where $SOC_{bat}(t+1)$ and $SOC_{bat}(t)$ are the SOC of battery at an instant $(t+1)$ and (t), respectively, σ is the self-discharge rate of the battery, η_{bdinv} is the efficiency of bi-directional inverter used, $P_{SPV\text{-}out}(t)$ and $P_{DG}(t)$ are the power output of SPV and DG, respectively, and η_{bat} is the round trip efficiency of the battery. The round-trip efficiency of the battery can be expressed using Equation (9):

$$\eta_{bat} = \eta_{bat}^{charge} \times \eta_{bat}^{discharge} \quad (9)$$

where η_{bat}^{charge} represents the charging efficiency of the battery, whereas $\eta_{bat}^{discharge}$ denotes the discharging efficiency of the battery. The round-trip efficiency of the battery bank is reported to be 92.2 percent. Furthermore, charging/discharging efficiencies are assumed to be 85/100 percent, respectively. SOC_{max} round-trip efficiency of the battery is the maximum value of SOC and is equivalent to the total energy of the battery bank C_{total}(Ampere hour), as follows:

$$C_{total}(\text{Ampere hour}) = \frac{N_{bat}}{N_{bat}^{series}} C_{sbat}(\text{Ampere hour}) \quad (10)$$

where C_{sbat} is the energy of a single battery, N_{bat} represents the total number of batteries, and N_{bat}^{series} represents the number of batteries connected in series. The battery bank is only allowed to discharge to a certain minimal level, known as SOC_{min}. This restriction can be used as a system constraint depending on how the battery bank is utilized. Batteries are arranged in series to get the desired bus voltage. The number of batteries in a series can be computed using the formula

$$N_{bat}^{series} = \frac{V_{bus}}{V_{bat}} \quad (11)$$

where V_{bat} is the voltage level of a single battery.

The maximum charge/discharge power at any given time is another important consideration in battery modeling. It is determined by the maximum charging current and may be computed using the equation below:

$$P_{bat}^{max} = \frac{N_{bat} \times V_{bat} \times I_{max}}{1000} \quad (12)$$

where I_{max} is the maximum charging current drawn by the battery in amperes, and P_{bat}^{max} is the maximum input/output power of the battery.

2.4. Bi-Directional Inverter Modeling

The bi-directional inverter performs the function of conversion of DC to AC power and vice versa. Solar panels provide DC electricity to EVs at the charging station. Hence, the bi-directional inverter is required to transform the DC power of SPV. The output of the bi-directional inverter can be computed as follows:

$$P_{AC} = \eta_{bdinv} \times P_{DC} \quad (13)$$

The proposed system takes into account a bi-directional inverter efficiency of 97 percent.

2.5. Utility Grid

In the first phase, when the demand exceeds the generated power from the different energy sources and BESS capacity, the energy shortage can be fulfilled by borrowing the energy from the grid network and may be represented using Equation (14):

$$\begin{aligned} P_P^{grid}(t) = P_D^{EV}(t) & \\ & -\left[P_{SPV\text{-}out}^{total}(t) + P_{DG}(t)\right. \\ & \left.+\left[\left(P_{SPV\text{-}out}^{total}(t) + \left(SOC_{bat}(t) - SOC_{bat}^{min}(t)\right)\right) \times \eta_{bdinv}\right]\right] \end{aligned} \quad (14)$$

where $P_P^{grid}(t)$ represents the energy to be borrowed from the grid in kWh, and $SOC_{bat}^{min}(t)$ denotes the minimum SOC of battery.

In the second phase, when the demand is less than the generated power from the different energy sources, and the battery is wholly charged, the extra energy is sold to the utility grid and may be represented using Equation (15):

$$\begin{aligned} P_S^{grid}(t) = & \left[P_{SPV\text{-}out}^{total}(t) + P_{DG}(t)\right. \\ & \left.+\left[\left(P_{SPV\text{-}out}^{total}(t) - (SOC_{bat}^{max}(t) - SOC_{bat}(t))\right) \times \eta_{bdinv}\right]\right] \\ & -P_D^{EV}(t) \end{aligned} \quad (15)$$

where $P_S^{grid}(t)$ represents the extra energy to be sold to the grid in kWh, and $SOC_{bat}^{max}(t)$ denotes the maximum SOC of the battery.

2.6. Modeling of EVCS Load

The main components of EVCS are dual converter, charging ports, and EVs. The charging station is coupled to a regulator, which facilitates controlling the direction of the power flow at a specific time. The charging of EV takes place based on its current SOC. SOC is defined as the ratio of available charging state to the maximum charging state, i.e., when the battery is fully charged. Hence, it describes how much battery needs to be charged.

The amount of power consumed by an EV is determined by the distance traveled, battery capacity, and driving mode. An EV uses power, which may be estimated using the equation

$$P_D^{EV} = \frac{D_{km} \times E_{req/km}}{T} \quad (16)$$

where D_{km} is the number of kilometers traveled, the power needed per kilometer is $E_{req/km}$, and T is the time to recharge the battery. T is the difference between the time of arrival and departure. T is determined by the SOC of the vehicle's battery. The power requirement of an EV can be illustrated using the capacity of battery, SOC, and time of charging:

$$P_D^{EV} = \frac{Q_{bat}^{EV} \times \left(SOC_{max}^{EV} - SOC_{min}^{EV}\right)}{T} \quad (17)$$

where Q_{bat}^{EV} represents the capacity of the EV battery, SOC_{max}^{EV} and SOC_{min}^{EV} are the maximum and minimum limit of SOC, and T denotes the duration of charging.

Therefore, the total power required for charging the N number of EVs is given as follows:

$$P_{total}^{EV} = \sum_{i=1}^{N_{EV}} P_D^i \quad (18)$$

The technical features of different components of a hybrid energy system are summarized in Table 2.

Table 2. Technical and economic parameters of different hybrid system components.

Components	Characteristics	Values
Solar PV panel (Canadian Solar Max Power CS6X-325P) [6]	Rated power	325 W
	Normal operating cell temperature	45 ± 2 °C
	Open circuit voltage (V_{OC})	45.5 Volts
	Short circuit current (I_{SC})	9.34 Amps
	Module efficiency:	16.94%
	Dimensions	76.90 × 38.70 × 1.57 inches
	Power tolerance	0/+5 W
	Initial cost	USD 950/kW
	Replacement cost	USD 900/kW
	Operation and maintenance cost	USD 10/kW
	Life span	25 years
	Derating factor	80%
Diesel generator (Generic 10 kW fixed capacity genset) [46]	Minimum load ratio	50%
	Heat recovery ratio	10%
	Operating time	15,000 h
	Fuel price	USD 1.21/L
	Density	820 Kg/m^3
	Initial capital cost	USD 500/kW
	Replacement cost	USD 500/kW
	Operation and maintenance cost	USD 10
Battery (Generic 1 kWh lead acid) [47]	Battery material	Lead acid
	Nominal voltage	12 V
	Nominal capacity	81–100 Ampere hour
	Lifetime	5–8 years
	Initial capital cost	USD 235/kW
	Replacement cost	USD 190/kW
	Operation and maintenance cost	USD 2/kW/year
Bi-directional converter [47]	Rated power	100 kW
	Initial capital cost	USD 171/kW
	Replacement cost	USD 171/kW
	Operation and maintenance cost	USD 4/kW/year
Utility grid [36]	Grid purchase price	USD 0.12/kWh
	Grid sell back price	USD 0.08/kWh
Others [47]	Nominal interest rate	3.75%
	Inflation rate	1.75%
	Lifetime of project	25 years
	SOC_{max}	0.95
	SOC_{min}	0.35
	Inverter efficiency (%)	0.95
	Rectifier efficiency (%)	0.95

3. Problem Formulation

In this paper, the problem formulation deals with the optimization of different configurations of energy system for satisfying the load demand of EVCS installed in the northwest region of Delhi, India, with real annual data of solar irradiance, temperature, and EV load. These configurations are stated as follows: (a) SPV/DG/battery-based EVCS, (b) SPV/battery-based EVCS, and (c) grid-and-SPV-based EVCS. The hybrid energy system is constructed with a maximum load of 25.45 kW, and the yearly average energy requirement is 257 kWh/day. The goal of optimization is to minimize the total net present cost (TNPC), keeping the reliability constraint in terms of loss of power supply probability within specified limits. Moreover, another parameter for analyzing the performance of the hybrid energy system is the levelized cost of energy (LCOE). The MSSA is implemented to improve its searching potential based on Levy flight distribution (LFD), enabling individuals to move to new locations and individual spiraling around the optimum solution to improve the exploitation phase. In this work, the energy management of the hybrid system is based on load following strategy (LFS). The numbers of SPV panels and batteries are chosen as the optimization variables, which are determined optimally using the MSSA. The primary goal of this research work is to perform a cost-effective analysis for all selected configurations with lowest TNPC, LCOE, and best reliability. The MSSA performance in the optimization of hybrid energy system is compared to conventional SSA and GWO. The objective functions, i.e., TNPC and LCOE and constraints, are coded in MATLAB environment considering the energy management strategy reported in Figures 4 and 5 for SPV/DG/battery-based EVCS and grid-and-SPV-based EVCS, respectively.

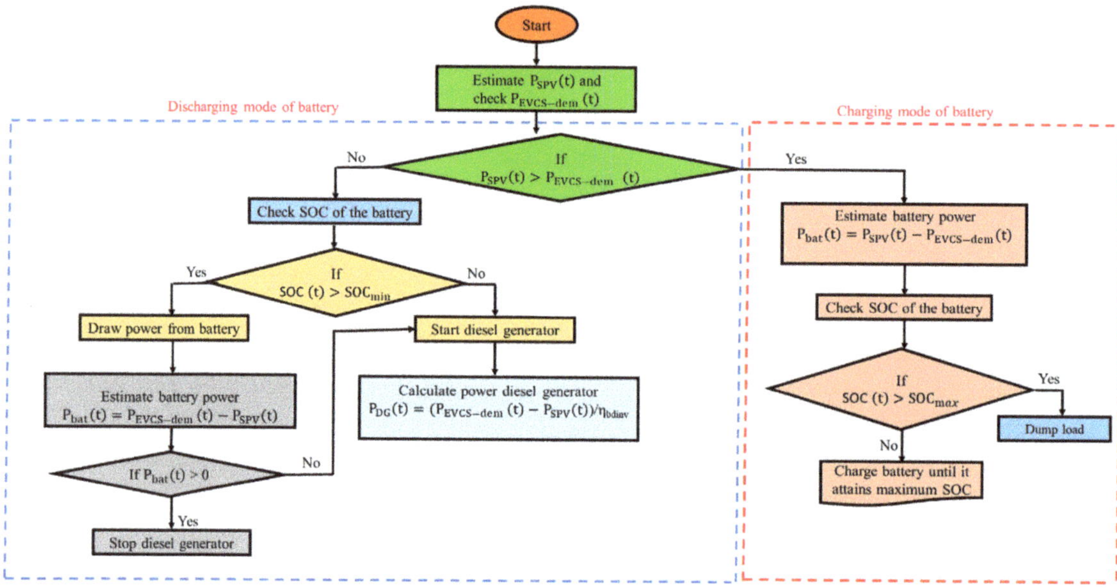

Figure 4. Flow chart of the energy management strategy of the SPV/DG/battery system.

3.1. Assumptions

- Charging stations operate for the whole day.
- Different number of EVs arrive at charging station on a daily basis.
- Only one charging station is considered.
- Limited output power of battery is taken into account.
- Load following strategy (LFS) is considered.
- Diesel generator is allowed to operate only when no power output from SPV is available.
- Cost of EV charger is around USD 50.

3.2. Objective Functions

The optimal number of solar panels, i.e., N_{SPV} and batteries, i.e., N_{bat}, must be determined to ensure that the system formulated in different configurations meets the charging demand of EVs. The objective function is formulated using the total net present cost (TNPC), levelized cost of energy (LCOE), and reliability in terms of loss of power supply probability (LPSP).

3.2.1. Total Net Present Cost (TNPC)

The total cost of a hybrid energy system is represented by the total net present cost (TNPC). It includes all costs and revenues incurred during the life cycle of the system, including system component capital costs, replacement costs incurred during the system's operation, and maintenance expenses. The numbers of SPV panels and batteries are chosen as the two primary decision variables for the optimum configuration. Hence, mathematical equations for the NPC of various system components are derived as follows:

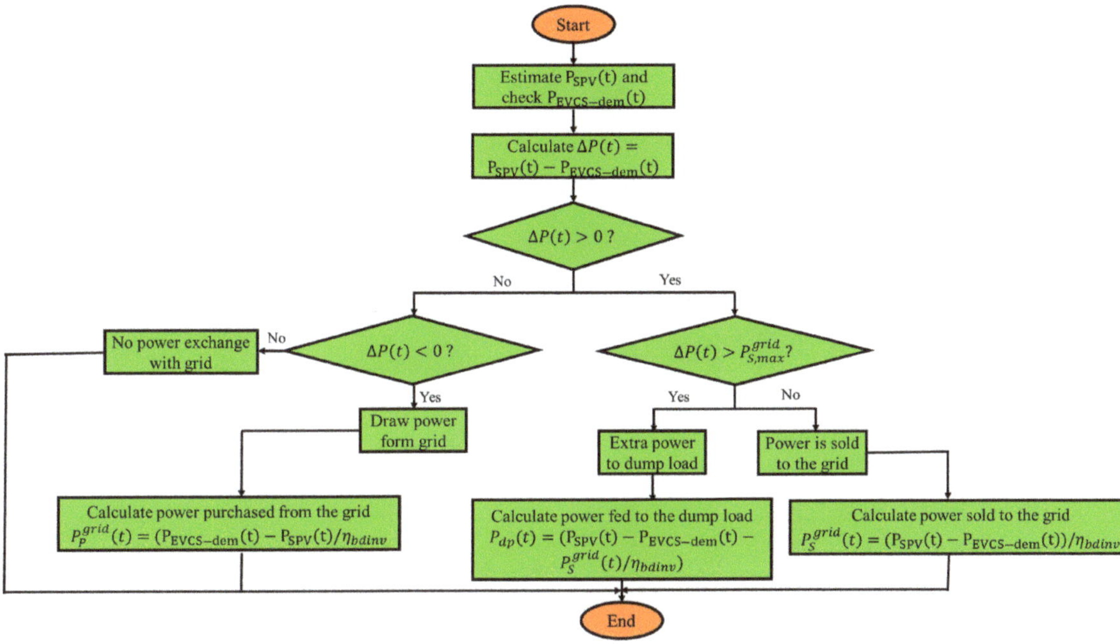

Figure 5. Flow chart of the energy management strategy in grid-and-solar PV-based EVCS.

Solar PV

The project lifetime is set to be equivalent to the lifetime of PV panels in this study. Furthermore, solar irradiance is used as a free fuel to generate power. Therefore, there will be no fuel costs. There, NPC of solar PV (NPC_{SPV}) consists of capital cost $\left(C^{CAP}_{SPV}\right)$, replacement cost $\left(C^{REP}_{SPV}\right)$, operation and maintenance cost $\left(C^{OM}_{SPV}\right)$, and salvage cost $\left(C^{SAL}_{SPV}\right)$ and can be given as follows

$$NPC_{SPV} = C^{CAP}_{SPV} + C^{REP}_{SPV} + C^{OM}_{SPV} - C^{SAL}_{SPV} \tag{19}$$

The capital cost of solar PV panels can be computed as

$$C^{CAP}_{SPV} = N_{SPV} \times \psi_{SPV} \times P_{SPV} \tag{20}$$

where ψ_{SPV} and P_{SPV} denote the initial cost and power rating of each SPV panel, respectively.

Replacement cost, i.e., $\left(C^{REP}_{SPV}\right)$ of SPV panel, can be determined using the following equation:

$$C^{REP}_{SPV} = \zeta^{rep}_{SPV} \times N_{SPV} \times P_{SPV} \times \frac{R(1+R)^{\Omega}}{\left((1+R)^{\Omega} - 1\right)(1+R)^{\Omega_{SPV}}} \tag{21}$$

where ζ^{rep}_{SPV} denotes the cost of replacing the SPV panel, R denotes the rate of interest, Ω indicates the life span of the project in years, and y represents the lifetime of the SPV panel.

Further, operation and maintenance cost $\left(C_{SPV}^{OM}\right)$ can be computed as follows:

$$C_{SPV}^{OM} = N_{SPV} \times \omega_{SPV} \times P_{SPV} \times \sum_{i=1}^{\Omega} \left(\frac{1+\mu_{SPV}}{1+R}\right)^{i} \quad (22)$$

where ω_{SPV} is the OM cost of each panel, and μ_{SPV} is the escalation rate.

Salvage value is the cost of the component's remaining life after the project is completed. The salvage cost of the SPV panel can be defined as

$$C_{SPV}^{SAL} = N_{SPV} \times \varepsilon_{SPV} \times \left(\frac{1+\partial}{1+R}\right)^{\Omega} \times P_{SPV} \quad (23)$$

where ε_{SPV} denotes the resale price of the solar panel after completing its life, and ∂ is the inflation rate.

Battery

The NPC of battery storage system, i.e., NPC_{bat}, involves capital cost $\left(C_{bat}^{CAP}\right)$, replacement cost $\left(C_{bat}^{REP}\right)$, operation and maintenance cost $\left(C_{bat}^{OM}\right)$, and salvage cost $\left(C_{bat}^{SAL}\right)$ and can be given as follows

$$NPC_{bat} = C_{bat}^{CAP} + C_{bat}^{REP} + C_{bat}^{OM} - C_{bat}^{SAL} \quad (24)$$

$$C_{bat}^{CAP} = N_{bat} \times \psi_{bat} \quad (25)$$

$$C_{bat}^{OM} = N_{bat} \times \omega_{bat} \times \sum_{i=1}^{\Omega} \left(\frac{1+\mu_{bat}}{1+R}\right)^{i} \quad (26)$$

$$C_{bat}^{SAL} = N_{bat} \times \varepsilon_{bat} \times \left(\frac{1+\partial}{1+R}\right)^{\Omega} \quad (27)$$

where ψ_{bat} is the cost of one battery in USD. ω_{bat} and μ_{bat} indicate annual OM cost in USD/year and escalation rate (0.075) of batteries, respectively. ε_{bat} denotes the resale value of one battery in USD.

The battery life (Ω_{bat}) has been considered to be 5 years in this study, which is shorter than the project lifetime of 25 years. Hence, the battery must be replaced every 5 years. The number of battery replacements (NBR) is determined as

$$NBR = \frac{\Omega}{\Omega_{bat}} - 1 \quad (28)$$

The replacement cost of the battery can be determined as

$$C_{bat}^{REP} = N_{bat} \times \zeta_{bat}^{rep} \times \sum_{i=5,\,10,\,15,\,20} \left(\frac{1+\mu_{bat}}{1+R}\right)^{i} \quad (29)$$

Bi-Directional Inverter

The NPC of the bi-directional inverter can be calculated as

$$NPC_{bdinv} = C_{bdinv}^{CAP} + C_{bdinv}^{REP} + C_{bdinv}^{OM} - C_{bdinv}^{SAL} \quad (30)$$

where C_{bdinv}^{CAP}, C_{bdinv}^{REP}, C_{bdinv}^{OM}, and C_{bdinv}^{SAL} represent the initial capital cost, replacement cost, operation and maintenance cost, and salvage cost of the bi-directional inverter, respectively.

$$C_{bdinv}^{CAP} = N_{bdinv} \times \psi_{bdinv} \quad (31)$$

$$C_{bdinv}^{OM} = N_{bdinv} \times \omega_{bdinv} \times \sum_{i=1}^{\Omega} \left(\frac{1 + \mu_{bdinv}}{1 + R} \right)^{i} \tag{32}$$

$$C_{bdinv}^{SAL} = N_{bdinv} \times \varepsilon_{bdinv} \times \left(\frac{1 + \partial}{1 + R} \right)^{\Omega} \tag{33}$$

The inverter's assumed lifetime is 10 years, which is shorter than the project's lifespan (25 years). As a result, the net present value of the replacement cost of the inverter is expressed as follows:

$$C_{bdinv}^{REP} = N_{bdinv} \times \zeta_{bdinv}^{rep} \times \sum_{i=10,20} \left(\frac{1 + \mu_{bdinv}}{1 + R} \right)^{i} \tag{34}$$

Sale and Purchase Capacity with Utility Grid

In an on-grid system, the NPC of selling electricity to the grid or purchasing electricity from the grid can be evaluated as follows:

$$C_{S}^{grid} = \xi_{S}^{grid} \times E_{S}^{grid} \times \left(\frac{1 + \partial_{grid}}{1 + R} \right)^{\Omega} \tag{35}$$

$$C_{P}^{grid} = \xi_{P}^{grid} \times E_{P}^{grid} \times \left(\frac{1 + \partial_{grid}}{1 + R} \right)^{\Omega} \tag{36}$$

where ξ_{S}^{grid} and ξ_{P}^{grid} indicate the cost for unit sale and of purchase of electricity to and from the utility grid, respectively. It is defined in USD per kWh.

Hence, one of the objectives of this research work is to minimize the TNPC of the system, which is described as follows:

$$TNPC(USD) = NPC_{SPV} + NPC_{DG} + NPC_{bat} + NPC_{bdinv} + C_{P}^{grid} - C_{S}^{grid} \tag{37}$$

3.2.2. Levelized Cost of Energy (LCOE)

Levelized cost of energy may be defined as the average cost incurred per kW of energy production by the system. LCOE is computed by dividing the annualized cost of energy generation by the total energy production per year. It can be computed with the help of capital recovery factor (CRF), using the following Equation (38) [10]:

$$LCOE = \frac{TNPC \times CRF}{\sum_{t=1}^{T} P_{gen}(t)} \ USD/kW \tag{38}$$

CRF can be evaluated using Equation (39), as follows:

$$CRF = \frac{R(1 + R)^{\Omega}}{\left((1 + R)^{\Omega} - 1\right)} \tag{39}$$

LCOE depends on numerous features, such as capital cost, solar radiation, lifetime, operation and maintenance cost, CRF and degradation of the SPV panels used, etc.

3.2.3. Renewable Fraction

The renewable fraction (RF) is an important criterion to consider when designing an energy system. The total energy produced via RE sources is divided by the total power drawn by the load to calculate the RF. RF can be calculated using Equation (40):

$$RF = \left(1 - \frac{\sum P_{DG}}{P_{SPV}} \right) \times 100\,\% \tag{40}$$

3.3. Decision Variables and Constraints

To optimize the hybrid system, a compromise must be made between the cost-based objectives and various technical constraints. However, the proposed optimization approach should be used to obtain the optimal decision variables. The decision variable of the suggested optimization procedure is subject to specific limits: the minimum limit put on the number of SPVs and battery, and the $N_{SPV}^{max} = 1000$, $N_{bat}^{max} = 600$.

- The following are the limits associated with the maximum and minimum sizes of decision variables:

$$N_{SPV}^{min} \leq N_{SPV} \leq N_{SPV}^{max} \tag{41}$$

$$N_{bat}^{min} \leq N_{bat} \leq N_{bat}^{max} \tag{42}$$

$$E_{bat}^{min} \leq N_{bat} \leq E_{bat}^{max} \tag{43}$$

- The charging state of the battery should be preserved using Equation (44). At any given time, the amount of energy stored in the battery $SOC_{bat}(t)$ is limited by the maximum and minimum quantities SOC_{bat}^{min} and SOC_{bat}^{max} as follows:

$$SOC_{bat}^{min} \leq SOC_{bat} \leq SOC_{bat}^{max} \tag{44}$$

SOC_{bat}^{max} takes the value of the theoretical capacity of the battery $C_{bus} \times V_{bus}$. The maximum depth of discharge, i.e., DOD and theoretical capacity $C_{bus} \times V_{bus}$ is used to calculate SOC_{bat}^{min}, as expressed in Equation (45):

$$SOC_{bat}^{min} = (1 - DOD) \times C_{bus} \times V_{bus} \tag{45}$$

DOD is assumed to be 50%.

- The maximum permissible lack of power supply should be considered for a reliable system:

$$LPSP \leq LPSP_{max} \tag{46}$$

where $LPSP_{max}$ represents the maximum allowable lack of power supply probability.

3.4. Operational Strategy

3.4.1. Solar PV and Diesel Generator-Based EVCS

To attain the reliability of the system in any hybrid energy system, optimal energy management is required. In this system, the DG is kept at the bottom of that list, which means it only needs to operate when the solar and battery systems are unable to supply the load requirement. The steps for the simplest implementation strategy for SPV/DG/battery system are as follows:

- If the total power generated by SPV panels is sufficient, demand can be met solely by solar power. After the load has been satisfied, excess electricity can be supplied to the BESS and is provided as follows:

$$P_{bat}(t) = P_{SPV\text{-}out}(t) - P_{EVCS\text{-}dem}(t) \tag{47}$$

- In the preceding case, if $P_{bat}(t)$ is greater than the maximum permissible capacity of the BESS, i.e., P_{bat}^{max}, then additional energy can be dumped or directed to deferrable loads. Excess or dump energy $(P_{dp}(t))$ can be given as

$$P_{dp}(t) = P_{bat}(t) - P_{bat}^{max}(t) \tag{48}$$

- If the SPV panels do not generate sufficient power, the battery can provide the remaining power, which can be calculated as

$$P_{bat}(t) = P_{EVCS\text{-}dem}(t) - P_{SPV\text{-}out}(t) \quad (49)$$

- If solar power is insufficient and batteries are unable to generate the required power to meet the power requirements, DG is used to energize the load. There are two ways to use DG.

(a) First, it employs a load-following strategy, which means that whenever it performs, it produces only the power necessary to satisfy the primary load requirements. The diesel generator's power output is calculated as

$$P_{DG}(t) = \frac{P_{EVCS\text{-}dem}(t) - P_{SPV\text{-}out}(t)}{\eta_{bdinv}} \quad (50)$$

(b) In the second strategy, it runs at maximum capacity or minimum load ratio. When the DG is fully operational, the surplus energy is used to charge batteries and is expressed as follows:

$$P_{bat}(t) = (P_{DG}(t) * \eta_{bdinv} - P_{EVCS\text{-}dem}(t)) + P_{SPV\text{-}out}(t) \quad (51)$$

The operational strategy of the SPV/DG/battery system can be understood with the help of the flow chart depicted in Figure 4.

3.4.2. Grid-and-Solar PV-Based EVCS

In this scenario, it is assumed that SPV is the only source to satisfy the load demand of EVs. The difference between power obtained by SPV and power demand by EVs can be determined as

$$\Delta P(t) = P_{SPV}(t) - P_{EVCS\text{-}dem}(t) \quad (52)$$

If solar panels are unable to fulfill the power requirements of EV load, the extra power is purchased from the grid network ($P_P^{grid}(t)$) to meet the requirements.

Furthermore, if more power is available from the SPV after meeting the requirements, it is sold to the grid ($P_S^{grid}(t)$). Nevertheless, there are some limitations on selling and purchasing power to and from the grid, which are defined as maximum purchase capacity of grid ($P_{P,\ max}^{grid}$) and maximum selling capacity of grid ($P_{S,\ max}^{grid}$). Beyond these limits, power cannot be purchased from or sold to the grid. The following cases are formed depending on the $\Delta P(t)$.

- When SPVs alone fulfill the power requirement of EV load, i.e., $\Delta P(t) > 0$, the extra power is sold to the grid network, which can be determined as

$$P_S^{grid}(t) = (P_{SPV}(t) - P_{EVCS\text{-}dem}(t))/\eta_{bdinv} \quad (53)$$

- When SPV power output is high enough, which fulfills the power requirement of EV load, as well as exceeds the maximum selling capacity of grid ($P_{S,\ max}^{grid}$), i.e., $\Delta P(t) > 0$ and $\Delta P(t) > P_{S,\ max}^{grid}$, the extra power is fed to the dump load, which can be determined as

$$P_{dp}(t) = (P_{SPV}(t) - P_{EVCS\text{-}dem}(t) - P_S^{grid}(t)/\eta_{bdinv}) \quad (54)$$

- When SPV power is unable to satisfy the charging demand of EVs, i.e., $\Delta P(t) < 0$, power is purchased from the grid which can be computed as

$$P_P^{grid}(t) = (P_{EVCS\text{-}dem}(t) - P_{SPV}(t)/\eta_{bdinv} \quad (55)$$

- If $\Delta P(t) = 0$, there is no exchange of power from grid, and SPV power fulfills the load requirement.

- When both SPV and grid are unable to satisfy the load demand, deficiency of power takes place, which can be given as

$$P_{def}(t) = P_{P}^{grid}(t) - P_{P,\ max}^{grid} \tag{56}$$

$P_{def}(t)$ must be zero to make sure the total power requirement of EV load demand is served in a reliable manner when minimizing the LPSP.

The LPSP can be kept within a specific tolerance band to solve the optimization problem. In this paper, the maximum limit of LPSP, i.e., $LPSP_{max}$ is assumed to be 1%. The energy management strategy for grid-and-solar PV-based EVCS can be demonstrated with the help of the flow chart shown in Figure 5.

4. Proposed Modified Salp Swarm Algorithm for Sizing of Hybrid Energy System

A brief discussion of the SSA and MSSA algorithms is presented in this section because they have yet to be examined for the application of optimal sizing of different energy sources for the designing of EVCS load in the northwest region of Delhi, India.

It is worth noting that the SSA is a robust heuristic algorithm that has been used to solve a variety of problems. However, it is susceptible to local optima in some instances and suffers from stagnation. To render and combat the shortcomings of the conventional SSA, an MSSA is presented. The suggested MSSA employs two different searching procedures to improve its searching potential. The first technique is to use Levy flight distribution (LFD) to improve the searching process, enabling individuals to move to new locations. The second technique relies on the individual spiraling around the optimum solution to improve the exploitation phase. Algorithm 1 [48] and Algorithm 2 [49] contain the pseudocode for the traditional SSA and the modified MSSA, respectively, whereas Figure 6 depicts the graphical representation of MSSA for SPV/DG/battery system optimization.

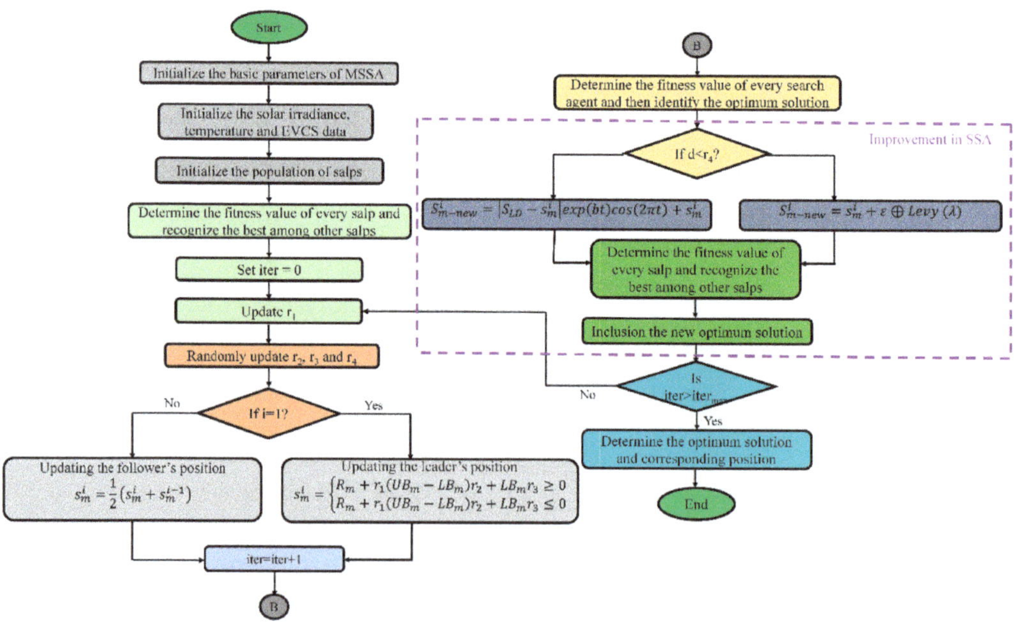

Figure 6. Graphical representation of MSSA for SPV/DG/battery system optimization.

4.1. Salp Swarm Algorithm (SSA)

In 2017, S. Mirjalili presented an individual-based algorithm known as salp swarm algorithm [48]. SSA models the mechanism of swarm salps, which will be scavenged in oceans. The salp at the front of the chain will be the leader, while the other salps will be considered followers in the ocean.

Like that of other meta-heuristic algorithms, the position of salp can be defined in n dimensions, where n is the number of variables for a specific problem. Hence, the positions of all salps are maintained in a two-dimensional matrix known as s. For example, it is assumed that the food source (R) is available in search space and considered as the target of all salps in a swarm. So, the describing equations of SSA are given as follows:

$$s_m^i = \begin{cases} R_m + r_1(UB_m - LB_m)r_2 + LB_m r_3 \geq 0 \\ R_m + r_1(UB_m - LB_m)r_2 + LB_m r_3 \leq 0 \end{cases} \tag{57}$$

where s_m^i indicates the position of leader, while R_m represents the position of the food source. UB_m and LB_m denote the upper and lower bound in m-th dimension of search space.

The coefficient r_1 plays a vital role in SSA because it maintains a proper balance between exploration and exploitation capability and can be expressed as follows:

$$r_1 = 2 \times \exp\left(-\left(\frac{4p}{P}\right)^2\right) \tag{58}$$

where p denotes the current iteration, while P represents the maximum iterations.

The parameters r_2 and r_3 are uniformly produced random values, which lie in the range of 0 and 1. They specify whether the next point in the m-th dimension should be toward positive or negative infinity, as well as the step size.

The positions of the follower salps are updated in accordance with the newton second law of motion, which is stated as follows:

$$s_m^i = \frac{1}{2}at^2 + u_o t \tag{59}$$

where $i \geq 2$ and s_m^i depict the position of i-th follower salp in m-th dimension, u_o denotes the initial velocity, and $a = \frac{u_{final}}{u_o}$, where $u = \frac{x - x_o}{t}$.

The simulation time is determined by the number of iterations, and if the conflict between iterations is equal to 1, and assuming $u_o = 0$, this equation will be as follows:

$$s_m^i = \frac{1}{2}\left(s_m^i + s_m^{i-1}\right) \tag{60}$$

The salp chains can be modeled with the help of Equations (57)–(60).

Algorithm 1: Pseudocode of conventional SSA method

Start SSA

Step 1: Fix the basic parameters of SSA algorithm, such as maximum iterations (P), population number, the upper and lower limits of the decision variables
Step 2: Randomly initialize the population of salps individuals
Step 3: Calculate the fitness function for each salp
Step 4: Set iteration p = 1
Step 5: Update r_1 according to Equation (58)
Step 6: For every member of salps, if i == 1, update the position of leading salp by Equation (57)
Otherwise, update the position of follower salp by Equation (60)
Fitness calculation of every salp : Update (S^*) if there is a superior solution
Increment t to 1
Repeat steps 5 to 7 until $p = P_{max}$ is satisfied
Step 7: Return the best solution (S^*) and its fitness value F (S^*)

End SSA

4.2. Modified Salp Swarm Algorithm (MSSA)

The MSSA is intended to address the shortcomings of the traditional SSA, such as its proclivity to local optima and stagnation in some circumstances [49]. To improve the capability of the traditional SSA, two search techniques are described. The first adjustment is used to speed up the exploration process by changing the positions of salps using the Levy flight distribution (LFD) to help the population find new places and break the cycle of stagnation. The second change entails changing and updating the locations of salps in the spiral path around the optimum solution to improve its exploitation. The first change, which is based on the LFD, is written as follows:

$$S^{i}_{m-new} = s^{i}_{m} + \varepsilon \oplus \text{Levy}\ (\lambda) \tag{61}$$

where ε denotes a step size parameter that is set as follows:

$$\varepsilon \oplus \text{Levy}\ (\lambda)\ \sim 0.01\frac{e}{|f|^{1/\lambda}}\left(S^{i}_{m} - S_{LD}\right) \tag{62}$$

where e and f can be obtained as follows:

$$e \sim N\left(0,\ \varnothing_e^2\right),\ f \sim N\left(0,\ \varnothing_f^2\right) \tag{63}$$

$$\varnothing_e = \left[\frac{\Gamma(1+\lambda)\times \sin(\pi\times\lambda/2)}{\Gamma[(1+\lambda)/2]\times\lambda}\right]^{1/\lambda},\ \varnothing_f = 1 \tag{64}$$

where Γ indicates the standard gamma function. The second modification is based on modifying the locations of salps in the spiral path around the best solution, which can be expressed as follows:

$$\varepsilon\varepsilon \oplus \text{Levy}\ (\lambda)\ \sim 0.01\frac{e}{|f|^{1/\lambda}}\left(S^{i}_{m} - S_{LD}\right) \tag{65}$$

The second adjustment is to change the salp locations in the spiral path around the optimum solution, which may be represented as follows:

$$S^{i}_{m-new} = \left|S_{LD} - s^{i}_{m}\right| \exp(bt)\cos(2\pi t) + s^{i}_{m} \tag{66}$$

where b is a number used to characterize the shape of a logarithmic spiral. The adaptive operator d is used to perform the job of balancing the exploration and exploitation processes:

$$d(t) = d_{min} + \left(\frac{d_{max} - d_{min}}{P_{max}}\right) \times p \tag{67}$$

where d_{max} and d_{min} are the maximum and minimum values of adaptive operator d. Lastly, the MSSA's exploring potential is enhanced by improving the exploration ability with the LFD by applying Equation (59) at the first iteration when the value of d is small, and the exploitation capability can be enhanced with the variable bandwidth transition by applying Equation (63) at the last iteration when the value of d is large.

Algorithm 2: Pseudocode of modified SSA method

Start MSSA

Step 1: Fix the basic parameters of SSA algorithm, such as maximum iterations (P), population number, the upper and lower limits of the decision variables
Step 2: Randomly initialize the population of salps individuals
Step 3: Calculate the fitness function for each salp
Step 4: Set iteration p = 1
Step 5: Update r_1 according to Equation (58)
Step 6: For every member of salps, if i == 1, update the position of leading salp by Equation (57)
Otherwise, update the position of follower salp by Equation (60)
Fitness calculation of every salp: Update (S^*), if there is a superior solution
Increment t to 1
Repeat steps 5 to 7 until $p = P_{max}$ is satisfied
Step 7: Arranging the salp according to their fitness values
Step 8: Update the d value according to Equation (67)
Step 9: Compare the value of d with a random value r_4
If the value of d is less than r_4
Update the salp positions based on the spiral orientation according to Equation (66)
Else
Update the salp positions based on Levy distribution according to Equation (61)
End
Step 10: Calculate the fitness function of the updated salps. Then, include the new solutions if these solutions are better than the solutions of the previous steps
Increment t to 1
Repeat steps 5 to 7 until ($p = P_{max}$) is satisfied
Step 11: Return the best solution (S^*) and its fitness value F (S^*)

End MSSA

5. Results and Discussion

5.1. Components and Methods

This section contains the important components, such as the place of the study area, resources available, and the load demand of the developed framework.

5.1.1. Area under Study

The proposed study is conducted in the northwest region of Delhi, India. The geographical coordinates of the studied area are 28.7408° N (latitude), 77.1126° E (longitude). Figure 7 depicts the study area from a geographical standpoint. The population of Rohini sector-17 is 21,460 individuals. Because the majority of citizens in this area are cultured and working, EVs are employed as a mode of transportation to get to their destinations on a day-to-day basis. This area has nearly 120–150 EVs, and there are 15 facilities in which these vehicles are kept and provided with electricity. As per the owner of different facilities, the amount required to charge the EVs is significantly high because these vehicles are charged using the grid.

5.1.2. Assessment of the Available Energy Sources

Solar radiation in Delhi is at its highest from March to October. It is available during the wintertime, but not at its high point. Additionally, it is regarded as an outstanding source of energy due to its practical features. Figure 8 shows the annual variation of irradiance at the chosen area. Figure 8 also depicts the clearness index, which is the percentage of solar radiation that is reflected by the atmosphere. It can also be defined as the ratio of surface irradiance and extraterrestrial irradiance. Figure 9 portrays the annual fluctuation of temperature for the study site. The DG is also incorporated as one of the energy providers in the proposed research work. Energy generation from DG is comparatively easy and cost effective as well. Thus, DG is used for charging EVs, especially during peak hours.

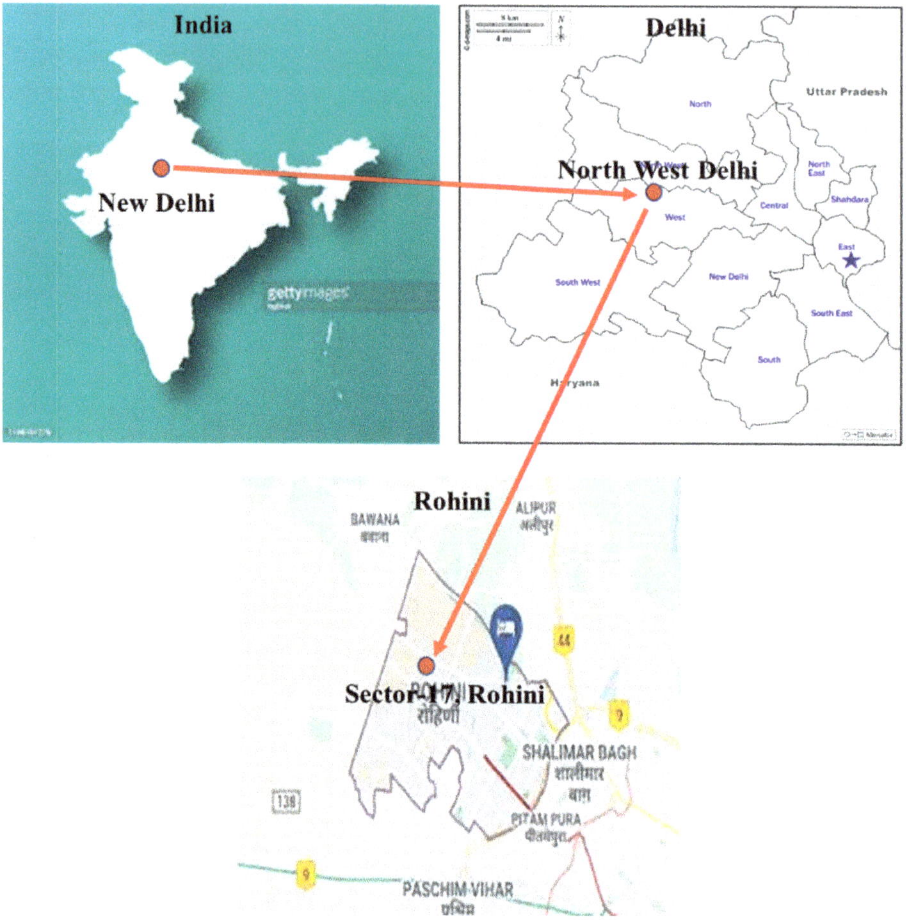

Figure 7. The geographical view of the study area.

Figure 8. Annual variation of solar insolation and clearness index of the selected area.

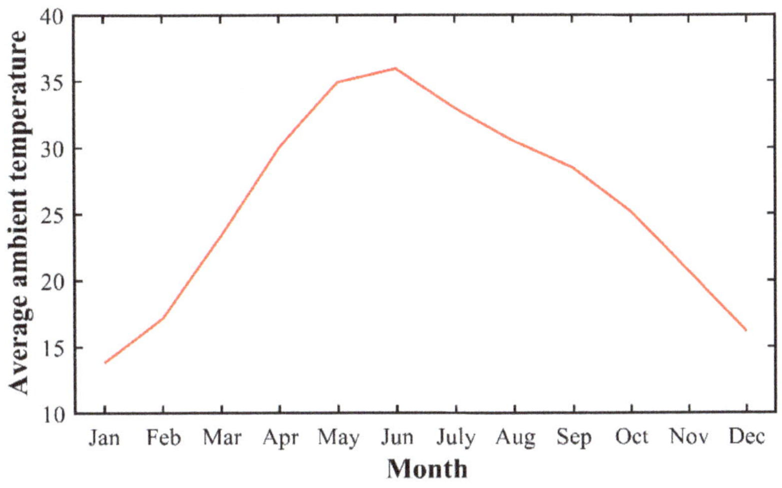

Figure 9. Annual variation of the mean temperature of the selected area.

5.1.3. Assessment of the EVCS Load in the Selected Area

The research study considers three-wheeler EVs, i.e., e-rickshaw, to be the system's load. Four 12-Volt lead-acid batteries with capacities of 160 to 180 Ampere hour are used in a three-wheeler. The theoretical battery energy is 2.2 kWh when the 12 V and 180 Ampere hour battery is considered. Generally, a 12 V battery is recharged to 90% of its design capacity. As a result, the battery's energy would be 1.94 kWh. The total energy consumption for four 12 V battery vehicles, such as e-rickshaws, would be 7.8 kWh. Taking the system's 15% system losses into account, the energy consumed is determined to be 9 kWh. Thus, the energy consumed by every e-rickshaw per day is 9 kWh. According to the research findings, the yearly average energy is 257 kWh/day, the average power is 10.71 kW, and the maximum load is 25.45 kW. A graphical representation of the EVCS load on an hourly and daily basis is shown in Figures 10 and 11, respectively.

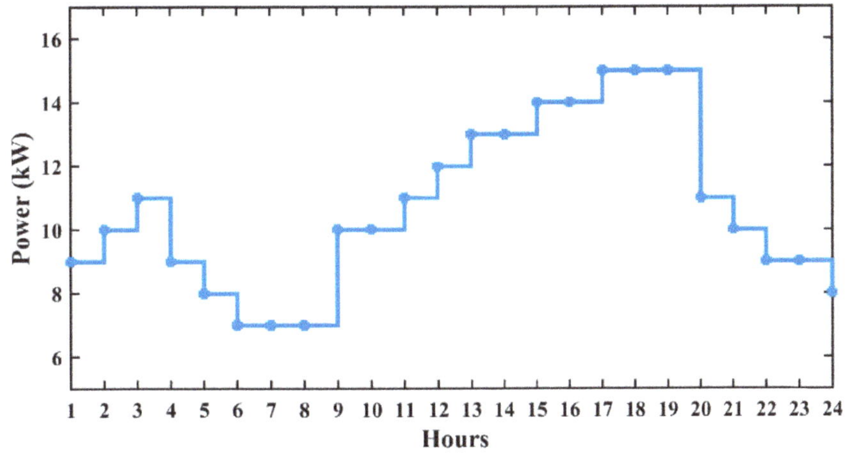

Figure 10. EVCS load profile on a daily basis.

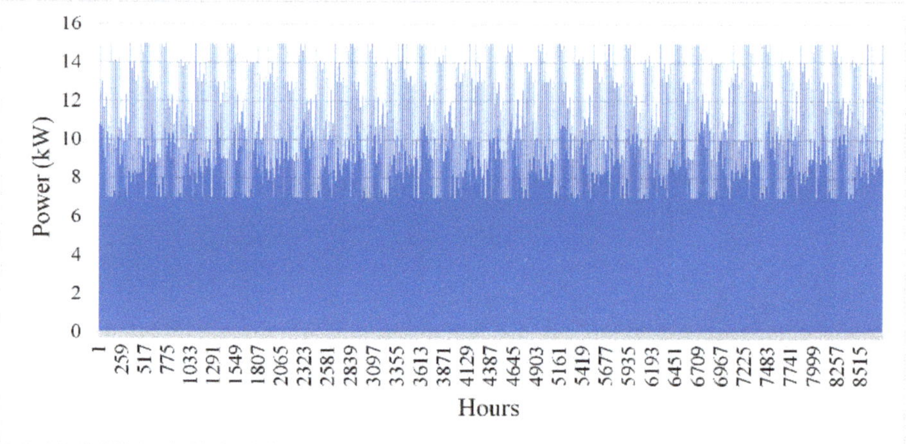

Figure 11. EVCS load profile on an hourly basis.

5.2. Implementation of Proposed MSSA Technique on Benchmark Functions

This section explores the application of the proposed MSSA technique on standard benchmark functions. Table 3 contains comprehensive descriptions of these functions. The proposed optimization strategy outperforms SSA and GWO in terms of maximum iterations and optimal solution of seven mathematical benchmark functions for 30 independent runs. MSSA adheres to a globally optimal solution without becoming stuck in a local optimal solution, resulting in faster convergence. Table 4 displays the numerical outcomes in terms of optimum solution and number of iterations.

Table 3. Standard benchmark functions.

S. No	Function Name	Mathematical Formulation	Dimension (D)	Search Range
1.	Sphere	$f(x) = \sum_{p=1}^{D} x_p^2$	30	[−100, 100]
2.	Rosenbrock	$f(x) = \sum_{p=1}^{D} \left[100 \times \left(x_p^2 - x_{p+1}\right)^2 + (1 - x_p)^2\right]$	30	[−2.048, 2.048]
3.	Rastragin	$f(x) = \sum_{p=1}^{D} \left[x_p^2 - 10\cos(2\pi x_p) + 10\right]$	30	[−5.12, 5.12]
4.	Greiwank	$f(x) = \frac{1}{4000} \sum_{p=1}^{D} x_p^2 - \prod_{p=1}^{D} \cos\left(\frac{x_p}{\sqrt{p}}\right) + 1$	30	[−600, 600]
5.	Schewefel	$f(x) = \sum_{p=1}^{D} \left(\sum_{q=1}^{p} x_q\right)^2$	30	[−100, 100]
6.	Ackley	$f(x) = -20e^{\left(-0.2\sqrt{\frac{1}{D}\sum_{p=1}^{D} x_p^2}\right)} - e^{\left(\frac{1}{D}\sum_{p=1}^{D} \cos(2\pi x_p)\right)} + 20 + e^1$	30	[−32.76, 32.76]
7.	Alpine	$f(x) = \sum_{p=1}^{D-1} \lvert x_p \sin x_p + 0.1 x_p \rvert$	30	[−10, 10]

Table 4. Numerical results obtained using MSSA and their comparison with SSA and GWO.

Function Name	MSSA		SSA		GWO	
	Maximum Iterations	Optimal Solution	Maximum Iterations	Optimal Solution	Maximum Iterations	Optimal Solution
Sphere	1000	1.55×10^{-23}	897	5.89×10^{-19}	1000	6.25×10^{-19}
Rosenbrock	415	2.72×10^{-3}	537	5.81×10^{1}	548	4.57×10^{2}
Rastragin	338	7.57×10^{-14}	458	5.53×10^{-3}	495	2.53×10^{-2}
Greiwank	973	4.25×10^{-1}	1000	3.45×10^{-3}	1000	4.87×10^{-1}
Schewefel	1000	8592.6	984	7956.4	1000	6085.8
Ackley	1000	7.74×10^{-7}	785	9.14×10^{-6}	1000	8.54×10^{-3}
Alpine	1000	−192.73	1000	−192.45	1000	−192.50

5.3. Simulation Outcomes and Main Findings

This section evaluates the technical and economic parameters for three different scenarios using HOMER software and the proposed MSSA technique. These scenarios are SPV- and DG-based EVCS, only SPV-based EVCS, and grid-and-SPV-based EVCS. The objective is to achieve the best EVCS design by reducing the total net present cost, levelized cost of energy while keeping the lack of power supply probability within limits, accounting for environmental emissions, and different energy supply options. The outcomes are estimated in terms of TNPC, LCOE, SPV energy, the contribution of DG, battery and converter capacity, additional electricity generation, and the proportion of renewables utilized. The study then compares a hybrid energy system with a grid connection to a standalone system. The analysis of the obtained results for different scenarios is summarized in the latter part of this section. Furthermore, to demonstrate the efficacy of meta-heuristic techniques, different algorithms are used to analyze the results and are compared and contrasted with HOMER results. The three different scenarios considered are as follows:

- Scenario 1: Solar PV and diesel generator–based EVCS

This scenario addresses "range anxiety," a prevalent consideration among EV users about the vehicle's range. EV users can schedule long trips with greater certainty if an EVCS is as easily accessible as a petrol pump. Hence, the layout of an EVCS on roadways with SPV, DG, and a BESS with the different electricity providers is investigated and considered as Scenario 1. The goal is to reduce the capital, operation and maintenance, replacement, and fuel cost of every component associated with the system. The size and number of the SPV, DG, battery, and converter are considered to be the decision variable of the optimization problem.

- Scenario 2: Only Solar PV–based EVCS

In some countries, such as Saudi Arabia, the DG acts as a primary source of energy. In such cases, the diesel option is a good example. On the other hand, DG units are prohibitively expensive due to the high costs of maintenance, fuel supply, and fuel transport. Furthermore, DGs emit a lot of pollution. Hence, SPV- and battery-based EVCS are also being investigated and considered as Scenario 2. In this scenario, the size and number of solar PVs and batteries are the only decision variables.

- Scenario 3: Grid-and-solar PV–based EVCS

In this scenario, the EVCS is supposed to be grid-connected, and solar PV, battery, and grid are the various energy supply sources. Because EVCS is grid connected, it can purchase/sell power from/to the grid network, and the design is improved to account for the EVCS's net costs. The price of electricity provided in the Feed-in Tariff program of New Delhi, India, is taken into account for drawing/selling power from/to the utility grid. Feed-in Tariffs are the rates charged to RE providers for the energy provided by their power plants. The evaluating model is distinguished by project length and technology method and offers a profitable return on investment. The Feed-in-Tariff is a standard rate for the buying and selling of power from and to the grid network throughout the agreement, whereas the Time-of-Use Tariff is only pertinent to buyers, in contrast to Feed-in Tariffs, and varies throughout the day over time. In this work, Feed-in Tariffs are considered.

5.3.1. Techno–Economic Analysis of Different Scenarios Using HOMER Software

HOMER is used to model the hybrid energy system to meet the load demand. HOMER defines weather information, different parametric values of components used, and EVCS load profile. Many different assumptions are made to attain the optimal plan that decreases the TNPC, LCOE, while maintaining restrictions, such as supply continuity. It is worth noting that the information related to cost is indicated in USD.

Table 5 shows the results obtained for Scenario 1, i.e., SPV/DG/battery system. SPV makes a significant contribution of 122,864 kWh per year and operates for 4369 h per year, while the DG constitutes 2767 kWh per year and functions for 364 h per year. To operate optimally, this system makes use of 590 batteries and a 12.1 kW bi-directional converter. The obtained system has a TNPC of USD 638,917.29, an initial capital cost of USD 375,855, and an operating cost of USD 16,144.69. The LCOE for this configuration is calculated to be USD 0.6328/kWh. Furthermore, the NPC of SPV and diesel generator is 42.8 percent and 11.1 percent of TNPC, respectively, whereas the NPC of the bi-directional converter is only 0.58 percent of TNPC.

Table 5. Results attained using HOMER software for Scenario 1.

SPV (kWh/Year)	DG (kWh/Year)	Batteries Used	Converter (kW)	Initial Capital Cost (USD)	Operating Cost (USD)	TNPC (USD)	LCOE (USD/kWh)	% of Renewables Utilized
122,864	2767	590	12.1	375,855	16,144.69	638,917.29	0.6328	96.5

Similarly, Tables 6 and 7 display the outcomes of Scenarios 2 and 3, respectively. When the only energy-producing source is an SPV and a battery, the LCOE is USD 0.7318/kWh, the TNPC is USD 875,481.24. In Scenario 3, when the grid is used to meet load demand, the LCOE is USD 0.587/kWh, and the TNPC is USD 275,349. In this scenario, the energy produced by SPV is 126,811 kWh per year, accounting for 71.3 percent of total energy production, while energy purchased from the grid is 51,155 kWh per year, accounting for 28.7 percent of total energy generation.

Table 6. Results attained using HOMER software for Scenario 2.

SPV (kWh/Year)	Batteries Used	Converter (kW)	Initial Capital Cost (USD)	Operating Cost (USD)	TNPC (USD)	LCOE (USD/kWh)	% of Renewables Utilized
298,911	590	-	647,261	14,079.41	875,481.24	0.7318	100

Table 7. Results attained using HOMER software for Scenario 3.

SPV (kWh/Year)	Grid Purchase (kWh/Year)	Batteries Used	Converter (kW)	Initial Capital Cost (USD)	Operating Cost (USD)	TNPC (USD)	LCOE (USD/kWh)	% of Renewables Utilized
126,811	51,155	590	49.4	228,402	2705	275,349	0.587	70.1

5.3.2. Techno-Economic Analysis of Different Scenarios Using Meta-Heuristic Algorithms

This section contains a summary of the optimization results obtained using various meta-heuristic techniques, as well as a comparison of those results for the various scenarios. The MSSA, SSA, and GWO are used in 30 separate runs with a population size of 50 and a maximum iteration of 100, with the best results cropped. The basic parameters used in the optimization problem are tabulated in Table 8. The three considered algorithms are executed for each scenario.

Table 8. Parameters used in MSSA.

Parameters	Value
Population size	50
Maximum iterations	100
d_{min}	0.34
d_{max}	0.79

- Techno–economic analysis for Scenario 1 (solar PV and diesel generator-based EVCS)

Table 9 shows a comparison of results for Scenario 1 using MSSA, SSA, and GWO in terms of LCOE, TNPC, and computational time. According to Table 9, MSSA has the lowest TNPC, LCOE, and computational time with USD 584,566.44, USD 0.4822/kWh, and 27,615.716 s, respectively, due to its lower capital, replacement, and operation and maintenance costs. MSSA takes less time than other compared techniques and gives minimum TNPC and LCOE as well. SSA comes in the second place, with high TNPC, LCOE, and taking more time to compute the results as compared to MSSA, whereas GWO gives a high value of TNPC, LCOE, and is slowest in comparison to the previous two meta-heuristic techniques. This system necessitates 232 SPV panels with 325 W energy, 3284 kWh of energy production per year from a diesel generator rated at 10 kW, 650 battery units, and 9 bi-directional converters rated at 100 kW. The maximum energy flow from the DC bus to the AC bus or vice versa determines the energy of a bi-directional converter.

Table 9. Comparative study of achieved outcomes using meta-heuristic techniques for Scenario 1.

Methods Used	Solar PV (325 W)	DG Power Production (kWh/Year)	Batteries	Converter (100 kW)	TNPC (USD)	LCOE (USD/kWh)	Computational Time (Sec)
MSSA	232	3284	650	9	5,84,566.44	0.4822	27,615.716
SSA	220	3056	580	9	6,04,482.82	0.5381	29,834.165
GWO	190	2859	610	9	6,77,615.38	0.6328	37,656.673

Figure 12 depicts hourly information about power generation for a three-day sample (1–3 October). It is observed that maximum EVs are charged during night hours, and there is no solar output available during that time. Hence, the energy requirement is fulfilled by DG and BESS using energy flow through the converter, as indicated in Figure 12. When PV power generates more than the load demand during the day, surplus electricity is used to charge the battery (Figure 12). Despite having a higher SPV energy, the converter energy is much lower in this context of system sizing. The amount of excess electricity produced in this scenario is 25,053 kWh per year, which is 19.2 % of total generated energy. It is worth noting that, despite the fact that the maximum energy shortage was set to zero, there was only a 23.7 kWh unmet load throughout the simulation model, accounting for 0.02 percent of the total load.

Figure 13 depicts a cost description of various system components for the SPV/DG/battery system. The SPV module and battery entail the most capital investment. Despite the fact that the BESS has a high cost of replacement over the project's lifespan, the resource cost is significantly more than the investment. As a result, this hybrid energy system needs constant economic infusion to keep the system running properly.

As shown in Figure 14, the DG contributes significantly when the load demand is high. The monthly average power production from the SPV module is lesser in summer weather in comparison to winter because of continuous rain and clouds in the summertime. Further, the effectiveness of the proposed algorithm, i.e., MSSA, can be proved using Figure 15. It has better convergence than other algorithms and results in a lower LCOE value.

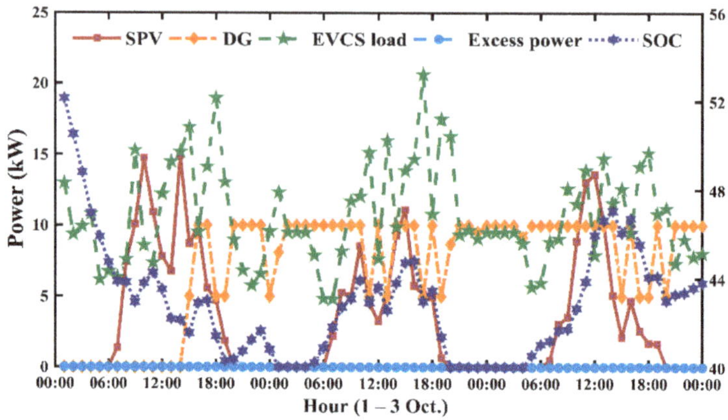

Figure 12. Power analysis on an hourly basis of SPV/DG/battery system.

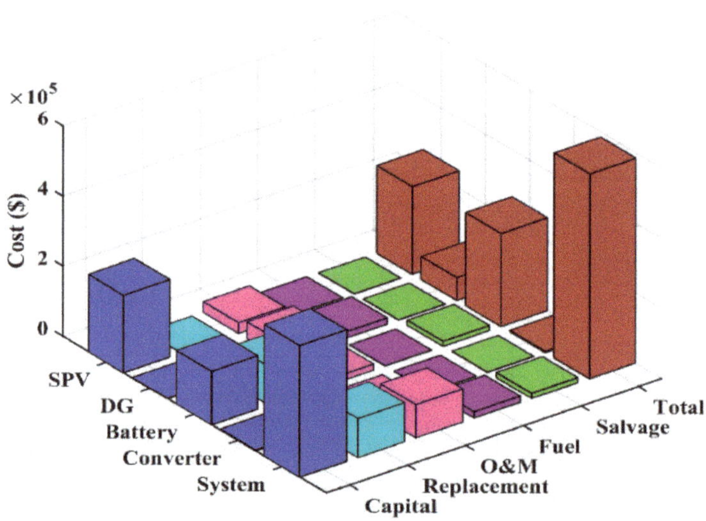

Figure 13. Cost description of the various system components for the SPV/DG/battery system.

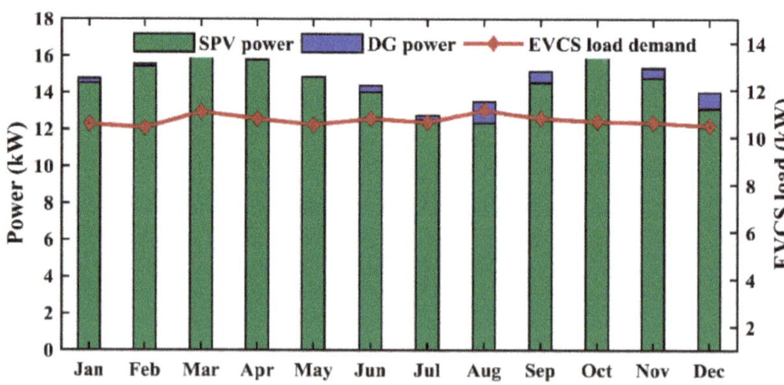

Figure 14. Monthly average power share for satisfying EVCS load demand in SPV/DG/battery system.

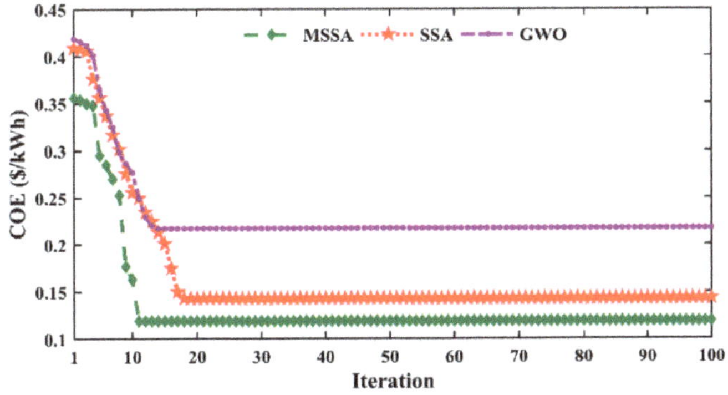

Figure 15. Convergence curves of different optimizers.

- Techno–economic analysis for Scenario 2 (only solar PV/battery-based EVCS)

In Scenario 2, diesel generator is not available, and solar PV is the only energy source available to meet the EVCS load requirement. This PV/battery system has the greatest LCOE and TNPC because it uses a larger PV module and battery storage to meet power demands, which increases the cost and produces more surplus power (194,140 kWh per year). Table 10 provides the different technical and economic parameters optimized by various algorithms for SPV/battery system.

Table 10. Comparative study of achieved outcomes using meta-heuristic techniques for Scenario 2.

Methods Used	Solar PV (325 W)	Batteries	Converter (100 kW)	TNPC (USD)	LCOE (USD/kWh)	Computational Time (Sec)
MSSA	545	548	12	843,461.28	0.6844	25,284.342
SSA	490	518	12	902,761.37	0.7045	29,451.984
GWO	415	476	12	981,736.38	0.7167	35,324.673

As in Scenario 1, computation time taken by MSSA is less as compared to other techniques. This system requires 545 solar PV panels with 325 W energy, 548 battery units, and 12 bi-directional converters rated at 100 kW. MSSA gives better results when equated with other techniques and, hence, proved its efficacy to solve various optimization problems. The hourly information about power generation for a three-day sample for SPV/battery system is depicted in Figure 16. The cost description of the various system components for the SPV/battery system is shown in Figure 17.

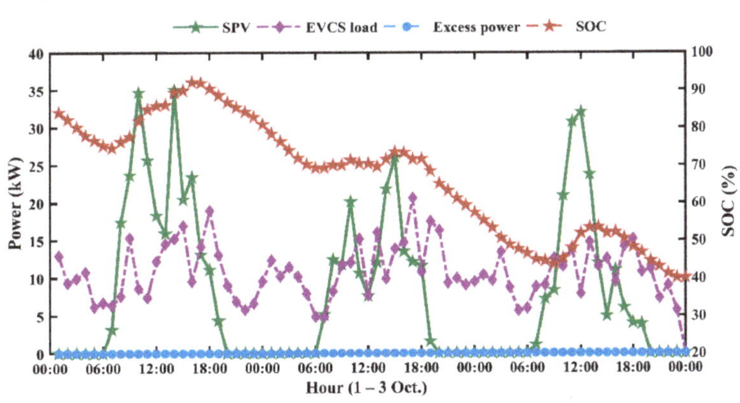

Figure 16. Power analysis on an hourly basis of SPV/battery system.

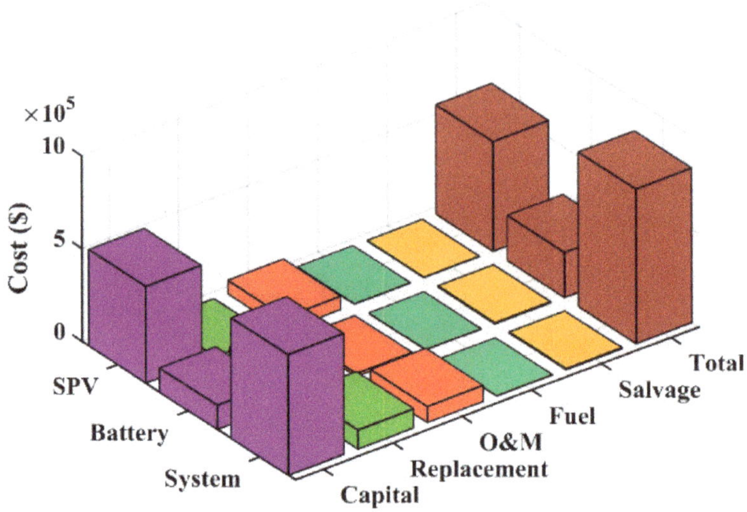

Figure 17. Cost description of the various system components for the SPV/battery system.

- Techno–economic analysis for Scenario 3 (grid-and-solar PV-based EVCS)

This scenario looks at the technological, economic, and environmental concerns of hybrid energy systems connected to the grid. In this strategy, electricity is purchased from the grid to meet power requirements when solar PV is unable to fulfill them. Excess power is sold back to the grid, requiring a small storage capacity while leveraging the monumental amount of additional energy dumped in hybrid energy options. In this respect, a fixed grid power price is chosen to be USD 0.12/kWh and a fixed grid sell back price is selected to be USD 0.08/kWh. According to the results in Table 11, the grid-connected solar PV-based EVCS has a significantly lower LCOE (0.119/kWh) than the solar PV and diesel generator-based EVCS (USD 0.482/kWh) and solar PV/battery-based EVCS (USD 0.684/kWh). The hybrid energy system in Scenarios 1 and 2 requires more resources and battery storage than the grid-connected system. whereas in the grid-connected system, net grid purchase is zero, i.e., the amount of energy purchased from the grid is equal to the amount of energy sold to the grid, and no battery storage is required. Despite having a 232 PV module of 325 W energy and a 49.4 kW bi-directional converter, the grid-connected hybrid system needs no battery storage and has insubstantial energy costs due to a significantly lower LCOE (USD 0.119/kWh) compared to a solar PV and diesel generator-based EVCS (USD 0.482/kWh) and solar PV/battery-based EVCS (USD 0.684/kWh). It is also worth noting that MSSA has a lower TNPC (USD 263377) and LCOE (USD 0.119/kWh) than SSA and GWO. More prominently, the extra electricity in the grid-connected system (408 kWh/year) is considerably lower than in the solar PV and diesel generator-based EVCS (25,053 kWh/year) and solar PV/battery-based EVCS (194,140 kWh/year). This is because a large portion of excess energy is sold back to the grid, and only a few storage devices are needed to meet the requirement when solar power is not available.

Table 11. Comparative study of achieved outcomes using meta-heuristic techniques for Scenario 3.

Methods Used	Solar PV (325 W)	Grid Purchase (kWh/Year)	Grid Sales (kWh/Year)	Converter (49.4 kW)	TNPC (USD)	LCOE (USD/kWh)	Computational Time (Sec)
MSSA	232	51,155	77,135	2	263,377	0.119	29,812.873
SSA	218	52,671	76,432	2	276,543	0.143	30,341.457
GWO	196	54,367	76,197	2	298,654	0.217	32,349.952

When solar power is not accessible, grid power meets the load requirements. According to Figure 18, the spring season necessitates a high amount of energy to be purchased from the grid due to higher power requirements. India, as a tropical country, has nearly equal solar exposure all year. Even though the summer season is supposed to generate more electricity from solar PV, rainy days lead to decreasing SPV production than the rest of the regular time of year.

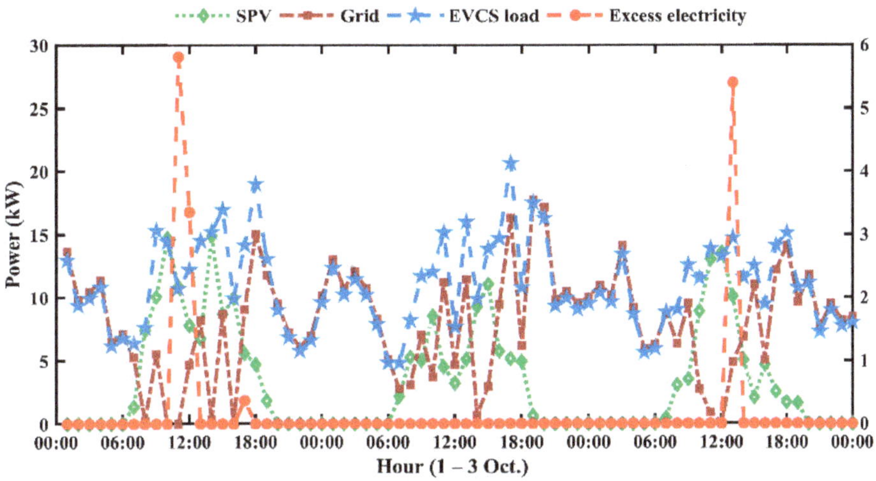

Figure 18. Power analysis on an hourly basis for grid-and-solar PV-based EVCS.

Solar energy accounts for approximately 71.3 percent of total energy demand, with grid power accounting for the remaining 28.7 percent. Figure 19 depicts the time series data for various components used to satisfy the load demand. The cost description of the various system components for the grid-and-solar PV-based EVCS is shown in Figure 20. The monthly average energy purchased/sold from/to the grid is portrayed in Figure 21.

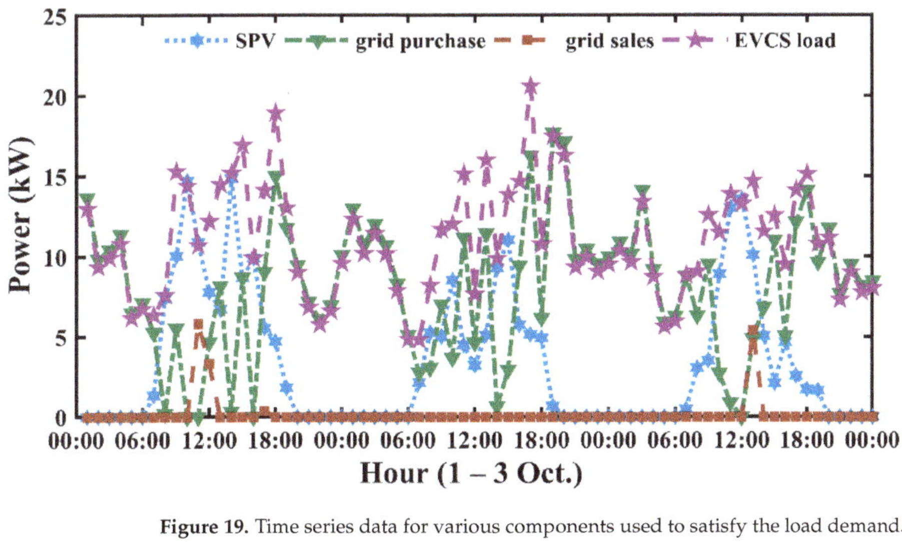

Figure 19. Time series data for various components used to satisfy the load demand.

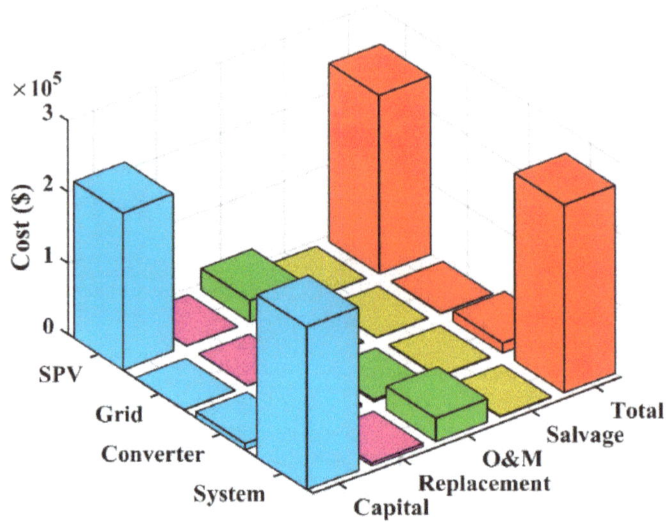

Figure 20. The cost-wise breakup of different components in grid-and-solar PV-based EVCS.

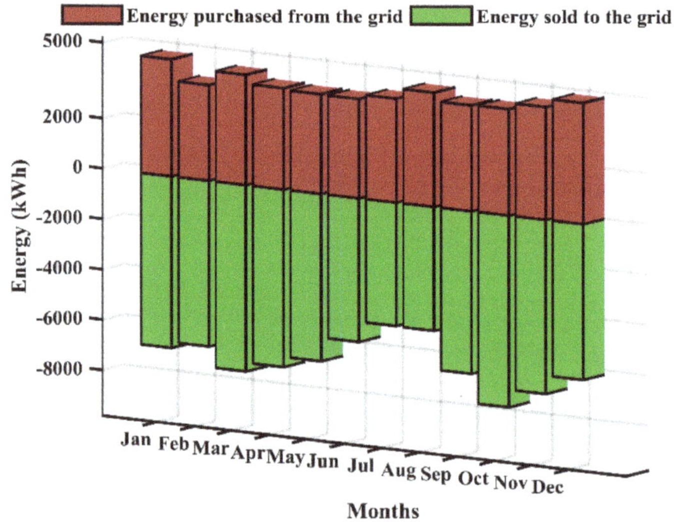

Figure 21. Monthly average energy purchased from the grid and sold to the grid.

- Impact of grid purchase and grid sales on LCOE

The implications of energy buying from grid infrastructure and the quantity of energy being sold to the grid on the LCOE are outlined and discussed in this section. A net grid procurement of 0% indicates that the quantity of energy buying from and sold back to the grid is the same. The value of LCOE increases more with increasing purchases from the grid than with increasing grid sell-back, as shown in Figure 22. According to Figure 22, an increase in grid energy to fulfill load demand leads to an increase in LCOE. For example, a 50% increase in grid energy purchase versus selling back to the grid results in a 36% increase in LCOE.

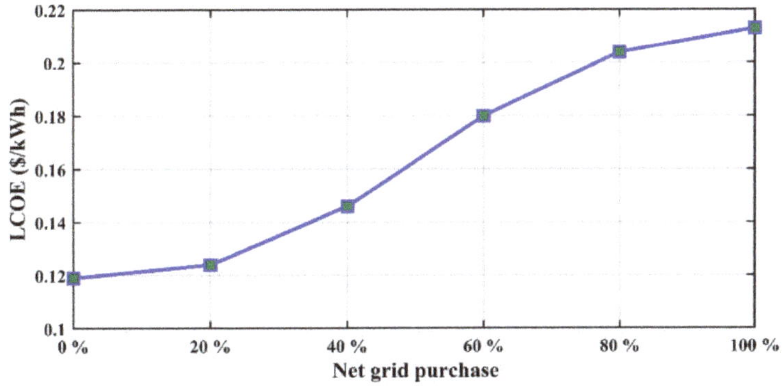

Figure 22. Impact of net grid purchase on LCOE.

5.3.3. Analysis of the Environmental Emissions in Different Scenarios

As previously stated, one of the primary goals of this work is to lower emissions by utilizing renewable technologies. The results displayed in Table 12 show that SPV/battery-based EVCS (Scenario 2) reduces the overall emissions by a significant amount when equated to all other cases. However, while the SPV/DG/battery-based EVCS (Scenario 1) produces higher emissions than SPV/battery-based EVCS, it will be far more ecologically friendly than the grid-and-SPV-based EVCS.

Table 12. Scenario-wise comparison of environmental emissions.

Pollutants	Scenario 1	Scenario 2	Scenario 3
Carbon dioxide (CO_2) (Kg/year)	2910	2417	32330
Carbon monoxide (CO) (Kg/year)	22	0	0
Unburned hydrocarbons (UHC) (Kg/year)	0.802	0	0
Particulate matter (PM) (Kg/year)	1.33	0	0
Sulphur dioxide (SO_2) (Kg/year)	7.14	4.94	140
Nitrogen dioxide (NO_x) (Kg/year)	25	0	68.5

6. Future Challenges and Opportunities

Along with industrial and organizational advancement to promote RE set up in India, identifying the potential obstacles to project implementation is important. The hybrid energy system's infrastructure and maintenance costs have hampered significant advances in RE technologies by increasing the COE. As a developing nation, most of the public are not able to avail high LCOE in comparison to traditional grid purchases. Financial constraints, such as ineffective economic pathways, intense competition with fossil fuels, a scarcity of shareholders, and very few public subsidies, all had a substantial influence on the growth of hybrid energy systems [50]. Furthermore, less economic investment and funding constraints as a result of the low lending charges and longer repayment time make RE investment dangerous. Although economic inflation and growing population necessarily dictate electricity requirements, substituting the current infrastructure with 100 percent renewable use is neither technically nor economically feasible due to ambiguity in energy price in the future [51]. There are still technical intricacies, such as specifications, indicators, processes, instructions, and power management systems for hybrid energy systems. In the meantime, research efforts to overcome such technological challenges are critical and must be taken into consideration when planning a hybrid energy system [52].

Despite their multiple issues, the hybrid energy systems provide lots of opportunities in the area of natural, cultural, financial, and energy security concerns. The hybrid energy system will reduce greenhouse gas emissions and protect the natural environment from deterioration. The steep decline in PV and wind prices will hasten the implementation of a hybrid energy system with a massive decrease in original investment. The International Renewable Energy Agency (IRENA) forecasted a 74% and a 24% decline in SPV and WT prices by 2020, respectively. Furthermore, continuing research and rapid technological advancements encourage RE systems to start competing with traditional energy sources. Effective hybrid technology selection, taking into account geographical region and environmental consequences, will broaden the existing energy resources and minimize reliance on non-renewable resources. Furthermore, these schemes will have an impact on employment options, the standard of living, business growth, health problems, economic status, women's rights, and economic development. RE is among the most basic needs for human progress, and it can help to generate earnings, improve healthcare and education performance, and poverty eradication. India is geologically favored for the spread of solar energy. Despite the numerous advantages of RE sources, the bulk of people are unwilling to accept the advanced technologies. The lack of knowledge shared among stakeholders is needed for the implementation of the RE-based hybridized micro-grid. In this respect, the government of India can introduce such projects through a public–private partnership framework, in which a group of developers, with support from the government, will participate in putting such projects in place.

7. Conclusions

Hybrid energy systems can deliver energy to the grid and off-grid sites cost effectively and reliably. The efficiency of the system improves when RE components are implemented with the grid. In this article, the complete mathematical modeling is presented for the diverse configurations of the hybrid energy system to meet the load requirements of EVCS situated in the northwest region of Delhi, India. The financial, technological, and ecologic implications of different configurations are also carefully investigated. The use of meta-heuristic methods, which include MSSA, SSA, and GWO, is carried out to optimize the objective functions. The obtained results using MSSA are compared with HOMER software and other techniques. According to comparisons, MSSA gives a more efficient solution than HOMER. Furthermore, the MSSA offers a robust framework, which aids model development. In all three configurations, the MSSA results in reduced LCOE values of USD 0.482/kWh, USD 0.684/kWh, and USD 0.119/kWh, respectively. According to the analysis, the grid-connected solar PV-based EVCS offers significant cost savings over the other two configurations. In addition, a hybrid energy system with equal grid purchase and sell-back can be a profitable choice. More significantly, the LCOE associated with the grid selling back schemes is lower than the grid purchase price. The results show that a 60% increase in grid sell-back price leads to a 46% decrease in LCOE.

Permitting tax-free imports can reduce the cost of hardware devices, lowering the COE. The government of India must enable full cooperation in research and technology by generating opportunities for cooperation and entrepreneurship communication between societies. Even though the hybrid micro-grid alternative is a lucrative, consistent, and long-term remedy, people in rural areas cannot accommodate it due to the increased investment. As a result, more alternative financial assistance from government and non-governmental organizations can address the underlying intricacy of the issue.

The feasibility analysis discussed in this work can be used to guide the development and operation of hybrid energy system applications in distant regions where grid utility is not accessible, as well as hybrid grid-connected systems in India. Future research is needed to investigate the impacts of charging/discharging cycles of a battery on its lifespan and cost of energy.

Author Contributions: Conceptualization, M.B. and I.A.; formal analysis, M.B., I.A., M.A. and F.M.A.; funding acquisition, I.A., M.A. and F.M.A.; investigation, M.B. and M.R.; methodology, M.B., I.A., M.A. and F.M.A.; project administration, M.R. and I.A.; supervision, M.R.; writing—original draft, M.B.; writing—review and editing, M.B., M.R., I.A., M.A. and F.M.A. All authors have read and agreed to the published version of the manuscript.

Funding: The research work is funded by the Chair of Prince Faisal for Artificial intelligence research (CPFAI) through project number QU-CPFAI-2-9-5.

Institutional Review Board Statement: Not applicable.

Informed Consent Statement: Not applicable.

Data Availability Statement: Not applicable.

Acknowledgments: The authors would like to thank the Chair of Prince Faisal for Artificial intelligence research (CPFAI) for funding this research work through project number QU-CPFAI-2-9-5. Additionally, the authors would like to extend their appreciation to the Deputyship for Research & Innovation, Ministry of Education, and the Deanship of Scientific Research, Qassim University for their support for this research.

Conflicts of Interest: The authors declare no conflict of interest.

Nomenclature

EV	Electric Vehicle
EVCS	Electric Vehicle Charging Station
SPV	Solar Photovoltaic
DG	Diesel Generator
HOMER	Hybrid Optimization of Multiple Energy Resources
GWO	Gray Wolf Optimization
SSA	Salp Swarm Algorithm
MSSA	Modified Salp Swarm Algorithm
RE	Renewable Energy
GHG	Green House Gases
WT	Wind Turbine
NBR	Number of Battery Replacements
RF	Renewable Fraction
GA	Genetic Algorithm
PSO	Particle Swarm Optimization
BESS	Battery Energy Storage System
NPC	Net Present Cost
TNPC	Total Net Present Cost
LCOE	Levelized Cost of Energy
LPSP	Lack of Power Supply Probability
MAS	Multi Agent System
MFO	Moth Flame Optimization
SOC	State of Charge
NOCT	Nominal Operating Cell Temperature
CRF	Capital Recovery Factor
DOD	Depth of Discharge

List of symbols

$P_{SPV\text{-}out}(t)$	Solar photovoltaic power output
η_{SPV}	Efficiency of solar panel
A_{SPV}	Surface area of the solar panel
$G_h(t)$	Hourly solar irradiance falling on the surface of solar panel
T_{cell}	Temperature of cell
T_a	Ambient temperature
P_{max}	Maximum power output of solar panel
V_{max}	Maximum value of voltage
I_{max}	Maximum value of current
N_{SPV}	Number of installed solar PV panel
$u(t)$	Fuel consumption in litres per hour
$P_{DG}(t)$	Power output of diesel generator in kW
$P_{rated\text{-}DG}$	Rated power of diesel generator
a	Constant value (0.246)

b	Constant value (0.08415)
$\eta_{overall}$	Overall efficiency of DG
$\eta_{brake\ thermal}$	Brake thermal efficiency of DG
$SOC_{bat}(t+1)$	State of charge of battery at time instant $(t+1)$
$SOC_{bat}(t)$	State of charge of battery at time instant (t)
σ	Self-discharge rate of battery
η_{bdinv}	Efficiency of bi-directional inverter
η_{bat}	Round trip efficiency of the battery
η_{bat}^{charge}	Charging efficiency of battery
$\eta_{bat}^{discharge}$	Discharging efficiency of battery
C_{total}	Total capacity of battery bank
N_{bat}	Total number of batteries
N_{bat}^{series}	Number of batteries connected in series
C_{sbat}	Capacity of a single battery
NPC_{bat}	Net present cost of battery
C_{bat}^{CAP}	Capital cost of battery
C_{bat}^{REP}	Replacement cost of battery
C_{bat}^{OM}	Operation and maintenance cost of battery
C_{bat}^{SAL}	Salvage cost of battery
ψ_{bat}	Cost of one battery in USD
ω_{bat}	OM cost of battery
μ_{bat}	Escalation rate of battery
ε_{bat}	Resale value of one battery in USD
ζ_{bat}^{rep}	Replacement cost of battery
NPC_{bdinv}	Net present cost of battery
C_{bdinv}^{CAP}	Capital cost of bi-directional inverter
$P_{S}^{grid}(t)$	Extra energy to be sold to the grid in kWh
$SOC_{bat}^{min}(t)$	Minimum SOC of battery
$SOC_{bat}^{max}(t)$	Maximum SOC of battery
D_{Km}	Number of kilometers traveled
$E_{req/km}$	Energy needed per kilometer
T	Duration of charging
Q_{bat}^{EV}	Capacity of EV battery
N_{SPV}	Number of solar panels
V_{bus}	Voltage level of bus
V_{bat}	Voltage level of single battery
P_{bat}^{max}	Maximum input/output power of battery.
$P_{P}^{grid}(t)$	Energy to be borrowed from the grid in kWh
$P_{D}^{EV}(t)$	Energy demand of EV
$P_{EVCS\text{-}dem}(t)$	Power demand of EVCS
NPC_{SPV}	Net present cost of solar panel
C_{SPV}^{CAP}	Capital cost of solar panel
C_{SPV}^{REP}	Replacement cost of solar panel
C_{SPV}^{OM}	Operation and maintenance cost of solar panel
C_{SPV}^{SAL}	Salvage cost of solar panel
ψ_{SPV}	Initial cost and of each SPV panel
P_{SPV}	Power rating of each SPV panel
ζ_{SPV}^{rep}	Cost of replacing the SPV panel
R	Rate of interest
Ω	Life span of project in years
ω_{SPV}	OM cost of each panel
μ_{SPV}	Escalation rate
ε_{SPV}	Resale price solar panel after completing their life
∂	Inflation rate
ε_{bdinv}	Resale value of bi-directional inverter in USD

ζ_{bdinv}^{rep}	Replacement cost of bi-directional inverter
ξ_{S}^{grid}	Cost for unit sale of electricity to the utility grid
ξ_{P}^{grid}	Cost for unit purchase of electricity from the utility grid
C_{P}^{grid}	Cost of purchasing electricity from the grid
C_{S}^{grid}	Cost of selling electricity to the grid
$P_{gen}(t)$	Total power generated by energy components
E_{bat}^{min}	Minimum energy of battery bank
E_{bat}^{max}	Maximum energy of battery bank
P_{def}	Deferrable load

References

1. Available online: https://mnre.gov.in/ (accessed on 2 January 2022).
2. Ashtari, A.; Bibeau, E.; Shahidinejad, S.; Molinski, T. PEV Charging Profile Prediction and Analysis Based on Vehicle Usage Data. *IEEE Trans. Smart Grid* **2012**, *3*, 341–350. [CrossRef]
3. Martinez, C.M.; Hu, X.; Cao, D.; Velenis, E.; Gao, B.; Wellers, M. Energy Management in Plug-in Hybrid Electric Vehicles: Recent Progress and a Connected Vehicles Perspective. *IEEE Trans. Veh. Technol.* **2017**, *66*, 4534–4549. [CrossRef]
4. Wu, X.; Hu, X.; Teng, Y.; Qian, S.; Cheng, R. Optimal integration of a hybrid solar-battery power source into smart home nanogrid with plug-in electric vehicle. *J. Power Sources* **2017**, *363*, 277–283. [CrossRef]
5. Karmaker, A.K.; Ahmed, M.R.; Hossain, M.A.; Sikder, M.M. Feasibility assessment & design of hybrid renewable energy based electric vehicle charging station in Bangladesh. *Sustain. Cities Soc.* **2018**, *39*, 189–202. [CrossRef]
6. Podder, A.K.; Supti, S.A.; Islam, S.; Malvoni, M.; Jayakumar, A.; Deb, S.; Kumar, N.M. Feasibility Assessment of Hybrid Solar Photovoltaic-Biogas Generator Based Charging Station: A Case of Easy Bike and Auto Rickshaw Scenario in a Developing Nation. *Sustainability* **2021**, *14*, 166. [CrossRef]
7. Fares, D.; Fathi, M.; Mekhilef, S. Performance evaluation of metaheuristic techniques for optimal sizing of a stand-alone hybrid PV/wind/battery system. *Appl. Energy* **2022**, *305*, 117823. [CrossRef]
8. Bernal-Agustín, J.L.; Dufo-López, R. Techno-economical optimization of the production of hydrogen from PV-Wind systems connected to the electrical grid. *Renew. Energy* **2010**, *35*, 747–758. [CrossRef]
9. Giallanza, A.; Porretto, M.; Puma, G.L.; Marannano, G. A sizing approach for stand-alone hybrid photovoltaic-wind-battery systems: A Sicilian case study. *J. Clean. Prod.* **2018**, *199*, 817–830. [CrossRef]
10. Halabi, L.M.; Mekhilef, S.; Olatomiwa, L.; Hazelton, J. Performance analysis of hybrid PV/diesel/battery system using HOMER: A case study Sabah, Malaysia. *Energy Convers. Manag.* **2017**, *144*, 322–339. [CrossRef]
11. Lian, J.; Zhang, Y.; Ma, C.; Yang, Y.; Chaima, E. A review on recent sizing methodologies of hybrid renewable energy systems. *Energy Convers. Manag.* **2019**, *199*, 112027. [CrossRef]
12. Al Busaidi, A.S.; Kazem, H.A.; Al-Badi, A.H.; Farooq Khan, M. A review of optimum sizing of hybrid PV–Wind renewable energy systems in oman. *Renew. Sustain. Energy Rev.* **2016**, *53*, 185–193. [CrossRef]
13. Zheng, X.; Lin, H.; Liu, Z.; Li, D.; Llopis-Albert, C.; Zeng, S. Manufacturing Decisions and Government Subsidies for Electric Vehicles in China: A Maximal Social Welfare Perspective. *Sustainability* **2018**, *10*, 672. [CrossRef]
14. Ye, B.; Jiang, J.; Miao, L.; Yang, P.; Li, J.; Shen, B. Feasibility Study of a Solar-Powered Electric Vehicle Charging Station Model. *Energies* **2015**, *8*, 13265–13283. [CrossRef]
15. Ilieva, L.M.; Iliev, S.P. Feasibility assessment of a solar-powered charging station for electric vehicles in the North Central region of Bulgaria. *Renew. Energy Environ. Sustain.* **2016**, *1*, 12. [CrossRef]
16. Filote, C.; Felseghi, R.; Raboaca, M.S.; Aşchilean, I. Environmental impact assessment of green energy systems for power supply of electric vehicle charging station. *Int. J. Energy Res.* **2020**, *44*, 10471–10494. [CrossRef]
17. Niccolai, A.; Bettini, L.; Zich, R. Optimization of electric vehicles charging station deployment by means of evolutionary algorithms. *Int. J. Intell. Syst.* **2021**, *36*, 5359–5383. [CrossRef]
18. Mohamed, A.; Salehi, V.; Ma, T.; Mohammed, O. Real-Time Energy Management Algorithm for Plug-In Hybrid Electric Vehicle Charging Parks Involving Sustainable Energy. *IEEE Trans. Sustain. Energy* **2014**, *5*, 577–586. [CrossRef]
19. Mozafar, M.R.; Moradi, M.H.; Amini, M.H. A simultaneous approach for optimal allocation of renewable energy sources and electric vehicle charging stations in smart grids based on improved GA-PSO algorithm. *Sustain. Cities Soc.* **2017**, *32*, 627–637. [CrossRef]
20. Xu, M.; Meng, Q. Optimal deployment of charging stations considering path deviation and nonlinear elastic demand. *Transp. Res. Part B Methodol.* **2020**, *135*, 120–142. [CrossRef]
21. Bhatti, A.R.; Salam, Z.; Sultana, B.; Rasheed, N.; Awan, A.B.; Sultana, U.; Younas, M. Optimized sizing of photovoltaic grid-connected electric vehicle charging system using particle swarm optimization. *Int. J. Energy Res.* **2019**, *43*, 500–522. [CrossRef]
22. Li, H.; Liu, H.; Ji, A.; Li, F.; Jia, Y. Design of a hybrid solar-wind powered charging station for electric vehicles. In Proceedings of the 2013 International Conference on Materials for Renewable Energy and Environment, Chengdu, China, 19–21 August 2013; IEEE: New York, NY, USA, 2013; pp. 977–981.

23. Hafez, O.; Bhattacharya, K. Optimal design of electric vehicle charging stations considering various energy resources. *Renew. Energy* **2017**, *107*, 576–589. [CrossRef]
24. Bouguerra, S.; Bhar Layeb, S. Determining optimal deployment of electric vehicles charging stations: Case of Tunis City, Tunisia. *Case Stud. Transp. Policy* **2019**, *7*, 628–642. [CrossRef]
25. Chandra Mouli, G.R.; Bauer, P.; Zeman, M. System design for a solar powered electric vehicle charging station for workplaces. *Appl. Energy* **2016**, *168*, 434–443. [CrossRef]
26. Sadeghian, O.; Nazari-Heris, M.; Abapour, M.; Taheri, S.S.; Zare, K. Improving reliability of distribution networks using plug-in electric vehicles and demand response. *J. Mod. Power Syst. Clean Energy* **2019**, *7*, 1189–1199. [CrossRef]
27. Zhang, Y.; Zhang, Q.; Farnoosh, A.; Chen, S.; Li, Y. GIS-Based Multi-Objective Particle Swarm Optimization of charging stations for electric vehicles. *Energy* **2019**, *169*, 844–853. [CrossRef]
28. Ray, P.; Bhattacharjee, C.; Dhenuvakonda, K.R. Swarm intelligence-based energy management of electric vehicle charging station integrated with renewable energy sources. *Int. J. Energy Res.* **2021**. [CrossRef]
29. Miceli, R.; Viola, F. Designing a Sustainable University Recharge Area for Electric Vehicles: Technical and Economic Analysis. *Energies* **2017**, *10*, 1604. [CrossRef]
30. Akbari, M.; Brenna, M.; Longo, M. Optimal Locating of Electric Vehicle Charging Stations by Application of Genetic Algorithm. *Sustainability* **2018**, *10*, 1076. [CrossRef]
31. Dai, Q.; Liu, J.; Wei, Q. Optimal Photovoltaic/Battery Energy Storage/Electric Vehicle Charging Station Design Based on Multi-Agent Particle Swarm Optimization Algorithm. *Sustainability* **2019**, *11*, 1973. [CrossRef]
32. Pan, A.; Zhao, T.; Yu, H.; Zhang, Y. Deploying Public Charging Stations for Electric Taxis: A Charging Demand Simulation Embedded Approach. *IEEE Access* **2019**, *7*, 17412–17424. [CrossRef]
33. Ekren, O.; Hakan Canbaz, C.; Güvel, Ç.B. Sizing of a solar-wind hybrid electric vehicle charging station by using HOMER software. *J. Clean. Prod.* **2021**, *279*, 123615. [CrossRef]
34. Ahmad, F.; Khalid, M.; Panigrahi, B.K. An enhanced approach to optimally place the solar powered electric vehicle charging station in distribution network. *J. Energy Storage* **2021**, *42*, 103090. [CrossRef]
35. Singh, S.; Chauhan, P.; Jap Singh, N. Feasibility of Grid-connected Solar-wind Hybrid System with Electric Vehicle Charging Station. *J. Mod. Power Syst. Clean Energy* **2021**, *9*, 295–306. [CrossRef]
36. Aldhanhani, T.; Al-Durra, A.; El-Saadany, E.F. Optimal design of electric vehicle charging stations integrated with renewable DG. In Proceedings of the 2017 IEEE Innovative Smart Grid Technologies-Asia (ISGT-Asia), Auckland, New Zealand, 4–7 December 2017; IEEE: New York, NY, USA, 2017; pp. 1–6.
37. Al Wahedi, A.; Bicer, Y. Techno-economic assessment of a renewable energy-based electric vehicle fast-charging station in Qatar. *Comput. Aided Chem. Eng.* **2021**, *50*, 1629–1634.
38. Abdelkader, A.; Rabeh, A.; Mohamed Ali, D.; Mohamed, J. Multi-objective genetic algorithm based sizing optimization of a stand-alone wind/PV power supply system with enhanced battery/supercapacitor hybrid energy storage. *Energy* **2018**, *163*, 351–363. [CrossRef]
39. Xu, X.; Hu, W.; Cao, D.; Huang, Q.; Chen, C.; Chen, Z. Optimized sizing of a standalone PV-wind-hydropower station with pumped-storage installation hybrid energy system. *Renew. Energy* **2020**, *147*, 1418–1431. [CrossRef]
40. Yan, D.; Ma, C. Stochastic planning of electric vehicle charging station integrated with photovoltaic and battery systems. *IET Gener. Transm. Distrib.* **2020**, *14*, 4217–4224. [CrossRef]
41. Bandopadhyay, J.; Roy, P.K. Application of hybrid multi-objective moth flame optimization technique for optimal performance of hybrid micro-grid system. *Appl. Soft Comput.* **2020**, *95*, 106487. [CrossRef]
42. Das, B.K.; Hassan, R.; Tushar, M.S.H.K.; Zaman, F.; Hasan, M.; Das, P. Techno-economic and environmental assessment of a hybrid renewable energy system using multi-objective genetic algorithm: A case study for remote Island in Bangladesh. *Energy Convers. Manag.* **2021**, *230*, 113823. [CrossRef]
43. Nishanthy, J.; Charles Raja, S.; Praveen, T.; Jeslin Drusila Nesamalar, J.; Venkatesh, P. Techno-economic analysis of a hybrid solar wind electric vehicle charging station in highway roads. *Int. J. Energy Res.* **2022**. [CrossRef]
44. Emad, D.; El-Hameed, M.A.; El-Fergany, A.A. Optimal techno-economic design of hybrid PV/wind system comprising battery energy storage: Case study for a remote area. *Energy Convers. Manag.* **2021**, *249*, 114847. [CrossRef]
45. Geleta, D.K.; Manshahia, M.S.; Vasant, P.; Banik, A. Grey wolf optimizer for optimal sizing of hybrid wind and solar renewable energy system. *Comput. Intell.* **2020**, *158*, 941–951. [CrossRef]
46. Borhanazad, H.; Mekhilef, S.; Gounder Ganapathy, V.; Modiri-Delshad, M.; Mirtaheri, A. Optimization of micro-grid system using MOPSO. *Renew. Energy* **2014**, *71*, 295–306. [CrossRef]
47. Kumar, A.; Rizwan, M.; Nangia, U. A new approach to design and optimize sizing of hybrid microgrid in deregulated electricity environment. *CSEE J. Power Energy Syst.* **2020**, 1–11. [CrossRef]
48. Mirjalili, S.; Gandomi, A.H.; Mirjalili, S.Z.; Saremi, S.; Faris, H.; Mirjalili, S.M. Salp Swarm Algorithm: A bio-inspired optimizer for engineering design problems. *Adv. Eng. Softw.* **2017**, *114*, 163–191. [CrossRef]
49. Khan, N.H.; Wang, Y.; Tian, D.; Jamal, R.; Kamel, S.; Ebeed, M. Optimal Siting and Sizing of SSSC Using Modified Salp Swarm Algorithm Considering Optimal Reactive Power Dispatch Problem. *IEEE Access* **2021**, *9*, 49249–49266. [CrossRef]
50. Nanaki, E.A.; Xydis, G.A. Deployment of Renewable Energy Systems: Barriers, Challenges, and Opportunities. In *Advances in Renewable Energies and Power Technologies*; Elsevier: Amsterdam, The Netherlands, 2018; pp. 207–229.

51. Sen, S.; Ganguly, S. Opportunities, barriers and issues with renewable energy development—A discussion. *Renew. Sustain. Energy Rev.* **2017**, *69*, 1170–1181. [CrossRef]
52. Dominković, D.F.; Bačeković, I.; Pedersen, A.S.; Krajačić, G. The future of transportation in sustainable energy systems: Opportunities and barriers in a clean energy transition. *Renew. Sustain. Energy Rev.* **2018**, *82*, 1823–1838. [CrossRef]

algorithms

MDPI

Article

Detection of Insulators on Power Transmission Line Based on an Improved Faster Region-Convolutional Neural Network

Haijian Hu [1], Yicen Liu [2] and Haina Rong [3,4,*]

[1] State Grid Sichuan Electric Power Company, Chengdu 610041, China; warshiphu@163.com
[2] State Grid Sichuan Electric Power Research Institute, Chengdu 610041, China; liuyc202201@163.com
[3] Chengdu Zhonglian Huarui Artificial Intelligence Technology Co., Ltd., Chengdu 610041, China
[4] School of Electrical Engineering, Southwest Jiaotong University, Chengdu 611756, China
* Correspondence: hainarong@swjtu.edu.cn

Abstract: Detecting insulators on a power transmission line is of great importance for the safe operation of power systems. Aiming at the problem of the missed detection and misjudgment of the original feature extraction network VGG16 of a faster region-convolutional neural network (R-CNN) in the face of insulators of different sizes, in order to improve the accuracy of insulators' detection on power transmission lines, an improved faster R-CNN algorithm is proposed. The improved algorithm replaces the original backbone feature extraction network VGG16 in faster R-CNN with the Resnet50 network with deeper layers and a more complex structure, adding an efficient channel attention module based on the channel attention mechanism. Experimental results show that the feature extraction performance has been effectively improved through the improvement of the backbone feature extraction network. The network model is trained on a training set consisting of 6174 insulator pictures, and is tested on a testing set consisting of 686 pictures. Compared with the traditional faster R-CNN, the mean average precision of the improved faster R-CNN increases to 89.37%, with an improvement of 1.63%.

Keywords: deep learning; insulator detection; target detection; faster R-CNN

Citation: Hu, H.; Liu, Y.; Rong, H. Detection of Insulators on Power Transmission Line Based on an Improved Faster Region-Convolutional Neural Network. *Algorithms* **2022**, *15*, 83. https://doi.org/10.3390/a15030083

Academic Editor: Mircea-Bogdan Radac

Received: 2 January 2022
Accepted: 25 February 2022
Published: 1 March 2022

1. Introduction

Insulators are widely used in power systems to provide electrical insulation and mechanical support for high-voltage transmission lines [1]. However, under the effects of long-term switching and lightening overvoltage, thermal strain, and natural aging, insulators will fail due to cracks or surface pollution which will hinder their safe operation and cause huge economic losses and casualties in the power transmission system. Therefore, it is crucial to detect insulators from complicated background to ensure the safe running of a power system. With the rapid development of Unmanned Aerial Vehicles (UAV), and since traditional manual inspection methods for detection are time-consuming and highly dangerous, the application of UAV inspection is becoming popular. As the massive aerial images are becoming increasingly easier to access, an accurate and real-time broken insulators location method is urgently needed.

At present, the methods for detecting insulators can be divided into two categories, according to the development stage. One is the traditional methods for detecting insulators, which combines human-designed features and classifiers [2–5]. Another is the the methods of detecting insulators that are based on deep learning networks [6–9]. For example, in [10], firstly, Xiaotong Yao et al. combined the Canny operator with the SURF algorithm to extract the edge feature points of the image insulator, then used the Haar wavelet to obtain the description information of the feature points and the Euclidean distance ratio to match the feature points of the target insulator, finally using the RANSAC algorithm to obtain the correct matching feature points for realizing the identification and detection of insulators. Although traditional methods provide a new idea for insulator detection, the drawbacks of traditional methods are gradually emerging. The detection results depend on the quality of image segmentation, which seriously affects the speed of subsequent feature extraction and classification and is not very good for insulator image segmentation with complex backgrounds in the traditional detecting methods of insulators.

In recent years, since deep learning networks have strong learning and expression ability, and feature extraction is generalized, experts and scholars at home and abroad have gradually applied deep learning networks to insulator detection. There are two main directions in the detecting methods of insulators based on deep learning networks: a single-stage target detection algorithm based on regression problems [5] and a two-stage target detection algorithm based on candidate regions. The main representatives of the single-stage target detection algorithm are the YOLO series algorithm and the SSD algorithm. For example, in [11], an insulator identification method combines traditional methods, such as edge detection and line detection, with the YOLO-V2 algorithm (You Only Look Once V2). However, the design of the loss function and anchor frame of the YOLO-V2 algorithm is far from that of the YOLO-V3 (You Only Look Once V3), and the detection performance is not as good as that of the YOLO-V3. Therefore, in [12], on the basis of the YOLO-V3 algorithm, the focal loss function and balanced cross-entropy function were introduced into YOLO-V3 for the imbalance of positive and negative samples in the training data set. However, since YOLO-V2 and YOLO-V3 are single-stage target detection algorithms, the detection accuracy of single-stage target detection algorithms is lower than of two-stage target detection algorithms. Liu et al. applied the YOLO-v3 target detection network to the localization and recognition of power insulator equipment, providing a new concept for power equipment inspection, but the recognition accuracy is still far from practical applications [13].

Compared with the single-stage target detection algorithm based on regression problems, the two-stage target detection algorithm based on candidate regions is superior in detection accuracy and positioning accuracy. Aiming at the low accuracy of YOLO-V3 in detecting insulator images, Yan Hongwen et al. used the focal loss function and the balanced cross-inheritance function to improve the loss function of the model for improving the insulator identification accuracy [14]. Ji Chao et al. proposed a saliency calculation method, which combines salient regions into the fast region candidate network Fast R-CNN, which avoids time-consuming candidate region traversal. In addition, a residual block is introduced based on the Fast R-CNN feature extraction network, which ensures the integrity of the feature information transfer and improves the detection efficiency [15]. Two convolutional neural networks on a limited insulator data set were discussed in [16]. Experimental results showed that the faster regional convolutional neural network (faster R-CNN) achieves a higher AP value (average precision) than the fully convolutional network (R-FCN). In [17], the faster R-CNN algorithm was used to extract features of insulators and the adaptive image preprocessing, area-based, non-maximum suppression and segmentation detection were introduced to effectively detect insulators, but the insulator positioning accuracy and model training efficiency were not high. In [18], an insulation detection method combining an attention mechanism and faster R-CNN was proposed. This method introduced a compressed excitation network structure in the convolutional feature extraction network.

In order to further improve the performance of faster R-CNN, we propose a faster R-CNN based on Resnet50 and efficient channel attention (ECA)-net. The contributions are summarized as follows.

(1) Resnet50 network is used as the backbone feature extraction network. The improved algorithm uses the Resnet50 network as the backbone feature extraction network to replace the original VGG16 network, which will result in more comprehensive features extracted;
(2) An efficient channel attention (ECA)-net based on the channel attention mechanism is added. The addition of ECA-net helps to extract useful information and suppress useless information, which helps to improve the overall performance of faster R-CNN.

The structure of the paper is arranged as follows. The development process from R-CNN to faster R-CNN and how to improve the faster R-CNN algorithm will be introduced in Section 2. In Section 3, experimental results and analysis are discussed. Finally, conclusions are described in Section 4.

2. Target Detection Method

This section, first, briefly describes the development process from R-CNN to faster R-CNN in Section 2.1, and then the network structure of faster R-CNN is described in Section 2.2. Finally, the focus of this part will be described in detail in Section 2.3, which outlines how to improve the faster R-CNN algorithm.

2.1. The Development History of Faster R-CNN

R-CNN is a milestone in the application of the CNN method to the target detection problem. It was proposed by RBG in 2014. Based on the good feature extraction and classification performance of CNN, the target detection problem is transformed through the method of Region Proposal. The steps of using R-CNN for target detection are listed as follows:

1. Input the image;
2. Use the selective search algorithm to extract about 2000 Region Proposals, from top to bottom, in the image;
3. Warp each Region Proposal to a size of 227×227 and input it to the CNN. The output of the fc7 layer is used as a feature;
4. Input the CNN features extracted by each Region Proposal into SVM for classification;
5. Perform border regression for the Region Proposal classified by SVM, and use the bounding box regression value to correct the original suggestion window. Generate prediction window coordinates.

On the Pascal VOC 2012 data set, R-CNN increased the verification index map for target detection to 53.3%, which is a full 30% improvement over the previous best results. However, R-CNN also has some obvious problems: fine-tuning the network + training SVM + training frame regressor, the steps are very cumbersome, and training is time-consuming, occupying a large space, using GPU to accelerate the VGG16 [19] model to process an image requires 47 s, and the CNN features are not learned and updated during the support vector machine (SVM) and regression process.

In response to these problems of R-CNN, in 2015, fast R-CNN [20] was improved on the basis of R-CNN. Compared with R-CNN, the main difference of fast R-CNN is that a RoI pooling layer is added after the last convolutional layer to make each suggestion window generate a fixed-size feature map. The loss function uses a multi-task loss function (multi-task loss). The bounding box regression is directly added to the CNN network for training, and the target classification and bounding box regression are corrected at the same time after the fully connected layer. Its target detection steps are listed as follows:

1. Input the image;
2. Use the selective search algorithm to extract 2000 or so proposal windows (Region Proposals), from top to bottom, in the image;
3. Input the entire picture into CNN for feature extraction;
4. Map the suggestion window to the last layer of the convolutional feature map of CNN;
5. Use the RoI pooling layer to generate a fixed-size feature map for each suggestion window;
6. Use Softmax Loss (probability of detection classification) and Smooth L1 Loss (bounding box regression) as a joint training for classification probability and bounding box regression.

Fast R-CNN normalizes the entire image, sends it directly to the CNN, and adds suggestion box information to the feature map output by the final convolutional layer, so that the previous CNN operations can be shared and the test speed is accelerated. During training, only one image needs to be sent to the network. Each image extracts

CNN features and suggested regions at one time. The training data is directly entered into the loss layer in the GPU memory, so that the first few layers of features in the candidate region do not need to be recalculated, thus improving the speed of training; fast R-CNN implements both category judgment and location regression using deep networks, so no additional storage is needed; thus, memory space is saved. Because of the proposal of RoI pooling, there is no need to input for corp and wrap operations, thereby avoiding pixel loss and cleverly solving the problem of scale scaling.

Although fast R-CNN has made a big leap in speed, it still takes more than 2 s to detect a picture, which cannot meet the requirements from time to time. This is mainly because it consumes a lot of time during the region proposal stage and does not run well on the GPU. For this reason, faster R-CNN [6] was proposed in 2017, and it has two main differences from fast R-CNN. One is to use a RPN (Region Proposal Network) instead of the original selective search method to generate the suggestion window; the other is to share the CNN that generates the suggestion window and the CNN for target detection. Faster R-CNN creatively uses the convolutional network to generate the suggestion frame by itself, and shares the convolutional network with the target detection network, so that the number of suggestion frames is reduced from about 2000 to 300, and the quality of the suggestion frame is also substantially improved. Moreover, faster R-CNN further introduces the RPN network on the basis of fast R-CNN, and proposes an anchor box to integrate the region proposal generation process into the network training process, which effectively reduces the time of RoIs, and the accuracy is also significantly improved.

2.2. Faster R-CNN Network Structure and Detection Steps

The network structure diagram of faster R-CNN is shown in Figure 1. The detection process is described as follows:

1. Input the image to be tested;
2. Use the VGG16 network to extract feature maps from the entire input image. The feature maps are shared for the subsequent RPN layer and fully connected layer;
3. The RPN network is used to generate region proposals. This layer uses softmax to determine whether the anchors are positive or negative, and then uses bounding box regression to correct the anchors to obtain accurate proposals.
4. The RoI pooling layer collects the input feature maps and proposals, combines the information to extract the proposal feature maps, and sends them to the subsequent fully connected layer to determine the target category;
5. Proposal feature maps are used to calculate the category of the proposal, and at the same time again, bounding box regression are used to obtain the final precise position of the detection frame.

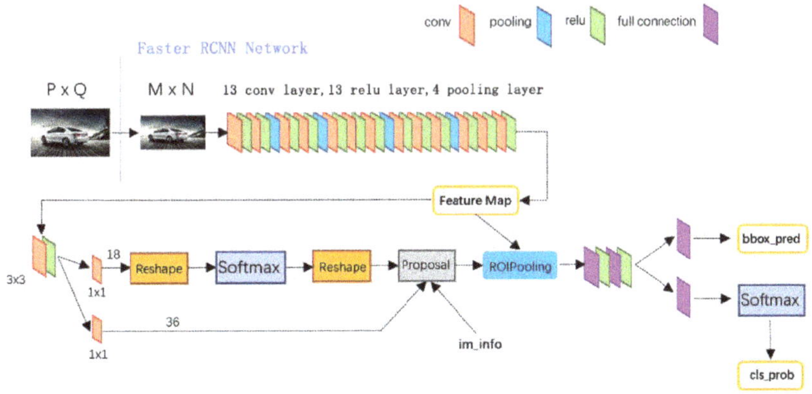

Figure 1. Faster R-CNN network structure diagram.

2.3. Improved Faster R-CNN

In order to improve the performance of the faster R-CNN network model for insulator recognition, this paper proposes an improved faster R-CNN based on the faster R-CNN network model, by replacing the VGG16 backbone feature extraction network with a deeper network and a more complex model structure, namely, the Resnet50 network. In addition, an ECA-net module based on the channel attention mechanism has been added, which makes the Resnet50 network much better at the extraction of insulator features.

2.3.1. Resnet50 Network Replacing VGG16 Network

The VGG16 network is used as the feature extraction backbone network by the original faster R-CNN model, and the features output by the final convolution output layer of VGG16 are used as the shared features of the RPN network and the RoI pooling layer. This algorithm is used mainly to detect everyday objects, and it is suitable for the coco data set. When this algorithm is used in the insulator data set, there are mainly the following shortcomings:

1. VGG16 uses a single-layer feature layer output to be suitable for the detection of single-sized targets. Because of the different sizes of insulators in the image, it is easy to cause missed detections and misjudgments;
2. Due to the different scales of aerial insulator images, many insulators have become small targets related to the entire picture. In order to identify the insulators more accurately, the feature extraction backbone network needs to be improved.

The Resnet50 network [21] consists of 49 convolutional layers and 1 fully connected layer. The innovative introduction of residual blocks makes the network depth increase, and deeper features can be extracted while effectively avoiding gradient disappearance and gradient explosion. The authors in [21] claimed that the residual network with a depth of up to 152 layers on the ImageNet data set has still lower complexity than the VGG16 network. Therefore, comparing with VGG16, Resnet50 increases the depth of the network but does not increase the complexity of the network.

2.3.2. ECA-Net Module Joining the Resnet50 Network

Recently, the channel attention mechanism has been shown to have great potential for improving the performance of convolutional neural networks (CNNs). In order to make CNNs have better performance, more complex attention modules have been developed, such as SANet [22], SKNet [23], and ResNeSt [24]. Although the performance of the network is improved, it also indirectly increases the complexity of the model. An ECA module to improve the channel attention mechanism has been proposed in [25]. It introduced a local cross-channel interaction strategy without dimensionality reduction and a method of adaptively selecting the size of the one-dimensional convolution kernel, thereby achieving performance improvement and balancing the contradiction between performance and complexity well. This module generates channel attention through fast one-dimensional convolution, and the size of the convolution kernel can be adaptively determined by the nonlinear mapping of the channel dimension. The complexity of the model added by this module is small, and the improvement effect is significant.

3. Experimental Results and Analysis

This section will elaborate on the three aspects of experimental configuration and data set, evaluation index, and result analysis.

3.1. Experiments

The hardware and software configuration is described as follows: AMD Ryzen 9 3950X 16-Core Processor 3.50 GHz, 64 GB RAM, GeForce RTX 3090, Windows 10. The environment configuration is set as follows: CUDA Version 11.1, Python3.7.10, PyTorch1.8.0. The selected insulator data set is a public one with 6860 images https://github.com/heitorcfelix/public-insulator-datasets (accessed on 12 April 2021). The data set occupies a storage space of 3.3 GB. The data set consists of real images and synthetic images, and has been enhanced. This experiment uses 6174 pictures as the training set and 686 pictures as the testing set. An example of the image is shown in Figure 2. Since the format of the network data set is a coco format, the first step is to write a script to convert it to a VOC2012 format. In the experiments, although the insulator data set used is public, no literature related to this data set is available. Therefore, we have to compare the performance of the three networks, the proposed faster R-CNN with Resnet50+ECA-net, the original faster R-CNN, and faster R-CNN with Resnet50, on the public data set mentioned in this paper.

Figure 2. Aerial images of insulators on different backgrounds.

3.2. Evaluation Index

Insulator detection belongs to single-label classification learning. This experiment uses mean average precision (MAP) as the performance evaluation index of the model. Before introducing MAP, we need to review the concepts of intersection over union (IOU), precision, recall, and average precision (AP) [21,26–30].

IOU is the intersection ratio, which measures the degree of overlap of two regions, and is the ratio of the overlapping area of the two regions to the total area of the two. As shown in Figure 3, the green box is the ground-truth box of the insulator; the red is the prediction box, and the IOU of the two rectangular boxes is the ratio of the cross area to the combined area.

Figure 3. Insulator true label frame and prediction frame.

Precision and recall is described as follows. In the field of target detection, we first assume that there is a set of pictures to be detected. Precision represents the proportion of real target objects in the targets detected by the model; recall is considered to be the proportion of real targets detected by the model to all target objects in the image to be detected, that is, how many out of all real targets are successfully detected by the model. The calculation formula of precision and recall is described as follows:

$$precision = \frac{TP}{TP + FP}, \tag{1}$$

$$precision = \frac{TP}{TP + FN}, \tag{2}$$

where TP, TN, FP, and FN represent a true positive, true negative, false positive, and false negative, respectively, and their meanings are described in Table 1.

Table 1. Meanings of TP, FP, TN and FN in target detection.

	The Detected Rectangular Box Is Greater than the Confidence Threshold (Positive)	The Detected Rectangular Box Is Less than the Confidence Threshold (Negative)
The IOU value of a target box in the data set is greater than 0.5 (True)	TP	TN
The IOU value of all target boxes in the data set is less than 0.5, repeated detection (False)	FP	FN

The precision–recall (PR) curve is shown in Figure 4. The higher the precision and recall, the better the model detection performance becomes, so we expect high values for both precision and recall, but in some cases they are contradictory. For example, in some extreme cases, only one result is detected and accurate; then, precision is 100%, and recall is very small. If all results are returned, then recall must be very large, but precision is very small. Therefore, it is necessary for different occasions to judge whether one wants higher precision or higher recall. Usually, the precision–recall curve is drawn to help understand them. As the name implies, AP is the average accuracy. Simply put, it is the average of the precision value on the PR curve, that is, the area under the PR curve. As shown in Figure 4, the area of the shaded area is the value of AP. AP is for a single category, and the average AP value of all categories is MAP.

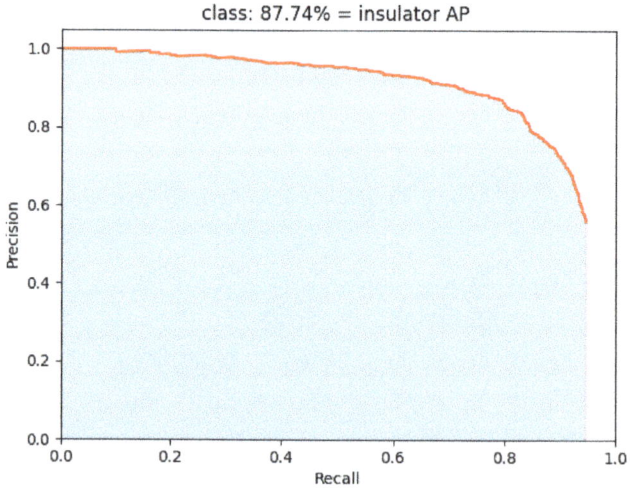

Figure 4. Precision–recall curve.

3.3. Analysis of Experimental Results

This article replaces faster R-CNN's backbone feature network VGG16 with a Resnet50 network that adds the ECA-net module based on the channel attention mechanism. After the three networks (original faster R-CNN with VGG16, faster R-CNN with Resnet50, and faster R-CNN with Resnet50+ECA-net) are independently trained on the same insulator data set, the performance of the three networks for insulator detection is compared under the same testing set. Experimental results show that the improved faster R-CNN is better than the faster R-CNN with respect to the MAP.

3.3.1. Comparisons of Insulator Target Detection Training Results

The model parameters are set as follows: the number of iterations, the initial size of the learning rate, and the batch size are set to 150, 0.00005, and 8, respectively. Figure 5 shows the train loss curve, validation loss curve, smooth train loss curve, and validation loss curve, respectively, of faster R-CNN with VGG16 network as the backbone feature extraction network. These loss curves refer to the training loss curves of migration learning under the pre-training model of imagenet. After 126 epochs, the training loss converges to around 0.535, and the validation loss converges to around 0.716.

In order to verify the effectiveness of Resnet50 and Resnet50+ECA-net, we compare the loss curve of faster R-CNN with Resnet50, and compare faster R-CNN with Resnet50+ECA-net with the loss curve of the original faster R-CNN with VGG16, respectively. Figures 6 and 7 show the train loss curve, validation loss curve, smooth train loss curve, and validation loss curve of the faster R-CNN with the Resnet50 backbone and the faster R-CNN with the Resnet50 backbone and ECA-net, respectively. In the faster R-CNN with the VGG16 backbone, when the number of epochs is 140, the train and val losses are 0.536 and 0.718, respectively. Compared with the faster R-CNN with the VGG16 backbone, the train loss of the faster R-CNN with the Resnet50 backbone and the faster R-CNN with the Resnet50 backbone and ECA-net decreases by 16.6% and 20.1%, respectively. The val loss of the faster R-CNN with the Resnet50 backbone and the faster R-CNN with the Resnet50 backbone and ECA-net decrease by 9.6% and 10.7%, compared to the faster R-CNN with the VGG16 backbone, respectively. Experimental results indicate that the addition of the Resnet50 backbone and ECA-net helps to reduce the loss value.

3.3.2. Comparisons of Insulator Target Detection Testing Results

The testing experiments are performed to compare the the improved faster R-CNN and the faster R-CNN. In order to show the performance before and after improvement more directly, experimental results are shown in Figures 8–10 and Table 2, where the final APs of the PR curve of the faster R-CNN, the first improved faster R-CNN, and the second improved faster R-CNN on the testing set are 87.74%, 87.91%, and 89.37%, respectively, with an improvement of 0.19% and 1.63%. In Table 2, comparing with the original faster R-CNN, the first improved faster R-CNN and the second improved faster R-CNN have fewer parameters, lower training and testing loss values, and higher accuracy. In summary, The addition of the Resnet50 backbone and ECA-net decreases losses and raises the AP, which leads to the improved performance of the faster R-CNN.

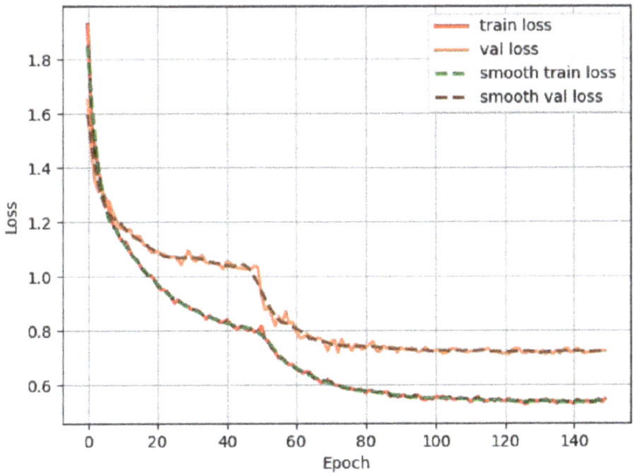

Figure 5. Training loss under the VGG16 backbone network.

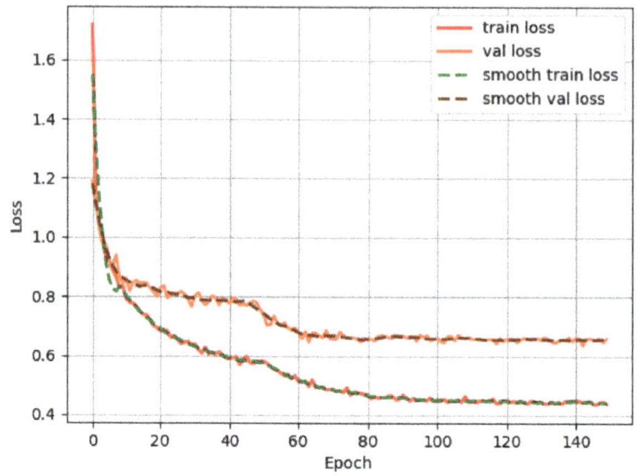

Figure 6. Training loss under the Resnet50 backbone network.

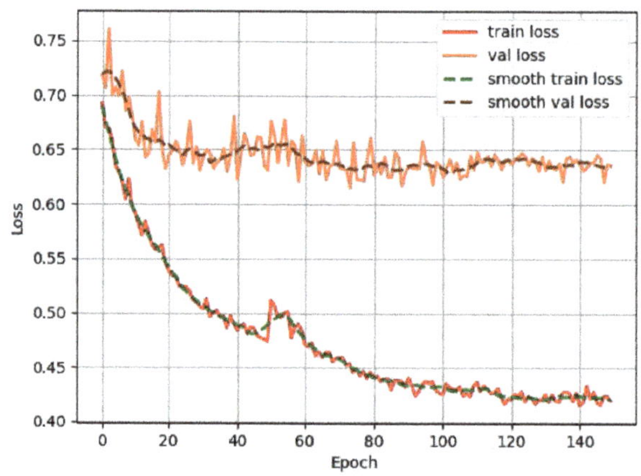

Figure 7. Training loss under the ECA-net+Resnet50 backbone network.

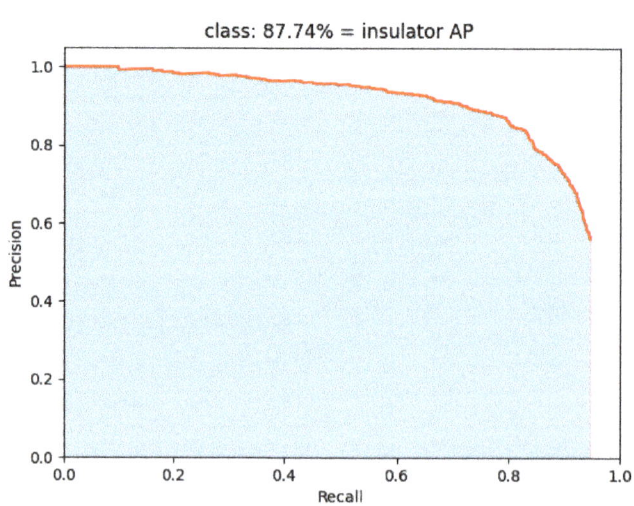

Figure 8. Precision–recall curve graph under the VGG16 backbone network.

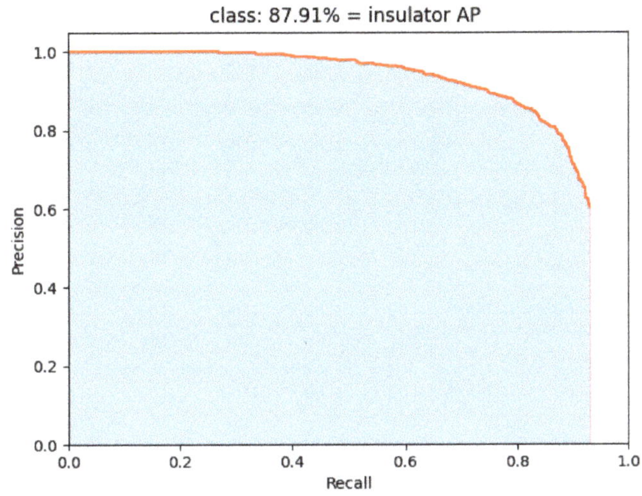

Figure 9. Precision–recall curve graph under the Resnet50 backbone network.

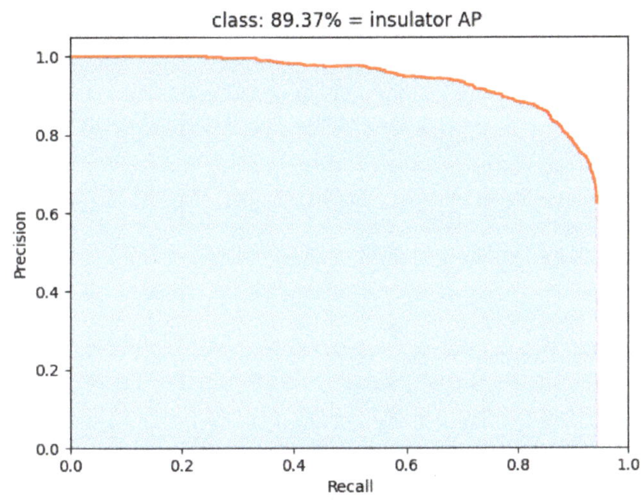

Figure 10. Precision–recall curve graph under the ECA-net+Resnet50 backbone network.

Table 2. Comparison of the original faster R-CNN and the improved faster R-CNN.

Model	Backbone	Parameters	Train Loss	Val Loss	AP
Original faster R-CNN	VGG16	136,689,024	0.535	0.716	87.74%
The first improved faster R-CNN	RESNET50	28,275,328	0.438	0.642	87.91%
The second improved faster R-CNN	ECA-net+RESNET50	28,275,376	0.412	0.612	89.37%

3.3.3. Display of Actual Test Results

The faster R-CNN target detection model, using the Resnet50 network with the ECA-net module as the backbone feature extraction network, is tested on the testing set, and the insulator detection results of aerial transmission lines are shown in Figure 11. It can be seen that good results can be achieved for the multi-target detection of insulators of different sizes.

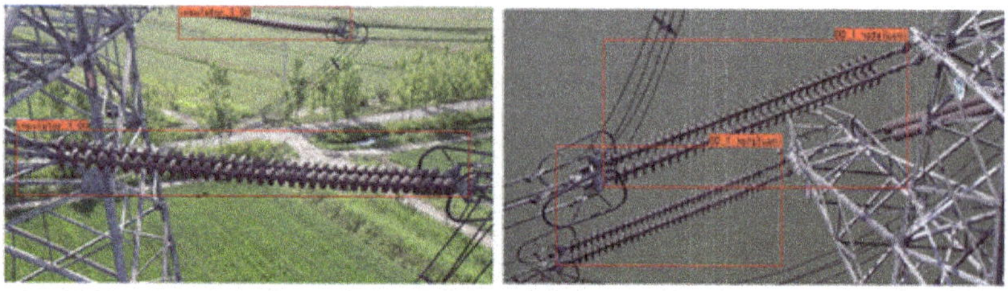

Figure 11. Insulator test results of the improved faster R-CNN.

4. Conclusions

This paper proposes an insulator detection method based on an improved faster R-CNN. Aiming at the problem of missed detections and misjudgments of the original feature extraction network VGG16 in the face of insulators with different sizes, the ECA-net+Resnet50 network is used to replace the original VGG16 backbone network. The deeper network has a larger field of view, so as to facilitate the detection of targets with different scales. Experimental results indicate the effectiveness of the introduced method for detecting insulators. Comparing with the original faster R-CNN, the improved faster R-CNN has fewer parameters, lower training and testing loss values, and higher accuracy.

Our future work will focus on the further improvement of insulator detection, e.g., the direction-aware defect detection network in [31] can be considered; optimization spiking neural P systems [32–34] can be introduced to optimize the structure and parameters of the ECA-net+Resnet50 network used in this paper; learning spiking neural P systems or fuzzy reasoning spiking neural P systems [35,36] can also be used to enhance the insulator detection [37,38].

Author Contributions: Conceptualization, H.H.; investigation, H.H., Y.L. and H.R.; software, Y.L. and H.R.; writing–original draft preparation H.H., Y.L. and H.R. All authors have read and agreed to the published version of the manuscript.

Funding: Science and Technology Project of State Grid Sichuan Electric Power Company (No. 521997180016).

Data Availability Statement: The selected insulator data set is a public one with 6860 images https://github.com/heitorcfelix/public-insulator-datasets (accessed on 12 April 2021).

Conflicts of Interest: The authors declare no conflict of interest.

References

1. Birlasekaran, S.; Li, H.J. Detection of insulators on power transmission line. *Power Eng. Soc. Winter Meet. IEEE* **2000**, *4*, 2817–2821.
2. Viola, P.; Jones, M. Rapid object detection using a boosted cascade of simple features. In Proceedings of the 2001 IEEE Computer Society Conference on Computer Vision and Pattern Recognition—CVPR 2001, Kauai, HI, USA, 8–14 December 2001.
3. Lienhart, R.; Maydt, J. An extended set of Haar-like features for rapid object detection. In Proceedings of the International Conference on Image Processing, Rochester, NY, USA, 22–25 September 2002; IEEE: Rochester, NY, USA, 2002.
4. Sun, D.; Wata, D. Detecting pedestrians and vehicles in traffic scene based on boosted HOG features and SVM. In Proceedings of the 2015 IEEE 9th International Symposium on Intelligent Signal Processing (WISP) Proceedings, Siena, Italy, 15–17 May 2015; IEEE: Siena, Italy, 2015; Volume 2015, pp. 1–4.
5. Zhang, D. Vehicle target detection methods based on color fusion deformable part model. *Eur. J. Wirel. Commun. Netw.* **2018**, *2018*, 94. [CrossRef]
6. Ren, S.; He, K.; Girshick, R. Faster R-CNN: Towards Real-Time Object Detection with Region Proposal Networks. *IEEE Trans. Pattern Anal. Mach. Intell.* **2017**, *39*, 1137–1149. [CrossRef] [PubMed]
7. Redmon, J.; Farhadi, A. YOLO9000: Better, Faster, Stronger. In Proceedings of the IEEE Computer Society Conference on Computer Vision and Pattern Recognition, Honolulu, HI, USA, 21–26 July 2017; Volume 2017, pp. 6517–6525.
8. Liu, S.; Lv, S.; Zhang, H. Pedestrian Detection Algorithm Based on the Improved SSD. In Proceedings of the 2019 Chinese Control And Decision Conference (CCDC), Nanchang, China, 3–5 June 2019; Volume 2019, p. 9.
9. Szegedy, C.; Reed, S.; Erhan, D. Scalable, High-Quality Object Detection. *arXiv* **2014**, arxiv:1412.1441.
10. Yao, X.T.; Liu, L.; Li, Z.Y.; Cheng, X. Identification Method of Railway Insulator Based on Edge Features. In *IOP Conference Series: Materials Science and Engineering*; IOP Publishing Ltd.: Bristol, UK, 2018; Volume 394.

11. Lai, Q.; Yang, J.; Tan, B. Automatic insulator recognition and defect diagnosis model based on YOLOv2 Network. *Electr. Power* **2019**, *52*, 31–39.
12. Yan, H.; Chen, J. Localization and state recognition of insulator string based on improved YOLOV3. *High Volt. Eng.* **2020**, *46*, 423–432.
13. Liu, Y.P.; Ji, X.X.; Pei, S.T.; Ma, Z.R.; Zhang, G.H.; Lin, Y.; Chen, Y.Y.F. Research onautomatic location and recognition of insulators in substation based on YOLOv3. *High Volt.* **2020**, *5*, 62–68. [CrossRef]
14. Yan, H.W.; Chen, J.X. Insulator String Positioning and State Recognition Method Based on Improved YOLOv3 Algorithm. *High Volt.* **2020**, *46*, 423–432.
15. Ji, C.; Huang, X.B.; Cao, W.; Zhu, Y.C.; Zhang, Y. Research on Infrared Insulator Detection Based on Improved Fast-CNN Mode. *Comput. Mod.* **2019**, *2019*, 59–64, 71.
16. Michal, T.; Pawel, M.; Jakub, O. Evaluation of Power Insulator Detection Efficiency with the Use of Limited Training Dataset. *Appl. Sci.* **2010**, *10*, 2104.
17. Bai, J.; Zhao, R.; Gu, F. Multi-target detection and fault recognition image processing method. *High Volt. Technol.* **2019**, *45*, 3504–3511.
18. Zhao, W.; Cheng, X.; Zhao, Z. Attentional mechanism combined with Faster RCNN for insulator recognition. *CAAI Trans. Intell. Syst.* **2020**, *15*, 92–98.
19. Simonyan, K.; Zisserman, A. Very Deep Convolutional Networks for Large-Scale Image Recognition. *arXiv* **2014**, arxiv:1409.1556.
20. Girshick, R. Fast R-CNN. In Proceedings of the 2015 IEEE International Conference on Computer Vision (ICCV), Santiago, Chile, 7–13 December 2015; pp. 1440–1448.
21. He, K.; Zhang, X.; Ren, S. Deep residual learning for image recognitiol. In Proceedings of the IEEE Conference on Computer Vision and Pattern Recognition, Las Vegas, NV, USA, 27–30 June 2016; pp. 770–778.
22. Jie, H.; Li, S.; Gang, S. Squeeze-and-Excitation Networks. In Proceedings of the Conference on Computer Vision and Pattern Recognition (CVPR), Salt Lake City, UT, USA, 18–22 June 2018; IEEE: Salt Lake City, UT, USA, 2018.
23. Li, X.; Wang, W.; Hu, X. Selective kernel networks. In Proceedings of the IEEE/CVF Conference on Computer Vision and Pattern Recognition, Long Beach, CA, USA, 15–20 June 2019; pp. 510–519.
24. Zhang, H.; Wu, C.; Zhang, Z. ResNeSt: Split-Attention Networks. *arXiv* **2018**, arxiv:2004.08955.
25. Wang, Q.; Wu, B.; Zhu, P. ECA-Net: Efficient Channel Attention for Deep Convolutional Neural Networks. In Proceedings of the Conference on Computer Vision and Pattern Recognition (CVPR), Seattle, WA, USA, 13–19 June 2020; IEEE: Seattle, WA, USA, 2020.
26. Zhao, Z.; Zhen, Z.; Zhang, L. Insulator Detection Method in Inspection Image Based on Improved Faster R-CNN. *Energies* **2019**, *12*, 1204. [CrossRef]
27. Zhang, D.; Fgvg, T. An accurate and real-time self-blast glass insulator location method based on faster R-CNN and U-net with aerial images. *CSEE J. Power Energy Syst.* **2019**, *5*, 474–482.
28. Huang, W.; Zhang, F.; Peng, L. Exploration and Application on Faster R-CNN Based Insulator Recognition. *South. Power Syst. Technol.* **2018**, *12*, 22–27.
29. Ma, L.; Xu, C.; Zuo, G. Detection Method of Insulator Based on Faster R-CNN. In Proceedings of the 2017 IEEE 7th Annual International Conference on CYBER Technology in Automation, Control, and Intelligent Systems (CYBER), Honolulu, HI, USA, 31 July–4 August 2017; IEEE: Honolulu, HI, USA, 2017.
30. Huo, H.; Min, Y.; Yang, K. Research on Detection Algorithm of Catenary Insulator Based on Improved Faster R-CNN. In Proceedings of the 2020 Chinese Automation Congress (CAC), Shanghai, China, 6–8 November 2020.
31. Liu, H.; Chang, Y.; Li, Z.; Zhong, S.; Yan, L. Directional-Aware Automatic Defect Detection in High-Speed Railway Catenary System. In Proceedings of the 2019 IEEE International Conference on Image Processing (ICIP), Taipei, Taiwan, 22–25 September 2019.
32. Zhang, G.; Rong, H.; Neri, F.; Pérez-Jiménez, M.J. An optimization spiking neural P system for approximately solving combinatorial optimization problems. *Int. J. Neural Syst.* **2014**, *24*, 1–16. [CrossRef]
33. Zhu, M.; Yang, Q.; Dong, J.; Zhang, G.; Gou, X.; Rong, H.; Paul, P.; Neri, F. An adaptive optimization spiking neural P system for binary problems. *Int. J. Neural Syst.* **2021**, *31*, 1–17. [CrossRef]
34. Zhang, G.; Pérez-Jiménez, M.J.; Gheorghe, M. *Real-Life Applications with Membrane Computing*; Springer: Berlin, Germany, 2017.
35. Rong, H.; Yi, K.; Zhang, G.; Dong, J.; Paul, P.; Huang, Z. Automatic implementation of fuzzy reasoning spiking neural P systems for diagnosing faults in complex power systems. *Complexity* **2019**, *2019*, 1–16. [CrossRef]
36. Wang, T.; Zhang, G.; Zhao, J.; He, Z.; Wang, J.; Pérez-Jiménez, M.J. Fault diagnosis of electric power systems based on fuzzy reasoning spiking neural P systems. *IEEE Trans. Power Syst.* **2015**, *30*, 1182–1194. [CrossRef]
37. Zhang, G.; Pérez-Jiménez, M.J.; Riscos-Núñez, A.; Verlan, S.; Konur, S.; Hinze, T.; Gheorghe, M. *Membrane Computing Models: Implementations*; Springer: Berlin, Germany, 2021.
38. Zhang, G.; Shang, Z.; Verlan, S.; Martínez-del-Amor, M.A.; Yuan, C.; Valencia-Cabrera, L.; Pérez-Jiménez, M.J. An overview of hardware implementation of membrane computing models. *ACM Comput. Surv.* **2020**, *53*, 1–38. [CrossRef]

mathematics

MDPI

Article

Stochastic Cognitive Dominance Leading Particle Swarm Optimization for Multimodal Problems

Qiang Yang [1], Litao Hua [1], Xudong Gao [1,*], Dongdong Xu [1], Zhenyu Lu [1], Sang-Woon Jeon [2] and Jun Zhang [2,3]

[1] School of Artificial Intelligence, Nanjing University of Information Science and Technology, Nanjing 210044, China; qiang_yang@nuist.edu.cn (Q.Y.); 201983460007@nuist.edu.cn (L.H.); 001601@nuist.edu.cn (D.X.); 001114@nuist.edu.cn (Z.L.)
[2] Department of Electrical and Electronic Engineering, Hanyang University, Ansan 15588, Korea; sangwoonjeon@hanyang.ac.kr (S.-W.J.); junzhang@ieee.org (J.Z.)
[3] Department of Computer Science and Information Engineering, Chaoyang University of Technology, Taichung 413310, Taiwan
* Correspondence: 003103@nuist.edu.cn

Abstract: Optimization problems become increasingly complicated in the era of big data and Internet of Things, which significantly challenges the effectiveness and efficiency of existing optimization methods. To effectively solve this kind of problems, this paper puts forward a stochastic cognitive dominance leading particle swarm optimization algorithm (SCDLPSO). Specifically, for each particle, two personal cognitive best positions are first randomly selected from those of all particles. Then, only when the cognitive best position of the particle is dominated by at least one of the two selected ones, this particle is updated by cognitively learning from the better personal positions; otherwise, this particle is not updated and directly enters the next generation. With this stochastic cognitive dominance leading mechanism, it is expected that the learning diversity and the learning efficiency of particles in the proposed optimizer could be promoted, and thus the optimizer is expected to explore and exploit the solution space properly. At last, extensive experiments are conducted on a widely acknowledged benchmark problem set with different dimension sizes to evaluate the effectiveness of the proposed SCDLPSO. Experimental results demonstrate that the devised optimizer achieves highly competitive or even much better performance than several state-of-the-art PSO variants.

Keywords: stochastic cognitive dominance leading; multimodal problems; particle swarm optimization; global optimization; evolutionary algorithm

Citation: Yang, Q.; Hua, L.; Gao, X.; Xu, D.; Lu, Z.; Jeon, S.-W.; Zhang, J. Stochastic Cognitive Dominance Leading Particle Swarm Optimization for Multimodal Problems. *Mathematics* **2022**, *10*, 761. https://doi.org/10.3390/math10050761

Academic Editor: Fabio Caraffini

Received: 30 January 2022
Accepted: 25 February 2022
Published: 27 February 2022

1. Introduction

Optimization problems widely exist in daily life and real-world engineering, such as resource allocation optimization [1], path planning optimization [2,3], and robot task allocation [4]. However, with the advent of Internet of Things and big data, optimization problems are becoming increasingly complicated [5,6], with many undesirable properties, such as non-differentiable, discontinuous, non-convex, non-linear and multimodal with many local areas [7]. In particular, these complex optimization problems tremendously challenge the effectiveness of traditional gradient descent based methods, or even make them infeasible [6,8,9]. Therefore, it is urgent to develop effective optimization algorithms to solve these complex problems, so as to boost the development of related areas.

Recently, heuristic algorithms, such as particle swarm optimization (PSO) [10,11] and differential evolution (DE) [12,13], have shown very promising performance in solving optimization problems. Unlike mathematical algorithms [14–17], which only maintain one solution to iteratively find the global optimum of the optimization problem, heuristic algorithms mainly maintain a population of feasible solutions to iteratively search the solution space to find the global optimum of the optimization problem. In particular, compared with traditional mathematical algorithms [14–17], heuristic algorithms preserve many unique advantages. (1) Heuristic algorithms usually have no requirements on the mathematic properties of the optimization problem, or even can optimize problems without mathematic models [18]. However, most mathematical algorithms, especially gradient decedent based optimization methods, usually require that the optimization problems be continuous, differentiable, and convex. Nevertheless, in the era of big data and Internet of Things (IoT), optimization problems usually have characteristics such as non-differentiable, discontinuous, non-convex, non-linear and multimodal [6,19]. Confronted with these optimization problems, heuristic algorithms such as PSO usually preserve many advantages over mathematical algorithms. In particular, heuristic algorithms, like PSO, preserve unique merits in solving NP-hard problems [20,21]. (2) Heuristic algorithms usually preserve strong global search ability due to the maintenance of a population to search the solution space. Mathematical algorithms usually maintain only one solution to iteratively find the global optimum of an optimization problem. In this case, they are at great risk of falling into local areas, especially when tackling multimodal problems with many wide and flat local regions, whereas heuristic algorithms could search the solution space in different directions by maintaining a population of individual solutions. (3) Heuristic algorithms usually preserve inherent parallelism to accelerate the iteration [18,22]. Specifically, at least, during the optimization, the fitness evaluation of each individual solution, which is usually the most time-consuming part in heuristic algorithms, could be separately executed, let alone that some parallel techniques could be designed and embedded into heuristic algorithms to accelerate the search process. However, most mathematical algorithms can only be executed sequentially because the current iteration usually relies on the results of the preceding iteration.

In recent years, PSO has received extensive attention from researchers since it was proposed by Eberhart and Kennedy in 1995 [10,11]. Therefore, many remarkable PSO variants have emerged [23–27], and PSO has been widely applied to solve various optimization problems [6,28], such as multimodal optimization [29–32] and multi-objective optimization [27,33].

In the literature of PSO, it is well recognized that the learning strategy of particles plays a key role in helping PSO achieve promising performance [8,34–37]. As a result, researchers have focused significant efforts in designing effective learning strategies for PSO to improve its performance, and thus many remarkable novel learning schemes have emerged, such as cooperative learning mechanisms [38,39], comprehensive learning strategies [8,40], and social learning methods [41,42]. In fact, the key to devising effective learning strategies is to select appropriate guiding exemplars for particles to update. In a broad sense, existing selection of guiding exemplars can be divided into the following two types.

The most popular way to select guiding exemplars for particles is to employ different topologies to communicate with other particles to find proper guiding exemplars [43,44]. In the earliest PSO [10,11], Eberhart and Kennedy utilized the global topology with full connections with all particles to choose the global best position in the whole swarm as one guiding exemplar to direct the update of particles. However, such a global topology leads to too greedy attraction, and thus the swarm usually falls into local areas when dealing with multimodal problems. To alleviate this predicament, researchers have attempted many other topologies, such as the ring topology [45], the star topology [46], the random topology [47,48], the wheel topology [46] and the dynamic topology [49]. In the early stage, researchers mainly employed the topologies on the personal best positions of particles to select proper exemplars. Since the obsolete historical evolutionary information may also contain useful information to guide the learning of particles, some researchers have attempted to deploy topologies on the personal best positions and the recorded historical best positions to find proper exemplars [35,50]. Nevertheless, these topology-based exemplar selection methods mainly find promising exemplars based on historical positions visited by particles.

To further improve the learning effectiveness of particles, researchers have sought new solutions in another direction, namely constructing guiding exemplars for particles. Unlike the former types of exemplar selection methods, this type of exemplar selection method mainly constructs new exemplars, which may not be visited by particles, to guide the evolution of the swarm. Consequently, many remarkable constructive learning strategies [8,51–53] have been devised for PSO to tackle complicated optimization problems. In this direction, the most representative method is the comprehensive learning PSO (CLPSO) [8], which constructs a guiding exemplar dimension by dimension for each particle based on the personal best positions of all particles. Inspired by this strategy, many other constructive approaches have been proposed, such as the orthogonal learning PSO (OLPSO) [51], which utilizes an orthogonal matrix to find suitable combinations of dimensions to construct the guiding exemplar for each particle, and the genetic learning PSO (GLPSO) [53], which adopts the operators in the genetic algorithm to construct the guiding exemplar for each particle.

Though most existing PSOs have shown promising performance on simple optimization problems, such as unimodal problems and simple multimodal problems [22,54], they are confronted with limitations when tackling complex optimization problems, such as multimodal problems with many interacted variables and an ocean of wide local regions, which are increasingly common in the era of big data and IoT. As a result, there is an increasing demand for effective PSO to solve emerging complicated optimization problems.

Inspired by the competition mechanisms in human society that groups of randomly assembled individuals spontaneously engage in costly group competition [55], this paper proposes a stochastic cognitive dominance leading particle swarm optimization algorithm (SCDLPSO) to improve the learning effectiveness and efficiency of particles when tackling complicated optimization problems. Instead of the competition between individuals, this paper employs the competition between the personal best positions of particles to select guiding exemplars for each updated particle. Specifically, for each particle to be updated, two different personal best positions are first randomly selected from the ones of the other particles. Then, the two selected personal best positions compete with the personal best position of the updated particle. Only when at least one randomly selected personal best position is better than the one of the particle to be updated, this particle is updated by cognitively learning from the two better personal best positions (either the two better randomly selected personal best positions or the better randomly selected one and its own one); otherwise, it is not updated, and thus directly enters the next generation. In this way, on the one hand, the swarm diversity could be largely promoted due to the random selection of the two competing personal best positions and the retention of some particles, preserving relatively good historical evolutionary information. On the other hand, the learning effectiveness of particles could also be largely improved because each updated particle learns from two better personal best positions. As a result, the proposed SCDLPSO is expected to balance search diversification and intensification well to explore and exploit the solution space appropriately.

To verify the effectiveness of the proposed SCDLPSO, comparative experiments are extensively conducted on the widely used CEC 2017 benchmark problem set [56] with different dimension sizes (namely 30, 50, and 100) by comparing SCDLPSO with seven state-of-the-art PSO methods.

The rest of this paper is organized as follows. Section 2 reviews the canonical PSO and the representative and latest PSO variants. Then, in Section 3, the devised SCDLPSO is elaborated in detail. This is followed in Section 4 by the verification of the effectiveness of SCDLPSO using extensive experiments. Finally, conclusions are given in Section 5.

2. Related Work

2.1. Canonical PSO

In the classical PSO [10,11], each particle is represented as two vectors, namely the position vector and the velocity vector. During the evolution, each particle memorizes its own personal best position found so far, while the whole swarm memorizes the global best position found so far by all particles. Then, each particle is updated by cognitively learning from its own search experience, namely the personal best position, and socially learning from the search experience of all particles, namely the global best position. Specifically, the velocity and the position of each particle are updated as follows:

$$v_i^{t+1} = wv_i^t + c_1r_1(\boldsymbol{pbest}_i^t - x_i^t) + c_2r_2(\boldsymbol{gbest}^t - x_i^t) \tag{1}$$

$$x_i^{t+1} = x_i^t + v_i^{t+1} \tag{2}$$

where x_i and v_i are the position vector and the velocity vector of the ith particle; $\boldsymbol{pbest}_i$ is the personal best position of the ith particle; $\boldsymbol{gbest}$ is the global best position of the whole swarm found by all particles; t denotes the generation index; $w \in [0, 1]$ is the inertia weight; r_1 and r_2 are two real random numbers uniformly sampled within [0, 1]; c_1 and c_2 are two acceleration factors in charge of the influence of the two guiding exemplars on the updated particle.

In the literature [8,11,57], a linearly decreased w defined as follows is usually adopted to alleviate the sensitivity of PSO to the inertia weight:

$$w = 0.9 - 0.5 \times \frac{t}{T_{max}} \tag{3}$$

where t is the number of generations used so far, while T_{max} stands for the maximum number of generations.

From Equation (1), it is found that in the classical PSO, all particles learn from one same guiding exemplar, namely the global best position $\boldsymbol{gbest}$ of the whole swarm. As a result, particles in the swarm converge very quickly to promising areas. However, once $\boldsymbol{gbest}$ falls into local areas, it is hard for particles to escape from the local basin, and thus the algorithm encounters premature convergence. Therefore, the classical PSO is very suitable for unimodal optimization problems, but unaccommodating for multimodal optimization problems [22,30,43].

2.2. *Advancement of Learning Strategies for PSO*

As shown in Equation (1), in the literature [8,22,26,45,51,52,58], the learning strategy used to update the velocity of particles plays the most crucial role in assisting PSO to achieve promising performance. Therefore, to further improve the optimization performance of PSO in solving various optimization problems, especially the complicated ones, such as multimodal problems, researchers have focused extensive attention on devising novel and effective learning strategies for PSO. As a consequence, an ocean of remarkable PSO variants have emerged [22,25,27,34,41]. Broadly speaking, existing learning strategies for PSO can be divided into two main categories, namely the topology-based learning strategies [43,48,59–61], and constructive learning strategies [8,35,51–53].

The topology-based learning strategies [23,43,45,49,62] mainly utilize different topologies to interact with particles to find proper guiding exemplars to direct the update of the associated particle. In fact, the learning strategy in the classical PSO [10,11] described above is also topology-based, where the topology is the full topology connecting all particles. However, such a topology leads to too greedy attraction of the second guiding exemplar in Equation (1), which results in premature convergence of PSO in solving multimodal problems. To alleviate this predicament, researchers have developed many novel topologies to find less greedy exemplars to improve the learning diversity of particles. To name a few representatives, in [45], a ring topology was utilized to organize particles into a ring, and each particle interacts with its left and right neighbors to select the locally best position (***lbest***) as the guiding exemplar to replace $\boldsymbol{gbest}$ in Equation (1). In addition, the star topology [46] and the wheel topology [46] were also utilized to organize particles and select the best position in the associated topology to replace $\boldsymbol{gbest}$. In [63], the cellular automata (CA) with the lattice and the "smart-cell" structures were integrated in PSO to select the second guiding exemplars for particles to update. In [60], the ring topology along with an elitist learning strategy was incorporated into PSO to maintain exploration and exploitation balance to properly search the solution space.

In the above studies, the adopted topologies are usually fixed during the whole optimization. This results in insufficient information exchange among particles. To further improve the learning effectiveness of particles, some researchers have attempted to adopt dynamic topologies to find promising guiding exemplars for particles. For instance, in [61], a dynamic-neighborhood-based switching PSO (DNSPSO) was proposed according to a distance-based dynamic topology along with a novel switching learning strategy to adaptively select the acceleration coefficients based on the searching state of the swarm. In this way, the evolutionary information of the swarm could be fully used to update particles. In [59], Gong and Zhang proposed a small-world network based topology to let each particle interact with its cohesive neighbors and by chance to communicate with some distant particles via small-world randomization with probability.

Except for the above researches that use only one topology on the whole swarm, some researchers have also been devoted to hybridizing different topologies based on subpopulation techniques to further improve the learning diversity of particles. For example, in [64], a fitness peak clustering (FPC) based dynamic multi-warm PSO with an enhanced learning strategy (FPCMSPSO) was designed. Specifically, this algorithm uses FPC to divide the swarm into several sub-swarms, and then evolves each sub-swarm independently based on the local topology. In [65], a PSO with double learning patterns (PSO-DLP) was developed. Specifically, this PSO variant adopts two swarms, namely the master swarm and the slave swarm, and employs two different learning patterns to update particles in the two swarms, so that a trade-off between the convergence speed and the swarm diversity can be achieved. In particular, in the slave swarm, a local topology is used to update its particles to explore the search space, while in the master swarm, a global topology is utilized to exploit the found promising areas. In [47], a memetic multi-topology particle swarm optimizer (MMTPSO) was devised by utilizing two different topologies and during the evolution, this algorithm biases to use the best performing topology to evolve the swarm.

Different from the topology-based learning strategies that select guiding exemplars from the historical best positions already found by particles, the constructive learning strategies mainly build new exemplars that may not appear during the evolution based on the historical best positions [8,35,51,53,66]. Specifically, these methods mainly aim to recombine dimensions of the historical best positions to try to generate promising guiding exemplars via some recombination techniques. In this direction, the most representative method is the comprehensive learning strategy [8], which constructs a new exemplar dimension by dimension for each particle based on the personal best positions of all particles. Since the advent of CLPSO, researchers have proposed many additional techniques to further improve its optimization ability. For instance, in [40], a local search method was incorporated into CLPSO, and an adaptive local search starting strategy was further put forward to adaptively trigger the local search by utilizing the quasi-entropy index. Instead of adopting fixed comprehensive learning (CL) probabilities in CLPSO, in [35], an adaptive CLPSO with cooperative archive (ACLPSO-CA) was developed by adaptively adjusting the CL probability along with a cooperative archive (CA). Specifically, this algorithm divides the CL probability into three levels and adjusts the CL probability level of each particle dynamically according to its performance during the evolution. In [66], a multi-leader CLPSO with adaptive mutation (ML-CLPSO-AM) was developed by incorporating a multi-leader (ML) strategy and an adaptive mutation (AM) strategy into CLPSO. In the ML strategy, a set of top-ranked particles form a pool of candidate leaders. During the update of each particle, a leader is randomly selected from the pool to guide the learning of this particle. In the AM strategy, the stagnated particles are adaptively mutated to restart their evolution. In [52], a heterogeneous CLPSO (HCLPSO) was devised by using the personal best positions of particles to generate guiding exemplars for the subpopulation responsible for exploration, while adopting the global best positions of the entire swarm to generate guiding exemplars for the subpopulation responsible for exploitation. In [50], a triple archives PSO (TAPSO) was designed by maintaining three archives. The first archive is used to store particles with better fitness, and the second archive is used to record the other particles. Then, similarly to CLPSO, this algorithm generates a new exemplar for each particle dimension by dimension by randomly choosing one particle from the first archive and one from the second archive based on the ordinary genetic operators. If the constructed exemplar has excellent performance, it will be saved in the third archive and then reused by inferior particles.

Although the above CLPSO variants have shown promising performance, the construction of new effective exemplars is inefficient because the recombination of dimension is thoroughly random. To improve the effectiveness of the recombination, in [51], an orthogonal learning PSO (OLPSO) was devised by orthogonal experimental design to discover useful information that lies in the historical positions found by particles. Specifically, this algorithm utilizes an orthogonal matrix to find the effective combinations of dimensions to construct a promising and efficient exemplar for each particle. However, this method is too resource-consuming because it needs many fitness evaluations in the orthogonal experimental design. To alleviate this issue, in [53], the genetic operators including crossover, mutation, and selection, were adopted to construct guiding exemplars based on the historical search information of particles. In this way, the generated exemplars are expected to be not only well diversified, but also of high quality. In [45], a global GLPSO with a ring topology (GGL-PSOD) was devised. In this PSO variant, the ring topology is used to breed diversified exemplars based on two directly connected neighbor particles, so that the exploration ability can be promoted.

Except for the learning strategies, the settings of key parameters in Equation (1) also play a crucial role in aiding PSO to achieve good performance when tackling optimization problems [19,57,67–69]. However, these parameter settings are usually different for different optimization problems. To alleviate this dilemma, researchers have designed many adaptive parameter adjustment strategies for PSO [70]. For instance, in [71], an adaptive PSO was proposed by adaptively adjusting the inertia weight and the acceleration coefficients. Specifically, a real-time evolutionary state estimation method was designed to classify the evolutionary states of the swarm into four types, namely exploration, exploitation, convergence, and jumping out in each generation. Then, the inertia weight and the acceleration coefficients are dynamically adjusted based on the estimated state. Taking inspiration from the activation function of neural networks, Liu et al. [57] proposed a sigmoid-function-based weighting strategy for PSO to update the acceleration coefficients by considering both the distances from the updated particle to ***gbest*** and from the particle to its ***pbest***. In [72], a self-adaptive parameter updating strategy based on success history information was proposed to automatically adjust the learning parameters to appropriate values.

Although PSO has been advanced to a large extent as briefly described above since its advent, its optimization ability is still challenged when confronted with complicated optimization problems with many interacting variables and wide and flat local areas. Therefore, how to improve the optimization performance of PSO is still an open issue and remains a hot and attractive topic in the evolutionary computation community. To this end, this paper proposes a stochastic cognitive dominance leading particle swarm optimization (SCDLPSO) to improve the optimization ability of PSO in solving complex optimization problems, which is elucidated in the next section.

3. Stochastic Cognitive Dominance Leading Particle Swarm Optimization

In human society, groups of randomly assembled individuals usually engage in costly group competition spontaneously [55]. Likewise, in the swarm of PSO, we can also randomly assemble particles and let them compete with each other based on their historical cognitive experience. Inspired by this, we propose a stochastic cognitive dominance leading particle swarm optimization algorithm (SCDLPSO) in this paper to improve the optimization ability of PSO in tackling optimization problems.

3.1. Stochastic Cognitive Dominance Leading Strategy

During the evolution, different particles usually preserve different historical experiences. These cognitive experiences preserve valuable information to guide the evolution of the swarm. In the classical PSO [10,11], each particle exchanges its own experience with all the other particles to find the best one to direct its learning. Such an exchange is too greedy and with low diversity, as all particles learn from the same best position, namely $\boldsymbol{gbest}$. To improve the learning diversity and effectiveness of particles, we propose a stochastic cognitive dominance leading strategy (SCDL) for PSO by imitating the stochastic competition mechanism in human society [55].

Specifically, given that NP particles are maintained in the swarm, during the evolution, for each particle to be updated (denoted as $\boldsymbol{x}_i$, $i \in [1, NP]$), we first randomly select two different personal best positions (denoted by $\boldsymbol{pbest}_{pr1}$ and $\boldsymbol{pbest}_{pr2}$) from those of the rest (NP-1) of the particles. Between the two selected personal best positions, we suppose $f(\boldsymbol{pbest}_{pr1}) \leq f(\boldsymbol{pbest}_{pr2})$. Then, the two selected best positions compete with the personal best position of the particle to be updated. Only when at least one of the two selected personal best positions shows dominance to the one of the particle, this particle is updated; otherwise, this particle is not updated and directly enters the next generation.

In particular, with the assumption that $f(\boldsymbol{pbest}_{pr1}) \leq f(\boldsymbol{pbest}_{pr2})$, in the competition between the two selected best positions ($\boldsymbol{pbest}_{pr1}$ and $\boldsymbol{pbest}_{pr2}$) and the personal best position ($\boldsymbol{pbest}_i$) of the particle to be updated, three cases may occur: namely, $\boldsymbol{pbest}_i$ is dominated by both of the two selected positions, $\boldsymbol{pbest}_i$ is dominated by only one of the two positions, or $\boldsymbol{pbest}_i$ dominates both of the two positions. In the three cases, the velocity of the associated particle is updated accordingly, as follows:

Case 1: $f(\boldsymbol{pbest}_{pr1}) \leq f(\boldsymbol{pbest}_{pr2}) \leq f(\boldsymbol{pbest}_i)$:

$$\boldsymbol{v}_i^{t+1} = w\boldsymbol{v}_i^t + \beta(r_1(\boldsymbol{pbest}_{pr1} - \boldsymbol{x}_i^t) + r_2(\boldsymbol{pbest}_{pr2} - \boldsymbol{x}_i^t)) \tag{4}$$

Case 2: $f(\boldsymbol{pbest}_{pr1}) \leq f(\boldsymbol{pbest}_i) < f(\boldsymbol{pbest}_{pr2})$:

$$\boldsymbol{v}_i^{t+1} = w\boldsymbol{v}_i^t + \beta(r_1(\boldsymbol{pbest}_{pr1} - \boldsymbol{x}_i^t) + r_2(\boldsymbol{pbest}_i - \boldsymbol{x}_i^t)) \tag{5}$$

Case 3: $f(\boldsymbol{pbest}_i) < f(\boldsymbol{pbest}_{pr1}) \leq f(\boldsymbol{pbest}_{pr2})$:

$$\boldsymbol{v}_i^{t+1} = \boldsymbol{v}_i^t \tag{6}$$

where $\boldsymbol{v}_i$ is the velocity vector of the ith particle; $\boldsymbol{pbest}_i$ is its personal best position; $\boldsymbol{pbest}_{pr1}$ and $\boldsymbol{pbest}_{pr2}$ are the two randomly selected personal best positions; $w \in [0, 1]$ is the inertia weight; r_1 and r_2 are two real random numbers uniformly sampled within [0, 1]; β is the acceleration parameter controlling the influence of the two guiding exemplars on the updated particle.

In Case 1, the randomly selected two personal best positions $\boldsymbol{pbest}_{pr1}$ and $\boldsymbol{pbest}_{pr2}$ are both superior to the one ($\boldsymbol{pbest}_i$) of the updated particle. In this situation, it is likely that the historical experience of the associated two particles is more valuable than the one of the particle to be updated. Therefore, to accelerate the learning efficiency of this particle, we let it cognitively learn from these two random best positions ($\boldsymbol{pbest}_{pr1}$ and $\boldsymbol{pbest}_{pr2}$) instead of learning from its own historical experience. In this way, it is expected that the updated particle could approach the promising areas quickly. It should be noticed that such a fast approach to promising areas is not at the expense of search diversity, because the two personal best positions are randomly selected from those of all particles.

In Case 2, only one (with the assumption that $f(\boldsymbol{pbest}_{pr1}) \leq f(\boldsymbol{pbest}_{pr2})$, it is actually $\boldsymbol{pbest}_{pr1}$) of the two selected personal best positions shows dominance to the personal best position ($\boldsymbol{pbest}_i$) of the particle to be updated. To enhance the probability of finding more promising areas, we let this particle cognitively learn from the better one ($\boldsymbol{pbest}_{pr1}$) of the two selected historical best positions and its own personal best position ($\boldsymbol{pbest}_i$). By this means, the updated particle also learns from good historical experience and thus likely approaches promising areas quickly.

In Case 3, both of the two randomly selected personal best positions are inferior to the one of the particle to be updated. In this situation, except for its own valuable experience, no extra useful experience is available for this particle to learn during this information exchange between the particle and the two particles associated with the two randomly selected personal best positions. Therefore, this particle is not updated and directly enters the next generation as shown in Equation (6), where the velocity of the particle remains unchanged.

On the whole, from the above three cases, we can see that the proposed SCDL strategy can assist PSO to maintain high search diversity from two perspectives. (1) The random selection of the two personal best positions allows different particles to learn from different exemplars. Therefore, the learning diversity of particles can be largely improved, which is beneficial for the promotion of the swarm diversity. (2) In Case 3, some particles with promising experience survive. Such an implicit retention mechanism affords great chances of improving the swarm diversity. Additionally, SCDL can also help PSO achieve fast convergence, because each updated particle in Case 1 and Case 2 learns from valuable cognitive experience. Such learning from elite experience offers high possibility for the updated particles to approach promising areas quickly. Consequently, based on the above investigation, it is expected that the proposed SCDLPSO can balance the search intensification and diversification well to explore and exploit the solution space appropriately to find high-quality solutions to optimization problems.

3.2. Difference between SCDL and Existing PSO Variants

In fact, the proposed SCDL strategy is a topology-based learning strategy. Specifically, the topology is a random triad topology connecting each particle to be updated with two random ones selected from the rest of the particles. Compared with existing studies adopting random topologies [47,48,59,73], the proposed SCDL distinguishes them from the following two perspectives:

(1) SCDL lets each particle learn from the best two personal best positions in the associated random triad topology. That is to say, the topology affects the selection of the two guiding exemplars in the velocity update. However, most existing random topology-based learning strategies [47,48,59,73] let each particle learn from its own personal best position and the best among the random topologies. In other words, the random topologies only influence the selection of the second guiding exemplar in the velocity update.

(2) In Case 3 in the proposed SCDL, some particles with promising historical experience are not updated and directly enter the next generation. With this retention mechanism, some promising historical experience is preserved from being attracted to local areas, which is beneficial for the swarm to escape from local areas. However, in most existing studies [47,48], all particles are updated, and thus there is no retention mechanism like Case 3 in SCDL.

Based on these two unique advantages, the proposed SCDL is expected to help PSO balance the swarm diversity and the convergence speed better to explore and exploit the solution space.

3.3. Overall Procedure

The overall procedure of the developed SCDLPSO is shown in Algorithm 1. Specifically, after NP particles are randomly initialized and evaluated as shown in Line 1, the algorithm goes to the main iteration of the evolution. During the update of the swarm (Lines 3–15), for each particle, two random personal best positions are first selected (Line 4), and then they compete with the personal best positions of the particle to be updated. With different competition cases, the particle is updated accordingly, as shown in Lines 9–13. After the particle is updated, it is reevaluated, and its personal best position is updated accordingly (Line 14). The above main iteration proceeds continuously until the maximum number of fitness evaluations is exhausted. At the end of the algorithm, the global best position found by the swarm is obtained as the final output (Line 17).

Algorithm 1 The pseudocode of SCDLPSO

Input: swarm size *NP*, maximum number of fitness evaluations *FEmax*

1: Initialize *NP* particles randomly and calculate their fitness, and set $fes = NP$;
2: **While** ($fes \leq FEmax$) **do**
3: **For** i = 1:*NP* **do**
4: Select two different exemplars randomly from the personal best positions of all particles: $\boldsymbol{pbest}_{pr1}$, $\boldsymbol{pbest}_{pr2}$;
5: **If** ($f(\boldsymbol{pbest}_{pr2}) < f(\boldsymbol{pbest}_{pr1})$) **then**
6: Swap *pr*2 and *pr*1;
7: **End If**
8: Compute the inertia weight w according to Equation (3);
9: **If** ($f(\boldsymbol{pbest}_{pr1}) \leq f(\boldsymbol{pbest}_{pr2}) \leq f(\boldsymbol{pbest}_i)$) **then**
10: Update the particle according to Equation (4) and Equation (2);
11: **Else If** ($f(\boldsymbol{pbest}_{pr1}) \leq f(\boldsymbol{pbest}_i) < f(\boldsymbol{pbest}_{pr2})$) **then**
12: Update the particle according to Equation (5) and Equation (2);
13: **End If**
14: Calculate the fitness of the updated particle: $f(\boldsymbol{x}_i)$, update its $\boldsymbol{pbest}_i$ and fes += 1;
15: **End For**
16: **End While**
17: Obtain the global best solution $\boldsymbol{gbest}$ and its fitness $f(\boldsymbol{gbest})$;

Output: $f(\boldsymbol{gbest})$ **and** $\boldsymbol{gbest}$

From Algorithm 1, it is found that at each generation, except for the fitness evaluation time, it takes O (*NP*) to select random personal best positions for all particles, and O (*NP*) to compete the selected personal best positions with the ones of the associated particles. Then, it takes O ($NP * D$) to update all particles. On the whole, it is found that the overall time complexity of SCDLPSO is $O(NP * D)$. As for the space complexity, the same with the classical PSO, SCDLPSO needs O ($NP * D$), O ($NP * D$) and O ($NP * D$) to store the velocity, the positions and the personal best positions of all particles, respectively.

In conclusion, we can see that the proposed SCDLPSO remains as efficient as the classical PSO in terms of both the time complexity and the space occupation.

4. Experiments

In this section, extensive experiments are carried out to verify the effectiveness of the proposed SCDLPSO on the widely used CEC 2017 benchmark problem set [56]. In particular, this benchmark set contains 29 optimization problems with four types, namely unimodal problems (F1 and F3), simple multimodal problems (F4–F10), hybrid problems (F11–F20) and composition problems (F21–F30). The latter two types of problems are much more difficult to optimize than the former two. For more detailed information about this benchmark set, please refer to [56].

4.1. Experimental Setup

First, to comprehensively verify the performance of SCDLPSO, we compare it with several state-of-the-art PSO algorithms. Specially, the selected representative and state-of-the-art PSO variants are XPSO [74], TCSPSO [75], DNSPSO [61], AWPSO [57], CLPSO_LS [40], GLPSO [53], and CLPSO [8]. Among these compared algorithms, XPSO and DNSPSO are topology-based learning PSO variants, while CLPSO_LS, GLPSO and CLPSO are constructive learning-based PSO methods.

Second, to make comprehensive comparisons between the proposed SCDLPSO and the compared PSO variants, we evaluate their performance on the CEC 2017 benchmark problems with three different dimension sizes, namely 30-D, 50-D, and 100-D. For fair comparisons, the maximum number of fitness evaluations (FE_{max}) is set as 10,000 $*$ D for all algorithms.

Third, to make fair comparisons, we fine-tune the population size for all algorithms on the CEC 2017 benchmark set with different dimension sizes. After preliminary experiments for the fine-tuning of the population size, the parameter settings of all algorithms were as listed in Table 1.

Table 1. Parameter settings of all comparative algorithms.

Algorithms	D		Parameter Settings
SCDLPSO	30	$NP = 100$	$w = 0.9–0.4$ $\beta = 0.5$
	50	$NP = 100$	
	100	$NP = 150$	
XPSO	30	$NP = 100$	$\eta = 0.2$ $p = 0.5$ $Stage_{max} = 5$
	50	$NP = 150$	
	100	$NP = 150$	
TCSPSO	30	$NP = 50$	$w = 0.9–0.4$ $c_1 = c_2 = 2$
	50	$NP = 50$	
	100	$NP = 50$	
DNSPSO	30	$NP = 50$	$w = 0.9–0.4$ $k = 5$ $F = 0.5$ $CR = 0.9$
	50	$NP = 50$	
	100	$NP = 50$	
AWPSO	30	$NP = 40$	$w = 0.9–0.4$
	50	$NP = 60$	
	100	$NP = 100$	
CLPSO_LS	30	$NP = 40$	$c = 1.4945$ $w = 0.9–0.4$ $\beta = 1/3$ $\theta = 0.94$ $Pc = 0.05–0.5$
	50	$NP = 50$	
	100	$NP = 50$	
GLPSO	30	$NP = 40$	$w = 0.7298$ $c_1 = c_2 = 1.49618$ $pm = 0.01$ $sg = 7$
	50	$NP = 40$	
	100	$NP = 50$	
CLSPO	30	$NP = 40$	$w = 0.9–0.2$ $c_1 = c_2 = 1.49445$ $Pc = 0.05–0.5$
	50	$NP = 60$	
	100	$NP = 60$	

Fourth, to comprehensively and fairly evaluate each algorithm, we run each algorithm independently for 30 times and utilize the median, the mean and the standard deviation over the 30 independent runs to evaluate its optimization performance. In addition, to identify the statistical significance, we conduct the Wilcoxon rank sum test at the significance level of $\alpha = 0.05$. Moreover, to investigate the overall performance of each algorithm on the whole CEC 2017 benchmark set, the Friedman test is also performed at the significance level of $\alpha = 0.05$ to obtain the average rank of each algorithm.

4.2. Parameter Sensitivity Analysis

In the proposed SCDLPSO, there are two parameters, namely the swarm size NP and the control parameter β, that need to be fine-tuned. Therefore, to investigate the sensitivity of SCDLPSO to the two parameters, we conduct experiments by varying NP from 50 to 200 and ranging β from 0.1 to 1.0 for SCDLPSO on the 50-D CEC 2017 benchmark set as a representative. Table 2 shows the comparison results for SCDLPSO with different settings of NP and β on the 50-D CEC 2017 benchmark problems. In this table, the best results are highlighted in bold, and the average rank of each setting of β under the same setting of NP, as presented in the last row of the table, is obtained by the Friedman test at the significance level of "$\alpha = 0.05$".

From Table 2, we can derive the following findings. (1) From the perspective of the Friedman test, when NP is fixed, the setting of β is neither too large nor too small, and the optimal setting is within [0.3, 0.5]. Specifically, when NP is 100 and 200, the optimal β is 0.5. When NP is set to 50, the optimal setting of β is 0.4. When NP is 150, the optimal setting of β is 0.3 and 0.4. (2) On closer observation, we can find that no matter what the swarm size is, the performance of SCDLPSO first improves with the increase in β at the beginning. After it reaches 0.5, the larger the setting of β is, the worse performance SCDLPSO achieves. (3) Taking comprehensive comparisons among all settings into consideration, we find that β is not so closely related to swarm size NP. Comprehensively, we recommend setting $\beta = 0.5$ for SCDLPSO to solve optimization problems.

To summarize, β is not so closely related to swarm size NP, and the optimal setting is generally within [0.3, 0.5]. In this paper, we recommend setting $\beta = 0.5$ for SCDLPSO to solve optimization problems.

4.3. Comparison with State-of-the-Art PSO Variants

In this section, we conduct extensive comparison experiments on the CEC 2017 benchmark set with different dimension sizes to compare the proposed SCDLPSO with the seven state-of-the-art and representative PSO variants. Tables 3–5 show the detailed comparison results for the 30-D, 50-D and 100-D CEC 2017 benchmark problems, respectively. In these tables, the symbols "+", "−" and "=" above the p-values mean that SCDLPSO is significantly superior, inferior, or equivalent to the compared algorithms on the associated problems. In the second to last rows of these tables, "$w/t/l$" count the number of problems where the devised SCDLPSO achieves significantly better, equivalent, or worse performance than the associated compared algorithms, respectively. They are actually the number of "+", "=" and "−", respectively. In the last rows of these tables, the average rank of each algorithm obtained by the Friedman test is presented. In addition, Table 6 summarizes the statistical comparison results between SCDLPSO and the seven state-of-the-art PSO variants on the CEC 2017 benchmark set with different dimensions in terms of "$w/t/l$".

Table 2. Comparison results for SCDLPSO with different settings of NP and β on the 50-D CEC 2017 problems.

F	$NP = 50$										$NP = 100$									
	$\beta = 0.1$	$\beta = 0.2$	$\beta = 0.3$	$\beta = 0.4$	$\beta = 0.5$	$\beta = 0.6$	$\beta = 0.7$	$\beta = 0.8$	$\beta = 0.9$	$\beta = 1.0$	$\beta = 0.1$	$\beta = 0.2$	$\beta = 0.3$	$\beta = 0.4$	$\beta = 0.5$	$\beta = 0.6$	$\beta = 0.7$	$\beta = 0.8$	$\beta = 0.9$	$\beta = 1.0$
F_1	8.01×10^{4}	2.21×10^{3}	**1.89×10^{3}**	5.02×10^{3}	5.38×10^{3}	8.37×10^{3}	1.16×10^{4}	1.36×10^{4}	2.17×10^{4}	2.24×10^{4}	2.39×10^{3}	7.77×10^{2}	**1.45×10^{3}**	2.32×10^{3}	2.32×10^{3}	6.33×10^{3}	9.69×10^{3}	1.40×10^{4}	1.76×10^{4}	2.69×10^{4}
F_3	7.85×10^{4}	2.94×10^{4}	1.52×10^{4}	8.64×10^{3}	**6.00×10^{3}**	6.07×10^{3}	8.68×10^{3}	1.81×10^{4}	2.98×10^{4}	5.09×10^{4}	7.04×10^{4}	3.65×10^{4}	2.02×10^{4}	1.46×10^{4}	**1.44×10^{4}**	1.77×10^{4}	2.50×10^{4}	2.89×10^{4}	5.06×10^{4}	8.59×10^{4}
F_4	1.82×10^{2}	1.19×10^{2}	**9.51×10^{1}**	1.23×10^{2}	1.29×10^{2}	1.55×10^{2}	1.42×10^{2}	1.59×10^{2}	1.67×10^{2}	1.83×10^{2}	1.83×10^{2}	1.38×10^{2}	**9.50×10^{1}**	1.02×10^{2}	1.23×10^{2}	1.34×10^{2}	1.57×10^{2}	1.73×10^{2}	1.81×10^{2}	1.97×10^{2}
F_5	1.73×10^{2}	7.94×10^{1}	3.03×10^{1}	2.11×10^{1}	**1.91×10^{1}**	1.99×10^{1}	2.42×10^{1}	1.70×10^{2}	3.34×10^{2}	3.62×10^{2}	1.58×10^{2}	6.59×10^{1}	2.43×10^{1}	1.22×10^{1}	**1.08×10^{1}**	1.08×10^{1}	5.79×10^{1}	3.15×10^{2}	3.39×10^{2}	3.63×10^{2}
F_6	3.02×10^{1}	4.17×10^{0}	2.68×10^{-1}	7.74×10^{-2}	**5.20×10^{-2}**	4.88×10^{-2}	7.08×10^{-2}	4.29×10^{-2}	3.97×10^{-2}	1.55×10^{-1}	2.45×10^{1}	2.80×10^{0}	9.15×10^{-2}	1.44×10^{-2}	**6.38×10^{-4}**	5.96×10^{-3}	2.48×10^{-3}	4.51×10^{-3}	2.74×10^{-3}	4.73×10^{-3}
F_7	2.77×10^{2}	1.24×10^{2}	8.19×10^{1}	6.99×10^{1}	**6.75×10^{1}**	7.56×10^{1}	1.61×10^{2}	3.65×10^{2}	3.83×10^{2}	4.04×10^{2}	2.18×10^{2}	1.01×10^{2}	7.33×10^{1}	6.38×10^{1}	**6.22×10^{1}**	9.15×10^{1}	3.32×10^{2}	3.70×10^{2}	3.84×10^{2}	4.15×10^{2}
F_8	1.81×10^{2}	7.74×10^{1}	2.97×10^{1}	1.98×10^{1}	**1.80×10^{1}**	2.01×10^{1}	2.31×10^{1}	1.22×10^{2}	3.36×10^{2}	3.60×10^{2}	1.53×10^{2}	6.98×10^{1}	2.36×10^{1}	1.41×10^{1}	**1.01×10^{1}**	1.14×10^{1}	4.48×10^{1}	3.03×10^{2}	3.34×10^{2}	3.65×10^{2}
F_9	4.48×10^{3}	2.62×10^{2}	2.40×10^{1}	**6.55×10^{0}**	1.13×10^{1}	1.91×10^{1}	2.36×10^{1}	1.37×10^{1}	1.55×10^{1}	3.55×10^{1}	2.80×10^{3}	1.11×10^{2}	7.89×10^{0}	2.07×10^{0}	1.57×10^{0}	5.58×10^{-1}	2.63×10^{0}	3.91×10^{0}	3.92×10^{-2}	**2.98×10^{-3}**
F_{10}	5.78×10^{3}	4.47×10^{3}	**4.25×10^{3}**	4.62×10^{3}	7.65×10^{3}	1.20×10^{4}	1.26×10^{4}	1.23×10^{4}	1.20×10^{4}	1.09×10^{4}	5.88×10^{3}	4.20×10^{3}	**4.04×10^{3}**	4.90×10^{3}	9.87×10^{3}	1.26×10^{4}	1.29×10^{4}	1.29×10^{4}	1.22×10^{4}	1.15×10^{4}
F_{11}	2.46×10^{2}	1.51×10^{2}	1.56×10^{2}	1.19×10^{2}	8.87×10^{1}	8.18×10^{1}	**6.36×10^{1}**	7.03×10^{1}	1.07×10^{2}	1.95×10^{2}	2.13×10^{2}	1.36×10^{2}	1.29×10^{2}	8.81×10^{1}	6.62×10^{1}	5.15×10^{1}	**3.88×10^{1}**	3.89×10^{1}	1.52×10^{2}	1.58×10^{2}
F_{12}	2.26×10^{6}	3.49×10^{5}	**2.15×10^{5}**	3.63×10^{5}	2.35×10^{5}	6.58×10^{7}	4.53×10^{5}	1.44×10^{6}	6.68×10^{6}	3.43×10^{6}	1.76×10^{6}	2.84×10^{5}	**1.99×10^{5}**	2.64×10^{5}	2.77×10^{5}	5.29×10^{5}	8.67×10^{5}	1.43×10^{6}	2.55×10^{6}	4.82×10^{6}
F_{13}	1.09×10^{4}	8.16×10^{3}	3.16×10^{5}	**8.62×10^{3}**	1.18×10^{4}	1.01×10^{6}	2.47×10^{4}	2.93×10^{4}	1.46×10^{8}	2.60×10^{7}	8.27×10^{3}	4.85×10^{3}	**4.00×10^{3}**	5.48×10^{3}	5.90×10^{3}	1.52×10^{4}	2.12×10^{4}	2.85×10^{4}	3.04×10^{4}	3.37×10^{4}
F_{14}	4.15×10^{4}	3.05×10^{4}	2.00×10^{4}	**1.64×10^{4}**	3.75×10^{4}	5.37×10^{4}	7.84×10^{4}	1.25×10^{5}	1.57×10^{5}	1.77×10^{5}	3.73×10^{4}	2.16×10^{4}	**1.74×10^{4}**	2.07×10^{4}	3.49×10^{4}	8.85×10^{4}	9.64×10^{4}	1.45×10^{5}	2.44×10^{5}	2.54×10^{5}

Table 2. *Cont.*

F_{15}	6.47×10^{3}	5.70×10^{3}	5.31×10^{3}	**5.87×10^{3}**	6.65×10^{3}	1.21×10^{4}	1.91×10^{4}	2.60×10^{4}	2.96×10^{4}	3.11×10^{4}	6.80×10^{3}	5.63×10^{3}	**5.96×10^{3}**	5.62×10^{3}	6.48×10^{3}	8.22×10^{3}	1.96×10^{4}	2.72×10^{4}	3.04×10^{4}	3.13×10^{4}
F_{16}	1.44×10^{3}	8.35×10^{2}	5.62×10^{2}	**4.44×10^{2}**	4.76×10^{2}	7.09×10^{2}	7.08×10^{2}	1.19×10^{3}	2.04×10^{3}	2.43×10^{3}	1.23×10^{3}	6.63×10^{2}	4.99×10^{2}	4.85×10^{2}	**4.29×10^{2}**	5.56×10^{2}	8.06×10^{2}	1.86×10^{3}	2.37×10^{3}	2.84×10^{3}
F_{17}	1.35×10^{3}	7.86×10^{2}	4.89×10^{2}	4.87×10^{2}	**4.29×10^{2}**	5.22×10^{2}	6.97×10^{2}	8.27×10^{2}	1.48×10^{3}	1.74×10^{3}	1.28×10^{3}	7.18×10^{2}	4.95×10^{2}	3.76×10^{2}	**3.70×10^{2}**	7.97×10^{2}	9.33×10^{2}	1.26×10^{3}	1.57×10^{3}	1.74×10^{3}
F_{18}	2.51×10^{5}	1.20×10^{5}	**7.21×10^{4}**	7.75×10^{4}	1.16×10^{5}	1.99×10^{5}	3.47×10^{5}	1.10×10^{6}	2.01×10^{6}	2.21×10^{6}	1.45×10^{5}	9.03×10^{4}	**6.69×10^{4}**	8.02×10^{4}	1.30×10^{5}	2.74×10^{5}	8.92×10^{5}	1.66×10^{6}	3.43×10^{6}	4.41×10^{6}
F_{19}	1.48×10^{4}	1.47×10^{4}	1.81×10^{4}	1.68×10^{4}	1.82×10^{4}	1.50×10^{4}	1.96×10^{4}	1.63×10^{4}	1.26×10^{4}	**3.73×10^{3}**	1.48×10^{4}	1.50×10^{4}	1.42×10^{4}	1.28×10^{4}	1.43×10^{4}	1.18×10^{4}	1.20×10^{4}	9.55×10^{3}	**4.58×10^{3}**	4.86×10^{3}
F_{20}	7.22×10^{2}	3.57×10^{2}	**2.50×10^{2}**	3.93×10^{2}	4.80×10^{2}	1.10×10^{3}	1.35×10^{3}	1.42×10^{3}	1.45×10^{3}	1.53×10^{3}	6.39×10^{2}	3.54×10^{2}	**2.37×10^{2}**	2.78×10^{2}	7.04×10^{2}	1.22×10^{3}	1.35×10^{3}	1.45×10^{3}	1.49×10^{3}	1.57×10^{3}
F_{21}	3.61×10^{2}	2.80×10^{2}	2.41×10^{2}	2.37×10^{2}	**2.31×10^{2}**	2.34×10^{2}	2.34×10^{2}	3.00×10^{2}	5.36×10^{2}	5.56×10^{2}	3.16×10^{2}	2.54×10^{2}	2.29×10^{2}	2.29×10^{2}	**2.19×10^{2}**	**2.19×10^{2}**	2.23×10^{2}	5.01×10^{2}	5.35×10^{2}	5.66×10^{2}
F_{22}	4.74×10^{3}	3.47×10^{3}	**3.57×10^{3}**	3.96×10^{3}	5.50×10^{3}	1.05×10^{4}	1.24×10^{4}	1.28×10^{4}	1.31×10^{4}	1.31×10^{4}	4.94×10^{3}	**2.19×10^{3}**	3.46×10^{3}	3.67×10^{3}	6.52×10^{3}	1.20×10^{4}	1.27×10^{4}	1.27×10^{4}	1.30×10^{4}	1.31×10^{4}
F_{23}	7.41×10^{2}	6.39×10^{2}	**5.95×10^{2}**	5.99×10^{2}	6.16×10^{2}	6.25×10^{2}	6.26×10^{2}	6.32×10^{2}	6.73×10^{2}	8.58×10^{2}	6.30×10^{2}	5.57×10^{2}	5.30×10^{2}	5.34×10^{2}	5.27×10^{2}	**5.24×10^{2}**	5.29×10^{2}	5.63×10^{2}	7.92×10^{2}	8.21×10^{2}
F_{24}	7.55×10^{2}	6.83×10^{2}	**6.76×10^{2}**	6.80×10^{2}	6.93×10^{2}	6.92×10^{2}	7.11×10^{2}	8.44×10^{2}	9.41×10^{2}	9.67×10^{2}	6.68×10^{2}	6.11×10^{2}	**5.97×10^{2}**	5.99×10^{2}	5.98×10^{2}	6.09×10^{2}	6.26×10^{2}	8.70×10^{2}	8.82×10^{2}	8.93×10^{2}
F_{25}	6.10×10^{2}	5.66×10^{2}	5.61×10^{2}	5.43×10^{2}	4.97×10^{2}	**4.81×10^{2}**	4.87×10^{2}	4.85×10^{2}	4.84×10^{2}	4.95×10^{2}	5.95×10^{2}	5.64×10^{2}	5.45×10^{2}	5.65×10^{2}	5.08×10^{2}	**4.80×10^{2}**	**4.80×10^{2}**	**4.80×10^{2}**	**4.80×10^{2}**	4.82×10^{2}
F_{26}	4.77×10^{3}	1.66×10^{3}	**1.35×10^{3}**	1.91×10^{3}	2.26×10^{3}	2.55×10^{3}	2.90×10^{3}	3.05×10^{3}	3.62×10^{3}	4.93×10^{3}	3.70×10^{3}	2.01×10^{3}	**1.15×10^{3}**	1.91×10^{3}	1.89×10^{3}	1.93×10^{3}	2.21×10^{3}	2.50×10^{3}	4.53×10^{3}	5.38×10^{3}
F_{27}	9.41×10^{2}	7.80×10^{2}	**7.27×10^{2}**	7.81×10^{2}	7.46×10^{2}	8.01×10^{2}	8.58×10^{2}	8.68×10^{2}	8.82×10^{2}	9.52×10^{2}	8.55×10^{2}	7.47×10^{2}	7.15×10^{2}	7.09×10^{2}	**6.79×10^{2}**	7.15×10^{2}	7.57×10^{2}	7.63×10^{2}	7.92×10^{2}	7.84×10^{2}
F_{28}	5.73×10^{2}	5.11×10^{2}	5.01×10^{2}	4.92×10^{2}	4.89×10^{2}	**4.79×10^{2}**	7.88×10^{2}	1.58×10^{3}	4.26×10^{3}	5.37×10^{3}	5.52×10^{2}	5.04×10^{2}	5.06×10^{2}	4.92×10^{2}	4.74×10^{2}	4.77×10^{2}	**4.73×10^{2}**	1.96×10^{3}	3.46×10^{3}	5.44×10^{3}
F_{29}	1.96×10^{3}	1.17×10^{3}	7.35×10^{2}	6.72×10^{2}	**6.62×10^{2}**	7.13×10^{2}	7.31×10^{2}	7.31×10^{2}	1.27×10^{3}	1.69×10^{3}	1.81×10^{3}	1.02×10^{3}	6.16×10^{2}	**5.05×10^{2}**	5.16×10^{2}	5.08×10^{2}	5.42×10^{2}	8.00×10^{2}	1.47×10^{3}	1.79×10^{3}

Table 2. *Cont.*

F_{30}	1.31×10^{6}	**8.35×10^{5}**	9.29×10^{5}	9.97×10^{5}	9.75×10^{5}	1.04×10^{6}	1.21×10^{6}	1.56×10^{6}	1.90×10^{6}	2.04×10^{6}	1.18×10^{6}	8.33×10^{5}	**8.12×10^{5}**	8.26×10^{5}	8.25×10^{5}	9.98×10^{5}	1.28×10^{6}	1.63×10^{6}	1.56×10^{6}	1.67×10^{6}
Rank	7.41	4.45	3.55	**3.14**	3.21	4.48	5.59	6.52	7.69	8.97	7.24	5.00	3.31	3.28	**2.93**	4.17	5.38	7.03	7.86	8.79
	$NP = 150$										$NP = 200$									
F	$\beta = 0.1$	$\beta = 0.2$	$\beta = 0.3$	$\beta = 0.4$	$\beta = 0.5$	$\beta = 0.6$	$\beta = 0.7$	$\beta = 0.8$	$\beta = 0.9$	$\beta = 1.0$	$\beta = 0.1$	$\beta = 0.2$	$\beta = 0.3$	$\beta = 0.4$	$\beta = 0.5$	$\beta = 0.6$	$\beta = 0.7$	$\beta = 0.8$	$\beta = 0.9$	$\beta = 1.0$
F_1	**1.34×10^{3}**	1.54×10^{3}	1.53×10^{3}	1.99×10^{3}	2.44×10^{3}	3.06×10^{3}	7.89×10^{3}	1.24×10^{4}	1.62×10^{4}	2.48×10^{4}	1.35×10^{3}	**9.97×10^{2}**	1.59×10^{3}	1.89×10^{3}	2.53×10^{3}	5.30×10^{3}	9.45×10^{3}	1.24×10^{4}	2.08×10^{4}	2.58×10^{4}
F_3	7.10×10^{4}	3.90×10^{4}	3.12×10^{4}	**2.62×10^{4}**	2.76×10^{4}	3.32×10^{4}	3.65×10^{4}	5.18×10^{4}	8.11×10^{4}	1.14×10^{3}	7.03×10^{4}	4.69×10^{4}	3.78×10^{4}	**3.60×10^{4}**	4.20×10^{4}	4.30×10^{4}	5.03×10^{4}	6.90×10^{4}	9.84×10^{4}	1.38×10^{3}
F_4	1.78×10^{2}	**1.11×10^{2}**	1.14×10^{2}	**1.11×10^{2}**	1.24×10^{2}	1.47×10^{2}	1.85×10^{2}	1.86×10^{2}	2.00×10^{2}	2.05×10^{2}	1.73×10^{2}	1.49×10^{2}	1.36×10^{2}	1.08×10^{2}	**1.06×10^{2}**	1.70×10^{2}	1.89×10^{2}	1.94×10^{2}	2.00×10^{2}	2.11×10^{2}
F_5	1.44×10^{2}	6.00×10^{1}	2.26×10^{1}	1.08×10^{1}	7.73×10^{0}	**7.36×10^{0}**	9.43×10^{1}	3.15×10^{2}	3.41×10^{2}	3.66×10^{2}	1.44×10^{2}	6.01×10^{1}	2.27×10^{1}	9.19×10^{0}	5.94×10^{0}	**5.31×10^{0}**	1.26×10^{2}	3.22×10^{2}	3.42×10^{2}	3.72×10^{2}
F_6	2.20×10^{1}	2.39×10^{0}	1.22×10^{-1}	1.04×10^{-2}	2.17×10^{-3}	2.13×10^{-4}	6.13×10^{-5}	5.61×10^{-3}	**3.41×10^{-6}**	1.46×10^{-3}	2.17×10^{1}	2.45×10^{0}	1.13×10^{-1}	1.00×10^{-2}	9.46×10^{-4}	1.30×10^{-3}	3.32×10^{-5}	3.78×10^{-4}	**3.00×10^{-6}**	1.43×10^{-2}
F_7	2.05×10^{2}	9.68×10^{1}	7.03×10^{1}	**6.22×10^{1}**	6.86×10^{1}	1.72×10^{2}	3.56×10^{2}	3.66×10^{2}	3.88×10^{2}	4.13×10^{2}	2.07×10^{2}	9.52×10^{1}	7.01×10^{1}	**6.20×10^{1}**	7.91×10^{1}	2.30×10^{2}	3.50×10^{2}	3.70×10^{2}	3.87×10^{2}	4.23×10^{2}
F_8	1.43×10^{2}	5.75×10^{1}	2.19×10^{1}	1.00×10^{1}	8.22×10^{0}	**7.50×10^{0}**	8.07×10^{1}	3.13×10^{2}	3.38×10^{2}	3.69×10^{2}	1.36×10^{2}	5.50×10^{1}	2.14×10^{1}	9.39×10^{0}	**5.44×10^{0}**	5.27×10^{0}	7.43×10^{1}	3.17×10^{2}	3.39×10^{2}	3.71×10^{2}
F_9	2.32×10^{3}	9.69×10^{1}	7.67×10^{0}	1.53×10^{0}	1.58×10^{0}	3.28×10^{-1}	1.18×10^{-1}	1.03×10^{-1}	**5.97×10^{-3}**	1.81×10^{-2}	1.96×10^{3}	1.24×10^{2}	6.87×10^{0}	2.01×10^{0}	7.63×10^{-1}	3.05×10^{-1}	1.02×10^{-1}	7.23×10^{-2}	**1.73×10^{-10}**	3.85×10^{-3}
F_{10}	5.76×10^{3}	4.79×10^{3}	**4.77×10^{3}**	4.94×10^{3}	1.10×10^{4}	1.25×10^{4}	1.25×10^{4}	1.29×10^{4}	1.21×10^{4}	1.21×10^{4}	6.06×10^{3}	4.73×10^{3}	**4.28×10^{3}**	4.76×10^{3}	1.12×10^{4}	1.27×10^{4}	1.28×10^{4}	1.29×10^{4}	1.29×10^{4}	1.19×10^{4}
F_{11}	1.86×10^{2}	1.25×10^{2}	1.23×10^{2}	1.00×10^{2}	6.69×10^{1}	5.12×10^{1}	**4.23×10^{1}**	4.64×10^{1}	1.43×10^{2}	1.79×10^{2}	1.74×10^{2}	1.20×10^{2}	1.15×10^{2}	9.74×10^{1}	7.41×10^{1}	5.93×10^{1}	**4.14×10^{1}**	5.23×10^{1}	1.52×10^{2}	1.78×10^{2}
F_{12}	1.33×10^{6}	3.23×10^{5}	**2.47×10^{5}**	**2.47×10^{5}**	3.06×10^{5}	6.99×10^{5}	1.40×10^{6}	2.33×10^{6}	3.90×10^{6}	7.79×10^{6}	1.70×10^{6}	4.03×10^{5}	**2.56×10^{5}**	3.24×10^{5}	4.74×10^{5}	7.62×10^{5}	1.44×10^{6}	2.12×10^{6}	3.75×10^{6}	8.57×10^{6}

Table 2. *Cont.*

F_{13}	7.45×10^3	4.81×10^3	**3.51×10^3**	3.87×10^3	7.54×10^3	1.12×10^4	1.63×10^4	2.94×10^4	3.22×10^4	3.16×10^4	7.92×10^3	4.21×10^3	3.44×10^3	**2.10×10^3**	3.24×10^3	1.14×10^4	1.59×10^4	2.77×10^4	3.06×10^4	3.30×10^4
F_{14}	3.24×10^4	2.16×10^4	**1.96×10^4**	1.95×10^4	3.79×10^4	6.97×10^4	1.10×10^5	1.85×10^5	2.21×10^5	3.14×10^5	2.69×10^4	**2.26×10^4**	2.31×10^4	2.54×10^4	3.01×10^4	1.05×10^5	1.44×10^5	2.20×10^5	2.38×10^5	3.00×10^5
F_{15}	6.25×10^3	5.30×10^3	**4.91×10^3**	5.64×10^3	4.94×10^3	7.13×10^3	1.32×10^4	2.59×10^4	3.10×10^4	3.13×10^4	6.34×10^3	5.72×10^3	5.92×10^3	**4.96×10^3**	5.17×10^3	5.59×10^3	1.42×10^4	2.56×10^4	3.10×10^4	3.12×10^4
F_{16}	1.19×10^3	7.21×10^2	5.07×10^2	4.95×10^2	**4.19×10^2**	4.90×10^2	1.32×10^3	2.26×10^3	2.63×10^3	2.88×10^3	1.09×10^3	7.40×10^2	**5.24×10^2**	5.39×10^2	5.39×10^2	7.37×10^2	1.45×10^3	2.21×10^3	2.65×10^3	2.88×10^3
F_{17}	1.19×10^3	7.65×10^2	5.64×10^2	4.37×10^2	**3.94×10^2**	6.73×10^2	1.13×10^3	1.42×10^3	1.52×10^3	1.71×10^3	1.27×10^3	7.66×102	4.54×10^2	3.86×10^2	**3.22×10^2**	8.72×10^2	1.12×10^3	1.43×10^3	1.56×10^3	1.76×10^3
F_{18}	1.59×10^5	7.85×10^4	**5.56×10^4**	8.38×10^4	1.18×10^5	4.23×10^5	1.27×10^6	2.43×10^6	3.90×10^6	5.93×10^6	2.21×10^5	7.01×10^4	6.44×10^4	7.45×10^4	**1.83×10^5**	5.85×10^5	1.27×10^6	3.08×10^6	5.30×10^6	4.23×10^6
F_{19}	1.51×10^4	1.51×10^4	1.42×10^4	1.50×10^4	1.54×10^4	1.34×10^4	7.49×10^3	7.54×10^3	4.91×10^3	**2.46×10^3**	1.50×10^4	1.48×10^4	1.42×10^4	1.45×10^4	1.29×10^4	1.18×10^4	1.03×10^4	7.09×10^3	5.41×10^3	**3.74×10^3**
F_{20}	6.32×10^2	3.88×10^2	**2.29×10^2**	3.94×10^2	9.29×10^2	1.24×10^3	1.31×10^3	1.38×10^3	1.51×10^3	1.56×10^3	6.62×10^2	4.50×10^2	**2.75×10^2**	5.11×10^2	9.92×10^2	1.24×10^3	1.35×10^3	1.41×10^3	1.53×10^3	1.53×10^3
F_{21}	2.95×10^2	2.44×10^2	2.23×10^2	2.17×10^2	**2.13×10^2**	2.14×10^2	2.36×10^2	5.14×10^2	5.38×10^2	5.68×10^2	2.94×10^2	2.41×10^2	2.20×10^2	2.14×10^2	2.14×10^2	**2.12×10^2**	2.98×10^2	5.10×10^2	5.43×10^2	5.73×10^2
F_{22}	3.14×10^3	**1.42×10^3**	3.18×10^3	4.06×10^3	7.58×10^3	1.19×10^4	1.29×10^4	1.29×10^4	1.31×10^4	1.32×10^4	2.59×10^3	**9.54×10^2**	3.28×10^3	5.62×10^3	9.13×10^3	1.25×10^4	1.28×10^4	1.29×10^4	1.32×10^4	1.31×10^4
F_{23}	6.06×10^2	5.29×10^2	5.11×10^2	5.06×10^2	**5.00×10^2**	5.06×10^2	5.01×10^2	6.23×10^2	7.89×10^2	8.17×10^2	5.77×10^2	5.19×10^2	5.05×10^2	5.05×10^2	4.96×10^2	**4.92×10^2**	5.16×10^2	7.62×10^2	7.85×10^2	8.13×10^2
F_{24}	6.20×10^2	5.79×10^2	**5.78×10^2**	5.84×10^2	5.83×10^2	5.86×10^2	7.15×10^2	8.53×10^2	8.65×10^2	8.80×10^2	6.02×10^2	5.78×10^2	**5.65×10^2**	5.75×10^2	5.72×10^2	5.77×10^2	7.64×10^2	8.55×10^2	8.65×10^2	8.70×10^2
F_{25}	5.97×10^2	5.71×10^2	5.66×10^2	5.48×10^2	4.96×10^2	**4.80×10^2**	**4.80×10^2**	**4.80×10^2**	**4.80×10^2**	5.05×10^2	5.93×10^2	5.73×10^2	5.66×10^2	5.40×10^2	5.12×10^2	**4.80×10^2**	**4.80×10^2**	**4.80×10^2**	4.81×10^2	5.20×10^2
F_{26}	3.05×10^3	1.59×10^3	**1.29×10^3**	1.57×10^3	1.67×10^3	1.89×10^3	2.02×10^3	3.08×10^3	4.92×10^3	5.14×10^3	3.27×10^3	2.48×10^3	**1.36×10^3**	1.59×10^3	1.70×10^3	1.79×10^3	1.93×10^3	3.23×10^3	4.87×10^3	5.12×10^3
F_{27}	8.33×10^2	7.36×10^2	6.80×10^2	**6.63×10^2**	6.76×10^2	7.16×10^2	7.22×10^2	7.26×10^2	7.36×10^2	7.69×10^2	8.45×10^2	7.15×10^2	6.68×10^2	6.64×10^2	**6.61×10^2**	7.07×10^2	7.17×10^2	7.24×10^2	7.61×10^2	7.14×10^2
F_{28}	5.41×10^2	5.07×10^2	4.97×10^2	4.91×10^2	4.73×10^2	**4.71×10^2**	4.75×10^2	7.48×10^2	3.35×10^3	5.10×10^3	5.47×10^2	5.07×10^2	5.06×10^2	5.06×10^2	4.85×10^2	**4.70×10^2**	4.77×10^2	7.11×10^2	3.78×10^3	5.14×10^3

Table 2. *Cont.*

F_{29}	1.77×10^3	1.09×10^3	6.71×10^2	5.90×10^2	**4.36×10^2**	4.51×10^2	4.74×10^2	1.06×10^3	1.44×10^3	1.88×10^3	1.74×10^3	1.10×10^3	6.93×10^2	5.79×10^2	**4.37×10^2**	4.61×10^2	4.93×10^2	1.06×10^3	1.68×10^3	1.99×10^3
F_{30}	1.14×10^6	8.25×10^5	**7.99×10^5**	8.63×10^5	8.17×10^5	9.02×10^5	1.39×10^6	1.51×10^6	1.54×10^6	1.55×10^6	1.21×10^6	**8.06×10^5**	8.14×10^5	8.35×10^5	8.25×10^5	8.55×10^5	1.21×10^6	1.52×10^6	1.56×10^6	1.51×10^6
Rank	6.52	4.72	**3.38**	**3.38**	3.45	4.21	5.48	7.14	7.86	8.86	6.66	4.72	3.41	3.34	**3.28**	4.07	5.66	7.03	8.10	8.72

Table 3. Comparison results between SCDLPSO and state-of-the-art PSO variants on the 30-D CEC 2017 benchmark functions.

F	Category	Quality	SCDLPSO	XPSO	TCSPSO	DNSPSO	AWPSO	CLPSO_LS	GLPSO	CLPSO
F_1	Unimodal Functions	Median	2.28×10^3	2.90×10^3	3.20×10^3	1.80×10^5	1.55×10^{10}	1.42×10^4	1.32×10^3	1.79×10^4
		Mean	3.11×10^3	4.27×10^3	3.66×10^3	2.35×10^5	1.63×10^{10}	1.67×10^4	**1.90×10^3**	1.82×10^4
		Std	3.15×10^3	4.47×10^3	4.08×10^3	2.18×10^5	5.52×10^9	7.78×10^3	2.23×10^3	5.65×10^3
		p-value	-	3.76×10^{-1} =	5.65×10^{-1} =	**3.88×10^{-7}** +	**7.28×23** +	**4.06×10^{-12}** +	9.86×10^{-2} =	**3.15×10^{-18}** +
F_3		Median	2.66×10^2	5.86×10^{-2}	9.94×10^3	1.55×10^5	5.25×10^4	6.80×10^{-11}	1.14×10^{-13}	6.68×10^3
		Mean	4.82×10^2	**5.45×10^{-1}**	1.15×10^4	1.56×10^5	5.20×10^4	4.28×10^3	2.58×10^{-13}	$6.20 \times 10_3$
		Std	5.86×10^2	1.57×10^0	3.68×10^3	2.67×10^4	3.62×10^4	1.61×10^4	6.22×10^{-13}	2.42×10^3
		p-value	-	4.39×10^{-5} −	**9.52×10^{-23}** +	**4.84×10^{-38}** +	**2.29×10^{-10}** +	2.10×10^{-1} =	4.28×10^{-5} −	**6.12×10^{-18}** +
$F_{1,3}$	*w/t/l*		-	0/1/1	1/1/0	2/0/0	2/0/0	1/1/0	0/1/1	2/0/0
F_4	Simple Multimodal Functions	Median	8.33×10^1	1.22×10^2	1.30×10^2	2.56×10^1	1.40×10^3	8.90×10^1	1.52×10^2	6.17×10^4
		Mean	7.66×10^1	1.17×10^2	1.32×10^2	**2.54×10^1**	1.70×10^3	8.90×10^1	1.60×10^2	5.87×10^4
		Std	1.11×10^1	2.67×10^1	4.84×10^1	8.61×10^{-1}	1.36×10^3	4.06×10^{-1}	6.32×10^1	1.06×10^4
		p-value	-	**8.13×10^{-11}** +	**1.25×10^{-7}** +	1.71×10^{-32} −	**2.38×10^{-8}** +	**1.18×10^{-7}** +	**2.68×10^{-9}** +	**9.78×10^{-37}** +
F_5		Median	4.97×10^0	4.18×10^1	8.56×10^1	1.96×10^2	1.89×10^2	2.17×10^2	5.67×10^1	9.58×10^1
		Mean	**5.14×10^0**	4.34×10^1	8.92×10^1	1.98×10^2	1.84×10^2	2.18×10^2	5.79×10^1	9.57×10^1
		Std	1.84×10^0	1.41×10^1	2.54×10^1	1.31×10^1	3.22×10^1	1.21×10^1	1.37×10^1	2.26×10^0
		p-value	-	**6.05×10^{-19}** +	**3.81×10^{-25}** +	**1.29×10^{-60}** +	**7.43×10^{-37}** +	**6.04×10^{-65}** +	**2.42×10^{-28}** +	**1.33×10^{-79}** +
F_6		Median	1.11×10^{-6}	2.02×10^{-3}	8.00×10^{-1}	1.48×10^{-1}	2.48×10^1	4.32×10^{-1}	6.21×10^{-3}	8.08×10^1
		Mean	**8.31×10^{-6}**	2.08×10^{-2}	1.04×10^0	1.48×10^{-1}	2.72×10^1	9.35×10^{-1}	1.36×10^{-2}	8.05×10^1
		Std	1.41×10^{-5}	6.31×10^{-2}	1.15×10^0	4.10×10^{-2}	1.04×10^1	1.04×10^0	1.60×10^{-2}	1.17×10^1
		p-value	-	**3.84×10^{-2}** +	**8.59×10^{-6}** +	**3.75×10^{-27}** +	**2.83×10^{-20}** +	**1.07×10^{-5}** +	**2.45×10^{-5}** +	**3.68×10^{-42}** +

Table 3. *Cont.*

F	Category	Quality	SCDLPSO	XPSO	TCSPSO	DNSPSO	AWPSO	CLPSO_LS	GLPSO	CLPSO
F_7		Median	3.48×10^{1}	7.70×10^{1}	1.45×10^{2}	2.34×10^{2}	3.24×10^{2}	2.36×10^{2}	1.06×10^{2}	1.43×10^{-3}
		Mean	3.94×10^{1}	8.06×10^{1}	1.42×10^{2}	2.32×10^{2}	3.09×10^{2}	2.33×10^{2}	1.07×10^{2}	**1.43×10^{-3}**
		Std	2.05×10^{1}	1.80×10^{1}	2.87×10^{1}	1.32×10^{1}	1.25×10^{2}	1.89×10^{1}	2.08×10^{1}	4.04×10^{-4}
		p-value	-	**1.82×10^{-11}** +	**1.68×10^{-22}** +	**2.15×10^{-45}** +	**1.43×10^{-16}** +	**2.80×10^{-42}** +	**3.76×10^{-18}** +	8.33×10^{-15} −
F_8		Median	3.98×10^{0}	4.18×10^{1}	9.55×10^{1}	2.04×10^{2}	1.81×10^{2}	2.25×10^{2}	6.47×10^{1}	1.03×10^{2}
		Mean	**4.58×10^{0}**	4.37×10^{1}	9.33×10^{1}	2.03×10^{2}	1.76×10^{2}	2.22×10^{2}	6.42×10^{1}	1.03×10^{2}
		Std	1.33×10^{0}	1.60×10^{1}	2.22×10^{1}	1.25×10^{1}	3.48×10^{1}	1.15×10^{1}	1.70×10^{1}	9.06×10^{0}
		p-value	-	**6.64×10^{-20}** +	**2.49×10^{-29}** +	**1.18×10^{-62}** +	**5.20×10^{-34}** +	**5.47×10^{-67}** +	**2.20×10^{-26}** +	**5.39×10^{-53}** +
F_9		Median	1.14×10^{-13}	1.72×10^{0}	3.01×10^{2}	1.61×10^{0}	4.44×10^{3}	1.90×10^{1}	5.56×10^{1}	9.06×10^{1}
		Mean	**2.11×10^{-2}**	3.10×10^{0}	3.85×10^{2}	2.32×10^{0}	4.40×10^{3}	2.29×10^{1}	6.39×10^{1}	9.10×10^{1}
		Std	8.34×10^{-2}	4.41×10^{0}	3.36×10^{2}	2.80×10^{0}	1.72×10^{3}	2.74×10^{1}	3.28×10^{1}	9.88×10^{0}
		p-value	-	**8.30×10^{-5}** +	**6.95×10^{-8}** +	**4.20×10^{-5}** +	**5.54×10^{-20}** +	**3.35×10^{-5}** +	**5.12×10^{-15}** +	**3.52×10^{-49}** +
F_{10}		Median	6.32×10^{3}	2.70×10^{3}	2.98×10^{3}	5.38×10^{3}	3.90×10^{3}	6.42×10^{3}	3.22×10^{3}	9.22×10^{2}
		Mean	5.97×10^{3}	2.61×10^{3}	2.97×10^{3}	5.24×10^{3}	4.00×10^{3}	6.26×10^{3}	3.46×10^{3}	**9.44×10^{2}**
		Std	1.35×10^{3}	6.38×10^{2}	4.22×10^{2}	1.01×10^{3}	5.96×10^{2}	6.36×10^{2}	8.35×10^{2}	2.89×10^{2}
		p-value	-	1.25×10^{-17} −	1.66×10^{-16} −	2.28×10^{-2} −	1.44×10^{-9} −	2.96×10^{-1} =	8.04×10^{-12} −	2.56×10^{-27} −
$F_{4\text{-}10}$	*w/t/l*		-	6/0/1	6/0/1	5/0/2	6/0/1	6/1/0	6/0/1	5/0/2
F_{11}		Median	1.28×10^{1}	8.06×10^{1}	1.16×10^{2}	8.95×10^{1}	1.34×10^{3}	1.82×10^{2}	1.01×10^{2}	3.15×10^{3}
		Mean	**3.41×10^{1}**	8.65×10^{1}	1.18×10^{2}	8.77×10^{1}	3.57×10^{3}	1.81×10^{2}	8.71×10^{1}	3.11×10^{3}
		Std	2.95×10^{1}	4.45×10^{1}	4.27×10^{1}	8.80×10^{0}	4.85×10^{3}	4.03×10^{1}	3.50×10^{1}	3.20×10^{2}
		p-value	-	**3.37×10^{-6}** +	**4.21×10^{-12}** +	**3.33×10^{-13}** +	**2.35×10^{-4}** +	**9.97×10^{-23}** +	**5.66×10^{-8}** +	**3.96×10^{-50}** +
F_{12}	Hybrid Functions	Median	2.68×10^{4}	2.48×10^{4}	1.86×10^{5}	5.51×10^{7}	1.04×10^{9}	4.68×10^{5}	1.08×10^{5}	1.42×10^{2}
		Mean	2.72×10^{4}	1.36×10^{5}	5.12×10^{5}	6.03×10^{7}	1.36×10^{9}	9.27×10^{5}	1.85×10^{6}	**1.48×10^{2}**
		Std	1.47×10^{4}	4.38×10^{5}	6.68×10^{5}	2.65×10^{7}	1.06×10^{9}	8.77×10^{5}	3.11×10^{6}	2.23×10^{1}
		p-value	-	1.15×10^{-1} =	**2.43×10^{-4}** +	**1.08×10^{-17}** +	**4.07×10^{-9}** +	**8.19×10^{-7}** +	**2.56×10^{-3}** +	4.27×10^{-14} −
F_{13}		Median	8.46×10^{3}	1.05×10^{4}	8.24×10^{3}	1.28×10^{6}	5.08×10^{6}	6.55×10^{3}	1.78×10^{4}	2.47×10^{6}
		Mean	1.80×10^{4}	**1.29×10^{4}**	3.28×10^{5}	1.37×10^{6}	5.11×10^{8}	1.74×10^{4}	1.10×10^{5}	2.83×10^{6}
		Std	1.79×10^{4}	1.22×10^{4}	1.14×10^{6}	5.04×10^{5}	9.07×10^{8}	2.18×10^{4}	4.13×10^{5}	1.19×10^{6}
		p-value	-	3.66×10^{-2} −	1.49×10^{-1} =	**7.39×10^{-21}** +	**3.59×10^{-3}** +	9.11×10^{-1} =	2.37×10^{-1} =	**2.38×10^{-18}** +

Table 3. *Cont.*

F	Category	Quality	SCDLPSO	XPSO	TCSPSO	DNSPSO	AWPSO	CLPSO_LS	GLPSO	CLPSO
F_{14}		Median	1.92×10^{3}	4.80×10^{3}	3.49×10^{4}	1.81×10^{2}	8.16×10^{4}	1.10×10^{5}	1.02×10^{3}	1.36×10^{4}
		Mean	3.19×10^{3}	6.46×10^{3}	5.20×10^{4}	$\mathbf{1.86 \times 10^{2}}$	5.84×10^{5}	1.06×10^{5}	9.06×10^{4}	1.32×10^{4}
		Std	3.20×10^{3}	5.50×10^{3}	7.90×10^{4}	2.27×10^{1}	1.71×10^{6}	5.32×10^{4}	1.62×10^{5}	4.83×10^{3}
		p-value	-	$\mathbf{3.55 \times 10^{-2}}$ +	$\mathbf{1.53 \times 10^{-3}}$ +	4.49×10^{-6} −	$\mathbf{7.26 \times 10^{-2}}$ +	$\mathbf{6.23 \times 10^{-15}}$ +	$\mathbf{5.30 \times 10^{-3}}$ +	$\mathbf{4.12 \times 10^{-13}}$ +
F_{15}		Median	6.60×10^{2}	2.13×10^{3}	1.08×10^{4}	3.58×10^{4}	1.55×10^{5}	4.13×10^{4}	1.92×10^{3}	4.39×10^{4}
		Mean	$\mathbf{1.70 \times 10^{3}}$	4.82×10^{3}	1.33×10^{4}	3.95×10^{4}	3.95×10^{7}	3.60×10^{4}	6.38×10^{3}	4.41×10^{4}
		Std	2.38×10^{3}	6.33×10^{3}	1.04×10^{4}	2.21×10^{4}	2.12×10^{8}	8.63×10^{3}	8.32×10^{3}	3.33×10^{4}
		p-value	-	6.02×10^{-2} =	$\mathbf{2.81 \times 10^{-7}}$ +	$\mathbf{7.51 \times 10^{-13}}$ +	3.20×10^{-1} =	$\mathbf{2.29 \times 10^{-28}}$ +	$\mathbf{5.06 \times 10^{-3}}$ +	$\mathbf{5.60 \times 10^{-9}}$ +
F_{16}		Median	2.21×10^{1}	5.70×10^{2}	8.65×10^{2}	1.90×10^{3}	1.38×10^{3}	1.29×10^{3}	7.00×10^{2}	7.82×10^{2}
		Mean	$\mathbf{9.21 \times 10^{1}}$	5.32×10^{2}	8.54×10^{2}	1.89×10^{3}	1.44×10^{3}	1.14×10^{3}	7.01×10^{2}	8.34×10^{2}
		Std	1.21×10^{2}	2.31×10^{2}	2.56×10^{2}	1.70×10^{2}	3.66×10^{2}	4.10×10^{2}	2.74×10^{2}	3.73×10^{2}
		p-value	-	$\mathbf{1.04 \times 10^{-14}}$ +	$\mathbf{6.95 \times 10^{-21}}$ +	$\mathbf{1.43 \times 10^{-47}}$ +	$\mathbf{2.11 \times 10^{-26}}$ +	$\mathbf{3.91 \times 10^{-19}}$ +	$\mathbf{9.54 \times 10^{-16}}$ +	$\mathbf{1.62 \times 10^{-14}}$ +
F_{17}		Median	5.11×10^{1}	1.56×10^{2}	3.18×10^{2}	8.59×10^{2}	6.29×10^{2}	6.85×10^{2}	1.82×10^{2}	6.32×10^{2}
		Mean	$\mathbf{5.80 \times 10^{1}}$	1.47×10^{2}	2.96×10^{2}	8.58×10^{2}	6.84×10^{2}	9.36×10^{2}	2.35×10^{2}	6.20×10^{2}
		Std	2.44×10^{1}	9.61×10^{1}	1.44×10^{2}	9.65×10^{1}	3.22×10^{2}	6.39×10^{2}	1.54×10^{2}	1.42×10^{2}
		p-value	-	$\mathbf{6.92 \times 10^{-7}}$ +	$\mathbf{3.09 \times 10^{-12}}$ +	$\mathbf{7.78 \times 10^{-46}}$ +	$\mathbf{6.20 \times 10^{-15}}$ +	$\mathbf{6.52 \times 10^{-10}}$ +	$\mathbf{8.63 \times 10^{-8}}$ +	$\mathbf{8.03 \times 10^{-29}}$ +
F_{18}		Median	8.54×10^{4}	9.98×10^{4}	1.41×10^{5}	2.40×10^{5}	6.70×10^{5}	6.81×10^{5}	1.53×10^{4}	2.00×10^{2}
		Mean	1.16×10^{5}	1.45×10^{5}	2.78×10^{5}	2.45×10^{5}	4.11×10^{6}	2.44×10^{6}	6.82×10^{4}	$\mathbf{1.91 \times 10^{2}}$
		Std	9.74×10^{4}	1.15×10^{5}	2.93×10^{5}	9.96×10^{4}	1.18×10^{7}	4.12×10^{6}	1.99×10^{5}	7.60×10^{1}
		p-value	-	3.26×10^{-1} =	$\mathbf{6.61 \times 10^{-3}}$ +	$\mathbf{5.90 \times 10^{-6}}$ +	7.34×10^{-2} =	$\mathbf{3.64 \times 10^{-3}}$ +	2.47×10^{-1} =	2.69×10^{-8} −
F	Category	Quality	SCDLPSO	XPSO	TCSPSO	DNSPSO	AWPSO	CLPSO_LS	GLPSO	CLPSO
F_{19}	Hybrid Functions	Median	1.72×10^{3}	2.28×10^{3}	7.66×10^{3}	1.75×10^{3}	1.80×10^{7}	3.51×10^{4}	6.16×10^{3}	2.00×10^{5}
		Mean	$\mathbf{3.58 \times 10^{3}}$	4.05×10^{3}	1.49×10^{4}	2.29×10^{3}	7.38×10^{7}	3.51×10^{4}	1.25×10^{4}	2.34×10^{5}
		Std	3.83×10^{3}	4.55×10^{3}	1.58×10^{4}	1.43×10^{3}	2.46×10^{8}	1.94×10^{4}	1.41×10^{4}	1.14×10^{5}
		p-value	-	3.98×10^{-1} =	$\mathbf{3.83 \times 10^{-4}}$ +	9.22×10^{-2} =	1.12×10^{-1} =	$\mathbf{6.65 \times 10^{-12}}$ +	$\mathbf{1.81 \times 10^{-3}}$ +	$\mathbf{1.32 \times 10^{-15}}$ +
F_{20}		Median	4.23×10^{1}	1.73×10^{2}	3.84×10^{2}	4.06×10^{2}	5.19×10^{2}	5.98×10^{2}	2.77×10^{2}	1.94×10^{2}
		Mean	$\mathbf{6.38 \times 10^{1}}$	1.85×10^{2}	3.70×10^{2}	4.07×10^{2}	5.38×10^{2}	5.87×10^{2}	2.72×10^{2}	2.30×10^{2}
		Std	7.20×10^{1}	7.16×10^{1}	1.39×10^{2}	1.05×10^{2}	1.97×10^{2}	1.55×10^{2}	1.19×10^{2}	1.71×10^{2}
		p-value	-	$\mathbf{1.55 \times 10^{-8}}$ +	$\mathbf{4.21 \times 10^{-15}}$ +	$\mathbf{5.22 \times 10^{-21}}$ +	$\mathbf{1.29 \times 10^{-17}}$ +	$\mathbf{1.39 \times 10^{-23}}$ +	$\mathbf{5.06 \times 10^{-11}}$ +	$\mathbf{9.95 \times 10^{-6}}$ +
$F_{11\text{-}20}$	*w/t/l*		-	5/4/1	9/1/0	8/1/1	7/3/0	9/1/0	8/2/0	8/0/2

Table 3. *Cont.*

F	Category	Quality	SCDLPSO	XPSO	TCSPSO	DNSPSO	AWPSO	CLPSO_LS	GLPSO	CLPSO
F_{21}		Median	2.09×10^{2}	2.38×10^{2}	2.80×10^{2}	3.98×10^{2}	3.74×10^{2}	4.00×10^{2}	2.62×10^{2}	2.02×10^{2}
		Mean	**2.09×10^{2}**	2.41×10^{2}	2.84×10^{2}	3.97×10^{2}	3.80×10^{2}	4.02×10^{2}	2.66×10^{2}	**2.09×10^{2}**
		Std	2.82×10^{0}	1.22×10^{1}	2.25×10^{1}	1.43×10^{1}	4.55×10^{1}	7.78×10^{0}	2.17×10^{1}	6.88×10^{1}
		p-value	-	**1.06×10^{-20} +**	**3.60×10^{-25} +**	**1.70×10^{-57} +**	**6.97×10^{-28} +**	**2.69×10^{-72} +**	**2.78×10^{-20} +**	9.99×10^{-1} =
F_{22}		Median	1.00×10^{2}	1.00×10^{2}	1.04×10^{2}	6.64×10^{3}	4.16×10^{3}	6.99×10^{3}	1.02×10^{2}	2.89×10^{2}
		Mean	2.73×10^{2}	2.95×10^{2}	1.70×10^{3}	6.52×10^{3}	4.04×10^{3}	6.95×10^{3}	2.06×10^{2}	2.78×10^{2}
		Std	9.33×10^{2}	7.77×10^{2}	1.75×10^{3}	5.90×10^{2}	1.03×10^{3}	3.10×10^{2}	5.56×10^{2}	4.38×10^{1}
		p-value	-	5.37×10^{-1} =	**2.63×10^{-4} +**	**2.23×10^{-37} +**	**4.31×10^{-21} +**	**9.29×10^{-42} +**	7.40×10^{-1} =	9.80×10^{-1} =
F_{23}		Median	3.92×10^{2}	3.97×10^{2}	4.44×10^{2}	5.68×10^{2}	6.64×10^{2}	5.61×10^{2}	4.26×10^{2}	3.05×10^{2}
		Mean	**3.91×10^{2}**	3.96×10^{2}	4.46×10^{2}	5.72×10^{2}	6.80×10^{2}	5.57×10^{2}	4.33×10^{2}	5.22×10^{2}
		Std	9.71×10^{0}	1.38×10^{1}	2.85×10^{1}	2.29×10^{1}	9.92×10^{1}	1.39×10^{1}	2.73×10^{1}	7.83×10^{2}
		p-value	-	**1.68×10^{-2} +**	**4.26×10^{-14} +**	**2.28×10^{-43} +**	**1.92×10^{-22} +**	**9.96×10^{-51} +**	**1.29×10^{-10} +**	3.70×10^{-1} =
F_{24}		Median	4.66×10^{2}	4.63×10^{2}	5.37×10^{2}	6.74×10^{2}	7.32×10^{2}	6.26×10^{2}	4.94×10^{2}	4.51×10^{2}
		Mean	4.70×10^{2}	4.68×10^{2}	5.38×10^{2}	6.97×10^{2}	7.51×10^{2}	6.20×10^{2}	5.14×10^{2}	**4.51×10^{2}**
		Std	1.60×10^{1}	2.19×10^{1}	5.08×10^{1}	7.02×10^{1}	8.86×10^{1}	1.02×10^{1}	5.44×10^{1}	9.85×10^{0}
	Composition Functions	*p*-value	-	5.75×10^{-1} =	**5.54×10^{-9} +**	**3.73×10^{-24} +**	**5.97×10^{-24} +**	**2.12×10^{-45} +**	**9.84×10^{-5} +**	7.08×10^{-7} −
F_{25}		Median	3.88×10^{2}	3.91×10^{2}	4.14×10^{2}	3.78×10^{2}	8.68×10^{2}	3.88×10^{2}	4.10×10^{2}	5.60×10^{2}
		Mean	3.88×10^{2}	3.95×10^{2}	4.12×10^{2}	3.78×10^{2}	1.13×10^{3}	3.88×10^{2}	4.11×10^{2}	5.62×10^{2}
		Std	5.00×10^{-1}	1.00×10^{1}	1.56×10^{1}	1.12×10^{0}	7.54×10^{2}	4.27×10^{-1}	2.15×10^{1}	1.54×10^{1}
		p-value	-	**7.63×10^{-6} +**	**6.91×10^{-12} +**	1.20×10^{-44} −	**1.64×10^{-6} +**	3.94×10^{-3} =	**2.73×10^{-7} +**	**2.56×10^{-54} +**
F_{26}		Median	1.33×10^{3}	3.00×10^{2}	2.32×10^{3}	3.28×10^{3}	4.30×10^{3}	3.16×10^{3}	1.94×10^{3}	**3.90×10^{2}**
		Mean	1.33×10^{3}	7.57×10^{2}	2.23×10^{3}	3.27×10^{3}	4.20×10^{3}	3.14×10^{3}	1.92×10^{3}	3.90×10^{2}
		Std	1.15×10^{2}	6.08×10^{2}	6.97×10^{2}	2.16×10^{2}	9.58×10^{2}	1.08×10^{2}	4.77×10^{2}	9.49×10^{-1}
		p-value	-	7.35×10^{-7} −	**5.77×10^{-9} +**	**1.68×10^{-45} +**	**5.88×10^{-23} +**	**1.31×10^{-54} +**	**3.08×10^{-8} +**	2.65×10^{-46} −
F_{27}		Median	5.14×10^{2}	5.36×10^{2}	5.61×10^{2}	5.00×10^{2}	6.85×10^{2}	5.18×10^{2}	5.48×10^{2}	1.99×10^{3}
		Mean	5.15×10^{2}	5.36×10^{2}	5.61×10^{2}	**5.00×10^{2}**	6.88×10^{2}	5.22×10^{2}	5.50×10^{2}	1.88×10^{3}
		Std	9.77×10^{0}	1.18×10^{1}	1.95×10^{1}	0.00×10^{0}	8.94×10^{1}	1.52×10^{1}	1.28×10^{1}	2.71×10^{2}
		p-value	-	**7.65×10^{-10} +**	**3.60×10^{-16} +**	1.08×10^{-11} −	**9.24×10^{-15} +**	6.01×10^{-2} =	**6.66×10^{-17} +**	**1.09×10^{-34} +**
F_{28}		Median	4.08×10^{2}	4.03×10^{2}	4.40×10^{2}	5.00×10^{2}	1.52×10^{3}	3.50×10^{3}	4.72×10^{2}	5.13×10^{2}
		Mean	4.15×10^{2}	**3.78×10^{2}**	4.51×10^{2}	5.00×10^{2}	1.98×10^{3}	3.07×10^{3}	4.51×10^{2}	5.13×10^{2}
		Std	3.61×10^{1}	6.75×10^{1}	5.21×10^{1}	0.00×10^{0}	1.25×10^{3}	9.06×10^{2}	7.00×10^{1}	3.62×10^{0}
		p-value	-	5.03×10^{-2} =	**3.32×10^{-3} +**	**2.27×10^{-18} +**	**7.59×10^{-9} +**	**1.24×10^{-22}+**	**1.74×10^{-2} +**	**6.55×10^{-21} +**

Table 3. *Cont.*

F	Category	Quality	SCDLPSO	XPSO	TCSPSO	DNSPSO	AWPSO	CLPSO_LS	GLPSO	CLPSO
F_{29}		Median	4.80×10^{2}	5.53×10^{2}	8.62×10^{2}	1.58×10^{3}	1.34×10^{3}	9.33×10^{2}	7.74×10^{2}	4.90×10^{2}
		Mean	$\mathbf{4.80 \times 10^{2}}$	5.73×10^{2}	9.05×10^{2}	1.60×10^{3}	1.38×10^{3}	1.11×10^{3}	8.17×10^{2}	5.00×10^{2}
		Std	2.24×10^{1}	8.08×10^{1}	1.86×10^{2}	2.07×10^{2}	4.13×10^{2}	6.02×10^{2}	2.37×10^{2}	2.44×10^{1}
		p-value	-	$\mathbf{5.04 \times 10^{-8}}$+	$\mathbf{1.21 \times 10^{-17}}$ +	$\mathbf{3.97 \times 10^{-36}}$ +	$\mathbf{5.93 \times 10^{-17}}$ +	$\mathbf{4.95 \times 10^{-7}}$ +	$\mathbf{2.52 \times 10^{-10}}$ +	$\mathbf{2.04 \times 10^{-3}}$ +
F_{30}		Median	3.90×10^{3}	8.08×10^{3}	1.20×10^{4}	5.00×10^{4}	8.80×10^{6}	1.43×10^{4}	9.44×10^{3}	6.47×10^{2}
		Mean	5.05×10^{3}	9.04×10^{3}	1.80×10^{4}	6.79×10^{4}	1.65×10^{7}	1.38×10^{4}	2.08×10^{4}	$\mathbf{6.46 \times 10^{2}}$
		Std	2.81×10^{3}	5.57×10^{3}	1.77×10^{4}	5.66×10^{4}	2.11×10^{7}	1.51×10^{3}	2.96×10^{4}	6.63×10^{1}
		p-value	-	$\mathbf{9.15 \times 10^{-4}}$ +	$\mathbf{2.58 \times 10^{-4}}$ +	$\mathbf{1.51 \times 10^{-7}}$ +	$\mathbf{8.77 \times 10^{-5}}$ +	$\mathbf{2.13 \times 10^{-21}}$ +	$\mathbf{5.93 \times 10^{-3}}$ +	1.07×10^{-11} −
$F_{21\text{-}30}$	*w/t/l*		-	6/3/1	10/0/0	8/0/2	10/0/0	8/2/0	9/1/0	5/3/2
	w/t/l		-	17/8/4	26/2/1	23/1/5	25/3/1	24/5/0	23/4/2	20/3/6
	rank		**1.93**	2.66	4.83	5.38	7.28	5.86	3.69	4.38

Table 4. Comparison results between SCDLPSO and state-of-the-art PSO variants on the 50-D CEC 2017 benchmark functions.

F	Category	Quality	SCDLPSO	XPSO	TCSPSO	DNSPSO	AWPSO	CLPSO_LS	GLPSO	CLPSO
F_1	Unimodal Functions	Median	1.01×10^{3}	1.84×10^{3}	5.46×10^{3}	3.32×10^{3}	6.18×10^{10}	4.59×10^{7}	1.64×10^{3}	2.15×10^{4}
		Mean	$\mathbf{2.32 \times 10^{3}}$	4.71×10^{3}	2.84×10^{6}	5.78×10^{3}	6.31×10^{10}	1.29×10^{8}	1.25×10^{4}	2.45×10^{4}
		Std	2.71×10^{3}	6.06×10^{3}	1.52×10^{7}	7.38×10^{3}	1.08×10^{10}	3.33×10^{8}	4.75×10^{4}	1.39×10^{4}
		p-value	-	2.20×10^{-1} =	3.21×10^{-1} =	$\mathbf{2.10 \times 10^{-2}}$ +	$\mathbf{4.04 \times 10^{-38}}$ +	$\mathbf{4.08 \times 10^{-2}}$ +	2.53×10^{-1} =	$\mathbf{1.23 \times 10^{-11}}$ +
F_3		Median	1.38×10^{4}	4.36×10^{3}	5.54×10^{4}	3.70×10^{5}	1.06×10^{5}	5.19×10^{-10}	5.68×10^{-13}	3.02×10^{4}
		Mean	1.44×10^{4}	4.62×10^{3}	5.83×10^{4}	3.70×10^{5}	1.19×10^{5}	2.37×10^{4}	$\mathbf{3.31 \times 10^{-12}}$	3.33×10^{4}
		Std	3.94×10^{3}	1.56×10^{3}	9.39×10^{3}	5.40×10^{4}	6.45×10^{4}	6.07×10^{4}	1.20×10^{-11}	1.89×10^{4}
		p-value	-	1.26×10^{-17} −	$\mathbf{4.84 \times 10^{-31}}$ +	$\mathbf{5.92 \times 10^{-41}}$ +	$\mathbf{3.50 \times 10^{-12}}$ +	4.11×10^{-1} =	2.70×10^{-27} −	$\mathbf{2.07 \times 10^{-6}}$ +
$F_{1,3}$	*w/t/l*		-	0/1/1	1/1/0	2/0/0	2/0/0	1/1/0	1/0/1	2/0/0

Table 4. *Cont.*

F	Category	Quality	SCDLPSO	XPSO	TCSPSO	DNSPSO	AWPSO	CLPSO_LS	GLPSO	CLPSO
F_4	Simple Multimodal Functions	Median	1.33×10^{2}	2.45×10^{2}	2.88×10^{2}	4.57×10^{1}	8.68×10^{3}	2.38×10^{2}	2.96×10^{2}	1.73×10^{5}
		Mean	1.23×10^{2}	2.34×10^{2}	2.93×10^{2}	**5.50×10^{1}**	8.83×10^{3}	2.40×10^{2}	3.03×10^{2}	1.75×10^{5}
		Std	5.20×10^{1}	5.08×10^{1}	9.09×10^{1}	2.53×10^{1}	3.63×10^{3}	3.13×10^{1}	6.31×10^{1}	1.50×10^{4}
		p-value	-	**3.54×10^{-11}** +	**3.48×10^{-12}** +	4.12×10^{-8} −	**1.08×10^{-18}** +	**6.79×10^{-15}** +	**4.07×10^{-17}** +	**5.95×10^{-55}** +
F_5		Median	1.04×10^{1}	8.16×10^{1}	1.87×10^{2}	4.10×10^{2}	4.26×10^{2}	4.40×10^{2}	1.46×10^{2}	2.03×10^{2}
		Mean	1.08×10^{1}	8.53×10^{1}	1.91×10^{2}	4.13×10^{2}	4.35×10^{2}	4.40×10^{2}	1.49×10^{2}	2.01×10^{2}
		Std	3.61×10^{0}	2.02×10^{1}	3.82×10^{1}	1.82×10^{1}	6.20×10^{1}	2.10×10^{1}	2.97×10^{1}	1.52×10^{1}
		p-value	-	**5.21×10^{-33}** +	**4.60×10^{-33}** +	**1.38×10^{-70}** +	**6.58×10^{-42}** +	**9.51×10^{-69}** +	**1.09×10^{-32}** +	**3.71×10^{-56}** +
F_6		Median	5.14×10^{-4}	5.67×10^{-2}	3.00×10^{0}	9.77×10^{-2}	4.25×10^{1}	5.71×10^{0}	1.75×10^{-2}	2.17×10^{2}
		Mean	**6.37×10^{-4}**	1.53×10^{-1}	3.93×10^{0}	1.03×10^{-1}	4.29×10^{1}	6.01×10^{0}	2.06×10^{-2}	2.16×10^{2}
		Std	5.24×10^{-4}	2.87×10^{-1}	3.68×10^{0}	2.85×10^{-2}	1.07×10^{1}	1.22×10^{0}	1.48×10^{-2}	2.07×10^{1}
		p-value	-	**2.41×10^{-3}** +	**3.51×10^{-7}** +	**5.50×10^{-27}** +	**2.56×10^{-29}** +	**4.11×10^{-34}** +	**9.84×10^{-10}** +	**2.85×10^{-52}** +
F_7		Median	6.18×10^{1}	1.60×10^{2}	3.18×10^{2}	4.72×10^{2}	1.04×10^{3}	5.15×10^{2}	2.38×10^{2}	9.44×10^{-5}
		Mean	6.22×10^{1}	1.64×10^{2}	3.35×10^{2}	4.71×10^{2}	1.02×10^{3}	5.15×10^{2}	2.34×10^{2}	**1.04×10^{-4}**
		Std	2.31×10^{0}	3.63×10^{1}	6.19×10^{1}	1.80×10^{1}	3.13×10^{2}	3.49×10^{1}	3.79×10^{1}	3.41×10^{-5}
		p-value	-	**1.27×10^{-26}** +	**1.44×10^{-31}** +	**1.70×10^{-71}** +	**1.85×10^{-23}** +	**1.30×10^{-57}** +	**4.20×10^{-32}** +	5.65×10^{-76} −
F_8		Median	9.45×10^{0}	8.66×10^{1}	1.96×10^{2}	4.13×10^{2}	4.35×10^{2}	4.44×10^{2}	1.40×10^{2}	2.39×10^{2}
		Mean	**1.01×10^{1}**	8.93×10^{1}	2.09×10^{2}	4.07×10^{2}	4.34×10^{2}	4.45×10^{2}	1.41×10^{2}	2.36×10^{2}
		Std	3.34×10^{0}	2.30×10^{1}	6.12×10^{1}	2.12×10^{1}	6.72×10^{1}	1.33×10^{1}	3.04×10^{1}	1.74×10^{1}
		p-value	-	**2.19×10^{-26}** +	**8.08×10^{-25}** +	**1.69×10^{-66}** +	**5.43×10^{-40}** +	**5.55×10^{-80}** +	**6.67×10^{-31}** +	**2.43×10^{-57}** +
F_9		Median	5.89×10^{-1}	1.50×10^{1}	2.84×10^{3}	1.21×10^{1}	1.24×10^{4}	1.10×10^{3}	4.76×10^{2}	2.25×10^{2}
		Mean	**1.57×10^{0}**	4.69×10^{1}	3.34×10^{3}	2.24×10^{1}	1.44×10^{4}	1.13×10^{3}	7.01×10^{2}	2.23×10^{2}
		Std	3.67×10^{0}	7.80×10^{1}	1.87×10^{3}	2.85×10^{1}	5.97×10^{3}	3.00×10^{2}	5.49×10^{2}	1.79×10^{1}
		p-value	-	**5.42×10^{-3}** +	**1.30×10^{-13}** +	**2.50×10^{-4}** +	**7.15×10^{-19}** +	**6.34×10^{-28}** +	**4.97×10^{-9}** +	**5.22×10^{-56}** +
F_{10}		Median	1.20×10^{4}	5.22×10^{3}	5.56×10^{3}	1.20×10^{4}	7.82×10^{3}	1.31×10^{4}	5.26×10^{3}	5.04×10^{3}
		Mean	9.87×10^{3}	**5.11×10^{3}**	5.56×10^{3}	1.16×10^{4}	7.96×10^{3}	1.31×10^{4}	5.73×10^{3}	5.14×10^{3}
		Std	3.77×10^{3}	8.60×10^{2}	6.71×10^{2}	1.44×10^{3}	7.97×10^{2}	4.34×10^{2}	1.53×10^{3}	1.10×10^{3}
		p-value		7.01×10^{-9} −	1.05×10^{-7} −	**2.24×10^{-2}** +	9.81×10^{-3} −	**2.41×10^{-5}** +	9.39×10^{-7} −	2.09×10^{-8} −
$F_{4\text{-}10}$	*w/t/l*		-	6/0/1	6/0/1	6/0/1	6/0/1	7/0/0	6/0/1	5/0/2

Table 4. *Cont.*

F	Category	Quality	SCDLPSO	XPSO	TCSPSO	DNSPSO	AWPSO	CLPSO_LS	GLPSO	CLPSO
F_{11}		Median	6.05×10^{1}	1.49×10^{2}	2.15×10^{2}	2.05×10^{2}	5.50×10^{3}	2.74×10^{2}	5.73×10^{2}	7.35×10^{3}
		Mean	**6.61×10^{1}**	1.54×10^{2}	2.36×10^{2}	2.06×10^{2}	8.68×10^{3}	1.24×10^{3}	9.01×10^{2}	7.29×10^{3}
		Std	1.85×10^{1}	3.28×10^{1}	9.86×10^{1}	2.07×10^{1}	7.82×10^{3}	5.21×10^{3}	9.97×10^{2}	3.63×10^{2}
		p-value	-	**3.41×10^{-18}** +	**8.47×10^{-13}** +	**1.12×10^{-34}** +	**1.76×10^{-7}** +	2.29×10^{-1} =	**3.26×10^{-5}** +	**2.54×10^{-68}** +
F_{12}		Median	2.22×10^{5}	3.96×10^{5}	1.83×10^{6}	2.87×10^{7}	1.58×10^{10}	2.56×10^{7}	1.98×10^{6}	2.07×10^{2}
		Mean	2.78×10^{5}	9.37×10^{5}	8.82×10^{6}	3.46×10^{7}	1.66×10^{10}	2.70×10^{7}	8.69×10^{6}	**2.15×10^{2}**
		Std	1.61×10^{5}	1.32×10^{6}	2.40×10^{7}	2.24×10^{7}	8.14×10^{9}	1.13×10^{7}	1.46×10^{7}	3.49×10^{1}
		p-value	-	**1.42×10^{-3}** +	6.01×10^{-2} =	**2.16×10^{-11}** +	**9.48×10^{-16}** +	**1.93×10^{-18}** +	**3.05×10^{-3}** +	5.24×10^{-13} −
F_{13}		Median	3.29×10^{3}	2.14×10^{3}	3.80×10^{3}	2.26×10^{6}	7.08×10^{9}	3.78×10^{4}	3.56×10^{3}	2.79×10^{7}
		Mean	5.90×10^{3}	**4.53×10^{3}**	7.76×10^{3}	2.84×10^{6}	7.30×10^{9}	1.03×10^{6}	1.82×10^{5}	3.05×10^{7}
		Std	6.31×10^{3}	4.68×10^{3}	9.15×10^{3}	1.77×10^{6}	5.13×10^{9}	5.34×10^{6}	9.49×10^{5}	1.21×10^{7}
		p-value	-	3.11×10^{-1} =	3.71×10^{-1} =	**5.67×10^{-12}** +	**2.35×10^{-10}** +	3.07×10^{-1} =	3.23×10^{-1} =	**1.14×10^{-19}** +
F_{14}		Median	2.21×10^{4}	2.99×10^{4}	4.09×10^{4}	7.80×10^{3}	1.33×10^{6}	3.32×10^{5}	1.94×10^{4}	3.01×10^{4}
		Mean	3.49×10^{4}	3.58×10^{4}	2.30×10^{5}	**8.34×10^{3}**	3.25×10^{6}	3.92×10^{5}	2.16×10^{5}	3.26×10^{4}
	Hybrid Functions	Std	2.94×10^{4}	2.68×10^{4}	5.29×10^{5}	2.55×10^{3}	6.94×10^{6}	3.44×10^{5}	5.80×10^{5}	1.10×10^{4}
		p-value	-	5.11×10^{-1} =	5.25×10^{-2} =	1.02×10^{-5} −	**1.57×10^{-2}** +	**6.69×10^{-7}** +	9.80×10^{-2} =	7.02×10^{-1} =
F_{15}		Median	7.16×10^{3}	2.71×10^{3}	7.12×10^{3}	4.18×10^{5}	7.14×10^{7}	3.16×10^{4}	3.08×10^{3}	4.58×10^{5}
		Mean	6.48×10^{3}	**4.02×10^{3}**	1.46×10^{4}	4.47×10^{5}	4.23×10^{8}	2.35×1007	6.26×10^{3}	5.01×10^{5}
		Std	4.35×10^{3}	4.07×10^{3}	2.54×10^{4}	2.19×10^{5}	8.11×10^{8}	7.71×1007	1.22×10^{4}	2.50×10^{5}
		p-value	-	4.79×10^{-2} −	9.31×10^{-2} =	**1.48×10^{-15}** +	**6.78×10^{-3}** +	1.07×10^{-1} =	9.29×10^{-1} =	**2.69×10^{-15}** +
F_{16}		Median	4.14×10^{2}	9.22×10^{2}	1.62×10^{3}	3.74×10^{3}	2.62×10^{3}	3.20×10^{3}	1.54×10^{3}	1.72×10^{3}
		Mean	**4.29×10^{2}**	9.45×10^{2}	1.70×10^{3}	3.78×10^{3}	2.68×10^{3}	3.21×10^{3}	1.59×10^{3}	1.81×10^{3}
		Std	2.14×10^{2}	3.49×10^{2}	4.35×10^{2}	2.22×10^{2}	6.31×10^{2}	1.96×10^{2}	4.61×10^{2}	7.35×10^{2}
		p-value	-	**1.09×10^{-9}** +	**1.80×10^{-20}** +	**2.88×10^{-53}** +	**1.15×10^{-25}** +	**3.58×10^{-50}** +	**9.35×10^{-18}** +	**8.37×10^{-14}** +
F_{17}		Median	2.48×10^{2}	8.63×10^{2}	1.15×10^{3}	2.34×10^{3}	2.61×10^{3}	2.06×10^{3}	9.67×10^{2}	1.34×10^{3}
		Mean	**3.70×10^{2}**	8.40×10^{2}	1.17×10^{3}	2.34×10^{3}	2.58×10^{3}	2.20×10^{3}	1.00×10^{3}	1.33×10^{3}
		Std	2.98×10^{2}	2.46×10^{2}	2.97×10^{2}	1.76×10^{2}	4.68×10^{2}	7.09×10^{2}	2.66×10^{2}	2.29×10^{2}
		p-value	-	**1.11×10^{-8}** +	**1.59×10^{-14}** +	**1.82×10^{-37}** +	**2.93×10^{-29}** +	**1.49×10^{-18}** +	**8.43×10^{-12}** +	**7.20×10^{-20}** +
F_{18}		Median	9.30×10^{4}	1.67×10^{5}	3.12×10^{6}	3.39×10^{6}	4.48×10^{6}	5.68×10^{6}	1.36×10^{6}	1.05×10^{3}
		Mean	1.30×10^{5}	3.41×10^{5}	5.58×10^{6}	3.69×10^{6}	9.85×10^{6}	8.35×10^{6}	2.45×10^{6}	**1.03×10^{3}**
		Std	8.55×10^{4}	4.26×10^{5}	5.71×10^{6}	1.28×10^{6}	1.88×10^{7}	6.36×10^{6}	2.70×10^{6}	1.59×10^{2}
		p-value	-	**5.67×10^{-3}** +	**3.38×10^{-6}** +	**1.35×10^{-21}** +	**7.13×10^{-3}** +	**3.35×10^{-9}** +	**2.08×10^{-5}** +	4.10×10^{-11} −

Table 4. *Cont.*

F	Category	Quality	SCDLPSO	XPSO	TCSPSO	DNSPSO	AWPSO	CLPSO_LS	GLPSO	CLPSO
F_{19}	Hybrid Functions	Median	1.43×10^4	1.06×10^4	1.30×10^4	3.06×10^4	2.16×10^7	2.52×10^3	8.92×10^3	1.33×10^6
		Mean	1.43×10^4	1.19×10^4	1.51×10^4	3.56×10^4	2.27×10^8	$\mathbf{2.51 \times 10^3}$	1.18×10^4	1.37×10^6
		Std	7.16×10^3	8.67×10^3	1.38×10^4	1.83×10^4	4.59×10^8	1.72×10^1	9.93×10^3	6.37×10^5
		p-value	-	1.23×10^{-1} =	7.70×10^{-1} =	$\mathbf{2.57 \times 10^{-7}}$ +	$\mathbf{9.88 \times 10^{-3}}$ +	2.17×10^{-12} −	2.81×10^{-1} =	$\mathbf{1.40 \times 10^{-16}}$ +
F_{20}		Median	8.70×10^2	4.75×10^2	9.58×10^2	1.56×10^3	1.36×10^3	1.72×10^3	7.44×10^2	3.53×10^2
		Mean	7.04×10^2	$\mathbf{4.82 \times 10^2}$	9.01×10^2	1.56×10^3	1.32×10^3	1.70×10^3	7.43×10^2	5.48×10^2
		Std	5.18×10^2	2.07×10^2	2.96×10^2	3.18×10^2	3.37×10^2	1.35×10^2	2.66×10^2	4.61×10^2
		p-value	-	2.91×10^{-2} −	7.94×10^{-2} =	$\mathbf{2.92 \times 10^{-10}}$ +	$\mathbf{1.66 \times 10^{-6}}$ +	$\mathbf{2.49 \times 10^{-14}}$ +	7.17×10^{-1} =	2.33×10^{-1} =
$F_{11\text{-}20}$	*w/t/l*		-	5/3/2	4/6/0	9/0/1	10/0/0	6/3/1	5/5/0	6/2/2
F_{21}	Composition Functions	Median	2.18×10^2	2.81×10^2	3.77×10^2	5.96×10^2	5.98×10^2	6.34×10^2	3.36×10^2	6.29×10^2
		Mean	$\mathbf{2.19 \times 10^2}$	2.82×10^2	4.02×10^2	5.98×10^2	6.13×10^2	6.36×10^2	3.46×10^2	6.08×10^2
		Std	4.32×10^0	1.93×10^1	6.26×10^1	1.81×10^1	6.90×10^1	1.65×10^1	4.16×10^1	1.50×10^2
		p-value	-	$\mathbf{3.92 \times 10^{-23}}$ +	$\mathbf{1.48 \times 10^{-22}}$ +	$\mathbf{6.05 \times 10^{-69}}$ +	$\mathbf{1.63 \times 10^{-37}}$ +	$\mathbf{1.39 \times 10^{-73}}$ +	$\mathbf{2.85 \times 10^{-23}}$ +	$\mathbf{3.56 \times 10^{-20}}$ +
F_{22}		Median	3.76×10^3	5.54×10^3	6.39×10^3	1.30×10^4	8.15×10^3	1.33×10^4	6.48×10^3	4.48×10^2
		Mean	6.52×10^3	4.61×10^3	6.14×10^3	1.29×10^4	8.12×10^3	1.32×10^4	5.88×10^3	$\mathbf{4.46 \times 10^2}$
		Std	4.14×10^3	2.32×10^3	1.52×10^3	7.68×10^2	1.12×10^3	3.66×10^2	3.61×10^3	1.99×10^1
		p-value	-	5.39×10^{-2} =	6.51×10^{-1} =	$\mathbf{3.68 \times 10^{-11}}$ +	$\mathbf{4.85 \times 10^{-2}}$ +	$\mathbf{3.82 \times 10^{-12}}$ +	5.33×10^{-1} =	8.94×10^{-11} −
F_{23}		Median	5.21×10^2	5.21×10^2	6.42×10^2	8.94×10^2	1.22×10^3	8.58×10^2	6.65×10^2	7.71×10^3
		Mean	5.27×10^2	$\mathbf{5.19 \times 10^2}$	6.50×10^2	9.04×10^2	1.25×10^3	8.56×10^2	6.80×10^2	7.19×10^3
		Std	3.04×10^1	2.93×10^1	5.85×10^1	5.59×10^1	1.82×10^2	1.38×10^1	8.50×10^1	1.70×10^3
		p-value	-	7.27×10^{-1} =	$\mathbf{3.26 \times 10^{-14}}$ +	$\mathbf{1.86 \times 10^{-38}}$ +	$\mathbf{9.39 \times 10^{-29}}$ +	$\mathbf{7.16 \times 10^{-51}}$ +	$\mathbf{7.53 \times 10^{-13}}$ +	$\mathbf{7.69 \times 10^{-29}}$ +
F_{24}		Median	5.91×10^1	6.03×10^2	7.06×10^2	1.14×10^3	1.30×10^3	9.06×10^2	7.38×10^2	6.69×10^2
		Mean	$\mathbf{5.98 \times 10^2}$	6.28×10^2	7.09×10^2	1.15×10^3	1.32×10^3	9.06×10^2	7.49×10^2	6.66×10^2
		Std	3.21×10^1	8.09×10^1	7.38×10^1	1.38×10^2	1.26×10^2	1.22×10^1	9.38×10^1	1.51×10^1
		p-value	-	5.13×10^{-1} =	$\mathbf{4.81 \times 10^{-10}}$ +	$\mathbf{1.30 \times 10^{-28}}$ +	$\mathbf{6.07 \times 10^{-37}}$ +	$\mathbf{1.44 \times 10^{-48}}$ +	$\mathbf{2.54 \times 10^{-11}}$ +	$\mathbf{6.76 \times 10^{-15}}$ +
F_{25}		Median	4.80×10^2	5.95×10^2	6.74×10^2	4.31×10^2	4.84×10^3	5.58×10^2	6.61×10^2	8.25×10^2
		Mean	5.08×10^2	5.93×10^2	6.76×10^2	4.33×10^2	5.22×10^3	5.59×10^2	6.66×10^2	8.20×10^2
		Std	3.86×10^1	2.38×10^1	6.58×10^1	8.36×10^0	2.83×10^3	9.38×1000	7.00×10^1	3.20×10^1
		p-value	-	$\mathbf{4.38 \times 10^{-15}}$ +	$\mathbf{3.72 \times 10^{-17}}$ +	1.39×10^{-14} −	$\mathbf{1.51 \times 10^{-12}}$ +	$\mathbf{4.36 \times 10^{-9}}$ +	$\mathbf{3.17 \times 10^{-15}}$ +	$\mathbf{1.24 \times 10^{-39}}$ +

Table 4. *Cont.*

F	Category	Quality	SCDLPSO	XPSO	TCSPSO	DNSPSO	AWPSO	CLPSO_LS	GLPSO	CLPSO
F_{26}		Median	1.84×10^{3}	1.73×10^{3}	3.98×10^{3}	7.50×10^{3}	1.05×10^{4}	5.62×10^{3}	2.96×10^{3}	5.37×10^{2}
		Mean	1.89×10^{3}	1.28×10^{3}	4.07×10^{3}	7.72×10^{3}	1.04×10^{4}	5.60×10^{3}	3.04×10^{3}	**5.37×10^{2}**
		Std	1.65×10^{2}	8.75×10^{2}	1.02×10^{3}	2.13×10^{3}	1.97×10^{3}	1.86×10^{2}	6.50×10^{2}	4.38×10^{0}
		p-value	-	3.58×10^{-5} −	**2.43×10^{-16}** +	**3.29×10^{-21}** +	**3.59×10^{-31}** +	**3.21×10^{-61}** +	**6.05×10^{-13}** +	2.20×10^{-46} −
F_{27}		Median	6.66×10^{2}	6.99×10^{2}	9.01×10^{2}	5.00×10^{2}	1.39×1003	7.18×10^{2}	5.48×10^{2}	3.65×10^{3}
		Mean	6.79×10^{2}	7.17×10^{2}	9.06×10^{2}	5.00×10^{2}	1.44×1003	7.32×10^{2}	5.50×10^{2}	3.58×10^{3}
		Std	6.07×10^{1}	8.03×10^{1}	9.24×10^{1}	0.00×10^{0}	2.58×10^{2}	7.84×10^{1}	1.28×10^{1}	2.08×10^{2}
		p-value	-	**7.25×10^{-3}** +	**7.07×10^{-16}** +	8.02×10^{-23} −	**2.86×10^{-22}** +	**6.22×10^{-3}** +	3.93×10^{-16} −	**1.94×10^{-58}** +
F_{28}		Median	4.59×10^{2}	5.39×10^{2}	6.68×10^{2}	5.00×10^{2}	7.56×10^{3}	5.44×10^{3}	4.72×10^{2}	6.54×10^{2}
		Mean	4.74×10^{2}	5.43×10^{2}	6.77×10^{2}	5.00×10^{2}	7.80×10^{3}	5.34×10^{3}	**4.51×10^{2}**	6.53×10^{2}
		Std	2.69×10^{1}	3.56×10^{1}	6.99×10^{1}	0.00×10^{0}	1.47×10^{3}	4.48×10^{2}	7.00×10^{1}	2.16×10^{1}
		p-value	-	**3.50×10^{-11}** +	**4.61×10^{-21}** +	**2.37×10^{-6}** +	**2.41×10^{-34}** +	**3.07×10^{-53}** +	1.02×10^{-1} =	**2.69×10^{-35}** +
F_{29}		Median	4.98×10^{2}	8.54×10^{2}	1.43×10^{3}	3.26×10^{3}	3.73×10^{3}	2.10×10^{3}	7.74×10^{2}	1.69×10^{3}
		Mean	**5.17×10^{2}**	8.60×10^{2}	1.43×10^{3}	3.25×10^{3}	4.01×10^{3}	2.24×10^{3}	8.17×10^{2}	1.78×10^{3}
		Std	1.27×10^{2}	1.93×10^{2}	2.49×10^{2}	2.65×10^{2}	1.07×10^{3}	5.86×10^{2}	2.37×10^{2}	4.53×10^{2}
		p-value	-	**1.67×10^{-11}** +	**7.15×10^{-25}** +	**2.16×10^{-49}** +	**9.40×10^{-25}** +	**2.91×10^{-22}** +	**1.26×10^{-7}** +	**6.43×10^{-21}** +
F_{30}		Median	7.96×10^{5}	1.82×10^{6}	2.00×10^{6}	2.39×10^{6}	3.41×10^{8}	1.46×10^{6}	2.00×10^{6}	1.04×10^{3}
		Mean	8.26×10^{5}	1.86×10^{6}	2.36×10^{6}	2.51×10^{6}	4.80×10^{8}	1.16×10^{7}	2.40×10^{6}	**1.03×10^{3}**
		Std	1.11×10^{5}	3.34×10^{5}	8.05×10^{5}	9.62×10^{5}	7.04×10^{8}	5.37×10^{7}	1.44×10^{6}	1.62×10^{2}
		p-value	-	**4.17×10^{-24}** +	**1.69×10^{-14}** +	**3.12×10^{-13}** +	**5.36×10^{-4}** +	2.87×10^{-1} =	**2.25×10^{-7}** +	6.00×10^{-44} −
$F_{21\text{-}30}$	*w/t/l*		-	6/3/1	9/1/0	8/0/2	10/0/0	9/1/0	7/2/1	7/0/3
	w/t/l		-	17/7/5	20/8/1	25/0/4	28/0/1	23/5/1	19/7/3	20/2/7
	rank		**2.14**	2.45	4.72	5.07	7.41	6.17	3.52	4.52

Table 5. Comparison results between SCDLPSO and state-of-the-art PSO variants on the 100-D CEC 2017 benchmark functions.

F	Category	Quality	SCDLPSO	XPSO	TCSPSO	DNSPSO	AWPSO	CLPSO_LS	GLPSO	CLPSO
F_{1}	Unimodal Functions	Median	1.94×10^{3}	3.97×10^{3}	2.62×10^{3}	4.08×10^{3}	2.15×10^{11}	7.93×10^{9}	1.46×10^{4}	2.71×10^{4}
		Mean	**5.40×10^{3}**	7.78×10^{3}	6.22×10^{3}	7.17×10^{3}	2.19×10^{11}	8.00×10^{9}	3.49×10^{4}	3.69×10^{8}
		Std	4.91×10^{3}	8.48×10^{3}	7.24×10^{3}	8.26×10^{3}	3.66×10^{10}	1.30×10^{9}	5.38×10^{4}	4.72×10^{8}
		p-value	-	**4.65×10^{-2}** +	2.14×10^{-1} =	9.87×10^{-2} =	**1.07×10^{-38}** +	**1.98×10^{-39}** +	**3.36×10^{-3}** +	**8.88×10^{-5}** +

Table 5. *Cont.*

F	Category	Quality	SCDLPSO	XPSO	TCSPSO	DNSPSO	AWPSO	CLPSO_LS	GLPSO	CLPSO
F_3		Median	1.74×10^{5}	7.15×10^{4}	2.54×10^{5}	1.04×10^{6}	4.94×10^{5}	2.72×10^{-8}	1.92×10^{5}	3.53×10^{4}
		Mean	1.72×10^{5}	7.11×10^{4}	2.53×10^{5}	1.03×10^{6}	5.19×10^{5}	2.51×10^{9}	1.98×10^{5}	**4.60×10^{4}**
		Std	2.09×10^{4}	1.09×10^{4}	3.05×10^{4}	1.18×10^{5}	1.67×10^{5}	1.34×10^{10}	2.70×10^{4}	2.70×10^{4}
		p-value	-	3.66×10^{-32} −	**1.15×10^{-16} +**	**4.26×10^{-43} +**	**7.31×10^{-16} +**	3.16×10^{-1} =	**3.34×10^{-4} +**	7.13×10^{-28} −
$F_{1,3}$	$w/t/l$		-	1/0/1	1/1/0	1/1/0	2/0/0	1/1/0	2/0/0	1/0/1
F_4	Simple Multimodal Functions	Median	2.18×10^{2}	4.89×10^{2}	6.28×10^{2}	2.02×10^{2}	5.03×10^{4}	7.20×10^{2}	1.62×10^{3}	5.67×10^{5}
		Mean	2.18×10^{2}	4.88×10^{2}	7.03×10^{2}	**2.05×10^{2}**	5.16×10^{4}	7.50×10^{2}	1.58×10^{3}	5.66×10^{5}
		Std	1.98×10^{1}	6.07×10^{1}	1.82×10^{2}	5.34×10^{1}	1.47×10^{4}	1.11×10^{2}	4.02×10^{2}	5.55×10^{4}
		p-value	-	**1.65×10^{-30} +**	**1.21×10^{-20} +**	2.21×10^{-1} =	**2.41×10^{-26} +**	**4.60×10^{-33} +**	**1.37×10^{-25} +**	**1.09×10^{-51} +**
F_5		Median	2.59×10^{1}	2.15×10^{2}	5.36×10^{2}	1.03×10^{3}	1.24×10^{3}	1.07×10^{3}	4.44×10^{2}	2.74×10^{2}
		Mean	**2.59×10^{1}**	2.22×10^{2}	5.55×10^{2}	1.03×10^{3}	1.23×10^{3}	1.07×10^{3}	4.51×10^{2}	2.82×10^{2}
		Std	3.52×10^{1}	4.44×10^{1}	1.09×10^{2}	4.69×10^{1}	1.57×10^{2}	2.54×10^{1}	8.62×10^{1}	2.03×10^{1}
		p-value	-	**2.07×10^{-30} +**	**8.72×10^{-34} +**	**5.07×10^{-70} +**	**1.11×10^{-44} +**	**2.54×10^{-86} +**	**4.04×10^{-34} +**	**1.48×10^{-56} +**
F_6		Median	3.73×10^{-2}	4.61×10^{0}	1.85×10^{1}	2.36×10^{-1}	7.13×10^{1}	2.92×10^{1}	2.42×10^{0}	7.74×10^{2}
		Mean	**5.58×10^{-2}**	5.06×10^{0}	1.78×10^{1}	2.78×10^{-1}	7.14×10^{1}	2.92×10^{1}	3.03×10^{0}	7.72×10^{2}
		Std	5.01×10^{-2}	3.54×10^{0}	6.21×10^{0}	2.20×10^{-1}	9.32×10^{0}	1.54×10^{0}	1.97×10^{0}	2.83×10^{1}
		p-value	-	**1.43×10^{-9} +**	**3.54×10^{-22} +**	**1.86×10^{-06} +**	**1.16×10^{-44} +**	**3.41×10^{-67} +**	**3.67×10^{-11} +**	**2.52×10^{-76} +**
F_7		Median	1.38×10^{2}	4.31×10^{2}	1.22×10^{3}	1.12×10^{3}	4.02×10^{3}	1.45×10^{3}	7.43×10^{2}	8.98×10^{-7}
		Mean	1.39×10^{2}	4.49×10^{2}	1.23×10^{3}	1.13×10^{3}	4.13×10^{3}	1.52×10^{3}	7.62×10^{2}	**7.95×10^{-3}**
		Std	5.87×10^{0}	8.30×10^{1}	1.99×10^{2}	3.75×10^{1}	7.71×10^{2}	1.72×10^{2}	1.19×10^{2}	2.08×10^{-2}
		p-value	-	**7.75×10^{-28} +**	**1.26×10^{-36} +**	**3.60×10^{-75} +**	**2.91×10^{-35} +**	**7.31×10^{-46} +**	**1.79×10^{-35} +**	1.07×10^{-72} −
F_8		Median	2.89×10^{1}	2.13×10^{2}	5.29×10^{2}	1.02×10^{3}	1.22×10^{3}	1.06×10^{3}	4.68×10^{2}	7.71×10^{2}
		Mean	**2.93×10^{1}**	2.13×10^{2}	5.54×10^{2}	1.03×10^{3}	1.25×10^{3}	1.06×10^{3}	5.10×10^{2}	7.69×10^{2}
		Std	4.21×10^{0}	3.84×10^{1}	8.00×10^{1}	3.98×10^{1}	1.43×10^{2}	2.61×10^{1}	1.53×10^{2}	5.38×10^{1}
		p-value	-	**1.95×10^{-32} +**	**7.17×10^{-41} +**	**5.08×10^{-74} +**	**3.44×10^{-47} +**	**2.73×10^{-85} +**	**3.96×10^{-24} +**	**5.03×10^{-59} +**
F_9		Median	1.89×10^{1}	4.19×10^{2}	1.38×10^{4}	1.16×10^{3}	4.41×10^{4}	1.32×10^{4}	1.28×10^{4}	7.89×10^{2}
		Mean	**2.59×10^{1}**	5.62×10^{2}	1.40×10^{4}	2.29×10^{3}	4.65×10^{4}	1.33×10^{4}	1.41×10^{4}	7.88×10^{2}
		Std	2.49×10^{1}	4.12×10^{2}	4.05×10^{3}	2.95×10^{3}	9.32×10^{3}	1.58×10^{3}	7.09×10^{3}	3.20×10^{1}
		p-value	-	**4.91×10^{-11} +**	**4.57×10^{-26} +**	**1.13×10^{-4} +**	**2.03×10^{-34} +**	**5.89×10^{-47} +**	**2.21×10^{-15} +**	**5.66×10^{-67} +**

Table 5. *Cont.*

F	Category	Quality	SCDLPSO	XPSO	TCSPSO	DNSPSO	AWPSO	CLPSO_LS	GLPSO	CLPSO
F_{10}		Median	1.05×10^4	1.22×10^4	1.36×10^4	3.03×10^4	1.87×10^4	3.01×10^4	1.98×10^4	2.65×10^4
		Mean	1.55×10^4	$\mathbf{1.24 \times 10^4}$	1.34×10^4	3.02×10^4	1.86×10^4	3.01×10^4	2.07×10^4	2.58×10^4
		Std	8.91×10^3	1.36×10^3	1.08×10^3	8.44×10^2	2.19×10^3	4.22×10^2	3.69×10^3	3.20×10^3
		p-value	-	4.95×10^{-2} −	1.67×10^{-1} =	$\mathbf{4.46 \times 10^{-12}}$ +	9.35×10^{-2} =	$\mathbf{4.84 \times 10^{-12}}$ +	$\mathbf{7.40 \times 10^{-3}}$ +	$\mathbf{3.53 \times 10^{-7}}$ +
$F_{4\text{-}10}$	*w/t/l*		-	-	6/0/1	6/0/1	6/1/0	6/1/0	7/0/0	7/0/0
F_{11}	Hybrid Functions	Median	8.01×10^2	1.15×10^3	2.74×10^3	2.24×10^4	1.42×10^5	1.44×10^3	2.51×10^4	2.26×10^4
		Mean	8.31×10^2	1.18×10^3	3.42×10^3	2.46×10^4	1.56×10^5	3.89×10^3	2.38×10^4	2.26×10^4
		Std	1.52×10^2	2.31×10^2	1.92×10^3	9.75×10^3	6.76×10^4	5.95×10^3	8.81×10^3	5.20×10^2
		p-value	-	$\mathbf{5.03 \times 10^{-9}}$ +	$\mathbf{1.18 \times 10^{-9}}$ +	$\mathbf{4.85 \times 10^{-19}}$ +	$\mathbf{7.25 \times 10^{-18}}$ +	$\mathbf{7.63 \times 10^{-3}}$ +	$\mathbf{2.52 \times 10^{-20}}$ +	$\mathbf{5.26 \times 10^{-86}}$ +
F_{12}		Median	5.91×10^5	1.18×10^7	5.19×10^7	1.51×10^7	6.93×10^{10}	6.02×10^8	2.31×10^8	2.94×10^3
		Mean	6.04×10^5	1.81×10^7	8.56×10^7	1.65×10^7	7.75×10^{10}	6.53×10^8	3.13×10^8	$\mathbf{3.00 \times 10^3}$
		Std	3.25×10^5	1.87×10^7	9.57×10^7	7.77×10^6	2.82×10^{10}	4.55×10^8	2.64×10^8	7.24×10^2
		p-value	-	$\mathbf{6.65 \times 10^{-6}}$ +	$\mathbf{1.22 \times 10^{-5}}$ +	$\mathbf{7.71 \times 10^{-16}}$ +	$\mathbf{2.51 \times 10^{-21}}$ +	$\mathbf{1.82 \times 10^{-10}}$ +	$\mathbf{3.40 \times 10^{-8}}$ +	2.77×10^{-14} −
F_{13}		Median	2.46×10^3	3.11×10^3	4.13×10^3	6.18×10^3	1.06×10^9	1.24×10^4	1.18×10^4	1.02×10^8
		Mean	$\mathbf{4.14 \times 10^3}$	4.42×10^3	6.81×10^3	1.11×10^4	9.85×10^9	1.28×10^8	2.56×10^7	1.01×10^8
		Std	4.23×10^3	3.81×10^3	5.71×10^3	1.41×10^4	4.78×10^9	2.63×10^8	1.36×10^8	2.98×10^7
		p-value	-	6.57×10^{-1} =	$\mathbf{4.76 \times 10^{-2}}$ +	$\mathbf{1.32 \times 10^{-2}}$ +	$\mathbf{5.55 \times 10^{-16}}$ +	$\mathbf{1.13 \times 10^{-2}}$ +	3.15×10^{-1} =	$\mathbf{1.22 \times 10^{-25}}$ +
F_{14}		Median	7.92×10^4	2.26×10^5	9.20×10^5	2.10×10^6	2.37×10^7	3.82×10^6	5.15×10^5	5.23×10^4
		Mean	8.59×10^4	3.69×10^5	1.48×10^6	2.38×10^6	3.30×10^7	5.14×10^6	2.70×10^6	$\mathbf{5.43 \times 10^4}$
		Std	3.53×10^4	4.94×10^5	1.37×10^6	8.62×10^5	2.89×10^7	6.48×10^6	3.96×10^6	2.17×10^4
		p-value	-	$\mathbf{3.25 \times 10^{-3}}$ +	$\mathbf{9.95 \times 10^{-7}}$ +	$\mathbf{1.18 \times 10^{-20}}$ +	$\mathbf{8.20 \times 10^{-8}}$ +	$\mathbf{9.52 \times 10^{-5}}$ +	$\mathbf{7.51 \times 10^{-4}}$ +	8.07×10^{-5} −
F_{15}		Median	1.06×10^3	1.54×10^3	2.32×10^3	5.08×10^4	4.72×10^9	7.66×10^3	2.72×10^3	5.12×10^6
		Mean	$\mathbf{2.64 \times 10^3}$	2.78×10^3	4.88×10^3	7.18×10^4	5.09×10^9	2.96×10^7	1.18×10^4	5.21×10^6
		Std	4.67×10^3	3.21×10^3	5.47×10^3	6.60×10^4	2.66×10^9	1.11×10^8	5.98×10^5	1.52×10^6
		p-value	-	6.26×10^{-1} =	9.89×10^{-2} =	$\mathbf{5.50 \times 10^{-7}}$ +	$\mathbf{9.75 \times 10^{-15}}$ +	1.55×10^{-1} =	3.06×10^{-1} =	$\mathbf{5.68 \times 10^{-26}}$ +
F_{16}		Median	1.26×10^3	2.90×10^3	3.71×10^3	8.82×10^3	8.86×10^3	8.44×10^3	4.64×10^3	6.39×10^3
		Mean	$\mathbf{1.19 \times 10^3}$	2.88×10^3	3.82×10^3	8.79×10^3	8.79×10^3	8.43×10^3	4.96×10^3	6.36×10^3
		Std	4.58×10^2	4.97×10^3	6.74×10^2	3.41×10^2	1.19×10^3	3.33×10^2	1.66×10^3	1.85×10^3
		p-value	-	$\mathbf{1.89 \times 10^{-19}}$ +	$\mathbf{1.09 \times 10^{-24}}$ +	$\mathbf{2.53 \times 10^{-58}}$ +	$\mathbf{1.46 \times 10^{-38}}$ +	$\mathbf{2.54 \times 10^{-57}}$ +	$\mathbf{5.24 \times 10^{-17}}$ +	$\mathbf{4.67 \times 10^{-21}}$ +

Table 5. *Cont.*

F	Category	Quality	SCDLPSO	XPSO	TCSPSO	DNSPSO	AWPSO	CLPSO_LS	GLPSO	CLPSO
F_{17}		Median	1.31×10^{3}	2.45×10^{3}	3.22×10^{3}	5.94×10^{3}	2.20×10^{4}	5.62×10^{3}	3.42×10^{3}	4.13×10^{3}
		Mean	**1.34×10^{3}**	2.43×10^{3}	3.07×10^{3}	5.97×10^{3}	1.41×10^{5}	6.07×10^{3}	3.35×10^{3}	4.09×10^{3}
		Std	5.07×10^{2}	4.69×10^{2}	5.17×10^{2}	2.44×10^{2}	3.96×10^{5}	1.62×10^{3}	1.01×10^{3}	3.01×10^{2}
		p-value	-	**1.68×10^{-11}** +	**1.30×10^{-18}** +	**2.09×10^{-46}** +	6.30×10^{-02} =	**1.39×10^{-21}** +	**1.74×10^{-13}** +	**8.14×10^{-33}** +
F_{18}		Median	2.80×10^{5}	3.87×10^{5}	2.76×10^{6}	2.92×10^{7}	4.18×10^{7}	1.88×10^{7}	6.31×10^{5}	3.37×10^{3}
		Mean	2.91×10^{5}	4.72×10^{5}	3.38×10^{6}	3.20×10^{7}	4.62×10^{7}	2.62×10^{7}	1.32×10^{6}	**3.32×10^{3}**
		Std	1.25×10^{5}	3.09×10^{5}	2.09×10^{6}	1.06×10^{7}	3.72×10^{7}	2.16×10^{7}	1.70×10^{6}	3.55×10^{2}
		p-value	-	**7.87×10^{-3}** +	**7.90×10^{-11}** +	**4.09×10^{-23}** +	**1.21×10^{-8}** +	**2.32×10^{-8}** +	**1.93×10^{-3}** +	2.68×10^{-18} −
F_{19}	Hybrid Functions	Median	8.30×10^{2}	3.33×10^{3}	2.30×10^{3}	5.82×10^{3}	3.08×10^{9}	2.60×10^{4}	1.37×10^{3}	8.17×10^{6}
		Mean	**2.07×10^{3}**	4.55×10^{3}	4.86×10^{3}	8.89×10^{3}	3.87×10^{9}	8.66×10^{7}	9.63×10^{5}	8.39×10^{6}
		Std	2.56×10^{3}	5.32×10^{3}	6.41×10^{3}	8.15×10^{3}	3.13×10^{9}	2.13×10^{8}	5.08×10^{6}	2.39×10^{6}
		p-value		5.23×10^{-2} =	**3.35×10^{-2}** +	**6.55×10^{-5}** +	**1.04×10^{-8}** +	**3.30×10^{-2}** +	3.12×10^{-1} =	**1.78×10^{-26}** +
F_{20}		Median	1.18×10^{3}	2.16×10^{3}	2.88×10^{3}	5.70×10^{3}	3.86×10^{3}	4.78×10^{3}	3.46×10^{3}	3.42×10^{3}
		Mean	**1.87×10^{3}**	2.19×10^{3}	2.83×10^{3}	5.58×10^{3}	3.75×10^{3}	4.74×10^{3}	3.59×10^{3}	4.15×10^{3}
		Std	1.32×10^{3}	4.39×10^{2}	4.93×10^{2}	5.74×10^{2}	6.26×10^{2}	2.62×10^{2}	9.00×10^{2}	2.05×10^{3}
		p-value		4.84×10^{-1} =	**1.96×10^{-3}** +	**1.72×10^{-19}** +	**2.00×10^{-8}** +	**7.71×10^{-16}** +	**1.26×10^{-6}** +	**1.24×10^{-5}** +
$F_{11\text{-}20}$	*w/t/l*		-	6/4/0	9/1/0	10/0/0	9/1/0	9/1/0	7/3/0	7/0/3
F_{21}		Median	2.88×10^{2}	4.54×10^{2}	7.91×10^{2}	1.22×10^{3}	1.62×10^{3}	1.30×10^{3}	7.46×10^{2}	2.21×10^{3}
		Mean	**2.94×10^{2}**	4.59×10^{2}	7.87×10^{2}	1.23×10^{3}	1.62×10^{3}	1.29×10^{3}	7.89×10^{2}	2.20×10^{3}
		Std	1.83×10^{1}	5.25×10^{1}	8.43×10^{1}	3.66×10^{1}	1.70×10^{2}	3.07×10^{1}	1.81×10^{2}	2.25×10^{2}
		p-value	-	**6.13×10^{-24}** +	**1.27×10^{-37}** +	**9.30×10^{-72}** +	**5.17×10^{-45}** +	**7.16×10^{-77}** +	**3.28×10^{-21}** +	**4.79×10^{-47}** +
F_{22}	Composition Functions	Median	9.97×10^{3}	1.38×10^{4}	1.46×10^{4}	3.09×10^{4}	1.98×10^{4}	3.08×10^{4}	2.30×10^{4}	1.02×10^{3}
		Mean	1.27×10^{4}	1.27×10^{4}	1.46×10^{4}	3.07×10^{4}	1.99×10^{4}	3.07×10^{4}	2.35×10^{4}	**1.02×10^{3}**
		Std	6.78×10^{3}	4.40×10^{3}	1.35×10^{3}	8.81×10^{2}	2.37×10^{3}	3.97×10^{2}	4.24×10^{3}	2.84×10^{1}
		p-value	-	7.26×10^{-1} =	1.44×10^{-1} =	**1.53×10^{-20}** +	**1.44×10^{-6}** +	**1.28×10^{-20}** +	**1.02×10^{-9}** +	4.54×10^{-13} −
F_{23}		Median	7.60×10^{2}	8.16×10^{2}	1.04×10^{3}	1.70×10^{3}	2.58×10^{3}	1.51×10^{3}	1.39×10^{3}	2.33×10^{4}
		Mean	**7.62×10^{2}**	8.19×10^{2}	1.07×10^{3}	1.83×10^{3}	2.53×10^{3}	1.51×10^{3}	1.39×10^{3}	2.34×10^{4}
		Std	4.79×10^{1}	5.65×10^{1}	1.09×10^{2}	3.60×10^{2}	2.47×10^{2}	2.84×10^{1}	1.92×10^{2}	5.35×10^{2}
		p-value	-	**6.24×10^{-4}** +	**2.35×10^{-20}** +	**8.10×10^{-23}** +	**1.46×10^{-42}** +	**1.22×10^{-58}** +	**2.38×10^{-24}** +	**3.44×10^{-87}** +

Table 5. *Cont.*

F	Category	Quality	SCDLPSO	XPSO	TCSPSO	DNSPSO	AWPSO	CLPSO_LS	GLPSO	CLPSO
F_{24}		Median	1.24×10^{3}	1.20×10^{3}	1.52×10^{3}	3.08×10^{3}	4.08×10^{3}	1.87×10^{3}	2.05×10^{3}	9.38×10^{2}
		Mean	1.24×10^{3}	1.24×10^{3}	1.55×10^{3}	3.14×10^{3}	4.12×10^{3}	1.88×10^{3}	2.03×10^{3}	**9.36×10^{2}**
		Std	7.89×10^{1}	1.18×10^{2}	1.46×10^{2}	6.53×10^{2}	4.76×10^{2}	7.17×10^{1}	1.32×10^{2}	2.01×10^{1}
		p-value	-	9.47×10^{-1} =	**4.48×10^{-14} +**	**2.23×10^{-22} +**	**1.11×10^{-38} +**	**1.37×10^{-38} +**	**5.80×10^{-35} +**	4.00×10^{-28} –
F_{25}		Median	8.22×10^{2}	1.08×10^{3}	1.29×10^{3}	7.34×10^{2}	2.32×10^{4}	2.82×10^{3}	1.79×10^{3}	1.51×10^{3}
		Mean	8.00×10^{2}	1.09×10^{3}	1.35×10^{3}	**7.54×10^{2}**	2.31×10^{4}	2.80×10^{3}	1.77×10^{3}	1.51×10^{3}
		Std	5.76×10^{1}	7.59×10^{1}	2.93×10^{2}	5.64×10^{1}	6.09×10^{3}	2.29×10^{2}	2.76×10^{2}	2.84×10^{1}
		p-value	-	**1.71×10^{-24} +**	**3.47×10^{-14} +**	3.03×10^{-3} –	**2.13×10^{-27} +**	**3.78×10^{-47} +**	**4.16×10^{-26} +**	**8.76×10^{-54} +**
F_{26}		Median	6.20×10^{3}	5.36×10^{3}	1.06×10^{4}	2.54×10^{4}	3.86×10^{4}	1.43×10^{4}	1.04×10^{4}	9.32×10^{2}
		Mean	6.16×10^{3}	4.18×10^{3}	1.14×10^{4}	2.62×10^{4}	3.74×10^{4}	1.45×10^{4}	1.10×10^{4}	**9.37×10^{2}**
		Std	4.81×10^{2}	2.45×10^{3}	2.40×10^{3}	7.04×10^{3}	4.48×10^{3}	6.46×1002	2.29×10^{3}	3.87×10^{1}
		p-value	-	2.09×10^{-5} –	**1.28×10^{-16} +**	**4.66×10^{-22} +**	**3.15×10^{-42} +**	**5.88×10^{-52} +**	**3.45×10^{-16} +**	3.19×10^{-53} –
F_{27}		Median	7.49×10^{2}	8.71×10^{2}	1.12×10^{3}	5.00×10^{2}	3.20×10^{3}	7.61×10^{2}	1.23×10^{3}	1.12×10^{4}
		Mean	7.66×10^{2}	8.83×10^{2}	1.12×10^{3}	**5.00×10^{2}**	3.32×10^{3}	7.61×10^{2}	1.26×10^{3}	1.12×10^{4}
		Std	6.46×10^{1}	7.08×10^{1}	1.75×10^{2}	0.00×10^{0}	8.62×10^{2}	4.86×10^{1}	9.86×10^{1}	3.26×10^{2}
		p-value	-	**2.15×10^{-7} +**	**1.23×10^{-14} +**	5.36×10^{-30} –	**7.09×10^{-23} +**	7.67×10^{-1} =	**3.18×10^{-30} +**	**6.83×10^{-80} +**
F_{28}		Median	5.64×10^{2}	8.31×10^{2}	1.33×10^{3}	5.00×10^{2}	2.88×10^{4}	1.34×10^{4}	2.20×10^{3}	7.68×10^{2}
		Mean	5.72×10^{2}	8.32×10^{2}	1.37×10^{3}	**5.00×10^{2}**	2.81×10^{4}	1.34×10^{4}	2.29×10^{3}	7.68×10^{2}
		Std	3.12×10^{1}	4.26×10^{1}	3.31×10^{2}	0.00×10^{0}	4.47×10^{3}	2.57×10^{2}	5.64×10^{2}	2.39×10^{1}
		p-value	-	**1.11×10^{-31} +**	**1.18×10^{-18} +**	6.71×10^{-18} –	**2.05×10^{-39} +**	**3.12×10^{-91} +**	**1.72×10^{-23} +**	**2.04×10^{-34} +**
F_{29}		Median	1.77×10^{3}	3.02×10^{3}	3.90×10^{3}	6.93×10^{3}	1.08×10^{4}	6.16×10^{3}	4.38×10^{3}	1.28×10^{4}
		Mean	**1.82×10^{3}**	3.03×10^{3}	3.91×10^{3}	6.87×10^{3}	1.71×10^{4}	6.13×10^{3}	4.43×10^{3}	1.28×10^{4}
		Std	3.99×10^{2}	4.92×10^{2}	5.41×10^{2}	3.37×10^{2}	1.70×10^{4}	8.58×10^{2}	6.82×10^{2}	3.67×10^{1}
		p-value	-	**2.42×10^{-15} +**	**6.71×10^{-24} +**	**2.20×10^{-50} +**	**1.00×10^{-5} +**	**2.60×10^{-32} +**	**3.49×10^{-25} +**	**2.66×10^{-76} +**
F_{30}		Median	5.68×10^{3}	2.87×10^{4}	1.04×10^{5}	7.89×10^{2}	7.81×10^{9}	1.26×10^{4}	1.84×10^{6}	3.45×10^{3}
		Mean	7.04×10^{3}	3.55×10^{4}	1.46×10^{5}	**7.99×10^{2}**	8.40×10^{9}	1.57×10^{8}	3.54×10^{6}	3.39×10^{3}
		Std	3.52×10^{3}	2.37×10^{4}	1.34×10^{5}	2.13×10^{2}	4.21×10^{9}	3.12×10^{8}	3.84×10^{6}	2.84×10^{2}
		p-value	-	**2.97×10^{-7} +**	**6.76×10^{-7} +**	1.83×10^{-13} –	**2.04×10^{-15} +**	**8.68×10^{-3} +**	**6.47×10^{-6} +**	6.99×10^{-7} –
$F_{21\text{-}30}$	*w/t/l*		-	7/2/1	9/1/0	6/0/4	10/0/0	9/1/0	10/0/0	6/0/4
	w/t/l		-	20/6/3	25/3/1	23/2/4	26/3/0	26/3/0	26/3/0	20/0/9
	rank		**1.69**	2.66	3.97	4.86	7.45	5.93	4.97	4.48

Table 6. Statistical comparison results between SCDLPSO and the seven state-of-the-art PSO variants on the CEC 2017 benchmark set with different dimensions in terms of "$w/t/l$".

Category	D	XPSO	TCSPSO	DNSPSO	AWPSO	CLPSO_LS	GLPSO	CLPSO
Unimodal Functions	30	0/1/1	1/1/0	2/0/0	2/0/0	1/1/0	0/1/1	2/0/0
	50	0/1/1	1/1/0	2/0/0	2/0/0	1/1/0	1/0/1	2/0/0
	100	1/0/1	1/1/0	1/1/0	2/0/0	1/1/0	2/0/0	1/0/1
Simple Multimodal Functions	30	6/0/1	6/0/1	5/0/2	6/0/1	6/1/0	6/0/1	5/0/2
	50	6/0/1	6/0/1	6/0/1	6/0/1	7/0/0	6/0/1	5/0/2
	100	6/0/1	6/0/1	6/1/0	6/1/0	7/0/0	7/0/0	6/0/1
Hybrid Functions	30	5/4/1	9/1/0	8/1/1	7/3/0	9/1/0	8/2/0	8/0/2
	50	5/3/2	4/6/0	9/0/1	10/0/0	6/3/1	5/5/0	6/2/2
	100	6/4/0	9/1/0	10/0/0	9/1/0	9/1/0	7/3/0	7/0/3
Composition Functions	30	6/3/1	10/0/0	8/0/2	10/0/0	8/2/0	9/1/0	5/3/2
	50	6/3/1	9/1/0	8/0/2	10/0/0	9/1/0	7/2/1	7/0/3
	100	7/2/1	9/1/0	6/0/4	10/0/0	9/1/0	10/0/0	6/0/4
Whole Set	30	17/8/4	26/2/1	23/1/5	25/3/1	24/5/0	23/4/2	20/3/6
	50	17/7/5	20/8/1	25/0/4	28/0/1	23/5/1	19/7/3	20/2/7
	100	20/6/3	25/3/1	23/2/4	26/3/0	26/3/0	26/3/0	20/0/9

As shown in Table 3, the comparison results on the 30-D CEC 2017 benchmark problems can be summarized as follows:

(1) As shown in the last row of Table 3, the proposed SCDLPSO achieves the lowest rank in terms of the Friedman test, and its rank is much smaller than those of the other algorithms. This demonstrates that SCDLPSO achieves the best overall performance on the 30-D CEC 2017 benchmark set and its overall performance is much superior to the compared algorithms.

(2) The second to last row of Table 3 shows that SCDLPSO performs much better than the seven compared algorithms from the perspective of the Wilcoxon rank sum test. Specifically, compared with TCSPSO, DNSPSO, AWPSO, CLPSO_LS, GLPSO, and CLPSO, SCDLPSO achieves significantly superior performance to the other algorithms on at least 20 problems, and displays inferiority to them on at most six problems. Compared with XPSO, the proposed SCDLPSO shows significant superiority on 17 problems and is worse than XPSO on only four problems.

(3) In terms of the comparison results on different types of optimization problems, on the two unimodal problems, SCDLPSO outperforms DNSPSO, AWPSO, and CLPSO, while it achieves competitive performance with XPSO, TCSPSO, CLPSO_LS, and GLPSO. On the seven simple multimodal problems, SCDLPSO presents significant superiority to the seven compared algorithms on at least five problems. As for the 10 hybrid problems, except for XPSO, SCDLPSO performs significantly better than the other compared PSO variants on at least seven problems. In comparison with XPSO, SCDLPSO shows great superiority on five problems and is defeated by XPSO on only one problem. In terms of the 10 composition problems, SCDLPSO is significantly better than AWPSO and TCSPSO on all of these problems. It achieves better performance than DNSPSO, CLPSO_LS, and GLPSO on eight, eight, and nine problems, respectively. In comparison with XPSO and CLPSO, SCDLPSO outperforms them on six and five problems respectively, and shows worse performance on only one and two problems, respectively.

Subsequently, from Table 4, we can draw the following conclusions in regard to the comparison results between SCDLPSO and the compared state-of-the-art PSO variants on the 50-D CEC 2017 benchmark problems:

(1) According to the last row of Table 4, SCDLPSO still achieves the lowest rank among all algorithms. This demonstrates that SCDLPSO still obtains the best overall performance on the 50-D CEC 2017 problems.

(2) From the perspective of the Wilcoxon rank sum test, as shown in the second to last row of Table 4, SCDLPSO achieves better performance than TCSPSO, DNSPSO, AWPSO, CLPSO_LS, and CLPSO on 20, 25, 28, 23 and 20 problems, respectively. In comparison with XPSO and GLPSO, SCDLPSO outperforms them on 17 and 19 problems, respectively.

(3) As for the comparison results on different types of optimization problems, on the two unimodal problems, SCDLPSO is significantly superior to DNSPSO, AWPSO, and CLPSO on both problems, while it obtains very competitive performance with the other compared algorithms. As for the seven simple multimodal problems, SCDLPSO significantly outperforms the seven compared PSO variants on at least five problems. On the 10 hybrid problems, SCDLPSO achieves significantly better performance than DNSPSO and AWPSO on 9 and 10 problems, respectively. In competition with the other five compared PSO variants, SCDLPSO shows no inferiority to them on at least eight problems. Confronted with the 10 composition problems, SCDLPSO presents great superiority to the seven compared PSO variants on at least six problems, and displays inferiority to them on at most three problems.

At last, from Table 5, the following conclusions can be drawn from the comparison results between SCDLPSO and the compared state-of-the-art PSO variants on the 100-D CEC 2017 benchmark problems:

(1) According to the averaged rank obtained from the Friedman test, SCDLPSO still ranks first among all algorithms. This verifies that SCDLPSO consistently achieves the best overall performance on the 100-D CEC 2017 benchmark set.

(2) From the results of the Wilcoxon rank sum test, SCDLPSO presents significant dominance to the seven compared algorithms on at least 20 problems. In particular, competed with AWPSO, CLPSO_LS, and GLPSO, SCDLPSO significantly outperforms them on 26 problems, and shows no inferiority on any of the 29 problems.

(3) Regarding the comparison results for different types of optimization problems, on the two unimodal problems, SCDLPSO is significantly superior to AWPSO and GLPSO on both problems and achieves competitive performance with the other algorithms. On the seven simple multimodal problems, SCDLPSO outperforms all seven compared algorithms on at least six problems. When it comes to the 10 hybrid problems, SCDLPSO achieves significantly better performance than all compared PSO variants on at least six problems. Particularly, SCDLPSO significantly beats TCSPSO, DNSPSO, AWPSO, and CLPSO_LS on at least nine problems, and shows no worse performance than them on all 10 problems. As for the 10 composition problems, SCDLSPO outperforms AWPSO and GLPSO on all of these problems, and wins the competition with TCSPSO and CLPSO_LS on nine. When compared with XPSO, DNSPSO, and CLPSO, SCDLPSO is superior on at least six problems and shows inferiority on at most four problems.

Overall, as shown in Table 6, it is found that the proposed SCDLPSO consistently shows great superiority to the compared state-of-the-art and representative PSO variants on the CEC 2017 benchmark problem set with different dimension sizes. On the one hand, from a comprehensive perspective, the above comparative experiments validate that SCDLPSO preserves good scalability to solve optimization problems. On the other hand, after deep investigation of the comparison results for different types of optimization problems, we find that SCDLPSO performs particularly much better than the compared algorithms on complicated problems, such as the multimodal problems, the hybrid problems and the composition problems. This demonstrates that SCDLPSO preserves the capability to solve complex optimization problems. All of these examples of SCDLPSO superiority mainly benefit from the proposed SCDL strategy, which affords high learning diversity and effectiveness for particles. With this strategy, SCDLPSO can balance search intensification and diversification well to explore and exploit the solution space to find high-quality solutions.

5. Conclusions

This paper has proposed a stochastic cognitive dominance leading particle swarm optimization (SCDLPSO) algorithm to tackle complicated optimization problems. Specifically, in this optimizer, a random triad topology is employed for each particle to communicate with two other particles randomly selected from the swarm. Then, the historical cognitive experiences of the three particles are allowed to compete with each other. Only when at least one of the personal best positions of the two randomly selected particles shows dominance to the personal best position of the particle to be updated, this particle is updated by learning from the two best cognitive experiences; otherwise this particle is not updated and directly enters the next generation. In this way, high learning diversity can be maintained due to the random interaction, and at the same time, high learning effectiveness can be preserved because each updated particle learns from the two best experiences. Therefore, the proposed SCDLPSO is expected to explore and exploit the solution space appropriately during the evolution.

Extensive comparative experiments were conducted on the CEC 2017 benchmark problem set with three different dimension sizes by comparing SCDLPSO with seven state-of-the-art and representative PSO variants. The experimental results demonstrate that SCDLPSO consistently achieves great superiority to the compared algorithms on most problems with the three dimension sizes. In particular, it was verified that SCDLPSO performs much better than the compared algorithms on complicated problems, such as multimodal problems, hybrid problems and composition problems. Therefore, SCDLPSO can be considered as a promising optimizer for optimization problems, especially the complicated ones.

Author Contributions: Q.Y.: Conceptualization, supervision, methodology, formal analysis, and writing—original draft preparation. L.H.: Implementation, formal analysis, and writing—original draft preparation. X.G.: Methodology, and writing—review and editing. D.X.: Writing—review and editing. Z.L.: Writing—review and editing, and funding acquisition. S.-W.J.: Writing—review and editing. J.Z.: Conceptualization and writing—review and editing. All authors have read and agreed to the published version of the manuscript.

Funding: This work was supported in part by the National Natural Science Foundation of China under Grant 62006124 and U20B2061, in part by the Natural Science Foundation of Jiangsu Province under Project BK20200811, in part by the Natural Science Foundation of the Jiangsu Higher Education Institutions of China under Grant 20KJB520006, in part by the National Research Foundation of Korea (NRF-2021H1D3A2A01082705), and in part by the Startup Foundation for Introducing Talent of NUIST.

Institutional Review Board Statement: Not applicable.

Informed Consent Statement: Not applicable.

Data Availability Statement: Not applicable.

Conflicts of Interest: The authors declare no conflict of interest.

References

1. Lu, W.; Xu, X.; Huang, G.; Li, B.; Wu, Y.; Zhao, N.; Yu, F.R. Energy Efficiency Optimization in SWIPT Enabled WSNs for Smart Agriculture. *IEEE Trans. Ind. Inform.* **2021**, *17*, 4335–4344. [CrossRef]
2. Zhou, J.; He, R.; Wang, Y.; Jiang, S.; Zhu, Z.; Hu, J.; Miao, J.; Luo, Q. Autonomous Driving Trajectory Optimization with Dual-Loop Iterative Anchoring Path Smoothing and Piecewise-Jerk Speed Optimization. *IEEE Robot. Autom. Lett.* **2021**, *6*, 439–446. [CrossRef]
3. Zhang, L.; Zhang, Y.; Li, Y. Mobile Robot Path Planning Based on Improved Localized Particle Swarm Optimization. *IEEE Sensors J.* **2021**, *21*, 6962–6972. [CrossRef]

4. Huang, L.; Ding, Y.; Zhou, M.; Jin, Y.; Hao, K. Multiple-Solution Optimization Strategy for Multirobot Task Allocation. *IEEE Trans. Syst. Man, Cybern. Syst.* **2018**, *50*, 4283–4294. [CrossRef]
5. Ghorpade, S.; Zennaro, M.; Chaudhari, B. Survey of Localization for Internet of Things Nodes: Approaches, Challenges and Open Issues. *Futur. Internet* **2021**, *13*, 210. [CrossRef]
6. Zhan, Z.-H.; Shi, L.; Tan, K.C.; Zhang, J. A survey on evolutionary computation for complex continuous optimization. *Artif. Intell. Rev.* **2021**, *55*, 59–110. [CrossRef]
7. Yang, Q.; Chen, W.-N.; Zhang, J. Probabilistic Multimodal Optimization. In *Metaheuristics for Finding Multiple Solutions*; Springer: Berlin/Heidelberg, Germany, 2021; pp. 191–228.
8. Liang, J.J.; Qin, A.K.; Suganthan, P.N.; Baskar, S. Comprehensive learning particle swarm optimizer for global optimization of multimodal functions. *IEEE Trans. Evol. Comput.* **2006**, *10*, 281–295. [CrossRef]
9. Yang, Q.; Li, Y.; Gao, X.-D.; Ma, Y.-Y.; Lu, Z.-Y.; Jeon, S.-W.; Zhang, J. An Adaptive Covariance Scaling Estimation of Distribution Algorithm. *Mathematics* **2021**, *9*, 3207. [CrossRef]
10. Kennedy, J.; Eberhart, R. Particle Swarm Optimization. In Proceedings of the International Conference on Neural Networks, Perth, Australia, 27 November–1 December 1995; pp. 1942–1948.
11. Eberhart, R.; Kennedy, J. A New Optimizer Using Particle Swarm Theory. In Proceedings of the International Symposium on Micro Machine and Human Science, Nagoya, Japan, 4–6 October 1995; pp. 39–43.
12. Yang, Q.; Xie, H.-Y.; Chen, W.-N.; Zhang, J. Multiple parents guided differential evolution for large scale optimization. In Proceedings of the Congress on Evolutionary Computation, Vancouver, BC, Canada, 24–29 July 2016; pp. 3549–3556.
13. Yu, W.-J.; Ji, J.-Y.; Gong, Y.-J.; Yang, Q.; Zhang, J. A tri-objective differential evolution approach for multimodal optimization. *Inf. Sci.* **2018**, *423*, 1–23. [CrossRef]
14. Zhigljavsky, A.; Žilinskas, A. Bi-objective Decisions and Partition-Based Methods in Bayesian Global Optimization. In *Bayesian and High-Dimensional Global Optimization*; Springer International Publishing: Cham, Switzerland, 2021; pp. 41–88.
15. Xue, Y.; Wang, Y.; Liang, J. A self-adaptive gradient descent search algorithm for fully-connected neural networks. *Neurocomputing* **2022**, *478*, 70–80. [CrossRef]
16. Žilinskas, A.; Calvin, J. Bi-Objective Decision Making in Global Optimization Based on Statistical Models. *J. Glob. Optim.* **2019**, *74*, 599–609. [CrossRef]
17. Pepelyshev, A.; Zhigljavsky, A.; Žilinskas, A. Performance of global random search algorithms for large dimensions. *J. Glob. Optim.* **2018**, *71*, 57–71. [CrossRef]
18. Zelinka, I. A Survey on Evolutionary Algorithms Dynamics and its Complexity–Mutual Relations, Past, Present and Future. Swarm and Evolutionary Computation. *Swarm Evol. Comput.* **2015**, *25*, 2–14. [CrossRef]
19. Bonyadi, M.R. A Theoretical Guideline for Designing an Effective Adaptive Particle Swarm. *IEEE Trans. Evol. Comput.* **2020**, *24*, 57–68. [CrossRef]
20. Mor, B.; Shabtay, D.; Yedidsion, L. Heuristic algorithms for solving a set of NP-hard single-machine scheduling problems with resource-dependent processing times. *Comput. Ind. Eng.* **2021**, *153*, 107024. [CrossRef]
21. Anbuudayasankar, S.P.; Ganesh, K.; Mohapatra, S. Survey of Methodologies for TSP and VRP. In *Models for Practical Routing Problems in Logistics: Design and Practices*; Springer: Berlin/Heidelberg, Germany, 2014; pp. 11–42.
22. Tang, J.; Liu, G.; Pan, Q. A Review on Representative Swarm Intelligence Algorithms for Solving Optimization Problems: Applications and Trends. *IEEE/CAA J. Autom. Sin.* **2021**, *8*, 1627–1643. [CrossRef]
23. Yang, Q.; Chen, W.-N.; Gu, T.; Jin, H.; Mao, W.; Zhang, J. An Adaptive Stochastic Dominant Learning Swarm Optimizer for High-Dimensional Optimization. *IEEE Trans. Cybern.* **2020**, 1–17. [CrossRef]
24. Ji, X.; Zhang, Y.; Gong, D.; Sun, X. Dual-Surrogate-Assisted Cooperative Particle Swarm Optimization for Expensive Multimodal Problems. *IEEE Trans. Evol. Comput.* **2021**, *25*, 794–808. [CrossRef]
25. Yang, Q.; Chen, W.-N.; Da Deng, J.; Li, Y.; Gu, T.; Zhang, J. A Level-Based Learning Swarm Optimizer for Large-Scale Optimization. *IEEE Trans. Evol. Comput.* **2017**, *22*, 578–594. [CrossRef]
26. Lan, R.; Zhu, Y.; Lu, H.; Liu, Z.; Luo, X. A Two-Phase Learning-Based Swarm Optimizer for Large-Scale Optimization. *IEEE Trans. Cybern.* **2020**, *51*, 6284–6293. [CrossRef]
27. Qu, B.; Li, G.; Yan, L.; Liang, J.; Yue, C.; Yu, K.; Crisalle, O.D. A grid-guided particle swarm optimizer for multimodal multi-objective problems. *Appl. Soft Comput.* **2022**, *117*, 108381. [CrossRef]
28. Wei, F.-F.; Chen, W.-N.; Yang, Q.; Deng, J.; Luo, X.-N.; Jin, H.; Zhang, J. A Classifier-Assisted Level-Based Learning Swarm Optimizer for Expensive Optimization. *IEEE Trans. Evol. Comput.* **2021**, *25*, 219–233. [CrossRef]
29. Jang-Ho, S.; Chang-Hwan, I.; Chang-Geun, H.; Jae-Kwang, K.; Hyun-Kyo, J.; Cheol-Gyun, L. Multimodal Function Optimization Based on Particle Swarm Optimization. *IEEE Trans. Magn.* **2006**, *42*, 1095–1098. [CrossRef]
30. Zou, J.; Deng, Q.; Zheng, J.; Yang, S. A close neighbor mobility method using particle swarm optimizer for solving multimodal optimization problems. *Inf. Sci.* **2020**, *519*, 332–347. [CrossRef]
31. Yang, Q.; Chen, W.-N.; Yu, Z.; Gu, T.; Li, Y.; Zhang, H.; Zhang, J. Adaptive Multimodal Continuous Ant Colony Optimization. *IEEE Trans. Evol. Comput.* **2016**, *21*, 191–205. [CrossRef]
32. Yang, Q.; Chen, W.-N.; Li, Y.; Chen, C.L.P.; Xu, X.-M.; Zhang, J. Multimodal Estimation of Distribution Algorithms. *IEEE Trans. Cybern.* **2017**, *47*, 636–650. [CrossRef]

33. Tanabe, R.; Ishibuchi, H. A Review of Evolutionary Multimodal Multiobjective Optimization. *IEEE Trans. Evol. Comput.* **2020**, *24*, 193–200. [CrossRef]
34. Molaei, S.; Moazen, H.; Najjar-Ghabel, S.; Farzinvash, L. Particle swarm optimization with an enhanced learning strategy and crossover operator. *Knowl.-Based Syst.* **2021**, *215*, 106768. [CrossRef]
35. Lin, A.; Sun, W.; Yu, H.; Wu, G.; Tang, H. Adaptive comprehensive learning particle swarm optimization with cooperative archive. *Appl. Soft Comput.* **2019**, *77*, 533–546. [CrossRef]
36. Yang, Q.; Chen, W.-N.; Gu, T.; Zhang, H.; Deng, J.D.; Li, Y.; Zhang, J. Segment-Based Predominant Learning Swarm Optimizer for Large-Scale Optimization. *IEEE Trans. Cybern.* **2016**, *47*, 2896–2910. [CrossRef]
37. Yang, Q.; Chen, W.-N.; Gu, T.; Zhang, H.; Yuan, H.; Kwong, S.; Zhang, J. A Distributed Swarm Optimizer with Adaptive Communication for Large-Scale Optimization. *IEEE Trans. Cybern.* **2020**, *50*, 3393–3408. [CrossRef] [PubMed]
38. Zhang, X.; Du, K.-J.; Zhan, Z.-H.; Kwong, S.; Gu, T.-L.; Zhang, J. Cooperative Coevolutionary Bare-Bones Particle Swarm Optimization with Function Independent Decomposition for Large-Scale Supply Chain Network Design with Uncertainties. *IEEE Trans. Cybern.* **2019**, *50*, 4454–4468. [CrossRef]
39. Song, X.-F.; Zhang, Y.; Guo, Y.-N.; Sun, X.-Y.; Wang, Y.-L. Variable-Size Cooperative Coevolutionary Particle Swarm Optimization for Feature Selection on High-Dimensional Data. *IEEE Trans. Evol. Comput.* **2020**, *24*, 882–895. [CrossRef]
40. Cao, Y.; Zhang, H.; Li, W.; Zhou, M.; Zhang, Y.; Chaovalitwongse, W.A. Comprehensive Learning Particle Swarm Optimization Algorithm with Local Search for Multimodal Functions. *IEEE Trans. Evol. Comput.* **2019**, *23*, 718–731. [CrossRef]
41. Zhang, X.; Wang, X.; Kang, Q.; Cheng, J. Differential mutation and novel social learning particle swarm optimization algorithm. *Inf. Sci.* **2019**, *480*, 109–129. [CrossRef]
42. Liang, X.; Li, W.; Liu, P.; Zhang, Y.; Agbo, A.A. Social Network-based Swarm Optimization algorithm. In Proceedings of the International Conference on Networking, Sensing and Control, Taipei, Taiwan, 9–11 April 2015; pp. 360–365.
43. Blackwell, T.; Kennedy, J. Impact of Communication Topology in Particle Swarm Optimization. *IEEE Trans. Evol. Comput.* **2019**, *23*, 689–702. [CrossRef]
44. Kennedy, J.; Mendes, R. Population Structure and Particle Swarm Performance. In Proceedings of the IEEE Congress on Evolutionary Computation, Honolulu, HI, USA, 12–17 May 2002; pp. 1671–1676.
45. Lin, A.; Sun, W.; Yu, H.; Wu, G.; Tang, H. Global genetic learning particle swarm optimization with diversity enhancement by ring topology. *Swarm Evol. Comput.* **2019**, *44*, 571–583. [CrossRef]
46. Kennedy, J. Small worlds and mega-minds: Effects of neighborhood topology on particle swarm performance. In Proceedings of the 1999 Congress on Evolutionary Computation-CEC99 (Cat. No. 99TH8406), Washington, DC, USA, 6–9 July 1999; Volume 3, p. 1931.
47. Elsayed, S.M.; Sarker, R.A.; Essam, D.L. Memetic multi-topology particle swarm optimizer for constrained optimization. In Proceedings of the IEEE Congress on Evolutionary Computation, Brisbane, Australia, 10–15 June 2012; pp. 1–8.
48. Li, F.; Guo, J. Topology Optimization of Particle Swarm Optimization. In Proceedings of the Advances in Swarm Intelligence, Hefei, China, 17–20 October 2014; pp. 142–149.
49. Bonyadi, M.R.; Li, X.; Michalewicz, Z. A hybrid particle swarm with a time-adaptive topology for constrained optimization. *Swarm Evol. Comput.* **2014**, *18*, 22–37. [CrossRef]
50. Xia, X.; Gui, L.; Yu, F.; Wu, H.; Wei, B.; Zhang, Y.-L.; Zhan, Z.-H. Triple Archives Particle Swarm Optimization. *IEEE Trans. Cybern.* **2019**, *50*, 4862–4875. [CrossRef]
51. Zhan, Z.; Zhang, J.; Li, Y.; Shi, Y. Orthogonal Learning Particle Swarm Optimization. *IEEE Trans. Evol. Comput.* **2011**, *15*, 832–847. [CrossRef]
52. Lynn, N.; Suganthan, P. Heterogeneous comprehensive learning particle swarm optimization with enhanced exploration and exploitation. *Swarm Evol. Comput.* **2015**, *24*, 11–24. [CrossRef]
53. Gong, Y.-J.; Li, J.-J.; Zhou, Y.; Li, Y.; Chung, H.S.-H.; Shi, Y.-H.; Zhang, J. Genetic Learning Particle Swarm Optimization. *IEEE Trans. Cybern.* **2016**, *46*, 2277–2290. [CrossRef] [PubMed]
54. Osaba, E.; Yang, X.-S. Applied Optimization and Swarm Intelligence: A Systematic Review and Prospect Opportunities. In *Applied Optimization and Swarm Intelligence*; Osaba, E., Yang, X.-S., Eds.; Springer: Singapore, 2021; pp. 1–23. [CrossRef]
55. Puurtinen, M.; Heap, S.; Mappes, T. The joint emergence of group competition and within-group cooperation. *Ethol. Sociobiol.* **2015**, *36*, 211–217. [CrossRef]
56. Wu, G.; Mallipeddi, R.; Suganthan, P.N. *Problem Definitions and Evaluation Criteria for the CEC 2017 Competition on Constrained Real-Parameter Optimization*; Technical Report; National University of Defense Technology: Changsha, China; Kyungpook National University: Daegu, Korea; Nanyang Technological University: Singapore, 2017; pp. 1–16.
57. Liu, W.; Wang, Z.; Yuan, Y.; Zeng, N.; Hone, K.; Liu, X. A Novel Sigmoid-Function-Based Adaptive Weighted Particle Swarm Optimizer. *IEEE Trans. Cybern.* **2021**, *51*, 1085–1093. [CrossRef]
58. Xie, H.-Y.; Yang, Q.; Hu, X.-M.; Chen, W.-N. Cross-generation Elites Guided Particle Swarm Optimization for large scale optimization. In Proceedings of the Symposium Series on Computational Intelligence, Athens, Greece, 6–9 December 2016; pp. 1–8.
59. Gong, Y.-j.; Zhang, J. Small-world Particle Swarm Optimization with Topology Adaptation. In Proceedings of the Annual Conference on Genetic and Evolutionary Computation, Amsterdam, The Netherlands, 6–10 July 2013; pp. 25–32.

60. Xu, G.; Zhao, X.; Wu, T.; Li, R.; Li, X. An Elitist Learning Particle Swarm Optimization with Scaling Mutation and Ring Topology. *IEEE Access* **2018**, *6*, 78453–78470. [CrossRef]
61. Zeng, N.; Wang, Z.; Liu, W.; Zhang, H.; Hone, K.; Liu, X. A Dynamic Neighborhood-Based Switching Particle Swarm Optimization Algorithm. *IEEE Trans. Cybern.* **2020**, 1–12. [CrossRef] [PubMed]
62. Tanweer, M.; Suresh, S.; Sundararajan, N. Dynamic mentoring and self-regulation based particle swarm optimization algorithm for solving complex real-world optimization problems. *Inf. Sci.* **2016**, *326*, 1–24. [CrossRef]
63. Shi, Y.; Liu, H.; Gao, L.; Zhang, G. Cellular particle swarm optimization. *Inf. Sci.* **2011**, *181*, 4460–4493. [CrossRef]
64. Tao, X.; Guo, W.; Li, X.; He, Q.; Liu, R.; Zou, J. Fitness peak clustering based dynamic multi-swarm particle swarm optimization with enhanced learning strategy. *Expert Syst. Appl.* **2020**, 116301. [CrossRef]
65. Shen, Y.; Wei, L.; Zeng, C.; Chen, J. Particle Swarm Optimization with Double Learning Patterns. *Comput. Intell. Neurosci.* **2016**, *2016*, 1–19. [CrossRef]
66. Lin, A.; Sun, W. Multi-Leader Comprehensive Learning Particle Swarm Optimization with Adaptive Mutation for Economic Load Dispatch Problems. *Energies* **2019**, *12*, 116. [CrossRef]
67. Wang, Z.-J.; Zhan, Z.-H.; Kwong, S.; Jin, H.; Zhang, J. Adaptive Granularity Learning Distributed Particle Swarm Optimization for Large-Scale Optimization. *IEEE Trans. Cybern.* **2021**, *51*, 1175–1188. [CrossRef] [PubMed]
68. Feng, Q.; Li, Q.; Wang, H.; Feng, Y.; Pan, Y. Two-Stage Adaptive Constrained Particle Swarm Optimization Based on Bi-Objective Method. *IEEE Access* **2020**, *8*, 150647–150664. [CrossRef]
69. Wang, R.; Hao, K.; Chen, L.; Wang, T.; Jiang, C. A novel hybrid particle swarm optimization using adaptive strategy. *Inf. Sci.* **2021**, *579*, 231–250. [CrossRef]
70. Song, G.-W.; Yang, Q.; Gao, X.-D.; Ma, Y.-Y.; Lu, Z.-Y.; Zhang, J. An Adaptive Level-Based Learning Swarm Optimizer for Large-Scale Optimization. In Proceedings of the International Conference on Systems, Man, and Cybernetics, Melbourne, Australia, 17–20 October 2021; pp. 152–159. [CrossRef]
71. Zhan, Z.; Zhang, J.; Li, Y.; Chung, H.S. Adaptive Particle Swarm Optimization. *IEEE Trans. Syst. Man Cybern.* **2009**, *39*, 1362–1381. [CrossRef]
72. Tao, X.; Li, X.; Chen, W.; Liang, T.; Li, Y.; Guo, J.; Qi, L. Self-Adaptive two roles hybrid learning strategies-based particle swarm optimization. *Inf. Sci.* **2021**, *578*, 457–481. [CrossRef]
73. Sun, W.; Lin, A.; Yu, H.; Liang, Q.; Wu, G. All-dimension neighborhood based particle swarm optimization with randomly selected neighbors. *Inf. Sci.* **2017**, *405*, 141–156. [CrossRef]
74. Xia, X.; Gui, L.; He, G.; Wei, B.; Zhang, Y.; Yu, F.; Wu, H.; Zhan, Z.-H. An expanded particle swarm optimization based on multi-exemplar and forgetting ability. *Inf. Sci.* **2020**, *508*, 105–120. [CrossRef]
75. Zhang, X.; Liu, H.; Zhang, T.; Wang, Q.; Wang, Y.; Tu, L. Terminal crossover and steering-based particle swarm optimization algorithm with disturbance. *Appl. Soft Comput.* **2019**, *85*, 105841. [CrossRef]

information

MDPI

Article

Learning Static-Adaptive Graphs for RGB-T Image Saliency Detection

Zhengmei Xu [1,†], Jin Tang [2,†], Aiwu Zhou [3,†] and Huaming Liu [1,*]

1 School of Computer and Information Engineering, Fuyang Normal University, Fuyang 236037, China; 200107032@fynu.edu.cn
2 Key Laboratory of Intelligent Computing and Signal Processing of Ministry of Education, Anhui University, Hefei 230601, China; tj@ahu.edu.cn
3 School of Computer Science and Technology, Anhui University, Hefei 230601, China; jyang@iim.ac.cn
* Correspondence: 200806004@fynu.edu.cn
† These authors contributed equally to this work.

Abstract: Many works have been proposed on image saliency detection to handle challenging issues including low illumination, cluttered background, low contrast, and so on. Although good performance has been achieved by these algorithms, detection results are still poor based on RGB modality. Inspired by the recent progress of multi-modality fusion, we propose a novel RGB-thermal saliency detection algorithm through learning static-adaptive graphs. Specifically, we first extract superpixels from the two modalities and calculate their affinity matrix. Then, we learn the affinity matrix dynamically and construct a static-adaptive graph. Finally, the saliency maps can be obtained by a two-stage ranking algorithm. Our method is evaluated on RGBT-Saliency Dataset with eleven kinds of challenging subsets. Experimental results show that the proposed method has better generalization performance. The complementary benefits of RGB and thermal images and the more robust feature expression of learning static-adaptive graphs create an effective way to improve the detection effectiveness of image saliency in complex scenes.

Keywords: RGB-thermal; static-adaptive graph; manifold ranking; saliency detection

Citation: Xu, Z.; Tang, J.; Zhou, A.; Liu, H. Learning Static-Adaptive Graphs for RGB-T Image Saliency Detection. *Information* **2022**, *13*, 84. https://doi.org/10.3390/info13020084

Academic Editor: Gholamreza Anbarjafari

Received: 25 December 2021
Accepted: 10 February 2022
Published: 12 February 2022

1. Introduction

Image saliency detection aims to quickly capture the most important and useful information from a scene by using the human visual attention mechanism, which can reduce the complexity of subsequent image processing, and has been applied to numerous vision problems including image classification [1], image retrieval [2], image encryption [3,4], video summary [5], and so on. In the past few decades, researchers have proposed many saliency detection algorithms, which can be divided into bottom-up data-driven models and top-down task-driven methods. Bottom-up models [6–9] take the underlying image features and some priors into consideration, such as color, texture, orientation, and brightness. Itti et al. [10] proposed a visual attention mechanism, which opened research on saliency detection in the field of computer vision. Cheng et al. [11] introduced a regional contrast-based salient object detection algorithm, which simultaneously evaluates global contrast differences and spatial weighted coherence scores. Wang et al. [12] improved the detection effect of image saliency by optimizing seeds. Top-down models [13,14] are task driven. They use a large amount of training data with category labels and supervised learning to conduct a task-oriented analysis. Recently, most of these methods are based on deep learning, they have better performance, but their training processes are time-consuming. We focus on the bottom-up models. Many scholars have made many attempts to improve image saliency detection and have obtained good performance in simple scenes. However, the effectiveness of traditional RGB saliency detection methods decreases sharply in complex scenes, such as poor lighting or saliency objects that have the same color and texture as the background.

In recent years, multi-source sensor technology has become popular in image processing. Li et al. [15–17] simultaneously extracted RGB and thermal features for tracking, which effectively improved the effect of video target tracking at night or in rainy, hazy, and foggy weather. Zhang et al. [18] extracted the depth features of RGB images and thermal images, and then fused the two extracted features for saliency detection, which greatly improved the detection effectiveness in the case of poor illumination or similar color and texture to the background. The fusion of RGB and thermal images is proven to be effective in image saliency detection. RGB images can provide texture details with high definition in a manner consistent with the human visual system in simple scenes. By contrast, thermal images can work well in low illumination, and also have good discrimination when the target and the background have similar colors or shapes RGB-T saliency detection algorithms can obtain better results by handling challenging issues including low illumination, cluttered background, low contrast, and so on. Graph-based models [19–21] use pixels or superpixels as nodes and the similarity weight between nodes as the edge to generate the graph, which can achieve a great structure character from initial input images for RGB-T saliency detection results. However, the existing graph-based fusion models only use the static graph. The limitation of this kind of method is that it cannot explore the relationship between nodes at the target level and gain better fusion of multi-modality information. Inspired by these methods, we consider the spatial connectivity feature of graph nodes to learn a static-adaptive graph, and propose a novel RGB-thermal saliency detection algorithm to obtain more effective results, as in Figure 1.

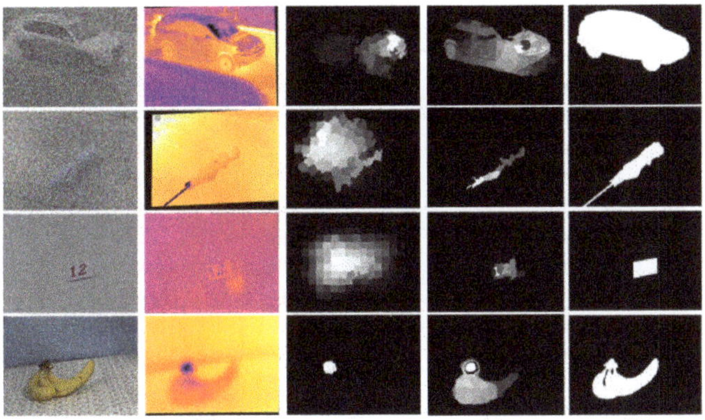

Figure 1. Comparative results of static-adaptive graph-based method with traditional static graph model. (**a**) RGB image; (**b**) thermal image; (**c**) the saliency map generated by the static graph-based model; (**d**) the saliency map generated by our model; (**e**) ground truth.

Specifically, we first extract superpixels from the two modalities and calculate their affinity matrix. Then, we learn the affinity matrix dynamically and construct a static-adaptive graph. Finally, the saliency maps can be obtained by a two-stage ranking algorithm. The contributions of this paper are summarized as follows.

- We construct an adaptive graph by sparse representation and carry out the optimization solution;
- We learn a novel static-adaptive graph model to increase the fusion capacity by considering the spatial connectivity features of graph nodes in RGB-T saliency detection;
- We compare our method with the state-of-the-art methods on an RGB-T dataset with 11 kinds of challenging subsets. The experimental results verify the effectiveness of our method.

2. Related Work

In this section, we give a brief review of methods closely related to our work. The relevant work in this paper mainly includes the graph-based method, multi-modality fusion method, and subspace-based method.

Graph-based method. In the past few decades, graph-based models have been successfully used for saliency detection problems. Harel et al. [19] proposed a graph model. The algorithm takes the pixel points as the graph nodes, constructs edges between the pixel points according to the spatial distance and feature distance, and uses Markov random field for feature fusion. Yang et al. [20] proposed a manifold ranking algorithm based on a static graph, which is a typical two-stage model to gain more accurate saliency maps. Jiang et al. [21] calculated a preliminary saliency map by Markov absorption probability on a weighted graph via partial image borders as prior background. Zhang et al. [22] used multi-scale to improve the manifold ranking algorithm. Xiao et al. [23] proposed a prior regularized multi-layer graph ranking model in which they used the prior calculating by boundary connectivity. Aytekin et al. [24] proposed a graph model that uses a convolution kernel function network to learn the connection weight coefficients.

Multi-modality fusion method. In recent years, with the development of multi-sensors, multi-modality fusion has become a new effective means to improve computer vision tasks. Li et al. [25] combined gray and thermal information to deal with target tracking in complex scenes. Li et al. [15] used multispectral (RGB and thermal) data to improve visual tracking effectiveness. Li et al. [26] established a unified RGB-T dataset and proposed a new algorithm to fuse RGB and thermal images for saliency detection, which incorporates the cross-modality consistent constraints to integrate different modalities collaboratively. RGB-D is an effective multi-modal fusion method in many aspects, such as manufacturing [27], semantic segmentation [28–30], and saliency detection [31,32]. Liu et al. [33] used three transformer encoders with shared weights to enhance multi-level features, and the algorithm they proposed effectively improves the effectiveness of saliency detection.

Subspace-based method. Subspace-based methods represent high-dimensional data in low-dimensional subspace. The purpose of subspace representation is to obtain a similarity matrix in the basic subspace of the original data. In a dataset, each data point can be reconstructed by an effective combination of other points, which are often helpful for data processing, because data can better reflect the characters of data in its low-dimensional subspace. Guo et al. [34] proposed a subspace segmentation method to jointly learn data representation and affinity matrix relationships simultaneously in a model. Li et al. [35] represented each patch with a linear combination of the remaining ones and learned the weights of the global and local features of the detection object, achieving good effectiveness in the application field of video tracking.

We learn static-adaptive graphs for saliency detection. The static graph is the traditional graph. Its structure is fixed, and it only considers the relationship between adjacent nodes. The adaptive graph is obtained by the subspace method to mine the internal relationship between superpixels. Therefore, our algorithm considers both local and global features, and has better effectiveness than the saliency detection algorithm which is only based on the static graph. In multi-modality selection, we fuse RGB image and thermal image, because RGB and thermal images have natural complementarity. Compared with the RGB-D saliency detection algorithm, the RGB-T saliency detection algorithm has much lower hardware requirements for computers, and can run well on computer with an i3 3.3G CPU and 4GB RAM.

3. Brief Review of Manifold Ranking

A manifold ranking (MR) model [20] is a typical graph-based method for saliency detection. For an image, simple linear iterative clustering (SLIC) [36] is always used to obtain n superpixels as graph nodes in most of these models. Take a graph $G = (V, E)$, where V is a node set. Some of nodes are labeled as queries and the rest need to be ranked according to their relevance to the queries. Let $\mathbf{X} = [x_1, x_2, \ldots, x_n] \in \mathbb{R}^{d \times n}$ be the character matrix, d the dimensionality of the feature vector, and n the number of the superpixels. E is the set of undirected edges and $\mathbf{W}_{ij}$ is the edge weight between node i and node j that can be calculated by feature vectors of two nodes. Let $\mathbf{q} = [q_1, q_2, \ldots, q_n]^T$ denote an indication vector, where $q_i = 1$ if node i is a labeled query, otherwise, $q_i = 0$. The aim of MR is to gain a ranking value f_i for each graph node, which can be computed by solving Equation (1),

$$\min_f \frac{1}{2}\left(\sum_{i,j=1}^{n} \mathbf{W}_{ij} \left\| \frac{f_i}{\sqrt{\mathbf{D}_{ii}}} - \frac{f_j}{\sqrt{\mathbf{D}_{jj}}} \right\|^2 + \mu \sum_{i=1}^{n} \|f_i - q_i\|^2\right) \tag{1}$$

where $\mathbf{D}_{ii} = \sum_{j=1}^{n} \mathbf{W}_{ij}$.

To obtain more effective results, Yang et al. [20] obtained the ranking value by using the un-normalized Laplacian matrix in Equation (2),

$$\mathbf{f} = (\mathbf{D} - \lambda \mathbf{W})^{-1}\mathbf{q}, \tag{2}$$

where $\mathbf{D}$ is a degree matrix, $\mathbf{D} = diag\{\mathbf{D}_{11}, \ldots, \mathbf{D}_{nn}\}$, $\lambda = 1/(1+\mu)$.

4. Static-Adaptive Graph Learning

4.1. Static-Adaptive Graph Construction

The graph of traditional models is static; most of them only consider adjacent nodes and boundary nodes. The limitation of this kind of method is that it cannot explore the relationship between nodes at the target level. Therefore, we consider the spatial connectivity features of graph nodes to construct a static-adaptive graph, in which superpixels with similar features in the region are also connected. Take multiple graphs $G^m = (V^m, E^m), m = 1, 2, \ldots, M$, where V^m is a node set, and E^m is the set of undirected edges. Let $\mathbf{X}^m = [x_1^m, x_2^m, \ldots, x_N^m] \in \mathbb{R}^{d \times N}, m = 1, 2, \ldots, M$ be the character matrix of the m-th modality. N is the number of graph nodes. d is the dimensionality of the feature vector. As in traditional static graphs [20], when two nodes meet one of the following three conditions, they are considered to have edges.

(1) Two nodes are directly adjacent;
(2) There is a common edge between two nodes;
(3) Superpixels are on the four boundaries.

If there is an edge between two nodes, the weight of the edge is calculated by Equation (3).

$$\mathbf{W}_{i,j}^m = e^{-\gamma_0 \|x_i^m - x_j^m\|}, m = 1, 2, \ldots, M, \tag{3}$$

where x_i^m denotes the mean of the i-th superpixel in the m-th modality, and γ_0 is a parameter.

We add the adaptive graph weight matrix to gain the weight matrix of the static-adaptive graph as in Figure 2, which can be calculated by Equation (4).

$$\mathbf{W} = \mathbf{W}^a + \sum_{m=1}^{M} t_m \mathbf{W}^m, \tag{4}$$

where $\mathbf{W}^a$ is the weight matrix of adaptive graph, which can be obtained by adaptive graph learning.

$\mathbf{W}^m = [\mathbf{W}_{ij}^m]_{N \times N}, m = 1, 2, \ldots, M$ is the initial weight matrix of the m-th modality. t_m can indicate the importance of different modalities of static and adaptive graphs.

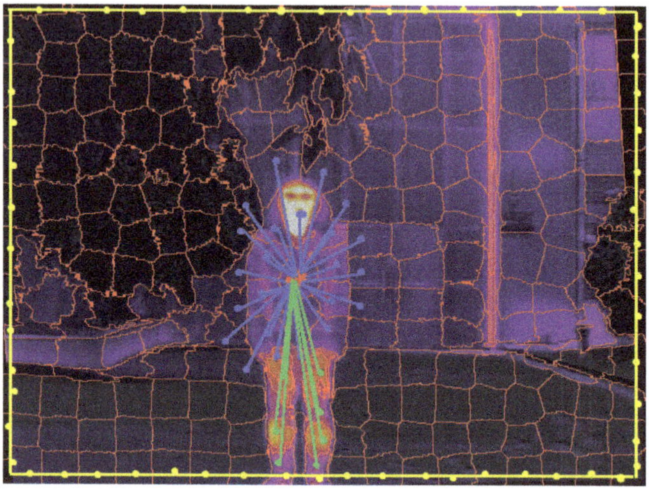

Figure 2. The general view of the static-adaptive graph on the multi-modality fusion image. The blue edges are obtained by the traditional static graph. The green edges are obtained by our adaptive graph learning model.

4.2. Adaptive Graph Learning Model Formulation

For M graphs $G^m = (V^m, E^m), m = 1, 2, \ldots, M$, we assume that all nodes in each graph belong to the same sparse subspace, in which each node can be sparsely represented by the remaining nodes. We can obtain $\mathbf{X}^m = \mathbf{X}^m\mathbf{Z}^m, m = 1, 2, \ldots, M$, where $\mathbf{Z}^m \in \mathbb{R}^{N\times N}$ is the sparse coefficient matrix. Sparse constraints can automatically select most informative neighbor nodes for each node, and make the graph more powerful. Since the nodes are often disturbed by noises, we introduce a noise matrix $\mathbf{E}^m \in \mathbb{R}^{d\times N}$ to improve the robustness. The joint sparse representation with the convex relaxation for all modalities can be written as,

$$\min_{\mathbf{Z},\mathbf{E}^m} \alpha\|\mathbf{Z}\|_1 + \beta\sum_{m=1}^{M}\|\mathbf{E}^m\|_{2,1}, \;\; s.t.\; \mathbf{X}^m = \mathbf{X}^m\mathbf{Z}^m + \mathbf{E}^m. \tag{5}$$

where α and β are balanced parameters. $\mathbf{Z} = [\mathbf{Z}^1; \cdots; \mathbf{Z}^M] \in \mathbb{R}^{N\times(M*N)}$ is the joint sparse representation coefficient matrix.

We consider the spatial connectivity feature of graph nodes and use $\mathbf{C} \in \mathbb{R}^{N\times N}$ to indicate the spatial connections of neighboring nodes.

If node i and j are 8-neighboring, $\mathbf{C}_{ij} = 1$; otherwise $\mathbf{C}_{ij} = 0$.

$$\mathbf{C}_{ij} = \begin{cases} 1, & if\ i\ and\ j\ are\ 8\text{-}neighboring, \\ 0, & else. \end{cases} \tag{6}$$

The closer the distance, the greater the relevance. Inspired by [35], to capture the global and local structure information, we employ Equation (7) to learn the adaptive graph affinity matrix.

$$\begin{aligned} &\min_{\mathbf{W}^a} \frac{\gamma}{2}\sum_{i,j=1}^{N}\|\mathbf{Z}_i - \mathbf{Z}_j\|_F^2\mathbf{W}^a_{ij} + \frac{\delta}{2}\sum_{i,j=1}^{N}\mathbf{C}_{ij}\|\mathbf{Z}_i - \mathbf{Z}_j\|_F^2 + \lambda_1\|\mathbf{W}^a\|_F^2, \\ &s.t.\; {\mathbf{W}^a}^T\mathbf{1} = \mathbf{1}, \mathbf{W}^a \geq \mathbf{0}. \end{aligned} \tag{7}$$

where γ and δ are the balancing parameters. The first item reflects the probability $\mathbf{W}^a_{ij}$ from the same cluster based on the distance between their representations $\mathbf{Z}_i$ and $\mathbf{Z}_j$. The second item indicates that two close nodes will have similar representations. $\lambda_1\|\mathbf{W}^a\|_F^2$ is to avoid over-fitting of $\mathbf{W}^a$. $\mathbf{1}$ denotes a unit vector. $\mathbf{W}^{a^T}\mathbf{1} = \mathbf{1}, \mathbf{W}^a \geq \mathbf{0}$ are constraints to guarantee the probability property of $\mathbf{W}^a_{ij}$. We combine the Equations (5) and (7) and obtain the following optimal function,

$$\begin{aligned} &\min_{\mathbf{Z},\mathbf{E}^m,\mathbf{W}^a} \alpha\|\mathbf{Z}\|_1 + \frac{\gamma}{2}\sum_{i,j=1}^{N}\|\mathbf{Z}_i - \mathbf{Z}_j\|_F^2 \mathbf{W}^a_{ij} + \frac{\delta}{2}\sum_{i,j=1}^{N}\mathbf{C}_{ij}\|\mathbf{Z}_i - \mathbf{Z}_j\|_F^2, \\ &+ \lambda_1\|\mathbf{W}^a\|_F^2 + \beta\sum_{m=1}^{M}\|\mathbf{E}^m\|_{2,1}, \\ &s.t.\ \mathbf{X}^m = \mathbf{X}^m\mathbf{Z}^m + \mathbf{E}^m, \mathbf{W}^{a^T}\mathbf{1} = \mathbf{1}, \mathbf{W}^a \geq \mathbf{0}. \end{aligned} \tag{8}$$

In order to solve the problem easily, let $\mathbf{D}^a_{ii} = \sum_{j=1}^{N}\mathbf{W}^a_{ij}$, $\mathbf{D}^c_{ii} = \sum_{j=1}^{N}\mathbf{C}_{ij}$. Equation (8) is a slightly algebraic transformation to,

$$\begin{aligned} &\min_{\mathbf{Z},\mathbf{E}^m,\mathbf{W}^a} \alpha\|\mathbf{Z}\|_1 + \gamma\, tr(\mathbf{Z}\mathbf{L}^a\mathbf{Z}^T) + \delta\, tr(\mathbf{Z}\mathbf{L}^c\mathbf{Z}^T) \\ &+ \lambda_1\|\mathbf{W}^a\|_F^2 + \beta\sum_{m=1}^{M}\|\mathbf{E}^m\|_{2,1}, \\ &s.t.\ \mathbf{X}^m = \mathbf{X}^m\mathbf{Z}^m + \mathbf{E}^m, \mathbf{W}^{a^T}\mathbf{1} = 1, \mathbf{W}^a \geq \mathbf{0}. \end{aligned} \tag{9}$$

where $\mathbf{L}^a = \mathbf{D}^a - \mathbf{W}^a$ and $\mathbf{L}^c = \mathbf{D}^c - \mathbf{C}$ are Laplacian matrices of $\mathbf{W}^a$ and $\mathbf{C}$, respectively.

4.3. Optimization

The variables in Equation (9) are not jointly convex; they are convex with respect to the subproblem of each variable when others are fixed and have a close form solution. We introduce two auxiliary variables, $\mathbf{P}^m$ and $\mathbf{Q}^m$, to make Equation (9) separable and then use the alternating direction multiplier (ADMM) algorithm [37] for optimization iteration. Then, we can obtain Equation (10).

$$\begin{aligned} &\min_{\mathbf{Z},\mathbf{E}^m,\mathbf{W}^a} \alpha\|\mathbf{Z}\|_1 + \gamma\, tr(\mathbf{Z}\mathbf{L}^a\mathbf{Z}^T) + \delta\, tr(\mathbf{Z}\mathbf{L}^c\mathbf{Z}^T) \\ &+ \lambda_1\|\mathbf{W}^a\|_F^2 + \beta\sum_{m=1}^{M}\|\mathbf{E}^m\|_{2,1}, \\ &s.t.\ \mathbf{P}^m = \mathbf{Z}^m, \mathbf{Q}^m = \mathbf{Z}^m, \mathbf{X}^m = \mathbf{X}^m\mathbf{Z}^m + \mathbf{E}^m, \mathbf{W}^{a^T}\mathbf{1} = 1, \mathbf{W}^a \geq \mathbf{0}. \end{aligned} \tag{10}$$

Thus, we obtain the Lagrange function [38] as Equation (11),

$$\begin{aligned} &\min_{\mathbf{Z},\mathbf{E}^m,\mathbf{W}^a,\mathbf{P},\mathbf{Q}} \alpha\|\mathbf{Q}\|_1 + \gamma\, tr(\mathbf{P}\mathbf{L}^a\mathbf{P}^T) + \delta\, tr(\mathbf{P}\mathbf{L}^c\mathbf{P}^T) + \lambda_1\|\mathbf{W}^a\|_F^2 \\ &+ \sum_{m=1}^{M}(\beta\|\mathbf{E}^m\|_{2,1} + \frac{\mu}{2}\|\mathbf{X}^m - \mathbf{X}^m\mathbf{Z}^m - \mathbf{E}^m + \frac{\mathbf{Y}_1^m}{\mu}\|_F^2 + \frac{\mu}{2}\|\mathbf{P}^m - \mathbf{Z}^m + \frac{\mathbf{Y}_2^m}{\mu}\|_F^2 \\ &+ \frac{\mu}{2}\|\mathbf{Q}^m - \mathbf{Z}^m + \frac{\mathbf{Y}_3^m}{\mu}\|_F^2 - \frac{1}{2\mu}(\|\mathbf{Y}_1^m\|_F^2 + \|\mathbf{Y}_2^m\|_F^2 + \|\mathbf{Y}_3^m\|_F^2)). \end{aligned} \tag{11}$$

where $\mathbf{P} = [\mathbf{P}^1; \mathbf{P}^2; \ldots; \mathbf{P}^M]$ and $\mathbf{Q} = [\mathbf{Q}^1; \mathbf{Q}^2; \ldots; \mathbf{Q}^M]$. μ is a penalty parameter; $\mathbf{Y}_1^m$, $\mathbf{Y}_2^m$, and $\mathbf{Y}_3^m$ are Lagrange multipliers.

There are five variables, $\mathbf{Z}$, $\mathbf{E}^m$, $\mathbf{W}^a$, $\mathbf{P}$, and $\mathbf{Q}$, needed to solve in Equation (11), The solver iteratively updates one variable at a time by fixing the others.

Z-subproblem: In order to calculate $\mathbf{Z}$, we fix other variables in Equation (11); the **Z**-subproblem can be written as Equation (12). Then, we divide $\mathbf{Z}$ and set it to 0 to obtain Equation (13),

$$\begin{aligned} &\min_{\mathbf{Z}} \sum_{m=1}^{M}(\frac{\mu}{2}\|\mathbf{X}^m - \mathbf{X}^m\mathbf{Z}^m - \mathbf{E}^m + \frac{\mathbf{Y}_1^m}{\mu}\|_F^2 \\ &+ \frac{\mu}{2}\|\mathbf{P}^m - \mathbf{Z}^m + \frac{\mathbf{Y}_2^m}{\mu}\|_F^2 + \frac{\mu}{2}\|\mathbf{Q}^m - \mathbf{Z}^m + \frac{\mathbf{Y}_3^m}{\mu}\|_F^2), \end{aligned} \tag{12}$$

$$\begin{aligned}\mathbf{Z}^{m,k+1} &= (\mu(\mathbf{X}^m)^T\mathbf{X}^m + 2\mu_k\mathbf{I})^{-1}(\mu_k(\mathbf{X}^m)^T\mathbf{X}^m \\ &- \mu_k(\mathbf{X}^m)^T\mathbf{E}^{m,k} + (\mathbf{X}^m)^T\mathbf{Y}_1^{m,k} + \mu_k\mathbf{P}^{m,k} + \mu_k\mathbf{Q}^{m,k} - \mathbf{Y}_2^{m,k} - \mathbf{Y}_3^{m,k}).\end{aligned} \tag{13}$$

P-subproblem: In order to calculate **P**, we fix other variables in Equation (11); the **P**-subproblem can be written as Equations (14) and (15), then dividing **P** and setting it to 0 to obtain Equation (16),

$$\min_{\mathbf{P}} \gamma\, tr(\mathbf{P}\mathbf{L}^a\mathbf{P}^T) + \delta\, tr(\mathbf{P}\mathbf{L}^c\mathbf{P}^T) + \sum_{m=1}^{M} \frac{\mu}{2}\|\mathbf{P}^m - \mathbf{Z}^m + \frac{\mathbf{Y}_2^m}{\mu}\|_F^2, \tag{14}$$

$$\min_{\mathbf{P}} \gamma\, tr(\mathbf{P}\mathbf{L}^a\mathbf{P}^T) + \delta\, tr(\mathbf{P}\mathbf{L}^c\mathbf{P}^T) + \frac{\mu}{2}\|\mathbf{P} - \mathbf{Z} + \frac{\mathbf{Y}_2}{\mu}\|_F^2, \tag{15}$$

$$\mathbf{P}^{k+1} = (\mu\mathbf{Z}^{k+1} - \mathbf{Y}_2^k)(\gamma(\mathbf{L}^a)^k + \gamma((\mathbf{L}^a)^k)^T + \delta(\mathbf{L}^a)^k + \delta((\mathbf{L}^c)^k)^T + \mu\mathbf{I})^{-1}. \tag{16}$$

Q-subproblem: In order to calculate **Q**, we fix other variables in Equation (11), then the **Q**-subproblem can be written as Equations (17) and (18). Then, divide **Q** and set it to 0, which is computed by the soft-thresholding (or shrinkage) method [39] as Equation (19),

$$\min_{\mathbf{Q}} \alpha\|\mathbf{Q}\|_1 + \sum_{m=1}^{M} \frac{\mu}{2}\|\mathbf{Q}^m - \mathbf{Z}^m + \frac{\mathbf{Y}_3^m}{\mu}\|_F^2, \tag{17}$$

$$\min_{\mathbf{Q}} \alpha\|\mathbf{Q}\|_1 + \frac{\mu}{2}\|\mathbf{Q} - \mathbf{Z} + \frac{\mathbf{Y}_3}{\mu}\|_F^2, \tag{18}$$

$$\mathbf{Q}^{k+1} = soft\_thr(\mathbf{Z}^{k+1} - \frac{\mathbf{Y}_3^k}{\mu}, \frac{\alpha}{\mu_k}). \tag{19}$$

$\mathbf{E}^m$-subproblem: In order to calculate $\mathbf{E}^m$, we fix other variables in Equation (11); then the **Q**-subproblem can be written as Equation (20). Then, by dividing **E** and setting it to 0, which is computed by the soft-thresholding (or shrinkage) method [39], we obtain Equation (21),

$$\min_{\mathbf{E}^m} \sum_{m=1}^{M} \beta\|\mathbf{E}^m\|_{2,1} + \frac{\mu}{2}\|\mathbf{X}^m - \mathbf{X}^m\mathbf{Z}^m - \mathbf{E}^m + \frac{\mathbf{Y}_1^m}{\mu}\|_F^2 \tag{20}$$

$$\mathbf{E}^{m,k+1} = S_{\frac{\beta}{\mu}}(\mathbf{X}^m\mathbf{Z}^{m,k+1} - \mathbf{X}^m - \frac{\mathbf{Y}_1^{m,k}}{\mu_k}) \tag{21}$$

$\mathbf{W}^a$-subproblem: In order to calculate $\mathbf{W}^a$, we fix other variables in Equation (11), then the $\mathbf{W}^a$-subproblem can be written as Equation (22). Then dividing $\mathbf{W}^a$ and set it to 0 obtains Equation (23),

$$\min_{\mathbf{W}^a} \gamma tr(\mathbf{P}\mathbf{L}^a\mathbf{P}^T) + \lambda_1\|\mathbf{W}^a\|_F^2 + \gamma\|\mathbf{P}_i - \mathbf{P}_j\|_F^2\mathbf{W}_{ij}^a \tag{22}$$

$$(\mathbf{W}^a)_i^{k+1} = (\frac{1 + \sum_{j=1}^{s}\hat{\mathbf{U}}_j}{s}\mathbf{1} - \mathbf{U}_{ij})_+ \tag{23}$$

where $\mathbf{U}_j \in \mathbb{R}^{N\times 1}$ is a vector whose i-th element is $\mathbf{U}_{ij} = \frac{\gamma\|\mathbf{P}_i - \mathbf{P}_j\|_F^2}{\lambda_1}$.

The Lagrange multiplier can be updated by Equation (24),

$$
\begin{aligned}
\mathbf{Y}_1^{m,k+1} &= \mathbf{Y}_1^{m,k} + \mu^k(\mathbf{X}^m - \mathbf{X}^m\mathbf{Z}^{m,k+1} - \mathbf{E}^{m,k+1}) \\
\mathbf{Y}_2^{m,k+1} &= \mathbf{Y}_2^{m,k} + \mu^k(\mathbf{Z}^{m,k+1} - \mathbf{P}^{m,k+1}) \\
\mathbf{Y}_3^{m,k+1} &= \mathbf{Y}_3^{m,k} + \mu^k(\mathbf{Z}^{m,k+1} - \mathbf{Q}^{m,k+1})
\end{aligned} \tag{24}
$$

5. RGB-T Salient Detection

Given a pair of RGB-T images, considering that the thermal image has stronger anti-interference ability in complex scenes, we first fuse the RGB and the thermal images at a ratio of 1:4. To generate N non-overlapping superpixels, we use a simple linear iterative clustering (SLIC) algorithm in the fused image. A two-stage ranking model is adapted to calculate the final saliency map. In the first stage, we take the boundary as a prior and select the nodes around the image as background seed queries. We use the top, bottom, left, and right sides of the image as four kinds of queries, $\mathbf{q}^t, \mathbf{q}^b, \mathbf{q}^l, \mathbf{q}^r$, which are selected separately to obtain four different detection results, $\mathbf{f}^t, \mathbf{f}^b, \mathbf{f}^l, \mathbf{f}^r$, by Equation (2). Considering that the symmetry of the image and saliency objects are often cross-left boundary and cross-bottom boundary, we select the large class nodes as queries by using the k-means method to obtain two clusters on the left and bottom boundaries separately. Then, we normalize $\mathbf{f}^k$ $(k = t, p, l, r)$ to the range between 0 and 1. The saliency value vector of N nodes $\mathbf{s}^k$ can be obtained by $\mathbf{s}^k = \mathbf{1} - \hat{\mathbf{f}}^k$ $(k = t, p, l, r)$. The saliency ranking value vector of all nodes $\mathbf{s}^1$ in the first stage can be calculated by Equation (25).

$$\mathbf{s}^1 = \mathbf{s}^t \times \mathbf{s}^b \times \mathbf{s}^l \times \mathbf{s}^r \tag{25}$$

By using the object characteristics, secondary ranking is performed to improve the first-stage saliency value. Given $\mathbf{s}^1$, we set an adaptive threshold to generate foreground as queries $\mathbf{q}_2$. Then, the Equation (2) is used to obtain the second ranking results $\mathbf{s}_2$, which are normalized to the range of 0 and 1 as $\hat{\mathbf{s}}_2$. In order to further reduce the background noise, we let $\mathbf{s} = \mathbf{s}^1 \times \mathbf{s}^2$ be the final saliency value and obtain the final salient map $\mathbf{S}$. The main steps of the two-stage RGB-T salient object detection algorithm are summarized in Algorithm 1.

Algorithm 1 The Static-Adaptive Graph based RGB-T Salient Detection Produce.

Require: The static-adaptive graph weight matrix $\mathbf{W}$, the indicator vectors of the four boundaries queries $\mathbf{q}^t, \mathbf{q}^b, \mathbf{q}^l, \mathbf{q}^r$.

1: Use Equation (2) to obtain $\mathbf{f}^t, \mathbf{f}^b, \mathbf{f}^l, \mathbf{f}^r$ separately;
2: $\mathbf{f}^t$, $\mathbf{f}^b$, $\mathbf{f}^l$ and $\mathbf{f}^r$ are normalized to 0 and 1;
3: Set $\mathbf{s}^t$=$\mathbf{1} - \hat{\mathbf{f}}^t$, $\mathbf{s}^b$=$\mathbf{1} - \hat{\mathbf{f}}^b$, $\mathbf{s}^l$=$\mathbf{1} - \hat{\mathbf{f}}^l$, $\mathbf{s}^r$=$\mathbf{1} - \hat{\mathbf{f}}^r$;
4: Obtain the first saliency value vector $\mathbf{s}^1 = \mathbf{s}^t \times \mathbf{s}^b \times \mathbf{s}^l \times \mathbf{s}^r$;
5: $\mathbf{s}^1$ is normalized to 0 and 1, and obtain $\hat{\mathbf{s}}^1$;
6: Use an adaptive threshold to binary $\hat{\mathbf{s}}^1$ and obtain foreground query $\mathbf{q}^2$;
7: Use Equation (2) to obtain the second saliency value vector $\mathbf{s}^2$;
8: $\mathbf{s}^2$ is normalized to 0 and 1 $\hat{\mathbf{s}}^2$;
9: Set $\mathbf{s} = \hat{\mathbf{s}}^1 \times \hat{\mathbf{s}}^2$ to suppress the background noise of image;
10: Set all superpixels value $\mathbf{s}_i$ to each pixel and obtain the final saliency map $\mathbf{S}$.

Ensure: $\mathbf{S}$ is the saliency map of the static-adaptive graph model for RGB-T saliency detection.

6. Experiment

6.1. Datasets and Experimental Settings

The RGBT-Saliency dataset [26] includes 821 pairs images with ground truth, in which the images with high diversity are recorded under different scenes and environmental conditions.

The datasets can be download from the address http://chenglongli.cn/people/lcl/journals.html (accessed on 20 December 2021).

The initial segmentation number of the superpixel N is set to 250. The edge weight coefficient θ is set to 29. Other parameters in this paper are set to $\alpha = 0.11$, $\beta = 0.15$, $\gamma = 0.04$, $\delta = 0.3$, and $\lambda_1 = 0.6$.

6.2. Measuring Standard

To verify the effectiveness of our algorithm, we compared with other methods with precision, recall, and F-measure (PRF) values, mean absolute error (MAE) values, and PR curve.

PR (*Precision, Recall*) curve. The PR curve is a curve with the "precision rate" as the ordinate and the "recall rate" as the abscissa. We binarize the original image S to obtain M, and then calculate the precision value and recall value by comparing M and G (ground truth) pixel by pixel in the following formula,

$$Precision = \frac{|M \cap G|}{|M|} \tag{26}$$

$$Recall = \frac{|M \cap G|}{|G|} \tag{27}$$

PRF (precision, recall, F-measure). Sometimes, the P and R indicators are contradictory, so they need to be considered comprehensively. The most common method is F-measure (also known as f-score). F-measure is the weighted average of precision and recall:

$$F_{\beta^2} = \frac{(1+\beta^2) \times P \times R}{\beta^2 \times P + R}, \tag{28}$$

where $\beta^2 = 0.3$.

MAE (mean absolute error). *MAE* is the direct calculation of the average absolute error between the salience map and the ground truth of the model output. It first binarizes them and then calculates them with the following formula:

$$MAE = \frac{1}{W \times H} \sum_{x=1}^{W} \sum_{y=1}^{H} |\overline{S}(x,y) - \overline{G}(x,y)| \tag{29}$$

where W is the width of the salient map S and the ground truth map G; H is the height of the salient map S and the ground truth map G.

6.3. Comparison Results

We compared our model with eight methods including BR [40], CA [41], MCI [42], NFI [43], SS-KDE [44], GMR [20], GR [45], and MTMR [26] on the RGBT-Saliency dataset.

We generated PR curves for 11 challenging subsets and the entire dataset, and listed their F values. The eleven subsets are eleven different challenges, which are: big salient object (BSO), bad weather (BW), center bias (CB), cross image boundary (CIB), image clutter (IC), low illumination (LI), multiple salient objects (MSO), out of focus (OF), similar appearance (SA), small salient object (SSO), and thermal crossover (TC). In Table 1, we describe in detail the division method of the eleven subsets [26].

Table 1. List of the 11 challenging subsets of RGBT-Saliency-Dataset.

Challenge	Description
BSO	The radio of ground truth salient objects over the image is more than 0.26.
BW	The image pairs are recorded in bad weather, such as snowy, rainy, hazy, or cloudy weather.
CB	The centers of salient objects are far away from the image center.
CIB	The salient objects cross the image boundaries.
IC	The image is cluttered.
LI	The environmental illumination is low.
MSO	The number of the salient objects in the image is more than one.
OF	The image is out of focus.
SA	The salient objects have similar color or shape to the background.
SSO	The radio of ground truth salient objects over the image is less the 0.05.
TC	The salient objects have similar temperature to the background.

As can be seen from Figure 3, only in the "BSO" and "CIB" subsets was our F-Measures slightly lower than the best detection result, and they were the best in the other nine subsets. Especially in the CB subset, the detection result has obvious advantages. Our detection curve has no crossover with other curves.

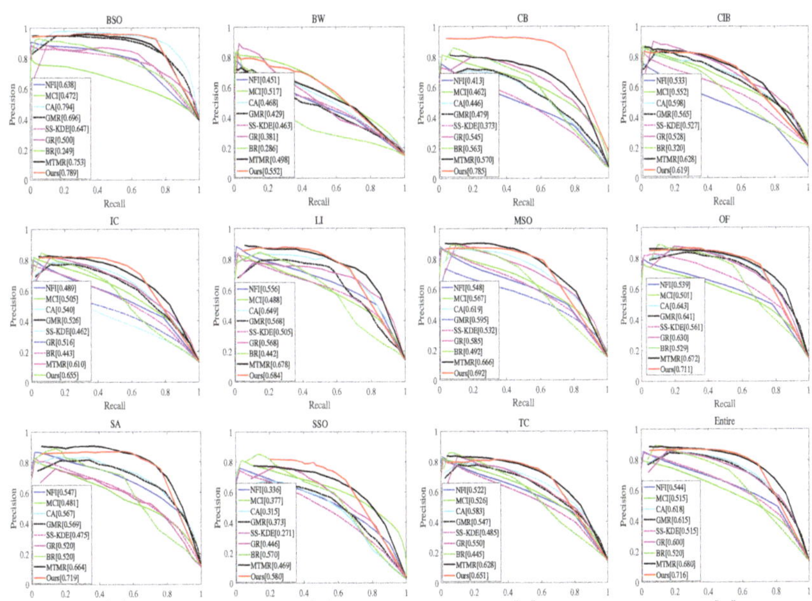

Figure 3. PR curves of the proposed approach with other baseline methods with RGB-T input on eleven subsets and the entire dataset. The $F_{0.3}$ values are shown in the legend.

The comparison results of the precision, recall, and F-measure values with other methods in different modalities as shown in Table 2. We only provide the detection results of MTMR [26] after multi-modality fusion because this model proposes to integrate multi-modal information and use multi-modal adaptive weights to detect image saliency objects. From the Table 2, we can see that the proposed algorithm is better than other methods in terms of P value and the comprehensive measure F-measure.

Table 2. Average precision (P), recall (R), F-measure (F) and mean absolute error (MAE) of our method against different kinds of methods on the RGBT-Saliency dataset. In the evaluation parameters, the larger the value of P, R, and F, the better the detection effect, while the smaller the value of MAE, the better the effect. The red font indicates the best performance. The green is second best.

Algorithm	RGB (P↑, R↑, F↑, MAE↓)	Thermal (P↑, R↑, F↑, MAE↓)	RGB-T (P↑, R↑, F↑, MAE↓)
BR [40]	0.724, 0.260, 0.411, 0.269	0.648, 0.413, 0.488, 0.323	0.804, 0.366, 0.520, 0.297
CA [41]	0.592, 0.667, 0.568, 0.163	0.623, 0.607, 0.573, 0.225	0.648, 0.697, 0.618, 0.195
MCI [42]	0.526, 0.604, 0.485, 0.211	0.445, 0.585, 0.435, 0.176	0.547, 0.652, 0.515, 0.195
NFI [43]	0.557, 0.639, 0.532, 0.126	0.581, 0.599, 0.541, 0.124	0.564, 0.665, 0.544, 0.125
SS-KDE [44]	0.581, 0.554, 0.532, 0.122	0.510, 0.635, 0.497, 0.132	0.528, 0.656, 0.515, 0.127
GMR [20]	0.644, 0.603, 0.587, 0.172	0.700, 0.574, 0.603, 0.232	0.694, 0.624, 0.615, 0.202
GR [45]	0.621, 0.582, 0.534, 0.197	0.639, 0.544, 0.545, 0.199	0.705, 0.593, 0.600, 0.199
MTMR [26]	-, -, -, -	-, -, -, -	0.716, 0.713, 0.680, 0.107
ours	0.697, 0.536, 0.603, 0.107	0.715, 0.569, 0.629, 0.112	0.804, 0.627, 0.716, 0.095

Sample Results. From the dataset, we extracted four photos with various challenges as the data source and compared the detection results of our algorithm with other algorithms for salient detection. It can be seen from the Figure 4 that our algorithm has a very robust detection effectiveness in challenging scenes such as fuzzy images, large targets, small targets, complex background, and center bias.

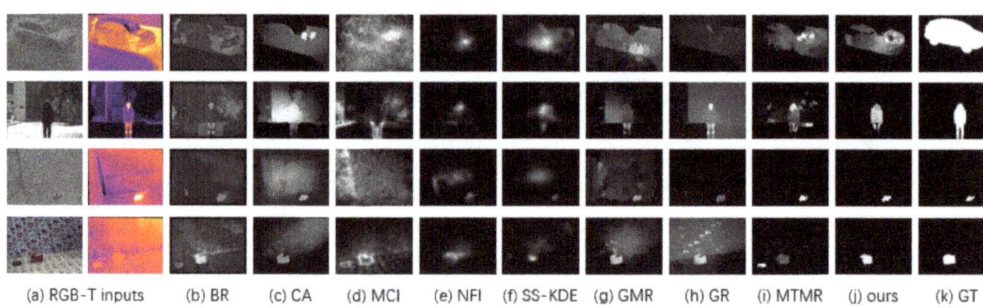

Figure 4. Sample results of the proposed approach and other baseline methods with the fusion of RGB and thermal inputs. (**a**) The first two columns are the origin RGB images and thermal images. (**b–i**) The results of the baseline methods with RGB and thermal inputs; (**j**) the result of our approach. (**k**) ground truth.

Runtime Results. All results were obtained on a Windows 10 64-bit operating system running Matlab 2014b with an i3 3.3G CPU and 4GB RAM. We compared the average running time with other algorithms in Table 3. Compared with the algorithm in [20], we spent more time mainly on the learning of the adaptive graph.

Table 3. Average runtime comparison on the RGBT-Saliency dataset.

Method	BR [40]	CA [41]	MCI [42]	NFI [43]	SS-KDE [44]	GMR [20]	GR [45]	MTMR [26]	Ours
Runtime(s)	21.95	3.13	58.37	33.16	2.51	2.96	6.48	3.71	5.18

6.4. Analysis of Our Approach

In our method, we compared the following four combinations of image salient detection results: (1) learning static-adaptive graphs for RGB image salient detection, called our1; (2) learning static-adaptive graph for thermal image salient detection, called our2; (3) not learning static-adaptive graphs and only fusing RGB and thermal image to detect the salient, called our3; (4) learning static-adaptive graphs for RGB-T image salient detection, called our4. It can be seen from Figure 5 that the fusion of multi-modality and the use of learning static-adaptive graphs are both effective methods to improve the salient detection.

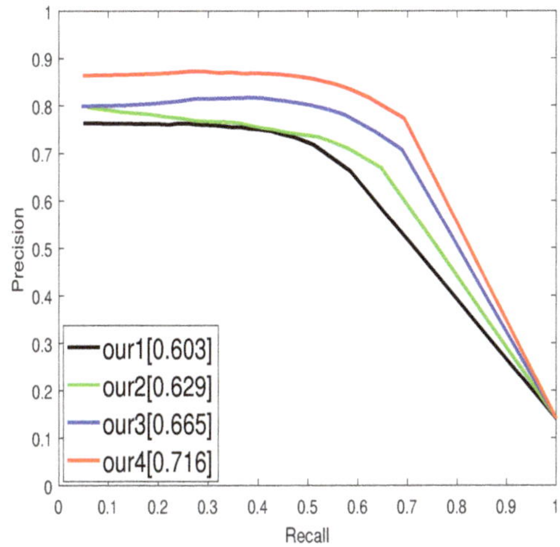

Figure 5. PR curves of our approach with its variants on the entire dataset.

Advantages. We fused thermal and RGB images for image salient detection, which can overcome the limitations of light, ambient temperature, background clutter, and color similarity in single mode. By learning the static-adaptive method, we not only retained the local features of superpixels, but also learned to mine their internal relations to obtain a better affinity matrix of superpixels and greatly improve the detection accuracy of image saliency.

Limitations. Through the experiment, we found that under complex scenes, multi-modality fusion can effectively improve the image in general. However, in some cases, the single-modality has better detection accuracy. Our future work will set the modality weight according to the image characteristics and further improve the detection effect of image saliency in complex scenes.

7. Conclusions

In this paper, we combine RGB-thermal modality information for image salient detection, which effectively improves the detection performance of single-modality RGB images under poor illumination and when the background and foreground colors are similar. At the same time, our method improves the detection accuracy of thermal images under normal lighting conditions, especially in the case of small temperature differences between the environment and the target. The image is dynamically learned, taking both global and local cues into account, and thus our method is capable of capturing the intrinsic relationship of superpixels. In the future, we will assign different weights to different modality images according to the characteristics of different modality images.

Author Contributions: Z.X. and J.T. proposed the idea, designed and performed the simulations, and wrote the paper; A.Z. and H.L. analyzed the data. All authors have read and agreed to the published version of the manuscript.

Funding: This paper is funded by the following foundations: National Natural Science Foundation of China (61906044), Natural Science Foundation of Anhui Higher Education Institution of China (KJ2019A0536, KJ2019A0529, KJ2020ZD46), Natural Science Foundation of Fuyang Normal University (2019FSKJ02ZD), Fuyang Normal University Scientific Research Project(2020KYQD0032), the Young Talents Projects of Fuyang Normal University(rcxm202001, 2021FSKJ01ZD), Fuyang City School Cooperation Project (SXHZ202103).

Institutional Review Board Statement: Not applicable.

Informed Consent Statement: Not applicable.

Data Availability Statement: The datasets can be download from the address http://chenglongli.cn/people/lcl/journals.html (accessed on 20 December 2021).

Conflicts of Interest: The authors declare no conflict of interest.

References

1. Wang, Q.; Lin, J.; Yuan, Y. Salient band selection for hyperspectral image classification via manifold ranking. *IEEE Trans. Neural Netw. Learn. Syst.* **2016**, *27*, 1279–1289. [CrossRef]
2. Yang, X.; Qian, X.; Xue, Y. Scalable mobile image retrieval by exploring contextual saliency. *IEEE Trans. Image Process.* **2015**, *24*, 1709–1721. [CrossRef]
3. Wen, W.; Zhang, Y.; Fang, Y.; Fang, Z. A novel selective image encryption method based on saliency detection. In Proceedings of the Visual Communications and Image Processing (VCIP), Chengdu, China, 27–30 November 2016; pp. 1–4.
4. Wen, W.; Zhang, Y.; Fang, Y.; Fang, Z. Image salient regions encryption for generating visually meaningful ciphertext image. *Neural Comput. Appl.* **2018**, *29*, 653–663. [CrossRef]
5. Jacob, H.; Padua, F.L.C.; Lacerda, A.; Pereira, A.C.M. A video summarization approach based on the emulation of bottom-up mechanisms of visual attention. *J. Intell. Inf. Syst.* **2017**, *49*, 193–211. [CrossRef]
6. Zhang, L.; Ai, J.; Jiang, B.; Lu, H.; Li, X. Saliency Detection via Absorbing Markov Chain With Learnt Transition Probability. *IEEE Trans. Image Process.* **2018**, *27*, 987–998. [CrossRef] [PubMed]
7. Borji, A.; Cheng, M.M.; Jiang, H.; Li, J. Salient object detection: A benchmark. *IEEE Trans. Image Process.* **2015**, *24*, 5706–5722. [CrossRef] [PubMed]
8. Tong, N.; Lu, H.; Ruan, X.; Yang, M.H. Salient object detection via bootstrap learning. In Proceedings of the Computer Vision and Pattern Recognition (CVPR), Boston, MA, USA, 7–12 June 2015; pp. 1884–1892.
9. Zhou, X.; Liu, Z.; Sun, G.; Wang, X. Adaptive saliency fusion based on quality assessment. *Multimed. Tools Appl.* **2017**, *76*, 23187–23211. [CrossRef]
10. Itti, L.; Koch, C.; Niebur, E. A model of saliency-based visual attention for rapid scene analysis. *IEEE Trans. Pattern Anal. Mach. Intell.* **1998**, *20*, 1254–1259. [CrossRef]
11. Cheng, M.M.; Mitra, N.J.; Huang, X.; Torr, P.H.; Hu, S.M. Global contrast based salient region detection. *IEEE Trans. Pattern Anal. Mach. Intell.* **2015**, *37*, 569–582. [CrossRef]
12. Wang, H.; Xu, L.; Wang, X.; Luo, B. Learning Optimal Seeds for Ranking Saliency. *Cogn. Comput.* **2018**, *10*, 347–358. [CrossRef]
13. Hou, Q.; Cheng, M.M.; Hu, X.; Borji, A.; Tu, Z.; Torr, P.H. Deeply supervised salient object detection with short connections. In Proceedings of the IEEE Conference on Computer Vision and Pattern Recognition, Honolulu, HI, USA, 21–26 July 2017; pp. 3203–3212.
14. Han, J.; Zhang, D.; Cheng, G.; Liu, N.; Xu, D. Advanced deep-learning techniques for salient and category-specific object detection: A survey. *IEEE Signal Process. Mag.* **2018**, *35*, 84–100. [CrossRef]
15. Li, C.; Zhao, N.; Lu, Y.; Zhu, C.; Tang, J. Weighted Sparse Representation Regularized Graph Learning for RGB-T Object Tracking. In Proceedings of the 25th ACM International Conference on Multimedia, Mountain View, CA, USA, 23–27 October 2017; pp. 1856–1864.
16. Li, C.; Zhu, C.; Huang, Y.; Tang, J.; Wang, L. Cross-modal ranking with soft consistency and noisy labels for robust rgb-t tracking. In Proceedings of the European Conference on Computer Vision (ECCV), Munich, Germany, 8–14 September 2018; pp. 808–823.
17. Li, C.; Liang, X.; Lu, Y.; Zhao, N.; Tang, J. RGB-T object tracking: Benchmark and baseline. *Pattern Recognit.* **2019**, *96*, 106977. [CrossRef]
18. Zhang, Q.; Huang, N.; Yao, L.; Zhang, D.; Shan, C.; Han, J. RGB-T salient object detection via fusing multi-level CNN features. *IEEE Trans. Image Process.* **2019**, *29*, 3321–3335. [CrossRef] [PubMed]
19. Harel, J.; Koch, C.; Perona, P. Graph-Based Visual Saliency. *Adv. Neural Inf. Process. Syst.* **2006**, *19, 545-552.*
20. Yang, C.; Zhang, L.; Lu, H.; Ruan, X.; Yang, M.H. Saliency detection via graph-based manifold ranking. In Proceedings of the IEEE Conference on Computer Vision and Pattern Recognition (CVPR), Washington, DC, USA, 23–28 June 2013; pp. 3166–3173.
21. Sun, J.; Lu, H.; Liu, X. Saliency region detection based on Markov absorption probabilities. *IEEE Trans. Image Process.* **2015**, *24*, 1639–1649. [CrossRef]
22. Zhang, L.; Yang, C.; Lu, H.; Ruan, X.; Yang, M.H. Ranking saliency. *IEEE Trans. Pattern Anal. Mach. Intell.* **2017**, *39*, 1892–1904. [CrossRef]
23. Xiao, Y.; Wang, L.; Jiang, B.; Tu, Z.; Tang, J. A global and local consistent ranking model for image saliency computation. *J. Vis. Commun. Image Represent.* **2017**, *46*, 199–207. [CrossRef]
24. Aytekin, Ç; Iosifidis, A.; Kiranyaz, S.; Gabbouj, M. Learning graph affinities for spectral graph-based salient object detection. *Pattern Recognit. J. Pattern Recognit. Soc.* **2017**, *64*, 159–167. [CrossRef]
25. Li, C.; Cheng, H.; Hu, S.; Liu, X.; Tang, J.; Lin, L. Learning collaborative sparse representation for grayscale-thermal tracking. *IEEE Trans. Image Process.* **2016**, *25*, 5743–5756. [CrossRef]
26. Li, C.; Wang, G.; Ma, Y.; Zheng, A.; Luo, B.; Tang, J. A Unified RGB-T Saliency Detection Benchmark: Dataset, Baselines, Analysis and A Novel Approach. *arXiv* **2017**, arXiv:1701.02829.

27. Giacomo, C.; Grazia, L.S.; Christian, N.; Rafi, S.; Marcin, W. Optimizing the Organic Solar Cell Manufacturing Process by Means of AFM Measurements and Neural Networks. *Energies* **2018**, *11*, 1221.
28. Huang, Z.; Wang, X.; Huang, L.; Huang, C.; Wei, Y.; Liu, W. CCNet: Criss-Cross Attention for Semantic Segmentation. In Proceedings of the IEEE/CVF International Conference on Computer Vision, Seoul, Korea, 27–28 October 2019.
29. Hu, X.; Yang, K.; Fei, L.; Wang, K. ACNet: Attention Based Network to Exploit Complementary Features for RGBD Semantic Segmentation. In Proceedings of the 2019 IEEE International Conference on Image Processing (ICIP), Taipei, China, 22–25 September 2019; pp. 1440–1444.
30. Zhang, J.; Yang, K.; Constantinescu, A.; Peng, K.; Müller, K.; Stiefelhagen, R. Trans4Trans: Efficient Transformer for Transparent Object Segmentation to Help Visually Impaired People Navigate in the Real World. In Proceedings of the IEEE/CVF International Conference on Computer Vision, Nashville, TN, USA, 19–25 June 2021; pp.1760-1770.
31. Liu, N.; Han, J.; Yang, M.H. PiCANet: Learning Pixel-wise Contextual Attention for Saliency Detection. In Proceedings of the IEEE Conference on Computer Vision and Pattern Recognition, Salt Lake City, UT, USA, 18–23 June 2018; pp. 3089–3098.
32. Liu, Z.; Tan, Y.; He, Q.; Xiao, Y. SwinNet: Swin Transformer drives edge-aware RGB-D and RGB-T salient object detection. *IEEE Trans. Circuits Syst. Video Technol.* **2021**. [CrossRef]
33. Liu, Z.; Wang, Y.; Tu, Z.; Xiao, Y.; Tang, B. TriTransNet: RGB-D Salient Object Detection with a Triplet Transformer Embedding Network. In Proceedings of the 29th ACM International Conference on Multimedia, New York, NY, USA, 20–24 October 2021; pp. 4481-4490.
34. Guo, X. Robust Subspace Segmentation by Simultaneously Learning Data Representations and Their Affinity Matrix. In Proceedings of the Twenty-Fourth International Joint Conference on Artificial Intelligence (IJCAI 2015), Buenos Aires, Argentina, 25–31 July 2015; AAAI Press: Palo Alto, CA, USA, 2015; pp. 3547–3553.
35. Li, C.; Wu, X.; Bao, Z.; Tang, J. ReGLe: Spatially Regularized Graph Learning for Visual Tracking. In Proceedings of the 25th ACM International Conference on Multimedia, Mountain View, CA, USA, 23–27 October 2017; pp. 252–260.
36. Achanta, R.; Shaji, A.; Smith, K.; Lucchi, A.; Fua, P.; Süsstrunk, S. SLIC superpixels compared to state-of-the-art superpixel methods. *IEEE Trans. Pattern Anal. Mach. Intell.* **2012**, *34*, 2274–2282. [CrossRef] [PubMed]
37. Stephen, B.; Neal, P.; Chu, E.; Borja, P.; EcKstein, J. *Distributed Optimization and Statistical Learning via the Alternating Direction Method of Multipliers*; Now Publishers Inc.: Hanover, MA, USA, 2010; Volume 3, pp. 1–122.
38. Lin, Z.; Chen, M.; Ma, Y. The Augmented Lagrange Multiplier Method for Exact Recovery of Corrupted Low-Rank Matrices. *arXiv* **2010**, arXiv:1009.5055.
39. Chen, M.; Ganesh, A.; Lin, Z.; Ma, Y.; Wright, J.; Wu, L. *Fast Convex Optimization Algorithms for Exact Recovery of a Corrupted Low-Rank Matrix*; Report No. UILU-ENG-09-2214; Coordinated Science Laboratory: Urbana, IL, USA, 2009.
40. Rahtu, E.; Kannala, J.; Salo, M.; Heikkilä, J. Segmenting salient objects from images and videos. In Proceedings of the European Conference on Computer Vision, Heraklion, Crete, Greece, 5–11 September 2010; Springer: Berlin/Heidelberg, Germany, 2010; pp. 366–379.
41. Qin, Y.; Lu, H.; Xu, Y.; Wang, H. Saliency detection via cellular automata. In Proceedings of the 2015 IEEE Conference on Computer Vision and Pattern Recognition (CVPR), Boston, MA, USA, 7–12 June 2015; pp. 110–119.
42. Goferman, S.; Zelnik-Manor, L.; Tal, A. Context-aware saliency detection. *IEEE Trans. Pattern Anal. Mach. Intell.* **2012**, *34*, 1915–1926. [CrossRef] [PubMed]
43. Erdem, E.; Erdem, A. Visual saliency estimation by nonlinearly integrating features using region covariances. *J. Vis.* **2013**, *13*, 11. [CrossRef]
44. Tavakoli, H.R.; Rahtu, E.; Heikkilä, J. Fast and efficient saliency detection using sparse sampling and kernel density estimation. In Proceedings of the Scandinavian Conference on Image Analysis, Ystad, Sweden, 23–25 May 2011; Springer: Berlin/Heidelberg, Germany, 2011; pp. 666–675.
45. Yang, C.; Zhang, L.; Lu, H. Graph-regularized saliency detection with convex-hull-based center prior. *IEEE Signal Process. Lett.* **2013**, *20*, 637–640. [CrossRef]

information

MDPI

Article

Security Graphics with Multilayered Elements in the Near-Infrared and Visible Spectrum

Jana Žiljak Gršić [1,2,*], Denis Jurečić [3], Lidija Tepeš Golubić [1,*] and Silvio Plehati [1]

1 Department of Informatics & Computing, Zagreb University of Applied Sciences, 10000 Zagreb, Croatia; splehati@tvz.hr
2 Department of Multimedia, Design and Application, University North, 48000 Koprivnica, Croatia
3 Faculty of Graphic Arts, University of Zagreb, 10000 Zagreb, Croatia; djurecic@grf.hr
* Correspondence: jana@tvz.hr (J.Ž.G.); ltepes2@tvz.hr (L.T.G.)

Abstract: In this paper, the fusion of four graphics into one integrated graphic is selectively observed in the visible and infrared spectrum. Each graphic represents its own information derived from the following sources: vector graphics, drawing, photograph and textual information. One graphic will be visible to the naked eye after the print. The other graphics will be observed with an NIR surveillance camera. These other graphics are nested into the selected visible graphics. All the graphics together make up a security print product with the characteristics of an individual solution with multilayered elements. Reprinting is possible only for the person in possession of the solution created according to the algorithm based on the INFRAREDESIGN® method. When these graphics are printed on paper, it is impossible to produce an identical graphic prepress (C, M, Y, K) to produce forgery with the same dual properties in the visible and NIR spectrum.

Keywords: NIR spectrum; dual graphics; spectroscopy of twin dyes; nested picture; CMYKIR VZ separation

Citation: Žiljak Gršić, J.; Jurečić, D.; Tepeš Golubić, L.; Plehati, S. Security Graphics with Multilayered Elements in the Near-Infrared and Visible Spectrum. *Information* **2022**, *13*, 47. https://doi.org/10.3390/info13020047

Academic Editor: Gholamreza Anbarjafari (Shahab)

Received: 10 November 2021
Accepted: 15 January 2022
Published: 20 January 2022

1. Introduction

In this paper, we extend the domain of computer graphics to include two light spectra—the visible and the near-infrared spectrum. The final solution and dyes are visually observed with a double ZRGB camera (Figure 1) in the "Z"-near-infrared and "V"-visible (RGB) spectrum [1]. In this paper, we develop and present the procedure of merging computer graphics with the INFRAREDESIGN® idea that will manifest itself separately in two light spectra: the visible (V) and near-infrared (NIR-Z).

State-of-the-art INFRAREDESIGN® is a field of security graphics with the idea of fusing two pictures that will manifest themselves separately in the near-infrared (NIR) and visible spectrum after integration. A one-way hierarchy of pictures was published, where the first picture hides the second picture. Previous solutions have shown the merging of only two graphics for two light spectra [2]: a dress with a hidden portrait or code and clothes with hidden textual information, printed with inkjet inks on a plotter for textiles. The clothing was designed with computer graphics, with the theme "camouflage clothing". ZRGB cameras are used to observe and photograph double solutions [1]. The Z camera, which photographs the NIR wavelength range, has a filter for 1000 nm. The visible and NIR condition were recorded in daylight without an additional IR source. Spectral analyses of light absorbance for process dyes in two light spectra—the visible (400 to 750 nm) and the beginning of the near-infrared (750 to 1000 nm)—have been performed for several printing techniques [3]. Since our eye does not register the NIR range, it can be said that the graphics in this range are "hidden from the naked eye" [4]. Several models have been published that combine the colors and dyes in the visible spectrum and the coverage of the dyes for those parts of the image that present two spectral ranges. The implementation of the dual image is suitable for the fast creation of highly protected individualized documents [5].

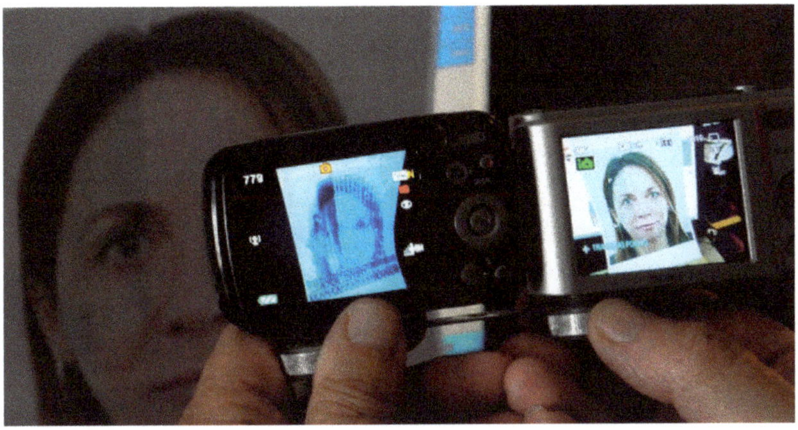

Figure 1. Print studied with a ZRGB camera (Z blockage at 1000 nm and in the visible spectrum; blockade at 400 nm).

The idea is to introduce INFRAREDESIGN® in different fields of security graphics. Motivation for this paper is the merging of three and four images whose parts are sorted for detection in the visible and near-infrared spectra. Printing is performed on a toner printer (OKI) for which spectrograms were performed and first published in this paper. These twin dyes are shown in two spectra with an emphasis on equality in the range of 400 to 700 nm. The separation of the dye twins when the measurements are shifted towards the NIR spectrum is also shown. The original digital record is attached as a JPG and PDF four-color graphic preparation for printing on a digital printer. Individual channels for process colors are available with an emphasis on the carbon black channel, which will be viewed separately with an NIR camera. Analysis of the transformation from V to Z spectra is attached as a continuous animated change into MP4 and SWF format.

2. Methodology: Twin Dyes

The merging of different images is based on the idea of twin dyes for dual graphics. Experimental work is performed on an OKI printer with the corresponding set of toners: cyan, magenta, yellow, and black. The first three toners (C, M, Y) absorb light only in the visible spectrum. Toner carbon black (K) absorbs it in two spectra, visible and near-infrared spectrum. By mixing toners, twins of colors and dyes are achieved, with which separate information is created for the visible and NIR spectrum.

A twin dye group consists of several twin dyes (pigments) made up of different components that visually manifest themselves as the same color [6,7]. These twin dyes will absorb light differently only in the NIR wavelength region, which is why they differ from each other.

In the printing practice, prepress for printing is performed in some phases with the use of GCR (Gray Component Replacement), while maintaining the same values of L*a*b*, HSB and RGB [8]. CMYKIR is a special GCR method designed with expansion to NIR spectrum. Near-infrared graphics are designed to be nested in a carbon black channel with the reduction of C, M, Y dyes. Therefore, our CMYKIR (VZ) method is also based on that same GCR theory. From there, routines are generated for our "VZGCR" software written in PostScript code. The focus of the software is data processing, presented in Table 1.

There are many different cameras around us: security cameras in the streets and in protected areas, IR money detectors in banks, IR reflectography in galleries. A, B, C and D spectra of dyes (Table 1) are derived with the forensic camera Projectina, model PAG [9] and with "XRITE–iONE". The light absorbance value at 1000 nm is called "Z". Our experimental work is based on process colorants C, M, Y and K is thus marked as "VZ". For example, the IR technology has been applied in the process of designing banknotes, but only with spot dyes whose material formulation is not published. On today's banknotes, the visible and NIR graphics are printed separately, one next to the other. There is no "hiding" of one image within the other, which is the innovation and purpose of this paper.

Table 1. CMYK formulations of twin dyes.

	Measured Data C, M, Y (K_0, K_{20}, K_{40})			
	C, M, Y (K = 0%)	**C, M, Y (K = 20%)**	**C, M, Y (K = 40%)**	**L*a*b***
A	99, 99, 59	88, 85, 41	74, 68, 19	1, 44, −45
B	53, 98, 73	38, 86, 56	20, 70, 36	25, 56, 9
C	70, 46, 99	58, 29, 84	44, 9, 67	45, −25, 43
D	40, 50, 40	28, 40, 24	14, 28, 5	57, 14, 6
	40, 40, 40	28, 30, 25	15, 18, 6	62, 5, 7
	37, 82, 36	26, 71, 23	7, 63, 0	42, 49, −1
	88, 35, 70	70, 24, 65	68, 1, 36	44, −41, 6
	30, 30, 35	19, 21, 20	5, 10, 2	71, 4, 11

2.1. Twin Dyes Components

The recipes (%) for eight groups of twin dyes have been given in Table 1. The A dyes are the same color but with three completely different formulations. The same happens with all other groups of dyes.

2.2. Spectroscopy of Twin Dyes

Figures 2 and 3 show spectrograms of A, B, C, and D dyes (Table 1) in the area from 400 to 900 nm. The measured reflected light is presented inversely as the absorbance of light in the process printing dyes: cyan, magenta, yellow and black [9]. The graphs of light absorbance show the pigment in its corresponding nanometer range. The maximum, "saddle" and "back" are displayed, which, in addition to the spectrograms of the components, shows the direction of the color twin repair [10].

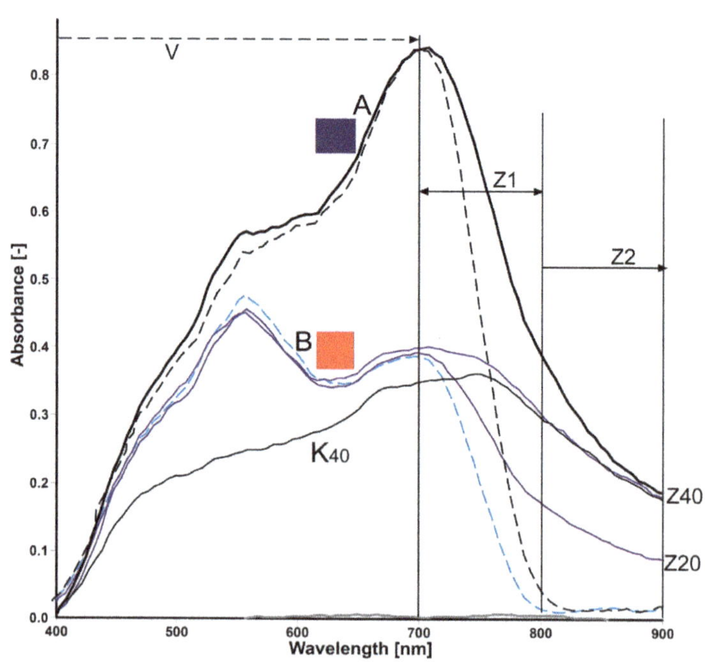

Figure 2. Twins A, B, K40.

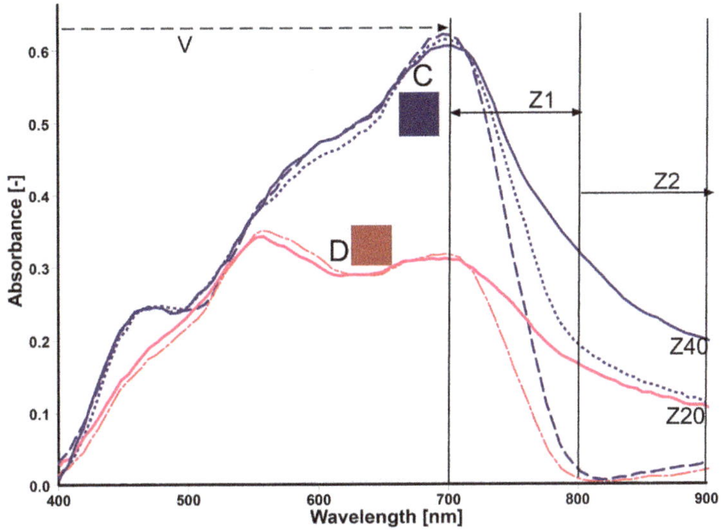

Figure 3. Twins C, D.

Conventionally, the visible range is defined as the range from 260 to 760 nm and it follows the near-infrared spectrum. In this paper, the range "V" is marked as a part of the visible spectrum from 400 to 700 nm. The dyes made up from twin dyes have the same values of light absorbance in the V area. The range between 700 and 800 nm is called Z1, and the values of light absorbance in it differ exceedingly. Z2 refers to the range from 800 to 900 nm. This range of the near-infrared spectrum is crucial for the clarity of the appearance of the hidden image detected by the NIR camera.

In this paper, the discussion of twin dyes is presented with the spectroscopy of the first four dyes from Table 1. The correct definition of the range between 800 and 1000 nm is the main subject of dual design and dual recognition of different graphics in the same place.

3. Results

This chapter demonstrates the fusion of different images as a "picture within the picture" technology. The result of printing is interpreted as a unique preparation for four-color printing. Some images were created by photographing, and some of them are computer graphics from the field of security graphics.

3.1. Experiment Plan for Creating Security Graphics

We present four graphics in color in order to demonstrate new ideas in the field of security graphics. The first graphic is a portrait *en face* (Figure 4). The second graphic is a vignette (Figure 5). These elements are usually found in securities. The lines are generated as vector Bézier lines. In our examples, the vignette is prepared in high resolution, but it is equalized with pixel graphics, such as images in Figures 4, 6 and 7. However, the completely different images are mutually connected through content: *en face*, profile and the text in the vignette.

Figure 6 is composed of alphanumeric characters. Each letter and number is a separate font that has 30 different values of thickness. An alphanumeric character is a simulation of a raster (screen) form, and it simulates the portrait *en face* from Figure 4. The text creates a surface with micro-text that can often be found in the design of securities. Here the text is connected to the information about the person: name and surname, date, place and country of birth (Figure 6). The fourth picture (Figure 7) is a profile of that same person.

The pictures are diverse; they derive from vector graphics, pixel graphics or a combination. All the colors are set in the RGB system of colors (only the region visible to our eyes).

Two solutions were conducted for the same group of images. It is shown that the solution is independent of which images are assigned to the visible spectrum, and which to the NIR spectrum.

Design plan no. 1: Figures 5 and 7 will be joined in the visible spectrum. Figure 4 (*en face*) will be joined in the near-infrared spectrum. After merging and printing, the security graphic becomes visible to the naked eye, and the hidden graphic will be recognized with an NIR (near- infrared) security camera.

Design plan no. 2: the portrait *en face* (Figure 4) will be joined in the visible spectrum. Three images will be joined to the near-infrared spectrum: the profile (Figure 7), vignette (Figure 5) and micro-text (Figure 6). A separate region is planned for each of these three images. Since there are four different images that will manifest themselves in two ways and in two schedules, the graphic size in pixels of each image was additionally adjusted.

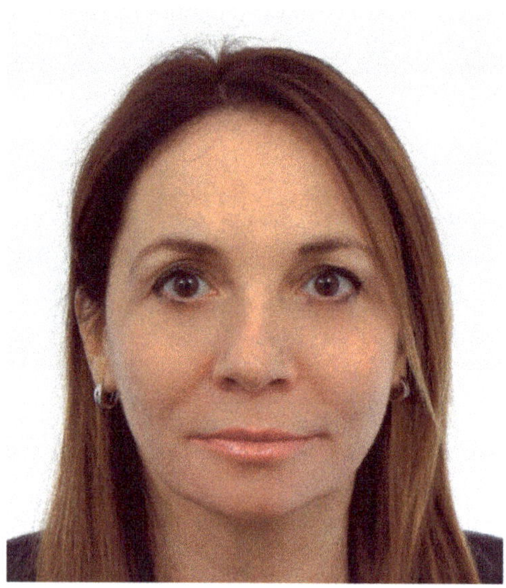

Figure 4. En face.

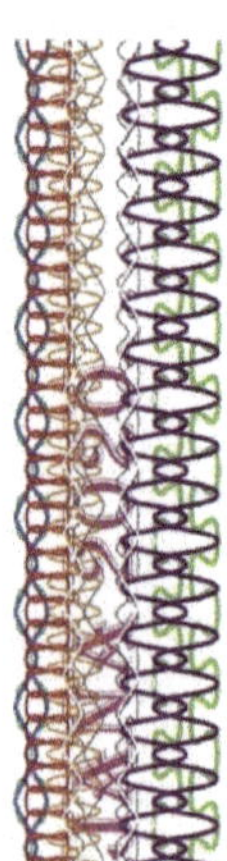

Figure 5. Vignette, security graphic.

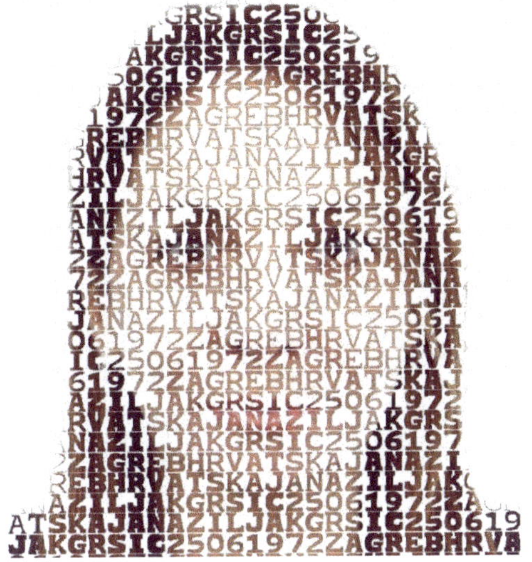

Figure 6. Portrait with a letter raster.

Figure 7. Profile.

For each image a (geometry) range is specified to be included in the common, nested Z picture. The plan is to print the merged picture with process CMYK dyes.

3.2. *A Security Graphic with Three Nested Images*

In the first example for the visible spectrum, two graphics were highlighted: the profile of the person and protective vignette. Those two graphics hide the portrait, which is captured by the NIR (Z) camera. The VZ merging and separation for the duality of the visible and near-infrared spectra have been limited according to the conventional printing separation practice–GCR. In Figure 8, gray scale is added. This image will be the final solution for the visible spectrum. Individuality of the graphic solution is introduced through different algorithms of gray tone replacement.

Figure 8. Visible presentation CMYK.

Figure 8 was prepared and printed. After the printing of Figure 8, the visible graphic is checked with an NIR–Z camera and the portrait *en face* in gray tone appears on the screen of the Z camera (Figure 9).

Figure 9. NIR graphic (1000 nm).

If, for some reason, Figure 8 does not get published in the article as a CMYK model, we are providing the link of the original (CMYK) Figure 8 on the author's website: http://www.jana.ziljak.hr/portret82.jpg (accessed on 10 January 2022).

The digital record was created as a four-color (CMYK) print. All images for INFRAREDESIGN® have a special (their own) procedure, from preparing the images, connecting the computer with the printer, to printing an image with NIR content. If such a record is sent to a digital printer, it is necessary to set the printing parameters. Printing (on the printer from, for example, Photoshop) is performed with the following options:

P1: No color management (to disable the internal process of image separation);
P2: In Page setup/Color, select: CMYK link profile—True Black.

The discernment of pictures in four channels, prepared in this way, with CMYKIR–VZ separation, is impossible even with a 24-filter forensic system [9]. This means that only the one who has the original image (C, M, Y, K) can reproduce this CMYKIR image.

3.3. A Security Graphic with Four Nested Images

A security graphic is composed with four merged color graphics. The portrait *en face* (Figure 4) will be visible to the naked eye, while the remaining three graphics are hidden and can be recognized with an NIR camera. The vignette (Figure 5) is intended for the horizontally lower part of the graphic.

The visible portrait *en face* (Figure 4—RGB) remains in color, with the goal that the initial portrait is visually the same as the portrait that has three nested images after VZ separation (CMYK). The photograph shows the eye color, hair color and skin tone. The advantage of the picture is the recognizability of the person. Eye color and skin tone are important elements of the portrait *en face*. Image Z is shown as a black-and-white micro-text, protective vignette and a part of the portrait in profile. After printing, our eyes see only Figure 4 although the print also contains the remaining three graphics.

In this example, it was decided that, after merging, the color of the portrait *en face* would be visible to the naked eye (in the visible spectrum, 400 nm blockage). After the merging of all four gray channels (Figures 10–13) in, for example, Photoshop, a pure color picture equal to the picture in Figure 4 can be seen. The check is performed by inserting the corresponding gray graphic into each of the C, M, Y, K channels (Figures 10–13).

In Photoshop (for example), each CMYK channel is written as a shade of gray. Together, they give a color solution for the visible graphic (identical to the picture in Figure 4) that hides the planned Z picture.

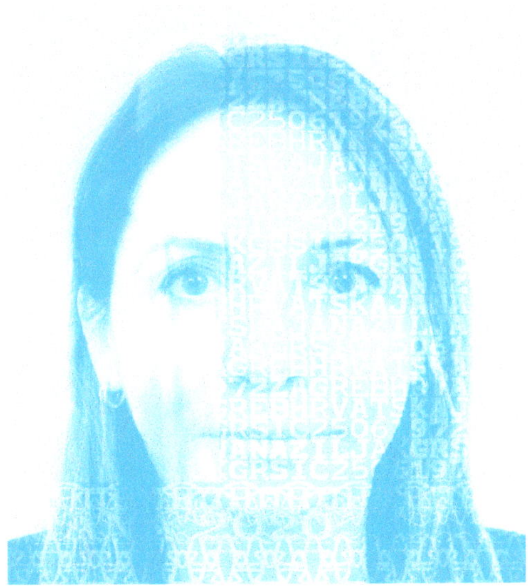

Figure 10. Cyan channel (C).

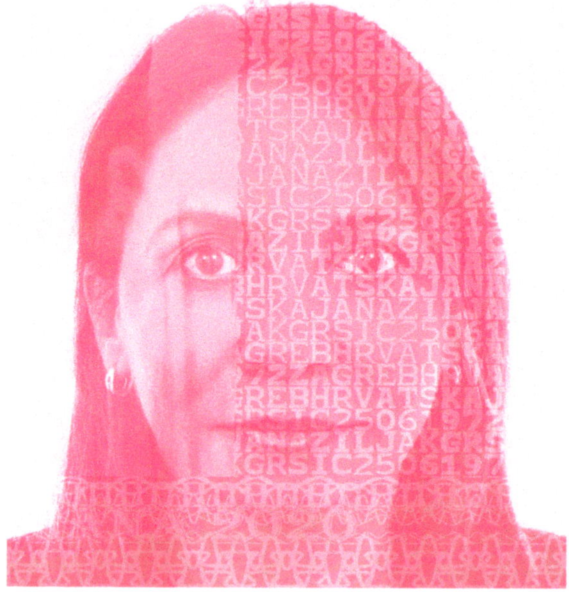

Figure 11. Magenta channel (M).

Figure 12. Yellow channel (Y).

The final, hierarchical graphic and print is a simulation of the color portrait (Figure 4), while the other images (Figures 5–7) are hidden within that portrait. The portrait, after the four images have been merged, is given in a C, M, Y, K, record.

Figure 13. Black channel (K).

The IRD security graphic represents the merging of four gray images to be printed with C, M, Y, K (1) colors, Figures 10–13. With this kind of graphic prepress, color printing is performed in an unlimited number of prints. The print on the paper is a color graphic that contains two different images: one for observation with the naked eye and one for observation with an NIR camera. It is not possible to use the color imprint on the paper and go back to creating four C, M, Y, K (2) channels that are equal to the channels (1) at the beginning of the paragraph using any existing scanning techniques. The intention of such a process in the opposite direction would be to make counterfeits. New colors (2) will not give the same visual and NIR image. The portrait in Figure 14 and the original merged image are recorded at the following address: http://www.jana.ziljak.hr/portret81.jpg (accessed on 10 January 2022).

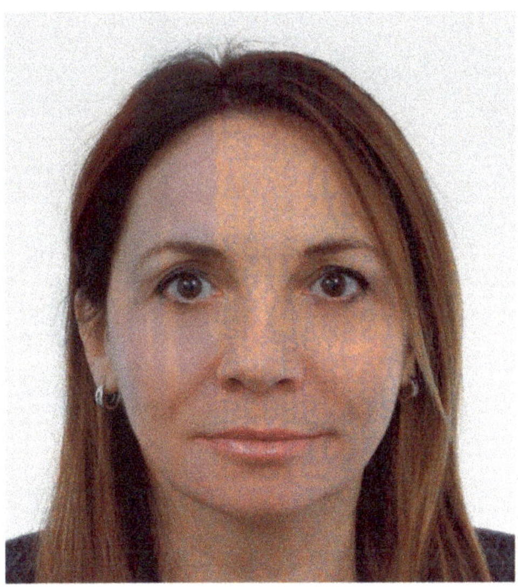

Figure 14. Picture for print (C, M, Y, K), all four channels merged, for the visible and NIR spectrum.

3.4. Blockage in the Visible and Near-Infrared Spectrum

The discernment of the merged picture is performed with light blockages (filters in cameras). Blockages at 400, 600, 700 and 850 nanometers are shown (Figures 15–18).

Animations of portrait changes were taken with a PAG camera that shoots in 24 light blocks. The display is located at the addresses: http://www.jana.ziljak.hr/portret81.mp4 (accessed on 10 January 2022). http://www.jana.ziljak.hr/portret82.mp4 (accessed on 10 January 2022).

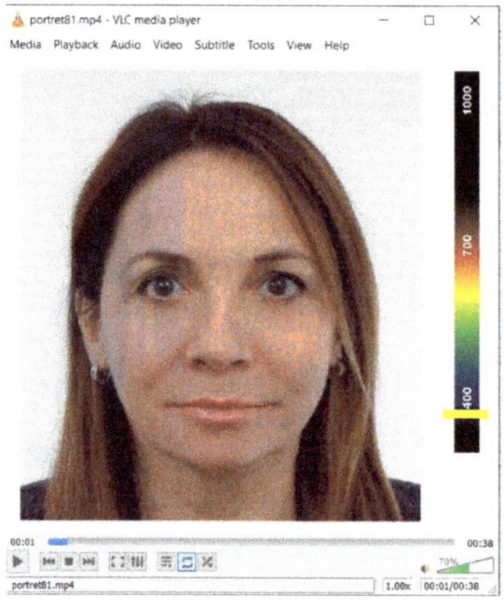

Figure 15. Blockage at 400 nm.

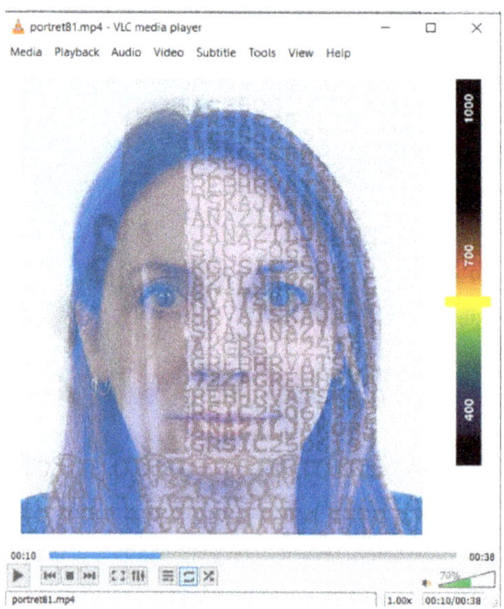

Figure 16. Blockage at 600 nm.

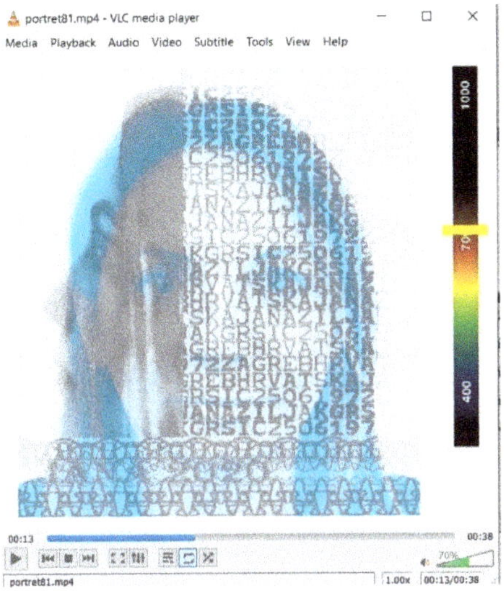

Figure 17. Blockage at 700 nm.

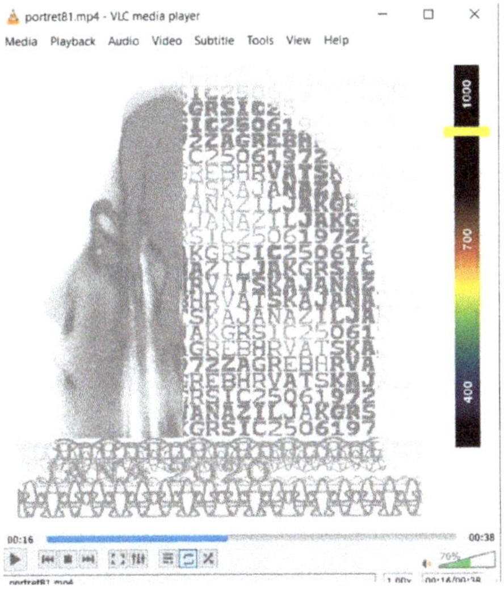

Figure 18. Blockage at 850 nm.

4. Conclusions

Control systems, such as NIR cameras, are all around us: cameras in road traffic, security night cameras, IR money detectors in banks, etc.

Outside the visible color range (400 to 750 nm), the graphic is shown instrumentally. Already after the first light blockage (shown in multilayer light presentation video at 600 nm) in the visible spectrum, images appear partially as "hidden".

Multiple pictures that are shown in the end make one picture and significantly raise the security of personal portraits and biometric portraits, as well as other pictures that require a high level of protection against forgery. Portrait images prepared in this manner cannot be subject to photo manipulation that is nowadays omnipresent. The image prepared in this way is designed to be implemented in extremely protected documents and identity cards.

The massiveness of NIR cameras in our environment initiated the idea for the launch of IRD design in new areas with materials: textiles, polypropylene, cardboard, postage stamps, new paper money design. A new information area of duality is opening up, such as the expansion of data on clothing, packaging, scenography, and costume design in the film and theater industry. Each color for artistic painting has its Z value. A new look at the information provided to us by fine artists is opening up. InfraRedArt is becoming a new "multilayeredness" in the creative field of multi-media art.

Author Contributions: Conceptualization: J.Ž.G. and L.T.G.; Methodology: J.Ž.G. and S.P.; Software: S.P. and J.Ž.G.; Validation: L.T.G. and D.J.; Formal analysis: D.J. and L.T.G.; Investigation: J.Ž.G. and D.J.; Resources: J.Ž.G. and D.J.; Data curation: J.Ž.G. and S.P.; Writing—original draft preparation: J.Ž.G.; Writing—review and editing: L.T.G., S.P.; Visualization: J.Ž.G., S.P.; Supervision: L.T.G., D.J. All authors have read and agreed to the published version of the manuscript.

Funding: This research received no external funding.

Institutional Review Board Statement: Not applicable.

Informed Consent Statement: Not applicable.

Data Availability Statement: Data is contained within the article.

Conflicts of Interest: The authors declare no conflict of interest.

References

1. Žiljak, V.; Pap, K.; Žiljak-Stanimirović, I. Development of a prototype for ZRGB INFRAREDESIGN device. *Tech. Gaz.* **2011**, *18*, 153–160.
2. Zhu, M.Z.; Chen, Z.; Liu, H.X. The Research on Special Printing Effects Based on Gray Component Replacement. *Adv. Mater. Res.* **2010**, *174*, 251–254. [CrossRef]
3. Wang, C.Y.; Li, C.; Huo, L.J. A Security Printing Method by Black Ink Hiding Infrared Image. *Appl. Mech. Mater.* **2012**, *200*, 730–733. [CrossRef]
4. Žiljak, V.; Pap, K.; Žiljak, I. CMYKIR security graphics separation in the infrared area. *Infrared Phys. Technol.* **2009**, *52*, 62–69. [CrossRef]
5. Friščić, M.; Agić, A.; Žiljak Stanimirović, I. Visual and infrared graphic applied through dedicated halftoning for transparent polypropylene packaging. *Tech. Gaz.* **2017**, *24*, 225–230.
6. Li, C.; Wang, C.Y.; Wang, S.J. A Black Generation Method for Black Ink Hiding Infrared Security Image. *Appl. Mech. Mater.* **2012**, *262*, 9–12. [CrossRef]
7. Pogarčić, I.; Agić, A.; Matas, M. Evaluation of the colorant twins for the neutral grey spectra in infrared graphic procedure. *Tech. Gaz.* **2016**, *23*, 1659–1664.
8. Glogar, M.I.; Parac-Osterman, Đ. Achromatic hues matching in graphic printing. *Acta Graph.* **2016**, *26*, 36–45.
9. Projectina Docucenter 4500, SP-2000 Color Spectroscopy Module & PAG B50 Custom Designed with 24 Barrier Filters, Projectina AG, Dammstrasse 2, P.O. Box CH-9435 Heerbrugg, Switzerland. Available online: https://www.ultra-forensictechnology.com/en/our-products/ (accessed on 2 November 2021).
10. Žiljak Gršić, J. Near infrared spektroskopy in print technology. *Polytech. Des.* **2017**, *5*, 32–36.

applied sciences

MDPI

Article

Procedural Video Game Scene Generation by Genetic and Neutrosophic WASPAS Algorithms

Aurimas Petrovas and Romualdas Bausys *

Department of Graphical Systems, Vilnius Gediminas Technical University, Sauletekio al. 11, LT-10223 Vilnius, Lithuania; aurimasp.petrovas@vilniustech.lt
* Correspondence: romualdas.bausys@vilniustech.lt

Abstract: The demand for automated game development assistance tools can be fulfilled by computational creativity algorithms. The procedural generation is one of the topics for creative content development. The main procedural generation challenge for game level layout is how to create a diverse set of levels that could match a human-crafted game scene. Our game scene layouts are created randomly and then sculpted using a genetic algorithm. To address the issue of fitness calculation with conflicting criteria, we use weighted aggregated sum product assessment (WASPAS) in a single-valued neutrosophic set environment (SVNS) that models the indeterminacy with truth, intermediacy, and falsehood memberships. Results are presented as an encoded game object grid where each game object type has a specific function. The algorithm creates a diverse set of game scene layouts by combining game rules validation and aesthetic principles. It successfully creates functional aesthetic patterns without specifically defining the shapes of the combination of games' objects.

Keywords: genetic algorithm; procedural generation; game scene; multicriteria decision making; WASPAS-SVNS

Citation: Petrovas, A.; Bausys, R. Procedural Video Game Scene Generation by Genetic and Neutrosophic WASPAS Algorithms. *Appl. Sci.* **2022**, *12*, 772. https://doi.org/10.3390/app12020772

Academic Editor: Rui Araújo

Received: 5 December 2021
Accepted: 7 January 2022
Published: 13 January 2022

1. Introduction

Today, researchers are discovering more and more new results in the artificial intelligence domain [1]. Increasing computing power, storage, and volumes of data creates new approaches to use Artificial Intelligence. Computational Creativity is one of the approaches that is gaining traction. It is used to solve creativity problems and is realized through computation-based systems that attempt to simulate creative work. Creativity definitions vary, and widely agreed upon definitions of creativity in engineering are not defined, but it is necessary to understand the essence of the concept to model a system according to creativity principles. The definition of creativity can be split into a few parts depending on how creative work is rated or created. Usually, there are four types of creativity modeling targets: person, process, product, and press. The most common machine learning targets for creative tasks are product and process [2]. The product target evaluates a completed work and attempts to replicate it by combining and expanding elements of the previous work. The process target tries to simulate logical loops, which are used to create work. The person target is rarely used, as it requires simulation of the creative agents or person. Press target is quite common when trying to filter creative and impactful work (i.e., Internet content scans). The product target is widely used in machine learning tasks, as most training data sets are made up of the available creative work—these systems usually create an independent logic loop of creativity directly unrelated to the original work process [3]. We are focusing on the process-related target, as it usually generates more example-independent results, which is one of computational creativity tasks. This means that the generated work differs more from the training data set.

To understand the structure of creativity, we can break it down into different classifications, which are important to understand when building a model. Creative value can be defined by these key terms: usefulness, aesthetics, originality, relevance to the task, and surprise [4]. These points are often referred to when trying to identify what makes the creative result valuable. Creativity involves a combination of expertise, chance, and intuition; adding these traits to a system generally makes the system more likely for its result to have more creative value [2]. Example approaches to cognitive creativity include concept combination, concept expansion, imagery, metaphor, and divergent thinking [5]. The cognitive approach is often compared to a heuristic search. Evaluation of results is an important task, as it defines creativity. Common evaluation methods are Classification, Regression, Predictive models, Generative models. These methods usually try to replicate creative content rather than exploring new spaces. Transformational creativity systems try to decide what is creative by themselves, autonomously using more abstract evaluation methods. The most effective evaluation is usually outside the system, feedback from other creative systems [6]. Transformational systems are not effectively realized or used today. Common challenges for creativity evaluation can be grouped into two categories: how to generate transformational creative content, which can add new value to existing parts of the results; another common challenge is that generated results may be quite similar between a few iterations of the result running on the same model. There are not that many creativity-oriented models. Creative models can generate artwork but lack contextual creative value [7].

There are a lot of possible criteria in the generative content ranging from functional to aesthetic; however, it is difficult to choose the criteria list for each task, and there is not a lot of research work done on creative fitness modeling. Ratios of selected criteria have a huge impact on the final results, and improper ratios may easily break the final result. Another important choice is the selection of the criteria itself. They can range from low to high levels. The low-level criteria define basic building block rules, and the high-level criteria define more abstract and specific tasks. Lower-level criteria usually increase the variety of generated results, while higher-level criteria can generate a specific result with fewer calculations. It is important to select criteria or create a criteria manipulation system to form a fitness function. There are various ways to realize a model using various algorithmic approaches. We are approaching our problem with a combination of procedural generation and machine learning methods.

One of the methods of content generation problem is procedural game content generation using machine learning models on existing content (PCGML) [8]. The use of procedural content generation is increasing in the game industry, and researchers are trying to find new ways to generate high-quality content. Generation assist levels can be categorized as partial, complete, autonomous, interactive, or guided. Game content is classified as functional and cosmetic. The main problems with the procedural generation machine learning approach include training on small datasets, lack of suitable data, parameter adjustment, and others [9]. Procedural content generation methods (PCG) usually lack evaluation, and objectives are created by designers. Use cases for PCGML are autonomous generation, artificial intelligence-assisted design, repair, analysis, and data compression. The proposed research is focused on autonomous generation, which creates game content without human interaction by combining the algorithm and the fitness function. Video games are a widely used form of multimedia that requires a broad scope of machine learning approaches. Game design generally requires the level of the game to be both playable and aesthetic [10]. At the same time, there is no common way to standardize datasets and evaluate performance for game design problems [8,11,12]. The objectives of PCG in game level generation are to make games more replayable, less demanding for creator time, reduce storage space, or enable particular aesthetics [10].

The fitness function for game design and computational creativity usually contains a subjective combination of criteria and is still in the early research state of its quantification [13]. There are no widely agreed upon definitions of how results should be compared. Game design can be broken into several parts, categorizing games by their objectives. Conversion to fitness criteria varies depending on the type of game. Patterns are elements that are present in levels across multiple games, rather than being a feature that is recurrent on the same game title. Patterns are usually categorized into several types. Guidance guides players in an intended direction. The safe zone is the area where players are not exposed to negative interactions. The foreshadowing hints at something that will happen later. Layering is the pattern of combining multiple objects to create a new experience. Branching provides players with multiple paths to reach their objectives. Pace breaking refers to the purpose of changing elements of the game to achieve a creative objective [14].

Most automated game design approaches follow reverse engineering principles, usually by using datasets generated by analyzing other games. Using this approach, fitness criteria for generations do not have to be defined [15]. The main problem with this approach is the lack of new concepts in the generated content [6]. The goal is to derive objective formulas from game design principles to generate game levels. Game flow strategy is one of the propositions to measure game design in the literature. This concept combines concentration, challenge, player skills, control, clear goals, feedback, immersion, and social interaction criteria [16]. Some authors try to measure game engagement by analyzing difficulty and applying constraints to make levels playable [17]. Current research is trying to quantify abstract creativity criteria so that they can be used in real-world digital applications. There are some examples in the field that use fuzzy logic to express criteria lists [18]. There is also a new emerging usage of neutrosophic sets combined with multicriteria decision making (MCDM), but this approach is not widely explored in the field of machine learning, but it can benefit the creativity of such models [19–22]. There are also not many MCDM algorithms used together with iterative optimization algorithms [23,24]. Some SVNS applications in the literature emphasize a greater focus on uncertainty [25]. In the following paragraphs, we add a more detailed explanation of related work, methodology, created framework, results, and conclusions.

2. Related Work

There is a rising interest in automated game level generation. Machine learning algorithms are masters of specific computational tasks, but there is no perfect solution to mimic human creativity. The primary goal of this type of research is to identify creativity measurements and apply them to automated content generation. The current stage of results in this field is mostly exploratory and does not substitute for creative work in most cases, but it gradually increases assistance levels for the creator by overtaking simple creative tasks. PCGML is one of the assistance tools for work generation. There are 4 modeling steps: problem identification, solution, results, and application of generated results.

There are research examples that tackle the problem of computed creativity in the game design field. One of the examples in the industry is the generation of physical puzzle game levels with the objective of fitness of the feasibility and stability of objects [26]. Final fitness is calculated using an agent that plays the game. This method reduces the computational cost for this problem and adds new solutions to calculate the rewards of the genetic algorithm, not focusing too much on the penalty. Another example is a level generator for a Lode Runner-type game. It assesses playability and connectivity using the 'A*' algorithm. Generator uses an autoencoder with a multi-channel approach, analyzes 150 pre-made levels, and uses evolutionary algorithms. Levels are encoded into multichannel strings. This solution adds some unpredictability. Performance evaluation compares similarity to the original game levels [27]. There is also a framework for general 2D games (mostly top-down adventures) [28]. It evaluates levels for symmetry, balance, density, and reachability with a focus on aesthetics and difficulty. For final fitness, it derives 3 different fitness values and calculates the average value (Score Difference Fitness, Unique Rule Fitness, and Metric Based Fitness). It tries to apply the procedural video game generation problem to a variety of games with differing rules. Another example is focused on creative patterns [29]. The match 3 type game is used as a base for the generator. For evaluation, it uses visual pattern recognition and line symmetry. The results are judged by expert study analysis. It learns from existing content and uses pattern-aware PCGML, random Markov fields with symmetric positional information, and visual analysis. This research tries to generate larger structural patterns. The Pac-Man arcade-type game [30] evaluates playability, the spread of objects, ratios, and evolving levels using a genetic algorithm. It tries to generate unique levels with each iteration of the algorithm. Another more generalized example is the generation of verticality for mostly flat surfaces on grid-based surfaces [31].

Most of the examples in the literature uses 2D space for experimentation and games from the 1980s or simple game levels made specifically for the selected tasks. Usual objects for game levels are empty space, wall, player, goal, collectibles, and hazards. General evaluation criteria are guidance, progression, aesthetics, safe zones, and pace breaking. The current state of the computational creativity field in video games is quite young and has not yet been applied to a game structure for large and complex games. It is also difficult to model systems that can fully replicate manual creative work. As the result becomes more complex, it is easier to distinguish synthetic creativity.

3. Scene Layout Modeling and Optimization Algorithm

We propose a PCGML framework for automated game scene layout generation. Our mathematical model consists of the fitness function, which is used by a genetic algorithm to evaluate the population. The MCDM utility function is used as a genetic algorithm fitness function. We chose fixed criteria parameters for difficulty, playability, and size adjustments. One algorithm iteration populates a game level grid, which is also used for further evaluation. Evaluation calculates fitness for each game level grid and selects the best performing grids. This model generates varied and unexpected results because the generation seed is randomly selected and fine-tuned by the algorithm.

We combine the level design criteria measurements into a multi-criteria decision-making table to formulate the problem. The final fitness for the value of the game scene can be measured by combining the scores for each criterion. It can evaluate different generated scenes by using generated alternatives for one table axis and fitness scores for another table axis. Based on the table results, we can then choose which alternatives should be used as a base for further scene generations. By calculating the fitness score for each criterion and combining them into a table, we can assess and evolve a combined fitness score to generate game levels. Proposed research is focused on the process evaluation type, which studies what types of actions are made that make results creative. We combine the criteria values by converting them to fuzzy sets. Our approach is to use the weighted aggregated sum product assessment with a single-valued neutrosophic sets (WASPAS-SVNS) method to find solutions when multiple conflicting criteria are present [19–24]. Calculations are made with fuzzy logic using neutrosophic sets [32].

From a computational creativity standpoint, we are using a creative process approach combining usefulness, aesthetics, and chance to create our model. These aspects form the constraints and criteria set for the mathematical model. This framework designs and generates video game level layouts. It generates random levels, modifies them with a genetic algorithm, and evaluates them with weighted aggregated sum product assessment to find the best alternatives. The framework can also be expanded with additional requirements and fitness criteria, and most of the parameters can be altered to suit needs. We explain our realization in detail next, broken into four chapters: game scene modeling methodology, game scene procedural generation criteria list, proposed extension of genetic algorithm by WASPAS-SVNS, and application of WASPAS-SVNS utility function to calculate fitness function.

3.1. Game Scene Encoding Modeling

A common set of game objects is applied, which is selected based on game level design principles. There are several possible object types encoded in the matrix. Each number represents a different object type. Game scene layout is discretized into a grid, and one object can occupy one cell. The single-scene layout forms a single genetic algorithm chromosome. These are:

- Player (number 0)—represents the starting position of the subject, which is intended to play the game;
- Exit (number 1)—marks the location that the player should reach to finish the game;
- Empty space (number 2)—traversable and empty space, which can be used to navigate by the player;
- Wall (number 3)—object that blocks player navigation;
- Hazzard or enemy (number 4)—a traversable object, which is dangerous for the player;
- Collectible (number 5)—a desirable object that can be collected by the player;
- Ground—this object is not encoded in the chromosome matrix but is used during the 3D projection visualization step as a floor layer.

The data of a single chromosome is stored in a two-dimensional number grid (Figure 1). We use a 10 unit wide and 10 unit long matrix for our experiments. Each object type is encoded as a different number. The final results are projected into the 3D space by adding a ground layer beneath the grid and converting numbers into 3D objects on the main grid.

3	3	3	3	3	3	3	3	3	3
3	0	2	2	2	2	2	2	2	3
3	2	2	2	2	2	2	2	2	3
3	2	2	2	2	2	2	2	2	3
3	2	2	2	2	2	2	2	2	3
3	2	2	2	2	2	2	2	2	3
3	2	2	2	2	2	2	2	2	3
3	2	2	2	2	2	2	2	2	3
3	2	2	2	2	2	2	2	1	3
3	3	3	3	3	3	3	3	3	3

Figure 1. Single-chromosome data example.

3.2. Game Scene Procedural Generation Criteria List

After discretization, we choose a set of criteria that define our game layout requirements. We proposed to use 4 fitness criteria functions and 3 constraint functions, which use function results to calculate the total fitness value with the WASPAS-SVNS algorithm for each iteration of the genetic algorithm. Aesthetics are defined by the symmetry and empty-space balance criteria. Usefulness is defined by the safe zone and player exit distance criteria. Criteria were selected based on recurrence in the literature [26–30], game design principles, and creativity definitions [2,4,5]. If one of the constraint functions does not pass, the total fitness is multiplied by zero.

The results of the criteria are normalized to fit in the 0 to 1 range to have a reference point for different criteria metrics [17]. Zero represents the worst possible value, and 1 represents the best possible value. The final values for each criterion are also multiplied by 0.9 so they will not get too close to 1, as it may skew the results in the evaluation steps using neutrosophic sets. Scalar values are converted to single-valued neutrosophic sets during evaluation. The fitness functions are as follows:

- Symmetry calculation for aesthetic purposes. The chromosome grid is crossed with a horizontal and vertical slice to form 4 smaller 5×5 grids. Each object is checked to determine if it has an identical vertically and horizontally symmetrically matching object (Figures 2 and 3) in the 5×5 grid. The final results are calculated by dividing the symmetrical matches by the maximal possible matches (each object has two matching objects with touching 5×5 grids) (1). x and y represent the size of the grid, s is a binary value, the value of which is 0 if the object does not have a matching pair. Each object is measured twice for each axis.

$$m = \frac{\sum_{i=0}^{2xy} s}{2xy} \tag{1}$$

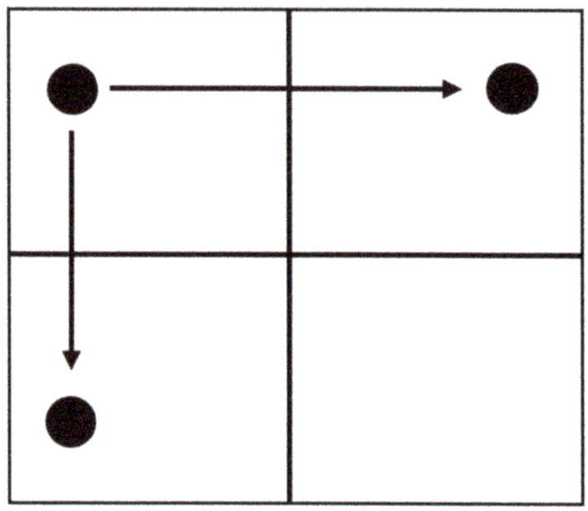

Figure 2. Symmetry calculation.

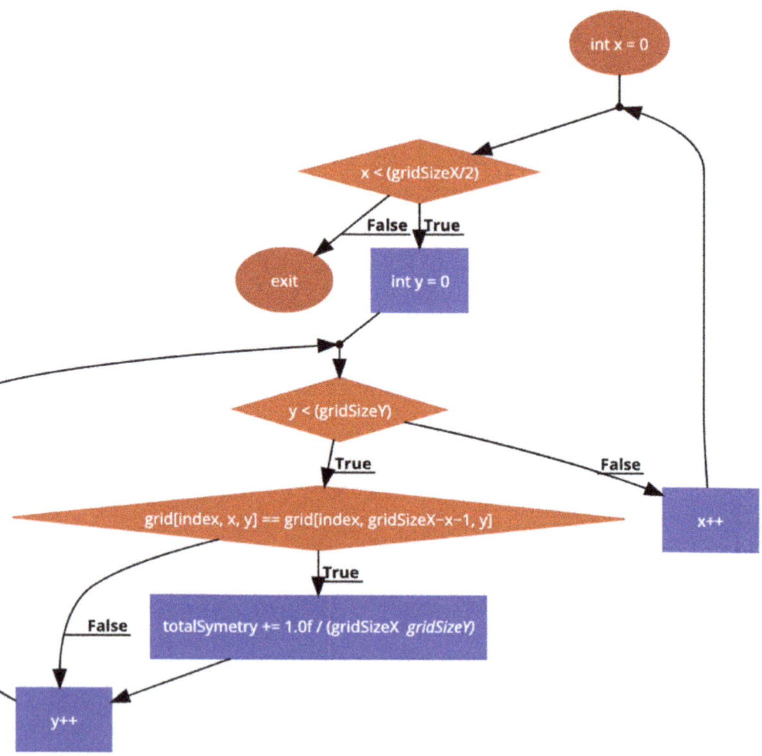

Figure 3. Symmetry calculation for single grid axis.

- Balance criteria for aesthetic purposes. Calculate how close to 50% is the ratio between empty game object count and total object count (Figure 4).

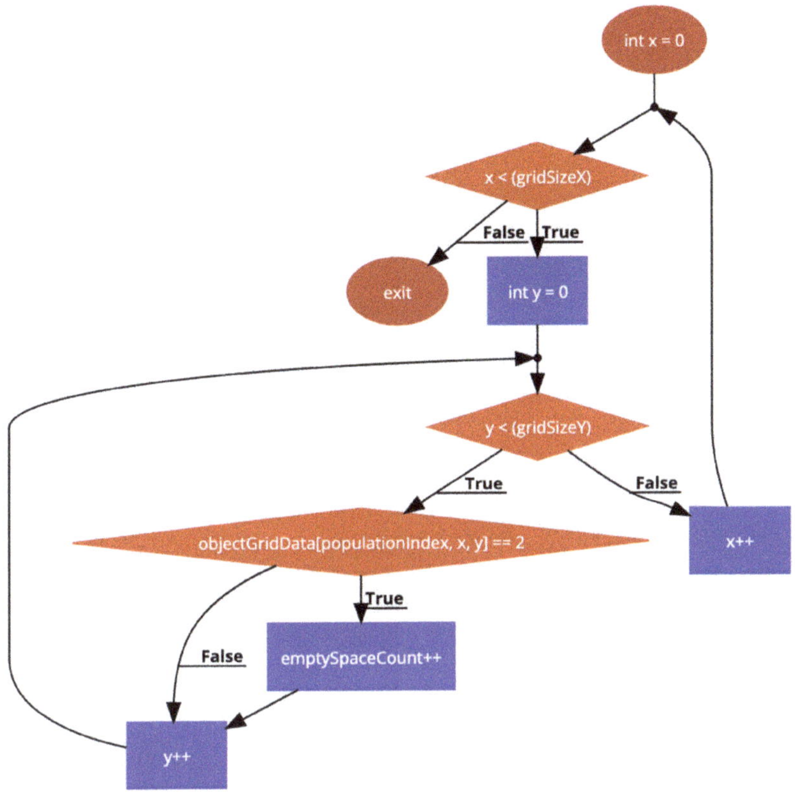

Figure 4. Empty-space balance.

Mathematically, it can be expressed in these steps (2), where e is the total empty space ratio normalized from 0 to 1, t is the sum of the empty objects (they are reversed if they exceed 50% of the grid), x and y represent the grid size, and s is a binary value, which value is 1 if the object is empty. t can be calculated by counting all empty space objects (3) and then reversing the value if it exceeds 50%.

$$e = \frac{t}{\frac{t}{\frac{1}{2}xy}} \tag{2}$$

$$\begin{cases} t = \sum_{i=0}^{n} s \\ t = \frac{1}{2}xy - t_1 - \frac{1}{2}xy \end{cases} \tag{3}$$

- Distance between player and exit game objects. x and y represent the coordinates of the player and exit (4). This rule makes sure that the player can see as much of the generated scene as possible while traveling to the exit point;

$$d = \sqrt{(x2 - x1) * (x2 - x1) + (y2 - y1) * (y2 - y1)} \tag{4}$$

- The safe zone criteria calculate the amount of Hazzard-type objects in a defined square around the Player and divide the result by the total area of this square (5).

$$z = \frac{x_1 y_1}{x_2 y_2} \quad (5)$$

Criteria are calculated for each member of the population (Figure 5) and can be modified on demand. The criteria calculations are the building blocks of the fitness function.

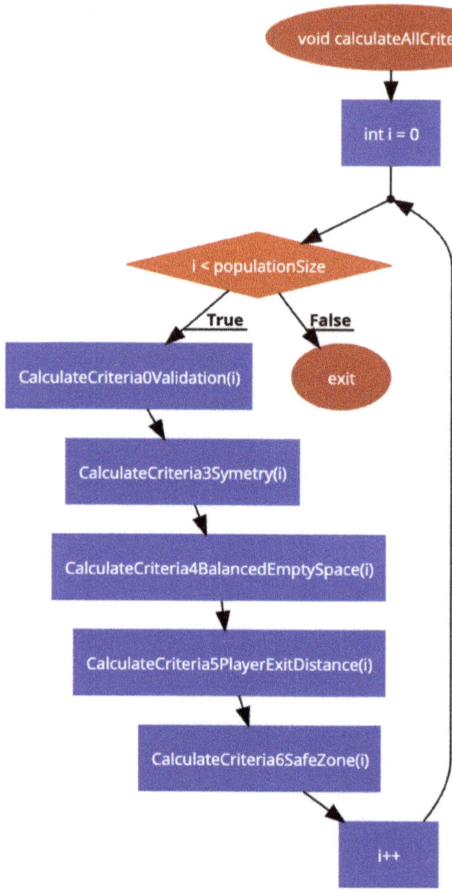

Figure 5. Criteria list.

Constraint functions are:

- Scan the chromosome grid and check if Player object exists;
- Scan chromosome grid and check if an exit object exists;
- Pathfinding algorithm to check if there is a passable way between Player and Exit.

3.3. *Application of WASPAS-SVNS in Genetic Algorithm*

In the evaluation step, we combine all the fitness results of the criteria functions using the modified WASPAS-SVNS algorithm (Figure 6) [19]. Most previous use cases for this algorithm were tested with single iterations [19–22]. This research focuses on an iterative process with WASPAS-SVNS, so there were tweaks made for it to work together with the genetic algorithm. These are the main steps of the final evaluation and explanations about how it was joined with our procedural generator:

1. Combining criteria evaluation data into matrix X where one dimension represents the index of a chromosome, and another dimension represents the index of the criteria (6);

$$X = \begin{bmatrix} x_{11} & x_{12} & \cdots & x_{1n} \\ x_{21} & x_{22} & \cdots & x_{2n} \\ \vdots & \vdots & \ddots & \vdots \\ x_{m1} & x_{m2} & \cdots & x_{mn} \end{bmatrix} \tag{6}$$

2. The original algorithm normalizes the data here, inside the WASPAS-SVNS algorithm, but for the iterative process it does not work because the local min–max and global min–max values are not the same, so we need to define boundaries before this step [33]. Normalization is made in the criteria functions to fit in the range of 0 to 1 (7). v represents current criteria value and v_{max} is the highest possible value for that criterion for the selected matrix size. $\widetilde{x}_{ij}$ is a normalized index ij criteria value of matrix X;

$$\widetilde{x}_{ij} = \frac{v}{v_{max}} \tag{7}$$

3. Neutrosophication step. In this step, we convert results from our normalized criteria function results into neutrosophic sets. The neutrosophic set consists of three numbers: truth (t), intermediary (i), and falsehood (f). For this, we map criteria results with neutrosophic numbers, but we do a linear conversion as even the slightest non-proportional shifts can make a huge error in the long evolutionary run. N represents a neutrosophic number and S represents a scalar number (8);

$$N(t,i,f) = \begin{cases} S \\ 1-S \\ 1-S \end{cases} \tag{8}$$

4. Sum of the total relative importance of the alternative (single evolutionary iteration chromosome);
5. Total relative importance of the product of the alternative;
6. A joint generalized criterion for the ranking alternatives (step 4 and step 5) (9);

$$\widetilde{Q}_i = 0.5\widetilde{Q}_i^{(1)} + 0.5\widetilde{Q}_i^{(2)} \tag{9}$$

7. Neutrosophic numbers (truth, intermediacy, and falsehood) are converted to scalar numbers using this formula and then used for chromosome evaluation in the genetic algorithm (10);

$$s\left(\widetilde{Q}_i\right) = \frac{3 + t_i - 2i_i - f_i}{4} \tag{10}$$

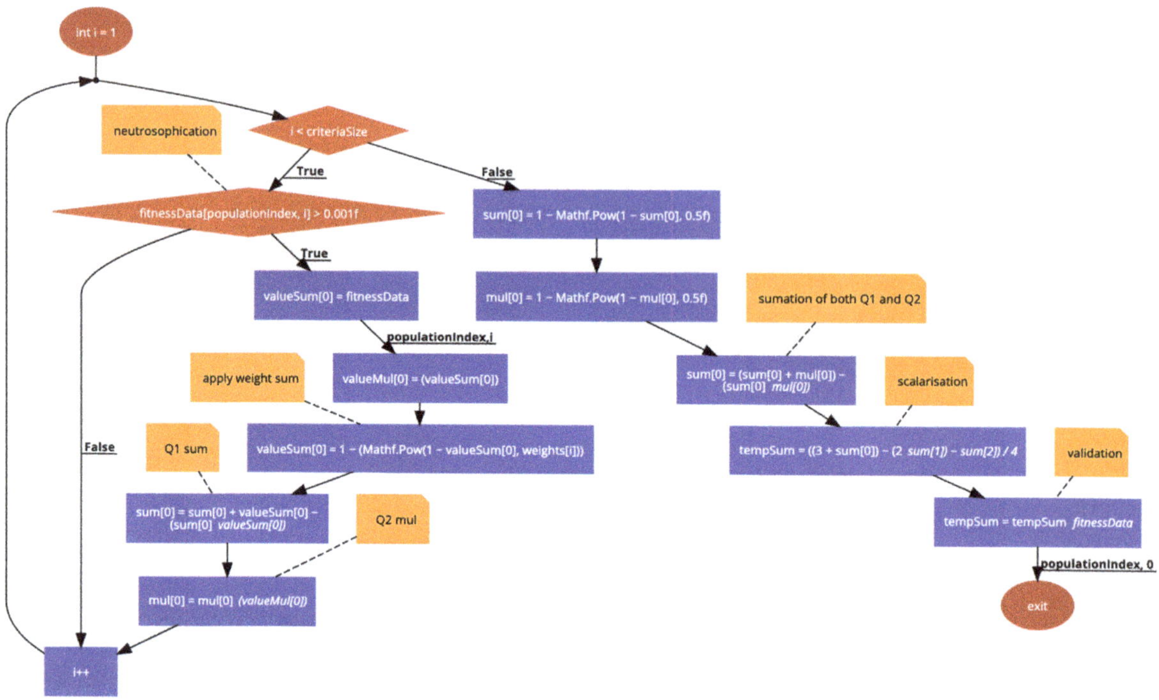

Figure 6. Fitness algorithm by the MCDM approach.

3.4. Proposed Extension of Genetic Algorithm

Genetic algorithms are used to solve optimization problems using natural selection [34]. It is convenient for us, as it can iterate in many different local maximums because of its random nature. It iteratively modifies the population of individual game level grids. At each generation, random grids are selected and modified. With time, the population evolves toward aesthetic and functional solutions. Genetic algorithms allow for finding good solutions without designing them. Transformative genetic algorithm operators are applied to a small set of grid cells. The main advantages from other optimization algorithms are: non-linear convergence, more than one solution can evolve in parallel and best solutions are kept, uses a lot of random numbers so it is not deterministic, and each run of the algorithm proposes different solutions. The fitness function is based on the WASPAS-SVNS algorithm and is used to find the best individuals using several criteria. One grid represents a single solution to a problem. Population defines the total concurrent grid pool. Best fitness shows the best grid designs in the current population. Two concurrent snapshots are used during calculations: parent and child generation.

The level layout is trained with the genetic algorithm (Algorithm 1), and the evaluation criteria for each iteration are combined with the WASPAS-SVNS algorithm to calculate the single fitness value. The population size is set to 50, and the algorithm runs for 2000 iterations. We are using selection and mutation operators to filter and repopulate the population. For initial data, we create empty chromosomes and fill them with random data where each object is coded with integer numbers from 2 to 6 (all possible objects except player (number 0) and exit (number 1)), and then add 1 Player and 1 Exit object. For each iteration, we calculate the median fitness value for all populations and split chromosomes into two temporal arrays, which store chromosomes below and above the median value. Chromosomes below the median value are replaced with chromosomes from the above median array, and then 5% of this new array data is mutated with new random values (Figure 7).

Algorithm 1. Genetic algorithm.

```
InitializeRandomPopulation:
DoFullEvolution:
  for amountOfEvolutionCycles
    CalculateAllCriteria
      for populationSize
        Validation
          PlayerExists
          ExitExists
          PathBetweenPlayer-ExitExists
        Symetry
        EmptySpaceBalance
        Player-ExitDistance
        SafeZone
    FindUnderperformersAndPerformers
      for populationSize calculateFitness
          WASPAS-SVNS
    EvolveUnderperformersWithGeneticAlgorithm
DrawGrid(best fitness):
```

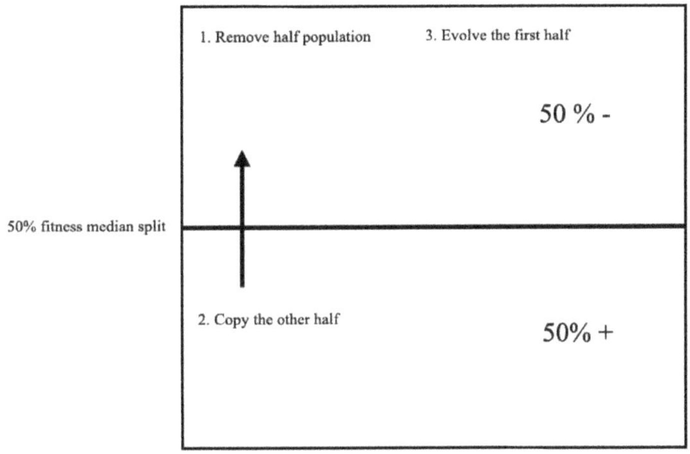

Figure 7. Data evolution.

4. Results

For this research, we developed a framework from scratch using the Unity game engine and visual game object assets from the Unity Asset Store. The results are generated with a custom C# script. Tests were performed with a 2.4 GHz 8-Core Intel Core i9 CPU. Procedural generator with neutrosophic evaluation generates quickly rising scores for the first 100–200 generations under current conditions compared to summation of individual criteria fitness scores, but generator usually requires more time to make symmetrical and visually balanced scene layouts while making sure that game rules apply. The final fitness score usually sets at around 0.75–0.85. It is important to have lots of local maxima for the game scene generation, as results must be unique and differ from each other. There are many possible solutions based on the random initial seed and mutations. Fitness examples with different seeds of random initial data and 500 generations (Figure 8). Note that close to 1 fitness is not possible, as the criteria conflict with each other. Fitness usually starts to converge after about 500–2000 generations. It takes about 21 s, on average, to calculate one 10×10 grid level with 2000 generations. As the initial population (50) is relatively low compared to the total possible scene layouts (7 to power of 100) and mutations are set to

5%, initial seeds usually define how wide the final fitness range is. We choose the lower population to have a wider number of possible solutions. The goal is not to optimize the algorithm for one solution, but to generate a diverse set of level layouts that satisfy the creativity and game design criteria.

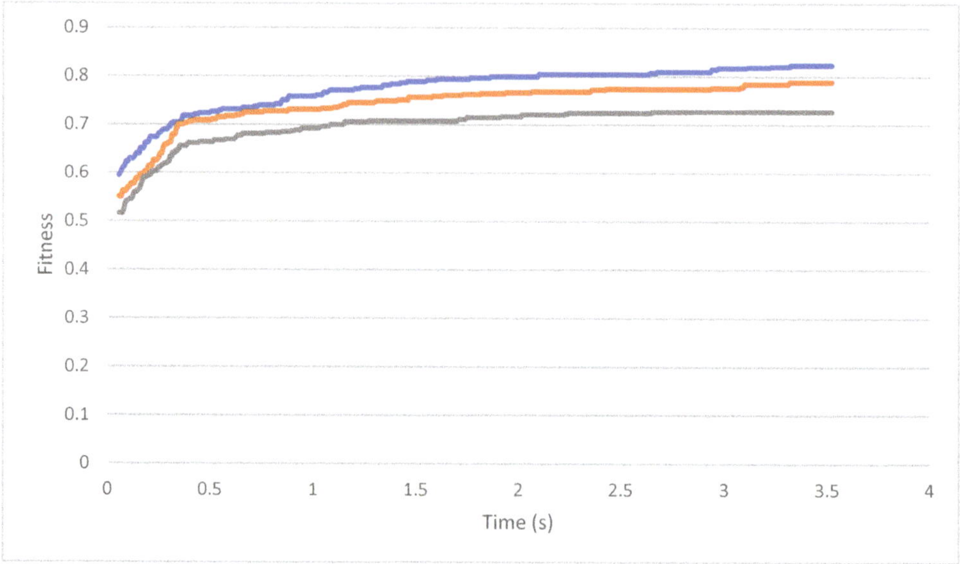

Figure 8. Fitness evolution examples.

We can also observe visual results, which generate aesthetically appealing game scenes. It has many elements of symmetry and space balance. Some examples are: room-like shape without a specific code that defines what a room is (Figure 9), as symmetry is conflicting with other criteria and is not strictly defined, we can also observe semi-symmetric shapes (Figure 10), a smaller room inside the scene with lots of coins/rewards (Figure 11), game scene without lots of walls (Figure 12). We can see that the generator can create many different aesthetic shapes.

Figure 9. Room-like generated scene.

Figure 10. Semi-symmetric results.

Figure 11. Small room inside the scene.

Figure 12. Game scene with almost no walls.

An example of intermediate evolution results can be observed (Figure 13). The grid is printed every 100 generations. We can observe a chaotic layout and quick progression early on and fine-tuning in the later generations.

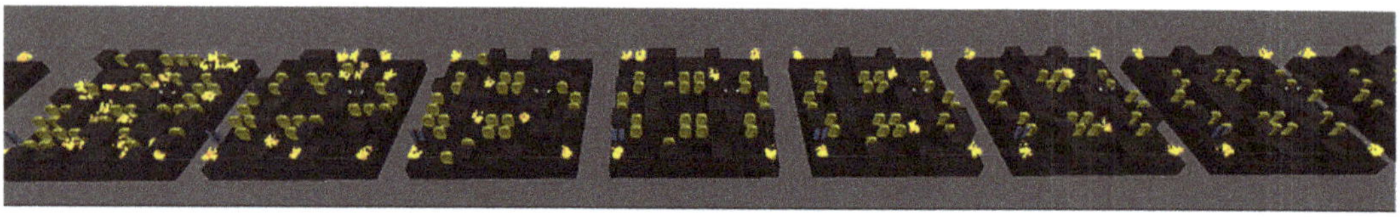

Figure 13. Intermediate evolution results (left to right).

On closer inspection, the realization of the aesthetic criterion can be seen in the visual examples (Figure 14) (symmetry and balance of the empty space balance). At the same time, game design requirements, such as pathfinding, are realized.

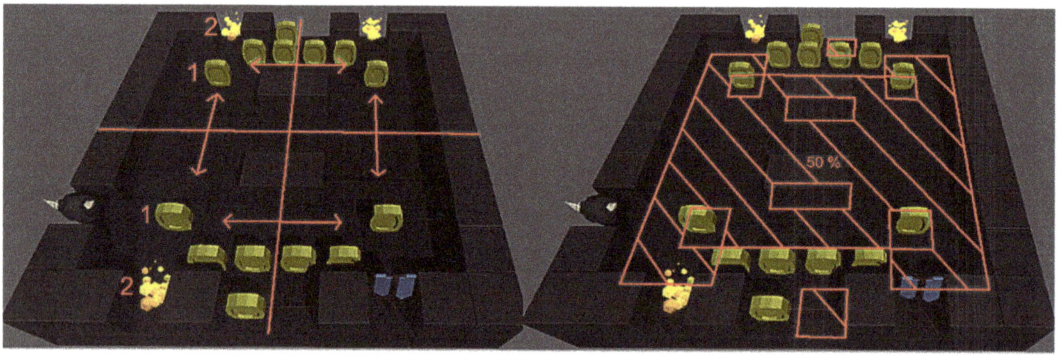

Figure 14. Symmetry and empty space balance. The numbers '1' and '2' represent the corresponding symmetrical objects. '50%' represents the balance between the number of objects and the empty space.

Each result is a different local maximum from many possible final results. It is a very small chance to generate an identical game level given the nature of the algorithm randomness and the number of possible solutions. It would also be possible to draw a crude initial room and then let the algorithm fine-tune it to satisfy aesthetic and playability criteria. The optimal solution for this task is total satisfaction with the proposed criterion under a given random initial seed. Compared to other similar research in the field, the proposed framework generates more visually noticeable aesthetic traits on small object resolutions while maintaining an above-average object pool. This approach to procedural generation has the potential to make creative work faster and easier.

5. Conclusions

The main problem that the proposed method solves is how to increase unique and not repetitive amounts of levels with several runs of the same algorithm. Observing the presented results, it can be concluded that our levels generate interesting game scene layouts, which differ with each run. It can also generate both aesthetic- and functional-level layouts at the same time. Visual representations of game assets can also be interchanged by a developer. The WASPAS-SVNS algorithm enables the evaluation of conflicting criteria. The proposed approach is realized by breaking down design principles into primary elements and defining them with the proposed criteria list. The algorithm generates a random shape and then sculpts a functional and aesthetic game level around that shape. The random nature of the genetic algorithm ensures surprise elements for the levels. It is also important to find a balance between different criteria weights and number of criteria that defines a certain objective to generate a coherent final level. Overabundance or lack of features may numb some of the game design elements. Creativity assistance algorithms can save time for game designers and developers, but at the moment, most commercial games use only light game design assistance tools, seeded procedurally generated or handcrafted levels. This work can be expanded by combining it with an algorithm, which can break down design elements from hand-crafted game levels and then use it as a base of criteria list.

Author Contributions: Conceptualization, A.P. and R.B.; methodology, A.P. and R.B.; software, A.P.; validation, A.P. and R.B.; formal analysis, A.P. and R.B.; investigation, A.P.; resources, A.P.; data curation, A.P.; writing—original draft preparation, A.P.; writing—review and editing, A.P. and R.B.; visualization, A.P.; supervision, R.B.; project administration, R.B. All authors have read and agreed to the published version of the manuscript.

Funding: This research received no external funding.

Institutional Review Board Statement: Not applicable.

Informed Consent Statement: Not applicable.

Data Availability Statement: The data used to support the findings of this research are available upon request to the authors.

Conflicts of Interest: The authors declare no conflict of interest.

References

1. Dick, S. Artificial Intelligence. *Harv. Data Sci. Rev.* **2019**, 1–8. [CrossRef]
2. Lamb, C.; Brown, D.G.; Clarke, C.L.A. Evaluating Computational Creativity: An Interdisciplinary Tutorial. *ACM Comput. Surv.* **2018**, *51*, 1–34. [CrossRef]
3. Carballal, A.; Fernandez-Lozano, C.; Rodriguez-Fernandez, N.; Castro, L.; Santos, A. Avoiding the Inherent Limitations in Datasets Used for Measuring Aesthetics When Using a Machine Learning Approach. *Complexity* **2019**, *2019*, 4659809. [CrossRef]
4. Ventura, D. Mere generation: Essential barometer or dated concept? In Proceedings of the Seventh International Conference on Computational Creativity ICCC 2016, Paris, France, 27 June–1 July 2016. Available online: http://www.computationalcreativity.net/iccc2016/wp-content/uploads/2016/01/Mere-Generation.pdf (accessed on 7 July 2021).
5. Cook, M.; Colton, S.; Gow, J. *Nobody's A Critic: On the Evaluation of Creative Code Generators—A Case Study in Video Game Design*; ICCC: Sydney, Australia, 2013.
6. Toivonen, H.; Gross, O. Data mining and machine learning in computational creativity. *Wiley Interdiscip. Rev. Data Min. Knowl. Discov.* **2015**, *5*, 265–275. [CrossRef]
7. Franceschelli, G.; Musolesi, M. Creativity and Machine Learning: A Survey. *arXiv* **2021**, arXiv:2104.02726 2021.
8. Summerville, A.; Snodgrass, S.; Guzdiaz, M.; Holmgård, C.; Hoover, A.K.; Isaksen, A.; Nealen, A.; Togelius, J. Procedural content generation via machine learning (PCGML). *IEEE Trans. Games* **2018**, *10*, 257–270. [CrossRef]
9. Togelius, J.; Champandard, A.J.; Lanzi, P.L.; Mateas, M.; Paiva, A.; Preuss, M.; Stanley, K.O. Procedural content generation: Goals, challenges and actionable steps. *Dagstuhl Follow-Ups* **2013**, *6*, 61–75.
10. Risi, S.; Togelius, J. Increasing generality in machine learning through procedural content generation. *Nat. Mach. Intell.* **2020**, *2*, 428–436. [CrossRef]
11. Isaksen, A.; Gopstein, D.; Togelius, J.; Nealen, A. Discovering Unique Game Variants Computational. In Proceedings of the Creativity and Games Workshop at the 2015 International Conference on Computational Creativity, Park City, UT, USA, 29 June–2 July 2015.
12. Liapis, A.; Yannakakis, G.N.; Togelius, J. *Limitations of Choice-Based Interactive Evolution for Game Level Design*; AAAI: Stanford, CA, USA, 2012.
13. Cook, M.; Colton, S.; Pease, A.; Llano, M.T. *Framing in Computational Creativity—A Survey and Taxonomy*; ICCC: Charlotte, NC, USA, 2019.
14. Khalifa, A.; de Mesentier Silva, F.; Togelius, J. Level Design Patterns in 2D Games. In Proceedings of the 2019 IEEE Conference on Games (CoG), London, UK, 20–23 August 2019; pp. 1–8. [CrossRef]
15. Togelius, J.; Schmidhuber, J. An experiment in automatic game design. In Proceedings of the 2008 IEEE Symposium on Computational Intelligence and Games, Perth, WA, Australia, 15–18 December 2008; pp. 111–118.
16. Sweetser, P.; Wyeth, P. GameFlow: A model for evaluating player enjoyment in games. *Comput. Entertain.* **2005**, *3*, 3. [CrossRef]
17. Sorenson, N.; Pasquier, P. Towards a Generic Framework for Automated Video Game Level Creation. In *European Conference on the Applications of Evolutionary Computation*; Springer: Berlin/Heidelberg, Germany, 2010; pp. 131–140.
18. Lara-Cabrera, R.; Cotta, C.; Fernández-Leiva, A.J. On balance and dynamism in procedural content generation with self-adaptive evolutionary algorithms. *Nat. Comput.* **2014**, *13*, 157–168. [CrossRef]

19. Lescauskiene, I.; Bausys, R.; Zavadskas, E.K.; Juodagalviene, B. VASMA weighting: Survey-based criteria weighting methodology that combines ENTROPY and WASPAS-SVNS to reflect the psychometric features of the VAS scales. *Symmetry* **2020**, *12*, 1641. [CrossRef]
20. Morkunaite, Z.; Bausys, R.; Zavadskas, E.K. Contractor Selection for Sgraffito Decoration of Cultural Heritage Buildings Using the WASPAS-SVNS Method. *Sustainability* **2019**, *11*, 6444. [CrossRef]
21. Bausys, R.; Juodagalviene, B.; Žiuriene, R.; Pankrasovaite, I.; Kamarauskas, J.; Usovaite, A.; Gaižauskas, D. *The Residence Plot Selection Model for Family House in Vilnius by Neutrosophic WASPAS Method*; VGTU Press: Vilnius, Lithuania, 2020; Volume 24, pp. 182–196.
22. Bausys, R.; Kazakeviciute-Januskeviciene, G. Qualitative rating of lossy compression for aerial imagery by neutrosophic WASPAS method. *Symmetry* **2021**, *13*, 273. [CrossRef]
23. Semenas, R.; Bausys, R.; Zavadskas, E.K. A novel environment exploration strategy by m-generalised q-neutrosophic WASPAS. In *Studies in Informatics and Control*; National Institute for R&D in Informatics, ICI Bucharest: Bucharest, Romania, 2021; Volume 30, pp. 19–28.
24. Bausys, R.; Kazakeviciute-Januskeviciene, G.; Cavallaro, F.; Usovaite, A. Algorithm Selection for Edge Detection in Satellite Images by Neutrosophic WASPAS Method. *Sustainability* **2020**, *12*, 548. [CrossRef]
25. Ali, J.; Bashir, Z.; Rashid, T. WASPAS-based decision making methodology with unknown weight information under uncertain evaluations. *Expert Syst. Appl.* **2020**, *168*, 114143. [CrossRef]
26. Pereira, L.T.; Toledo, C.; Ferreira, L.N.; Lelis, L.H. Learning to Speed up Evolutionary Content Generation in Physics-Based Puzzle Games. In Proceedings of the 2016 IEEE 28th International Conference on Tools with Artificial Intelligence (ICTAI), San Jose, CA, USA, 6–8 November 2016; pp. 901–907.
27. Thakkar, S.; Cao, C.; Wang, L.; Choi, T.J.; Togelius, J. Autoencoder and Evolutionary Algorithm for Level Generation in Lode Runner. In Proceedings of the IEEE Conference on Games (CoG), London, UK, 20–23 August 2019; pp. 1–4.
28. Zafar, A.; Mujtaba, H.; Beg, M.O. Search-based procedural content generation for GVG-LG. *Appl. Soft Comput.* **2020**, *86*, 105909. [CrossRef]
29. Volz, V.; Justesen, N.; Snodgrass, S.; Asadi, S.; Purmonen, S.; Holmgård, C.; Togelius, J.; Risi, S. Capturing Local and Global Patterns in Procedural Content Generation via Machine Learning. In Proceedings of the 2020 IEEE Conference on Games (CoG), Osaka, Japan, 24–27 August 2020; pp. 399–406.
30. Safak, A.B.; Bostanci, E.; Soylucicek, A.E. Automated Maze Generation for Ms. Pac-Man Using Genetic Algorithms. *Int. J. Mach. Learn. Comput.* **2016**, *6*, 226–230. [CrossRef]
31. Petrovas, A.; Bausys, R. Automated Digital Terrain Elevation Modification by Procedural Generation Approach. In Proceedings of the 2019 Open Conference of Electrical, Electronic and Information Sciences (eStream), Vilnius, Lithuania, 25 April 2019; pp. 1–5. [CrossRef]
32. Stanujkić, D.; Karabašević, D.; Popović, G.; Pamučar, D.; Stević, Ž.; Zavadskas, E.K.; Smarandache, F.A. Single-Valued Neutrosophic Extension of the EDAS Method. *Axioms* **2021**, *10*, 245. [CrossRef]
33. Padhye, N.; Deb, K. Multi-objective optimisation and multi-criteria decision making in SLS using evolutionary approaches. *Rapid Prototyp. J.* **2011**, *17*, 458–478. [CrossRef]
34. Whitley, D. A genetic algorithm tutorial. *Stat. Comput.* **1994**, *4*, 65–85. [CrossRef]

applied sciences

MDPI

Article

Exploring Bidirectional Performance of Hotel Attributes through Online Reviews Based on Sentiment Analysis and Kano-IPA Model

Yanyan Chen, Yumei Zhong, Sumin Yu *, Yan Xiao and Sining Chen

College of Management, Shenzhen University, Shenzhen 518061, China; csucyy@163.com (Y.C.); ymzhong7@163.com (Y.Z.); 2060133024@email.szu.edu.cn (Y.X.); 2100133003@email.szu.edu.cn (S.C.)
* Correspondence: yusumin@szu.edu.cn

Abstract: As people increasingly make hotel booking decisions relying on online reviews, how to effectively improve customer ratings has become a major point for hotel managers. Online reviews serve as a promising data source to enhance service attributes in order to improve online bookings. This paper employs online customer ratings and textual reviews to explore the bidirectional performance (good performance in positive reviews and poor performance in negative reviews) of hotel attributes in terms of four hotel star ratings. Sentiment analysis and a combination of the Kano model and importance-performance analysis (IPA) are applied. Feature extraction and sentiment analysis techniques are used to analyze the bidirectional performance of hotel attributes in terms of four hotel star ratings from 1,090,341 online reviews of hotels in London collected from TripAdvisor.com (accessed on 4 January 2022). In particular, a new sentiment lexicon for hospitality domain is built from numerous online reviews using the PolarityRank algorithm to convert textual reviews into sentiment scores. The Kano-IPA model is applied to explain customers' rating behaviors and prioritize attributes for improvement. The results provide determinants of high/low customer ratings to different star hotels and suggest that hotel attributes contributing to high/low customer ratings vary across hotel star ratings. In addition, this paper analyzed the Kano categories and priority rankings of six hotel attributes for each star rating of hotels to formulate improvement strategies. Theoretical and practical implications of these results are discussed in the end.

Keywords: online reviews; hotel attribute; attribute bidirectional performance; sentiment analysis; Kano model; importance-performance analysis

Citation: Chen, Y.; Zhong, Y.; Yu, S.; Xiao, Y.; Chen, S. Exploring Bidirectional Performance of Hotel Attributes through Online Reviews Based on Sentiment Analysis and Kano-IPA Model. *Appl. Sci.* **2022**, *12*, 692. https://doi.org/10.3390/app12020692

Academic Editor: Rui Araújo

Received: 12 November 2021
Accepted: 7 January 2022
Published: 11 January 2022

1. Introduction

Unlike using recommendations of relatives and friends in the past, people increasingly make hotel booking decisions relying on online reviews on various online travel platforms in the modern era. Hotel online reviews are posted by numerous customers according to their experiences in hotels, which are perceived as more objective, trustworthy and helpful than information provided by hotels [1,2]. Online reviews generally consist of online ratings and textual reviews. Online ratings signal customer satisfaction or dissatisfaction with hotels. Textual reviews contain customers' actual expectations, feelings and perceptions about hotel services. According to bounded rationality model, customers are unable to elaborate and extract useful information from numerous and heterogeneous data, thus driving them to prefer and rely more on ratings than on textual reviews [3]. As more and more potential customers regard the online ratings as one of the direct references of hotel quality when selecting hotels, it is crucial for hotels to obtain high customer ratings to achieve the goal of improving online bookings [4,5]. Therefore, exploring what contributes to the difference in online ratings between satisfied and dissatisfied customers is particularly important for hotels. In other words, for the purpose of being competitive sustainably in the hospitality industry, it is critical for hotels to understand the determinants of customer satisfaction and dissatisfaction which are proxied by online ratings [6,7].

Existing studies have proved that the performance of multiple hotel attributes is strongly correlated with customer satisfaction [8–10]. Most studies have investigated the hotel attributes that lead to customer satisfaction and dissatisfaction through surveys [11–13]. Recently, with the development of data mining techniques, online reviews serve as the promising data source for customer satisfaction analysis. Several scholars have analyzed the attribute performance through online reviews using sentiment analysis methods, and hence found the determinants of customer satisfaction in the hotel industry [14,15]. However, these studies processed hotel reviews as a whole dataset, neglecting discriminating positive and negative reviews. Processing hotel reviews as a whole can compare the overall performance of multiple attributes from the perspective of all customers but could not distinguish between contributors of customer satisfaction and factors resulting in customer dissatisfaction. Previous studies have found that dual-valence (that is, featuring both positive and negative sentiment) reviews existing in hotels of one–five-star ratings [16,17]. The presence of negative sentiment toward attributes in positive reviews and positive sentiment toward attributes in negative reviews was observed [18–20]. In other words, even if the performance of several hotel attributes does not meet customer expectations, customers are still satisfied with the hotel and give high ratings to the hotel because of the good performance of other hotel attributes. Meanwhile, customers can be very dissatisfied with the hotel and give low ratings to hotels when the performance of some certain hotel attributes is poor, even though they think other hotel attributes perform well. Therefore, it's necessary to investigate the following question:

Research Question 1 (RQ1). *Which hotel attribute with good performance contributes to high customer ratings and which hotel attribute with poor performance causes low customer ratings?*

In fact, it should be pointed out that customers' expectations and perceptions vary across different market segments, such as different hotel star ratings [14,21]. Exploring the determinants of customer satisfaction and dissatisfaction of each market segment is beneficial for making more appropriate and precise strategies [10]. Moreover, it is helpful for hotel managers to understand customer demands for different star hotels in the decision-making of marching into new markets through comparing the difference of attribute performance in different star hotels. However, whether the hotel attribute contributing to high/low customer ratings varies across different star hotels has not been verified. Therefore, this study intends to investigate the following question:

Research Question 2 (RQ2). *Does the hotel attribute contributing to high/low customer ratings vary across different star hotels?*

To answer the above two questions, it is necessary to analyze the effect of attribute performance on customer satisfaction. Customers' preferences, expectations and perceptions on each hotel attribute are influenced by comprehensive factors, thus driving positive and negative customer evaluations toward the bidirectional (good and poor) performance of hotel attributes [22]. Traditionally, one unit increase in good performance and one unit decrease in poor performance concerning a certain hotel attribute should cause the same change of customer satisfaction, thus the relationship between attribute performance and customer satisfaction is assumed to be linear or symmetric [23]. However, some studies have demonstrated that some attributes provide more satisfaction than dissatisfaction [24–26]. In other words, hotel attributes can have asymmetric effects on customer satisfaction [24]. The Kano model was proposed by Kano et al. (1984) to identify these non-linear or asymmetric relationships between attribute performance and customer satisfaction. The Kano model is often applied to classify hotel attributes into different categories in terms of customer demands, which is helpful for hotel managers to better understand customer expectations and perceptions [27,28]. Meanwhile, considering the limited hotel resources, it is critical to determine attribute priority to maximize customer satisfaction through service improvement. Many studies have shown that applying a combination of the Kano model and importance-performance analysis (IPA) in customer satisfaction analysis can not only analyze customer requirements toward service attributes, but also determine attribute priority [10,29–33]. The IPA is a common and effective technique to formulate improvement strategies according to the importance and performance of the attribute [34]. However, existing studies concerning the Kano-IPA model are mainly based on surveys, and few studies use online reviews as a data source for the Kano-IPA model. There are two main reasons limiting the application of the Kano-IPA model in online reviews. On the one hand, online textual reviews are unstructured and therefore need to be processed before they can be converted into usable structured data. On the other hand, there is a question of how to apply the processed data to the customer satisfaction model to obtain different Kano categories. Considering online reviews serving as promising data source for analyzing and improving hotel services, this study intends to apply feature extraction and natural language processing (NLP) techniques to conduct Kano-IPA model through online reviews.

In summary, this study aims to identify the well-performed attributes contributing to high customer ratings and poorly performed attributes causing low customer ratings for different star hotels. For this, firstly, we distinguish between positive and negative reviews for different star hotels according to online ratings. Next, we apply feature extraction and sentiment analysis techniques to explore bidirectional performance of hotel attributes. In particular, a new sentiment lexicon for hospitality domain was built from numerous online reviews using the PolarityRank algorithm. To further understand customers' rating behaviors and demands for hotel service, this study intends to conduct the Kano-IPA model through online reviews for attribute classification and prioritizing. We propose an approach to classify attributes into the Kano model, which provides convenience for the application of the Kano model in textual reviews. Lastly, the comparative analysis of attribute performance and priority rankings is carried out to enhance the understanding of customers' demands for different star hotels.

The remainder of this paper is organized as follows. Section 2 briefly reviews the relevant literature to provide the motivation for this study. Section 3 presents the framework and methodology employed in this study. Section 4 presents the results and provides some discussion of this study. Section 5 concludes and offers theoretical and practical implications, limitations, and directions for future research.

2. Literature Review

2.1. Studies on Hotel Online Reviews

Hotel online reviews, in the form of online ratings and textual reviews, represent customers' emotions and experience toward service quality based on their expectations against their actual experience. In general, most websites collect customer ratings and opinions on hotels by offering several critical attributes for evaluation. Many researchers have used these hotel attributes to explore customer behaviors. For example, Wang et al. [35] investigated the importance of six attributes including value, location, service, room, cleanliness and sleep quality offered by TripAdvisor.com during the process of hotel selection decision-making. Liu at al. [9] verified the differences of these six hotel attributes' preferences between domestic and international tourists. Bi et al. [10] also used online reviews from TripAdvisor.com to analyze the asymmetric effect of the performance of these six attributes on overall customer ratings. Nicolau et al. [36] analyzed the influence of the variations in the ratings of hotel attributes (comfort, staff, services, value for money and cleanliness) on the variation in the ratings of location to test the halo effect, where these attributes are offered by Booking.com (accessed on 4 January 2022) for evaluation. Evidently, online reviews contain various information of service quality concerning hotel attributes. Thus, it's significant to extract useful information from massive online reviews to help hotel management to improve service quality.

Many previous studies focused on the analysis of numerical ratings, including the overall and multi-attribute ratings. Though overall ratings can indicate customers' overall satisfaction in a straightforward way [37], multi-attribute ratings can obtain a better understanding of factors driving customer satisfaction for different segments within hotels [38]. Sharma et al. [39] classified the multi-attribute ratings into positive, neutral and negative sentiments and then applied interval-valued neutrosophic TOPSIS for ranking hotels. Bi et al. [10] used both overall and multi-attribute ratings to explore the asymmetric effects of attribute performance on customer satisfaction. However, multi-attribute ratings are usually incomplete, limiting utilizing multi-attribute ratings to obtain more information about customers' feelings on service quality of each attribute.

Online textual reviews, a kind of unstructured data source, contain a wealth of information, including customers' preferences, expectations, feelings and perceptions toward hotels [6,38], have gained growing interest among scholars. Especially, with the advance in NLP techniques, more and more studies based on text analysis have been conducted. In the current hospitality research, topic analysis that aims to extract the review's important aspects has been popular and a number of topic mining algorithms have been applied. Guo et al. [40] used Latent Dirichlet Allocation (LDA) topic modeling tool to analyze customers' preferences of hotel attributes. Hu et al. [41] adopted a structural topic model text analysis method to analyze the causes of customers' complaints for hotel service improvement. Wang et al. [35] extracted key factors of different attributes for ranking hotels using the term frequency-inverse document frequency (TF-IDF) and Word2Vec methods. While topic mining is useful to identify key service factors, it cannot reflect whether customers are satisfied with hotel service quality. Sentiment analysis of textual data, which focuses on extracting a review's sentiment polarity, such as positive, negative and neutral, can indicate customers' real emotions and satisfaction toward hotel services [42,43]. Therefore, some hospitality scholars have gradually applied both topic mining and sentiment analysis techniques to capture customers' concerns, emotions, satisfaction and complaints toward hotel services. Bi et al. [44] applied LDA and IOVO-SVM algorithms to identify hotel attributes and their sentiment strengths to conduct IPA plotting for attribute improvement strategies. Al-Smadi et al. [45] used the bidirectional Long Short-Term Memory (LSTM) to extract opinionated aspects and their polarity from Arabic hotels' reviews. Nie et al. [46] applied a semantic partitioned sentiment dictionary to obtain sentiment values of different attributes to rank hotels.

Existing studies suggested that online textual reviews indicate details on customers' demands and perceptions of hotel attributes [6,9]. Although numerical textual reviews have been studied widely to extract critical hotel attributes and their sentiment [14,15,46], few studies have distinguished between positive reviews and negative reviews to identify well-performed attributes contributing to satisfied customers and poorly performed attributes resulting in dissatisfied customers. Therefore, this study attempts to explore the difference between the good performance and poor performance of the same attribute with respect to positive reviews and negative reviews concerning different star hotels through sentiment analysis.

2.2. Studies on Sentiment Analysis

Sentiment analysis has emerged as an important aspect of NLP. Sentiment analysis leverages a variety of NLP techniques to extract the sentiment expressed in texts and determine whether they are positive, negative or neutral [47,48]. Analysis of text sentiments has spread across many fields such as consumer information, marketing, books, application, social media, tourism destination and hotels [49–55]. The approaches to sentiment analysis can be mainly divided into two types, namely, machine learning and lexicon-based methods.

Machine learning methods represent documents as vectors in a feature space and classify them into predefined sentiment categories [56]. There are several machine learning methods for sentiment classification, such as naive Bayes (NB), maximum entropy and support vector machine (SVM). These classifiers usually use the bag-of-words (BoW) method comprising unigrams or n-grams to determine how the documents are represented [57], resulting in high dimensionality of the feature space. With the help of feature selection techniques, such as part-of-speech (POS) tagging which aims to disambiguate text sense based on a lexical category, machine learning algorithms can reduce high-dimensional feature space by eliminating the noisy and irrelevant features. Existing studies concluded that some machine learning classifier algorithms have better performance than lexicon-based [58,59], but these methods have some certain defects: (1) classifiers trained for a domain-specific problem do not perform well in other domains [58]; (2) feature construction is critical but can hardly implement [60] and (3) these methods usually rely on a great volume of manually labeled training data [61]. Given these drawbacks, unsupervised methods like lexicon-based methods are applied.

Lexicon-based methods are based on the assumption that the contextual sentiment orientation is the sum of the sentiment orientation of each word or phrase by matching a word or phrase with words from sentiment lexicon and their associated sentiment scores [62]. In general, adjectives are used as indicators of the semantic orientation of a text [58]. More recently, verbs and nouns are also used to compile into a sentiment dictionary [63]. Such a lexicon or dictionary can be created manually, or automatically, using seed words to expand the list of words [64]. Abdulla et al. [65] built a lexicon for the Arabic language and the proposed approach gained better accuracy than other methods. Taboada et al. [58] constructed a dictionary incorporating intensification and negation to compute text sentiment scores, which is called the semantic orientation calculator (SO-CAL) approach. Dey et al. [66] developed an n-gram sentiment dictionary called Senti-N-Gram for automatic score calculation. Compared to machine learning methods, sentiment lexicons learned from a certain domain preserve the domain-based orientation of words, which provides greater accuracy for sentiment analysis tasks [67]. Furthermore, lexicon-based methods take the lexical and syntactical information in linguistic content into account in order to revise the sentiment valence [56]. That is, in the sentiment scoring process, negation, intensification, and the rhetorical roles of text segments are taken into account as well. The language-dependent features can also be considered in lexicon-based model [68].

In summary, machine learning sentiment analysis trained on a particular dataset by using features, which may reach quite high accuracy in detecting the polarity of a text. However, this is highly dependent on labeled data, limiting its application. Unsupervised lexicon-based methods, such as knowledge-graph propagation and seed word-based methods, not only overcome the absence of labeled data, but are able to extract domain-specific sentiment words [69,70]. Thus, lexicon-based methods are considered preferable for sentiment analysis in a certain domain, and in this study an unsupervised lexicon-based sentiment method is used for sentiment analysis to explore the bidirectional performance of hotel attributes.

2.3. The Kano Model

Traditionally, customer satisfaction has been regarded as one-dimensional or symmetric: the higher the perceived product/service quality is, the higher the customer's satisfaction is and vice versa [23]. However, continuous improvements in product/service attributes without considering what customers actually want may not engender a higher level of customer satisfaction. Some researchers argued that the relationship between attribute performance and customer satisfaction is nonlinear or asymmetric [23,24,71]. Consequently, Kano et al. (1984) introduced a two-dimensional model, called the Kano model, that clarifies the asymmetric and nonlinear relationship between product/service attribute performance and customer satisfaction, and classifies the attributes into five categories, namely, must-be factors, one-dimensional factors, attractive factors, indifferent factors and reverse factors [72]. Later, the simplified Kano model classifies attributes into the following three factors, basic, performance and excitement factors corresponding to must-be, one-dimensional, attractive factors [73], which has been widely used in different research domains, as shown in Figure 1.

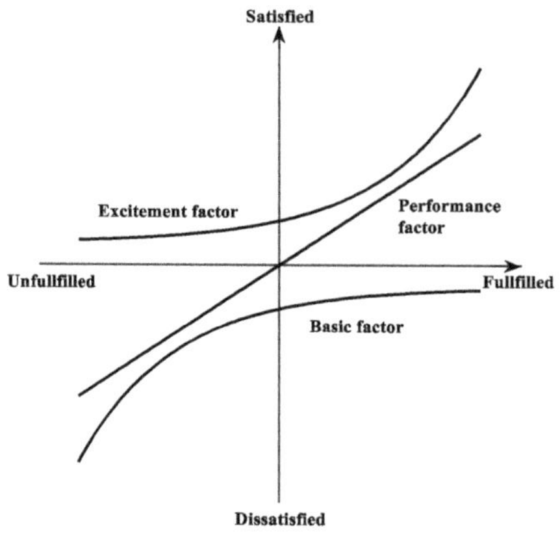

Figure 1. The Kano model.

1. Basic: These attributes are the basic requirements of product/service. Customers are extremely dissatisfied when these attributes don't meet their expectations. However, when customer expectations are exceeded, customers are just neutral since they take it for granted.
2. Performance: The performance of these attributes is positively and linearly related to customer satisfaction. In other words, the customer satisfaction increases with the increase in attribute performance, and vice versa.
3. Excitement: When the performance of these attributes exceeds customer expectations, customers are satisfied, but they are not dissatisfied when these attributes are absent. Therefore, good performance of this category has a stronger impact on customer satisfaction than its poor performance.

Identifying different categories of attributes is beneficial for hotels to understand the determinants of customer satisfaction and dissatisfaction, and hence improving service attributes effectively [29,74]. Many scholars have applied the Kano model to understand customer expectations and perceptions toward different service attributes in hospitality research, as shown in Table 1.

Table 1. Related studies of Kano model in hospitality research.

Literature	Objective	Classifying Method	Sample Source
Yang et al., (2009) [75]	Offer enhanced value to the hotel customer through low prices while meeting appropriate features using refined Kano model and a strategic price model.	Kano's method	Questionnaire
Chang and Chen (2011) [27]	Use the Kano model and quality function deployment (QFD) to explore hotel brand contact elements perceived by customers.	Kano's method	Questionnaire
Tontini et al., (2017) [76]	Explore nonlinear effects of service quality on customers' evaluation of three-star hotels in Rio de Janeiro, Brazil.	CIT and PRCA	TripAdvisor.com
Lai and Hitchcock (2016) [30]	Integrate IPA with the Kano three-factor theory to examine the difference of service attribute importance in different market segments using the case of luxury hotels in Macau.	Importance grid	Questionnaire
Beheshtinia and Farzaneh Azad (2017) [12]	Identify customer needs for the hotels and prioritize them using a combination of the SERVQUAL and Kano approaches.	Kano's method	Questionnaire

Table 1. *Cont.*

Literature	Objective	Classifying Method	Sample Source
Cheng and Chen (2018) [11]	Analyze competitive qualities required for improvement to enhance service quality of motels in Taiwan.	Moderated regression	Questionnaire
Bi et al., (2020) [10]	Explore the asymmetric effects of attribute performance on customer satisfaction with respect to different market segments.	PRCA	TripAdvisor.com

Most of these studies are based on a questionnaire survey, scarcely extracting key attributes from big data such as online reviews. One of problems limiting its application to big data is that it is hard to classify service attributes into Kano categories using existing methods. In current Kano model analysis, several methods are introduced to classify quality attributes. Kano et al. (1984) provided an approach using a structured questionnaire with functional and dysfunctional questions for each attribute [73]. The penalty-reward contrast analysis (PRCA) has been used widely to classify quality attributes by regression analysis with two sets of dummy variables for each attribute [25,73]. The moderated regression approach based on a five-point Likert scale, proposed by Lin et al. [77], uses regression coefficients to classify attributes. Another quantitative method called the "importance grid" has also been applied to a variety of studies, which compares explicit and implicit importance of each attribute to category in three factors [30,78]. Qualitative data methods including critical incident technique (CIT) and the "analysis of complaints and compliments" (ACC) have been applied to category attributes by comparing the difference in attribute frequency mentioned by customers in a positive context or a negative context [76,79,80]. In conclusion, these methods distinguish between different types of attributes by comparing the impacts of good performance and poor performance of the attribute on customer satisfaction.

Most studies relied on questionnaire survey when using the above Kano classifying methods to category attributes, which indicates existing classifying methods may not be suitable for the application of the Kano model in numerous textual reviews because of unstructured feature. Following the Kano model classifying principle, this study aims to propose a novel approach to classify hotel attributes into the Kano model in text analysis. This new approach will provide support for the application of the Kano model in numerous unstructured data to explore customer satisfaction.

2.4. Importance-Performance Analysis

Importance-performance analysis (IPA), proposed by Martilla and James (1977), is a graphical tool to classify attributes for improvement and rank their priority based on the importance and performance of each product/service attribute [34,81]. This approach constructs a plot with two dimensions, importance and performance of product/service attribute perceived by consumers, and classifies attributes into four quadrants equipped with different management strategies. An example of IPA is given in Figure 2. Quadrant I is labeled 'Keep up the good work', where attributes are considered highly important, and their performance is high. Attributes located in quadrant I can be considered as the major strengths of the product/service and should be maintained. Quadrant II is labeled 'Possible overkill', where attributes have low importance but high performance. The resources dedicated to these attributes may be excessive, so reallocating limited resources to address other more important attributes is proper. Quadrant III is labeled 'low priority', where attributes have both low importance and performance. Attributes in this quadrant are regarded as the minor weakness and have a low priority for improvement. Quadrant IV is labeled 'Concentrate here', where attributes are considered highly important but are poorly performed. Attributes in this quadrant are regarded as the major weaknesses and should be given a high priority for improvement.

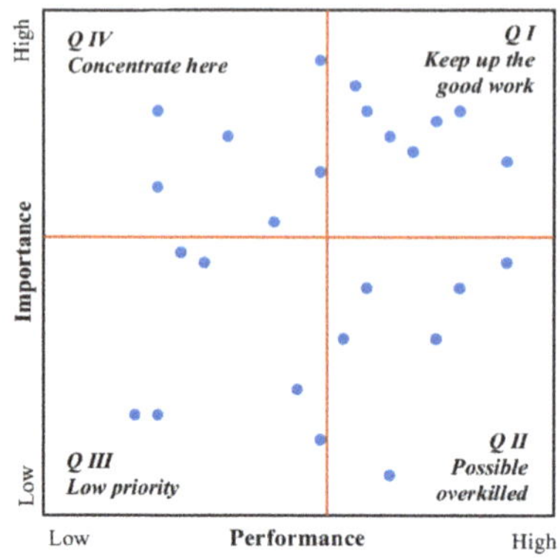

Figure 2. An example of IPA plot.

IPA is applied in a wide variety of research domains, partly due to the clear managerial strategies it provides on how to allocate resources and efforts [82], and also due to its ability to identify the strengths and weaknesses of product/service to guide management in taking effective measures to keep competitive [83]. In hospitality research, IPA is commonly integrated with other techniques, such as SERVQUAL [84,85], data envelopment analysis [86,87], partial least squares path modeling [88], the Kano model [10,30–33]. Many scholars have applied a combination of the Kano model and IPA on customer satisfaction analysis. For example, Bi et al. [10] applied the Kano model and asymmetric IPA to explore the asymmetric effects of hotel attribute performance on customer satisfaction through online ratings. Jou and Day [31] integrated the Kano model and IPA into a three-dimensional IPA approach to identify the critical service attributes for hotel online booking through survey. Tseng [32] constructed an IPA-Kano model for classifying and diagnosing service attributes at the TPE airport. Pai et al. [33] combined the Kano model and IPA to investigate the critical service quality attributes to enhance customer satisfaction in the chain restaurant industry.

However, few hotel studies conducted Kano-IPA analysis using online textual review. Furthermore, few studies have applied Kano-IPA model to obtain the hotel attribute priority ranking for resource allocation to get improved across different hotel star ratings. These literature gaps need to be dealt with. Thus, considering the effectiveness of the Kano model and IPA for providing constructive guidelines to hotels to enhance customer satisfaction, it is of great significance to explore the application of Kano-IPA model in hotel textual reviews across different hotel star ratings.

3. Materials and Methods

The main objective of this study is to explore what contributes to the difference in hotel customer ratings for different star hotels. Specifically, this study identifies well-performed attributes contributing to high customer ratings and poorly performed attributes causing low customer ratings in terms of hotel star ratings by exploring the bidirectional performance of hotel attributes. This study also aims to apply the Kano-IPA model in online textual reviews for a better understanding of customers' rating behaviors and demands, and hence provides effective attribute improvement strategies for different star hotels.

In this section, we propose a methodology to realize the above objectives and the structure of this methodology framework is shown in Figure 3. First, we collected data from TripAdvisor.com and processed the data according to hotel star ratings and customer ratings. Second, sub-attributes of six hotel attributes (value, location, service, room, cleanliness and sleep quality) that customers mentioned in online reviews were extracted. Specifically, similar terms and the similarity under each attribute are identified through the Word2Vec algorithm. Third, a sentiment lexicon for the hospitality domain to obtain sentiment values of each attribute and sub-attribute was obtained through the PolarityRank algorithm. Fourth, well-performed attributes that contribute to customer satisfaction and poorly performed attributes that cause customer dissatisfaction were identified for hotels of different star ratings through sentiment analysis. Finally, the above results by text mining were applied to conduct the Kano-IPA analysis for different star hotels. In particular, a novel approach for Kano model classification is proposed. Thus, the improvement strategies and priority of attributes are provided for different star hotels.

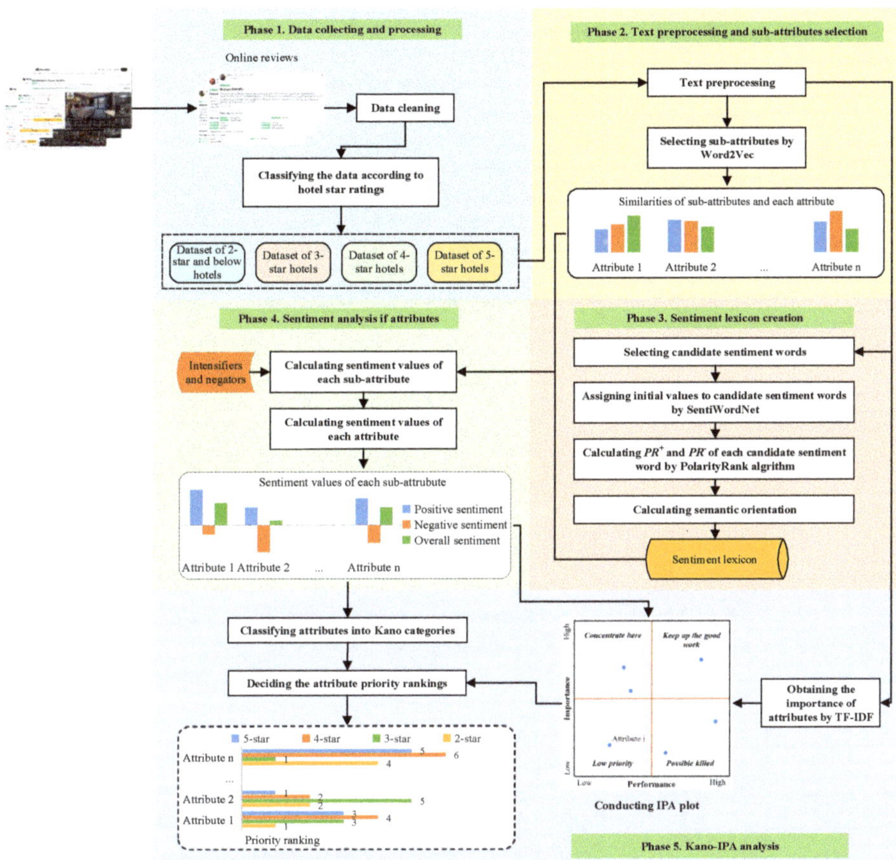

Figure 3. Framework of the proposed methodology.

3.1. Data Collecting and Processing

We collected hotel online reviews in London from TripAdvisor.com, which is the world's largest travel-sharing website. TripAdvisor.com contains millions of unbiased user-generated reviews from customers worldwide; thus, it's feasible to collect a large volume of online reviews. The data collection and processing steps in this paper are as follows.

First, hotels in London were selected as data source for this research. London is one of the largest financial centers in Europe, as well as one of the world's most famous tourist attractions. It attracts millions of customers across the world. Statistically, London recorded 28.47 million bed nights of domestic tourists and 118.9 million nights of international visitors in 2019 [89,90].

Second, we crawled all available information at both hotel-level and review-level in London using a Python program. The hotels with fewer than 400 reviews in English were removed to ensure the credibility of this research sample. A total of 640 hotels with 1,090,341 reviews in English satisfied our requirements. Hotel-level information contains hotel name, star, rating, number of reviews and address. Each review-level data contains reviewer, travel type, posting time, stay time, textual review, overall rating and ratings on six hotel attributes (value, location, service, room, cleanliness and sleep quality).

Finally, we classified the hotel reviews into different datasets according to the hotel star ratings and review overall ratings. Following the studies in [9,91], we categorized the online reviews into four datasets, namely, two-star and below hotels (1-, 1.5-, 2- and 2.5-stars hotels), three-star hotels (3- and 3.5-stars hotels), four-star hotels (4- and 4.5-stars hotels) and five-star hotels according to the hotel star ratings. Review classification based on the review overall rating is controversial [18]. The main argument is whether the 3-score rating reviews should be classified as neutral or negative. Studies have shown that a 3-score evaluation is close to the service failure for most of potential customers [18,92]. Therefore, in this study, according to review overall ratings, online reviews of each hotel star were divided into two sub-datasets respectively, 1–3-score rating reviews as negative reviews and 4–5-score rating reviews as positive reviews. Let D_t^{neg} and D_t^{pos} respectively indicate the negative and positive dataset of t-star hotels, $t = 2,3,4,5$. The final distribution of sub-datasets is shown in Table 2.

Table 2. Distribution of sub-datasets according to the hotel star and overall customer ratings.

	Hotel Star Ratings			
	Two-Star and Below	**Three-Star**	**Four-Star**	**Five-Star**
Total number of hotels	72	231	237	100
negative reviews	21,771	86,720	106,585	21,997
positive reviews	41,486	231,938	408,553	171,291
Total	63,257	318,658	515,138	193,288

3.2. *Text Preprocessing and Sub-Attributes Selection*

3.2.1. Text Preprocessing

Several standard steps were adopted to complete the text preprocessing task by using modules of the Natural Language Toolkit in Python programming environment, including:

- Correcting spelling errors and transforming words with variant spellings (e.g., isn't and is not);
- Sentence segmentation and word tokenization;
- Transforming capital letters to lowercases;
- Removing non-English characters, punctuations and stopwords (an existing stopwords list from https://www.ranks.nl/stopwords) (accessed on 30 October 2021);
- POS tagging;
- Lemmatization (reduce the inflectional forms to their root forms, e.g., rooms and room).

3.2.2. Sub-Attributes Selection

In this study, six key attributes including value, location, service, room, cleanliness and sleep quality [9,10,35,36,46] are selected to explore the role of their bidirectional performance on customer overall ratings. These six attributes are provided by TripAdvisor.com as significant factors for customers to review [35]. Hotel customers use a variety of elements to evaluate the performance of the same attribute [46,93,94]. For example, customers may use "locate", "place" and "distance" to describe the attribute "location". Therefore, extracting words that are semantically similar to each hotel attribute is essential to comprehensively understand customers' opinions.

In this study, we use Word2Vec algorithm to extract words semantically similar to each hotel attribute from textual reviews. Word2Vec is a generative similarity analysis method used to compare the degrees of semantic similarity between two words or two texts. Given a text corpus, Word2Vec learns a vector for each word in the vocabulary using the Continuous Bag-of-Words or the Skip-Gram neural network architectures [95]. Continuous Bag-of-Words is suitable for a small corpus, while Skip-Gram performs better in a large corpus. After training the word vector model, the similarity of the words can be obtained. For this study, gensim is used as library which provides ready-made implementation of Word2Vec algorithm. We trained word vectors from each dataset of different hotel star using the Skip-Gram model. With the pre-trained Word2Vec model for each dataset, the similarity value between attribute A_i and each word in dataset is calculated, where $i = 1, 2, 3, 4, 5, 6$. The words with similarity value under attribute A_i greater than 0.5 are selected as the sub-attribute B_{ij} of attribute A_i, where $j = 1, 2, \ldots, P$ is the number of sub-attributes. Let SS_{ij} denoted the similarity value between sub-attribute B_{ij} and attribute A_i.

3.3. Sentiment Lexicon Creation

We used the PolarityRank algorithm to create a sentiment lexicon from hotel reviews, which has achieved reasonable accuracy without training for domain-specific sentiment analysis [63,96]. The PolarityRank algorithm is a non-supervised sentiment analysis method based on PageRank, with the ability to consider the relevance between nodes, and spread both positive PolarityRank (PR^+) and negative PolarityRank (PR^-) of one node to other nodes through the relevance by edges of weights in a graph [63,96,97]. The main idea behind PolarityRank is to calculate two measures of relevance, the positive and the negative for each node in the graph [63].

Given a text, a graph can be built based on lexical and syntactical dependency, which is named a dependency-based parse tree in NLP. The lexical graph is defined as $G = (N, E)$, where $N = \{g_x\}$ is a set of nodes and E is a set of bidirectional edges between pairs of nodes according to the syntactic dependencies and between all nodes contained in descendant branches. The edge E between node g_x and g_y contains an associated weight denoted by w_{xy}. An example of lexical graph is given, shown in Figure 4. After generating the graph, propagation process with PR^+ and PR^- of each node begins. The detailed descriptions of implementation steps are given in Sections 3.3.1–3.3.3

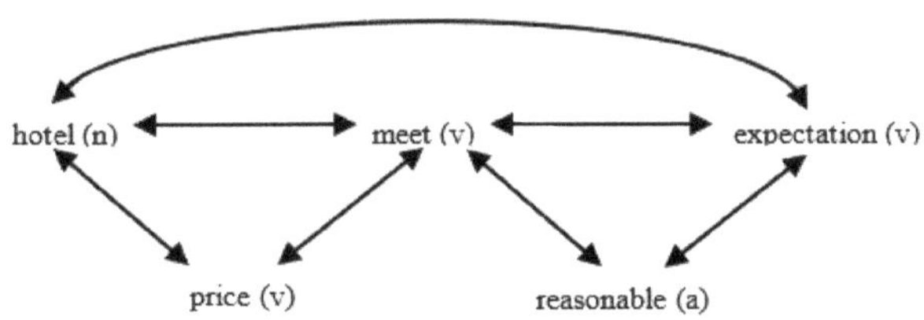

Figure 4. An example of a lexical graph.

3.3.1. Selecting Candidate Sentiment Words

After text preprocessing, the words were lemmatized as nouns, verbs, adjectives, adverbs, pronouns, etc. Previous studies selected the lemmatized nouns, adjectives and verbs as candidate sentiment words, discarding adverbs for they merely alter the degree of the polarity of the words they modify, but do not carry an inherent sentiment polarity [96,98]. Actually, many adverbs carry sentiment polarity, such as the adverb "luckily" in sentence "Luckily, there was one room available" expresses positive emotion.

To accurately analyze customers' feelings, we used all the lemmatized nouns (n), verbs (v), adjectives (a) and adverbs (ad) as candidate sentiment words. The nodes of the graph corresponding to candidate sentiment words from hotel reviews are connected by the bidirectional edges. Following the study of Fernández-Gavilanes et al. [96], the co-occurrence frequency of node g_x and g_y in the whole dataset is assigned to the weight w_{xy} of edge E joining node g_x and g_y.

3.3.2. Assigning Initial Values to Candidate Sentiment Words

In this section, the candidate sentiment words are assigned initial positive value e^+ and negative value e^- by SentiWordNet 3.0 through encoding a Python program. SentiWordNet 3.0 is a general sentiment lexicon publicly available for researchers, with three sentiment scores for each word, namely positive, negative and objective scores [99]. For each candidate sentiment word, we assigned the positive value from SentiWordNet 3.0 to e^+ and the negative value from SentiWordNet 3.0 to e^-. For the words excluded in SentiWordNet 3.0, the e^+ and e^- are equal to zero.

3.3.3. Calculation of PR^+ and PR^-

With weights for edges and pairs of initial sentiment values for nodes, calculation of PR^+ and PR^- could commence. Let $E(g_x)$ be a set of indices y of the nodes for which there exists an edge to node g_x. Then, suppose e_x^+ and e_x^- be the initial positive and negative values of node g_x respectively. The parameter α is set to 0.85 based on the original definition of PageRank, which is a damping factor to ensure convergence [63,97]. The PR^+ and PR^- are estimated as follows:

$$PR^+(g_x) = (1-\alpha)e_x^+ + \alpha \sum_{y \in E(g_x)} \frac{w_{xy}}{\sum_{z \in E(g_y)} w_{xz}} PR^+(g_y) \tag{1}$$

$$PR^-(g_x) = (1-\alpha)e_x^- + \alpha \sum_{y \in E(g_x)} \frac{w_{xy}}{\sum_{z \in E(g_y)} w_{xz}} PR^-(g_y) \tag{2}$$

The propagation process is stopped until the calculation converges or iteration times reach a fixed approximation threshold. In this study, after testing this process, a maximum of 300 iterations is set as the stopping criterion.

3.3.4. Calculation of Semantic Orientation

With the final values PR^+ and PR^-, referred to Cruz et al. [63], semantic orientation SO of each candidate sentiment word is normalized as:

$$SO(g_x) = 5 \cdot \frac{PR^+(g_x) - PR^-(g_x)}{PR^+(g_x) + PR^-(g_x)} \tag{3}$$

Finally, we dropped the candidate sentiment words with a zero SO. Thus, the sentiment lexicon from hotel reviews consists of the words with nonzero SO. Let two-tuple (g_k, SO_k) denote sentiment word g_k and the corresponding sentiment value SO_k, where $SO_k \in [-5, 5]$ and $k = 1, 2, \ldots, m$, with m representing the number of words in the lexicon.

3.4. *Sentiment Analysis of Attributes*

3.4.1. Calculation of Sub-Attribute Sentiment Values

According to the principle of Lexicon-based methods to sentiment analysis, the polarity of a sentence can be obtained from the polarities of words in that sentence [62]. To obtain the sentiment value of each sub-attribute from different sub-datasets, we calculate the sentiment value of each sentence in different sub-datasets and record whether sub-attribute B_{ij} exists in that sentence. For a single dataset, let $\beta_q^l = \left(G_q^l, SO_q^l\right)$ be a two-tuple consisting of the qth sentiment word G_q^l and corresponding sentiment value SO_q^l of the lth sentence, where $l = 1, 2, \ldots, L$, with L denoting the number of sentences in the dataset, $q = 1, 2, \ldots, Q$, with Q denoting the number of sentiment words, and G_q^l belongs to the sentiment lexicon we created. Then, let $\beta^l = \left\{\beta_1^l, \beta_2^l, \ldots, \beta_q^l, \ldots, \beta_Q^l\right\}$ be a set of pairs of sentiment words and the corresponding sentiment values in the lth sentence. For sub-attribute B_{ij} existing in the lth sentence, the sentiment value of B_{ij} in the lth sentence is calculated by the following Equation (4):

$$S_{ij}^l = \begin{cases} \left| \sum_{q=1}^{Q} SO_q^l, \right. & \beta^l \neq \varnothing \\ \left| 0, \right. & \beta^l = \varnothing \end{cases} \tag{4}$$

where $l = 1, 2, \ldots, L$, with L denoting the number of sentences in the dataset.

To improve the accuracy of sub-attributes sentiment polarities, it is important to take the intensifiers and negators into account since these words can affect the sentiment values [46,56]. The sentiment propagation for intensification and negation is described as follows.

Propagation 1: Intensification.

Intensifiers are linguistic terms that primarily combine with adjectives, as well as modify nouns, adverbs and verbs. These words serve to influence the strength of the sentiment word, enhancing or diminishing the sentiment strength. The most common way of identifying these valence shifters is using a list of words, such as adverbs and adjectives, associated with fixed values for intensifiers [100,101]. In this study, we used a list of intensifiers, adapted from Brooke, where each element is a modifier that emphasizes or attenuates words [102]. Let γ_r represent the shift value of intensifier r, where $r = 1, 2, \ldots, R$. Following the above description of sentiment calculation, for a dataset, if there're intensifiers existing in lth sentence, the sum of these shift values γ_l^{sum} is calculated. If not, the γ_l^{sum} is assigned zero. The propagation of S_{ij}^l is represented as:

$$\overline{S}_{ij}^l = S_{ij}^l + S_{ij}^l \, \gamma_l^{sum} \tag{5}$$

where $l = 1, 2, \ldots, L$, with L denoting the number of sentences in the dataset.

Propagation 2: Negation.

In sentiment analysis, negators are the words like "not" that cause negation. Negators could alter the meaning of a word, sentence or provide a negation context, like converting an affirmative statement into a negative statement. The most common way to process negators is attaching these terms to the nearest words [96]; i.e., in "This story is not interesting", the word "interesting" is converted into "NOT-interesting". In this processing method, negators are considered as polarity shifters of polar expressions that produce the opposite polarity. In other words, the polarity value was simply inverted if a polar expression fell within the negation scope [101]. Thus, as the term "perfect" assigned a positive sentiment value of +4, "NOT-perfect" has the sentiment value of −4. However, some researchers hold the opinion that it is more reasonable to decrease the strength of sentiment words rather than directly invert them [96,102]. We use a list of negators, adapted from Brooke, where the negators are used as sentiment shifter with a default shift value of 4 [102]. If there's at least one negator existing in the lth sentence, the negation propagation begins and is represented by Equation (6):

$$\hat{S}_{ij}^l = \begin{cases} |\overline{S}_{ij}^l + 4, & \overline{S}_{ij}^l < 0 \\ |0, & \overline{S}_{ij}^l = 0 \\ |\overline{S}_{ij}^l - 4, & \overline{S}_{ij}^l > 0 \end{cases} \tag{6}$$

where $l = 1, 2, \ldots, L$, with L denoting the number of sentences in the dataset.

3.4.2. Calculation of Attribute Sentiment Values

For the purpose of ensuring that we get the pure positive sentiment value of attribute A_i in each positive dataset, only the positive sentiment value of each sub-attribute under attribute A_i is retained. In other words, the negative sub-attribute sentiment values in the positive dataset are re-assigned to zero, i.e., in the 6th sentence of five-star positive dataset D_5^{pos}, the sentiment value of sub-attribute B_{12} is equal to −3 denoting $\hat{S}_{12}^6 = -3$, and then it should be re-assigned to zero. Similarly, the positive sub-attribute sentiment values in each negative dataset are re-assigned to zero. Let $\widetilde{S}_{ij}^{posl}$ indicate the re-assigned sentiment value of sub-attribute B_{ij} in the lth sentence of the positive dataset, and $\widetilde{S}_{ij}^{negl}$ indicate the re-assigned sentiment value of sub-attribute B_{ij} in the lth sentence of the negative dataset. These two concepts can be computed as follows:

$$\widetilde{S}_{ij}^{posl} = \begin{cases} |0, & \hat{S}_{ij}^{l} < 0 \\ |\hat{S}_{jl}^{l}, & \hat{S}_{ij}^{l} \geq 0 \end{cases} \quad (7)$$

$$\widetilde{S}_{ij}^{negl} = \begin{cases} |\hat{S}_{jl}^{l}, & \hat{S}_{ij}^{l} < 0 \\ |0, & \hat{S}_{ij}^{l} \geq 0 \end{cases} \quad (8)$$

where $l = 1, 2, \ldots, L$, with L denoting the number of sentences in the dataset.

Given that the sub-attribute B_{ij} is the homonymsemantic similar word of attribute A_i but not exactly equal to A_i, it's necessary to consider the semantic similarity between sub-attribute B_{ij} and attribute A_i. Let SC_{ij}^{pos} indicate the positive sentiment value of sub-attribute B_{ij} under attribute A_i in the positive dataset, and SC_{ij}^{neg} indicate the negative sentiment value of sub-attribute B_{ij} under attribute A_i in the negative dataset. Considering the semantic similarity between sub-attribute B_{ij} and attribute A_i, SC_{ij}^{pos} and SC_{ij}^{neg} are estimated as follows:

$$SC_{ij}^{pos} = SS_{ij} \sum_{l=1}^{L} \widetilde{S}_{ij}^{posl} \quad (9)$$

$$SC_{ij}^{neg} = SS_{ij} \sum_{l=1}^{L} \widetilde{S}_{ij}^{negl} \quad (10)$$

Finally, with sub-attribute overall sentiment values, the sentiment values of each attribute in different datasets can be calculated. The sentiment values of attribute A_i in the positive dataset and negative dataset are calculated respectively, as shown in Equations (11) and (12):

$$SC_{i}^{pos} = \sum_{j=1}^{P} SC_{ij}^{pos} \quad (11)$$

$$SC_{i}^{neg} = \sum_{j=1}^{P} SC_{ij}^{neg} \quad (12)$$

In addition, we also calculate the sentiment values of each attribute without re-assigned propagation for the following studies. The positive and negative datasets of the same hotel star are merged, and let SC_i represent the overall sentiment value of attribute A_i without discriminating positive and negative reviews, which is estimated as:

$$SC_i = \sum_{j=1}^{P} SS_{ij} \sum_{l=1}^{L'} \hat{S}_{ij}^{l} \quad (13)$$

where $l = 1, 2, \ldots, L'$, with L' indicating the total number of sentences in the review datasets of different hotel star ratings.

3.5. Kano-IPA Analysis

In this study, the Kano-IPA analysis contains three relevant parts. First, the six hotel attributes of each hotel star rating are classified into different categories in order to understand the effect of attribute performance on customer satisfaction. Second, we construct the IPA plot for hotels of different star ratings through analyzing the attributes' importance and performance. Finally, the attribute priority rankings for improvement and resource allocation are given, so the different improvement strategies are provided for hotels of different star ratings. A detailed description of the Kano-IPA analysis is given as below.

3.5.1. Classifying Attributes into Kano Categories

In this study, a new approach to classify hotel attributes into Kano categories is proposed. As the above descriptions in our study, the positive sentiment value SC_i^{pos} of attribute A_i is obtained from customers whose expectations toward hotel attribute A_i has been met or even exceeded. So SC_i^{pos} of attribute A_i indicates the customer satisfaction that attribute A_i can bring when it performs well. Likewise, the negative sentiment value SC_i^{neg} of attribute A_i is obtained from customers who think the attribute realistic performance hasn't met their expectations, which represents customer dissatisfaction that attribute A_i causes when its performance is poor. The overall sentiment value SC_i of attribute A_i is obtained from all customers stayed in the hotels of the same star. Thus, the SC_i is regarded as the expectant customer satisfaction that attribute A_i should generate. In accordance with the obtained SC_i^{pos}, SC_i^{neg} and SC_i, following the previous index value classifying methods [10,24], here we define an index SI to compare the effects of the attributes' good performance and poor performance on customer satisfaction in hotels of the same star rating, and the SI index of attribute A_i can be calculated as:

$$SI_i = \frac{SC_i^{pos} - SC_i}{SC_i - SC_i^{neg}} \tag{14}$$

Obviously, $SI_i \in [0, +\infty]$. The SI index indicates the ratio of the customer satisfaction of good performance to the customer dissatisfaction of poor performance comparing with the expectant customer satisfaction of the overall performance concerning attribute A_i. To determine the Kano category of each hotel attributes, a cut-off point θ is defined subjectively. According to the testing results based on different assignment methods in these review datasets, we define $\theta = \left(SI^{MAX} - SI^{MIN}\right)/6$, where SI^{MAX} and SI^{MIN} represent the largest and smallest values of the SI index among the six hotel attributes. Moreover, the mean of the SI index among the six hotel attributes is calculated, denoting $\overline{SI}$. Hence, hotel attributes can be classified into Kano categories as follows:

If $0 \le SI_i < \overline{SI} - \theta$, attribute A_i is regarded as basic factor, indicating attributes in this category bring more customer dissatisfaction compared to other attributes.

If $\overline{SI} - \theta \le SI_i \le \overline{SI} + \theta$, attribute A_i is regarded as performance factor, indicating attributes in this category bring equal or approximate customer satisfaction and dissatisfaction compared to other attributes.

If $SI_i > \overline{SI} + \theta$, attribute A_i is regarded as excitement factor, indicating attributes in this category bring more customer satisfaction compared to other attributes.

3.5.2. Constructing the IPA Plot

In this section, we try to construct an IPA plot of the six attributes. From Section 4.4, SC_i indicating the overall performance of each attribute A_i, so our next task is to estimate the importance of each attribute. In this study, the term frequency-inverse document frequency (TF-IDF) algorithm is utilized to estimate the importance of each sub-attribute. TF-IDF is a statistical method, which is widely used to evaluate the relative importance of a word to a particular document in a set of documents or a corpus [35,103]. The term's importance increases as it appears more frequently in the document, but at the same time, its importance decreases as the frequency it appears increases in the whole corpus. Based on TF-IDF algorithm, we defined u_{ij} indicating the weight of sub-attribute B_{ij}. As mentioned above, the sub-attribute B_{ij} is semanticly similar to the attribute A_i and the similarity SS_{ij} indicating the degree of semantic proximity. Therefore, we adopted the processing method of attribute importance from the study of Wang et al. [35], and the attribute importance is calculated as follows:

$$u_i = \frac{\sum_j^P u_{ij} * SS_{ij}}{\sum_i^6 \sum_j^P u_{ij} * SS_{ij}} \tag{15}$$

With the performance and importance of each attribute, the IPA plot can be constructed. The IPA plot is drawn with importance on the vertical axis and performance on the horizontal axis, with the crosshair located inside based on the data-centered method [104], as shown in Figure 2. According to IPA, hotel managers should improve the attributes in Q IV and Q III in that order, maintain the attributes in Q I, and finally consider reducing investment for attributes in Q II [10,29].

3.5.3. Analyzing the Attribute Priority Rankings

Due to the limitation of hotel resource and efforts, the detailed priority rankings for resource allocation in the same quadrant still need to be determined. The Kano model indicates that the effect of attribute performance on customer satisfaction varies from different Kano categories. According to product lifecycle, the attributes of a product or service are regarded as excitement, performance and basic factors [32], which provides a guideline for resource allocation. Specifically, the basic factors should be given the first priority to fulfill, the performance factors should be put in the second order to fulfill, and the excitement factors are given the lower priority to fulfill [10,29]. Therefore, based on the integrated Kano-IPA model, the attribute priority rankings for resource allocation are as shown in Table 3.

Table 3. The attribute priority rankings for resource allocation.

Kano Category	IPA Strategy			
	Concentrate Here	Low Priority	Keep Up the Good Work	Possible Overkill
Basic	1	4	7	10
Performance	2	5	8	11
Excitement	3	6	9	12

4. Results and Discussion

4.1. Results of Sub-Attributes Selection

According to the process described in Section 3.2, sub-attributes and the corresponding similarity under each attribute are obtained from online reviews through Word2Vec algorithm. The sub-attributes of each attribute are sorted by the similarity values. Due to space limitations, we only show the top 10 similar sub-attributes with respect to the six attributes extracted from the five-star hotel reviews in Table 4. In Table 4, "Similarity" indicates the similarity between sub-attributes and the corresponding attribute. Considering the six attribute terms also appear in textual reviews, the six attribute terms are also considered as sub-attributes of themselves. For example, room is a sub-attribute of attribute room, and the similarity is 1. As results shown in Table 4, we find that some terms may be the sub-attributes of two or more attributes. For example, the similarity between the term "bed" and attribute room is 0.5867, meanwhile the similarity between the term "bed" and attribute sleep quality is 0.6537. That is, term "bed" is a sub-attribute of both attributes room and sleep quality. This observation is similar to the sub-attributes (or key factors) selection findings of Wang et al. [35] and Nie et al. [46], indicating that the scopes of different attributes may overlap.

4.2. Results of Sentiment Lexicon from Hotel Reviews

According to the process given in Section 3.3, the PolarityRank algorithm is employed to create a sentiment lexicon from the corpus composed of all textual reviews after preprocessing.

Based on the selecting criteria, the nouns, adjectives, verbs and adverbs with POS are selected as candidate sentiment words. To ensure the efficiency of the PolarityRank algorithm, the final list of candidate sentiment words is composed of words that exist in at least 30 reviews. A total of 13,933 candidate sentiment words and the co-occurrence frequency of any two nodes in the whole dataset are obtained. Subsequently, the initial positive and negative sentiment values of each candidate word are assigned based on SentiWordNet. The results of candidate sentiment words with POS, frequency and initial sentiment value are shown in Table 5. In Table 5, "Tag" indicates the POS of each candidate sentiment words, and "Number of Words" indicates the number of times that candidate sentiment words appear in the whole corpus.

Table 4. Top 10 similar sub-attributes with respect to the six attributes from five-star hotels.

Value		Location		Service	
Sub-Attributes	**Similarity**	**Sub-Attributes**	**Similarity**	**Sub-Attributes**	**Similarity**
represent	0.6609	position	0.7390	middling	0.6122
competitively	0.6373	localisation	0.6764	exemplar	0.6041
comparative	0.6291	harrow	0.6467	fatless	0.5972
money	0.6283	fatherland	0.6444	outstanding	0.5942
ratio	0.6070	breckenridge	0.6380	approachability	0.5908
introductory	0.6064	geest	0.6367	topnotch	0.5858
comparably	0.6055	truistic	0.6341	cleanness	0.5787
reasonable	0.5948	locality	0.6308	tentativeness	0.5771
affordability	0.5733	heartland	0.6254	exceptional	0.5751
inline	0.5720	kenton	0.6245	staff	0.5668
Room		**Cleanliness**		**Sleep Quality**	
Sub-Attributes	**Similarity**	**Sub-Attributes**	**Similarity**	**Sub-Attributes**	**Similarity**
plushly	0.6636	spotless	0.7262	soundly	0.8287
suite	0.6609	immaculate	0.6335	undisturbed	0.7309
uprate	0.6463	tidy	0.6275	hypnos	0.6987
handspring	0.6425	spacey	0.6169	insomniac	0.6950
furbished	0.6340	furbished	0.6079	restless	0.6841
cubit	0.6321	conformable	0.6057	soundproofed	0.6737
spacious	0.6289	appoint	0.6056	silent	0.6722
pristinely	0.6284	spacious	0.6055	pillow	0.6691
spacey	0.6214	equipped	0.5960	uninterrupted	0.6646
luminous	0.6207	scrupulously	0.5927	blackout	0.6643

Table 5. Candidate sentiment words with tag, frequency and initial sentiment value.

Terms	Tag	Number of Words	e^{+}	e^{-}
adequate	a	27,210	0.0795	0.0682
close	a	2257	0.1810	0.0697
abusive	a	120	0	0.8750
affable	a	115	0.6250	0
inspirational	a	75	0.6250	0
trusty	a	52	0.5000	0
accomplished	a	51	0.4432	0
mannerly	a	44	0.7500	0
stubborn	a	40	0	0.6667
immune	a	38	0.0900	0.0950
scramble	n	213	0	0.0833
defect	n	95	0	0.0950
blind	n	68	0	0.0300
despair	n	44	0.0833	0.3750
easiness	n	38	0.0568	0.2045
unwillingness	n	33	0	0.5000
enhancement	n	32	0.3750	0
prejudice	n	30	0	0.8750
horribly	r	508	0	0.7500
painfully	r	506	0	0.0833
marvellously	r	70	0.5000	0.1250

Table 5. *Cont.*

Terms	Tag	Number of Words	e^+	e^-
pathetically	r	51	0.3333	0.1667
guiltily	r	38	0.1250	0
responsibly	r	31	0.5000	0
stop	v	2146	0	0.0183
hate	v	283	0	0.7500
free	v	269	0.2523	0.0103
adore	v	97	0.5000	0.1250
desire	v	44	0.1705	0.0341
respect	v	34	0.4583	0
...	...	...	...	...

Based on the PolarityRank algorithm, PR^+ and PR^- of each candidate sentiment word can be estimated by Equations (1) and (2). The PolarityRank algorithm propagation process stopped until convergence. Additionally, *SO* of each candidate sentiment word can be calculated by Equation (3). According to the results, we can see the *SO* of some candidate sentiment words is equal to zero. The word with a zero *SO* is dropped because it is neutral without sentiment polarity. Finally, the sentiment lexicon composed of 5837 sentiment words with nonzero *SO* is created for attribute sentiment analysis. Due to space limitations, only the results of top 10 positive and negative sentiment words are shown in Table 6.

Table 6. TOP 10 positive and negative sentiment words.

Sentiment Words	Tag	Number of Words	PR^+	PR^-	*SO*
deferentially	r	33	0.8250	0.0051	4.9382
upbeat	n	45	1.2375	0.0077	4.9380
mannerly	a	44	1.8582	0.0134	4.9282
glowing	a	32	1.8582	0.0135	4.9279
topnotch	a	33	1.5494	0.0121	4.9224
uxorious	a	35	1.2398	0.0099	4.9205
fortune	n	41	1.5504	0.0132	4.9157
maestro	n	43	1.5505	0.0132	4.9156
amusingly	r	32	1.8606	0.0159	4.9153
rosy	a	48	1.5389	0.0139	4.9103
mustiness	n	31	0.0128	2.0339	−4.9372
untrustworthy	a	33	0.0119	1.7442	−4.9321
unemployed	a	34	0.0104	1.4540	−4.9290
untypical	a	31	0.0129	1.7452	−4.9265
malodorous	a	42	0.0163	2.0373	−4.9206
bogus	a	32	0.0140	1.7464	−4.9203
prejudice	n	30	0.0170	2.0380	−4.9174
damage	v	30	0.0149	1.7472	−4.9155
ungraded	a	52	0.0124	1.4560	−4.9154
egregious	a	33	0.0177	2.0387	−4.9141

From the results of sentiment lexicon, we find that some words that may not be used in daily life, but express emotions are identified. These less-common words are identified from numerous user-generated data of hotel domain through PolarityRank algorithm. This sentiment lexicon preserves some terms particular to the hotel domain, and hence it is a preferable choice to be used for sentiment analysis of hotel attributes to ensure greater accuracy [67].

4.3. Results and Discussions of Attribute Bidirectional Performance

4.3.1. Results of Attribute Bidirectional Performance

With the obtained sentiment lexicon, the sentiment values of sub-attributes under each attribute in t-star negative dataset D_t^{neg} and positive D_t^{pos} are respectively obtained by Equations (4)–(10). Each sub-attribute has bidirectional performance, represented by positive sentiment value and negative sentiment value. The top 20 well-performed sub-attributes with the strongest positive sentiment polarity and top 20 poorly performed sub-attributes with the strongest negative sentiment polarity with respect to six attributes in five-star hotel reviews are shown in Tables 7 and 8. The results indicate that the bidirectional performance of the same sub-attributes may affect customer satisfaction differently. For example, considering the sub-attribute "decorate" under attribute room, its positive sentiment value from positive reviews is 18,064.37, ranked 3, but on the other hand, the negative sentiment value from negative reviews is −61.57, ranked 17. That is, for "decorate", customers tend to give much more praises when it performs well, whereas customers are probably not sensitive to its poor performance. The observation provides support for the existence of asymmetric relationship between attribute performance and customer satisfaction [24,72].

Table 7. Top 20 well-performed sub-attributes with the strongest positive sentiment polarity with respect to six attributes in five-star hotels.

Value		Location		Service	
Sub-Attributes	**Sentiment Values**	**Sub-Attributes**	**Sentiment Values**	**Sub-Attributes**	**Sentiment Values**
quality	9287.06	hotel	127,281.60	service	83,160.98
price	7723.98	location	51,065.79	staff	77,198.91
value	7457.58	great	47,273.65	excellent	19,643.26
overall	4718.63	London	43,130.88	food	18,286.84
cheap	3431.94	place	15,842.34	experience	17,114.62
money	3167.76	perfect	11,355.46	quality	9531.05
rate	3115.13	close	6832.03	attentive	8627.21
deal	2744.77	station	6794.12	superb	7623.67
compare	1427.76	locate	6678.65	outstanding	5455.66
reasonable	1384.81	central	5362.67	attention	4843.12
opinion	874.96	tube	5162.28	exceptional	3612.60
discount	441.41	overall	5036.61	deliver	2642.43
represent	344.97	near	3440.62	impeccable	1426.53
inexpensive	282.49	distance	3431.87	exemplary	1093.08
bargain	226.16	heart	2727.13	cleanliness	1029.09
affordable	176.04	attraction	2529.99	presentation	415.14
favourably	169.36	convenient	2336.89	unmatched	230.21
competitive	164.94	boutique	2066.52	noteworthy	157.20
ratio	73.53	ideal	1999.84	commendable	155.89
phenomenally	52.47	spot	1966.29	incomparable	70.66
Room		**Cleanliness**		**Sleep Quality**	
Sub-Attributes	**Sentiment Values**	**Sub-Attributes**	**Sentiment Values**	**Sub-Attributes**	**Sentiment Values**
room	215,490.83	room	112,559.79	bed	21,150.78
bed	20,678.19	very	87,177.71	comfortable	18,250.67
decorate	18,064.37	well	29,182.25	sleep	8989.04
bathroom	16,396.52	clean	26,561.49	comfy	2552.32
clean	13,874.17	comfortable	21,693.18	pillow	2535.55
suite	13,413.97	bed	20,495.64	linen	2019.79
spacious	9802.36	nice	20,308.36	peaceful	1063.44
large	9007.25	excellent	18,263.84	bedding	1058.61
although	5901.86	decorate	17,812.75	wake	913.49
size	5253.50	bathroom	17,204.46	restful	606.27
bedroom	4876.74	amenity	15,456.15	mattress	499.52
appoint	4173.68	modern	10,517.20	duvet	301.12

Table 7. *Cont.*

deluxe	3075.58	spacious	10,358.82	heavenly	255.83
apartment	2444.80	extremely	9127.39	sleeper	250.90
tastefully	2220.49	luxurious	6218.94	silent	177.54
spotless	1873.90	appoint	5207.32	soundly	176.52
superior	1617.76	super	4743.52	blackout	162.35
furnish	1525.96	elegant	4253.82	dreamy	107.29
spotlessly	1203.26	nicely	3671.61	supremely	104.01
junior	1133.58	polite	3554.43	soundproof	103.71

Table 8. Top 20 poorly performed sub-attributes with the strongest negative sentiment polarity with respect to six attributes in five-star hotels.

Value		Location		Service	
Sub-Attributes	**Sentiment Values**	**Sub-Attributes**	**Sentiment Values**	**Sub-Attributes**	**Sentiment Values**
price	−3459.05	hotel	−24,483.27	service	−14,546.04
money	−2399.80	location	−5800.25	staff	−8881.41
value	−1547.67	London	−3815.88	experience	−3077.38
overall	−1298.75	great	−3269.14	food	−2274.30
quality	−1208.69	place	−2614.52	quality	−1240.45
rate	−1145.92	overall	−1386.27	excellent	−708.58
cheap	−861.07	close	−1289.30	deliver	−629.82
deal	−662.34	near	−720.90	attention	−493.67
opinion	−291.08	locate	−560.99	attentive	−261.83
discount	−265.94	station	−509.16	exceptional	−141.31
compare	−265.43	central	−462.92	superb	−138.37
reasonable	−218.69	tube	−332.92	cleanliness	−134.34
represent	−69.34	convenient	−303.54	outstanding	−113.61
bargain	−42.48	base	−302.06	presentation	−50.10
competitive	−27.97	perfect	−287.74	impeccable	−40.52
advertised	−19.33	ideal	−255.15	cleanness	−15.81
disproportionate	−16.96	position	−198.63	exemplary	−10.32
ratio	−16.09	spot	−166.93	servicing	−8.97
commensurate	−13.71	distance	−162.97	middling	−8.62
inexpensive	−12.85	boutique	−145.95	tentativeness	−7.46
Room		**Cleanliness**		**Sleep Quality**	
Sub-Attributes	**Sentiment Values**	**Sub-Attributes**	**Sentiment Values**	**Sub-Attributes**	**Sentiment Values**
room	−53,448.27	room	−27,918.25	bed	−3801.36
bed	−3716.42	very	−11,076.76	sleep	−3001.91
bathroom	−3220.28	clean	−3866.42	comfortable	−886.78
clean	−2019.59	bed	−3683.61	wake	−581.08
although	−1719.86	nice	−3652.38	pillow	−428.58
suite	−1527.98	bathroom	−3378.96	mattress	−300.63
large	−1048.14	well	−3181.78	comfy	−188.85
size	−800.25	comfortable	−1054.04	bedding	−145.26
bedroom	−783.50	extremely	−1009.08	duvet	−136.03
spacious	−417.60	excellent	−658.82	linen	−131.69
deluxe	−364.08	spacious	−441.31	soundproofing	−70.98
superior	−270.01	modern	−437.53	sleepless	−65.18
apartment	−211.28	polite	−300.16	proofing	−51.64
junior	−176.58	luxurious	−235.98	peaceful	−44.81
appoint	−160.08	super	−234.58	insulation	−44.18
furnish	−65.87	amenity	−229.05	soundproof	−40.43
decorate	−61.57	appoint	−199.73	restless	−39.13
spotless	−28.02	comfy	−198.85	comforter	−37.62
furnished	−26.66	maintain	−141.21	disturbed	−34.95
airy	−17.59	functional	−140.51	sleeper	−28.19

By Equations (11)–(13), the sentiment values of each attribute in hotel reviews of different star ratings are calculated. The positive, negative and overall sentiment values of each attribute are given in Table 9 according to the hotel star ratings. According to the negative sentiment values of six attributes in negative reviews, the ranking of poorly performed attributes is derived for hotels of each star rating. Hotels of three-stars, two stars and below have the same poorly performed attribute ranking, while hotels of four stars and five stars also have the same poorly performed attribute ranking. That is, *Room* < *Cleanliness* < *Location* < *Value* < *Service* < *Sleep quality* in negative reviews of three-star, two-star and below hotels, and *Room* < *Cleanliness* < *Location* < *Service* < *Value* < *Sleep quality* in negative reviews of four-star and five-star hotels. Similarly, according to the positive sentiment values of six attributes in positive reviews, the ranking of well-performed attributes is derived for hotels of each star rating. That is, *Cleanliness* > *Location* > *Room* > *Value* > *Service* > *Sleep quality* in positive reviews of three-star, two-star and below hotels, *Cleanliness* > *Room* > *Location* > *Service* > *Value* > *Sleep quality* in positive reviews of four-star hotels, and *Cleanliness* > *Room* > *Location* > *Service* > *Sleep quality* > *Value* in positive reviews of five-star hotels.

Table 9. The negative, positive and overall sentiment values of each attribute according to the hotel star ratings.

Hotel Star Ratings	Sentiment Values	Attribute						
		Value	Location	Service	Room	Cleanliness	Sleep Quality	Mean
Two-star and below	Negative	−44,402.05	−77,480.13	−34,111.69	−113,060.25	−88,818.79	−7304.89	−60,862.97
	Positive	75,603.2	141,847.69	60,384.86	136,324.38	147,678.6	5874.08	94,618.8
	Overall	15,868.76	35,830.05	17,723.35	9693.39	31,797.07	490.12	18,567.12
Three-star	Negative	−157,906.9	−277,368.47	−119,342.87	−371,405.85	−318,294	−43,190.77	−214,584.81
	Positive	369,114.39	707,692.72	294,129.77	606,484.11	777,154.88	65,084.39	469,943.38
	Overall	84,434.11	205,247.57	90,848.91	96,265.24	238,493.17	12,018.31	121,217.88
Four-star	Negative	−102,202.45	−242,558.76	−192,196.55	−409,184.1	−364,975.08	−46,462.47	−226,263.23
	Positive	296,090.96	931,501.55	630,134.85	1,058,169.12	1,384,338.72	81,519.98	730,292.53
	Overall	92,485.82	352,708.09	244,041.18	362,762.47	622,552.97	18,611.68	282,193.7
Five-star	Negative	−13,991.48	−48,218.33	−32,808.41	−70,719.73	−63,615.54	−10,346.97	−39,950.08
	Positive	47,803.81	369,732.22	263,145.74	358,219.8	482,105.65	62,616.95	263,937.36
	Overall	14,494.87	164,098.46	128,769.04	169,039.22	274,199.1	28,743.64	129,890.72

4.3.2. Comparative Analysis of Attributes' Bidirectional Performance

To better analyze the antecedents of both high and low customer ratings, the percentages of negative sentiment values and positive sentiment values concerning six attributes in terms of hotel star ratings are respectively calculated, shown in Figures 5 and 6. Results show that room, cleanliness and location account for about 75% of the sum of attribute negative sentiment values, meanwhile these three attributes also account for about 75% of the sum of attribute positive sentiment values for hotels of each star rating. Room, cleanliness and location are core attributes of hotels, in line with some prior research [35,94,105,106]. This finding also implies the main contributors to high customer ratings and causes of low customer ratings are the same for hotels of each star rating, similar to the studies of Berezina et al. [94] and Kitsios et al. [107]. Value, service and sleep quality have less impact on both high and low customer ratings, contrary to some previous research. For instance, the study of Ban et al. [105] implied that intangible service has the greatest impact on customer satisfaction.

The results also imply that the percentage of positive/negative sentiment values concerning location, value and service fluctuate with hotel star rating. Thus, the effect of good/poor performance concerning location, value, service on high/low customer ratings varies across hotel star ratings. For location, its good/poor performance contributes less to customer satisfaction/dissatisfaction in four-star hotels than in other star hotels. For value, the percentages of both positive and negative sentiment values gradually drop with the increase in hotel star ratings above three-star hotels. That is, for value, poor performance in high-star (four-star and five-star) hotels does not cause as many complaints as in low-star (three-star and below) hotels, and good performance brings less satisfaction for customers of high-star hotels. This finding is consistent with common sense that customers who choose low-star hotels lay greater emphasis on value for money [108], and customers in high-star hotels may take it for granted when value performs well because they spend more [10]. On the contrary, for service, good/poor performance contributes markedly more to high/low customer ratings in high-star hotels than in low-star hotels. The results show the same finding as earlier studies which showed that the effect of service's poor performance on customer dissatisfaction increases with the improvement of hotel level and luxury (i.e., four–five-star ratings) hotel customers emphasize good service [10,21]. Moreover, it is observed that the good performance of sleep quality has the potential to be the incentive for high customer ratings in five-star hotels.

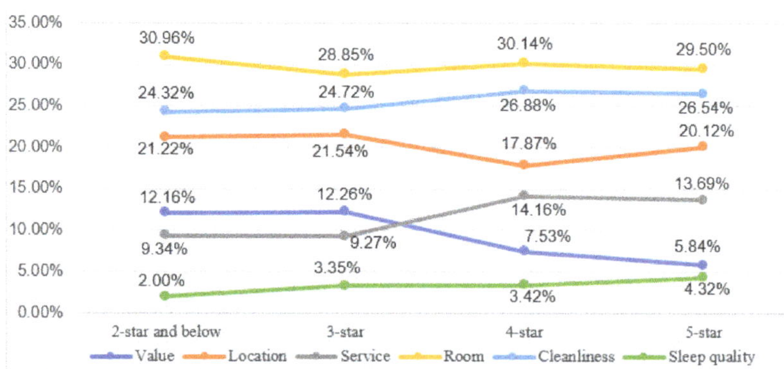

Figure 5. The percentage of negative sentiment values concerning six attributes.

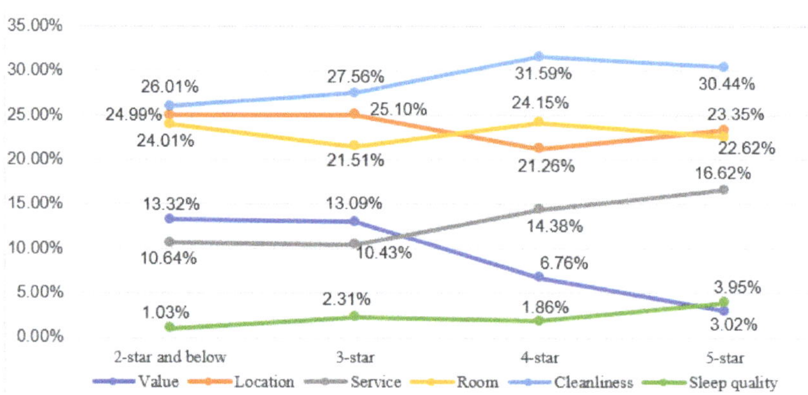

Figure 6. The percentage of positive sentiment values concerning six attributes.

By comparing the lines in Figures 5 and 6 for one certain attribute, it can be found that the impact of the bidirectional performance concerning one attribute on high/low customer ratings is different. For room and sleep quality, the effect of their poor performance on low customer ratings is stronger than the effect of their good performance on high customer ratings. On the contrary, for cleanliness, location and service, the effect of their good performance on high customer ratings is stronger than the effect of their poor performance on high customer ratings. For value, the effect of its good performance on high customer ratings is stronger than the effect of its poor performance on low customer ratings in low-star hotels, while quite the opposite is true for high-star hotels. Therefore, the results indicate that the effect of attribute performance on customer ratings is asymmetric, consistent with many previous studies [8,10,23,24,71,76]. Furthermore, the asymmetric effect of values' performance on customer ratings is different between high-star and low-star hotels.

4.4. Results and Discussions of Kano-IPA Analysis

4.4.1. Attribute Classification Based on the Kano Model

According to the obtained sentiment values and Equation (14), the SI values of six attributes concerning four hotel star ratings can be calculated and further the six attributes are classified into three Kano categories, as shown in Table 10. The final classification of attribute categories is basically consistent with the relative effect of each attribute on customer satisfaction for hotels of the same star rating. On the whole, the categories of all attributes except value vary across different hotel star ratings.

Table 10. The three Kano categories of six attributes concerning four hotel star ratings.

Attribute	Two Stars and Below Hotels		Three-Star Hotels		Four-Star Hotels		Five-Star Hotels	
	SI	Category	*SI*	Category	*SI*	Category	*SI*	Category
Value	0.9911	Excitement	1.1747	Excitement	1.0458	Excitement	1.1693	Excitement
Location	0.9356	Performance	1.0411	Performance	0.9723	Excitement	0.9685	Excitement
Service	0.8230	Basic	0.9671	Basic	0.8851	Performance	0.8317	Performance
Room	1.0316	Excitement	1.0910	Excitement	0.9008	Performance	0.7890	Performance
Cleanliness	0.9607	Performance	0.9674	Basic	0.7714	Basic	0.6154	Basic
Sleep quality	0.6907	Basic	0.9612	Basic	0.9667	Performance	0.8665	Performance
Mean	0.9055		1.0338		0.9237		0.8734	
θ	0.0568		0.0356		0.0457		0.0923	
Upper threshold	0.9623		1.0693		0.9694		0.9657	
Lower threshold	0.8487		0.9982		0.8780		0.7811	

Specifically, value is always classified as an excitement factor, indicating that value can bring more satisfaction when it performs well regardless of hotel star ratings. Unlike the study of Bi et al. [10], this study shows that value is an excitement factor, providing support for the finding (value and price is the attractive factor for four–five-star hotels) of Chiang et al. [13]. Location is classified as a performance factor in hotels of three stars and below and is classified as an excitement factor in hotels of four stars and five stars. Compared with hotels of three stars and below, the good performance of location can bring more customer satisfaction for four-star and five-star hotels. Luxury hotel customers are willing to pay more for a convenient location [21]. Thus, customers in high-star hotels will be very satisfied when the performance of location, which is the core requirement, exceeds their expectations. Service and sleep quality, showing the same change with the increase in hotel star, are classified as basic factors in hotels of three stars and below, and are classified as performance factors in hotels of four and five stars. Thus, customers in hotels of three stars and below may not be sensitive to the good performance of service and sleep quality, but they are dissatisfied when the performance of service and sleep quality is poor. Meanwhile, customers in hotels of four and five stars are sensitive to the bidirectional performance of service and sleep quality. Room is classified as an excitement factor in hotels of three stars and below and is classified as a performance factor in hotels of four and five stars. The good performance of room can bring customer satisfaction in hotels of each star rating, while poor performance of room can bring more customer satisfaction in hotels of four and five stars than in hotels of three stars and below. Cleanliness is classified as a performance factor in hotels of two stars and below and is classified as a basic factor in hotels of three, four and five stars. This result indicates that customers in two-star and below hotels may feel satisfied when the room is clean, but customers in other star hotels take the good performance of cleanliness for granted. These findings are different to the study of Bi et al. [10] who classified service, sleep quality, room and cleanliness as basic factors in hotels of each type.

4.4.2. The IPA Plot

Based on the TF-IDF algorithm, we obtained the weight of each sub-attribute with respect to the six attributes concerning four hotel star ratings. Then the importance of the six attributes concerning four hotel star ratings was calculated respectively by Equation (15), as shown in Table 11. On the whole, the importance of value, service and sleep quality varies across the hotel star ratings, while other attributes' importance fluctuates slightly and are considered as very important for all hotels. This finding is consistent with previous research that revealed that customers of high-star hotels are more likely to value some intangible attributes (i.e., service and sleep quality) [40]. Specifically, with the increase in hotel stars, the importance of value decreases, while the importance of service and sleep quality increases. In other words, customers who select high-star hotels pay more attention to service and sleep quality, and consider value as less important. On the contrary, customers who choose low-star hotels highly emphasize value, but consider service and sleep quality as less important. The importance of value shows a significant downward trend with the improvement in hotel stars, coinciding with Zhao's [108] research.

Table 11. The importance of the six attributes concerning four hotel star ratings.

Hotel Star Rating	Attribute					
	Value	Location	Service	Room	Cleanliness	Sleep Quality
Two-star and below	0.1354	0.2412	0.0990	0.2566	0.2574	0.0104
Three-star	0.1328	0.2446	0.0962	0.2337	0.2690	0.0238
Four-star	0.0692	0.2267	0.1353	0.2508	0.2978	0.0202
Five-star	0.0315	0.2626	0.1545	0.2315	0.2812	0.0387

With the obtained importance and performance of each attribute concerning four hotel star ratings, the IPA plots can be constructed, as shown in Figure 7. Location and cleanliness are located in Q I in hotels of all stars, which indicates that location and cleanliness should be well remained for their high importance and performance. Value, service and sleep quality are located in Q III in hotels of all stars, with low importance and performance, so they are of low priority for improvement. In contrast, room is located in Q IV in hotels of three stars, two stars and below, urgent to be improved, while it is located in Q I in four and five star hotels, indicating it is the hotels' strength.

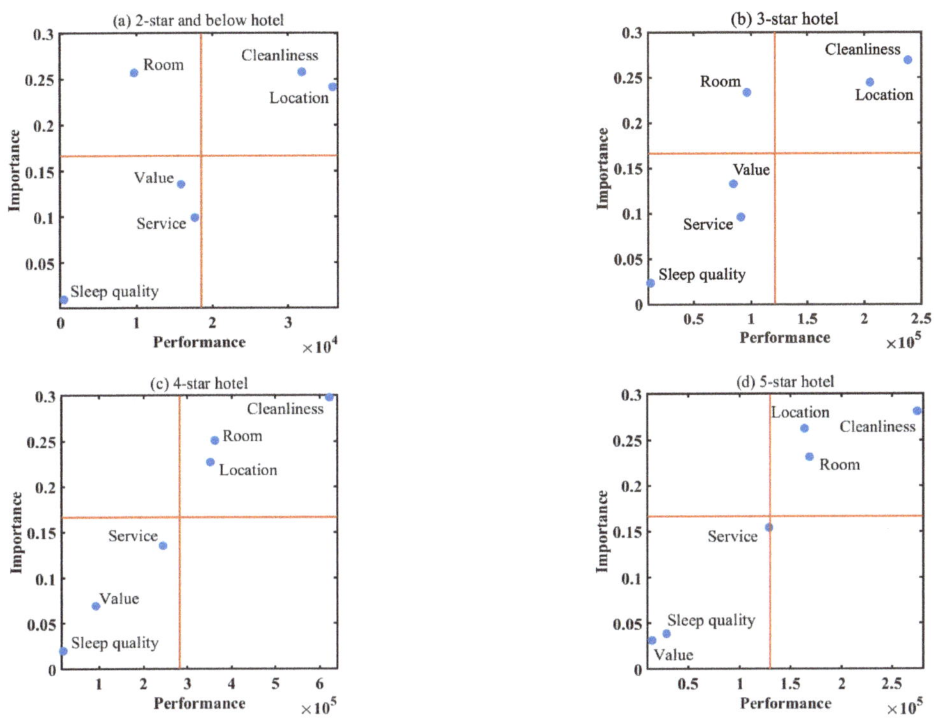

Figure 7. The IPA plots concerning four hotel star ratings.

4.4.3. Suggestions for Attribute Improvement and Priority

With the obtained performance and importance of the six attributes, the attribute priority rankings for resource allocation concerning four hotel star ratings are obtained by integrating the Kano categories of six attributes with the IPA plot, as shown in Table 12. The attribute priority rankings are divided into two groups, namely, *Room* > *Service* > *Sleep quality* > *Value* > *Cleanliness* > *Location* for low-star (three stars, two stars and below) hotels, and *Service* > *Sleep quality* > *Value* > *Cleanliness* > *Room* > *Location* for high-star (five-star and four-star) hotels.

Table 12. The attribute priority rankings of the six attributes for resource allocation concerning four hotel star ratings.

Attribute	Hotel Star Rating			
	2-Star and Below	3-Star	4-Star	5-Star
Value	4	4	3	3
Location	6	6	6	6
Service	2	2	1	1
Room	1	1	5	5
Cleanliness	5	5	4	4
Sleep quality	3	3	2	2

For low-star hotels, room (an excitement factor) is of the first priorities to get improved since it is very important, but it performs poorly from the perspective of customers. According to sub-attributes results, some effective measures can be taken to improve the room performance, such as paying attention to improving the facilities, tidiness, room size and soundproofing. The importance and performance of service, sleep quality and value are low. Service and sleep quality are basic factors which can cause numerous complaints when they perform poorly, while value is an excitement factor which can generate more customer satisfaction when it performs well. Service's importance is higher than sleep quality, so service and sleep quality are respectively given the second and third priority for resource allocation for improvement. Regarding service, improving staff's skill and attitude is important, and more professional, friendly and polite staff are needed. Moreover, hotel managers should pay attention to beds, pillows and soundproofing facilities to improve customers' sleep quality. Value is given the fourth priority for improvement. The importance of value is significantly higher in low-star hotels than in high-star hotels, which indicates that customers who choose low-star hotels are more likely to emphasize value for money. Offering a variety of discounts, reasonable price and member reward is helpful to enhance customer satisfaction. Lastly, cleanliness and location are the strengths of hotels, which should be well maintained. Considering cleanliness is considered more important than location, cleanliness is given the fifth priority for resource allocation to get improved.

For high-star hotels, there are no attributes that should be improved urgently. However, it is still necessary to invest resources and effort in service, sleep quality and value. Service is given the highest priority for resource allocation for improvement. From Figure 7, it can be concluded that service performs much better in five-star hotels than other hotels and its importance gradually increases, but it has not been the strength of five-star hotels yet. Unlike common service aspects which should be strengthened for low-star hotels, some advanced service aspects need to be improved. For example, improving staffing levels, providing more proactive, pet-friendly and infant-related service, multilingual receptionists, etc., are preferable ways to obtain more customer satisfaction. Sleep quality (a performance factor) and value (an excitement factor) are given the second and third priorities, respectively, and their importance is very low. If possible, investing resources in improving sleep quality and value (i.e., improving bedding and soundproofing facilities, offering discounts and reasonable prices) can also improve hotel customer ratings. Cleanliness, room and location are the strengths of high-star hotels, and these attributes should be well maintained. Especially, room is a unique strength for high-star hotels, while it is the weakness of low-star hotels. This result is consistent with the hotel star rating system offered by the Automobile Association that good performance of room is a must-be requirement for hotels to be rated as high-star [109]. According to attribute categories, cleanliness, room and location are prioritized in order for resource allocation because they are basic, performance and excitement factors, respectively.

5. Conclusions

5.1. Theoretical Implications

This study explored the attribute bidirectional performance by dividing online reviews into positive reviews and negative reviews. The Kano-IPA model was used for further understanding of customer's rating behaviors and demands for hotel service. The proposed methodology in five phases of sentiment analysis and Kano-IPA model enriches the research on online hotel reviews. The main theoretical contributions introduced are as follows:

First, this study explores the well-performed attributes contributing to high customer ratings and the poorly performed attributes causing low customer ratings. By dividing 1,090,341 online reviews into positive and negative reviews, the six attributes' good performance (positive sentiment values) in positive reviews and poor performance (negative sentiment values) in negative reviews are calculated through sentiment analysis. Our findings suggest that room, cleanliness and location are the most crucial determinants of both high and low customer ratings for hotels of these four levels. By contrast, other attributes, including value, service and sleep quality, have less impact on customers' rating behaviors. Therefore, the most crucial hotel attributes influencing customer satisfaction and dissatisfaction are exactly the same. Focusing on improving service quality of these general attributes including room, cleanliness and location is the key to win high customer ratings for all hotels. Thus, the effect of good/poor performance concerning location, value, service on high/low customer ratings varies across hotel star ratings.

Second, comparative analysis of attribute bidirectional performance concerning four hotel star ratings was conducted to verify the difference of hotel attributes contributing to high/low customer ratings among different hotel star ratings. This study indicates that the impact of several attributes on high/low customer ratings varies across different star hotels. On one hand, the impact of value and service's poor performance on low customer ratings varies across hotel star ratings. With the improvement in the hotel level, the impact of value's poor performance on low customer ratings shows a downward trend, while the impact of service's poor performance on low customer ratings shows an upward trend. For three-star and below hotels, value's poor performance contributes more to low customer ratings than service's poor performance. In contrast, for four and five star hotels, service's poor performance has greater impact on low customer ratings. On the other hand, the good performance in room, location, value, service and sleep quality contributes to high customer ratings differently among different star hotels, where the impact of value and service's good performance on high customer ratings shows a larger range of changes. Interestingly, for value and service, with the improvement in the hotel level, the impact of their good performance on high customer ratings shows the same trend as the impact of their poor performance on low customer ratings. These findings indicate that customers' expectations and perceptions on the good/poor performance of each attribute may vary across hotel star ratings. Thus, it is necessary to take hotel star ratings into consideration on customer satisfaction research.

Third, this study suggests that the effect of good performance on high customer ratings may not be equal to the effect of poor performance on low customer ratings for the same hotel attribute. In other words, the effect of attribute performance on customer satisfaction is asymmetric. For this reason, the Kano-IPA model was applied to better understand customer's rating behaviors and demands for hotel service. The Kano categories of five attributes (location, service, room, cleanliness and sleep quality) vary across different hotel star ratings. Furthermore, suggestions on priority for attribute improvement are formulated for hotels of the four star ratings according to the results of Kano-IPA model.

Fourth, this study proposes a methodology for hotel attribute sentiment analysis based on the automated textual analysis techniques including the Word2Vec and PolarityRank algorithms. A new sentiment lexicon was created from user generated reviews based on the PolarityRank algorithm, contributing to sentiment analysis in the hotel domain. The advance in the sentiment lexicon creation contains the following two points. On the one hand, we adopted more words (i.e., adverbs) than existing studies for PolarityRank propagation [63,96], which avoids missing some important sentiment words. On the other hand, initial both positive and negative sentiment values of each candidate sentiment word are assigned by a function from SentiWordNet instead of assigning positive seed words and negative seed words sentiment values manually, which is considered more objective and trustworthy. In addition, to our best knowledge, our sentiment lexicon built from the 1,090,341 textual reviews is the instructive application of the PolarityRank algorithm in million-level datasets. Thus, the comprehensive and complete sentiment propagation provides a guarantee of more precise sentiment calculation.

Lastly, this study proposed a novel index approach for Kano model classification and further makes it possible to apply the Kano-IPA model to numerous textual reviews. The SI index is defined to represent the satisfaction-stimulating ability of any one hotel attribute. Then the six attributes are classified into three Kano categories by comparing each SI index with the average index value for hotels of each star rating. The proposed approach enriches the existing research on the classification of the Kano model. Additionally, based on the TF-IDF algorithm, the importance of each attribute is obtained to construct the IPA plot. This study is a preferable attempt to apply online reviews to explore the effects of attribute performance on customer satisfaction to understand customers' rating behaviors.

5.2. Practical Implications

As consumers' reliance on the Internet grows, online reviews are increasingly important since customers usually browse a lot of hotel reviews when making hotel choices. It is important to analyze how hotel attributes contribute to high and low customer ratings. This study enables hotel managers and hotel online platforms to understand customers' rating behaviors, expectations and perceptions on hotel attributes. Furthermore, our findings and discussions provide a reference for hotel managers to allocate resources for attribute improvement and prioritization to achieve higher customer ratings.

First, due to the findings that the final attribute priority rankings for improvement are divided into two groups, two strategies for attribute improvement are given to low-star (three stars and below) and high-star (four- and five-star) hotels, respectively. For low-star hotels, room, which is an excitement factor, should be given the highest priority for resource allocation for improvement. Effective measures such as refurnishing, renovating, providing tidy and spacious rooms and proper decoration could be taken to improve room's performance in order to enhance customer satisfaction. Service, sleep quality and value are of lower priority for improvement, and they are basic, basic, and excitement factors, respectively. Some effective measures should be taken to enhance the performance of service and sleep quality in order to reduce customer dissatisfaction, which might include, for instance, staff training for work skill and attitude improvement, quality improvement in beds, pillows and soundproofing. With sufficient resource, low-star hotel managers should also provide attractive discounts or reasonable prices to customers since value for money is highly important for them. For high-star hotels, though nothing calls for urgent improvement, there still a need for better performance in service, sleep quality and value. Service and sleep quality are performance factors, and their importance is significantly higher for customers in high-star hotels. Service improvement (i.e., higher staffing levels, proactive, pet-friendly and infant-related services and multilingual receptionists) and providing better sleeping conditions (i.e., better bedding and soundproofing) are preferable methods to enhance customer satisfaction. Moreover, providing proper discounts and price for customers is also needed.

Second, some strengths should be well maintained for different star hotels. For low-star hotel managers, cleanliness and location are the strengths to win customer satisfaction. Since cleanliness and location are performance or basic factors and of high importance for customers in low-star hotels, it is necessary to invest sufficient resource to ensure their high quality. For high-star hotel managers, cleanliness, room and location are the strengths that need to be well maintained. In contrast to customers in low-star hotels, cleanliness and location are, respectively, basic and excitement factors for customers in high-star hotels. Investing more in hotel location is a preferable way for high-star hotels to enhance customer satisfaction. While it is hard to transform the existing locations, some convenient transportation services can be offered to improve access to attractions or traffic stations, such as free shuttles, attraction brochures. Additionally, room is a unique strength for high-star hotels, while it is a weakness of low-star hotels. These findings are in line with the hotel star rating system offered by the Automobile Association that room is a basic and quantitative indicator for hotel star rating [109]. Therefore, hotel managers should pay great attention to room improvement for higher star ratings.

Third, this study indicates that attribute improvement priorities are the same for hotels of three stars, two stars and below. However, compared with two-star and below hotels, service and sleep quality's importance is higher but performance is worse in three-star hotels. Service and sleep quality are basic factors, so their poor performance is more likely to cause great customer dissatisfaction. Customers pay more for a better hotel, so their expectations increase [110]. Thus, three-star hotel managers should pay more attention to improve performance in service and sleep quality to reduce customer dissatisfaction, and further enhance the competitive strengths against two-star and below hotels. Similarly, five-star hotel managers should keep alert for the pursuit of higher service quality since the SI index values of location, service, room, cleanliness and sleep quality show a downward trend compared to four-star hotels. This can be explained as follows: customers place much higher expectations on five-star hotels than four-star hotels, so very minor service failures can also cause great complaints. Compared to four-star hotels, investing resources to provide customers more attentive service and better sleep quality is necessary for five-star hotels.

Last but not least, for hotel online platforms, two aspects of practical significance are as follows. On the one hand, this study serves as references for online websites to recommend hotels to customers when they filter hotel star ratings. Our findings imply that customers have different expectations, preferences and demands for the six attributes when they choose hotels of different star ratings. Thus, different weights assigned to each hotel attribute according to hotel star ratings can be considered when designing the hotel recommendation system. On the other hand, we suggest that the six evaluation dimensions on the website should be upgraded. For example, considering the sub-attribute lists of room and cleanliness are similar, they can be merged into one dimension or given some notes for each attribute to help customers to distinguish between them.

5.3. Limitations

This study also has several limitations, which might serve as avenues for future research. First, the data were collected from one online travel website, which may not provide the complete information about customers' opinions. In addition, not all customers write textual reviews and give ratings to the hotels after leaving. Therefore, hotel reviews can be collected from multiple online websites and customers who book hotels offline. Second, although this study explores the differences in the categories and performance of six attributes across four hotel star ratings, attribute differences between different traveling purposes or different regions may exist. Customers with different traveling purposes and from different districts have different preferences on hotel attributes, which may influence attribute performance and further influence attribute classification in the Kano model. In the future, classifying online reviews based on other methods involves complex research. Third, for each hotel, its star rating may move up or down when the hotel makes some changes such as redecoration, management mode upgrades or becoming run-down. Although the cost of improving hotel star ratings is very high, some hotels may attempt to make efforts for higher star ratings. As a result, for some hotels, earlier online reviews may not reveal their quality appropriately in the current star ratings. This will affect the attribute bidirectional performance analysis results among different hotel star ratings. Thus, it is preferable to select online reviews during the current star rating period or exclude the hotels with changes in star ratings in the future research. Additionally, exploring the difference in determinants of customer satisfaction and dissatisfaction between the previous and current hotel star ratings is a future research direction. Finally, the attributes used in this study are the six evaluation dimensions on TripAdvisor.com, which may not include all topics expressed in textual reviews. To comprehensively understand customer demands, different categories of attributes can be extracted from textual reviews in future research.

Author Contributions: Supervision, conceptualization, project administration, resources, writing—review and editing, Y.C.; conceptualization, methodology, software, investigation, data curation, visualization, writing—original draft preparation, Y.Z.; supervision, conceptualization, methodology, formal analysis, validation, funding acquisition, S.Y.; investigation, data curation, visualization, Y.X.; investigation, data curation, software, S.C. All authors have read and agreed to the published version of the manuscript.

Funding: This work was supported by the National Natural Science Foundation of China (No. 71901151).

Institutional Review Board Statement: Not applicable.

Informed Consent Statement: Not applicable.

Data Availability Statement: Not applicable.

Conflicts of Interest: The authors declare no conflict of interest.

References

1. Ahani, A.; Nilashi, M.; Yadegaridehkordi, E.; Sanzogni, L.; Tarik, A.R.; Knox, K.; Samad, S.; Ibrahim, O. Revealing customers' satisfaction and preferences through online review analysis: The case of Canary Islands hotels. *J. Retail. Consum. Serv.* **2019**, *51*, 331–343. [CrossRef]
2. Schuckert, M.; Liu, X.; Law, R. Hospitality and Tourism Online Reviews: Recent Trends and Future Directions. *J. Travel. Tour. Mark.* **2015**, *32*, 608–621. [CrossRef]
3. Yang, Y.; Park, S.; Hu, X. Electronic word of mouth and hotel performance: A meta-analysis. *Tour. Manag.* **2018**, *67*, 248–260. [CrossRef]
4. Ye, Q.; Law, R.; Gu, B.; Chen, W. The influence of user-generated content on traveler behavior: An empirical investigation on the effects of e-word-of-mouth to hotel online bookings. *Comput. Hum. Behav.* **2011**, *27*, 634–639. [CrossRef]
5. Gavilan, D.; Avello, M.; Martinez-Navarro, G. The influence of online ratings and reviews on hotel booking consideration. *Tour. Manag.* **2018**, *66*, 53–61. [CrossRef]
6. Gao, B.; Li, X.; Liu, S.; Fang, D. How power distance affects online hotel ratings: The positive moderating roles of hotel chain and reviewers' travel experience. *Tour. Manag.* **2018**, *65*, 176–186. [CrossRef]
7. Chatterjee, S. Drivers of helpfulness of online hotel reviews: A sentiment and emotion mining approach. *Int. J. Hosp. Manag.* **2020**, *85*, 102356. [CrossRef]
8. Chen, L.-F. Exploring asymmetric effects of attribute performance on customer satisfaction using association rule method. *Int. J. Hosp. Manag.* **2015**, *47*, 54–64. [CrossRef]
9. Liu, Y.; Teichert, T.; Rossi, M.; Li, H.; Hu, F. Big data for big insights: Investigating language-specific drivers of hotel satisfaction with 412,784 user-generated reviews. *Tour. Manag.* **2017**, *59*, 554–563. [CrossRef]
10. Bi, J.-W.; Liu, Y.; Fan, Z.-P.; Zhang, J. Exploring asymmetric effects of attribute performance on customer satisfaction in the hotel industry. *Tour. Manag.* **2020**, *77*, 104006. [CrossRef]
11. Cheng, C.-C.; Chen, C.-T. Creating excellent and competitive motels services. *Int. J. Contemp. Hosp. Manag.* **2018**, *30*, 836–854. [CrossRef]
12. Beheshtinia, M.A.; Farzaneh Azad, M. A fuzzy QFD approach using SERVQUAL and Kano models under budget constraint for hotel services. *Total Qual. Manag. Bus. Excell.* **2017**, *30*, 808–830. [CrossRef]
13. Chiang, C.-F.; Chen, W.-Y.; Hsu, C.-Y. Classifying technological innovation attributes for hotels: An application of the Kano model. *J. Travel. Tour. Mark.* **2019**, *36*, 796–807. [CrossRef]
14. Xu, X.; Li, Y. The antecedents of customer satisfaction and dissatisfaction toward various types of hotels: A text mining approach. *Int. J. Hosp. Manag.* **2016**, *55*, 57–69. [CrossRef]
15. Zhao, M.; Zhang, C.; Hu, Y.; Xu, Z.; Liu, H. Modelling Consumer Satisfaction Based on Online Reviews Using the Improved Kano Model from the Perspective of Risk Attitude and Aspiration. *Technol. Econ. Dev. Econ.* **2021**, *27*, 550–582. [CrossRef]
16. Fong, L.H.N.; Lei, S.S.I.; Law, R. Asymmetry of Hotel Ratings on TripAdvisor: Evidence from Single-Versus Dual-Valence Reviews. *J. Hosp. Mark. Manag.* **2016**, *26*, 67–82. [CrossRef]
17. Filieri, R.; McLeay, F.; Tsui, B.; Lin, Z. Consumer perceptions of information helpfulness and determinants of purchase intention in online consumer reviews of services. *Inf. Manag.* **2018**, *55*, 956–970. [CrossRef]
18. Kirilenko, A.P.; Stepchenkova, S.O.; Dai, X. Automated topic modeling of tourist reviews: Does the Anna Karenina principle apply? *Tour. Manag.* **2021**, *83*, 104241. [CrossRef]
19. Dinçer, M.Z.; Alrawadieh, Z. Negative Word of Mouse in the Hotel Industry: A Content Analysis of Online Reviews on Luxury Hotels in Jordan. *J. Hosp. Mark. Manag.* **2017**, *26*, 785–804. [CrossRef]
20. Vásquez, C. Complaints online: The case of TripAdvisor. *J. Pragmat.* **2011**, *43*, 1707–1717. [CrossRef]
21. Zhang, Z.; Ye, Q.; Law, R. Determinants of hotel room price: An exploration of travelers' hierarchy of accommodation needs. *Int. J. Contemp. Hosp. Manag.* **2011**, *23*, 972–981. [CrossRef]
22. Francesco, G.; Roberta, G. Cross-country analysis of perception and emphasis of hotel attributes. *Tour. Manag.* **2019**, *74*, 24–42. [CrossRef]
23. Slevitch, L.; Oh, H. Asymmetric relationship between attribute performance and customer satisfaction: A new perspective. *Int. J. Hosp. Manag.* **2010**, *29*, 559–569. [CrossRef]
24. Albayrak, T.; Caber, M. The symmetric and asymmetric influences of destination attributes on overall visitor satisfaction. *Curr. Issues Tour.* **2013**, *16*, 149–166. [CrossRef]
25. Alegre, J.; Garau, J. The Factor Structure of Tourist Satisfaction at Sun and Sand Destinations. *J. Travel. Res.* **2009**, *50*, 78–86. [CrossRef]
26. Füller, J.; Matzler, K. Customer delight and market segmentation: An application of the three-factor theory of customer satisfaction on life style groups. *Tour. Manag.* **2008**, *29*, 116–126. [CrossRef]
27. Chang, K.-C.; Chen, M.-C. Applying the Kano model and QFD to explore customers' brand contacts in the hotel business: A study of a hot spring hotel. *Total Qual. Manag. Bus. Excell.* **2011**, *22*, 1–27. [CrossRef]
28. Kuo, C.-M.; Chen, H.-T.; Boger, E. Implementing City Hotel Service Quality Enhancements: Integration of Kano and QFD Analytical Models. *J. Hosp. Mark. Manag.* **2015**, *25*, 748–770. [CrossRef]

29. Kuo, Y.-F.; Chen, J.-Y.; Deng, W.-J. IPA–Kano model: A new tool for categorising and diagnosing service quality attributes. *Total Qual. Manag. Bus. Excell.* **2012**, *23*, 731–748. [CrossRef]
30. Lai, I.K.W.; Hitchcock, M. A comparison of service quality attributes for stand-alone and resort-based luxury hotels in Macau: 3-Dimensional importance-performance analysis. *Tour. Manag.* **2016**, *55*, 139–159. [CrossRef]
31. Jou, R.-C.; Day, Y.-J. Application of Revised Importance–Performance Analysis to Investigate Critical Service Quality of Hotel Online Booking. *Sustainability* **2021**, *13*, 2043. [CrossRef]
32. Tseng, C.C. An IPA-Kano model for classifying and diagnosing airport service attributes. *Res. Transp. Bus. Manag.* **2020**, *37*, 100499. [CrossRef]
33. Pai, F.-Y.; Yeh, T.-M.; Tang, C.-Y. Classifying restaurant service quality attributes by using Kano model and IPA approach. *Total Qual. Manag. Bus. Excell.* **2016**, *29*, 301–328. [CrossRef]
34. Martilla, J.A.; James, J.C. Importance-performance analysis. *J. Mark* **1977**, *41*, 77–79. [CrossRef]
35. Wang, L.; Wang, X.; Peng, J.; Wang, J. The differences in hotel selection among various types of travellers: A comparative analysis with a useful bounded rationality behavioural decision support model. *Tour. Manag.* **2020**, *76*, 103961. [CrossRef]
36. Nicolau, J.L.; Mellinas, J.P.; Martín-Fuentes, E. The halo effect: A longitudinal approach. *Ann. Tour. Res.* **2020**, *83*, 102938. [CrossRef]
37. Zhao, Y.; Xu, X.; Wang, M. Predicting overall customer satisfaction: Big data evidence from hotel online textual reviews. *Int. J. Hosp. Manag.* **2019**, *76*, 111–121. [CrossRef]
38. Jannach, D.; Zanker, M.; Fuchs, M. Leveraging multi-criteria customer feedback for satisfaction analysis and improved recommendations. *J. Inf. Technol. Tour.* **2014**, *14*, 119–149. [CrossRef]
39. Sharma, H.; Tandon, A.; Kapur, P.K.; Aggarwal, A.G. Ranking hotels using aspect ratings based sentiment classification and interval-valued neutrosophic TOPSIS. *Int. J. Syst. Assur. Eng. Manag.* **2019**, *10*, 973–983. [CrossRef]
40. Guo, Y.; Barnes, S.J.; Jia, Q. Mining meaning from online ratings and reviews: Tourist satisfaction analysis using latent Dirichlet allocation. *Tour. Manag.* **2017**, *59*, 467–483. [CrossRef]
41. Hu, N.; Zhang, T.; Gao, B.; Bose, I. What do hotel customers complain about? Text analysis using structural topic model. *Tour. Manag.* **2019**, *72*, 417–426. [CrossRef]
42. Philander, K.; Zhong, Y. Twitter sentiment analysis: Capturing sentiment from integrated resort tweets. *Int. J. Hosp. Manag.* **2016**, *55*, 16–24. [CrossRef]
43. Yadav, M.L.; Roychoudhury, B. Effect of trip mode on opinion about hotel aspects: A social media analysis approach. *Int. J. Hosp. Manag.* **2019**, *80*, 155–165. [CrossRef]
44. Bi, J.-W.; Liu, Y.; Fan, Z.-P.; Zhang, J. Wisdom of crowds: Conducting importance-performance analysis (IPA) through online reviews. *Tour. Manag.* **2019**, *70*, 460–478. [CrossRef]
45. Al-Smadi, M.; Talafha, B.; Al-Ayyoub, M.; Jararweh, Y. Using long short-term memory deep neural networks for aspect-based sentiment analysis of Arabic reviews. *Int. J. Mach. Learn. Cybern.* **2018**, *10*, 2163–2175. [CrossRef]
46. Nie, R.; Tian, Z.; Wang, J.; Chin, K.S. Hotel selection driven by online textual reviews: Applying a semantic partitioned sentiment dictionary and evidence theory. *Int. J. Hosp. Manag.* **2020**, *88*, 102495. [CrossRef]
47. Ravi, K.; Ravi, V. A survey on opinion mining and sentiment analysis: Tasks, approaches and applications. *Knowl. Based. Syst.* **2015**, *89*, 14–46. [CrossRef]
48. Fang, X.; Zhan, J. Sentiment analysis using product review data. *J. Big Data* **2015**, *2*, 1–14. [CrossRef]
49. Jansen, B.J.; Zhang, M.; Sobel, K.; Chowdury, A. Twitter power: Tweets as electronic word of mouth. *J. Am. Soc. Inf. Sci. Technol.* **2009**, *60*, 2169–2188. [CrossRef]
50. Bowden, J.; Mirzaei, A. Consumer engagement within retail communication channels: An examination of online brand communities and digital content marketing initiatives. *Eur. J. Mark.* **2021**, *55*, 1411–1439. [CrossRef]
51. Almjawel, A.; Bayoumi, S.; Alshehri, D.; Alzahrani, S.; Alotaibi, M. Sentiment analysis and visualization of amazon books' reviews. In Proceedings of the 2019 2nd International Conference on Computer Applications & Information Security (ICCAIS), Riyadh, Saudi Arabia, 19–21 March 2019; pp. 1–6.
52. Liang, T.-P.; Li, X.; Yang, C.-T.; Wang, M. What in Consumer Reviews Affects the Sales of Mobile Apps: A Multifacet Sentiment Analysis Approach. *Int. J. Electron. Commer.* **2015**, *20*, 236–260. [CrossRef]
53. Kharde, V.A.; Sonawane, S.S. Sentiment Analysis of Twitter Data: A Survey of Techniques. *Int. J. Comput. Appl.* **2016**, *139*, 5–15.
54. González-Rodríguez, M.R.; Martínez-Torres, R.; Toral, S. Post-visit and pre-visit tourist destination image through eWOM sentiment analysis and perceived helpfulness. *Int. J. Contemp. Hosp. Manag.* **2016**, *28*, 2609–2627. [CrossRef]
55. Geetha, M.; Singha, P.; Sinha, S. Relationship between customer sentiment and online customer ratings for hotels—An empirical analysis. *Tour. Manag.* **2017**, *61*, 43–54. [CrossRef]
56. Araque, O.; Corcuera-Platas, I.; Sánchez-Rada, J.F.; Iglesias, C.A. Enhancing deep learning sentiment analysis with ensemble techniques in social applications. *Expert. Syst. Appl.* **2017**, *77*, 236–246. [CrossRef]
57. Yoshikawa, Y.; Iwata, T.; Sawada, H. Latent support measure machines for bag-of-words data classification. In *Advances in Neural Information Processing Systems 27*; Ghahramani, Z., Welling, M., Cortes, C., Lawrence, N.D., Weinberger, K.Q., Eds.; Curran Associates, Inc.: Red Hook, NY, USA, 2014; pp. 1961–1969.
58. Taboada, M.; Brooke, J.; Tofiloski, M.; Voll, K.; Stede, M. Lexicon-Based Methods for Sentiment Analysis. *Comput. Linguist.* **2011**, *37*, 267–307. [CrossRef]

59. Basarslan, M.S.; Kayaalp, F. Sentiment Analysis with Machine Learning Methods on Social Media. *ADCAIJ Adv. Dist. Comput. Artif. Intell. J.* **2020**, *9*, 5–15. [CrossRef]
60. Bengio, Y.; Courville, A.; Vincent, P. Representation learning: A review and new perspectives. *IEEE Trans. Pattern. Anal. Mach. Intell.* **2013**, *35*, 1798–1828. [CrossRef]
61. Ghiassi, M.; Lee, S. A domain transferable lexicon set for Twitter sentiment analysis using a supervised machine learning approach. *Expert. Syst. Appl.* **2018**, *106*, 197–216. [CrossRef]
62. Hogenboom, A.; Heerschop, B.; Frasincar, F.; Kaymak, U.; de Jong, F. Multi-lingual support for lexicon-based sentiment analysis guided by semantics. *Decis. Support Syst.* **2014**, *62*, 43–53. [CrossRef]
63. Cruz, F.L.; Vallejo, C.G.; Enrı´quez, F.; Troyano, J.A. PolarityRank: Finding an equilibrium between followers and contraries in a network. *Inf. Process Manag.* **2012**, *48*, 271–282. [CrossRef]
64. Turney, P.D.; Littman, M.L. Measuring praise and criticism. *ACM Trans. Inf. Syst.* **2003**, *21*, 315–346. [CrossRef]
65. Abdulla, N.A.; Ahmed, N.A.; Shehab, M.A.; Al-Ayyoub, M. Arabic sentiment analysis: Lexicon-based and corpus-based. In Proceedings of the 2013 IEEE Jordan conference on applied electrical engineering and computing technologies (AEECT), Amman, Jordan, 3–5 December 2013.
66. Dey, A.; Jenamani, M.; Thakkar, J.J. Senti-N-Gram: An ngram lexicon for sentiment analysis. *Expert. Syst. Appl.* **2018**, *103*, 92–105. [CrossRef]
67. Sanagar, S.; Gupta, D. Unsupervised Genre-Based Multidomain Sentiment Lexicon Learning Using Corpus-Generated Polarity Seed Words. *IEEE Access.* **2020**, *8*, 118050–118071. [CrossRef]
68. Sánchez-Rada, J.F.; Iglesias, C.A. Social context in sentiment analysis: Formal definition, overview of current trends and framework for comparison. *Inf. Fusion.* **2019**, *52*, 344–356. [CrossRef]
69. Fares, M.; Moufarrej, A.; Jreij, E.; Tekli, J.; Grosky, W. Unsupervised word-level affect analysis and propagation in a lexical knowledge graph. *Knowl. Based Syst.* **2019**, *165*, 432–459. [CrossRef]
70. Qiu, G.; Liu, B.; Bu, J.; Chen, C. Expanding domain sentiment lexicon through double propagation. In Proceedings of the 21st International Joint Conference on Artificial Intelligence, Pasadena, CA, USA, 11–17 July 2009.
71. Mittal, V.; Ross, W.T., Jr.; Baldasare, P.M. The asymmetric impact of negative and positive attribute-level performance on overall satisfaction and repurchase intentions. *J. Mark.* **1998**, *62*, 33–47. [CrossRef]
72. Kano, N.; Seraku, N.; Takahashi, F.; Tsuji, S. Attractive Quality and Must-Be Quality. *J. Japan. Soc. Qual. Control.* **1984**, *14*, 39–48.
73. Matzler, K.; Sauerwein, E. The factor structure of customer satisfaction. *Int. J. Serv. Ind. Manag.* **2002**, *13*, 314–332. [CrossRef]
74. Chen, L.-F. A novel approach to regression analysis for the classification of quality attributes in the Kano model: An empirical test in the food and beverage industry. *Omega* **2012**, *40*, 651–659. [CrossRef]
75. Yang, C.-C.; Jou, Y.-T.; Cheng, L.-Y. Using integrated quality assessment for hotel service quality. *Qual. Quant.* **2009**, *45*, 349–364. [CrossRef]
76. Tontini, G.; Bento, G.d.S.; Milbratz, T.C.; Volles, B.K.; Ferrari, D. Exploring the nonlinear impact of critical incidents on customers' general evaluation of hospitality services. *Int. J. Hosp. Manag.* **2017**, *66*, 106–116. [CrossRef]
77. Lin, S.-P.; Yang, C.-L.; Chan, Y.; Sheu, C. Refining Kano's 'quality attributes–satisfaction' model: A moderated regression approach. *Int. J. Prod. Econ.* **2010**, *126*, 255–263. [CrossRef]
78. Busacca, B.; Padula, G. Understanding the relationship between attribute performance and overall satisfaction. *Mark. Intell. Plan.* **2005**, *23*, 543–561. [CrossRef]
79. Friman, M.; Edvardsson, B. A content analysis of complaints and compliments. *Manag. Serv. Qual. Int. J.* **2003**, *13*, 20–26. [CrossRef]
80. Mikulić, J.; Prebežac, D. A critical review of techniques for classifying quality attributes in the Kano model. *Manag. Serv. Qual. Int. J.* **2011**, *21*, 46–66. [CrossRef]
81. Sampson, S.E.; Showalter, M.J. The Performance-Importance Response Function: Observations and Implications. *Serv. Ind. J.* **1999**, *19*, 1–25. [CrossRef]
82. Sever, I. Importance-performance analysis: A valid management tool? *Tour. Manag.* **2015**, *48*, 43–53. [CrossRef]
83. Sörensson, A.; von Friedrichs, Y. An importance–performance analysis of sustainable tourism: A comparison between international and national tourists. *J. Dest. Mark. Manage* **2013**, *2*, 14–21. [CrossRef]
84. Parasuraman, A.P.; Zeithaml, V.A.; Berry, L.L. A Conceptual Model of Service Quality and Its Implications for Future Research. *J. Mark* **1985**, *49*, 41–50. [CrossRef]
85. Cheng, Y.-S.; Kuo, N.-T.; Chang, K.-C.; Hu, S.-M. Integrating the Kano model and IPA to measure quality of museum interpretation service: A comparison of visitors from Taiwan and Mainland China. *Asia Pac. J. Tour. Res.* **2019**, *24*, 483–500. [CrossRef]
86. Dabestani, R.; Shahin, A.; Saljoughian, M.; Shirouyehzad, H. Importance-performance analysis of service quality dimensions for the customer groups segmented by DEA. *Int. J. Qual. Reliab. Manage* **2016**, *33*, 160–177. [CrossRef]
87. Yin, P.; Chu, J.; Wu, J.; Ding, J.; Yang, M.; Wang, Y. A DEA-based two-stage network approach for hotel performance analysis: An internal cooperation perspective. *Omega* **2020**, *93*, 102035. [CrossRef]
88. Phillips, P.; Barnes, S.; Zigan, K.; Schegg, R. Understanding the Impact of Online Reviews on Hotel Performance. *J. Travel. Res.* **2016**, *56*, 235–249. [CrossRef]
89. Domestic Overnight Tourism Trips to London. Available online: https://data.london.gov.uk/dataset/domestic-overnight-tourism-trips-to-london (accessed on 1 August 2020).

90. Number of International Visitors to London. Available online: https://data.london.gov.uk/dataset/number-international-visitors-london (accessed on 1 August 2021).
91. Keller, D.; Kostromitina, M. Characterizing non-chain restaurants' Yelp star-ratings: Generalizable findings from a representative sample of Yelp reviews. *Int. J. Hosp. Manag.* **2020**, *86*, 102440. [CrossRef]
92. Local Consumer Review Survey. 2015. Available online: https://www.brightlocal.com/research/local-consumer-review-survey-2015/ (accessed on 19 October 2015).
93. Ban, H.-J.; Choi, H.; Choi, E.-K.; Lee, S.; Kim, H.-S. Investigating Key Attributes in Experience and Satisfaction of Hotel Customer Using Online Review Data. *Sustainability* **2019**, *11*, 6570. [CrossRef]
94. Berezina, K.; Bilgihan, A.; Cobanoglu, C.; Okumus, F. Understanding Satisfied and Dissatisfied Hotel Customers: Text Mining of Online Hotel Reviews. *J. Hosp. Mark. Manag.* **2015**, *25*, 1–24. [CrossRef]
95. Zhang, D.; Xu, H.; Su, Z.; Xu, Y. Chinese comments sentiment classification based on word2vec and SVMperf. *Expert. Syst. Appl.* **2015**, *42*, 1857–1863. [CrossRef]
96. Fernández-Gavilanes, M.; Álvarez-López, T.; Juncal-Martínez, J.; Costa-Montenegro, E.; Javier González-Castaño, F. Unsupervised method for sentiment analysis in online texts. *Expert. Syst. Appl.* **2016**, *58*, 57–75. [CrossRef]
97. Page, L.; Brin, S.; Motwani, R.; Winograd, T. The PageRank Citation Ranking: Bringing Order to the Web. *Tech. Rep. Stanf. Digit. Libr. Technol. Proj.* **1998**.
98. Dragut, E.C.; Fellbaum, C. The role of adverbs in sentiment analysis. In Proceedings of the Frame Semantics in NLP: A Workshop in Honor of Chuck Fillmore (1929–2014), Baltimore, MD, USA, 27 June 2014; pp. 38–41.
99. Baccianella, S.; EsuliAndrea, A.; Sebastiani, F. SentiWordNet 3.0: An enhanced lexical resource for sentiment analysis and opinion mining. In Proceedings of the International Conference on Language Resources and Evaluation, Valletta, Malta, 17–23 May 2010.
100. Kennedy, A.; Inkpen, D. Sentiment Classication of Movie and Product Reviews Using Contextual Valence Shifters. *Comput. Intell.* **2006**, *22*, 110–125. [CrossRef]
101. Polanyi, L.; Zaenen, A. Contextual valence shifters. In *Computing Attitude and Affect in Text: Theory and Applications*; Shanahan, J.G., Qu, Y., Wiebe, J., Eds.; The Information Retrieval Series; Springer: Dordrecht, The Netherlands, 2006; Volume 20, pp. 1–10.
102. Brooke, J. A Semantic Approach to Automatic Text Sentiment Analysis. Master's Thesis, Simon Fraser University, Burnaby, BC, Canada, 31 March 2009.
103. Kim, D.; Seo, D.; Cho, S.; Kang, P. Multi-co-training for document classification using various document representations: TF–IDF, LDA, and Doc2Vec. *Inf. Sci.* **2019**, *477*, 15–29. [CrossRef]
104. Azzopardi, E.; Nash, R. A critical evaluation of importance–performance analysis. *Tour. Manag.* **2013**, *35*, 222–233. [CrossRef]
105. O'Connor, P. Managing a Hotel's Image on TripAdvisor. *J. Hosp. Mark. Manag.* **2010**, *19*, 754–772. [CrossRef]
106. Manolitzas, P.; Glaveli, N.; Palamas, S.; Talias, M.; Grigoroudis, E. Hotel guests' demanding level and importance of attribute satisfaction ratings: An application of MUltiplecriteria Satisfaction Analysis on TripAdvisor's hotel guests ratings. *Curr. Issues Tour.* **2021**, *24*, 1–6. [CrossRef]
107. Kitsios, F.; Kamariotou, M.; Karanikolas, P.; Grigoroudis, E. Digital Marketing Platforms and Customer Satisfaction: Identifying eWOM Using Big Data and Text Mining. *Appl. Sci.* **2021**, *11*, 8032. [CrossRef]
108. Zhao, S. Thumb Up or Down? A Text-Mining Approach of Understanding Consumers through Reviews. *Decis. Sci.* **2021**, *52*, 699–719. [CrossRef]
109. UK Hotel Star Rating System. Available online: https://www.narehotel.co.uk/uk-hotel-star-rating-system (accessed on 4 January 2022).
110. Kashyap, R.; Bojanic, D.C. A Structural Analysis of Value, Quality, and Price Perceptions of Business and Leisure Travelers. *J. Travel. Res.* **2000**, *39*, 45–51. [CrossRef]

applied sciences

MDPI

Article

A Note on "Wiener Index of a Fuzzy Graph and Application to Illegal Immigration Networks"

Hoon Lee [1], Xue-gang Chen [2] and Moo Young Sohn [3,*]

1 Information and Communications Engineering, Changwon National University, Changwon 51140, Korea; hoony@changwon.ac.kr
2 Mathematics, North China Electric Power University, Beijing 102206, China; gxcxdm@163.com
3 Mathematics, Changwon National University, Changwon 51140, Korea
* Correspondence: mysohn@changwon.ac.kr

Abstract: Connectivity parameters have an important role in the study of communication networks. Wiener index is such a parameter with several applications in networking, facility location, cryptology, chemistry, and molecular biology, etc. In this paper, we show two notes related to the Wiener index of a fuzzy graph. First, we argue that Theorem 3.10 in the paper "Wiener index of a fuzzy graph and application to illegal immigration networks, Fuzzy Sets and Syst. 384 (2020) 132–147" is not correct. We give a correct statement of Theorem 3.10. Second, by using a new operator on matrix, we propose a simple and polynomial-time algorithm to compute the Wiener index of a fuzzy graph.

Keywords: fuzzy graph; Wiener index; algorithm

Citation: Lee, H.; Chen, X.-g.; Sohn, M.Y. A Note on "Wiener Index of a Fuzzy Graph and Application to Illegal Immigration Networks". *Appl. Sci.* **2022**, *12*, 304. https://doi.org/10.3390/app12010304

Academic Editor: Ugo Vaccaro

Received: 19 November 2021
Accepted: 22 December 2021
Published: 29 December 2021

1. Introduction

In many real world problems we get only partial information about the problem, and the vagueness in the description and uncertainty has led to the growth of fuzzy graph theory. A mathematical framework to describe uncertainty in real life situation was first suggested by A.L. Zadeh [1]. Rosenfeld [2] introduced the notion of fuzzy graph and several fuzzy analogs of graph theoretic concepts such as paths, cycles and connectedness. Wiener index of graphs has been studied in the field of mathematics, chemistry, and molecular biology [3–5].

There are many situations which are modeled by a connected fuzzy graph. Wiener index is such an accepted index used in various fields like communication networks, facility location, crytopology, medicine, etcs. Let us start with a basic definition and concepts of fuzzy graphs; most of them can be found in [6].

Let S be a set. A *fuzzy graph* $G = (\sigma, \mu)$ is a pair of membership functions on fuzzy sets $\sigma : S \to [0,1]$ and $\mu : S \times S \to [0,1]$ such that $\mu(u,v) \le \sigma(u) \wedge \sigma(v)$. Here $\wedge$ represents the minimum. Throughout the paper, we assume that S is finite and nonempty, μ is reflexive and symmetric. We denote the underlying crisp graph by $G^* = (\sigma^*, \mu^*)$ where $\sigma^* = \{u \in V : \sigma(u) > 0\}$ and $\mu^* = \{(u,v) \in V \times V : \mu(u,v) > 0\}$. We denote an element (x,y) of μ^* by xy and call it an *edge* of G. Elements of σ^* are called *vertices* of the fuzzy graph G. A fuzzy graph $H = (\tau, \nu)$ is called a *partial fuzzy subgraph* of $G = (\sigma, \mu)$ if $\tau(v) = \sigma(v)$ for all $vs. \in \tau^*$ and $\nu(uv) = \mu(uv)$ for all $uv \in \nu^*$. Note that $G - uv$ denotes the fuzzy subgraph of G in which $\mu(uv) = 0$ and $G - u$ is used for the fuzzy subgraph of G in which $\sigma(u) = 0$.

In a fuzzy graph $G = (\sigma, \mu)$, a *path* P of length n is a sequence of distinct vertices $u_0, u_1, \ldots, u_n$ such that $\mu(u_{i-1}, u_i) > 0$, $i = 1, 2, \ldots, n$ and the degree of membership of a weakest edge is defined to be the *strength* of the path P. A path P is called a *cycle* if $u_0 = u_n$.

For any two vertices x and y, let $d(x,y)$ denotes the length of the shortest path between x and y. The diameter of G, denoted by $diam(G)$, is the maximum distance $d(x,y)$ for any two vertices x, y in G. The *strength of connectedness* between two vertices x and y is defined as the maximum of the strengths of all paths between x and y and is denoted by $Conn_G(x,y)$. If the strength of a path P is equal to $Conn_G(x,y)$, then a path P is called a *strongest* $x-y$ path. A fuzzy graph $G=(\sigma,\mu)$ is *connected* if for every $u,v\in\sigma^*$, $Conn_G(u,v)>0$. An edge xy of a fuzzy graph $G=(\sigma,\mu)$ is called α-strong if $\mu(xy)>Conn_{G-xy}(x,y)$. An edge xy of a fuzzy graph $G=(\sigma,\mu)$ is called β-strong if $\mu(xy)=Conn_{G-xy}(x,y)$. An edge xy of a fuzzy graph $G=(\sigma,\mu)$ is called δ-strong if $\mu(xy)<Conn_{G-xy}(x,y)$. An edge is called strong if it is either α-strong or β-strong. A path P is called a *strong path* if all of its edges are strong. Let $G=(\sigma,\mu)$ be a fuzzy graph and $x,y\in\sigma^*$. A strong path P from x to y is called *geodesic* if there is no shorter strong path from x to y. The weight of a geodesic is the sum of membership values of all edges in the geodesic. Let $G=(\sigma,\mu)$ be a fuzzy graph. The *Wiener index (WI)* of G is defined by $WI(G)=\sum_{u,v\in\sigma^*}\sigma(u)\sigma(v)d_S(u,v)$, where $d_S(u,v)$ is the minimum sum of weights of geodesic from u to v. In this paper, it is assumed that $\sigma(u)=1$ for $u\in\sigma^*$ in all examples of fuzzy graphs $G=(\sigma,\mu)$, for convenience. The outline of this paper is organized as follows. In Section 2, it is shown that Theorem 3.10 in the paper "Wiener index of a fuzzy graph and application to illegal immigration networks, Fuzzy Sets and Syst. 384 (2020) 132–147" is not correct. A corrected statement of Theorem 3.10 is given. In Section 3, we present a simple algorithm to compute the wiener index of a fuzzy graph by using a new operator on matrix.

2. Counterexamples and Revision

At first, we recall the Theorem 3.10 of [7] and give two counterexamples to it.

Theorem 1 (Theorem 3.10 of [7]). *Let $G=(\sigma,\mu)$ be a fuzzy graph. For $s,t\in\sigma^*$, let $P_{s,t}$ denote the path which has the minimum sum of membership values among all shortest strong paths between s and t. Let $uv\in\mu^*$. If uv is an α or β-strong edge and uv is not a part of any $P_{s,t}$ for $s,t\in\sigma^*$ with $\{s,t\}\neq\{u,v\}$, then $WI(G-uv)\neq WI(G)$.*

Theorem 3.10 of [7] is not correct as shown in the following two counterexamples. In the following two graphs, uv is an α-strong edge and β-strong edge, respectively.

Example 1. *Let $G=(\sigma,\mu)$ be the fuzzy graph shown in Figure 1 with vertex set $\{a,b,u,v\}$ and $\sigma(x)=1$ for any $x\in\sigma^*$, $\mu(ua)=0.1$, $\mu(ab)=0.4$, $\mu(bv)=0.1$, $\mu(uv)=0.6$. Then each edge of the graph G is strong. So $d_S(u,a)=0.1$, $d_S(a,b)=0.4$, $d_S(b,v)=0.1$, $d_S(u,v)=0.6$, $d_S(u,b)=0.5$ and $d_S(a,v)=0.5$. Therefore, $WI(G)=\sum_{x,y\in\sigma^*}\sigma(x)\sigma(y)d_S(x,y)=2.2$. It is obvious that uv is an α-strong edge and uv is not a part of any $P_{s,t}$ for $s,t\in\sigma^*$ with $\{s,t\}\neq\{u,v\}$. For any two vertices $x,y\in\sigma^*$, the number $d_S^{G-uv}(x,y)$ in $G-uv$ is equal to the number $d_S(x,y)$ in G. Then $WI(G-uv)=\sum_{x,y\in\sigma^*}\sigma(x)\sigma(y)d_S^{G-uv}(x,y)=2.2$ and $WI(G-uv)=WI(G)$.*

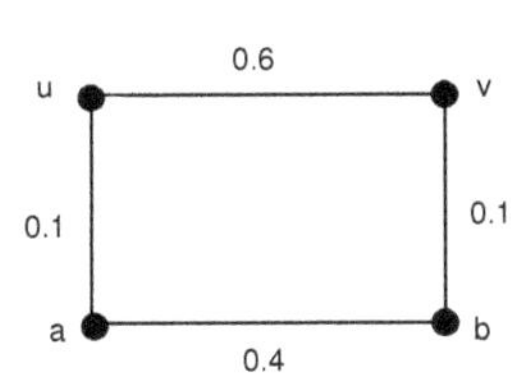

Figure 1. uv is an α-strong edge and $WI(G)=WI(G-uv)$.

Example 2. *Let $G=(\sigma,\mu)$ be the fuzzy graph shown in Figure 2 with vertex set $\{a,b,c,u,v\}$ and $\sigma(x)=1$ for any $x\in\sigma^*$, $\mu(au)=\mu(av)=0.2$, $\mu(uv)=\mu(uc)=\mu(bv)=\mu(bc)=0.4$, $\mu(ub)=0.5$, $\mu(cv)=0.6$. Then each edge of the graph G is strong. So $d_S(a,u)=d_S(a,v)=0.2$, $d_S(u,v)=d_S(u,c)=d_S(b,v)=d_S(b,c)=0.4$, $d_S(u,b)=0.5$, $d_S(c,v)=0.6$, $d_S(a,b)=0.6$ and $d_S(a,c)=0.6$. Therefore, $WI(G)=\sum_{x,y\in\sigma^*}\sigma(x)\sigma(y)d_S(x,y)=4.3$. It is obvious that uv is an β-strong edge and uv is not a part of any $P_{s,t}$ for $s,t\in\sigma^*$ with $\{s,t\}\neq\{u,v\}$. For any two vertices $x,y\in\sigma^*$, the number $d_S^{G-uv}(x,y)$ in $G-uv$ is equal to the number $d_S(x,y)$ in G. So $WI(G-uv)=\sum_{x,y\in\sigma^*}\sigma(x)\sigma(y)d_S^{G-uv}(x,y)=4.3$ and $WI(G-uv)=WI(G)$.*

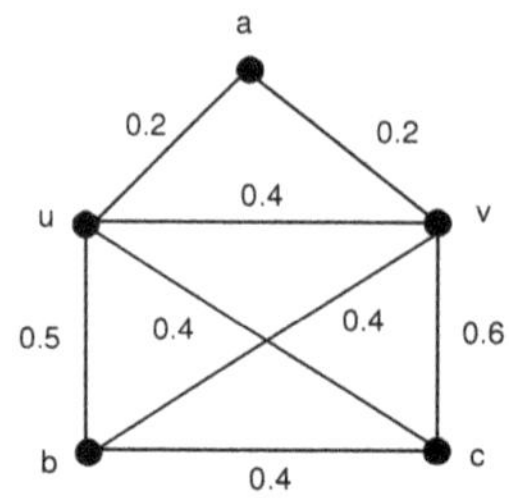

Figure 2. uv is an β-strong edge and $WI(G) = WI(G - uv)$.

Therefore, Theorem 3. 10 of [7] can be changed as follows

Theorem 2. *Let $G = (\sigma, \mu)$ be a fuzzy graph with each edge being strong. For $s, t \in \sigma^*$, let $P_{s,t}$ denote the path which has the minimum sum of membership values among all shortest strong paths between s and t. Suppose that uv is not a part of any $P_{s,t}$ for $s, t \in \sigma^*$ with $\{s,t\} \neq \{u,v\}$. Then*
(1) If $d_S^{G-uv}(u,v) > \mu(uv)$, then $WI(G - uv) > WI(G)$.
(2) If $d_S^{G-uv}(u,v) = \mu(uv)$, then $WI(G - uv) = WI(G)$.
(3) If $d_S^{G-uv}(u,v) < \mu(uv)$, then $WI(G - uv) < WI(G)$.

Proof. Since each edge of G is strong, it follows that each edge in $G - uv$ is also strong edge. Owing to uv being a strong edge in G, $d_S^G(u,v) = \mu(uv)$. Let $\{a,b\} \neq \{u,v\}$. Since uv is not part of any $P_{a,b}$, $d_S^{G-uv}(a,b) = d_S^G(a,b)$. Thus, $WI(G - uv) = \sum_{x,y \in \sigma^*} \sigma(x)\sigma(y) d_S^{G-uv}(x,y) = \sum_{x,y \in \sigma^*} \sigma(x)\sigma(y) d_S^G(x,y) + (d_S^{G-uv}(u,v) - d_S^G(u,v)) = WI(G) + (d_S^{G-uv}(u,v) - \mu(uv))$. So, if $d_S^{G-uv}(u,v) > \mu(uv)$, then $WI(G - uv) > WI(G)$. If $d_S^{G-uv}(u,v) = \mu(uv)$, then $WI(G - uv) = WI(G)$. If $d_S^{G-uv}(u,v) < \mu(uv)$, then $WI(G - uv) < WI(G)$. □

Note: The condition "each edge is strong" is necessary in Theorem 2. For example, let $G = (\sigma, \mu)$ be the fuzzy graph shown in Figure 3 with vertex set $\{a, b, c, u, v\}$ and $\sigma(x) = 1$ for any $x \in \sigma^*$, $\mu(uv) = 0.95$, $\mu(au) = \mu(ab) = 0.3$, $\mu(cu) = \mu(cb) = 0.1$, $\mu(vb) = 0.5$, $\mu(ub) = 0.4$. Then each edge of the graph G except edge ub is strong. Edge ub is a weak edge in G. So $d_S(u,v) = 0.95$, $d_S(a,u) = d_S(a,b) = 0.3$, $d_S(c,u) = d_S(c,b) = 0.1$, $d_S(v,b) = 0.5$, $d_S(a,c) = 0.4$, $d_S(a,v) = 0.8$, $d_S(u,b) = 0.2$ and $d_S(c,v) = 0.6$. Therefore, $WI(G) = \sum_{x,y \in \sigma^*} \sigma(x)\sigma(y) d_S(x,y) = 4.25$. It is obvious that uv is a strong edge and uv is not a part of any $P_{s,t}$ for $s, t \in \sigma^*$ with $\{s,t\} \neq \{u,v\}$. It is obvious that ub is a strong edge in $G - uv$. Hence, $d_S^{G-uv}(u,v) = 0.9$, $d_S^{G-uv}(a,u) = d_S^{G-uv}(a,b) = 0.3$, $d_S^{G-uv}(c,u) = d_S^{G-uv}(c,b) = 0.1$, $d_S^{G-uv}(v,b) = 0.5$, $d_S^{G-uv}(u,b) = 0.4$, $d_S^{G-uv}(a,c) = 0.4$, $d_S^{G-uv}(a,v) = 0.8$ and $d_S^{G-uv}(c,v) = 0.6$. So $WI(G - uv) = \sum_{x,y \in \sigma^*} \sigma(x)\sigma(y) d_S^{G-uv}(x,y) = 4.4$. Though $d_S^{G-uv}(u,v) < \mu(uv)$, $WI(G - uv) > WI(G)$.

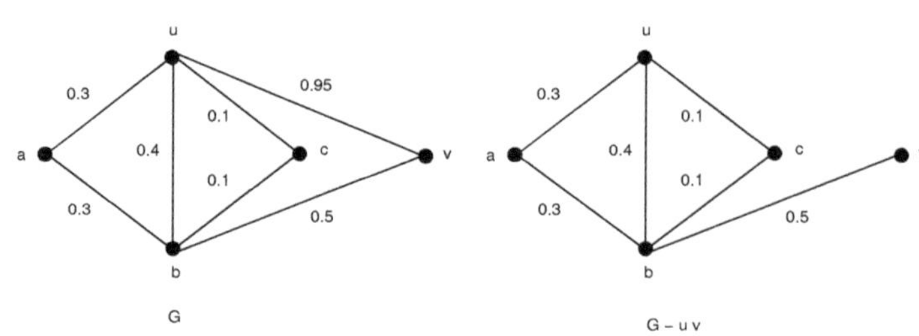

Figure 3. Fuzzy graphs G and $G - uv$.

3. A New Algorithm to Compute Wiener Index of a Fuzzy Graph

Let $G = (\sigma, \mu)$ be a fuzzy graph with $|\sigma^*| = n$. The underlying graph of $G = (\sigma, \mu)$ is denoted by $G^* = (\sigma^*, \mu^*)$, where $\sigma^* = \{v|\sigma(v) > 0\}$ and $\mu^* = \{uv|\mu(uv) > 0\}$. M. Binu et al. in [7] give an Algorithm 1 to compute Wiener index of a fuzzy graph as follows.

Algorithm 1: Computing Wiener index of a fuzzy graph [7].

Step 1. Identify strong edges of G using the algorithm in [8].
Step 2. Let $G' = (\sigma', \mu')$ be the fuzzy subgraph of G obtained by deleting the δ-edges of G.
Step 3. Use Dijkstra's algorithm to identify geodesics between u and v in G'^*, for each $u, v \in \sigma^*$. Let $P^1, P^2, \cdots, P^k$ be the geodesics connecting u and v in G'^*.
Step 4. Calculate S_{p^i} for $i = 1, 2, \cdots, k$, where S_{p^i} is the sum of membership values of edges of P^i.
Step 5. Let $d_S(u, v) = \wedge\{S_{pi}|i = 1, 2, \cdots, k\}$.
Step 6. Construct an $n \times n$ matrix D corresponding to $G = (\sigma, \mu)$ with the following properties. Each row and column corresponds to vertices in σ^*. If row i corresponds to vertex u and column j corresponds to vertex v, then $d_S(u, v)$ is the entry corresponds to row i and column j.
Step 7. Calculate $WI(G) = \sum_{u,v\in\sigma^*} \sigma(u)\sigma(v)d_S(u, v)$.

The main drawback of the Algorithm 1 in [7] is as follows.

Dijkstra's algorithm is only used to identify the length of the shortest path between u and v in G'^*. However, it can not be used to identify the number k of the shortest path. For any big graph, the total number of the shortest path between u and v can be very high. In such case, it is difficult to perform Step 3 and Step 4.

In order to give a simple algorithm to compute Wiener index of a fuzzy graph $G = (\sigma, \mu)$, we define an operator $\circledast$ on matrix as follows. Let $G' = (\sigma', \mu')$ be the fuzzy subgraph of G obtained by deleting the δ-edges of G. Let $A^1 = (a^1_{ij})_{n\times n}$ be the adjacent matrix of the fuzzy graph G', where $a^1_{ii} = 0$ and $a^1_{ij} = \mu'(v_iv_j)$ for $i, j \in \{1, 2, \cdots, n\}$. Let $A^k = (a^k_{ij})_{n\times n}$ for $i = 2, 3, \cdots, diam(G'^*)$, where $diam(G'^*)$ denote the diameter of G'^*. Define $A^{k+1} = A^k \circledast A^1$ as follows:

$$a^{k+1}_{ij} = \begin{cases} 0, & \text{if } i = j \\ a^k_{ij}, & \text{if } i \neq j \text{ and } a^k_{ij} \neq 0 \\ \min_{1\leq t\leq n}\{a^k_{it} + a^1_{tj} | a^k_{it} \neq 0, a^1_{tj} \neq 0\}, & \text{if } i \neq j \text{ and } a^k_{ij} = 0 \end{cases}$$

Theorem 3. *Let $G' = (\sigma', \mu')$ be a fuzzy graph such that every edge is strong edge. If $a^k_{ij} = 0$ and $a^{k+1}_{ij} \neq 0$, then $d(v_i, v_j) = k + 1$ and $d_S(v_i, v_j) = a^{k+1}_{ij}$, where $d(v_i, v_j)$ is the distance between v_i and v_j in G'^*.*

Proof. We will prove it by induction on the number k. Suppose $k = 1$. If $a^1_{ij} \neq 0$, then $v_iv_j \in \mu'$ and $a^1_{ij} = \mu'(v_iv_j)$. Since every edge in G' is strong edge, then $d(v_i, v_j) = 1$ and $d_S(v_i, v_j) = a^1_{ij}$. Suppose that $a^1_{ij} = 0$ and $a^2_{ij} \neq 0$. Then $v_iv_j \notin \mu'$. Since $a^2_{ij} = \min_{1\leq t\leq n}\{a^1_{it} + a^1_{tj} | a^1_{it} \neq 0, a^1_{tj} \neq 0\}$, it follows that $d(v_i, v_j) = 2$ and $d_S(v_i, v_j) = a^2_{ij}$.

Assume that the theorem holds for $k < l$. Suppose that $a^l_{ij} = 0$ and $a^{l+1}_{ij} \neq 0$. By definition, $a^{l+1}_{ij} = \min_{1\leq t\leq n}\{a^l_{it} + a^1_{tj} | a^l_{it} \neq 0, a^1_{tj} \neq 0\}$. For any $a^l_{it} \neq 0$ and $a^1_{tj} \neq 0$, if there exists $l' < l$ such that $a^{l'}_{it} \neq 0$, then $a^{l'+1}_{ij} \neq 0$. Since $l' + 1 \leq l$, it follows that $a^l_{ij} \neq 0$, which is a contradiction. Hence for any $l' < l$, $a^{l'}_{it} = 0$. That is $a^{l-1}_{it} = 0$ and $a^l_{it} \neq 0$. By inductive hypotheses on k, it follows that $d(v_i, v_t) = l$ and $d_S(v_i, v_t) = a^l_{it}$. Since

$a_{ij}^{l+1} = \min_{1\leq t\leq n}\{a_{it}^{l} + a_{tj}^{1} | a_{it}^{l} \neq 0, a_{tj}^{1} \neq 0\}$, there exists t such that $a_{ij}^{l+1} = a_{it}^{l} + a_{tj}^{1}$, where $a_{it}^{l} \neq 0$ and $a_{tj}^{1} \neq 0$. Since $d(v_i, v_t) = l$ and $d_S(v_i, v_t) = a_{it}^{l}$, it follows that $d(v_i, v_j) = l + 1$ and $d_S(v_i, v_j) = a_{ij}^{l+1}$. □

Corollary 1. *Let $G' = (\sigma', \mu')$ be a fuzzy graph such that every edge is strong edge. Let $diam(G'^*)$ be the diameter of G'^*. Then for any two vertices $v_i, v_j \in \sigma'$, $d_S(v_i, v_j) = a_{ij}^{diam(G'^*)}$, where $i, j \in \{1, 2, \cdots, n\}$.*

Algorithm 2: A new algorithm to compute Wiener index of a fuzzy graph.

Step 1. Identify strong edges of G using the algorithm in [8].
Step 2. Let $G' = (\sigma', \mu')$ be the fuzzy subgraph of G obtained by deleting the δ-edges of G.
Step 3. Calculate $A^1, A^2, \cdots, A^{diam(G'^*)}$, where $d_S(v_i, v_j) = a_{ij}^{diam(G'^*)}$ for $1 \leq i < j \leq n$.
Step 4. Calculate $WI(G) = \sum_{u,v\in\sigma^*} \sigma(u)\sigma(v)d_S(u, v)$.

Obviously, it is a polynomial-time algorithm. The correctness of the Algorithm 2 follows from Theorem 3 and Corollary 1. So we have the following:

Theorem 4. *Let $G = (\sigma, \mu)$ be a fuzzy graph. Let $A^1, A^2, \cdots, A^{diam(G'^*)}$ be defined as in the Algorithm 2. Then $WI(G) = \sum_{1\leq i<j\leq n} \sigma(v_i)\sigma(v_j)a_{ij}^{diam(G'^*)}$.*

Proof. By Theorem 3, let $G' = (\sigma', \mu')$ be a fuzzy graph such that every edge is strong edge. If $a_{ij}^{k} = 0$ and $a_{ij}^{k+1} \neq 0$, then $d(v_i, v_j) = k + 1$ and $d_S(v_i, v_j) = a_{ij}^{k+1}$. By the definition on the new operator on matrix, for any two vertices $v_i, v_j \in \sigma'$, $d_S(v_i, v_j) = a_{ij}^{diam(G'^*)}$, where $i, j \in \{1, 2, \cdots, n\}$. So, $WI(G) = \sum_{1\leq i<j\leq n} \sigma(v_i)\sigma(v_j)a_{ij}^{diam(G'^*)}$. □

Example 3. *Let $G = (\sigma, \mu)$ be the fuzzy graph shown in Figure 4 with vertex set $\{a, b, c, d, e\}$ and $\sigma(v) = 1$ for any $v \in \sigma^*$, $\mu(ab) = 0.2$, $\mu(ac) = 0.2$, $\mu(bc) = 0.3$, $\mu(cd) = 0.4$, $\mu(de) = 0.5$. Then each edge of the graph G is strong.*

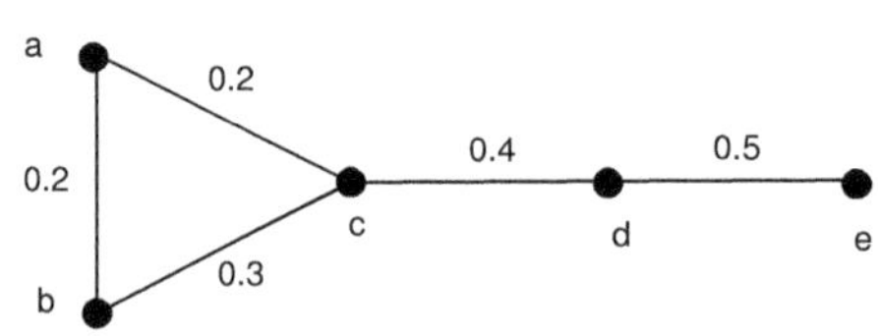

Figure 4. Fuzzy graph G with $diam(G'^*) = 3$.

By using Algorithm 2 and $diam(G'^*) = 3$, we can compute A^1, A^2, and A^3 as follows:

A^1	a	b	c	d	e
a	0	0.2	0.2	0	0
b	0.2	0	0.3	0	0
c	0.2	0.3	0	0.4	0
d	0	0	0.4	0	0.5
e	0	0	0	0.5	0

A^2	a	b	c	d	e
a	0	0.2	0.2	0.6	0
b	0.2	0	0.3	0.7	0
c	0.2	0.3	0	0.4	0.9
d	0.6	0.7	0.4	0	0.5
e	0	0	0.9	0.5	0

A^3	a	b	c	d	e
a	0	0.2	0.2	0.6	1.1
b	0.2	0	0.3	0.7	1.2
c	0.2	0.3	0	0.4	0.9
d	0.6	0.7	0.4	0	0.5
e	1.1	1.2	0.9	0.5	0

As membership values of all vertices are one, the sum of all upper triangular entries of A^3 will be the WI of G. Hence $WI(G) = 6.1$.

4. Conclusions

In this work, we discussed two problems related to the Wiener index of a fuzzy graph. First, we argued that Theorem 3.10 in the paper "Wiener index of a fuzzy graph and application to illegal immigration networks, Fuzzy Sets and Syst. 384 (2020) 132–147" is not correct. We gave a correct statement of Theorem 3.10, where a different result is given for the same conditions. Second, by using a new operator on matrix, we proposed a simple algorithm to compute the wiener index of a fuzzy graph. The main contribution of the proposed algorithm is as follows: First, for a general fuzzy graph, computation of the Wiener index by hand is possible. At the same time, the algorithm is easily realized in the computer. Furthermore, the new algorithm is simpler and more efficient, which is a polynomial-time algorithm. The property on Wiener index can help us to understand the critical property on the communication network. That is, when some edge is deleted, the Wiener index in communication network may be changed.

Author Contributions: The contributions of all authors are equal. All authors have read and agreed to the published version of the manuscript.

Funding: This research was supported by the Basic Science Research Program through the National Research Foundation of Korea (NRF) funded by the Ministry of Education (2020R1I1A3A04036669).

Conflicts of Interest: The authors declare no conflict of interest.

References

1. Zadeh, A.L. Fuzzy sets. *Inf. Sci.* **1965**, *8*, 338–353. [CrossRef]
2. Rosenfeld, A. Fuzzy graphs. In *Fuzzy Sets and Their Applications to Cognitive and Decision Processes*; Zadeh, L.A., Fu, K.S., Tanaka, K., Shimura, M., Eds.; Academic Press: New York, NY, USA, 1975; pp. 77–95.
3. Dobrynin, A.A.; Gutman, I.; Klavžar, S.; Žigert, P. Wiener index of hexagonal systems. *Acta Appl. Math.* **2002**, *72*, 247–294. [CrossRef]
4. Knor, M.; Skrekovski, R.; Tepeh, A. Mathematical aspects of Wiener index. *Ars Math. Contemp.* **2016**, *2*, 327–352. [CrossRef]
5. Wiener, H. Structural determination of paraffin points. *J. Am. Chem. Soc.* **1947**, *69*, 17–20. [CrossRef]
6. Mathew, S.; Mordeson, J.N.; Malik, D. *Fuzzy Graph Theory*; Springer: Berlin/Heidelberg, Germany, 2018.
7. Binu, M.; Mathew, S.; Mordeson, J.N. Wiener index of a fuzzy graph and application to illegal immigration networks. *Fuzzy Sets Syst.* **2020**, *384*, 132–147.
8. Mathew, S.; Sunitha, M.S. Types of arcs in a fuzzy graph. *Inf. Sci.* **2009**, *179*, 1760–1768. [CrossRef]

applied sciences

MDPI

Article

Improving the Energy Efficiency of an Existing Building by Dynamic Numerical Simulation

Lelia Letitia Popescu *, Razvan Stefan Popescu and Tiberiu Catalina

Faculty of Building Services, Technical University of Construction of Bucharest, 66 Pache Protopopescu Blvd., RO-021414 Bucharest, Romania; razvan.popescu@utcb.ro (R.S.P.); tiberiu.catalina@utcb.ro (T.C.)
* Correspondence: lelia.popescu@utcb.ro

Abstract: Nowadays, the enhancement of the existing building stock energy performance is a priority. To promote building energy renovation, the European Committee asks Member States to define retrofit strategies, finding cost-effective solutions. This research aims to investigate the relationship between the initial characteristics of an existing residential buildings and different types of retrofit solutions in terms of final/primary energy consumption and CO_2 emissions. A multi-objective optimization has been carried out using experimental data in DesignBuilder dynamic simulation tool.

Keywords: energy efficiency; renewable energy; dynamic simulation; thermal image; thermal conductivity

Citation: Popescu, L.L.; Popescu, R.S.; Catalina, T. Improving the Energy Efficiency of an Existing Building by Dynamic Numerical Simulation. *Appl. Sci.* **2021**, *11*, 12150. https://doi.org/10.3390/app112412150

Academic Editors: Luisa F. Cabeza and Cesare Biserni

Received: 17 November 2021
Accepted: 16 December 2021
Published: 20 December 2021

1. Introduction

Today's global concern is major climate change, which affects the entire planet, and therefore a strong emphasis on reducing greenhouse gases has been established. The rising of global temperatures rise has disastrous effects on nature, leading to irreversible changes in many ecosystems, thus affecting the biodiversity. All this will have enormous costs for the EU economy and will dramatically affect countries' ability to produce food.

Recently constructed buildings in the EU under energy efficiency criteria represent only a small amount of the total stock of buildings, which is mainly comprised of buildings that are several decades old, long before the first EU energy efficiency directives. Therefore, energy rehabilitation work on existing buildings is of great interest nowadays, as it is an important way to reduce energy consumption, which is currently mainly supplied from conventional sources (fossil fuels). It is already observed at the European level that renovation and energy rehabilitation works represent twice the value obtained by the construction of new buildings.

The construction sector is the largest consumer of energy in Europe: 40% of final energy use. According to European data [1], 75% of the buildings are not energy efficient and out of the total of these buildings, only 0.4–1.2% of stock is renovated each year. Data are collected in 2016 in support of the proposal to revise Directive 2010/31/EU, on the occasion of the publication of "Clean Energy for All Europeans", a package of measures to accelerate the transition to clean energy, in line with the commitment in terms of reducing CO_2 emissions by at least 40% by 2030, modernizing the economy and creating the conditions for sustainable jobs and growth [1].

The need for the Union to achieve its energy efficiency targets at the Union level—expressed in primary and/or final energy consumption—should be clearly set out in the form of a target of at least 32.5% for 2030 [2].

Nowadays, many studies were focused on the retrofitting of the existing building stock due to the climate change issue, but this process is also affected by the cost–benefit analysis of energy efficiency methods [3], or indoor comfort of retrofitted buildings [4]. The effects of energy efficient measures on climate change due to CO_2 emissions from the point of view of buildings' service performance is of interest [5,6], as it is an important parameter.

Existing HVAC (heating, ventilation, and air conditioning) systems using conventional fuels can be partially or completely replaced with 'green energy' HVAC systems.

As a function of climate change over time, it is a great challenge for both architects and engineers to obtain an optimal solution which can achieve minimal energy consumption while also ensuring indoor comfort for new buildings. This goal is even more difficult to achieve for existing buildings, where the degree of freedom is considerably lower. Therefore, a way to approach the accomplishment of degree of energy consumption decrease through different methods is with the help of dynamic simulation tools.

A good building refurbishment requires a proper combination between thermal insulation for different buildings' envelope zones (walls, ceiling, floor, ...), better thermal resistance windows, but also renewable energy systems to make the building sustainable within its environment, a graphical representation of passive house's principles is shown in Figure 1.

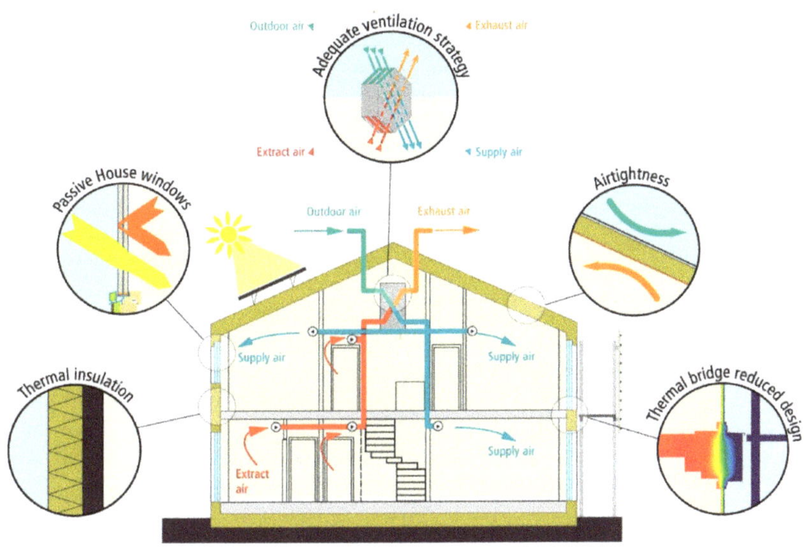

Figure 1. Basic structural principles for the construction of passive houses (with permission from [7]).

In order to be considered a passive house [7,8], a new or retrofitted house should fulfill certain criteria, such as:

- All opaque building components of the exterior envelope of the house must be very well-insulated, this means a heat transfer coefficient (U-value) of 0.15 W/(m^{2}K) at the most;
- Window frames must be well insulated and fitted with low-e glazing filled with argon or krypton to prevent heat transfer, this means a U-value of 0.80 W/(m^{2}K) or less, with g-values around 50% (g-value = total solar transmittance, proportion of the solar energy available for the room);
- At least 75% of the heat from the exhaust air should be transferred to the fresh air by means of a heat exchanger, allowing a good indoor air quality same time with saving energy;
- All edges, corners, connections, and penetrations must be planned and executed with great care, so that thermal bridges can be avoided. Thermal bridge care should be considered, as they cannot be completely excluded, but they must be reduced as far as possible;
- The space heating energy demand should not exceed 15 kWh/m^2 year (considering square meter of net living space);

- The total energy to be used for all domestic applications (heating, hot water, and domestic electricity) must not exceed 60 kWh/m^2 year (considering square meter of net living space);
- A maximum of 0.6 air changes per hour at 50 Pascal pressure should be achieved, and verified with on-site pressure testing (in both pressurized and depressurized states);
- Thermal comfort must be met for all living areas during winter as well as in summer, with not more than 10% of the hours in a given year over 25 °C.

Zero, net zero, nearly zero, passive, plus [9,10] which is the optimal balance between energy efficient and economical solution, but also possible for an existing house by means of dynamical simulation, is our purpose for this study. Residential and tertiary ground source heat pumps are widely used and many studies are found in the literature, with sensitivity analysis [10,11]. Another very important issue in sizing a heat pump system is also to calculate the investment, so a technical–economic analysis is crucial [12].

Romanian regulations followed the European trend in the last decade, concerning energy savings and greenhouse gases (GHG) emissions for the building sector [13–16]. Additionally, as a Member State of the European Union since early 2007, our country developed its own methodology for buildings energy performance calculation [17], which is, at present, in a recast stage. According to new regulation [14], starting from January 2021, all new buildings or existing ones which are to undergo major repairs (works performed on the building envelope and/or building's installations whose costs exceed 25% of the tax value of the building, exclusively the value of the land on which the building is located) have to be near zero energy buildings, but also to be provided with at least 30% of their consumption from renewable energy. The near zero energy concept in Romania, as well as Romania's climate zones for the winter, were recently re-defined [13]. Romania has five climate zones for winter time, the mildest one, "the climate zone I" defined by a conventional exterior temperature of −12 °C and the coldest one, "the climate zone IV" defined by a conventional exterior temperature of −24 °C. The temperature difference between two consecutive climate zones is of −3 °C. The climate of Romania is temperate continental, with four seasons. Summers are generally very warm to hot, depending on the climate zone. Winters are frosty or very frosty, with snow precipitation for most climatic zones.

2. Case Study

The objectives of the present study are to study by means of dynamical simulation with DesignBuilder tool, an existing residential building, situated in the countryside zone of Romania, district of Prahova, having a −15 °C conventional temperature for winter time. Our aim is to determine the optimal balance between energy efficient and economical solutions possible for an existing house by means of dynamic simulation. The energy efficiency of existing house envelope is experimentally fully determined.

The net living space of the house is of 214 m^2, between the ground floor and first floor. The studied house was built in the year 2008, in concordance with the Romanian regulations from that period, making it a real case study house. For indoor heating, the building is equipped with wood fuel type boiler. The hot water supply is represented by an electrical boiler and no cooling system is provided, as the building is located near the mountains, having a mild summer period. The studied house is inhabited by a family of four, a couple of retirees and a couple of young people. Due to the fact that the retired couple stay home all day long, the heating system works 24 h and seven days a week. It should also be mentioned that there are unheated/unoccupied spaces, namely: the garage, the technical room, the attic of the house, and the basement. The domestic hot water demand was calculated according to the Romania legislation [17], considering a 50 L per person and day hot water demand.

Our goal is to apply as many of the passive house requirements, as the existing building allows in order to reduce the building's energy consumption. As already described, the study was conducted on dynamical simulation level using DesignBuilder tool [18], which allows the design of a perfect match, between the architecture of the existing building and the simulated model, as can be seen in Figure 2.

(**a**) (**b**)

Figure 2. (**a**) Real building photo; (**b**) DesignBuilder model.

3. Simulations and Scenarios

The studied house's envelope has the following component description: the walls are made of bricks and thermally insulated with 10 cm polystyrene (partially insulated), the ground floor upon the basement is made of concrete, the ceiling of the first floor is made of concrete also, and no thermal insulation. The windows have a U-value of 0.5 W/(m^{2}K), according to Romanian Energy Standards from the period when they were bought. Measurements were conducted for the thermal performance of walls using TESTO 435 equipment for thermal resistance or U-values, as well as infrared TESTO 885-2 for thermal bridges discovery. Figure 3 shows measured U-value for different walls of the house, which were used to determine the mean U-value used in the dynamical simulations.

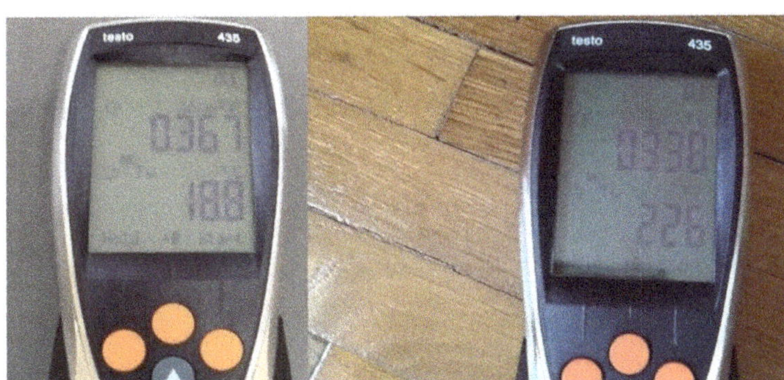

U- Value surface to surface (W/m^2 K):	**0.380**
R- Value (m^2 K/W):	**2.800**
U- Value (W/m^2 K):	**0.357**

Figure 3. Measured U-value for different walls of the house with TESTO 435.

Being an existing house, the thermal bridges were also our concern, so thermal imager TESTO 885-2 equipment was used to determine their presence. Both thermal images presented in Figures 4 and 5 show that thermal bridges are presented and should be taken into consideration in further rehabilitation.

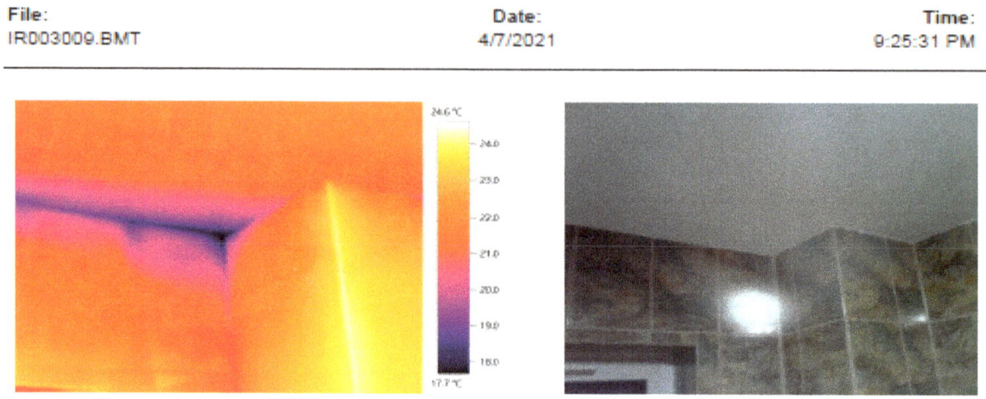

Figure 4. Thermal image from the living room, from the inside of the house.

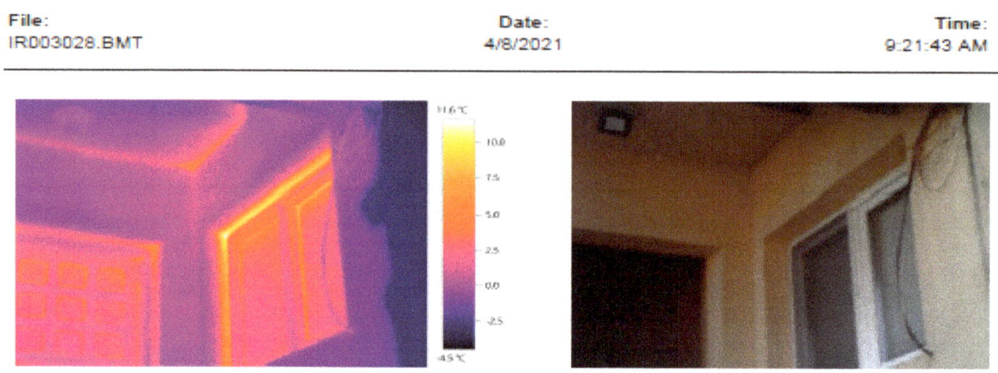

Figure 5. Thermal image from outside, ground floor area.

Both thermal images and measurements of the U-value for building's exterior walls concluded, that house retrofit is needed from the energy consumption point of view, as well as monthly paid bills for energy.

Concerning the building's installations, the heating source is represented by a standard wood fuel boiler (a 70% boiler efficiency), made of cast iron, with natural circulation of evacuated gases. The hot water is obtained with an electric boiler, for the four people who presently occupy the house.

4. Simulations and Scenarios

Designbuilder is a powerful tool for building simulation, using EnergyPlus software for energetic simulations, natural lighting, comfort, and CO_2 emissions. The architecture is built in 3D with a high matching level with the real building with its envelope (walls, windows, basement), as well as its interior zones. Introducing the building thermal zones, it is a very important step because we can introduce different temperatures for every created zone.

After the construction of the architecture, presented in Figure 2, the next step is to choose the HVAC scheme's components from Designbuilder library [18] and to correlate then in order to create the real existing scheme, from the initial studied case. Figure 6 shows the heating and hot water supply systems scheme used in Designbuilder for the initial case.

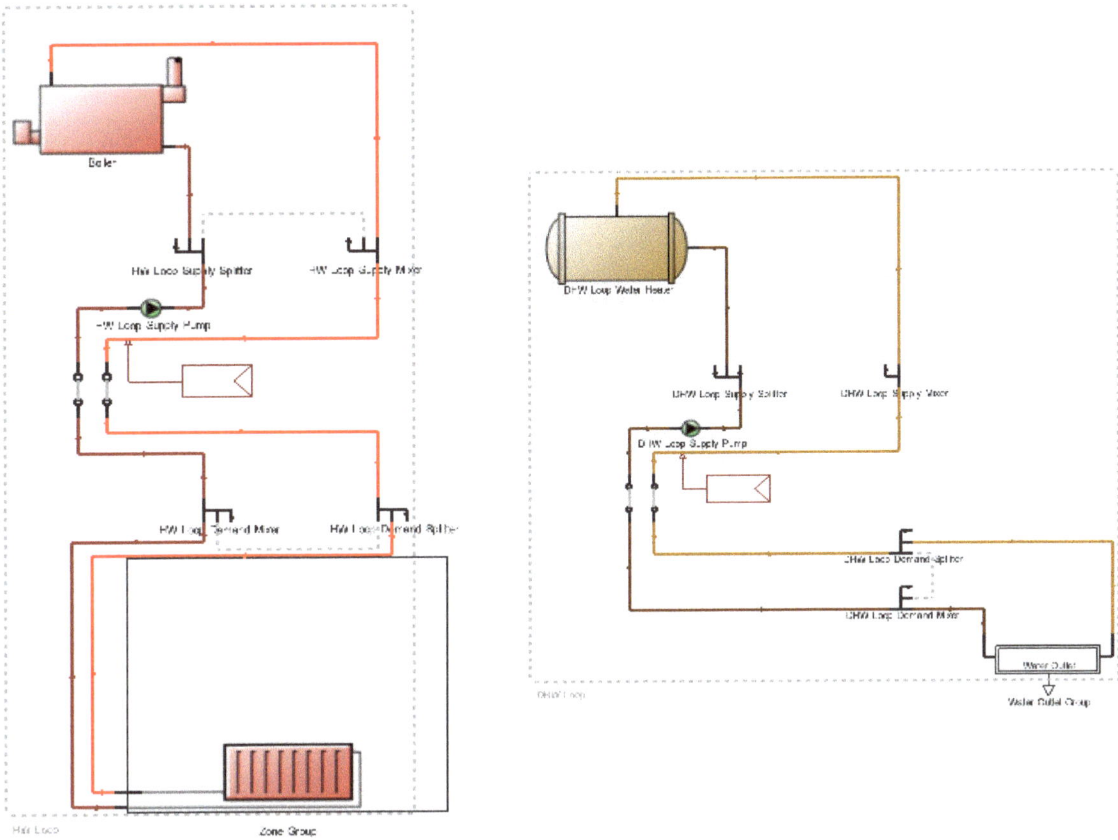

(a) (b)

Figure 6. (**a**) Heating of the initial case with wood boiler and radiators in Designbuilder; (**b**) Domestic hot water initial scheme in Designbuilder.

For the retrofit case of the house, the following scenarios were simulated using the Designbuilder tool:

Scenario 1: The retrofit of the building's envelope opaque components. The 'before' and 'after' of the envelope's retrofit thermal performances are listed in Table 1. The exterior walls are partially insulated, so a mean thermal resistance is listed in the table; but in the simulation measured, the mean for each element was considered.

Table 1. Thermal resistance for exterior walls.

Envelope's Component	Thermal Resistance [m^2K/W]	
	Initial Case	Retrofit Case
Insulated wall	2.81	5.63
Not insulated wall	0.91	5.63
Last floor ceiling	0.92	5.11

Scenario 2: As in Scenario 1, adding the exchange of well-insulated windows presented in Figure 7. For the heating system, there were not changes made. The wood fueled boiler was used to cover the hot water demand.

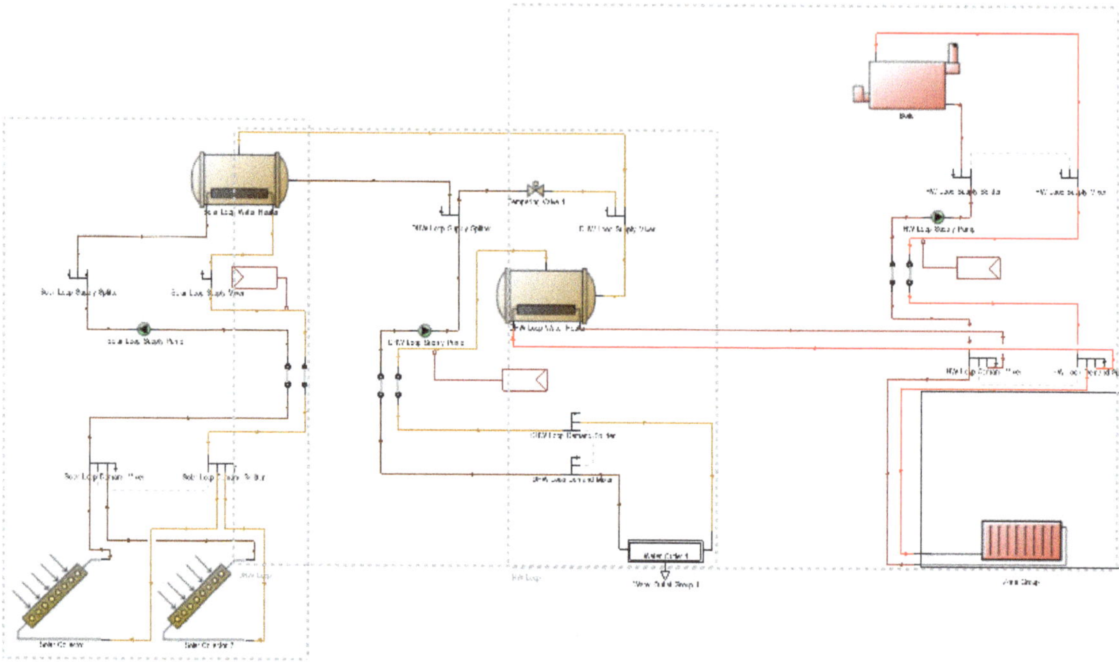

Figure 7. Heating and hot water supply systems schema used in Designbuilder for the third and fourth scenarios.

Scenario 3: Same as above—adding two flat solar panels, of 2 m^2 surface each, were used for hot water supply.

Scenario 4: Adding thermal solar panels and photovoltaic panels (monocrystalline of 1.5 kW_p installed power, on grid type), to cover total on-site electric sources, for the retrofit envelope's case. Additionally, the envelope of the house is thermally insulated, the purpose being to obtain low energy consumption for the studied house.

These scenarios were taken into consideration, to be able to apply for the funding of the National Environmental Program (AFM), to improve energy efficiency of buildings. These funds are obtained only by reducing energy consumption and CO_2 emissions, according to fixed levels imposed. Therefore, exploratory simulations using dynamic software are needed; later on, economic analysis for the investment is easier to obtain.

5. Results

The simulations using DesignBuilder software show all energy consumption—heating, cooling, lighting, domestic hot water, as well as renewable energy production—for the four presented rehabilitation solutions.

For the presented scenarios results are being presented in the table below:

Table 2 presents the results for the energy consumption decrease for every presented scenario analyzed. If only the envelope is thermally insulated, 11.63% of final energy is decreased; then, by adding to Scenario 1, the window replacement a 24.35% specific final energy decrease is obtained. Scenario 3 adds the thermal solar panel implementation to Scenario 2 and determines a 40.78% decrease of the specific final energy consumption. Scenario 4, the most complete one, brings a 44.84% decrease of specific final energy if added PV collectors are added. Concerning primary energy, 55.33% can be achieved for the best studied scenario. Carbon dioxide emissions are decreasing accordingly with maximum 56% for the last case study.

Table 2. Final energy, primary energy, and CO_2 emissions.

	Initial Case	Scenario 1	Scenario 2	Scenario 3	Scenario 4	NZEB [13] Romanian Climatic Zone II
Thermal energy consumption (kWh/m^2 an)	163.6	140.8	116.0	108.0	108.0	
Thermal energy decrease compared to initial case (%)	-	14%	29%	33.95%	33.95%	
Electrical energy consumption (kWh/m^2 year)	32.36	32.29	32.24	7.99	0.04	
Electrical energy decrease compared to initial case (%)	-	0.21%	0.37%	75.32%	99.88%	
Final energy specific economy (%)	-	11.63%	24.35%	40.78%	44.84%	
Conversion to primary energy Fossil fuel (1) (kWh/m^2 year)	176.7	152.1	125.3	114.1	116.7	
Conversion to primary energy Electric energy (2) (kWh/m^2 year)	84.8	84.6	84.5	46.3	0.1	
Yearly primary energy consumption Total (1) + (2) (kWh/m^2 year)	261.4	236.7	209.7	160.4	116.8	111.0
Primary energy economy compared to initial case (%)	-	9.46%	19.78%	38.66%	55.33%	
CO_2 emission (kg CO_2/year)	1892	1714	1521	990	837	
Specific CO_2 emission (kg CO_2/m^2 year)	8.80	7.98	7.08	4.61	3.89	30
CO_2 emission decrease (%)	-	9%	20%	48%	56%	

The PV yearly coverage of the electrical energy consumption is 99.5%, a good coverage degree being obtained.

The total energy class of the initial building is B, and the chosen retrofit scenario has an A energy class, according to Romanian methodology existing in the legislation [17]. According to Romanian Legislation [13] the NZEB (Near Zero Energy Building) situated in the same climate as our building (−15 °C conventional temperature for wintertime, climate zone II) should have a yearly primary energy consumption less than 111.0 kWh/m^2 year and specific CO_2 emission of 30 kg CO_2/m^2 year. For our best scenario, the yearly primary energy target could not be reached by almost 5%, but the specific yearly CO_2 emissions were 7.7 times decreased because of the fact that renewable energies were used (biomass and solar). As a perspective maybe a heat pump should be investigated and taken into consideration; additionally, PV solar collectors can be added for heating and DHW (domestic hot water) purposes, with this renewable energy source.

An investment of 28 k euro was taken into account which includes building envelope complete rehabilitation and renewable energy systems implementation for Scenario 4. Concerning the building envelope rehabilitation, we took into account the external walls and windows, roof. The floors could not be isolated as it is an existing building and the free internal level high does not allow it. The economic viability depends on the local costs of energy for internal use and, in some countries, the often higher value per kW to sell back to the power grid. Without the benefits of the different 'efficient energy' programs in progress in Romania, the amortization is around 13 years, less than their life time, but this period can be significantly reduced by applying for one of the National Agency of Environmental Protection programs. The payback time was calculated according to Romanian legislation [17].

6. Conclusions

The initial case of the simulated building had a poorly thermal insulation envelope with high monthly energy consumption. The total energy class of this building according to Romanian legislation of buildings' energy efficiency [17] was labeled "B". The purpose of this research is to find measures to improve energy efficiency considering also renewable energy sources. The initial aim was to obtain a Near Zero Energy House (NZEB) but only some few percentage points are delimiting from this purpose, so we can conclude that a low energy house is simulated for the fourth scenario analyzed, the building being classed in energy label 'A'. The Romanian Government's target for the existing houses in the year 2020 was to improve energy efficiency by 40%, 50%, or 60% primary energy decrease founding by specific programs that are even now available [19]. Starting from the first of January 2021, the Romanian Government's target changed and only NZEB buildings are allowed to be built.

A primary energy reduction of 55% is an acceptable savings, attained only by using PV and solar thermal collectors, as well as a thicker thermal insulation for the envelope. The final energy reduction between the initial case and the fourth scenario was 44.85% and the CO_2 emissions decreased by 56%. The PV panels ensure 99.9% of the house's electric energy consumption and the solar thermal collectors determine a 32% decrease in specific final energy consumption. Generally speaking, both the building's envelope and use of renewable energy sources should be taken into account in order to achieve a near NZEB house, according to the dynamical simulations. These results should be further considered by the policy makers for future eligible funding of the government's energy efficiency programs.

This study combined experimental work with numerical simulations, proving the importance of both sides in developing an energy strategy. The experimental data obtained by means of thermal images and building's envelope components' conductivity measurements describe the stage of the initial case and represent the incomes for the numerical simulations. Without the help of a dynamical simulation, the real energy consumption cannot be revealed, as the use of stationary heat transfer equations give a maximum instant energy consumption, which is reached few days over the year.

Further studies will take place to achieve a passive house with a lower energy consumption, with this case being a close perspective. In order to achieve that goal, ground source heat pumps, with active and/or passive cooling and a heat recovery unit for ventilation, should further studied.

Author Contributions: Conceptualization, L.L.P.; methodology, all authors; software, L.L.P.; validation, R.S.P.; formal analysis, T.C.; writing—original draft preparation, L.L.P.; writing—review and editing, L.L.P., R.S.P. and T.C. All authors have read and agreed to the published version of the manuscript.

Funding: This research was funded by The Technical University of Construction of Bucharest, Romania.

Institutional Review Board Statement: Not applied.

Informed Consent Statement: Not applied.

Data Availability Statement: Not applied.

Acknowledgments: The paper was elaborated within the research activities of the research grants: UTCB-CDI-2021-004 "The study of the indoor air quality- particle pollution" and UTCB-CDI-2021-005 "The study of the outdoor air quality from Bucharest, Romania based on pollution measurements".

Conflicts of Interest: The authors declare no conflict of interest.

References

1. *Directive of the European Parliament and of the Council COM (2016) 765 Amending Directive 2010/31/EU on the Energy Performance of Buildings*; European Commission: Brussels, Belgium, 2016.
2. Directive (EU) 2018/2002 of the European Parliament and of the Council of 11 December 2018 amending Directive 2012/27/EU on Energy Efficiency. Available online: http://data.europa.eu/eli/dir/2018/2002/oj (accessed on 1 December 2021).
3. Liu, Y.; Liu, T.; Ye, S.; Liu, Y. Cost-benefit analysis for Energy Efficiency Retrofit of existing buildings: A case study in China. *J. Clean. Prod.* **2018**, *177*, 493–506. [CrossRef]
4. Penna, P.; Prada, A.; Cappelletti, F.; Gasparella, A. Multi-objectives optimization of Energy Efficiency Measures in existing buildings. *Energy Build.* **2015**, *95*, 57–69. [CrossRef]
5. Pylsy, P.; Lylykangas, K.; Kurnitski, J. Buildings' energy efficiency measures effect on CO_2 emissions in combined heating, cooling and electricity production. *Renew. Sustain. Energy Rev.* **2020**, *134*, 110299. [CrossRef]
6. Popescu, R.S.; Popescu, L.; Preda, A.; Limam, K. Using a geothermal heat pump for an office building with low energy consumption. *Balt. J.* **2020**, *33*, 39–45.

7. Passive House Requirements. Available online: https://passiv.de/en/02_informations/02_passive-house-requirements/02_passive-house-requirements.htm (accessed on 1 December 2021).
8. Primary Energy Renewable. Available online: https://passipedia.org/basics/energy_and_ecology/primary_energy_renewable_per (accessed on 1 December 2021).
9. Molinari, M.; Lazzarotto, A.; Bjork, F. *The Application of the Parametric Analysis for Improved Energy Design of a Ground Source Heat Pump for Residential Buildings*; Elsevier: Amsterdam, The Netherlands, 2013; Volume 63.
10. Casasso, A.; Sethi, R. Sensitivity Analysis on the Performance of a Ground Source Heat Pump Equipped with a Double U-Pipe Borehole Heat Exchanger. *Energy Procedia* **2014**, *59*, 301–308. [CrossRef]
11. Michopoulos, A.; Kyriakis, N. The influence of a vertical ground heat exchanger length on the electricity consumption of the heat pumps. *Renew. Energy* **2010**, *35*, 1403–1407. [CrossRef]
12. Blum, P.; Campillo, G.; Kölbel, T. Techno-economic and spatial analysis of vertical ground source heat pump systems in Germany. *Energy* **2011**, *36*, 3002–3011. [CrossRef]
13. DesignBuilder Simulation Software. Available online: http://www.designbuilder.co.uk (accessed on 1 December 2021).
14. *ORDIN nr. 386 din 28 Martie 2016 Pentru Modificarea şi Completarea Reglementării Tehnice "Normativ Privind Calculul Termotehnic al Elementelor de Construcţie ale Clădirilor", Indicativ C 107-2005, Aprobată Prin Ordinul Ministrului Transporturilor, Construcţiilor şi Turismului nr. 2.055/2005*; Ministry of Transport, Constructions and Tourism: Bucharest, Romania, 2016.
15. *The Law Concerning the Buildings Energy Performance, no.372/2005 Re-Casted by the Law 101/2020, nr.868/23.09.2020*; Romanian Official Monitor Part I: Bucharest, Romania, 2020.
16. *The Law no.121/2014 Concerning the Energy Efficiency, Recasted by the Law no.160/2016*; Ministry of Transport, Constructions and Tourism: Bucharest, Romania, 2016.
17. *Order no.2641/2017 Concerning the Modification and Completion of the "Romanian Methodology for the Calculation of Buildings Energy Performance"—Mc001/1,2,3-2007*; Ministry of Transport, Constructions and Tourism: Bucharest, Romania, 2017.
18. *Romanian Methodology for the Calculation of Buildings Energy Performance"—Mc001/1,2,3-2007, Approved by the Order of the Ministry of Transport, Constructions and Tourism no.157/2007*; Ministry of Transport, Constructions and Tourism: Bucharest, Romania, 2007.
19. Efficient Energy House Guide. Available online: https://www.afm.ro/casa_eficienta_energetic_ghid_finantare.php (accessed on 1 December 2021).

mathematics

MDPI

Article

An Adaptive Covariance Scaling Estimation of Distribution Algorithm

Qiang Yang [1,*], Yong Li [1], Xu-Dong Gao [1], Yuan-Yuan Ma [2], Zhen-Yu Lu [1], Sang-Woon Jeon [3] and Jun Zhang [3,4]

[1] School of Artificial Intelligence, Nanjing University of Information Science and Technology, Nanjing 210044, China; 20201249539@nuist.edu.cn (Y.L.); 003103@nuist.edu.cn (X.-D.G.); 001114@nuist.edu.cn (Z.-Y.L.)
[2] College of Computer and Information Engineering, Henan Normal University, Xinxiang 453007, China; 121100@htu.edu.cn
[3] Department of Electrical and Electronic Engineering, Hanyang University, Ansan 15588, Korea; sangwoonjeon@hanyang.ac.kr (S.-W.J.); junzhang@ieee.org (J.Z.)
[4] Department of Computer Science and Information Engineering, Chaoyang University of Technology, Taichung 413310, Taiwan
* Correspondence: qiang_yang@nuist.edu.cn or mmmyq@126.com

Citation: Yang, Q.; Li, Y.; Gao, X.-D.; Ma, Y.-Y.; Lu, Z.-Y.; Jeon, S.-W.; Zhang, J. An Adaptive Covariance Scaling Estimation of Distribution Algorithm. *Mathematics* **2021**, *9*, 3207. https://doi.org/10.3390/math9243207

Academic Editor: Simeon Reich

Received: 8 November 2021
Accepted: 7 December 2021
Published: 11 December 2021

Publisher's Note: MDPI stays neutral with regard to jurisdictional claims in published maps and institutional affiliations.

Abstract: Optimization problems are ubiquitous in every field, and they are becoming more and more complex, which greatly challenges the effectiveness of existing optimization methods. To solve the increasingly complicated optimization problems with high effectiveness, this paper proposes an adaptive covariance scaling estimation of distribution algorithm (ACSEDA) based on the Gaussian distribution model. Unlike traditional EDAs, which estimate the covariance and the mean vector, based on the same selected promising individuals, ACSEDA calculates the covariance according to an enlarged number of promising individuals (compared with those for the mean vector). To alleviate the sensitivity of the parameters in promising individual selections, this paper further devises an adaptive promising individual selection strategy for the estimation of the mean vector and an adaptive covariance scaling strategy for the covariance estimation. These two adaptive strategies dynamically adjust the associated numbers of promising individuals as the evolution continues. In addition, we further devise a cross-generation individual selection strategy for the parent population, used to estimate the probability distribution by combing the sampled offspring in the last generation and the one in the current generation. With the above mechanisms, ACSEDA is expected to compromise intensification and diversification of the search process to explore and exploit the solution space and thus could achieve promising performance. To verify the effectiveness of ACSEDA, extensive experiments are conducted on 30 widely used benchmark optimization problems with different dimension sizes. Experimental results demonstrate that the proposed ACSEDA presents significant superiority to several state-of-the-art EDA variants, and it preserves good scalability in solving optimization problems.

Keywords: estimation of distribution algorithm; covariance scaling; gaussian distribution; meta-heuristic algorithm; problem optimization

1. Introduction

Optimization problems are ubiquitous in daily life and industrial engineering [1,2], such as protein structure prediction [3], community detection [4], control of pollutant spreading [5] and multi-compartment electric vehicle routing [6]. These optimization problems often preserve characteristics such as non-convex, discontinuous, and non-differentiable [7–10], which greatly challenge the effectiveness of traditional gradient-based optimization algorithms or even make them infeasible [11]. In particular, in the era of big data and the Internet of Things, optimization problems are becoming more and more complex due to the increase in dimensionality [12–14]. For instance, some unimodal problems become multimodal with many local optima [15], while some multimodal problems become more complicated with an increasing number of wide and flat local areas [16–18]. Such complicated optimization problems are becoming more and more common nowadays, and thus, it is urgent to develop effective optimization algorithms to solve them, so as to promote the development of related fields.

As a kind of gradient-free meta-heuristic algorithm, estimation of distribution algorithm (EDA) mainly maintains a population of individuals to iteratively search the solution space, with each individual representing a feasible solution [19]. During each generation, it selects a number of promising individuals to estimate the probability distribution of the population, and it then randomly samples a new population of solutions based on the estimated probability distribution [20,21]. Due to the randomness in sampling the offspring, EDA preserves high diversity and strong global search ability [22]. Therefore, a lot of researchers have paid extensive attention to developing effective EDAs, and, consequently, not only have EDAs been applied to solve various optimization problems, such as multimodal optimization problems [23] and multi-objective optimization problems [24], but also they have been employed to solve many real-world problems, such as multi-policy insurance investment planning [25] and multi-source heterogeneous user-generated content-driven interactive [22].

In the literature, most EDAs utilize the Gaussian distribution model to evaluate the probability distribution of the population, which is then adopted to sample new solutions [26,27]. During the estimation of the distribution, based on whether the correlation between variables is considered, the current Gaussian estimations of distribution algorithms (GEDAs) are mainly divided into two categories [20,27,28], namely univariate GEDAs (UGEDAs) [29–31] and multivariate GEDAs (MGEDAs) [32–36].

UGEDAs [30] consider that each variable is independent on each other. Therefore, the probability distribution of each variable is estimated individually. The most advantageous property of UGEDAs is that the computational cost of the distribution estimation and the offspring sampling is low [31]. However, their effectiveness deteriorates drastically when confronted with optimization problems with interacted variables [29].

Different from UGEDAs, MGEDAs take the correlation among variables into consideration [34], which is realized by estimating the covariance among all variables. With the covariance matrix, MGEDAs could capture the structure of the optimization problem and thus implicitly offer useful information to direct the search of the population [35]. Due to this advantage, MGEDAs achieve much better performance than UGEDAs, especially on problems with many interactive variables [20]. As a result, MGEDAs have been extensively researched in the literature [37–39]. However, such superiority of MGEDAs is at the sacrifice of efficiency, as calculating the covariance among all variables is very time-consuming [32].

In MGEDAs, the parameters (i.e., the mean vector and the covariance matrix) of the probability distribution are usually estimated based on a certain number of promising individuals [28]. Specifically, in the probability distribution, the mean vector plays a key role in controlling the center of the offspring to be sampled, while the covariance takes charge of the range of the offspring around the center. In other words, the mean vector makes a crucial influence on the convergence of the population to the optimal areas, while the covariance affects the population diversity [40]. Therefore, to maintain high search diversity for EDAs, many researchers have designed variance or covariance scaling methods [37,41–43] to enlarge the sampling range of the estimated probability distribution. However, on the one side, the basic covariance (variance) is still estimated on the same promising individuals selected for the estimation of the mean vector in most existing GEDAs. Therefore, after the covariance (variance) scaling, implemented by multiplying a scaling factor on the estimated covariance (variance) in most covariance scaling methods, the learned structure of the optimization problem, by them, remains unchanged; on the other side, most existing covariance (variance) scaling methods enlarge the estimated covariance (variance) the same degree in different directions.

To remedy the above shortcomings, this paper devises an adaptive covariance scaling method for MGEDAs, leading to an adaptive covariance scaling estimation of distribution algorithm (ACSEDA). Specifically, for the estimation of the mean vector in the probability distribution, it is the same as existing MGEDAs, namely estimating it based on a certain number of promising individuals. However, for the estimation of the covariance in the probability distribution, different from existing MGEDAs, ACSEDA first adaptively enlarges the number of promising individuals, and then, it calculates the covariance on the basis of the scaled promising individuals. In this way, the sampling range of the estimated probability distribution could be enlarged, which is helpful for sampling diversified offspring. As a result, the search diversity of EDA could be amplified, and thus, the chance of falling into local areas could be declined.

As a whole, the main contributions of this paper are summarized as follows:

(1) An adaptive covariance scaling method is proposed to adaptively enlarge the sampling range of the estimated probability distribution. Different from most existing covariance scaling methods, we scale the covariance by calculating the covariance based on an amplified number of promising individuals. As a result, not only could the learned structure of the optimization problem captured by the algorithm be improved but the covariance in different directions is also scaled differently. In this way, it is expected that the sampled offspring are not only of high quality, but they are also diversified in different areas.
(2) An adaptive selection of promising individuals for the estimation of the mean vector is further designed by adaptively decreasing the selection ratio, which is the number of the selected promising individuals out of the whole population. In this way, the estimated mean vector, namely the center of the offspring to be sampled, is gradually close to the promising areas that the current population covers. Therefore, the search process is gradually biased toward exploiting the solution space to refine the solution accuracy. However, it should be mentioned that such a bias is not greedy and not at the serious sacrifice of the population diversity because of the aforementioned covariance scaling technique.
(3) A cross-generation individual selection scheme, for the parent population to estimate the probability distribution, is devised by combining the sampled offspring in the last generation and the one in the current generation to select parent individuals for the next generation. Instead of directly utilizing the sampled offspring as the parent population for the next generation in most existing MGEDAs, the proposed ACSEDA combines the sampled offspring in the last generation and the one in the current generation to select the top half best individuals to form the parent population for the next generation. In this way, the parent population formed is neither too crowded nor too scattered, and thus, the estimated probability distribution is of high quality to sample slightly diversified offspring to approach the optimal areas.
(4) With the above mechanisms, the proposed ACSEDA is expected to compromise intensification and diversification of the search process well to explore and exploit the solution space and could thus achieve promising performance.

To verify the effectiveness of the proposed ACSEDA, this paper conducts extensive experiments on the widely used CEC2014 [44] benchmark optimization problems, with different dimension sizes, by comparing ACSEDA to 7 state-of-the-art GEDAs. In addition, deep investigations on the components of ACSEDA are also taken to observe what contributes to its promising performance in solving optimization problems.

The remainder of this study is organized as follows: Section 2 reviews related works on GEDAs; then, the proposed ACSEDA is elucidated in detail in Section 3; in Section 4, extensive experiments are conducted to verify the effectiveness of the developed ACSEDA; last, in Section 5, conclusions are presented.

2. Related Work

2.1. Basic GEDA

The overall framework of a general GEDA is outlined in Algorithm 1. As a whole, the basic principle of GEDA is to iteratively build a Gaussian probability distribution model, based on a certain number of promising individuals selected from the current population, and then sample new individuals based on the built probability model for the next generation [21].

Algorithm 1: The Procedure of GEDA.

Input: population size *PS*, selection ratio *sr*;
1: Set $g = 0$, and randomly initialize the population $\boldsymbol{P^g}$;
2: Obtain the global best solution ***Gbest***;
3: **Repeat**
4: Select $\lceil sr * PS \rceil$ promising solutions $\boldsymbol{S^g}$ from $\boldsymbol{P^g}$;
5: Build a Gaussian probability distribution model $\boldsymbol{G^g}$ based on $\boldsymbol{S^g}$;
6: Randomly generate a new population $\boldsymbol{P^{g+1}}$ by sampling from $\boldsymbol{G^g}$;
7: Update the global best solution ***Gbest***;
8: $g = g + 1$;
9: **Until** *the stopping criterion is met.*
Output: the global best solution ***Gbest***;

Specifically, as shown in Algorithm 1, given that the population size is *PS* and the selection ratio is *sr* (which is the number of selected promising individuals out of the whole population), the number of selected promising individuals in each generation is $s = \lceil sr * PS \rceil$. After *PS* individuals are initialized randomly and evaluated accordingly, as shown in Line 1, the global best solution found so far is obtained (as shown in Line 2). Subsequently, it comes to the main iteration of the algorithm. First, a set (**S**) of *s* promising individuals are selected from the current population (Line 4). Then, a Gaussian probability distribution model is estimated based on the selected individuals (Line 5). After that, *PS* new individuals are randomly sampled based on the estimated Gaussian distribution to form a new population (Line 6). Subsequently, the newly generated individuals are evaluated, and the global best solution is updated. The above process proceeds repeatedly until the termination condition is met. At last, the found global best solution is output.

In GEDAs, the key component is the way to estimate the probability distribution. Different manners of probability distribution estimation result in different kinds of GEDAs. In the literature, based on whether the linkage between variables is considered, existing GEDAs are mainly classified into two categories [20,27,28], namely univariate GEDAs (UGEDAs) [29–31,45,46] and multivariate GEDAs [27,32–36].

(1) *UGEDAs*: In UGEDAs [29,45,46], each variable is considered to be separable and independent on each other. As a result, the probability distribution of *D* variables can be estimated separately, and the joint probability distribution of *D* variables is computed as follows:

$$P(x^1, x^2, \ldots, x^D) = \prod_{i=1}^{D} P(x^i). \tag{1}$$

where $P(x^i)$ is the probability distribution of the *i*th variable, which is estimated as:

$$P(x^i) = \frac{1}{\sigma^i \sqrt{2\pi}} e^{-\frac{(x^i - \mu^i)^2}{2\sigma^i}} \tag{2}$$

where u^i and σ^i are the mean value and the variance of the *i*th variable respectively, which are calculated as follows:

$$\mu^i = \frac{1}{s} \sum_{j=1}^{s} S_j^{\,i} \tag{3}$$

$$\sigma^i = \sqrt{\left(\frac{1}{s-1}\sum_{j=1}^{s}(S_j{}^i - \mu^i)^2\right)} \tag{4}$$

where S is the set of the selected promising individuals, $S_j{}^i$ is the ith dimension of the jth promising individual in S, and D denotes the dimension size of the optimization problem. Based on the estimated probability distribution of each variable, a new solution can be constructed by randomly sampling a new value of each variable separately, based on the associated probability distribution.

(2) *MGEDAs*: In MGEDAs [33–36], the correlations between variables are taken into consideration to estimate the probability distribution. Consequently, different from UGEDAs, the probability distribution of D variables in MGEDAs is estimated together, and the joint probability distribution of D variables is computed as follows:

$$P(x^1, x^2, \ldots, x^D) = \frac{1}{\sqrt{(2\pi)^D \left|C\right|}} e^{\left(-\frac{1}{2}(X-\mu)^T C^{-1}(X-\mu)\right)} \tag{5}$$

where u is the mean vector of the multivariate Gaussian distribution, which is calculated by Equation (3). C is the covariance matrix, which is calculated as follows:

$$C = \frac{1}{s-1}(S-\mu)(S-\mu)^T \tag{6}$$

Based on the estimated joint probability distribution, a new solution is constructed by jointly sampling values for all variables, randomly, from the multivariate Gaussian distribution model. In general, to make the sampling of new solutions simple, a modified version presented below is usually utilized to generate the offspring in most MGEDAs [32,35]:

$$\begin{aligned} X &= \mu + A\Lambda Z \\ Z &\sim N(0,1) \end{aligned} \tag{7}$$

$$C = A\Lambda^2 A^T \tag{8}$$

where A is the eigenvector matrix of C, and Λ is the diagonal matrix whose entries are the square root of the eigenvalues of C. Z is a real number vector, each value of which is randomly sampled from a standard normal distribution separately.

With respect to the computational cost, UGEDAs are less time-consuming, while MGEDAs take more computational cost due to the calculation of the covariance matrix [33–35]. However, in terms of the optimization performance, MGEDAs show much better performance, especially on problems with many interacted variables, while UGEDAs only present promising performance on separable optimization problems [29,30,47]. This is because MGEDAs could capture the interaction between variables and thus evolve the population more effectively than UGEDAs [27,37,48].

2.2. Recent Advance of GEDAs

During the optimization, one crucial challenge that most existing GEDAs encounter is the rapid shrinkage of the variance (or the covariance) [20,42,43], which leads to the quickly narrowed sampling range of the probability distribution. This may lead to a quick loss of the search diversity and thus may result in premature convergence and falling into local areas. To remedy this shortcoming, researchers have devoted plenty of attention to designing novel mechanisms to improve the quality of the probability distribution in GEDAs [38,49–51].

In [51], the authors demonstrated empirically that high diversity maintenance is very crucial for EDAs to achieve satisfactory performance. Then, based on the findings, they further developed a novel three-step method by combining clustering methods with EDAs to search for the optimal areas with high diversity. To prevent premature convergence in EDAs, Pošík [52] directly multiplied a constant factor on the estimated variance of the Gaussian distribution in each generation to enlarge the sampling range. In [43], Grahl et al. proposed a correlation-triggered adaptive variance scaling strategy to reduce the risk of premature convergence and then embedded it into the iterated density–estimation evolutionary algorithm (IDEA). Specifically, similar to [52], the proposed method multiplies a factor to the estimated variance. The difference lies in such a factor not being constant but dynamically adjusted during the evolution, based on whether the global best solution is improved or not. In addition, such adjustment is triggered based on the correlation between the ranks of the normal density and the fitness of the selected solutions. To further trigger the dynamic adjustment of the scaling factor properly, in [42], Bosman et al. proposed a novel indicator, named Standard–Deviation Ratio (SDR), to trigger the adjustment adaptively. Specifically, based on this indicator, the variance scaling is triggered only when the improvements are found to be far away from the mean vector. In [53], a cross-entropy based adaptive variance scaling method was proposed. In this method, the difference between the sampled population and the prediction of the probabilistic model is first measured, and the scaling factor on the variance is then computed by minimizing the cross-entropy between the two distributions.

Different from the above variance scaling methods that directly multiply a scaling factor on the estimated variance, in [49], the authors proposed a novel probability density estimator based on the new mean vector obtained by the anticipated mean shift strategy. Then, once the new mean vector gets better, the variance estimator adaptively enlarges the variance without using an explicit factor, but rather, by using the new better mean vector to calculate the variance. Furthermore, they also developed a reflecting sampling strategy to further improve the search efficiency of GEDA. Accompanied with these two schemes, a new GEDA variant named EDA_{VERS} [49] was developed. Subsequently, a novel anisotropic adaptive variance scaling (AAVS) method was proposed in [41], and a new GEDA named AAVS-EDA was designed. Specifically, in this algorithm, a topology-based detection method was devised to detect the landscape characteristics of the optimization problems, and then, based on the captured characteristics, the variances along different eigen-directions are anisotropically scaled. In this way, the variances and the main search direction of GEDA could be simultaneously adjusted. Recently, Liang et al. proposed a new GEDA variant, named EDA2 [37], to improve the optimization performance of EDA. Specifically, instead of only utilizing promising individuals in the current generation to estimate the Gaussian model, this algorithm stores historical high-quality individuals generated in the previous generations into an archive and adopts these individuals to collaboratively estimate the covariance of the Gaussian model. In this manner, valuable historical evolution information could be integrated into the estimated model.

The above mentioned MGEDAs usually adopt the full rank covariance matrix to estimate the covariance. Since the calculation of the full-rank covariance matrix is very time-consuming, the computational complexity of most MGEDAs is usually high. To alleviate this shortcoming, researchers turn to seeking efficient covariance matrix adaption (CMA) techniques for EDA [54]. As for the covariance matrix, a direct and simple method to speed up its computation is to reduce the degrees of freedom. To this end, Ros and Hansen [55] proposed to only update the elements in the diagonal of the covariance matrix, leading to a de-randomized evolution strategy, named sep-CMA-ES. This method reduces the updating time and space complexity of the covariance matrix from quadratic to linear. In [56], the authors devised an adaptive diagonal decoding scheme to accelerate covariance matrix adaptation. Further, ref. [57] developed a matrix-free CMA strategy by employing combinations of difference vectors between archived individuals and random vectors, generated by the univariate Gaussian distribution along directions of the past shifts of the mean vector. In [58], Beyer and Sendhoff proposed a matrix adaptation evolution strategy (MA-ES) by removing one evolution path in the calculation of the covariance matrix, leading to that the covariance update is no longer needed. In [59], Li and Zhang first designed a rank one evolution strategy by using a single principal search direction, which is of linear complexity. Then, they developed a rank-m evolution strategy by employing multiple search directions. In particular, these two evolution strategies mainly adopt principal search directions to seek for the optimal low rank approximation to the covariance matrix. In [60], He et al. put forward a search direction adaptation evolution strategy (SDA-ES) with linear time and space complexity. Specifically, this algorithm first models the covariance matrix with an identity matrix along with multiple search directions. Then, it uses a heuristic to update the search directions such as the principal component analysis.

Besides the advance of GEDAs in covariance (or variance) scaling and adaption, some researchers have also attempted to design new ways to shift the mean vector of the Gaussian distribution model. For instance, in [61], Bosman et al. proposed an anticipated mean shift to update the mean vector and then used the updated mean vector to calculate the variance. In addition, some researchers have also attempted to adopt other distribution models, instead of the Gaussian distribution model, to estimate the probability distribution. For example, in [62], a probabilistic graphical model was designed to consider the dependencies between multivariate variables. Specifically, in this algorithm, a parallel of a certain number of subgraphs, with a smaller number of variables, is estimated separately to capture the dependencies among variables in each subgraph. Then, each estimated graph model associated with the subgraph samples new values for the associated variables separately. In [39], the authors utilized the Boltzmann distribution to build the probability distribution model in EDA, leading to BUMDA. In particular, the distribution parameters are derived from the analytical minimization of the Kullback–Leibler divergence. In [50], the authors devised a novel multiple sub-models maintenance technique for EDA, leading to a new EDA variant, named maintaining and processing sub-models (MAPS). Specifically, this algorithm maintains multiple sub-models to detect promising areas.

Since EDAs utilize the estimated probability distribution model to sample new solutions, they generally lack subtle refinement to improve the solution accuracy [63]. To fill this gap, local search methods are commonly accompanied with EDAs to refine the found promising solutions [38,49,50]. For instance, in [64], simulated annealing (SA) based local search operator was incorporated into EDA to balance the exploration and exploitation to search the solution space properly. Specifically, the SA-based local search is probabilistically executed on some good solutions to improve their accuracy. To improve the solution accuracy, Zhou et al. [38] developed cheap and expensive local search methods for EDA, leading to a new EDA variant named EDA/LS. In particular, this EDA variant adopts a modified univariate histogram probabilistic model to sample a part of individuals, and it then utilizes a cheap local search method to sample the rest of the individuals. Besides, it also employs an expensive local search method to refine the found promising solutions. Along this direction, an extension of EDA/LS, named EDA/LS-MS, was developed in [65] by introducing a mean shift strategy to replace the cheap local search method in EDA/LS to refine some good parent solutions.

Though a lot of remarkable GEDA variants have emerged and shown promising performance in solving optimization problems, they still encounter limitations, such as falling into local areas and premature convergence. In particular, it is found that most existing GEDAs estimate the variance (or covariance) based on the same selected promising individuals used for the estimation of the mean vector. Although various variance (or covariance) scaling methods [37,43,49,53,66] and covariance matrix adaption methods [56–60] have been proposed to improve the sampling range of the estimated probability distribution model, on the one hand, the structure of the optimization problem captured by most existing GEDA variants remains unchanged after the scaling; on the other hand, most existing variance (covariance) scaling methods scale the estimated variance (covariance) equally in different directions. This is actually not beneficial to effectively sample new promising individuals.

To alleviate the above concern, this paper devises an adaptive covariance scaling EDA by adaptively enlarging the number of promising individuals (as compared with those for the mean vector estimation) to estimate the covariance. In this way, not only does the structure of the optimization problem captured by the algorithm become better, but also the covariance is scaled differently in different directions.

3. Proposed ACSEDA

To improve the effectiveness of EDA in solving optimization problems, this paper proposes an adaptive covariance scaling EDA (ACSEDA) by introducing more promising individuals to calculate the covariance. Furthermore, to alleviate the sensitivity of the proposed ACSEDA to parameters, this paper further devises two adaptive strategies for the two key parameters in ACSEDA. The components of ACSEDA are elucidated as follows.

3.1. Adaptive Covariance Scaling

In traditional GEDAs [27], both the mean vector and the covariance of the multivariate Gaussian distribution model are estimated based on the selected promising individuals. Then, on the basis of the estimated probability distribution model, the offspring are sampled randomly. In particular, we can see that the mean vector has a great influence on the convergence speed of GEDAs to promising areas, while the covariance mainly takes charge of the sampling range of the distribution model, which plays a significant role in high diversity maintenance.

During the evolution, the population gradually approaches the promising areas and the selected promising individuals used for probability distribution estimation are gradually aggregated together as well. In this situation, the estimated covariance would become smaller and smaller. Once the estimated mean vector falls into local areas, the sampled offspring could hardly escape from local areas. As a consequence, the population falls into local areas, and premature convergence occurs. Such a predicament is encountered by many existing GEDAs [20].

To alleviate this issue, this paper proposes a covariance scaling strategy to enlarge the covariance by introducing more promising individuals on the basis of the selected individuals for the estimation of the mean vector. Specifically, given the population size is PS, $s = \lceil sr * PS \rceil$ promising individuals are first selected from the population to estimate the mean vector of the probability distribution, where sr is the selection ratio, defined as the number of selected individuals out of the population. Then, different from most existing GEDAs [67], which estimate the covariance based on the $sc = \lceil cs * PS \rceil$ individuals as well, this paper selects sc promising individuals to estimate the covariance, where cs is the covariance scaling parameter, which is the number of the promising individuals out of the population and is usually larger than sr. In this way, more promising individuals are selected to participate in the estimation of the covariance and thus, the covariance is enlarged.

As shown in Figure 1, after the population is sorted from the best to the worst with respect to the fitness, s best individuals are selected to form the promising individual set $\mathbf{S}$, and then, the mean vector $\boldsymbol{u}$ is estimated based on $\mathbf{S}$ according to Equation (3). Subsequently, different from most existing GEDAs, the proposed covariance scaling method selects sc best individuals to form the promising individual set $\mathbf{S_C}$ to estimate the covariance. It should be mentioned that cs is usually larger than sr, which also indicates that $\mathbf{S_C}$ is larger than $\mathbf{S}$. In this way, $\mathbf{S}$ is a subset of $\mathbf{S_C}$. Subsequently, instead of using $\mathbf{S}$ to calculate the covariance according to Equation (6), the proposed method utilizes $\mathbf{S_C}$, namely an enlarged individual set, to estimate the covariance as follows:

$$\mathbf{C} = \frac{1}{sc-1}(\mathbf{S_C}-\boldsymbol{\mu})(\mathbf{S_C}-\boldsymbol{\mu})^T \tag{9}$$

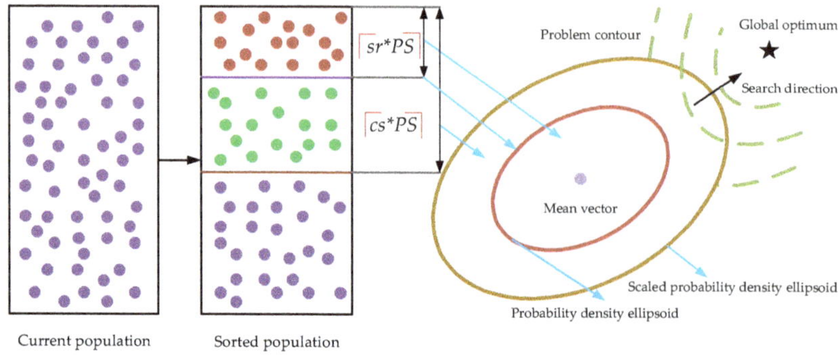

Figure 1. The visual structure of the covariance scaling strategy.

As shown in Figure 1, since more promising individuals participate in the estimation of the covariance, the sampling range of the estimated probability distribution model is enlarged. On the one hand, the sampled offspring based on this model are more diversified, which is very beneficial for the population to avoid falling into local areas. On the other hand, it might also be likely to generate more promising individuals close to the promising areas, and thus, the convergence could also be strengthened to some extent.

Remark 1. *It deserves attention that different from existing covariance scaling methods, which scales the covariance directly with a fixed scalar, the proposed covariance scaling method estimates the covariance based on an enlarged number of promising individuals as compared to those for the estimation of the mean vector. This brings the following two benefits for the estimated probability distribution model:*

(1) *By introducing more promising individuals, the proposed scaling method enlarges the covariance differently in different directions between variables and thus it implicitly takes the difference between variables into consideration. However, existing scaling methods [37] enlarge the covariance with a same scalar and hence they do not consider the difference between variables.*
(2) *The proposed scaling method is likely to better capture the structure of the optimization problem with respect to the correlations between variables by introducing more promising individuals. Nevertheless, the structure captured by existing scaling methods remains unchanged after the scaling.*

Taking a deep investigation on the parameter cs, in the proposed covariance scaling method, we find that neither a too large cs, nor a too small cs are proper to aid EDA to achieve promising performance. On the one side, a too-large cs may lead to a too large sampling range of the probability distribution. This may result in too diversified offspring sampled from the distribution model. In particular, it is found that, in the early stage of the evolution, a large cs may be beneficial to maintain a large sampling range and thus sample diversified offspring. This is helpful for EDA to explore the solution space in very different directions, whereas, in the late stage, such a setting of cs is not appropriate because it is not beneficial for the population to extensively exploit the found promising areas to refine the solution accuracy. On the other side, a too-small cs may bring in a too-small sampling range of the distribution model, which may sample concentrated offspring. Though it is desirable in the late stage of the evolution, it is not suitable during the whole evolution, because it may increase the risk of EDA in falling into local areas. Consequently, based on the above analysis, it is found that cs should not be fixed, but dynamically adjusted during the evolution process.

To the above end, this paper further designs an adaptive strategy for cs as follows:

$$cs = 1 - (1 - sr_{\min})(\frac{FEs}{FEs_{\max}})^2 \tag{10}$$

where $sr_{\min}$ denotes the lower bound of the selection ratio sr used for the estimation of the mean vector of the multivariate Gaussian distribution model, $FEs_{\max}$ is the maximum number of fitness evaluations, while FEs denotes the used number of fitness evaluations up to the current generation.

From Equation (10), we can see that cs decreases from 1.0 to $sr_{\min}$ as the evolution continues. Specifically, it is found that, in the early stage, most individuals in the population are used to estimate the covariance. This brings two benefits for EDA: (1) the sampling range of the probability distribution model is large and thus the sampled offspring are diversified and scatter dispersedly to explore the solution space. It is not only beneficial for the population to find more promising areas, but it is also very profitable for the population to avoid falling into local areas. (2) The captured structure of the optimization problem tends to be global and accurate with a large number of promising individuals. In the early stage, the individuals are usually scattered diversely in the solution space. In this situation, the captured structure of the optimization problem is usually global. Therefore, to accurately capture the correlations between variables globally, a large number of promising individuals are usually needed. Consequently, in the early stage, it is helpful to capture an accurate structure of the optimization problem when cs is large.

Conversely, in the late stage, from Equation (10), it is found that cs becomes smaller and smaller. This leads to a narrow sampling range of the probability distribution. Therefore, the sampled offspring are concentrated and surrounded around the mean vector. In this situation, the population exploits the found promising areas, and thus, the accuracy of the solution can be improved.

To summarize, with the above adaptive covariance scaling scheme, the proposed EDA variant is expected to obtain a promising balance between diversification and intensification of the population. Therefore, the algorithm could explore and exploit the complicated solution space properly to obtain promising performance in solving complicated optimization problems.

3.2. *Adaptive Promising Individuals Selection*

In GEDAs, the number ($s = \lceil sr * PS \rceil$) of selected promising individuals, for the estimation of the mean vector, makes a significant influence on the convergence speed of EDAs. As shown in Figure 1, the mean vector mainly takes control of the center of the sampled offspring. A too-large sr may lead to that a large number of promising individuals being used to estimate the mean vector. As a result, the estimated mean vector may be too far away from the promising areas. In the early stage of the evolution, this is beneficial for EDAs to maintain high search diversity. Nevertheless, in the late stage of the evolution, a large sr may slow down the convergence of the population to find high-quality solutions. On the contrary, a too-small sr may result in the estimated mean vector being too close to the promising areas. This case is suitable, in the late stage of the evolution, to exploit the found promising areas. However, it may lead to premature convergence if we keep sr small during the whole evolution, especially when the selected promising individuals all fall into local areas.

Based on the above analysis, it might as well dynamically adjust sr during the evolution. To this end, this paper devises a simple adaptive strategy for sr as follows:

$$sr = sr_{\max} - (sr_{\max} - sr_{\min})\left(\frac{FEs}{FEs_{\max}}\right)^{0.1} \quad (11)$$

where $sr_{\max}$ and $sr_{\min}$ represent the maximum selection ratio and the minimum selection ratio, which accordingly determine the maximum number ($s_{\max} = \lceil sr_{\max} * PS \rceil$) and the minimum number ($s_{\min} = \lceil sr_{\min} * PS \rceil$) of the selected promising individuals. In this paper, we set them as 0.35 and 0.05, respectively.

From Equation (11), we can see that in the early stage, sr is large, and then, it decreases gradually as the evolution goes. This indicates that during the evolution, the mean vector of the estimated probability distribution is becoming closer and closer to the promising areas. In this way, the population gradually tends to exploit the found promising areas.

Remark 2. *In particular, compared Equation (11) with Equation (10), as shown in Figure 2, the following findings can be obtained:*

(1) *sr decreases dramatically in the early stage, and mildly in the late stage, while cs decreases mildly in the early stage, and dramatically in the late stage. This actually matches the expectation that the proposed ACSEDA should explore the solution space in the early stage without serious loss of convergence, while it should exploit the search space in the late stage without serious sacrifice of search diversity. For one thing, in the early stage, sr decreases rapidly and thus the estimated mean vector is close to the promising areas that the current population lies. However, it should be mentioned that in such a situation, the sampling diversity of the estimated probability distribution is not declined, because the estimated covariance is large due to the large cs. On the contrary, in this situation, the sampling quality of the estimated probability distribution could be improved due to the high-quality mean vector and thus the population could effectively explore the search space to find promising areas in the early stage. In the late stage, sr decreases mildly, while cs descends quickly. In this situation, the quality of the mean vector is gradually promoted by approaching the promising areas closer and closer. At the same time, the sampling range of the estimated distribution gradually shrinks due to the covariance estimated on the reduced number of promising individuals. Therefore, in the late stage, ACSEDA gradually biases to exploiting the found promising areas to improve the solution quality. However, it should be mentioned that such a bias is not at the serious sacrifice of the search diversity because of the proposed covariance scaling technique.*

(2) *cs is always larger than sr during the evolution and the gap between cs and sr gradually shrinks as the evolution goes. This indicates that during the evolution, compared with traditional GEDAs, the covariance is always amplified, so that the estimated probability distribution could sample diversified offspring around the estimated mean vector with high quality. In addition, the gradually narrowed difference between sr and cs indicates that the scaling of the covariance is gradually declined. This implies that the proposed ACSEDA gradually concentrates on exploiting the solution space to refine the solution accuracy.*

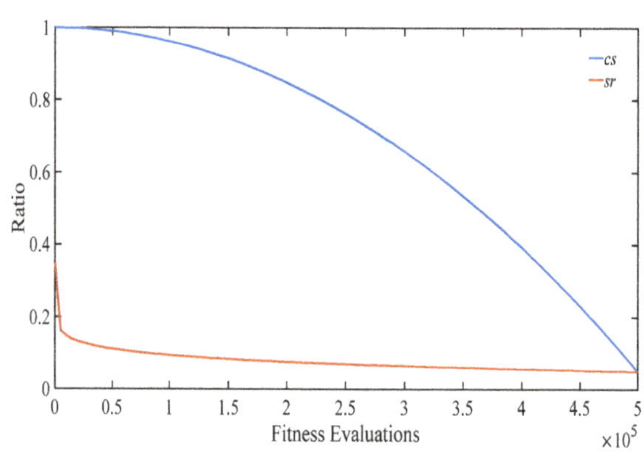

Figure 2. The change curves of *cs* and *sr* with the proposed two adaptive strategies.

3.3. Cross-Generation Individual Selection for Parent Population

In traditional EDAs [28], the offspring is directly utilized as the parent population for the next generation to estimate the probability distribution model. Since the quality of the sampled offspring is uncertain, the quality of the estimated probability distribution model may not be improved or even degrade compared with the estimated probability distribution in the last generation. This may slow down the convergence of the population to promising areas. Therefore, to remedy this shortcoming, some EDA variants [68] combine the offspring and the parent population together and then select the best *PS* individuals as the parent population for the next generation. However, such selection is too greedy and thus may lead to premature convergence and falling into local areas.

To alleviate the above predicament, and to further make a promising compromise between convergence and diversity, this paper further devises a cross-generation individual selection strategy for the parent population.

As Figure 3 shows, this paper combines the sampled offspring in the last generation and the sampled offspring in the current generation and then selects the best *PS* individuals as the parent population for the next generation to estimate the probability distribution model. In this way, the historical information in the last generation can be utilized to build the probability distribution model.

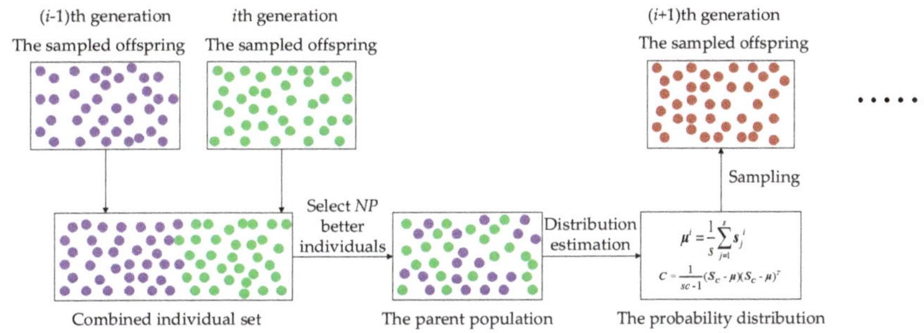

Figure 3. The process of the cross-generation individual selection strategy.

Remark 3. *Different from existing individual selection methods [26,48,69], the proposed cross-generation individual selection strategy takes advantage of the sampled offspring in the two consecutive generations to select individuals for the parent population in the next generation. Such a selection strategy brings the following benefits to ACSEDA:*

(1) *Individuals in the parent population are diversified. The sampled offspring in the last generation usually have big difference with the offspring in the current generation. Combining them together to select individuals is less likely to generate crowded individuals for the parent population. As a result, the estimated probability distribution model is less likely to fall into local areas and at the same time, has a wide sampling range to generate diversified offspring. In this way, ACSEDA could preserve high search diversity during the evolution.*

(2) *With this strategy, the latest historical promising individuals in the last generation could be integrated with those in the current offspring. As a consequence, individuals in the parent population are not only diversified, but also of high quality. Hence, the estimated probability distribution is of high quality to generate more promising offspring. By this means, the convergence of ACSEDA could be guaranteed.*

(3) *With this selection strategy, ACSEDA is further expected to preserve a good compromise between exploration and exploitation to search the solution space effectively. Experiments conducted in Section 4 will demonstrate the effectiveness of the proposed cross-generation individual selection strategy.*

3.4. Overall Procedure of ACSEDA

Combining the above three schemes together, the proposed ACSEDA is outlined in Algorithm 2. Specifically, after the initialization of the population (Line 2), the algorithm goes to the main iteration loop for evolution (Lines 5–16). In the main loop, it first executes the proposed adaptive promising individual selection strategy for the estimation of the mean vector (Lines 6 and 7). Then, it comes to the proposed adaptive covariance scaling strategy to estimate the covariance of the probability distribution model (Lines 8 and 9).

Subsequently, the offspring are randomly sampled based on the estimated probability distribution model (Line 10). Hereafter, it arrives at the proposed cross-generation individual selection strategy to select individuals for the parent population in the next generation (Line 12). At last, a local search method is conducted on the global best solution to refine its accuracy (Line 14).

Algorithm 2: The Procedure of ACSEDA.

Input: population size PS;
1: Set $FEs = 0$;
2: Initialize PS individuals randomly and evaluate their fitness;
3: $FEs = FEs + PS$;
4: Obtain the global best solution ***Gbest*** and store the current population;
5: **While** ($FEs < FEs_{\max}$)
6: Calculate the selection ratio sr according to Equation (11);
7: Select $\lceil sr * PS \rceil$ promising solutions from the population and calculate the mean value $\boldsymbol{\mu}$ using Equation (3);
8: Calculate the covariance scaling parameter cs according to Equation (10);
9: Estimate the covariance matrix $\mathbf{C}$ according to Equation (9);
10: Randomly sample PS new individuals based on the estimated multivariate Gaussian model, evaluate their fitness and store them;
11: $FEs = FEs + PS$;
12: Combine the offspring in the last generation and the offspring in the current generation to select PS better individuals to form the parent population for the next generation;
13: Update the global best solution ***Gbest***;
14: Execute local search *2* times on ***Gbest***;
15: $FEs = FEs + 2$;
16: **End While**
Output: the global best solution ***Gbest***;

In Algorithm 2, it should be noticed that a local search strategy is additionally added to improve the solution accuracy of the global best solution. This is because EDAs are probability distribution model based optimization algorithms, and as a consequence, EDAs usually lack strong local exploitation [26,38,70]. Therefore, in the literature [23,38,70], local search methods are generally accompanied by EDAs to improve the solution quality. Hence, the same as most existing EDA variants [23,26,38,70,71], ACSEDA also adopts a local search method to refine the global best solution, as shown in Line 14.

For simplicity and keeping consistent with the probability distribution model in ACSEDA, this paper applies the univariate Gaussian distribution with a small variance to execute the local search on the global best solution. In this paper, the small variance is set as 1.0×10^{-4}. In addition, for saving computational resources, we execute the local search method on the global best solution only two times, as shown in Line 14.

As a whole, with the proposed three main techniques and the local search method, ACSEDA is expected to explore and exploit the solution space properly to locate the optima of optimization problems.

4. Experimental Studies

This section mainly conducts extensive experiments to verify the effectiveness of the proposed ACSEDA. Specifically, the commonly used CEC 2014 benchmark problem set [44] is adopted in this paper. This benchmark set contains 30 various complicated optimization problems, such as unimodal problems, multimodal problems, hybrid problems, and composition problems. For detailed information on this benchmark set, please refer to [44].

4.1. Experimental Settings

First, to comprehensively demonstrate the effectiveness of the proposed ACSEDA, we select several state-of-the-art EDA variants to make comparisons in solving the complicated CEC 2014 benchmark problems. Specifically, the selected state-of-the-art EDA variants are

EDA2 [37], EDA_{VERS} [49], EDA/LS [38], EDA/LS-MS [65], MA-ES [58], and BUMDA [39]. In addition, as a baseline method, the traditional multivariate Gaussian model based EDA [28] is also utilized as a compared method. To tell it apart from the others, we denote it as TRA-EDA in the experiments.

Second, to make comprehensive comparisons between the proposed ACSEDA and the above compared EDA variants, we compare their optimization performance in solving the CEC 2014 problems with three different dimension sizes, namely 30-D, 50-D, and 100-D. For fairness, the maximum number of fitness evaluations ($FEs_{\max}$) is set as 10,000 $*$ D for all algorithms.

Third, for fair comparisons, the key parameter settings of the compared algorithms are set, as recommended, in the associated papers. For the population size, we tune the settings of all algorithms on the CEC 2014 benchmark set with different dimension sizes. Specifically, after preliminary experiments, the parameter settings of all algorithms are shown in Table 1.

Table 1. Parameter settings of ACSEDA and the compared algorithms.

Algorithm		ACSEDA	EDA/LS					EDA2		EDA/LS-MS					EDA_{VERS}		BUMDA	TRA-EDA		MA-ES	
Parameter		PS	PS	M	P_b	P_c	θ	PS	l	PS	M	P_a	P_b	θ	PS	sr	PS	PS	sr	PS	mu
D	30	1300	150					100		150					500		900	2500			
	50	1800	150	15	0.2	0.2	0.1	200	20	1000	15	0.2	0.2	0.1	600	0.35	1100	4700	0.2	$4+\lfloor 3*\ln D\rfloor$	$\lfloor PS/2\rfloor$
	100	3200	150					300		2000					700		1500	6000			

Fourth, to comprehensively evaluate the optimization performance of each algorithm, we execute each algorithm independently for 30 runs and utilize the median, mean, and standard deviation values over the 30 independent runs to evaluate its optimization performance. Furthermore, to tell the statistical significance, we conduct the Wilcoxon rank-sum test, at the significance level of $\alpha = 0.05$, to compare the proposed ACSEDA with each associated EDA variant. In addition, to compare the overall optimization performance of all algorithms on the whole CEC 2014 benchmark set, we further conduct the Friedman test at the significance level of $\alpha = 0.05$ by taking advantage of the mean value of each algorithm on each function in the benchmark set.

At last, it deserves attention that all algorithms are programmed under MATLAB R2018a, and they are run on the same computer with Intel(R) Core(TM) i7-10700T CPU @ 2.90 GHz 2.90 GHz and 8 G RAM.

4.2. Comparison with State-of-the-Art EDAs

Table 2, Table 3, and Table 4 display the comparison results between ACSEDA and the compared EDA variants on the 30-D, 50-D, and 100-D CEC 2014 benchmark problems, respectively. In these tables, the symbols "+", "−" and "=" represent that ACSEDA is significantly better than, significantly worse than, and equivalent to the associated compared algorithms on the associated problems, respectively. Besides, "$w/t/l$" denotes the numbers of the problems where ACSEDA achieves significantly better performance, equivalent performance, and significantly worse performance than the compare algorithms, respectively. Actually, "$w/t/l$" is equal to the numbers of "+", "=" and "−", respectively. Additionally, in the last rows of this table, the averaged rank of each algorithm obtained from the Friedman test is presented.

Table 2. Comparison between ACSEDA and the compared state-of-the-art EDA variants on the 30-*D* CEC2014 benchmark problems. The bold results indicate that ACSEDA is significantly better than the compared methods.

F	Category	Quality	ACSEDA	EDA2	EDA_{VERS}	EDA/LS	EDA/LS-MS	TRA-EDA	BUMDA	MA-ES
F_1	Unimodal Prob-lems	Median	0.00×10^{0}	0.00×10^{0}	6.48×10^{4}	1.90×10^{-9}	1.83×10^{-9}	6.94×10^{7}	1.15×10^{8}	1.42×10^{-14}
		Mean	0.00×10^{0}	0.00×10^{0}	7.76×10^{4}	9.09×10^{1}	1.98×10^{1}	6.80×10^{7}	1.16×10^{8}	1.37×10^{-14}
		Std	0.00×10^{0}	0.00×10^{0}	4.85×10^{4}	4.89×10^{2}	1.07×10^{2}	1.17×10^{7}	2.12×10^{7}	2.55×10^{-15}
		p-value	-	NaN=	$\mathbf{1.21 \times 10^{-12}}$+	$\mathbf{1.21 \times 10^{-12}}$+	$\mathbf{1.21 \times 10^{-12}}$+	$\mathbf{1.21 \times 10^{-12}}$+	$\mathbf{1.21 \times 10^{-12}}$+	$\mathbf{1.17 \times 10^{-13}}$+
F_2		Median	0.00×10^{0}	0.00×10^{0}	2.54×10^{3}	1.48×10^{-11}	4.59×10^{-11}	6.61×10^{9}	1.80×10^{4}	2.84×10^{-14}
		Mean	0.00×10^{0}	0.00×10^{0}	2.36×10^{3}	2.52×10^{-10}	1.45×10^{-10}	6.58×10^{9}	5.15×10^{5}	2.65×10^{-14}
		Std	0.00×10^{0}	0.00×10^{0}	8.94×10^{2}	5.04×10^{-10}	2.43×10^{-10}	1.05×10^{9}	2.45×10^{6}	7.09×10^{-15}
		p-value	-	NaN=	$\mathbf{1.21 \times 10^{-12}}$+	$\mathbf{1.21 \times 10^{-12}}$+	$\mathbf{1.21 \times 10^{-12}}$+	$\mathbf{1.21 \times 10^{-12}}$+	$\mathbf{1.21 \times 10^{-12}}$+	$\mathbf{7.15 \times 10^{-13}}$+
F_3		Median	0.00×10^{0}	0.00×10^{0}	1.41×10^{3}	8.42×10^{-9}	8.85×10^{-10}	9.83×10^{3}	6.49×10^{3}	5.68×10^{-14}
		Mean	0.00×10^{0}	0.00×10^{0}	1.44×10^{3}	3.10×10^{-9}	2.34×10^{-9}	1.00×10^{4}	6.52×10^{3}	5.68×10^{-14}
		Std	0.00×10^{0}	0.00×10^{0}	5.54×10^{2}	4.33×10^{-9}	4.21×10^{-9}	1.97×10^{3}	1.85×10^{3}	5.05×10^{-29}
		p-value	-	NaN=	$\mathbf{1.21 \times 10^{-12}}$+	$\mathbf{1.21 \times 10^{-12}}$+	$\mathbf{1.21 \times 10^{-12}}$+	$\mathbf{1.21 \times 10^{-12}}$+	$\mathbf{1.21 \times 10^{-12}}$+	$\mathbf{1.69 \times 10^{-14}}$+
F_{1-3}	$w/t/l$		-	0/3/0	3/0/0	3/0/0	3/0/0	3/0/0	3/0/0	3/0/0
F_4	Simple Multi-modal Prob-lems	Median	3.25×10^{0}	0.00×10^{0}	6.39×10^{-1}	1.02×10^{-9}	4.75×10^{-9}	7.93×10^{2}	9.74×10^{1}	5.68×10^{-14}
		Mean	3.15×10^{0}	0.00×10^{0}	6.14×10^{-1}	1.33×10^{-1}	2.66×10^{-1}	7.72×10^{2}	9.88×10^{1}	2.66×10^{-1}
		Std	1.02×10^{0}	0.00×10^{0}	2.76×10^{-1}	7.16×10^{-1}	9.94×10^{-1}	1.27×10^{2}	2.69×10^{1}	9.94×10^{-1}
		p-value	-	1.21×10^{-12}−	3.82×10^{-10}−	2.15×10^{-10}−	1.41×10^{-9}−	$\mathbf{3.02 \times 10^{-11}}$+	$\mathbf{3.02 \times 10^{-11}}$+	1.69×10^{-10}−
F_5		Median	2.09×10^{1}	2.10×10^{1}	2.10×10^{1}	2.00×10^{1}	2.00×10^{1}	2.09×10^{1}	2.10×10^{1}	2.00×10^{1}
		Mean	2.09×10^{1}	2.09×10^{1}	2.09×10^{1}	2.00×10^{1}	2.00×10^{1}	2.09×10^{1}	2.10×10^{1}	2.04×10^{1}
		Std	4.96×10^{-2}	4.86×10^{-2}	7.13×10^{-2}	5.47×10^{-6}	3.43×10^{-2}	5.47×10^{-2}	5.00×10^{-2}	6.93×10^{-1}
		p-value	-	7.01×10^{-2}=	7.48×10^{-2}=	1.17×10^{-11}−	9.37×10^{-12}−	8.77×10^{-1}=	$\mathbf{2.32 \times 10^{-2}}$+	3.96×10^{-4}−
F_6		Median	1.99×10^{-6}	0.00×10^{0}	1.77×10^{-5}	4.09×10^{1}	3.81×10^{1}	4.58×10^{-1}	1.50×10^{0}	2.25×10^{1}
		Mean	2.07×10^{-6}	2.09×10^{-1}	4.76×10^{-1}	4.05×10^{1}	3.84×10^{1}	5.16×10^{-1}	1.61×10^{0}	2.24×10^{1}
		Std	8.29×10^{-7}	5.63×10^{-1}	8.19×10^{-1}	2.14×10^{0}	2.75×10^{0}	4.43×10^{-1}	1.12×10^{0}	4.08×10^{0}
		p-value	-	$\mathbf{3.82 \times 10^{-5}}$+	$\mathbf{6.61 \times 10^{-1}}$=	$\mathbf{3.02 \times 10^{-11}}$+	$\mathbf{3.02 \times 10^{-11}}$+	$\mathbf{3.02 \times 10^{-11}}$+	$\mathbf{1.07 \times 10^{-7}}$+	$\mathbf{3.02 \times 10^{-11}}$+
F_7		Median	0.00×10^{0}	0.00×10^{0}	1.14×10^{-13}	2.21×10^{-2}	3.69×10^{-2}	9.12×10^{1}	3.24×10^{-2}	1.14×10^{-13}
		Mean	0.00×10^{0}	0.00×10^{0}	1.06×10^{-13}	3.52×10^{-2}	4.84×10^{-2}	9.01×10^{1}	9.35×10^{-2}	2.79×10^{-3}
		Std	0.00×10^{0}	0.00×10^{0}	2.84×10^{-14}	4.56×10^{-2}	6.23×10^{-2}	1.25×10^{1}	1.41×10^{-1}	4.40×10^{-3}
		p-value	-	NaN=	$\mathbf{7.15 \times 10^{-13}}$+	$\mathbf{1.20 \times 10^{-12}}$+	$\mathbf{1.19 \times 10^{-12}}$+	$\mathbf{1.21 \times 10^{-12}}$+	$\mathbf{1.21 \times 10^{-12}}$+	$\mathbf{3.85 \times 10^{-13}}$+
F_8		Median	9.95×10^{-1}	6.47×10^{0}	1.52×10^{2}	2.27×10^{2}	2.06×10^{2}	1.49×10^{1}	0.00×10^{0}	1.68×10^{2}
		Mean	1.16×10^{0}	6.60×10^{0}	1.32×10^{2}	2.38×10^{2}	2.13×10^{2}	1.54×10^{1}	0.00×10^{0}	1.68×10^{2}
		Std	8.93×10^{-1}	1.81×10^{0}	5.09×10^{1}	6.69×10^{1}	6.08×10^{1}	3.39×10^{0}	0.00×10^{0}	3.98×10^{0}
		p-value	-	$\mathbf{1.92 \times 10^{-11}}$+	$\mathbf{3.86 \times 10^{-11}}$+	$\mathbf{2.13 \times 10^{-11}}$+	$\mathbf{2.13 \times 10^{-11}}$+	$\mathbf{2.13 \times 10^{-11}}$+	$\mathbf{4.26 \times 10^{-9}}$−	$\mathbf{2.02 \times 10^{-11}}$+

Table 2. *Cont.*

F	Category	Quality	ACSEDA	EDA2	EDA$_{VERS}$	EDA/LS	EDA/LS-MS	TRA-EDA	BUMDA	MA-ES
F_9		Median	9.95×10^{-1}	5.47×10^{0}	1.48×10^{2}	3.33×10^{2}	3.20×10^{2}	1.10×10^{1}	9.95×10^{-1}	1.87×10^{2}
		Mean	1.16×10^{0}	5.80×10^{0}	1.31×10^{2}	3.28×10^{2}	3.18×10^{2}	1.13×10^{1}	1.41×10^{0}	1.88×10^{2}
		Std	9.29×10^{-1}	1.69×10^{0}	4.96×10^{1}	6.66×10^{1}	4.01×10^{1}	2.68×10^{0}	1.05×10^{0}	5.42×10^{0}
		p-value	-	**2.85×10^{-11}+**	**1.13×10^{-11}+**	**2.40×10^{-11}+**	**2.40×10^{-11}+**	**2.40×10^{-11}+**	**2.03×10^{-2}+**	**2.10×10^{-11}+**
F_{10}		Median	3.59×10^{0}	6.33×10^{0}	5.69×10^{3}	4.76×10^{3}	4.04×10^{3}	2.39×10^{2}	2.84×10^{1}	3.88×10^{3}
		Mean	1.13×10^{1}	5.28×10^{1}	5.62×10^{3}	4.76×10^{3}	4.11×10^{3}	2.86×10^{2}	3.81×10^{1}	3.90×10^{3}
		Std	2.89×10^{1}	8.32×10^{1}	3.90×10^{2}	1.24×10^{3}	8.86×10^{2}	1.98×10^{2}	5.51×10^{1}	3.24×10^{2}
		p-value	-	**2.10×10^{-3}+**	**3.02×10^{-11}+**	**3.02×10^{-11}+**	**3.02×10^{-11}+**	**1.09×10^{-10}+**	**8.77×10^{-2}=**	**3.01×10^{-11}+**
F_{11}		Median	8.36×10^{0}	1.52×10^{1}	6.55×10^{3}	4.64×10^{3}	4.69×10^{3}	2.63×10^{2}	2.26×10^{2}	4.31×10^{3}
		Mean	2.80×10^{1}	7.94×10^{1}	6.53×10^{3}	5.04×10^{3}	4.99×10^{3}	2.58×10^{2}	2.43×10^{2}	4.31×10^{3}
		Std	4.53×10^{1}	1.19×10^{2}	2.54×10^{2}	1.36×10^{3}	1.07×10^{3}	1.53×10^{2}	2.10×10^{2}	2.21×10^{2}
		p-value	-	2.87×10^{-2}+	**3.02×10^{-11}+**	**3.02×10^{-11}+**	**3.02×10^{-11}+**	**2.44×10^{-9}+**	**8.84×10^{-7}+**	**2.98×10^{-11}+**
F_{12}		Median	2.35×10^{0}	2.46×10^{0}	2.56×10^{0}	6.89×10^{-1}	6.67×10^{-1}	2.47×10^{0}	2.51×10^{0}	3.34×10^{-2}
		Mean	2.32×10^{0}	2.49×10^{0}	2.56×10^{0}	9.21×10^{-1}	7.84×10^{-1}	2.39×10^{0}	2.47×10^{0}	3.68×10^{-2}
		Std	2.41×10^{-1}	2.71×10^{-1}	1.98×10^{-1}	8.73×10^{-1}	4.11×10^{-1}	3.49×10^{-1}	2.56×10^{-1}	2.29×10^{-2}
		p-value	-	**1.76×10^{-2}+**	**1.89×10^{-4}+**	**5.57×10^{-10}−**	**4.50×10^{-11}−**	**1.45×10^{-1}=**	**1.70×10^{-2}+**	**3.02×10^{-11}−**
F_{13}		Median	9.34×10^{-2}	4.40×10^{-2}	1.85×10^{-1}	5.86×10^{-1}	7.04×10^{-1}	3.24×10^{0}	4.27×10^{-2}	3.16×10^{-1}
		Mean	9.35×10^{-2}	4.39×10^{-2}	1.84×10^{-1}	1.85×10^{0}	2.57×10^{0}	3.20×10^{0}	4.25×10^{-2}	3.40×10^{-1}
		Std	1.50×10^{-2}	1.08×10^{-2}	2.68×10^{-2}	2.53×10^{0}	2.87×10^{0}	2.03×10^{-1}	8.97×10^{-3}	9.18×10^{-3}
		p-value	-	6.07×10^{-11}−	**3.34×10^{-11}+**	**3.02×10^{-11}+**	**3.02×10^{-11}+**	**3.02×10^{-11}+**	3.69×10^{-11}-	**3.02×10^{-11}+**
F_{14}		Median	2.59×10^{-1}	4.10×10^{-1}	3.10×10^{-1}	3.18×10^{-1}	3.27×10^{-1}	4.63×10^{1}	3.89×10^{-1}	3.39×10^{-1}
		Mean	2.62×10^{-1}	4.06×10^{-1}	3.11×10^{-1}	3.36×10^{-1}	3.42×10^{-1}	4.68×10^{1}	3.89×10^{-1}	3.81×10^{-1}
		Std	5.24×10^{-2}	3.87×10^{-2}	3.61×10^{-2}	9.64×10^{-2}	9.18×10^{-2}	5.15×10^{0}	3.91×10^{-2}	1.37×10^{-1}
		p-value	-	**1.78×10^{-10}+**	**3.01×10^{-4}+**	**7.66×10^{-5}+**	**9.51×10^{-6}+**	**3.02×10^{-11}+**	**5.07×10^{-11}+**	**1.61×10^{-6}+**
F_{15}		Median	4.16×10^{0}	4.24×10^{0}	1.30×10^{1}	9.26×10^{1}	8.31×10^{1}	3.60×10^{0}	3.46×10^{0}	3.24×10^{0}
		Mean	4.21×10^{0}	4.26×10^{0}	1.30×10^{1}	9.75×10^{1}	8.43×10^{1}	4.69×10^{0}	4.68×10^{0}	3.40×10^{0}
		Std	1.43×10^{0}	1.08×10^{0}	1.04×10^{0}	3.23×10^{1}	2.67×10^{1}	3.53×10^{0}	5.52×10^{0}	9.00×10^{-1}
		p-value	-	7.39×10^{-1}=	**3.02×10^{-11}+**	**3.02×10^{-11}+**	**3.02×10^{-11}+**	9.82×10^{-1}=	3.11×10^{-1}=	3.51×10^{-2}−
F_{16}		Median	8.81×10^{0}	1.24×10^{1}	1.11×10^{1}	1.36×10^{1}	1.34×10^{1}	1.06×10^{1}	1.09×10^{1}	1.37×10^{1}
		Mean	8.74×10^{0}	1.23×10^{1}	1.11×10^{1}	1.35×10^{1}	1.34×10^{1}	1.05×10^{1}	1.09×10^{1}	1.37×10^{1}
		Std	5.75×10^{-1}	1.92×10^{-1}	3.95×10^{-1}	3.73×10^{-1}	2.65×10^{-1}	3.39×10^{-1}	4.89×10^{-1}	9.40×10^{-2}
		p-value	-	**3.02×10^{-11}+**	**3.34×10^{-11}+**	**3.02×10^{-11}+**	**3.02×10^{-11}+**	**4.08×10^{-11}+**	**3.34×10^{-11}+**	**3.02×10^{-11}+**

Table 2. *Cont.*

F	Category	Quality	ACSEDA	EDA2	EDA_{VERS}	EDA/LS	EDA/LS-MS	TRA-EDA	BUMDA	MA-ES
$F_{4\text{–}16}$	$w/t/l$		-	8/3/2	10/2/1	10/0/3	10/0/3	10/3/0	9/2/2	9/0/4
F_{17}	Hybrid Problems	Median	1.45×10^{1}	1.16×10^{1}	3.19×10^{4}	2.13×10^{3}	6.76×10^{4}	9.97×10^{4}	3.05×10^{7}	1.76×10^{3}
		Mean	2.17×10^{1}	2.00×10^{1}	3.58×10^{4}	4.10×10^{7}	2.38×10^{7}	1.36×10^{5}	2.98×10^{7}	1.77×10^{3}
		Std	2.98×10^{1}	2.80×10^{1}	1.97×10^{4}	7.15×10^{7}	5.06×10^{7}	1.05×10^{5}	7.53×10^{6}	4.44×10^{2}
		p-value	-	$4.16 \times 10^{-1=}$	**$3.02 \times 10^{-11+}$**	**$3.02 \times 10^{-11+}$**	**$3.02 \times 10^{-11+}$**	**$3.02 \times 10^{-11+}$**	**$3.02 \times 10^{-11+}$**	**$3.02 \times 10^{-11+}$**
F_{18}		Median	4.71×10^{-1}	1.33×10^{0}	6.85×10^{1}	1.61×10^{2}	1.56×10^{2}	2.28×10^{1}	1.38×10^{2}	8.24×10^{1}
		Mean	8.77×10^{-1}	1.14×10^{0}	1.13×10^{2}	9.57×10^{8}	5.05×10^{8}	2.57×10^{1}	1.79×10^{2}	8.36×10^{1}
		Std	6.39×10^{-1}	7.04×10^{-1}	9.31×10^{1}	1.82×10^{9}	1.32×10^{9}	9.48×10^{0}	1.36×10^{2}	3.01×10^{1}
		p-value	-	$6.79 \times 10^{-2=}$	**$3.02 \times 10^{-11+}$**	**$3.02 \times 10^{-11+}$**	**$3.02 \times 10^{-11+}$**	**$3.02 \times 10^{-11+}$**	**$3.02 \times 10^{-11+}$**	**$3.02 \times 10^{-11+}$**
F_{19}		Median	2.78×10^{0}	3.66×10^{0}	4.86×10^{0}	1.56×10^{1}	1.44×10^{1}	4.28×10^{1}	2.10×10^{1}	9.97×10^{0}
		Mean	2.78×10^{0}	3.72×10^{0}	4.89×10^{0}	6.83×10^{1}	5.28×10^{1}	4.36×10^{1}	2.42×10^{1}	9.86×10^{0}
		Std	5.02×10^{-1}	4.02×10^{-1}	4.92×10^{-1}	1.85×10^{2}	1.33×10^{2}	7.25×10^{-1}	1.00×10^{1}	1.52×10^{0}
		p-value	-	**$7.12 \times 10^{-9+}$**	**$3.34 \times 10^{-11+}$**	**$3.02 \times 10^{-11+}$**	**$3.02 \times 10^{-11+}$**	**$2.87 \times 10^{-11+}$**	**$3.02 \times 10^{-11+}$**	**$3.02 \times 10^{-11+}$**
F_{20}		Median	1.25×10^{0}	1.57×10^{0}	3.95×10^{0}	1.52×10^{4}	1.62×10^{4}	7.85×10^{2}	3.29×10^{4}	2.31×10^{2}
		Mean	1.33×10^{0}	1.76×10^{0}	4.14×10^{0}	2.10×10^{4}	2.61×10^{4}	1.03×10^{3}	3.18×10^{4}	2.72×10^{2}
		Std	1.98×10^{-1}	5.96×10^{-1}	1.26×10^{0}	1.89×10^{4}	2.55×10^{4}	6.34×10^{2}	5.47×10^{3}	1.43×10^{2}
		p-value	-	**$7.70 \times 10^{-4+}$**	**$3.02 \times 10^{-11+}$**	**$3.02 \times 10^{-11+}$**	**$3.02 \times 10^{-11+}$**	**$3.02 \times 10^{-11+}$**	**$3.02 \times 10^{-11+}$**	**$3.02 \times 10^{-11+}$**
F_{21}		Median	1.92×10^{0}	1.85×10^{0}	2.65×10^{4}	2.23×10^{4}	3.49×10^{4}	2.29×10^{2}	3.40×10^{6}	8.32×10^{2}
		Mean	1.45×10^{1}	3.35×10^{1}	2.93×10^{4}	5.96×10^{6}	1.05×10^{7}	2.86×10^{2}	3.38×10^{6}	8.98×10^{2}
		Std	3.40×10^{1}	5.26×10^{1}	9.29×10^{3}	2.53×10^{7}	2.38×10^{7}	1.94×10^{2}	1.58×10^{6}	3.58×10^{2}
		p-value	-	$8.88 \times 10^{-1=}$	**$3.02 \times 10^{-11+}$**	**$3.02 \times 10^{-11+}$**	**$3.02 \times 10^{-11+}$**	**$3.02 \times 10^{-11+}$**	**$3.02 \times 10^{-11+}$**	**$3.02 \times 10^{-11+}$**
F_{22}		Median	2.67×10^{1}	2.81×10^{1}	2.46×10^{2}	1.15×10^{3}	1.24×10^{3}	3.67×10^{1}	1.53×10^{2}	4.15×10^{2}
		Mean	3.46×10^{1}	3.63×10^{1}	2.54×10^{2}	1.28×10^{3}	1.41×10^{3}	8.88×10^{1}	1.57×10^{2}	4.95×10^{2}
		Std	2.97×10^{1}	3.17×10^{1}	4.84×10^{1}	6.04×10^{2}	7.89×10^{2}	6.13×10^{1}	1.16×10^{2}	1.63×10^{2}
		p-value	-	**$2.77 \times 10^{-5+}$**	**$3.02 \times 10^{-11+}$**	**$3.02 \times 10^{-11+}$**	**$3.02 \times 10^{-11+}$**	**$6.72 \times 10^{-10+}$**	**$9.76 \times 10^{-10+}$**	**$3.02 \times 10^{-11+}$**
$F_{17\text{–}22}$	$w/t/l$		-	3/3/0	6/0/0	6/0/0	6/0/0	6/0/0	6/0/0	6/0/0
F_{23}	Composition Problems	p-value	3.15×10^{2}	3.15×10^{2}	3.15×10^{2}	3.20×10^{2}	3.20×10^{2}	3.37×10^{2}	3.19×10^{2}	2.00×10^{2}
		Mean	3.15×10^{2}	3.15×10^{2}	3.15×10^{2}	4.20×10^{2}	4.87×10^{2}	3.38×10^{2}	3.20×10^{2}	2.00×10^{2}
		Std	5.68×10^{-14}	5.68×10^{-14}	5.68×10^{-14}	2.59×10^{2}	2.96×10^{2}	3.78×10^{0}	1.92×10^{0}	0.00×10^{0}
		p-value	-	NaN$^{=}$	NaN$^{=}$	**$1.21 \times 10^{-12+}$**	**$1.21 \times 10^{-12+}$**	**$1.21 \times 10^{-12+}$**	**$1.21 \times 10^{-12+}$**	$1.69 \times 10^{-14-}$

Table 2. *Cont.*

F	Category	Quality	ACSEDA	EDA2	EDA$_{VERS}$	EDA/LS	EDA/LS-MS	TRA-EDA	BUMDA	MA-ES
F_{24}		Median	2.00×10^{2}	2.00×10^{2}	2.23×10^{2}	2.63×10^{2}	2.56×10^{2}	2.27×10^{2}	2.30×10^{2}	2.00×10^{2}
		Mean	2.04×10^{2}	2.00×10^{2}	2.23×10^{2}	2.80×10^{2}	2.65×10^{2}	2.27×10^{2}	2.30×10^{2}	2.00×10^{2}
		Std	8.66×10^{0}	0.00×10^{0}	7.69×10^{-1}	5.55×10^{1}	3.19×10^{1}	1.50×10^{0}	2.92×10^{0}	1.31×10^{-4}
		p-value	-	$1.10 \times 10^{-2-}$	**$9.93 \times 10^{-12+}$**	**$6.48 \times 10^{-12+}$**	**$6.48 \times 10^{-12+}$**	**$6.48 \times 10^{-12+}$**	**$6.48 \times 10^{-12+}$**	$2.03 \times 10^{-1=}$
F_{25}		Median	2.03×10^{2}	2.03×10^{2}	2.00×10^{2}	3.06×10^{2}	2.82×10^{2}	2.07×10^{2}	2.06×10^{2}	2.00×10^{2}
		Mean	2.03×10^{2}	2.03×10^{2}	2.00×10^{2}	2.97×10^{2}	2.78×10^{2}	2.07×10^{2}	2.06×10^{2}	2.00×10^{2}
		Std	1.33×10^{2}	2.63×10^{-2}	0.00×10^{0}	4.49×10^{2}	2.76×10^{1}	3.90×10^{-1}	1.24×10^{0}	0.00×10^{0}
		p-value	-	$8.27 \times 10^{-1=}$	$5.91 \times 10^{-13-}$	**$1.74 \times 10^{-11+}$**	**$1.74 \times 10^{-11+}$**	**$1.74 \times 10^{-11+}$**	**$1.74 \times 10^{-11+}$**	$5.91 \times 10^{-13-}$
F_{26}		Median	1.00×10^{2}	1.00×10^{2}	1.00×10^{2}	1.01×10^{2}	1.00×10^{2}	1.00×10^{2}	1.00×10^{2}	2.00×10^{2}
		Mean	1.00×10^{2}	1.00×10^{2}	1.00×10^{2}	1.05×10^{2}	1.04×10^{2}	1.01×10^{2}	1.01×10^{2}	2.00×10^{2}
		Std	1.84×10^{-2}	1.25×10^{-2}	1.06×10^{2}	2.38×10^{2}	1.82×10^{2}	1.05×10^{0}	9.10×10^{-1}	0.00×10^{0}
		p-value	-	$2.37 \times 10^{-1-}$	**$3.34 \times 10^{-11+}$**	**$3.02 \times 10^{-11+}$**	**$3.02 \times 10^{-11+}$**	**$3.02 \times 10^{-11+}$**	**$3.02 \times 10^{-11+}$**	**$3.02 \times 10^{-11+}$**
F_{27}		Median	3.00×10^{2}	3.41×10^{2}	3.00×10^{2}	1.58×10^{3}	1.46×10^{3}	4.24×10^{2}	3.00×10^{2}	2.00×10^{2}
		Mean	3.00×10^{2}	3.38×10^{2}	3.00×10^{2}	1.49×10^{3}	1.21×10^{3}	4.14×10^{2}	3.17×10^{2}	2.00×10^{2}
		Std	0.00×10^{0}	4.02×10^{1}	0.00×10^{0}	2.92×10^{2}	4.53×10^{2}	3.09×10^{2}	2.56×10^{2}	0.00×10^{0}
		p-value	-	**$2.20 \times 10^{-6+}$**	NaN$^{=}$	**$1.21 \times 10^{-12+}$**	**$1.21 \times 10^{-12+}$**	**$1.21 \times 10^{-12+}$**	**$1.21 \times 10^{-12+}$**	$1.69 \times 10^{-14-}$
F_{28}		Median	8.41×10^{2}	8.19×10^{2}	8.16×10^{2}	5.49×10^{3}	4.98×10^{3}	7.10×10^{2}	4.78×10^{2}	2.00×10^{2}
		Mean	8.40×10^{2}	8.21×10^{2}	8.22×10^{2}	5.56×10^{3}	5.11×10^{3}	7.03×10^{2}	4.76×10^{2}	2.00×10^{2}
		Std	2.05×10^{1}	4.03×10^{1}	2.02×10^{1}	1.59×10^{3}	1.06×10^{3}	6.79×10^{1}	5.60×10^{0}	0.00×10^{0}
		p-value	-	$2.15 \times 10^{-2-}$	$3.34 \times 10^{-3-}$	**$3.01 \times 10^{-11+}$**	**$3.01 \times 10^{-11+}$**	$7.09 \times 10^{-9-}$	$3.01 \times 10^{-11-}$	$1.20 \times 10^{-12-}$
F_{29}		Median	7.21×10^{2}	7.15×10^{2}	1.24×10^{3}	4.17×10^{8}	4.37×10^{8}	1.00×10^{3}	5.88×10^{2}	2.00×10^{2}
		Mean	7.28×10^{2}	1.40×10^{6}	1.28×10^{3}	4.14×10^{8}	4.40×10^{8}	1.04×10^{3}	8.95×10^{2}	2.00×10^{2}
		Std	2.71×10^{1}	3.59×10^{6}	1.39×10^{2}	1.18×10^{8}	1.53×10^{8}	1.60×10^{2}	8.34×10^{2}	0.00×10^{0}
		p-value	-	**$1.56 \times 10^{-2+}$**	**$3.02 \times 10^{-11+}$**	**$3.02 \times 10^{-11+}$**	**$3.02 \times 10^{-11+}$**	**$3.34 \times 10^{-11+}$**	$6.57 \times 10^{-2=}$	$1.21 \times 10^{-12-}$
F_{30}		Median	1.35×10^{3}	8.73×10^{2}	2.13×10^{3}	3.40×10^{6}	3.33×10^{6}	9.55×10^{2}	4.69×10^{2}	2.00×10^{2}
		Mean	1.57×10^{3}	1.00×10^{3}	2.03×10^{3}	3.60×10^{6}	3.41×10^{6}	1.02×10^{3}	4.82×10^{2}	2.00×10^{2}
		Std	5.78×10^{2}	4.70×10^{2}	5.37×10^{2}	1.45×10^{6}	1.62×10^{6}	2.69×10^{2}	2.16×10^{2}	0.00×10^{0}
		p-value	-	$1.43 \times 10^{-5-}$	**$5.56 \times 10^{-4+}$**	**$3.02 \times 10^{-11+}$**	**$3.02 \times 10^{-11+}$**	$5.46 \times 10^{-6-}$	$3.02 \times 10^{-11-}$	$1.21 \times 10^{-12-}$
$F_{23\text{–}30}$	*w/t/l*		-	2/2/4	4/2/2	8/0/0	8/0/0	6/0/2	5/1/2	1/1/6
w/t/l			-	13/11/6	23/4/3	27/0/3	27/0/3	25/3/2	23/3/4	19/1/10
Rank			**2.25**	2.97	4.77	6.43	6.10	5.03	4.83	3.62

Table 3. Comparison between ACSEDA and the compared state-of-the-art EDA variants on the 50-*D* CEC2014 benchmark problems. The bold results indicate that ACSEDA is significantly better than the compared methods.

F	Category	Quality	ACSEDA	EDA2	EDA_{VERS}	EDA/LS	EDA/LS-MS	TRA-EDA	BUMDA	MA-ES
F_1	Unimodal Problems	Median	1.42×10^{-14}	7.11×10^{-14}	4.09×10^{5}	6.84×10^{-9}	1.28×10^{-8}	3.42×10^{8}	1.61×10^{7}	1.42×10^{-14}
		Mean	1.14×10^{-14}	7.06×10^{-14}	4.13×10^{5}	6.41×10^{2}	9.33×10^{7}	3.50×10^{8}	1.63×10^{7}	1.80×10^{-14}
		Std	6.77×10^{-15}	1.00×10^{-14}	7.25×10^{4}	2.32×10^{3}	5.02×10^{8}	5.08×10^{7}	3.06×10^{6}	6.28×10^{-15}
		p-value	-	**5.50×10^{-12}+**	**9.04×10^{-12}+**	**9.04×10^{-12}+**	**9.04×10^{-12}+**	**9.04×10^{-12}+**	**9.04×10^{-12}+**	**5.37×10^{-4}+**
F_2		Median	5.97×10^{-13}	1.23×10^{-11}	3.30×10^{3}	1.77×10^{-11}	3.67×10^{-11}	3.27×10^{10}	3.27×10^{3}	2.84×10^{-14}
		Mean	7.40×10^{-13}	1.23×10^{-11}	3.49×10^{3}	7.56×10^{-10}	1.59×10^{-9}	3.32×10^{10}	1.56×10^{4}	3.60×10^{-14}
		Std	3.83×10^{-13}	1.92×10^{-12}	1.76×10^{3}	2.98×10^{-9}	3.81×10^{-9}	1.79×10^{9}	2.60×10^{4}	1.26×10^{-14}
		p-value	-	**2.98×10^{-12}+**	**2.98×10^{-12}+**	**8.43×10^{-7}+**	**1.48×10^{-6}+**	**2.98×10^{-11}+**	**2.98×10^{-11}+**	8.74×10^{-12}−
F_3		Median	0.00×10^{0}	0.00×10^{0}	4.85×10^{3}	1.10×10^{-8}	1.11×10^{-8}	2.93×10^{4}	1.25×10^{4}	1.14×10^{-13}
		Mean	0.00×10^{0}	0.00×10^{0}	5.10×10^{3}	1.74×10^{1}	4.35×10^{-5}	2.87×10^{4}	1.35×10^{4}	1.00×10^{-13}
		Std	0.00×10^{0}	0.00×10^{0}	1.03×10^{3}	9.36×10^{1}	1.72×10^{-4}	2.52×10^{3}	5.05×10^{3}	2.40×10^{-14}
		p-value	-	NaN=	**1.21×10^{-12}+**	**1.21×10^{-12}+**	**1.21×10^{-12}+**	**1.21×10^{-12}+**	**1.21×10^{-12}+**	**1.97×10^{-13}+**
$F_{1\text{-}3}$	$w/t/l$		-	2/1/0	3/0/0	3/0/0	3/0/0	3/0/0	3/0/0	2/0/1
F_4	Simple Multimodal Problems	Median	9.81×10^{1}	5.35×10^{0}	8.16×10^{1}	5.71×10^{-9}	1.29×10^{-8}	4.47×10^{3}	1.25×10^{2}	1.14×10^{-13}
		Mean	9.25×10^{1}	3.93×10^{1}	7.80×10^{1}	1.04×10^{3}	3.86×10^{3}	4.46×10^{3}	1.28×10^{2}	3.99×10^{-1}
		Std	7.64×10^{0}	4.51×10^{1}	2.13×10^{1}	5.11×10^{3}	1.27×10^{4}	4.24×10^{2}	1.57×10^{1}	1.20×10^{0}
		p-value	-	1.93×10^{-3}−	4.80×10^{-4}−	**5.29×10^{-9}+**	**7.12×10^{-8}+**	**1.62×10^{-11}+**	**1.10×10^{-10}+**	8.76×10^{-12}−
F_5		Median	2.11×10^{1}	2.11×10^{1}	2.11×10^{1}	2.00×10^{1}	2.00×10^{1}	2.11×10^{1}	2.11×10^{1}	2.00×10^{1}
		Mean	2.11×10^{1}	2.11×10^{1}	2.11×10^{1}	2.00×10^{1}	2.00×10^{1}	2.11×10^{1}	2.11×10^{1}	2.05×10^{1}
		Std	4.14×10^{-2}	2.54×10^{-2}	3.32×10^{-2}	0.00×10^{0}	2.21×10^{-2}	4.83×10^{-2}	3.07×10^{-2}	7.59×10^{-1}
		p-value	-	8.30×10^{-1}=	2.51×10^{-2}−	1.21×10^{-12}−	4.10×10^{-12}−	1.45×10^{-1}=	1.19×10^{-1}=	1.95×10^{-3}−
F_6		Median	5.22×10^{-5}	1.39×10^{-4}	5.51×10^{-1}	7.19×10^{1}	6.91×10^{1}	6.96×10^{0}	3.64×10^{-1}	4.21×10^{1}
		Mean	1.74×10^{-2}	2.52×10^{-1}	8.59×10^{-1}	7.17×10^{1}	6.93×10^{1}	7.18×10^{0}	9.50×10^{-1}	4.15×10^{1}
		Std	9.31×10^{-2}	5.81×10^{-1}	9.72×10^{-1}	3.34×10^{0}	3.41×10^{0}	2.00×10^{0}	1.23×10^{0}	6.17×10^{0}
		p-value	-	**3.16×10^{-10}+**	**1.09×10^{-10}+**	**3.02×10^{-11}+**	**3.02×10^{-11}+**	**3.02×10^{-11}+**	**1.41×10^{-9}+**	**3.02×10^{-11}+**
F_7		Median	0.00×10^{0}	0.00×10^{0}	3.41×10^{-13}	1.72×10^{-2}	1.97×10^{-2}	2.98×10^{2}	5.73×10^{-3}	2.27×10^{-13}
		Mean	0.00×10^{0}	0.00×10^{0}	9.86×10^{-4}	2.39×10^{-2}	2.18×10^{-2}	2.99×10^{2}	2.20×10^{-2}	2.47×10^{-4}
		Std	0.00×10^{0}	0.00×10^{0}	2.96×10^{-3}	2.69×10^{-2}	2.33×10^{-2}	1.73×10^{1}	4.80×10^{-2}	1.33×10^{-3}
		p-value	-	NaN=	**6.50×10^{-13}+**	**1.20×10^{-12}+**	**1.20×10^{-12}+**	**1.21×10^{-12}+**	**1.21×10^{-12}+**	**5.02×10^{-13}+**
F_8		Median	2.98×10^{0}	7.96×10^{0}	3.09×10^{2}	4.69×10^{2}	4.35×10^{2}	4.93×10^{1}	0.00×10^{0}	3.10×10^{2}
		Mean	2.98×10^{0}	7.89×10^{0}	2.96×10^{2}	4.83×10^{2}	4.57×10^{2}	4.82×10^{1}	3.32×10^{-2}	3.10×10^{2}
		Std	1.23×10^{0}	2.56×10^{0}	5.42×10^{1}	1.07×10^{2}	9.68×10^{1}	6.28×10^{0}	1.79×10^{-1}	5.00×10^{0}
		p-value	-	**7.55×10^{-10}+**	**2.51×10^{-11}+**	**2.51×10^{-11}+**	**2.51×10^{-11}+**	**2.51×10^{-11}+**	1.69×10^{-12}−	**2.47×10^{-11}+**

Table 3. *Cont.*

F	Category	Quality	ACSEDA	EDA2	EDA$_{VERS}$	EDA/LS	EDA/LS-MS	TRA-EDA	BUMDA	MA-ES
F_9		Median	2.49×10^{0}	6.96×10^{0}	3.11×10^{2}	7.00×10^{2}	6.23×10^{2}	3.80×10^{1}	1.99×10^{0}	4.37×10^{2}
		Mean	2.49×10^{0}	6.96×10^{0}	2.82×10^{2}	6.95×10^{2}	6.54×10^{2}	3.75×10^{1}	2.14×10^{0}	4.35×10^{2}
		Std	1.35×10^{0}	1.87×10^{0}	8.86×10^{1}	1.20×10^{2}	9.46×10^{1}	4.75×10^{0}	2.74×10^{0}	9.41×10^{0}
		p-value	-	**1.59×10^{-10}+**	**2.61×10^{-11}+**	**2.61×10^{-11}+**	**2.61×10^{-11}+**	**2.61×10^{-11}+**	2.95×10^{-1}=	**2.59×10^{-11}+**
F_{10}		Median	8.44×10^{0}	1.30×10^{2}	1.18×10^{4}	7.57×10^{3}	7.42×10^{3}	1.06×10^{3}	5.79×10^{1}	7.87×10^{3}
		Mean	1.07×10^{2}	1.99×10^{2}	1.18×10^{4}	8.30×10^{3}	8.24×10^{3}	1.04×10^{3}	6.25×10^{1}	7.84×10^{3}
		Std	1.65×10^{2}	1.80×10^{2}	4.28×10^{2}	2.40×10^{3}	2.07×10^{3}	4.09×10^{2}	4.11×10^{1}	3.26×10^{2}
		p-value	-	**5.44×10^{-3}+**	**3.02×10^{-11}+**	**3.02×10^{-11}+**	**3.02×10^{-11}+**	**9.91×10^{-11}+**	6.95×10^{-1}=	**3.02×10^{-11}+**
F_{11}		Median	1.36×10^{2}	1.42×10^{2}	1.25×10^{4}	8.39×10^{3}	8.52×10^{3}	2.57×10^{3}	4.52×10^{2}	8.44×10^{3}
		Mean	1.18×10^{2}	1.63×10^{2}	1.25×10^{4}	9.46×10^{3}	9.43×10^{3}	2.52×10^{3}	4.58×10^{2}	8.47×10^{3}
		Std	9.29×10^{0}	1.65×10^{2}	3.67×10^{2}	2.78×10^{3}	3.02×10^{3}	9.47×10^{2}	2.82×10^{2}	2.77×10^{2}
		p-value	-	2.84×10^{-1}=	**3.02×10^{-11}+**	**3.02×10^{-11}+**	**3.02×10^{-11}+**	**3.02×10^{-11}+**	**3.65×10^{-11}+**	**3.02×10^{-11}+**
F_{12}		Median	3.35×10^{0}	3.30×10^{0}	3.20×10^{0}	7.43×10^{-1}	6.52×10^{-1}	3.27×10^{0}	3.38×10^{0}	2.85×10^{-2}
		Mean	3.35×10^{0}	3.27×10^{0}	3.25×10^{0}	8.15×10^{-1}	6.83×10^{-1}	3.26×10^{0}	3.35×10^{0}	3.26×10^{-2}
		Std	2.52×10^{-1}	2.71×10^{-1}	3.05×10^{-1}	3.04×10^{-1}	2.57×10^{-1}	2.56×10^{-1}	1.99×10^{-1}	1.70×10^{-2}
		p-value	-	2.84×10^{-1}=	1.96×10^{-1}=	3.02×10^{-1}−	3.02×10^{-11}−	2.17×10^{-1}=	8.77×10^{-1}=	3.02×10^{-11}−
F_{13}		Median	1.40×10^{-1}	7.28×10^{-2}	2.46×10^{-1}	4.98×10^{-1}	4.50×10^{-1}	3.71×10^{0}	6.67×10^{-2}	3.63×10^{-1}
		Mean	1.40×10^{-1}	7.48×10^{-2}	2.44×10^{-1}	1.24×10^{0}	4.97×10^{-1}	3.72×10^{0}	6.38×10^{-2}	3.63×10^{-1}
		Std	1.89×10^{-2}	1.03×10^{-2}	2.22×10^{-2}	2.15×10^{0}	1.24×10^{-1}	1.31×10^{-1}	1.03×10^{-2}	6.78×10^{-2}
		p-value	-	3.02×10^{-11}−	**4.08×10^{-11}+**	**3.02×10^{-11}+**	**3.02×10^{-11}+**	**3.02×10^{-11}+**	3.02×10^{-11}−	**3.02×10^{-11}+**
F_{14}		Median	2.57×10^{-1}	4.01×10^{-1}	3.73×10^{-1}	3.82×10^{-1}	3.97×10^{-1}	6.42×10^{1}	4.36×10^{-1}	3.37×10^{-1}
		Mean	2.40×10^{-1}	3.93×10^{-1}	3.76×10^{-1}	4.58×10^{-1}	4.95×10^{-1}	6.40×10^{1}	4.27×10^{-1}	4.23×10^{-1}
		Std	6.75×10^{-2}	4.80×10^{-2}	2.87×10^{-2}	1.86×10^{-1}	1.86×10^{-1}	3.95×10^{0}	3.49×10^{-2}	2.25×10^{-1}
		p-value	-	**5.07×10^{-10}+**	**1.61×10^{-10}+**	**2.87×10^{-10}+**	**3.16×10^{-10}+**	**3.02×10^{-11}+**	**4.08×10^{-11}+**	**1.87×10^{-7}+**
F_{15}		Median	4.87×10^{0}	1.23×10^{1}	2.74×10^{1}	1.84×10^{2}	1.73×10^{2}	2.39×10^{3}	5.30×10^{0}	6.61×10^{0}
		Mean	4.81×10^{0}	1.19×10^{1}	2.64×10^{1}	1.84×10^{2}	1.82×10^{2}	2.45×10^{2}	5.27×10^{0}	6.61×10^{0}
		Std	4.93×10^{-1}	1.79×10^{0}	4.02×10^{0}	4.72×10^{1}	5.73×10^{1}	9.48×10^{2}	6.27×10^{-1}	1.37×10^{0}
		p-value	-	**3.02×10^{-11}+**	**3.34×10^{-11}+**	**3.02×10^{-11}+**	**3.02×10^{-11}+**	**3.02×10^{-11}+**	**2.38×10^{-3}+**	**7.69×10^{-8}+**
F_{16}		Median	1.80×10^{1}	2.18×10^{1}	2.07×10^{1}	2.28×10^{1}	2.25×10^{1}	2.01×10^{1}	1.98×10^{1}	2.24×10^{1}
		Mean	1.80×10^{1}	2.18×10^{1}	2.07×10^{1}	2.28×10^{1}	2.26×10^{1}	2.00×10^{1}	1.97×10^{1}	2.24×10^{1}
		Std	6.67×10^{-1}	2.58×10^{-1}	3.75×10^{-1}	5.14×10^{-1}	4.27×10^{-1}	2.92×10^{-1}	4.87×10^{-1}	1.40×10^{-1}
		p-value	-	**3.02×10^{-11}+**	**3.02×10^{-11}+**	**3.02×10^{-11}+**	**3.02×10^{-11}+**	**3.02×10^{-11}+**	**1.09×10^{-10}+**	**3.02×10^{-11}+**

Table 3. *Cont.*

F	Category	Quality	ACSEDA	EDA2	EDA_{VERS}	EDA/LS	EDA/LS-MS	TRA-EDA	BUMDA	MA-ES
$F_{4\text{–}16}$	$w/t/l$		-	7/4/2	11/1/1	11/0/2	11/0/2	11/2/0	07/4/2	10/0/3
F_{17}	Hybrid Problems	Median	7.79×10^{1}	3.16×10^{1}	3.45×10^{4}	5.74×10^{7}	1.49×10^{4}	1.96×10^{7}	1.69×10^{7}	2.45×10^{3}
		Mean	1.16×10^{2}	3.77×10^{1}	3.62×10^{4}	3.42×10^{8}	1.04×10^{8}	2.02×10^{7}	1.76×10^{7}	2.49×10^{3}
		Std	7.34×10^{1}	2.47×10^{1}	1.57×10^{4}	4.09×10^{8}	1.77×10^{8}	4.76×10^{6}	3.45×10^{6}	5.06×10^{2}
		p-value	-	1.10×10^{-11}−	**3.02×10^{-11}+**	**3.02×10^{-11}+**	**3.02×10^{-11}+**	**3.02×10^{-11}+**	**3.02×10^{-11}+**	**3.02×10^{-11}+**
F_{18}		Median	7.84×10^{0}	1.40×10^{0}	1.58×10^{2}	2.88×10^{2}	2.98×10^{2}	5.85×10^{8}	1.02×10^{3}	1.66×10^{2}
		Mean	8.47×10^{0}	1.35×10^{0}	2.13×10^{2}	3.49×10^{9}	1.34×10^{9}	6.42×10^{8}	1.08×10^{3}	1.65×10^{2}
		Std	3.23×10^{0}	6.37×10^{-1}	1.59×10^{2}	7.03×10^{9}	3.62×10^{9}	1.93×10^{8}	5.06×10^{2}	4.18×10^{1}
		p-value	-	3.02×10^{-11}−	**4.98×10^{-11}+**	**3.02×10^{-11}+**	**3.02×10^{-11}+**	**3.02×10^{-11}+**	**3.02×10^{-11}+**	**3.02×10^{-11}+**
F_{19}		Median	1.09×10^{1}	7.36×10^{0}	3.50×10^{1}	2.53×10^{1}	2.55×10^{1}	5.99×10^{1}	6.30×10^{1}	1.85×10^{1}
		Mean	1.08×10^{1}	7.46×10^{0}	3.48×10^{1}	2.59×10^{1}	9.34×10^{1}	5.81×10^{1}	6.07×10^{1}	1.84×10^{1}
		Std	7.71×10^{-1}	7.65×10^{-1}	4.99×10^{0}	2.77×10^{0}	3.64×10^{2}	9.80×10^{0}	9.75×10^{0}	2.52×10^{0}
		p-value	-	4.50×10^{-11}−	**4.50×10^{-11}+**	**3.02×10^{-11}+**	**3.02×10^{-11}+**	**3.02×10^{-11}+**	**3.02×10^{-11}+**	**4.08×10^{-11}+**
F_{20}		Median	1.96×10^{0}	2.52×10^{0}	2.33×10^{3}	6.41×10^{4}	8.45×10^{4}	5.01×10^{2}	3.41×10^{4}	3.01×10^{2}
		Mean	1.95×10^{0}	2.50×10^{0}	2.37×10^{3}	9.01×10^{4}	7.81×10^{4}	5.08×10^{2}	3.35×10^{4}	3.27×10^{2}
		Std	2.27×10^{-1}	3.48×10^{-1}	7.57×10^{2}	1.47×10^{5}	4.91×10^{4}	2.60×10^{2}	6.10×10^{3}	1.35×10^{2}
		p-value	-	**1.16×10^{-7}+**	**3.02×10^{-11}+**	**3.02×10^{-11}+**	**3.02×10^{-11}+**	**3.02×10^{-11}+**	**3.02×10^{-11}+**	**3.02×10^{-11}+**
F_{21}		Median	2.47×10^{2}	1.23×10^{2}	7.13×10^{4}	2.35×10^{7}	1.23×10^{6}	1.64×10^{3}	8.45×10^{6}	1.75×10^{3}
		Mean	1.99×10^{2}	9.74×10^{1}	7.27×10^{4}	8.60×10^{7}	1.83×10^{7}	1.98×10^{3}	8.53×10^{6}	1.76×10^{3}
		Std	8.11×10^{1}	7.43×10^{1}	1.94×10^{4}	1.08×10^{8}	3.03×10^{7}	1.49×10^{3}	2.01×10^{6}	4.26×10^{2}
		p-value	-	8.15×10^{-5}−	**3.02×10^{-11}+**	**3.02×10^{-11}+**	**3.02×10^{-11}+**	**3.34×10^{-11}+**	**3.02×10^{-11}+**	**3.02×10^{-11}+**
F_{22}		Median	2.72×10^{1}	3.02×10^{1}	9.51×10^{2}	1.89×10^{3}	2.11×10^{3}	9.87×10^{1}	4.69×10^{1}	7.94×10^{2}
		Mean	3.54×10^{1}	3.02×10^{1}	9.06×10^{2}	9.23×10^{3}	3.33×10^{3}	1.18×10^{2}	1.37×10^{2}	7.58×10^{2}
		Std	3.02×10^{1}	6.70×10^{-1}	1.90×10^{2}	2.74×10^{4}	5.36×10^{3}	6.15×10^{1}	1.46×10^{2}	2.84×10^{2}
		p-value	-	1.36×10^{-7}−	**3.02×10^{-11}+**	**3.02×10^{-11}+**	**3.02×10^{-11}+**	**2.23×10^{-9}+**	**1.17×10^{-9}+**	**3.02×10^{-11}+**
$F_{17\text{–}22}$	$w/t/l$		-	1/0/5	6/0/0	6/0/0	6/0/0	6/0/0	6/0/0	6/0/0
F_{23}	Composition Problems	Median	3.44×10^{2}	3.44×10^{2}	3.44×10^{2}	2.48×10^{3}	1.70×10^{3}	4.17×10^{2}	3.66×10^{2}	2.00×10^{2}
		Mean	3.44×10^{2}	3.44×10^{2}	3.44×10^{2}	2.20×10^{3}	1.52×10^{3}	4.21×10^{2}	3.66×10^{2}	2.00×10^{2}
		Std	1.71×10^{-13}	1.71×10^{-13}	1.71×10^{-13}	9.29×10^{2}	6.07×10^{2}	1.44×10^{1}	2.74×10^{0}	0.00×10^{0}
		p-value	-	NaN=	NaN=	**1.21×10^{-12}+**	**1.21×10^{-12}+**	**1.21×10^{-12}+**	**1.21×10^{-12}+**	1.69×10^{-14}−
F_{24}		Median	2.68×10^{2}	2.68×10^{2}	2.67×10^{2}	4.09×10^{2}	3.60×10^{2}	2.72×10^{2}	2.78×10^{2}	2.00×10^{2}
		Mean	2.68×10^{2}	2.68×10^{2}	2.64×10^{2}	4.78×10^{2}	4.44×10^{2}	2.72×10^{2}	2.78×10^{2}	2.00×10^{2}
		Std	1.33×10^{0}	1.10×10^{0}	5.66×10^{0}	1.69×10^{2}	1.40×10^{2}	8.52×10^{-1}	1.11×10^{0}	0.00×10^{0}
		p-value	-	9.17×10^{-1}=	3.00×10^{-1}=	**2.88×10^{-11}+**	**2.88×10^{-11}+**	**5.25×10^{-11}+**	**2.88×10^{-11}+**	1.14×10^{-12}−

Table 3. *Cont.*

F	Category	Quality	ACSEDA	EDA2	EDA_{VERS}	EDA/LS	EDA/LS-MS	TRA-EDA	BUMDA	MA-ES
F_{25}		Median	2.05×10^2	2.05×10^2	2.00×10^2	5.77×10^2	4.32×10^2	2.17×10^2	2.14×10^2	2.00×10^2
		Mean	2.05×10^2	2.05×10^2	2.00×10^2	5.57×10^2	4.29×10^2	2.17×10^2	2.15×10^2	2.00×10^2
		Std	1.49×10^{-1}	1.58×10^{-1}	0.00×10^0	8.87×10^1	5.87×10^1	1.56×10^0	5.84×10^0	0.00×10^0
		p-value	-	$4.12 \times 10^{-1=}$	$1.19 \times 10^{-12-}$	$\mathbf{2.98 \times 10^{-11+}}$	$\mathbf{2.98 \times 10^{-11+}}$	$\mathbf{2.98 \times 10^{-11+}}$	$\mathbf{2.98 \times 10^{-11+}}$	$1.19 \times 10^{-12-}$
F_{26}		Median	1.00×10^2	1.00×10^2	2.00×10^2	1.01×10^2	1.01×10^2	1.06×10^2	1.74×10^2	2.00×10^2
		Mean	1.00×10^2	1.00×10^2	1.96×10^2	2.90×10^2	1.72×10^2	1.06×10^2	1.58×10^2	2.00×10^2
		Std	4.78×10^{-2}	1.35×10^{-2}	1.64×10^1	2.27×10^2	1.18×10^2	1.35×10^0	4.00×10^1	0.00×10^0
		p-value	-	$8.98 \times 10^{-11-}$	$\mathbf{2.80 \times 10^{-11+}}$	$\mathbf{6.06 \times 10^{-11+}}$	$\mathbf{1.09 \times 10^{-10+}}$	$\mathbf{2.98 \times 10^{-11+}}$	$\mathbf{2.98 \times 10^{-11+}}$	$\mathbf{1.21 \times 10^{-11+}}$
F_{27}		Median	3.00×10^2	4.19×10^2	3.52×10^2	2.68×10^3	2.51×10^3	7.00×10^2	3.68×10^2	2.00×10^2
		Mean	3.17×10^2	4.23×10^2	3.48×10^2	2.67×10^3	2.49×10^3	7.00×10^2	3.74×10^2	2.00×10^2
		Std	2.20×10^1	4.98×10^1	4.35×10^1	1.79×10^2	1.16×10^2	5.13×10^1	2.72×10^1	0.00×10^0
		p-value	-	$\mathbf{5.37 \times 10^{-11+}}$	$\mathbf{5.79 \times 10^{-6+}}$	$\mathbf{2.95 \times 10^{-11+}}$	$\mathbf{2.95 \times 10^{-11+}}$	$\mathbf{2.95 \times 10^{-11+}}$	$\mathbf{4.11 \times 10^{-9+}}$	$1.18 \times 10^{-12-}$
F_{28}		Median	1.16×10^3	1.16×10^3	1.15×10^3	1.15×10^4	1.02×10^4	1.92×10^3	4.29×10^2	2.00×10^2
		Mean	1.16×10^3	1.19×10^3	1.15×10^3	1.15×10^4	9.91×10^3	2.01×10^3	4.31×10^2	2.00×10^2
		Std	4.37×10^1	1.68×10^2	3.88×10^1	2.14×10^3	2.17×10^3	4.59×10^2	6.89×10^0	0.00×10^0
		p-value	-	$9.35 \times 10^{-1=}$	$1.81 \times 10^{-1=}$	$\mathbf{3.02 \times 10^{-11+}}$	$\mathbf{3.02 \times 10^{-11+}}$	$\mathbf{1.33 \times 10^{-11+}}$	$\mathbf{3.02 \times 10^{-11+}}$	$1.21 \times 10^{-12-}$
F_{29}		Median	8.14×10^2	7.39×10^2	1.66×10^3	1.57×10^9	1.05×10^9	1.23×10^3	7.29×10^2	2.00×10^2
		Mean	8.34×10^2	4.24×10^6	1.67×10^3	1.57×10^9	1.06×10^9	2.65×10^3	9.82×10^2	2.00×10^2
		Std	5.15×10^1	1.63×10^7	1.50×10^2	3.30×10^8	2.06×10^8	2.81×10^3	7.50×10^2	0.00×10^0
		p-value	-	$\mathbf{6.77 \times 10^{-5+}}$	$\mathbf{3.02 \times 10^{-11+}}$	$\mathbf{3.02 \times 10^{-11+}}$	$\mathbf{3.02 \times 10^{-11+}}$	$\mathbf{9.92 \times 10^{-11+}}$	$1.33 \times 10^{-1=}$	$1.21 \times 10^{-12-}$
F_{30}		Median	8.56×10^3	9.37×10^3	9.02×10^3	2.49×10^7	1.83×10^7	1.74×10^5	8.19×10^2	2.00×10^2
		Mean	8.57×10^3	9.50×10^3	9.20×10^3	2.66×10^7	1.92×10^7	1.67×10^5	8.17×10^2	2.00×10^2
		Std	3.37×10^2	7.30×10^2	3.67×10^2	1.14×10^7	6.42×10^6	2.95×10^4	1.82×10^2	0.00×10^0
		p-value	-	$\mathbf{1.07 \times 10^{-7+}}$	$\mathbf{7.77 \times 10^{-9+}}$	$\mathbf{3.02 \times 10^{-11+}}$	$\mathbf{3.02 \times 10^{-11+}}$	$\mathbf{3.02 \times 10^{-11+}}$	$\mathbf{3.02 \times 10^{-11+}}$	$1.21 \times 10^{-12-}$
$F_{23–30}$	*w/t/l*		-	3/4/1	4/3/1	8/0/0	8/0/0	8/0/0	5/1/2	1/0/7
w/t/l			-	13/9/8	24/4/2	28/0/2	28/0/2	28/2/0	21/5/4	19/0/11
Rank			**2.33**	2.93	4.52	6.80	6.27	5.70	4.30	3.15

Table 4. Comparison between ACSEDA and the compared state-of-the-art EDA variants on the 100-*D* CEC2014 benchmark problems. The bold results indicate that ACSEDA is significantly better than the compared methods.

F	Category	Quality	ACSEDA	EDA2	EDA_{VERS}	EDA/LS	EDA/LS-MS	TRA-EDA	BUMDA	MA-ES
F_1	Unimodal Prob-lems	Median	1.80×10^{-9}	6.82×10^{-13}	1.25×10^{6}	1.87×10^{6}	1.39×10^{3}	4.74×10^{8}	8.33×10^{7}	4.26×10^{-14}
		Mean	1.90×10^{-9}	7.24×10^{-13}	1.22×10^{6}	8.24×10^{9}	4.41×10^{8}	4.62×10^{8}	8.48×10^{7}	1.46×10^{-13}
		Std	7.43×10^{-9}	1.70×10^{-13}	1.68×10^{5}	8.95×10^{9}	1.03×10^{9}	6.07×10^{7}	8.05×10^{6}	5.46×10^{-13}
		p-value	-	2.99×10^{-11}−	**3.02×10^{-11}+**	**3.02×10^{-11}+**	**3.02×10^{-11}+**	**3.02×10^{-11}+**	**3.02×10^{-11}+**	1.20×10^{-11}−
F_2		Median	3.84×10^{-7}	9.36×10^{-11}	9.84×10^{3}	1.88×10^{-10}	1.14×10^{-10}	9.69×10^{10}	4.27×10^{4}	8.53×10^{-14}
		Mean	3.94×10^{-7}	9.67×10^{-11}	1.12×10^{4}	1.54×10^{-9}	3.87×10^{1}	9.74×10^{10}	7.15×10^{4}	7.48×10^{-14}
		Std	2.02×10^{-7}	2.78×10^{-11}	5.42×10^{3}	3.04×10^{-9}	2.08×10^{2}	3.46×10^{9}	9.73×10^{4}	1.55×10^{-14}
		p-value	-	**3.02×10^{-11}−**	**3.02×10^{-11}+**	**3.02×10^{-11}+**	**5.57×10^{-10}+**	**3.02×10^{-11}+**	**3.02×10^{-11}+**	1.48×10^{-11}−
F_3		Median	5.40×10^{-13}	0.00×10^{0}	6.11×10^{3}	2.13×10^{-8}	2.57×10^{-8}	1.11×10^{5}	2.39×10^{4}	1.71×10^{-13}
		Mean	7.09×10^{-13}	0.00×10^{0}	6.25×10^{3}	9.02×10^{-1}	4.18×10^{2}	1.10×10^{5}	2.34×10^{4}	2.01×10^{-13}
		Std	4.71×10^{-13}	0.00×10^{0}	1.48×10^{3}	4.31×10^{0}	1.13×10^{3}	5.11×10^{3}	3.80×10^{3}	4.08×10^{-14}
		p-value	-	1.18×10^{-12}−	**2.96×10^{-11}+**	**2.96×10^{-11}+**	**2.96×10^{-11}+**	**2.96×10^{-11}+**	**2.96×10^{-11}+**	4.68×10^{-9}−
F_{1-3}		$w/t/l$	-	0/0/3	3/0/0	2/0/1	3/0/0	3/0/0	3/0/0	0/0/3
F_4	Simple Multi-modal Prob-lems	Median	1.85×10^{2}	2.01×10^{2}	1.43×10^{2}	3.99×10^{0}	1.37×10^{-7}	1.27×10^{4}	1.91×10^{2}	1.71×10^{-13}
		Mean	1.81×10^{2}	1.92×10^{2}	1.40×10^{2}	1.04×10^{4}	4.09×10^{3}	1.27×10^{4}	1.82×10^{2}	1.20×10^{0}
		Std	2.93×10^{1}	2.69×10^{1}	7.29×10^{0}	3.91×10^{4}	1.31×10^{4}	7.87×10^{2}	2.44×10^{1}	1.83×10^{0}
		p-value	-	3.76×10^{-1}=	1.83×10^{-9}−	**3.51×10^{-9}+**	**1.45×10^{-7}+**	**2.61×10^{-11}+**	9.70×10^{-1}=	1.14×10^{-11}−
F_5		Median	2.13×10^{1}	2.13×10^{1}	2.13×10^{1}	2.00×10^{1}	2.00×10^{1}	2.13×10^{1}	2.13×10^{1}	2.00×10^{1}
		Mean	2.13×10^{1}	2.13×10^{1}	2.13×10^{1}	2.00×10^{1}	2.00×10^{1}	2.13×10^{1}	2.13×10^{1}	2.03×10^{1}
		Std	2.74×10^{-2}	2.56×10^{-2}	2.46×10^{-2}	1.47×10^{-2}	0.00×10^{0}	2.28×10^{-2}	3.62×10^{-2}	6.25×10^{-1}
		p-value	-	1.70×10^{-2}−	1.86×10^{-1}=	3.16×10^{-12}−	1.21×10^{-12}−	1.69×10^{-1}=	5.44×10^{-1}=	9.19×10^{-6}−
F_6		Median	4.10×10^{-3}	1.94×10^{0}	1.01×10^{1}	1.55×10^{2}	6.88×10^{1}	3.37×10^{1}	4.93×10^{0}	1.19×10^{2}
		Mean	3.83×10^{-1}	2.43×10^{0}	1.06×10^{1}	1.56×10^{2}	6.96×10^{1}	3.39×10^{1}	5.05×10^{0}	1.20×10^{2}
		Std	5.70×10^{-1}	1.85×10^{0}	3.51×10^{0}	5.33×10^{0}	2.75×10^{0}	3.54×10^{0}	1.58×10^{0}	7.18×10^{0}
		p-value	-	**9.51×10^{-6}+**	**3.02×10^{-11}+**	**3.02×10^{-11}+**	**3.02×10^{-11}+**	**3.02×10^{-11}+**	**3.34×10^{-11}+**	**3.02×10^{-11}+**
F_7		Median	2.27×10^{-13}	0.00×10^{0}	1.02×10^{-12}	1.90×10^{-8}	1.72×10^{-2}	1.10×10^{3}	5.50×10^{-2}	3.41×10^{-13}
		Mean	2.31×10^{-13}	0.00×10^{0}	1.00×10^{-12}	9.27×10^{-3}	2.65×10^{-2}	1.10×10^{3}	7.16×10^{-2}	4.93×10^{-4}
		Std	1.19×10^{-13}	0.00×10^{0}	1.84×10^{-13}	2.21×10^{-2}	2.82×10^{-2}	3.23×10^{1}	5.06×10^{-2}	1.84×10^{-3}
		p-value	-	8.27×10^{-13}−	**1.98×10^{-11}+**	**2.22×10^{-11}+**	**2.25×10^{-11}+**	**2.25×10^{-11}+**	**2.25×10^{-11}+**	**4.97×10^{-6}+**
F_8		Median	9.45×10^{0}	2.39×10^{1}	7.33×10^{2}	1.56×10^{3}	4.64×10^{2}	2.71×10^{2}	0.00×10^{0}	6.15×10^{2}
		Mean	9.22×10^{0}	2.37×10^{1}	5.79×10^{2}	1.48×10^{3}	4.93×10^{2}	2.75×10^{2}	1.33×10^{-1}	6.13×10^{2}
		Std	2.52×10^{0}	4.46×10^{0}	2.94×10^{2}	3.00×10^{2}	1.31×10^{2}	1.62×10^{1}	3.38×10^{-1}	1.08×10^{1}
		p-value	-	**3.11×10^{-11}+**	**2.88×10^{-11}+**	**2.88×10^{-11}+**	**2.88×10^{-11}+**	**2.88×10^{-11}+**	3.86×10^{-12}−	**2.86×10^{-11}+**

Table 4. *Cont.*

F	Category	Quality	ACSEDA	EDA2	EDA_{VERS}	EDA/LS	EDA/LS-MS	TRA-EDA	BUMDA	MA-ES
F_9		Median	7.96×10^{0}	2.14×10^{1}	7.43×10^{2}	1.68×10^{3}	5.92×10^{2}	2.77×10^{2}	2.98×10^{0}	8.09×10^{2}
		Mean	8.06×10^{0}	2.08×10^{1}	6.50×10^{2}	1.68×10^{3}	6.27×10^{2}	2.78×10^{2}	3.26×10^{0}	8.06×10^{2}
		Std	2.29×10^{0}	4.72×10^{0}	2.40×10^{2}	2.62×10^{2}	1.31×10^{2}	1.83×10^{1}	1.46×10^{0}	1.52×10^{1}
		p-value	-	**$4.34 \times 10^{-11+}$**	**$2.78 \times 10^{-11+}$**	**$2.78 \times 10^{-11+}$**	**$2.78 \times 10^{-11+}$**	**$2.78 \times 10^{-11+}$**	$2.22 \times 10^{-9-}$	**$2.77 \times 10^{-11+}$**
F_{10}		Median	6.11×10^{2}	1.63×10^{3}	2.79×10^{4}	1.78×10^{4}	7.72×10^{3}	6.51×10^{3}	1.01×10^{2}	1.64×10^{4}
		Mean	6.60×10^{2}	1.79×10^{3}	2.78×10^{4}	2.20×10^{4}	8.58×10^{3}	6.55×10^{3}	1.17×10^{2}	1.65×10^{4}
		Std	3.85×10^{2}	7.29×10^{2}	7.77×10^{2}	7.13×10^{3}	2.27×10^{3}	7.46×10^{2}	1.16×10^{2}	4.17×10^{2}
		p-value	-	**$1.55 \times 10^{-9+}$**	**$3.02 \times 10^{-11+}$**	**$3.02 \times 10^{-11+}$**	**$3.02 \times 10^{-11+}$**	**$3.02 \times 10^{-11+}$**	$1.20 \times 10^{-8-}$	**$3.02 \times 10^{-11+}$**
F_{11}		Median	1.43×10^{3}	1.81×10^{3}	2.89×10^{4}	2.17×10^{4}	7.99×10^{3}	5.73×10^{3}	1.25×10^{3}	1.31×10^{4}
		Mean	1.44×10^{3}	1.85×10^{3}	2.87×10^{4}	2.29×10^{4}	8.34×10^{3}	5.66×10^{3}	1.23×10^{3}	1.31×10^{4}
		Std	3.41×10^{2}	4.59×10^{2}	9.63×10^{2}	6.21×10^{3}	1.97×10^{3}	4.87×10^{2}	5.13×10^{2}	4.55×10^{2}
		p-value	-	**$1.24 \times 10^{-3+}$**	**$3.02 \times 10^{-11+}$**	**$3.02 \times 10^{-11+}$**	**$3.02 \times 10^{-11+}$**	**$3.02 \times 10^{-11+}$**	$6.57 \times 10^{-2=}$	**$3.02 \times 10^{-11+}$**
F_{12}		Median	3.80×10^{0}	3.97×10^{0}	3.99×10^{0}	1.04×10^{0}	6.78×10^{-1}	4.00×10^{0}	3.93×10^{0}	2.03×10^{-2}
		Mean	3.85×10^{0}	3.99×10^{0}	3.97×10^{0}	1.19×10^{0}	7.28×10^{-1}	3.93×10^{0}	3.96×10^{0}	2.03×10^{-2}
		Std	2.16×10^{-1}	2.21×10^{-1}	2.00×10^{-1}	5.49×10^{-1}	2.81×10^{-1}	3.04×10^{-1}	2.10×10^{-1}	6.72×10^{-3}
		p-value	-	**$3.15 \times 10^{-2+}$**	**$3.92 \times 10^{-2+}$**	$3.02 \times 10^{-11-}$	$3.02 \times 10^{-11-}$	$5.75 \times 10^{-2=}$	$7.98 \times 10^{-2=}$	$3.02 \times 10^{-11-}$
F_{13}		Median	2.38×10^{-1}	1.33×10^{-1}	3.15×10^{-1}	4.35×10^{-1}	4.99×10^{-1}	5.54×10^{0}	5.77×10^{-2}	5.49×10^{-1}
		Mean	2.36×10^{-1}	1.35×10^{-1}	3.23×10^{-1}	2.39×10^{0}	8.72×10^{-1}	5.54×10^{0}	5.85×10^{-2}	5.57×10^{-1}
		Std	2.01×10^{-2}	1.85×10^{-2}	2.55×10^{-2}	3.94×10^{0}	1.17×10^{0}	7.78×10^{-2}	9.94×10^{-3}	7.13×10^{-2}
		p-value	-	$3.02 \times 10^{-11-}$	**$3.69 \times 10^{-11+}$**	**$3.02 \times 10^{-11+}$**	**$2.15 \times 10^{-10+}$**	**$3.02 \times 10^{-11+}$**	$3.02 \times 10^{-11-}$	**$3.02 \times 10^{-11+}$**
F_{14}		Median	2.74×10^{-1}	4.00×10^{-1}	3.59×10^{-1}	1.44×10^{2}	3.77×10^{-1}	3.29×10^{2}	4.56×10^{-1}	2.92×10^{-1}
		Mean	2.70×10^{-1}	3.86×10^{-1}	3.63×10^{-1}	2.40×10^{2}	4.26×10^{-1}	3.27×10^{2}	4.50×10^{-1}	3.66×10^{-1}
		Std	4.12×10^{-2}	4.86×10^{-2}	1.83×10^{-2}	2.63×10^{2}	1.50×10^{-1}	7.41×10^{0}	2.05×10^{-2}	2.36×10^{-1}
		p-value	-	**$3.82 \times 10^{-9+}$**	**$4.98 \times 10^{-11+}$**	**$3.02 \times 10^{-11+}$**	**$1.29 \times 10^{-9+}$**	**$3.02 \times 10^{-11+}$**	**$3.02 \times 10^{-11+}$**	**$3.02 \times 10^{-11+}$**
F_{15}		Median	1.04×10^{1}	1.04×10^{1}	6.55×10^{1}	5.57×10^{2}	1.93×10^{2}	1.90×10^{5}	1.20×10^{1}	1.38×10^{1}
		Mean	1.06×10^{1}	1.03×10^{1}	6.01×10^{1}	2.48×10^{7}	2.48×10^{5}	1.90×10^{5}	1.22×10^{1}	1.45×10^{1}
		Std	8.45×10^{-1}	8.26×10^{-1}	1.66×10^{1}	7.66×10^{7}	1.33×10^{6}	3.06×10^{4}	1.07×10^{0}	2.62×10^{0}
		p-value	-	$3.33 \times 10^{-1=}$	**$1.29 \times 10^{-9+}$**	**$3.02 \times 10^{-11+}$**	**$3.02 \times 10^{-11+}$**	**$3.02 \times 10^{-11+}$**	**$1.29 \times 10^{-6+}$**	**$1.69 \times 10^{-9+}$**
F_{16}		Median	4.12×10^{1}	4.60×10^{1}	4.53×10^{1}	4.65×10^{1}	2.26×10^{1}	4.33×10^{1}	4.26×10^{1}	4.59×10^{1}
		Mean	4.12×10^{1}	4.59×10^{1}	4.53×10^{1}	4.64×10^{1}	2.26×10^{1}	4.33×10^{1}	4.26×10^{1}	4.58×10^{1}
		Std	5.74×10^{-1}	2.25×10^{-1}	3.31×10^{-1}	9.07×10^{-1}	4.77×10^{-1}	4.56×10^{-1}	5.66×10^{-1}	1.42×10^{-1}
		p-value	-	**$3.02 \times 10^{-11+}$**	**$3.02 \times 10^{-11+}$**	**$3.02 \times 10^{-11+}$**	$3.02 \times 10^{-11-}$	**$3.69 \times 10^{-11+}$**	**$1.07 \times 10^{-9+}$**	**$3.02 \times 10^{-11+}$**

Table 4. *Cont.*

F	Category	Quality	ACSEDA	EDA2	EDA_{VERS}	EDA/LS	EDA/LS-MS	TRA-EDA	BUMDA	MA-ES
$F_{4–16}$	*w*/*t*/*l*		-	8/2/3	11/1/1	11/0/2	10/0/3	11/2/0	5/4/4	10/0/3
F_{17}	Hybrid Problems	Median	7.10×10^{2}	8.16×10^{2}	2.78×10^{5}	1.29×10^{9}	1.33×10^{7}	9.57×10^{7}	3.87×10^{7}	5.48×10^{3}
		Mean	7.27×10^{2}	7.72×10^{2}	2.74×10^{5}	1.32×10^{9}	1.19×10^{8}	9.54×10^{7}	3.85×10^{7}	5.59×10^{3}
		Std	1.93×10^{2}	2.45×10^{2}	5.52×10^{4}	9.08×10^{8}	1.52×10^{8}	1.15×10^{7}	3.48×10^{6}	6.64×10^{2}
		p-value	-	3.26×10^{-1}=	**3.02×10^{-11}+**	**3.02×10^{-11}+**	**3.02×10^{-11}+**	**3.02×10^{-11}+**	**3.02×10^{-11}+**	**3.02×10^{-11}+**
F_{18}		Median	5.80×10^{1}	2.62×10^{1}	2.02×10^{2}	5.64×10^{8}	2.47×10^{2}	3.04×10^{9}	1.36×10^{2}	3.63×10^{2}
		Mean	5.89×10^{1}	2.60×10^{1}	2.54×10^{2}	1.97×10^{10}	1.41×10^{9}	3.01×10^{9}	1.73×10^{2}	3.63×10^{2}
		Std	1.37×10^{1}	7.82×10^{0}	1.87×10^{2}	2.58×10^{10}	3.71×10^{9}	4.29×10^{8}	1.43×10^{2}	6.79×10^{1}
		p-value	-	7.39×10^{-11}−	**7.12×10^{-9}+**	**3.02×10^{-11}+**	**3.02×10^{-11}+**	**3.02×10^{-11}+**	**2.50×10^{-3}+**	**3.02×10^{-11}+**
F_{19}		Median	9.09×10^{1}	9.40×10^{1}	9.44×10^{1}	5.32×10^{1}	2.61×10^{1}	5.03×10^{2}	6.97×10^{1}	6.03×10^{1}
		Mean	9.10×10^{1}	9.40×10^{1}	9.44×10^{1}	2.43×10^{3}	2.57×10^{1}	5.14×10^{2}	7.21×10^{1}	6.28×10^{1}
		Std	1.21×10^{0}	1.38×10^{0}	1.41×10^{0}	4.59×10^{3}	2.12×10^{0}	6.67×10^{1}	1.71×10^{1}	1.41×10^{1}
		p-value	-	**2.67×10^{-9}+**	**3.16×10^{-10}+**	1.86×10^{-1}=	3.02×10^{-11}−	**3.02×10^{-11}+**	9.51×10^{-6}−	8.48×10^{-9}−
F_{20}		Median	6.20×10^{0}	9.31×10^{0}	8.31×10^{3}	2.43×10^{5}	6.28×10^{4}	2.68×10^{4}	1.08×10^{5}	6.01×10^{2}
		Mean	6.68×10^{0}	9.56×10^{0}	8.43×10^{3}	2.33×10^{6}	6.75×10^{4}	2.60×10^{4}	1.08×10^{5}	6.36×10^{2}
		Std	2.34×10^{0}	2.40×10^{0}	2.00×10^{3}	5.76×10^{6}	3.67×10^{4}	2.96×10^{3}	1.89×10^{4}	1.92×10^{2}
		p-value	-	**8.88×10^{-6}+**	**3.02×10^{-11}+**	**3.02×10^{-11}+**	**3.02×10^{-11}+**	**3.02×10^{-11}+**	**3.02×10^{-11}+**	**3.02×10^{-11}+**
F_{21}		Median	2.23×10^{2}	1.83×10^{2}	1.59×10^{5}	9.21×10^{8}	5.22×10^{6}	3.03×10^{6}	2.42×10^{7}	3.58×10^{3}
		Mean	2.51×10^{2}	1.98×10^{2}	1.51×10^{5}	8.09×10^{8}	2.02×10^{7}	3.18×10^{6}	2.40×10^{7}	3.51×10^{3}
		Std	8.04×10^{1}	6.34×10^{1}	3.04×10^{4}	5.30×10^{8}	3.12×10^{7}	1.09×10^{6}	2.95×10^{6}	5.75×10^{2}
		p-value	-	1.08×10^{-2}−	**3.02×10^{-11}+**	**3.02×10^{-11}+**	**3.02×10^{-11}+**	**3.02×10^{-11}+**	**3.02×10^{-11}+**	**3.02×10^{-11}+**
F_{22}		Median	3.89×10^{1}	3.75×10^{1}	3.54×10^{3}	4.09×10^{3}	1.86×10^{3}	7.75×10^{1}	2.58×10^{2}	1.20×10^{3}
		Mean	4.62×10^{1}	6.56×10^{1}	3.49×10^{3}	1.07×10^{5}	2.43×10^{3}	9.21×10^{1}	2.90×10^{2}	1.26×10^{3}
		Std	3.11×10^{1}	5.21×10^{1}	2.48×10^{2}	2.59×10^{5}	1.73×10^{3}	4.21×10^{1}	2.09×10^{2}	3.12×10^{2}
		p-value	-	6.00×10^{-1}=	**3.02×10^{-11}+**	**3.02×10^{-11}+**	**3.02×10^{-11}+**	**5.00×10^{-9}+**	**1.43×10^{-8}+**	**3.02×10^{-11}+**
$F_{17–22}$	*w*/*t*/*l*		-	2/2/2	6/0/0	5/1/0	5/0/1	6/0/0	5/0/1	5/0/1
F_{23}	Composition Problems	Median	3.48×10^{2}	3.48×10^{2}	3.48×10^{2}	6.57×10^{3}	1.64×10^{3}	4.43×10^{2}	3.59×10^{2}	2.00×10^{2}
		Mean	3.48×10^{2}	3.48×10^{2}	3.48×10^{2}	6.72×10^{3}	1.44×10^{3}	4.41×10^{2}	3.59×10^{2}	2.00×10^{2}
		Std	0.00×10^{0}	0.00×10^{0}	0.00×10^{0}	1.60×10^{3}	6.17×10^{2}	1.26×10^{1}	2.91×10^{0}	0.00×10^{0}
		p-value	-	NaN=	NaN=	**1.21×10^{-12}+**	**3.36×10^{-11}+**	**1.21×10^{-12}+**	**1.21×10^{-12}+**	1.69×10^{-14}−
F_{24}		Median	3.72×10^{2}	3.72×10^{2}	3.62×10^{2}	5.05×10^{2}	3.39×10^{2}	3.82×10^{2}	3.82×10^{2}	2.00×10^{2}
		Mean	3.73×10^{2}	3.73×10^{2}	3.61×10^{2}	5.07×10^{2}	4.29×10^{2}	3.82×10^{2}	3.82×10^{2}	2.00×10^{2}
		Std	2.22×10^{0}	1.90×10^{0}	4.63×10^{0}	2.74×10^{1}	1.32×10^{2}	1.41×10^{0}	1.59×10^{0}	0.00×10^{0}
		p-value	-	7.73×10^{-1}=	4.08×10^{-11}−	**3.02×10^{-11}+**	1.91×10^{-1}=	**3.02×10^{-11}+**	**3.02×10^{-11}+**	1.21×10^{-12}−

Table 4. *Cont.*

F	Category	Quality	ACSEDA	EDA2	EDA_{VERS}	EDA/LS	EDA/LS-MS	TRA-EDA	BUMDA	MA-ES
F_{25}		Median	2.15×10^{2}	2.16×10^{2}	2.00×10^{2}	1.35×10^{3}	4.27×10^{2}	2.42×10^{2}	2.22×10^{2}	2.00×10^{2}
		Mean	2.15×10^{2}	2.16×10^{2}	2.00×10^{2}	1.37×10^{3}	4.20×10^{2}	2.42×10^{2}	2.23×10^{2}	2.00×10^{2}
		Std	3.90×10^{-1}	8.62×10^{-1}	0.00×10^{0}	2.04×10^{2}	5.42×10^{1}	3.06×10^{0}	9.36×10^{0}	0.00×10^{0}
		p-value	-	**1.11×10^{-4}+**	1.21×10^{-12}−	**3.02×10^{-11}+**	**3.02×10^{-11}+**	**3.02×10^{-11}+**	**2.39×10^{-4}+**	1.21×10^{-12}−
F_{26}		Median	2.00×10^{2}	2.00×10^{2}	2.00×10^{2}	1.07×10^{3}	1.01×10^{2}	2.00×10^{2}	2.00×10^{2}	2.00×10^{2}
		Mean	2.00×10^{2}	2.00×10^{2}	2.00×10^{2}	8.64×10^{2}	1.25×10^{2}	2.00×10^{2}	2.00×10^{2}	2.00×10^{2}
		Std	2.50×10^{-4}	3.97×10^{-4}	7.67×10^{-3}	4.77×10^{2}	6.80×10^{1}	2.31×10^{-3}	4.88×10^{-2}	0.00×10^{0}
		p-value	-	3.16×10^{-1}=	**1.96×10^{-12}+**	**9.65×10^{-8}+**	1.02×10^{-6}−	**9.17×10^{-6}+**	**2.59×10^{-11}+**	9.91×10^{-13}−
F_{27}		Median	3.00×10^{2}	4.65×10^{2}	3.98×10^{2}	5.45×10^{3}	2.42×10^{3}	1.78×10^{3}	3.13×10^{2}	2.00×10^{2}
		Mean	3.04×10^{2}	4.96×10^{2}	4.15×10^{2}	5.50×10^{3}	2.43×10^{3}	1.78×10^{3}	3.23×10^{2}	2.00×10^{2}
		Std	7.91×10^{0}	1.13×10^{2}	5.12×10^{1}	3.00×10^{2}	1.17×10^{2}	1.05×10^{2}	2.48×10^{1}	0.00×10^{0}
		p-value	-	**3.02×10^{-11}+**	**3.02×10^{-11}+**	**3.02×10^{-11}+**	**3.02×10^{-11}+**	**3.02×10^{-11}+**	**7.60×10^{-7}+**	1.21×10^{-12}−
F_{28}		Median	2.31×10^{3}	2.79×10^{3}	2.17×10^{3}	2.99×10^{4}	1.07×10^{4}	3.99×10^{3}	8.14×10^{2}	2.00×10^{2}
		Mean	2.31×10^{3}	2.74×10^{3}	2.17×10^{3}	2.98×10^{4}	1.05×10^{4}	4.45×10^{3}	8.14×10^{2}	2.00×10^{2}
		Std	1.10×10^{2}	5.59×10^{2}	5.72×10^{1}	5.35×10^{3}	1.89×10^{3}	1.10×10^{3}	1.85×10^{1}	0.00×10^{0}
		p-value	-	**1.17×10^{-2}+**	2.38×10^{-7}−	**3.02×10^{-11}+**	**3.02×10^{-11}+**	**3.02×10^{-11}+**	3.02×10^{-11}−	1.21×10^{-12}−
F_{29}		Median	1.13×10^{3}	7.44×10^{2}	1.79×10^{3}	5.47×10^{9}	9.15×10^{8}	1.49×10^{5}	3.32×10^{2}	2.00×10^{2}
		Mean	1.26×10^{3}	7.85×10^{2}	1.80×10^{3}	5.60×10^{9}	9.62×10^{8}	1.69×10^{5}	4.54×10^{2}	2.00×10^{2}
		Std	2.79×10^{2}	1.55×10^{2}	4.30×10^{1}	1.23×10^{9}	2.57×10^{8}	1.26×10^{5}	3.73×10^{2}	0.00×10^{0}
		p-value	-	3.16×10^{-10}−	**3.82×10^{-10}+**	**3.02×10^{-11}+**	**3.02×10^{-11}+**	**3.02×10^{-11}+**	5.00×10^{-9}−	1.21×10^{-12}−
F_{30}		Median	9.17×10^{3}	4.37×10^{3}	8.66×10^{3}	3.02×10^{8}	1.31×10^{7}	7.14×10^{4}	1.92×10^{3}	2.00×10^{2}
		Mean	8.91×10^{3}	4.71×10^{3}	8.82×10^{3}	3.11×10^{8}	1.38×10^{7}	8.01×10^{4}	1.88×10^{3}	2.00×10^{2}
		Std	1.03×10^{3}	1.21×10^{3}	6.54×10^{2}	1.03×10^{8}	5.63×10^{6}	2.63×10^{4}	4.86×10^{2}	0.00×10^{0}
		p-value	-	9.92×10^{-11}−	3.87×10^{-1}=	**3.02×10^{-11}+**	**3.02×10^{-11}+**	**3.90×10^{-10}+**	3.02×10^{-11}−	1.21×10^{-12}−
$F_{23–30}$	*w/t/l*		-	3/3/2	3/2/3	8/0/0	6/1/1	8/0/0	5/0/3	0/0/8
w/t/l			-	13/7/10	23/3/4	26/1/3	24/1/5	28/2/0	18/4/8	15/0/15
Rank			2.70	3.37	4.58	7.07	5.27	5.83	4.03	3.15

From Table 2, the comparison results between ACSEDA and the compared state-of-the-art EDAs on the 30-*D* CEC 2014 benchmark problems can be summarized as follows:

(1) As shown in the last row of Table 2, in view of the Friedman test, it is found that the proposed ACSEDA obtains the smallest rank and this rank value is much smaller than those of the other algorithms. This indicates that ACSEDA achieves the best overall performance on the 30-*D* CEC 2014 benchmark set and obtains significant superiority to the compared algorithms.

(2) As shown in the second-to-last row of Table 2, from the perspective of the Wilcoxon rank-sum test, we can see that ACSEDA achieves significantly better performance than the compared algorithms on at least 23 problems, except for EDA2 and MA-ES. Compared with EDA2, ACSEDA shows significant superiority on 13 problems, and only presents inferiority on 6 problems. Competing with MA-ES, ACSEDA presents significant dominance on 19 problems, and it loses the competition on 10 problems.

(3) With respect to the optimization performance on different kinds of problems, on the three unimodal problems, both ACSEDA and EDA2 achieve the true global optima of these three problems and thus show significantly better performance than the other 6 EDA variants. On the 13 simple multimodal problems, ACSEDA is significantly superior to EDA_{VESR}, EDA/LS, EDA/LS-MS, and TRA-EDA on 10 problems, and it also beats EDA2, BUMDA, and MA-ES down on 8, 9, and 9 problems, respectively. In terms of the six hybrid problems, the optimization performance of ACSEDA is significantly better than the compared EDA variants on all these functions, except for EDA2. In comparison with EDA2, ACSEDA shows great superiority on three problems and achieves equivalent performance with EDA2 on three problems. In particular, on these six hybrid problems, ACSEDA shows no inferiority to all the compared EDA variants. As for the eight composition problems, it is observed that ACSEDA is significantly better than EDA/LS and EDA/LS-MS on all these problems. In comparison with TRA-EDA and BUMDA, it outperforms them on six and five problems, respectively. Particularly, on this kind of problem, ACSEDA is a litter inferior to EDA2 and MA-ES.

(4) To sum up, it is observed that ACSEDA shows very competitive, or even significantly better, performance in solving the 30-*D* CEC 2014 benchmark problems, as compared with the selected state-of-the-art EDA variants. In particular, encountered with complicated optimization problems, such as multimodal problems, hybrid problems, and composition problems, the proposed ACSEDA shows great superiority to the compared algorithms, which indicates that it is very promising for complicated problems.

Subsequently, from Table 3, we can get the following findings, with respect to the comparison results between ACSEDA and the compared state-of-the-art EDA variants, on the 50-*D* CEC 2014 benchmark problems:

(1) In terms of the Friedman test, as shown in the last row of Table 3, it is found that ACSEDA still achieves the lowest rank among all algorithms. This verifies that ACSEDA still obtains the best overall performance on the 50-*D* CEC 2014 problems.

(2) With respect to the Wilcoxon rank sum test, as shown in the second last row of Table 3, ACSEDA outperforms EDA_{VERS}, EDA/LS, EDA/LS-MS, TRA-EDA, and BUMDA significantly on 24, 28, 28, 28, and 21 problems, respectively. Compared with EDA2, ACSEDA attains much better performance on 13 problems and equivalent performance on 9 problems. Competing with MA-ES, ACSEDA significantly outperforms it on 19 problems and only shows inferiority on 11 problems.

(3) Regarding the performance on different kinds of optimization problems, on the three unimodal problems, except for EDA2 and MA-ES, ACSEDA still presents great superiority to the other compared EDA variants on all the three problems. In particular, both ACSEDA and EDA2 locate the true global optimum of F_3, while ACSEDA displays great dominance over EDA2 on the other two problems. Compared with MA-ES, ACSEDA is much better on two problems, and it obtains worse performance on only

one problem. On the 13 simple multimodal functions, ACSEDA significantly outperforms EDA_{VERS}, EDA/LS, EDA/LS-MS, and TRA-EDA on 11 problems, performs significantly better than MA-ES on 10 problems, and wins the competition on 7 problems, as competed with both EDA2 and BUMDA. On the 6 hybrid problems, ACSEDA is significantly superior to the compared EDA variants on all the six problems, except for EDA2. In competition with EDA2, ACSEDA loses the competition on five problems. On the 8 composition problems, ACSEDA is better than EDA/LS, EDA/LS-MS, TRA-EDA on all eight problems. At the same time, it achieves equivalent or even much better performance than EDA2, EDA_{VERS}, and BUMDA on at least six problems. However, ACSEDA shows inferior performance to MA-ES on seven problems.

(4) In summary, encountered with the 50-*D* CEC 2014 problems, ACSEDA still exhibits significantly better performance than the compared EDA variants. This further demonstrates that ACSEDA is promising for both simple optimization problems, such as unimodal problems, and complicated optimization problems, such as hybrid problems and composition problems.

At last, from Table 4, the following observations can be achieved from the comparison results between ACSEDA and the compared state-of-the-art EDA variants on the 100-*D* CEC 2014 benchmark problems:

(1) From the averaged rank obtained from the Friedman test, it is observed that ACSEDA still obtains the smallest rank value among all algorithms. This means that ACSEDA consistently achieves the best overall optimization performance on the 100-*D* CEC 2014 benchmark set.

(2) According to the results of the Wilcoxon rank sum test, ACSEDA presents great dominance to EDA_{VERS}, EDA/LS, EDA/LS-MS, and TRA-EDA on 23, 26, 24, and 28 problems, respectively. In comparison with EDA2 and BUMDA, ACSEDA obtains competitive or even better performance on 20 and 22 problems, respectively. Compared with MA-ES, ACSEDA achieves much better performance on 15 problems and presents inferiority on 15 problems as well. This indicates that ACSEDA is very competitive to MA-ES on the 100-*D* CEC2014 benchmark problems.

(3) Concerning the optimization performance on different kinds of optimization problems, on the three unimodal problems, ACSEDA outperforms EDA_{VERS}, EDA/LS-MS, TRA-EDA, and BUMDA on all these three problems, and it performs much better than EDA/LS on two problems. However, it loses the competition on these three problems to both EDA2 and MA-ES. When it comes to the 13 simple multimodal functions, ACSEDA shows significantly better performance than EDA_{VERS}, EDA/LS, EDA/LS-MS, TRA-EDA, and MA-ES on at least 10 problems, and presents great superiority to EDA2 on 8 problems. On these 13 problems, ACSEDA and BUMDA achieve very similar performance. Encountered with the six hybrid problems, ACSEDA exhibits much better performance than the compared EDA variants on at least five problems, except for EDA2. Faced with the eight composition problems, ACSEDA is better than EDA/LS, EDA/LS-MS, TRA-EDA, and BUMDA on at least five problems, presents very competitive performance with EDA2 and EDA_{VERS}, and is only inferior to MA-ES.

(4) To conclude, encountered with the 100-*D* CEC 2014 problems, ACSEDA still shows great superiority to the compared state-of-the-art EDA variants in solving such high-dimensional problems. In particular, on the complicated problems with such high dimensionality, such as the hybrid problems and the composition problems, ACSEDA still presents significant dominance to most of the compared algorithms. This further demonstrates that the proposed ACSEDA is promising for optimization problems.

Comprehensively speaking, from the above comparisons, we can see that ACSEDA consistently exhibits great superiority to the compared state-of-the-art EDA variants on the CEC 2014 benchmark problem set with different dimension sizes. This demonstrates that ACSEDA is promising for both simple unimodal problems and complicated multimodal problems. Besides, it also preserves good scalability in solving optimization problems.

The above demonstrated superiority of the proposed ACSEDA mainly benefits from the proposed three techniques. With the cohesive cooperation of them, ACSEDA could strike a promising balance between exploration and exploitation to search the complicated solution space properly.

4.3. Deep Investegation on ACSEDA

From the above comparison experiments, we can see that ACSEDA shows great dominance over the compared state-of-the-art EDA variants. In this section, we take a deep observation on ACSEDA to investigate the influence of each component, so it is clear to see what contributes to the promising performance of ACSEDA.

4.3.1. Effectiveness of the Covariance Scaling Strategy

First, we conduct experiments to verify the effectiveness of the proposed covariance scaling strategy, which is realized by setting a larger cs (to estimate the covariance) than the selection ratio sr (to estimate the mean vector). To this end, first, we fix different sr. Then, based on each fixed sr, we set different cs, each of which is larger than the associated sr. Subsequently, based on the above settings of sr and cs, we conduct experiments on the 50-D CEC 2014 benchmark problems. Table 5 shows the comparison results among ACSEDA with different settings of the selection ratio (sr) and different settings of the covariance scaling (cs) parameter on the 50-D CEC 2014 benchmark problems. In this table, the best results are highlighted in bold in each part associated with each fixed sr. In addition, the averaged ranks of each cs, in each part obtained from the Friedman test, are listed in the last row of the table.

From Table 5, we can get the following findings:

(1) With respect to the comparison results of each part, it is found that a larger cs than sr helps ACSEDA achieve much better performance than the one with $cs = sr$. In particular, it is interesting to find that the superiority of the ACSEDA with a larger cs (than sr) to the one without the covariance scaling ($cs = sr$) is particularly significant in solving complicated problems such as F_{20}-F_{30}. This demonstrates the covariance scaling technique is helpful for ACSEDA to obtain promising performance in solving optimization problems, especially on complicated problems.

(2) It is also interesting to find that neither a too small cs, nor a too large cs (compared with sr) are suitable for ACSEDA to achieve promising performance. For instance, when $sr = 0.1$, we find that, though ACSEDA with a too-small cs ($cs \leq 0.4$) achieves much better performance than the one with $cs = sr$ (namely without covariance scaling), its performance is much worse than the ones with a larger cs ($0.4 < cs < 0.8$). On the contrary, when ACSEDA has a too large cs ($cs \geq 0.8$), its performance degrades dramatically, as compared with the ones with a proper cs. A similar situation occurs to ACSEDA with the other settings of sr when cs is either too large or too small. Such experimental results verify the analysis presented in Section 3.1.

To sum up, it is found that the proposed covariance scaling strategy is effective to help EDA achieve promising performance in solving optimization problems, especially complicated problems.

Table 5. Comparison among ACSEDA, with different settings of the selection ratio (*sr*) and different settings of the covariance scaling (*cs*) parameter, on the 50-*D* CEC 2014 benchmark problems. The best results in each part are highlighted in bold in this table.

F	$sr = 0.1$										$sr = 0.2$					$sr = 0.3$				$sr = 0.4$			$sr = 0.5$	
	$cs = 0.1$, $cs/sr = 1.0$	$cs = 0.2$, $cs/sr = 2.0$	$cs = 0.3$, $cs/sr = 3.0$	$cs = 0.4$, $cs/sr = 4.0$	$cs = 0.5$, $cs/sr = 5.0$	$cs = 0.6$, $cs/sr = 6.0$	$cs = 0.7$, $cs/sr = 7.0$	$cs = 0.8$, $cs/sr = 8.0$	$cs = 0.9$, $cs/sr = 9.0$	$cs = 1.0$, $cs/sr = 10.0$	$cs = 0.2$, $cs/sr = 1.0$	$cs = 0.4$, $cs/sr = 2.0$	$cs = 0.6$, $cs/sr = 3.0$	$cs = 0.8$, $cs/sr = 4.0$	$cs = 1.0$, $cs/sr = 5.0$	$cs = 0.3$, $cs/sr = 1.0$	$cs = 0.6$, $cs/sr = 2.0$	$cs = 0.9$, $cs/sr = 3.0$	$cs = 1.0$, $cs/sr = 3.3$	$cs = 0.4$, $cs/sr = 1.0$	$cs = 0.8$, $cs/sr = 2.0$	$cs = 1.0$, $cs/sr = 2.5$	$cs = 0.5$, $cs/sr = 1.0$	$cs = 1.0$, $cs/sr = 2.0$
F_1	2.32×10^{8}	2.33×10^{7}	1.44×10^{5}	3.43×10^{3}	**5.87×10^{-14}**	5.74×10^{-10}	4.28×10^{-6}	2.57×10^{-2}	5.70×10^{2}	1.03×10^{8}	4.00×10^{8}	3.98×10^{6}	4.67×10^{1}	**1.87×10^{-2}**	9.68×10^{7}	6.93×10^{8}	9.64×10^{4}	**2.52×10^{3}**	1.02×10^{8}	1.05×10^{9}	**3.46×10^{3}**	1.26×10^{8}	1.44×10^{9}	**1.47×10^{8}**
F_2	3.39×10^{10}	1.41×10^{10}	5.17×10^{8}	9.41×10^{1}	**9.35×10^{-12}**	1.08×10^{-7}	7.97×10^{-4}	5.14×10^{0}	4.17×10^{4}	2.97×10^{9}	3.94×10^{10}	7.22×10^{8}	**1.87×10^{-2}**	4.16×10^{0}	2.34×10^{9}	4.71×10^{10}	3.71×10^{8}	**2.33×10^{4}**	2.15×10^{9}	5.63×10^{10}	**9.39×10^{2}**	2.09×10^{9}	6.48×10^{10}	**2.01×10^{9}**
F_3	3.47×10^{4}	1.30×10^{4}	4.03×10^{2}	**0.00×10^{0}**	**0.00×10^{0}**	2.80×10^{-13}	2.45×10^{-9}	1.77×10^{-5}	1.75×10^{-1}	1.13×10^{5}	3.72×10^{4}	5.53×10^{3}	**1.71×10^{-13}**	1.02×10^{-5}	1.13×10^{5}	4.33×10^{4}	2.81×10^{2}	**8.26×10^{-2}**	1.17×10^{5}	4.81×10^{4}	**9.73×10^{-6}**	1.17×10^{5}	**5.47×10^{4}**	1.17×10^{5}
F_4	4.69×10^{3}	1.46×10^{3}	1.36×10^{2}	9.02×10^{1}	**8.96×10^{1}**	9.26×10^{1}	9.42×10^{1}	9.65×10^{1}	9.73×10^{1}	3.38×10^{2}	5.58×10^{3}	6.39×10^{2}	**9.31×10^{1}**	9.53×10^{1}	3.31×10^{2}	6.81×10^{3}	1.33×10^{2}	**9.55×10^{1}**	3.52×10^{2}	8.82×10^{3}	**9.32×10^{0}**	3.76×10^{2}	1.06×10^{4}	**4.20×10^{2}**
F_5	**2.11×10^{1}**	**2.11×10^{1}**	**2.11×10^{1}**	**2.11×10^{1}**	**2.11×10^{1}**	**2.11×10^{1}**	**2.11×10^{1}**	**2.11×10^{1}**	**2.11×10^{1}**	**2.11×10^{1}**	**2.11×10^{1}**	**2.11×10^{1}**	**2.11×10^{1}**	**2.11×10^{1}**	**2.11×10^{1}**	**2.11×10^{1}**	**2.11×10^{1}**	**2.11×10^{1}**	**2.11×10^{1}**	**2.11×10^{1}**	**2.11×10^{1}**	**2.11×10^{1}**	**2.11×10^{1}**	**2.11×10^{1}**
F_6	1.91×10^{1}	8.04×10^{0}	1.36×10^{0}	1.65×10^{-1}	6.99×10^{-2}	**1.89×10^{-2}**	3.50×10^{-2}	3.66×10^{-1}	1.80×10^{0}	3.38×10^{1}	1.39×10^{1}	2.32×10^{0}	**1.21×10^{-1}**	1.46×10^{-1}	3.12×10^{1}	1.44×10^{1}	**1.06×10^{-1}**	1.51×10^{0}	3.06×10^{1}	1.58×10^{1}	**1.50×10^{-1}**	3.06×10^{1}	**1.76×10^{1}**	3.13×10^{1}
F_7	3.19×10^{2}	1.29×10^{2}	6.81×10^{0}	**0.00×10^{0}**	**0.00×10^{0}**	1.17×10^{-13}	1.03×10^{-9}	6.92×10^{-6}	6.71×10^{-2}	3.39×10^{1}	3.68×10^{2}	6.40×10^{1}	**1.14×10^{-13}**	4.14×10^{-6}	2.64×10^{1}	4.33×10^{2}	7.17×10^{0}	**3.47×10^{-2}**	2.46×10^{1}	5.12×10^{2}	**4.55×10^{-6}**	2.45×10^{1}	5.92×10^{2}	**2.44×10^{1}**
F_8	1.61×10^{2}	8.34×10^{1}	3.44×10^{1}	1.63×10^{1}	8.86×10^{0}	6.00×10^{0}	4.88×10^{0}	**3.39×10^{0}**	1.71×10^{1}	3.02×10^{2}	1.26×10^{2}	4.67×10^{1}	1.19×10^{1}	**4.02×10^{0}**	2.97×10^{2}	1.14×10^{2}	2.81×10^{1}	**1.09×10^{1}**	2.96×10^{2}	1.15×10^{2}	**1.56×10^{1}**	2.96×10^{2}	**1.29×10^{2}**	2.96×10^{2}
F_9	1.60×10^{2}	7.09×10^{1}	2.67×10^{1}	9.24×10^{0}	6.53×10^{0}	4.11×10^{0}	**3.68×10^{0}**	3.79×10^{0}	1.15×10^{2}	3.32×10^{2}	1.09×10^{2}	3.34×10^{1}	6.37×10^{0}	**2.70×10^{0}**	3.27×10^{2}	1.02×10^{2}	**1.93×10^{1}**	3.32×10^{1}	3.26×10^{2}	1.07×10^{2}	**9.42×10^{0}**	3.21×10^{2}	**1.23×10^{2}**	3.26×10^{2}
F_{10}	4.63×10^{3}	2.46×10^{3}	1.25×10^{3}	8.25×10^{2}	5.36×10^{2}	3.14×10^{2}	2.76×10^{2}	**1.30×10^{2}**	3.55×10^{2}	8.99×10^{3}	3.26×10^{3}	1.37×10^{3}	7.34×10^{2}	**2.96×10^{2}**	8.83×10^{3}	2.79×10^{3}	1.05×10^{3}	**9.53×10^{2}**	9.21×10^{3}	2.70×10^{3}	**1.23×10^{3}**	9.28×10^{3}	**2.86×10^{3}**	1.01×10^{4}
F_{11}	4.85×10^{3}	2.46×10^{3}	1.31×10^{3}	8.47×10^{2}	6.18×10^{2}	4.83×10^{2}	3.77×10^{2}	**3.00×10^{2}**	3.22×10^{3}	1.05×10^{4}	3.24×10^{3}	1.33×10^{3}	7.81×10^{2}	**5.04×10^{2}**	1.05×10^{4}	2.87×10^{3}	**1.26×10^{3}**	3.65×10^{3}	1.05×10^{4}	2.73×10^{3}	**1.23×10^{3}**	1.08×10^{4}	**2.66×10^{3}**	1.10×10^{4}
F_{12}	**1.14×10^{0}**	3.17×10^{0}	3.21×10^{0}	3.20×10^{0}	3.15×10^{0}	3.28×10^{0}	3.23×10^{0}	3.18×10^{0}	3.23×10^{0}	3.42×10^{0}	3.21×10^{0}	3.28×10^{0}	**3.20×10^{0}**	3.30×10^{0}	**3.20×10^{0}**	3.27×10^{0}	3.28×10^{0}	**3.22×10^{0}**	3.35×10^{0}	3.26×10^{0}	3.28×10^{0}	**3.25×10^{0}**	**3.25×10^{0}**	3.33×10^{0}
F_{13}	3.81×10^{0}	1.53×10^{0}	1.78×10^{-1}	1.43×10^{-1}	1.40×10^{-1}	1.30×10^{-1}	1.08×10^{-1}	8.13×10^{-2}	**6.16×10^{-2}**	4.44×10^{-1}	4.01×10^{0}	3.50×10^{-1}	1.33×10^{-1}	**9.14×10^{-2}**	3.78×10^{-1}	4.38×10^{0}	1.63×10^{-1}	**6.22×10^{-2}**	4.16×10^{-1}	4.74×10^{0}	**1.22×10^{-1}**	4.24×10^{-1}	5.13×10^{0}	**4.27×10^{-1}**
F_{14}	6.72×10^{1}	2.05×10^{1}	2.85×10^{-1}	3.37×10^{-1}	3.30×10^{-1}	3.15×10^{-1}	2.93×10^{-1}	**2.61×10^{-1}**	2.69×10^{-1}	8.46×10^{0}	8.23×10^{1}	8.36×10^{-1}	3.20×10^{-1}	**2.98×10^{-1}**	5.38×10^{0}	9.72×10^{1}	**3.02×10^{-1}**	3.06×10^{-1}	3.74×10^{0}	1.14×10^{2}	**3.26×10^{-1}**	4.16×10^{0}	1.37×10^{2}	**3.64×10^{0}**
F_{15}	5.07×10^{3}	5.31×10^{1}	5.65×10^{0}	**4.75×10^{0}**	4.94×10^{0}	4.77×10^{0}	7.47×10^{0}	1.38×10^{1}	2.54×10^{1}	8.19×10^{1}	7.77×10^{3}	6.69×10^{0}	**5.02×10^{0}**	1.30×10^{1}	6.93×10^{1}	1.20×10^{4}	**5.41×10^{0}**	2.34×10^{1}	6.93×10^{1}	2.26×10^{4}	**1.27×10^{1}**	7.29×10^{1}	4.24×10^{4}	**8.86×10^{1}**
F_{16}	1.72×10^{1}	**1.58×10^{1}**	1.84×10^{1}	1.91×10^{1}	2.03×10^{1}	2.02×10^{1}	2.02×10^{1}	2.01×10^{1}	2.02×10^{1}	2.03×10^{1}	**1.74×10^{1}**	1.97×10^{1}	2.07×10^{1}	2.08×10^{1}	2.05×10^{1}	**2.00×10^{1}**	2.13×10^{1}	2.12×10^{1}	2.10×10^{1}	**2.12×10^{1}**	2.20×10^{1}	2.16×10^{1}	2.24×10^{1}	**2.24×10^{1}**
F_{17}	1.18×10^{7}	6.54×10^{2}	3.78×10^{2}	2.87×10^{2}	2.51×10^{2}	2.42×10^{2}	**2.30×10^{2}**	3.50×10^{2}	7.20×10^{2}	5.30×10^{6}	2.58×10^{7}	3.90×10^{2}	**3.20×10^{2}**	4.51×10^{2}	6.01×10^{6}	4.60×10^{7}	**4.82×10^{2}**	9.18×10^{2}	6.59×10^{6}	8.43×10^{7}	**6.54×10^{2}**	7.59×10^{6}	1.23×10^{8}	**9.22×10^{6}**
F_{18}	2.44×10^{8}	1.30×10^{2}	9.18×10^{1}	4.83×10^{1}	2.76×10^{1}	2.44×10^{1}	**2.42×10^{1}**	2.63×10^{1}	1.63×10^{2}	1.72×10^{8}	8.31×10^{8}	8.80×10^{1}	**3.95×10^{1}**	4.11×10^{1}	1.40×10^{8}	1.77×10^{9}	**7.36×10^{1}**	1.66×10^{2}	1.32×10^{8}	3.13×10^{9}	**7.62×10^{1}**	1.27×10^{8}	4.56×10^{9}	**1.22×10^{8}**
F_{19}	8.07×10^{1}	3.42×10^{1}	2.69×10^{1}	1.45×10^{1}	1.23×10^{1}	**1.19×10^{1}**	1.22×10^{1}	1.23×10^{1}	1.31×10^{1}	8.38×10^{1}	9.25×10^{1}	3.61×10^{1}	1.44×10^{1}	**1.23×10^{1}**	8.69×10^{1}	1.34×10^{2}	3.42×10^{1}	**1.30×10^{1}**	8.90×10^{1}	2.11×10^{2}	**1.62×10^{1}**	9.02×10^{1}	3.22×10^{2}	**9.37×10^{1}**
F_{20}	7.88×10^{2}	9.08×10^{1}	3.64×10^{1}	9.82×10^{0}	5.40×10^{0}	3.19×10^{0}	**2.71×10^{0}**	4.25×10^{0}	3.90×10^{1}	2.17×10^{4}	1.21×10^{3}	4.07×10^{1}	7.49×10^{0}	**5.36×10^{0}**	2.97×10^{4}	1.69×10^{3}	**2.34×10^{1}**	3.38×10^{1}	3.55×10^{4}	2.51×10^{3}	**2.10×10^{1}**	3.88×10^{4}	**3.84×10^{3}**	3.82×10^{4}
F_{21}	1.22×10^{4}	9.05×10^{2}	4.28×10^{2}	3.15×10^{2}	**2.61×10^{2}**	2.62×10^{2}	2.84×10^{2}	3.24×10^{2}	5.40×10^{2}	3.39×10^{6}	1.45×10^{4}	4.69×10^{2}	**3.00×10^{2}**	3.99×10^{2}	4.06×10^{6}	2.85×10^{4}	**4.04×10^{2}**	7.46×10^{2}	4.15×10^{6}	5.22×10^{4}	**5.38×10^{2}**	4.96×10^{6}	**1.19×10^{5}**	5.55×10^{6}
F_{22}	1.48×10^{2}	6.22×10^{1}	5.03×10^{1}	**3.77×10^{1}**	3.87×10^{1}	5.43×10^{1}	5.87×10^{1}	6.81×10^{1}	8.39×10^{1}	8.07×10^{2}	1.43×10^{2}	7.32×10^{1}	**6.23×10^{1}**	7.91×10^{1}	8.02×10^{2}	3.79×10^{2}	**4.38×10^{1}**	1.20×10^{2}	8.64×10^{2}	1.12×10^{3}	**8.50×10^{1}**	8.52×10^{2}	3.10×10^{3}	**8.66×10^{2}**
F_{23}	4.62×10^{2}	3.93×10^{2}	3.52×10^{2}	**3.44×10^{2}**	**3.44×10^{2}**	**3.44×10^{2}**	**3.44×10^{2}**	**3.44×10^{2}**	**3.44×10^{2}**	**3.94×10^{2}**	4.53×10^{2}	3.75×10^{2}	**3.44×10^{2}**	**3.44×10^{2}**	4.02×10^{2}	4.62×10^{2}	3.52×10^{2}	**3.44×10^{2}**	4.14×10^{2}	4.94×10^{2}	**3.44×10^{2}**	4.24×10^{2}	5.24×10^{2}	**4.44×10^{2}**
F_{24}	2.80×10^{2}	2.74×10^{2}	2.73×10^{2}	2.72×10^{2}	2.72×10^{2}	2.70×10^{2}	2.70×10^{2}	2.69×10^{2}	2.70×10^{2}	3.27×10^{2}	2.76×10^{2}	2.73×10^{2}	2.73×10^{2}	**2.71×10^{2}**	3.22×10^{2}	2.75×10^{2}	**2.73×10^{2}**	**2.73×10^{2}**	3.21×10^{2}	**2.74×10^{2}**	**2.74×10^{2}**	3.21×10^{2}	**2.72×10^{2}**	3.22×10^{2}
F_{25}	2.22×10^{2}	2.15×10^{2}	2.09×10^{2}	**2.05×10^{2}**	**2.05×10^{2}**	**2.05×10^{2}**	**2.05×10^{2}**	**2.05×10^{2}**	**2.05×10^{2}**	**2.36×10^{2}**	2.20×10^{2}	2.13×10^{2}	**2.05×10^{2}**	**2.05×10^{2}**	2.47×10^{2}	2.21×10^{2}	2.09×10^{2}	**2.05×10^{2}**	2.51×10^{2}	2.20×10^{2}	**2.05×10^{2}**	2.51×10^{2}	**2.19×10^{2}**	2.51×10^{2}
F_{26}	1.27×10^{2}	1.11×10^{2}	1.04×10^{2}	1.01×10^{2}	1.03×10^{2}	1.00×10^{2}	**1.00×10^{2}**	**1.00×10^{2}**	**1.00×10^{2}**	**1.62×10^{2}**	1.33×10^{2}	1.17×10^{2}	**1.05×10^{2}**	**1.05×10^{2}**	1.85×10^{2}	1.38×10^{2}	1.19×10^{2}	**1.06×10^{2}**	1.91×10^{2}	1.64×10^{2}	**1.08×10^{2}**	1.99×10^{2}	**1.71×10^{2}**	1.80×10^{2}
F_{27}	1.05×10^{3}	6.94×10^{2}	4.08×10^{2}	3.37×10^{2}	**3.29×10^{2}**	3.36×10^{2}	3.30×10^{2}	3.34×10^{2}	3.81×10^{2}	1.08×10^{3}	9.37×10^{2}	5.39×10^{2}	**3.22×10^{2}**	3.29×10^{2}	1.03×10^{3}	9.48×10^{2}	3.93×10^{2}	**3.62×10^{2}**	1.02×10^{3}	1.02×10^{3}	**3.24×10^{2}**	1.02×10^{3}	1.10×10^{3}	**1.04×10^{3}**
F_{28}	2.40×10^{3}	1.73×10^{3}	1.51×10^{3}	**1.16×10^{3}**	1.17×10^{3}	1.18×10^{3}	1.21×10^{3}	1.23×10^{3}	1.28×10^{3}	1.47×10^{3}	2.34×10^{3}	1.47×10^{3}	**1.16×10^{3}**	1.25×10^{3}	1.51×10^{3}	2.43×10^{3}	1.48×10^{3}	**1.30×10^{3}**	1.56×10^{3}	2.15×10^{3}	**1.23×10^{3}**	1.56×10^{3}	2.33×10^{3}	**1.56×10^{3}**
F_{29}	1.24×10^{4}	2.66×10^{3}	1.97×10^{3}	1.20×10^{3}	**8.88×10^{2}**	9.91×10^{2}	1.29×10^{3}	1.71×10^{3}	4.11×10^{3}	1.07×10^{6}	1.16×10^{4}	1.64×10^{3}	**9.69×10^{2}**	1.62×10^{3}	1.70×10^{6}	3.79×10^{4}	**1.59×10^{3}**	4.49×10^{3}	5.00×10^{6}	4.45×10^{4}	**1.24×10^{3}**	4.60×10^{6}	**1.15×10^{5}**	7.17×10^{6}
F_{30}	1.68×10^{5}	3.69×10^{4}	1.66×10^{4}	1.06×10^{4}	9.03×10^{3}	**8.77×10^{3}**	9.08×10^{3}	9.47×10^{3}	9.29×10^{3}	3.70×10^{4}	2.36×10^{5}	2.54×10^{4}	9.83×10^{3}	**8.95×10^{3}**	5.48×10^{4}	3.11×10^{5}	1.73×10^{4}	**9.12×10^{3}**	5.35×10^{4}	5.07×10^{5}	**9.99×10^{3}**	6.20×10^{4}	6.99×10^{5}	**7.81×10^{4}**
Rank	8.83	7.43	6.03	4.53	3.25	3.25	**3.22**	3.65	5.73	9.07	4.20	3.00	**1.62**	1.98	4.20	3.30	1.70	**1.60**	3.40	2.43	**1.20**	2.37	1.57	**1.43**

4.3.2. Effectiveness of the Proposed Adaptive Covariance Scaling Strategy

Then, we conduct experiments to verify the effectiveness of the proposed adaptive covariance scaling method, which is realized by dynamically adjusting *cs* according to Equation (10). Since the proposed adaptive covariance scaling method is related to *sr*, as can be seen from Equation (10), we first fix *sr* as 0.1 and 0.2 to investigate the effectiveness of the proposed adaptive method. These two settings of *sr* are utilized because, from the experimental result in the last subsection as shown in Table 5, when *sr* is larger than 0.3, ACSEDA achieves much worse performance and thus, it is meaningless to investigate the effectiveness of the proposed adaptive *cs* when *sr* is larger than 0.3. Subsequently, for each set of *sr*, we set different fixed *cs* for ACSEDA and then compare them with the ACSEDA with the adaptive *cs* strategy.

Table 6 presents the comparison between ACSEDA with the adaptive covariance scaling method and the ones with different fixed settings of *cs* on the 50-*D* CEC 2014 benchmark problems. From this table, we attain the following observations:

(1) As a whole, no matter if it is from the perspective of the averaged rank obtained from the Friedman test or from the perspective of the number of problems where the algorithm achieves the best results, the ACSEDA with the proposed adaptive *cs* obtains much better performance than those with different fixed *cs*. This verifies that the proposed adaptive *cs* strategy is effective to help ACSEDA achieve promising performance.

(2) Under the same set of *sr*, we find that the optimal fixed *cs* for ACSEDA to achieve the best performance is different on different problems. This indicates that the optimal setting of *cs* is not consistent for all problems. With the proposed adaptive strategy, we can see that not only is the sensitivity of ACSEDA to *cs* alleviated, but its optimization performance is also largely promoted.

In summary, based on the above experiments, we can see that the proposed covariance scaling strategy is very beneficial for ACSEDA to achieve promising performance. This is mainly because the proposed adaptive strategy helps ACSEDA bias to explore the solution space in the early stage and, gradually, bias to exploit the found promising areas without serious loss of diversity as the evolution iterates. As a result, with this adaptive strategy, ACSEDA could explore and exploit the solution space properly to find the optima of optimization problems.

4.3.3. Effectiveness of the Proposed Adaptive Promising Selection Strategy

Subsequently, we conduct experiments to verify the effectiveness of the proposed promising individual selection strategy, which is realized by dynamically adjusting the parameter *sr*, based on Equation (11). Since the setting of *sr* influences the covariance scaling parameter *cs*, we first fix *cs* as 0.6 and then accordingly set different fixed *sr*. It should be noticed that $cs = 0.6$ is adopted here because, in the last subsection, as shown in Table 6, ACSEDA obtains the best overall performance when $cs = 0.6$ under the two settings of *sr*. Then, we compare the ACSEDA with the proposed adaptive *sr* and the ones with different fixed *sr* under the same set of *cs* (namely $cs = 0.6$).

Table 6. Comparison between ACSEDA, with and without the adaptive covariance scaling method, on the 50-*D* CEC 2014 benchmark problems. The best results in each part are highlighted in bold in this table.

	$sr = 0.1$											$sr = 0.2$					
F	$cs = 0.1$	$cs = 0.2$	$cs = 0.3$	$cs = 0.4$	$cs = 0.5$	$cs = 0.6$	$cs = 0.7$	$cs = 0.8$	$cs = 0.9$	$cs = 1.0$	Adaptive $- cs$	$cs = 0.2$	$cs = 0.4$	$cs = 0.6$	$cs = 0.8$	$cs = 1.0$	Adaptive $- cs$
F_1	2.32×10^{8}	2.33×10^{7}	1.44×10^{5}	3.43×10^{3}	5.87×10^{-14}	5.74×10^{-10}	4.28×10^{-6}	2.57×10^{-2}	5.70×10^{2}	1.03×10^{8}	**1.47×10^{-14}**	4.00×10^{8}	3.98×10^{6}	4.67×10^{1}	1.87×10^{-2}	9.68×10^{7}	**2.53×10^{-13}**
F_2	3.39×10^{8}	1.41×10^{10}	5.17×10^{8}	9.41×10^{1}	9.35×10^{-12}	1.08×10^{-7}	7.97×10^{-4}	5.14×10^{0}	4.17×10^{4}	2.97×10^{9}	**2.04×10^{-12}**	3.94×10^{10}	7.22×10^{9}	1.87×10^{-2}	4.16×10^{0}	2.34×10^{9}	**3.83×10^{-11}**
F_3	3.47×10^{4}	1.30×10^{4}	4.03×10^{2}	**0.00×10^{0}**	**0.00×10^{0}**	2.80×10^{-13}	2.45×10^{-9}	1.77×10^{-5}	1.75×10^{-1}	1.13×10^{5}	**0.00×10^{0}**	3.72×10^{4}	5.53×10^{3}	1.71×10^{-14}	1.02×10^{-5}	1.13×10^{5}	**0.00×10^{0}**
F_4	4.69×10^{3}	1.46×10^{3}	1.36×10^{2}	9.02×10^{1}	**8.96×10^{1}**	9.26×10^{1}	9.42×10^{1}	9.65×10^{1}	9.73×10^{1}	3.38×10^{2}	9.31×10^{1}	5.58×10^{3}	6.39×10^{2}	**9.31×10^{1}**	9.53×10^{1}	3.31×10^{2}	9.35×10^{1}
F_5	**2.11×10^{1}**	**2.11×10^{1}**	**2.11×10^{1}**	**2.11×10^{1}**	**2.11×10^{1}**	**2.11×10^{1}**	**2.11×10^{1}**	**2.11×10^{1}**	**2.11×10^{1}**	**2.11×10^{1}**	**2.11×10^{1}**	**2.11×10^{1}**	**2.11×10^{1}**	**2.11×10^{1}**	**2.11×10^{1}**	**2.11×10^{1}**	**2.11×10^{1}**
F_6	1.91×10^{1}	8.04×10^{0}	1.36×10^{0}	1.65×10^{-1}	6.99×10^{-2}	1.89×10^{-2}	3.50×10^{-2}	3.66×10^{-1}	1.80×10^{0}	3.38×10^{1}	**1.74×10^{-2}**	1.39×10^{1}	2.32×10^{0}	1.21×10^{-1}	1.46×10^{-1}	3.12×10^{1}	**2.20×10^{-4}**
F_7	3.19×10^{2}	1.29×10^{2}	6.81×10^{0}	0.00×10^{0}	**0.00×10^{0}**	1.17×10^{-13}	1.03×10^{-9}	6.92×10^{-6}	6.71×10^{-2}	3.39×10^{1}	**0.00×10^{0}**	3.68×10^{2}	6.40×10^{1}	1.14×10^{-13}	4.14×10^{-6}	2.64×10^{1}	**0.00×10^{0}**
F_8	1.61×10^{2}	8.34×10^{1}	3.44×10^{1}	1.63×10^{1}	8.86×10^{0}	6.00×10^{0}	4.88×10^{0}	3.39×10^{0}	1.71×10^{1}	3.02×10^{2}	**3.05×10^{0}**	1.26×10^{2}	4.67×10^{1}	1.19×10^{1}	4.02×10^{0}	2.97×10^{2}	**2.35×10^{0}**
F_9	1.60×10^{2}	7.09×10^{1}	2.67×10^{1}	9.24×10^{0}	6.53×10^{0}	4.11×10^{0}	3.68×10^{0}	3.79×10^{0}	1.15×10^{2}	3.32×10^{2}	**2.06×10^{0}**	1.09×10^{2}	3.34×10^{1}	6.37×10^{0}	2.70×10^{0}	3.27×10^{2}	**1.86×10^{0}**
F_{10}	4.63×10^{3}	2.46×10^{3}	1.25×10^{3}	8.25×10^{2}	5.36×10^{2}	3.14×10^{2}	2.76×10^{2}	1.30×10^{2}	3.55×10^{2}	8.99×10^{3}	**7.00×10^{1}**	3.26×10^{3}	1.37×10^{3}	7.34×10^{2}	2.96×10^{2}	8.83×10^{3}	**1.12×10^{2}**
F_{11}	4.85×10^{3}	2.46×10^{3}	1.31×10^{3}	8.47×10^{2}	6.18×10^{2}	4.83×10^{2}	3.77×10^{2}	3.00×10^{2}	3.22×10^{3}	1.05×10^{4}	**1.22×10^{2}**	3.24×10^{3}	1.33×10^{3}	7.81×10^{2}	5.04×10^{2}	1.05×10^{4}	**1.72×10^{2}**
F_{12}	**1.14×10^{0}**	3.17×10^{0}	3.21×10^{0}	3.20×10^{0}	3.15×10^{0}	3.28×10^{0}	3.23×10^{0}	3.18×10^{0}	3.23×10^{0}	3.42×10^{0}	3.27×10^{0}	3.21×10^{0}	3.28×10^{0}	3.20×10^{0}	3.30×10^{0}	3.20×10^{0}	**3.16×10^{0}**
F_{13}	3.81×10^{0}	1.53×10^{0}	1.78×10^{-1}	1.43×10^{-1}	1.40×10^{-1}	1.30×10^{-1}	1.08×10^{-1}	8.13×10^{-2}	**6.16×10^{-2}**	4.44×10^{-1}	1.47×10^{-1}	4.01×10^{0}	3.50×10^{-1}	1.33×10^{-1}	**9.14×10^{-2}**	3.78×10^{-1}	1.67×10^{-1}
F_{14}	6.72×10^{1}	2.05×10^{1}	2.85×10^{-1}	3.37×10^{-1}	3.30×10^{-1}	3.15×10^{-1}	2.93×10^{-1}	**2.61×10^{-1}**	2.69×10^{-1}	8.46×10^{0}	2.81×10^{-1}	8.23×10^{1}	8.36×10^{-1}	3.20×10^{-1}	2.98×10^{-1}	5.38×10^{0}	**2.96×10^{-1}**
F_{15}	5.07×10^{3}	5.31×10^{1}	5.65×10^{0}	**4.75×10^{0}**	4.94×10^{0}	4.77×10^{0}	7.47×10^{0}	1.38×10^{1}	2.54×10^{1}	8.19×10^{1}	4.78×10^{0}	7.77×10^{3}	6.69×10^{0}	**5.02×10^{0}**	1.30×10^{1}	6.93×10^{1}	5.04×10^{0}
F_{16}	1.72×10^{1}	**1.58×10^{1}**	1.84×10^{1}	1.91×10^{1}	2.03×10^{1}	2.02×10^{1}	2.02×10^{1}	2.01×10^{1}	2.02×10^{1}	2.03×10^{1}	1.80×10^{1}	**1.74×10^{1}**	1.97×10^{1}	2.07×10^{1}	2.08×10^{1}	2.05×10^{1}	1.86×10^{1}
F_{17}	1.18×10^{7}	6.54×10^{2}	3.78×10^{2}	2.87×10^{2}	2.51×10^{2}	2.42×10^{2}	2.30×10^{2}	3.50×10^{2}	7.20×10^{2}	5.30×10^{6}	**1.28×10^{2}**	2.58×10^{7}	3.90×10^{2}	3.20×10^{2}	4.51×10^{2}	6.01×10^{6}	**1.40×10^{2}**
F_{18}	2.44×10^{8}	1.30×10^{2}	9.18×10^{1}	4.83×10^{1}	2.76×10^{1}	2.44×10^{1}	2.42×10^{1}	2.63×10^{1}	1.63×10^{2}	1.72×10^{8}	**7.69×10^{0}**	8.31×10^{8}	8.80×10^{1}	3.95×10^{1}	4.11×10^{1}	1.40×10^{8}	**1.40×10^{1}**
F_{19}	8.07×10^{1}	3.42×10^{1}	2.69×10^{1}	1.45×10^{1}	1.23×10^{1}	1.19×10^{1}	1.22×10^{1}	1.23×10^{1}	1.31×10^{1}	8.38×10^{1}	**1.11×10^{1}**	9.25×10^{1}	3.61×10^{1}	1.44×10^{1}	1.23×10^{1}	8.69×10^{1}	**1.12×10^{1}**
F_{20}	7.88×10^{2}	9.08×10^{1}	3.64×10^{1}	9.82×10^{0}	5.40×10^{0}	3.19×10^{0}	2.71×10^{0}	4.25×10^{0}	3.90×10^{1}	2.17×10^{4}	**2.04×10^{0}**	1.21×10^{3}	4.07×10^{1}	7.49×10^{0}	5.36×10^{0}	2.97×10^{4}	**2.17×10^{0}**
F_{21}	1.22×10^{4}	9.05×10^{2}	4.28×10^{2}	3.15×10^{2}	2.61×10^{2}	2.62×10^{2}	2.84×10^{2}	3.24×10^{2}	5.40×10^{2}	3.39×10^{6}	**2.12×10^{2}**	1.45×10^{4}	4.69×10^{2}	3.00×10^{2}	3.99×10^{2}	4.06×10^{6}	**2.13×10^{2}**
F_{22}	1.48×10^{2}	6.22×10^{1}	5.03×10^{1}	3.77×10^{1}	3.87×10^{1}	5.43×10^{1}	5.87×10^{1}	6.81×10^{1}	8.39×10^{1}	8.07×10^{2}	**3.70×10^{1}**	1.43×10^{2}	7.32×10^{1}	6.23×10^{1}	7.91×10^{1}	8.02×10^{2}	**3.70×10^{1}**
F_{23}	4.62×10^{2}	3.93×10^{2}	3.52×10^{2}	**3.44×10^{2}**	**3.44×10^{2}**	**3.44×10^{2}**	**3.44×10^{2}**	**3.44×10^{2}**	**3.44×10^{2}**	3.94×10^{2}	**3.44×10^{2}**	4.53×10^{2}	3.75×10^{2}	**3.44×10^{2}**	**3.44×10^{2}**	4.02×10^{2}	**3.44×10^{2}**
F_{24}	2.80×10^{2}	2.74×10^{2}	2.73×10^{2}	2.72×10^{2}	2.72×10^{2}	2.70×10^{2}	2.70×10^{2}	2.69×10^{2}	2.70×10^{2}	3.27×10^{2}	**2.68×10^{2}**	2.76×10^{2}	2.73×10^{2}	2.73×10^{2}	2.71×10^{2}	3.22×10^{2}	**2.69×10^{2}**
F_{25}	2.22×10^{2}	2.15×10^{2}	2.09×10^{2}	**2.05×10^{2}**	**2.05×10^{2}**	**2.05×10^{2}**	**2.05×10^{2}**	**2.05×10^{2}**	**2.05×10^{2}**	2.36×10^{2}	**2.05×10^{2}**	2.20×10^{2}	2.13×10^{2}	**2.05×10^{2}**	**2.05×10^{2}**	2.47×10^{2}	**2.05×10^{2}**
F_{26}	1.27×10^{2}	1.11×10^{2}	1.04×10^{2}	1.01×10^{2}	1.03×10^{2}	**1.00×10^{2}**	**1.00×10^{2}**	**1.00×10^{2}**	**1.00×10^{2}**	1.62×10^{2}	**1.00×10^{2}**	1.33×10^{2}	1.17×10^{2}	1.05×10^{2}	1.05×10^{2}	1.85×10^{2}	**1.00×10^{2}**
F_{27}	1.05×10^{3}	6.94×10^{2}	4.08×10^{2}	3.37×10^{2}	**3.29×10^{2}**	3.36×10^{2}	3.30×10^{2}	3.34×10^{2}	3.81×10^{2}	1.08×10^{3}	3.31×10^{2}	9.37×10^{2}	5.39×10^{2}	3.22×10^{2}	3.29×10^{2}	1.03×10^{3}	**3.21×10^{2}**
F_{28}	2.40×10^{3}	1.73×10^{3}	1.51×10^{3}	**1.16×10^{3}**	1.17×10^{3}	1.18×10^{3}	1.21×10^{3}	1.23×10^{3}	1.28×10^{3}	1.47×10^{3}	1.18×10^{3}	2.34×10^{3}	1.47×10^{3}	1.16×10^{3}	1.25×10^{3}	1.51×10^{3}	**1.14×10^{3}**
F_{29}	1.24×10^{4}	2.66×10^{3}	1.97×10^{3}	1.20×10^{3}	8.88×10^{2}	9.91×10^{2}	1.29×10^{3}	1.71×10^{3}	4.11×10^{3}	1.07×10^{6}	**8.77×10^{2}**	1.16×10^{4}	1.64×10^{3}	9.69×10^{2}	1.62×10^{3}	1.70×10^{6}	**9.02×10^{2}**
F_{30}	1.68×10^{5}	3.69×10^{4}	1.66×10^{4}	1.06×10^{4}	9.03×10^{3}	8.77×10^{3}	9.08×10^{3}	9.47×10^{3}	9.29×10^{3}	3.70×10^{4}	**8.60×10^{3}**	2.36×10^{5}	2.54×10^{4}	9.83×10^{3}	8.95×10^{3}	5.48×10^{4}	**8.51×10^{3}**
Rank	9.70	8.40	7.13	4.98	4.05	3.95	4.08	4.55	6.55	10.18	**2.42**	5.22	3.97	2.47	2.83	5.20	**1.32**

Table 7 shows the comparison results between the ACSEDA with the proposed adaptive sr and the ones with different fixed settings of sr on the 50-D CEC 2014 benchmark problems. From this table, the following findings can be attained:

(1) In view of the averaged rank obtained from the Friedman test, the ACSEDA with the adaptive sr achieves the best overall performance than the ones with different fixed sr. This verifies the effectiveness of the proposed adaptive promising individual selection strategy for the estimation of the mean vector.

(2) For different problems, the optimal sr is different for ACSEDA to achieve the best performance. In particular, we find that a small sr tends to help ACSEDA obtain better performance than a large sr. The proposed adaptive strategy, based on Equation (11), matches this observation that sr is dramatically decreased to a small value in the early stage, and then, it mildly declines as the evolution goes as stated in Section 3.2.

Table 7. Comparison between ACSEDA, with and without the adaptive promising individual selection method, on the 50-D CEC 2014 benchmark problems. The best results are highlighted in bold in this table.

				cs = 0.6				
F	sr = 0.05	sr = 0.10	sr = 0.15	sr = 0.20	sr = 0.25	sr = 0.30	sr = 0.35	Adaptive − sr
F_1	1.61×10^{-14}	1.42×10^{-14}	$\mathbf{1.28 \times 10^{-14}}$	7.55×10^{1}	2.52×10^{4}	4.67×10^{5}	1.75×10^{7}	1.42×10^{-14}
F_2	2.91×10^{-12}	1.08×10^{-12}	$\mathbf{9.56 \times 10^{-13}}$	2.02×10^{0}	1.65×10^{7}	2.91×10^{9}	1.20×10^{10}	2.22×10^{-12}
F_3	$\mathbf{0.00 \times 10^{0}}$	$\mathbf{0.00 \times 10^{0}}$	$\mathbf{0.00 \times 10^{0}}$	$\mathbf{0.00 \times 10^{0}}$	2.28×10^{0}	1.93×10^{3}	8.44×10^{3}	$\mathbf{0.00 \times 10^{0}}$
F_4	9.07×10^{1}	8.88×10^{1}	9.14×10^{1}	9.26×10^{1}	1.04×10^{2}	3.45×10^{2}	1.14×10^{3}	$\mathbf{8.67 \times 10^{1}}$
F_5	$\mathbf{2.11 \times 10^{1}}$	$\mathbf{2.11 \times 10^{1}}$	$\mathbf{2.11 \times 10^{1}}$	$\mathbf{2.11 \times 10^{1}}$	$\mathbf{2.11 \times 10^{1}}$	$\mathbf{2.11 \times 10^{1}}$	$\mathbf{2.11 \times 10^{1}}$	$\mathbf{2.11 \times 10^{1}}$
F_6	1.23×10^{-4}	7.35×10^{-5}	5.20×10^{-2}	$\mathbf{6.25 \times 10^{-5}}$	1.74×10^{-2}	1.64×10^{-2}	2.24×10^{-1}	9.75×10^{-5}
F_7	$\mathbf{0.00 \times 10^{0}}$	$\mathbf{0.00 \times 10^{0}}$	$\mathbf{0.00 \times 10^{0}}$	$\mathbf{0.00 \times 10^{0}}$	6.22×10^{-1}	3.07×10^{1}	1.13×10^{2}	$\mathbf{0.00 \times 10^{0}}$
F_8	7.36×10^{0}	$\mathbf{6.80 \times 10^{0}}$	8.66×10^{0}	1.43×10^{1}	1.95×10^{1}	2.63×10^{1}	3.48×10^{1}	9.78×10^{0}
F_9	6.20×10^{0}	$\mathbf{4.91 \times 10^{0}}$	6.10×10^{0}	7.16×10^{0}	1.20×10^{1}	1.71×10^{1}	2.40×10^{1}	6.70×10^{0}
F_{10}	$\mathbf{2.52 \times 10^{2}}$	3.05×10^{2}	4.49×10^{2}	5.11×10^{2}	7.11×10^{2}	8.08×10^{2}	7.93×10^{2}	3.22×10^{2}
F_{11}	$\mathbf{2.30 \times 10^{2}}$	3.36×10^{2}	4.32×10^{2}	4.96×10^{2}	5.80×10^{2}	7.30×10^{2}	6.84×10^{2}	3.57×10^{2}
F_{12}	3.27×10^{0}	3.27×10^{0}	3.21×10^{0}	3.20×10^{0}	$\mathbf{3.20 \times 10^{0}}$	3.30×10^{0}	3.31×10^{0}	3.25×10^{0}
F_{13}	$\mathbf{1.39 \times 10^{-1}}$	1.55×10^{-1}	1.60×10^{-1}	1.57×10^{-1}	1.53×10^{-1}	2.45×10^{-1}	8.50×10^{-1}	1.51×10^{-1}
F_{14}	2.85×10^{-1}	2.95×10^{-1}	3.17×10^{-1}	3.11×10^{-1}	3.16×10^{-1}	$\mathbf{2.77 \times 10^{-1}}$	1.25×10^{1}	3.09×10^{-1}
F_{15}	4.82×10^{0}	4.78×10^{0}	$\mathbf{4.77 \times 10^{0}}$	5.04×10^{0}	5.07×10^{0}	5.10×10^{0}	8.61×10^{0}	4.77×10^{0}
F_{16}	1.86×10^{1}	1.85×10^{1}	$\mathbf{1.84 \times 10^{1}}$	1.87×10^{1}	1.90×10^{1}	1.91×10^{1}	1.93×10^{1}	1.87×10^{1}
F_{17}	1.80×10^{2}	1.75×10^{2}	1.62×10^{2}	1.97×10^{2}	1.70×10^{2}	2.37×10^{2}	2.28×10^{2}	$\mathbf{1.44 \times 10^{2}}$
F_{18}	$\mathbf{1.35 \times 10^{1}}$	1.85×10^{1}	2.08×10^{1}	2.98×10^{1}	3.97×10^{1}	5.17×10^{1}	5.58×10^{1}	1.61×10^{1}
F_{19}	1.17×10^{1}	$\mathbf{1.15 \times 10^{1}}$	1.22×10^{1}	1.33×10^{1}	2.17×10^{1}	2.81×10^{1}	2.23×10^{1}	1.17×10^{1}
F_{20}	$\mathbf{2.58 \times 10^{0}}$	2.71×10^{0}	3.76×10^{0}	6.29×10^{0}	9.09×10^{0}	1.50×10^{1}	2.34×10^{1}	3.14×10^{0}
F_{21}	2.29×10^{2}	2.11×10^{2}	2.28×10^{2}	2.47×10^{2}	2.72×10^{2}	3.10×10^{2}	3.15×10^{2}	$\mathbf{1.98 \times 10^{2}}$
F_{22}	4.16×10^{1}	$\mathbf{3.13 \times 10^{1}}$	4.45×10^{1}	3.27×10^{1}	4.57×10^{1}	5.78×10^{1}	5.63×10^{1}	3.18×10^{1}
F_{23}	$\mathbf{3.44 \times 10^{2}}$	$\mathbf{3.44 \times 10^{2}}$	$\mathbf{3.44 \times 10^{2}}$	$\mathbf{3.44 \times 10^{2}}$	3.50×10^{2}	3.64×10^{2}	3.80×10^{2}	$\mathbf{3.44 \times 10^{2}}$
F_{24}	$\mathbf{2.69 \times 10^{2}}$	2.71×10^{2}	2.72×10^{2}	2.72×10^{2}	2.72×10^{2}	2.72×10^{2}	2.71×10^{2}	2.70×10^{2}
F_{25}	2.05×10^{2}	2.05×10^{2}	2.05×10^{2}	2.05×10^{2}	2.08×10^{2}	2.10×10^{2}	2.12×10^{2}	$\mathbf{2.05 \times 10^{2}}$
F_{26}	$\mathbf{1.00 \times 10^{2}}$	$\mathbf{1.00 \times 10^{2}}$	$\mathbf{1.00 \times 10^{2}}$	1.02×10^{2}	1.04×10^{2}	1.06×10^{2}	1.09×10^{2}	$\mathbf{1.00 \times 10^{2}}$
F_{27}	3.33×10^{2}	3.23×10^{2}	$\mathbf{3.11 \times 10^{2}}$	3.12×10^{2}	3.17×10^{2}	4.09×10^{2}	5.12×10^{2}	3.13×10^{2}
F_{28}	1.15×10^{3}	1.13×10^{3}	1.15×10^{3}	1.20×10^{3}	1.46×10^{3}	1.40×10^{3}	1.40×10^{3}	$\mathbf{1.11 \times 10^{3}}$
F_{29}	8.43×10^{2}	8.32×10^{2}	8.30×10^{2}	1.07×10^{3}	1.50×10^{3}	1.40×10^{3}	1.34×10^{3}	$\mathbf{8.24 \times 10^{2}}$
F_{30}	$\mathbf{8.92 \times 10^{3}}$	9.05×10^{3}	9.18×10^{3}	1.14×10^{4}	1.54×10^{4}	2.05×10^{4}	3.29×10^{4}	9.20×10^{3}
Rank	2.98	2.70	3.25	4.53	5.80	6.67	7.43	2.63

Based on the above experiments, it is demonstrated that the proposed adaptive promising individual selection strategy, for the estimation of the mean vector, is very useful for ACSEDA to not only achieve promising performance but also alleviate the sensitivity to the parameter sr.

4.3.4. Effectiveness of the Proposed Cross-Generation Individual Selection Strategy

At last, we conduct experiments to verify the usefulness of the proposed cross-generation individual selection strategy for the parent population. To this end, we first develop three other ACSEDA variants by using some existing typical selection strategies for the parent population. The first is to directly utilize the generated offspring as the parent population, such as in some traditional EDAs [20,21,28,30]. This variant of ACSEDA is denoted as "ACSEDA-O". The second is to combine the parent population in the last generation and the generated offspring and then, select the best half of the combined population as the parent population for the next generation, as in some EDA variants [51,68]. This variant of ACSEDA is represented as "ACSEDA-OP". The last one is to maintain an archive l in some EDA variants [37] to store the historical useful individuals and then they are combined with the generated offspring to select the best half of the combined population as the parent population. This ACSEDA variant is denoted as "ACSEDA-OA".

After the preparation of the compared methods, we conduct experiments on the 50-*D* CEC 2015 benchmark problems to compare the ACSEDA with the proposed cross-generation individual selection strategy and the ones with the above mentioned three compared strategies. Table 8 presents the comparison results among these different variants of ACSEDA.

Table 8. Comparison among ACSEDA with different selection strategies for the parent population on the 50-*D* CEC 2014 benchmark problems. The best results are highlighted in bold in this table.

F	ACSEDA	ACSEDA-O	ACSEDA-OP	ACSEDA-OA
F_1	1.14×10^{-14}	9.80×10^{1}	5.68×10^{-15}	$\mathbf{5.21 \times 10^{-15}}$
F_2	7.40×10^{-2}	6.04×10^{2}	$\mathbf{1.04 \times 10^{-13}}$	8.62×10^{-13}
F_3	$\mathbf{0.00 \times 10^{0}}$	4.29×10^{-11}	$\mathbf{0.00 \times 10^{0}}$	$\mathbf{0.00 \times 10^{0}}$
F_4	9.25×10^{1}	9.59×10^{1}	$\mathbf{9.12 \times 10^{1}}$	9.68×10^{1}
F_5	$\mathbf{2.11 \times 10^{1}}$	$\mathbf{2.11 \times 10^{1}}$	$\mathbf{2.11 \times 10^{1}}$	$\mathbf{2.11 \times 10^{1}}$
F_6	1.74×10^{-2}	$\mathbf{4.15 \times 10^{-3}}$	1.74×10^{-2}	3.47×10^{-2}
F_7	$\mathbf{0.00 \times 10^{0}}$	8.03×10^{-12}	1.14×10^{-14}	$\mathbf{0.00 \times 10^{0}}$
F_8	$\mathbf{2.98 \times 10^{0}}$	4.84×10^{0}	3.13×10^{2}	5.14×10^{1}
F_9	$\mathbf{2.49 \times 10^{0}}$	4.88×10^{0}	3.16×10^{2}	1.23×10^{2}
F_{10}	1.07×10^{2}	$\mathbf{4.99 \times 10^{1}}$	1.22×10^{4}	6.17×10^{3}
F_{11}	$\mathbf{1.18 \times 10^{2}}$	2.84×10^{2}	1.27×10^{4}	9.74×10^{3}
F_{12}	3.35×10^{0}	$\mathbf{3.19 \times 10^{0}}$	3.23×10^{0}	3.20×10^{0}
F_{13}	1.40×10^{-1}	$\mathbf{8.34 \times 10^{-2}}$	2.66×10^{-1}	2.21×10^{-1}
F_{14}	2.40×10^{-1}	2.91×10^{-1}	$\mathbf{2.28 \times 10^{-1}}$	2.35×10^{-1}
F_{15}	$\mathbf{4.81 \times 10^{0}}$	5.42×10^{0}	2.81×10^{1}	2.43×10^{1}
F_{16}	$\mathbf{1.80 \times 10^{1}}$	1.88×10^{1}	2.15×10^{1}	1.96×10^{1}
F_{17}	$\mathbf{1.16 \times 10^{2}}$	7.10×10^{2}	1.54×10^{3}	1.42×10^{2}
F_{18}	$\mathbf{8.47 \times 10^{0}}$	7.40×10^{1}	9.10×10^{1}	1.15×10^{1}
F_{19}	$\mathbf{1.08 \times 10^{1}}$	1.22×10^{1}	1.15×10^{1}	1.14×10^{1}
F_{20}	$\mathbf{1.95 \times 10^{0}}$	1.54×10^{1}	6.69×10^{1}	4.34×10^{0}
F_{21}	$\mathbf{1.99 \times 10^{2}}$	4.93×10^{2}	9.70×10^{2}	2.59×10^{2}
F_{22}	$\mathbf{3.54 \times 10^{1}}$	8.56×10^{1}	9.26×10^{2}	5.28×10^{1}
F_{23}	$\mathbf{3.44 \times 10^{2}}$	$\mathbf{3.44 \times 10^{2}}$	$\mathbf{3.44 \times 10^{2}}$	$\mathbf{3.44 \times 10^{2}}$
F_{24}	2.68×10^{2}	2.72×10^{2}	2.68×10^{2}	$\mathbf{2.67 \times 10^{2}}$
F_{25}	$\mathbf{2.05 \times 10^{2}}$	$\mathbf{2.05 \times 10^{2}}$	$\mathbf{2.05 \times 10^{2}}$	$\mathbf{2.05 \times 10^{2}}$
F_{26}	$\mathbf{1.00 \times 10^{2}}$	1.01×10^{2}	$\mathbf{1.00 \times 10^{2}}$	$\mathbf{1.00 \times 10^{2}}$
F_{27}	$\mathbf{3.17 \times 10^{2}}$	3.44×10^{2}	3.25×10^{2}	3.30×10^{2}
F_{28}	1.16×10^{3}	1.36×10^{3}	$\mathbf{1.14 \times 10^{3}}$	1.19×10^{3}
F_{29}	8.34×10^{2}	1.81×10^{4}	$\mathbf{8.17 \times 10^{2}}$	9.04×10^{2}
F_{30}	$\mathbf{8.57 \times 10^{3}}$	1.07×10^{4}	8.60×10^{3}	8.74×10^{3}
Rank	**1.77**	2.92	2.82	2.50

From Table 8, we can see that, from the perspective of the averaged rank obtained from the Friedman test and the number of problems where the algorithm achieves the best results, the ACSEDA with the proposed cross-generation individual selection strategy obtains the best overall performance. In particular, not only is the averaged rank is much smaller than those of the compared methods but the number of problems where the proposed ACSEDA achieves the best results is also much larger than those of the compared methods.

The above observations demonstrate that the proposed cross-generation individual selection strategy for the parent population is very helpful for ACSEDA to obtain promising performance. This is because, by combining the generated offspring in the last generation and in the current generation, this strategy is less likely to generate crowded individuals for the parent population in the next generation and could, thus, aid ACSEDA to preserve high search diversity during the evolution, as analyzed in Section 3.3.

5. Conclusions

This paper has proposed an adaptive covariance scaling estimation of distribution algorithm (ACSEDA) to solve optimization problems. First, instead of estimating the mean vector and the covariance, based on the same selected promising individuals like traditional EDAs, the proposed ACSEDA estimates the covariance based on an enlarged number of promising individuals. In this way, the sampling range of the estimated probability distribution model is enlarged and thus, the estimated model could generate more diversified offspring, which is helpful to avoid falling into local areas. To alleviate the sensitivity of the associated parameter, we further devise an adaptive covariance scaling method to dynamically adjust the covariance scaling parameter during the evolution. Second, to further help ACSEDA to explore and exploit the solution space properly, this paper further devises an adaptive promising individual selection strategy for the estimation of the mean vector. By dynamically adjusting the selection ratio parameter related to the estimation of the mean vector, the proposed ACSEDA gradually biases to exploit the found promising areas without serious loss of diversity as the evolution goes. At last, to further promote the diversity of the proposed ACSEDA, we develop a cross-generation individual selection strategy for the parent population. Different from existing selection methods, the proposed selection method combines the randomly sampled offspring in the last generation and the one in the current generation, together, and then selects the best half of the combined population as the parent population to estimate the probability distribution model. With the cohesive collaboration among the three devised techniques, the proposed ACSEDA is expected to explore and exploit the solution space appropriately and thus, is likely to achieve promising performance.

Extensive comparison experiments have been conducted on the widely used CEC 2014 benchmark problem set with different dimension sizes (30-*D*, 50-*D*, and 100-*D*). Experimental results have demonstrated that the proposed ACSEDA achieves very competitive, or even much better, performance than several state-of-the-art EDA variants. The comparison results also show that ACSEDA preserves good scalability to solve higher-dimensional optimization problems. In addition, deep investigations on the effectiveness of the three proposed techniques have also been performed. The investigation results have demonstrated that the three proposed mechanisms make great contributions to helping ACSEDA to achieve promising performance.

Author Contributions: Q.Y.: Conceptualization, supervision, methodology, formal analysis, and writing—original draft preparation. Y.L.: Implementation, formal analysis, and writing—original draft preparation. X.-D.G.: Methodology, and writing—review and editing. Y.-Y.M.: Writing—review and editing. Z.-Y.L.: Writing—review and editing, and funding acquisition. S.-W.J.: Writing—review and editing. J.Z.: Conceptualization and writing—review and editing. All authors have read and agreed to the published version of the manuscript.

Funding: This work was supported in part by the National Natural Science Foundation of China under Grant 62006124, U20B2061, 62002103, and 61873097, in part by the Natural Science Foundation of Jiangsu Province under Project BK20200811, in part by the Natural Science Foundation of the Jiangsu Higher Education Institutions of China under Grant 20KJB520006, in part by the National Research Foundation of Korea (NRF-2021H1D3A2A01082705), and in part by the Startup Foundation for Introducing Talent of NUIST.

Conflicts of Interest: The authors declare no conflict of interest.

References

1. Hasan, M.Z.; Al-Rizzo, H. Optimization of Sensor Deployment for Industrial Internet of Things Using a Multiswarm Algorithm. *IEEE Internet Things J.* **2019**, *6*, 10344–10362. [CrossRef]
2. Li, H.; Yu, J.; Yang, M.; Kong, F. Secure Outsourcing of Large-scale Convex Optimization Problem in Internet of Things. *IEEE Internet Things J.* **2021**, 1. [CrossRef]
3. Zhou, X.G.; Peng, C.X.; Liu, J.; Zhang, Y.; Zhang, G.J. Underestimation-Assisted Global-Local Cooperative Differential Evolution and the Application to Protein Structure Prediction. *IEEE Trans. Evol. Comput.* **2020**, *24*, 536–550. [CrossRef]
4. Zeng, X.; Wang, W.; Chen, C.; Yen, G.G. A Consensus Community-Based Particle Swarm Optimization for Dynamic Community Detection. *IEEE Trans. Cybern.* **2020**, *50*, 2502–2513. [CrossRef]
5. Chen, W.N.; Tan, D.Z.; Yang, Q.; Gu, T.; Zhang, J. Ant Colony Optimization for the Control of Pollutant Spreading on Social Networks. *IEEE Trans. Cybern.* **2020**, *50*, 4053–4065. [CrossRef]
6. Shen, Y.; Li, W.; Li, J. An Improved Estimation of Distribution Algorithm for Multi-compartment Electric Vehicle Routing Problem. *J. Syst. Eng. Electron.* **2021**, *32*, 365–379.
7. Li, X.; Tang, K.; Omidvar, M.N.; Yang, Z.; Qin, K.; China, H. Benchmark Functions for the CEC 2013 Special Session and Competition on Large-scale Global Optimization. 2013. Available online: https://www.tflsgo.org/assets/cec2018/cec2013-lsgo-benchmark-tech-report.pdf (accessed on 6 December 2021).
8. Wu, G.; Mallipeddi, R.; Suganthan, P.N. Problem Definitions and Evaluation Criteria for the CEC 2017 Competition on Constrained Real-parameter Optimization. 2017. Available online: https://moam.info/problem-definitions-and-evaluation-criteria-for-the-_5bad2530097c479e798b46a8.html (accessed on 6 December 2021).
9. Wei, F.F.; Chen, W.N.; Yang, Q.; Deng, J.; Luo, X.N.; Jin, H.; Zhang, J. A Classifier-Assisted Level-Based Learning Swarm Optimizer for Expensive Optimization. *IEEE Trans. Evol. Comput.* **2021**, *25*, 219–233. [CrossRef]
10. Yang, Q.; Chen, W.N.; Gu, T.; Jin, H.; Mao, W.; Zhang, J. An Adaptive Stochastic Dominant Learning Swarm Optimizer for High-Dimensional Optimization. *IEEE Trans. Cybern.* **2020**, 1–17. [CrossRef]
11. Tanweer, M.R.; Suresh, S.; Sundararajan, N. Dynamic Mentoring and Self-regulation Based Particle Swarm Optimization Algorithm for Solving Complex Real-world Optimization Problems. *Inf. Sci.* **2016**, *326*, 1–24. [CrossRef]
12. Wu, X.; Zhao, J.; Tong, Y. Big Data Analysis and Scheduling Optimization System Oriented Assembly Process for Complex Equipment. *IEEE Access* **2018**, *6*, 36479–36486. [CrossRef]
13. Yang, Q.; Chen, W.; Deng, J.D.; Li, Y.; Gu, T.; Zhang, J. A Level-Based Learning Swarm Optimizer for Large-Scale Optimization. *IEEE Trans. Evol. Comput.* **2018**, *22*, 578–594. [CrossRef]
14. Yang, Q.; Chen, W.; Gu, T.; Zhang, H.; Deng, J.D.; Li, Y.; Zhang, J. Segment-Based Predominant Learning Swarm Optimizer for Large-Scale Optimization. *IEEE Trans. Cybern.* **2017**, *47*, 2896–2910. [CrossRef]
15. Yang, Q.; Chen, W.; Yu, Z.; Gu, T.; Li, Y.; Zhang, H.; Zhang, J. Adaptive Multimodal Continuous Ant Colony Optimization. *IEEE Trans. Evol. Comput.* **2017**, *21*, 191–205. [CrossRef]
16. Tanabe, R.; Ishibuchi, H. A Review of Evolutionary Multimodal Multiobjective Optimization. *IEEE Trans. Evol. Comput.* **2020**, *24*, 193–200. [CrossRef]
17. Yang, Q.; Chen, W.; Zhang, J. Evolution Consistency Based Decomposition for Cooperative Coevolution. *IEEE Access* **2018**, *6*, 51084–51097. [CrossRef]
18. Yang, Q.; Chen, W.N.; Gu, T.; Zhang, H.; Yuan, H.; Kwong, S.; Zhang, J. A Distributed Swarm Optimizer with Adaptive Communication for Large-Scale Optimization. *IEEE Trans. Cybern.* **2020**, *50*, 3393–3408. [CrossRef]
19. Doerr, B.; Krejca, M.S. Significance-Based Estimation-of-Distribution Algorithms. *IEEE Trans. Evol. Comput.* **2020**, *24*, 1025–1034. [CrossRef]
20. Hauschild, M.; Pelikan, M. An Introduction and Survey of Estimation of Distribution Algorithms. *Swarm Evol. Comput.* **2011**, *1*, 111–128. [CrossRef]
21. Larrañaga, P.; Lozano, J.A. *Estimation of Distribution Algorithms: A New Tool for Evolutionary Computation*; Springer: Berlin/Heidelberg, Germany, 2001.
22. Bao, L.; Sun, X.; Gong, D.; Zhang, Y. Multi-source Heterogeneous User Generated Contents-driven Interactive Estimation of Distribution Algorithms for Personalized Search. *IEEE Trans. Evol. Comput.* **2021**, 1. [CrossRef]
23. Yang, Q.; Chen, W.; Li, Y.; Chen, C.L.P.; Xu, X.; Zhang, J. Multimodal Estimation of Distribution Algorithms. *IEEE Trans. Cybern.* **2017**, *47*, 636–650. [CrossRef] [PubMed]
24. Shao, W.; Pi, D.; Shao, Z. A Pareto-Based Estimation of Distribution Algorithm for Solving Multiobjective Distributed No-Wait Flow-Shop Scheduling Problem with Sequence-Dependent Setup Time. *IEEE Trans. Autom. Sci. Eng.* **2019**, *16*, 1344–1360. [CrossRef]
25. Shi, W.; Chen, W.; Lin, Y.; Gu, T.; Kwong, S.; Zhang, J. An Adaptive Estimation of Distribution Algorithm for Multipolicy Insurance Investment Planning. *IEEE Trans. Evol. Comput.* **2019**, *23*, 1–14. [CrossRef]
26. Wang, X.; Han, T.; Zhao, H. An Estimation of Distribution Algorithm with Multi-Leader Search. *IEEE Access* **2020**, *8*, 37383–37405. [CrossRef]
27. Krejca, M.S.; Witt, C. Theory of Estimation-of-distribution Algorithms. In *Theory of Evolutionary Computation*; Springer: Berlin/Heidelberg, Germany, 2020; pp. 405–442.

28. Pelikan, M.; Hauschild, M.W.; Lobo, F.G. Estimation of Distribution Algorithms. In *Springer Handbook of Computational Intelligence*; Springer: Berlin/Heidelberg, Germany, 2015; pp. 899–928.
29. Montaño, O.D.L.; Gómez-Castro, F.I.; Gutierrez-Antonio, C. Design and Optimization of a Shell-and-tube Heat Exchanger Using the Univariate Marginal Distribution Algorithm. In *Computer Aided Chemical Engineering*; Elsevier: Amsterdam, The Netherlands, 2021; Volume 50, pp. 43–49.
30. Muelas, S.; Mendiburu, A.; LaTorre, A.; Peña, J.-M. Distributed Estimation of Distribution Algorithms for Continuous Optimization: How Does the Exchanged Information Influence Their Behavior? *Inf. Sci.* **2014**, *268*, 231–254. [CrossRef]
31. Zhang, Q. On Stability of Fixed Points of Limit Models of Univariate Marginal Distribution Algorithm and Factorized Distribution Algorithm. *IEEE Trans. Evol. Comput.* **2004**, *8*, 80–93. [CrossRef]
32. Dong, W.; Yao, X. Unified Eigen Analysis on Multivariate Gaussian Based Estimation of Distribution Algorithms. *Inf. Sci.* **2008**, *178*, 3000–3023. [CrossRef]
33. Gao, B.; Wood, I. TAM-EDA: Multivariate T Distribution, Archive and Mutation Based Estimation of Distribution Algorithm. *Anziam J.* **2012**, *54*, C720–C746. [CrossRef]
34. Gao, K.; Harrison, J.P. Multivariate Distribution Model for Stress Variability Characterisation. *Int. J. Rock Mech. Min. Sci.* **2018**, *102*, 144–154. [CrossRef]
35. Gao, Y.; Hu, X.; Liu, H. Estimation of Distribution Algorithm Based on Multivariate Gaussian Copulas. In Proceedings of the IEEE International Conference on Progress in Informatics and Computing, Shanghai, China, 10–12 December 2010; pp. 254–257.
36. Yang, G.; Li, H.; Yang, W.; Fu, K.; Sun, Y.; Emery, W.J. Unsupervised Change Detection of SAR Images Based on Variational Multivariate Gaussian Mixture Model and Shannon Entropy. *IEEE Geosci. Remote. Sens. Lett.* **2019**, *16*, 826–830. [CrossRef]
37. Liang, Y.; Ren, Z.; Yao, X.; Feng, Z.; Chen, A.; Guo, W. Enhancing Gaussian Estimation of Distribution Algorithm by Exploiting Evolution Direction with Archive. *IEEE Trans. Cybern.* **2020**, *50*, 140–152. [CrossRef]
38. Zhou, A.; Sun, J.; Zhang, Q. An Estimation of Distribution Algorithm with Cheap and Expensive Local Search Methods. *IEEE Trans. Evol. Comput.* **2015**, *19*, 807–822. [CrossRef]
39. Valdez, S.I.; Hernández, A.; Botello, S. A Boltzmann Based Estimation of Distribution Algorithm. *Inf. Sci.* **2013**, *236*, 126–137. [CrossRef]
40. Ceberio, J.; Irurozki, E.; Mendiburu, A.; Lozano, J.A. A review on estimation of distribution algorithms in permutation-based combinatorial optimization problems. *Prog. Artif. Intell.* **2012**, *1*, 103–117. [CrossRef]
41. Ren, Z.; Liang, Y.; Wang, L.; Zhang, A.; Pang, B.; Li, B. Anisotropic Adaptive Variance Scaling for Gaussian Estimation of Distribution Algorithm. *Knowl.-Based Syst.* **2018**, *146*, 142–151. [CrossRef]
42. Bosman, P.A.; Grahl, J.; Rothlauf, F. SDR: A Better Trigger for Adaptive Variance Scaling in Normal EDAs. In Proceedings of the 9th Annual Conference on Genetic and Evolutionary Computation, New York, NY, USA, 7–11 July 2007; pp. 492–499. [CrossRef]
43. Grahl, J.; Bosman, P.A.; Rothlauf, F. The Correlation-Triggered Adaptive Variance Scaling IDEA. In Proceedings of the Annual Conference on Genetic and Evolutionary Computation, Seattle, WA, USA, 8–12 July 2006; pp. 397–404.
44. Liang, J.J.; Qu, B.Y.; Suganthan, P.N. Problem Definitions and Evaluation Criteria for the CEC 2014 Special Session and Competition on Single Objective Real-parameter Numerical Optimization. In *Computational Intelligence Laboratory, Zhengzhou University, Zhengzhou China and Technical Report*; Nanyang Technological University: Singapore, 2013.
45. Bronevich, A.G.; De Oliveira, J.V. On the Model Updating Operators in Univariate Estimation of Distribution Algorithms. *Nat. Comput.* **2016**, *15*, 335–354. [CrossRef]
46. Rastegar, R. On the Optimal Convergence Probability of Univariate Estimation of Distribution Algorithms. *Evol. Comput.* **2011**, *19*, 225–248. [CrossRef]
47. Krejca, M.S. *Theoretical Analyses of Univariate Estimation-of-Distribution Algorithms*; Universität Potsdam: Potsdam, Germany, 2019.
48. Wang, X.; Zhao, H.; Han, T.; Wei, Z.; Liang, Y.; Li, Y. A Gaussian Estimation of Distribution Algorithm with Random Walk Strategies and Its Application in Optimal Missile Guidance Handover for Multi-UCAV in Over-the-Horizon Air Combat. *IEEE Access* **2019**, *7*, 43298–43317. [CrossRef]
49. Ren, Z.; He, C.; Zhong, D.; Huang, S.; Liang, Y. Enhance Continuous Estimation of Distribution Algorithm by Variance Enlargement and Reflecting Sampling. In Proceedings of the IEEE Congress on Evolutionary Computation, Vancouver, BC, Canada, 24–29 July 2016; pp. 3441–3447.
50. Yang, P.; Tang, K.; Lu, X. Improving Estimation of Distribution Algorithm on Multimodal Problems by Detecting Promising Areas. *IEEE Trans. Cybern.* **2015**, *45*, 1438–1449. [CrossRef]
51. Yuan, B.; Gallagher, M. On the Importance of Diversity Maintenance in Estimation of Distribution Algorithms. In Proceedings of the Annual Conference on Genetic and Evolutionary Computation, Washington, DC, USA, 25–29 June 2005; pp. 719–726.
52. Pošík, P. Preventing Premature Convergence in A Simple EDA Via Global Step Size Setting. In Proceedings of the International Conference on Parallel Problem Solving from Nature, Lecture Notes in Computer Science, Technische Universität, Dortmund, Germany, 13–17 September 2008; pp. 549–558.
53. Cai, Y.; Sun, X.; Xu, H.; Jia, P. Cross Entropy and Adaptive Variance Scaling in Continuous EDA. In Proceedings of the Annual Conference on Genetic and Evolutionary Computation, New York, NY, USA, 7–11 July 2007; pp. 609–616.
54. Emmerich, M.; Shir, O.M.; Wang, H. Evolution Strategies. In *Handbook of Heuristics*; Martí, R., Panos, P., Resende, M.G.C., Eds.; Springer International Publishing: Cham, Switzerland, 2018; pp. 1–31.

55. Ros, R.; Hansen, N. A Simple Modification in CMA-ES Achieving Linear Time and Space Complexity. In Proceedings of the Parallel Problem Solving from Nature–PPSN X, Berlin/Heidelberg, Germany, 13–17 September 2008; pp. 296–305.
56. Akimoto, Y.; Hansen, N. Diagonal Acceleration for Covariance Matrix Adaptation Evolution Strategies. *Evol. Comput.* **2020**, *28*, 405–435. [CrossRef]
57. Arabas, J.; Jagodziński, D. Toward a Matrix-Free Covariance Matrix Adaptation Evolution Strategy. *IEEE Trans. Evol. Comput.* **2020**, *24*, 84–98. [CrossRef]
58. Beyer, H.; Sendhoff, B. Simplify Your Covariance Matrix Adaptation Evolution Strategy. *IEEE Trans. Evol. Comput.* **2017**, *21*, 746–759. [CrossRef]
59. Li, Z.; Zhang, Q. A Simple Yet Efficient Evolution Strategy for Large-Scale Black-Box Optimization. *IEEE Trans. Evol. Comput.* **2018**, *22*, 637–646. [CrossRef]
60. He, X.; Zhou, Y.; Chen, Z.; Zhang, J.; Chen, W.N. Large-Scale Evolution Strategy Based on Search Direction Adaptation. *IEEE Trans. Cybern.* **2021**, *51*, 1651–1665. [CrossRef]
61. Bosman, P.A.; Grahl, J.; Thierens, D. Enhancing the Performance of Maximum–likelihood Gaussian EDAs Using Anticipated Mean Shift. In Proceedings of the International Conference on Parallel Problem Solving from Nature, Technische Universität, Dortmund, Germany, 13–17 September 2008; pp. 133–143.
62. PourMohammadBagher, L.; Ebadzadeh, M.M.; Safabakhsh, R. Graphical Model Based Continuous Estimation of Distribution Algorithm. *Appl. Soft Comput.* **2017**, *58*, 388–400. [CrossRef]
63. Yang, Q.; Chen, W.-N.; Zhang, J. Probabilistic Multimodal Optimization. In *Metaheuristics for Finding Multiple Solutions;* Preuss, M., Epitropakis, M.G., Li, X., Fieldsend, J.E., Eds.; Springer International Publishing: Cham, Switzerland, 2021; pp. 191–228.
64. Huang, X.; Jia, P.; Liu, B. Controlling Chaos by an Improved Estimation of Distribution Algorithm. *Math. Comput. Appl.* **2010**, *15*, 866–871. [CrossRef]
65. Fang, H.; Zhou, A.; Zhang, G. An Estimation of Distribution Algorithm Guided by Mean Shift. In Proceedings of the IEEE Congress on Evolutionary Computation, Vancouver, BC, Canada, 24–29 July 2016; pp. 3268–3275.
66. Liu, J.; Wang, Y.; Teng, H. Variance Analysis and Adaptive Control in Intelligent System Based on Gaussian Model. *Int. J. Model. Identif. Control* **2013**, *18*, 26–33. [CrossRef]
67. Santana, R.; Larranaga, P.; Lozano, J.A. Adaptive Estimation of Distribution Algorithms. In *Adaptive and Multilevel Metaheuristics;* Springer: Berlin/Heidelberg, Germany, 2008; Volume 136, pp. 177–197.
68. Dong, W.; Chen, T.; Tiňo, P.; Yao, X. Scaling Up Estimation of Distribution Algorithms for Continuous Optimization. *IEEE Trans. Evol. Comput.* **2013**, *17*, 797–822. [CrossRef]
69. Hansen, N. Towards a New Evolutionary Computation. *Stud. Fuzziness Soft Comput.* **2006**, *192*, 75–102.
70. Hedar, A.-R.; Allam, A.A.; Fahim, A. Estimation of Distribution Algorithms with Fuzzy Sampling for Stochastic Programming Problems. *Appl. Sci.* **2020**, *10*, 6937. [CrossRef]
71. Maza, S.; Touahria, M. Feature Selection for Intrusion Detection Using New Multi-objective Estimation of Distribution Algorithms. *Appl. Intell.* **2019**, *49*, 4237–4257. [CrossRef]

applied sciences

MDPI

Article

Application of Hierarchical Agglomerative Clustering (HAC) for Systemic Classification of Pop-Up Housing (PUH) Environments

Thomas Märzinger, Jan Kotík * and Christoph Pfeifer

Department of Material Sciences and Process Engineering, Institute for Chemical and Energy Engineering (IVET), University of Natural Resources and Life Sciences, Peter-Jordan Strasse 82, 1190 Vienna, Austria; thomas.maerzinger@boku.ac.at (T.M.); christoph.pfeifer@boku.ac.at (C.P.)
* Correspondence: jan.kotik@boku.ac.at; Tel.: +43-1-47654-89312

Abstract: This paper is the result of the first-phase, inter-disciplinary work of a multi-disciplinary research project ("Urban pop-up housing environments and their potential as local innovation systems") consisting of energy engineers and waste managers, landscape architects and spatial planners, innovation researchers and technology assessors. The project is aiming at globally analyzing and describing existing pop-up housings (PUH), developing modeling and assessment tools for sustainable, energy-efficient and socially innovative temporary housing solutions (THS), especially for sustainable and resilient urban structures. The present paper presents an effective application of hierarchical agglomerative clustering (HAC) for analyses of large datasets typically derived from field studies. As can be shown, the method, although well-known and successfully established in (soft) computing science, can also be used very constructively as a potential urban planning tool. The main aim of the underlying multi-disciplinary research project was to deeply analyze and structure THS and PUE. Multiple aspects are to be considered when it comes to the characterization and classification of such environments. A thorough (global) web survey of PUH and analysis of scientific literature concerning descriptive work of PUH and THS has been performed. Moreover, out of several tested different approaches and methods for classifying PUH, hierarchical clustering algorithms functioned well when properly selected metrics and cut-off criteria were applied. To be specific, the 'Minkowski'-metric and the 'Calinski-Harabasz'-criteria, as clustering indices, have shown the best overall results in clustering the inhomogeneous data concerning PUH. Several additional algorithms/functions derived from the field of hierarchical clustering have also been tested to exploit their potential in interpreting and graphically analyzing particular structures and dependencies in the resulting clusters. Hereby, (math.) the significance 'S' and (math.) proportion 'P' have been concluded to yield the best interpretable and comprehensible results when it comes to analyzing the given set (objects $n = 85$) of researched PUH-objects together with their properties ($n > 190$). The resulting easily readable graphs clearly demonstrate the applicability and usability of hierarchical clustering- and their derivative algorithms for scientifically profound building classification tasks in Urban Planning by effectively managing huge inhomogeneous building datasets.

Keywords: temporary environments; pop-up housings; building classification; Hierarchical Agglomerative Clustering (HAC) algorithms; cluster proportion/significance

Citation: Märzinger, T.; Kotík, J.; Pfeifer, C. Application of Hierarchical Agglomerative Clustering (HAC) for Systemic Classification of Pop-Up Housing (PUH) Environments. *Appl. Sci.* **2021**, *11*, 11122. https://doi.org/10.3390/app112311122

Academic Editors: Rui Araújo, Nuno Lau, António Pedro Aguiar, Rodrigo Ventura and João Fabro

Received: 8 October 2021
Accepted: 18 November 2021
Published: 24 November 2021

1. Introduction

1.1. Description of PUH-Environments–Web Search and Literature Survey

Pop-up housings (PUH) and temporary housing solutions (THS), respectively, can be characterized based on multiple aspects in terms of technical, architectural, infrastructural, economic, ecological, temporal and socio-cultural considerations as well as local conditions and contextual circumstances.

The following introductory statements are based on the general analysis of over 85 PUH/THS structures (see list 'Collection of PUH-examples' in Supplementary Materials)

that have been researched and recorded/documented for later classification of PUH, which is the main aim of this paper. Moreover, the principal methodology of clustering, applicable to numerous similarly structured problems, is presented.

On the one hand, a rising need for affordable and sustainable THS, mainly in urban regions of developed countries, can be observed. This refers to a trend in the housing demand of young people, such as singles, students, workers and professionals, artists as well as young families. Rising requirements in the working environment concerning flexibility and mobility/stand-by duty, potentially lower-income situations and changing social developments/living conditions, among others, are triggering the subtle change in housing needs, particularly of younger generations. Such THS might play a role in the decarbonization of the building sector as well. Tumminia et al. [1] showed through LCA- and energy consumption simulations that for a high energy efficiency prefabricated building module for temporary housing, mainly consisting of fiber-reinforced plastic (FRP), the pre-use phase causes the most (72%) total environmental impacts compared to the use-phase (23%). Generally speaking, it is vital to use LCA methods to assess the overall energy and environmental performances of such temporary housing units [1].

Apart from housing units of solely architectural experiments and design studies, or artists' visualization projects, there is also a market for (fairly extravagant) micro or tiny homes, ashore and afloat. These units are often single and provide (temporary) housing solutions for users who are seeking exclusive, often self-sustaining housing conditions, frequently being rather highly priced.

On the other hand, there are THS for people and user groups with a more 'urgent or critical' housing demand, such as socially underprivileged (local) people, lowest-income households, asylum seekers, refugees and homeless people, to name a few. These THS aim to provide more 'basic' housing/shelter options with high affordability and ease of build possibilities in mind.

In general, the global need for temporary housing is rising through the increasing severity of natural disasters, increase in number of climate refugees due to climate change and rapid urban population growth in developing countries [2].

THS for post-disaster situations seem to be analyzed most in the scientific literature [3–6]. Concerning a post-disaster temporary housing settlement, several contextual factors have to be considered, such as physical characteristics of the settlement, availability of vital services such as education, health and work, infrastructure services, accessibility to the temporary houses within the settlement, economic prospect of the temporary settlement as well as socio-cultural, educational and financial standards of the occupants, as pointed out by Abulnour [3]. Temporary accommodation in post-disaster situations is an issue that goes beyond the simple provisioning of housings since the whole space for the temporary settlement is important [4].

A considerable part of the provided THS is not produced in the affected region but in a different country, often neglecting local resources, such as building materials and labor workforce. Therefore, social problems and environmental degradation are often the consequence of false planning strategies (i.e., not considering the option of prolonged usage) and commonly require transition to permanent housing [4]. The lack of environmental, economic and social sustainability of temporary housing environments in post-disaster situations stems from poor governmental decision-making, lack of understanding of user's needs and lack of realization and adaptability to local conditions, as discussed by Perrucci et al. [2].

As a consequence of the need to capture the great diversity of and to classify the researched and analyzed THS, a method being able to 'cope' with the heavy inter-disciplinary characteristics of the THS (i.e., mixed categorical and numerical data) had to be found. Therefore, a high amount of time has been spent for selecting, testing and refining a wide variety of already known clustering algorithms to come up with the best suitable clustering method in order to solve the aforementioned THS classification/structuring problem.

1.2. Data/Buildings Classification and Applied Clustering Algorithms

Throughout the past decade, data classification and clustering algorithms have been applied extensively in the fields of biology, engineering, information science, medical sciences as well as behavioral and social sciences, to name a few. Furthermore, cluster analysis has been widely used in many applications such as business intelligence, image pattern recognition, web search, biology and security [7]. The need for, and advantageous uses of, cluster analysis probably stem from the rising amount of readily available and processable big data and the urge to generate interpretable structures and usable classifications and correlations out of seemingly unstructured or highly heterogeneous datasets.

Concerning the classification and (technical) evaluation of buildings, various scientific studies have already been performed, mainly for energy consumption profiling [8–14] and assessing/improving energy efficiencies of different buildings [15,16] as well as benchmarking buildings [17–23]. A comprehensive review of performed cluster analyses in the building sector can be found in [15].

Throughout these performed studies (covering building topics), two methods for classification and clustering of data have proven themselves to be highly suitable and appropriate: hierarchical and k-means clustering algorithms. K-means is an iterative algorithm that divides a given dataset into k clusters by minimizing the sum of all squared Euclidian distances to the respective cluster centers (centroids to be chosen). Different 'cluster quality indices' are applied herein to obtain the optimal number of clusters, e.g., the 'Davies and Bouldin' and the 'Silhouette' indices. Grouping data into clusters so that objects within each specific cluster are very similar to each other but also very dissimilar to objects in other clusters requires distance measures (metrics based on indices of proximity) so that similarities and dissimilarities can be assessed correctly. Hierarchical data clustering results in a so-called *'dendrogram'* (graphical clustering tree structure), which represents groupings of objects at specific (dis-) similarity levels. Which clustering algorithm to choose is a matter of specific properties, i.e., type and scale of the given dataset, for example. As a matter of fact, HAC can yield impressive and rather easily readable results concerning clustering and classification of general building information/datasets. In the following, the effective application of HAC on inhomogeneous data of PUH-environments is presented, leading to the obvious suggestion that HAC could be far more intensely used for building classification tasks in urban planning. For a detailed description of a novel clustering algorithm (*k*-CMM), especially suited to mixed numerical and categorical data, see e.g., Dinh et al. [24].

2. Methodology

2.1. Data Preparation–Data Type and Scale

Firstly, before initiating a cluster analysis through a specific clustering algorithm, an examination of the given dataset has to be performed in order to specify general clusterability (is cluster structure present?), data quality and data structure, as pointed out in Jain and Dubes [25]. The result of this first data analysis helps in choosing an appropriate distance metric, which is crucial for proper clustering results. In the present study, an analysis of over 85 PUH/THS (from now on referring to 'objects') has been performed, and a representative table with all (n) objects and over a hundred (d) attributes, describing them as holistically as possible, has been generated. This dataset is viewed as the (d x n) pattern matrix, where each row of this matrix represents a pattern, and each column denotes a feature/property. Features are categorized as *binary*; thus, each individual object does (*Yes*) or does not (*No*) contain a specific attribute. Binary features are best coded on a qualitative, nominal scale, e.g., (0, 1). Therefore, a suitable distance metric, measuring proximity between objects, must be chosen according to the data type and scale. The implemented metric for the (d x n) pattern matrix of PUH/THS results in the so-called *proximity matrix* $[d(i, j)]$, containing the pairwise indices of proximity between the ith and jth object.

2.2. Cluster Analysis

2.2.1. Descriptions and Definitions of Application-Oriented Hierarchical Cluster Analysis (HCA)

As mentioned before, the *proximity matrix* describes the proximity between objects to be clustered using a problem-specific metric. It is the only, and therefore crucial, input to a clustering algorithm. The following application-oriented hierarchical clustering represents an *exclusive* (each object belongs to exactly one cluster with no overlapping), *intrinsic* (unsupervised, no a priori partition) and *hierarchical* (interlaces sequence of partitions) form of *object classification*. The following mathematical notations are explicitly taken from Jain et al. [26].

Hierarchical cluster analysis (HCA) is a procedure for transforming a defined proximity matrix into a sequence of interlaced partitions.

The n objects to be clustered are denoted by the set χ where x_i is the ith object:

$$\chi = \{x_1, x_2, \; x_3 \ldots . x_n \} \tag{1}$$

A partition, Γ of χ, divides χ into subsets $\{C_1, C_2 \ldots . \; C_m\}$, which satisfy the following expressions:

$$C_i \cap C_j = \varnothing \tag{2}$$

for all i and j from 1 to m, $i \neq j$, where $\emptyset$ is the empty set,

$$C_1 \cup C_2 \cup \ldots \cup C_m = \chi \tag{3}$$

the union of all clusters results in the total quantity of all n objects.

Thus, the hierarchical clustering algorithm yields a sequence of partitions in which each partition is nested into the next partition in the sequence. The process is repeated to form a sequence of nested clusters until a single cluster containing all n objects remains [26]. This approach is also called hierarchical *agglomerative* clustering (HAC). The visualization of the clustering results is obtained through a special type of tree structure called a *dendrogram*. A dendrogram consists of layers of nodes, each representing a cluster. Cutting a dendrogram horizontally creates a clustering [26]. Therefore, a suitable 'cut-off' criterion, described later in detail in the present chapter, must be defined in order to achieve an optimum number of clusters for a given set of objects.

Two specific hierarchical clustering methods can be distinguished, the *single-link* and the *complete-link* form. The sequences of clusterings created by these two methods depend on the proximities only through their rank order. *Single-link* clusters are characterized as *maximally connected* subgraphs, whereas *complete-link* clusters are *maximally complete* subgraphs [26]. In single-link clustering, the similarity of two clusters stems from the similarity of their most proximate similar objects. In complete-link clustering, the similarity of two clusters stems from the similarity of their most dissimilar objects. This diverse approach induces a different dendrogram; thus, a conscientious interpretation and analysis of the resulting clusterings are necessary.

2.2.2. Applied Clustering Technique

Figure 1 illustrates the step-by-step procedure performed to obtain the most suitable metric and cut-off criterion for the optimum number of clusters, thus calculating the final proximity matrix and displaying the obtained clusters through dendrograms. Accordingly, a final table with all objects (listed PUH) clustered by inconsistency coefficient and weighted match of columns has been generated. The following distance measures were included in this selection algorithm as depicted in Figure 1: basic, squared and standardized *Euclidean distances*, the *City Block-*, *Minkowski-*, *Chebychev-*, *Mahalanobis-*, *Cosine-*, *Correlation-*, *Spearman-*, *Hamming-* and *Jaccard distances*.

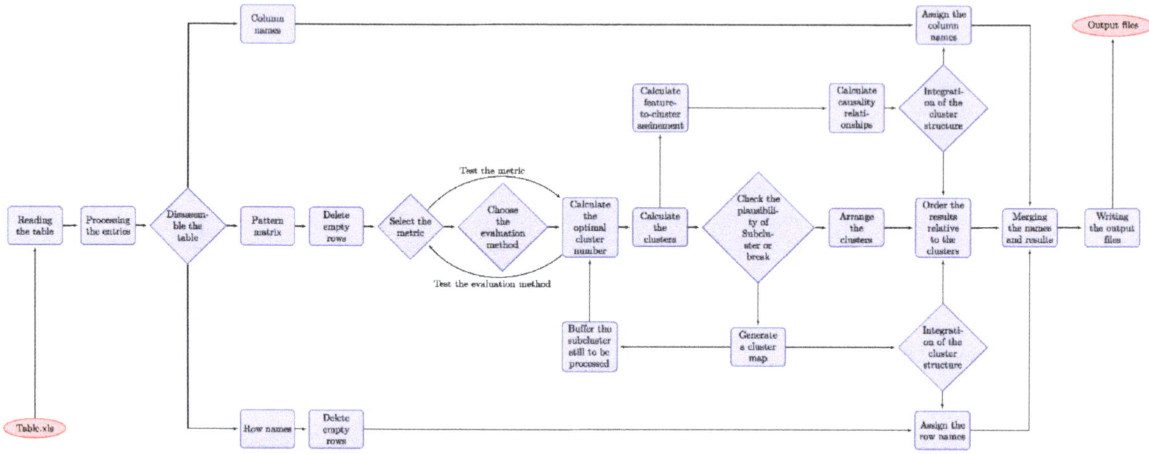

Figure 1. Algorithm for selecting the best suitable metric and writing the final clustered results.

2.2.3. Measuring Dissimilarity: Proximity Index 'Hamming Distance' (Minkowski Metric)

Because of the underlying data nature and, as a result of the conducted selection algorithm, the most suitable and comprehensibly applicable metric out of all investigated proximity indices was chosen as the Minkowski proximity index $d(i, k)$ between the ith and kth patterns, measuring dissimilarity $d(i, i) = 0$ for all i:

$$d(i,\ k) = \left(\sum_{j=1}^{d} \left| x_{ij} - x_{kj} \right|^{r} \right)^{\frac{1}{r}} \tag{4}$$

where $r \geq 1$.

Choosing $r = 2$ results in the *Euclidean* distance, the most common of the *Minkowski* metrics. For $r = 1$ and exclusively for binary features, the values are either 0 or 1. In this particular case, the *Minkowski* metric corresponds to the so-called '*Hamming distance*'. This metric is particularly useful for measuring differences and similarities as:

$$d(i,\ k) = \sum_{j=1}^{d} \left| x_{ij} - x_{kj} \right| \tag{5}$$

with $r = 1$, see Equation (4).

2.2.4. Determining the Number of Clusters: Clustering Index 'Calinski-Harabasz'-Criterion

In order to find the structure contained in the data, it is necessary to specify a suitable and ideal number of clusters for hierarchical clustering. To determine this number, the following evaluation criteria were analyzed in more detail, the Calinsky-Harabasz, Davies-Bouldin, Gap, Silhouette and Cut-off criteria for inconsistent data. A detailed description of the application of the Silhouette coefficient can be found in Dinh et al. [27].

The chosen Calinski-Harabasz index (known as Variance-Ratio criterion) is defined as the quotient of the degree of separation and compactness, using a normalization factor to limit the monotonous increase of the index with an increasing number of clusters. The Calinski-Harabasz index (CH-i) can therefore be written as:

$$CH(n) = \frac{SS_B}{SS_W} \times \frac{n-k}{k-1} \tag{6}$$

where n is the total number of data points (observed objects), k is the number of clusters, SS_B is the overall between (inter)-cluster variance or separation degree and SS_W is the overall within (intra)-cluster variance or compactness degree. Accordingly, the optimum number of clusters is achieved with the highest value of CH-i.

2.2.5. Deep Search

A depth-first search is used to identify any deeper structure in the data. The structure of the data is exposed recursively. In the first step, the optimal clustering is calculated. This step is called Layer 1. In the next step, the same procedure is applied again to each of the calculated clusters. After each new application, the calculated clusters are checked for meaningfulness with respect to the resulting clusters. If the result of the decomposition is mostly a cluster with only one element, this clustering step is discarded, and the search for this branch is stopped. If a cluster contains too few objects at the beginning of the new step, the process is stopped for this branch. Each new cycle is assigned to a new layer. The notation for this is Cluster-number_layer1-number_layer 2- . . .

For a more distinctive cluster analysis, and in order to obtain a deeper interpretation of the clustering results (assignments of an object (PUH) to a specific cluster), Bayesian rules are applied, and possible conclusions are drawn from there. Figure 2 shows the respective flow chart of the applied depth-first-search algorithm (showing procedure exemplary for the first two layers).

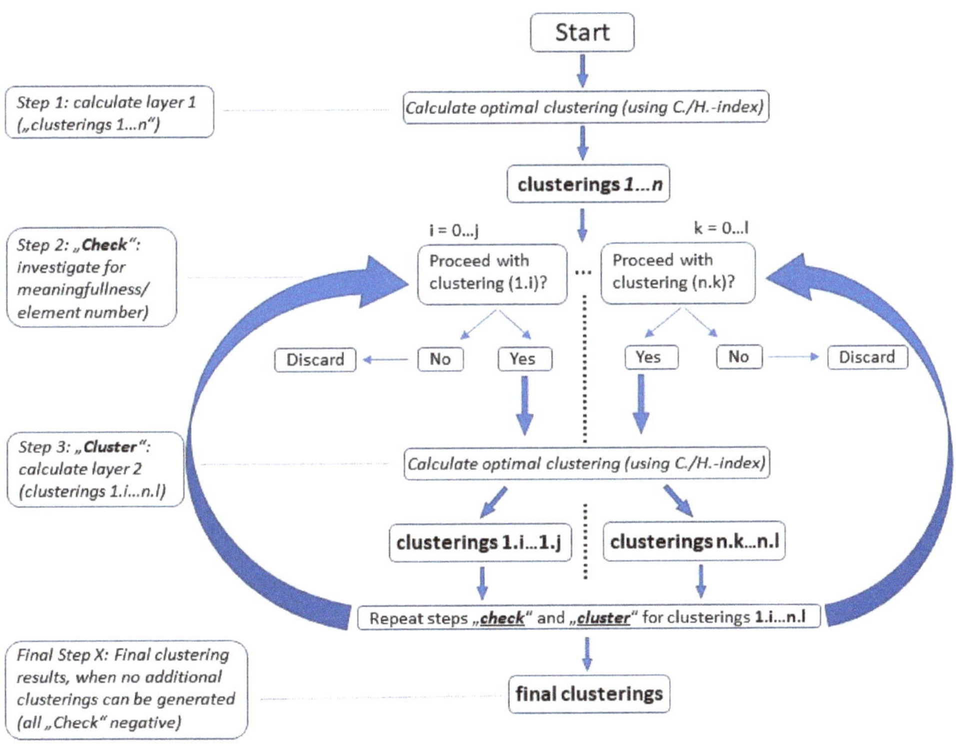

Figure 2. Algorithm for depth-first-search to identify any deeper clustering structure.

2.2.6. Limitations of Applied Method

As mentioned in the introductory section, there exist several proven distance (dissimilarity) measures as well as various clustering indices (cut-off criteria) to provide the optimum number of clusters for scattered, inhomogeneous data. The authors tried to conscientiously identify the most suitable metric and the best applicable clustering index according to the given datasets' characteristics. Overall, several Matlab simulation test runs with different metrics and cut-off criteria were performed prior to settling on the mentioned Minkowski metric together with the 'Calinski-Harabasz'-criterion. This choice, although evaluable and logically deducted, confines the possible clustering results, which potentially differ from each other when applying a specific metric and cut-off criterion. As a consequence, the provided clusterings in this paper must be considered as specific results from the application of an individually chosen and user-specific clustering technique, which might, albeit not heavily, differ from the results of other applied clustering algorithms.

2.3. Assessment

Two values were defined to evaluate whether a characteristic is meaningful with regard to a cluster. The first number evaluates how large the proportion of objects with this property is in the cluster. This value is referred to as P (proportion). The second number indicates how meaningful a characteristic of a cluster is in relation to all other clusters in the same level. This number is referred to as S (significance).

2.3.1. Proportion P_i

P_i describes the proportion of objects which possess the characteristic in relation to the number of objects in the considered cluster i.

$$P_i := \frac{\#(Objects\ in\ the\ cluster\ which\ have\ the\ characteristic)}{\#(Objects\ in\ cluster)} \tag{7}$$

Equation (7) defines the proportion of the considered characteristic in the cluster with #, being the number of elements that have the described property and i being the cluster number.

2.3.2. Significance S_i

Significance S_i describes the likelihood that an object with an observed characteristic is in a particular cluster on the level under consideration. The Bayes theorem is used to define this value.

In the Bayesian approach, an observation is not allocated to a cluster with probability 1. The Bayesian approach generates cluster probabilities for each object. This is especially important for observations close to cluster boundaries [27].

Bayes' theorem can be stated mathematically as follows, hereby defining the significance S_i:

$$S_i = P(A|B) = \frac{P(B|A)\,P(A)}{P(B)}, \tag{8}$$

where A and B are events and $P(B) \neq 0$.

$P(A|B)$ and $P(B|A)$ respectively denote the likelihood of event A (B) occurring, given that B (A) is true, called the conditional probability. $P(A)$ and $P(B)$ are the probabilities of observing events A and B independently of each other, called the marginal probability. Based on the definition, S_i describes a measure that, if an object has a characteristic, it is located in cluster i.

3. Results

3.1. Input Data for Clustering Algorithm

The presented methodology can be applied to all systems following the below-listed criteria. The input data for the verification of the model has been chosen from an evaluation of case examples for temporary living. The types of housing examples are called 'objects' within this publication. The objects have been selected within a scientific study on a global basis, whereas only objects in Vienna were physically visited. All other objects were only selected if comprehensive information could be accessed online. The described criteria were developed and selected within an interdisciplinary, scientific team of energy engineers, architects, waste management experts, sanitary engineers, landscape architects and special planning experts. Each criterion had to fulfill the condition that only 'yes' or 'no', and 'existing' or 'not existing' could be selected, e.g., one floor (yes/no), private kitchen (existing/not existing); where yes and no (as well as existing and not existing) were coded as 1 and 0, respectively.

Pattern and Proximity Matrix

As mentioned in the Introduction, 54 out of more than 85 analyzed objects (type of housing) were put together, i.e., listed as a complete dataset in a representative table with all (*n*) objects and nearly two hundred (*d*) properties, viewed as the (*d x n*) pattern matrix. Features were categorized as *binary*, and thus, each individual object did (*marked with an* X and being represented as 1) or did not (*empty field being represented as* 0) contain a specific property.

The data for the remaining objects were too incomplete, scattered or unclear to be included in the final table. The complete matrix is shown in the Supplementary Materials.

The pattern matrix represents the starting basis for all further clustering considerations and algorithmic calculations. Being rather large (see complete table with all objects and properties in 'attachments' to this paper), only a fragment of the matrix is depicted at this point, solely for illustrative purposes, showing all final objects, but only a fraction of all listed properties, see Table 1.

A fragment of the *proximity matrix [d(i, j)]*, containing the pairwise indices of proximity, i.e., the dissimilarity between the *i*th and *j*th object/feature, is presented in Table 2 for illustrative purposes only. As can be seen in Table 2, the diagonal entries of the proximity matrix can be ignored since all patterns are assumed to have the same degree of dissimilarity (namely zero) with themselves. One can also identify the symmetry of the proximity matrix since all pairs of objects have the same proximity index, independent of the order in which they are stated.

3.2. *Visualization of Clustering Results*

3.2.1. Overview of the Resulting Structuring

Table 3 gives a graphical overview of the resulting clustering of the PUH environment. The distribution of clusters in the individual layers, as well as the assignment of the THS to the individual clusters, are clearly visible. Moreover, it can be seen that Cluster 1 within Layer 1 could not be broken down further. However, enlarged sample sizes could result in a meaningful clustering of the THS samples in Layer 2. The former Cluster 1 could be implemented into a new cluster structure. Finally, the number of layers is not limited to three. In the corresponding case study, the mathematical modeling did not result in any further, meaningful sub-clusters. Again, an enlarged sample size could lead to different results, eventually leading to a higher number of layers. Summarizing, after a correct selection of states for each criterion, the mathematical solution shows the clusters as well as the number of layers.

Table 1. Extraction of (d x n)-pattern matrix as input table for further clustering calculations.

(dxn)-pattern matrix

		Object name / Properties	Weaving a Home	Base camp field hospital in Balukhali II	Emergency (..) center Hamburg	Cuvrybrache Berlin	Refugee Shelter Balukhali II	Instant (..) light weight/..) Munich	Better Shelter	Lifeshelter	Habitat for humanity	instant home	Wagendorf Lohmühle	Little Home Köln e.V.	Getaway House	Wohnwagon	FEMA Temporary housing units
		Overall table (pattern matrix) of pop-up examples	A	A	A	A	A	A	A	A	A	A	B	B	B	B	C
		Consecutive Number	1	2	3	4	5	6	7	8	9	10	11	12	13	14	15
		ID-Number	43	51	19	58	45	9	32	44	6	49	61	37	16	10	62
1	6	free standing	x	x	x	x		x	x	x	x	x	x	x	x	x	x
2	3	newly built accomodation	x	x	x	x	x	x	x	x	x	x		x	x	x	x
3	180	unit transportable	x	x	x	x	x	x	x	x		x	x	x	x	x	x
4	4	prefabricated elements or modules	x	x	x		x	x	x	x		x	x	x	x	x	x
5	12	one floor	x	x	x	x	x	x	x	x	x	x	x	x	x	x	x
6	29	collectively used open space	x	x	x	x	x	x	x	x	x		x				
7	31	access to shared toilet or similar	x	x	x			x	x	x	x		x				
8	10	industrial kitchen/field kitchen		x				x	.	.					x		
9	32	access to shared bathroom or washing area	.	x	x			x	x	x	x		x				
10	25	green land (former land use)		x	x		x		x	x							
11	8	lightweight (rigid) materials				x		x	x	x	x	x	x	x	x	x	x
12	49	private kitchen/cooking facility	.				x				x		x	x	x	x	
13	50	residential area (zoning plan)	.					x		.	.			x		x	x
14	52	one bed unit				x						x		x	x	x	x
15	54	primary material: wood				x							x	x	x	x	
16	184	complete unit in more than one dwelling	x	x	x	x	x	x	x	x		x	x	x			
17	185	one room only	x	x	x	x	x	x	x	x		x					
18	183	frameconstruction with lightweight walls	x	x	x	x	x	x	x	x	x						
19	182	supported by poles and ropes		x	x	x	x	x									

		Object name / Properties	EXO Emergency Shelters	Seven Skies Pop Up Hotel	micro compact home (m-ch)	Heijmans ONE	Sliding hub	Nader Khalili & Cal Earth Earth Build-	BaleHouse	3D Printed home	3D Housing 05	BigDelta project	Prototype ME-01	Meme Meadows Experimental House	YMCA's Y:Cube house	The Blob	Tanzania Rural Housing	Abaton casa transportable	Arc House	microHOME	Home Box	Passive house pop-up	Airclad Suffolk Pod	Concrete shelter Ahrweiler
		Overall table (pattern matrix) of pop-up examples	C	C	C	C	C	C	C	C	C	C	C	C	C	C	C	C	C	C	C	C	C	C
		Consecutive Number	16	17	18	19	20	21	22	23	24	25	26	27	29	30	31	32	33	34	35	36	28	37
		ID-Number	46	63	31	5	13	47	28	40	38	39	12	11	34	27	15	29	30	33	23	4	64	17
1	6	free standing	x	x	x	x	x	x	x	x	x	x	x	x	x	x	x	x	x	x	x	x	x	
2	3	newly built accomodation	x	x	x	x	x	x	x	x	x	x	x	x	x	x	x	x	x	x	x	x	x	x
3	180	unit transportable	x	x	x	x	x		x				x		x	x	x	x	x	x	x	x	x	x
4	4	prefabricated elements or modules	x	x	x	x	x		x				x	x	x	x	x	x	x	x	x	x	x	x
5	12	one floor	x	x	x		x	x		x	x	x	x	x	x	x	x	x	x	x		x	x	x
6	29	collectively used open space	x	x	x	x	x	.									x		x					x
7	31	access to shared toilet or similar	x	x																			x	x
8	10	industrial kitchen/field kitchen		x	x		x	.									x	x					x	
9	32	access to shared bathroom or washing area	x	x																			x	x
10	25	green land (former land use)								x		x												x
11	8	lightweight (rigid) materials	x	x	x	x	x		x	x	x	x	x	x	x	x	x	x	x	x	x	x	x	
12	49	private kitchen/cooking facility	.		x	x	x		x	x	x	+	x	x	x	x		x	x	x	x	x		
13	50	residential area (zoning plan)	.		x	x				x					x					x	x	x	x	
14	52	one bed unit		x	x	x				x	x	x	x	x	x	x	x	x		x		x		
15	54	primary material: wood					x		x					x	x			x	x	x	x	x		
16	184	complete unit in more than one dwelling	x	x								x											x	x
17	185	one room only	x	x								x											x	
18	183	frameconstruction with lightweight walls																						
19	182	supported by poles and ropes																						

		Object name / Properties	Urban Rigger	Feel Home Taufkirchen	Pop-up dorms Seestadt Aspern	Wohncontainer Pfarre Rodaun	City Container Zürich	NYC Emergency Housing Prototype	Onagawa Temporary Container Housing	Startblok Riekerhaven	Cubity	Slim Building Siemensstraße	Holzmodule Podhaskygasse	Haus Liebhartstal	HAWI	Notunterkunft Primavesigasse	Pop-down Hotel	Sleepbox mobile hotel	Dwelling with SHED
		Overall table (pattern matrix) of pop-up examples	D	D	D	D	D	D	D	D	D	D	D	E	E	E	E	E	E
		Consecutive Number	38	39	40	41	42	43	44	45	46	47	48	49	50	51	52	53	54
		ID-Number	52	22	25	50	8	14	53	60	18	41	42	54	59	48	21	65	3
1	6	free standing	x	x	x	x	x	x	x	x	x	x	x	x	x	x	x		
2	3	newly built accomodation	x	x	x	x	x	x	x	x	x	x	x						
3	180	unit transportable	x	x	x	x	x	x	x	x	x	x	x					x	
4	4	prefabricated elements or modules	x	x	x	x	x	x	x	x	x	x	x					x	
5	12	one floor				x					x								
6	29	collectively used open space	x	x	x	x	x		x	x	x	x	x	x	x		x		x
7	31	access to shared toilet or similar		x	x	x	x							x	x	x		x	
8	10	industrial kitchen/field kitchen														x	x	x	
9	32	access to shared bathroom or washing area		x	x	x	x							x	x	x		x	
10	25	green land (former land use)		x									x						
11	8	lightweight (rigid) materials	x	x	x	x	x		x	x	x		x					x	
12	49	private kitchen/cooking facility	x					x	x	x		x	x						
13	50	residential area (zoning plan)		x		x		x		x	x			x				x	
14	52	one bed unit	x		x					x	x				x			x	x
15	54	primary material: wood		x	x						x		x		x			x	.
16	184	complete unit in more than one dwelling																	
17	185	one room only																x	
18	183	frameconstruction with lightweight walls																	
19	182	supported by poles and ropes																	

Table 1. *Cont.*

		(dxn)-pattern matrix																																																							
20	41	lightweight-textile (flexible) materials	x	x	x	x	x	x			x																																																
21	139	innovative structure	x																																		x																							
22	121	rely on public toilette				x	x					x																																															
23	122	rely on public bathroom				x	x					x																																															
24	177	unit accessed by footpath		x	x	x	x	x	x	x	x	x							x	x	x	x																	x	x																			
25	187	unit with different qualities in the same PUE				x	x						x																										x											x									x	
26	188	selfmade improvised dwelling				x							x																																															
27	125	appropriation of open space (other than private)				x	x				x		x																																														.
28	126	expansion of unit during PUE life span					x				x		x																																															
29	35	≥100 units on-site	.				x		x	.																																						x	x		x	x							
30	186	tent units in a row (connected to each other)					x																																																				
31	39	connection to "buildings" on both sides					x																																																			x	
32	142	expansion of PUE during PUE life span				x	x																																						x														
33	84	mass shelter						x																																															x			
34	175	lightweight ridig plastic							x	x																											x																				
35	16	primary material: coated paper										x																																														
36	191	earth cellar (as refridgerator)											x																																													
37	192	vermicomosting for organic waste streams											x																																													
38	57	unit on wheels											x	x	x	x	x																																										x
39		one mobile unit											x	x	x	x	x																																										
40	70	PUE consists of 1 unit (one unit on site)												x	x	x						x	x		x	x	x	x	x	x	x	x	x	x	x	x	x	x																					
41	178	Unit accessed on private plot												x	x	x				x			x	x	x	x	x	x	x	x	x	x	x	x	x	x	x	x	x																				
42	7	lockable unit(s)	.			.			x	x	x			x	x	x	x		x	x	x	x	x	x	x	x	x	x	x	x	x	x	x	x	x	x	x	x	x	x		x	x	x	x	x	x	x	x	x	x	x		x	x		x	x	
43	63	unit has private toilet	.								x			x	x	x	x		.		x	x	x		x	x	x		x	x	x	x	x	x	x	x	x	x				x					x	x	x	x	x	x					x		.
44	62	unit has private bathroom	.												x	x	x		.		x	x	x		x	x	x		x	x	x	x	x	x	x	x	x	x				x					x	x	x	x	x	x					x		.
45	5	reversible (reusable) foundation															x		x	x	x		x	x	x	x	x	x	x	x	x	x	x	x	x	x	x	x	x	x			x		x			x		x									
46	13	house with one unit per floor																	x	x	x	x	x	x	x	x	x	x	x	x	x	x	x	x	x	x	x	x	x	x																			
47	118	constructed on site only	x			x					x													x		x	x	x																											x	x	x		
48	112	building ban with temporary building permit								.												x		.	x		x	x			x		x									x	.				x					x							
49	128	3d printed																								x	x	x																															

Table 2. Proximity matrix $[d(i, j)]$ (extr.), based on previously mentioned dissimilarity index/Minkowski metric.

	1	2	3	4	5	6	7	8	9	10	11	12	13	14	15
1	0.0000	0.1907	0.1753	0.2371	0.2165	0.2062	0.2165	0.1959	0.1649	0.2062	0.3041	0.2577	0.2474	0.3093	0.2680
2	0.1907	0.0000	0.1289	0.2732	0.1289	0.1289	0.1495	0.1495	0.1392	0.2113	0.3093	0.2113	0.2526	0.3144	0.2526
3	0.1753	0.1289	0.0000	0.2577	0.1753	0.0825	0.1340	0.1237	0.1443	0.1959	0.3247	0.2371	0.2268	0.2990	0.2371
4	0.2371	0.2732	0.2577	0.0000	0.2371	0.2577	0.2990	0.2887	0.2474	0.2474	0.2216	0.2371	0.2990	0.3196	0.2887
5	0.2165	0.1289	0.1753	0.2371	0.0000	0.1959	0.1856	0.1959	0.1546	0.2371	0.3041	0.2371	0.2680	0.3093	0.2887
6	0.2062	0.1289	0.0825	0.2577	0.1959	0.0000	0.1443	0.1237	0.1649	0.2165	0.3247	0.2371	0.2577	0.2887	0.2062
7	0.2165	0.1495	0.1340	0.2990	0.1856	0.1443	0.0000	0.0722	0.1443	0.1753	0.3144	0.2165	0.2165	0.2784	0.2474
8	0.1959	0.1495	0.1237	0.2887	0.1959	0.1237	0.0722	0.0000	0.1237	0.1856	0.3041	0.1959	0.2268	0.2577	0.2268
9	0.1649	0.1392	0.1443	0.2474	0.1546	0.1649	0.1443	0.1237	0.0000	0.2062	0.2835	0.1753	0.2062	0.2577	0.2165
10	0.2062	0.2113	0.1959	0.2474	0.2371	0.2165	0.1753	0.1856	0.2062	0.0000	0.3351	0.1753	0.1546	0.2784	0.2268
11	0.3041	0.3093	0.3247	0.2216	0.3041	0.3247	0.3144	0.3041	0.2835	0.3351	0.0000	0.2732	0.3144	0.2938	0.3144
12	0.2577	0.2113	0.2371	0.2371	0.2371	0.2371	0.2165	0.1959	0.1753	0.1753	0.2732	0.0000	0.1237	0.1546	0.1649
13	0.2474	0.2526	0.2268	0.2990	0.2680	0.2577	0.2165	0.2268	0.2062	0.1546	0.3144	0.1237	0.0000	0.1753	0.1856
14	0.3093	0.3144	0.2990	0.3196	0.3093	0.2887	0.2784	0.2577	0.2577	0.2784	0.2938	0.1546	0.1753	0.0000	0.1856
15	0.2680	0.2526	0.2371	0.2887	0.2887	0.2062	0.2474	0.2268	0.2165	0.2268	0.3144	0.1649	0.1856	0.1856	0.0000
16	0.1907	0.2062	0.1804	0.3041	0.2629	0.2113	0.1804	0.2010	0.2113	0.1804	0.3299	0.2629	0.2216	0.3247	0.2113
17	0.2423	0.2165	0.1701	0.2835	0.2938	0.1804	0.2010	0.1804	0.2216	0.2113	0.2990	0.2216	0.2320	0.2938	0.1907
18	0.2732	0.2268	0.1907	0.3144	0.2835	0.2113	0.2216	0.2010	0.2010	0.2216	0.3299	0.1907	0.1907	0.2113	0.1186
19	0.2990	0.2629	0.2474	0.3093	0.3093	0.2474	0.2474	0.2268	0.2268	0.2268	0.3351	0.1959	0.2165	0.2062	0.1649
20	0.2577	0.2526	0.2474	0.3196	0.3093	0.2474	0.2371	0.2371	0.2474	0.2062	0.3557	0.2062	0.1443	0.2165	0.1546
21	0.1856	0.2320	0.2474	0.2474	0.2165	0.2577	0.2268	0.2062	0.1546	0.2474	0.2835	0.1856	0.2474	0.2577	0.2062
22	0.2784	0.2835	0.2680	0.3402	0.3093	0.2887	0.2474	0.2474	0.2371	0.1959	0.3763	0.1959	0.1340	0.2062	0.1856
23	0.2938	0.2990	0.2526	0.3454	0.3351	0.2629	0.2526	0.2423	0.2320	0.1907	0.3711	0.2113	0.1598	0.2113	0.1907
24	0.2629	0.2990	0.2526	0.3454	0.3351	0.2835	0.2526	0.2423	0.2320	0.1804	0.3814	0.2320	0.1598	0.2423	0.1907
25	0.2629	0.2680	0.2216	0.3247	0.3041	0.2629	0.2113	0.2216	0.2113	0.1598	0.3711	0.2113	0.1495	0.2732	0.2320
26	0.2216	0.2577	0.2526	0.3454	0.2835	0.2732	0.2320	0.2113	0.1907	0.2010	0.3505	0.1907	0.1495	0.2216	0.1907
27	0.2680	0.3041	0.2784	0.2990	0.3299	0.2887	0.2577	0.2371	0.2474	0.2474	0.3247	0.1856	0.1753	0.1856	0.2165
28	0.2990	0.2526	0.2268	0.3196	0.2887	0.2165	0.2268	0.2165	0.2165	0.2062	0.3557	0.1443	0.1546	0.1546	0.1134
29	0.2938	0.2577	0.2320	0.2835	0.3041	0.2526	0.2423	0.2216	0.2320	0.1907	0.3299	0.1598	0.1701	0.1701	0.1392
30	0.2680	0.2320	0.2371	0.3093	0.2887	0.2474	0.2165	0.1753	0.2062	0.2165	0.3454	0.1546	0.1753	0.2062	0.2165
31	0.2577	0.2526	0.2268	0.2784	0.2680	0.2577	0.2062	0.2062	0.2062	0.1856	0.3247	0.1340	0.0722	0.1753	0.1753
32	0.2629	0.2474	0.2629	0.3041	0.2732	0.2629	0.2010	0.2216	0.2216	0.2216	0.3196	0.1495	0.1495	0.1598	0.1907
33	0.2938	0.3299	0.2732	0.3454	0.3454	0.2938	0.2835	0.2526	0.2732	0.2629	0.3711	0.2113	0.2113	0.1804	0.1804
34	0.3041	0.2990	0.2629	0.3041	0.3247	0.2629	0.2835	0.2526	0.2423	0.2423	0.3402	0.1495	0.1907	0.1598	0.1495
35	0.2990	0.2835	0.2371	0.3299	0.2990	0.2474	0.2577	0.2268	0.2268	0.2371	0.3454	0.1649	0.1649	0.1340	0.1340
36	0.2577	0.2320	0.1856	0.2680	0.2990	0.1959	0.2371	0.2062	0.2371	0.2371	0.2938	0.2268	0.2474	0.2371	0.1753
37	0.2526	0.1753	0.1289	0.3351	0.2526	0.1392	0.1495	0.1289	0.2010	0.2320	0.3299	0.2526	0.2423	0.3041	0.2113
38	0.3351	0.3093	0.3041	0.3557	0.3557	0.2938	0.3144	0.2835	0.2835	0.3144	0.3299	0.2732	0.2835	0.2835	0.2113
39	0.3093	0.2423	0.1753	0.3402	0.2784	0.1753	0.2062	0.1649	0.2268	0.2680	0.3454	0.2371	0.2474	0.2474	0.1856
40	0.3093	0.2216	0.2371	0.3196	0.3093	0.2165	0.2268	0.2062	0.2371	0.2680	0.3247	0.2268	0.2577	0.2474	0.1959
41	0.3196	0.2216	0.2062	0.3299	0.2887	0.1753	0.2165	0.1856	0.2474	0.2887	0.3144	0.2577	0.2887	0.2887	0.1959
42	0.2887	0.2423	0.1856	0.3299	0.2784	0.1753	0.2165	0.1753	0.2165	0.2577	0.3454	0.2577	0.2474	0.2784	0.2165
43	0.2835	0.2474	0.2320	0.3660	0.2835	0.2423	0.2320	0.2113	0.2216	0.2629	0.3814	0.2216	0.2320	0.2320	0.1701
44	0.2887	0.2629	0.2371	0.3505	0.2680	0.2474	0.2371	0.2062	0.2062	0.2680	0.3454	0.2371	0.2371	0.2371	0.1340
45	0.3505	0.3041	0.2680	0.3402	0.3093	0.2577	0.2784	0.2577	0.2784	0.2990	0.3351	0.2577	0.2577	0.2784	0.1649
46	0.3247	0.2784	0.2629	0.3247	0.3351	0.2629	0.2629	0.2526	0.2835	0.2835	0.3196	0.2423	0.2423	0.2423	0.1598
47	0.3196	0.2835	0.2680	0.3402	0.2990	0.2577	0.2784	0.2474	0.2577	0.2990	0.3866	0.2474	0.2577	0.2371	0.1959
48	0.3196	0.2732	0.2577	0.3505	0.2990	0.2474	0.2680	0.2371	0.2268	0.2990	0.3557	0.2371	0.2474	0.2165	0.1856
49	0.3454	0.2887	0.2526	0.3660	0.3351	0.2320	0.2835	0.2526	0.2732	0.3247	0.3711	0.3041	0.3247	0.3247	0.2526
50	0.3505	0.2938	0.2887	0.3608	0.3608	0.2474	0.2887	0.2680	0.2990	0.3505	0.3454	0.3093	0.3299	0.3711	0.2887
51	0.2835	0.2577	0.2113	0.3454	0.3247	0.2113	0.2732	0.2526	0.2423	0.2526	0.3814	0.2732	0.2629	0.3247	0.2629
52	0.3041	0.2990	0.2526	0.3557	0.3351	0.2732	0.2835	0.2526	0.2526	0.2938	0.4124	0.2835	0.2423	0.2629	0.2320
53	0.3093	0.2629	0.2268	0.3402	0.3196	0.2371	0.2784	0.2371	0.2784	0.2784	0.3144	0.2577	0.2680	0.3093	0.2474
54	0.2938	0.2680	0.2320	0.3351	0.3454	0.2526	0.2629	0.2526	0.2526	0.2835	0.3402	0.2629	0.2732	0.3144	0.2216

3.2.2. Proximity Dendrogram

As mentioned above, a dendrogram represents the graphical form of clusterings, either in *single-link* or *complete-link* form. Moreover, a dendrogram yields the nested grouping of patterns and similarity levels at which groupings change [26].

Figure 3 represents the complete-link dendrogram for the input data. Two major clusters can be identified (differentiated by color); The length of the various bars describes the dissimilarity or proximity between various objects using the specified metric. As can be seen in Figure 3 (marked by dotted blue line), several sub-clusters can be formed and visualized when using a specific clustering index as 'stop criterion' (e.g., 0.2).

Table 3. Overview of the resulting PUH-environment.

<table>
<tr><th>Layer 1</th><th>Layer 2</th><th>Layer 3</th><th>Name of the PUH-Environment</th></tr>
<tr><td rowspan="22">Layer I Cluster 2</td><td rowspan="10">Layer II Cluster 2-1</td><td rowspan="4">C2-1-3</td><td>Haus Liebhartstal</td></tr>
<tr><td>HAWI</td></tr>
<tr><td>Dwelling with SHED</td></tr>
<tr><td>Sleepbox mobile hotel</td></tr>
<tr><td rowspan="2">C2-1-2</td><td>Notunterkuft Primavesigasse</td></tr>
<tr><td>Pop-down Hotel</td></tr>
<tr><td rowspan="4">C2-1-1</td><td>Cubity</td></tr>
<tr><td>Pop-up dorms Seestadt Aspern</td></tr>
<tr><td>City Container Zürich</td></tr>
<tr><td>Wohncontainer Pfarre Rodaun</td></tr>
<tr><td rowspan="12">Layer II Cluster 2-2</td><td rowspan="7">C2-2-3</td><td>Slim Building Siemensstraße</td></tr>
<tr><td>Holzmodule Podhaskygasse</td></tr>
<tr><td>Onagawa Temporary Container Housing</td></tr>
<tr><td>NYC Emergency Housing Prototype</td></tr>
<tr><td>Urban Rigger</td></tr>
<tr><td>Startblok Rikerhaven</td></tr>
<tr><td>Feel Home Taufkirchen</td></tr>
<tr><td rowspan="5">C2-2-2</td><td>Concrete shelter Ahrweiler</td></tr>
<tr><td>Airclade Suffolk Pod</td></tr>
<tr><td>FEMA Temporary housing units</td></tr>
<tr><td>Micro compact home (m-ch)</td></tr>
<tr><td>Heijmans One</td></tr>
</table>

Table 3. *Cont.*

Layer 1	Layer 2	Layer 3	Name of the PUH-Environment
			Sliding hub
			Bale House
			The Blob
			YMCA's: Cube house
			Passive house pop-up
			Microhome
			Home Box
			Wohnwagon
			Little Home Köln e.V.
			Getaway House
			Abaton casa transportable
			Meme meadows Experimental House
			Arc House
			Tanzania Rural Housing
			Prototype ME-01
			Seven Skies Pop Up Hotel
		C2-2-1	3D Housing 05
			3D Printed home
			Big Delta project
			Instant home
			EXO Emergency Shelters
			Weaving a Home
			Nader Khalili & Cal Earth Buildings
			Habitat for humanity
			Lifeshelter
Layer I Cluster 1			Better Shelter
			Emergency (…) center Hamburg
			Instant (…) lightweight...) Munich
			Base camp field hospital in Balukhali II
			Refugee Shelter Balukhali II
			Cuvrybrache Berlin
			Wagendorf Lohnmühle

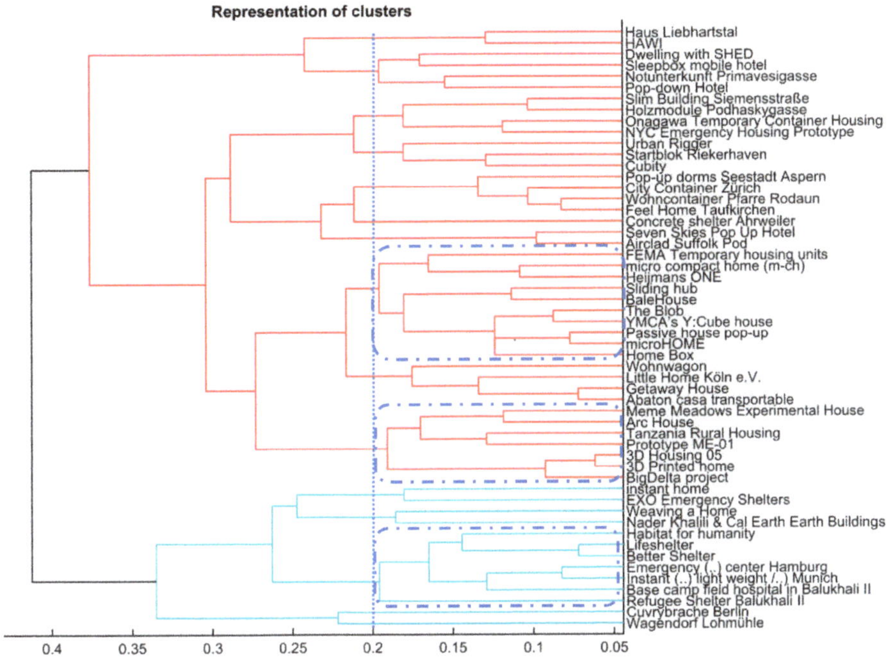

Figure 3. Complete-link dendrogram for final PUE-objects ($n = 54$).

4. Discussion

4.1. Bar Plots–Feature-To-Cluster Assignment

In the following, typical results of the mathematical modeling are presented. The selected types of illustration are typical examples of how those results can be presented. Figure 4 shows the proportion as calculated according to the above-derived equation for each characteristic element. The frequency of occurrence is normalized to one, which means that one represents 100% of examples in the cluster having the considered characteristic, whereas 0 means 0% of the examples show this specific characteristic. Thus, the values, and especially the shape, of the graphs indicate the importance of specific characteristics for each cluster. It is straightforward to create illustrations, as shown in Figure 4 for each cluster, to compare the significance of each group of characteristics with each other. These illustrations help the experts who provide the initial data to neutrally evaluate the results.

In Figure 5, the significance of each characteristic in a cluster in relation to the so-called parent cluster is shown. Based on the investigated cluster, a specific number 'i' of sub-clusters results from the calculations. For simplicity, again, only two sub-clusters are used in the explanatory illustrations. A significance of '1' means that the specific characteristic is only present in one sub-cluster of all sub-clusters extracted out of one parent cluster. A significance of '0' means that this characteristic is not existing in the sub-cluster 'i'. Again, this type of illustration helps to group the classifications and allows a graphical evaluation of the results based on the mathematical modeling. This allows a dispassionate evaluation of the results of the initial classification of the example cases. In comparison to the classical evaluation of large groups of objects, these findings and the selected methods of presentation are big advantages to avoid misjudgments by the experts.

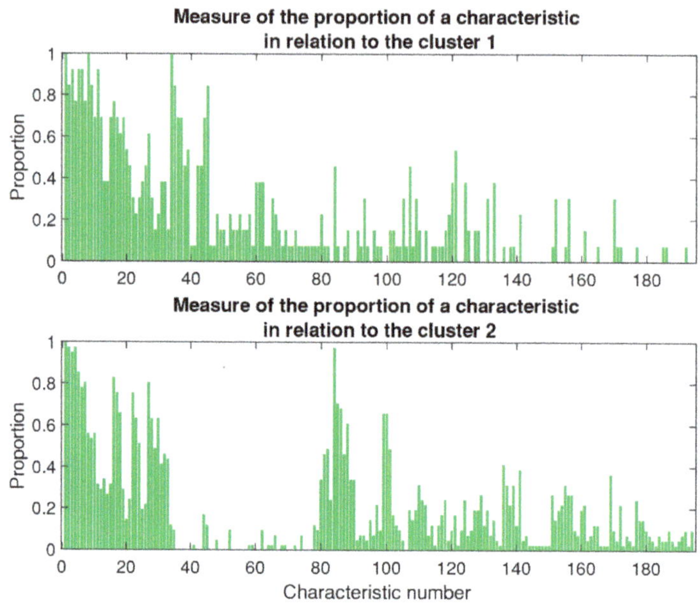

Figure 4. The illustration shows how present a feature is in the determining cluster.

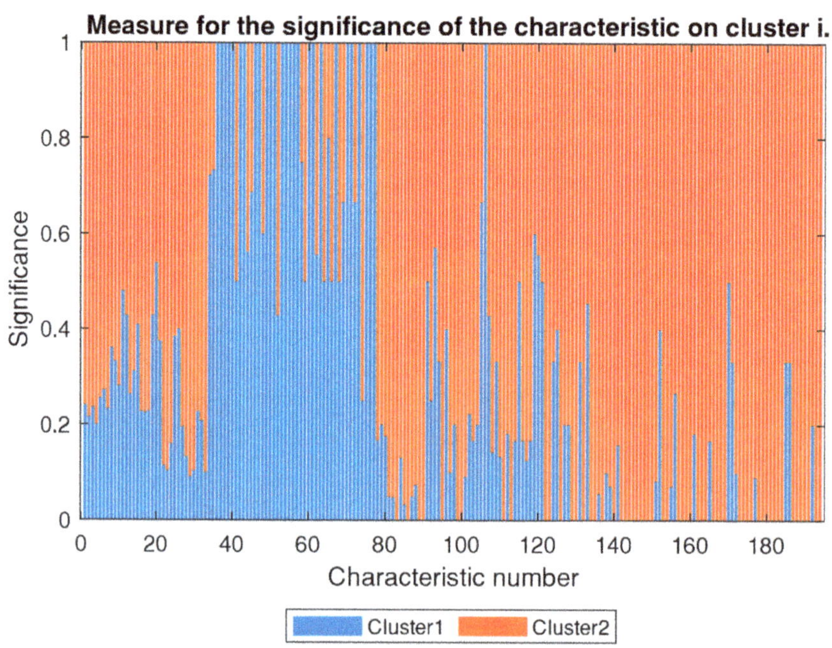

Figure 5. The significance of each characteristic to the clusters.

Figure 6 gives a graphical overview of the whole assessment of data, including the conciseness, given as proportion, and significances of each characteristic number for several layers of classes (parent cluster down to each layer of sub-clusters). Again, the values are normalized to '1', representing all cases in the corresponding cluster, and '0', meaning no proportion/significance. The visual presentation of the results of mathematical modeling can be used to distinguish relevance. If both criteria (proportion and significance) are high for a specific characteristic within a specific cluster, it can be stated that this characteristic is relevant for the assignment of objects into the cluster. If only the proportion is high, this means that the characteristic criteria are relevant for the cluster but should not be exclusively assigned. In other words, many objects within the observed cluster possess specific characteristics, but other objects in other clusters possess this characteristic as well. Vice versa, it can be stated if only the significance is high, this means that the characteristic is the reason for assigning the object to the cluster, but many other objects in the same cluster do not possess this characteristic.

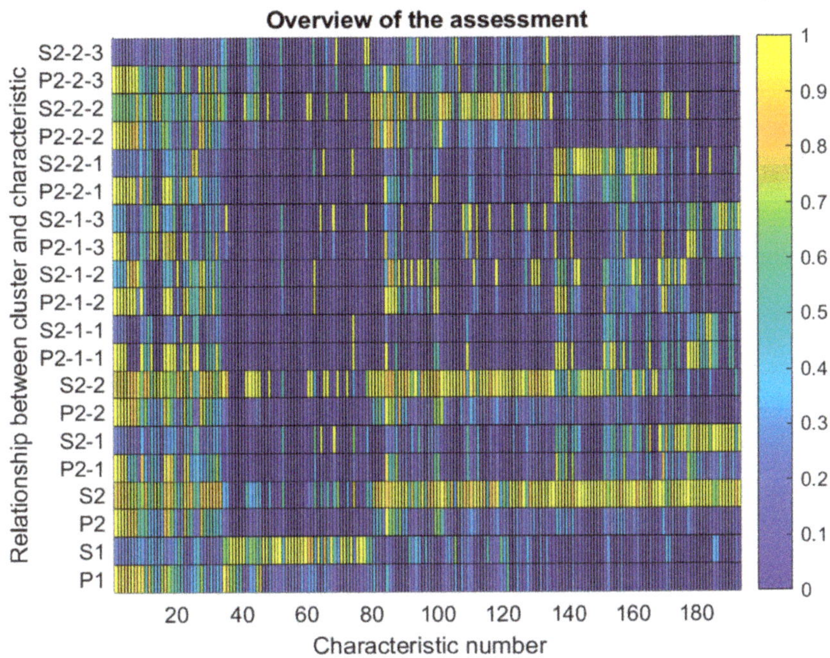

Figure 6. The relation of proportion and significance of characteristics to the clusters.

Summarizing, the combination of proportion and significance, and especially their gradient between the different layers of clusters (parent cluster, sub-clusters, sub-sub-clusters, …), allows a description for each specific cluster. Based on the graphical presentation, which is created automatically out of the results of the mathematical modeling, the experts can easily determine the relations between the list of objects.

4.2. Interpretation

As an example, properties Nr. 1 ('residential use') and Nr. 2 ('embedded in formal setting') generally represented descriptive characteristics. Both exhibited a proportion of $p = 1$ in all clusters and layers. Whereas the significance S was clearly below 1 ($S << 1$), meaning that the specific PUH-objects, namely dwelling forms, are very homogeneously distributed in the particular clusters.

It is possible to identify properties that are representative of a particular cluster, i.e., properties that are relevant for the assignment of a PUH-object to a particular cluster (classification). Properties Nr. 36 ('space-saving transport'), Nr. 37 ('frame construction with lightweight walls'), Nr. 38 ('supported by poles and ropes') and Nr. 58 ('embedded in informal setting') exhibited relevance for their specific clusters, i.e., their significance S was $S = 1$, and their proportion P was rather high.

Properties that are strongly descriptive, i.e., relevant for classification/assignment to particular clusters, can be explicitly found for Cluster 2 and its subclusters. Property Nr. 87

('thermal insulation'), Nr. 88 ('reversible/reusable foundation'), Nr. 99 ('unit connected to centralized water supply'), 100 ('unit has sewage connection') and Nr. 101 ('house with one unit per floor') were the most significant characteristics.

5. Conclusions

5.1. Final Classification Remarks

5.1.1. General

At least 80% of all analyzed objects (n = 54) were free-standing, all-year-round habitable, newly built and transportable structures made of prefabricated elements in modular design. Nearly all (95%) were officially built and positioned according to legal zoning formalities, although 'illegally' built pop-up housings are obviously less likely to be registered and/or well-documented in scientific literature.

5.1.2. Building Flexibility

Very few (~7%) objects exhibited appropriation of public open space, while, simultaneously, these pop-up housings seemed to be readily expandable during their specific life span. The majority (93%) of the analyzed objects were not designed with a potential expansion in mind as one could suggest regarding the changing needs of their users.

5.1.3. Environment

A vast majority (~75%) of the investigated pop-up housings were inhabited all-year-round, excluding emergency, short-term and basic shelters, as well as solely experimental setups. Those objects that are being used throughout the year were most commonly based in urban environments. It is essentially the micro-compact and 3D-printed housings, as well as structures on wheels, such as mobile homes, that are not necessarily built or placed in urban regions, although the target area for compact (3D build) housing concepts is definitely the urban region as well. To date, micro-compact and 3D-printed housing concepts remain as prototypes without real-life applications, although this could change rather quickly depending on progress in technology and material advances.

5.1.4. Infrastructure/Technical Layout

Although some (~10%) of the investigated objects, like refugee and emergency shelters, only have access to a public bathroom and toilet, most of the analyzed pop-up housing structures do incorporate in-house sanitary facilities, even if about 40% of these objects have only shared access to a toilet/bathroom. Those housings with private sanitary rooms (toilet/bathroom) typically also incorporate a private kitchen or cooking facility. A tendency towards shared cooking facilities can be observed in shared housing, e.g., student dorms/halls and flat-sharing solutions for migrants and asylum seekers.

Apparently, it is very difficult to obtain detailed and reliable data on energy supply, i.e., thermal energy for heating/DHW purposes, as well as supply of electrical energy and sufficient data on (waste) water infrastructure. However, the results from the performed cluster analysis reveal that pop-up housings claiming energy autarchy also (have to?) operate water purification plants. This suggests that 'overall autarchy' is basically a need in remote areas where public energy and water supply are not easily accessible.

The more basic pop-up structures are seldom connected to the power grid or centralized water supply and have a sewage connection or even municipal waste collection. These infrastructural attributes, in most cases, only apply to more solidly built and profoundly positioned housings, which are often owned by their inhabitants.

5.1.5. Construction

The need and justification for building an underground cellar is found only in a few (~10%) permanently built objects. Above all, these structures are mostly used for temporary housing under rental conditions (e.g., temporary hotels).

Nearly 60% of the analyzed objects, particularly micro-compact and 3D-printed housings, are based on a reversible, i.e., reusable foundation on a private plot. Apparently, because of this, these typically modular structures are also built according to the one-unit-per-floor concept and often, due to the static structure and used materials, do not incorporate a second floor or multi-unit-layouts.

The more urgent the need for accommodation, as is the case for migrants and refugees, the more straightforward and faster it is to build or prepare the pop-up housings. Therefore 'simple and quick' solutions like basic shelters, modified containers or converted multi-use buildings are preferred. Shelters and 'camps' are necessarily realized by incorporating lightweight walls for frame construction, often supported by poles/ropes and by using lightweight materials, i.e., textile and/or flexible, which can be transported easily and in a space-saving way.

About 75% of all investigated objects used lightweight, rigid materials for construction, whereas two-thirds of the analyzed structures consisted of a single housing unit with one floor. Sixty percent of the documented pop-up housings incorporated units that can be reused or recycled, which implies simple modular construction, removable foundations and recyclable materials. About 37% of the documented objects use wood as the primary building material and thermal insulation with conventional insulators.

5.2. Final Conclusions

This paper intends to showcase the structured and very effective application of hierarchical clustering algorithms to a set of seemingly inhomogeneous objects of pop-Up environments/temporary housing solutions from around the globe under interdisciplinary consideration. It tries to identify the necessary steps, i.e., the inherent procedure to be followed for performing effective and productive clustering of data. Furthermore, the authors intended to illustrate several possible options for analyzing and interpreting scattered information/data in detail by using hierarchical clustering algorithms and their derivative considerations and calculations, respectively. Finally, conclusive classification remarks are given, pointing out in detail the similarities and differences of the clustered set of PUE objects.

This procedure can be applied to all kinds of data to allow a neutral presentation of the connections between datasets. Thus, interpretation of connection clustering within large datasets can be performed objectively.

Supplementary Materials: The following are available online at https://www.mdpi.com/article/10.3390/app112311122/s1, Table S1: complete pattern matrix, Table S2: complete proximity matrix.

Author Contributions: Conceptualization, T.M., J.K. and C.P.; methodology, T.M., J.K. and C.P.; software, T.M.; validation, T.M. and J.K.; formal analysis, T.M. and J.K.; investigation, T.M., J.K. and C.P.; resources, T.M., J.K. and C.P.; data curation, T.M. and J.K.; writing—original draft preparation, T.M. and J.K.; writing—review and editing, J.K. and C.P.; visualization, T.M.; supervision, J.K. and C.P.; project administration, C.P.; funding acquisition, C.P. All authors have read and agreed to the published version of the manuscript.

Funding: This research has been funded by the Vienna Science and Technology Fund (WWTF) through project ESR17-010.

Institutional Review Board Statement: Not applicable.

Informed Consent Statement: Not applicable.

Data Availability Statement: Data supporting reported results can be found at: https://popupenvironments.boku.ac.at/ (accessed on 11 November 2021).

Conflicts of Interest: The authors declare no conflict of interest.

References

1. Tumminia, G.; Guarino, F.; Longo, S.; Ferraro, M.; Cellura, M.; Antonucci, V. Life cycle energy performances and environmental impacts of a prefabricated building module. *Renew. Sustain. Energy Rev.* **2018**, *92*, 272–283. [CrossRef]
2. Perrucci, D.V.; Vazquez, B.A.; Aktas, C.B. Sustainable Temporary Housing: Global Trends and Outlook. *Procedia Eng.* **2016**, *145*, 327–332. [CrossRef]
3. Abulnour, A.H. The post-disaster temporary dwelling: Fundamentals of provision, design and construction. *HBRC J.* **2014**, *10*, 10–24. [CrossRef]
4. Félix, D.; Monteiro, D.; Feio, A. Estimating the needs for temporary accommodation units to improve pre-disaster urban planning in seismic risk cities. *Sustain. Cities Soc.* **2020**, *61*, 102276. [CrossRef]
5. Arslan, H. Re-design, re-use and recycle of temporary houses. *Build. Environ.* **2007**, *42*, 400–406. [CrossRef]
6. Potangaroa, R. Sustainability by Design: The Challenge of Shelter in Post Disaster Reconstruction. *Procedia—Soc. Behav. Sci.* **2015**, *179*, 212–221. [CrossRef]
7. Mora, D.; Fajilla, G.; Austin, M.C.; de Simone, M. Occupancy patterns obtained by heuristic approaches: Cluster analysis and logical flowcharts. A case study in a university office. *Energy Build.* **2019**, *186*, 147–168. [CrossRef]
8. Wei, Y.; Zhang, X.; Shi, Y.; Xia, L.; Pan, S.; Wu, J.; Han, M.; Zhao, X. A review of data-driven approaches for prediction and classification of building energy consumption. *Renew. Sustain. Energy Rev.* **2018**, *82*, 1027–1047. [CrossRef]
9. Deb, C.; Lee, S.E. Determining key variables influencing energy consumption in office buildings through cluster analysis of pre- and post-retrofit building data. *Energy Build.* **2018**, *159*, 228–245. [CrossRef]
10. Yang, J.; Ning, C.; Deb, C.; Zhang, F.; Cheong, D.; Lee, S.E.; Sekhar, C.; Tham, K.W. K-Shape clustering algorithm for building energy usage patterns analysis and forecasting model accuracy improvement. *Energy Build.* **2017**, *146*, 27–37. [CrossRef]
11. Li, K.; Ma, Z.; Robinson, D.; Ma, J. Identification of typical building daily electricity usage profiles using Gaussian mixture model-based clustering and hierarchical clustering. *Appl. Energy* **2018**, *231*, 331–342. [CrossRef]
12. Heidarinejad, M.; Dahlhausen, M.; McMahon, S.; Pyke, C.; Srebric, J. Cluster analysis of simulated energy use for LEED certified U.S. office buildings. *Energy Build.* **2014**, *85*, 86–97. [CrossRef]
13. Pieri, S.P.; Tzouvadakis, I.; Santamouris, M. Identifying energy consumption patterns in the Attica hotel sector using cluster analysis techniques with the aim of reducing hotels' CO_2 footprint. *Energy Build.* **2015**, *94*, 252–262. [CrossRef]
14. Ma, Z.; Yan, R.; Li, K.; Nord, N. Building energy performance assessment using volatility change based symbolic transformation and hierarchical clustering. *Energy Build.* **2018**, *166*, 284–295. [CrossRef]
15. Yoganathan, D.; Kondepudi, S.; Kalluri, B.; Manthapuri, S. Optimal sensor placement strategy for office buildings using clustering algorithms. *Energy Build.* **2018**, *158*, 1206–1225. [CrossRef]
16. Huang, P.; Sun, Y. A clustering based grouping method of nearly zero energy buildings for performance improvements. *Appl. Energy* **2019**, *235*, 43–55. [CrossRef]
17. Schaefer, A.; Ghisi, E. Method for obtaining reference buildings. *Energy Build.* **2016**, *128*, 660–672. [CrossRef]
18. Tardioli, G.; Kerrigan, R.; Oates, M.; O'Donnell, J.; Finn, D.P. Identification of representative buildings and building groups in urban datasets using a novel pre-processing, classification, clustering and predictive modelling approach. *Build. Environ.* **2018**, *140*, 90–106. [CrossRef]
19. Owen, S.M.; MacKenzie, A.R.; Bunce, R.G.H.; Stewart, H.E.; Donovan, R.G.; Stark, G.; Hewitt, C.N. Urban land classification and its uncertainties using principal component and cluster analyses: A case study for the UK West Midlands. *Landsc. Urban Plan.* **2006**, *78*, 311–321. [CrossRef]
20. Santamouris, M.; Mihalakakou, G.; Patargias, P.; Gaitani, N.; Sfakianaki, K.; Papaglastra, M.; Pavlou, C.; Doukas, P.; Primikiri, E.; Geros, V.; et al. Using intelligent clustering techniques to classify the energy performance of school buildings. *Energy Build.* **2007**, *39*, 45–51. [CrossRef]
21. Ghiassi, N.; Mahdavi, A. Reductive bottom-up urban energy computing supported by multivariate cluster analysis. *Energy Build.* **2017**, *144*, 372–386. [CrossRef]
22. Gao, X.; Malkawi, A. A new methodology for building energy performance benchmarking: An approach based on intelligent clustering algorithm. *Energy Build.* **2014**, *84*, 607–616. [CrossRef]
23. Xinyi, L.; Runming, Y.; Meng, L.; Costanzo, V.; Wei, Y.; Wenbo, W.; Short, A.; Baizhan, L. Developing urban residential reference buildings using clustering analysis of satellite images. *Energy Build.* **2018**, *169*, 417–429. [CrossRef]
24. Dinh, D.-T.; Huynh, V.-N.; Sriboonchitta, S. Clustering mixed numerical and categorical data with missing values. *Inf. Sci.* **2021**, *571*, 418–442. [CrossRef]
25. Jain, A.K.; Dubes, R.C. *Algorithms for Clustering Data*; Prentice Hall-Inc.: Hoboken, NJ, USA, 1988; p. 334.
26. Jain, A.K.; Murty, M.N.; Flynn, P.J. Data Clustering: A Review. *ACM Comput. Surv.* **1999**, *31*, 60. [CrossRef]
27. Dinh, D.-T.; Fujinami, T.; Huynh, V.N. Estimating the Optimal Number of Clusters in Categorical Data Clustering by Silhouette Coefficient. In *Knowledge and Systems Sciences Communications in Computer and Information Science, Proceedings of the KSS 2019, Da Nang, Vietnam, 29 November–1 December 2019*; Springer: Singapore, 2019; Volume 1103. [CrossRef]

algorithms

MDPI

Article

Path Planning of a Mechanical Arm Based on an Improved Artificial Potential Field and a Rapid Expansion Random Tree Hybrid Algorithm

Qingni Yuan, Junhui Yi *, Ruitong Sun and Huan Bai

Key Laboratory of Modern Manufacturing Technology of Ministry of Education, Guizhou University, Guiyang 550025, China; qnyuan@gzu.edu.cn (Q.Y.); sunruitong2018@163.com (R.S.); baihuan2010@163.com (H.B.)
* Correspondence: junhuiyi163@163.com

Abstract: To improve the path planning efficiency of a robotic arm in three-dimensional space and improve the obstacle avoidance ability, this paper proposes an improved artificial potential field and rapid expansion random tree (APF-RRT) hybrid algorithm for the mechanical arm path planning method. The improved APF algorithm (I-APF) introduces a heuristic method based on the number of adjacent obstacles to escape from local minima, which solves the local minimum problem of the APF method and improves the search speed. The improved RRT algorithm (I-RRT) changes the selection method of the nearest neighbor node by introducing a triangular nearest neighbor node selection method, adopts an adaptive step and generates a virtual new node strategy to explore the path, and removes redundant path nodes generated by the RRT algorithm, which effectively improves the obstacle avoidance ability and efficiency of the algorithm. Bezier curves are used to fit the final generated path. Finally, an experimental analysis based on Python shows that the search time of the hybrid algorithm in a multi-obstacle environment is reduced to 2.8 s from 37.8 s (classic RRT algorithm), 10.1 s (RRT* algorithm), and 7.4 s (P_RRT* algorithm), and the success rate and efficiency of the search are both significantly improved. Furthermore, the hybrid algorithm is simulated in a robot operating system (ROS) using the UR5 mechanical arm, and the results prove the effectiveness and reliability of the hybrid algorithm.

Keywords: mechanical arm; path planning; artificial potential field method; rapid expansion random tree algorithm; virtual new node

Citation: Yuan, Q.; Yi, J.; Sun, R.; Bai, H. Path Planning of a Mechanical Arm Based on an Improved Artificial Potential Field and a Rapid Expansion Random Tree Hybrid Algorithm. *Algorithms* **2021**, *14*, 321. https://doi.org/10.3390/a14110321

Academic Editor: Rui Araújo

Received: 10 October 2021
Accepted: 31 October 2021
Published: 1 November 2021

1. Introduction

In recent years, China's logistics industry has developed rapidly to meet the increasing demand for e-commerce in response to the internet economy and the rapid development of warehousing automation technology [1]. The intelligent robotic arm industry is developing rapidly. The robotic arm, as its name implies, is designed to imitate a human arm for moving, grasping, lifting, and loading objects, among other operations [2]. Therefore, robotic arms are widely used in the logistics and warehousing industry. To grasp a specified object and place it in the specified position, it is necessary to bypass complex obstacles and find an efficient and collision-free path, which is very simple for humans but presents many technical problems that need to be considered for the robotic arm. Thus, for the robotic arm, its path planning is one of the most important technical problems. Successful path planning can greatly improve the storage and grabbing efficiency and has long been a research hotspot in robotic applications.

2. Related Research

Among the existing path planning algorithms, the artificial potential field (APF) method for path planning was first proposed by Khatib [3] in 1986. The idea is to use virtual force to make the robot navigate obstacles. The disadvantages of this algorithm are as follows: The algorithm easily falls into local minima or oscillates, and it is difficult to reach the target point when there are obstacles nearby. The rapid expansion random tree (RRT) algorithm was first proposed by the American professor LaValle [4] in 1998. The RRT method is a path planning algorithm based on sampling and an efficient multi-dimensional space with complete probability but is not optimal. However, the randomness of the RRT method causes it to be blind and exhibit a low efficiency; in addition, the resulting path is tortuous and not smooth enough, and the search speed is slow in a narrow area. A large number of scholars have made different improvements to these two algorithms. Zheng et al. [5] proposed a new minimum criterion and designed an improved virtual obstacle local path planning method to overcome the tendency of the APF algorithm to easily fall into local minima and other shortcomings. Sun et al. [6] proposed the use of dynamic windows to improve the APF method to solve the problem of being trapped in local minima. Zhang et al. [7] proposed a curved path planning algorithm for overtaking cars based on an improved APF method, and an optimal guaranteed performance control strategy for tracking the curved paths for overtaking cars based on linear robust control theory was proposed. Han et al. [8] proposed an improved APF method to solve the problems of large swinging trajectories and easily falling into local minima that are encountered by the classic APF method in unmanned aerial vehicle (UAV) trajectory planning. Zhang et al. [9] proposed a path planning method for multiple underwater unmanned vehicles (UUVs) in a three-dimensional environment based on the "domain", which solves the disadvantages of unreachable targets near obstacles, local minima, and oscillations encountered in the classic APF method. Tian et al. [10] proposed an overall configuration planning method of continuum hyper-redundant manipulators (CHRMs) based on an improved APF method that avoids complicated inverse kinematics and vastly reduces the computational complexity. Li et al. [11] proposed a path planning method for mobile mechanical arms based on the sparse node RRT algorithm that solves the problem of excessively searching in the local space and reduces the number of invalid nodes. Ge et al. [12] proposed a free-floating space robot (FFSR) trajectory planning method based on the dynamic RRT* algorithm, which can rapidly generate a feasible robot movement trajectory. Gan et al. [13] proposed a 1–0 goal-bias RRT algorithm to reduce the computational time and complexity, even in complex environments. Qureshi et al. [14] proposed the potential function-based RRT* (P-RRT*) method by incorporating the APF algorithm into the RRT* method. The proposed algorithm allows a considerable decrease in the number of iterations and thus leads to more efficient memory utilization and an accelerated convergence rate. Jeong et al. [15] proposed Quick-RRT* (Q-RRT*), a modified RRT* algorithm that generates a better initial solution and converges to the optimal solution faster than RRT*. Q-RRT* enlarges the set of possible parent vertices by considering not only a set of vertices contained in a hypersphere, as in RRT*, but also their ancestry up to a user-defined parameter, thus resulting in paths with less cost than those of RRT*. Wang et al. [16] proposed a novel learning-based multi-RRT (LM-RRT) approach for robot path planning in narrow passages. The LM-RRT approach models the tree selection process as a multi-armed bandit problem and uses a reinforcement learning algorithm that learns action values and selects actions with an improved epsilon-greedy (epsilon (t)-greedy) strategy. Lee et al. [17] proposed a motion planning algorithm by exploiting RRT stars (RRT stars) and dynamic movement primitives (DMPs). Hao et al. [18] proposed a Dubins-RRT* algorithm to involve the construction of a recovery path for an agricultural mobile robot (AMR) under kinematic constraints. The planned path avoids obstacles and incurs the minimum cost from a rendezvous point to the recovery platform. However, in general, the problems in the above research are as follows. (1) The improved APF algorithms cannot readily solve the problem of local minima in the search process, and the local minima cannot be adjusted in a timely manner. In addition, when there are obstacles at both the target point and the starting point, the obstacles cannot be effectively avoided to reach the target point quickly, and the generated path is relatively tortuous. (2) The improved RRT algorithms cannot quickly find a reliable path in a complex, multi-obstacle environment. Moreover, the algorithm cannot be adjusted well for different environmental conditions, which is not conducive to improving the algorithm efficiency, and the generated path is not sufficiently smooth, causing the mechanical arm to undergo impacts and become damaged in actual operation and severely shortening the service life of the mechanical arm.

In response to the above problems, the present paper proposes an improved hybrid three-dimensional path planning algorithm for mechanical arms that combines the APF method and the RRT algorithm. The proposed algorithm is used to solve the path planning problem of the manipulator in an environment with complex obstacles. Compared with the existing path planning algorithms, the main contributions of this article are as follows:

1. In the I-APF (I-APF) method, a heuristic method based on the number of adjacent obstacles to break away from the local minimum is proposed so that the algorithm can quickly eliminate local minima and break away from obstacles.
2. In the I-RRT (I-RRT) algorithm, a triangular nearest neighbor node selection method is proposed, which improves the convergence of the algorithm.
3. Based on the I-APF algorithm and the I-RRT algorithm, a hybrid algorithm is proposed that combines these two improved algorithms to search for the optimal path. First, the distance between the nearest neighbor node and the obstacle is judged. According to the different distances, the I-APF method and the I-RRT algorithm are used to expand the path, which improves the search speed and obstacle avoidance ability of the algorithm.

The rest of this article is organized as follows. Section 3 introduces the classic APF method and the classic RRT algorithm. Then, Section 4 introduces the improved APF and RRT hybrid algorithm. In Section 5, to smooth the path, the path is fitted with a Bezier curve. In Section 6, the planned path is verified by an experimental simulation using Python language development tools and robot operating system (ROS) tools.

3. Background

3.1. Principles of the Classic APF Method

The APF method makes the object move and reach the target point under the action of a force field, which includes a gravitational field and a repulsive field [19–21].

The gravitational field function $U_{att}(q)$ is as follows:

$$U_{att}(q) = \frac{1}{2}\varepsilon\rho^2(q, q_{goal}) \tag{1}$$

where ε is the scale factor, $\rho(q,q_{goal})$ represents the Euclidean distance between the target object q_{goal} and the current position q, and the gravitation $F_{att}(q)$ of the corresponding target object is the derivative of the gravitational field:

$$F_{att}(q) = -\nabla(U_{att}(q)) = \varepsilon(q_{goal} - q) \tag{2}$$

The repulsive field $U_{rep}(q)$ function is as follows:

$$U_{\mathrm{rep}}(q) = \begin{cases} \frac{1}{2}\eta\left(\frac{1}{\rho(q,\, q_{obs})} - \frac{1}{\rho_0}\right)^2 & ,if\ \rho(q, q_{obs}) \le \rho_0 \\ 0 & ,if\ \rho(q, q_{obs}) > \rho_0 \end{cases} \tag{3}$$

where η is the repulsion scale factor, $\rho(q,q_{obs})$ represents the Euclidean distance between obstacle q_{obs} and current position q, and ρ_0 represents the influence radius of each obstacle.

Then, the repulsion $F_{rep}(q)$ is the derivative of the repulsive field:

$$F_{rep}(q) = -\nabla U_{rep}(q) = \begin{cases} \eta(\frac{1}{\rho(q,q_{obs})} - \frac{1}{\rho_0})\frac{1}{\rho^2(q,q_{obs})}\nabla\rho(q, q_{obs}) & ,if\ \rho(q, q_{obs}) \le \rho_0 \\ 0 & ,if\ \rho(q, q_{obs}) > \rho_0 \end{cases} \tag{4}$$

where $\nabla\rho(q,q_{obs})$ represents the derivative of $\rho(q,q_{obs})$, the target object generates gravitation, the object is guided to move towards the target object, and the obstacle generates repulsion to avoid the obstacle, as shown in Figure 1. The resultant force field $U(q)$ (resultant force $F(q)$) that an object receives at any point in the field is equal to the vector sum of the target object's gravitational field (gravity) and the total repulsive field (repulsion) of obstacles encountered. The formula is as follows:

$$U(q) = U_{att}(q) + U_{rep}(q) \tag{5}$$

$$F(q) = -\nabla U(q) \tag{6}$$

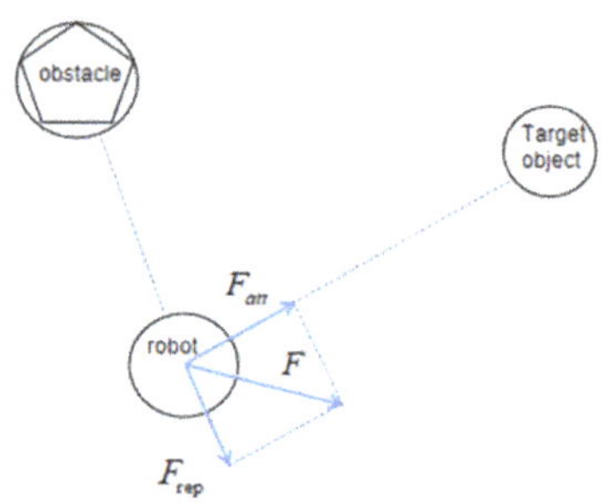

Figure 1. Diagram illustrating the principle of the APF method.

3.2. Principles of the Classic RRT Algorithm

The principle of the classic RRT algorithm is to use an initial point as the root node q_{init}. In each subsequent expansion, a random node q_{rand} is randomly generated in the map. The nearest function selects a node $q_{nearest}$ to q_{rand} based on the Euclidean distance on the existing random tree and expands this node in the direction of q_{rand} with a step distance through the extend function to obtain a new node q_{new}. Evaluating whether q_{new} collides with an obstacle proceeds as follows. If it collides, growth is abandoned, q_{new} is removed, and a random node is regenerated; if there is no collision, q_{new} is added to the random tree, and q_{new}'s parent node is assigned to $q_{nearest}$, as shown in Figure 2. When the distance between the node of the random tree and the target point is less than a specific value m, the program is terminated, as a collision-free path from the starting point to the target point has been obtained. The above steps are followed until reaching the target point, as shown in Figure 3.

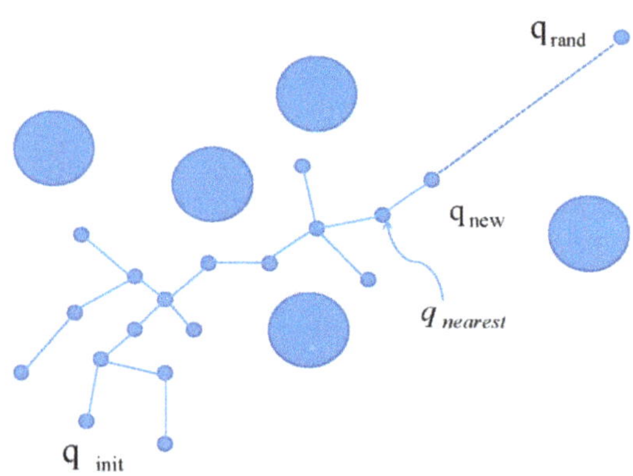

Figure 2. Classic RRT mind map.

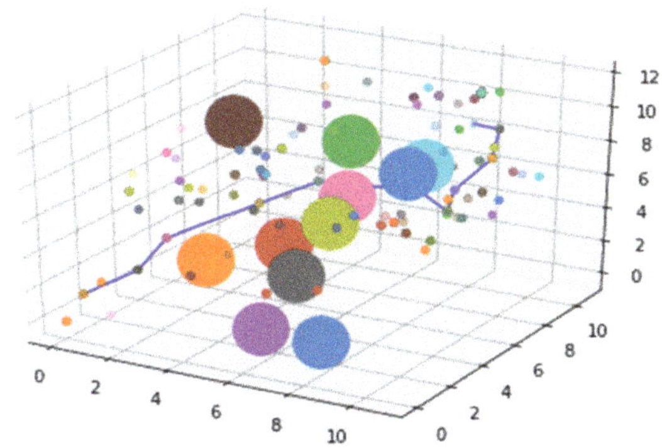

Figure 3. Path planning diagram of the classic RRT algorithm.

4. Improved Algorithms

This section first conducts a collision detection analysis of obstacles, and the results are used to plan the path of the UR5 manipulator. An I-APF method is proposed to avoid local minima of APF, and an I-RRT algorithm is proposed to improve upon the slow convergence speed and poor search efficiency of the classic RRT algorithm. However, the I-APF method leads to tortuous paths and unreachable targets when there are obstacles at the start and end points, and the I-RRT algorithm still has a low search efficiency in an environment with complex obstacles. Notably, the hybrid algorithm that combines the I-APF method and the I-RRT algorithm overcomes the shortcomings of these respective algorithms. Finally, the principles and implementation steps of the hybrid algorithm are given below.

4.1. Collision Detection

A collision detection analysis is carried out. This article uses the universal robot UR5 mechanical arm for the research, uses the geometric envelope in space to simplify the obstacles and mechanical arm model, establishes an environment perception model through sensors, establishes an environmental model map, and divides the map into an obstacle space and obstacle-free space [22]. Usually, obstacles are irregular. To facilitate the calculation, the obstacles are usually idealized as an enclosed ball, and the joints of the robotic arm are idealized as cylindrical. In this way, the computational load can be reduced [23]. Suppose the coordinates of the center of the sphere are (x, y, z), the radii of the spheres are R_1 and R_2, and the radius of the cylinder is r; then, the distances between the coordinates of the centers of the two spheres and the central axis of the cylinder are calculated, denoted d_1 and d_2. As shown in Figure 4, when $d > r + R$, the robotic arm does not collide with the obstacle; otherwise, the arm collides with the obstacle. This method can greatly improve the computational efficiency.

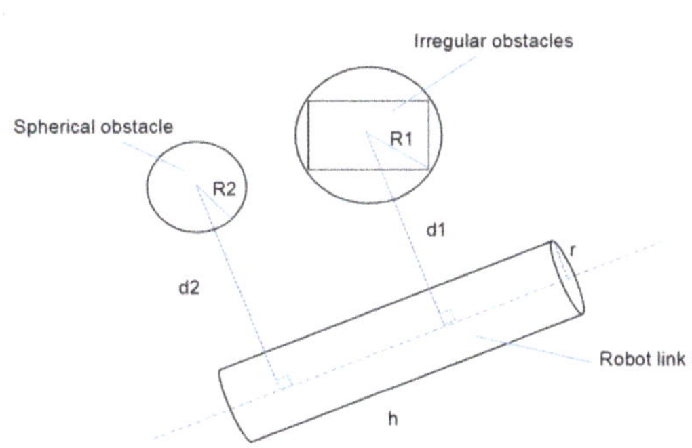

Figure 4. Collision detection model.

4.2. I-APF Algorithm

According to the principles of the classic APF method, the algorithm has some shortcomings. When there is an obstacle at the starting point, the repulsive force causes the path to be tortuous. When the target point is near an obstacle, the obstacle makes it difficult for the object to reach the target point. When the gravitational force and the repulsive force on the object are equal, objects fall into local minima.

In view of the local minima problem of the classic algorithm, some scholars have proposed the idea of using virtual sub-targets [24] to make the robot escape from the local minimum. However, the location of the virtual sub-target is random, which inevitably leads to the blindness of the algorithm; that is, the robot deviates from the target or enters an obstacle area. This paper proposes a heuristic method for deviating from local minima based on the number of adjacent obstacles. The method includes three steps. Step 1: Make a judgement on the local minimum. Step 2: If the object is stuck, take the local minimum as the center, draw a circle with twice the step length in the RRT algorithm as the radius, record the number of obstacles in the circle as O, and record the total number of obstacles in the space as S. Step 3: Introduce a heuristic function to calculate the new potential field force. The specific process is as follows.

When the object falls into a local minimum, the gravitational field and repulsive force field received by the robotic arm are equal in size but in opposite directions. By judging the size and direction of the gravitational field and repulsive force field of the robotic arm, whether the robotic arm falls into a local minimum can be evaluated. When the robotic arm falls into a local minimum, the number of nearby obstacles is obtained by using twice the step length in the RRT algorithm as the radius, and then a heuristic function is established to calculate the new force field. The heuristic function F is as follows:

$$F = \alpha(O/S)M + \beta(1 - O/S)N \tag{7}$$

where α and β are the repulsion scale factor and the gravitational scale factor, respectively, O is the number of obstacles, S is the total number of obstacles in space, M represents the repulsive force of the obstacles in the circle on the robotic arm, and N represents the gravitational force of the target point on the robotic arm. According to the heuristic function, the more obstacles there are in the circle, the greater the repulsive force of the robot arm will be, and the path will expand in the direction deviating from the obstacle to the target point, which is beneficial for the robot arm to quickly avoid the obstacle. Figure 5 shows a path planning diagram of the I-APF method.

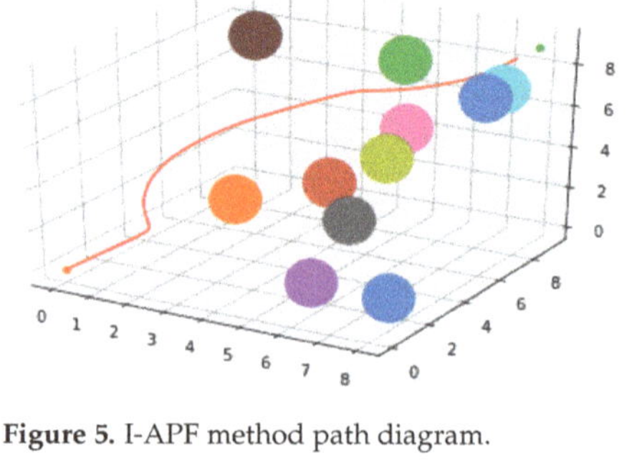

Figure 5. I-APF method path diagram.

4.3. I-RRT Algorithm

The principle of the classic RRT algorithm shows that the algorithm has randomness and a strong obstacle avoidance ability, but this inevitably increases the search time and reduces the efficiency, and the search path is long and tortuous, which takes up a relatively large amount of memory. When there are many obstacles, the efficiency of the algorithm is greatly reduced. Therefore, this paper improves on the classic RRT algorithm and proposed the I-RRT algorithm. The specific improvements are as follows.

4.3.1. Path Collision Detection

The classic RRT algorithm performs collision detection for nodes but does not conduct collision detection along the path. If the path results in a collision with obstacles, the generated path does not meet the actual requirements. In this paper, collision detection is performed on the path in the Collision_check_line() function. The specific method is to pass between two nodes in the random tree, calculate the Euclidean distance between the two nodes in the three-dimensional space, and set the division rate (Discretepoint). The number of divided points is equal to the distance between the two points divided by the division rate. The path is divided into many points, and whether these points collide with obstacles is evaluated.

4.3.2. Goal-Bias Strategy

According to the classic algorithm, the random tree is expanded by selecting random points, resulting in a low path search efficiency. Therefore, this paper adopts the idea of target bias [25]. Random point generation is used to select the target point with a certain probability $P_{goalDampleRate}$ and can effectively reduce the blindness of the algorithm. When the random probability is greater than $P_{goalDampleRate}$, a random point is generated, and when the random probability is less than $P_{goalDampleRate}$, the target point is taken as the random point, which speeds up the convergence rate.

4.3.3. Triangular Nearest Neighbor Node Selection Strategy

The traditional nearest neighbor node selection strategy uses the node in the tree with the closest Euclidean distance to the random sampling point as the nearest neighbor node. This paper proposes a triangular nearest neighbor node selection method, and the specific steps are as follows: If a random sampling point is not the target point, the connections between the random sampling point, the node in the tree, and the target point are established to form a triangular area; then, the sum of the distances of the three sides of the triangle is calculated as the cost function of the nodes. Then, the cost function can be expressed as:

$$\text{Cost}\ (q) = \left(||q_{goal} - q_{rand}||^2 + ||q_{rand} - q||^2 + ||q_{goal} - q||^2\right) \tag{8}$$

Then, the nearest neighbor node in the triangle is:

$$q_{nearest} = \{q \in T_{node} \mid \text{Cost}(q_i) = \min\{\text{Cost}\ (q)\}\} \tag{9}$$

The triangular nearest neighbor node selection can combine the nearest Euclidean distance to the random point and the nearest Euclidean distance to the target point to select the nearest neighbor node, which improves the convergence efficiency of the algorithm and further reduces the blindness of the algorithm.

4.3.4. Adaptive Step Size

The classic algorithm has a fixed step length for expansion; approaching obstacles cannot be avoided well, causing collisions and occupying a large amount of memory, and the generated step length grows towards a random point without directionality. Both of these issues lead to a reduction in the efficiency of the algorithm, so an adaptive step size strategy is adopted [26]. When the minimum distance between $q_{nearest}$ and the obstacle is greater than the step size, the obstacle is marked as s, the idea of gravity in the APF method is introduced into the adaptive step size of the RRT algorithm [27], and the random tree is guided to grow towards the target. On the basis of the original RRT algorithm growing towards random points, the step component G(n) towards the target is added so that the new node has a tendency to deviate towards the target point. The formula of step F(n) is defined as follows:

$$\mathrm{F(n)} = \mathrm{R(n)} + \mathrm{G(n)} \tag{10}$$

where F(n) represents the traction step length of the nth node, R(n) represents the traction step length of the nth node by a random point, and G(n) represents the traction step length of the nth node by the target object. The gravitational potential energy U of the target point to the nearest node is known by the APF method and is defined as follows:

$$U = \frac{1}{2}K_p * \left\| x_{goal} - x_{near} \right\|^2 \tag{11}$$

where k_p represents the gravitational coefficient and $\|x_{goal} - x_{near}\|$ represents the absolute value of the Euclidean distance between the position vector x_{goal} of the target point and the position vector x_{near} of the nearest node. Then, the gravitational force G is the derivative of the gravitational potential energy U, namely,

$$G = K_p * \left\| x_{goal} - x_{near} \right\| \tag{12}$$

Then

$$G(n) = \rho * k_p \frac{x_{goal} - x_{near}}{\left\| x_{goal} - x_{near} \right\|} \tag{13}$$

where ρ represents the step length of random growth; then, $R(n)$ can also be deduced as

$$R(n) = \rho * \frac{x_{rand} - x_{near}}{\| x_{rand} - x_{near} \|} \tag{14}$$

where $\|x_{rand} - x_{near}\|$ represents the absolute value of the Euclidean distance between the position vector $\mathrm{x_{rand}}$ of the random point and the position vector x_{near} of the nearest node. By inserting Equations (13) and (14) into Equation (10) to obtain the traction step length of the nearest node as Equation (15), defined as $F_1(n)$, the following formula is obtained:

$$F_1(n) = \rho * \frac{x_{rand} - x_{near}}{\| x_{rand} - x_{near} \|} + \rho * k_p \frac{x_{goal} - x_{near}}{\left\| x_{goal} - x_{near} \right\|} \tag{15}$$

When the minimum distance to the obstacle is less than the step length, two situations are possible.

Case 1: First, a virtual new node is generated according to the above steps, and whether the distance between the virtual new node and obstacle s is less than the distance between the nearest node and the obstacle is evaluated. If the distance is less than the distance between the nearest node and the obstacle, then the virtual new node has a tendency to approach the obstacle is proven. Then, the virtual new node is removed, the step length of the new node growth is changed, and the step length is reduced on the basis of the step length $F_1(n)$, which is defined as the step length $F_2(n)$. The formula is as follows

$$F_2(n) = \frac{\mathrm{dist}_2}{\mathrm{dist}_1}\left(\rho * \frac{x_{rand} - x_{near}}{\| x_{rand} - x_{near} \|} + \rho * k_p \frac{x_{goal} - x_{near}}{\left\| x_{goal} - x_{near} \right\|}\right) \tag{16}$$

where dist_1 represents the distance between the nearest node and obstacles and dist_2 represents the distance between the virtual new node and obstacles. Whether the node collides with an obstacle is evaluated. If it collides, a random node is regenerated; if there is no collision, a new node is added to the random tree.

Case 2: If the distance between the virtual new node and obstacles is greater than the distance between the nearest node and obstacles, it proves that the new node has a tendency to grow away from the obstacle. Then, the step length of the new node generation adopts the step length of the classic RRT algorithm, defined as the step length $F_3(n)$, and the formula is as follows:

$$F_3(n) = \rho * \frac{x_{rand} - x_{near}}{\|x_{rand} - x_{near}\|} \quad (17)$$

Then, whether the node collides with an obstacle is evaluated; if it collides, a random node is regenerated, and if there is no collision, a new node is added to the random tree.

4.3.5. Removing Redundant Nodes

Due to the randomness of the classic RRT algorithm, the path may oscillate. Redundant nodes are removed to process the path from the starting point to the target point generated by the random tree [28]. Starting from the first node q_{init}, the subsequent path nodes are connected in turn, the second node is ignored, and the third node is connected. If the object does not collide with the obstacle, the second node on the path is deleted. If there is a collision, then the node is retained, the parent node of the collision point is used as the new evaluation node, and the above steps are repeated until the target point is reached. The final series of reserved nodes is saved into Path 2 and connected to obtain the path after removing redundant nodes, as shown in Figure 6. The collision here refers to the collision detection of the path mentioned in Section 4.3.1. Figure 7 shows the path diagram generated by the I-RRT algorithm. Algorithm 1 shows the pseudocode of the I-RRT algorithm.

Algorithm 1: I-RRT

```
initialization
S ← q_init;
while true do
    q(rand)=random() or end();
    if q(rand) = random()
        q(nearest) ← Triangle_nearest_list_index(node_list, q(rand));
    dist1 ← Nearest distance(q(nearest),obstacles);
    if dist1 > expandDis
        q(new) ← Extend(expandDis,q(nearest), Direction Angle)
    else
        q(new)=virtual(Extend(expandDis,q(nearest),DirectionAngle)
        dist2 ← Nearest_distance(q(new), obstacles)
        if dist1 > dist2
            q(new) ← Extend(expanddis,q(nearest),Direction Angle)
        else
            q(new) ← Extend(expandDis,q(nearest),Direction Angle)
    if collision_check(q(new))
        Tree.add(q(new))
    if dist(q(new),q(end))< expandDis
        return Goal
    else
        continue
return path
path2 ← remove_redundant_nodes(path)
return Graph
```

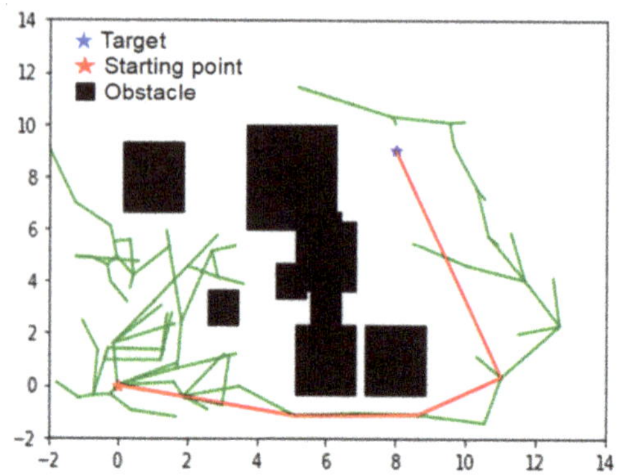

Figure 6. Diagram of the I-RRT algorithm implementation to remove redundant nodes.

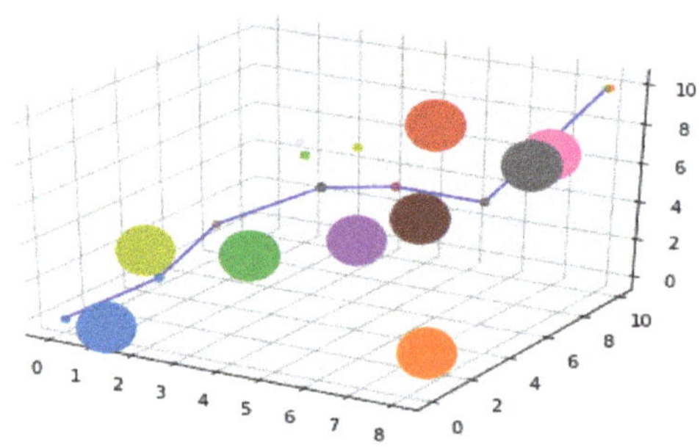

Figure 7. I-RRT algorithm generation path diagram.

4.4. *Improved Hybrid Algorithm of the APF and RRT*

The I-APF method deals with the local minimum problem in the classic algorithm, but when there are obstacles at the starting point, the repulsion causes a tortuous path. When there are obstacles at the target point, the target is unreachable, and vibration occurs. The I-RRT algorithm significantly improves the search efficiency, but its efficiency is still low in the case of multiple obstacles. This article adopts the strategy of combining the two improved algorithms (I-APF, I-RRT) to give full play to the advantages of the two algorithms to avoid defects.

4.4.1. Principle of the Improved Hybrid Algorithm

The principle of the hybrid algorithm is as follows. First, the tree node is initialized, and the distance between the nearest node and the obstacle is continuously detected. If it is detected that the minimum distance between the current node and the obstacle is greater than twice the step length, it means that there is no obstacle near the current node, and the I-APF method is used for rapid expansion. If the minimum distance between the current node and the obstacle is less than twice the step length, the I-RRT algorithm is adopted to make full use of the efficient obstacle avoidance ability of the RRT algorithm, and the above steps are repeated until the target point is reached. The hybrid algorithm can effectively improve the efficiency of path searching and resolve the following problem of the APF method: When there are obstacles at the starting point, the repulsive force causes a tortuous path, and the target is unreachable when there is an obstacle at the target point. Furthermore, the hybrid algorithm also solves the problem of the classic RRT algorithm, which has a significantly lower efficiency when there are more obstacles, and the generated path is shorter and smoother.

4.4.2. Improved Hybrid Algorithm Implementation

According to the principle of the improved APF and RRT hybrid algorithm, the specific implementation steps are described as follows.

Step 1: Initialize the parameters, and define the obstacle environment, starting point, target point, step length, and target sampling rate.

Step 2: Determine the distance between the current node and the obstacle. If the distance between the current node and the nearest obstacle is greater than twice the step length, execute Step 3. If the distance between the current node and the nearest obstacle is less than twice the step length, execute Step 4.

Step 3: Use the I-APF method to search and move forward under the combined force of the target and obstacle. (1) Calculate the gravitational and repulsive forces. (2) Determine whether the gravitational and repulsive forces experienced by the current node are equal and opposite. If they are equal, the object falls into a local minimum, the heuristic method based on the number of adjacent obstacles is used to escape from the local minimum so that the algorithm escapes from the local minimum, and the end effector of the robotic arm is guided to continue to move. If they are not equal, proceed to Step (3). (3) If the distance between the current node and the target point is less than the step length or if the distance between the nearest node and the nearest obstacle is less than twice the step length, then the I-APF method search process is ended, the path node obtained by the APF method is added to the Pathpath, and the Pathpath and the latest node q(new) are returned. Otherwise, jump to Step (1).

Step 4: If the distance between the current node and the nearest obstacle is less than twice the step length, the I-RRT algorithm is used to search for the path. (1) Initialize the tree, set the initial node and target point, and define the step size, target sampling rate, and segmentation rate. (2) Start the iteration and sample the state space. When the random probability is less than the target sampling rate, the sampling point selects the target point. If it is greater than the target sampling rate, random sampling points in the space are selected. (3) Select the nearest neighbor node according to the triangle nearest neighbor node selection method and calculate the distance dist_1 between this node and the nearest obstacle. (4) Determine whether the distance dist_1 is greater than the step length. If so, the step length $F_1(n)$ is used for expansion. If not, generate a virtual new node according to the step length $F_1(n)$ and calculate the distance dist_2 between the virtual new node and the nearest obstacle. (5) If $\text{dist}_2 < \text{dist}_1$, the new node has a tendency to move towards obstacles. Remove the virtual new node and use step $F_2(n)$ to expand. If $\text{dist}_2 > \text{dist}_1$, the new node has a tendency to move away from obstacles; remove the virtual new node and use the step size $F_3(n)$ to expand. (6) Determine whether the new node collides with obstacles; if there is a collision, skip to Step (2). Re-sample random points; if there is no collision, add the new node to the tree and assign the parent node of the new node to the nearest node (q_{nearest}). (7) When the distance between the new node and the target point is less than the step length or the distance between the nearest node and the nearest obstacle is greater than twice the step length, end the iterative process; otherwise, skip to Step (2). (8) Use the collision detection method of the path to perform the process of removing redundant nodes on the generated path to obtain the processed path. (9) Add the processed path node to the Pathpath and return the Pathpath and the latest node q(new).

Step 5: Determine whether the distance between the new node and the target point is less than the step length. If so, reach the target point, connect the new node and the target point, output the path graph, obtain a collision-free path from the start point to the end point, and exit the program. Otherwise, skip to Step 2.

According to the specific implementation steps of the hybrid algorithm, Figure 8 shows the flowchart of the hybrid algorithm and Algorithm 2 shows the pseudocode of the hybrid algorithm.

Algorithm 2: APF-RRT

```
Initialization;
While True do
  if distance(q(new),obstacle_list)>2*expandDis
      q(new) ← q(recent_node);
      q(new) ← Extend(F_att, F_rep)
      Path.append(q(new))
      if dist(q(new),q(end))< expandDis
          return Goal
      else
          Continue
  return Path recent_node
  Pathpath.append(path)
  else
      q(new) ← q(recent_node);
      q(rand)= random() or end();
      if q(rand)= random()
        q(nearest) ← Triangle_nearest_list_index(node_list,q(rand));
      dist1 ← Nearest distance(q(nearest), obstacles);
      if dist1 > expandDis
        q(new) ← Extend(expandDis,q(nearest), Direction Angle)
      else
        q(new)=virtual(Extend(expandDis,q(nearest),Direction Angle))
        dist2 ← Nearest_distance(q(new), obstacles)
        if dist1 > dist2
          q(new) ← Extend(expanddis, q(nearest), Direction Angle)
        else
          q(new) ← Extend(expandDis,q(nearest),Direction Angle)
      if collision_check(q(new))
        Tree.add(q(new))
      if dist(q(new),q(end))< expandDis
        return Goal path
      else
        continue
  path2 ← remove_redundant_nodes(path)
  return Path2 recent_node
  Pathpath.append(path2)
return Pathpath Graph
```

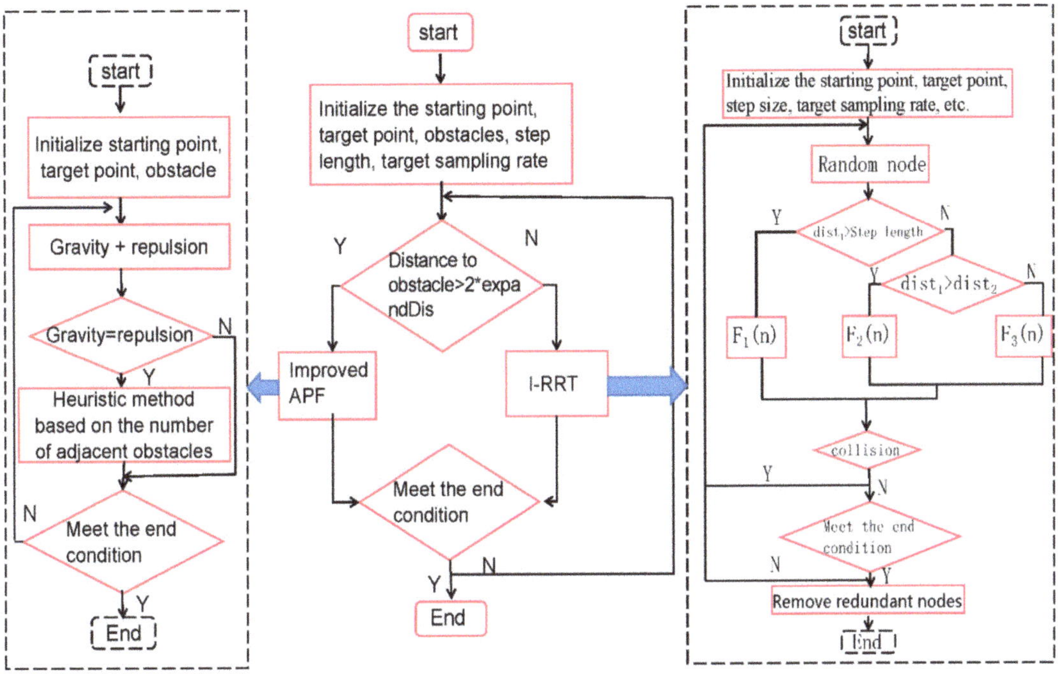

Figure 8. APF-RRT algorithm flow chart.

5. Bezier Curve Path Smoothing

Aiming at the phenomenon that the path generated by the algorithm has turning points and is not sufficiently smooth, reducing the performance of the robot arm due to its acceleration in actual operation, this paper uses Bezier curves to smooth the path [17,29]. Smoothing is realized on the basis of the original path, and $n + 1$ nodes obtain the formula of an n-order Bezier curve:

$$C(u) = \sum_{i=0}^{n} B_{n,i}(u)p_i, \qquad u \in [0,1] \tag{18}$$

where P_i represents $n + 1$ points in space and the weight coefficient $B_{n,i}(u)$ with the parameter u is the Bernstein basis function. The calculation method is as follows:

$$B_{n,i}(u) = \frac{n!}{i!(n-i)!}u^i(1-u)^{n-i} \tag{19}$$

The final generated curve has a relationship with each of the $n + 1$ points. These points determine the final direction of the curve and are called control points. The Bessel order in Equation (18) is n and is controlled by the $n + 1$ control points. The start point and end point correspond to $u = 0$ and $u = 1$, respectively. The curve after Bezier fitting is shown in Figure 9a. The slight gap between the fitted curve and the original path may risk collision between the path and the obstacle. It can be seen from the figure that the fitting curve is likely to result in a collision with the obstacle only when the obstacle environment is very complex. For the simulation experiment, the success rate of the 200-path fitting experiment in this paper is 100%.

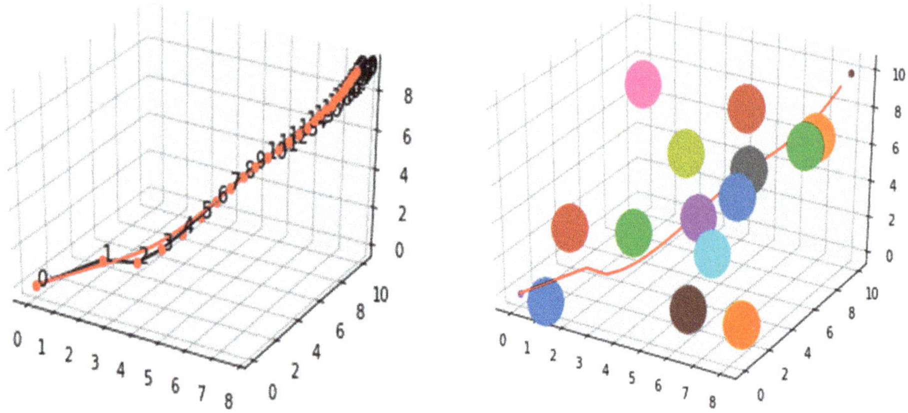

Figure 9. Comparison of the Bezier fitted curve and original path.

6. Simulation and Experiment (Python and ROS Simulation)

In this section, the improved APF and RRT hybrid algorithm is verified experimentally. To verify that the algorithm can maintain excellent results in a multi-obstacle environment, this hybrid algorithm is compared with the RRT, RRT*, and P_RRT* algorithms. The RRT* algorithm is a landmark algorithm among the improved RRT algorithms. It has a higher convergence rate and has been widely studied by scholars, while the P_RRT* algorithm introduces the idea of the APF method on the basis of RRT*, making the P_RRT* algorithm one of the path planning algorithms with the highest convergence efficiency. Experiments were carried out on different obstacle environments to verify the effectiveness of the algorithm. This experiment used Python language development tools on a laboratory desktop HP computer with 4-GB memory and an Intel(R) Core (TM)i5-6500 CPU @3.20 GHz–3.19 GHz to find a smooth and collision-free path. Combining theory with reality, this work chose a simulation robot, the UR5 of the Danish UAO Company, for simulation experiments. The UR5 is a six-degree-of-freedom manipulator. Table 1 shows the motion control parameters of the UR5 robotic arm.

Table 1. Manipulator motion control parameter table.

Parameter	Value
Search step ρ	1.0 dm
Starting point	(0,0,0)/dm
Ending point	(8,10,10)/dm
Gravitational coefficient ε, K_p	0.05
Repulsion coefficient η	100.0
α	0.4
β	0.6
Obstacle influence radius ρ_0	0.3 dm
Target sampling rate	0.1

6.1. Experiments and Analysis in Python

Experiment 1. A comparison among the algorithms for different numbers of obstacles was carried out to verify that the improved APF and RRT hybrid algorithm maintained a better search effect in the case of multiple obstacles. With a gradually increasing number of obstacles, the advantages and disadvantages of the proposed path planning algorithm were compared with those of the classic RRT algorithm, the RRT* method, and the P_RRT* algorithm. Each group of experiments was performed 200 times, and the average of the results is shown in Table 2, where the search success rate was that the search was within 100 s of a successful search, and the search was unsuccessful outside of 100 s.

Table 2. Comparison of various algorithms for different obstacle numbers.

Obstacle	Average Search Time				Average Path Length				Average Number of Sampling Nodes				Search Success Rate			
	RRT				RRT*				P_RRT*				APF-RRT			
10	20.0	22.5	70.4	100%	7.7	21.0	20.4	100%	6.9	20.6	20.5	100%	2.0	18.0	10.3	100%
12	21.0	22.8	74.1	100%	8.13	21.2	20.6	100%	7.3	20.7	20.8	100%	2.1	18.0	10.4	100%
14	37.8	23.5	95.7	95%	10.1	21.8	21.9	100%	7.4	20.7	21.0	100%	2.8	18.0	11.5	100%
16	40.9	23.6	111.2	90%	10.3	21.8	23.8	100%	7.8	21.1	21.3	100%	3.2	19.1	11.9	100%

According to Table 2, the search time, path length, number of sampling nodes, and search success rate of the algorithms are compared. The search time of the classic RRT algorithm was relatively short when there are few obstacles. When the number of obstacles gradually increased, the search time greatly increased, and the search success rate decreased. Compared with the classic RRT algorithm, the RRT* method was superior. The average search time and the average number of sampling nodes were greatly improved, and the algorithm maintained good results when the number of obstacles gradually increased. Compared with the RRT* algorithm, the P_RRT* algorithm exhibited further improvements, and the search efficiency was higher. With an increasing number of obstacles, stable search results can be maintained, but compared with the improved APF and RRT hybrid algorithm in this article, it still had the disadvantages of a low search efficiency, tortuous paths, and high average number of sampling nodes, which consumed more computing memory. The hybrid algorithm in this paper showed a better effect when the number of obstacles gradually increased.

Experiment 2. To analyze the effectiveness of the improved APF and RRT hybrid algorithm, this work compared various algorithms under the same conditions of obstacles and step lengths. Each set of experiments was performed 200 times, and the average of the results is shown in Table 3.

Table 3. Comparison of various algorithms under the same conditions.

	Average Search Time	Average Number of Sampling Nodes	Average Path Length	Search Success Rate
RRT	37.8	95.7	23.5	95%
RRT*	10.1	21.9	21.8	100%
P_RRT*	7.4	21.0	20.7	100%
APF-RRT	2.8	11.5	18.0	100%

According to the data in Table 3, the classic RRT algorithm had a slow search speed, a large number of sampling nodes, and tortuous paths when there were many obstacles, as shown in Figure 10a. Compared with the classic RRT algorithm, the RRT* algorithm had a great improvement in the average search time, the path planning efficiency was higher, and the generated path was smoother, as shown in Figure 10b. The average search time of the P_RRT* algorithm was shorter than that of the RRT* algorithm. However, compared with the improved APF and RRT hybrid algorithm in this paper, there were still shortcomings, such as a low search efficiency, tortuous paths, and a large demand on the computing memory. The improved hybrid algorithm in this paper still showed excellent results in the case of many obstacles, the search efficiency was higher, the path was shorter and smoother, and it overcame the phenomenon of tortuous paths and unreachable targets in the APF method when there were obstacles near the starting point and target point. To a certain extent, the blindness of the RRT algorithm was reduced, and the efficient obstacle avoidance ability of the RRT algorithm was fully utilized, as shown in Figure 10d.

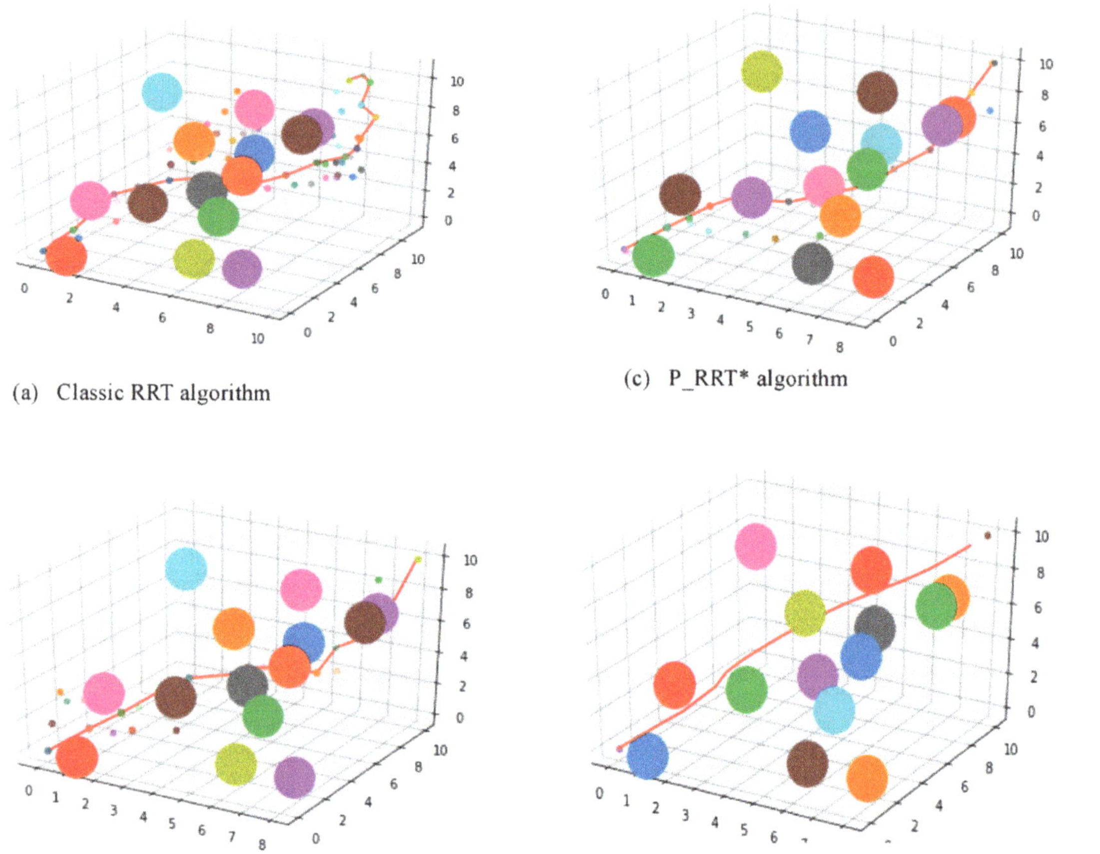

(a) Classic RRT algorithm
(c) P_RRT* algorithm
(b) RRT* algorithm
(d) APF-RRT

Figure 10. Comparison of paths generated by the different algorithms under the same conditions, such as the number of obstacles and step lengths.

The comparison and analysis with the classic RRT algorithm, RRT* algorithm, and P_RRT* algorithm verified the effectiveness of the improved hybrid algorithm in this paper.

Experiment 3. The reliability of the algorithm was evaluated by changing the step length of the RRT algorithm in the hybrid algorithm. Each step was carried out 200 times. The experimental results are shown in Table 4, and the trajectory of different step lengths is shown in Figure 11.

Table 4. Comparison of hybrid algorithms with different step sizes.

Step	Average Search Time	Average Number of Sampling Nodes	Average Path Length	Search Success Rate
0.6	6.2	13.8	19.4	100%
0.8	5.9	12.6	19.1	100%
1.0	3.2	11.9	19.1	100%
1.2	5.6	12.0	19.1	100%
1.4	6.7	12.7	19.2	100%

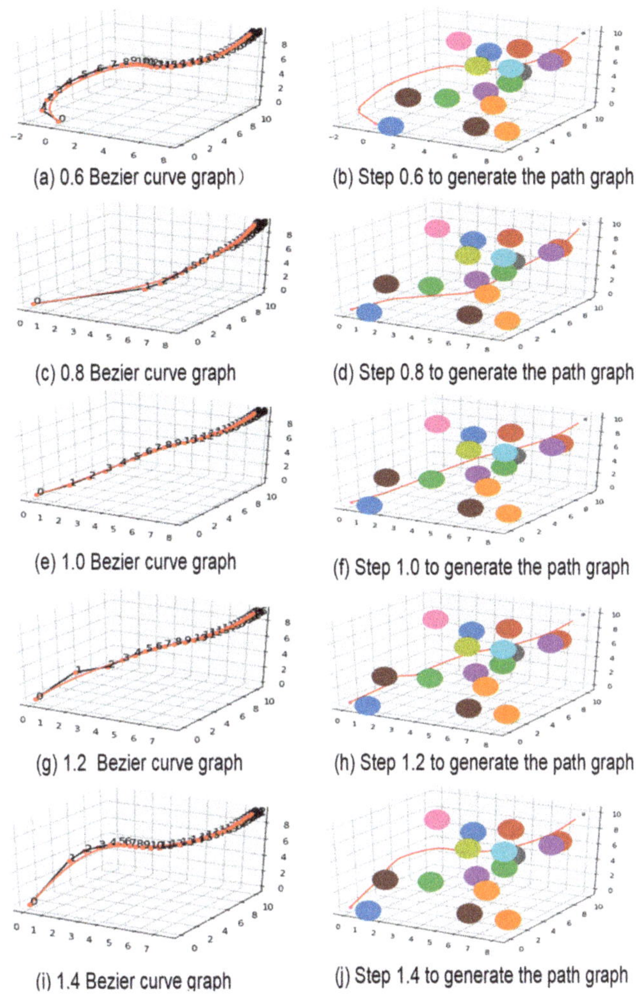

Figure 11. Comparison of hybrid algorithms with different step sizes.

As shown in Table 4, the search time was the smallest when the step size was 1.0 (42.8~52.2% shorter than at the other step sizes). The search time at the left end with a step size of 1.0 gradually decreased, while the search time at the right end had an increasing trend due to the excessive step size. The search time of the path did not decrease with increasing step size, and when the step size was 1.0, the search path was the shortest and the time was the shortest. In summary, when the step size was 1.0, the algorithm achieved the optimal effect. Therefore, the selection of the step size is still very important for the hybrid algorithm, and the experiment verified the reliability of the algorithm.

Figure 11 shows the trajectory diagrams of different step lengths fitted with Bezier curves. The effect is best when the step length is 1.0 in the figure: The trajectory is smooth, and the path is the shortest.

6.2. ROS Simulation Experiment

This section took the UR5 mechanical arm as the research object, conducted a simulation analysis in the ROS, and set up the scenes required for robotic arm motion planning in MoveIt. The objects were added in MoveIt by creating a topic publisher, setting the basic shape and position of required obstacles and target objects, and publishing object information. The experiment was demonstrated by the visualization tool Rviz in the ROS. First, the UR5 robotic arm model was loaded, the simulation function was enabled, and the start point and end point of the robotic arm were set. As shown in Figure 12, the gray robotic arm was the pose of the starting point, and the yellow robotic arm was the pose of the ending point. The obstacles in the picture are a table and eight regular-shaped cubes, and the green cuboid is the grasping target. This scenario was a locally restricted test scenario with obstacle constraints and platform constraints. Before motion planning, the error transformation matrix was used to compensate for the parking error of the manipulator, and the positions of the start point and end point of the manipulator were the postures after compensation. The improved APF and RRT hybrid algorithm was added to the Open Motion Planning Library (OMPL), and the corresponding YAML Ain't Markup Language (YMAL) was modified. The Kinematics and Dynamics Library (KDL) solver that comes with MoveIt was used to solve the angle changes of each joint during the movement from the starting point to the ending point. A smooth and collision-free path from the starting point to the target point was obtained, as shown in Figure 13. The motion trajectory was smooth and did not collide with obstacles.

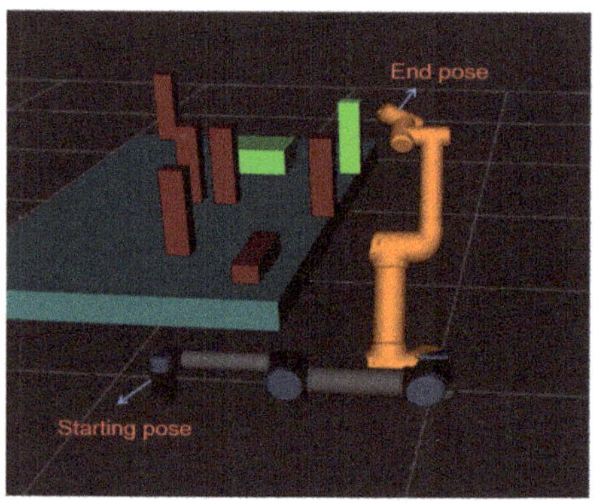

Figure 12. The poses of the start and end points of the UR5.

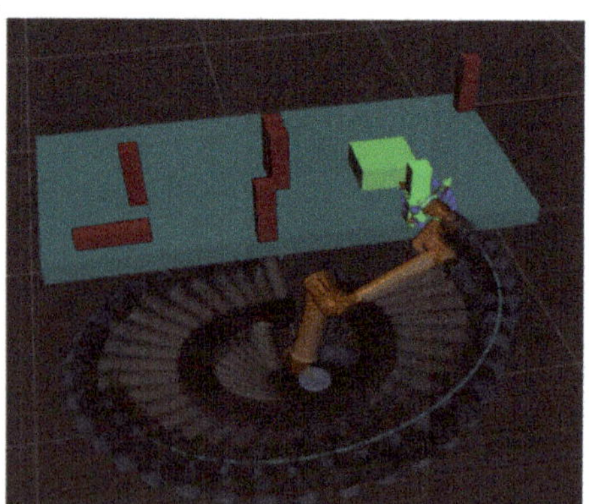

Figure 13. The trajectory diagram of the UR5 robotic arm.

Figure 14 shows the changes in the six joints during the movement of the robotic arm. The position change of each joint was relatively stable and met the real movement needs of the robotic arm. The position of the joints at the start and end positions are shown in Table 5. Table 6 shows the average search time and search success rate of different algorithms in the same obstacle environment. Each set of experiments was carried out 20 times. Simulation experiments verified the feasibility of the algorithm.

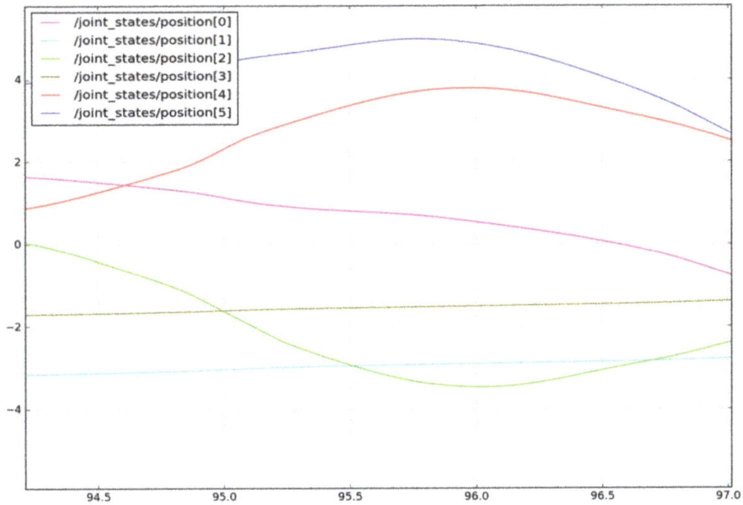

Figure 14. The position change of each joint of the robotic arm.

Table 5. Initiation and termination of the robotic arm joint.

	Joint 0	Joint 1	Joint 2	Joint 3	Joint 4	Joint 5
The starting position	0.00112	0.00322	−0.00114	5.793×10^{-5}	4.422×10^{-6}	5.521×10^{-6}
The ending position	−1.42552	−0.67803	−0.27888	4.10375	−0.70497	−3.14746

Table 6. Search time and search success rate of different algorithms.

	RRT	RRT*	P_RRT*	APF-RRT
Search time	19.1	9.3	6.2	3.0
Search success rate	90%	100%	100%	100%

7. Conclusions and Discussions

This paper improves the classic RRT algorithm and the classic APF algorithm and combines the two improved algorithms. The combined hybrid algorithm made full use of the efficient obstacle avoidance ability of the RRT algorithm and the efficient guidance ability of the APF method. Moreover, it avoided the disadvantages of each algorithm, and the comparison and analysis with the other three algorithms verified the effectiveness of the improved algorithm in this paper.

7.1. Discussions

To resolve the shortcomings of the classic APF method and the classic RRT algorithm, this paper proposes an improved path planning method that combines the APF method and the RRT algorithm for the path planning of the manipulator. First, the distance between the obstacle and the nearest node was evaluated. Through the distance value, the I-APF and I-RRT algorithms were used to explore the path. This simultaneously solved the shortcoming that the APF method cannot reach the target point when there are obstacles at the target point and made full use of the efficient obstacle avoidance ability of the RRT algorithm. The two algorithms alternated exploring the path until the target point was reached. The I-APF method introduced the heuristic method of breaking away from the local minimum based on the number of adjacent obstacles to solve the local minimum problem in the algorithm. The I-RRT algorithm included the triangular nearest neighbor node selection strategy, which effectively improved the obstacle avoidance ability and efficiency of the algorithm. In the same obstacle environment, compared with the classic RRT algorithm, the search time of the improved APF and RRT hybrid algorithm was reduced by 92.5%, the number of sampling nodes was reduced by 87.9%, and the path length was reduced by 23.4%. Compared with the RRT* algorithm, the search time of the APF and RRT hybrid algorithm was reduced by 72.2%, the number of sampling nodes was reduced by 47.4%, and the path length was reduced by 17.4%. Compared with the P_RRT* algorithm, the search time of the APF and RRT hybrid algorithm was reduced by 62.1%, the number of sampling nodes was reduced by 45.2%, and the path length was reduced by 13.0%. With an increase in the number of obstacles, the improved hybrid algorithm also showed the excellent effect of steady increases in the search time, number of sampling nodes, and path length.

7.2. Conclusions

The improved hybrid algorithm gives full play to the advantages of the individual algorithms while avoiding the disadvantages of the individual algorithms and is more suitable for the path planning of robotic arms. However, in a complex environment with more obstacles, the number of sampling nodes of the improved hybrid algorithm increases significantly. This is because the number of nodes randomly sampled by the RRT algorithm in the improved hybrid algorithm increases, increasing the consumption of computational memory and reducing the search efficiency. Therefore, it is recommended that future work focus on how to reduce the number of sampling nodes of the improved hybrid algorithm to reduce memory consumption and improve the efficiency of the algorithm.

This article focuses on the research of robotic arm path planning in a three-dimensional environment, which can be used in a variety of unstructured environments, such as warehouse automation and handling on production lines. Future research will focus on adaptive path planning in a dynamic environment.

Author Contributions: Conceptualization, Q.Y., J.Y., R.S. and H.B.; Data curation, Q.Y., J.Y., R.S. and H.B.; Formal analysis, Q.Y., J.Y. and H.B.; Funding acquisition, Q.Y.; Investigation, R.S.; Methodology, Q.Y. and J.Y.; Project administration, Q.Y. and J.Y.; Supervision, Q.Y., J.Y., R.S. and H.B.; Visualization, Q.Y. and J.Y.; Writing—original draft, J.Y.; Writing—review and editing, J.Y. All authors have read and agreed to the published version of the manuscript.

Funding: This research was funded by the National Natural Science Foundation of China (No. 51865004), Guizhou Province Education Department Science and Technology Talents Support Project (Branch Support KY [2017]062), and Horizontal Topic (K19-0204-001).

Institutional Review Board Statement: Not applicable.

Informed Consent Statement: Not applicable.

Data Availability Statement: Not applicable.

Conflicts of Interest: The authors declare that they have no known competing financial interests or personal relationships that could have appeared to influence the work reported in this paper.

References

1. Zeng, A.; Yu, K.-T.; Song, S.; Suo, D.; Walker, E.; Rodriguez, A.; Xiao, J. Multi-view self-supervised deep learning for 6D pose estimation in the Amazon Picking Challenge. In Proceedings of the 2017 IEEE International Conference on Robotics and Automation (ICRA), Singapore, 29 May–3 June 2017; IEEE: Piscataway, NJ, USA, 2017; pp. 1386–1393.
2. Wang, M.; Hou, Z. Continuous Trajectory Point Control Research of Six Degree of Freedom Mechanical Intelligent Arm Position. *Int. J. Precis. Eng. Manuf.* **2018**, *19*, 221–226. [CrossRef]
3. Khatib, O. Real-time obstacle avoidance for manipulators and mobile robots. In *Autonomous Robot Vehicles*; Springer: New York, NY, USA, 1986; pp. 396–404.
4. LaValle, S.M. Rapidly-Exploring Random Trees: A New Tool for Path Planning. *Computer Science Dept.* **1998**, *11*, 4.

5. Zheng, Y.; Shao, X.; Chen, Z.; Zhang, J. Improvements on the virtual obstacle method. *Int. J. Adv. Robot. Syst.* **2020**, *17*, 1729881420911763. [CrossRef]
6. Sun, J.; Liu, G.; Tian, G.; Zhang, J. Smart Obstacle Avoidance Using a Danger Index for a Dynamic Environment. *Appl. Sci.* **2019**, *9*, 1589. [CrossRef]
7. Zhang, J.; Wang, C.; Zhao, J. Overtaking path planning and tracking control of automobile curve based on improved artificial potential field method. *Automob. Eng.* **2021**, *43*, 546–552.
8. Han, Y.; Li, S. UAV trajectory planning based on improved artificial potential field method. *Syst. Eng. Electron. Technol.* **2021**, 1–9. Available online: https://scjg.cnki.net/kcms/detail/detail.aspx?filename=XTYD20210528005&dbcode=CJFQ&dbname=CAPJ2021&v= (accessed on 20 September 2021). (In Chinese).
9. Zhang, W.; Wei, S.; Zeng, J.; Wang, N. Multi-UUV path planning based on improved artificial potential field method. *Int. J. Robot. Autom.* **2021**, *36*. [CrossRef]
10. Tian, Y.; Zhu, X.; Meng, D.; Wang, X.; Liang, B. An Overall Configuration Planning Method of Continuum Hyper-Redundant Manipulators Based on Improved Artificial Potential Field Method. *IEEE Robot. Autom. Lett.* **2021**, *6*, 4867–4874. [CrossRef]
11. Li, Y.; Wang, S.; Jiang, L. Motion planning of mobile manipulator based on RRT with sparse nodes. *China Mech. Eng.* **2021**, *32*, 1462.
12. Ge, J.; Liu, L.; Dong, X.; Tian, W.; Lu, T. Free-floating space robot trajectory planning based on dynamic RRT*. *J. Aeronaut.* **2021**, *42*, 176–185.
13. Gan, Y.; Zhang, B.; Ke, C.; Zhu, X.; He, W.; Ihara, T. Research on Robot Motion Planning Based on RRT Algorithm with Nonholonomic Constraints. *Neural Process. Lett.* **2021**, *53*, 3011–3029. [CrossRef]
14. Qureshi, A.H.; Ayaz, Y. Potential functions based sampling heuristic for optimal path planning. *Auton. Robot.* **2015**, *40*, 1079–1093. [CrossRef]
15. Jeong, I.-B.; Lee, S.-J.; Kim, J.-H. Quick-RRT*: Triangular inequality-based implementation of RRT* with improved initial solution and convergence rate. *Expert Syst. Appl.* **2019**, *123*, 82–90. [CrossRef]
16. Wang, W.; Zuo, L.; Xu, X. A Learning-based Multi-RRT Approach for Robot Path Planning in Narrow Passages. *J. Intell. Robot. Syst.* **2017**, *90*, 81–100. [CrossRef]
17. Lee, H.; Kim, H.; Kim, H.J. Planning and Control for Collision-Free Cooperative Aerial Transportation. *IEEE Trans. Autom. Sci. Eng.* **2016**, *15*, 189–201.
18. Hao, B.; Yan, Z. Recovery path planning for an agricultural mobile robot by Dubins-RRT* algorithm. *Int. J. Robot. Autom.* **2018**, *33*. [CrossRef]
19. Zhou, H.; Zhou, S.; Yu, J.; Zhang, Z.; Liu, Z. Zhou Trajectory Optimization of Pickup Manipulator in Obstacle Environment Based on Improved Artificial Potential Field Method. *Appl. Sci.* **2020**, *10*, 935. [CrossRef]
20. Yao, Q.; Zheng, Z.; Qi, L.; Yuan, H.T.; Guo, X.; Zhao, M.; Liu, Z.; Yang, T. Path Planning Method With Improved Artificial Potential Field—A Reinforcement Learning Perspective. *IEEE Access* **2020**, *8*, 135513–135523. [CrossRef]
21. Wang, H.; Hao, C.; Zhang, P.; Zhang, M. Mobile robot path planning based on A* algorithm and artificial potential field method. *China Mech. Eng.* **2019**, *30*, 2489–2496.
22. Li, Y.; Wei, W.; Gao, Y.; Wang, D.; Fan, Z. PQ-RRT*: An improved path planning algorithm for mobile robots. *Expert Syst. Appl.* **2020**, *152*, 113425. [CrossRef]
23. Chen, M.; Zhang, Q.; Zhang, G. Research on obstacle avoidance path planning of manipulator under multi-obstacle environment. *Comput. Integr. Manuf. Syst.* **2021**, *27*, 990–998.
24. Long, Z. Virtual target point-based obstacle-avoidance method for manipulator systems in a cluttered environment. *Eng. Optim.* **2019**, *52*, 1–17. [CrossRef]
25. Jia, Z.; Lin, P.; Liu, J.; Liang, L. Online cooperative path planning for multi-quadrotors in an unknown dynamic environment. *Proc. Inst. Mech. Eng. Part G: J. Aerosp. Eng.* **2021**, *2021*, 09544100211016615. [CrossRef]
26. Zhang, Z.; Wu, D.; Gu, J. A Path-Planning Strategy for Unmanned Surface Vehicles Based on an Adaptive Hybrid Dynamic Step size and Target Attractive Force-RRT Algorithm. *J. Mar. Sci. Eng.* **2019**, *7*, 132.
27. Li, Y.; Xu, D. Cooperative path planning of dual-arm robot based on gravitational adaptive step size RRT. *Robot* **2020**, *42*, 606–616.
28. Shi, Y.; Li, Q.; Bu, S.; Yang, J.; Zhu, L. Research on Intelligent Vehicle Path Planning Based on Rapidly-Exploring Random Tree. *Math. Probl. Eng.* **2020**, *2020*, 5910503. [CrossRef]
29. Yuan, J.; Yang, L.; Tang, X.; Chen, A. Intelligent vehicle motion planning based on improved RRT* and driving trajectory optimization. *Acta Autom. Sin.* **2020**, *46*, 1–10. (In Chinese)

MDPI AG
Grosspeteranlage 5
4052 Basel
Switzerland
Tel.: +41 61 683 77 34

MDPI Books Editorial Office
E-mail: books@mdpi.com
www.mdpi.com/books

www.ingramcontent.com/pod-product-compliance
Lightning Source LLC
Chambersburg PA
CBHW061447220926
51850CB00004B/35
9783725818914